Handbook of Research on Microbial Remediation and Microbial Biotechnology for Sustainable Soil

Junaid Ahmad Malik
Government Degree College, Bijbehara, India

A volume in the Advances in Environmental Engineering and Green Technologies (AEEGT) Book Series

IGI Global
PUBLISHER of TIMELY KNOWLEDGE

Published in the United States of America by
IGI Global
Engineering Science Reference (an imprint of IGI Global)
701 E. Chocolate Avenue
Hershey PA, USA 17033
Tel: 717-533-8845
Fax: 717-533-8661
E-mail: cust@igi-global.com
Web site: http://www.igi-global.com

Library of Congress Cataloging-in-Publication Data

Names: Malik, Junaid Ahmad, 1987- editor.
Title: Handbook of research on microbial remediation and microbial biotechnology for sustainable soil / Junaid Ahmad Malik, editor.
Description: Hershey, PA : Engineering Science Reference, [2021] | Includes bibliographical references and index. | Summary: "This book highlights some of the significantly important microbial species involved in remediation, the physiology, biochemistry and the mechanisms of remediation by various microbes, and suggestions for future improvement of bioremediation technology"-- Provided by publisher.
Identifiers: LCCN 2020046702 (print) | LCCN 2020046703 (ebook) | ISBN 9781799870623 (h/c) | ISBN 9781799870647 (eISBN)
Subjects: LCSH: Soil remediation. | In situ bioremediation. | Soil microbiology.
Classification: LCC TD878 .M53 2021 (print) | LCC TD878 (ebook) | DDC 628.5/5--dc23
LC record available at https://lccn.loc.gov/2020046702
LC ebook record available at https://lccn.loc.gov/2020046703

This book is published in the IGI Global book series Advances in Environmental Engineering and Green Technologies (AEEGT) (ISSN: 2326-9162; eISSN: 2326-9170)

British Cataloguing in Publication Data
A Cataloguing in Publication record for this book is available from the British Library.

For electronic access to this publication, please contact: eresources@igi-global.com.

Advances in Environmental Engineering and Green Technologies (AEEGT) Book Series

Sang-Bing Tsai
Zhongshan Institute, University of Electronic Science and Technology of China, China & Wuyi University, China
Ming-Lang Tseng
Lunghwa University of Science and Technology, Taiwan
Yuchi Wang
University of Electronic Science and Technology of China Zhongshan Institute, China

ISSN:2326-9162
EISSN:2326-9170

Mission

Growing awareness and an increased focus on environmental issues such as climate change, energy use, and loss of non-renewable resources have brought about a greater need for research that provides potential solutions to these problems. Research in environmental science and engineering continues to play a vital role in uncovering new opportunities for a "green" future.

The **Advances in Environmental Engineering and Green Technologies (AEEGT)** book series is a mouthpiece for research in all aspects of environmental science, earth science, and green initiatives. This series supports the ongoing research in this field through publishing books that discuss topics within environmental engineering or that deal with the interdisciplinary field of green technologies.

Coverage

- Cleantech
- Policies Involving Green Technologies and Environmental Engineering
- Sustainable Communities
- Water Supply and Treatment
- Industrial Waste Management and Minimization
- Electric Vehicles
- Biofilters and Biofiltration
- Waste Management
- Green Transportation
- Contaminated Site Remediation

IGI Global is currently accepting manuscripts for publication within this series. To submit a proposal for a volume in this series, please contact our Acquisition Editors at Acquisitions@igi-global.com or visit: http://www.igi-global.com/publish/.

The Advances in Environmental Engineering and Green Technologies (AEEGT) Book Series (ISSN 2326-9162) is published by IGI Global, 701 E. Chocolate Avenue, Hershey, PA 17033-1240, USA, www.igi-global.com. This series is composed of titles available for purchase individually; each title is edited to be contextually exclusive from any other title within the series. For pricing and ordering information please visit http://www.igi-global.com/book-series/advances-environmental-engineering-green-technologies/73679. Postmaster: Send all address changes to above address.

Titles in this Series

For a list of additional titles in this series, please visit: http://www.igi-global.com/book-series/advances-environmental-engineering-green-technologies/73679

Improving Integrated Pest Management for Crop Protection and Food Security
Muhammad Haseeb (Florida A&M University, USA) Lambert H.B. Kanga (Florida A&M University, USA) and Jawwad A. Qureshi (University of Florida, USA)
Engineering Science Reference • © 2021 • 315pp • H/C (ISBN: 9781799841111) • US $195.00

Role of IoT in Green Energy Systems
Vasaki Ponnusamy (Universiti Tunku Abdul Rahman, Malaysia) Noor Zaman (Taylor's University, Malaysia) Low Tang Jung (Universiti Teknologi PETRONAS, Malaysia) and Anang Hudaya Muhamad Amin (Higher Colleges of Technology, UAE)
Engineering Science Reference • © 2021 • 405pp • H/C (ISBN: 9781799867098) • US $195.00

Combating and Controlling Nagana and Tick-Borne Diseases in Livestock
Caleb Oburu Orenge (Department of Veterinary Anatomy and Physiology, Faculty of Veterinary Medicine and Surgery, Egerton University, Kenya)
Engineering Science Reference • © 2021 • 440pp • H/C (ISBN: 9781799864332) • US $185.00

Artificial Intelligence and IoT-Based Technologies for Sustainable Farming and Smart Agriculture
Pradeep Tomar (Gautam Buddha University, India) and Gurjit Kaur (Delhi Technological University, India)
Engineering Science Reference • © 2021 • 400pp • H/C (ISBN: 9781799817222) • US $215.00

Handbook of Research on Waste Diversion and Minimization Technologies for the Industrial Sector
Ashok K. Rathoure (M/s Akone Services, Lucknow, India)
Engineering Science Reference • © 2021 • 455pp • H/C (ISBN: 9781799849216) • US $295.00

Smart Farming Techniques and Knowledge-Based Agriculture Emerging Research and Opportunities
Petraq Papajorgji (European University of Tirana, Albania) and Francois Pinet (Irstea Centre de Clermont-Ferrand, France)
Engineering Science Reference • © 2021 • 230pp • H/C (ISBN: 9781799845409) • US $195.00

Solar Concentrating Modules With Louvered Heliostats Emerging Research and Opportunities
Dmitry Strebkov (Federal State Budget Scientific Institution, Russia & Federal Scientific Agroengineering Center VIM, Russia) Natalya Filippchenkova (Federal State Budget Scientific Institution, Russia & Federal Scientific Agroengineering Center VIM, Russia) and Anatoly Irodionov (Federal State Budget Scientific Institution, Russia & Federal Scientific Agroengineering Center VIM, Russia)
Engineering Science Reference • © 2021 • 267pp • H/C (ISBN: 9781799842767) • US $190.00

IGI Global
PUBLISHER of TIMELY KNOWLEDGE

701 East Chocolate Avenue, Hershey, PA 17033, USA
Tel: 717-533-8845 x100 • Fax: 717-533-8661
E-Mail: cust@igi-global.com • www.igi-global.com

List of Contributors

Table of Contents

Section 2
Microbes and Sustainable Development

Detailed Table of Contents

The soil is considered to be one of the most important substances for the existence of the biotic community. The quality of the soil is continually degrading due to the continuous exploitation of human activity. The superiority of a soil is rated on the basis of its chemical and physical characteristics. The contaminants added to the soil mainly because of human activity change the usual function and ecological properties and cause of negative impacts on agricultural productivity and soil health. The property of the soil is potentially affected by urban wastes, industrial wastes, sewage water, mining wastes, oil, radioactive wastes, deforestation, and massive use of fertilizers and pesticides. Heavy metal contamination of the soil is a vital environmental problem because it is the cause of adverse effects on the biological community through the contamination of the food chain. A continuous exposure of municipal solid waste (MSW) in the landfill sites causes leachate formation; this is percolated inside the soil leading to the change in properties.

Industrialization led to the release of synthetic and toxic compounds. Partial or improper treatment increases environmental pollution. Conventional methods possess more disadvantages, such as increased duration of degradation and release of secondary pollutants. The drawbacks paved the way for the significant bioremediation perspective. The ubiquitous nature of microbes enables it to utilize toxic compounds, which attracted the focus of treatment towards the biological and eco-friendly methods. The recent decade has shown interest in the application of indigenous microbes in the polluted environment. Apart from the microbial application, phytoremediation is an emerging tool for treating soil contaminated

Section 2
Microbes and Sustainable Development

Beneficial microbes are used as the best alternative against the synthetic fertilizers and pesticides. The beneficial microbes not only help with plant growth, nutrition uptake, nitrogen fixation, but also help in acquiring the ions, not freely available to plants to uptake; these microbes also guard the plants by secreting toxic chemicals by inducing defense systems against pathogens. These microbes can provide best choice to look forward to sustainable agriculture and sustainable ecosystem. The addition of soil inoculants in the form of microorganisms or bio stimulants promise more environmentally friendly approaches for augmenting crop yields. The crop becomes less reliant on chemical fungicides and herbicides as many strains of microorganism have abilities of controlling pests. In this chapter, the interaction of beneficial plant bacteria, bio stimulants, effects on native microbial communities, and bacteria influencing economically important crops are discussed.

Terrestrial soil is a complex part of the ecosystem hosting bacteria, fungi, protists, animals, and huge source of nutrients to plants. These soil-dwelling organisms exhibit an array of interactions with plants to span the full range of ecological possibilities. In the 19th century, many different bacterial strains were described as having plant growth favouring potential like Pseudomonas, Azospirillum, and even crop seeds were coated with bacterial cultures to improve growth and yield. The soil microbial community also recognized their considerable role to improve the soil health via energy transfer, catalyzing reactions, and nutrient mineralization. Thus, soil microorganisms and enzymatic process are generally regarded as rate-limiting steps in decomposition and nutrient cycling.

Microbial products are being used from ages in known as well as unknown forms. Some common products harvested from microbes include proteins, amino acids, antibiotics, antibodies, secondary metabolites, organic acids, lipids, and so on. It also includes antivirals, polymers, surfactants, enzyme inhibitors, nutraceuticals, and many industrial and agricultural products. Moreover, sometimes the whole single

is a critical concern in today's world. The use of plant growth-promoting rhizobacteria (PGPR) has been found to be an environmentally sound way of increasing agricultural productivity by promoting plant growth either through a direct or indirect mechanism. PGPRs are commonly occurring soil microbes that colonize the root system, which is an ideal location for interactions with plant microbes. PGPRs can provide an enticing way of reducing the use of toxic chemicals and can affect plant growth and development, either through releasing plant growth regulators or other bioactive stimulants and by taking up nutrients through fixation and mobilization, minimizing adverse effects of microbial pathogens on crops by using numerous mechanisms. In addition, they also play a significant role in soil fertility. This chapter aims to explore the diversified plant growth mechanisms that promote rhizobacteria in fostering crop yields and promoting sustainable agriculture.

Chapter 13
An Account on Mycoviruses and Their Applications... 358
Md. Idrish Raja Khan, College of Fisheries, Central Agricultural University, Tripura, India

Mycoviruses are obligate parasites of fungi and can infect the majority of the fungal groups. They remain mysterious to various communities throughout the globe. Mycoviruses are responsible for certain changes in fungal hyphae, which could be asymptomatic and may cause a reduction or elimination of the virulence capacity of fungal hosts by the process called hypovirulence. Such fungal-virus system could be valuable for the development of novel biocontrol approaches against fungal pathogens for the development of a sustainable environment. There are adequate reports where mycovirus has been employed as a biocontrol approach against the pathogenic fungi in the fields of agriculture and other allied sciences. The prime focus of this review is to emphasize naturally available mycoviruses and strategies to adopt the mycovirus therapy which could serve as an excellent alternative strategy against chemical prophylactic and therapeutic approaches.

Section 3
Microbes and Site Remediation

Chapter 14
Application of Dehalogenase Enzymes in Bioremediation of Halogenated Pollutants: A Short Review... 373
Raghunath Satpathy, Gangadhar Meher University, Samabalpur, India

The halogenated hydrocarbons have been widely used by human beings. They are xenobiotic and toxic. The microbes having a specific group of hydrolase enzymes, known as dehalogenases, that actually break the carbon-halogen bonds of the halogenated substances and subsequently convert them into their non-toxic forms. In this chapter, the categories of dehalogenase enzymes possessed by microorganisms are narrated. The overall source, mechanism of catalysis, and structural aspects of the haloalkane dehalogenase enzymes have been discussed with special focus to the bioremediation of 1, 2 dichloroethane.

Quorum quenching is the process that prevents quorum sensing through the disruption of signalling cascade and bacterial communication among themselves mediated by the degradation of the signalling molecules. Therefore, quorum quenching has a considerable contribution in the negative regulation of threatening diseases and eventually increasing soil reclamation through different mechanism mediated by microorganisms in reclamation of soil. Quorum sensing has a significant contribution in enhancement of soil quality through microbial-based enzymes and mechanism in the versatile fields which are a component of the environment. The current chapter discusses the details of various direct and indirect mechanisms mediated by microbial systems that have a significant role in soil reclamation for the sustenance of the environment.

Industrialization led to an increase in chemicals in the environment. The soil absorbs these chemicals and holds them for years until treated. The action of bacteria, fungi, and algae utilize the pollutants and generate energy. The bioremediation contains a diverse treatment process, but the effectiveness of the bioremediation increases by the enzymatic action. Laccase, a copper-containing enzyme, is versatile and oxidizes complex organic compounds without generating reactive oxygen species (ROS). This process is carried by laccase-mediated systems (LCMs) controlled by low redox potential. The presence of redox mediators oxidizes the chemical compounds at the higher rate, making laccase degradation of the pollutants effectively. The chapter provides a glimpse of soil bioremediation by bacteria and fungi as individual species and symbiotic species, the production of laccase enzyme by bacteria and fungi, methods adopted to enhance the enzyme activity, and degradation of pollutants in soil.

Soils are a vital part of agricultural production. Soil health plays a significant role in the best crop production. Nowadays, our lands are under immense pressure. This pressure may be in the form of climatic changes that affect crop productivity or may be due to population increment that forces our current food system to produce more food to meet consumer needs. Climatic changes affect soil sustainability in the wrong way. Salinity, drought, and heavy metals disturb land structure badly. As the population increases, it dramatically impacts the current production system to fulfill the present needs. In all these situations, agricultural soil sustainability is a challenging factor for soil scientists to make our agriculture sustainable because agricultural sustainability couldn't be possible without maintaining soil health. Many approaches are available to improve soil structure and health. Among these, plant growth-promoting rhizobacterium is a good option. It not only improves soil structure but also helps the plants under abiotic stress conditions.

Preface

Contamination is creating a significant health and environmental crisis owing to its unhealthy origin and has become a major issue for environmentalists. A variety of carcinogenic compounds are released by different ways into soil and water every day. Not only does the presence of these substances in water and soil damage our earth's environment, but it also poses a danger to the welfare of living organisms.

Advances in microbiology and biotechnology have led to the launch of microbial biotechnology as a distinct research field, leading significantly to the advancement of fields such as agriculture, the environment, biopharmaceuticals, fermented foods, etc. Latest innovations in biotechnology for the environment have been influential in solving numerous environmental issues and global problems. For the science and academic fraternity, this frontier branch of biotechnology has always been a fascinating field that continuously draws the scientific community to delineate intricate pathways to address global environmental issues. In terms of our climate, which is mainly influenced by extensive human activities, accelerated urbanisation, and industrial development, we face many challenges today. Therefore, understanding nuanced information regarding this interaction of humans with the world is becoming exceedingly important. As a result of excessive extraction and an increased consumption rate, the bulk of natural resources have reached the stage of extinction.

Sustainable tools and technologies that are environmentally friendly are used to dramatically mitigate emissions. As historically hired, less environmentally friendly chemical means have many pitfalls associated with them, prompting the research fraternity to search for sustainable new biotechnological alternatives. Modern environmental biotechnology procedures could provide us with a stronger forum, at least for enhancing our quality standards and the environment as a whole, and for exploring the use of sustainable raw materials. Latest developments require cost-effective, eco-friendly, safe technologies for the manufacturing of goods, retaining high quality standards and more recycling and emission control of waste products.

Humans are known to be the world's most highly advanced organisms. Microbes are, possibly, placed in the lowest strata of evolution. Planet, however, is useless, non-functional and uninhabitable without these 'plain, small, invisible' beings. Soil microbiologists and microbial ecologists have for many years distinguished soil microbes according to their role as positive, harmful or neutral, and how they impact soil quality, plant growth and yield, and plant health. It is important to actively promote the use of microbial communities to degrade or replace environmental toxins such as heavy metals, chemicals, dyes, etc. There is a need to consider the metal-microbial relationship for the safe use of the microbial population in the bioremediation process, so that an in-depth mechanism of bioremediation and biodegradation can be delineated. In view of the above issues, the writers have progressed by this book under the banner of IGI Global Publishers to make important developments in the use of new environmental technology. The book seeks to offer a detailed review of innovative environmental approaches for wastewater disposal, elimination of heavy metals, oxidation of chemicals, removal of dyes, waste control, environmental contaminants' microbial transformation, etc.

The book is divided into four major sections: (1) 'Fundamentals and Approaches', (2) 'Microbes for Sustainable Development', (3) 'Microbes and Site Remediation', and (4) 'Microbial Bioremediation: Tools and Technologies'. These sections cover the recent developments in the applications of microorganism in various fields such as agriculture and environment. The first section of the book covers some of the fundamental areas and potential applications of microbial biotechnology emphasizing its current application and future prospects, promising applications of microbiome. Chapters cover the main potential pollutants and their remediation process. The principles of bioremediation and phytoremediation along with the role of microbes as biofertlizers are also discussed. The second section elaborates the role of microbes in sustainable development and agricultural production. The perspectives of plant growth through various microbial enzymatic processes are discussed. Articles are addressing the role of *Bacillus* sp., plant growth-promoting rhizobacteria and bacterial siderophore as promoters of plant growth, and investigating the use of fungi and actinobacteria for sustainable agriculture. The third section covers the fate of microbes and their products in various types of decontamination pathways for frequent pollutants of different composition and origin. As a component of sustainable agriculture practice, the use of microbes in detoxifying the petroleum hydrocarbons, halogenated contaminants, dyes, heavy metals and phenolic compounds is also discussed. The fourth and last section includes the articles discussing the fate of microbial biotechnology for various agronomic improvements and soil reclamation procedures. This section also covers the role of nanotechnology and rhizosphere engineering for sustainable soil and environment.

The new developments and reach by using different state-of-the-art methods to clean up and save our world have been well briefed by our readers. The continuing popularity of the books published under the banner of IGI Global Publishers is the result of a collaborative endeavour by a devoted editorial and publishing team, and to the benefit of our contributors and readers, we will continue to grow gradually. I have taken advantage of the support and recommendations of a wide number of biotechnology researchers around the globe in preparation for this book. I gratefully accept my debt to the reviewers, who, at different times, offered constructive feedback and useful recommendations. I affirm my contribution to the ethical and quality work published in this book, *Handbook of Research on Microbial Remediation and Microbial Biotechnology for Sustainable Soil*, while thanking all the authors. I hope this book to be able to provide detailed, available, up-to-date information on sustainable approaches to developing an environmentally healthy climate. In addition, this book gives environmental biotechnologists and microbial and biochemical technologists immediate access to a wealth of data along with student fraternity from diverse streams of environmental engineering and industrial biotechnology.

I would like to thank my spouse and parents, who in the editing process have given invaluable continuous encouragement.

I hope that this volume will support a large audience of scholars, educators, students, and others who are searching for technically sound and workable alternatives to today's agricultural systems' heavy chemicalization.

Junaid Ahmad Malik
Government Degree College, Bijbehara, India
Editor

Section 1
Fundamentals and Approaches

Chapter 1
Effect of Pollution on Physical and Chemical Properties of Soil

Trinath Biswal
VSS University of Technology, India

Junaid Ahmad Malik
https://orcid.org/0000-0003-4411-2015
Government Degree College, Bijbehara, India

ABSTRACT

The soil is considered to be one of the most important substances for the existence of the biotic community. The quality of the soil is continually degrading due to the continuous exploitation of human activity. The superiority of a soil is rated on the basis of its chemical and physical characteristics. The contaminants added to the soil mainly because of human activity change the usual function and ecological properties and cause of negative impacts on agricultural productivity and soil health. The property of the soil is potentially affected by urban wastes, industrial wastes, sewage water, mining wastes, oil, radioactive wastes, deforestation, and massive use of fertilizers and pesticides. Heavy metal contamination of the soil is a vital environmental problem because it is the cause of adverse effects on the biological community through the contamination of the food chain. A continuous exposure of municipal solid waste (MSW) in the landfill sites causes leachate formation; this is percolated inside the soil leading to the change in properties.

INTRODUCTION

The soil science is started from beginning of the human civilization. All kinds of soil contain minerals, air, water, nutrients, inorganic and organic matter. The proportion of all these parameters determines the soil properties such as soil texture, porosity, structure, color and chemistry. Soil is formed with various particle size associated with different percentage composition of silt, sand, clay and organic matter, which is commonly termed as soil texture. The soil containing mixture of sand, clay and silt is said to be loam. Soil is a significant parameter in order to sustain life on the surface of the earth and is solely responsible

DOI: 10.4018/978-1-7998-7062-3.ch001

for the production of various kinds of crops that are required for fulfilling the demand of food of plants, animals and human beings (Oliver et al., 2013; Kothawala et al., 2012). It is the base of the establishment of thousands of industries and settlements. Soils are solely responsible for the creation of forests in our globe and maintain the ecological balance and biodiversity. Soil health, its physical and chemical properties play a vital role for not only fulfilling the food demand, but also greatly responsible to sustain life in the globe in every aspect. The soil quality means it's capability to sustain life (plants, animals, human beings and microorganisms) on the earth's surface in a sustainable way (Quesada et al., 2020; Bünemann et al., 2018). The extensive agricultural activities and land conservation drastically retards the organic carbon with simultaneous addition of toxins in the form of chemical fertilizers, pesticides, insecticides and finally changing the mineralization composition of the soil. The sustainable land management and agriculture can retain the soil properties and quality with food security and ecological balance. The soil health and properties normally referred to as the quality of the soil, which can change, due to different anthropogenic and natural activities like agriculture, mining, addition of inorganic and organic substances, heavy metals, industrial wastes, urban wastes etc. (Qaswar et al., 2019; Stewart et al., 2020). Hence, to get optimum output and productivity, we have to use the soil in eco-friendly and sustainable manner. The property of soil and soil health depends upon soil texture, soil structure, composition, soil organic carbon, macro and micro nutrients present in the soil. The physical properties of the soil generally include soil structure (blocky, columnar, platy prismatic), soil texture (sand, clay and silt), particle density, bulk density; void space, permeability and colour. The chemical property of the soil includes ion exchange capacity, electrical conductivity, pH etc. (Biesalski et al., 2018). The alternation of land use pattern is highly influenced to the supply of nutrients to the ecosystem; therefore the nutrient status of the soil is comparatively better in the uninhibited land than on plantation land. The soil contamination is the degradation of the soil with change in soil properties mainly due to anthropogenic activities and sometimes changes in the natural environmental condition. In many cases the properties of the soil changes due to agricultural activity (addition of fertilizers, insecticides and pesticides), industrial operations (swage and solid waste), open cast mining (solid mining wastes, mining effluents), plastic and polymeric materials, heavy metals and other non-biodegradable wastes. Hence finally we can say that the properties of soil and soil quality degrades due to industrialization, population growth, deforestation, industrialization, extensive mining, and expansion of agricultural land and use of nonbiodegrdable materials in our day to day life (Ye et al., 2020; Carvalho, 2017; Dwivedi et al., 2019).

CLASSIFICATION OF SOIL

Soil can be classified as sand, silt, peat, clay, chalk and loam on the basis of the size of the soil particles (Hartemink, 2015; Schoonover & Crim, 2015; Azuka et al., 2015);

Sandy Soil

This kind of soil is warm, dry, lightweight and relatively acidic in nature having less concentration of nutrients. The sandy soil is normally light because of the presence of more void space, sand and is comparatively lighter than clay. These soils have the capability of speedy water drainage and are easily warmed up in the spring as compared to clay soils. It can be dried out in summer having low nutrients and can be easily washed away by rainwater because of less cohesive force of attraction. The organic matter present in the soil can facilitate plant growth and extra boosting of nutrients in order to improve water holding capacity and nutrient concentration of the soil.

Clay Soil

This category of soil is comparatively denser having high nutrient content. This kind of soil is cold in the winter season, whereas dried out in the summer season. This kind of soil contains about 25% clay and, due to spaces in between clay particles, it has high value of water holding capacity. Hence, these kinds of soils can drain slowly and takes a longer time period to warm up especially in summer.

Silt Soil

This kind of soil is of comparatively less density having high moisture retaining capacity and higher rates of fertility because of the high level of nutrients. The soil of these kinds consists of particles of medium size and due to medium sized particles; it can be easily compacted and susceptible to washing by rain water.

Peat Soil

This kind of soil contains a high percentage of organic matter and holds the maximum amount of moisture, therefore shows high fertility. This is an important kind of soil for purpose of gardening and plantation.

Chalk Soil

This kind of soil is highly alkaline in nature because of the presence of more amounts of lime and $CaCO_3$ in its chemical composition. Due to its alkaline nature the growth of ericaceous plants is inhibited in this soil, because these kinds of plants only grow in acidic soils.

Loam Soil

This soil is only a combination of sand, clay and silt. These kinds of soils are highly fertile, easy for agricultural activity and make available good drainage. On the basis of high fraction composition, it is considered as either clay loam or sandy. Since loam soil is a perfect composition of soil particles, therefore, is best considered for gardening and its performance in gardening is further enhanced by the addition of any extra organic matter in it.

SOIL CONTAMINATION

The soil contamination may be defined as the addition of persistent toxic substances, heavy metals, salts, organic chemicals, radioactive wastes, pathogens in the soil, which is the cause of adverse effects for plant growth, health hazard effect on animals and human beings. Soil is a layer of inorganic and organic substance cover over the rocky surface of the earth. The organic fraction of the soil is the decomposition products of animals and plants and rich in uppermost layer of the soil or topsoil. The inorganic fraction is due to the weathering of rocks and its formation is an extremely slow process and takes thousands of years. The top layer of the soil is called as protective layer, which is used for agriculture and fulfills the demand of food.

Soil contamination and the change in properties of the soil are due to different factors, which are illustrated as follows;

1. Percolation of liquid leachate from landfill sites.
2. Direct dumping of industrial wastes on the soil.
3. Percolation of industrial effluents inside the soil.
4. Addition of mining waste directly on the surrounding land.
5. Leakage of underground storage tanks.
6. Excessive use of pesticides, fertilizer and herbicides.
7. Seepage of solid waste.

The important soil pollutants causing soil contamination include pesticides, petroleum hydrocarbons, heavy metals, solvents, fertilizers, organic chemicals, chlorinated compounds, radioactive materials etc. (Yao et al., 2012; Balseiro-Romero, & Baveye, 2018).

Classification of Soil Contamination

The soil contamination is classified into the following categories (Steffan et al., 2017);

1. Soil Pollution due to agriculture
2. Contamination of surface layer soil
3. Soil contamination through solid wastes and industrial effluents
4. Change in soil profile
5. Soil contamination owing to urban activities
6. Contamination of underground soil
7. Soil pollution due to air pollution

PHYSICAL PROPERTIES OF SOIL

Some important physical properties of soil are as follows;

Soil Texture

The soil texture determines the relative percentage of sand, clay and silt in the soil and the soil is named on the basis of textural class as sandy clay, loam or sandy clay loam. The soil containing 70% or more sand is called as sandy soil or loamy sand soil. The soil containing more than 35% clay is termed as clay soil. The soil containing almost 40% sand, 20% clay and 40% silt is considered as an ideal loam soil, which is suitable for agronomic crops.

The soil containing a higher percentage of clay and silt normally retards the branching and growth of plant roots. The nutrients extracted by the plants from the soil are adsorbed by the colloidal particles present in the soil; hence the soil having fine structure exhibits higher fertility and higher rate of water infiltration. The infiltration rate of water for soil having a higher percentage of clay is less, whereas the surface runoff is more. The soil having fine texture prevents the flow of water, whereas the water movement is high in case of sandy soil.

Soil texture is more significant because of the following reasons;

- It determines the water holding capacity of the soil.
- It estimates the rate of water flow through the soil.
- It determines the fertility and usability of soil for various purposes.

Example: Although sandy soil is well aerated, but is able to hold less water and less quantity of nutrients. The soil containing a high percentage of clay can hold more water with more nutrient content.

Soil texture changes in accordance with the depth of the soil, therefore roots of the plant cope itself with different soil texture having the different penetrating ability inside the soil. On the basis of change in soil texture with depth, soil is classified into three profiles.

1. **Uniform:** Same soil texture all over the soil profile.
2. **Texture-contrast:** The soil texture changes unexpectedly in between the top soil and sub soil.
3. **Gradational:** The soil texture progressively increases down the soil profile.

The soil particles having size < 2 mm spherical diameter include in the explanations of soil texture because almost all physicochemical activity is carried out in this fraction of fine-size particles of the soil. The particle sizes > 2 mm spherical diameter are called as "skeletal grains" due to their less water absorption capacity. The different classes of soil texture are mainly important because it determines the water-holding capacity along with the base saturation of the soil and these factors are interlinked with the agronomic productivity (Seema et al., 2019; Paz-Ferreiro et al., 2018).

Soil Structure

The soil structure normally denotes the grouping or arrangement of the soil particles such as silt, clay, sand, fertilizer particles, organic matter into an aggregation of porous compounds termed as soil aggregates. The smallest aggregate is called as ped. The soil structure is generally a perfect arrangement of these soil aggregates and each soil aggregate contains pore or void space and solid particles. The soil aggregated due to different natural factors such as freezing, thawing, drying, wetting and microbial activity is the cause of decay of biological matter or organic matter, soil organisms and adsorption of cations. The cations adsorbed by the soil support to form aggregates because the cations present in the soil can bind two or more particles of the soil. The aggregates are classified on the basis of shape, volume, size and stability.

Soil structures significantly influence the down flow of water after saturation infiltration. The soil having a well-ordered structured will have a comparatively higher drainage capacity in irrigation or during heavy rain and minimizes the temporary water logging of that land. Although well-ordered structured soil has higher capacity of water drainage, but still it can hold an adequate quantity of water for plant growth. The soil having disordered structure or less ordered structured is the cause of water logging in especially rainy season and inhabits the plant growth and damages the root system of the plants.

Soil having one horizon exhibits exclusively one structure, whereas the soil having different structure exhibits different horizons. The basic kind of arrangement of aggregate includes granular, prismatic, blocky and massive structures. If the massive structure is present at the top layer soil, it prevents the infiltration of water in the soil and the germination of seeds is difficult due to less aeration in soil. If the top layer soil is granular kind, water can easily enter into the soil and germination of seed is easier.

Like soil texture, the soil structure also changes naturally by change in weathering condition through the penetration of roots of the plants, which is important for agriculture. The soil having a high percentage of organic matter and clay possesses relatively more stable structure as compared to the soil containing a higher percentage of silt and sand. The soil containing free $CaCO_3$ and small amount of iron oxide exhibits highly stable structure.

The soil structure serves a significant role for the estimation of nutrients and its availability for acceptance by the roots. The change in climate specially affects the kinds of structure of the soil. The crumb and granular structure normally found at the soil surface of A-horizon. The blocky, columnar, prismatic structure and sub-angular blocky structure is predominantly found in the subsoil of B-horizon. The platy structure is present at subsoil or surface layer, whereas single grain and no structure are found at C-horizon.

If the soil aggregates are more developed in the horizontal zone than vertical zone, then it is called plate like soil, whereas if it is developed more in vertical axis than horizontal, then the soil is called as prism –like soil. If the dimension of the aggregate of the soil is of equal size, the soil is called as block –like soil, whereas if aggregates are almost roundish, then it is called sphere – like soil (Wankhade & Ghugal, 2016; Guéguen & Bard, 2005).

Soil Density

Density can be defined as the ratio M/V to any kind of monophasic homogeneous material having mass (M) and volume (V). The density of the soil can be measured by means of either particle density or true density and bulk density or apparent density. The particle density depends upon only the percentage of solid soil mass and is defined as the mass per unit volume of the solids. The average particle density normally ranges from 2.4 to 2.75 g/cc. The particle density only estimates the volumetric mass of the solid soil and the volume used for measurement does not consider the available pore spaces in the soil.

$$\text{Particle density} = \frac{Over\ dry\ soil\ mass}{Volume\ of\ the\ solid\ soil}$$

The particle density indicates all the various kinds of minerals and its composition present in the soil. For many soils its value is very close to 2.65 g/cm^3, because the density of quartz is 2.65 g/cm^3 and normally quartz is the leading mineral in most of the soils. The particle density varies very less in between different minerals and possesses less practical significance.

The bulk density is defined as the ratio between the total volume of solid to the volume of both pores and solids.

$$\text{Bulk density} = \frac{Total\ volume\ occupied\ by\ the\ solid}{Volume\ of\ both\ the\ pores\ and\ solid} = \frac{Dry\ weight\ of\ the\ solid}{Total\ volume\ of\ the\ solid}$$

The bulk density of soil is always less than the particle density. The bulk density is considered as the indicator of the compaction of the soil. The bulk density of the soil is solely influenced by densities of the soil mineral (such as silt, clay, sand and particles of organic matter), soil texture and the packing of the soil solids. Most of the soils have the bulk density of 2.65 g/cm^3. Normally, soil having loose compaction and more porous containing higher % of organic matter possesses less bulk density.

The sandy soils are comparatively found to be having a high volume of bulk density, because the total pore space present in the sand is always less than clay or silt soils. The soil having fine texture (clay loams and silt) possesses an adequate structure having more pore space and less bulk density relative to the sandy soil. The bulk density of the soil normally increases with an increase in the depth of the soil because the subsurface layer of the soil have less aggregation, organic matter and less capability of root penetration than the surface layers soil, therefore comprising less pore space (Håkansson & Lipiec, 2000; Demattê et al., 2010).

Soil Porosity

The pore space or porosity is defined as the volume of the pores in the soil, which can be filled by air or water or both and is inversely proportional to the bulk density

Bulk density x 100 = % solid space particle density

% of pore space = 100% – % Solid space

$$\text{Porosity} = \frac{Total\ pore\ volume}{Total\ volume\ of\ the\ soil\ sample}$$

Soil is generally composed of particles of different dimensions and the space in between the particles is termed as pore space. The pore space regulates the quantity of water that can be held by a particular volume of soil. The porosity normally referred to as how many void pores or holes are present in a given volume of soil. The soil porosity is normally expressed as the percentage of the entire volume of the soil substances. Porosity is an important factor of the soil, where the ground water can be purified, which is the major source of drinking water (Liu et al., 2017; Emerson & McGarry, 2003).

Soil Colour

Soil colour is one of the physical properties of the soil, which does not influence the behavior, properties and use of the soils, but however it can give the composition of a soil and predicts its usability. It has no direct influence on the soil fertility and growth of the plants, but indirectly its influence can be observed in soil moisture and soil temperature.

There are a number of different colours exhibited by different soils such as black, grey, brown, brick red, rust colour, yellow and sometimes green colour, but the principal colours are grey, rust and brown. The colour of the soil depends upon different factors. Clay is the major component impacting colour in the soil; humus is either black or dark brown if rich in oxides of iron. Quartz is an important component of soil which impacts greyish white or grey colour to the soil. The colour of the soil also changes with the change in concentration of the moisture. The soils having a higher percentage of moisture are generally darker because of the change in refractive properties and when light is incident on the surface of the dark soil a major fraction of it is reflected back. Hence the most suitable environment for determination of soil colour is clear sky, because the wavelength of incident light is in consistence. More is the darkness of the soil, the higher is its productivity because of more organic matter and vice-versa. The light colour of the soil is due to the percentage of quartz present in it. Black and dark colour soils absorb more heat energy of radiation than other coloured soil and comparatively more warmer (Moritsuka et al., 2014; Owens & Rutledge, 2005).

Soil Temperature

Soil temperature plays a key role in the biological interactions and chemical reactions in the soil. It directly influences the soil fertility, plant growth, % of moisture, microbial activity, soil structure, aeration in soil, decomposition of the plant residues, availability of nutrients and enzyme activities. The hot climatic condition is the cause of retarding the crop production because of decrease in the metabolic activities, retardation in root elongation and seed germination. There are many factors affecting soil temperature, such as environmental factors including solar radiations, soil air, soil water, cloud, snow and plants. Rainfall may be the cause of warming or cooling of the soil. The soil temperature changes daily and seasonally due to variation in the energy of radiation and change in energy that is absorbed through the surface of the soil.

The temperature is normally associated with the biological and physiochemical processes and also impacts the procedures of gas exchange between the soil and earth's atmosphere. The soil temperature can be controlled and balanced by the environmental factors such as quantity of heat absorbed by the soil and the heat dissipated from the surface of the soil. Soil temperature also influences the soil properties due to change in decomposition of organic matter and mineralization of various forms of biological waste products. It also affects the rate of water absorption, soil nutrients, electrical conductivity (EC) of the soil, plant growth, soil water transmission and retention.

Soil is considered as the major storage basin for heat and act as a reservoir of energy source during the day time as well as the heat source for the soil surface at night time. The soil also stores heat energy in summer season and releases this heat energy in winter. The soil temperature depends on the ratio of the energy absorbed to that of energy lost from the soil and fluctuates due to variations in temperature of the air and intensity of solar radiation. Soil temperature is one of the important factors that vastly influences both physical, chemical properties of the soil, seedling emergence, plant growth, nutrient availability and root growth. Hence the soil temperature is treated as the heat flux of the soil and also heat exchanger in between the troposphere and soil (Onwuka, 2016; Heinze et al., 2016).

CHEMICAL PROPERTIES OF SOIL

Cation Exchange

The organic materials and clay particles are present in the soil in the form of colloidal particles and possess negative charges, which obviously attacks the cations such as Ca^{2+} (from lime), K^+ ions (as potassium fertilizer), and NH_4^+ ions (from ammonium fertilizer) towards the surface of the soil. The cation adsorption normally happens at the surface of the colloid micelle and is associated with the release of one or more ions, which are present at the colloid micelle. This is popularly called as cationic exchange.

Example: Suppose in a colloid micelle 50% of its capacity is fulfilled with Ca^{2+} ions, 25% with K^+ ions and the remaining 25% is by H^+ ions. The H^+ and Ca^{2+} ions may combine with Cl^- ions of KCl and forms HCl and $CaCl_2$ respectively.

The exchange reaction in soil is very rapid, reversible and ion exchanges generally carry on till equilibrium is attained. All cations do not have equal capability of adsorption; some of these can be adsorbed very slowly, whereas some are substituted rarely. The cation exchange is generally carried out in between the following;

- The cations of the soil solution and the cations previously present on the surface of the soil colloids.
- The cations previously present on the surface of the soil colloids and the cations released from the plant roots.
- The cations in between a clay colloid and an organic colloid or cations in between two organic colloids.

Therefore, two different kinds of colloids are found in the soil, such as;

1. Clay colloids or mineral colloids
2. Humus colloids or organic colloids

These two kinds of colloidal particles present in the soil are strongly interlinked with each other and cannot be separated easily. The organic colloids are normally found in the form of humus particles, whereas the inorganic colloids are found in the form of very tiny particles. The cation exchange ability is important because of the following two reasons;

- The exchangeable cations (Ca, Mg, K are easily available for plant consumption).
- The exchange sites, where the cations are adsorbed are highly resistant to leaching or it can move downward in the soil with water.

The cations like Ca, Mg, K and Na are the cause of alkaline reaction in the water and are said to be the basic cations. The cations like H^+ ions in the water are said to be as acidic cations. The percentage of the cation exchange capacity occupied through the basic cations is said to be percentage of base saturation. More is the percentage of base saturation, higher is the pH of the soil (Sorkau et al., 2017; Olorunfemi et al., 2016).

Soil pH

The pH of the soil is considered as the most basic chemical property of the soil and is one of the most informative soil parameters. The soil pH indicates many properties connected with the soil and increases with decrease in acidity. The soil having pH>7 is treated as alkaline, pH< 7 is treated as acidic and pH=7 is called as neutral soil. But normally the pH of the soil varies from 4 to 8.5. The pH of the soil influences the soil property, quality, kinds and concentration of microorganisms present in soils, the decomposition of waste crop residues, sludges, organic matter, nutrients and manures. It influences the plant growth, percentage of phosphorus, solubility and transformation of the nutrients.

Example: Al, Mn, Fe is highly soluble in water at pH < 5.5 and toxic to the plants. The essential bacteria present in the soil, which serve as mediators for the number of nutrient transformation mechanisms from soil to plants, are normally most efficient in slightly alkaline medium.

The soil pH is normally affected by mineral concentration, soil texture and climatic condition. The pH of natural soil is due to the combined influence of different soil forming factors such as original materials, topography, time, climatic condition and kind of organisms. The pH of newly shaped soils only depends upon the concentration of the minerals present in the original materials. The intensity of rainfall and temperature influences the weathering of soil minerals and leaching capacity of the soil. Hence, in humid and warm environmental condition the soil pH declines because of more leaching, due to higher fraction of rainfall, whereas in the dry environment, the intensity of leaching and weathering is less. Hence the soil may be alkaline or neutral having pH $\geq$ 7. Soil having a higher percentage of organic matter and clay are normally more resistant to pH change and exhibits good buffering capacity as compared to sandy soils. The sandy soils contain less percentage of organic matter, therefore, possess low buffering capacity along with high infiltration and percolation rate for water (Penn, & Camberato, 2019; Gentili et al., 2018).

Soil Inorganic Matters

There are a number of inorganic constituents of soil such as the compounds of Al, Ca, Si, Fe, Na, K, Mn, which are added into the soil because of the weathering process. Besides these some other different inorganic compounds of less concentration like Cu, B, Mg, Zn, Mo, Co, I, F etc. are existing in the soil due to natural process and contribute inorganic material to the soil. The concentration of minerals or inorganic matter present in the soil varies from place to place and depends upon the quality of soil. The composition of the inorganic matter present in the soil varies greatly from one horizon to another horizon (Liu et al., 2017).

Soil Organic Matters

The most common source of organic matter present in the soil is from the dead bodies of plants and animals. Its percentage is very less in the sandy soil of dry zone, whereas highest in peaty soil. The decomposition of the dead bodies of plants and animals contributes a major percentage of organic matter to the soil. The decomposition of a number of kinds of organic compounds is converted into an organic complex of dark colour called as humus. Living organisms in some cases contribute enough amounts of organic matters in the soil due to wastes from metabolic activities. The presence of excessive amounts of organic matter affects both the physical and chemical properties of soil and also the soil health. The different properties affected by the organic matters include the function of soil microorganisms, moisture holding capacity, diversity of soil organisms and availability of soil nutrients. The organic matter also greatly influences soil fertility, alterations of chemicals, pesticides and herbicides. The presence of organic matter may influence the physical properties of the soil in different ways. The plant residues usually act as a protective layer over the surface of the soil and protect the soil from crusting and sealing from rain water, which is the cause of increasing the rate of infiltration of rainwater and reduces the surface runoff. The rate of infiltration of rain water normally depends upon various factors such as aggregation of the particles and stability, pore continuousness and stability, presence of cracks and condition of the surface of the soil.

The increase in the percentage of organic matter indirectly contributes the porosity of the soil through the enhancement of soil fauna activity. The newly formed organic matter supports the macrofauna activity mainly of earthworms, which is the cause of holes lined with the glue-like matter secreted from the bodies of earthworms and are intermittently occupied with this larva cast material (Tully & McAskill, 2020; Smith et al., 2017).

Plant Nutrients

There are 16 different nutrients required for the growth of plants and existence of living organisms in the soil. All these 16 kinds of nutrients are broadly divided into two categories such as micronutrients and macronutrients. The macronutrients include Hydrogen (H), Carbon (C), Nitrogen (N), Oxygen (O), Calcium (Ca), Sulphur (S), Magnesium (Mg), Potassium (K) and Phosphorus (P), whereas the micronutrients are Iron (Fe), Manganese (Mn), Zinc (Zn), Copper (Cu), Boron (B), Chlorine (Cl) and Molybdenum (Mo). Both macro and micro nutrients are essential for enhancing soil fertility and plant growth. Normally the plant receives both macro and micro nutrients in the ionic form (both cationic and anionic forms) from the soil solution. Plant requires large amount of macronutrients and very small amounts of micronutrients. The optimum health and fertility of the soil is the combination of micronutrients and macronutrients. The macronutrient, Nitrogen is the cause of greenery of the plants, growth of plant leave and impacts green colour to the plants on support with the production of the chlorophyll. Phosphorous is the cause of healthy growth of plant roots, flower germination and plant can also survive in drastic climatic conditions. Potassium impacts strength to the plants, promotes for early growth, helps the plants in retaining water and preventing the plants from insects and diseases. Magnesium is the cause of impacting green colour to the plants. Sulphur prevents plant diseases, supports seed growth and cause of the production of proteins, amino acids, vitamins and enzymes. Calcium is the cause of growth of the cell walls of the plants and prevents attacks of disease, helps cell metabolism and acceptance of NO_3^-.

Micronutrients also offer some major benefits to the soil such as Fe is necessary for the synthesis of chlorophyll in plants. Manganese assists Fe in the synthesis of chlorophyll and act as an activator for the enzymes during the growth process of plants. Zinc is highly essential for plant and root growth. Boron controls the metabolism of the carbohydrates in the plants and promotes fertilization and pollination. Copper stimulates enzymes in plants. Chlorine is necessary for growth of roots and activates photosynthesis. Molybdenum is necessary for the conversion of nitrate nitrogen into the amino acids. Nickel is necessary for feasible seed and to complete the life cycle of the plants (Bashagaluke et al., 2018; Singh & Sood, 2016).

Soil Salinity

The salt can be moved into the surface of the soil horizon due to capillary action from the salt laden water table and deposited on the soil surface because of evaporation. Normally, soil salinization happens, if the extensive irrigation is carried out without leaching the salts from the soil and proper drainage system. Salts can be accumulated on the soil surface because of seawater intrusion or the soil may be salted naturally. The soil salinity is the cause of degradation of the soil and affects vegetation. The most common salts are the combination of cations such as Ca^{2+}, Na^+, K^+, Mg^{2+} with the anions such as SO_4^{2-}, Cl^-, CO_3^{2-}.

Soil salinity is nothing, but the increase in the concentration of the salts in the soil and the process of increase is known as salinization. The increase in salt concentration in the soil is caused by both natural process and man-made activities. In natural process it is due to weathering of minerals, continuous withdrawal of salts from the ocean. In manmade activity it is due to road salt and irrigation practices. Salinity is the cause of damaging plant growth, infrastructure such as roads, corrosion of pipes, bricks. Excessive salt concentration is the cause of sedimentation problems, degradation in the quality of the water, increased leaching of metals (especially Cu, Cd, Zn and Mn) and ultimately promotes soil erosion (Shrivastava & Kumar, 2015; Huang, 2018).

Electrical Conductivity

Electrical conductivity (EC) in soil is generally a measure of soil salinity and is an important parameter affecting soil health. It influences the percentage of yield of the crops, availability of plant nutrients, suitability of the crops, and function of soil microorganisms, which is the vital factor in influencing the soil process such as production of greenhouse gases (NO_X, CO_2, and CH_4). The presence of excess amounts of salts is the cause of hindrance of the plant growth and affects the balance between soil and water. Electrical conductivity does not show a direct measurement of some specific ions or salts. It is always interlinked with the K, Na, Cl^-, NO_3^-, NH_3, and SO_4^{2-}. The water used for irrigation contains at least some salts and the concentration of salts over the agricultural land periodically increases, which is the cause of damage of plant roots, reduced crop productivity, fertility and also cause of change in composition and structure of the soil. Hence, in order to protect the agricultural land from excessive salt, we have to take steps for proper management of the salts (Shannon et al., 2020; Lech et al., 2020).

Buffer Action

Buffer action means the resistance of changing the pH of the soil. There are some solutions, which possesses reasonably constant pH even after addition of little acid or alkali or on dilution is said to have reserved alkalinity or acidity or commonly called as buffer solutions. If small amount of acid is added with the suspension of neutral soil, the change in pH is negligible and this property of the soil to resist the pH change is termed as buffer action. Hence soil possesses strong buffering capacity; therefore in order to change the pH of a soil, large quantities of alkalis or acids has to be added in the soil. The buffering action of a soil is mainly owing to the presence of excessive amounts of weak acids and their corresponding salts in the soil. The salts of PO_4^{3-}, HCO_3^-, CO_3^{2-} and salts of some weak inorganic acids and their corresponding salts serves as significant buffering agents present in the soils. Besides these salts and weak acids colloids linked with cations also act as significant buffering agents. The buffering action of the soil directly depends upon the presence of quantity and nature of clay particles and organic matter or humus colloids present in the soil (Wei et al., 2019; Shi et al., 2017).

EFFECT OF HEAVY METAL CONTAMINATION ON PHYSICAL AND CHEMICAL PROPERTIES OF THE SOIL

The contamination of the soil is due to the presence of excessive amount of toxic heavy metals and is now considered as one of the important issues in our present industrialized world. This is partly due to anthropogenic activities and partly from geologic factors. The soil containing excessive amounts of heavy metals is the cause of retardation in the growth of the plants, soil quality, changes the soil composition and activity of microbial community present in the soil. Hence the heavy metals are the significant source of soil pollution, which is caused by various metals such as Cu, Ni, Hg, As, Cd, Fe, Zn, Pb, Mg and Cr. The properties of soil including the presence of clay, organic matter and pH are greatly affected by the presence of toxic heavy metals. The presences of heavy metals affect indirectly the enzymatic activities of the soil due to shifting of the microbial community, which are produced by enzymes. The heavy metal strongly influences the soil biotic community and retards the population and activity of essential soil microorganisms due to the negative impact towards the vital microbial processes, because of its toxicity. If soil is long-term exposed by toxic metallic contaminants, the tolerance limit of the bacterial community like arbuscular mycorrhizal (AM) fungi are also increasing, which can play a vital role in the refurbishment of contaminated soil ecosystems.

The presence of excessive amount of heavy metals is the cause of retardation in the population of the bacterial species with subsequent increase in the soil actinomycetes. It is also the cause of destroying the biodiversity of the bacterial communities in the metal contaminated soil and decreases the biomass. Different enzymes function differently in the soil depending upon their nature and toxicity of metal because of variation in the chemical affinities by the enzymes present in the contaminated soil system. Cd is comparatively more toxic towards the enzymes than Pb due to its less affinity to the colloidal particles of the soil and higher mobility. Cu prevents the activity of b-glucosidase more relative to the cellulose activity. Pb significantly decreases the activity of some important soil microorganisms such as acid phosphatase, urease, catalase and invertase. Although sulfatase and phosphatase are significantly inhibited due to the presence of As (V), but urease is not affected. The contamination of soil by Cd exhibits negative impact on the activities of many soil microorganisms such as protease, alkaline phosphatase, urease and arylsulfatase, but no noticeable effect is observed on invertase. Every soil enzyme shows a different activity towards different heavy metals. The order of retarding activity of the enzyme urease normally decreases in the order of Cr > Cd > Zn > Mn > Pb. Since the activity, population and diversity of soil microorganisms plays a vital role in the physical and chemical properties of soil including the plant nutrients recycling, change in soil structure, change in soil organic matter, detoxification of toxic chemicals controlling of pests, soil structure, soil density, which ultimately impacts the plant growth.

Cr is the most common and highly toxic heavy metal present in the soil in the form of Cr (III) and Cr (VI) and both of these two states of Cr show different properties and toxic effects. Cr (VI) is acting as a strong oxidizing agent and extremely harmful and toxic in nature, but Cr (III) is a micronutrient comparatively non-toxic in nature and its toxicity is almost 10 to 100 times less than Cr (VI). The presence of Cr (VI) at higher concentration is the cause of highly negative impact on the population of soil microbes and unfavorably acting on the metabolism of microbial cell and cause of change in soil density with other physical and chemical properties of soil. The increase in the concentration of heavy metals than the permissible limit affects negatively towards the properties of the soil microbial community, including enzyme activity and rate of respiration, which is the important indicator for soil pollution and properties of the soil. The presence of metals in the soil affects the soil properties in a diversified way. pH is the major factor that affects significantly the availability of metals in soil.

Example: The accessibility of Zn and Cd surrounding the roots of *Thlaspi caerulescens* is inhabited with the increase in pH of the soil. The hydrous ferric oxide and organic matter has the capability of retarding the availability of heavy metals by the process of immobilization of Zn and Cd.

There is an important positive correlation found between some of the physical properties (including water holding capacity, % of moisture) of the soil and heavy metals. Some other properties such as soil density, charge of the soil colloidal particles, complex formation and relative surface area of the soil also affects significantly due to the availability excess metals in the soil. Soil colloidal particles provides particular surface areas and large interface, which helps in regulating the amount of heavy metals present in the natural soils. The soluble state of metal in the contaminated soil can be decreased through the soil particles having a large surface area. The soil aeration, composition of minerals and microbial activity is also changed due to change in availability of heavy metals in the soil (Alamgir, 2016; Bhatti et al., 2016; Wang et al., 2007). Metals may change the properties of the soil by monitoring the biochemical and microbiological changes in the soil. The toxicity due to heavy metals on soil microbes depends upon a number of factors, including soil temperature, clay minerals, pH, inorganic anions, organic matter, cations and chemical forms of the metals. Normally metals are present in the soil either in the uncombined state or on combination with some other components. These other components include exchangeable ions (cations & anions) sorbed at the surfaces of inorganic solids, nonexchangeable ions (cations & anions) and insoluble state of inorganic metal compounds mainly CO_3^{2-} & PO_4^{3-}, free metal ions in the soil solution, metals on combination with the silicate bases minerals, metal complex of organic substances.

The metals bonded with the silicate minerals show the contextual soil metal concentration and are not the cause of any contamination issue as compared with the metals in the free state. The natural resource of metal in soil is called rock bed. Heavy metals are the underlying forces in the soil and the plant consumption of it depends on the properties of soil and plays a crucial role in the bioavailability of metals. The level of the metal compounds builds up in the plant depends on various soil properties including soil texture, soil pH, humidity, and micronutrient concentration.

The soil having pH <7 or in acidic soil, the capacity of absorption of heavy metals are more in vegetables, whereas in alkaline soil pH>7 the leaching of heavy metals and their bioavailability to the plants is less. The solubility of metal in the soil determines the soil properties because the heavy metals present in the soil might be immobilized in its solid state. The concentration of organic and inorganic substances in soil influences the bioavailability and metal mobility). Different sources of heavy metals in the soil are shown in Figure 1.

Figure 1. Sources of heavy metals

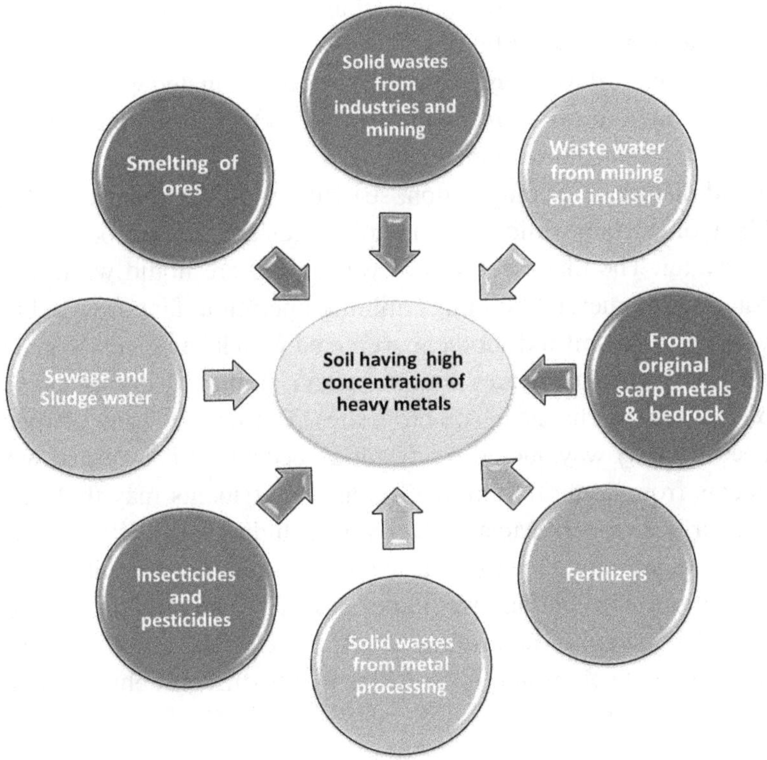

The addition of PO_4^{3-} in the soil enhances the soil permeability and is the main cause of arsenic migration into the deep soil profile, and then finally mixed with the ground water sources. The age of the soil plays a crucial role in the variability of metal bioavailability in the plant. The highest concentration of heavy metals is identified in the old flat wetlands because of long-term discharges of industrial and municipal wastewater. The natural bedrock predominantly is the source of Mn and Fe, whereas the existence of other heavy metals (Cu, Zn, Cr, Cd, Ni, and Pb) in the soil is due to anthropogenic origin (Su et al., 2014; Kapahi & Sachdeva, 2019).

IMPACT OF WATER POLLUTION ON PHYSICAL AND CHEMICAL PROPERTIES OF SOIL

In the urban areas of the developing countries the deficiency of water is the cause of encouraging people for water storage and use of this polluted water in agriculture, gardening and some other purposes. The river water affected by chemicals, fertilizer and textile industries gets polluted. Among the different soil properties, the electrical conductivity, pH, exchangeable cations and nutrient availability are strongly influenced by waste water added to the soil. Fresh uncontaminated water is the stock resource of our globe and due to continuous increase in population, the demand of fresh drinking water is highly increased. In many regions of our globe life of plants, animals and human beings are now in risk due to the scarcity of fresh uncontaminated water. The major sources of water contamination are the commercial areas, chemical plants, agricultural runoff, mining, industrial and domestic sewage. Now to prevent this global issue, the waste effluent water from various sources is properly treated before adding into the land surface and nearest sources of water bodies.

Although in some parts of the developing countries the contaminated water is treated properly before addition to soil and water, but in most of the countries the discharged polluted water is commonly used for irrigation in agriculture, aquaculture, constructional activities and some cases such as for rejuvenation of groundwater and disturbs the composition, structure with the change in physical and chemical properties of soil. Many of the large cities and municipalities are facing trouble for proper disposal of sewage water or wastewater. The major sources of wastewater are liquid wastewater from residential areas, hospitals, institutes, commercial buildings, mining operation, factories and industries. All these are partially directly added to the soil and another part mixed with lakes, rivers, canals, streams and a few % in the ground water. These water resources after use of agriculture, fisheries and some other activities again mixed with soil leads to the change in soil properties. These wastewater effluents can also be used in many ways in an eco-friendly way, including irrigation because it is enriched with nutrient content. The wastewater effluents from domestic, mining, industrial effluents may undergo leaching on addition with rain water or surface runoffs. Most of the rivers in India are heavily polluted due to industrial, mining and municipal sewage water. Due to extensive increase in the agricultural, urban and industrial sector, environmental pollution is continuously increasing. This unfavorable activity of human beings is not only the cause of destruction of natural resources, but also the cause of ever-increasing demand of healthy and clean resource of soil, water. Figure 2 shows the different physical and chemical changes due to water pollution.

Figure 2. Change in physical and chemical properties of soil due to contaminated water

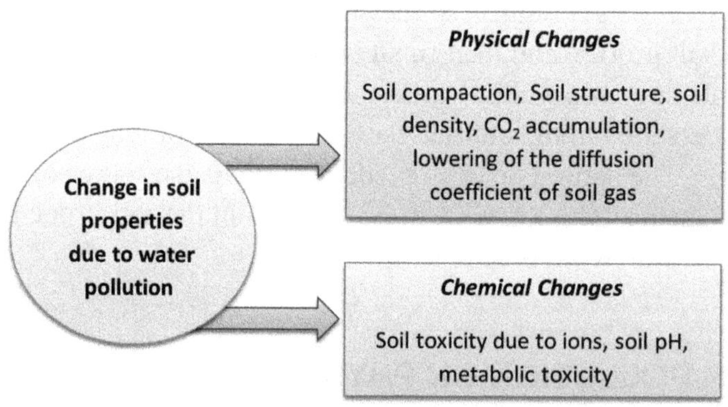

Only 60% of the industrial and mining wastewater is treated and the remaining is added directly in the surface water, soil and causes the change in chemical composition, properties of the soil which proves dangerous to soil ecology. Usually the waste water effluents contain organic, inorganic materials along with toxic metallic contaminants and pathogens. The soil consists of organic materials, including plant materials, paper, feces, ceramic materials, various kinds of salts along with toxic chemicals (cleaners, dyes, detergents pesticides, soap solution/surfactants) and millions of bacteria, protozoa, virus, fungi and some other soil microorganisms. The macro-particles are mostly accommodated only at the surface of the soil, whereas the micro-particles are found in the subsurface layers of the soil. The nutrients leach from the topmost layer of the soil and distributed in the inner layers, then reacts or combines with the nutrients already existing in the soil of that layer and this combination or reaction leads to the formation of hard pans which change the structure of the soil. The change in the chemical structure is the cause of the decrease in fertility and porosity, acidification or salinization (Dunne et al., 2011; Keller & Fox, 2019; Khaledian et al., 2012).

Impact of Waste Water or Sewage Water Pollution on Soil pH

The Soil pH directly influences the plant growth because it can change the availability of all the nutrients present in the soil. The soil pH decreases with increase in salinity. The decrease in pH of the soil is possibly due to the addition of acidic components in the sewage water or wastewater and converts it into acidic matter leading to the decrease in soil pH. Normally the best choice of using wastewater is irrigation and by the addition of wastewater or sewage water in agricultural field, there is a drastic change in pH of the soil. The waste water discharged from pharmaceutical plants, packaging plants, textile industries, fertilizer plants, dyeing and textile industries, etc. contain a noticeable amount of Ca^{2+} and Mg^{2+} and increases the pH of the soil, if it is added or irrigated. On the other hand the presence of some acidic substances along with the nitrification of NH_3 and oxidation of organic matter is the cause of retardation in the pH of the soil, if irrigated in the agricultural land. The use of sewage water in irrigation of distillery units and urban sewage is also the cause of acidic nature of the soil. The generation of organic acids by anaerobic degradation of organic matter is due to the use of wastewater in agriculture and gardening, which is one of the major causes of decrease in soil pH (Khalid et al., 2018; Jaramillo & Restrepo, 2017).

Impact of Waste Water on Concentration of Metallic Contaminates in Soil

The wastewater and sewage water from different industries used for agriculture and gardening contains different categories of heavy metals such as Cu, Mn, Zn, Fe and micronutrients. The presence of the micronutrients significantly retards the growth of the plants in the soil exposed to the industrial effluents or sewage water for prolonged time periods (5-10 years). The concentration of Pb & Cd significantly enhances, if the application of wastewater or sewage water is for 10 years. If waste water effluents or sewage water is continually added to the agricultural land for more than 20 years, the heavy metal accumulation in the soil becomes much higher than the permissible limit and the presence of Cd, Ni and Pb are reflected in the crops or vegetables cultivated in these agricultural fields and become toxic to human health. The percentage contribution of Cu, Pb, Ni and Cr in fruit was observed to be higher than the green vegetables, but the reverse is true in case of heavy metals Zn and Cd. The presence of excessive amounts of toxic metals in soils can inhibit shoot and root growth, nutrient consumption, homeostasis and found that the heavy metals accumulated regularly in agriculturally significant crops.

The addition of sewage sludge with the inorganic fertilizer is the cause of accumulation of most of the micronutrients like Fe, Zn, Mn, Cu, etc. and crops; vegetables harvested from these agricultural lands contains a higher percentage of the micronutrients leading to the health hazard effect to both human beings and animals. In spite of these, the addition of sewage water is also the cause of the accumulation of Pb; therefore the sewage water should be disposed in a safe manner without exposing to the agricultural fields. If sewage water is used for irrigation purpose, it affects the physicochemical properties of soil along with the accumulation of many heavy metals (Pb, Cr, Cd). There is a positive impact of the sewage sludge (increase in concentration of micronutrients such as Zn, Fe, Cu and Ni) observed, which is the cause of accelerating the electrical conductivity of the treated soil.

The change in metal concentration is directly proportional to the application of sewage sludge. The application of barueri sewage sludge is the cause of the increase in the metal concentration, such as Cu, Ni, Mn, Pb, and Zn, which is extracted through diethylenetriaminepentaacetic acid (DTPA) or Pentetic acid solution. It was found that the total % of heavy metals present in the soil is more in the land fertilized by the barueri sewage sludge or domestic and industrial wastewater effluents as compared to the soil fertilized by Franca sludge or domestic waste. It was experimented that the solution of Mehlich-1 and DTPA are quite useful in estimating the availability of Zn in the corn plants by analyzing grains and leaves. The various toxic effects of heavy metals are represented in Figure 3.

The concentration of metals such as Cu, Ni, Mn, Pb, and Zn present in the grains and leaves of the corn can be estimated from the extracts got from nitric-perchloric digestion. The concentration of metals like Mn, Zn, Fe, and Cu in the leaves and grains of corn changes significantly with the change in dose of application of sewage sludge (Gola et al., 2016; Kinuthia et al., 2020).

Figure 3. Toxic effects of heavy metals on human health

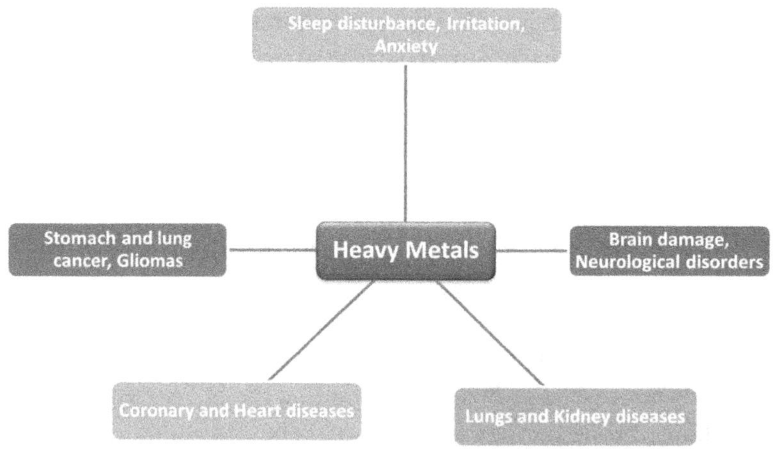

Impact of Wastewater Pollution on the Electrical Conductivity of Soil

It was found that the soils irrigated with contaminated water possess electrical conductivity of about 0.2 dSm^{-1} at a depth of around 61 cm, which is comparatively higher than the soil irrigated with regular water. The soluble salt of metals present in the municipal wastewater or sewage water is accumulated in the upper layer of the soil up to 60cm depth due to leaching effect. This is the cause of comparatively higher electrical conductivity (EC) in the soils irrigated with waste contaminated water. The soil exposed to the wastewater or sewage water traps the nutrients at its surface and sub-layer, which increases the EC of the soil to a noticeable level. This increase in EC is the cause of change in the growth rate of the plants and the percentage yield of the agricultural products. The soil irrigated with wastewater possesses the EC of 893 to 943 μS/cm and the soil irrigated with normal ground water possesses the EC of 600 to 705 μS/cm. The EC value of soil is the measure of soil salinity and is considered as a significant indicator to the agricultural land irrigated with wastewater.

The soil EC indirectly shows a very significant correlation between several kinds of soil with their chemical and physical properties. Hence the EC of soil or any material is the capability of the material to transmit electrical current, which is due to the presence of ions in the soil. Now a day farmers are using the technologies of EC to know about the soil variability in the agricultural land efficiently, cost effectively and in a sustainable manner. The mapping of EC is directly related to the maps of the percentage of crop yield. Especially the land having contrasting drainage categories, the correlations between the percentage of yield and electrical conductivity not significantly varied depending on whether the year is a dry or wet year. Sand possesses a low electrical conductivity, whereas clay possesses high EC and the EC of the soil correlates very strongly with the soil texture and size of the particles. The soils prone to excessive water or drought exhibit the variations in soil texture and can be defined by using soil EC. The water holding capability of soil is extensively related to percentage of yield of the crops.

The EC of the soil can explain the distinction between the concentration of the organic matter and the cation exchange capacity. The major problem for estimation of the EC of the soil is a variation of nutrient availability and soil texture. The soil management of partitioning zones on the basis of EC is a unique strategy, which is now more popular throughout the world (Chaoua et al., 2019; Shakir et al., 2017).

Impact of Wastewater Pollution on the Soil Organic Carbon (SOC)

The soil carbon present in soil can mainly be divided into two categories; soil inorganic carbon and soil organic carbon. The soil inorganic carbon generally contains the carbon in the form of minerals, which is obtained either from reaction of soil minerals with tropospheric CO_2 or weathering of rocks (original material). In the desert area or dry climate, the carbonate minerals are the leading form of soil carbon. The soil organic carbon existing in the soil as soil organic matter includes humus matter, available carbon from fresh plant residues, comparatively inert carbon in substances derived from plant residues and charcoal.

The SOC can be obtained from the biomass of dead biotic materials and from the living soil biota. Both of these comprise food web in the soil, where the living biota sustains through the biotic material constituents. There are different kinds of soil biota found such as earthworms, protozoa, nematodes, bacteria, fungi, and various forms of arthropods. Detritus organisms obtained from plant senescence are considered as the primary source of SOC. The biomass products of plants having the cell walls containing a higher percentage of lignin and cellulose are disintegrating and the carbon which is not breathed, but retained on the soil surface as humus. Starch and cellulose are easily degraded in a short residence time, therefore higher persistent forms of organic carbon such as humus, lignin, and organic matter captured in the form of charcoal and soil aggregates with more residence times.

Normally, soil carbons are more concentrated in the topmost layer of the soil which ranges from 0.5% to 3.0% in most of the upland soils. Less than 0.5% organic carbon is only found in the soil of deserts. The soil having 12 - 18% organic carbon is normally termed as organic soil. The soil having a higher percentage of organic carbon promotes wetland ecology, fire ecology, flood deposition, and human activity. The soil organic carbon by 5 - 50% is obtained from char, whereas the level more than 50% is found in mollisol, terra preta soils and chernozem.

Root exudates are also an important source of soil carbon. Almost 5 - 20% of the entire plant carbon fixed during the process of photosynthesis is normally supplied in the form of root exudates in favor of rhizospheric mutualistic biota. Hence the population of the microbial community is specifically higher in the rhizosphere relative to the nearest bulk soil. The soil organic carbon is considered as a quantifiable constituent of soil organic matter. The organic matter present in the soil is in the range of only 2–10%

of most of the soil mass and is found to be playing a significant role in the chemical, biological and physical function of the soil of agricultural lands. The contribution of organic matter in the soil is the cause of retention of soil nutrients and turnover, moisture retention, soil structure, moisture availability, carbon sequestration, degradation of soil contaminants and soil resilience. Confiscating the carbon in SOC is found to be a path for mitigating the climate change by decreasing the atmospheric CO_2. The increase in less percentage of SOC even in a large area of pastoral lands and agricultural land will drastically decrease the atmospheric CO_2. The SOC is less than 0.3% in desert soils and the highest 14% of intensive dairy soils. Maximum percentage of organic matter is found on the nearest part of the soil surface and was estimated to be almost about 60% of organic matter present within 30cm depth from the surface of the soil. Hence the SOC signifies the quantity and quality of nutrients consumed and stored by plants. Carbon is a unique element, which is the part of every living organism and everything derived from it contains carbon as a major constituent. Hence all kinds of pollution (air, water, soil) are due to the addition of some extra carbon. The different kinds of wastewater are enriched with various kinds of organic matter. Hence the use of wastewater in irrigation or agriculture increases the SOC in a noticeable quantity than the soil irrigated with usual water.

If the industrial effluents are frequently added to the soil surface, it becomes enriched with SOC. The addition of waste effluents of paper mills leads to the increase in pH, organic matter, EC, percentage of total N and similarly by addition of biomedical and pharmaceutical effluents is the cause of the increase in SOC in the range of 1 to 1.08%. Hence, it was concluded that the addition of wastewater effluents in the agricultural lands continuously for a longer time period is the cause of accumulation of appreciable amount of organic carbon in the soil surface, which is although some way helpful for better agricultural productivity. In some cases the excessive accumulation of organic material from sewage water for irrigation purposes is the cause of creating an anaerobic environment, which subsequently decreases the rate of decomposition, particularly organic carbon because excessive amount of organic matter will get deposited in the soil layers (Elcossy et al., 2020; Friedel et al., 2000; Lal, 2016).

Impact of Wastewater Pollution on Soil Nutrients

Healthy soil is the key factor for the production of a higher percentage of yield and good quality of food grains. The composition of soil nutrients plays a vital role for determination of soil health or quality of the soil. A good quality of the soil or healthy soil must have to contain all the required elements that are necessary for healthy growth of plants, however, additional inorganic or organic fertilizers are added for higher percentage of yield, because these are providing extra nutrient for growth of plants. The nutrients supplied by the soil to the plants are called as *mineral nutrients*. The non-mineral nutrients, including carbon (C), oxygen (O) and hydrogen (H) come from water and air during the process of photosynthesis. The mineral nutrients in the soil are divided into two groups such as micronutrients and macronutrients. The macronutrients are further divided into two classes such as intermediate nutrients and primary nutrients. The primary nutrients are necessary for the plants in large quantities and included nitrogen (N), potassium (K), phosphorus (P) and commonly termed as NPK. The necessity of intermediate nutrients are comparatively less than primary nutrients and includes calcium (Ca) sulphur (S) and magnesium (Mg). The requirement of micronutrients to the plants is comparatively less. The micronutrients include Fe, B, Mn, Cu, Zn, Mo, Ni and Cl. Both the groups of nutrients are equally important for plant growth and soil fertility.

The different sources of nutrients in soil include the decomposition of organic matter, biological fixation of nitrogen, precipitation, application of fertilizers, weathering of minerals and rocks. There are a number of factors influencing the availability of soil nutrients which include erosion of soil, leaching, soil pH, denitrification, volatilization, immobilization, nitrogen and consumption of crop nutrient. Nitrogen promotes foliage growth, increases the quality of the leaf and impacts green colour to the plants, due to stimulating the production of chlorophyll. Phosphorous promotes the growth of plant roots, flower production and assist the plants surviving in harsh climatic condition. Potassium is the cause of retaining water in the plants, protects the plants from insects, diseases, helps in early growth and strengthening plants. Magnesium impacts green coloration to the plants. Sulphur protects the plants from diseases, helps in healthy growth of seeds, helps in producing amino acids, enzymes proteins and vitamins. Calcium helps in growth of the cell walls of the plants, assist in cell metabolism and nitrate consumption. Fe helps the formation of chlorophyll. Mg acts as an activator for the enzymes and promotes Fe for chlorophyll synthesis. Zn is necessary for the growth of plants and plant roots. Boron promotes fertilization, pollination and controls the metabolism of carbohydrates in plants. Cu is essential for the growth of plant roots and process of photosynthesis, whereas Chlorine supports the growth of plant roots. Mo is necessary for nitrogen utilization by the plants, whereas Ni is necessary for feasible seed and the whole the life cycle of the plants.

During cultivation some areas of soils are frequently irrigated with wastewater of industries or urban sewage effluents throughout the world and the concentration of micronutrients in the soil is enhanced because of the addition of extra metals from sewage or wastewater. The agricultural land irrigated with sewage wastewater becomes enriched with extra nutrients, including Pb, Cd, Ni, Zn, and Fe in a noticeable concentration at the top layer of the soil than the ordinary water. The volcanic soils are always enriched with Cu & Zn. The concentration of Zn, Cu, Ni, Fe is continuously increased to whooping 208%, 170%, 63%, 170%, respectively, if sewage wastewater is used in irrigation continuously for 20 years. The contamination of green vegetables with Cu, Ni, Cd, Zn, Pb, and Cr are generally observed, if the waste sewage water is used for cultivation of vegetables.

The treatment of domestic wastewater affects the soil properties in various ways such as;

- Sludge formation of the upper most layer of the soil.
- Retardation of the fertility of the soil.
- Decrease in soil pH or cause of soil acidification.
- Nitrification of the soil.
- Enhancement of soil salinity.

The wastewater effluents from diary plants, textile and paper industries is the cause of adverse impact on the properties of the soil, if irrigated for cultivation purposes and frequent irrigation is the cause of the decrease in soil fertility. The wastewater used in irrigation also partly leached inside the soil and degrades the quality of ground water. Due to the presence of higher concentration of salts, it affects the growth of vegetation. Over irrigation of wastewater or sewage water leads to the rise in underlying water table and cause deterioration of the soil surface and vegetation. There are a number of factors that affect the availability of micronutrients in the soil and increase the disease resistance properties, which are reflected in Figure 4 (Lal, 2016; Biswas et al., 2018; Olowoyo, & Mugivhisa, 2019).

Figure 4. Different factors affecting the improvement of plant nutrients and protection of diseases

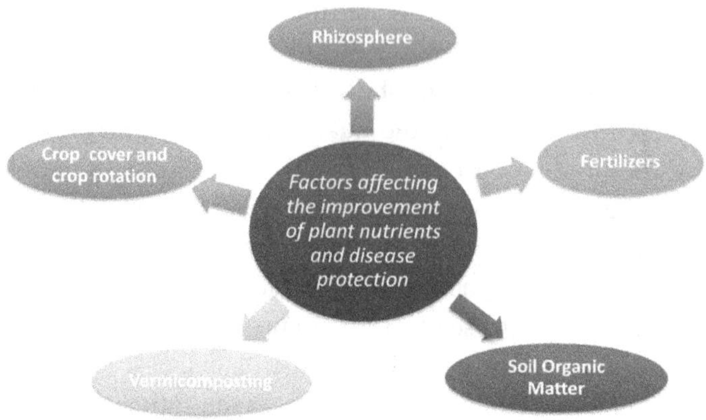

Carbon Sequestration and Soil Properties

Carbon sequestration defines long-term storage of CO_2 or any other states of carbon to either accept global warming or alleviate and avoid a drastic change in climatic condition. It is the process of slow accumulation of the several greenhouse gases in the troposphere and marine ecosystem, due to combustion of fossil fuels. The CO_2 is captured naturally from the troposphere through physical, biological and chemical processes or some artificial processes may be used for capturing CO_2. Soil organic carbon is treated as the major carbon pool on the surface of the earth and serves as either source or a sink of tropospheric CO_2. The SOC also impacts the soil fertility and causes better yield of crops and vegetables. But, the knowledge of using the techniques of cropping on the long-term performance of soil carbon is rare.

In many studies it was found that continuous cropping in a particular agricultural land is the cause of retardation of the SOC. In the starting year the decrease is more rapid, whereas it decreases slowly in the subsequent years and equilibrium is established after 30 to 50 years. The rigorous corn cropping in a particular agricultural field for about 35 years on temperate soils exhibits a 50% retardation of SOC. Carbon is the major source of all biotic communities and is considered as the major building block for the existence of life on the surface of the earth. In nature, carbon is found in many forms. The predominant form of carbon in natural environment is the biomass of plants, SOC, atmospheric CO_2 and dissolved in seawater. The carbon sequestration is mainly the long term storage of carbon in soil. The major percentage of carbon is stored by the oceans and the soil contains almost 75% of the carbon pool on the land, which is almost three times more as compared to the quantity stored in animals and living plants. Hence the soil plays a vital role in maintaining equilibrium in the global carbon cycle.

At the time of photosynthesis, plants consume carbon and again returned back to the atmosphere during the process of respiration. The carbon stored in the plant tissues is taken by animals and human beings and later on added to the soil in the form of litter, after the death of plants and animals. The primary method of storage of carbon is in the form of SOM, which is a complex form of carbon compounds containing the decomposed products of animals, plants, microbial community (protozoa, fungi, nematodes, and bacteria) and the soil minerals enriched with carbon.

Carbon can be deposited in soils for a longer period of time or rapidly released back into the atmosphere. The environmental condition, soil texture, natural vegetation, and drainage are influenced by the length and quantity of carbon stored. Removal of CO_2 from the troposphere is the major advantage of increasing carbon storage in the soils. The improvement in the quality of the soil by adopting different steps such as minimal loss of nutrients, reduction in erosion of the soil, increase in water holding capacity and cause the increase in the percentage of crop production due to increase in the quantity of carbon stored in the soil of the agricultural lands. The proper techniques of management offer a net carbon sink in soils as follows;

- **Conservation tillage**: It eliminates or reduces manipulation of the soil for the production of crops and reduces SOC. Proper procedure of these normally improves the efficiency of water utilization, increases carbon at the topmost layer of soil, prevention of soil erosion. The carbon sequestration is the cause of a decrease in the quantity of fossil fuel consumption.
- **Cover cropping**: It is the process of utilization of crops, including small grains and clover for soil improvement in soil quality and soil protection during the regular period of crop cultivation. The cover crops increases carbon sequestration by improving the soil structure, and adding some organic substances to the soil.
- **Crop rotation**: The crop rotation is the order of crop production or cultivation frequently in recurring succession within the same agricultural land. The diversity of the natural ecosystems is more or less related with the mono-cropping agriculture. The rotation of crops leads to the increase in the percentage of SOC and soil organic matter however the effectiveness of the rotation of crops mainly depends upon the time or month of rotation and kinds of crops (Arunrat et al., 2020; Gmach et al., 2020; Dignac et al., 2017).

Effect of Pollution due to Agricultural Activity on Physical and Chemical Properties of Soil

Agricultural activity is developed from the beginning of the human civilization and the agricultural activities are modified, developed from time to time according to the necessity. Till now many hybrid and industrialized seeds are developed, various techniques are investigated and many management techniques are used for qualitative and quantitative agriculture. The advanced technique of agricultural activity, although to a major extent fulfilling the whole demand of food in our globe, simultaneously causes the addition of different toxic fertilizer, pesticides, insecticides in the soil, leading to the degradation of soil health, which changes the physical, biological and chemical properties of the soil. The natural ecosystem is a proper balance or equilibrium in between the fertility and plant growth. Because of extensive human need, natural ecosystem is gradually changing due to its negative impact on anthropogenic activities especially in soil. For increasing the percentage of yield the ecological function of the living biota gradually made trouble with the excessive use of chemical fertilizers, insecticides, and pesticides. These chemicals in water soluble form are added to the soil surface as agricultural runoff. The nutrients present in the applied fertilizers are either added to the nearest water bodies or may leach inside the soil and mixed with the groundwater.

The land use and management impacts the supply of nutrients into the ecosystem, therefore the nutrient grade is comparatively better in planed agricultural lands than orchards. The land use and agricultural activity are the cause of the retardation of SOC, due to the decrease in the input of SOC, mineralization change and soil redistribution. Therefore, starting from the implementation of agriculture upto the tilling and clearing the agricultural land, many potential components of the soil like soil structure, soil density, soil fertility and health are appreciably changed. Since the top surface layer of the soil is in direct exposure to air and sunlight, therefore by tilling actions and the passage of time soil got tossed with the major nutrients. Due to periodic farming of particular crops the soil fertility declines, which is again the cause of continuous decrease in nutrients and organic matter in the soil. Hence, to get continuous quality and quantity of crop production, some extra nutrients in the form of chemical fertilizer must have to be added. In the agricultural methods like slash-and-burn system, a patch of the forest land is converted into open land by burning and the desired crops are cultivated continuously until the land gets turned into a barren land or empty of nutrients. After that, the land is left and new land is selected, whereby more and more land becomes barren, which severely affects the soil health globally in a long term basis.

Since the percentage of organic matter present in the soil decreases gradually, subsequently the ion exchange capacity of the soil also simultaneously retarded due to the less binding of the nutrients causing more leaching towards ground water and inside the soil. Hence, the disturbances of the soil due to anthropogenic activities are the cause of affecting macro and microscopic fauna and flora. The agricultural activity disturbs the ecological cycle, which is due to digging, fertilizers, and pesticides. The tillage is the cause of damaging soil fauna and microflora, soil erosion and has either indirect or direct negative impacts on the environment and soil. Zero tillage is the cause of developing water retaining capacity and soil health. The intensive and overuse of agricultural activity retards the concentration of SOM and is the cause of degradation of the soil quality and again related to the depletion of soil microbial population and activity (Bai et al., 2018; Tesfahunegn & Gebru, 2020; Jat et al., 2017).

Impact on Soil Carbon

The extensive use of agricultural land is the cause of modification of natural ecosystems and greatly changes the soil carbon regionally, and also in the whole globe. The extensive reduction of total C of the soil is owing to strengthening of agricultural practice and proper land use management. This can change the croplands to worst unused land due to extensive agricultural practices with reduction in soil organic material and affects other factors such as mineralization and soil erosion. In an analysis, it was found that the change of pastures and native forests to croplands is the cause of reduction in stock soil C in the range of 42 - 59%. The percentage of total C is less in agricultural soils as compared to the natural soils, whereas no appreciable change is observed in other regions. The change in land-use pattern, addition of litter due to new vegetation causes the increase in soil carbon, because the litters are mineralized and decomposed by some soil microorganisms leading to the addition of some extra carbon to the soil.

In the tropical and continental regions the soil carbon is always less than natural soil. This inconsistency may be due to the changes in soil management practices and can trouble regimes such as residue retention, grazing, tillage and the period of change in land use. Hence, it was concluded that agriculture is the cause of strong reduction in soil carbon because tillage may stimulate the generation of organo-mineral and change of soil C with the depth of the soil and might decrease decomposition. There is no appreciable change in soil C, due to cultivation in tropical or arid regions because of stimulating the production and greater rates of turnover. This increase in concentration of carbon in soil is due to application of manure or fertilizer and residues of crops (Morais et al., 2019; Jat et al., 2019).

Impact on Soil Nitrogen

The change of natural soil into arable soil not only losses the stock carbon in the soil in our ecosystem, but also causes a significant loss in stock nitrogen along with gaseous volatilization, soil erosion and hydrological cycle. From meta-analysis, it was found that the average decrease in soil N due to conversion of forest lands to croplands was almost about 15%. There is a substantial difference in percentage of soil N between the natural soil and agricultural soil from the continental, tropical, and temperate regions. Hence the widespread application of chemical Nitrogenous fertilizer in the agricultural land will compensate the loss of Nitrogen due to natural process and maintains equilibrium or balance in the soil N concentration. In addition to that the rotation of crops, addition of organic fertilizers can promote the increase in stock of the Nitrogen. The total soil N is appreciably more in agricultural land in arid regions than natural systems. In arid regions the N stocks in soil and SOC are significantly dependent upon the kinds of soils and a strong interaction between the land use patterns exist. In the arid regions the enhancement of soil N may be due to growth of leguminous crops, because this crop can grow in less water (LI et al., 2014; Singh, 2018; Nendel et al., 2019).

Impact on Soil pH

In the similar climatic condition of continental, tropical, and temperate regions, soils of the agricultural land are found to be more alkaline as compared to natural soils. The addition of lime in the agricultural field is considered as one of the important factors to enhance the pH of the soil. In arid regions a positive impact in soil pH is observed even in acidic soil. The pH of the soil tends to become alkaline and no appreciable difference is observed in between natural soils vs. agricultural soil, which indicates that the agricultural carry out was soil dependent.

Normally, natural ecosystems are found to be more acidic (lower pH) than agricultural systems. This difference in pH value or acidity is due to different mechanisms such as an increase in the generation of some organic acids or production of H_2CO_3 from the autotrophic respiration in the natural soils at higher rates. The decrease in pH value or an increase in acidity in forest lands is due to more acceptances of cations by trees with the resulting change in the percentage of cations adsorbed to the soil exchange complex. The high value of acidity (lower pH) can be created by the change in concentration of cations such as Mg, Ca, K and Na (Goulding, 2016; Bünemann et al., 2006).

Effect of Pollution Due to Solid Waste on Physical and Chemical Properties of Soil

The solid waste generally contains a mixture of garbage waste, twaddle waste, industrial wastes, mining wastes and biomedical wastes. Solid waste pollution is considered as the third most important kind of pollution that impacts immediately to our ecosystem. The dumping of solid wastes in open land or open space is a normal practice in most of the developing countries; where proper procedures are not followed for dumping of solid wastes and these open dumping solid wastes become nastier in rainfall. This is because the toxins from the solid wastes are mixed with rain water and percolated inside the soil surface and finally goes deeper inside the soil and cause of contamination of both soil and ground water.

The average composition of urban solid wastes is 43% moisture, 46.5% organic matter, 3.2% rags, 6% paper wastes, 0.7% broken glass, and 1.1% plastic wastes. The open dumping of solid wastes both directly and indirectly affects the biotic community, particularly plants and the main cause of deterioration of our ecosystem. This process is irreversible in nature. The urban solid waste consists of various kinds of unused materials such as nutrition litters, paper packing materials, waste plastics, broken glass, garden wastes, rejected dress, radioactive waste, medical wastes, and dangerous leftover. The poor inefficiency of waste management is the cause of increasing solid wastes in a rapid rate worldwide (Sharma et al., 2018; Abdel-Shafy, & Mansour, 2018).

Effect on Physical Properties of the Soil Due to Urban Solid Wastes

The lots of solid waste and huge quantities of litters are carelessly dumped in the urban or intense population areas and these solid wastes occupied a huge land area because of inefficient and unplanned management. This is a challenge to the proper waste disposal and management. The accumulation of huge amounts of undiscriminating wastes along the roadside, drainage, gutters, open spaces, etc. is the cause of drastic nuisance towards the ecosystem and cause of destruction of properties and lives.

The household wastes contain a large percentage of non-biodegradable materials like plastics, metals or biodegradable materials. The composted form of organic domestic wastes on an addition to soil surface is the cause of change in chemical, biological and physical properties of the soil and this soil is the cause of increase in plant growth due to excessive nutrients. But, the addition of high carbon content materials such as compost, organic municipal solid waste (MSW), influence of physical, biological, and chemical properties, changes the soil quality and this change depends upon sources of organic wastes, kinds and climatic condition (Ali et al., 2014; Przydatek, & Kanownik, 2019).

Impact of Bulk Density and Soil Moisture Content

There is a high coefficient of estrangement in bulk density among landfill or dumping yard or any non-dumping sites. The dumping site having a high percentage of garbage wastes had 9-13% less bulk density in their second and first horizons, whereas very fewer alterations in bulk density were found in the third horizon of the soil. The bulk density of the soil of the dumping yard is always less than soil in a non-dumped sites because of enrichment of inorganic and organic materials in the MSW. Since non-dumpsite soils have less capacity to retain water, therefore it possesses lowest percentage of moisture and total porosity content. The removal of MSW is the cause of decrease in bulk density. In the upper layer of the soil the decrease in bulk density is due to biological activity at the surface of the soil, whereas at certain depth, soil texture having a high percentage of the gravel particles possesses increase in bulk density of the soil. In addition to bulk density the percentage of moisture in the soil and water penetration capacity is greatly influenced by the addition of compost form the MSW (AL-Shammary et al., 2018; Guo & Liu, 2019).

Impact on Soil Texture

The sites selected for waste dumping should be sandy-loam kind of soil in the second and first horizons, whereas sandy-clay-loam soil in the third horizon having predominant clayey texture. The difference in the soil texture in between non-dump and dumping yard sites are mainly due to the differences in the percentage of clayey soils and sandy clay soil. It was observed that the percentage of sand in the soil reduced slightly with the increase in depth of the soil. The top layer soil of the dumping yards enriched with garbage wastes was the cause of alteration in the soil texture in different horizons. The samples of soil collected at horizons 0-15 and 30-45 were exceptionally sandy in nature with the combination of clay and silt, where the class of soil texture was sandy loam class. The soil of dumping yard possesses higher cation exchange capacity, which leads to the increase in the resilience of the soil and soil fertility (Lin & Cheng, 2016; Colazo & Buschiazzo, 2014).

Effect on Chemical Properties of the Soil Due to Urban Solid Wastes

Macro Minerals

The pH of the soil, concentration of organic matter and some other properties of the soil are influenced by the quantity of leaching. The compost of MSW contains hazardous metals like Cd, Hg and Pd and the compost when added on the agricultural land, a small concentration of these toxic metals can be dangerous to human beings and animals. The deposition of mining wastes, industrial waste, urban wastes, atmospheric deposition, agricultural wastes and chemicals are various sources of toxic heavy metals causing soil contamination. The pH of the soil at the horizon of 0-15cm is decreased from 6.94 to 6.03, because the decomposed form of scrapheap or the landfill sites wastes are the cause of decrease in acidity in the soil of dumping yards. The addition of garbage in the soil in the horizon of 0–15cm increases SOC, organic matter in soil, total N, Ca, K, Mg, Na and befits the cause of percentage of base saturation. The organic matter from the compost of the MSW improves soil structure along with the development of water holding capacity of the soil. There is no change in SO_4^{2-} concentration observed in the contaminated soil. The change in NO_3^- concentration is very less because in anaerobic condition the microorganisms consume O_2 from the existing NO_3^- and the concentration of NO_3^-, whereas in the aerobic condition there is no change in the concentration of NO_3^- observed. Excessive deficiency of O_2 avoids denitrification and nitrification, therefore the concentration of NH_3 is not changed through the soil reactions. The landfill sites enriched with garbage wastes, soil contains higher percentage of organic matter as compared to the soil of non-dump sites. The increase in % of organic matter is approximately 701–743% in the top layer of soil and then rapidly decreases by an average of 731% in the 2nd and 3rd horizon of the soil of dumping yards. In connection with this the concentration of the organic matter in soil at the upper horizon of the non-dumping sites is less in 2nd and 3rd layer as compared to 1st one on an average of 73%. The urban solid waste normally contains refining soil products, organic matter, and microbial action, which is very closely related to fertility of the soil. The biological wastes and food wastes present in MSW is the cause of the increase in nitrogen content, cation exchange capacity, pH, microbial biomass in soil and water holding capacity. There is a high percentage of organic matter and plant nutrients present in the manure sludge. Compost is considered as a common waste, which is the cause of modification of soil properties due to the extra addition of nutrients and increase in the enzyme and microbial activity in soil (Kanmani and Gandhimathi 2013; Srivastava et al., 2016).

Micro Minerals

The concentration of some heavy metals such as Pb, Fe, Cu, and Zn is more in the soil of dumping sites of MSW as compared to non-dump sites. The increase in the heavy metal concentration in the soil is the main cause for the increase in metals consumed by the plants, which is dangerous to health effect of both animals and human beings. The compost of animate litters is the most common source which can change the chemical and physical properties of the soil along with the biological movement and withstands soil health. The animal wastes are the cause of improving the percentage of organic matter in the soil, infiltration capacity, water holding capacity, porosity, hydraulic conductivity and decreases surface crusting, bulk density. The compost of animal wastes is the cause of the increased microbial property; the manures from animal wastes improve enzyme actions, metabolism of microbial communities and thus increase the availability of nutrient that required for effective agriculture. The addition of animal waste on agricultural fields improves the chemical and physical properties of the soil and helps to stimulate the fertility level of the soil and soil microorganisms, which offers cementing polysaccharides that serve as the binding agents for the mineral particles leading to the increase in aggregation along with the modification of soil structure. This improvement in the structure of the soil facilities the penetration of plant roots in the soil with subsequent plant growth. The dumping of solid wastes are the cause of the increase in Pb, Cd, Cu, NH_4^+, SO_4^{2-} and NO_3^- in the soil, but it is almost within the acceptable levels.

The soil contamination, due to heavy metals is owing to both natural and anthropogenic activities. The anthropogenic actions include mining, agriculture and smelting process and increase the level of the heavy metals such as Cd, Co, As, Pd, Cr, and Ni in the soil leading to the hazardous effect to our biological community. Among the 19 different heavy metals, Cd, Pd and Hg have not any valuable use or biological significance, therefore extremely toxic in nature. The metals like Cr, Mg, Cu, Sn, Zn and Ni, if added to the biosphere persist for a longer time period without any biodegradation; hence these metals are toxic to both plants and animals. The heavy metals including Pb, Cd, Co and Hg are highly toxic and are different than other pollutants, because these are not biodegradable and accumulated in the living organisms leading to the cause of various dangerous diseases and health hazard effect even at low concentration. The concentrations of lead in the soil of dumping yards are found to be higher, whereas Cd is detected to be lowest in concentration. The leachate generated from the landfill sites reacts with the different parameters of soil and changes the pH and also affects the percentage organic matter of the soil. The biodegradable wastes such as paper, kitchen waste, agricultural waste, cardboard and wood are not destructive to the soil, whereas the decomposition of these waste materials increases the organic matter of the soil leading to the improvement in porosity and soil structure. The significant amounts of non-biodegradable wastes like as metals, plastics, e-wastes and electronic articles leaches a number of elements, including Fe, Cr, Zn, Mn, Cu, Pb, Hg, and Ar, which are added to soil and surface water directly and percolating inside the soil surface causing ground water pollution (Michaud et al.,2020; Jacoby et al., 2017; Ndukwu et al., 2016).

CONCLUSION

Soil is considered as one of the primary requirements for the existence of life on the surface of the earth. The soil is used by human beings in a variety of ways, including agriculture, mining, industrial activity, construction, etc. The overuse of the soil without proper management and sustainability, not only causes the health hazard effect to the biological community, but also imparts a negative impact to our whole ecosystem and biodiversity. The change in soil properties mainly affects the soil health and agricultural productivity and is the major cause of food scarcity. The irrigation of wastewater effluents and sewage water in the agricultural field is one of a prominent cause of change in soil properties inhibiting plant growth, agricultural productivity. The extensive addition of synthetic fertilizers, insecticides and pesticides decreases the quality of the soil with change in both chemical and physical properties. The open dumping of garbage wastes, MSW, industrial and mining wastes are the cause of leaching some toxic elements into the soil, which is irreversible and significantly changes both chemical and physical properties of the soil. Heavy metals from industrial and mining wastes are one of the significant causes for the change in the properties of soil. Hence it is concluded that mankind is the sole cause of change in properties of the soil and soil quality degradation and threaten the existence of life on the surface of the earth.

REFERENCES

Abdel-Shafy, H. I., & Mansour, M. S. M. (2018). Solid waste issue: Sources, composition, disposal, recycling, and valorization. *Egyptian Journal of Petroleum*, *27*(4), 1275–1290. doi:10.1016/j.ejpe.2018.07.003

Alamgir, M. (2016). The Effects of Soil Properties to the Extent of Soil Contamination with Metals. Environmental Remediation Technologies for Metal-Contaminated Soils. *Chapter*, *1*, 1–19.

Ali, S. M., Pervaiz, A., Afzal, B., Hamid, N., & Yasmin, A. (2014). Open dumping of municipal solid waste and its hazardous impacts on soil and vegetation diversity at waste dumping sites of Islamabad city. *Journal of King Saud University. Science*, *26*(1), 59–65. doi:10.1016/j.jksus.2013.08.003

Arunrat, N., Pumijumnong, N., Sereenonchai, S., & Chareonwong, U. (2020). Factors Controlling Soil Organic Carbon Sequestration of Highland Agricultural Areas in the Mae Chaem Basin, Northern Thailand. *Agronomy (Basel)*, *10*(2), 305–330. doi:10.3390/agronomy10020305

Azuka, C. V., Igu eacute, A. M., Diekkr uuml ger, B., & Igwe, C. A. (2015). Soil survey and soil classification of the Koupendri catchment in Benin, West Africa. *African Journal of Agricultural Research*, *10*(42), 3938–3951. doi:10.5897/AJAR2015.9904

Bai, Z., Caspari, T., Gonzalez, M. R., Batjes, N. H., Mäder, P., Bünemann, E. K., & Tóth, Z. (2018). Effects of agricultural management practices on soil quality: A review of long-term experiments for Europe and China. *Agriculture, Ecosystems & Environment*, *265*, 1–7. doi:10.1016/j.agee.2018.05.028

Balseiro-Romero, M., & Baveye, P. C. (2018). Book Review: Soil Pollution: A Hidden Danger Beneath our Feet. *Frontiers in Environmental Science*, *6*, 130–134. doi:10.3389/fenvs.2018.00130

Bashagaluke, J. B., Logah, V., Opoku, A., Sarkodie-Addo, J., & Quansah, C. (2018). Soil nutrient loss through erosion: Impact of different cropping systems and soil amendments in Ghana. *PLoS One*, *13*(12), 1–17. doi:10.1371/journal.pone.0208250 PMID:30566517

Bhatti, S. S., Kumar, V., Singh, N., Sambyal, V., Singh, J., Katnoria, J. K., & Nagpal, A. K. (2016). Physico-chemical Properties and Heavy Metal Contents of Soils and Kharif Crops of Punjab, India. *Procedia Environmental Sciences*, *35*, 801–808. doi:10.1016/j.proenv.2016.07.096

Biesalski Hans, K., & Jana, T. (2018). Micronutrients in the life cycle: Requirements and sufficient supply. *NFS Journal*, *11*, 1–11. doi:10.1016/j.nfs.2018.03.001

Biswas, B., Qi, F., Biswas, J., Wijayawardena, A., Khan, M., & Naidu, R. (2018). The Fate of Chemical Pollutants with Soil Properties and Processes in the Climate Change Paradigm—A Review. *Soil Systems*, *2*(3), 51–71. doi:10.3390oilsystems2030051

Bünemann, E. K., Bongiorno, G., Bai, Z., Creamer, R. E., De Deyn, G., de Goede, R., & Brussaard, L. (2018). Soil quality – A critical review. *Soil Biology & Biochemistry*, *120*, 105–125. doi:10.1016/j.soilbio.2018.01.030

Bünemann, E. K., Schwenke, G. D., & Van Zwieten, L. (2006). Impact of agricultural inputs on soil organisms—A review. *Australian Journal of Soil Research*, *44*(4), 379. doi:10.1071/SR05125

Carvalho, F. P. (2017). Mining industry and sustainable development: Time for change. *Food and Energy Security*, *6*(2), 61–77. doi:10.1002/fes3.109

Chaoua, S., Boussaa, S., El Gharmali, A., & Boumezzough, A. (2019). Impact of irrigation with wastewater on accumulation of heavy metals in soil and crops in the region of Marrakech in Morocco. *Journal of the Saudi Society of Agricultural Sciences*, *18*(4), 429–436. doi:10.1016/j.jssas.2018.02.003

Colazo, J. C., & Buschiazzo, D. (2014). The Impact of Agriculture on Soil Texture Due to Wind Erosion. *Land Degradation & Development*, *26*(1), 62–70. doi:10.1002/ldr.2297

Demattê, J. A. M., Nanni, M. R., da Silva, A. P., de Melo Filho, J. F., Dos Santos, W. C., & Campos, R. C. (2010). Soil density evaluated by spectral reflectance as an evidence of compaction effects. *International Journal of Remote Sensing*, *31*(2), 403–422. doi:10.1080/01431160902893469

Dignac, M., Derrien, D., Barré, P., Barot, S., Cécillon, L., Chenu, C., Chevallier, T., Freschet, G. T., Garnier, P., Guenet, B., Hedde, M., Klumpp, K., Lashermes, G., Maron, P.-A., Nunan, N., Roumet, C., & Basile-Doelsch, I. (2017). Increasing soil carbon storage: Mechanisms, effects of agricultural practices and proxies A review. *Agronomy for Sustainable Development*, *37*(2), 14–44. doi:10.100713593-017-0421-2

Dunne, E. J., Clark, M. W., Corstanje, R., & Reddy, K. R. (2011). Legacy phosphorus in subtropical wetland soils: Influence of dairy, improved and unimproved pasture land use. *Ecological Engineering*, *37*(10), 1481–1491. doi:10.1016/j.ecoleng.2011.04.003

Dwivedi, P., Mishra, P. K., Mondal, M. K., & Srivastava, N. (2019). Non-biodegradable polymeric waste pyrolysis for energy recovery. *Heliyon*, *5*(8), 1–15. doi:10.1016/j.heliyon.2019.e02198 PMID:32368634

Elcossy, S. A. E., Abbas, M. H. H., Farid, I. M., Beheiry, G. G. S., Abou Yuossef, M. F., Abbas, H. H., Abdelhafez, A. A., & Mohamed, I. (2020). Dynamics of soil organic carbon in Typic Torripsamment soils irrigated with raw effluent sewage water. *Environmental Science and Pollution Research International*, *27*(8), 8188–8198. doi:10.100711356-019-07526-4 PMID:31900766

Emerson, W. W., & McGarry, D. (2003). Organic carbon and soil porosity. *Australian Journal of Soil Research*, *41*(1), 107–118. doi:10.1071/SR01064

Friedel, J. K., Langer, T., Siebe, C., & Stahr, K. (2000). Effects of long-term waste water irrigation on soil organic matter, soil microbial biomass and its activities in central Mexico. *Biology and Fertility of Soils*, *31*(5), 414–421. doi:10.1007003749900188

Gentili, R., Ambrosini, R., Montagnani, C., Caronni, S., & Citterio, S. (2018). Effect of Soil pH on the Growth, Reproductive Investment and Pollen Allergenicity of Ambrosia artemisiifolia L. *Frontiers in Plant Science*, *9*, 1335–1347. doi:10.3389/fpls.2018.01335 PMID:30294333

Gmach, M. R., & Cherubin, M. R. (2020). Processes that influence dissolved organic matter in the soil: A review. *Scientia Agrícola*, *77*(3), 1–10. doi:10.1590/1678-992X-2018-0164

Gola, D., Malik, A., Shaikh, Z. A., & Sreekrishnan, T. R. (2016). Impact of Heavy Metal Containing Wastewater on Agricultural Soil and Produce: Relevance of Biological Treatment. *Environ. Process.*, *3*(4), 1063–1080. doi:10.100740710-016-0176-9

Goulding, K. W. T. (2016). Soil acidification and the importance of liming agricultural soils with particular reference to the United Kingdom. *Soil Use and Management*, *32*(3), 390–399. doi:10.1111um.12270 PMID:27708478

Guéguen, P., & Bard, P.-Y. (2005). Soil-Structure and Soil-Structure-Soil Interaction: Experimental Evidence at the Volvi Test Site. *Journal of Earthquake Engineering*, *9*(5), 657–693. doi:10.1080/13632460509350561

Guo, K., & Liu, X. (2019). Effect of initial soil water content and bulk density on the infiltration and desalination of melting saline ice water in coastal saline soil. *European Journal of Soil Science*, *70*(6), 1249–1266. doi:10.1111/ejss.12816

Håkansson, I., & Lipiec, J. (2000). A review of the usefulness of relative bulk density values in studies of soil structure and compaction. *Soil & Tillage Research*, *53*(2), 71–85. doi:10.1016/S0167-1987(99)00095-1

Hartemink, A. E. (2015). The use of soil classification in journal papers between 1975 and 2014. *Geoderma Regional*, *5*, 127–139. doi:10.1016/j.geodrs.2015.05.002

Heinze, J., Gensch, S., Weber, E., & Joshi, J. (2016). Soil temperature modifies effects of soil biota on plant growth. *Journal of Plant Ecology*, *10*(5), 898–821. doi:10.1093/jpe/rtw097

Huang, R. (2018). Research progress on plant tolerance to soil salinity and alkalinity in sorghum. *Journal of Integrative Agriculture*, *17*(4), 739–746. doi:10.1016/S2095-3119(17)61728-3

Jacoby, R., Peukert, M., Succurro, A., Koprivova, A., & Kopriva, S. (2017). The Role of Soil Microorganisms in Plant Mineral Nutrition—Current Knowledge and Future Directions. *Frontiers in Plant Science*, *8*, 1617–1636. doi:10.3389/fpls.2017.01617 PMID:28974956

Jaramillo, M. F., & Restrepo, I. (2017). Wastewater Reuse in Agriculture: A Review about Its Limitations and Benefits. *Sustainability*, *9*(10), 1734–1753. doi:10.3390u9101734

Jat, H. S., Datta, A., Choudhary, M., Yadav, A. K., Choudhary, V., Sharma, P. C., & McDonald, A. (2019). Effects of tillage, crop establishment and diversification on soil organic carbon, aggregation, aggregate associated carbon and productivity in cereal systems of semi-arid Northwest India. *Soil & Tillage Research*, *190*, 128–138. doi:10.1016/j.still.2019.03.005 PMID:32055081

Jat, H. S., Datta, A., Sharma, P. C., Kumar, V., Yadav, A. K., Choudhary, M., & McDonald, A. (2017). Assessing soil properties and nutrient availability under conservation agriculture practices in a reclaimed sodic soil in cereal-based systems of North-West India. *Archives of Agronomy and Soil Science*, *64*(4), 531–545. doi:10.1080/03650340.2017.1359415 PMID:30363929

Kanmani, S., & Gandhimathi, R. (2013). Assessment of heavy metal contamination in soil due to leachate migration from an open dumping site. *Applied Water Science*, *3*(1), 193–205. doi:10.100713201-012-0072-z

Kapahi, M., & Sachdeva, S. (2019). Bioremediation Options for Heavy Metal Pollution. *Journal of Health & Pollution*, *9*(24), 1–20. doi:10.5696/2156-9614-9.24.191203 PMID:31893164

Keller, A. A., & Fox, J. (2019). Giving credit to reforestation for water quality benefits. *PLoS One*, *14*(6), 1–18. doi:10.1371/journal.pone.0217756 PMID:31163057

Khaledian, Y., Kiani, F., & Ebrahimi, S. (2012). The effect of land use change on soil and water quality in northern Iran. *Journal of Mountain Science*, *9*(6), 798–816. doi:10.100711629-012-2301-1

Khalid, S., Shahid, M., Natasha, Bibi, I., Sarwar, T., Shah, A., & Niazi, N. (2018). A Review of Environmental Contamination and Health Risk Assessment of Wastewater Use for Crop Irrigation with a Focus on Low and High-Income Countries. *International Journal of Environmental Research and Public Health*, *15*(5), 895–931. doi:10.3390/ijerph15050895 PMID:29724015

Kinuthia, G. K., Ngure, V., Beti, D., Lugalia, R., Wangila, A., & Kamau, L. (2020). Levels of heavy metals in wastewater and soil samples from open drainage channels in Nairobi, Kenya: Community health implication. *Scientific Reports*, *10*(1), 8434–8447. doi:10.103841598-020-65359-5 PMID:32439896

Kothawala, D. N., Roehm, C., Blodau, C., & Moore, T. R. (2012). Selective adsorption of dissolved organic matter to mineral soils. *Geoderma*, *189-190*, 334–342. doi:10.1016/j.geoderma.2012.07.001

Lal, R. (2016). Soil health and carbon management. *Food and Energy Security*, *5*(4), 212–222. doi:10.1002/fes3.96

Lech, M., Skutnik, Z., Bajda, M., & Markowska-Lech, K. (2020). Applications of Electrical Resistivity Surveys in Solving Selected Geotechnical and Environmental Problems. *Applied Sciences (Basel, Switzerland)*, *10*(7), 2263–2282. doi:10.3390/app10072263

Li, S., Wang, Z., Miao, Y., & Li, S. (2014). Soil Organic Nitrogen and Its Contribution to Crop Production. *Journal of Integrative Agriculture*, *13*(10), 2061–2080. doi:10.1016/S2095-3119(14)60847-9

Lin, T.-S., & Cheng, F.-Y. (2016). Impact of Soil Moisture Initialization and Soil Texture on Simulated Land–Atmosphere Interaction in Taiwan. *Journal of Hydrometeorology*, *17*(5), 1337–1355. doi:10.1175/JHM-D-15-0024.1

Liu, Z., Dugan, B., Masiello, C. A., & Gonnermann, H. M. (2017). Biochar particle size, shape, and porosity act together to influence soil water properties. *PLoS One*, *12*(6), 1–28. doi:10.1371/journal.pone.0179079 PMID:28598988

Liu, Z., Rong, Q., Zhou, W., & Liang, G. (2017). Effects of inorganic and organic amendment on soil chemical properties, enzyme activities, microbial community and soil quality in yellow clayey soil. *PLoS One*, *12*(3), 1–20. doi:10.1371/journal.pone.0172767 PMID:28263999

Michaud, A. M., Cambier, P., Sappin-Didier, V., Deltreil, V., Mercier, V., Rampon, J.-N., & Houot, S. (2020). Mass balance and long-term soil accumulation of trace elements in arable crop systems amended with urban composts or cattle manure during 17 years. *Aurélia Marcelline Michaud*, *27*(5), 5367–5386. doi:10.100711356-019-07166-8 PMID:31848970

Morais, T. G., Teixeira, R. F. M., & Domingos, T. (2019). Detailed global modelling of soil organic carbon in cropland, grassland and forest soils. *PLoS One*, *14*(9), 1–27. doi:10.1371/journal.pone.0222604 PMID:31536571

Moritsuka, N., Matsuoka, K., Katsura, K., Sano, S., & Yanai, J. (2014). Soil color analysis for statistically estimating total carbon, total nitrogen and active iron contents in Japanese agricultural soils. *Soil Science and Plant Nutrition*, *60*(4), 475–485. doi:10.1080/00380768.2014.906295

Ndukwu, B. N., Osujieke, D. N., Ahukaemere, C. M., & Umeh, M. O. (2016). Micronutrients and Physicochemical Properties Of Soils Affected By Municipal Solid Wastes In Ekwulobia Southeastern Nigeria. *Journal of Multidisciplinary Engineering Science Studies*, *2*(11), 1032–1040.

Nendel, C., Melzer, D., & Thorburn, P. J. (2019). The nitrogen nutrition potential of arable soils. *Scientific Reports*, *9*(1), 5851–5860. doi:10.103841598-019-42274-y PMID:30971710

Oliver, D. P., Bramley, R. G. V., Riches, D., Porter, I., & Edwards, J. (2013). Review: Soil physical and chemical properties as indicators of soil quality in Australian viticulture. *Australian Journal of Grape and Wine Research*, *19*(2), 129–139. doi:10.1111/ajgw.12016

Olorunfemi, I., Fasinmirin, J., & Ojo, A. (2016). Modeling cation exchange capacity and soil water holding capacity from basic soil properties. *Eurasian Journal of Soil Science*, *5*(4), 266–274. doi:10.18393/ejss.2016.4.266-274

Olowoyo, J. O., & Mugivhisa, L. L. (2019). Evidence of uptake of different pollutants in plants harvested from soil treated and fertilized with organic materials as source of soil nutrients from developing countries. *Chemical and Biological Technologies in Agriculture*, *6*(1), 28–39. doi:10.118640538-019-0165-0

Onwuka, B. M. (2016). Effects of soil temperature on Some Soil properties and plant growth. *The Journal of Agricultural Science*, *6*(3), 89–93.

Owens, P. R., & Rutledge, E. M. (2005). Morphology. Encyclopedia of Soils in the Environment, 511–520. doi:10.1016/B0-12-348530-4/00002-3

Paz-Ferreiro, J., Gascó, G., Méndez, A., & Reichman, S. (2018). Soil Pollution and Remediation. *International Journal of Environmental Research and Public Health*, *15*(8), 1657–1660. doi:10.3390/ijerph15081657 PMID:30081583

Al-Shammary, A. A. G., Kouzani, A. Z., Kaynak, A., Khoo, S. Y., Norton, M., & Gates, W. (2018). Soil Bulk Density Estimation Methods: A Review. *Pedosphere*, *28*(4), 581–596. doi:10.10161002-0160(18)60034

Penn, C. J., & Camberato, J. J. (2019). A Critical Review on Soil Chemical Processes that Control How Soil pH Affects Phosphorus Availability to Plants. *Agriculture*, *9*(6), 120–138. doi:10.3390/agriculture9060120

Przydatek, G., & Kanownik, W. (2019). Impact of small municipal solid waste landfill on groundwater quality. *Environmental Monitoring and Assessment*, *191*(3), 169–183. doi:10.100710661-019-7279-5 PMID:30778777

Qaswar, M., Ahmed, W., Jing, H., Hongzhu, F., Xiaojun, S., Xianjun, J., & Zhang, H. (2019). Soil carbon (C), nitrogen (N) and phosphorus (P) stoichiometry drives phosphorus lability in paddy soil under long-term fertilization: A fractionation and path analysis study. *PLoS One*, *14*(6), 1–20. doi:10.1371/journal.pone.0218195 PMID:31233510

Quesada, C. A., Paz, C., Oblitas Mendoza, E., Phillips, O. L., Saiz, G., & Lloyd, J. (2020). Variations in soil chemical and physical properties explain basin-wide Amazon forest soil carbon concentrations. *Soil (Göttingen)*, *6*(1), 53–88. doi:10.5194oil-6-53-2020

Schoonover, J. E., & Crim, J. F. (2015). An Introduction to Soil Concepts and the Role of Soils in Watershed Management. *Journal of Contemporary Water Research & Education*, *154*(1), 21–47. doi:10.1111/j.1936-704X.2015.03186.x

Seema, D., Dahiya, R., Phogat, V. K., & Sheoran, H. S. (2019). Hydraulic Properties and Their Dependence on Physico-chemical Properties of Soils: A Review. *Current Journal of Applied Science and Technology*, *38*(2), 1–7. doi:10.9734/cjast/2019/v38i230355

Shakir, E., Zahraw, Z., & Al-Obaidy, A. H. M. J. (2017). Environmental and health risks associated with reuse of wastewater for irrigation. *Egyptian Journal of Petroleum*, *26*(1), 95–102. doi:10.1016/j.ejpe.2016.01.003

Shannon, T. P., Ahler, S. J., Mathers, A., Ziter, C. D., & Dugan, H. A. (2020). Road salt impact on soil electrical conductivity across an urban landscape. *Journal of Urban Economics*, *6*(1), juaa006. Advance online publication. doi:10.1093/jue/juaa006

Sharma, A., Gupta, A. K., & Ganguly, R. (2018). Impact of open dumping of municipal solid waste on soil properties in mountainous region. *Journal of Rock Mechanics and Geotechnical Engineering*, *10*(4), 725–739. doi:10.1016/j.jrmge.2017.12.009

Shi, R., Hong, Z., Li, J., Jiang, J., Baquy, M. A.-A., Xu, R., & Qian, W. (2017). Mechanisms for Increasing the pH Buffering Capacity of an Acidic Ultisol by Crop Residue-Derived Biochars. *Journal of Agricultural and Food Chemistry*, *65*(37), 8111–8119. doi:10.1021/acs.jafc.7b02266 PMID:28846405

Shrivastava, P., & Kumar, R. (2015). Soil salinity: A serious environmental issue and plant growth promoting bacteria as one of the tools for its alleviation. *Saudi Journal of Biological Sciences*, *22*(2), 123–131. doi:10.1016/j.sjbs.2014.12.001 PMID:25737642

Singh, B. (2018). Are Nitrogen Fertilizers Deleterious to Soil Health? *Agronomy (Basel)*, *8*(4), 48–67. doi:10.3390/agronomy8040048

Singh, V., & Sood, A. K. (2016). Plant Nutrition: A tool for the management of hemipteran insect-pests-A review. *Agricultural Reviews (Karnal)*, *38*(04), 260–270. doi:10.18805/ag.R-1637

Smith, P., Lutfalla, S., Riley, W. J., Torn, M. S., Schmidt, M. W. I., & Soussana, J. F. (2017). The changing faces of soil organic matter research. *European Journal of Soil Science*, *69*(1), 23–30. doi:10.1111/ejss.12500

Sorkau, E., Boch, S., Boeddinghaus, R. S., Bonkowski, M., Fischer, M., Kandeler, E., & Oelmann, Y. (2017). The role of soil chemical properties, land use and plant diversity for microbial phosphorus in forest and grassland soils. *Journal of Plant Nutrition and Soil Science*, *181*(2), 185–197. doi:10.1002/jpln.201700082

Srivastava, V., de Araujo, A. S. F., Vaish, B., Bartelt-Hunt, S., Singh, P., & Singh, R. P. (2016). Biological response of using municipal solid waste compost in agriculture as fertilizer supplement. *Reviews in Environmental Science and Biotechnology*, *15*(4), 677–696. doi:10.100711157-016-9407-9 PMID:32214923

Steffan, J. J., Brevik, E. C., Burgess, L. C., & Cerdà, A. (2017). The effect of soil on human health: An overview. *European Journal of Soil Science*, *69*(1), 159–171. doi:10.1111/ejss.12451 PMID:29430209

Stewart, Z. P., Pierzynski, G. M., Middendorf, B. J., & Vara Prasad, P. V. (2020). Approaches to improve soil fertility in sub-Saharan Africa. *Journal of Experimental Botany*, *71*(2), 632–641. doi:10.1093/jxb/erz446 PMID:31586430

Su, C., LiQin, J., Zhang, W. (2014). A review on heavy metal contamination in the soil worldwide: Situation, impact and remediation techniques. *Environmental Skeptics and Critics*, *3*(2), 24–38.

Tesfahunegn, G. B., & Gebru, T. A. (2020). Variation in soil properties under different cropping and other land-use systems in Dura catchment, Northern Ethiopia. *PLoS One*, *15*(2), 1–27. doi:10.1371/journal.pone.0222476 PMID:32023243

Tully, K. L., & McAskill, C. (2020). Promoting soil health in organically managed systems: A review. *Organic Agriculture*, *10*(3), 339–358. doi:10.100713165-019-00275-1

Wang, Y., Shi, J., Lin, Q., Chen, X., & Chen, Y. (2007). Heavy metal availability and impact on activity of soil microorganisms along a Cu/Zn contamination gradient. *Journal of Environmental Sciences (China)*, *19*(7), 848–853. doi:10.1016/S1001-0742(07)60141-7 PMID:17966873

Wankhade, R., & Ghugal, Y. M. (2016). Study on Soil-Structure Interaction: A Review. *International Journal of Engine Research*, *5*(3), 737–741. doi:10.17950/ijer/v5i3/047

Wei, H., Liu, Y., Xiang, H., Zhang, J., Li, S., & Yang, J. (2019). Soil pH Responses to Simulated Acid Rain Leaching in Three Agricultural Soils. *Sustainability*, *12*(1), 280–292. doi:10.3390u12010280

Yao, Z., Li, J., Xie, H., & Yu, C. (2012). Review on Remediation Technologies of Soil Contaminated by Heavy Metals. *Procedia Environmental Sciences*, *16*, 722–729. doi:10.1016/j.proenv.2012.10.099

Ye, L., Zhao, X., Bao, E., Li, J., Zou, Z., & Cao, K. (2020). Bio-organic fertilizer with reduced rates of chemical fertilization improves soil fertility and enhances tomato yield and quality. *Scientific Reports*, *10*(1), 177. Advance online publication. doi:10.103841598-019-56954-2 PMID:31932626

ADDITIONAL READING

Bhattacharyya, R., Prakash, V., Kundu, S., Srivastva, A. K., & Gupta, H. S. (2009). Soil properties and their relationships with crop productivity after 30 years of different fertilization in the Indian Himalayas. *Archives of Agronomy and Soil Science*, *55*(6), 641–661. doi:10.1080/03650340902718615 de Souza Machado, A. A., Lau, C. W., Kloas, W., Bergmann, J., Bachelier, J. B., Faltin, E., Becker, R., Görlich, A. S., & Rillig, M. C. (2019). Microplastics can change soil properties and affect plant performance. *Environmental Science & Technology*, *53*(10), 6044–6052. doi:10.1021/acs.est.9b01339 PMID:31021077

Gregory, A. S., Ritz, K., McGrath, S. P., Quinton, J. N., Goulding, K. W. T., Jones, R. J. A., & Whitmore, A. P. (2015). A review of the impacts of degradation threats on soil properties in the UK. *Soil Use and Management*, *31*, 1–15. doi:10.1111um.12212 PMID:27667890

Jahn, L. G., Polen, M. J., Jahl, L. G., Brubaker, T. A., Somers, J., & Sullivan, R. C. (2020). Biomass combustion produces ice-active minerals in biomass-burning aerosol and bottom ash. *Proceedings of the National Academy of Sciences of the United States of America*, *117*(36), 21928–21937. doi:10.1073/pnas.1922128117 PMID:32839314

Kalbitz, K., Solinger, S., Park, J.-H., Michalzik, B., & Matzner, E. (2000). Controls on the Dynamics of Dissolved Organic Matter in Soils: A Review. *Soil Science*, *165*(4), 277–304. doi:10.1097/00010694-200004000-00001

Kim, M. S., Min, H. G., Lee, S. H., & Kim, J. G. (2016). The effects of various amendments on trace element stabilization in acidic, neutral and alkali soil with similar pollution index. *Plos ONE, 11*(11), 1-12. doi:.pone.0166335 doi:10.1371/journal

Savci, S. (2012). An Agricultural Pollutant: Chemical Fertilizer. *International Journal of Environmental Sciences and Development*, *3*(1), 77–81.

Xiaoyan, L., Shuwen, Z., Zongming, W., & Huilin, Z. (2004). Spatial variability and pattern analysis of soil properties in Dehui city, Jilin province. *Journal of Geographical Sciences*, *14*(4), 503–511. doi:10.1007/BF02837495

KEY TERMS AND DEFINITIONS

Bioavailability: On the basis of soil and environmental sciences, bioavailability may be defined as the quantity of compound or element that can be available to an organism for adsorption or consumption across its cellular membrane.

Colloidal Particles: The microscopic form of solid particles that are suspended in a fluid are termed as colloidal particles. The size of the colloidal particles ranges from 1 nanometer to 1 micrometer.

Effluent: According to the US Environmental protection agency, effluent may be defined as the untreated or treated wastewater that has capability of flow and flows from the treatment plant, industrial outfall, or sewer system.

Enzyme: The substance that functions as a catalyst in living organisms and controls the rate of growth at which the chemical reaction occurs without the use of itself in the chemical reaction and altering the process.

Heavy Metal: Heavy metals are the metals having relatively high densities, atomic numbers or atomic weights.

Leaching: Leaching is primarily defined as one of the methods of carrying small particles and soluble substances through rock or soil.

Suspension: Suspension may be defined as a heterogeneous mixture in which the solute particles cannot be dissolved, but suspended and floated freely in any direction throughout the medium.

Water Holding Capacity: The term water holding capacity is defined as the quantity of water that a particular soil can hold for crop use. Larger is the surface area of the soil, easier is to hold the water and thereby having higher water holding capacity.

Chapter 2
Bioremediation and Phytoremediation:
The Remedies for Xenobiotics

Leena Merlin Biju
Kumararani Meena Muthiah College of Arts and Science, University of Madras, Chennai, India

Veena Gayathri Krishnaswamy
https://orcid.org/0000-0002-3012-8561
Stella Maris College (Autonomous), University of Madras Chennai, India

ABSTRACT

Industrialization led to the release of synthetic and toxic compounds. Partial or improper treatment increases environmental pollution. Conventional methods possess more disadvantages, such as increased duration of degradation and release of secondary pollutants. The drawbacks paved the way for the significant bioremediation perspective. The ubiquitous nature of microbes enables it to utilize toxic compounds, which attracted the focus of treatment towards the biological and eco-friendly methods. The recent decade has shown interest in the application of indigenous microbes in the polluted environment. Apart from the microbial application, phytoremediation is an emerging tool for treating soil contaminated with hazardous pollutants. Technological advancement in biotechnology ensures a safe and healthy environment for a better future.

INTRODUCTION

Rapid urbanization has led to the development of transport facilities and industries, destroying productive land ecosystems (Nawarot et al., 2006). Anthropogenic activities contribute the majority to pollution. The massive concern is soil contamination and emerging it as a "universal sink" (Doran, 1996). The recalcitrant substances resistant to degradation are known as xenobiotic compounds. The term xenobiotic is derived from the Greek word '*xenos*', which means foreign or strange, and '*bios*,' which means life. Chemicals beyond its threshold cause pollution and harm to humanity (Embrandiri et al., 2016).

DOI: 10.4018/978-1-7998-7062-3.ch002

The accumulation of these compounds will be ascending as it passes through various levels of the food chain (Dubey et al., 2014). Pollutants are merely classified as biodegradable, partially degradable, and non-biodegradable compounds based on the degradable nature. The toxicity and concentration of these substances in the environment alter the ecosystem (Mishra et al., 2019). Bharadwaj (2018) has reported bacteria and fungi in converting problematic pollutants into simpler non-toxic forms. Junghare et al. (2019) have reported the role of *Syntrophorhabdus aromaticivorans* in the anaerobic remediation of isophthalate, a xenobiotic compound. Bioremediation using microbes employs diverse metabolic pathways for generating enzymes (Sharma et al., 2018; Dangi et al., 2019). The current scenario focus on overcoming the time consumption, minimal removal of hazardous toxicants, loss of ecological balance, and off-odours generated in the environment during the conventional treatment methods (Barghava et al. 2019; Dangi et al. 2019; Kumar and Femina Carolin, 2019).

Havugimana et al. (2015) reported on the wide range of organic pollutants such as PolyChlorinated Biphenyls (PCBs), Polybrominated biphenyls, PolyChlorinated DibenzoFurans (PCDFs), Polycyclic Aromatic Hydrocarbons (PAHs), organophosphorus, carbamate insecticides/ pesticides, herbicides, organic fuels and pharmaceuticals, and their metabolites. Pesticides applied to increase the yield have turned out to be a bane disturbing flora and fauna of the natural habitat (Nishimoto, 2019; Tuomisto et al., 2017). The application has created a negative impact in neurological, reproductive, and oncogenic effect on children (Cognitive development) and pregnant women (Fetal death/anomalies) (Ward et al., 2006; Bouchard et al., 2011; Carmichael et al., 2016; Rahbar et al., 2016).

Studies focus on green remediation of soil pollutants using microbes like bacteria, fungi, algae, and plants.Bioremediation has gained a lot of attention for remediating different contaminants, especially for compounds like volatile organic compounds - benzene, toluene, ethylbenzene, xylene (BTEX) compounds, phenolic compounds, PAHs, petroleum hydrocarbons, nitroaromatic compounds, metals, complex (high molecular weight) PAHs, and chlorinated hydrocarbons- (Kumar and Femina Carolin, 2019). Based on the nature of pollutants, various methods for removing xenobiotics are practised, like *in-situ* and *ex-situ* bioremediation processes. *In-Situ* methods are cost-effective and economical compared with Ex-Situ procedures (Azubuike et al., 2016). Studies are focused on contemporary heterogeneous bioremediation approaches to accelerate the degradation rate and enable it to be cost-effective (Cassidy et al., 2015; Garcı´a-Delgado et al., 2015; Martı´nez-Pascual et al. 2015).

The impact of recalcitrant compounds on the environment and living systems are explored in the past and recent decades. The present chapter attempts to review the varied bioremediation and phytoremediation strategies employed from past decades to treat various xenobiotic compounds posing an offensive threat to the ecosystem. It gives an overview of the different sources of xenobiotic pollutants; the hazardous effects generated due to the presence of these recalcitrant compounds, the *in-situ* and *ex-situ* micro remediation techniques, the eco-friendly phytoremediation approaches and integrated degradation approaches using plant and microbes.

WASTES GENERATED FROM DIFFERENT SOURCES

Pollutants vary in their toxicity based on their chemical nature. Surface water and groundwater are the victims of these pollutants but not the cause of pollution in the environment. The organic and inorganic pollutants resemble the indispensable compounds required for life (Esumang 2013). They enter into the environment from different sources. Amid various organic pollutants, a few are indicated as Persistent

Organic Pollutants (POPs) and challenging due to its recalcitrant nature. The fate of soil and indigenous microbes has been altered due to the impact of industries with a wide range of applications (Reineke and Knackmuss 1988). The source of pollutants from different industrial sources is shown in Figure 1.

Figure 1. Source of Pollutants

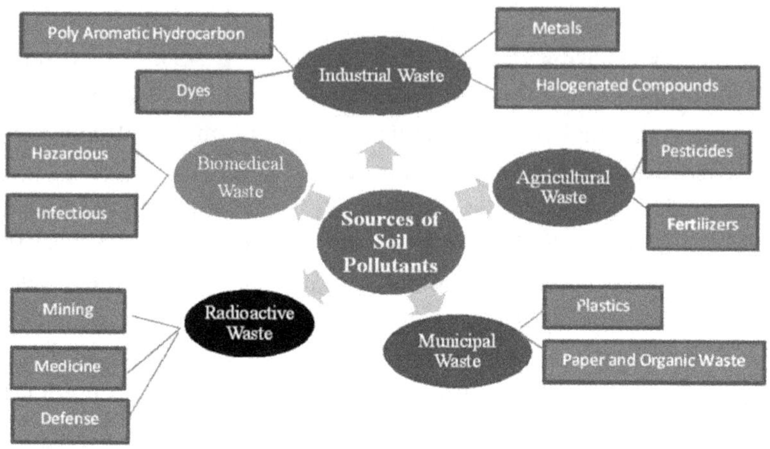

Industrial Waste

Imprudent disposal of waste from these industries is a threat to the living beings and the natural ecosystem. The waste products generated can be in gaseous, liquid, or solid form (Havugimana et al.,2015). Studies suggest that rivers near textile industries are more prone to heavy metal pollution by discharged sewage sludge (Kurnia et al., 2000). The accumulation of metals like Cu, Na, K, Ca, Mn, Se, Fe, and Zn beyond the threshold damages the aquatic ecosystem (Chakraborty, 2019).

Agricultural Waste

The critical source of soil pollution is the immense and inappropriate use of fertilizers and pesticides. Agricultural waste chiefly comprises organic or natural pesticides and animal wastes (Havugimana et al., 2015). Kasno et al. (2000) reported that lead and cadmium present in soils would have been originated due to phosphate fertilizer application. Based on lead and cadmium levels in rice, researchers categorized the ground as high, medium and unpolluted polluted soils. The toxicity of Metribuzin application (synthetic organic compound), a selective herbicide applied to restrict the growth of weeds in wheat, potato, tomato, was explored (Samir et al., 2020).

Municipal Waste

Municipal waste released by homes and industries comprising plastics, paper, and organic waste. Disposal of this kind of waste results in contamination of the surrounding environment. Processing of this generated wastes result in sewage sludge production further contaminating the groundwater. Some of these are reclaimed as compost or discarded in landfills (Havugimana et al., 2015).

Biomedical Waste

Health care industries generate wastes that are contagious, hazardous, and lethal in nature (Burke, 1994; Klangsin and Harding, 1998; Pruss et al., 1999). Many researchers have highlighted the negative impact and bioactive nature of biomedical waste in aquatic ecosystems at low concentrations (Purdom et al., 1994; Daughton and Ternes, 1999; Jobling et al., 1998). Bonjoko (2014), studied the presence of pharmaceutical preparations in detectable concentrations in foods and the aquatic ecosystem. The heavy metal leaching to groundwater is a matter of concern in the treatment process (Al Raisi et al., 2014). Heera and Rajor (2014) highlighted enormous amounts of heavy metals and PAHs from incinerated biomedical waste in the surface and groundwater. Mansoor and Sharma (2019) reported the risk factors associated with biomedical waste depends on the hospitals and medical centers' standard hygienic practices.

Radioactive Waste

Radioactive wastes are substances released during - nuclear fuel production, application of radioisotopes in research, medicine, agriculture, as a by-product in mining, combustion of fossil fuels, natural gas, and oil - (Rahman et al., 2011). In earlier times, radioactive waste was classified based on safety aspects, characteristics, engineering processes, and regulatory issues as reported by International Atomic Energy Agency (1970), International Atomic Energy Agency (1999). Radioactive waste is currently being classified based on activity level and half-life (Rahman et al., 2011). Appleton (2007) reported on the presence of natural gas Radon in underground basements. Islami et al. (2015) featured that exposure to the naturally present gas can result in lung cancer. Sufficient aeration in these systems is required to minimize the pollutant gases' accumulation (Khan and Gomes 2018).

LETHAL EFFECTS OF POLLUTANTS ON ECOSYSTEM

The effects of hazardous, recalcitrant pollutants in the ecosystem and its living organisms triggered young minds to find a solution. Embrandiri et al. (2016) reported the effect of xenobiotic compounds on the ecosystem. The impact of the recalcitrant molecules on plants, animals, and humans are highlighted in many events and social gatherings. The presence of xenobiotic compounds in the aquatic environment can disturb the balance of the ecosystem and cause ill health to mankind. (Fatta Kassinos et al., 2011). Studies suggest that xenobiotic compounds' outcomes on animals (insects and fish) and trophic levels of the food chain (Bhat, 2013; Rosi-Marshall, 2013). The persistence of recalcitrant molecules in aquatic bodies is focused (Essumang, 2010). Living beings of the marine ecosystem are considered indicators of these recalcitrant compounds (Fent et al, 2006). Xenobiotic combinations in soil result in incomplete mineralization, the formation of toxic metabolites, assimilation, accumulation in plants, and leading to lethal consequences on the biosphere (Mishra et al., 2019). The presence of xenobiotic compounds in the environment has negatively influenced the diversity of fauna and flora (Zacharia, 2011). The sustained accumulation of these hazardous pollutants in food and water resources produces autoimmune disorders, liver and kidney ailments, cardiac and carcinogenic issues in humans (Embrandiri et al. 2016).

BIOREMEDIATION METHODS IN REMOVAL OF XENOBIOTICS

Bioremediation is an emerging and effective tool for remediating contaminated sites. Joutey et al. (2013) defined biodegradation as a method that decreased the complexity of a compound and converted into simpler forms by living microbes. Comprehensive research has been performed for biotechnology using distinct microbes. Remediation using microflora and plants has been employed for the degradation of the pollutants. The eco-friendly and economical approach has taken over the conventional physicochemical processes. Current scenario opts for newer bioremediation methods with integrated approach (Gong et al. 2017). Monica et al. (2011) defined Effective Microorganism (EM) as the consortia of microbes capable of synthesizing enzymes and organic acids to suppress xenobiotic compounds. Various biotic and abiotic factors determine the rate of degradation. Abatenh et al. (2017) has highlighted that rate of degradation depends on the bioavailability of pollutants. Gaur and Narasimhulu (2018), described the factors affecting biodegradation, such as accessibility, bioavailability, and microbial metabolism to transform the pollutant into less toxic forms. Extensive *ex-situ* (Biopile, Composting, Bioreactor, and Land farming) and *in-situ* methods (Bioslurping, Bioventing, Biosparging, and Phytoremediation) are adapted for treating the polluted sites (Azubuike et al. 2016).

Biopile comprises of piling up of excavated contaminant soil with adequate aeration, nutrition, and temperature for enhanced degradation by microorganisms (Whelan et al. 2015; Azubuike et al. 2016). Composting is a *ex – situ* remediation applied in the transformation of organic pollutants. The technique involves mixing polluted soil with organic substances and setting it as piles or windrows under controlled aeration and temperature in the presence of indigenous microorganisms (Kastner and Miltner,2016; Choudhary and Kim,2019). A bioreactor is used to convert raw materials into metabolic products in a biological system under optimized physical and biological conditions (Tomei and Daugulis, 2013; Choudhary and Kim, 2019). Different bioreactors are available based on its mode of operation (Batch, fed-batch, continuous, and multistage (Azubuike et al. 2016). Land farming is a simple, economical, and widely employed *ex-situ* method of remediation. In this method, the pollutant soil is excavated are arranged uniformly in a bed for treatment (Choudhary and Kim, 2019). Jeong et al. (2015) reported the significance of bioaugmentation and biosurfactants in bettering Land farming strategies.

Bioventing is a process of aerating the polluted site to enhance *in situ* degradation of organic pollutants and improve bioremediation (Hinchee and Leeson, 1996; Hyman and Dupont, 2001). Hohener and Ponsin (2014) reported on accepting this technique for treating light spilled petroleum products. Bioslurping involves vacuum-stimulated pumping, soil vapor extraction, and bioventing for remediating soil pollutants. The technique is used to eradicate volatile and semi-volatile organic compounds (Gidarakos and Aivalioti, 2007; Choudhary and Kim, 2019). The technique of biosparging involves aerating the subsurface of soil, facilitating the degradation at the pollution site by enhancing microbial activity (Azubuike et al. 2016). Philp and Atlas (2005) discussed that the efficacy of biosparging is dependable on the bioavailability and biodegradability of the contaminants. Phytoremediation is the process of remediation applying plants and their enzymes for extraction, accumulation, degradation, stabilization, filtration and volatilization of recalcitrant compounds from soil and water is known as phytoremediation (Kabra et al., 2011; Azubuike et al. 2016; Choudhary and Kim, 2019). Biostimulation helps in transforming the polluted environment by providing nutrition to the microbes, altering the environmental parameters, and adding limited nutrients to promote the C:N::P ratio (Singh et al., 2011; Choudhary and Kim,2019). Bioaugmentation involves the inclusion of microbiota (bacteria, fungi) and biostimulant (enzyme and gene) to degrade organic and inorganic contaminants (Stroo et al., 2012). Choudhary and Kim (2019) highlighted different strategies of bioaugmentation, such as consortia - bioaugmentation (Mixed microflora), recombinant- bioaugmentation (Genetically modified organisms), and biosurfactant-bioaugmentation.

Microorganisms employ two methods for degradation of these recalcitrant compounds. These include aerobic mode and anaerobic mode of degradation. The aerobic technique utilizes the presence of oxygen for the treatment of contaminants. The end products in aerobic degradation are CO2, water, and residual metabolites (Kumar et al., 2017). An anaerobic mode is a process that converts the complex organic molecules into simpler molecules by a distinct group of microbes such as bacteria and archaea in the absence of oxygen, evolving a mixture of gases (chiefly CH_4 and CO_2). The key steps involved in anaerobic degradation are hydrolysis, acidogenesis, acetogenesis, and methanogenesis (Reyes et al. 2015)

Biodegradation of xenobiotic compounds through endo and exoenzymes with specific metabolic pathways were explored (Singh et al., 2014). Recently, composting is found to be a favorite *in-situ* remediation tool for contaminants because of its efficacy and easiness in handling (Cerda et al. 2017). Theerachat et al. (2018) discussed the ability of laccases derived from marine microorganisms in bioremediation. Strategies like biopiling and compiling are cost-effective, eco-friendly techniques for treating hazardous organic pollutants (Singh et al., 2017; Kaewlaoyoong et al., 2020). Reports of the past decade elaborate on biodegradation of Total Petroleum Hydrocarbon (TPH) or related hydrocarbons by composting. After treatment, the fate of these compounds was also highlighted by Kastner and Miltner (2016); Aguelmous et al. (2019). Tran et al. (2020) has explored the degradation of TPHs in polluted soil through aerobic composting under optimized environmental conditions. Ismail et al., (2014) reported on the biodegradation of spent engine oil by rhizosphere isolated *Pseudomonas putrefacience* CR33, *Klebsiella pneumonia* CR23, *Pseudomonas alcaligenes* LR14, *Klebsiella aerogenes* CR21, and *Bacillus coagulans* CR31 at the rate of 68%, 62%, 59%, 58%and 45% respectively after 21days of incubation.

Like other microbes, cyanobacteria also have the innate ability to degrade limited amounts of organic pollutants (Safari et al. 2016; Pimda and Bunnag 2012). Attempts have been made on the uranium sorption by the algal cell wall. Further, the algal biomass was fed to the heterotrophic bacterial strains, which reduced hexavalent uranium to tetravalent uranium. (Kalin et al., 2004). Blending bioaugmentation with biostimulation was proven to be effective in treating the Atrazine polluted environment. Bacterial species belonging to Enterobacter, Pseudomonas, Bacillus, and Providencia genera were resistant towards Atrazine and effective in the degradation of the same (El-Bestawy et al., 2014). Jesubunmi (2014) isolated five bacterial species (*Pseudomonas* sp., *Klebsiella* sp., *Bacillus* sp., *Micrococcus* sp., and *Proteus* sp.) and four fungal isolates (*Streptomyces* sp., *Penicillium* sp., *Cheatomium* sp., and *Aspergillus* sp.) from engine oil contaminated soil and reported on the degradation ability of the isolates. Abbes et al. (2018) reported on the degradation efficacy of *Advenella Kashmirensis* MB-PR to utilize DDT as a unique carbon source and produce intermediate metabolites such as DDD, DDE, and DBH. The role of humin, a humic substance (HS), was studied to treat wastewater in aerobic and anaerobic conditions (Lipczynska-Kochany., 2018).

Synthetic dyes released from the textile industry also serve as a critical xenobiotic source, causing environmental pollution. Research is being carried out in the field of biotechnology to remediate these toxic pollutants. Both aerobic and anaerobic methods are applied for the degradation studies. Table 1, gives an outline on different bioremediation methods used for textile dye removal.

Table 1. Bioremediation of textile dyes

S.No	Mode of Degradation	Microorganisms Involved	Name/Con. of Dyes	% of Decolourization	Reference
1.	Aerobic Mode	*B. cereus AZ27* *A. faecalis AZ26* *Bacillus sp.AZ28*	Novacron Super BlackG/ 200mg/L	*B. cereus AZ27*-93% *A.faecalisAZ26* - 92% *Bacillus sp.AZ28*-91%,	Hossen *et al.*, (2019)
2	Aerobic Mode	*Scheffersomyces spartinae*	Acid Scarlet 3RDye/20mg/L & 100mg/L	90%- 20mg/L 80%- 100mg/	Tan *et al.*, (2016)
3	Aerobic Mode	*Bacillus stratosphericus*	Methyl Orange/150mg/L	100%- 150mg/L	Akansha *et al*, (2019)
4.	Aerobic & Anaerobic Mode (Combined	Psychrotrophic bacterial consortia *Stenotrophomonas Sphingomonas*,(StSp) & mesophilic bacterial consortia *Pseudoarthrobacter& Gordonia*, (PsGo)	RB-5azo dye/50mg/L	50mg/L-PsGo- 54% StSp- 34%	Eskandari *et al.*, (2019)
5.	Aerobic Mode	Mixed bacterial Culture	Reactive Brilliant *R*edX-3 B, Direct Blue-6 & Direct Black-19 20-100 mg/L	RBRX-3B-31.2% DB-6- 71.5% DB19- 87.6%	Krishnan *et al* (2016)
6.	Aerobic Mode Static&Shaking	Halotolerant *Nesterenkonia lacusekhoensis EMLA3*:	Azodye Methyl Red/ 50mg/L	97%-50mg/L Methyl red	*Bhattacharya et al* (2017)
7.	Aerobic Mode Static&Shaking	*Aeromonas hydrophila*	Reactive Black 5 /100mg/L	76%- (Static) 56%-(Shaking)	El Bouraie, M. and El Din, W.S, (2016)
8	Aerobic mode (Shaking)	Mixed alkaliphilic bacterial consortium-*B.cereus B.cytotoxicusBacillus sp. L10*, and *B. flexus*	Azo dyes - Direct Blue 151and Direct Red 31/100-300mg/l	200mg/L DB151-97.57% DR 31-95.25% *B.cereus*-Mixture of Dyes-93.37% *B. cytotoxicus, Bacillus sp. L10*, and *B. flexus* -mixture of dyes- 92.77%, 86.86%, and 85% respectively	Sylvine Lalnunhlimi, and Veenagayathri Krishnaswamy (2016)
9.	Aerobic mode (Shaking)	*Halomonas* strain IP8	Toludine Red Dye/10mg/L 25mg/L	67%- 10mg/L 70%- 25mg/L	Moharrey *et al* (2018)
10.	Aerobic mode Static & Shaking	*E.feacalis* *Klebsiella varicola*	Reactive Red 198/10-100mg/L	98%-10-25mg/L 55.62%-50mg/L 25.82%-75mg/L 15.42%100mg/L	*Eslami et al* (2019)

PHYTOREMEDIATION OF XENOBIOTICS

Phytoremediation is a looming technology that is applied in environmental biotechnology for the treatment of hazardous pollutants. This technology involves plants to remediate toxic compounds like heavy metals, synthetic dyes, pesticides, polyaromatic hydrocarbons, chlorinated solvents, and polychlorinated biphenyls (Susarala et al., 2002; PilonSmits 2005; Nwoko 2010). The pollutants' hazardous nature can suppress plant metabolism and retards tissues' growth (Smith et al., 2006; Meudec et al., 2007; Euliss et al., 2008). Accordingly, it is preferred for plants with rising resistance against toxicants, competitive pollutant uptake, and relevant abilities to metabolize the organic pollutants (Wenzel, 2009). Of late, researchers are exploring Phytoextraction and phytomining of metals and metalloids using hyperaccumulator plants (Srivastava. N., 2020). Studies have pinpointed the significance of plants in the treatment of textile dye effluents over microbial application (Chandanshive et al., 2017). The ability of plants to naturally filter the environment's pollutants by varied absorption, accumulation, extraction, immobilization, volatilization and degradation enables phytoremediation as a useful tool for biodegradation (Kabra et al., 2011). The technology succeeds as a remediation tool in recent decades as it demands no chemicals, simple nutrient intake, advantageous, and eco-friendly (Adki et al., 2013; Srikantan et al., 2018). The phytoremediation potential of Bacopa monnieri, a perennial marshy plant, was proven in the decolorization of azo-dyes (Shanmugham et al., 2020).

Phytoremediation is one of the *in-situ* environment-friendly methods, diminishing soil erosion, improving soil fertility by increasing organic matters in soil, and utilizes plants for degradation, extraction, transformation or detoxification of chemical contamination (Organum and Bacon 2006). Phytoremediation methods can remove dominant pollutants such as metals, solvents, pesticides, explosives, crude oil, landfill leachates and hydrocarbons. Alberto and Sigua (2013) reported on the plants' genetic adaptation for enduring the toxic contaminants. During the photosynthesis process, phytoremediation can also assist in the eradication of carbon dioxide from the air. Structure and biological functions of the environments can be persevered by phytoremediation. Contaminants from the accumulated site can be efficiently removed by excluding the plant used in the process, which is the main advantage of phytoremediation (Balarak *et al.,* 2015). Phytoremediation technology has been employed using growing aquatic plants like *Phagmites australis* (Hussein and Scholz 2017) and using free-floating plants like *L. minor* (Uysal *et al.,* 2014). Researchers explore newer approaches of finding suitable trees for practical bioremediation.

Phytoremedial Mechanism

Eradication of contaminants from its sites using plants are applied in several possible ways. Plants can eliminate pollutants by acting as filters or traps from the soil, sediment, and water. The root system of plants up-take the pollutants present in the site and protect the environment from the hazardous contaminants. Therefore, contaminants from various sources must be ready to be absorbed by roots, and this phenomenon facilitates the removal of dyes combined with the bioavailability of dyes. The plants' root systems absorb essential nutrients for their growth and other contaminants due to adaptation. More toxic complex organic pollutants are degraded into simple complex, non-toxic forms of carbon, oxygen, and hydrogen (Mahar *et al.,* 2016). In contrast, inorganic chemical substances change their chemical structure or are transferred from one medium to another, resulting in the withdrawal of toxic chemical elements (Dickinson 2017). The supply of textile wastewater and canal water mixing enhanced the growth and yield of field mustard (*Brassica campestris L.*) (Yaseen *et al.,* 2017). An integrated system of using *the Prescaria barbata* plant inoculated with microbes and supplemented with agricultural rice waste showed effective removal of Reactive Black 5 dyes (Beenish *et al.,* 2015).

PHYTOREMEDIATION APPROACHES

The phytoremediation process can be classified principally based on techniques involved, application, and kind of pollutant. (Wang et al., 2017)

Figure 2. Phytoremediation methods

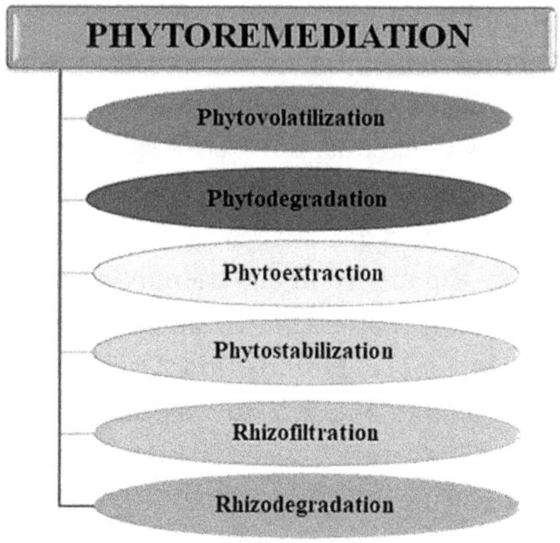

The process that involves the absorption of contaminants from the polluted site and releasing it as volatilized compounds into the atmosphere is known as phytovolatilization (Wiszniewska et al., 2016). Phytovolatalization is carried out through stem, leaves (direct volatilization), and roots (indirect volatilization) (Matt Limmer and Joel Burken, 2016). This method is employed for the treatment of metals and organic compounds. (Bharathiraja et al., 2018). The methylation and de-methylation in rice plants resulted in volatilization and conversion of 2, 4-dibromophenol, and 2, 4-dibromoanisole (Zhang et al., 2020). Bioconversion in various plants resulting in methylation and demethylation of organic and inorganic pollutants were extensively studied (Fu et al., 2018; Hou et al., 2018; Li et al., 2018; Sun et al., 2016; Xu et al., 2016; Zhang et al., 2019).

The phytodegradation technique is also known as phytotransformation, involving the breakdown or mineralization of organic substances by the plant's unique enzymes. The remediation potential of a diverse group of enzymes nitroreductases (reduction of nitroaromatic compounds), laccases (degradation of anilines), and dehalogenases (break down of chlorinated solvents and pesticides) were proven decades ago and implemented in treating contaminated sites. *Populus* species is a crucial plant employed for the degradation of organic pollutants. (Schnoor et al., 1995; Elizabeth L Rylott and Neil C. Bruce.,2008).

Heavy metals present in the environment that includes As, Cu, Cd, Pb, Cr, Ni, Hg and Zn pose a threat to human beings as well the environment. Uptake of these toxic metals may interfere in human health and cause diseases. Excessive concentration of these metals also produces toxic effects on plants by affecting the growth and metabolic functions of plants and sometimes lead to the death of plants. These metals present in soil also harm the microbial community of soil and characteristics of soil. (Garbisu and Alkorta, 2001; Schmidt, 2003; Schwartz *et* al., 2003). Different approaches have been developed to remove heavy metal contaminants from soil and water and most methods that have been used are expensive, time consuming and may have side effects. Phytoremediation is a new approaching technology that to reclaim polluted and contaminated soils and environment (McGrath et al., 2001). In phytoremediation, plants will uptake and accumulate the heavy metals in their cells, shoot and roots and remove the contaminants from environment. This heavy metal accumulation may cause toxic effects to the plants and they should be tested. Plants that grow in contaminated areas can also be tested for heavy metal toxicity.

The suction and accumulation of hazardous pollutants in the plant leaves, stem, and roots are known as phytoextraction. This technique is applied to remove metal and organic pollutants (Cd. Ni, Cu, Zn, Pb, Se, As) (Favas et al., 2014). Srivastava (2020) has affirmed the efficacy of Brassicaceae plants in permanent removal of toxic metals and metalloids. The response of specific plants in absorption, transfer, and compiling of aromatic hydrocarbons from soil to their roots and shoots is a promising phenomenon employed for remediating xenobiotic compounds (Wild et al., 2005; Lu et al., 2010). The phytoremediation ability of switchgrass (Panicum virgatum) in removing toxic heavy metals like Zinc, Cd, Pb, Co, and Ni was proven in in-vitro studies (Shrestha et al., 2019). Table 2 shows the effects of metals on plants.

Table 2. Effects of heavy metals on the plants

S.No	Plant Name	Heavy Metal	Effect on Plants	Reference
1.	*Hordeum vulgare*	Cd, Hg	Overexpression of gene – dehydration stress.	Tamás et al. (2010)
2	*Medicago sativa*	Cd, Hg	oxidative stress & glutathione depletion	Hernandez et al. (2012)
3.	*Triticum sp*	Cd	Inhibition of seed germination and seedling growth	Zhang et al. (2002)
4.	*Helainthus annuus*	As	reduction in plumule and radicle length	Imran et al. (2013)
5.	*Oryza sativa*	As	Reduced seed germination and decreased seedling height	Abedin et al. (2002)
6.	*Brassica napus*	As	Stunted growth; chlorosis; wilting	Cox et al. (1996)
7.	*Allium cepa*	Cr	Inhibition of germination process; reduction of plant biomass	Nematshahi et al. (2012)
8.	*Lycopersicon esculentum*	Hg	Reduction in germination percentage; reduced plant height; reduction in flowering and fruit weight; chlorosis	Shekar et al. (2011)
9.	*Zea mays*	Pb	Suppressed growth; reduced plant biomass; decrease in plant protein content	Hussain et al. (2013)

Phytostabilization involves the immobilization of pollutants in soil. This process is applicable in the remediation of heavy metals (Cunningham et al., 1995) and minimizes the communication with associated microbes. The process retains the contaminants harmless and maintains the ecological balance (Carlos and Alkorta., 2001). Helianthus petiolaris, an aromatic plant species, was proven to tolerate 50mg/kg of Cd (Cadmium) up to 1000mg/kg of Pb (Lead). This plant's capability has enabled it to be a favourite choice among researchers in phytostabilization (Saran et al., 2019). Researchers reported the exploitation of fast-growing non-food crop aromatic plants in phytostabilization and yielding essential oils without cross-contamination of remediating pollutants (Croes et al. 2015; Pandey et al. 2015, 2019). The enhanced anti-oxidant system of free-floating aquatic plants was demonstrated on the addition of benoxacor, suggesting phytoremediation efficacy to treat xenobiotic polluted sites (Panfili, Bartucca, Del Buono 2019). Del Buono et al. (2020) discussed the application of safeners to enable plants to tolerate the effect of hazardous pollutants like heavy metals and herbicides. Visconti et al. (2020) reported the immobilization of potentially toxic elements (PTEs) like Cd, Pb, and Zn in a polluted mine environment using *Brassica juncea Dactylis glomerata*. The process was enhanced with the addition of compost and biochar.

The Rhizofiltration method applies to both aquatic and terrestrial plants. This method is used to eradicate radioactive substances from groundwater and wastewater (Ibrahim et al., 2015). Panfili et al. (2017) reported the combination of aquatic plants with safeners as a dependable remedy for the `phytofiltration of Copper metal-contaminated water. Plants are grown in a hydroponic system to remove pollutants from the contaminated environment (Shanmugham et al., 2020). The technique helps in the concentration or precipitation of toxicants by which the roots of the plant act as filters. (Ali et al., 2013). The method of remediating the pollutant by rhizosphere microbes is known as rhizodegradation. The process uses the plant metabolites and exudates as a source of carbon and energy and releases enzymes capable of bioremediation (Favas et al., 2014). The heterogenous microflora of the rhizosphere region is a boon for the bioremediation of hazardous toxicants.

Role of Plant Enzymes in Phytoremediation

Vasavi et al., (2010), discussed on the enzymes involved in phytodegradation and phytotransformation of organic pollutants. Studies reported on the role of plant enzymes (Phosphatase, Aromatic dehalogenases, Cytochrome 450 peroxidases, Peroxygenases, Glutathione, S-transferase, O-glucosyltransferases, O-malonyltransferases B-cyanoalanine synthase) in the degradation of the recalcitrant compounds (Organo Phosphates, Chlorinated aromatic Compounds (DDT, PCH's), PCB's, Xenobiotics and Cyanide). Phytotransformation of organochlorine pesticides by various enzymes such as phenoloxidases, peroxidases, cytochrome P450s, monooxygenases through oxidation process was reported (Miguel et al. 2013; Kurasvili et al. 2014). Kurasvili et al., (2014), highlighted the role of monooxygenases, cytochromes, phenoloxidases or peroxidases in the initial oxidation of DDT and lindane and further conjugation by glutathione S-transferases. The role of different oxidoreductase enzymes of plants (laccase, veratryl alcohol oxidase, lignin peroxidase, tyrosinase, azo reductase, DCIP reductase) and stress diminishing enzymes (superoxide dismutase and catalase) involved in activation of phytoremediation of textile dyes are studied (Kabra et al., 2013; Watharkar and Jadhav, 2014; Kagalkar et al., 2015; Rane et al., 2016).

The advantages and disadvantages of phytoremediation are tabulated in Table 3

Table 3. Phytoremediation advantages and disadvantages

S. No	Advantages	Disadvantages
1.	Applicable for varied organic and inorganic compounds	Limited within the rhizosphere region of plants used, in case of negligible contamination.
2.	The best remedy for *In Situ* and *Ex-Situ* application of effluents and sludge.	Requires a longer duration for remediation of polluted sites.
3.	*In-Situ* mode diminishes the soil interruption compared with routine techniques employed.	Confined to the least polluted sites.
4	Landfilling of waste can be decreased (up to 95%), and can be exploited as bio-ore of heavy metals.	Proper disposal of biomass harvested is required due to the generation of hazardous waste in phytoextraction.
5.	*In-Situ* methods are cost-effective and require no specific device or specialized personnel.	The influx of unnatural species may affect biodiversity.
6.	*In-Situ* modes are efficient to minimize the spread of pollutants through air and water.	Environmental factors like climatic conditions serve as a limiting factor.
7.	Energy stored in large scale applications can be used to generate thermal energy.	The exploitation of contaminated plant biomass is a task for ecologists.

(Ghosh and Singh, 2005)

INTERACTION OF PLANTS AND MICROBES IN REMEDIATION

The synergistic association of plants and microbes is a recently explored field in bioremediation. The microbes are associated with plants in many ways, rhizosphere microbes, endophytic bacteria, and mycorhizal association. The association of plant and microbes is highly explored in the field of phytoremediation (Doty et al. 2017; Deng and Cao, 2017; Feng et al. 2017; Jambon et al. 2018). Aransiola *et al.,* (2019), reviewed the inputs of plant and microbial association. The role of plants in accumulating and segregating the pollutants like heavy metals was highlighted. The microbial aid in transforming the pollutants into less toxic forms and uptake of pollutants by plant roots was discussed. Dai et al., (2020) reported on the role of Fire Phoenix plants remediation of PAH-Cd co- contaminated soil in association with rhizosphere bacteria. Many researchers report the mutualistic association of plants and fungi as a model of remediation. Hakeem et al., (2015), reported on the absorption capacity of roots of associated mycorrhizal plants. Endophytic bacteria are known for its ability to detoxify the hazardous pollutants that are accumulated in the plant tissues. The textile effluent degradation efficacy of *Typha domingensis* in a constructed wetland system was stimulated by inoculation of endophytic bacterial strains *Microbacterium arborescens* TYSI04 and *Bacilluspumilus PIRI30* (Shehzadi et al., 2014). Datta et al. (2020) reviewed the strategies utilizing endophytic bacteria for the break-down of recalcitrant compounds as well as being employed as a plant growth-promoting factor. Researchers have studied the various methods adopted to lessen metal toxicity in plants and trees for Cd toxicity, repression of metal toxicity by *Lonicera japonica,* stimulated tolerance towards heavy metals in *Eucalyptus tereticornis,* and reduction in root to shoot translocation of pollutants in *Betula pubescens* (Jiang *et al.*, 2016; Reddy *et al.*, 2016; Ferna´ndez-Fuego *et al.*, 2017). Khandre *et al.*, (2013) reported on the plant-bacteria consortium of *Portulaca grandiflora* and *Pseudomonas putida* in 100% decolourization of diazo dye Direct Red 5B. The researcher highlighted the synergistic enzymatic activity of both plant and bacteria in the degradation of the synthetic dye.

The plant-associated microbes can improve the efficacy of phytoremediation. The endurance and strength of plants towards heavy metals are enhanced by rhizosphere microorganisms (Gupta et al., 2013; Fasani et al., 2018). Ma et al., (2011), highlighted the ability of plant growth-promoting rhizobacteria (PGPR) in stimulating plant growth, resistance towards heavy metals, absorption of nutrients, heavy metals and its translocation.

PGPR produced IAA (Indole Acetic Acid) can induce improved lateral root initiation and root hair development (Glick, 2010; DalCorso et al., 2019). The presence of Arbuscular Mycorrhizal Fungi (AMF) in rhizosphere region increases the rate of absorption of heavy metals by the massive hyphal growth and promotes phytoremediation by enhanced water and nutrient uptake (Göhre and Paszkowski, 2006).

CONCLUSION

The chapter is an attempt to review the past and current methods used to remediate the xenobiotic compounds. The toxic effects of the recalcitrant compounds and the outcome of bioaccumulation of these pollutants were discussed. The release of toxic metabolites during physicochemical treatments paved the way for eco-friendly and cost-effective bioremediation approaches. Remediation using microbes in treating the xenobiotic compounds are much studied and discussed in this era of industrialization. Apart from the limitation of time consumption and a chance of mutation for the native or induced microbes, bioremediation is a proven tool in treating the hazardous compounds. The recent decade has attracted the focus on phytoremediation as an alternative to degradation using microbes. The treatment of pollutant compounds using different varieties of plants and the microbes associated with them are studied for better results. The methods of improving the tolerance level of plants using PGPRs and towards these toxicants is also focus of study among environmentalists. Hence the integrated approaches are preferable than single approaches in detoxifying the xenobiotic compounds. The researchers are keeping an eye on science and technology's progress and inspire the future generation by implementing varied green technologies for a better tomorrow.

REFERENCES

Abbes, C., Mansouri, A., Werfelli, N., & Landoulsi, A. (2018). Aerobic Biodegradation of DDT by *Advenella kashmirensis* and Its Potential Use in Soil Bioremediation. *Soil and Sediment Contamination: An International Journal, 27*(6), 455–468. doi:10.1080/15320383.2018.1485629

Abedin, M. J., Cotter-Howells, J., & Meharg, A. A. (2002). Arsenic uptake and accumulation in rice (*Oryza sativa* L.) irrigated with contaminated water. *Plant and Soil, 240*(2), 311–319. doi:10.1023/A:1015792723288

Adki, V. S., Jadhav, J. P., & Bapat, V. A. (2012). Exploring the phytoremediation potential of cactus (Nopalea cochenillifera Salm. Dyck) cell cultures for textile dye degradation. *International Journal of Phytoremediation, 14*(6), 554–569. doi:10.1080/15226514.2011.619226 PMID:22908626

Aguelmous, A., El Fels, L., Souabi, S., Zamama, M., & Hafidi, M. (2019). The fate of total petroleum hydrocarbons during oily sludge composting: A critical review. *Reviews in Environmental Science and Biotechnology, 18*(3), 1–21. doi:10.100711157-019-09509-w

Akansha, K., Chakraborty, D., & Sachan, S. G. (2019). Decolorization and degradation of methyl orange by *Bacillus stratospherericus* SCA1007. *Biocatalysis and Agricultural Biotechnology*, *18*, 101044. doi:10.1016/j.bcab.2019.101044

Al Raisi, S. A. H., Sulaiman, H., Suliman, F. E., & Abdallah, O. (2014). Assessment of heavy metals in leachate of an unlined landfill in the Sultanate of Oman. *International Journal of Environmental Sciences and Development*, *5*(1), 60.

Ali, H., Khan, E., & Sajad, M. A. (2013). Phytoremediation of heavy metals—Concepts and applications. *Chemosphere*, *91*(7), 869–881. doi:10.1016/j.chemosphere.2013.01.075 PMID:23466085

Appleton, J. D. (2007). Radon: Sources, health risks, and hazard mapping. *Ambio*, *36*(1), 85–89. doi:10.1579/0044-7447(2007)36[85:RSHRAH]2.0.CO;2 PMID:17408197

Aransiola, S. A., Ijah, U. J. J., Abioye, O. P., & Bala, J. D. (2019). Microbial-aided phytoremediation of heavy metals contaminated soil: A review. *European Journal of Biological Research*, *9*(2), 104–125.

Azubuike, C. C., Chikere, C. B., & Okpokwasili, G. C. (2016). Bioremediation techniques–classification based on site of application: Principles, advantages, limitations and prospects. *World Journal of Microbiology & Biotechnology*, *32*(11), 180. doi:10.100711274-016-2137-x PMID:27638318

Balarak, D., Jaafari, J., Hassani, G., Mahdavi, Y., Tyagi, I., Agarwal, S., & Gupta, V. K. (2015). The use of low-cost adsorbent (Canola residues) for the adsorption of methylene blue from aqueous solution: Isotherm, kinetic and thermodynamic studies. *Colloid and Interface Science Communications*, *7*, 16–19. doi:10.1016/j.colcom.2015.11.004

Bharadwaj, A. (2018). Green Chemistry in Environmental Sustainability and Chemical Education. In *Green Chemistry in Environmental Sustainability and Chemical Education*. Springer Singapore.

Bharagava, R. N., Purchase, D., Saxena, G., & Mulla, S. I. (2019). Applications of metagenomics in microbial bioremediation of pollutants: from genomics to environmental cleanup. In *Microbial diversity in the genomic era* (pp. 459–477). Academic Press. doi:10.1016/B978-0-12-814849-5.00026-5

Bharathiraja, B., Jayamuthunagai, J., Praveenkumar, R., & Iyyappan, J. (2018). Phytoremediation techniques for the removal of dye in wastewater. In *Bioremediation: applications for environmental protection and management* (pp. 243–252). Springer. doi:10.1007/978-981-10-7485-1_12

Bhat, S. (2013). Ecotoxicology & Impact on Biodiversity. *Journal of Pharmacognosy and Phytochemistry*, *2*(2).

Bhattacharya, A., Goyal, N., & Gupta, A. (2017). Degradation of azo dye methyl red by alkaliphilic, halotolerant *Nesterenkonia lacusekhoensis* EMLA3: Application in alkaline and salt-rich dyeing effluent treatment. *Extremophiles*, *21*(3), 479–490. doi:10.100700792-017-0918-2 PMID:28255636

Bouchard, M. F., Chevrier, J., Harley, K. G., Kogut, K., Vedar, M., Calderon, N., Trujillo, C., Johnson, C., Bradman, A., Barr, D. B., & Eskenazi, B. (2011). Prenatal exposure to organophosphate pesticides and IQ in 7-year-old children. *Environmental Health Perspectives*, *119*(8), 1189–1195. doi:10.1289/ehp.1003185 PMID:21507776

Brahushi, F., Kengara, F. O., & Yang, S. O. N. G. (2017). Fate processes of chlorobenzenes in soil and potential remediation strategies: A review. *Pedosphere*, *27*(3), 407–420. doi:10.1016/S1002-0160(17)60338-2

Burke, E. L. (1994). A survey of recent literature on medical waste. *Journal of Environmental Health*, 11–14.

Carmichael, S. L., Yang, W., Roberts, E., Kegley, S. E., Brown, T. J., English, P. B., & Shaw, G. M. (2016). Residential agricultural pesticide exposures and risks of selected birth defects among offspring in the San Joaquin Valley of California. *Birth Defects Research. Part A, Clinical and Molecular Teratology*, *106*(1), 27–35. doi:10.1002/bdra.23459 PMID:26689858

Cassidy, D. P., Srivastava, V. J., Dombrowski, F. J., & Lingle, J. W. (2015). Combining in situ chemical oxidation, stabilization, and anaerobic bioremediation in a single application to reduce contaminant mass and leachability in soil. *Journal of Hazardous Materials*, *297*, 347–355. doi:10.1016/j.jhazmat.2015.05.030 PMID:26093352

Cerda, A., Artola, A., Font, X., Barrena, R., Gea, T., & Sánchez, A. (2018). Composting of food wastes: Status and challenges. *Bioresource Technology*, *248*, 57–67. doi:10.1016/j.biortech.2017.06.133 PMID:28693949

Chakraborty, S. K. (2019). Bioinvasion and Environmental Perturbation: Synergistic Impact on Coastal–Mangrove Ecosystems of West Bengal, India. In *Impacts of Invasive Species on Coastal Environments* (pp. 171–245). Springer. doi:10.1007/978-3-319-91382-7_6

Chandanshive, V. V., Rane, N. R., Tamboli, A. S., Gholave, A. R., Khandare, R. V., & Govindwar, S. P. (2017). Co-plantation of aquatic macrophytes *Typha angustifolia* and *Paspalum scrobiculatum* for effective treatment of textile industry effluent. *Journal of Hazardous Materials*, *338*, 47–56. doi:10.1016/j.jhazmat.2017.05.021 PMID:28531658

Chaudhary, D. K., & Kim, J. (2019). New insights into bioremediation strategies for oil-contaminated soil in cold environments. *International Biodeterioration & Biodegradation*, *142*, 58–72. doi:10.1016/j.ibiod.2019.05.001

Cox, M. S., Bell, P. F., & Kovar, J. L. (1996). Differential tolerance of canola to arsenic when grown hydroponically or in soil. *Journal of Plant Nutrition*, *19*(12), 1599–1610. doi:10.1080/01904169609365224

Croes, S., Weyens, N., Colpaert, J., & Vangronsveld, J. (2015). Characterization of the cultivable bacterial populations associated with field grown *Brassica napus L.*: An evaluation of sampling and isolation protocols. *Environmental Microbiology*, *17*(7), 2379–2392. doi:10.1111/1462-2920.12701 PMID:25367683

Dai, Y., Liu, R., Zhou, Y., Li, N., Hou, L., Ma, Q., & Gao, B. (2020). Fire Phoenix facilitates phytoremediation of PAH-Cd co-contaminated soil through promotion of beneficial rhizosphere bacterial communities. *Environment International*, *136*, 105421. doi:10.1016/j.envint.2019.105421 PMID:31884414

DalCorso, G., Fasani, E., Manara, A., Visioli, G., & Furini, A. (2019). Heavy metal pollutions: State of the art and innovation in phytoremediation. *International Journal of Molecular Sciences*, *20*(14), 3412. doi:10.3390/ijms20143412 PMID:31336773

Dangi, A. K., Sharma, B., Hill, R. T., & Shukla, P. (2019). Bioremediation through microbes: Systems biology and metabolic engineering approach. *Critical Reviews in Biotechnology*, *39*(1), 79–98. doi:10.1080/07388551.2018.1500997 PMID:30198342

Datta, S., Singh, S., Kumar, V., Dhanjal, D. S., Sidhu, G. K., Amin, D. S., & Singh, J. (2020). Endophytic bacteria in xenobiotic degradation. In *Microbial Endophytes* (pp. 125–156). Woodhead Publishing. doi:10.1016/B978-0-12-818734-0.00006-1

Daughton, C. G., & Ternes, T. A. (1999). Pharmaceuticals and personal care products in the environment: Agents of subtle change? *Environmental Health Perspectives*, *107*(suppl 6), 907–938. doi:10.1289/ehp.99107s6907 PMID:10592150

Del Buono, D., Terzano, R., Panfili, I., & Bartucca, M. L. (2020). Phytoremediation and detoxification of xenobiotics in plants: Herbicide-safeners as a tool to improve plant efficiency in the remediation of polluted environments. A mini-review. *International Journal of Phytoremediation*, *22*(8), 789–803. doi:10.1080/15226514.2019.1710817 PMID:31960714

Deng, Z., & Cao, L. (2017). Fungal endophytes and their interactions with plants in phytoremediation: A review. *Chemosphere*, *168*, 1100–1106. doi:10.1016/j.chemosphere.2016.10.097 PMID:28029384

Doran, J. W., Sarrantonio, M., & Liebig, M. A. (1996). Soil health and sustainability. *Advances in Agronomy*, *56*, 1–54. doi:10.1016/S0065-2113(08)60178-9

Doty, S. L., Freeman, J. L., Cohu, C. M., Burken, J. G., Firrincieli, A., Simon, A., & Blaylock, M. J. (2017). Enhanced degradation of TCE on a superfund site using endophyte-assisted poplar tree phytoremediation. *Environmental Science & Technology*, *51*(17), 10050–10058. doi:10.1021/acs.est.7b01504 PMID:28737929

Dubey, K. K., Kumar, P., Singh, P. K., & Shukla, P. (2014). Exploring Prospects of Monooxygenase-Based Biocatalysts in Xenobiotics. In *Microbial Biodegradation and Bioremediation* (pp. 577–614). Elsevier. doi:10.1016/B978-0-12-800021-2.00026-1

El-Bestawy, E., Sabir, J., Mansy, A. H., & Zabermawi, N. (2014). Comparison among the efficiency of different bioremediation technologies of Atrazine-contaminated soils. *Journal of Bioremediation & Biodegradation*, *5*(5), 237.

El Bouraie, M., & El Din, W. S. (2016). Biodegradation of Reactive Black 5 by *Aeromonas hydrophila* strain isolated from dye-contaminated textile wastewater. *Sustainable Environment Research*, *26*(5), 209–216. doi:10.1016/j.serj.2016.04.014

Embrandiri, A., Kiyasudeen, S. K., Rupani, P. F., & Ibrahim, M. H. (2016). Environmental xenobiotics and its effects on natural ecosystem. In *Plant Responses to Xenobiotics* (pp. 1–18). Springer. doi:10.1007/978-981-10-2860-1_1

Eskandari, F., Shahnavaz, B., & Mashreghi, M. (2019). Optimization of complete RB-5 azo dye decolorization using novel cold-adapted and mesophilic bacterial consortia. *Journal of Environmental Management*, *241*(March), 91–98. doi:10.1016/j.jenvman.2019.03.125 PMID:30986666

Eslami, H., Shariatifar, A., Rafiee, E., Shiranian, M., Salehi, F., Hosseini, S. S., & Ebrahimi, A. A. (2019). Decolorization and biodegradation of reactive Red 198 Azo dye by a new *Enterococcus faecalis–Klebsiella variicola* bacterial consortium isolated from textile wastewater sludge. *World Journal of Microbiology & Biotechnology, 35*(3), 38. doi:10.100711274-019-2608-y PMID:30739299

Essumang, D. K. (2010). Distribution, levels, and risk assessment of polycyclic aromatic hydrocarbons (PAHs) in some water bodies along the coastal belt of Ghana. *TheScientificWorldJournal, 10*, 972–985. doi:10.1100/tsw.2010.96 PMID:20526527

Essumang, D. K. (2013). *Environmental Xenobiotics: PAHs In Soil (Heavy Metals), Indoor Air And Water Environment, Case Studies Of Ghana And Denmark* (Doctoral Dissertation). Luma Print.

Euliss, K., Ho, C. H., Schwab, A. P., Rock, S., & Banks, M. K. (2008). Greenhouse and field assessment of phytoremediation for petroleum contaminants in a riparian zone. *Bioresource Technology.*

Fasani, E., Manara, A., Martini, F., Furini, A., & DalCorso, G. (2018). The potential of genetic engineering of plants for the remediation of soils contaminated with heavy metals. *Plant, Cell & Environment, 41*(5), 1201-1232.

Fatta-Kassinos, D., Meric, S., & Nikolaou, A. (2011). Pharmaceutical residues in environmental waters and wastewater: Current state of knowledge and future research. *Analytical and Bioanalytical Chemistry, 399*(1), 251–275. doi:10.100700216-010-4300-9 PMID:21063687

Favas, P. J., Pratas, J., Varun, M., D'Souza, R., & Paul, M. S. (2014). Phytoremediation of soils contaminated with metals and metalloids at mining areas: potential of native flora. *Environmental Risk Assessment of Soil Contamination, 3*, 485-516..

Feng, N. X., Yu, J., Zhao, H. M., Cheng, Y. T., Mo, C. H., Cai, Q. Y., & Wong, M. H. (2017). Efficient phytoremediation of organic contaminants in soils using plant–endophyte partnerships. *The Science of the Total Environment, 583*, 352–368. doi:10.1016/j.scitotenv.2017.01.075 PMID:28117167

Fernández-Fuego, D., Bertrand, A., & González, A. (2017). Metal accumulation and detoxification mechanisms in mycorrhizal *Betula pubescens. Environmental Pollution, 231*, 1153–1162. doi:10.1016/j.envpol.2017.07.072 PMID:28941719

Fu, Q., Liao, C., Du, X., Schlenk, D., & Gan, J. (2018). Back conversion from product to parent: Methyl triclosan to triclosan in plants. *Environmental Science & Technology Letters, 5*(3), 181–185. doi:10.1021/acs.estlett.8b00071

Gadd, G. M. (2010). Metals, minerals and microbes: Geomicrobiology and bioremediation. *Microbiology, 156*(3), 609–643. doi:10.1099/mic.0.037143-0 PMID:20019082

Garbisu, C., & Alkorta, I. (2001). Phytoextraction: A cost-effective plant-based technology for the removal of metals from the environment. *Bioresource Technology, 77*(3), 229–236. doi:10.1016/S0960-8524(00)00108-5 PMID:11272009

Garbisu, C., & Alkorta, I. (2001). Phytoextraction: A cost effective plant-based technology for the removal of metals from the environment. *Bioresource Technology, 77*(3), 229–236. doi:10.1016/S0960-8524(00)00108-5 PMID:11272009

García-Delgado, C., Alfaro-Barta, I., & Eymar, E. (2015). Combination of biochar amendment and mycoremediation for polycyclic aromatic hydrocarbons immobilization and biodegradation in creosote-contaminated soil. *Journal of Hazardous Materials, 285*, 259–266. doi:10.1016/j.jhazmat.2014.12.002 PMID:25506817

Gaur, N., Narasimhulu, K., & y, P. S. (2018). Recent advances in the bio-remediation of persistent organic pollutants and its effect on environment. *Journal of Cleaner Production, 198*, 1602–1631. doi:10.1016/j.jclepro.2018.07.076

Ghosh, M., & Singh, S. P. (2005). Comparative uptake and phytoextraction study of soil induced chromium by accumulator and high biomass weed species. *Applied Ecology and Environmental Research, 3*(2), 67–79. doi:10.15666/aeer/0302_067079

Gidarakos, E., & Aivalioti, M. (2007). Large scale and long term application of bioslurping: The case of a Greek petroleum refinery site. *Journal of Hazardous Materials, 149*(3), 574–581. doi:10.1016/j.jhazmat.2007.06.110 PMID:17709182

Glick, B. R. (2010). Using soil bacteria to facilitate phytoremediation. *Biotechnology Advances, 28*(3), 367–374. doi:10.1016/j.biotechadv.2010.02.001 PMID:20149857

Göhre, V., & Paszkowski, U. (2006). Contribution of the arbuscular mycorrhizal symbiosis to heavy metal phytoremediation. *Planta, 223*(6), 1115–1122. doi:10.100700425-006-0225-0 PMID:16555102

Gong, X., Huang, D., Liu, Y., Peng, Z., Zeng, G., Xu, P., Cheng, M., Wang, R., & Wan, J. (2018). Remediation of contaminated soils by biotechnology with nanomaterials: Bio-behavior, applications, and perspectives. *Critical Reviews in Biotechnology, 38*(3), 455–468. doi:10.1080/07388551.2017.1368446 PMID:28903604

Gupta, D. K., Huang, H. G., & Corpas, F. J. (2013). Lead tolerance in plants: Strategies for phytoremediation. *Environmental Science and Pollution Research International, 20*(4), 2150–2161. doi:10.100711356-013-1485-4 PMID:23338995

Hakeem, K. R., Sabir, M., Ozturk, M., Mermut, A. R., Surriya, O., Saleem, S. S., & Kazi, A. G. (2015). *Soil remediation and plants*. Soil Remediation. Plants.

Havugimana, E., Bhople, B. S., Kumar, A., Byiringiro, E., Mugabo, J. P., & Kumar, A. (2015). Soil pollution–major sources and types of soil pollutants. *Environmental Science and Engineering, 11*, 53-86.

Heera, S., & Rajor, A. (2014). Bacterial treatment and metal characterization of biomedical waste ash. *Journal of Waste Management*.

Hernández, L. E., Garate, A., & Carpena-Ruiz, R. (1997). Effects of cadmium on the uptake, distribution and assimilation of nitrate in *Pisum sativum*. *Plant and Soil, 189*(1), 97–106. doi:10.1023/A:1004252816355

Hinchee, R. E., & Leeson, A. (1996). *Soil bioventing: Principles and practice*. CRC Press.

Höhener, P., & Ponsin, V. (2014). In situ vadose zone bioremediation. *Current Opinion in Biotechnology, 27*, 1–7. doi:10.1016/j.copbio.2013.08.018 PMID:24863890

Hossen, M. Z., Hussain, M. E., Hakim, A., Islam, K., Uddin, M. N., & Azad, A. K. (2019). Biodegradation of reactive textile dye Novacron Super Black G by free cells of newly isolated *Alcaligenes faecalis AZ26* and *Bacillus spp* obtained from textile effluents. *Heliyon*, *5*(7), e02068. doi:10.1016/j.heliyon.2019.e02068 PMID:31338473

Hou, X., Yu, M., Liu, A., Li, Y., Ruan, T., Liu, J., Schnoor, J. L., & Jiang, G. (2018). Biotransformation of tetrabromobisphenol A dimethyl ether back to tetrabromobisphenol A in whole pumpkin plants. *Environmental Pollution*, *241*, 331–338. doi:10.1016/j.envpol.2018.05.075 PMID:29843015

Hussain, A., Abbas, N., Arshad, F., Akram, M., Khan, Z. I., Ahmad, K., Mansha, M., & Mirzaei, F. (2013). Effects of diverse doses of lead (Pb) on different growth attributes of *Zea mays* L. *Agricultural Sciences*, *4*(5), 262–265. doi:10.4236/as.2013.45037

Hyman, M., & Dupont, R. R. (2001). *Groundwater and soil remediation: process design and cost estimating of proven technologies* (Vol. 137). ASCE Press. doi:10.1061/9780784404270

IAEA. (1999). *Review of the factors affecting the selection and implementation of waste management technologies; IAEA-TECDOC-1096*. IAEA.

IAEA (International Atomic Energy Agency). (1970). *Standardization of radioactive waste categories; TRS no. 101*. IAEA.

Imran, M. A., Ch, M. N., Khan, R. M., Ali, Z., & Mahmood, T. (2013). Toxicity of arsenic (As) on seed germination of sunflower (*Helianthus annuus L.*). *International Journal of Physical Sciences*, *8*(17), 840–847. doi:10.5897/IJPS2013.3894

Islami, F., Torre, L. A., & Jemal, A. (2015). Global trends of lung cancer mortality and smoking prevalence. *Translational Lung Cancer Research*, *4*(4), 327. PMID:26380174

Ismail, H. Y., Ijah, U. J. J., Riskuwa, M. L., & Allamin, I. I. (2014). Biodegradation of spent engine oil by bacteria isolated from the rhizosphere of legumes grown in contaminated soil. *International Journal of Environment*, *3*(2), 63–75. doi:10.3126/ije.v3i2.10515

Jambon, I., Thijs, S., Weyens, N., & Vangronsveld, J. (2018). Harnessing plant-bacteria-fungi interactions to improve plant growth and degradation of organic pollutants. *Journal of Plant Interactions*, *13*(1), 119–130. doi:10.1080/17429145.2018.1441450

Jeong, S. W., Jeong, J., & Kim, J. (2015). Simple surface foam application enhances bioremediation of oil-contaminated soil in cold conditions. *Journal of Hazardous Materials*, *286*, 164–170. doi:10.1016/j.jhazmat.2014.12.058 PMID:25577318

Jesubunmi, C. O. (2014). Isolation of oil-degrading microorganisms in spent engine oil-contaminated soil. *Journal of Biology, Agriculture and Healthcare*, *4*(24), 191–195.

Jiang, Q. Y., Zhuo, F., Long, S. H., Zhao, H. D., Yang, D. J., Ye, Z. H., & Jing, Y. X. (2016). Can arbuscular mycorrhizal fungi reduce Cd uptake and alleviate Cd toxicity of *Lonicera japonica* grown in Cd-added soils? *Scientific Reports*, *6*(1), 1–9. doi:10.1038rep21805 PMID:26892768

Jobling, S., Nolan, M., Tyler, C. R., Brighty, G., & Sumpter, J. P. (1998). Widespread sexual disruption in wild fish. *Environmental Science & Technology*, *32*(17), 2498–2506. doi:10.1021/es9710870

Joutey, N. T., Bahafid, W., Sayel, H., & El Ghachtouli, N. (2013). Biodegradation: involved microorganisms and genetically engineered microorganisms. *Biodegradation-Life of Science*, 289-320.

Junghare, M., Spiteller, D., & Schink, B. (2019). Anaerobic degradation of xenobiotic isophthalate by the fermenting bacterium *Syntrophorhabdus aromaticivorans. The ISME Journal*, *13*(5), 1252–1268. doi:10.103841396-019-0348-5 PMID:30647456

Kabra, A. N., Khandare, R. V., & Govindwar, S. P. (2013). Development of a bioreactor for remediation of textile effluent and dye mixture: A plant–bacterial synergistic strategy. *Water Research*, *47*(3), 1035–1048. doi:10.1016/j.watres.2012.11.007 PMID:23245543

Kabra, A. N., Khandare, R. V., Kurade, M. B., & Govindwar, S. P. (2011). Phytoremediation of a sulphonated azo dye Green HE4B by *Glandularia pulchella* (Sweet) Tronc.(Moss Verbena). *Environmental Science and Pollution Research International*, *18*(8), 1360–1373. doi:10.100711356-011-0491-7 PMID:21465161

Kaewlaoyoong, A., Cheng, C. Y., Lin, C., Chen, J. R., Huang, W. Y., & Sriprom, P. (2020). White rot fungus Pleurotus pulmonarius enhanced bioremediation of highly PCDD/F-contaminated field soil via solid state fermentation. *The Science of the Total Environment*, *738*, 139670. doi:10.1016/j.scitotenv.2020.139670 PMID:32534283

Kagalkar, A. N., Khandare, R. V., & Govindwar, S. P. (2015). Textile dye degradation potential of plant laccase significantly enhances upon augmentation with redox mediators. *RSC Advances*, *5*(98), 80505–80517. doi:10.1039/C5RA12454A

Kasno, A., Adiningsih, S., & Subowo, S. (2000). Pollution status of Lead and Cadmium in lowland rice intensifies the Pantura route of West Java. *Journal of Soil and Environmental Sciences*, *3*(2), 25–32.

Kästner, M., & Miltner, A. (2016). Application of compost for effective bioremediation of organic contaminants and pollutants in soil. *Applied Microbiology and Biotechnology*, *100*(8), 3433–3449. doi:10.100700253-016-7378-y PMID:26921182

Khan, S. M., & Gomes, J. (2018). *An interdisciplinary population health approach to the radon health risk management in Canada*. Academic Press.

Khandare, R. V., Kabra, A. N., Awate, A. V., & Govindwar, S. P. (2013). Synergistic degradation of diazo dye Direct Red 5B by *Portulaca grandiflora* and *Pseudomonas putida. International Journal of Environmental Science and Technology*, *10*(5), 1039–1050. doi:10.100713762-013-0244-x

Klangsin, P., & Harding, A. K. (1998). Medical waste treatment and disposal methods used by hospitals in Oregon, Washington, and Idaho. *Journal of the Air & Waste Management Association*, *48*(6), 516–526. doi:10.1080/10473289.1998.10463706 PMID:9949738

Krishnan, J., Kishore, A. A., Suresh, A., Madhumeetha, B., & Prakash, D. G. (2017). Effect of pH, inoculum dose and initial dye concentration on the removal of azo dye mixture under aerobic conditions. *International Biodeterioration & Biodegradation*, *119*, 16–27. doi:10.1016/j.ibiod.2016.11.024

Kuhad, R. C., Singh, S., & Singh, A. (2011). Phosphate-solubilizing microorganisms. In *Bioaugmentation, Biostimulation and Biocontrol* (pp. 65–84). Springer. doi:10.1007/978-3-642-19769-7_4

Kumar, M., Prasad, R., Goyal, P., Teotia, P., Tuteja, N., Varma, A., & Kumar, V. (2017). Environmental biodegradation of xenobiotics: role of potential microflora. In *Xenobiotics in the Soil Environment* (pp. 319–334). Springer. doi:10.1007/978-3-319-47744-2_21

Kumar, P. S. (2019). Soil bioremediation techniques. In *Advanced Treatment Techniques for Industrial Wastewater* (pp. 35–50). IGI Global. doi:10.4018/978-1-5225-5754-8.ch003

Kurasvili, M. V., Adamia, G. S., Ananiasvili, T. I., Varazi, T. G., Pruidze, M. V., Gordeziani, M. S., & Khatisashvili, G. A. (2014). Plants as tools for control and remediation of the environment polluted by organochlorine toxicants. *Annals of Agrarian Science*, *12*(3), 84–87.

Kurnia, Sutono, Anda, Sulaeman, Kurniawansyah, & Tala'ohu. (2000). *Assessment of soil quality standards on agricultural land.* Final Report on Research Collaboration between Bapedal and Puslitbangtanak.

Lalnunhlimi, S., & Krishnaswamy, V. (2016). Decolorization of azo dyes (Direct Blue 151 and Direct Red 31) by moderately alkaliphilic bacterial consortium. *Brazilian Journal of Microbiology, 47*(1), 39-46.

Li, Y., Chuang, Y. H., Sallach, J. B., Zhang, W., Boyd, S. A., & Li, H. (2018). Potential metabolism of pharmaceuticals in radish: Comparison of in vivo and in vitro exposure. *Environmental Pollution*, *242*, 962–969. doi:10.1016/j.envpol.2018.07.060 PMID:30373041

Limmer, M., & Burken, J. (2016). Phytovolatilization of organic contaminants. *Environmental Science & Technology*, *50*(13), 6632–6643. doi:10.1021/acs.est.5b04113 PMID:27249664

Lipczynska-Kochany, E. (2018). Humic substances, their microbial interactions and effects on biological transformations of organic pollutants in water and soil: A review. *Chemosphere*, *202*, 420–437. doi:10.1016/j.chemosphere.2018.03.104 PMID:29579677

Lu, S., Teng, Y., Wang, J., & Sun, Z. (2010). Enhancement of pyrene removed from contaminated soils by *Bidens maximowicziana. Chemosphere*, *81*(5), 645–650. doi:10.1016/j.chemosphere.2010.08.022 PMID:20832842

Ma, Y., Prasad, M. N. V., Rajkumar, M., & Freitas, H. (2011). Plant growth promoting rhizobacteria and endophytes accelerate phytoremediation of metalliferous soils. *Biotechnology Advances*, *29*(2), 248–258. doi:10.1016/j.biotechadv.2010.12.001 PMID:21147211

Mahar, A., Wang, P., Ali, A., Awasthi, M. K., Lahori, A. H., Wang, Q., & Zhang, Z. (2016). Challenges and opportunities in the phytoremediation of heavy metals contaminated soils: A review. *Ecotoxicology and Environmental Safety*, *126*, 111–121. doi:10.1016/j.ecoenv.2015.12.023 PMID:26741880

Manzoor, J., & Sharma, M. (2019). Impact of Biomedical Waste on Environment and Human Health. *Environmental Claims Journal*, *31*(4), 311–334. doi:10.1080/10406026.2019.1619265

Martínez-Pascual, E., Grotenhuis, T., Solanas, A. M., & Viñas, M. (2015). Coupling chemical oxidation and biostimulation: Effects on the natural attenuation capacity and resilience of the native microbial community in alkylbenzene-polluted soil. *Journal of Hazardous Materials*, *300*, 135–143. doi:10.1016/j.jhazmat.2015.06.061 PMID:26177489

McGrath, S. P., Zhao, F. J., & Lombi, E. (2001). Plant and rhizosphere process involved in phytoremediation of metal-contaminated soils. *Plant and Soil*, *232*(1/2), 207–214. doi:10.1023/A:1010358708525

Meudec, A., Poupart, N., Dussauze, J., & Deslandes, E. (2007). Relationship between heavy fuel oil phytotoxicity and polycyclic aromatic hydrocarbon contamination in Mishra, V. K., Singh, G., & Shukla, R. (2019). Impact of Xenobiotics under a Changing Climate Scenario. In *Climate Change and Agricultural Ecosystems* (pp. 133–151). Woodhead Publishing.

Moharrery, L., Otadi, M., Miraly, N., Rezaei Zangeneh, M. M., & Amiri, R. (2019). Degradation of toluidine red, an oil soluble azo dye by *Halomonas strain IP8* at alkaline condition. *Chemical Engineering Communications*, *206*(1), 61–68. doi:10.1080/00986445.2018.1472587

Monica, S., Karthik, L., Mythili, S., & Sathiavelu, A. (2011). Formulation of effective microbial consortia and its application for sewage treatment. *J Microbial Biochem Technol, 3*, 51-55.

Nawrot, T., Plusquin, M., Hogervorst, J., Roels, H. A., Celis, H., Thijs, L., & Staessen, J. A. (2006). Environmental exposure to cadmium and risk of cancer: A prospective population-based study. *The Lancet. Oncology*, *7*(2), 119–126. doi:10.1016/S1470-2045(06)70545-9 PMID:16455475

Nematshahi, N., Lahouti, M., & Ganjeali, A. (2012). Accumulation of chromium and its effect on growth of (*Allium cepa* cv. Hybrid). *European Journal of Experimental Biology*, 2(4), 969–974.

Nishimoto, R. (2019). Global trends in the crop protection industry. *Journal of pesticide science*, D19-101.Nwoko, C. O. (2010). Trends in phytoremediation of toxic elemental and organic pollutants. *African Journal of Biotechnology*, *9*(37), 6010–6016.

Organum, N., & Bacon, F. (2006). Bioremediation technologies. In P. J. J. Alvarez & W. A. Illman (Eds.), *Bioremediation and natural attenuation* (pp. 351–455). John Wiley & Sons.

Pandey, J., Verma, R. K., & Singh, S. (2019). Suitability of aromatic plants for phytoremediation of heavy metal contaminated areas: A review. *International Journal of Phytoremediation*, *21*(5), 405–418. doi:10.1080/15226514.2018.1540546 PMID:30656974

Pandey, V. C., Pandey, D. N., & Singh, N. (2015). Sustainable phytoremediation based on naturally colonizing and economically valuable plants. *Journal of Cleaner Production*, *86*, 37–39. doi:10.1016/j.jclepro.2014.08.030

Panfili, I., Bartucca, M. L., Ballerini, E., & Del Buono, D. (2017). Combination of aquatic species and safeners improves the remediation of copper polluted water. *The Science of the Total Environment*, *601*, 1263–1270. doi:10.1016/j.scitotenv.2017.06.003 PMID:28605844

Panfili, I., Bartucca, M. L., & Del Buono, D. (2019). The treatment of duckweed with a plant biostimulant or a safener improves the plant capacity to clean water polluted by terbuthylazine. *The Science of the Total Environment*, *646*, 832–840. doi:10.1016/j.scitotenv.2018.07.356 PMID:30064109

Paz-Alberto, A. M., & Sigua, G. C. (2013). Phytoremediation: A green technology to remove environmental pollutants. *American Journal of Climate Change*, 2(01), 71–86. doi:10.4236/ajcc.2013.21008

Pereda Reyes, I., & Sárvári Horváth, I. (2015). *Anaerobic Biodegradation of Solid Substrates from Agroindustrial Activities—Slaughterhouse Wastes and Agrowastes*. Academic Press.

Philp, J. C., & Atlas, R. M. (2005). Bioremediation of contaminated soils and aquifers. In *Bioremediation* (pp. 139–236). American Society of Microbiology.

Pilon-Smits, E. A. H. (2005). Phytoremediation. *Annual Review of Plant Biology, 56*(1), 15–39. doi:10.1146/annurev.arplant.56.032604.144214 PMID:15862088

Pimda, W., & Bunnag, S. (2012). Biodegradation of used motor oil by single and mixed cultures of cyanobacteria. *African Journal of Biotechnology, 11*(37), 9074–9078.

Prüss-Üstün, A., Giroult, E., Rushbrook, P., & World Health Organization. (1999). *Safe management of wastes from health-care activities*. World Health Organization.

Purdom, C. E., Hardiman, P. A., Bye, V. V. J., Eno, N. C., Tyler, C. R., & Sumpter, J. P. (1994). Estrogenic effects of effluents from sewage treatment works. *Chemistry and Ecology, 8*(4), 275–285. doi:10.1080/02757549408038554

Rahbar, M. H., Samms-Vaughan, M., Hessabi, M., Dickerson, A. S., Lee, M., Bressler, J., & Shakespeare-Pellington, S. (2016). Concentrations of polychlorinated biphenyls and organochlorine pesticides in umbilical cord blood serum of newborns in Kingston, Jamaica. *International Journal of Environmental Research and Public Health, 13*(10), 1032. doi:10.3390/ijerph13101032 PMID:27775677

Rahman, R. O. (2011). Abdel et al., "Liquid Radioactive Wastes Treatment: A Review". *Water*, 551-565.

Rane, N. R., Patil, S. M., Chandanshive, V. V., Kadam, S. K., Khandare, R. V., Jadhav, J. P., & Govindwar, S. P. (2016). Ipomoea hederifolia rooted soil bed and Ipomoea aquatica rhizofiltration coupled phytoreactors for efficient treatment of textile wastewater. *Water Research, 96*, 1–11. doi:10.1016/j.watres.2016.03.029 PMID:27016633

Reddy, M. S., Kour, M., Aggarwal, S., Ahuja, S., Marmeisse, R., & Fraissinet-Tachet, L. (2016). Metal induction of a P isolithus albus metallothionein and its potential involvement in heavy metal tolerance during mycorrhizal symbiosis. *Environmental Microbiology, 18*(8), 2446–2454. doi:10.1111/1462-2920.13149 PMID:26626627

Reineke, W., & Knackmuss, H. J. (1988). Microbial degradation of haloaromatics. *Annual Review of Microbiology, 42*(1), 263–287. doi:10.1146/annurev.mi.42.100188.001403 PMID:3059995

Rosi-Marshall, E. (2013). *Streams stressed by pharmaceutical pollution*. www.environmentalchange.nd.edu/events/2

Rylott, E. L., & Bruce, N. C. (2009). Plants disarm soil: Engineering plants for the phytoremediation of explosives. *Trends in Biotechnology, 27*(2), 73–81. doi:10.1016/j.tibtech.2008.11.001 PMID:19110329

Saba, B., Jabeen, M., Khalid, A., Aziz, I., & Christy, A. D. (2015). Effectiveness of rice agricultural waste, microbes and wetland plants in the removal of reactive black-5 azo dye in microcosm constructed wetlands. *International Journal of Phytoremediation, 17*(11), 1060–1067. doi:10.1080/15226514.2014.1003787 PMID:25849115

Safari, M., Ahmadi, A. S., & Soltani, N. (2016). *The Potential Of Cyanobacterium Fischerella Ambigua Isc67 in Biodegradation Of Crude Oil*. Academic Press.

Samir, D., Mohcem, R., Selma, O., & Asma, S. (2020). The Effect of Herbicide Metribuzin on Environment and Human: A Systematic Review. *Pharmaceutical and Biosciences Journal*. Available at www.ukjpb.com

San Miguel, A., Ravanel, P., & Raveton, M. (2013). A comparative study on the uptake and translocation of organochlorines by *Phragmites australis*. *Journal of Hazardous Materials*, *244*, 60–69. doi:10.1016/j.jhazmat.2012.11.025 PMID:23246941

Saran, A., Fernandez, L., Cora, F., Savio, M., Thijs, S., Vangronsveld, J., & Merini, L. J. (2020). Phytostabilization of Pb and Cd polluted soils using *Helianthus petiolaris* as pioneer aromatic plant species. *International Journal of Phytoremediation*, *22*(5), 459–467. doi:10.1080/15226514.2019.1675140 PMID:31602996

Schmidt, U. (2003). Enhancing phytoremediation: The effect of chemical soil manipulation on mobility, plant accumulation, and leaching of heavy metals. *Journal of Environmental Quality*, *32*, 1939–1954. doi:10.2134/jeq2003.1939 PMID:14674516

Schnoor, J. L., Light, L. A., McCutcheon, S. C., Wolfe, N. L., & Carreia, L. H. (1995). Phytoremediation of organic and nutrient contaminants. *Environmental Science & Technology*, *29*(7), 318A–323A. doi:10.1021/es00007a747 PMID:22667744

Schwartz, C., Echevarria, G., & Morel, J. L. (2003). Phytoextraction of cadmium with *Thlaspi caerulescens*. *Plant and Soil*, *249*(1), 27–35. doi:10.1023/A:1022584220411

Shanmugam, L., Ahire, M., & Nikam, T. (2020). *Bacopa monnieri (L.) Pennell*, a potential plant species for degradation of textile azo dyes. *Environmental Science and Pollution Research International*, *27*(9), 9349–9363. doi:10.100711356-019-07430-x PMID:31912399

Sharma, B., Dangi, A. K., & Shukla, P. (2018). Contemporary enzyme based technologies for bioremediation: A review. *Journal of Environmental Management*, *210*, 10–22. doi:10.1016/j.jenvman.2017.12.075 PMID:29329004

Shehzadi, M., Afzal, M., Khan, M. U., Islam, E., Mobin, A., Anwar, S., & Khan, Q. M. (2014). Enhanced degradation of textile effluent in constructed wetland system using *Typha domingensis* and textile effluent-degrading endophytic bacteria. *Water Research*, *58*, 152–159. doi:10.1016/j.watres.2014.03.064 PMID:24755300

Shekar, C. H. C., Sammaiah, D., Shasthree, T., & Reddy, K. J. (2011). Effect of mercury on tomato growth and yield attributes. *International Journal of Pharma and Bio Sciences*, *2*(2), B358–B364.

Shrestha, P., Bellitürk, K., & Görres, J. H. (2019). Phytoremediation of heavy metal-contaminated soil by switchgrass: A comparative study utilizing different composts and coir fiber on pollution remediation, plant productivity, and nutrient leaching. *International Journal of Environmental Research and Public Health*, *16*(7), 1261. doi:10.3390/ijerph16071261 PMID:30970575

Singh, P., Jain, R., Srivastava, N., Borthakur, A., Pal, D. B., Singh, R., & Mishra, P. K. (2017). Current and emerging trends in bioremediation of petrochemical waste: A review. *Critical Reviews in Environmental Science and Technology*, *47*(3), 155–201. doi:10.1080/10643389.2017.1318616

Singh, R. (2014). Microorganism as a tool of bioremediation technology for cleaning environment: A review. *Proceedings of the International Academy of Ecology and Environmental Sciences*, *4*(1), 1.

Smith, M. J., Flowers, T. H., Duncan, H. J., & Alder, J. (2006). Effects of polycyclic aromatic hydrocarbons on germination and subsequent growth of grasses and legumes in freshly contaminated soil and soil with aged PAHs residues. *Environmental Pollution*, *141*(3), 519–525. doi:10.1016/j.envpol.2005.08.061 PMID:16246476

Srikantan, C., Suraishkumar, G. K., & Srivastava, S. (2018). Effect of light on the kinetics and equilibrium of the textile dye (Reactive Red 120) adsorption by *Helianthus annuus* hairy roots. *Bioresource Technology*, *257*, 84–91. doi:10.1016/j.biortech.2018.02.075 PMID:29486410

Srivastava, N. (2020). Phytoremediation of Toxic Metals/Metalloids and Pollutants by *Brassicaceae* Plants. In *The Plant Family Brassicaceae* (pp. 409–435). Springer. doi:10.1007/978-981-15-6345-4_14

Stroo, H. F., Leeson, A., & Ward, C. H. (Eds.). (2012). *Bioaugmentation for groundwater remediation* (Vol. 5). Springer Science & Business Media.

Sun, J., Chen, Q., Qian, Z., Zheng, Y., Yu, S., & Zhang, A. (2018). Plant uptake and metabolism of 2, 4-dibromophenol in carrot: In vitro enzymatic direct conjugation. *Journal of Agricultural and Food Chemistry*, *66*(17), 4328–4335. doi:10.1021/acs.jafc.8b00543 PMID:29656645

Susarla, S., Medina, V. F., & McCutcheon, S. C. (2002). Phytoremediation: An ecological solution to organic chemical contamination. *Ecological Engineering*, *18*(5), 647–658. doi:10.1016/S0925-8574(02)00026-5

Tamás, L., Mistrík, I., Huttová, J., Halusková, L., Valentovicová, K., & Zelinová, V. (2010). Role of reactive oxygen species-generating enzymes and hydrogen peroxide during cadmium, mercury and osmotic stresses in barley root tip. *Planta*, *231*(2), 221–231. doi:10.100700425-009-1042-z PMID:19898864

Tan, L., He, M., Song, L., Fu, X., & Shi, S. (2016). Aerobic decolorization, degradation and detoxification of azo dyes by a newly isolated salt-tolerant yeast *Scheffersomyces spartinae TLHS-SF1*. *Bioresource Technology*, *203*, 287–294. doi:10.1016/j.biortech.2015.12.058 PMID:26744802

Tano, Z. J. (2011). *7 Ecological Effects of Pesticides*. Academic Press.

Theerachat, M., Guieysse, D., Morel, S., Remaud-Siméon, M., & Chulalaksananukul, W. (2019). Laccases from marine organisms and their applications in the biodegradation of toxic and environmental pollutants: A review. *Applied Biochemistry and Biotechnology*, *187*(2), 583–611. doi:10.100712010-018-2829-9 PMID:30009326

Tomei, M. C., & Daugulis, A. J. (2013). Feasibility of operating a solid–liquid bioreactor with used automobile tires as the sequestering phase for the biodegradation of inhibitory compounds. *Journal of Environmental Management*, *125*, 7–11. doi:10.1016/j.jenvman.2013.03.047 PMID:23629012

Tran, H. T., Lin, C., Bui, X. T., Ngo, H. H., Cheruiyot, N. K., Hoang, H. G., & Vu, C. T. (1920). Aerobic composting remediation of petroleum hydrocarbon-contaminated soil. Current and future perspectives. *The Science of the Total Environment*, *753*, 142250. doi:10.1016/j.scitotenv.2020.142250 PMID:33207468

TuomistoH. L.ScheelbeekP. F.ChalabiZ.GreenR.SmithR. D.HainesA.DangourA. D. (2017). Effects of environmental change on agriculture, nutrition and health: A framework with a focus on fruits and vegetables. Wellcome open research, *2*. doi:10.12688/wellcomeopenres.11190.2

Uysal, A. K., & Gunal, S. (2014). The impact of preprocessing on text classification. *Information Processing & Management, 50*(1), 104–112. doi:10.1016/j.ipm.2013.08.006

Vasavi, A., Usha, R., & Swamy, P. M. (2010). Phytoremediation–an overview review. *J Ind Pollut Control, 26*(1), 83–88.

Visconti, D., Álvarez-Robles, M. J., Fiorentino, N., Fagnano, M., & Clemente, R. (2020). Use of *Brassica juncea* and *Dactylis glomerata* for the phytostabilization of mine soils amended with compost or biochar. *Chemosphere, 260*, 127661. doi:10.1016/j.chemosphere.2020.127661 PMID:32688327

Wang, L., Ji, B., Hu, Y., Liu, R., & Sun, W. (2017). A review on in situ phytoremediation of mine tailings. *Chemosphere, 184*, 594–600. doi:10.1016/j.chemosphere.2017.06.025 PMID:28623832

Ward, M. H., Lubin, J., Giglierano, J., Colt, J. S., Wolter, C., Bekiroglu, N., Camann, D., Hartge, P., & Nuckols, J. R. (2006). Proximity to crops and residential exposure to agricultural herbicides in Iowa. *Environmental Health Perspectives, 114*(6), 893–897. doi:10.1289/ehp.8770 PMID:16759991

Watharkar, A. D., & Jadhav, J. P. (2014). Detoxification and decolorization of a simulated textile dye mixture by phytoremediation using Petunia grandiflora and, Gailardia grandiflora: A plant–plant consortial strategy. *Ecotoxicology and Environmental Safety, 103*, 1–8. doi:10.1016/j.ecoenv.2014.01.033 PMID:24561240

Wenzel, W. W. (2009). Rhizosphere processes and management in plant-assisted bioremediation (phytoremediation) of soils. *Plant and Soil, 321*(1-2), 385–408. doi:10.100711104-008-9686-1

Whelan, M. J., Coulon, F., Hince, G., Rayner, J., McWatters, R., Spedding, T., & Snape, I. (2015). Fate and transport of petroleum hydrocarbons in engineered biopiles in polar regions. *Chemosphere, 131*, 232–240. doi:10.1016/j.chemosphere.2014.10.088 PMID:25563162

Wild, E., Dent, J., Thomas, G. O., & Jones, K. C. (2005). Direct observation of organic contaminant uptake, storage, and metabolism within plant roots. *Environmental Science & Technology, 39*(10), 3695–3702. doi:10.1021/es048136a PMID:15952374

Wiszniewska, A., Hanus-Fajerska, E., Muszyńska, E., & Ciarkowska, K. (2016). Natural organic amendments for improved phytoremediation of polluted soils: A review of recent progress. *Pedosphere, 26*(1), 1–12. doi:10.1016/S1002-0160(15)60017-0

Xu, X., Wen, B., Huang, H., Wang, S., Han, R., & Zhang, S. (2016). Uptake, translocation and biotransformation kinetics of BDE-47, 6-OH-BDE-47 and 6-MeO-BDE-47 in maize (Zea mays L.). *Environmental Pollution, 208*(Pt B), 714–722.

Zhang, Q., Kong, W., Wei, L., Wang, Y., Luo, Y., Wang, P., & Jiang, G. (2020). Uptake, phytovolatilization, and interconversion of 2, 4-dibromophenol and 2, 4-dibromoanisole in rice plants. *Environment International, 142*, 105888. doi:10.1016/j.envint.2020.105888 PMID:32593840

Zhang, Q., Liu, Y., Lin, Y., Kong, W., Zhao, X., Ruan, T., & Jiang, G. (2019). Multiple metabolic pathways of 2, 4, 6-tribromophenol in rice plants. *Environmental Science & Technology, 53*(13), 7473–7482. doi:10.1021/acs.est.9b01514 PMID:31244074

Zhang, Y., Yu, Z., Fu, X., & Liang, C. (2002). Noc3p, a bHLH protein, plays an integral role in the initiation of DNA replication in budding yeast. *Cell, 109*(7), 849–860. doi:10.1016/S0092-8674(02)00805-X PMID:12110182

Chapter 3
Microbes and Their Role in Bioremediation of Soil:
A Detailed Review

A. Madhavi
Sri Krishnadevaraya University, India

M. Srinivasulu
Yogi Vemana University, India

V. Rangaswamy
Sri Krishnadevaraya University, India

ABSTRACT

Soil is the Earth's shell and is getting polluted in a number of ways in the present scenario. Human activities are the root cause of different types of soil pollution, which is an alarming issue and has become a major obstacle that needs to be overcome to build a cleaner environment. The area of polluted soil is widening day by day by virtue of a sharp increase in people from all over the world. It has been expected that the global population will continue to increase up to 9 billion by 2050, and such prodigious population may be in need of advanced agricultural and industrial systems, which may inevitably cause soil pollution. Therefore, it is essential to control soil pollution, and fortunately, the solution for this is microbes that are the real creatures of life on Earth. In fact, microorganisms play a unique role in the detoxification of polluted soil environments, and in the last several years, this process has been called bioremediation. Remediation of polluted soils is necessary, and research continues to develop novel, science-based remediation methods.

DOI: 10.4018/978-1-7998-7062-3.ch003

INTRODUCTION

Soil on the surface of the earth is a diverse natural entity which is home to a large amount of living elements, including plants, animals and microbes that communicate with each other (Dwivedi, 1997). Soil filters water, decomposes waste, stores heat and exchanges the gases and therefore have great bearing on environmental balance. As the life on earth mainly concentrates on the top of soil, hence, it is extremely important to pay attention on pollutants or hazardous substances affecting predominantly the soil ecosystems. In the past few years an estimated 12.6 million people have lost their lives worldwide from more than 100 diseases resulting from unhealthy environments such as contaminated soils (WHO, 2016). The formation of 1 cm top layer of soil requires 100-400 years (Chandra & Singh, 2009). Soil is the layer of mixture of inorganic and organic material, where inorganic part is composed of fine rock particles produced as a result of weathering and the organic part is produced by decay of plants and animals. Life is believed to emerge from the soil and is an integral part of the environment, ecosystem and also an important natural resource for plant growth, and is a repository for biogeochemical cycle. Soil is highly susceptible to environmental transformations (Yu, 2016) and is often the most important sink for environmental pollution due to its strong binding capacity (Sun et al., 2017). According to Rodriguez et al. (2018), soil pollution is defined as the presence of chemicals or substances in the soil that are inappropriate or at an increased concentration than normal with deleterious effects on any non-target organism. A contaminant is an unwanted substance introduced into the environment. Harmful effects by contaminants lead to pollution, a process by which a resource (natural or man-made) is rendered unsuitable for use. Plants, animals and aquatic life depend on soil for their survival. Plants relay upon soil for anchorage, nutrients, water and even oxygen. The soil influences the distribution of plant species and provides a habitat for a large number of organisms such as both micro and macro organisms. Soils are essential for biodiversity conservation above and below the ground. Huge amount of chemicals employed in day to day lives and excessive amounts of urban, industrial and agricultural wastes, mining etc., have all led to soil contamination across the planet and also leaving it barren and deteriorated.

Industrialization and extensive use of chemical compounds such as petroleum products, hydrocarbons (aliphatic, aromatic, polycyclic aromatic hydrocarbons (PAHs), BTEX (benzene, toluene, ethylbenzene and xylene), chlorinated hydrocarbons such as polychlorinated biphenyls (PCBs), trichloroethylene (TCE) and perchloroethylene, nitroaromatic compounds, organophosphorus compounds) pose an alarming threat to crop production, food safety, and for the health of citizens. Since soil quality is directly linked to food security, human health and sustainable economic and social progress, soil pollution management is important (Esmaeili et al., 2013; Wan et al., 2018). Biological life prevailing in a gram of soil includes tiny microbes such as algae, actinomycetes, bacteria, bacteriophages, protozoa, nematodes and fungi. The role of these organisms is highly complex and form an integral part of cycling the nutrients through the environment and they drive the processes such as decomposition, mineralization, storage and release of nutrients, breakdown of pollutants before they reach groundwater or surface water, carbon cycling, carbon sequestration, and soil organic matter transformations, nitrogen cycling (N fixation, denitrification, nitrification).

The biological transformation by the action of microorganisms led to development of abundant nutrients (Kiflu & Beyene, 2013). Soil microbes are the principal participants of all the soil biochemical processes. These biochemical processes are devices for soil quality stabilization, soil organic matter production, hazardous material decomposition, soil structure formation and physiological cycles. Soil degradation by harmful metals reduces the microbial properties of the soil, such as soil respiration and enzymatic processes. One of the reasons that impact life in soils is the degradation of soils by highly poisonous materials attributable to multiple anthropogenic activities (Prajapati & Meravi, 2014; Zojiali et al., 2014; Baishya & Samra, 2014). Elements with high density and high relative atomic weight are inherently poisonous elements, exhibiting metallic properties such as ductility, malleability, conductivity and specificity of the ligand (Algreen et al., 2012). Especially zinc, cadmium and copper are the potentially toxic elements that may alter the microbiological equilibrium of soil (Olaniran et al., 2013; Liu et al., 2013; Markowicz et al., 2016; Shi & Ma, 2017). Finally, soil contaminated by such potentially toxic elements (PTEs) has led to negative impact on the environment. In the soil microbes are the first to react to PTEs and microbial metabolisms can interfere PTE speciation change (Bolan et al., 2013).

SOURCES OF SOIL POLLUTION

Agricultural Practices

Agriculture is one of the main pillars of economy and principal productive sectors, and the main land use activity in many countries. Agriculture is a basic industry, which provides endless power for the development of national economy and it is also the foundation for human survival and development. Agricultural wastes are those produced by agricultural and livestock practices such as fertilizer containers, agricultural pesticides, feed, harvest residues, and manure. In soils and sediments, the prolonged application of pesticides persists where they can directly penetrate the food chain or percolate down to the water table. Not only in farming areas, but also in schools, parks, highways, houses, buildings and trees, pesticides are used almost everywhere and it is impossible to find any location where pesticides are not used - from the can of bug spray under the kitchen sink to the aircraft crop dusting acres of farmland. The farming activities contribute to the soil pollution with harmful substances such as cadmium by the use of mineral phosphate fertilizers or organic pollutants due to application of pesticides (Kanianska, 2016). Exploitation of chemical fertilizers and pesticides in crop production brought about soil pollution. Soil pollution is a result of long-term accumulation and a large number of pollutants accumulated in the soil, which inturn lead to the extension of pollution, such as ground water pollution. It appears to be difficult to control soil pollution. Contaminants from agrochemical sources include pesticides, fertilizers and manure. The crop protection products and fertilizers are chemicals that are manufactured synthetically and broken down into numerous soil components and they gradually bring down the fertility and quality of the soil (Usman, 2018). Pesticide is a generic term that comprises of all the chemicals used to kill or control pests either in farming sector or in different settings such as store rooms, human houses and gardens as noted by the Food and Agricultural Organization (FAO) of the United Nation (FAO, 2002). The pesticide formulations were utilized to control, eliminate and in preventing any pests, which includes rodents, nematodes, weeds, birds, insects and microbes. These chemicals are classified into herbicides, insecticides, fungicides, nematicides and rodenticides. The annual increase in worldwide pesticides production is 11% from 0.2 million tons in the 1950s and exceeding 5 million tons by 2000 (Carvalho, 2017). The chemical pesticides applied to farm field in 2012; on an average is around 3.8 million tons (FAO, 2020). About two million people chiefly, livings in the developing economies are at an elevated health risks because of pesticides utilization (Hicks, 2019). Pesticides cause damage to soil biomass and microorganisms such as bacteria, fungi, and earthworms. The labile component of organic matter in soil is microbial biomass which plays a significant role in soil nutrient element cycle

(Azam et al., 2003). Quality of the soil is a major factor for the growth of crop plants and the deciding factor for the availability of plant nutrients. Microorganisms present in the soil are able to metabolize and degrade plenty of pollutants and pesticides. Healthy levels of soil microbes are essential for preserving soil structure and soil fertility. The soil fungi, algae, cyanobacteria and actinomycetes are mainly involve in the decomposition of organic residues and release the nutrients including phosphorus, which enhance plant growth and contribute to the pollution control. The biological transformation by the action of microorganisms led to accumulation (develop) of abundant nutrients in the soil (Kiflu & Beyene, 2013). Pesticides may cause considerable changes in the composition, diversity and basic functioning of important soil microflora (Ahemad & Khan, 2013; Yousaf et al., 2013; Riah et al., 2014). Soil enzymes help in speedup chemical reactions in soils, regulate cellular metabolism of soil organisms, participate in the decomposition of organic matter and also play a key role in the formation of humus. The quality and fertility of soil depend to a great extent on the activity of soil enzymes. Soil enzyme activities (SEA) are sensitive to management practices (Medeiros et al., 2015).

Industrial Wastes

Heavy Metals

The disposal of industrial wastes is a serious issue for soil pollution. About 90% of soil pollution is caused by industrial waste products. Industrial wastes may be liquid, solid, or gaseous. The types of industrial waste that are discharged into the environment and which are harmful are the scrap lumber, scrap metal dirt and gravel, plastics, oil, solvents, masonry and concrete, chemicals, even vegetable matters from restaurants (Awuchi & Awuchi, 2019a, 2019b). Heavy metals are the life threatening group of soil pollutants produced from natural processes and anthropogenic sources such as industrial, agricultural, military activities, tannery, dyeing, mining, sewage sludge, electroplating and waste water treatment plants. Soil pollution with heavy metals is a major challenge because of their adverse consequences on the living biota. Metals and metalloids are heavy metals and have a greater density relative to water (Dotaniya et al., 2016). Heavy metal is called a metal with a specific gravity greater than 5.0 or an atomic number greater than 20 (Dotaniya et al., 2013).

The metals are most widely distributed pollutants, because of their toxicity, metal pollution represent a potential threat for the soil microorganisms (Singh et al., 2014). Hydroxides, oxides, sulphides, sulphates, phosphates, silicates and organic compounds can be found in heavy metals. Agricultural crop processing inputs are the main cause of heavy metal pollution in the soil and water bodies. The use of rock phosphates or their products to improve crop production always implies the addition of a large amount of Pb and Cd into soils. During the course of phosphate fertilizer application for crop production, the accumulated heavy metals on the surface of soil are instantly accessible to plants (Meena et al., 2015; Dominguez-Nunezet al., 2016; Dotaniya et al., 2016). Heavy metals are a set of components with relatively higher densities, atomic numbers, and atomic weights. The heavy metals are classified into two types i.e., essential and non-essential heavy metals. Cu, Fe, Mn, Ni, and Zn are the essential heavy metals that are required for physiological and biochemical processes during plant life cycle (Cempel & Nikel, 2006) and on the other hand they become toxic when present in excess. The non-essential heavy metals such as Pb, Cd, As, and Hg are extremely toxic and their function was unknown in plants (Fasani et al., 2018) leading to environmental pollution and have a serious impact on a variety of physiological and biochemical processes in crop plants thereby, lowering the agricultural productivity (Clemens,

2006). These heavy metals/metalloids come from natural and anthropogenic sources, like wastewater produced in the oil and gas industries (Neff et al., 2011; Pichtel, 2016), usage of fertilizers of phosphates in agriculture (Hamzah et al., 2016; Rafique & Tariq, 2016), sludge from wastewater (Farahat & Linderholm, 2015), metal mining and smelting (Chen et al., 2016), pesticide application (Iqbal et al., 2016), electroplating, and fossil fuel burning (Muradoglu et al., 2015). Suman et al. (2018) reported that heavy metals remain in the soil for prolonged period of time, which pose a lasting threat on the environment and also they are non-degradable by any biological or physical process.

Plant tissues can absorb heavy metals that remain in the soil, penetrate the biosphere and accumulate trophic levels in the food web (Liu et al., 2005; Clemens, 2006). Mercury is regarded as extremely dangerous of them (Polak-Juszczak, 2009). Mercury toxicity depends on the form in which it exists. The deadly type of mercury is methylmercury, which is produced by the ionic mercury methylation process (Boeing, 2000). The aggregation and translocation of heavy metals in the soil environment cause destruction to the environment (Chang et. al. 2014). Observations made by Juwarkar et al. (2007) reveals that the microbial populations are very low in Cd and Pb polluted soils compared to non-polluted soils. Xie et al. (2016) reported that the levels of heavy metals in soils have major impacts not only on the population size, but also the physiological activity of soil microorganisms. Heavy metals may have the lowest level of soil concentration, but they have a huge effect on the biotic life cycle (Dotaniya et al., 2014; Meena et al., 2013). They are carcinogenic in nature and thus need to be detoxified from a method for the safe development of crops or a healthier climate.

Dyes

Pollutants from the industries of dyeing, printing and finishing have become a troubling concern. The textile industry is liable for the extensive environmental effects of toxicants (Muthu, 2017). The textile industry is one of the biggest industries in the world and is one of the most polluting industries in the country and consumes water for several processes such as scouring, sizing and bleaching, dying and other related methods. Dyeing is the process to color textile material and these dyes are recalcitrant to microbial degradation because they contain substitutions such as azo, nitro, or sulpho groups. The waste water released from the textile industry is a combination of a variety of contaminating materials, which include synthetic dyes and other chemical substances used in washing and colour stripping of over or unequal dyed cotton and during the dying of coloured fabric. The color used in the textile industry is the key attraction of any textile material that may cause potential risk to environment and living organisms. In ancient times, natural dye was the primary substance for dyeing cloth fabrics. But natural dyes are unable to satisfy the necessary demand for dyed and printed textiles because of the inadequacy of natural dyes and the growing demand for dyed and printed textiles. At present, synthetic dye has taken over the natural dye business, and people have also begun to use synthetic dye in both textile dyeing and printing segments. Moreover, the problem with synthetic dye is that it highly toxic and hazardous to our environment. Dyes are the soluble organic compounds which are classified as reactive, direct, basic and acids (Mahapatra, 2016) and are extremely water soluble which becomes hard to remove by the conventional methods (Hassan & Carr, 2018). With over 7,107 tonnes of dye stuff processed annually worldwide, there are more than 1,00,000 commercially available dyes. Owing to the prevalence of multiple contaminants in the water system, cloth waste water produces heavy colour, high humidity, high turbidity, broadly fluctuating pH, high COD and BOD concentration, significant amounts of suspended solids and overall dissolved solids. In addition, these effluents are directly or indirectly toxic and unfavourable to the ecosystem and human health (Elango et. al. 2016). Central Pollution Control Board included the dyeing industry as severly polluting industries (Rajan, 2014).

Every year over one million tons of synthetic dyes are manufacturing across the globe for use in the plastic, food, pharmaceutical, textile, cosmetic, paint, leather and paper industries (Shamraiz et al., 2016), of which, almost 60% are of azo dyes (Shah, 2014; Gürses et al., 2016). On the other hand, azo dyes are poisonous, carcinogenic and mutagenic in nature. They depict a pollution hazard because they contain components such as benzidine and aromatic compounds in their structure. Their degradation products (colorless amines) are also detrimental and/or mutagenic to living organisms (Xu et al., 2016). In the textile processing, a large amounts of textile effluents produced, contain organic and inorganic compounds (Elliott et al., 1954). Distribution of these substances in the environment can cause major negative implications on the environment (Islam et al., 2011). Toxic nature of the dyes causes death to the soil microorganisms which ultimately affect the agricultural productivity (Savin & Butnaru, 2008). Cotton fibres are predominantly treated with azo dyes, one of the main classes of synthetic dyes used in the industry (Mohan et al., 2002). These dyes are capable of modifying the soil's physical and chemical properties, degrading water sources, and damaging the environment's flora and fauna (Mohamed et al, 2017).

Urban Wastes

By 2050, the urban population is expected to rise by 2.5 billion, accounting for 66% of the global population (UN DESA, 2014). Solid waste, such as organic and inorganic waste, is the main source of pollution in urban society. The municipal solid wastes (MSWs) are unwanted materials mainly consisting of household wastes called as household garbage. Municipal Solid Waste (MSW) is a major concern particularly in urban areas and this problem has worsened due to the improper disposal plans. Nowadays, cities throughout the world generate around 1.3 billion tons of solid waste per year (Orhorhoro & Oghoghorie, 2019). The increasing population has contributed to an increase in the production of municipal solid waste in urban areas, resulting in hundreds of tonnes of waste per day. Municipal solid waste such as paper, plastic, metal, glass discards, clothing, etc. exert influence on soil properties. Around 60-75 percent of the global population will live in metropolitan areas in a more globalized and urban world during 2025-2050, as well as in an environmentally degraded world (UN DESA, 2014). The emissions in urban areas are formed during the transport (fossil fuel combustion, petrol and engine oil leaks, attrition of parts and tyres), industrial activities (metallurgy, mining and chemical engineering), building and waste disposal, incineration contaminate the soils and ecosystems, coal combustion (power plants and heating) (Cachada et al., 2012; Luo et al., 2012). Janas & Zawadzka (2017) noticed that the operation of mining and metallurgical plants within the urban areas may have huge impact on soil quality and the aquatic environments and also influence the human health and life in an indirect manner. Prolonged period of disposal of biowaste and municipal waste affects the physico-chemical properties of soil (Anikwe & Nowobodo, 2002; Yuksel et al., 2004; Montemurro et al., 2005) and also contains heavy metals (Lisk, 1988; Zhang et al., 2006; Pasquini & Alexander, 2004).

Radioactive Waste

Various types of radionuclides or radioisotopes are found in the environment. In the biosphere, radioactive compounds appear to be widespread and they can be synthesized spontaneously or actively.

Radioactivity occurs as a result of the accidental disintegration of the parent radionuclide and the creation of a daughter nuclide by releasing gamma, beta and/or alpha radiation in the phase. The natural radionuclides are primitive, secondary or cosmogenic in origin. The artificial radionuclides are formed by nuclear explosions, nuclear reactors or radionuclide generators. Soil is a medium of migration and transfer of radionuclides to biological systems (Badhan et al., 2017; Manigandan and Shekar, 2014). Usually, radionuclides are dispersed in nature in little concentrations. The natural radioactivity in soil varies depending on soil type, mineral make up and density. In India, naturally occurring radionuclides are ^{238}U, ^{232}Th, ^{4}K, ^{226}Ra, Ca. The flow of primordial radionuclides into the soil may be intensified by man-made operations, such as mining. Generally, radioactive wastes are the by-products of nuclear power generation and other applications of nuclear technology. A recent research has established an effective tool to classify and measure the concentration of natural radioactive material (NORM) in soil horizons (Michalik, 2017). The artificial radionuclides are emitted into the atmosphere by the open tests of nuclear warheads, approved discharges from the nuclear reprocessing plants, and incidents at businesses using nuclear energy (Aarkrog, 1994; 2003). Radionuclides are found in all parts of the earth's surface and are present in air, soil and water and enter the soil either directly by the introduction of liquid wastes or indirectly by water infiltrating through the soil. For the last ten years, military operations, uranium mining, and failures at nuclear power plants have released anthropogenic radioactive materials into the setting (Hu et al., 2008). In the northeastern Indian state of Jharkhand, uranium mining and milling from the Jaduguda uranium mine in the Bay of Bengal was found to emit alpha particles that influenced the local microbial communities (Dhal & Sar, 2014). In several ways, such as an oxide, organic or inorganic complex, and occasionally as a free metallic ion, uranium is present in the world. The free forms of elemental uranium mostly exist in higher states of oxidation and are normally bound to oxygen. In the aqueous stage, cationic uranium rapidly mixes with oxygen and forms incredibly mobile and highly reactive uranium oxy-cations (uranyl ions) e.g., U(VI) is highly soluble in water in the form of (UO_2^{2+}). But, the reduced form of U(IV), existing as uraninite (UO_2) is less soluble and therefore indicate a lower risk in the environment. Most of the key microbial interactions with radionuclides have been studied using uranium as a model.

Sources of Radiation in the Environment

The major sources of radiation are the cosmic radiation; nuclear power production; nuclear fuel cycle activities; the mining and chemical processing connected with impurities of U and Th; production and use of radioactive substances for the medical, research and industrial purposes; military activities, production, testing and use of nuclear weapons.

BIOREMEDIATION

New technologies have been introduced in the past few years to increase the efficiency for the removal of pollutants. Among them, bioremediation techniques have been proven to be a new and effective process for cleaning up contaminants in a variety of environments and a quite flexible management option to be implemented, also at a large scale (Azubuike et al., 2016). To recover the functions of the contaminated environment, for both environmental preservation and urban development, the remediation of contaminated sites is essential. Bioremediation is defined as the technique through which living organisms such as plants, algae and microorganisms are utilized to clean-up, decrease or eliminate contamination from the environment (Saxena & Bharagava, 2020; Kuppusamy et al., 2020). Several processes such as physical, chemical and biological have already been implemented to remediate contaminated soils (Smith et al., 1995; Mulligan et al., 2001). These methods either decontaminate the soil or stabilize the contaminant within it. Bioremediation is considered as an "environmentally-friendly" soil clean-up technology which has a mild impact on soil functional properties, and the environment in general, and employs soil organisms (including plants, bacteria, and/or fungi) to breakdown impurities in soil (Pilon-Smits, 2005). Bioremediation of contaminated soil is defined as the application of living organisms (especially bacteria, fungi, algae) to make environment free from toxicity of contaminant by means of transformation, degradation and mineralization of the contaminants to less harmful compound. It is the technology of eliminating pollutants from the environment to restoring the original natural environment and preventing further pollution (Gallego et al., 2001; Ubani et al., 2013; Sasikumar & Papinazath, 2003). For bioremediation purposes biological agents particularly microbes (microremediation), plants (phytoremediation) or both (rhizoremediation) were utilized. In the bioremediation process different microbes are employed for the degradation or detoxification of xenobiotic compounds, volatile organic compounds, aromatic hydrocarbons, herbicides, pesticides, heavy metals, radionuclides, jet fuels, crude oil, explosives and petroleum products (Gaur et al., 2014). Bioremediation techniques were widely applied for the detoxification of soils from a broad range of environments using laboratory or in situ approaches (Azubuike et al., 2016; Margesin, 2007; Varjani, 2017). Globally, soil remediation is among the most expensive treatments throughout the world (Agamuthu et al., 2013). The bioremediation methods are mainly depedning on the enzymatic activities of microorganisms for conversion and degradation of environmental contaminants or wastes into less toxic or non-toxic constituents like carbon dioxide and water (Das & Dash, 2014). The bioremediation approach may be either aerobic (Bedard & May, 1995; Wiegel & Wu, 2000) or anaerobic (Komancova et al., 2003). Microorganisms can degrade pollutants under aerobic and anaerobic conditions. In aerobic degradation, microbes use oxygen as final electron acceptor to convert organic and inorganic pollutants into harmless products such as carbon dioxide and water. In anaerobic degradation microbes utilize other electron acceptor such as sulphate, iron, nitrate, manganese etc. to degrade organic compounds into carbon dioxide and methane. Mostly, aerobic microorganisms have the potential to degrade the contaminants at a faster rate than anaerobic organisms.

Principle of Bioremediation

The secret to effective bioremediation is to leverage the inherent catabolic capacity of species to catalyse environmental pollutant transformations (Vidali, 2001; Chakraborty et al., 2012). The principle behind bioremediation is the application of microbes to destroy the hazardous pollutants or transform them into less harmful forms. The microorganisms act against the pollutants, only when they have gain access to a different materials and compounds to help them generate energy and nutrients to build more cells. Bioremediation technology uses the physiological potential of microbes and plants for the degradation of pollutants (Odukkathil & Vasudevan, 2013). While choosing any bioremediation technique, some of the measures to be taken into account are the nature of contaminant, depth and degree of pollution, type of environment, location, cost, and environmental policies (Frutos et al., 2012; Smith et al., 2015). According to Dua et al. (2002), the following are the three basic principles of bioremediation

1. Availability of pollutant to biological transformation
2. Availability of the pollutant to microbes
3. Optimization of biological activity

TYPES OF BIOREMEDIATION

There are two types of bioremediation techniques;

1. In-situ bioremediation
2. Ex-situ bioremediation

In-Situ Bioremediation

In Latin, '*in situ*' means in the original place. This method is carried out at the original contaminated site or this approach implies treating contaminated substances at the site of pollution. The in situ technology option is good as unearthing, and transfer of contaminated materials is prevented, but achieving uniform remediation is challenging because of soil heterogeneity (Simarro et al., 2013; Vogt & Richnow, 2014). In *In situ* process, the breakdown of pollutants is brought about by the stimulating naturally occurring bacteria with the supply of oxygen and nutrients by circulating aqueous solutions through contaminated soils. *In situ* bioremediation is inexpensive and more feasible to carry out than ex situ bioremediation and is suitable for ecological rejuvenation (Megharaj et al., 2011). Utilization of in situ method is depends on the oxygen supply, soil nature, and the depth of penetration of the contaminants into the soil (Angelucci & Tomei, 2016). *In-situ* method has proved to be successful in groundwater remediation as well as surface soils and sub soils polluted with petroleum hydrocarbons (Pierzynski et al., 1994). In situ bioremediation techniques are successful in treating the sites contaminated with heavy metals, chlorinated solvents, dyes and hydrocarbons (Folch et al., 2013; Kim et al., 2014; Frascari et al., 2015; Roy et al., 2015). The significant benefits of in situ bioremediation are low cost, (having no excavation), minimal site disruption, minimal dust production, and the possibility of simultaneous treatments of soil and groundwater in the future. However, the main disadvantage of this method is running out of time, seasonal fluctuations in the microbial activity and troublesome application of treatment additives in the natural habitats (Rayu et al., 2012). In-situ remediation process is the only solution to treat a huge contaminated site, when considering the scale of the area and the relationship to the cost benefit. The *in-situ* process is applied for wide area of soil and/or contaminated sediments as it causes little disturbance to the site in which, the operation is quite easy, and the expendature is lower than the *ex-situ* treatment process (Song et al., 2017). In site bioremediation techniques include Biosparging, Bioventing, Bioaugmentation and Bioslurping;

Biosparging

Biosparging is an efficient in situ approach, which is carried out by treating the soil with air by applying pressure to intensify the activity of microbes for the degradation of hazardous substances at the contaminated sites. Efficacy of biosparging is dependent on soil permeability that greatly determines the pollutants availability to microorganisms in addition to the biodegradability of pollutants (Godheja et al., 2019). The injection of air triggers the movement of organic compounds of volatile nature in the unsaturated zone of soils for strengthening biodegradation. The biosparging is maninly used to reduce the concentration of the petroleum constituents, which dissolve in ground water, adsorbed to soil.

Bioventing

Bioventing is the most prevalent in situ treatment that is carried out by the controlled stimulation of the air flow, supplying oxygen to augment microbial activity and therefore, enhancing the bioremediation (Brown et al., 2017). It requires providing oxygen and nutrients to a polluted soil by wells to activate the indigenous aerobic bacteria. In the remediation of phenanthrene-contaminated soil, after seven months, 93% contaminant elimination was accomplished (Frutos et al., 2010). Among other *in situ* approaches, bioventing gained immense popularity mainly in restoring polluted sites with light spilled petroleum products (Hohener & Ponsin, 2014). Using bioventing technique, aerobically degradable contaminants such as fuels are treated along with contaminants such as non-halogenated solvents (e.g., benzene, acetone, toluene, and phenol), lightly halogenated solvents (e.g., 1,2-dichloroethane, dichloromethane, and chlorobenzene), and some semi-volatile organic compounds (SVOCs) (e.g., lighter polycyclic aromatic hydrocarbons (PAHs) (EPA, 2006).

Bioaugmentation

The bioaugmentation is introduction of the specific competent strains of microbes that can biotransform or biodegrade a particular pollutant in a particular environment. In this method, exogenous microbial population is introduced to the polluted environment (Thierry et al., 2008, Łebkowska et al., 2011; Taccari et al., 2012; Wu et al., 2013). The contaminated soils are inoculated with specially cultivated microbes with capabilities for the degradation of certain pollutants. Alternative technique of bioremediation is bioaugmentation in which, hydrocarbon degrading microbes are introduced into the waste (D'Annibale et al., 2006; Thompson et al., 2005). Bioaugmentation is effective, in the case of inadequate or absence of native microbial community (Wu et al. 2017; Liu et al. 2014). Several years ago, the advantage of bioaugmentation with pesticide-degrading microorganisms in cleaning the polluted soil was demonstrated (Bidlan et al., 2004; Karpouzas & Walker, 2000; Li et al., 2007; Singh et al., 2006; Yang et al., 2010; Zhang et al., 2006). Over the past few years, bioaugmentation was put into operation in the remediation of soil contaminated with diverse organochlorinated pesticides such as endosulfan, DDT and lindane (Abhilash et al., 2011; Kataoka et al., 2011; Saez et al., 2014), organophosphorus pesticides (OPPs) (chlorpyrifos, fenitrothion, methyl parathion) (Aceves-Diez et al., 2015; Cycon et al., 2013; Wang et al., 2014), triazines (atrazine, simazine, terbuthylazine) (Sagarkar et al., 2013; Silva et al., 2015; Wang et al., 2013), pyrethroids (bifenthrin, cypermethrin, fenvalerate, deltamethrin) (Chen et al., 2011, 2012, 2014; Cycon et al., 2014; Liu et al., 2014) and other such as carbamate (carbofuran) (Pimmata et al., 2013), chloroacetamide (butachlor) (Zheng et al., 2012), benzimidazole (carbendazim) (Wang et al., 2010) and derivatives of phenoxyacetic acid (Onneby et al., 2014). In fact, bioaugmentation seems to be effective for the removal of polycyclic aromatic hydrocarbon (PAH) compounds at contaminated sites (Wu et al., 2016) or for the remediation of pesticides and their residues from soil (Cycon et al., 2017). With respect to efficiency and economy, this strategy for treating contaminated sites gives best outcome compared to chemical and/or physical methods (Isaac et al., 2017).

Bioslurping

Bioslurping is also known as vacuum-enhanced extraction or dual-phase extraction (DPE). In this procedure, very high vacuum pumps are used to remove further variations from the subsurface of toxic water, hydrocarbon vapour and distinct phase petroleum products. Usually, in this method the pollutant extraction rates are increased particularly in layered and fine grained soils. This process is powerful for moderate to low permeable soils. Bioslurping, also known as multiphase extraction, is a combination of intensified vacuum pumping, soil vapor extraction and bioventing and it is carried out for the remediation of groundwater and soil by indirect O_2 supply thereby enhancing the degradation of pollutant (Azubuike et al., 2016).

Ex Situ Bioremediation

The ex situ bioremediation involves the excavation or elimination of the polluted materials from the affected sites. These treatment processes were conducted in a container or away from the original sites. More commonly used *ex situ* techniques were physical separation and solidification/stabilization (USEPA, 2006). Ex situ therapies are easier to manage and track, but at the excavated site they create high-cost health risks, waste generation, and habitat destruction. *Ex-situ* processes of bioremediation include Biopiling, Landfarming, Composting and Bioreactor.

Biopiling

This technique has been widely used for remediating a wide range of petrochemical contaminants in soils. Biopiles are also known as biocells, bioheaps, biomounds, and compost piles. Biopiling involves the accumulation of contaminated soils into piles and stimulating the biodegrading activity of microbial populations by setting up near optimum growth conditions. Bio-piles are elevated (in mountain-like structures) of contaminant soils and are constantly aerated by an injector pump by mechanical injection of oxygen into soil mounds (USEPA, 2006).

Landfarming

Landfarming can, depending on where the treatment takes place, be categorised as ex-situ or in-situ technology. The principle of this process involves the use of microbial communities to extract organic contaminants mainly through their conversion into CO_2 and water (Straube et al., 2003; Maila & Cloete, 2004). This process is carried out by the application on the ground surface of a treatment site of excavated polluted soils in a thin layer and by inducing aerobic microbial activity within the soils to accelerate the natural methods of biodegradation (Brown et al., 2017; Vidali, 2001). In fact, this method was widely used by the mineral oil processing industry, as it is easier and cost-effective method to remediate accidentally contaminated soils by oil spills. Maila & Cloete (2004) reported that successful landfarming relies on certain conditions, like well-drained soil, biodegradability of contaminants by the existing microorganisms, abundant presence of microorganisms, and a closed greenhouse needed to minimize soil erosion and runoff from rain and to control air emissions.

Composting

Composting is an ex-situ aerobic mechanism by which thermophilic biological agents decay agricultural waste to obtain a humic amendment known as compost (Narayan Chadar, 2018), that is applied as fertilizer to the soil (Rosca et al., 2019). In order to attain high degradation rates, it is extremely important to maintain easy availability of nutrients and oxygen, a temperature range of 40 and 70° C along with neutral pH rates (Macaulay & Rees, 2014). While composting is mostly used for organic waste disposal, it is often used for the bioremediation of degraded soil or sludge. Microbial activity is capable of biodegrading harmful organic compounds in such a process, thus minimising metal bioavailability (Aguelmous et al., 2019). When waste or final compost is combined with the soil, soil microorganisms are added (Kumar et al., 2018). As this process turned out to be more efficient in the degradation of several chemical contaminants such as pesticides, chlorophenols, explosives and petroleum hydrocarbons, it received more focus over the past decades (Sayara et al., 2011). While it is possible to use composting itself as part of other processes (e.g. phytoremediation and landfarming), its individual use is becoming more common (Aguelmous et al., 2019; Grasserová et al., 2020; Loick et al., 2009; Sayara et al., 2011) as part of other processes (e.g., phytoremediation and landfarming).

Bioreactor

Tough glass or stainless steel bioreactors are typically cylindrical in shape and have a volume varying from a few litres to cubic metres. As a dry product or suspension, the polluted material should be supplied to the reactor. The use of bioreactorsis considered among the best ways to treat polluted soil, as the operating conditions can be controlled, therefore, enabling the increase in microbial biodegradation activity (Dzionek et al., 2016). Bioreactors are utilized to treat the soil and other materials which are contaminated with petroleum residues (Sardrood et al., 2013). Fulekar (2009) showed remediation of fenvalerate by means of *Pseudomonas aeruginosa* in an upgraded bioreactor and proven that this process shall be advantageous to fenvalerate detoxification.

Characteristics of Microorganisms Involved in Bioremediation

Natural organisms, either indigenous or extraneous, are the important agents used for bioremediation (Prescott et al., 2002). According to Alexander (1994) the following requirements are fulfilled by the organisms which were employed in bioremediation.

1. The organisms, which have the potent enzymes that are crucial in bio-remediation.
2. The organisms must have the ability to survive and exhibits bioactivity under conditions of pollution.
3. The organisms must be accessible towards the contaminant that is insoluble in aqueous environments or strongly adsorbed to hard surfaces.
4. Substrate site of the contaminant should be available for the active site of the enzyme that plays an important role in bioremediation.
5. The contaminant and the enzymatic system should come in close contact at somewhere inside or outside of the cell.
6. There must be favorable environmental conditions for the population to develop potential bioremediant.

ROLE OF MICROORGANISMS IN BIOREMEDIATION

Microorganisms have been one of the promising instruments that maintain natural sources' self-cleaning ability. Microorganisms are ideal for the removal of pollutants because they have enzymes that allow them to use organic pollutants as a source of food and energy. Employing microorganisms in the bioremediation approach would enzymatically degrade the hazardous organic pollutants and transform them into CO_2, CH_4, and H_2O without adversely affecting the environment (Ron & Rosenberg 2014; Yuniati, 2018). The bioremediation phase is quite sluggish. Only some species of bacteria and fungi have demonstrated their potential as potent degraders of contaminants.

Microorganisms that carry out biodegradation are known as active members of microbial consortiums in many different ecosystems, and they include: *Acinetobacter, Actinobacter, Alcaligenes, Arthrobacter, Bacillus, Beijerinckia, Flavobacterium, Methylosinus, Mycobacterium, Mycococcus, Nitrosomonas, Nocardia, Penicillium, Phanerochaete, Pseudomonas, Rhizoctonia, Serratia, Trametes and Xanthobacter* etc.

Bacteria

Bioremediation has been widely used to convert hazardous heavy metals into a less harmful state by employing microorgansism (Ndeddy Aka & Babalola, 2016; Akcil et al., 2015; Abbas et al., 2014) or their enzymes to clear the polluted environment (Okoduwa et al., 2017). To clean-up heavy metal polluted surroundings, numerous microorganisms like bacteria, fungi, and algae have been utilized (Kim et al., 2015; Neha et al., 2013). The detoxification of metals has been carried out by microorganisms by means of processes such as valence conversion, volatilization, or extracellular chemical precipitation (Ramasamy et al., 2006). Microorganisms possess negative charge on their cell surface due to the presence of anionic structures which allows the microbes to adhere to metal cations (Gavrilescu, 2004). Bacteria use metals and metalloids as electron donors or acceptors for energy generation. Heavy metals transfigure the physiological and biochemical properties of microorganisms. Chromium (Cr) and cadmium (Cd) are capable of inducing oxidative damage and denaturation of microorganisms as well as weakening the bioremediation capacity of microbes.

The heavy metal tolerant microorganisms such as *Proteobacteria, Geobacter, Edaphobacter, Pseudomonas, Gemmatiomonas, Nitrosomonas, Xanthobacter, Sphingomonas, Pedobacter, Ktedonobacter, Thermotogales, Enterobacter, Polyangium* sp., *Stenotrophomonas* sp., *Variovorax* sp., *Hafinaa* sp., *Clostridia, Spingomaonsa* sp., *Acidobacteria, Acinetobacter*, etc. were isolated from a number of industrial waste polluted soils (Hemmat-Jou, 2018; Chien et al., 2008; Sandaa et al., 1999) and the bacterial species that breakdown the pesticides belongs to genera *Flavobacterium, Arthobacter, Aztobacter, Burkholderia,* and *pseudomonas* (Glazer & Nikaido, 2007). Diverse strains of *Pseudomonas putida, Escherichia coli, Ralstonia eutropha, Sphingomonas desiccabilis, Mycobacterium marinum, Bacillus idriensis*, etc., have genes in their genomes, which allow them to selectively bioremediate toxic metal compounds (Valls et al., 2000; Ackerley et al., 2004; Kube et al., 2005; Parnell et al., 2006; Schue et al., 2009; Liu et al., 2011). *Lysinibacillus sphaericus* CBAM5 is a gram positive, spore forming bacteria found in soil. It is a heavy metal tolerant stain isolated from Easter Planes of *Colombia*. L. *sphaericus* biomass has been known to bioremediate metals such as cobalt, copper, chromium and lead (Peña-Montenegro et al., 2015).

Bacilli have been characterized for the bioreduction of chromium from Cr(VI) to Cr(III) - *Bacillus* sp. strain KSUCr9a (Ibrahim et al., 2012), *Bacillus* sp. FY1 (Xiao et al., 2017). The uranium transformation from U(VI) into nanouramphite was studied in two B. *Thuringiensis* strains isolated from uranium mine

(Pan et al., 2015). The studies carried out by Zolgharnein et al. (2010), have assessed that the heavy metals such as copper, lead, zinc, and cadmium were extracted by the *Delfetia* and *Methylobacter* spp. It has become clear that the bacteria, *Methylococcus capsulatus* has the ability to decontaminate lead and zinc at low pH (Chen et al., 2013) and chromium (Al Hasin et al., 2010). As reported by Youssef et al. (2009), both bacteria, *Neisseria mucosa* and *Rahnella aquatilis* are capable of reducing arsenate and selenate. The bacteria used for the bioremediation of various heavy metals are presented in Table 1.

Table 1. Bacteria used in bioremediation of heavy metals

Name of the Bacteria	Name of the Heavy Metal	Reference
Lysinibacillus sphaericus CBAM5	Copper, chromium, cobalt and lead	Peña-Montenegro et al., 2015
Rhizobiales and *Proteobacteria*	Strontium and cesium	Quero et al., 2015
Acinetobacter sp. and *Klebsiella*	Mercury	Jan et al., 2016
Delfetia and *Methylobacter* spp.	Copper, lead, zinc, and cadmium	Zolgharnein et al., 2010
Methylococcus capsulatus	Lead and zinc	Chen et al., 2013
Neisseria mucosa and *Rahnella aquatilis*	Arsenate and selenate.	Youssef et al., 2009
Bacillus sp. strain KSUCr9a *Bacillus* sp. FY1 *Bacillus* sp. MH778713 *Bacillus cereus* TN10 *Bacillus thuringiensis*	Chromium Chromium Chromium Chromium Uranium	Ibrahim et al., 2012 Xiao et al., 2017 Ramírez et al., 2019 Hossain et al., 2020 Pan et al., 2015

The microbial system is well - suited for the synthetic pyrethroids biodegradation (Bhatt et al., 2019). *Bacillus, Pseudomonas, Raoultella, Achromobacter, Acidomonas, Brevibacterium, Pseudomonas, Streptomyces, Serratia, Sphingobium, Clostridium, Klebsiella*, and *Lysini bacillus* are the bacterial genera involved for pyrethroid degradation (Cycon & Piotrowska-Seget, 2016; Birolli et al., 2019; Hu et al., 2019; Zhao et al., 2019). Over the past several decades, the wide-spread use of lindane caused a high distribution that negatively affects the biota and lindane is now believed to be one of the major hazardous persistant organic pollutants (Tsygankov et al., 2019). Degradation of lindane and other xenobiotics by the bacteria and other xenobiotics have been reported (Chen et al., 2012, 2013; Yang et al., 2010; Zhang et al., 2018; Bhatt et al., 2020). Bacteria play a fundamental role in lindane biodegradation through chemical and physical interactions resulting in the structural changes or total degradation of the target molecule. Various lindane-degrading bacteria include *Streptomyces* (Sineli et al., 2016), *Paracoccus* (Sahoo et al., 2019), *Achromobacter* (Singh and Singh, 2019), *Burkholderia* (Kumar, 2018), *Rhodococcus* (Egorova et al., 2017), *Kocuria* and *Staphylococcus* (Kumar et al., 2016), *Chromohalobacter* (Bajaj et al., 2017).

In the breakdown of thiamethoxam, Rana et al. (2015), employed *Bacillus aerophilus* and Sharma et al. (2014), applied imidachlorpid for the degradation of *Bacillus alkalinitrilicus*. Pailan et al. (2015) reported that *Bacillus aryabhattai* isolated from agricultural soil of West Bengal, India, is highly efficient in chlorpyriphos degradation in addition to parathion at an optimal concentration of 200 mg mL^{-1}. Several *Pseudomonas* species such as *Pseudomonas putida, Pseudomonas stutzeri, Pseudomonas aeruginosa, Pseudomonas nitroreducens* and *Pseudomonas fluorescence*, isolated from agricultural soils and polluted effluents across several areas revealed to be effective enough in the biodegradation of chlorpyriphos (Bhagobaty and Malik, 2008; Vidya Lakshmi et al., 2008; Maya et al., 2011; Latifi et al., 2012; Sasikala et al., 2012). *Rhizobiales* and *Proteobacteria* are found in coastal areas polluted with strontium and cesium, are the potentially toxic metals (Quero et al., 2015). A study conducted by Jan et al. (2016), showed that *Acinetobacter* sp. and *Klebsiella* sp. were able to tolerate 148 ppm of mercury. Methyl violet or Paraquat (1, 10 -dimethyl-4, 40-bipyridinium dichloride) which is a broad-spectrum cationic contact herbicide has wide applications above 100 countries (Rashidipour et al., 2019). *Oscillospira* sp. BCK1, *Clostridium prazmowski* BCK-2, and *Sporohalobacter orenetal* BCK-3 are the bacterial strains are found to be proficient in degrading the paraquat up to 79.35, 80.26, and 86.22%, respectively, after 3 days of treatment (Han et al., 2014). Observations made by Li et al. (2017), showed that the employment of four microorganisms such as *Escherichia coli*, *Roseateles terrae*, *Bacillus* sp. and *Pseudomonas fluorescens*, in a mixed culture for the degradation of paraquat, resulted in 97% degradation of initial paraquat dose (100 mg/L) over 7 days. *Psedomonas alcaligens* bacterial strain is capable of degrading the herbicide butachlor and proved that it attained 75% of degradation at 50 mg kg^{-1} of soil over 21 days (Abd-Alrahman & Salem-Bekhit, 2013). The degradation of lot of xenobiotic compounds including cypermethrin, chlorate and cyhalothrin etc. were carried out by the widely available bacterium such as *Ochrobactrum anthropic* which lives in natural habitats, has broad catabolic competency (Zhai et al., 2012; Chudasama et al., 2017; Chen et al., 2019). The bacteria used for the bioremediation of different pesticides are presented in Table 2.

Table 2. Bacteria used in bioremediation of pesticides

Name of the Bacteria	Name of the Pesticide	Reference
Streptomyces *Paracoccus* *Achromobacter* *Burkholderia* *Rhodococcus* *Kocuria* and *Staphylococcus* *Microbacterium* *Chromohalobacter*	Lindane	Sineli et al., 2016 Sahoo et al., 2019 Singh and Singh, 2019a Kumar, 2018 Egorova et al., 2017 Kumar et al., 2016 Singh and Singh, 2019b Bajaj et al., 2017
Bacillus aerophilus *Bacillus alkalinitrilicus* *Bacillusaryabhattai*	Thiamethoxam Imidachlorpid Chlorpyriphos and parathion	Rana et al., 2015 Sharma et al., 2014 Pailan et al., 2015
Pseudomonas. putida *Pseudomonas stutzeri* *Pseudomonas aeruginosa* *Pseudomonas nitroreducens* *Pseudomonas fluorescence,*	Chlorpyriphos	Bhagobaty and Malik, 2008 Vidya Lakshmi et al., 2008 Maya et al., 2011 Latifi et al., 2012 Sasikala et al., 2012
Ochrobactrum anthropic	Cypermethrin chlorate cyhalothrin	Zhai et al., 2012 Chudasama et al., 2017 Chen et al., 2019
Escherichia coli, Roseateles terrae, Bacillus sp. and *Pseudomonas fluorescens* *Psedomonas alcaligens*	Paraquat butachlor	Li et al., 2017 Abd-Alrahman & Salem-Bekhit, 2013

New inquires about the use of autochthonous bacterial communities in bioremediation of azo dyes has been implemented currently in which indigenous microorganisms are used. These microbes have the capacity to degrade azo dyes both aerobically and anaerobically (Knapp & Newby, 1995). On the other hand, consortia of *Pseudomonas spp*, *Proteus spp*.,. and *Acinetobacter* spp. were being capable of degrading the dyes including methyl red and carbol fuchsin (Joshi et al., 2015). According to the reports published by Mahbub et al.(2011), *Staphylococcus aureus* which was formerly isolated from textile effluent carries out the degradation of cibacron blue FN-R, cibacron orange FN-R, cibacron yellow F-4G, cibacron navy FN-B, terasil black WNS and terasil red W-FS dyes. Azo dyes decolorization was brought about by few bacterial strains including *Proteus mirabilis*, *Pseudomonas aeruginosa*, *Bacillus licheniformis*, *Serratia marcescens* and *Alcaligenes faecalis* at static and shaking conditions (Neelambari et al., 2013). The observations made by Jadhav et al. (2010) showed that *Pseudomonas aeruginosa* will detoxify the dye, Direct Orange 39 (1,000 ppm each day) effectively. The decolorization of turquoise blue dye (Remazol Blue BB) by *Bacillus megaterium* isolated from a sample gathered from the dye industry was documented by Joshi et al. (2013). This species will decolorize up to a level of 5 mg/ml of turquoise blue dye. Aerobic decolorization of the textile azo dye Direct Red-22 by an obligate alkaliphilic bacterium *B. cohnii* MTCC 3616 has been documented by Prasad and Rao (2013). In static conditions, this strain was able to decolorize Direct Red-22 (5,000 mg/l) with 95 percent efficiency at 37 ^{0}C and pH 9 in 4 hours. The bacteria used for the bioremediation of various dyes are represented in Table 3.

Table 3. Bacteria used in bioremediation of dyes

Name of the Bacteria	Name of the Dye	Reference
Pseudomonas spp., *Proteus* spp. and *Acinetobacter* spp.	Methyl red and Carbol fuchsin	Joshi et al., 2015
Staphylococcus aureus	Cibacron blue FN-R, Cibacron orange FN-R Cibacron yellow F-4G Cibacron navy FN-B Terasil black WNS Terasil red W-FS	Mahbub et al., 2011
Proteus mirabilis, *Pseudomonas aeruginosa*, *Bacillus licheniformis*, *Serratia marcescens* and *Alcaligenes faecalis*	Azo dyes	Maharani et al., 2013
Bacillus megaterium	Turquoise blue dye	Joshi et al. (2013
Bacillus cohnii MTCC 3616	Direct Red-22	Prasad and Rao, 2013

Algae

Phycoremediation or algal bioremediation is the use of algae to expel contaminants from the environment or to transform them into nontoxic form. Algae are highly adaptive and can grow autotrophically, heterotrophically or mixotrophically in any setting. Algae serve as a substantial constituent of soil microflora and are present everywhere, accounting up-to 27% of the whole microbial biomass in the soil (McCann & Cullimore, 1979). These microorganisms are engaged in preserving soil fertility and oxygen production. They are more widespread than the other free-living microorganisms capable of dinitrogen fixation (Burns & Hardy, 1975) and thus are very important for the nitrogen economy of soils. Algae are used as indicators to assess the ecotoxicity, genotoxicity and environmental risk of pollutants, both in soil and sediments due to their sensitivity to the presence of the toxic chemicals (Subash chandrabose et al., 2013; Tigini et al., 2011) and were also utilized as a soil conditioners, or as biofertilizers (Metting, 1981). Different microalgae showed enormous benefits over bacteria and fungi in the breakdown of organic pollutants.

Blue green alga, *Anabaena sphaerica* extracts cadmium (II) and lead(II) from aqueous solution (Abdel et al., 2013). The other study showed that the ability of chemically altered *Cystoseira indica* in the biosorption of uranium (VI) oxide and lead(II) cations (Moghaddam et al., 2013). Some researchers reported that, pharmaceuticals were degraded by the microalgae (Leng et al., 2020; Gentili & Fick, 2017). For eg: *Chlamydomonas oblonga* (Chlorophyta) was exposed to 200 mg/L of an emergent contaminant, carbamazepine, a pharmaceutical used in the treatment of epilepsy, and a 30% growth inhibition and a 35% degradation rate of the pollutant was registered (Derakhshan et al., 2019). The green microalgae such as *Fischerella* sp., *Scenedesmus* sp., *Chlorella vulgaris* and *Chlorococcum* sp. and cyanobacteria such as *Lyngbya spiralis Geitler*, *Tolypothrix tenuis Kützing*, *Stigonema* sp. and *Phormidium molle* has greater efficiency in eliminating Pb (II), Cd (II) and Hg (II) ions (Inthorn et al., 2002). The algae used for the bioremediation of heavy metals are represented in Table 4.

Table 4. Algae used in bioremediation of heavy metals

Name of the Algae	Name of the Heavy Metal	Reference
Fischerella sp., *Scenedesmus* sp., *Chlorella vulgaris* and *Chlorococcum* sp.	Pb (II), Cd (II) and Hg (II) ions	Inthorn et al., 2002
Blue green alga and *Anabaena sphaerica*	Cadmium(II) and Lead(II)	Abdel et al., 2013
Chlamydomonas oblonga	carbamazepine	Derakhshan et al., 2019

Chlorella vulgaris and *Scenedesmus bijugatus* are the microgreen algae which breakdown the organophosphorus insecticides such as monocrotophos and quinalphos and the cyanobacteria such as *Synechococcus elongatus*, *Phormidium tenue* and *Nostoc linckia* convert these pesticides at 5 to 50 ppm concentration via 30 days (Megharaj et al., 1987). El-Bestawy et al.(2007) noted that the cyanobacterial strains such as *Nostoc, Nodularia, Oscilatoria, Cyanothece*and *Synechococcus*are capable of degrading the pesticide lindane very quickly. *Chlamydomonas reinhardtii,* green alga has a great potential to accumulate and degrade the herbicide prometryne (Jin et al., 2012). The algae used for the bioremediation of different pesticides are represented in Table 5.

Table 5. Algae used in bioremediation of pesticides

Name of the Fungi	Name of the Heavy Metal	Reference
Chlorella vulgaris and *Scenedesmus bijugatus*	Monocrotophos and Quinalphos	Megharaj et al., 1987
Nostoc, Nodularia, Oscilatoria, Cyanothece and *Synechococcus*	Lindane	El-Bestawy et al., 2007
Chlamydomonas reinhardtii	Prometryne	Jin et al., 2012

Table 6. Fungi used in bioremediation of pesticides

Name of the Fungi	Name of the Pesticide	Reference
Cunninghamella elegans	Cyhalothrin	Palmer-Brown, 2018
Phlebiaacanthocystis, Phlebia brevispora, and *Phlebia aurea*	Aldrin and dieldrin	Bhalerao and Puranik, 2007
Aspergillus niger, Aspergillus terreus, Cladosporium oxysporum, Mucor thermohyalospora, Fusarium ventricosum, Phanerochaetechrysosporium, and *Trichoderma harzianum*	Endosulfan	Jin et al., 2012
Hypholoma dispersum ECS-705	Paraquat	Camachomorales et al., 2017
Aspergillus niger ARIFCC 1053	Endosulfan	Bhalerao, 2012
Antracophyllum discolour, Phanerochaete chrysosporium, Trametes versicolor, Ganoderma lucidum, Armillaria mellea, and *Gloeophyllum striatum*	Pentachlorophenol	Bosso et al., 2015
Eurotiumcristatum ET1	β-cypermethrin and 3-phenoxybenzaldehye	Hu et al., 2018

Fungi

Since they are the ultimate decomposers of waste products, fungi are among the successful candidates for bioremediation. In general, fungal biodegradation of organic compounds, including pesticides, has been observed as a comparatively slower process and also does not contribute to the complete elimination of pollutants (Sasec & Cajthaml, 2014). It was also noted that tolerance to the polluted atmosphere and elimination of the toxins take longer periods of time for the fungi (Kulshreshtha et al., 2014). Fungi was naturally considered to be more beneficial for the degradation of recalcitrant compounds than bacteria (Gangola et al., 2019; Yu et al., 2011). The capacity of *Phlebia acanthocystis*, *Phlebia brevispora*, and *Phlebia aurea* to degrade pesticides with aldrin and dieldrin was reported by Xiao et al. (2011). *Cunninghamella elegans* are the Pyrethroid-degrading fungi that degrade the pesticide cyhalothrin (Palmer-Brown, 2018). Large number of fungi such as *Aspergillus niger*, *Aspergillus terreus*, *Cladosporium oxysporum*, *Mucor thermohyalospora*, *Fusarium ventricosum*, *Phanerochaete chrysosporium*, and *Trichodermaharzianum* were cross checked for their ability to degrade endosulfan (Bhalerao & Puranik, 2007). Camachomorales et al.(2017b), found that the percentage elimination of paraquat- herbicide by *Hypholoma dispersum* ECS-705 strain was around 70.7% in 12 days. It became evident that the strain of *Aspergillus niger* ARIFCC 1053 turned out to be very effective in endosulfan degradation (Bhalerao, 2012). The research reports of Bosso et al.(2015) has shown that the fungal species such as *Antracophyllum discolor*, *Antracophyllum* discolor, *Phanerochaete chrysosporium*, *Trametes versicolor*, *Ganoderma lucidum*, *Armillaria mellea*, and *Gloeophyllum striatum* are invoved in Pentachlorophenol (PCP) degradation. According to the investigation reports of Sharma and Adholeya (2011), *Paecilomyces lilacinus* fungi accumulate only 24% of chromium from spent chrome effluent supplemented with cane sugar, whereas, 100% removal was noted from a synthetic medium. The assessment carried out by Akar et al.(2005), showed that the fungi, *Botrytis cinerea* is able to remove Pb removal in a batch reactor. The filamentous fungi have greater adsorption potential for the removal of heavy metal (Singh & Gauba, 2014). The *Trichoderma* and *Mortierella* sps., isolated from the soil and *Aspergillus* and *Penicillium* sps., isolated from marine and terrestrial environments have large capability to clean up the polluted environment (Thenmozhi et al., 2013). Reports indicated that the pyrethroids degradation was brought about by *Candida*, *Trichoderma*, *Eurotium*, *Phanerochaete* and *Aspergillus* (Chen et al., 2011; Birolli et al., 2016; Birolli et al., 2018; Palmer-Brown et al., 2018; Hu et al., 2018). β-cypermethrin and 3-phenoxybenzaldehye degradation was carried out by the fungus, *Eurotiumcristatum* ET1, from fu brick tea in China (Hu et al., 2018). Many fungal enzymes have low specificity, permitting the fungal strain to metabolize multiple compounds of diverse pollutants even dissimilar in structure simultaneously, like *Phanerochaete chrysosporium* is a crust fungi, which degrades several hazardous chemicals including benzene, ethylbenzene, xylene, toluene, organochlorines, N-heterocyclic explosives, polycyclic aromatic hydrocarbons (PAHs), nitroaromatic compounds, pesticides, synthetic dyes, polychlorinated dibenzo-p-dioxins, and synthetic polymers simultaneously even in mixture of all (Kues, 2015). The fungi used for the bioremediation of various pesticides are represented in Table 6.

Table 7. Fungi used in bioremediation of dyes

Name of the Fungi	Name of the Dye	Reference
White-rot fungi	Reactive Green 19	Sari et al., 2016
Trichoderma harzianum	Congo red, Acid red, Basic blue and Bromophenol blue, Direct green	Singh and Singh, 2010
Aspergillus fumigatus	Malchite green, Trypan Blue Viscose Orange-A dyes	Kalyani et al., 2017; Madhuri et al., 2014; Saranraj et al., 2010
Cyberlindnera samutprakarnen sis	Acid Red B	Song et al., 2018
Trametes versicolor *Aspergillus niger* and *Aspergillus terreus*	Red dye 27 Red azo dye MX-5	Rekik et al., 2019 Almeida and Corso, 2014.

Earlier studies revealed the ability of wood-rotting fungi *Antrodia xanthan* and *Fomitopsis palustris* in remediating the copper accumulation in the wood (Deshmukh et al., 2016; Voberkova et al., 2017). *Pleurotus ostreatus*, *Aspergillus nidulans*, *Funalia trogii* and *Irpex lacteus* etc., are few plant-associated fungi that are capable to survive in and decolorize textile industry effluents. *Suillus bovinus* and *Rhizopogon roseolus* are the ectomycorrhizal fungi, coupled with pinus are involved in the removal of cadmium that is also subjected to the effect of other environmental factors such as types of nutrients and pH (Mao & Guan, 2016). Perissini-Lopes et al.(2016), had studied the potential of 14 different fungal strains involved in the degradation of diuron (*Absidia* sp., *Aspergillus spp.*, *Cunninghamella* spp., *Fusarium* spp., *Mucor* sp., *Paecilomyces* sp., *Trichoderma* spp. and *Verticillium* sp.)

Fungi have the ability to degrade a broad spectrum of pollutants and are attracting wide-spread use in bioremediation. Numerous bioreactors such as fluidized beds and rotating biological contactors have been designed for the remediation of pollutants with fungi (Gautam et al., 2012). The novel bioreactor systems have been set up for the removal of dyes such as Reactive Green 19 by white-rot fungi (Sari et al., 2016). The microbial degradation of different hazardous dyes like, congo red, acid red, basic blue and bromophenol blue, Direct green by the fungus *Trichoderma harzianum* (Singh and Singh, 2010) by using different fungal strains has been investigated earlier). The discoloration of Malchite green, Trypan Blue and Viscose Orange-A dyes is brought about by *Aspergillus fumigatus* (Kalyani et al., 2017, Madhuri et al., 2014, Saranraj et al., 2010). Similarly, *Cyberlindnera samutprakarnensis* decolourise Acid Red B (ARB) dye within 18 h with 97% efficacy under optimal conditions (Song et al., 2018). *Trametes versicolor* degrades the red dye 27 by means of lignins peroxidases (Rekik et al., 2019). *Aspergillus niger* and *Aspergillus terreus* degrade and absorb the red azo dye MX-5 by lowering its toxicity (Almeida & Corso, 2014). The fungi used for the bioremediation of different dyes are represented in Table 7.

MICROPLASTICS POLLUTION AND BIOREMEDIATION

The plastic materials are synthetic polymer compounds which contain many other chemicals to increase the efficiency. The majority of plastics synthesized from petrochemical sources, which have high molecular mass and plasticity (Costa et al., 2016). The microplastics (MPs) are the emerging contaminants that exist in various environmental media. The particles and fibres smaller than five millimetres are usually referred to as microplastics (GESAMP, 2015; Bertling et al., 2018; Galgani et al., 2010; De Souza Machado et al., 2018). The microplastics are frequently abundant in soils than in ocean waters (He et al., 2018; Rezania et al., 2018). The world's agricultural soils alone might contain a lot more microplastic mass than oceanic surface waters (Nizzetto et al., 2016). Reportedly, plenty of microplastics were found in Vembanad Lake, Kerala, with a mean abundance of 252.80 ± 25.76 particles m^{-2}. Low-density polyethene is the essential polymer compound (Sruthy and Ramasamy, 2016). It was stated that, microplastics may exhibit an adverse consequences on soil biota, causing increase in mortality rate and lowering the growth and reproduction rates of soil life (Huerta Lwanga et al., 2016; Zhu et al., 2018). In extremely polluted top soils, the abundance of microplastics differs at different soil depths, reaching up to 7% by weight (Fuller & Gautam, 2016; Liu et al., 2018). The implications of microplastics of different form, density and chemical composition, bulk density, water holding capacity, water stable aggregates and soil microbial activities were investigated by De Souza Machado et al. (2018).

In India, the inquiries on microplastics was noted and presence of plastic waste (81 mg/kg) such as polyurethane, nylon, polystyrene, polyester particles in the marine sediments of Gujarat coast was detected by Reddy et al. (2006). Plentiful of plastics waste (7.49 gm^{-2} and 68.83 items m^{-2}) were found in Mumbai beach (Jayasiri et al., 2013). The Central Pollution Control board, New Delhi 2014 reported that India is one of the principal plastic consumers of the world with an average plastic generation of 5.6 million tons of plastic annually (Toxics link, 2014). Microplastics sources are generally categorised as (i) primary sources of microplastics produced directly in industry (Gregory, 2009) and (ii) secondary microplastics generated indirectly from the fragmentation of larger plastic residues. The low density polyethylene (LDPE) is the main source for microplastic pollution globally.

Microplastics may undergo degradation, normally by biodegradation, where the carbon in the polymer is converted into CO_2 and incorporated into the marine biomass by microbial colonies. The lack of knowledge on the mineralisation of microplastics in the atmosphere and the presence of nano-scale plastics in the ocean has been noted (Andrady, 2011; GESAMP, 2015). *Aspergillus* and *Bacillus* were found to be involved in the low density polyethylene degradation (Esmaeili et al., 2013). *Pseudomonas* species are most highly implicated in the biodegradation of LDPEs (Bhatia et al., 2014). They isolated *Pseudomonas citronellolis* EMBS027 strain which led to 17.8% weight reduction on polyethylene sheets. The *Brevibaccillus borstelensis* strain 707 was isolated after 30-day incubation at 50°C reduced the gravimetric and molecular weights of polyethylene sheets by 11 and 30% respectively (Hadad et al., 2005). *Bacillus subtilis* is capable of degrading polyethylene (Vimala & Mathew, 2016) in the presence and absence of bio-surfactants. Several fungal genera were isolated which degrades the polyethylene sheets with *Aspergillus niger* showing the maximum weight reduction by, 4.32% (Vinay et al., 2016). According-ing to the reports of source Hadad et al. (2005), the Gram-positive bacteria *Brevibacillus borstelensis* and *Rhodococcus ruber* were reported to have throughput to degrade the CH_2 backbone of plastics and utilize the polyethylene as their sole carbon source.

Table 8. Genetically modified bacteria used in bioremediation of heavy metals/OPs/PCBs

Name of the Genetically Modified Bacteria	Heavy Metals/OPs/PCBs	References
Rhodopseudomonas palustris (Recombinant photosynthetic bacterium)	Elimination of Hg^{2+} from the heavy metal wastewater	Xu & Pei, 2011
E. coli SE5000	Nickel	Sari et al., 2016
Sphingomonas desiccabilis, Bacillus idriensis	Arsenic	Singh and Singh, 2010
*Achromobacter*sp AO22	Mercury	Kalyani et al., 2017; Madhuri et al., 2014; Saranraj et al., 2010
Ralstoniaeutropha CH34	Cd^{2+}	Song et al., 2018
Alcaligenes eutrophus AE104 (pEBZ141)	Chromium expulsion from industrial wastewater	Srivastava et al., 2010
Stenotrophomonas sp. strain YC-1 [organophosphorus hydrolase (OPH)]	Mixture of organophosphates (OPs)	Yang et al., 2010
Rhodococcus sp. RHA1 (pRHD34) and *Burkholderia xenovorans* LB400 (pRO41)	Mixture of polychlorinated biphenyls (PCBs)	Rodrigues et al., 2006

GENETICALLY MODIFIED MICROBES

Genetic engineering is a new technique that enables microorganisms that are capable of degrading specific pollutants to be engineered. It creates an incentive to construct an artificial mix of genes in nature that do not occur together. Genetically engineered microorganisms (GEMs) are classified as bacteria, fungi or viruses in which the genetic material has been changed mainly by the use of recombinant DNA technology, i.e. through means which are not naturally occurring. GEMs showed potential applications for the bioremediation in soil, groundwater and activated sludge environments, because of their enhanced degradative capabilities of a wide range of pollutants (Menn et al., 2008). Several naturally occuring or genetically modified microbes have the capability to degrade, chelate or transform various toxic chemical compounds and therefore provide better strategies to combat environmental contamination. Great attention was paid to the applications for GEMs for bioremediation, but they were mainly confined to the laboratory area. This was attributed to questions regarding regulatory risk management and, to a large degree, the ambiguity of their realistic effect and delivery under field circumstances. At least four principal approaches are available for the development of GEMs for bioremediation application, which include:(a) modification of enzyme specificity and affinity (b) pathway construction and regulation (c) bioprocess development, monitoring and control and (d) bioaffinity bioreporter sensor applications for chemical sensing, toxicity reduction, and end point analysis. On the regular basis, scientists deploy either natural or modified microorganisms to eliminate pollutants, viz., radioactive waste, heavy metals, metalloids and oil products from contaminated sites (Dixit et al., 2015). Genetically engineered bacteria are an advanced technology that has attracted public attention when employed in cleaning up toxic waste and heavy metals from contaminated sites (Shukla et al., 2010; Liu et al., 2011). It has also contributed to the detoxification of heavy metals and other recalcitrant compounds (Muhammad et al., 2008). The genetically modified bacteria used for the bioremediation of heavy metals/OPs/PCBs are represented in Table 8.

The bacteria have a high potential force for the degradation of environmental contaminants. A study for expanding the substrate range of enzymes was reported by Yang et al. (2010). In this research, *Stenotrophomonas* sp. strain YC-1, a soil bacterium was genetically engineered to generate organophosphorus hydrolase (OPH) enzyme with wider substrate range for the organophosphates (OPs). A mixture of six Synthetic organophosphate pesticides could be degraded completely within 5 hours. The broader substrate specificity in combination with the rapid degradation rate makes this engineered strain a promising candidate for in situ remediation of OP-contaminated sites. Rodrigues et al. (2006) studied the ability of two genetically modified strains *Rhodococcus* sp. RHA1 (pRHD34) and *Burkholderiaxenovorans* LB400 (pRO41) to degrade mixture of PCBs in soil polluted with Aroclor 1242.

Bioremediation becomes successful with the application of recombinant DNA and RNA technologies and genetically modified microbes. Bioremediation approach is strengthened by altering the genes of microorganisms, that generate novel metabolic pathway.The transformation of hazardous heavy metals into non-toxic form is achieved by the bacterial metal regulatory genes (Bondarenko et al., 2008; Jan et al., 2009; Ng et al., 2009; Hasin et al., 2010). The expression of metallothioneins (MT) by the genetically engineered bacteria will hasten heavy metal deposition (Pazirandeh et al., 1995).

ADVANTAGES AND DISADVANTAGES OF BIOREMEDIATION

Advantages

1. Bioremediation is a natural process and requires acceptable waste treatment system for the polluted substance such as soil.
2. It is helpful for the total removal of pollutants.
3. It is inexpensive when compared with other mechanisms or technologies that are used for the elimination of hazardous waste.

Disadvantages

1. Bioremediation is restricted to those compounds, which are biodegradable.
2. It is hard to generalise from the bench and pilot scale studies to full scale field operations.
3. It require longer than other treatment options, like excavation and removal of soil or incineration.

CONCLUSION

Based on the above literature, it is clearly revealed that soil has been extensively polluting by various kinds of organic and inorganic compounds derived from various types of sources such as agricultural, industrial, radioactive and urban wastes. The organic compounds like pesticides, insecticides, fungicides, and herbicides etc., inorganic compounds such as heavy metals and radionuclides are dramatically polluting the soil environment. However, many researchers have been studied on various remedial methods like physical, chemical and bioremedial techniques. Literature survey clearly open that bioremediation of diverse pollutants by using potential microorganisms such as bacteria and fungi are proved to be more efficient, economical, eco-friendly. The information available on the diverse chemical pollutants, which adversely affects the environment, could be useful for the development of potential bioremedial techniques for the removal and/ or detoxification of pollutants. Bioremediation is safe and is an alternative to traditional physicochemical techniques for the remediation of organic pollutants at contaminated sites and removes the pollutants by speeding up the natural process of biodegradation. Engineering of useful microorganisms to improve detoxifying potential provides further betterment in pesticide decontamination. Microorganisms are predominantly responsible for biodegradation and the elimination of toxic chemicals. It seems that, microbial bioremediation is economical and the most efficient method for the removal of heavy metal pollutants from the environment. Both growth operations, such as agriculture, mining, manufacturing, manufacturing practices, power plants, transport services, consciously or indirectly, whether consciously or inadvertently, make a critical contribution to the processing of food, water and soil chemical pollutants. The best remediation methods must be identified after completing necessary laboratory studies, considering the specialities of pollutants. Majority of the successful applications of bioremediation involve combined remediation systems. The biological degradation or biodegradation of synthetic dyes using microbes is risk-free, economical and eco-friendly.

REFERENCES

Aarkrog, A. (1994). Source terms and inventories of anthropogenic radionuclides. In E. Holm (Ed.), *Radioecology. Lectures in Environmental Radioactivity* (pp. 21–38). World Scientific Publishing.

Aarkrog, A. (2003). Input of anthropogenic radionuclides into the World Ocean. *Deep-sea Research. Part II, Topical Studies in Oceanography*, *50*(17-21), 2597–2606. doi:10.1016/S0967-0645(03)00137-1

Abbas, S. H., Ismail, I. M., Mostafa, T. M., & Sulaymon, A. H. (2014). Biosorption of heavy metals: A review. *Journal of Chemical Science and Technology*, *3*, 74–102.

Abd-Alrahman, S. H., & Salem-Bekhit, M. M. (2013). Microbial biodegradation of butachlor pollution (obsolete pesticide Machete 60% EC). *Asian Journal of Multidimensional Research*, *7*(4), 330–335.

Abdel-Aty, A. M., Ammar, N. S., Abdel Ghafar, H. H., & Ali, R. K. (2013). Biosorption of cadmium and lead from aqueous solution by fresh water alga *Anabaena sphaerica* biomass. *Journal of Advanced Research*, *4*(4), 367–374. doi:10.1016/j.jare.2012.07.004 PMID:25685442

Abhilash, P. C., Srivastava, S., & Singh, N. (2011). Comparative bioremediation potential of four rhizospheric microbial species against lindane. *Chemosphere*, *82*(1), 56–63. doi:10.1016/j.chemosphere.2010.10.009 PMID:21044795

Aceves-Diez, A. E., Estrada-Castaneda, K. J., & Castaneda-Sandoval, L. M. (2015). Use of Bacillus thuringiensis supernatant from a fermentation process to improve bioremediation of chlorpyrifos in contaminated soils. *Journal of Environmental Management*, *157*, 213–219. doi:10.1016/j.jenvman.2015.04.026 PMID:25910975

Ackerley, D. F., Gonzalez, C. F., Keyhan, M., Blake, R., & Matin, A. (2004). Mechanism of chromate reduction by the Escherichia coli protein, NfsA, and the role of different chromate reductases in minimizing oxidative stress during chromate reduction. *Environmental Microbiology*, *6*(8), 851–860. doi:10.1111/j.1462-2920.2004.00639.x PMID:15250887

Agamuthu, P., Tan, Y. S., & Fauziah, S. H. (2013). Bioremediation of hydrocarbon contaminated soil using selected organic wastes. *Procedia Environmental Sciences*, *18*, 694–702. doi:10.1016/j.proenv.2013.04.094

Aguelmous, A., El Fels, L., Souabi, S., Zamama, M., & Hafidi, M. (2019). The fate of total petroleum hydrocarbons during oily sludge composting: A critical review. *Reviews in Environmental Science and Biotechnology, 18*, 473 - 493.

Ahemad Munees, M. (2013). Pesticides as Antagonists of Rhizobia and the Legume-Rhizobium Symbiosis: A Paradigmatic and Mechanistic Outlook. *Biochemistry and Molecular Biology*, *1*(4), 63–75. doi:10.12966/bmb.12.02.2013

Akar, T., Tunali, S., & Kiran, I. (2005). *Botrytis cinerea* as a new fungal biosorbent for removal of Pb(II) from aqueous solutions. *Biochemical Engineering Journal*, *5*(25), 227–235. doi:10.1016/j.bej.2005.05.006

Akcil, A., Erust, C., Ozdemiroglu, S., Fonti, V., & Beolchini, F. (2015). A review of approaches and techniques used in aquatic contaminated sediments: Metal removal and stabilization by chemical and biotechnological processes. *Journal of Cleaner Production*, *86*, 24–36. doi:10.1016/j.jclepro.2014.08.009

Al Hasin, A., Gurman, S. J., Murphy, L. M., Perry, A., Smith, T. J., & Gardiner, P. H. E. (2010). Remediation of Chromium (VI) by a Methane-Oxidizing Bacterium. *Environmental Science & Technology*, *44*(1), 400–405. doi:10.1021/es901723c PMID:20039753

Alexander, M. (1994). *Biodegradation and bioremediation*. Academic.

Algreen, M., Rein, A., Legind, C. N., Amundsen, C. E., Karlson, U. G., & Trapp, S. (2012). Test of tree core sampling for screening of toxic elements in soils from a Norwegian site. *International Journal of Phytoremediation*, *14*(4), 305–319. doi:10.1080/15226514.2011.620648 PMID:22567713

Almeida, E. J. R., & Corso, C. R. (2014). Comparative study of toxicity of azo dye procion red MX-5B following biosorption and biodegradation treatments with the fungi *Aspergillus niger* and *Aspergillus terreus*. *Chemosphere*, *112*, 317–322. doi:10.1016/j.chemosphere.2014.04.060 PMID:25048922

Andrady, A. L. (2011). Microplastics in the marine environment. *Marine Pollution Bulletin*, *62*(8), 1596–1605. doi:10.1016/j.marpolbul.2011.05.030 PMID:21742351

Angelucci, D., & Tomei, M. (2016). Ex-situ bioremediation of chlorophenol contaminated soil: Comparison of slurry and solid-phase bioreactors with the two-step polymer extraction-bioregeneration process. *Journal of Chemical Technology and Biotechnology (Oxford, Oxfordshire)*, *91*(6), 1577–1584. doi:10.1002/jctb.4882

Anikwe, M. A. N., & Nwobodo, K. C. A. (2002). Long-term effect of municipal waste disposal on soil properties and productivity of sites used for urban agriculture in Abakaliki, Nigeria. *Bioresource Technology*, *83*(3), 241–250. doi:10.1016/S0960-8524(01)00154-7 PMID:12094801

Awuchi Godswill, C., & Awuchi Gospel, C. (2019a). Physiological Effects of Plastic Wastes on the Endocrine System (Bisphenol A, Phthalates, Bisphenol S, PBDEs, TBBPA). *International Journal of Bioinformatics and Computational Biology*, *4*(2), 11–29.

Awuchi Godswill, C., & Awuchi Gospel, C. (2019b). Impacts of Plastic Pollution on the Sustainability of Seafood Value Chain and Human Health. *International Journal of Advanced Academic Research*, *5*(11), 46–138.

Azam, F., Farooq, S., & Lodhi, A. (2003). Microbial biomass in agricultural soils-determination, synthesis, dynamics and role in plant nutrition. *Pakistan Journal of Biological Sciences*, *6*(7), 629–639. doi:10.3923/pjbs.2003.629.639

Azubuike, C. C., Chikere, C. B., & Okpokwasili, G. C. (2016). Bioremediation techniques - classification based on site of application: Principles, advantages, limitations and prospects. *World Journal of Microbiology & Biotechnology*, *32*(11), 180. doi:10.100711274-016-2137-x PMID:27638318

Badhan, K., Mehra, R., & Sonkawade, R. G. (2017). Natural Radioactivity Measurements in Soils of Jalandhar and Hoshiarpur Districts of Punjab, India. *International Journal of Pure and Applied Physics*, *13*, 232–237.

Baishya, K., & Samra, H. P. (2014). Effect of agrochemicals application on accumulation of heavy metals on soil of different land uses with respect to nutrient status. *Journal of Environmental Science Toxicology and Food Technology*, *8*(7), 46–54. doi:10.9790/2402-08724654

Bajaj, S., Sagar, S., Khare, S., & Singh, D. K. (2017). Biodegradation of γhexachlorocyclohexane (lindane) by halophilic bacterium *Chromohalobacter* sp. LD2 isolated from HCH dumpsite. *International Biodeterioration & Biodegradation*, *122*, 23–28. doi:10.1016/j.ibiod.2017.04.014

Bedard, D. L., & May, R. J. (1995). Characterization of the polychlorinated biphenyls in the sediments of Woods Pond: Evidence for microbial dechlorination of Aroclor 1260 *in situ*. *Environmental Science & Technology*, *30*(1), 237–245. doi:10.1021/es950262e

Bertling, J., Bertling, R., & Hamann, L. (2018). Kunststoffe in der Umwelt. In Fraunhofer Institut für Umwelt, Sicherheits- und Energietechnik Umsicht (pp. 1–56). Oberhausen, Germany: Academic Press.

Bhagobaty, R. K., & Malik, A. (2008). Utilization of chlorpyrifos as a sole source of carbon by bacteria isolated from wastewater irrigated agricultural soils in an industrial area of western Uttar Pradesh, India. *Research Journal of Microbiology*, *3*(5), 293–307. doi:10.3923/jm.2008.293.307

Bhalerao, T. S. (2012). Bioremediation of endosulfancontaminated soil by using bioaugmentation treatment of fungal inoculant *Aspergillus niger*. *Turkish Journal of Biology*, *35*, 561–567.

Bhalerao, T. S., & Puranik, P. R. (2007). Biodegradation of organochlorine pesticide, endosulfan, by a fungal soil isolate, Aspergillus niger. *International Biodeterioration & Biodegradation*, *59*(4), 315–321. doi:10.1016/j.ibiod.2006.09.002

Bhatia, M., Girdhar, A., Tiwari, A., & Nayarisseri, A. (2014). Implications of a novel Pseudomonas species on low density polyethylene biodegradation: An in vitro to in silico approach. *SpringerPlus*, *3*(1), 497. doi:10.1186/2193-1801-3-497 PMID:25932357

Bhatt, P., Gangola, S., Chaudhary, P., Khati, P., Kumar, G., Sharma, A., & Srivastava, A. (2019). Pesticide induced up-regulation of esterase and aldehyde dehydrogenase in indigenous *Bacillus* spp. *Bioremediation Journal, 23*(1), 42–52. doi:10.1080/10889868.2019.1569586

Bhatt, P., Huang, Y., Zhang, W., Sharma, A., & Chen, S. (2020). Enhanced cypermethrin degradation kinetics and metabolic pathway in *Bacillus thuringiensis* strain SG4. *Microorganisms, 8*(2), 223. doi:10.3390/microorganisms8020223 PMID:32046050

Bidlan, R., Afsar, M., & Mononmani, H. K. (2004). Bioremediation of HCH-contaminated soil: Elimination of inhibitory effects of the insecticide on radish and green gram seed germination. *Chemosphere, 56*(8), 803–811. doi:10.1016/j.chemosphere.2004.01.015 PMID:15251295

Birolli, W. G., Alvarenga, N., Seleghim, M. H. R., & Porto, A. L. M. (2016). Biodegradation of the pyrethroid pesticide esfenvalerate by marine-derived fungi. *Marine Biotechnology (New York, N.Y.), 18*(4), 511–520. doi:10.100710126-016-9710-z PMID:27381569

Birolli, W. G., Arai, M. S., Nitschke, M., & Porto, A. L. M. (2019). The pyrethroid (±)-lambda-cyhalothrin enantioselective biodegradation by a bacterial consortium. *Pesticide Biochemistry and Physiology, 156*, 129–137. doi:10.1016/j.pestbp.2019.02.014 PMID:31027572

Boeing, D. W. (2000). Ecological effects, transport, and fate of mercury: A general review. *Chemosphere, 40*(12), 1335–1351. doi:10.1016/S0045-6535(99)00283-0 PMID:10789973

Bolan, N. S., Choppala, G., Kunhikrishnan, A., Park, J., & Naidu, R. (2013). Microbial transformation of trace elements in soils in relation to bioavailability and remediation. *Reviews of Environmental Contamination and Toxicology, 225*, 1–56. doi:10.1007/978-1-4614-6470-9_1 PMID:23494555

Bondarenko, O., Rolova, T., Kahru, A., & Ivask, A. (2008). Bioavailability of Cd, Zn and Hg in soil to nine recombinant luminescent metal sensor bacteria. *Sensors (Basel), 8*(11), 6899–6923. doi:10.33908116899 PMID:27873907

Bosso, L., Scelza, R., Testa, A., Cristinzio, G., & Rao, M. A. (2015). Depletion of pentachlorophenol contamination in an agricultural soil treated with Byssochlamys nivea, Scopulariopsis brumptii and urban waste compost: A laboratory microcosm study. *Water, Air, and Soil Pollution, 226*(6), 183. doi:10.100711270-015-2436-0

Brown, L. D., Cologgi, D. L., Gee, K. F., & Ulrich, A. C. (2017). Bioremediation of Oil Spills on Land. In Oil Spill Science and Technology (2nd ed., pp. 699 - 729). Gulf Professional Publishing. doi:10.1016/B978-0-12-809413-6.00012-6

Burns, R. C., & Hardy, R. W. F. (1975). *Nitrogen Fixation in Bacteria and Higher Plants*. Springer - Verlag. doi:10.1007/978-3-642-80926-2

Cachada, A., Pato, P., Rocha-Santos, T., Da Silva, E. F., & Duarte, A. C. (2012). Levels, sources and potential human health risks of organic pollutants in urban soils. *The Science of the Total Environment, 430*, 184–192. doi:10.1016/j.scitotenv.2012.04.075 PMID:22652008

Camachomorales, R. L., Karina, G. N., & José, E. S. (2017). Degradation of the herbicide paraquat by macromycetes isolated from southeastern Mexico. *Biotech, 7*, 324 - 334.

Carvalho, F. P. (2017). Pesticides, environment, and food safety. *Food and Energy Security*, *6*(2), 48–60. doi:10.1002/fes3.108

Cempel, M., & Nikel, G. (2006). Nickel: A review of its sources and environmental toxicology. *Polish Journal of Environmental Studies*, *15*, 375–382.

Chakraborty, R., Wu, C. H., & Hazen, T. C. (2012). Systems biology approach to bioremediation. *Current Opinion in Biotechnology*, *23*(3), 483–490. doi:10.1016/j.copbio.2012.01.015 PMID:22342400

Chandra, R., & Singh, S. k. (2009). *Fundamentals and management of soil quality*. Westville Publishing House.

Chang, C. Y., Yu, H. Y., Chen, J. J., Li, F. B., Zhang, H. H., & Liu, C. P. (2014). Accumulation of heavymetalsinleaf vegetables from agricultural soils and associated potential health risks in the Pearl River Delta, South China. *Environmental Monitoring and Assessment*, *186*(3), 1547–1560. doi:10.100710661-013-3472-0 PMID:24185814

Chen, B., Stein, A. F., Castell, N., Gonzalez-Castanedo, Y., De La Campa, A. S., & De La Rosa, J. (2016). Modeling and evaluation of urban pollution events of atmospheric heavy metals from a large Cu-smelter. *The Science of the Total Environment*, *539*, 17–25. doi:10.1016/j.scitotenv.2015.08.117 PMID:26352643

Chen, H. W., Xu, M., Ma, X. W., Tong, Z. H., & Liu, D. F. (2019). Isolation and characterization of a chlorate-reducing bacterium *Ochrobactrum anthropi* XM-1. *Journal of Hazardous Materials*, *380*, 120–873. doi:10.1016/j.jhazmat.2019.120873 PMID:31325697

Chen, S., Chang, C., Deng, Y., An, S., Dong, Y. H., Zhou, J., Hu, M., Zhong, G., & Zhang, L. H. (2014). Fenpropathrin biodegradation pathway in Bacillus sp. DG-02 and its potential for bioremediation of pyrethroid-contaminated soils. *Journal of Agricultural and Food Chemistry*, *12*, 147–157. PMID:24576059

Chen, S., Chao, L., Sun, L. N., & Sun, T. H. (2013). Plant-microorganism combined remediation for sediments contaminated with heavy metals. *Advanced Materials Research*, *123*, 610–613. doi:10.4028/www.scientific.net/AMR.726-731.610

Chen, S., Geng, P., Xiao, Y., & Hu, M. (2012). Bioremediation of b-cypermethrin and 3- phenoxybenzaldehyde contaminated soils using Streptomyces aureus HP-S- 01. *Applied Microbiology and Biotechnology*, *94*(2), 505–515. doi:10.100700253-011-3640-5 PMID:22038248

Chen, S., Yang, L., Hu, M., & Liu, J. (2011). Biodegradation of fenvalerate and 3- phenoxybenzoic acid by a novel *Stenotrophomonas* sp. strain ZS-S-01 and its use in bioremediation of contaminated soils. *Applied Microbiology and Biotechnology*, *90*(2), 755–767. doi:10.100700253-010-3035-z PMID:21184062

Chien, C., Yumei, K. U. O., Changchieh, C. H. E. N., Chunwei, H. U. N. G., Chihwei, Y. E. H., & Weijen, Y. E. H. (2008). Microbial diversity of soil bacteria in agricultural field contaminated with heavy metals. *Journal of Environmental Sciences (China)*, *20*(3), 359–363. doi:10.1016/S1001-0742(08)60056-X PMID:18595405

Chudasama, K. S., & Thaker, V. S. (2017). Genome sequence of *Ochrobactrum anthropi* strain SUBG007, a plant pathogen and potential xenobiotic compounds degradation bacterium. *Genomics Data*, *11*, 116–117. doi:10.1016/j.gdata.2017.01.001 PMID:28119820

Clemens, S. (2006). Toxic metal accumulation, responses to exposure and mechanisms of tolerance in plants. *Biochimie, 88*(11), 1707–1719. doi:10.1016/j.biochi.2006.07.003 PMID:16914250

Costa, J. P., Santos, P. S. M., Duarte, A. C., & Rocha-Santos, T. (2016). Nanoplastics in the environment - sources, fates and effects. *The Science of the Total Environment, 567*, 15–26. doi:10.1016/j.scitotenv.2016.05.041 PMID:27213666

Cycon, M., Mrozik, A., & Piotrowska-Seget, Z. (2017). Bioaugmentation as a strategy for the remediation of pesticide-polluted soil: A review. *Chemosphere, 172*, 52–71. doi:10.1016/j.chemosphere.2016.12.129 PMID:28061345

Cycon, M., & Piotrowska-Seget, Z. (2016). Pyrethroid-degrading microorganisms and their potential for the bioremediation of contaminated soils: A review. *Frontiers in Microbiology, 7*, 1463. doi:10.3389/fmicb.2016.01463 PMID:27695449

Cycon, M., Zmijowska, A., & Piotrowska-Seget, Z. (2014). Enhancement of deltamethrin degradation by soil bioaugmentation with two different strains of *Serratia marcescens. International Journal of Environmental Science and Technology, 11*(5), 1305–1316. doi:10.100713762-013-0322-0

Cycon, M. Z., Mijowska, A., Wojcik, M., & Piotrowska-Seget, Z. (2013). Biodegradation and bioremediation potential of diazinon-degrading Serratia marcescens to remove other organophosphorus pesticides from soils. *Journal of Environmental Management, 117*, 7–16. doi:10.1016/j.jenvman.2012.12.031 PMID:23333465

D'Annibale, A., Rosetto, F., Leonardi, V., Federici, F., & Petruccioli, M. (2006). Role of autochthonous filamentous fungi in bioremediation of a soil historically contaminated with aromatic hydrocarbons. *Applied and Environmental Microbiology, 72*(1), 28–36. doi:10.1128/AEM.72.1.28-36.2006 PMID:16391021

Das, S., & Dash, H. R. (2014). 1 - Microbial Bioremediation: *A Potential Tool for Restoration of Contaminated Areas*. In S. Das (Ed.), *Microbial Biodegradation and Bioremediation* (pp. 1–21). Elsevier. doi:10.1016/B978-0-12-800021-2.00001-7

De Souza Machado, A. A., Kloas, W., Zarfl, C., Hempel, S., & Rillig, M. C. (2018). Microplastics as an emerging threat to terrestrial ecosystems. *Global Change Biology, 24*(4), 1405–1416. doi:10.1111/gcb.14020 PMID:29245177

Derakhshan, Z., Ehrampoush, M. H., Mahvi, A. H., Dehghani, M., Faramarzian, M., & Eslami, H. (2019). A comparative study of hybrid membrane photobioreactor and membrane photobioreactor for simultaneous biological removal of atrazine and CNP from wastewater: A performance analysis and modeling. *Chemical Engineering Journal, 355*, 428–438. doi:10.1016/j.cej.2018.08.155

Deshmukh, R., Khardenavis, A. A., & Purohit, H. J. (2016). Diverse metabolic capacities of fungi for bioremediation. *Indian Journal of Microbiology, 56*(3), 247–264. doi:10.100712088-016-0584-6 PMID:27407289

Dhal, P. K., & Sar, P. (2014). Microbial communities in uranium mine tailings and mine water sediment from JadugudaU mine, India: A culture independent analysis. *Journal of Environmental Science and Health. Part A, Toxic/Hazardous Substances & Environmental Engineering, 49*(6), 694–709. doi:10.1080/10934529.2014.865458 PMID:24521415

Dixit, R., Wasiullah, E., Malaviya, D., Pandiyan, K., Singh, U., Sahu, A., Shukla, R., Singh, B., Rai, J., Sharma, P., Lade, H., & Paul, D. (2015). Bioremediation of heavy metals from soil and aquatic environment: An overview of principles and criteria of fundamental processes. *Sustainability*, *7*(2), 2189–2212. doi:10.3390u7022189

Dominguez-Nunez, J. A., Benito, B., Berrocal-Lobo, M., & Albanesi, A. (2016). Mycorrhizal fungi: role in the solubilization of potassium. In V. S. Meena, B. R. Maurya, J. P. Verma, & R. S. Meena (Eds.), *Potassium solubilizing microorganisms for sustainable agriculture* (pp. 77–98). Springer. doi:10.1007/978-81-322-2776-2_6

Dotaniya, M. L., Meena, H. M., & Lata, M. (2013b). Heavy metal toxicity: Pandora's box for human disease. *Read Shelf*, *9*(6), 5–6.

Dotaniya, M. L., Rajendiran, S., Meena, V. D., Saha, J. K., Coumar, M. V., Kundu, S., & Patra, A. K. (2016). Influence of chromium contamination on carbon mineralization and enzymatic activities in Vertisol. *Agricultural Research*, *6*(1), 91–96. doi:10.100740003-016-0242-6

Dotaniya, M. L., Thakur, J. K., Meena, D., Jajoria, D. K., & Rathor, G. (2014). Chromium pollution: A threat to environment. *Agricultural Reviews (Karnal)*, *35*(2), 153–157. doi:10.5958/0976-0741.2014.00094.4

Dua, M., & Singh, A. (2002). Biotechnology and bioremediation: Successes and limitations. *Applied Microbiology and Biotechnology*, *59*(2-3), 143–152. doi:10.100700253-002-1024-6 PMID:12111139

Dwivedi, R. S. (1997). Perspectives of Microbial interactions. *Journal of the Indian Botanical Society*, *76*, 145–156.

Dzionek, A., Wojcieszy, D., & Guzik, U. (2016). Natural carriers in bioremediation: A review. *Electronic Journal of Biotechnology*, *23*, 28–36. doi:10.1016/j.ejbt.2016.07.003

Egorova, D. O., Buzmakov, S. A., Nazarova, E. A., Andreev, D. N., Demakov, V. A., & Plotnikova, E. G. (2017). Bioremediation of hexachlorocyclohexane-contaminated soil by the new *Rhodococcuswratislaviensis* Strain Ch628. *Water, Air, and Soil Pollution*, *228*(5), 183. doi:10.100711270-017-3344-2

El-Bestawy, E., El-Salam, Z., & Mansy, E. R. H. (2007). Potential Use of Environmental Cyanobacterial Species in Bioremediation of Lindane-Contaminated Effluents. *International Biodeterioration & Biodegradation*, *59*(3), 180–192. doi:10.1016/j.ibiod.2006.12.005

Elango, G., Rathika, G., & Elango, S. (2016). Physico-Chemical Parameters of Textile Dyeing Effluent and Its Impacts with Case study. *International Journal of Research in Chemistry and Environment*, *7*(1), 17–24.

Elliott, A., Hanby, W., & Malcolm, B. (1954). The near infra-red absorption spectra of natural and synthetic fibres. *British Journal of Applied Physics*, *5*(11), 337. doi:10.1088/0508-3443/5/11/301

EPA. (2006). In Situ Treatment Technologies for Contaminated Soil. Engineering Issue. EPA 542-F-06-013.

Esmaeili, A., Pourbabaee, A. A., Alikhani, H. A., Shabani, F., & Esmaeili, E. (2013). Biodegradation of Low-Density Polyethylene (LDPE) by Mixed Culture of *Lysinibacillus xylanilyticus* and *Aspergillus niger* in Soil. *PlosOne*, 8. doi:10.1371/journal.pone.0071720

FAO. (2002). International code of conduct on the distribution and use of pesticides. In *Hundred and Twenty-third Session of the FAO Council in November 2002*. Publishing Management Service, Information Division, FAO.

FAO. (2020). *Agriculturalpollution: pesticides*. documents.worldbank.org/curated/en/689281521218090562/pdf/124345-BRI-p153343-PUBLIC-march-22-9-pm-WB-Knowledge-Pesticides.pdf

Farahat, E., & Linderholm, H. W. (2015). The effect of long-term wastewater irrigation on accumulation and transfer of heavy metals in Cupressus sempervirens leaves and adjacent soils. *The Science of the Total Environment*, *51*, 1–7. doi:10.1016/j.scitotenv.2015.01.032 PMID:25613764

Fasani, E., Manara, A., Martini, F., Furini, A., & DalCorso, G. (2018). The potential of genetic engineering of plants for the remediation of soils contaminated with heavy metals. *Plant, Cell & Environment*, *41*(5), 1201–1232. doi:10.1111/pce.12963 PMID:28386947

Folch, A., Vilaplana, M., Amado, L., Vicent, R., & Caminal, G. (2013). Fungal permeable reactive barrier to remediate groundwater in an artificial aquifer. *Journal of Hazardous Materials*, *262*, 554–560. doi:10.1016/j.jhazmat.2013.09.004 PMID:24095995

Frascari, D., Zanaroli, G., & Danko, A. S. (2015). *In situ* aerobic cometabolism of chlorinated solvents: A review. *Journal of Hazardous Materials*, *283*, 382–399. doi:10.1016/j.jhazmat.2014.09.041 PMID:25306537

Frutos, F. J. G., Escolano, O., Garcı'a, S., Mar Babın, M., & Ferna'ndez, M. D. (2010). Bioventing remediation and ecotoxicity evaluation of phenanthrene-contaminated soil. *Journal of Hazardous Materials*, *183*(1-3), 806–813. doi:10.1016/j.jhazmat.2010.07.098 PMID:20800967

Frutos, F. J. G., Pe'rez, R., Escolano, O., Rubio, A., Gimeno, A., Fernandez, M. D., Carbonell, G., Perucha, C., & Laguna, J. (2012). Remediation trials for hydrocarbon-contaminated sludge from a soil washing process: Evaluation of bioremediation technologies. *Journal of Hazardous Materials*, *199*, 262–271. doi:10.1016/j.jhazmat.2011.11.017 PMID:22118850

Fulekar, M. H. (2009). Bioremediation of fenvalerate by Pseudomonas aeruginosa in a scale upbioreactor. *Romanian Biotechnological Letters*, *14*(6), 4900–4905.

Fuller, S., & Gautam, A. (2016). A procedure for measuring microplasticsusing pressurized fluid extraction. *Environmental Science & Technology*, *50*(11), 5774–5780. doi:10.1021/acs.est.6b00816 PMID:27172172

Galgani, F., Fleet, J., Van Franeker, S., Kastanevakis, T., Maes, J., Mouat, L., & Oosterban, I. (2010). Marine Strategy Framework Directive. Task Group 10 Report. Marine Litter; EU Commission, Joint Research Centre.

Gallego, J. L., Loredo, J., Llamas, J., Vázquez, F., & Sánchez, J. (2001). Bioremediation of diesel-contaminated soils: Evaluation of potential in situ techniques by study of bacterial degradation. *Biodegradation*, *12*(5), 325–335. doi:10.1023/A:1014397732435 PMID:11995826

Gangola, S., Joshi, S., Kumar, S., & Pandey, S. C. (2019). Comparative analysis of fungal and bacterial enzymes in biodegradation of xenobiotic compounds. *Smart Bioremediation Technologies: Microbial Enzymes*, *10*, 169–189. doi:10.1016/B978-0-12-818307-6.00010-X

Gaur, N., Flora, G., Yadav, M., & Tiwari, A. (2014). A review with recent advancements on bioremediation-based abolition of heavy metals. *Environmental Science. Processes & Impacts*, *16*(2), 180–193. doi:10.1039/C3EM00491K PMID:24362580

Gautam, S. P., Bundela, P. S., Pandey, A. K., Jamaluddin, M. K., Awasthi, M. K., & Sarsaiya, S. (2012). Diversity of cellulolytic microbes and the biodegradation of municipal solid waste by a potential strain. *International Journal of Microbiology*, *2012*, 1–12. doi:10.1155/2012/325907 PMID:22518141

Gavrilescu, M. (2004). Removal of heavy metals from the environment by biosorption. *Engineering in Life Sciences*, *4*(3), 219–232. doi:10.1002/elsc.200420026

Gentili, F. G., & Fick, J. (2017). Algal cultivation in urban wastewater: An efficient way to reduce pharmaceutical pollutants. *Journal of Applied Phycology*, *29*(1), 255–262. doi:10.100710811-016-0950-0 PMID:28344390

GESAMP. (2015). Sources, fate and effects of microplastics in the marine environment: a global assessment. In IMO/FAO/ UNESCOIOC/ UNIDO/WMO/IAEA/UN/UNEP/UNDP Joint Group of Experts on the Scientific Aspects of Marine Environmental Protection. Rep Stud GESAMP.

Glazer, A. N., & Nikaido, H. (2007). *Microbial biotechnology: Fundamentals of applied Microbiology*. Cambridge University Press. doi:10.1017/CBO9780511811227

Godheja, J., Modi, D. R., Kolla, V., Pereira, A. M., Bajpai, R., Mishra, M., Sharma, S. V., Sinha, K., & Shekhar, S. K. (2019). Environmental remediation: Microbial and nonmicrobial prospects. In *Microbial Interventions in Agriculture and Environment* (Vol. 1, pp. 379–409). Springer. doi:10.1007/978-981-13-8383-0_13

Grasserová, A., Hanc, A., Innemanová, P., & Cajthaml, T. (2020). Composting and vermicomposting used to break down and remove pollutants from organic waste: A mini review. *European Journal of Environmental Sciences*, *10*(1), 9–14. doi:10.14712/23361964.2020.2

Gregory, M. R. (2009). Environmental implications of plastic debris in marine settings–entanglement, ingestion, smothering, hangers-on, hitch-hiking and alien invasions. *Philosophical Transactions of the Royal Society of London. Series B, Biological Sciences*, *364*(1526), 2013–2025. doi:10.1098/rstb.2008.0265 PMID:19528053

Gürses, A., Acikyildiz, M., Günes, K., & Gürses, M.S. (2016). Dyes and pigments VIII. *Springer Brief in Green Chemistry for Sustainability*, *83*, 23.

Hadad, D., Geresh, S., & Sivan, A. (2005). Biodegradation of polyethylene by the thermophilic bacterium *Brevibacillus borstelensis*. *Journal of Applied Microbiology*, *98*(5), 1093–1100. doi:10.1111/j.1365-2672.2005.02553.x PMID:15836478

Hamzah, A., Hapsari, R. I., & Wisnubroto, E. I. (2016). Phytoremediation of Cadmium-contaminated agricultural land using indigenous plants. *International Journal of Environmental & Agriculture Research, 2*, 8–14.

Han, X., Yuan, R., Wang, G. Q., & Zhang, C. J. (2014). Isolation of paraquat degrading bacteria and identification of degradation characteristics. *Anhui Agricultural Science Bulletin, 20*, 38–39.

Hasin, A. A., Gurman, S. J., & Murphy, L. M. (2010). Remediation of chromium (VI) by a methane-oxidizing bacterium. *Environmental Science & Technology, 44*(1), 400–405. doi:10.1021/es901723c PMID:20039753

Hassan, M. M., & Carr, C. M. (2018). A critical review on recentadvancements of the removal of reactive dyes from dyehouseeffluent by ion-exchange adsorbents. *Chemosphere, 209*(1), 201–219. doi:10.1016/j.chemosphere.2018.06.043 PMID:29933158

He, D., Luo, Y., Lu, S., Liu, M., Song, Y., & Lei, L. (2018). Microplasticsin soils: Analytical methods, pollution characteristics and ecologicalrisks. *Trends inAnalytical Chemistry, 109*, 163–172. doi:10.1016/j.trac.2018.10.006

Hassaan & Ahmed. (2017). Health and Environmental Impacts of Dyes: Mini Review. *American Journal of Environmental Science and Engineering, 1*(3), 64–67.

Hemmat-Jou, M. H., Safari-Sinegani, A. A., Mirzaie-Asl, A., & Tahmourespour, A. (2018). Analysis of microbial communities in heavy metals-contaminated soils using the metagenomic approach. *Ecotoxicology (London, England), 27*(9), 1281–1291. doi:10.100710646-018-1981-x PMID:30242595

Hicks, B. (2019). *Agricultural pesticides and human health, department of earth sciences*. Montana State University. serc.carleton.edu/NAGTWorkshops/health/case_studies/pesticides.html

Hohener, P., & Ponsin, V. (2014). *In situ* vadose zone bioremediation. *Current Opinion in Biotechnology, 27*, 1–7. doi:10.1016/j.copbio.2013.08.018 PMID:24863890

Hossain, A., Islam Masum, M. M., Wu, X., Abdallah, Y., Ogunyemi, S. O., Wang, Y., Sun, G., Li, B., & An, Q. (2020). Screening of Bacillus strains in biocontrol of pathogen Dickeya dadantii causing stem and root rot disease of sweet potato. *Biocontrol Science and Technology, 30*(11), 1180–1198. doi:10.1080/09583157.2020.1798356

Hu, K., Deng, W., Zhu, Y., Yao, K., Li, J., Liu, A., Ao, X., Zou, L., Zhou, K., & He, L. (2018). Simultaneous degradation of β-cypermethrin and 3-phenoxybenzoic acid by Eurotium cristatum ET1, a novel"golden flower fungus" strain isolated from Fu Brick Tea. *MicrobiologyOpen, 8*, 7. PMID:30548839

Hu, Q. H., Weng, J. Q., & Wang, J. S. (2008). Sources of anthropogenic radionuclides in the environment: A review. *Journal of Environmental Radioactivity, 101*(6), 426–437. doi:10.1016/j.jenvrad.2008.08.004 PMID:18819734

Hu, W., Lu, Q., Zhong, G., Hu, M., & Yi, X. (2019). Biodegradation of pyrethroids by a hydrolyzing carboxylesterase EstA from *Bacillus cereus* BCC01. *Applied Sciences (Basel, Switzerland), 9*(3), 477. doi:10.3390/app9030477

Huerta Lwanga, E., Gertsen, H., Gooren, H., Peters, P., Salánki, T., van der Ploeg, M., Besseling, E., Koelmans, A. A., & Geissen, V. (2016). Microplastics in the terrestrial ecosystem: Implications for Lumbricus terrestris (Oligochaeta, Lumbricidae). *Environmental Science & Technology*, *50*(5), 2685–2691. doi:10.1021/acs.est.5b05478 PMID:26852875

Ibrahim, A. S., Elbadawi, Y. B., El-Tayeb, M. A., & Al-Salamah, A. A. (2012). Hexavalent chromium reduction by novel chromate resistant alkaliphilic *Bacillus* sp. strain KSUCr9a. *African Journal of Biotechnology*, *11*, 3832–3841.

Inthorn, D., Sidtitoon, N., Silapanuntakul, S., & Incharoensakdi, A. (2002). Sorption of mercury, cadmium and lead by microalgae. *Science Asia*, *28*(3), 253–261. doi:10.2306cienceasia1513-1874.2002.28.253

Iqbal, M., Iqbal, N., Bhatti, I. A., Ahmad, N., & Zahid, M. (2016). Response surface methodology application in optimization of cadmium adsorption by shoe waste: A good option of waste mitigation by waste. *Ecological Engineering*, *88*, 265–275. doi:10.1016/j.ecoleng.2015.12.041

Isaac, P., Alessandrello, M. J., Macedo, A. J., Estévez, M. C., & Ferrero, M. A. (2017). Pre-exposition to polycyclic aromatic hydrocarbons (PAHs) enhance biofilm formation and hydrocarbon removal by native multi-species consortium. *Journal of Environmental Chemical Engineering*, *5*(2), 1372–1378. doi:10.1016/j.jece.2017.02.031

Islam, M. M., Mahmud, K., Faruk, O., & Billah, M. S. (2011). Textile Dyeing Industries in Bangladesh for Sustainable Development. *International Journal of Environmental Sciences and Development*, *2*(6), 428–436. doi:10.7763/IJESD.2011.V2.164

Islam, M. M., Mahmud, K., Faruk, O., & Billah, M. S. (2011). Textile Dyeing Industries in Bangladesh for Sustainable Development. *International Journal of Environmental Sciences and Development*, *2*(6), 428–436. doi:10.7763/IJESD.2011.V2.164

Jadhav, J. P., Phugare, S. S., Dhanve, R. S., & Jadhav, S. B. (2010). Rapid biodegradation and decolorization of Direct Orange 39 (Orange TGLL) by an isolated bacterium *Pseudomonas aeruginosa* strain BCH. *Biodegradation*, *21*(3), 453–463. doi:10.100710532-009-9315-6 PMID:19937265

Jan, A. T., Azam, M., Choi, I., Ali, A., & Haq, Q. M. R. (2016). Analysis for the presence of determinants involved in the transport of mercury across bacterial membrane from polluted water bodies of India. *Brazilian Journal of Microbiology*, *47*(1), 55–62. doi:10.1016/j.bjm.2015.11.023 PMID:26887227

Jan, A. T., Murtaza, I., Ali, A., & Rizwanul Haq, Q. M. (2009). Mercury pollution: An emerging problem and potential bacterial remediation strategies. *World Journal of Microbiology & Biotechnology*, *25*(9), 1529–1537. doi:10.100711274-009-0050-2

Janas, M., & Zawadzka, A. (2017). The impact of waste landfill on the environment. *Inzynieria Ekologiczna*, *18*(3), 64–73. doi:10.12912/23920629/70259

Jayasiri, H. B., Purushothaman, C. S., & Vennila, A. (2013). Plastic litter accumulation on high-water strandline of urban beaches in Mumbai, India. *Environmental Monitoring and Assessment*, *185*(9), 7709–7719. doi:10.100710661-013-3129-z PMID:23430068

Jin, Z. P., Luo, K., Zhang, S., Zheng, Q., & Yang, H. (2012). Bioaccumulation and catabolism of prometryne in green algae. *Chemosphere*, *87*(3), 278–284. doi:10.1016/j.chemosphere.2011.12.071 PMID:22273183

Joshi, B., Kabariya, K., Nakrani, S., Khan, A., Parabia, F. M., Doshi, H. V., & Thakur, M. C. (2013). Biodegradation of turquoise blue dye by *Bacillus megaterium* isolated from industrial effluent. *American Journal of Environmental Protection*, *1*(2), 41–46. doi:10.12691/env-1-2-5

Joshi, P. A., Jaybhaye, S., & Mhatre, K. (2015). Biodegradation of dyes using consortium of bacterial strains isolated from textile effluent. *European Journal of Experimental Biology*, *5*, 36–40.

Juwarkar, A. A., Nair, A., Dubey, K. V., Singh, S. K., & Devotta, S. (2007). Biosurfactant technology for remediation of cadmium and lead contaminated soils. *Chemosphere*, *68*(10), 1996–2002. doi:10.1016/j.chemosphere.2007.02.027 PMID:17399765

Kalyani, P., Sailaja, B., & Hemalatha, K. P. J. (2017). Degradation of textile dyes by *Aspergillus fumigatus* strain and their culture optimization. *International Journal of Current Research*, *9*(12), 62229–62232.

Kanianska, R. (2016). Agriculture and Its Impact on Land-Use, Environment,and Ecosystem Services. In A. Almusaed (Eds.), Landscape Ecology. The Influences of Land Use and Anthropogenic Impacts of Landscape Creation. IntechOpen. doi:10.5772/63719

Karpouzas, D. G., & Walker, A. (2000). Factors influencing the ability of Pseudomonas putida epI to degrade ethoprophos in soil. *Soil Biology & Biochemistry*, *32*(11-12), 1753–1762. doi:10.1016/S0038-0717(00)00093-6

Kataoka, R., Takagi, K., & Sakakibara, F. (2011). Biodegradation of endosulfan by *Mortieralla* sp. strain W8 in soil: Influence of different substrates on biodegradation. *Chemosphere*, *85*(3), 548–552. doi:10.1016/j.chemosphere.2011.08.021 PMID:21893334

Kiflu, A., & Beyene, S. (2013). Effects of different land use systems on selected soil properties in south Ethiopia. *Journal of Soil Science and Environmental Management*, *4*(5), 100–107. doi:10.5897/JSSEM2013.0380

Kim, I. H., Choi, J. H., Joo, J. O., Kim, Y. K., Choi, J. W., & Oh, B. K. (2015). Development of a microbe-zeolite carrier for the efective elimination of heavy metals from seawater. *Journal of Microbiology and Biotechnology*, *25*(9), 1542–1546. doi:10.4014/jmb.1504.04067 PMID:26032363

Kim, S., Krajmalnik-Brown, R., Kim, J. O., & Chung, J. (2014). Remediation of petroleum hydrocarbon-contaminated sites by DNA diagnosis-based bioslurping technology. *The Science of the Total Environment*, *497*, 250–259. doi:10.1016/j.scitotenv.2014.08.002 PMID:25129160

Knapp, J. S., & Newby, P. S. (1995). The microbiological decolorization of an industrial effluent containing a diazo-linked chromophore. *Water Research*, *29*(7), 1807–1809. doi:10.1016/0043-1354(94)00341-4

Komancová, M., Jurčová, I., Kochánková, L., & Burkhard, J. (2003). Metabolic pathways of polychlorinated biphenyls degradation by *Pseudomonas* sp. 2. *Chemosphere*, *50*(4), 537–543. doi:10.1016/S0045-6535(02)00374-0 PMID:12685753

Kube, M., Beck, A., Zinder, S. H., Kuhl, H., Reinhardt, R., & Adrian, L. (2005). Genome sequence of the chlorinated compound respiring bacterium *Dehalococcoides* species strain CBDB1. *Nature Biotechnology*, *23*(10), 1269–1273. doi:10.1038/nbt1131 PMID:16116419

Kues, U. (2015). Fungal enzymes for environmental management. *Current Opinion in Biotechnology*, *33*, 268–278. doi:10.1016/j.copbio.2015.03.006 PMID:25867110

Kulshreshtha, S., Mathur, N., & Bhatnagar, P. (2014). Mushroom as a product and their role in mycoremediation. *AMB Express*, *4*(1), 29. doi:10.118613568-014-0029-8 PMID:24949264

Kumar, D. (2018). Biodegradation of γ-Hexachlorocyclohexane by Burkholderia sp. IPL04. *Biocatalysis and Agricultural Biotechnology*, *16*, 331–339. doi:10.1016/j.bcab.2018.09.001

Kumar, D., Kumar, A., & Sharma, J. (2016). Degradation study of lindane by novel strains Kocuria sp. DAB-1Y and *Staphylococcus* sp. DAB-1W. *Bioresources and Bioprocessing*, *3*(1), 53. doi:10.118640643-016-0130-8 PMID:28090433

Kuppusamy, S., Maddela, N. R., Megharaj, M., Venkateswarlu, K., Kuppusamy, S., Megharaj, M., & Venkateswarlu, K. (2020). Approaches for remediation of sites contaminated with total petroleum hydrocarbons. In *Total Petroleum Hydrocarbons* (pp. 167–205). Springer International Publishing. doi:10.1007/978-3-030-24035-6_7

Latifi, A. M., Khodi, S., Mirzaei, M., Miresmaeili, M., & Babavalian, H. (2012). Isolation and characterization of five chlorpyrifos degrading bacteria. *African Journal of Biotechnology*, *11*, 3140–3146.

Łebkowska, M., Zborowska, E., Karwowska, E., Mia'skiewicz-Peska, E., Muszy'nski, A., Tabernacka, A., Naumczyk, J., & Jeczalik, M. (2011). Bioremediation of soil polluted with fuels by sequential multiple injection of native microorganisms: Field-scale processes in Poland. *Ecological Engineering*, *37*(11), 1895–1900. doi:10.1016/j.ecoleng.2011.06.047

Leng, L., Wei, L., Xiong, Q., Xu, S., Li, W., Lv, S., Lu, Q., Wan, L., Wen, Z., & Zhou, W. (2020). Use of microalgae based technology for the removal of antibiotics from wastewater: A review. *Chemosphere*, *238*, 124680. Advance online publication. doi:10.1016/j.chemosphere.2019.124680 PMID:31545213

Li, X., He, J., & Li, S. (2007). Isolation of a chlorpyrifos-degrading bacterium, *Sphingomonas* sp. strain Dsp-2, and cloning of the mpd gene. *Research in Microbiology*, *158*(2), 143–149. doi:10.1016/j.resmic.2006.11.007 PMID:17306510

Li, Y., Ge, X. Z., Wang, X. Y., & Gao, R. (2017). *The Invention Discloses a Compound Bacterial Agent used to Degrade Paraquat and a Preparation Method. China.* Patent No CN 106520618 A. Beijing: National Intellectual Property Administration.

Lisk, D. J. (1988). Environmental implications of incineration of municipal solid waste and ash disposal. *The Science of the Total Environment*, *74*, 39–66. doi:10.1016/0048-9697(88)90128-3 PMID:3065938

Liu, A., Wang, H., Gao, P., & Xu, H. (2013). Chemical fractionation of Cu and Zn and their impacts on microbial properties in slightly contaminated soils. *International Journal of Agricultural Research, Innovation and Technology*, *3*(1), 20–25. doi:10.3329/ijarit.v3i1.16045

Liu, F., Chi, Y., Wu, S., Jia, J., & Yao, K. (2014). Simultaneous degradation of cypermethrin and its metabolite, 3-phenoxybenzoic acid, by the cooperation of *Bacillus licheniformis* B-1 and *Sphingomonas* sp. SC-1. *Journal of Agricultural and Food Chemistry*, *62*(33), 8256–8262. doi:10.1021/jf502835n PMID:25068244

Liu, H., Probst, A., & Liao, B. (2005). Metal contamination of soils and crops affected by the Chenzhoulead/zincminespill (Hunan, China). *The Science of the Total Environment*, *339*(1-3), 153–166. doi:10.1016/j.scitotenv.2004.07.030 PMID:15740766

Liu, M., Lu, S., Song, Y., Lei, L., Hu, J., Lv, W., & He, D. (2018). Microplasticand mesoplastic pollution in farmland soils in suburbs of Shanghai,China. *Environmental Pollution*, *242*, 855–862. doi:10.1016/j.envpol.2018.07.051 PMID:30036839

Liu, S., Zhang, F., Chen, J., & Sun, G. X. (2011). Arsenic removal from contaminated soil via biovolatilization by genetically engineered bacteria under laboratory conditions. *Journal of Environmental Sciences (China)*, *23*(9), 1544–1550. doi:10.1016/S1001-0742(10)60570-0 PMID:22432292

Loick, N., Hobbs, P. J., Hale, M. D. C., & Jones, D. L. (2009). Bioremediation of poly-aromatic hydrocarbon (PAH)-contaminated soil by composting. *Critical Reviews in Environmental Science and Technology*, *39*(4), 271–332. doi:10.1080/10643380701413682

Luo, X. S., Yu, S., Zhu, Y. G., & Li, X. D. (2012). Trace metal contamination in urban soils of China. *The Science of the Total Environment*, *421*, 17–30. doi:10.1016/j.scitotenv.2011.04.020 PMID:21575982

Macaulay, B. M., & Rees, D. (2014). Bioremediation of oil spills. A review of challenges for resesrch advancement. *Annals of Environmental Science (Boston, Mass.)*, *8*, 9–37.

Madhuri, R. J., & Vijayalakshmi, G. (2014). Biodegradation of diazodye, trypan blue by Aspergillus species from dye contaminated sites. *International Journal of Research Studies in Biosciences*, *2*(11), 49–61.

Mahapatra, N. N. (2016). Textile dyes. Boca Raton: CRC Press.

Maharani, V., Vijayalakshmi, S., & Balasubramanian, T. (2013). Degradation and detoxification of reactive azo dyes by native bacterial communities. *African Journal of Microbiological Research*, *7*(20), 2274–2282. doi:10.5897/AJMR12.1539

Mahbub, K. R., Ferdouse, J., & Anwar, M. N. (2011). Demonstration of Decolorization of Various Dyes by Some Bacterial Isolates Recovered from Textile Effluents. *Bangladesh Journal of Scientific and Industrial Research*, *46*(3), 323–328. doi:10.3329/bjsir.v46i3.9037

Maila, M. P., & Cloete, T. E. (2004). Bioremediation of petroleum hydrocarbons through landfarming: Are simplicity and costeffectiveness the only advantages? *Reviews in Environmental Science and Biotechnology*, *3*(4), 49–360. doi:10.100711157-004-6653-z

Manigandan, P. K., & Shekar, C. (2014). Evaluation of radionuclides in the terrestrial environment of Western Ghats. *Journal of Radiation Research and Applied. The Sciences*, *7*, 310–316.

Mao, J., & Guan, W. (2016). Fungal degradation of polycyclic aromatic hydrocarbons (PAHs) by Scopulariopsis brevicaulis and its application in bioremediation of PAH-contaminated soil. *Acta Agriculturæ Scandinavica. Section B, Soil and Plant Science*, *66*(5), 399–405. doi:10.1080/09064710.2015.1137629

Margesin, R., Hammerle, M., & Tscherko, D. (2007). Microbial activity and community composition during bioremediation of diesel-oilcontaminated soil: Effects of hydrocarbon concentration, fertilizers, and incubation time. *Microbial Ecology*, *53*(2), 259–269. doi:10.100700248-006-9136-7 PMID:17265002

Markowicz, A., Plaza, G., & Piotrowskaseget, Z. (2016). Activity and functional diversity of microbial communities in long-term hydrocarbon and potentially toxic element contaminated soils. *Archives of Environmental Protection*, *42*, 3–11. doi:10.1515/aep-2016-0041

Maya, K., Singh, R. S., Upadhyay, S. N., & Dubey, S. K. (2011). Kinetic analysis reveals bacterial efficacy for biodegradation of chlorpyrifos and its hydrolyzing metabolite TCP. *Process Biochemistry*, *46*(11), 2130–2136. doi:10.1016/j.procbio.2011.08.012

McCann, A. E., & Cullimore, D. R. (1979). Influence of pesticides on the soil algal flora. *Residue Reviews*, *72*, 1–31. doi:10.1007/978-1-4612-6214-5_1

Medeiros, E. V., Krystal, A. N., Jamilly, A. B., Wendson, S. M., Aline, O. S., & Keila, A. M. (2015). Absolute and specific enzymatic activities of sandy entisol from tropical dry forest, monoculture and intercropping areas. *Soil & Tillage Research*, *145*, 208–215. doi:10.1016/j.still.2014.09.013

Meena, R. S., Meena, V. S., Meena, S. K., & Verma, J. P. (2015b). The needs of healthy soils for a healthy world. *Journal of Cleaner Production*, *102*, 560–561. doi:10.1016/j.jclepro.2015.04.045

Meena, V. D., Dotaniya, M. L., Meena, B. P., & Das, H. (2013a). Organic food safer but not healthy: Truth in myth. *Indian Farmers Dig*, *46*(8), 43–44.

Megharaj, M., Ramakrishnan, B., Venkateswarlu, K., Sethunathan, N., & Naidu, R. (2011). Bioremediation approaches for organic pollutants: A critical perspective. *Environment International*, *37*(8), 1362–1375. doi:10.1016/j.envint.2011.06.003 PMID:21722961

Megharaj, M., Venkateswarlu, K., & Rao, A. S. (1987). Metabolism of monocrotophos and quinalphos by algae isolated from soil. *Bulletin of Environmental Contamination and Toxicology*, *39*(2), 251–256. doi:10.1007/BF01689414 PMID:3663978

Menn, F. M., Easter, J. P., & Sayler, G. S. (2008). Genetically engineered microorganisms and bioremediation. In H. J. Rehm & B. Reed (Eds.), *Biotechnology set* (pp. 441–463). Wiley. doi:10.1002/9783527620951.ch21

Metting, B. (1981). The systematics and ecology of soil algae. The Botanical Review, 47(2), 195 – 312.

Michalik, B. (2017). NORM contaminated area identification using radionuclides activity concentration pattern in a soil profile. *Journal of Environmental Radioactivity*, *169*, 9–18. doi:10.1016/j.jenvrad.2016.11.035 PMID:28408134

Moghaddam, M. R., Fatemi, S., & Keshtkar, A. (2013). Adsorption of lead (Pb2þ) and uranium (UO22þ) cations by brown algae; experimental and thermodynamic modeling. *Chemical Engineering Journal*, *231*, 294–303. doi:10.1016/j.cej.2013.07.037

Mohan, V., Rao, C., & Karthikeyan, J. (2002). Adsorptive Removal of Direct Azo Dye from Aqueous Phase into Coal Based Sorbents: A Kinetic and Mechanistic Study. *Journal of Hazardous Materials*, *90*(2), 189–204. doi:10.1016/S0304-3894(01)00348-X PMID:11827721

Montemurro, F., Maiorana, M., Convertini, G., & Fornaro, F. (2005). Improvement of soil properties and nitrogen utilization of sunflower by amending municipal solid waste compost. *Agronomy for Sustainable Development*, *25*(3), 369–375. doi:10.1051/agro:2005038

Muhammad, S., Muhammad, S., & Sarfraz, H. (2008). Perspectives of bacterial ACC deaminase in phytoremediation. *Trends in Biotechnology*, *25*, 356–362. PMID:17573137

Mulligan, C. N., Young, R. N., & Gibbs, B. F. (2001). Remediation technologies for metal-contaminated soils and groundwater: An evaluation. *Engineering Geology*, *60*(1-4), 193–207. doi:10.1016/S0013-7952(00)00101-0

Muradoglu, F., Gundogdu, M., Ercisli, S., Encu, T., Balta, F., Jaafar, H. Z., & Zia-Ul-Haq, M. (2015). Cadmium toxicity affects chlorophyll a and b content, antioxidant enzyme activities and mineral nutrient accumulation in strawberry. *Biological Research*, *48*(1), 11. doi:10.118640659-015-0001-3 PMID:25762051

Muthu, S. S. (2017). Introduction. In S. S. Muthu (Ed.), *Sustainabilityin the textile industry* (pp. 1–8). Springer. doi:10.1007/978-981-10-2639-3_1

Narayan Chadar, S. (2018). Composting as an eco-friendly method to recycle organic waste. *Progress in Petrochemical Science*, *2*(5), 252–254. doi:10.31031/PPS.2018.02.000548

Ndeddy Aka, R. J., & Babalola, O. O. (2016). Effect of bacterial inoculation of strains of *Pseudomonas aeruginosa, Alcaligenes feacalis* and *Bacillus subtilis* on germination, growth and heavy metal (Cd, Cr, and Ni) uptake of Brassica juncea. *International Journal of Phytoremediation*, *18*(2), 200–209. doi:10.1080/15226514.2015.1073671 PMID:26503637

Neff, J., Lee, K., & Deblois, E. M. (2011). Produced water: overview of composition, fates, and effects. In K. Lee & J. Neff (Eds.), *Produced Water: Environmental Risks and Advances in Mitigation Technologies* (pp. 3–54). Springer. doi:10.1007/978-1-4614-0046-2_1

Neha, S., Tuhina, V., & Rajeeva, G. (2013). Detoxifcation of hexavalent chromium by an indigenous facultative anaerobic *Bacillus cereus* strain isolated from tannery effluent. *African Journal of Biotechnology*, *12*(10), 1091–1103.

Ng, S. P., Davis, B., & Polombo, E. A. (2009). Tn5051 like mer containing transposon identified in a heavy metal tolerant strain*Achromobacter* sp. AO22. *BMC Research Notes*, *7*(1), 2–38. doi:10.1186/1756-0500-2-38 PMID:19284535

Nizzetto, L., Futter, M., & Langaas, S. (2016). Are agricultural soils dumpsfor microplastics of urban origin? *Environmental Science & Technology*, *50*(20), 10777–10779. doi:10.1021/acs.est.6b04140 PMID:27682621

Odukkathil, G., & Vasudevan, M. (2013). Toxicity and bioremediation of pesticides in agricultural soil. *Reviews in Environmental Science and Biotechnology*, *12*(4), 421–444. doi:10.100711157-013-9320-4

Okoduwa, S. I. R., Igiri, B., Udeh, C. B., Edenta, C., & Gauje, B. (2017). Tannery effluent treatment by yeast species isolates from watermelon. *Toxics*, *5*(1), 6. doi:10.3390/toxics5010006 PMID:29051437

Olaniran, A. O., Balgobind, A., & Pillay, B. (2013). Review bioavailability of heavy metal in soil: Impact on microbial biodegradation of organic compounds and possible improvement strategies. *International Journal of Molecular Sciences*, *14*(5), 10197–10228. doi:10.3390/ijms140510197 PMID:23676353

Onneby, K., Håkansson, S., Pizzul, L., & Stenstrom, J. (2014). Reduced leaching of the herbicide MCPA after bioaugmentation with formulated and stored *Sphingobium* sp. *Biodegradation*, *25*(2), 291–300. doi:10.100710532-013-9660-3 PMID:23982656

Orhorhoro, E., & Oghoghorie, O. (2019). Review on solid waste generation and management in sub-Saharan Africa: A case study of Nigeria. *Journal of Applied Science & Environmental Management*, *23*(9), 1729–1737. doi:10.4314/jasem.v23i9.19

Pailan, S., Gupta, D., Apte, S., Krishnamurthi, S., & Saha, P. (2015). Degradation of organophosphate insecticide by a novel Bacillus aryabhattai strain SanPS1, isolated from soil of agricultural field in Burdwan, West Bengal, India. *International Biodeterioration & Biodegradation*, *103*, 191–195. doi:10.1016/j.ibiod.2015.05.006

Palmer-Brown, W., de Melo Souza, P. L., & Murphy, C. D. (2018). Cyhalothrin biodegradation in Cunninghamella elegans. *Environmental Science and Pollution Research International*, *26*(2), 1414–1421. doi:10.100711356-018-3689-0 PMID:30426373

Pan, X., Chen, Z., Chen, F., Cheng, Y., Lin, Z., & Guan, X. (2015). The mechanism of uranium transformation from U (VI) into nano-uramphite by two indigenous *Bacillus thuringiensis* strains. *Journal of Hazardous Materials*, *297*, 313–319. doi:10.1016/j.jhazmat.2015.05.019 PMID:26026850

Parnell, J. J., Park, J., Denef, V., Tsoi, T., Hashsham, S., Quensen, J. I. III, & Tiedje, J. M. (2006). Coping with polychlorinated biphenyl (PCB) toxicity: Physiological and genomewide responses of Burkholderia xenovorans LB400 to PCB-mediated stress. *Applied and Environmental Microbiology*, *72*(10), 6607–6614. doi:10.1128/AEM.01129-06 PMID:17021212

Pasquini, M. W., & Alexander, M. J. (2004). Chemical properties of urban waste ash produced by open burning on the Jos Plateau: Implications for agriculture. *The Science of the Total Environment*, *319*(1-3), 225–240. doi:10.1016/S0048-9697(03)00434-0 PMID:14967513

Pazirandeh, M., Chrisey, L. A., Mauro, J. M., Campbell, J. R., & Gaber, B. P. (1995). Expression of the Neurospora crassa metallothionein gene in Escherichia coli and its effect on heavy-metal uptake. *Applied Microbiology and Biotechnology*, *43*(6), 1112–1117. doi:10.1007/BF00166934 PMID:8590662

Peña-Montenegro, T. D., Lozano, L., & Dussán, J. (2015). Genome sequence and description of the mosquitocidal and heavy metal tolerant strain *Lysinibacillussphaericus* CBAM5. *Standards in Genomic Sciences*, *10*(1), 2. doi:10.1186/1944-3277-10-2 PMID:25685257

Perissini-Lopes, B., Egea, T. C., Monteiro, D. A., Vici, A. C., & Da Silva, D. G. H. (2016). Lisboa DCO, De Almeida EA, Parsons JR, Da Silva R, Gomes E:Evaluation of diuron tolerance and biotransformation byfungi from a sugar cane plantation sandy-loam soil. *Journal of Agricultural and Food Chemistry*, *64*, 268–9275. doi:10.1021/acs.jafc.6b03247

Pichtel, J. (2016). Oil and gas production wastewater: Soil contamination and pollution prevention. *Applied and Environmental Soil Science*, *2707989*, 1–24. Advance online publication. doi:10.1155/2016/2707989

Pierzynski, G. M., Sims, J. T., & Vance, G. F. (1994). *Soils and environmental quality*. Lewis Publishers.

Pilon-Smits, E. (2005). Phytoremediation. *Annual Review of Plant Biology*, *56*(1), 15–39. doi:10.1146/annurev.arplant.56.032604.144214 PMID:15862088

Pimmata, P., Reungsang, A., & Plangklang, P. (2013). Comparative bioremediation of carbofuran contaminated soil by natural attenuation, bioaugmentation and biostimulation. *International Biodeterioration & Biodegradation*, *85*, 196–204. doi:10.1016/j.ibiod.2013.07.009

Polak-Juszczak, L. (2009). Temporal trends in the bioaccumulation of trace metals in herring, sprat, and cod from the southern Baltic Sea in the 1994-2003 period. *Chemosphere*, *76*(10), 1334–1339. doi:10.1016/j.chemosphere.2009.06.030 PMID:19580989

Prajapati, S. K., & Meravi, N. (2014). Potentially toxic element speciation of soil and calotropis procera from thermal power plant area. *Proceedings of the International Academy of Ecology and Environmental Sciences*, *4*, 68–71.

Prasad, M. P., Bhakat, P., & Chatterjee, S. (2013). Optimization of textile dye degradation by bacterial species isolated from natural sources. *Journal of Ecology and Environmental Sciences*, *4*(1), 97–99.

Prescott, M. I., Harle, J. D., & Klein, D. A. (2002). *Microbiology of Food* (5th ed.). McGraw-Hill Ltd.

Quero, G. M., Cassin, D., Botter, M., Perini, L., & Luna, G. M. (2015). Patterns of benthic bacterial-diversity in coastal areas contaminated by heavy metals,polycyclic aromatic hydrocarbons (PAHs) and polychlorinatedbiphenyls (PCBs). *Frontiers in Microbiology*, *6*, 1–15. doi:10.3389/fmicb.2015.01053 PMID:26528247

Rafique, N., & Tariq, S. R. (2016). Distribution and source apportionment studies of heavy metals in soil of cotton/wheat fields. *Environmental Monitoring and Assessment*, *188*(5), 309. doi:10.100710661-016-5309-0 PMID:27115422

Rajan, M. R. D. N. S. (2014). Impact of Dyeing Industry Effluent on Groundwater Quality by Water Quality Index and Correlation Analysis. *Journal of Pollution Effects and Control*, *2*(2), 1–4. doi:10.4172/2375-4397.1000126

Ramasamy, K., Kamaludeen, S., & Parwin, B. (2006). Bioremediation of metals microbial processes and techniques. In Environmental Bioremediation Technologies (pp. 173–187). Springer Publication.

Ramírez, V., Baez, A., López, P., Bustillos, R., Villalobos, M. Á., Carreño, R., Contreras, J. L., Muñoz-Rojas, J., Fuentes, L. E., Martínez, J., & Munive, J. A. (2019). Chromium hyper-tolerant Bacillus sp. MH778713 assists phytoremediation of heavy metals by mesquite trees (Prosopis laevigata). *Frontiers in Microbiology*, *10*, 1833. doi:10.3389/fmicb.2019.01833 PMID:31456770

Rana, S., Jindal, V., Mandal, K., Kaur, G., & Gupta, V. K. (2015). Thiamethoxam degradation by Pseudomonas and Bacillus strains isolated from agricultural soils. *Environmental Monitoring and Assessment*, *187*(5), 4532. doi:10.100710661-015-4532-4 PMID:25917187

Rashidipour, M., Maleki, A., Kordi, S., Birjandi, M., Pajouhi, N., Mohammadi, E., Heydari, R., Rezaee, R., Rasoulian, B., & Davari, B. (2019). Pectin/chitosan/tripolyphosphate nanoparticles: Efficient carriers for reducing soil sorption, cytotoxicity, and mutagenicity of paraquat and enhancing its herbicide activity. *Journal of Agricultural and Food Chemistry*, *67*, 5736–5745. doi:10.1021/acs.jafc.9b01106 PMID:31042035

Rayu, S., Karpouzas, D. G., & Singh, B. K. (2012). Emerging technologies in bioremediation: Constraints and opportunities. *Biodegradation*, *23*(6), 917–926. doi:10.100710532-012-9576-3 PMID:22836784

Reddy, S. M., Basha, S., Adimurthy, S., & Ramachandraiah, G. (2006). Description of small plastics fragments in marine sediments along the Alang–Sosiyaship-breaking yard, India. *Estuarine, Coastal and Shelf Science*, *68*(3-4), 656–660. doi:10.1016/j.ecss.2006.03.018

Rekik, H., Zaraî Jaouadi, N., Bouacem, K., Zenati, B., Kourdali, S., Badis, A., Annane, R., Bouanane-Darenfed, A., Bejar, S., & Jaouadi, B. (2019). Physical and enzymatic properties of a new manganese peroxidase from the white-rot fungus *Trametes pubescens* strain i8 for lignin biodegradation and textile-dyes biodecolorization. *International Journal of Biological Macromolecules*, *15*, 514–525. doi:10.1016/j.ijbiomac.2018.12.053 PMID:30528991

Rezania, S., Park, J., Md Din, M. F., Mat Taib, S., Talaiekhozani, A., Kumar Yadav, K., & Kamyab, H. (2018). Microplastics pollution in differentaquatic environments and biota: A review of recent studies. *Marine Pollution Bulletin*, *133*, 191–208. doi:10.1016/j.marpolbul.2018.05.022 PMID:30041307

Riah, W., Laval, K., Laroche-Ajzenberg, E., Mougin, C., Latour, X., & Trinsoutrot-Gattin, I. (2014). Effects of pesticides on soil enzymes: A review. *Environmental Chemistry Letters*, *12*(2), 257–273. doi:10.100710311-014-0458-2

Rodrigues, J. L. M., Kachel, C. A., Aiello, M. R., Quensen, J. F. III, Maltseva, O. V., Tsoi, T. V., & Tiedje, J. M. (2006). Degradation of Aroclor 1242 Dechlorination Products in Sediments by Burkholderiaxenovorans LB400(Ohb) and *Rhodococcus* sp. Strain RHA1(Fcb). *Applied and Environmental Microbiology*, *72*(4), 2476–2482. doi:10.1128/AEM.72.4.2476-2482.2006 PMID:16597946

Rodríguez Eugenio, N., McLaughlin, M., & Pennock, D. (2018). *Soil Pollution: a hidden reality. Food and Agriculture Organization of the United Nations edition*. FAO.

Ron, E. Z., & Rosenberg, E. (2014). Enhanced bioremediation of oil spills in the sea. *Current Opinion in Biotechnology*, *27*, 191–194. doi:10.1016/j.copbio.2014.02.004 PMID:24657912

Rosca, M., Hlihor, R. M., & Gavrilescu, M. (2019). Bioremediation of Persistent Toxic Substances: From conventional to new approaches in using microorganisms and plants. In *Microbial Technology for the Welfare of Society* (pp. 289–312). Springer. doi:10.1007/978-981-13-8844-6_14

Roy, M., Giri, A. K., Dutta, S., & Mukherjee, P. (2015). Integrated phytobial remediation for sustainable management of arsenic in soil and water. *Environment International*, *75*, 180–198. doi:10.1016/j.envint.2014.11.010 PMID:25481297

Saez, J. M., Alvarez, A., Benimelli, C. S., & Amorosso, M. J. (2014). Enhanced lindane removal &from soil slurry by immobilized Streptomyces consortium. *International Biodeterioration & Biodegradation*, *93*, 63–69. doi:10.1016/j.ibiod.2014.05.013

Sagarkar, S., Mukherjee, S., Nousiainen, A., Björklöf, K., Purohit, H. J., Jørgensen, K. S., & Kapley, A. (2013). Monitoring bioremediation of atrazine in soil microcosms using molecular tools. *Environmental Pollution*, *172*, 108–115. doi:10.1016/j.envpol.2012.07.048 PMID:23022948

Sahoo, B., Ningthoujam, R., & Chaudhuri, S. (2019). Isolation and characterization of a lindane degrading bacteria Paracoccus sp. NITDBR1 and evaluation of its plant growth promoting traits. *International Microbiology*, *22*(1), 155–167. doi:10.100710123-018-00037-1 PMID:30810939

Sandaa, R. A., Torsvik, V., Enger, O., Daae, F. L., Castberg, T., & Hahn, D. (1999). Analysis of bacterial communities in heavy metal-contaminated soils at different levels of resolution. *FEMS Microbiology Ecology*, *30*(3), 237–251. doi:10.1111/j.1574-6941.1999.tb00652.x PMID:10525180

Saranraj, P., Sumathi, V., Reetha, D., & Stella, D. (2010). Fungal decolourization of direct Azo dyes and biodegradation of textile dye effluent. *Journal of Ecobiotechnology*, *2*(7), 12–16.

Sardrood, B. P., Goltapeh, E. M., & Varma, A. (2013). An introduction to bioremediation. In Funghi as bioremediators (pp. 3-29). Springer Science and Business Media. doi:10.1007/978-3-642-33811-3_1

Sari, A. A., Tachibana, S., Muryanto, & Hadibarata, T. (2016). Development of bioreactor systems for decolorization of reactive green 19 using white rot fungus. *Desalination and Water Treatment*, *57*(15), 7029–7039. doi:10.1080/19443994.2015.1012121

Sasec, V., & Cajthaml, T. (2014). *Mycoremediation*: Current status and perspectives. *International Journal of Medicinal Mushrooms*, *7*(3), 360–361. doi:10.1615/IntJMedMushr.v7.i3.200

Sasikala, C., Jiwal, S., Rout, P., & Ramya, M. (2012). Biodegradation of chlorpyrifos by bacterial consortium isolated from agriculture soil. *World Journal of Microbiology & Biotechnology*, *28*(3), 1301–1308. doi:10.100711274-011-0879-z PMID:22805851

Sasikala, C., Jiwal, S., Rout, P., & Ramya, M. (2012). Biodegradation of chlorpyrifos by bacterial consortium isolated from agriculture soil. *World Journal of Microbiology & Biotechnology*, *28*(3), 1301–1308. doi:10.100711274-011-0879-z PMID:22805851

Sasikumar, C. S., & Papinazath, T. (2003). Environmental management: bioremediation of polluted environment. In *Proceedings of the third international conference on environment and health* (pp 465-469). Chennai, India: Department of Geography, University of Madras and Faculty of Environmental Studies, York University, Chennai.

Savin, I. I., & Butnaru, R. (2008). Wastewater Characteristics in Textile Finishing Mills. *Environmental Engineering and Management Journal*, *7*(6), 859–864. doi:10.30638/eemj.2008.113

Saxena, G., & Bharagava, R. N. (2020). Bioremediation of Industrial Waste for Environmental Safety: Volume I: Industrial Waste and Its Management. Springer.

Sayara, T., Borràs, E., Caminal, G., Sarrà, M., & Sánchez, A. (2011). Bioremediation of PAHs-contaminated soil through composting: Influence of bioaugmentation and biostimulation on contaminant biodegradation. *International Biodeterioration & Biodegradation*, *65*(6), 859–865. doi:10.1016/j.ibiod.2011.05.006

Schue, M., Dover, L. G., Besra, G. S., Parkhill, J., & Brown, N. L. (2009). Sequence and analysis of a plasmid encoded mercury resistance operon from *Mycobacterium marinum* identifies MerH, a new mercuric ion transporter. *Journal of Bacteriology*, *19*(1), 439–444. doi:10.1128/JB.01063-08 PMID:18931130

Shah, M. (2014). Effective treatment systems for azo dye degradation: A joint venture between physico – chemical and microbiological process. *International Journal of Environmental Bioremediation & Biodegradation*, *2*(5), 231–242.

Shamraiz, U., Hussain, R. A., Badshah, A., Raza, B., & Saba, S. (2016). Functional metal sulfides and selenides for the removal of hazardous dyes from water. *Journal of Photochemistry and Photobiology*, *159*, 33–41. doi:10.1016/j.jphotobiol.2016.03.013 PMID:27010842

Sharma, S., & Adholeya, A. (2011). Detoxification and accumulation of chromium from tannery effluent and spent chrome effluent by Paecilomyces lilacinus fungi. *International Biodeterioration & Biodegradation*, *65*(2), 309–317. doi:10.1016/j.ibiod.2010.12.003

Sharma, S., Singh, B., & Gupta, V. K. (2014). Assessment of imidacloprid degradation by soil isolated *Bacillus alkalinitrilicus*. *Environmental Monitoring and Assessment*, *186*(11), 7183–7193. doi:10.100710661-014-3919-y PMID:25052329

Shi, W., & Ma, X. (2017). Effects of heavy metal Cd pollution on microbial activities in soil. *Annals of Agricultural and Environmental Medicine*, *24*(4), 722–725. doi:10.26444/aaem/80920 PMID:29284254

Silva, V. P., Moreira-Santos, M., Mateus, C., Teixeira, T., Ribeiro, R., & Viegas, C. (2015). Evaluation of *Arthrobacter aurescens* strain TC1 as bioaugmentation bacterium in soils contaminated with the herbicidal substance terbuthylazine. *PLoS One*, *10*(12), e0144978. doi:10.1371/journal.pone.0144978 PMID:26662024

Simarro, R., González, N., Bautista, L., & Molina, M. (2013). Assessment of the efficiency of in situ bioremediation techniques in a creosote polluted soil: Change in bacterial community. *Journal of Hazardous Materials*, *262*, 158–167. doi:10.1016/j.jhazmat.2013.08.025 PMID:24025312

Sineli, P. E., Tortella, G., Costa, J. D., Benimeli, C. S., & Cuozzo, S. A. (2016). Evidence of α-β-and γ-HCH mixture aerobic degradation by the native actinobacteria *Streptomyces* sp. M7. *World Journal of Microbiology & Biotechnology*, *32*(5), 81. doi:10.100711274-016-2037-0 PMID:27038951

Singh, A., & Gauba, P. (2014). Mycoremediation: A treatment for heavy metal pollution of soil. *Journal of Civil and Engineeringand Environmental Technology*, *1*, 59–61.

Singh, B. K., Quince, C., Macdonald, C. A., Khachane, A., Thomas, N., AlSoud, W. A., Sørensen, S. J., He, Z., White, D., Sinclair, A., Crooks, B., Zhou, J., & Campbell, C. D. (2014). Loss of microbial diversity in soils is coincident with reductions in some specialized functions. *Environmental Microbiology*, *16*(8), 2408–2420. doi:10.1111/1462-2920.12353 PMID:24422656

Singh, B. K., Walker, A., & Wright, D. J. (2006). Bioremedial potential of fenamiphos and chlorpyrifos degrading isolates: Influence of different environmental conditions. *Soil Biology & Biochemistry*, *38*(9), 2682–2693. doi:10.1016/j.soilbio.2006.04.019

Singh, L., & Singh, V. P. (2010). Microbial degradation and decolourization of dyes in semi-solid medium by the fungus-*Trichoderma harzianum*. Environment &*We. International Journal of Science and Technology*, *5*(3), 147–153.

Singh, T., & Singh, D. K. (2019a). Lindane degradation by root epiphytic bacterium *Achromobacter* sp. strain A3 from Acorus calamus and characterization of associated proteins. *International Journal of Phytoremediation*, *21*(5), 419–424. doi:10.1080/15226514.2018.1524835 PMID:30648424

Smith, E., Thavamani, P., Ramadass, K., Naidu, R., Srivastava, P., & Megharaj, M. (2015). Remediation trials for hydrocarbon-contaminated soils in arid environments: Evaluation of bioslurry and biopiling techniques. *International Biodeterioration & Biodegradation*, *101*, 56–65. doi:10.1016/j.ibiod.2015.03.029

Smith, L. A., Means, J. L., Chen, A., Alleman, B., Chapma, C. C., Tixier, J. R., Brauning, S. E., Gavaskar, A. R., & Royer, M. D. (1995). Remedial options for metal contaminated sites. Academic Press.

Song, B., Zeng, G., Gong, J., Liang, J., Xu, P., Liu, Z., Zhang, Y., Zhang, C., Cheng, M., Liu, Y., Ye, S., Yi, H., & Ren, X. (2017). Evaluation methods for assessing effectiveness of in situ remediation of soil and sediment contaminated with organic pollutants and heavy metals. *Environment International*, *105*, 43–55. doi:10.1016/j.envint.2017.05.001 PMID:28500873

Song, Z., Song, L., Shao, Y., & Tan, L. (2018). Degradation and detoxification of azo dyes by a salt-tolerant yeast Cyberlindnera samutprakarnensis S4 under high-salt conditions. *World Journal of Microbiology & Biotechnology*, *34*(9), 131. doi:10.100711274-018-2515-7 PMID:30105649

Srivastava, N. K., Jha, M. K., Mall, I. D., & Singh, D. (2010). Application of Genetic Engineering for Chromium Removal from Industrial Wastewater. *International Journal of Chemical and Biological Engineering*, *3*, 3.

Sruthy, S., & Ramasamy, E. V. (2016). Microplastic pollution in Vembanad Lake, Kerala, India: The first report of microplastics in lake and estuarine sediments in India. *Environmental Pollution*, 1–8. PMID:28041839

Sruthy, S., & Ramasamy, E. V. (2017). *Microplastic pollution in Vembanad Lake, Kerala, India: The first report of microplastics in lake and estuarine sediments in India*. Academic Press.

Straube, W. L., Nestler, C. C., Hansen, L. D., Ringleberg, D., Pritchard, P. H., & Jones-Meehan, J. (2003). Remediation of polyaromatic hydrocarbons (PAHs) through landfarming with biostimulation and bioaugmentation. *Acta Biotechnologica*, *23*(23), 179–196. doi:10.1002/abio.200390025

Subashchandrabose, S. R., Ramakrishnan, B., Megharaj, M., Venkateswarlu, K., & Naidu, R. (2013). Mixotrophic Cyanobacteria and Microalgae as Distinctive Biological Agents for Organic Pollutant Degradation. *Environment International*, *51*, 59–72. doi:10.1016/j.envint.2012.10.007 PMID:23201778

Suman, J., Uhlik, O., Viktorova, J., & Macek, T. (2018). Phytoextraction of heavy metals: A promising tool for clean-up of polluted environment? *Frontiers in Plant Science*, *9*, 1476. doi:10.3389/fpls.2018.01476 PMID:30459775

Sun, S. L., Yang, W. L., Guo, J. J., Zhou, Y. N., Rui, X., Chen, C., Ge, F., & Dai, Y. J. (2017). Biodegradation of the neonicotinoid insecticide acetamiprid in surface water by the bacterium Variovorax boronicumulans CGMCC 4969 and its enzymatic mechanism. *Royal Society of Chemistry Advances*, *7*(41), 25387–25397. doi:10.1039/C7RA01501A

Taccari, M., Milanovic, V., Comitini, F., Casucci, C., & Ciani, M. (2012). Effects of biostimulation and bioaugmentation on diesel removal and bacterial community. *International Biodeterioration & Biodegradation*, *66*(1), 39–46. doi:10.1016/j.ibiod.2011.09.012

Thenmozhi, R., Arumugam, K., Nagasathya, A., Thajuddin, N., & Paneerselvam, A. (2013). Studies on mycoremediation of used engine oil contaminated soil samples. *Advances in Applied Science Research*, *4*, 110–118.

Thierry, L., Armelle, B., & Karine, J. (2008). Performance of bioaugmentation–assisted phytoextraction applied to metal contaminated soils: A review. *Environmental Pollution*, *153*(3), 497–522. doi:10.1016/j.envpol.2007.09.015 PMID:17981382

Thompson, I. P., & Christopher, J. (2005). Bioaugmentation for bioremediation: The challenge of strain selection. *Environmental Microbiology*, *7*(7), 909–915. doi:10.1111/j.1462-2920.2005.00804.x PMID:15946288

Tigini, V., Giansanti, P., Mangiavillano, A., Pannocchiam, A., & Varese, G. C. (2011). Evaluation of toxicity, genotoxicity and environmental risk of simulated textile and tannery wastewaters with a battery of biotests. *Ecotoxicology and Environmental Safety*, *74*(4), 866–873. doi:10.1016/j.ecoenv.2010.12.001 PMID:21176963

Toxics Link. (2014). *Plastics and the Environment Assessing the Impact of the Complete Ban on Plastic Carry Bag*. Central Pollution Control Board, CPCB.

Tsygankov, V. Y., Lukyanova, O., Boyarova, M., Gumovskiy, A., Donets, M., Lyakh, V., Korchagin, V. P., & Prikhodko, Y. V. (2019). Organochlorine pesticides in commercial Pacific salmon in the Russian Far Eastern seas: Food safety and human health risk assessment. *Marine Pollution Bulletin*, *140*, 503–508. doi:10.1016/j.marpolbul.2019.02.008 PMID:30803671

Ubani, O., Atagana, I. H., & Thantsha, S. M. (2013). Biological degradation of oil sludge: A review of the current state of development. *African Journal of Biotechnology*, *12*(47), 6544–6567. doi:10.5897/AJB11.1139

UN DESA (United Nations, Department of Economic and Social Affairs). (2014). World Urbanization Prospects. Population Department, United Nations.

USEPA (United States Environmental Protection Agency). (2006). *A Citizen's Guide to Bioremediation*. USEPA.

Usman, S. (2018). *Technology of bioorganic fertilizer production: Treasures for north-west Nigeria*. Furtunate Print.

Valls, M., Atrian, S., de Lorenzo, V., & Fernandez, L. A. (2000). Engineering a mouse metallothionein on the cell surface of Ralstonia eutropha CH34 for immobilization of heavy metals in soil. *Nature Biotechnology*, *18*, 661–665.

Varjani, S. J. (2017). Microbial degradation of petroleum hydrocarbons. *Bioresource Technology*, *223*, 277–286.

Vidali, M. (2001). Bioremediation. An overview. *Pure and Applied Chemistry*, *73*, 1163–1172.

Vidya Lakshmi, C., Kumar, M., & Khanna, S. (2008). Biotransformation of chlorpyrifos and bioremediation of contaminated soil. *International Biodeterioration & Biodegradation*, *62*, 204–209.

Vimala, P. P., & Mathew, L. (2016). Biodegradation of Polyethylene using *Bacillus subtilis*. *Procedia Technology*, *24*, 232–239.

Vinay, B. R., Uzma, M., Govindappa, M., Vasantha, R. A., & Lokesh, S. (2016). Screening and Identification of Polyurethane (PU) and low density poly-ethene (LDPE) degrading soil fungi isolated from municipal solid waste. *International Journal of Current Research*, *8*, 34752–34761.

Voberkova, S., Vaverkova, M. D., Buresova, A., Adamcova, D., Vrsanska, M., Kynicky, J., Brtnicky, M., & Adam, V. (2017). Effect of inoculation with white-rot fungi and fungal consortium on the composting efficiency of municipal solid waste. *Waste Management (New York, N.Y.)*, *61*, 157–164.

Vogt, C., & Richnow, H. (2014). Bioremediation via in situ microbial degradations of organic pollutants. *Advances in Biochemical Engineering/Biotechnology*, *142*, 123–146.

Wan, X., Yang, J., & Song, W. (2018). Pollution Status of Agricultural Land in China: Impact of Land Use and Geographical Position. *Soil and Water Research*, *13*, 234–242.

Wang, L., Chi, X. Q., Zhang, J. J., Sun, D. L., & Zhou, N. Y. (2014). Bioaugmentation of a methyl parathion contaminated soil with Pseudomonas sp. strain WBC-3. *International Biodeterioration & Biodegradation*, *87*, 116–121.

Wang, Q., Xie, S., & Hu, R. (2013). Bioaugmentation with Arthrobacter sp. strain DAT1 for remediation of heavily atrazine-contaminated soil. *International Biodeterioration & Biodegradation*, *77*, 63–67.

Wang, Y., Jiao, X., & Song, L. (2014). Soil and soil environmental quality monitoring in China: A review. *Environment International*, *69*, 177–199. doi:10.1016/j.envint.2014.04.014 PMID:24875802

Wang, Z., Wang, Y., Gong, F., Zhang, J., Hong, Q., & Li, S. (2010). Biodegradation of carbendazim by a novel actinobacterium *Rhodococcus* jialingiae djl-6-2. *Chemosphere*, *81*, 639–644.

WHO. (2016). *News Release*. https://www.who.int/news-room/detail/15-03-2016-an-estimated-12-6-million-deaths-each-year-are-attributable-to-unhealthyenvironments

Wiegel, J., & Wu, Q. (2000). Microbial reductive dehalogenation of polychlorinated biphenyls. *FEMS Microbiology Ecology*, *32*, 1–15.

Wu, M., Chen, L., Tian, Y., Ding, Y., & Dick, W. A. (2013). Degradation of polycyclic aromatic hydrocarbons by microbial consortia enriched from three soils using two different culture media. *Environmental Pollution*, *178*, 152–158.

Wu, M., Dick, W. A., Li, W., Wang, X., Yang, Q., & Wang, T. (2016). Bioaugmentation and biostimulation of hydrocarbon degradation and the microbial community in a petroleum-contaminated soil. *International Biodeterioration & Biodegradation, 107*, 158–164.

Wu, M., Li, W., Dick, W. A., Ye, X., Chen, K., Kost, D., & Chen, L. (2017). Bioremediation of hydrocarbon degradation in a petroleum-contaminated soil and microbial population and activity determination. *Chemosphere, 169*, 124–130. doi:10.1016/j.chemosphere.2016.11.059 PMID:27870933

Xiao, P., Mori, T., Kamei, I., Kiyota, H., Takagi, K., & Kondo, R. (2011). Novel metabolic pathways of organochlorine pesticides dieldrin and aldrin by the white rot fungi of the genus Phlebia. *Chemosphere, 85*(2), 218–224.

Xiao, W., Ye, X., Yang, X., Zhu, Z., Sun, C., & Zhang, Q. (2017). Isolation and characterization of chromium (VI)-reducing Bacillus sp. FY1 and Arthrobacter sp. WZ2 and their bioremediation potential. *Bioremediation Journal, 2*, 100–108.

Xie, Y., Fan, J., Zhu, W., Amombo, E., Lou, Y., Chen, L., & Fu, J. (2016). Effect of Heavy Metals Pollution on Soil Microbial Diversity and Bermudagrass Genetic Variation. *Frontiers in Plant Science, 7*, 755.

Xu, D., & Pei, J. (2011). Construction and characterization of a photosynthetic bacterium ge- netically engineered for Hg^{2+} uptake. *Bioresource Technology, 102*, 3083–3088.

Xu, F., Mou, Z., Geng, J., Zhang, X., & Li, C. (2016). Azo dye decolorization by a halotolerant exoelectrogenic decolorizer isolated frommarine sediment. *Chemosphere, 158*, 30–36.

Yang, C., Song, C., Mulchandani, A., & Qiao, C. (2010). Genetic engineering of Stenotrophomonas strain YC-1 to possess a broader substrate range for organophosphates. *Journal of Agricultural and Food Chemistry, 58*, 6762–6766.

Yousaf, S., Khan, S., & Aslam, M. T. (2013). Effect of pesticides on the soil microbial activity. *Pakistan Journal of Zoology, 45*(4).

Youssef, N., Sheik, C. S., Krumholz, L. R., Najar, F. Z., Roe, B. A., & Elshahed, M. S. (2009). Comparison of species richness estimates obtained using nearly complete fragments and simulated pyrosequencing-generated fragments in 16S rRNA gene-based environmental surveys. *Applied and Environmental Microbiology, 75*(16), 5227–5236.

Yu, Y. (2016). *Study on Rapid Screening of Pesticides and Antibiotics in Soil.* Chinese Academy of Agricultural Sciences.

Yu, Y. L., Fang, H., Wang, X., Wu, X. M., & Shan, M. (2011). Characteristics of fungal a fungal strain capable of degrading chlorpyrifos and its use in detoxification of the insecticide on vegetables. *Biodegradation, 17*, 487–494.

Yuksel, O., Kavdr, Y., & Bahtiyar, M. (2004). The effect of municipal waste compost on physical characteristics of clay soils. *Fresenius Environmental Bulletin, 13*(11a), 1094–1098.

Yuniati, M. D. (2018). Bioremediation of petroleum-contaminated soil: A Review. I*OP Conf. Series: Earth and Environmental Science, 118*. doi:10.1088/1755-1315/118/1/012063

Zhai, Y., Li, K., Song, J., Shi, Y., & Yan, Y. (2012). Molecular cloning, purification and biochemical characterization of a novel pyrethroid-hydrolyzing carboxylesterase gene from *Ochrobactrumanthropi* YZ-1. *Journal of Hazardous Materials*, *221*, 206–212.

Zhang, J., Lu, L., Chen, F., Chen, L., Yin, J., & Huang, X. (2018). Detoxification of diphenyl ether herbicide lactofen by *Bacillus* sp. Za and enantioselective characteristics of an esterase gene lacE. *Journal of Hazardous Materials*, *341*, 336–345.

Zhang, Z., Hong, Q., Xu, J., Zhang, X., & Li, S. (2006). Isolation of fenitrothion-degrading strain Burkholderia sp. FDS-1 and cloning of mpd gene. *Biodegradation*, *17*, 275–283.

Zhao, J., Jia, D., Du, J., Chi, Y., & Yao, K. (2019). Substrate regulation on cometabolic degradation of β-cypermethrin by *Bacillus licheniformis* B-1. *AMB Express*, *9*, 83.

Zheng, J., Li, R., Zhu, J., Zhang, J., He, J., Li, S., & Jiang, J. (2012). Degradation of the chloroacetamide herbicide butachlor by *Catellibacterium caeni* sp. nov DCA-1T. *International Biodeterioration & Biodegradation*, *73*, 16–22.

Zhu, D., Chen, Q. L., An, X. L., Yang, X. R., Christie, P., Ke, X., & Zhu, Y. G. (2018). Exposure of Soil collembolans to microplastics perturbs their gut microbiota and alters their isotopic composition. *Soil Biology & Biochemistry*, *116*, 302–310.

Zojiali, F., Hassani, A. H., & Sayedi, M. H. (2014). Bioaccumulation of chromium by Zea mays in a waste water-irrigated soil. An experimental study. *Proceeding of the International Academy of Ecology and Environmental Sciences*, *4*, 62-67.

Zolgharnein, H., Karami, K., Assadi, M.M., & Sohrab, A.D. (2010). Molecular characterization and phylogenetic analyses of heavy metal removal bacteria from the Persian *Gulf Biotechnology, 29*, 1-8.

KEY TERMS AND DEFINITIONS

Bioaugmentation: It requires the introduction of microbes that in a given ecosystem can biotransform or biodegrade a certain pollutants. Exogenous microbial populations are added to the contaminated ecosystem in this process.

Biopiling: It is a full-scale technology in which excavated soils are stacked and usually installed in a leachate storage and aeration facility that consists of a treatment field. It is widely implemented by using the biodegradation process to minimise concentrations of petroleum components in soils.

Bioreactor: As the term bioreactor means, is a vessel in which raw materials are transformed after a sequence of biological reactions to a particular product(s). Tough glass or stainless steel bioreactors are typically cylindrical in shape and have a volume varying from a few litres to cubic metres.

Bioventing: It is a method of promoting the natural biodegradation of pollutants in the soil in situ by supplying current soil microorganisms with air or oxygen.

Composting: Compost bioremediation refers to the use of a biological community of micro-organisms for sequestering or breaking down toxins in water or soil in mature and cured compost.

Genetically Engineered Microorganisms (GEMs): GEMs are classified as bacteria, fungi, or viruses in which the genetic material has been changed predominantly by recombinant DNA technology, i.e. by means which are not naturally occurring.

Land Farming: Typically, this technology entails distributing the excavated polluted soils on the ground surface in a thin layer and promoting aerobic microbial development by aeration and/or the incorporation of minerals, nutrients, and humidity within the soils. The increased microbial activity results in the degradation by microbial respiration of adsorbed petroleum product constituents.

Chapter 4
Microbes as Sustainable Biofertilizers:
Current Scenario and Challenges

Halima M.
B.S. Abdur Rahman Crescent Institute of Science and Technology, India

Sneha Unnikrishnan
B.S. Abdur Rahman Crescent Institute of Science and Technology, India

Karthikeyan Ramalingam
https://orcid.org/0000-0002-9334-427X
B.S. Abdur Rahman Crescent Institute of Science and Technology, India

ABSTRACT

Across the globe, in both developed and developing countries, wheat provides the fundamental support for all other important foods. However, due to climate change, environmental stress, soil infertility, etc., the yield of wheat is affected. To overcome these issues, biofertilizers are recommended. They are eco-friendly, cost-efficient, and affordable by marginal farmers too when compared with chemical fertilizers. Biofertilizers are made up of living microorganisms that colonize the rhizosphere to promote plant yield and prevent plant disease. Pesticide degrading strains of bacteria are emerging as the best technique to overcome the negative effect of pesticides. Due to insufficient awareness among farmers, agricultural land and crops are cultivated through chemical fertilizers, which became a major threat to human health and agriculture. On the other hand, the government is implementing several measures in marketing bio-fertilizers for the betterment of agriculture and human health. In this chapter, the significance and future perspectives of biofertilizers have been covered.

DOI: 10.4018/978-1-7998-7062-3.ch004

INTRODUCTION

The pivotal food consumed by millions of people in the world is wheat, which is one among the globally produced cereals. Wheat is cultivated all over the world but wheat origin was traced back to south- east Turkey (Morris et al., 2016). Archaeological research of wild emmer specifies that wheat was firstly grown in the Karacadag Mountains in southeastern Turkey. Through DNA analysis of wheat seeds, the oldest corroboration for hexaploidy wheat has been substantiated, dating around 6400-6200 BCE recovered by catalhoyuk. The Egyptians were the developers of bread and baking which made a huge revolution in the food production industries. Mean time wheat started to spread all over Europe to Asia. The northern region of India is traditionally dominated in wheat cultivation. The prolific producers are northern states of Punjab and Haryana plains in India. By today's date all types of extensive research efforts are been taken by India for improving the output in the years to come. It is said that wheat and wheat flour play the vital role in developing India's food economy (Michael et al., 2019; Bell, 1987).

In spite of hundreds of food group, the ultimate reason for choosing wheat for our study is that it is being grown in large scale over a huge range of soils and climatic conditions, along with a wide geographical distribution. Over 40 countries in the world, wheat has been declared as the national food over one third of the world's population. The annual production of wheat has been raised from 171 million metric tons to 308 million tons between the years 1948-1952 to 1966. During the same stretch the areas for cultivation were extended from 173 to 217 million hectors, and the world average yield became 900-1420 kg per hectare (Lupton, 1987). Factors such as decrease in crop yield; poor quality of land, loss of soil texture, animals and insects affects the crops which would simultaneously affect the growing population. To overcome this problem's fertilizer was introduced in markets. A fertilizer is a substance that is added to the soil to supply one or more plant nutrients needed for the growth of plant. It is of two types- inorganic fertilizer (made up of chemical products) and organic fertilizer (obtained from animal source) (International fertilizer development center 1980). Bio fertilizers are microbial inoculants consisting of living cells of microorganisms like bacteria, algae, fungi, or a combination which may help in increasing the crop productivity. Biofertilizers such as *Rhizobium, Azotobium, Azospirillium* and *Cyanobacterium* have been used for long time purposes. *Azolla* can be used for crops like wheat, maize, cotton and other vegetable crops. Phosphorous is the most important nutrient next to nitrogen for the growth of wheat. In case of inadequate phosphorous content, the crop resembles stunted growth, dark color over older leaves and inhibition of root and flowering system. To overcome this deficiency phosphorous rich bio fertilizers can be used. Generally, biofertilizers are suggested to be better than chemical fertilizers, as chemical fertilizers pollute ground water, affect soil health and soil fertility (Rai, 2006).

The growth of biofertilizer market is driven by growing organic food, initiative taken by government and organizing several awareness programs about the need for sustainability in modern agriculture (Ghosh, 2003). The foremost intention of the article is to describe the importance of wheat, concept of fertilizer, usage of bio-fertilizer on wheat, their benefits, limitation in modern agriculture and a portion of the contextual analyses applicable to the article are also discussed.

BACKGROUND

Fertilizers and Their Relation With Yield

Approximately 175 years ago, there was a scientific debate acquiring in Europe regarding the importance of nitrogen for the growth of plants. British scientist, Bennet Lawes and Joseph Henry Gilbert settled the debate declaring that nitrogen fertilizer gave massive increase of wheat yield in England. Through this discovery, fertilizer kept its foremost food print in the agricultural world (International fertilizer development center 1980). They are classified as single nutrient (K, P, N) called as straight fertilizer, if they tend to provide two or more nutrients (N and P) they are called as multi nutrient fertilizer, inorganic fertilizer and organic fertilizer. Generally, we use fertilizer with application rate depending on the soil fertility. Muriates of potash and potassium sulphates are the only potassic fertilizers available in the market. Both are considered to give the best results when used for wheat cultivation (Maguire et al., 2019). Fertilizer's consumption in India has increased significantly in last three decades. The total consumption of N, P, K fertilizers have been increased 9- fold between 1969/1970 and 1999/2000, it also showed increased per hectare from 11-95kg in the same period. After reaching a record level in 1999/2000, the consumption has been irregular and also fluctuated around 17 million tons since 2000. The soils of India were noted with less organic matter, nitrogen, phosphorous and zinc contents. To overcome inadequate nutrient supply for plant fertilizers were introduced (Lamb, 2003).

Though fertilizers developed the production of agriculture it ended up with its own negative impact on the crops. The major drawbacks of organic fertilizers are they do not contain primary nutrients called as NPK with an exception of manure-based fertilizers. Inorganic fertilizer also contains salts and other compounds which are difficult for the crops to absorb. These compounds settle in soil and damage the soil chemistry and natural agro ecosystem. If one uses excess of fertilizer without proper knowledge it may burn the plant too. When there is sudden rainfall or excess watering, the nutrients and other compounds present wash out into nearby water body causing water pollution. It affects the natural biotic environment; based on its usage it has the ability of destroying the soil organisms, natural weeds, and microorganism (Savci, 2012). It is said that many of the "quick release" fertilizer is the reason for oxygen loss in water ways. High amount of nitrogen presence in water ways may cause excess algae which simultaneously results in loss of oxygen thereby causing a negative impact on marine life. Most of the people are still unaware that these fertilizers are also made up of residuals of waste water treatment facilities that have high chances of testing positive for toxicity. To overcome these negativities and for betterment of agriculture, bio fertilizers were introduced in market (Tripathi et al., 2020).

Microbes as Fertilizers- Biofertilizers

The term bio fertilizer is defined in many ways, over past 20 years, which frequently highlighted the relationship between rhizosphere microorganisms and crops. The common definition of biofertilizer is the substance made up of living microorganisms that colonize the rhizosphere and promote the growth of plants by increasing the supply of primary nutrients to the targeted crops. The only requirements of crops for good yield are organic matter and the relationship between the soil and microorganisms. Biofertilizers are capable of mobilizing the nutritionally potent elements from a state of non-usable to usable, and restore the natural nutrient cycle of the soil and develop soil organic matter (Trujillo and Ramirez, 2016). When bio fertilizers are used in crops, they grow healthier by enhancing the sustainability and health of the soil. From this context we can summarize that they include products that mainly contain carrier-based living microorganisms, which are meant for phosphorous solubilizing, nitrogen fixation, and nutrient mobilizing to enhance soil or crop growth (Rai, 2006; Soumare et al., 2020). At present nitrogen and phosphorus are available, potassium and zinc are yet to be commercialized. Most of the farmers choose bio fertilizers to maintain soil health, minimize environmental pollution, reduce the use of chemicals in food crops and increase the availability of nutrients by 10% to 20% without causing adverse damage to the natural system (Panda, 2011). Figure 1 represents the potent role of biofertilizers in sustainable agriculture.

Figure 1. Biofertilizers potent role in sustainable agriculture.
(Source- Mahanty et al., 2017)

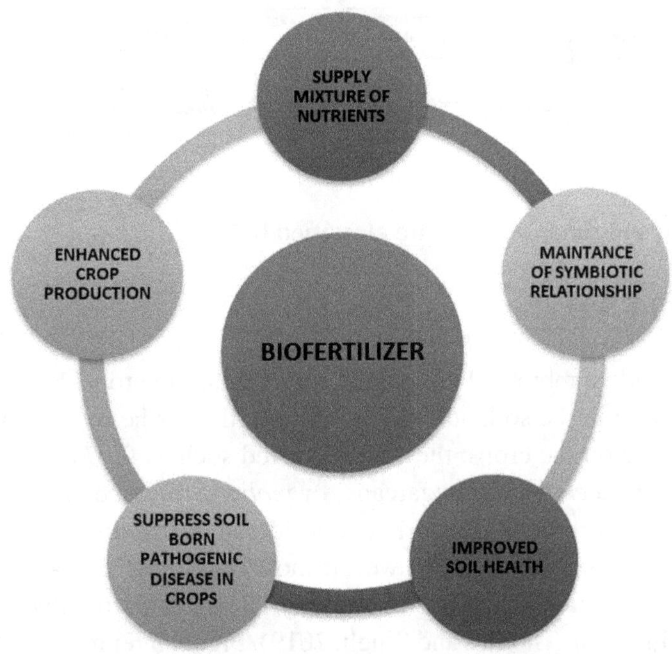

BIOFERTILIZERS: PRODUCTION AND TYPES

When biofertilizers are added to the crops, they initially look for phosphates in soil layer and solubilize them simultaneously. They also have the ability to promote atmospheric nitrogen fixation in root nodules of legume plants, making them available to the crops. By producing the needed anti- metabolites, they develop the root system for improving the yield (Itelima et al., 2018; Reddy et al., 2020). The flowchart of the steps involved in production of bacterial biofertilizer is given in Figure 2.

Figure 2. Isolated bacterial cultures are sub cultured in nutrient broth. They are allowed to grow under shaking condition at 30±2°C. They are incubated until it reaches its maximum cell population of 1010-1011 cfu/mL. At the same time for mass production, inoculum from the starter culture is transferred into fermenter and grown until required cell count.

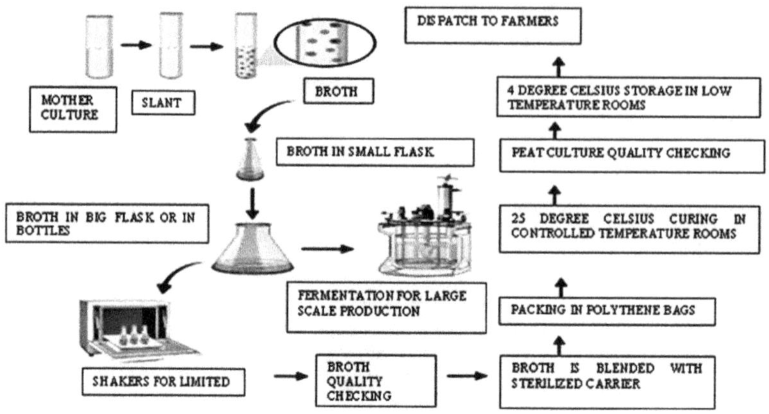

Based on their nature and function, they are classified below:

Nitrogen Fixing Microbes

Nitrogen bio fertilizers work symbiotically in fixing nitrogen to the crops. Every crop requires limited amount of nitrogen to thrive in the soil, adding these bio fertilizers helps to correct soil nitrogen level and promote growth. Based on the crops, they are preferred such as for *Rhizobia* -legume crops, *Azotobacte*r- non legume crop, *Acetobacter*- sugarcane, *cyanobacteria*- rice paddies (Macik et al., 2020). Nitrogen is infinite and omnipresent in the air, yet considered as limiting nutrient for plants due to its complexity in fixation and uptake process. However, most of the microorganisms are associated with plants and enable nitrogen fixation. This special property allows them to reduce the losses of leaching, denitrification and volatilization (Thomas and Singh, 2019). Figure 3 represents plant growth enhancing mechanisms of plant growth promoting rhizobacteria (PGPR). These microbes are;

Free Living Microbes

It is found that assessment of fixing nitrogen in plants is difficult but with exception to some plants like *Medicago sativa*, range of nitrogen fixation is noted between 3 kg N ha^{-1} to 10 kg N ha^{-1} (Roper et al., 1995). In arable soils, *Azotobacter chroococcum* can fix carbon source between 2-15 mg N g^{-1} in culture media and further initiate the production of lime thereby aggregating soil. *Frankia*, free living cultures of nodulating bacterial symbiont have been found in fixing atmospheric nitrogen for their host and non-host plant in the rhizosphere (Smolander and Sarsa, 1990). When cucumber and barley plants were interacted with *Beijerinckia mobilis* and *Clostridium* spp., through leaf spray and seed soaking methods resulted in stimulation of growth by sufficient nitrogen fixation and bacterial plant growth hormone synthesis mechanisms (Polyanskayart al., 2002). Under ideal condition, rice cultivation in

India is estimated to provide up to 20-30 kg N ha^{-1}, when cyanobacteria are harnessed along rice during cultivation (Kannaiyan, 2002).

Figure 3. Plant growth enhancing mechanisms of plant growth promoting rhizobacteria (PGPR).
(Source- Mahanty et al., 2017)

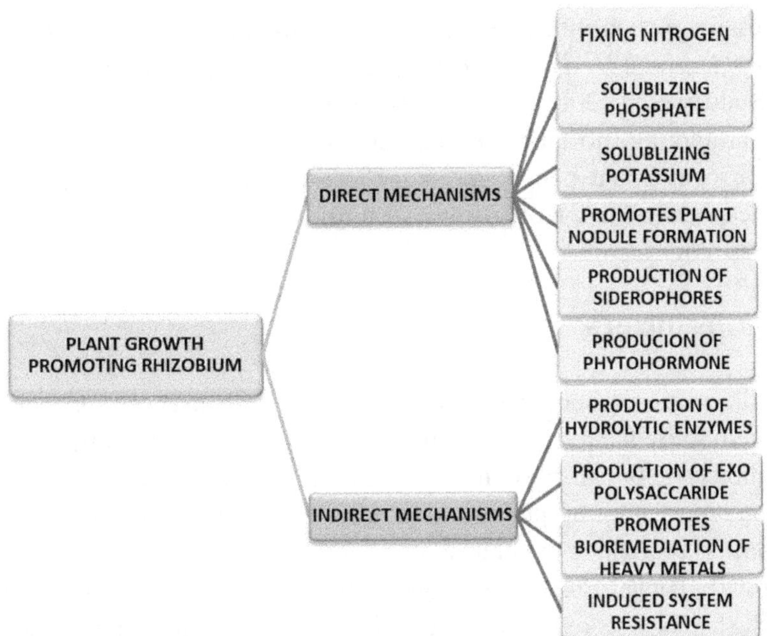

Symbiotic and Endophytic Association of Rhizobia, Cyanobacteria and Frankia With Plants

A Rhizobia bacterium is a potent group of biofertilizer that includes organisms like *Rhizobium, Bradyrhizobium, Sinorhizobium, Azorhizobium, Messorhizobium* and *Allorhizobium.* Their nitrogen fixing efficiency can vary up to 450 kg N ha^{-1} based on different strain and host legume species, were the root nodules are formed (Stamford et al., 1997; Unkovich et al., 1997; Spaink et al., 1998; Graham and Vance, 2000; Unkovich and Pate, 2000). Like *Rhizobium, Frankia* is an Actinomycete that can fix nitrogen among woody plants (Torrey, 1978; Dawson, 1986; Benson and Silvester, 1993). Non-leguminous plants such as *Casuarina, Rubus* etc. are associated with *Frankia.* Few plants such as *Ardisia* give space for symbiotic nitrogen fixing bacteria to settle by developing internal cavities. Such leaves are meant for nitrogen fertilizers constant source (Gentili and Jumpponen, 2006). Ecologically potent group among these microbes are Cyanobacteria called as blue green algae. This group includes symbiotic association with plants such as cycad roots, liverworts and *Azolla* (Thomas and Singh, 2019). *Azolla,* floating fern suitable for paddy cultivation allows this group of bacteria to settle in lower cavities and fix atmospheric nitrogen. Presence of *Anabaena azollae* in leaf cavities of fern fix nitrogen that are excreted through cavities and easily made available to ferns. During the decay period of ferns, the nitrogen is released and utilized by rice plants (Ali et al., 1998). *Trichodesmium, Nostoc* and *Anabaena* have been reported as the enhancers for rice field fertility in many parts of the world (Kundu and Ladha, 1995; Gallon, 2001). On the other side, the production and application of Cyanobacteria is poorly developed and it should be promoted as biofertilizer for sustainable agriculture (Hashem, 2001).

Living in Rhizosphere Without Endophytic Symbioses

The microbes show less interaction with roots when compared to endophytic symbionts. This group includes *Acetobacter diazotrophicus* and *Herbaspirillum* spp. for sugarcane, sorghum and maize (Triplett, 1996; James et al., 1997; Boddey et al., 2000). Several research studies have reported that due to the nitrogen fixation and growth promoting substance production, the growth and yield of food crops such as rice, wheat tomato, oak, carrot eggplant, pepper and sugar beet was increased with *Azospirillum* (Bashan and Holguin, 1997). Hence the production of nitrogen in substantial quantity makes these microbes suitable for their application as biofertilizer.

Phosphorous Biofertilizers

Phosphorous biofertilizers are similar to nitrogen biofertilizers with an exception to not showing a dependency on soil where crops are harvested. Their main function is to maintain the optimum phosphorous level. The concentration level of phosphorous is high in soil but it is not easily available for plants hence it is said as the second most limiting plant nutrient after nitrogen (Schachtman et al., 1998). *Bacillus* and *Pseudomonas* bacteria mobilize the unavailable state of phosphorous in the soil and provide them to plants for their growth (Richardson, 2001). Soil fungi such as *Aspergillus* and *Penicillum* along with phosphate solubilizing bacteria secrets organic acids for the dissolution of bound phosphate. When Rock phosphate along with *Bacillus megaterium var. phosphticum* were used in sugar cane, increase in yield and juice quality by 12.6% was noted and also reduced the used of phosphate by 25% (Sundara et al., 2002).

Compost Biofertilizers

Compost is a decomposing murky material, which contains potassium, phosphorus, and nitrogen along with microorganisms, earthworms and dung beetles. This compost aerates, aggregates, keeps soil moist, provides minerals and increases soil microbial activity through the formation of human- containing material which are formed by oxidation of microbial organic solid residue (Yu et al., 2016). Composts formed from materials such as straw, leaves, vegetable and fruit waste etc. give rise to the microorganisms like *Trichoderma viridae, Aspergillus, Bacillus* spp., gram negative bacteria etc. that have plant cell wall degrading cellulolytic enzymes which promotes the suppression of parasitic microorganisms. Vermicompost is otherwise known as organic fertilizers, which contain earthworm cocoons, excreta, microorganisms for example, bacteria, Actinomyctes, fungi and also different organic matter provides N, P, K and several other micronutrients. To overcome loss of soil fertility, quality/quantity of yield, and salinity, vermicompost is preferred. Bio compost was prepared from sugar industry waste materials, which are decomposed and enriched with plant and human friendly bacteria and fungi. Presence of N, phosphate solubilizing bacteria and other beneficial fungi like *Trichoderma viridae,* prevent soil borne disease and increases yield and produces quality products (Boulter et al., 2002).

Microphos Biofertilizers

The biofertilizers are used for releasing phosphate from bound and insoluble state e.g., *Bacillus polymyxa* (Thomas and Singh, 2019).

Mycorrhiza

Mycorrhiza is mutually beneficial fungus found on roots of higher plant. *Glomus* species are common mycorrhiza fungal partner, this single fungus have the ability of forming mycorrhizal association with several numbers of plants. Mycorrhiza associated biofertilizers are highly efficient in mobilizing nutrient elements such as P, Fe, Zn, B and other trace elements (Rao et al., 2020). Long duration crops are preferred for using this biofertilizer. They provide moisture from far of inches and 2 years of storage capacity. They are phosphorus mobilizing biofertilizers. The mycelium of these mycorrhizal fungi extends from root surface into soil, thereby extending the surface area for more efficient nutrient supply. Apart from this quality they are also known to increase soil quality, soil aeration, water dynamics and to make plants less sensitive to herbivores or root pathogen (Rillig et al., 2002; Thakur and Singh, 2018). This valuable property makes them more suitable for application in agriculture and land reclamation (Menge, 1983; Sylvia, 1990). Based on residence of fungus they are classified in two groups: Ectomycorrhiza and Endomycorrhiza (Smith and Read, 1997). Endomycorrhiza enables exchange of nutrients between the soil and host. Whereas Ectomycorrhiza are generally found in trees such as *Eucalyptus*, *Quercus*, peach, pine etc.; potent role of these fungi are absorption of water and minerals by increasing surface area of roots, secreting antimicrobial substance that protect plants from root pathogen and solubilizing soil humus organic matter to release and absorb inorganic nutrients (White, 1941; Wilde, 1944; Mikola, 1970; Smith and Read, 1997).

Vesicular Arbuscular Mycorrhizal (VAM)

Intercellular obligate endosymbionts, made up of special structures known as vesicles and arbuscular are said as VAM fungi. They produce micronutrients and supply it to the host plant; they also increase the availability and mobility of phosphorous and help up taking Zn, Cu, P and water (Douds et al., 2002). However, the obligate nature and uncultivability characteristic of these fungi had made the inoculation incompatible with large scale agriculture and thus it might require additional research for betterment (Wood and Cummings, 1992; Ryan and Graham, 2001).

BIOFERTILIZERS FOR WHEAT

Due to infertility agricultural soil and use of pesticides, million people's lives and livelihood are in threat. To overcome these problems advanced agricultural practice is advised to be followed. However, investing huge amount of money in buying expensive fertilizer and machines are not affordable by low-income farmers. For effective, economically friendly, eco-friendly practice of agriculture, use of biofertilizers can be followed (Macik et al., 2020). Many researchers are trying different types of biofertilizer on wheat to increase their yield, to protect them from disease or insects etc. Some of them are discussed in Table 1.

Table 1. Distinctive combination of biofertilizers used for wheat

S. No.	Biofertilizer Combination	Benefits in Wheat	Reference
1	A bacterial consortium (*Azospirillum spp.* + *Azoarcus spp.* + *Azorhizobium spp.*); and two mycorrhizal fungal-bacterial consortia, viz. *Rhizophagus irregularis* + *Azotobacter vinelandii*, *R. irregularis* + *Bacillus megaterium* + *Frateuria aurantia*	• Increased plant growth and nitrogen content. • Growth without harming and changing the resident microbiome.	Dal et al., 2020
2	Bio inoculants of Azotobacter and PSB	Use of nitrogen and phosphorous fertilizers were saved.	Khandare et al., 2020
3	Plant growth-promoting rhizobacteria (PGPR), arbuscular mycorrhizal fungi (AMF), and spectral properties for deriving need-based fertilizer N	• Improvement in rhizosphere mycorrhixation and mycorrhizal colonization. • Highest grain production and reduced use of fertilizer N and P • Improved PGPR population, dehydrogenase and alkaline phosphatase activities on soil.	Varinderpal et al., 2020
4	Azotobacter, phosphate solubilizing bacteria and potash mobilizing bacteria, in combinations with different doses of inorganic fertilizers.	• Increased nitrogen fixation, phosphate solubilization, Potash mobilization and efficiency of both organic and inorganic fertilizer. • Production of plant growth promoting substances promoted nutrient availability to plants.	Game et al., 2020
5	Consortium of endophyte and rhizosphere phosphate solubilizing bacteria	Rhizospheric and endophytic bacterial inoculation improved-Root and shoot dry matter, Grain yield, Length, Surface area, volume of the root and enhanced P efficiency of wheat cultivars.	Emami et al., 2020
6	Root-Associated Rock Phosphate Solubilizing Bacteria	Good resistance against different environmental stresses like antibiotics and fungi.	Rfaki et al., 2020
7	*Streptomyces* strains	• Salinity significantly decreased seedling fresh and dry weight, K^+ and chlorophyll content. and Glutathione S- transferase activity. • Increased chlorophyll and carotenoid simultaneously decreased Na^+ content.	Akbari et al., 2020
8	*Bacillus methylotrophicus M41*	• Attenuates salt stress injury in wheat under both low and high salt stress • Reduced soil pH, Electrical conductivity, content of Na^+ in leaves. • Increased the exchangeable K content and the uptake of Mg^{2+} by wheat roots.	Ji et al., 2020
9	PGPB and arbuscular mycorrhizal (AM) fungi	• Increased macronutrient and micronutrient content in wheat grains. • Increased total chlorophyll content in wheat leaves. • Improved soil health parameters.	Yadav et al., 2020
10	Diazotrophic bacterium (*Paenibacillus beijingensis BJ-18*) and a P-solubilizing bacterium (*Paenibacillus sp. B1*)	• Increased plant biomass, plant nitrogen content, P content • Increased N fixation in soils and the endosphere. • Improved soil available P and plant P uptake.	Li et al., 2020
11	Potassium and Residue Management Options	• Improved -crop and soil quality related parameters. • Increase in crop growth, physiological parameters, grain yield.	Madar et al., 2020
12	*Pseudomonas* strain	Increased wheat shoots, length, Root length, Fresh biomass, Dry biomass, Leaf greenness	Dar et al., 2020
13	*Penicilliumbilaiae* and *Bacillus simpex*	• Increased-P concentration in root mass at all P level, Mg, Mn and S concentration in shoot biomass in low P soil. • Increased- P uptake • Improved nutritional status of winter wheat at low P soil	Hansen et al., 2020
14	*Bacillus subtilis HG-15*	• Increased total N, Organic matter, K^+, Ca^{2+}, Mg^{2+} • Increased-Dry weight, Plant height- Root length, and obtained induced systematic tolerance.	Ji et al., 2020
15	*Bacillus halotolerns MSR- h4* and *Lelliottiaamnigena MSR-M49*	• Increased plant height, • Straw dry weight, Spike number and grain yield, N% and protein content in grains • sustainable approach to reduced salt effect on wheat production.	El- Akhdar et al., 2019
16	Sludge, compost and bio fertilizer under newly reclaimed soil	Increased spike length, weight of spike and grain and straw yield	Mohamed et al., 2019
17	*Chlorella sorokiniana*	• Increased plant length. • *Total dry biomass of above ground and below ground parts were improved.*	Mohamed et al., 2019
18	Nitrogen fixing cyanobacteria	Increased plant height, number of spikes/m^2, grain yield, straw yield and protein content% on wheat crop.	Joshi et al., 2019

Methods of Application

Two methods are commonly used for applying biofertilizers. They are seed treatment and seedling root dip.

Seed Treatment

Seed treatment is the most effective, economic and ordinary method used for all types of inoculants (Sethi et al., 2014). Generally, fungicide seed treatment is preferred for wheat crops to overcome soil born disease, fall season insects such as aphids (Mubeen at al., 2006). Initially, preparation of slurry is done by mixing one package of inoculants to 200 mL of rice kanji. Secondly the seeds required for the cultivation are mixed inside the slurry, and carefully noted whether all the seeds are uniformly covered with slurry. Finally slurry mixed seeds are dried for 30 minutes, after 24 hours the dried seeds are ready to sow (Chen, 2006). In case of liquid biofertilizer, the coating can be done in plastic bag if quantity is small or if quantity is large it can be done in buckets, thus it depends upon the quantity. Two or more bacteria can be used in seed treatment without any antagonistic effect, and the maximum quantity of each bacterium can be provided on individual seeds for better results (Chen, 2006). Generally, fungicide seed treatment is preferred for wheat crops to overcome soil born diseases, fall season insects such as aphids (Mubeen at al., 2006).

Seedling Root Dip

Seed root dipping method is common for plantation of crops such as vegetables, fruits, trees, cereals, sugarcane, cotton and tobacco. 1 to 2 kg of nitrogen fixing and phosphate solubilizing biofertilizers are mixed with water (quantity of water depends on quantity of seedling). The roots of seedling are mixed in the mixture for about 20-30 minutes before sowing (Barea and Brown, 1974). Root tip method for wheat can also be done with *F. graminearum.* It is done by filling 5 mL of macro conidia suspension in 12 chambers along with seedling placed onto small flat tray, upon the chamber (Mubeen et al., 2006).

COMBINATION OF BIOFERTILIZERS AND CHEMICAL FERTILIZERS

Chen (2006) combined both biofertilizer - mixture of *Bacillus* sp. *B. subtilis, B erythropolis, B. pumilus* and *P. rubiacearump* and 50% of chemical fertilizer- half of CF and biofertilizer on lettuce, which increased 25% of yield when it was treated with half of CF and biofertilizer when compared to half of CF alone. Likewise, when (i) Four levels of chemical nitrogen fertilizer (0, 100, 150 and 200 kg N ha–1), (ii) Two levels of biofertilizer (with and without inoculation) containing *Azotobacter* sp. *and Azospirillum* sp. and (iii) Two levels of weed interference, were tested in wheat showed increased plant height, spike number per unit of area, grains number per spike, grain yield and grain protein content (Namvar and Khandan, 2013). Mustard was cultivated with six different reduced doses of chemical fertilizer combined with biofertilizers and vermicompost resulted in optimum plant growth and enhanced plant defense system against insect and disease (Mondal et al., 2019). From the mentioned examples we can conclude that combination of two types of fertilizers were harmless and showed increase in yield.

BIOFERTILIZERS IN REMEDIATION OF PESTICIDES

The best technique to overcome the harmful effects of pesticides is to grow pesticide degrading strains of bacteria. The role of plant growth promoting *Rhizobium* in pesticide bioremediation had been subjected to several set of investigations. It has been concluded that microorganisms like *Azospirillum, Bacillus, Enterobacter, Gordonia, Pseudomonas, Serratia,* etc. have the ability to reduce pesticide toxicity and have the potential for biotransformation and biodegradation (Mondal et al., 2017). For the pesticide degradation process, the initial step taken by the microorganisms is enzymatic degradation three important systems involved in pesticides degradation are: hydrolases, esterase and mixed function of oxidase hydrolases (first metabolic stage) and glutathione (second stage) (Shaheen and Sundari, 2013). Based on several reports, it can be summarized that plant growth promoting rhizobium holds a strong approach in decreasing the pesticide contamination in soil.

DRAWBACKS OF BIOFERTILIZERS

Despite the fact that biofertilizer innovation is eco-friendly there are a few constraints and significant disadvantages of microbial biofertilizers which include; they require prompt consideration for additional examination and appropriate arranging, for example, plant explicitness, lower supplement thickness, prerequisite of independent apparatus and ability for creation, trouble away and lacking mindfulness about their advantages (Malusa et al., 2019). The various biofertilizer technology constraints are given in Table 2 along with examples.

Table 2. Biofertilizer Technology Constraints with examples.

Biofertilizer Technology Constraints	Examples
Technological	• Less efficient microbial strain and carrier materials • Low quality microbial inoculants
Infrastructural	• Non-availability of suitable production facilities
Financial and marketing	• Insufficient funds • Non availability of right inoculant • Lack of retail outlets
Environmental	• Seasonal bio fertilizer demand • Soil characteristics
Human resource	• Lack of training • Unawareness • Ignorance on the environmental problems caused by chemical fertilizer

(Source: Giri et al., 2019)

GOVERNMENT INTERVENTION IN BIOFERTILIZERS: INDIAN PERSPECTIVE

For the production, distribution and promotion of biofertilizers, the ninth plan (National project on development and use of bio fertilizer) was implemented by the Government of India. The NPDB, Central sector scheme was launched for organizing training courses, field demonstration and quality control service (Malusa and Vassilev, 2014).

Reach of biofertilizers among people was inefficient. Hence government took the opportunity to advertise biofertilizers in three ways (El-Akkdar et al., 2020):

1. State government by means of district level and village workers to farmers.
2. State marketing federation by means of cooperative bodies to farmers.
3. State agro industries cooperative by means of agro service center to farmers.

Government of India implemented the scheme for promoting biofertilizers since 7th five-year plan. One national centre and six regional centres have been established under his scheme. These centres organize training, demonstrate and also supply 10 efficient cultures for production of biofertilizers thereby promoting them. At Ghaziabad, National Biofertilizer Development Center was set up as a subordinate office of the Department of Agriculture and Cooperation with six centers. The significant point of this plan zeroed in on the broad laborers by giving secured association of instructional class and field showing alongside quality control administration. Appropriation of various biofertilizers were taken however stopped as the focuses reclassified their part towards R&D and H&D related exercises (Ghosh, 2003).

Case Studies

Biofertilizers were not being embraced on a wide scale, for dissecting this issue a field study was started in two regions in Haryana. The investigation essentially centered around two areas: karnal (speaks to serious horticulture with serious extent of water system) and Bhawani (speaks to dryland cultivating, with low level irrigation). This study found that biofertilizers were not broadly acknowledged by farmers in Haryana. This absence of acknowledgment was because of inaccessibility and their low quality. Then the examination additionally found that both the State Agriculture Department and business people are reluctant to stock and sell biofertilizers as they feel that their quality is inconsistent. Due to the low interest for biofertilizers, enormous interest in cutting edge creation and storerooms are forestalled (Alam et al., 2002).

To analyze the usage of biofertilizers and organic manures a study was directed in Durg locale arranged in Chhattisgarh, India. The investigation uncovered that there is a positive acknowledgment among ranchers for the Integrated Nutrient Management (INM) strategy. Investigation of the 10 years information (1995-96 to 2004-2005) presumed that biofertilizers use regarding land was similarly higher than organic components. Among these components, utilization of PSB (38.64) was greatest trailed by *Rhizobium* culture (20.48%), *Azotobacter* (10.02%) and vermicompost (0.16%) (Singh et al., 2007).

A case of biofertilizers production has been chosen from an agro- biotechnology firm located in Ichalkaranji, southern part of Maharashtra. Factors impacting innovation commercialization measure are distinguished utilizing content investigation. From the case examination three new factors were distinguished to be specific altered framework, infrastructure accessibility and government intercession. This examination will support associations, business people, and policy makers to devise approaches separately to improve biofertilizers commercialization in India. The current examination endeavored to comprehend the TC process in an Indian biofertilizer firm and to distinguish the variables affecting the sub-processes of TC. The utilization of biofertilizers improves soil just as farming yield quality by providing fundamental micronutrients. This, prompts secure a serious situation for Indian vegetables and natural products in global market particularly the nations following USFDA standards. The perspective like suitable determination of advances like fermentation and harvesting lead to improvement of excellent biofertilizers for the chose crops, in this way upgrading the notoriety of the firm. This further empowered the firm to increase serious edge by differentiating into other novel items and extending their limits to different countries like Sri Lanka and Bangladesh. The case features that facilities like client call focuses and utilization of advanced media empowered clients particularly famers to comprehend the present moment just as long-haul advantages of biofertilizer. Government uphold through single tax assessment conspire like merchandise and administration charges; online authority record freedom discovered fundamental for consolidating simplicity of business in Indian biological system. The examination implies the part of biofertilizer commercialization at firm level just as at nation level (Tawate et al., 2018).

FUTURE PRESPECTIVE OF BIOFERTILIZERS

These days biofertilizers have transformed into a basic part of agribusiness. Microorganisms utilized in biofertilizer are as of now being utilized in scarcely any creating nations and it is relied upon to spread with time (Weekley et al., 2012). Hence, it is sensible to expect that later on the tremendous utilization of biofertilizer will advance a few methodologies for the improvement of horticulture. Adjusting the use of biofertilizers, however, would require further thought and essential measures to be taken to assess (Gamalero et al., 2008);

- Multifunction biofertilizer with powerful and serious function over assortment of harvests ought to be chosen.
- Extending the utilization of biofertilizer from research center analysis to enormous scope business use will require progressed approaches for these microbes, for example, in their development, stockpiling, definition, and delivery.
- It is essential to instructed individuals about the impacts of delayed utilization of concoction compost and the misguided judgment about microorganisms that they can just aim's sicknesses should be revised.
- "Biofertilizer Act" and Quality control framework with severe guideline in business sectors and application ought to be set up.
- Under distressing conditions, the action of bio compost ought to be contemplated.
- Agronomic and monetary assessment for assortment of farming creation ought to be considered.

CONCLUSION

The use of chemical fertilizers oversupply has generated nutrients, such as phosphate, to accumulate in soil, triggering soil futility. Despite the fact that chemical fertilizers brought about upgrading the development of yields quicker, its cynicism indicated significant effects on horticultural land and plants. To survive and lessen these issues biofertilizers were presented in business sectors. Despite the fact that these natural composts are totally comprised of living microorganisms, they stay sheltered and innocuous to the harvests just as the buyer. Biofertilizers were utilized as single or even in blend to increment and make illness free wheat. From the investigation we can infer that bio manures, for example, *Rhizo-*

bium, *Azotobacter*, *Azospirillum* and *Cyanobacteria* growth can be utilized for long haul reason. When *Azotobacter* was utilized as bio manure in wheat crops, they brought about most elevated grain yield and overall gain. In the year 2003, Sikkim restricted the section of substance manure and pesticides for farmland and was proclaimed as India's first completely natural state by changing over around 75,000 hectares of agrarian land into feasible development. Consequently, ranchers had no other choice except for to follow natural development, which is liberated from substance pesticides and compost as it attempts an agreeable offset with the environment. At the point when natural cultivating is followed ceaselessly for an extensive stretch it prompts means of horticulture, biodiversity preservation and ecological insurance. Shockingly, because of ill-advised mindfulness among ranchers, the utilization of biofertilizers is less. Government had found a way to advance biofertilizers and substance free agribusiness among ranchers. Notwithstanding, with the assistance of Science and innovation, scientists are striving to beat not many downsides related with biofertilizer and make synthetic free rural items.

REFERENCES

Akbari, A., Gharanjik, S., Koobaz, P., & Sadeghi, A. (2020). Plant growth promoting Streptomyces strains are selectively interacting with the wheat cultivars especially in saline conditions. *Heliyon*, *6*(2), e03445. doi:10.1016/j.heliyon.2020.e03445 PMID:32095655

Alam, G., & Alam, G. (2000). *A study of biopesticides and biofertilisers in Haryana*. International Institute for Environment and Development.

Ali, S., Hamid, N., Khan, D., & Malik, K. A. (1998). Use of *Azolla* as biofertilizer to enhance crop yield in a rice—wheat cropping system under mild climate. In *Nitrogen Fixation with Non-Legumes* (pp. 353–357). Springer. doi:10.1007/978-94-011-5232-7_41

Barea, J. M., & Brown, M. E. (1974). Effects on plant growth produced by *Azotobacter paspali* related to synthesis of plant growth regulating substances. *The Journal of Applied Bacteriology*, *37*(4), 583–593. doi:10.1111/j.1365-2672.1974.tb00483.x PMID:4611996

Bashan, Y., & Holguin, G. (1997). Azospirillum-plant relationships: Environmental and physiological advances (1990–1996). *Canadian Journal of Microbiology*, *43*(2), 103–121. doi:10.1139/m97-015

Bell, G. D. H. (1987). The history of wheat cultivation. In *Wheat breeding* (pp. 31–49). Springer. doi:10.1007/978-94-009-3131-2_2

Benson, D. R., & Silvester, W. B. (1993). Biology of Frankia strains, actinomycete symbionts of actinorhizal plants. *Microbiological Reviews*, *57*(2), 293–319. doi:10.1128/MR.57.2.293-319.1993 PMID:8336669

Boddey, R. M., Da Silva, L. G., Reis, V., Alves, B. J. R., & Urquiaga, S. (2000). Assessment of bacterial nitrogen fixation in grass species. In Prokaryotic nitrogen fixation: a model system for analysis of a biological process. Horizon Scientific Press.

Boulter, J. I., Trevors, J. T., & Boland, G. J. (2002). Microbial studies of compost: Bacterial identification and their potential for turfgrass pathogen suppression. *World Journal of Microbiology & Biotechnology*, *18*(7), 661–671. doi:10.1023/A:1016827929432

Carvajal-Muñoz, J. S., & Carmona-Garcia, C. E. (2012). Benefits and limitations of biofertilization in agricultural practices. *Livestock Research for Rural Development*, *24*, 43.

Chen, J. H. (2006). The combined use of chemical and organic fertilizers and/or biofertilizer for crop growth and soil fertility. In *International workshop on sustained management of the soil-rhizosphere system for efficient crop production and fertilizer use* (*Vol. 16*, No. 20, pp. 1-11). Land Development Department.

Dal Cortivo, C., Ferrari, M., Visioli, G., Lauro, M., Fornasier, F., Barion, G., & Vamerali, T. (2020). Effects of seed-applied biofertilizers on rhizosphere biodiversity and growth of common wheat (*Triticum aestivum* L.) in the *Field. Frontiers in Plant Science*, *11*, 72. doi:10.3389/fpls.2020.00072 PMID:32174929

Dar, A., Zahir, Z. A., Asghar, H. N., & Ahmad, R. (2020). Preliminary screening of rhizobacteria for biocontrol of little seed canary grass (Phalaris minor Retz.) and wild oat (*Avena fatua* L.) in wheat. *Canadian Journal of Microbiology*, *66*(5), 368–376. doi:10.1139/cjm-2019-0427 PMID:32040347

Das, H. K. (2019). Azotobacters as biofertilizer. *Advances in Applied Microbiology*, *108*, 1–43. doi:10.1016/bs.aambs.2019.07.001 PMID:31495403

Dawson, J. O. (1986). Actinorhizal plants: Their use in forestry and agriculture. *Outlook on Agriculture*, *15*(4), 202–208. doi:10.1177/003072708601500406

Douds, D. D., Gadkar, V., & Adholeya, A. (2000). Mass production of VAM fungus biofertilizer. In *Mycorrhizal Biology* (pp. 197–215). Springer. doi:10.1007/978-1-4615-4265-0_13

El-Akhdar, I., Elsakhawy, T., & Abo-Koura, H. A. (2020). Alleviation of Salt Stress on Wheat (*Triticum aestivum* L.) by Plant Growth Promoting Bacteria strains *Bacillus halotolerans* MSR-H4 and *Lelliottia amnigena* MSR-M49. *Journal of Advances in Microbiology*, 44–58. doi:10.9734/jamb/2020/v20i130208

Emami, S., Alikhani, H. A., Pourbabaee, A. A., Etesami, H., Motasharezadeh, B., & Sarmadian, F. (2020). Consortium of endophyte and rhizosphere phosphate solubilizing bacteria improves phosphorous use efficiency in wheat cultivars in phosphorus deficient soils. *Rhizosphere*, *14*, 100196. doi:10.1016/j.rhisph.2020.100196

Gallon, J. R. (2001). N_2 fixation in phototrophs: Adaptation to a specialized way of life. *Plant and Soil*, *230*(1), 39–48. doi:10.1023/A:1004640219659

Gamalero, E., Berta, G., Massa, N., Glick, B. R., & Lingua, G. (2008). Synergistic interactions between the ACC deaminase-producing bacterium Pseudomonas putida UW4 and the AM fungus Gigasporarosea positively affect cucumber plant growth. *FEMS Microbiology Ecology*, *64*(3), 459–467. doi:10.1111/j.1574-6941.2008.00485.x PMID:18400004

Game, B. C., Ilhe, B. M., Pawar, V. S., & Khandagale, P. P. (2020). Effect of Azotobacter, Phosphate Solubilising Bacteria and Potash Mobilising Bacteria Inoculants on Productivity of Wheat (Triticum aestivum L.). *International Journal of Current Microbiology and Applied Sciences*, *9*(3), 2800–2807. doi:10.20546/ijcmas.2020.903.322

Gentili, F., & Jumpponen, A. (2006). Bacterial and Fungal Biofertilizers. Handbook of microbial biofertilizers, 1, 25-89.

Ghosh, N. (2004). Promoting biofertilisers in Indian agriculture. *Economic and Political Weekly*, 5617–5625.

Giri, B., Prasad, R., Wu, Q. S., & Varma, A. (Eds.). (2019). *Biofertilizers for sustainable agriculture and environment* (Vol. 55). Springer. doi:10.1007/978-3-030-18933-4

Graham, P. H., & Vance, C. P. (2000). Nitrogen fixation in perspective: An overview of research and extension needs. *Field Crops Research*, *65*(2-3), 93–106. doi:10.1016/S0378-4290(99)00080-5

Hansen, V., Bonnichsen, L., Nunes, I., Sexlinger, K., Lopez, S. R., van der Bom, F. J. T., & Jensen, L. S. (2020). Seed inoculation with *Penicillium bilaiae* and *Bacillus simplex* affects the nutrient status of winter wheat. *Biology and Fertility of Soils*, *56*(1), 97–109. doi:10.100700374-019-01401-7

Hashem, M. A. (2001). Problems and prospects of cyanobacterial biofertilizer for rice cultivation. *Australian Journal of Plant Physiology*, *28*, 881–888.

International Fertilizer Development Center (1980). *Fertilizer manual* (No. 13). UN.

Itelima, J. U., Bang, W. J., Onyimba, I. A., & Oj, E. (2018). A review: Biofertilizer; a key player in enhancing soil fertility and crop productivity. *Journal of Microbiology and Biotechnology Research*, *2*(1), 22–28.

James, E. K., Olivares, F. L., Baldani, J. I., & Döbereiner, J. (1997). Herbaspirillum, an endophytic diazotroph colonizing vascular tissue in leaves of Sorghum bicolor L. *Moench J Exp Bot*, *48*(3), 785–797. doi:10.1093/jxb/48.3.785

Ji, C., Tian, H., Wang, X., Hao, L., Wang, C., Zhou, Y., & Liu, X. (2020). Bacillus subtilis *HG-15, a Halotolerant Rhizoplane Bacterium, Promotes Growth and Salinity Tolerance in Wheat (*Triticum aestivum*)*. Preprint.

Ji, C., Wang, X., Tian, H., Hao, L., Wang, C., Zhou, Y., & Liu, X. (2020). Effects of Bacillus methylotrophicus M4-1 on physiological and biochemical traits of wheat under salinity stress. *Journal of Applied Microbiology*, *129*(3), 695–711. doi:10.1111/jam.14644 PMID:32215987

Joshi, H., Shourie, A., & Singh, A. (2020). Cyanobacteria as a source of biofertilizers for sustainable agriculture. In *Advances in Cyanobacterial Biology* (pp. 385–396). Academic Press. doi:10.1016/B978-0-12-819311-2.00025-5

Kannaiyan, S. (2002). Biotechnology of biofertilizers. Alpha Science Int'l Ltd.

Khandare, R. N., Chandra, R., Pareek, N., & Raverkar, K. P. (2020). Carrier-based and liquid bioinoculants of Azotobacter and PSB saved chemical fertilizers in wheat (Triticum aestivum L.) and enhanced soil biological properties in Mollisols. *Journal of Plant Nutrition*, *43*(1), 36–50. doi:10.1080/01904167.2019.1659333

Kholssi, R., Marks, E. A., Miñón, J., Montero, O., Debdoubi, A., & Rad, C. (2019). Biofertilizing effect of Chlorella sorokiniana suspensions on wheat growth. *Journal of Plant Growth Regulation*, *38*(2), 644–649. doi:10.100700344-018-9879-7

Kundu, D. K., & Ladha, J. K. (1995). Efficient management of soil and biologically fixed N2 in intensively cultivated rice fields. *Soil Biology & Biochemistry*, *27*(4-5), 431–439. doi:10.1016/0038-0717(95)98615-U

Lamb, R. L. (2003). Fertilizer use, risk, and off-farm labor markets in the semi-arid tropics of India. *American Journal of Agricultural Economics*, *85*(2), 359–371. doi:10.1111/1467-8276.00125

Li, Y., Li, Q., Guan, G., & Chen, S. (2020). Phosphate solubilizing bacteria stimulate wheat rhizosphere and endosphere biological nitrogen fixation by improving phosphorus content. *PeerJ*, *8*, e9062. doi:10.7717/peerj.9062 PMID:32411531

Lupton, F. G. H. (1987). History of wheat breeding. In Wheat breeding (pp. 51-70). doi:10.1007/978-94-009-3131-2_3

Mącik, M., Gryta, A., & Frą, M. (2020). Biofertilizers in agriculture: An overview on concepts, strategies and effects on soil microorganisms. *Advances in Agronomy*, *162*, 31–87. doi:10.1016/bs.agron.2020.02.001

Mącik, M., Gryta, A., & Frąc, M. (2020). Biofertilizers in agriculture: An overview on concepts, strategies and effects on soil microorganisms. *Advances in Agronomy*, *160*, 31–87. doi:10.1016/bs.agron.2020.02.001

Madar, R., Singh, Y. V., Meena, M. C., Das, T. K., Gaind, S., & Verma, R. K. (2020). Potassium and Residue Management Options to Enhance Productivity and Soil Quality in Zero till Maize–Wheat Rotation. *CLEAN–Soil, Air. Water (Basel)*, *48*(3), 1900316.

Maguire, R., Alley, M. M., & Flowers, W. (2019). *Fertilizer types and calculating application rates*. Academic Press.

Mahanty, T., Bhattacharjee, S., Goswami, M., Bhattacharyya, P., Das, B., Ghosh, A., & Tribedi, P. (2017). Biofertilizers: A potential approach for sustainable agriculture development. *Environmental Science and Pollution Research International*, *24*(4), 3315–3335. doi:10.100711356-016-8104-0 PMID:27888482

Malusá, E., & Vassilev, N. (2014). A contribution to set a legal framework for biofertilisers. *Applied Microbiology and Biotechnology*, *98*(15), 6599–6607. doi:10.100700253-014-5828-y PMID:24903811

Menge, J. A. (1983). Utilization of vesicular arbuscular mycorrhizal fungi in agriculture. *The New Phytologist*, *81*, 553–559. doi:10.1111/j.1469-8137.1978.tb01628.x

Mia, M. B., & Shamsuddin, Z. H. (2010). Rhizobium as a crop enhancer and biofertilizer for increased cereal production. *African Journal of Biotechnology*, *9*(37), 6001–6009.

Mohamed, M. F., Thalooth, A. T., Elewa, T. A., & Ahmed, A. G. (2019). Yield and nutrient status of wheat plants (Triticum aestivum) as affected by sludge, compost, and biofertilizers under newly reclaimed soil. *Bulletin of the National Research Center*, *43*(1), 31. doi:10.118642269-019-0069-y

Mondal, T., Datta, J. K., & Mondal, N. K. (2017). Chemical fertilizer in conjunction with biofertilizer and vermicompost induced changes in morpho-physiological and bio-chemical traits of mustard crop. *Journal of the Saudi Society of Agricultural Sciences*, *16*(2), 135–144. doi:10.1016/j.jssas.2015.05.001

Morris, C. F., Wrigley, C., Corke, H., Seetharaman, K., & Faubion, J. (2016). *Cereals: Overview of uses: Accent on wheat grain. Encyclopedia of food grains* (2nd ed.). Academic Press.

Mubeen, F., Aslam, A., Sheikh, M., Iqbal, T., Hameed, S., Malik, K. A., & Hafeez, F. Y. (2006). Response of wheat yield under combine use of Fungicide and Biofertilizer. *International Journal of Agriculture and Biology*, *8*(5), 580–582.

Namvar, A., & Khandan, T. (2013). Response of wheat to mineral nitrogen fertilizer and biofertilizer (Azotobacter sp. and Azospirillum sp.) inoculation under different levels of weed interference. *Ekologija (Lietuvos Mokslu Akademija)*, *59*(2), 2. doi:10.6001/ekologija.v59i2.2711

Ortiz-Hernández, M. L., Sánchez-Salinas, E., Dantán-González, E., & Castrejón-Godínez, M. L. (2013). Pesticide biodegradation: mechanisms, genetics and strategies to enhance the process. *Biodegradation-Life of Science*, 251-287.

Panda, H. (2011). *Manufacture of biofertilizer and organic farming*. ASIA PACIFIC BUSINESS PRESS Inc.

Polyanskaya, L. M., Vedina, O. T., Lysak, L. V., & Zvyagintsev, D. G. (2002). The growth-promoting effects of Beijerinckiamobilis and Clostridium sp. cultures on some agricultural crops. *Microbiology*, *71*(1), 109–115. doi:10.1023/A:1017914803544

Rai, M. (Ed.). (2006). *Handbook of microbial biofertilizers*. CRC Press. doi:10.1201/9781482277760

Rao, H. C. Y., Mohana, N. C., & Satish, S. (2020). *Biocommercial aspects of microbial endophytes for sustainable agriculture*. Microbial Endophytes, Functional Biology and Applications. doi:10.1016/B978-0-12-819654-0.00013-2

Reddy, G. C., Goyal, R. K., Puranik, S., Waghmar, V., Vikram, K. V., & Sruthy, K. S. (2020). *Biofertilizers Toward Sustainable Agricultural Development*. Plant Microbe Symbiosis. doi:10.1007/978-3-030-36248-5_7

Rfaki, A., Zennouhi, O., Aliyat, F. Z., Nassiri, L., & Ibijbijen, J. (2020). Isolation, selection and characterization of root-associated rock phosphate solubilizing bacteria in moroccan wheat (Triticum aestivum L.). *Geomicrobiology Journal*, *37*(3), 230–241. doi:10.1080/01490451.2019.1694106

Richardson, A. E. (2001). Prospects for using soil microorganisms to improve the acquisition of phosphorus by plants. *Australian Journal of Plant Physiology*, *28*, 897–906.

Rillig, M. C., Wright, S. F., & Eviner, V. T. (2002). The role of arbuscular mycorrhizal fungi and glomalin in soil aggregation: Comparing effects of five plant species. *Plant and Soil*, *238*(2), 325–333. doi:10.1023/A:1014483303813

Roper, M. M., Gault, R. R., & Smith, N. A. (1995). Contribution to the N status of soil by free-living N2-fixing bacteria in a Lucerne stand. *Soil Biology & Biochemistry*, *27*(4-5), 467–471. doi:10.1016/0038-0717(95)98621-T

Savci, S. (2012). An agricultural pollutant: Chemical fertilizer. *International Journal of Environmental Sciences and Development*, *3*(1), 73–80. doi:10.7763/IJESD.2012.V3.191

Schachtman, D. P., Reid, R. J., & Ayling, S. M. (1998). Phosphorus uptake by plants: From soil to cell. *Plant Physiology*, *116*(2), 447–453. doi:10.1104/pp.116.2.447 PMID:9490752

Scott, M. F., Botigué, L. R., Brace, S., Stevens, C. J., Mullin, V. E., Stevenson, A., & Mott, R. (2019). A 3,000-year-old Egyptian emmer wheat genome reveals dispersal and domestication history. *Nature Plants*, *5*(11), 1120–1128. doi:10.103841477-019-0534-5 PMID:31685951

Sethi, S. K., Sahu J. K., Adhikary S. P. (2014). Microbial biofertilizers and their pilot-scale production. *Microbial Biotechnol. Progr. Trends,* 297.

Shaheen, S., & Sundari, K. (2013). Exploring the applicability of PGPR to remediate residual organophosphate and carbamate pesticides used in agriculture fields. *International Journal of Agriculture and Food Science Technology*, *4*(10), 947–954.

Singh, S. P., Shrivastava, S. K., Kolhe, S. S., Patel, J. R., & Bargali, S. S. (2007). Prospects of biofertilizers and organic manure utilization: A case study in Durg district. *Agricultural Science Digest*, *27*(3), 157–161.

Smith, S. E., & Read, D. J. (1997). *Mycorrhizal symbiosis*. Academic Press.

Smolander, A., & Sarsa, M. L. (1990). Frankia strains of soil under Betula pendula: Behaviour in soil and in pure culture. *Plant and Soil*, *122*(1), 129–136. doi:10.1007/BF02851920

Soumare, A., Diedhiou, A. G., Thuita, M., Hafidi, M., Ouhdouch, Y., Gopalakrishnan, S., & Kouisni, L. (2020). Exploiting Biological Nitrogen Fixation: A Route Towards a Sustainable Agriculture. *Plants*, *9*(8), 10–11. doi:10.3390/plants9081011 PMID:32796519

Spaink, H. P., Kondorosi, A., & Hooykaas, P. J. J. (Eds.). (1998). *The Rhizobiaceae*. Kluwer Academic Publishers. doi:10.1007/978-94-011-5060-6

Stamford, N. P., Ortega, A. D., Temprano, F., & Santos, D. R. (1997). Effects of phosphorus fertilization and inoculation of Bradyrhizobium and mycorrhizal fungi on growth of Mimosacaesalpiniaefolia in an acid soil. *Soil Biology & Biochemistry*, *29*(5-6), 959–964. doi:10.1016/S0038-0717(96)00240-4

Sundara, B., Natarajan, V., & Hari, K. (2002). Influence of phosphorus solubilizing bacteria on the changes in soil available phosphorus and sugarcane and sugar yields. *Field Crops Research*, *77*(1), 43–49. doi:10.1016/S0378-4290(02)00048-5

Sylvia, D. M. (1990). Inoculation of native woody plants with vesicular–arbuscular fungi for phosphate mine land reclamation. *Agriculture, Ecosystems & Environment*, *31*(3), 847–897. doi:10.1016/0167-8809(90)90224-2

Tawate, S., Gupta, R., & Jain, K. (2018). Technology Commercialization in Bio-fertilizer Firm: An Indian Case. *International Journal of Global Business and Competitiveness*, *13*(1), 65–74.

Thakur, P., & Singh, I. (2018). Biocontrol of soilborne root pathogens: An Overview. *Root Biology*, 181-220.

Thomas, L., & Singh, I. (2019). Microbial biofertilizers: types and applications. In *Biofertilizers for Sustainable Agriculture and Environment* (pp. 1–19). Springer. doi:10.1007/978-3-030-18933-4_1

Torrey, J. G. (1978). Nitrogen fixation by actinomycete-nodulated angiosperms. *Bioscience*, *28*(9), 586–592. doi:10.2307/1307515

Tripathi, S., Srivastava, P., Devi, R. S., & Bhadouria, R. (2020). Influence of synthetic fertilizers and pesticides on soil health and soil microbiology. In Agrochemicals Detection, Treatment and Remediation (pp. 25-54). Butterworth-Heinemann. doi:10.1016/B978-0-08-103017-2.00002-7

Triplett, E. (1996). Diazotrophic endophytes: Progress and prospects for nitrogen fixation in monocots. *Plant and Soil*, *186*(1), 29–38. doi:10.1007/BF00035052

Trujillo-Tapia, M. N., & Ramírez-Fuentes, E. (2016). Bio-fertilizer: An alternative to reduce chemical fertilizer in agriculture. *Journal of Global Agriculture and Ecology*, *4*(2), 99–103.

Unkovich, M. J., & Pate, J. S. (2000). An appraisal of recent field measurements of symbiotic N_2 fixation by annual legumes. *Field Crops Research*, *65*(2-3), 211–228. doi:10.1016/S0378-4290(99)00088-X

Unkovich, M. J., Pate, J. S., & Sanford, P. (1997). Nitrogen fixation by annual legumes in Australian Mediterranean agriculture. *Australian Journal of Agricultural Research*, *48*(3), 267–293. doi:10.1071/A96099

Varinderpal-Singh, Sharma, S., Kunal, Gosal, S. K., Choudhary, R., Singh, R., Adholeya, A., & Bijay-Singh. (2020). Synergistic Use of Plant Growth-Promoting Rhizobacteria, Arbuscular Mycorrhizal Fungi, and Spectral Properties for Improving Nutrient Use Efficiencies in Wheat (*Triticum aestivum* L.). *Communications in Soil Science and Plant Analysis*, *51*(1), 14–27. doi:10.1080/00103624.2019.1689259

Weekley, J., Gabbard, J., & Nowak, J. (2012). Micro-level management of agricultural inputs: Emerging approaches. *Agronomy (Basel)*, *2*(4), 321–357. doi:10.3390/agronomy2040321

White, D. P. (1941). Prairie soil as a medium for tree growth. *Ecology*, *22*(4), 398–407. doi:10.2307/1930714

Wilde, S. A. (1944). Mycorrhizae and silviculture. *Journal of Forestry*, *42*, 290.

Wood, T., & Cummings, B. (1992). Biotechnology and the future of VAM commercialization. In M. F. Allen (Ed.), *Mycorrhizal functioning* (pp. 468–487). Chapman and Hall.

Yadav, R., Ror, P., Rathore, P., & Ramakrishna, W. (2020). Bacteria from native soil in combination with arbuscular mycorrhizal fungi augment wheat yield and biofortification. *Plant Physiology and Biochemistry*, *150*, 222–233. doi:10.1016/j.plaphy.2020.02.039 PMID:32155450

Yu, G., Ran, W., & Shen, Q. (2016). Compost process and organic fertilizers application in China. In *Organic fertilizers – From Basic concepts to applied outcomes*. IntechOpen. doi:10.5772/62324

Section 2

Microbes and Sustainable Development

Chapter 5
Favorable Soil Microbes for Sustainable Agriculture

Umair Riaz
Soil and Water Testing Laboratory for Research, Bahawalpur, Pakistan

Laila Shahzad
Sustainable Development Study Center, Government College University, Lahore, Pakistan

Wajiha Anum
Regional Agricultural Research Institute, Bahawalpur, Pakistan

Anam Waheed
Government College University, Lahore, Pakistan

ABSTRACT

Beneficial microbes are used as the best alternative against the synthetic fertilizers and pesticides. The beneficial microbes not only help with plant growth, nutrition uptake, nitrogen fixation, but also help in acquiring the ions, not freely available to plants to uptake; these microbes also guard the plants by secreting toxic chemicals by inducing defense systems against pathogens. These microbes can provide best choice to look forward to sustainable agriculture and sustainable ecosystem. The addition of soil inoculants in the form of microorganisms or bio stimulants promise more environmentally friendly approaches for augmenting crop yields. The crop becomes less reliant on chemical fungicides and herbicides as many strains of microorganism have abilities of controlling pests. In this chapter, the interaction of beneficial plant bacteria, bio stimulants, effects on native microbial communities, and bacteria influencing economically important crops are discussed.

BENEFICIAL MICROORGANISMS FOR CROP PLANTS

The use of chemical fertilizers and pesticides in agriculture unlocked the hurdles to combat food insecurity issues. With the advancement of science and research, it manifested that the use of these chemical inputs, in a long run, found as curse in disguise due to adverse health impacts to human and animals as well as

DOI: 10.4018/978-1-7998-7062-3.ch005

in degradation of soil ecosystem. In this situation, the use of bio inoculants (beneficial microorganisms) is a hopeful option for having properties to fight against diseases, caused by various viruses, bacteria, nematodes, etc., and aggrandize the growth of plant (Pereg & McMillan, 2015). The growth of plant is depending on the soil type and the healthy plant-roots relationship, containing the soil microbial community. However, their association may give different response with different crop type, existing soil micro biota, geography, climatic conditions, bacterial strain and soil health.

The beneficial microorganism create association with plant roots which help to elevate nutrient uptake from soil by nitrogen fixation and reduced chemical fertilizers demand (Naik et al., 2019), potassium and phosphate solubilization and clampdown of pathogens, hence increase the agronomic productivity (Romano et al., 2020). These associational benefits are naturally evolved and may get benefits from a single microbial community or a consortium (Naik et al., 2019). The use of beneficial microorganisms make rhizosphere good in physical and chemical composition with longstanding fertility, maintaining healthy soil, combating framing costs and ensures sustainability (Debnath et al., 2019). Many species and strains of bacteria, fungi, yeast and actinomycetes are used as bio-inoculants. The most widely used bioincoulants are plant growth-promoting (PGP) microbes and bio-control agents.

Plant Growth Promoting Bacteria (PGPBs)

The plant-growth promoting bacteria (PGPBs) are involved in such symbiotic association which produces that kind of metabolites which make the process of P solublization, N_2 fixation, pathogens inactivity, and siderophores and phytohormones production, effortless for plant. In return the only thing the PGPBs get from plant is their energy essential carbohydrates (Nion, 2015; Saad et al., 2020; Sivasakthi et al., 2014). The extracellular matrix, exopolysaccharide are released by bacteria for the protection of its cell from toxic components and dehydration. These exopolysaccharides have cementing characteristics. Plants take advantage by bio-filming these exopolysaccharides which bind soil aggregates, fix nutrients and support water movement through roots (Meneses et al., 2017; Saad et al., 2020). The success of applied PGPB in agricultural crops is affect by the bacterial colonization with root, soil health and diffusion across plant roots and soil (Beauregard et al., 2013; Carvalhais et al., 2013). However, to select applicable PGPBs, following points must be considered; PGPBs should be according to regulatory agencies of country to ensure safe release and environmental safety, confirm the fitness of PGPB strain to efficient biological activity, selection to their particular environmental conditions (like organism working well in cold environment versus those well adopted for hot environment), the relationship of endophytic and rhizospheric bacteria in soil and the detailed associational analysis of PGPB with existing soil microbes and mycorrhizae (Glick,2012). Looking forward to sustainable agriculture, many strains of *Bacillus* and *Pseudomonas* sp. are used and are under screening to be used in treatment of heavy metal eradication, salinity problem and enhancing the drought tolerance of plant under water stress regions (Singh et al., 2018).

Plant Growth Promoting Fungi (PGPFs)

Like PGPBs, the use of plant growth-promoting fungi (PGPFs) are the microbes to boost-up organ enlargement, hormonal indicators and growth, by indole-3-cetic acid (IAA), cytokinins, and gibberellins etc. hormones, and reduced disease-associated risks (Chaudhary et al., 2018). Initially, the PGPF are not recognized by plant as symbionts, the pattern recognition receptors (PRRs) detect the microbe associated molecular patterns (MAMPs) like chitin, peptidoglycans, lipopolysaccharides and elicit the plant root and activates the plant hormones like jasmonic acid, salicylic acid, ethylene etc. and immune response (Pieterse et al., 2014; Venturi & Keel, 2016). When PGPFs inhabited roots, the plant immune and/or defense system tempted, named as induced systematic resistance (ISR) activated by jasmonic acid and ethylene signaling (Contreras-Cornejo et al., 2014; Venturi & Keel, 2016; Vos et al., 2013). The ISR is mediated by PGPFs while the second defense system systematic acquired resistance (SAR) is induced by pathogens and activated by salicylic signaling (Romera et al., 2019). The PGPFs itself secrete defensive composites in battle against phyto-pathogens like harzianic acid, anti-fungal chitinases, hydrogen cyanide, etc. (El-Maraghy et al., 2020). *Trichoderma* sp. and many other species of PGPFs are found advantageous for plant in abiotic stress conditions like salinity, heat, cold, etc. (Contreras-Cornejo et al., 2014).

Bio Control Agent

Keeping the eye on sustainability and healthy environment, the microbes as bio control agents are extensively put to crop plants to cease or eliminate the invasive and pathogenic species (Amsellem et al., 2017). Micro-organisms as bio control agent nullify the chemical cost against bio-pesticide, bionematicides and bio-insecticides, and its negative impact on environment, however, the approach is host-specific (Bourguet & Guillemaud, 2016); Khasa, 2017). Antibiosis, enzyme production, hyper parasitism, nutrient and space competition, resistance induction, etc., are involved in bio control agents' mechanism (Khasa, 2017). Biotechnology and genetic engineering technique are widely used to enhance and promote the bio control agent activity (Amsellem et al., 2017). This host-specific approach is started by introducing the parasite that is native to the invasive species region. The risk behind the approach is the inception of another invasive species, may be detrimental for native population or if the parasite can fail to survive in the environment. Another risk is that apparently the control seems fruitful, but after few to many years, the negative impacts suddenly manifest (Middleton, 2008). In rhizosphere, the bio control agents are released by three strategies: classical, inductive and integrated management (Hillel & Hatfield, 2005). In classical approach, the release of native enemy is on targeted pests or host. The reason is to maintain the pest-prey relationship, which was lost due to some reasons (Kumar, 2016). While the inductive approach is not native to environment, so it is the introduction of host-targeted bio control agent strain in each season and integrated approach is to protect and/or increase natural enemies. The microbial-biological control agents are achieved via two methods; first is to isolate the natural enemy from soil, roots of native environment and second is to isolate pathogen specific from suppressant soils. The screening trials are done, and the successful reproducible agents are then applied to specific disease, crop and environment (Köhl et al., 2019). The time is very important in application of bio control agents as the early application will lead to vigorous microbial interaction, however the delayed application may cause the death or inactivity of bio control agent due to less nutrients (Postma et al., 2008). To treat the attack of post-harvest fungi the bacterial strain of *Pantoea agglomerans* EPS125/CPA-2 and *Pseudomonas syringae* ESC10/ESC11 and fungal strains of *Aureobasidium pullulans* CF10, *Aureobasidium pullulans* Ach1-1, *Candida oleophila* I-182/Q and *Candida sake* CPA-1 are used. The bacterial strain of *Streptomyces griseoviridis* K61 and fungal strains of *Gliocladium catenulatum* J1446, *Gliocladium virens* GL21 are used as bio control of soil-borne pathogens (Montesinos & Bonaterra, 2009).

STRUCTURE AND FUNCTIONING OF PLANT ASSOCIATED SOIL BACTERIA

The well-structure community of microbes is associated with plant performing different functions according to their association. Bacteria secrete polysaccharides to build well-organized structure along the soil by aggregate formation by utilizing more carbon from roots (Gliński et al., 2011). The functions of the soil-associated bacteria are differing according to their association with plants. The symbiotic association of bacteria with plant function as suppression of pathogens, nutrient acquisition, plant growth, soil respiration, stress tolerant, bio control agent and bio-fertilizer (Bulgarelli et al., 2013).

MECHANISM OF PLANT BACTERIA INTERACTION

Crowded with bacteria, fungi, algea, protozoa and other microbial life-forms; the microbes are the most commonly found micro-organsims in rhizosphere (Schoenborn et al., 2004). The soil microbes are unevenly distributed, the rhizospheric soil have more microbes than the surrounded soil to fix nutrients by roots (Badri et al., 2009). In all the micro biota of soil, the effective plant growth promoting microbes (PGPMs) provide productive, defensive and growth benefits to the plants either directly or indirectly (Sharma et al., 2011) . Not only this but also the PGPMs enhance the lignin catabolism and photosynthetic activity of the crop plants (Naik et al., 2019). The plant-bacteria interaction usually involves hormonal induction, resisting pathogenic strain activity, pathogen attack and nutrient absorption by plants (Mendes et al., 2013; Verbon & Liberman, 2016). The direct and indirect mechanism of plant bacterial interaction is depicted in Figure 1.

Figure 1. Indicating direct and indirect plant-bacteria interaction
(Shahzad et al., 2020)

Direct Mechanism

The direct mechanism is subcategorized into two, based on their performance; as bio fertilizers by fixing nitrogen, solubilizing K and P, root colonization and siderophore production and as phytohormone producers by production of IAA, cytokinin, ethylene, abscisic acid and GA_3 production.

Bio Fertilizer

The dependency on synthetic chemicals for crop protection and enhanced production provide benefits at the required time but are directly associated with the damages to environment, human health and animals. To avoid these damages, there is a need to shift towards more natural and nature friendly options. Microbes as bio-fertilizers are the best alternative of chemical fertilizers for nutrient capturization and mobilization as well as for the conversion of non-usable soil nutrient into usable form (Debnath et al., 2019).

To achieve maximum crop yield, the nitrogen is most important component of plant productivity and nutrition. The nitrogen in atmosphere and rhizosphere is not readily available for plant to uptake, the beneficial microbes, specifically *Rhizobia* sp. are the most common bacteria used for conversion of triple bonded N_2 compound into the usable form (Olanrewaju et al., 2017) . The bacteria work in both symbiotic (*Frankia* and *Rhizobium*) and free living (*Azospirillum* and *Azotobactor*) form (Debnath et al., 2019). The energy consuming nitrogenase activity uses 16 moles of ATP to reduce each mole of nitrogen. The diazotrophic bacteria fixes nitrogen of the soil by nitrogenase enzyme (*nif*, Nitrogen fixation) containing FeMo-protein (dinitrogenase) to reduce N_2 and NH_3, and Fe-protein (dinitrogenaseredutase) for high reducing power to convert gaseous nitrogen to ammonia (Souza et al., 2015). The structural gene in *nif* triggers molybdenum, Fe-protein and other regulatory genes, to functionally activate and synthesize the enzyme (Kundan et al., 2015). In this process plant give carbon to the bacteria via photosynthesis and bacteria fix nitrogen in the roots (Olanrewaju et al., 2017). The nitrogenase activity is inhibited by oxygen but for *Rhizobium* respiration, oxygen is necessary. For excess oxygen to bacteria and no oxygen for better nitrogenase activity it would be possible to introduce microbial hemoglobin to bind the oxygen (Glick, 2012).

The root colonization mechanism in rhizosphere is different from N_2 fixation. The plant first recognize the microbe via roots exudation (enrich in carbon nutrition) as an enticement for bacteria as a source of energy. After recognition, the microbes start penetrating around the roots; and start proliferation in, and around the roots. This colonization help in bio control tolerance against stresses and improve plant growth (Berg, 2009; Harman, 2011; Hermosa et al., 2012; Sachdev & Singh, 2018).

In natural soil ecosystem, the potassium and phosphate are less bioavailable to plant to uptake. The phosphate-solubilizing bacteria (PSB) and potassium-solubilizing bacteria (KSB) work on same principle. The P and K are present in abundance in soil but in non-usable or insoluble form. The phosphate present in insoluble inositol, phosphotriesters, etc., and potassium is present in mica, illite, etc., forms. The mechanism involves the utilization of mineral dissolving compounds; it will reduce pH by proton release or carboxyl and hydroxyl ion chelate cation. The bacteria secrete organic acid resulting in the acidification of microbial cell and its surrounding and then release phosphate and/or potassium ion. So the production of K and P ion is inversely related to pH, the lower the pH, the higher will be the ion liberation (Glick, 2012; Kalayu, 2019).

Another important process of plant-microbe interaction is siderophore production. The siderophores are iron-binding peptide molecules. The bacteria by producing high affinity for ferric ion siderophore lower the amount of ferric for pathogens hence make them unable to behave like pathogens (Shen et al., 2013). The siderophores, a low molecular iron chelates are secreted by bacteria which have high affinity for Fe^{+3} ion. After binding through plant membrane, plant gets iron rich siderophore by direct uptake of Fe-siderophore compound, by chelate formation or by ligand exchange method (Ahemad & Kibret, 2014).

Phyto-Hormones Production

In rhizosphere, the plant-bacteria interaction is playing important role in producing and inducing phytohormones. Indole-3-acetic acid, an auxin, is an important hormone for plant cell division, root development, seed stimulation, photosynthesis, resistance and many more (Spaepen & Vanderleyden, 2011). The auxin synthesis occurs in shoot apex and is transported to root apical meristem of plant. The rhizobacteria synthesize and release auxin as secondary metabolite, the endogenous plant auxin regulate by secreted microbial auxin and hence regulate the plant development.

Gibberellin and cytokinin are largely produced in plants however, the bacterial producing hormones provide purified exogenous hormones in addition to growing plants. The cytokinin is produced in root tip and help in root elongation, cell division, seed germination, etc. The bacteria producing cytokinin help plant to grow in dry seasons by closing stomata by accumulation of abscisic acid (ABA) in leaves (Mishra et al., 2017; Olanrewaju et al., 2017).

Indirect Mechanism

Stress Tolerance

Almost all plants produce ethylene as a hormone to enhance fruit ripening, root initiation, seed germination, synthesis of other plant hormones, etc. In environmental stresses like drought, salinity, organic pollution, high temperature, etc., the ethylene production is increased and negatively influences the plant health and reduces crop production (Glick et al., 2013). In that particular stress situation, bacteria produce ACC (1-Aminocyclopropane-1-carboxylate) deaminase to lower the ethylene concentration. The plant roots secrete ACC synthase that is taken up by bacteria and synthesize ACC. The plant acquires ACC from neighboring soil bacteria. On the same time, bacteria hydrolyze ACC into ammonia and 2-oxobutanoate that make the low concentration of ACC in bacterial cell and high concentration in plants (Kang et al., 2010).

Under drought stress, the electron transport system lead to the formation of reactive oxygen species (ROS) like OH^-, which start destructing chloroplast and chlorophyll and untimely start damaging the cells rapidly. The rhizobacteria produce antioxidant compounds like peroxidase (POX), superoxide dismutase (SOD) and catalase (CAT). These compounds reduce the effect of ROS and help plant to withstand in drought stress (Ghorbanpour et al., 2013).

In cold stress conditions, many fungal pathogens are at peak to attack the plant. The bacteria release anti-freeze protein to enhance cold tolerance of crops. At low temperature, outside the bacterial cell the ice-nucleation occurs by formation of ice crystals that helps to protect and prevent the bacterial cell at low temperature. The bacteria around the roots produce antifreeze protein to protect bacterial cell as well as roots from damage via freeze thaw (Garnham et al., 2008).

Biological Control

The bacteria as biological control agents are widely used. Induced systematic resistance, a plant immune system, is the phenomenon that is triggered by bacteria. It is not a process involving the pathogen killing by bacteria but involves the prime of plant immune system. The non-pathogenic bacteria signals the plant by outer lipopolysaccharide, salicylic acid, ethylene signaling, etc., the plant exudation tracked by bacteria and these signaling activate the ISR against non-pathogenic bacteria, and protect the plant from pathogenic attack (Bakker et al., 2007).

As an indirect biocontrol mechanism, competition is found to be useful for the plant protection. For nutrients, space, and energy or for binding to the roots, the pathogenic and non-pathogenic bacteria are in continuous competition. The competition between bacteria indirectly benefits the plant for growth, disease protection and other stress control mechanisms however, the chance is equal, that if pathogenic bacteria outcompete the non-pathogenic bacteria, the scenario will be opposite (Innerebner et al., 2011).

Bacteria to control root diseases of plant produce an anti-fungal, anti-biotic compound, hydrogen cyanide (HCN). Not only HCN but some cell wall- degrading enzymes are also produced to cease the phytopathogen activity by inhibiting cytochrome *c* oxidase and necessary metallo-enzymes (Michelsen et al., 2012).

CHARACTERISTICS OF BENEFICIAL SOIL BACTERIA

Following are the characteristics of beneficial soil bacteria:

1. The soil bacteria, either free-living or in symbiotic relationship, function in wide range of environmental conditions; improve nutrients, eradicate heavy metals and many more.
2. The beneficial bacteria are widely used as bio-inoculants to improve the plant growth.
3. The beneficial soil microbes help the plant to fight against pathogens as bio control agent by releasing various growth inhibiting hormones and enzymes.
4. Other than self-bio control agents, beneficial microbes also alerts the plant from pathogens by inducing plant's immunity system for example, induced systematic resistance (ISR).
5. In rhizosphere, the beneficial bacteria make the bio-availability of non-usable molecule into useable ions like phosphorus solubilization.
6. The bacteria also help in mobilizing iron, an essential nutrient for major plant processes like photosynthesis.
7. Beneficial bacteria also release hormones which help in plant organ development, like auxin.

AGRICULTURAL USE OF BENEFICIAL SOIL BACTERIA

The smooth ecosystem functioning is reliant upon numerous factors. One of them is the rhizosphere that embodies diverse habitats on the planet. Several developments are continuously in process in rhizosphere for sustaining its functioning. Few of them are; the presence of microbes and their activities, production of root exudates, transformation of nutrients, genetic exchange and gradient diffusion occurring through the substrates. Hence there is a need to know rhizosphere utilities in more depth for extra effectual management strategies and extract the likely benefits from rhizosphere. The anticipated benefits are agricultural and forest sustainability, preservation of biota, minimization of changing climate effects along with improved water quality which can be acquired by handling the rhizosphere accordingly (Trabelsi & Mhamdi, 2013).

As the agricultural soils are becoming deficits in nutrients and are losing their fertility status (Riaz et al., 2020a). There is a dire need to maintain the fertility by any means. Soil biota can render benefits which otherwise are not attainable by simply depending on the synthetic chemicals. The benefits are achievable by altering/adding substances/microbes in soil rhizosphere. The addition of beneficial rhizopheric or endophyto microbes in the soil for enhancing the plant growth and development is termed as plant bioinoculation. The use of inoculants/microorganisms in crops is attractive for farmers as they decrease the use of chemicals like synthetic fertilizers (Riaz et al., 2020b) and pesticides for adequate plant growth and disease control. Owing to their benefits, inoculants are being encouraged and commercialized for numerous crops (Berg, 2009). However any alteration (addition of microorganism for any purpose) may disturb the native microbial community, hence before selecting the type of inoculant, knowledge about its pros and cons is requisite.

Azospirillum is the most considered PGPR (Plant growth promoting bacteria) with abundant agronomic benefits (Dobbelaere et al., 2001; Okon & Labandera-Gonzalez, 1994). As a bio inoculant, it synthesizes phytohormones especially IAA, **3**-IAA (indole-3-acetic acid) (Spaepen et al., 2007). But its addition also impacts the native microbial community by mostly disturbing root development which consecutively influences the production and release of root exudates (Jacoud et al., 1998). A research indicated that under application of *Azospirillum lipoferum* CRT1 (a commercial *Azospirillum* strain) in maize field, a more variable and genetically assorted rhizobacterial community was observed amongst individual field-grown maize plants and between sampling times without modifying the total number of root bacteria (Baudoin et al., 2009; Schumpp & Deakin, 2010).

Similarly, in the phylum Glomeromycota, Arbuscular mycorrhizal fungi (AMF) are grouped which form a mutual relationship with plants and occupy a large soil volume (Schwarzott & Walker, 2001). They supply the plants with minerals and water and in return consume carbon produced by host plants (Smith & Read, 2010). The native microbial community that comes in contact with ERM (extraradical mycillium) of AMF forms a mycorrhizosphere and can be affected positively (Albertsen et al., 2006), negatively (Hang et al., 2005), or sometimes are not affected (Cavagnaro et al., 2006). The response depends on the nature and type of both AMF and the inherent microorganism (Artursson et al., 2006). In few examples, a more indirect synergic affect takes place between plant, AMF and native bacteria (Barea, 1997). For example, Gamalero et al. (2002) revealed that *P fluorescens* A6RI and *G mosseae* BEG12 affected the root morphogenesis under the different ratios of soil and sand mixtures. *P fluorescens* significantly increased the root characters and plant growth in more fertile soil while similar results were obtained by *G mosseae* in less fertile soil.

Plant Bio-Stimulation

Plant biostimulants are defined as the substance/microorganism intended at stimulating the natural processes of plants, boost the nutrient uptake and inclusive performance of the plant (Du Jardin, 2015). Bio stimulants are applied to rhizosphere or plants, while they indirectly reduce the use of chemical inputs for maintaining/recovering the natural equipoise in agro ecosystems (Woo & Pepe, 2018). They make the plant more tolerant to the diseases, surge nutrient uptake efficiency and yield a more quality crop.

Ahmad et al. (2008) revealed that microorganisms interrelate in behaviors with each other or with the host plants, for instance, the fungi may act as parasites or can be involved in mutualism. The bacteria are distributed in the rhizosphere and can be vertically distributed through seeds (Du Jardin, 2015).

Protein hydrolysates (PHs) are the mixtures of amino acids, polypeptides and oligopeptides and have gained popularity as plant bio stimulants in vegetable crop production (Colla et al. 2015a). Studies have revealed that the PHs (leaf or root application) can develop tolerance against environmental stresses like drought, thermal stress, nutrient deficiency, alkalinity and salinity (Botta,2012; Cerdán et al., 2008; Colla et al., 2013a; Colla et al., 2014; Lucini et al., 2015; Petrozza et al., 2014). The positive dealing of PHs with stresses is often linked to their ability in changing hormones networks, osmatic adjustments and to modify the oxidative stresses (Colla et al., 2015a; Lucini et al., 2015).

Under the alkaline and saline environments the crop growth and development is extremely exaggerated but under such cases the use of AMF fungi and *Trichoderma* spp. are helpful in incapacitating limitations/affects instigated by salt stress. For example, in vegetables, the use of AMF and *Trichderma* spp. upsurges nutrient uptake by greater and effective root area and enhance solubilisation of micronutrients. They also produce volatiles, small peptides and metabolites which have hormonal activity like auxin analogs or indole-3-acetic acid (for *Trichoderma* spp.) (Giovannetti et al., 2001 ; López-Bucio et al., 2015; Rouphael et al., 2015a; Rouphael et al., 2015b) . A research also revealed that under non-stressed conditions, the combined inoculation of AMF and *Trichoderma atroviride* improves the growth of certain vegetables (Colla et al., 2015b).

Beneficial Plant-Associated Microorganisms as Bio-Fungicides

The advancement in microbiological research has open ways to utilize the microorganisms for controlling plant diseases by colonizing around the plant pathogens and making them harmless to the plant. Plant growth promoting rhizobacteria (PGPR) also have fungicide characteristics and suppresses the fungal infections by limiting the activity of the pathogens (Ardakani et al., 2009). The *Pseudomonas* spp. and *Bacillus* spp. are amongst the imperative bio control agents (Chen et al., 2000). Both of them have been utilized in controlling the diseases caused by *Fusarium, Sclerotinia, Gaeumannomyces, Rhizoctonia* and *Pythium* (Bacon et al., 2001; Schmiedeknecht et al., 1998; Zhang et al., 1996) by producing antibiotic substances mostly cyclic peptides or dipeptides (Loeffler et al., 1990).

Botrytis is a type of fungus that grows as parasite or saprophytes on several economical important agriculture crops and forest plants. Bacterial genus (few of them) acts as bio control agent for controlling botrytis. Ge et al. (2016) reported that *B. cinerea* can be controlled through the anti-fungal activity of genus bacillus. Similarly a research claims that grey mould can be reduced upto 85% in strawberries by the action of *Bacillus subtilis* (S1-0210) (Hang et al., 2005). *Bacillus* spp. effectively control plant diseases because they have capability of producing broad spectrum antibiotics and have endosporic nature with extended shelf life (Emmert & Handelsman, 1999). *Bacillus (cereus)* induces resistance against grey mould (*B. cinerea*) (Nie et al., 2017). Some of the commercial products are Kodiak HB (from B. subtilis GB03) and Serenade (from *B. subtilis* QST-713) that have confirmed abilities to control grey mould (Mahaffee et al., 1993; Marrone, 2002; Percival et al., 2016). *Pseudomonas* is another bacterium that has ability to control grey mould (Gao et al., 2018; Wallace et al., 2018). Two antagonist bacterial isolates i.e. *Burkholderia cepacia*, T1A-2B, and *Pseudomonas* sp., T4B-2A), were tested against the disease caused by *Rhizoctonia solani* and *Sclerotium rolfsii* in tomato. Both strains were effective in reducing severity and incidence of the disease (De Curtis et al., 2010).

Beneficial Plant-Associated Microorganisms as Bio-Herbicides

Pesticide application in field crops is posing threat to the environment as well as agricultural sustainability is at the verge of damage. An alternative option is use of microbes for protecting plants against pests by enhancing their resistance to insect and diseases and weeds (Gadhave et al., 2016).

Weeds are a severe delinquent in field crop production and pose economical threats. Some weeds also release chemical compounds known as allelochemicals (Anum et al., 2016) that may influence the crop growth positively or negatively. Regardless of the situation, weeds have always been unwanted intruder. For their control certain selective and non-selective weedicides have been employed. However, the continuous application of weedicides can damage the ecological set up to irreversible way. A friendly approach is the use of microorganism for controlling weeds.

In this regard DRB (deleterious rhizobacteria) can be used for weed control, as they the bacteria that are non-parasitic in nature and they can slow down the plant growth without attacking the root tissues (Kremer, 2006). This character makes it attractive for investigation about its bio herbicidal potential (Kremer, 2005). *Pseudomonas fluorescens* D7 is a soil-applied DRB bio-herbicide formulated as a liquid suspension or encapsulated in clay that effectively suppresses downy brome (*Bromus tectorum*) in cereal grain crops (Kennedy et al., 1991).

FUTURE CHALLENGES

The use of beneficial bacteria is a good option for managing agricultural productivity without posing threats to the environments; however there exist few critical points that need to be addressed. First of all, the characteristics like molecular setup, chemical processes, mode of action, shelf life, genetic analysis and genomics of the specific strains needs to be carried out, secondly the movement and mechanism in rhizosphere and how symbiosis takes place. No doubt the ongoing investigation is identifying the microorganisms that may be beneficial but an extensive research is still missing for understanding their all dimensions. Moreover the information regarding the affects is haphazard in few cases. The strains caused positive as well as negative effects. This gap needs to be researched again for their confirmed usage in agriculture. Lastly, the commercialization and acceptance is a big complication in its implementation at farmers' level. In this regard the government, industrialists, microbiologists and extension workers should cooperate and set up a technique which can make the best possible use of microorganisms for promoting crop production.

SUMMARY

Microorganisms are crucial in almost every function of life and running the life cycle of living organisms. Similar is the case in plants and its allied environment. As soil contains a huge number of micro biota and microorganism, researchers are continuously exploring the major characteristics, their influence on the plants, role in soil make up, mode of action and shelf life of soil microorganisms. Owing to the factors, an extensive research is present and further needed to explore the role of beneficial bacteria in agriculture production. The addition of soil beneficial bacteria and those influencing the agricultural crops positively promise a rigorous and effective process in enhancing the growth and development of crop as well as soil community. Nowadays commercial soil inoculants are available in the market and recommended instead of chemical growth promoters, however they also alter the native microbial communities. Similar is the case of plant bio-stimulants, and their effectiveness crucially depends upon how they retaliate with the native ecosystem. Plant bio stimulation is successful if microorganism/bio stimulant makes an uninterrupted mutualism with the plant. Most importantly, use of beneficial bacteria is becoming popular as they can take place of the synthetic chemicals in near future. They act as bio fungicides, herbicides and control plant diseases and unwanted weeds through their natural apparatus. All of the above mentioned points ensure that plant beneficial bacteria are an emerging too for controlling many crucial problems in agriculture production. However, they synchronic relationship among the bacteria and targeted plant has to be understood more deeply. Such species should be researched and chosen that influence the non-targeted organisms (soil microbione). Their long term effect and sustainability and feasibility are also needed to research upon extensively so that maximum benefits can be extracted through natural means.

CONCLUSION

Soil microorganisms play a beneficial role for plant growth by developing associations with their roots in soils. Plant growth promoting bacteria and fungi are such associations which are significant for plants and overall ecosystem sustainability. Such favorable microbes provide long term solutions for sustainable agriculture and are more environment friendly than chemical fertilizers and pesticides.

REFERENCES

Ahemad, M., & Khan, M. S. (2012a). Effect of fungicides on plant growth promoting activities of phosphate solubilizing Pseudomonas putida isolated from mustard (*Brassica compestris*) rhizosphere. *Chemosphere*, *86*(9), 945–950. doi:10.1016/j.chemosphere.2011.11.013 PMID:22133911

Ahemad, M., & Khan, M. S. (2012b). Evaluation of plant-growth-promoting activities of rhizobacterium *Pseudomonas putida* under herbicide stress. *Annals of Microbiology*, *62*(4), 1531–1540. doi:10.100713213-011-0407-2

Ahemad, M., & Kibret, M. (2014). Mechanisms and applications of plant growth promoting rhizobacteria: Current perspective. *Journal of King Saud University. Science*, *26*(1), 1–20. doi:10.1016/j.jksus.2013.05.001

Ahmad, I., Pichtel, J., & Hayat, S. (Eds.). (2008). *Plant-bacteria interactions: strategies and techniques to promote plant growth*. John Wiley & Sons. doi:10.1002/9783527621989

Albertsen, A., Ravnskov, S., Green, H., Jensen, D. F., & Larsen, J. (2006). Interactions between the external mycelium of the mycorrhizal fungus *Glomus intraradices* and other soil microorganisms as affected by organic matter. *Soil Biology & Biochemistry*, *38*(5), 1008–1014. doi:10.1016/j.soilbio.2005.08.015

Amsellem, L., Brouat, C., Duron, O., Porter, S. S., & Facon, B. (2017). Networks of invasion: Empirical evidence and case studies. *Advances in Ecological Research*, *57*, 99–146. doi:10.1016/bs.aecr.2016.10.005

Antoun, H., Beauchamp, C. J., Goussard, N., Chabot, R., & Lalande, R. (1998). Potential of Rhizobium and Bradyrhizobium species as plant growth promoting rhizobacteria on non-legumes: effect on radishes (*Raphanus sativus* L.). In *Molecular microbial ecology of the soil* (pp. 57–67). Springer. doi:10.1007/978-94-017-2321-3_5

Anum, W., Naeem, M., Tanveer, A., Ali, H. H., Nazir, M. Q., Hanif, Z., & Kashif, M. S. (2016). Effects of African purslane (*Zaleya pentandra* L.) on germination and seedling growth of maize. *Allelopathy Journal, 39*(1), 83–91.

Ardakani, S. S., Heydari, A., Khorasani, N. A., Arjmandi, R., & Ehteshami, M. (2009). Preparation of new biofungicides using antagonistic bacteria and mineral compounds for controlling cotton seedling damping-off disease. *Journal of Plant Protection Research, 49*(1). Advance online publication. doi:10.2478/v10045-009-0007-3

Artursson, V., Finlay, R. D., & Jansson, J. K. (2006). Interactions between arbuscular mycorrhizal fungi and bacteria and their potential for stimulating plant growth. *Environmental Microbiology, 8*(1), 1–10. doi:10.1111/j.1462-2920.2005.00942.x PMID:16343316

Bacon, C. W., Yates, I. E., Hinton, D. M., & Meredith, F. (2001). Biological control of *Fusarium moniliforme* in maize. *Environmental Health Perspectives, 109*(suppl 2), 325–332. PMID:11359703

Badri, D. V., Weir, T. L., van der Lelie, D., & Vivanco, J. M. (2009). Rhizosphere chemical dialogues: Plant–microbe interactions. *Current Opinion in Biotechnology, 20*(6), 642–650. doi:10.1016/j.copbio.2009.09.014 PMID:19875278

Bakker, P. A., Pieterse, C. M., & Van Loon, L. C. (2007). Induced systemic resistance by fluorescent Pseudomonas spp. *Phytopathology, 97*(2), 239–243. doi:10.1094/PHYTO-97-2-0239 PMID:18944381

Barea, J. M. (1997). *Mycorrhiza/bacteria interactions on plant growth promotion. In Plant growth-promoting rhizobacteria, present status and future prospects*. OECD.

Baudoin, E., Nazaret, S., Mougel, C., Ranjard, L., & Moënne-Loccoz, Y. (2009). Impact of inoculation with the phytostimulatory PGPR *Azospirillum lipoferum* CRT1 on the genetic structure of the rhizobacterial community of field-grown maize. *Soil Biology & Biochemistry, 41*(2), 409–413. doi:10.1016/j.soilbio.2008.10.015

Beauregard, P. B., Chai, Y., Vlamakis, H., Losick, R., & Kolter, R. (2013). *Bacillus subtilis* biofilm induction by plant polysaccharides. *Proceedings of the National Academy of Sciences of the United States of America, 110*(17), E1621–E1630. doi:10.1073/pnas.1218984110 PMID:23569226

Berg, G. (2009). Plant–microbe interactions promoting plant growth and health: Perspectives for controlled use of microorganisms in agriculture. *Applied Microbiology and Biotechnology, 84*(1), 11–18. doi:10.100700253-009-2092-7 PMID:19568745

Boddey, R. M., Baldani, V. L., Baldani, J. I., & Döbereiner, J. (1986). Effect of inoculation of *Azospirillum* spp. on nitrogen accumulation by field-grown wheat. *Plant and Soil, 95*(1), 109–121. doi:10.1007/BF02378857

Boddey, R. M., Polidoro, J. C., Resende, A. S., Alves, B. J., & Urquiaga, S. (2001). Use of the15N natural abundance technique for the quantification of the contribution of N2 fixation to sugar cane and other grasses. *Functional Plant Biology*, *28*(9), 889–895. doi:10.1071/PP01058

Botta, A. (2012). Enhancing plant tolerance to temperature stress with amino acids: an approach to their mode of action. *I World Congress on the Use of Biostimulants in Agriculture 1009*, 29-35.

Bourguet, D., & Guillemaud, T. (2016). The hidden and external costs of pesticide use. In *Sustainable agriculture reviews* (pp. 35–120). Springer. doi:10.1007/978-3-319-26777-7_2

Bulgarelli, D., Schlaeppi, K., Spaepen, S., Van Themaat, E. V. L., & Schulze-Lefert, P. (2013). Structure and functions of the bacterial microbiota of plants. *Annual Review of Plant Biology*, *64*(1), 807–838. doi:10.1146/annurev-arplant-050312-120106 PMID:23373698

Carvalhais, L. C., Dennis, P. G., Fan, B., Fedoseyenko, D., Kierul, K., Becker, A., von Wiren, N., & Borriss, R. (2013). Linking plant nutritional status to plant-microbe interactions. *PLoS One*, *8*(7), e68555. doi:10.1371/journal.pone.0068555 PMID:23874669

Castellane, T. C. L., Otoboni, A. M. M. B., & Lemos, E. G. D. M. (2015). Characterization of exopolysaccharides produced by rhizobia species. *Revista Brasileira de Ciência do Solo*, *39*(6), 1566–1575. do i:10.1590/01000683rbcs20150084

Cavaglieri, L., Orlando, J. R. M. I., Rodriguez, M. I., Chulze, S., & Etcheverry, M. (2005). Biocontrol of Bacillus subtilis against Fusarium verticillioides in vitro and at the maize root level. *Research in Microbiology*, *156*(5-6), 748–754. doi:10.1016/j.resmic.2005.03.001 PMID:15950130

Cavagnaro, T. R., Jackson, L. E., Six, J., Ferris, H., Goyal, S., Asami, D., & Scow, K. M. (2006). Arbuscular mycorrhizas, microbial communities, nutrient availability, and soil aggregates in organic tomato production. *Plant and Soil*, *282*(1-2), 209–225. doi:10.100711104-005-5847-7

Cerdán, M., Sánchez-Sánchez, A., Oliver, M., Juárez, M., & Sánchez-Andreu, J. J. (2008). Effect of foliar and root applications of amino acids on iron uptake by tomato plants. *IV Balkan Symposium on Vegetables and Potatoes 830*, 481-488.

Chaudhary, D., Narula, N., Sindhu, S. S., & Behl, R. K. (2013). Plant growth stimulation of wheat (Triticum aestivum L.) by inoculation of salinity tolerant Azotobacter strains. *Physiology and Molecular Biology of Plants*, *19*(4), 515–519. doi:10.100712298-013-0178-2 PMID:24431520

Chaudhary, S., Shankar, A., Singh, A., & Prasad, V. (2018). Usefulness of Penicillium in Enhancing Plants Resistance to Abiotic Stresses: An Overview. In *New and Future Developments in Microbial Biotechnology and Bioengineering* (pp. 277–284). Elsevier. doi:10.1016/B978-0-444-63501-3.00017-X

Chen, C., Belanger, R. R., Benhamou, N., & Paulitz, T. C. (2000). Defense enzymes induced in cucumber roots by treatment with plant growth-promoting rhizobacteria (PGPR) and Pythium aphanidermatum. *Physiological and Molecular Plant Pathology*, *56*(1), 13–23. doi:10.1006/pmpp.1999.0243

Choudhary, D. K., Varma, A., & Tuteja, N. (Eds.). (2016). *Plant-microbe interaction: an approach to sustainable agriculture*. Springer Singapore. doi:10.1007/978-981-10-2854-0

Colla, G., Nardi, S., Cardarelli, M., Ertani, A., Lucini, L., Canaguier, R., & Rouphael, Y. (2015a). Protein hydrolysates as biostimulants in horticulture. *Scientia Horticulturae*, *196*, 28–38. doi:10.1016/j.scienta.2015.08.037

Colla, G., Rouphael, Y., Canaguier, R., Svecova, E., & Cardarelli, M. (2014). Biostimulant action of a plant-derived protein hydrolysate produced through enzymatic hydrolysis. *Frontiers in Plant Science*, *5*, 448. doi:10.3389/fpls.2014.00448 PMID:25250039

Colla, G., Rouphael, Y., Di Mattia, E., El-Nakhel, C., & Cardarelli, M. (2015b). Co-inoculation of Glomus intraradices and Trichoderma atroviride acts as a biostimulant to promote growth, yield and nutrient uptake of vegetable crops. *Journal of the Science of Food and Agriculture*, *95*(8), 1706–1715. doi:10.1002/jsfa.6875 PMID:25123953

Colla, G., Rouphael, Y., Jawad, R., Kumar, P., Rea, E., & Cardarelli, M. (2013a). The effectiveness of grafting to improve NaCl and CaCl2 tolerance in cucumber. *Scientia Horticulturae*, *164*, 380–391. doi:10.1016/j.scienta.2013.09.023

Collins, D. P., & Jacobsen, B. J. (2003). Optimizing a *Bacillus subtilis* isolate for biological control of sugar beet Cercospora leaf spot. *Biological Control*, *26*(2), 153–161. doi:10.1016/S1049-9644(02)00132-9

Contreras-Cornejo, H. A., Macías-Rodríguez, L., López-Bucio, J. S., & López-Bucio, J. (2014). Enhanced plant immunity using Trichoderma. In *Biotechnology and Biology of Trichoderma* (pp. 495–504). Elsevier. doi:10.1016/B978-0-444-59576-8.00036-9

Dal Cortivo, C., Barion, G., Visioli, G., Mattarozzi, M., Mosca, G., & Vamerali, T. (2017). Increased root growth and nitrogen accumulation in common wheat following PGPR inoculation: Assessment of plant-microbe interactions by ESEM. *Agriculture, Ecosystems & Environment*, *247*, 396–408. doi:10.1016/j.agee.2017.07.006

Danish, S., Zafar-ul-Hye, M., Mohsin, F., & Hussain, M. (2020). ACC-deaminase producing plant growth promoting rhizobacteria and biochar mitigate adverse effects of drought stress on maize growth. *PLoS One*, *15*(4), e0230615. doi:10.1371/journal.pone.0230615 PMID:32251430

De Curtis, F., Lima, G., Vitullo, D., & De Cicco, V. (2010). Biocontrol of Rhizoctonia solani and Sclerotium rolfsii on tomato by delivering antagonistic bacteria through a drip irrigation system. *Crop Protection (Guildford, Surrey)*, *29*(7), 663–670. doi:10.1016/j.cropro.2010.01.012

de Salamone, I. G., Döbereiner, J., Urquiaga, S., & Boddey, R. M. (1996). Biological nitrogen fixation in Azospirillum strain-maize genotype associations as evaluated by the 15 N isotope dilution technique. *Biology and Fertility of Soils*, *23*(3), 249–256. doi:10.1007/BF00335952

Debnath, S., Rawat, D., Mukherjee, A. K., Adhikary, S., & Kundu, R. (2019). Applications and Constraints of Plant Beneficial Microorganisms in Agriculture. In *Rhizosphere and Soil Microbes-Utilization in Agriculture and Industry Under Current Scenario*. IntechOpen.

Devescovi, G., Aguilar, C., Majolini, M. B., Marugg, J., Weisbeek, P., & Venturi, V. (2001). A siderophore peptide synthetase gene from plant-growth-promoting Pseudomonas putida WCS358. *Systematic and Applied Microbiology*, *24*(3), 321–330. doi:10.1078/0723-2020-00063 PMID:11822666

Di Benedetto, N. A., Corbo, M. R., Campaniello, D., Cataldi, M. P., Bevilacqua, A., Sinigaglia, M., & Flagella, Z. (2017). The role of plant growth promoting bacteria in improving nitrogen use efficiency for sustainable crop production: A focus on wheat. *AIMS Microbiology*, *3*(3), 413–434. doi:10.3934/microbiol.2017.3.413 PMID:31294169

Dobbelaere, S., Croonenborghs, A., Thys, A., Ptacek, D., Vanderleyden, J., Dutto, P., & Brener, S. (2001). Responses of agronomically important crops to inoculation with Azospirillum. *Functional Plant Biology*, *28*(9), 871–879. doi:10.1071/PP01074

Du Jardin, P. (2015). Plant biostimulants: Definition, concept, main categories and regulation. *Scientia Horticulturae*, *196*, 3–14. doi:10.1016/j.scienta.2015.09.021

El-Maraghy, S. S., Tohamy, T. A., & Hussein, K. A. (2020). Role of Plant-Growth Promoting Fungi (PGPF) in Defensive Genes Expression of Triticum aestivum against Wilt Disease. *Rhizosphere*, *15*, 100223. doi:10.1016/j.rhisph.2020.100223

Emmert, E. A., & Handelsman, J. (1999). Biocontrol of plant disease: A (Gram-) positive perspective. *FEMS Microbiology Letters*, *171*(1), 1–9. doi:10.1111/j.1574-6968.1999.tb13405.x PMID:9987836

Etesami, H., Emami, S., & Alikhani, H. A. (2017). Potassium solubilizing bacteria (KSB): Mechanisms, promotion of plant growth, and future prospects A review. *Journal of Soil Science and Plant Nutrition*, *17*(4), 897–911. doi:10.4067/S0718-95162017000400005

Gadhave, K. R., Hourston, J. E., & Gange, A. C. (2016). Developing soil microbial inoculants for pest management: Can one have too much of a good thing? *Journal of Chemical Ecology*, *42*(4), 348–356. doi:10.100710886-016-0689-8 PMID:27059329

Gamalero, E., Martinotti, M. G., Trotta, A., Lemanceau, P., & Berta, G. (2002). Morphogenetic modifications induced by Pseudomonas fluorescens A6RI and Glomus mosseae BEG12 in the root system of tomato differ according to plant growth conditions. *The New Phytologist*, *155*(2), 293–300. doi:10.1046/j.1469-8137.2002.00460.x

Gao, P., Qin, J., Li, D., & Zhou, S. (2018). Inhibitory effect and possible mechanism of a Pseudomonas strain QBA5 against gray mold on tomato leaves and fruits caused by Botrytis cinerea. *PLoS One*, *13*(1), e0190932. doi:10.1371/journal.pone.0190932 PMID:29320571

Garnham, C. P., Gilbert, J. A., Hartman, C. P., Campbell, R. L., Laybourn-Parry, J., & Davies, P. L. (2008). A Ca2+-dependent bacterial antifreeze protein domain has a novel β-helical ice-binding fold. *The Biochemical Journal*, *411*(1), 171–180. doi:10.1042/BJ20071372 PMID:18095937

Ge, B., Liu, B., Nwet, T. T., Zhao, W., Shi, L., & Zhang, K. (2016). Bacillus methylotrophicus strain NKG-1, isolated from Changbai Mountain, China, has potential applications as a biofertilizer or biocontrol agent. *PLoS One*, *11*(11), e0166079. doi:10.1371/journal.pone.0166079 PMID:27832162

Ghorbanpour, M., Hatami, M., & Khavazi, K. (2013). Role of plant growth promoting rhizobacteria on antioxidant enzyme activities and tropane alkaloid production of *Hyoscyamus niger* under water deficit stress. *Turkish Journal of Biology*, *37*(3), 350–360. doi:10.3906/biy-1209-12

Giovannetti, M., Fortuna, P., Citernesi, A. S., Morini, S., & Nuti, M. P. (2001). The occurrence of anastomosis formation and nuclear exchange in intact arbuscular mycorrhizal networks. *The New Phytologist, 151*(3), 717–724. doi:10.1046/j.0028-646x.2001.00216.x

Glick, B. R. (2012). Plant growth-promoting bacteria: Mechanisms and applications. *Scientifica, 2012*, 1–15. doi:10.6064/2012/963401 PMID:24278762

Glick, B. R., Cheng, Z., Czarny, J., & Duan, J. (2013). Promotion of plant growth by ACC deaminase-producing soil bacteria. In *New perspectives and approaches in plant growth-promoting Rhizobacteria research* (pp. 329–339). Springer.

Gliński, J., Horabik, J., & Lipiec, J. (Eds.). (2011). *Encyclopedia of agrophysics* (pp. 264–267). Springer. doi:10.1007/978-90-481-3585-1

Hang, N. T. T., Oh, S. O., Kim, G. H., Hur, J. S., & Koh, Y. J. (2005). Bacillus subtilis S1-0210 as a biocontrol agent against Botrytis cinerea in strawberries. *The Plant Pathology Journal, 21*(1), 59–63. doi:10.5423/PPJ.2005.21.1.059

Harman, G. E. (2011). Trichoderma—Not just for biocontrol anymore. *Phytoparasitica, 39*(2), 103–108. doi:10.100712600-011-0151-y

Hermosa, R., Viterbo, A., Chet, I., & Monte, E. (2012). Plant-beneficial effects of Trichoderma and of its genes. *Microbiology, 158*(1), 17–25. doi:10.1099/mic.0.052274-0 PMID:21998166

Hillel, D., & Hatfield, J. L. (Eds.). (2005). *Encyclopedia of Soils in the Environment* (Vol. 3). Elsevier.

Innerebner, G., Knief, C., & Vorholt, J. A. (2011). Protection of Arabidopsis thaliana against leaf-pathogenic Pseudomonas syringae by Sphingomonas strains in a controlled model system. *Applied and Environmental Microbiology, 77*(10), 3202–3210. doi:10.1128/AEM.00133-11 PMID:21421777

Jacoud, C., Faure, D., Wadoux, P., & Bally, R. (1998). Development of a strain-specific probe to follow inoculated Azospirillum lipoferum CRT1 under field conditions and enhancement of maize root development by inoculation. *FEMS Microbiology Ecology, 27*(1), 43–51. doi:10.1111/j.1574-6941.1998.tb00524.x

Kalayu, G. (2019). Phosphate solubilizing microorganisms: promising approach as biofertilizers. *International Journal of Agronomy*.

Kang, B. G., Kim, W. T., Yun, H. S., & Chang, S. C. (2010). Use of plant growth-promoting rhizobacteria to control stress responses of plant roots. *Plant Biotechnology Reports, 4*(3), 179–183. doi:10.100711816-010-0136-1

Khasa, Y. P. (2017). Microbes as biocontrol agents. In *Probiotics and Plant Health* (pp. 507–552). Springer.

Kisiala, A., Laffont, C., Emery, R. N., & Frugier, F. (2013). Bioactive cytokinins are selectively secreted by Sinorhizobium meliloti nodulating and nonnodulating strains. *Molecular Plant-Microbe Interactions, 26*(10), 1225–1231. doi:10.1094/MPMI-02-13-0054-R PMID:24001254

Köhl, J., Kolnaar, R., & Ravensberg, W. J. (2019). Mode of action of microbial biological control agents against plant diseases: Relevance beyond efficacy. *Frontiers in Plant Science, 10*, 845. doi:10.3389/fpls.2019.00845 PMID:31379891

Kumar, B. (2016). Biocontrol of insect pests. In *Ecofriendly pest management for food security* (pp. 25–61). Academic Press.

Kundan, R., Pant, G., Jadon, N., & Agrawal, P. K. (2015). Plant growth promoting rhizobacteria: Mechanism and current prospective. *Journal of Fertilizers and Pesticides*, *6*(2), 9. doi:10.4172/2471-2728.1000155

Liba, C. M., Ferrara, F. I. S., Manfio, G. P., Fantinatti-Garboggini, F., Albuquerque, R. C., Pavan, C., Ramos, P. L., Moreira-Filho, C. A., & Barbosa, H. R. (2006). Nitrogen-fixing chemo-organotrophic bacteria isolated from cyanobacteria-deprived lichens and their ability to solubilize phosphate and to release amino acids and phytohormones. *Journal of Applied Microbiology*, *101*(5), 1076–1086. doi:10.1111/j.1365-2672.2006.03010.x PMID:17040231

Liu, C. H., Chen, X., Liu, T. T., Lian, B., Gu, Y., Caer, V., Xue, Y. R., & Wang, B. T. (2007). Study of the antifungal activity of Acinetobacter baumannii LCH001 in vitro and identification of its antifungal components. *Applied Microbiology and Biotechnology*, *76*(2), 459–466. doi:10.100700253-007-1010-0 PMID:17534613

Loeffler, W., Katzer, W., Kremer, S., Kugler, M., Petersen, F., Jung, G., & Tschen, J. S. M. (1990). Gegen Pilze wirksame antibiotika der Bacillus subtilis-gruppe. Forum Mikrobiol, 90, 156-163.

López-Bucio, J., Pelagio-Flores, R., & Herrera-Estrella, A. (2015). Trichoderma as biostimulant: Exploiting the multilevel properties of a plant beneficial fungus. *Scientia Horticulturae*, *196*, 109–123. doi:10.1016/j.scienta.2015.08.043

Lucini, L., Rouphael, Y., Cardarelli, M., Canaguier, R., Kumar, P., & Colla, G. (2015). The effect of a plant-derived biostimulant on metabolic profiling and crop performance of lettuce grown under saline conditions. *Scientia Horticulturae*, *182*, 124–133. doi:10.1016/j.scienta.2014.11.022

Mahaffee, W. F., & Backman, P. A. (1993). Effects of seed factors on spermosphere and rhizosphere colonization of cotton by Bacillus subtilis GB03. *Phytopathology*, *83*(10), 1120–1125. doi:10.1094/Phyto-83-1120

Mahdi, I., Fahsi, N., Hafidi, M., Allaoui, A., & Biskri, L. (2020). Plant Growth Enhancement using Rhizospheric Halotolerant Phosphate Solubilizing Bacterium Bacillus licheniformis QA1 and Enterobacter asburiae QF11 Isolated from *Chenopodium quinoa* Willd. *Microorganisms*, *8*(6), 948. doi:10.3390/microorganisms8060948 PMID:32599701

Malik, K. A., Bilal, R., Mehnaz, S., Rasul, G., Mirza, M. S., & Ali, S. (1997). Association of nitrogen-fixing, plant-growth-promoting rhizobacteria (PGPR) with kallar grass and rice. In *Opportunities for Biological Nitrogen Fixation in Rice and Other Non-Legumes* (pp. 37–44). Springer. doi:10.1007/978-94-011-5744-5_5

Marrone, P. G. (2002). An effective biofungicide with novel modes of action. *Pesticide Outlook*, *13*(5), 193–194. doi:10.1039/b209431m

Mendes, R., Garbeva, P., & Raaijmakers, J. M. (2013). The rhizosphere microbiome: Significance of plant beneficial, plant pathogenic, and human pathogenic microorganisms. *FEMS Microbiology Reviews*, *37*(5), 634–663. doi:10.1111/1574-6976.12028 PMID:23790204

Meneses, C., Gonçalves, T., Alquéres, S., Rouws, L., Serrato, R., Vidal, M., & Baldani, J. I. (2017). Gluconacetobacter diazotrophicus exopolysaccharide protects bacterial cells against oxidative stress in vitro and during rice plant colonization. *Plant and Soil*, *416*(1-2), 133–147. doi:10.100711104-017-3201-5

Michelsen, C. F., & Stougaard, P. (2012). Hydrogen cyanide synthesis and antifungal activity of the biocontrol strain Pseudomonas fluorescens In5 from Greenland is highly dependent on growth medium. *Canadian Journal of Microbiology*, *58*(4), 381–390. doi:10.1139/w2012-004 PMID:22417387

Middleton, B. A. (2008). Invasive Species. In S. E. Jørgensen & B. D. Fath (Eds.), *Encyclopedia of Ecology* (pp. 2020–2028). Academic Press. doi:10.1016/B978-008045405-4.00060-4

Mishra, J., Singh, R., & Arora, N. K. (2017). Plant growth-promoting microbes: diverse roles in agriculture and environmental sustainability. In *Probiotics and plant health* (pp. 71–111). Springer. doi:10.1007/978-981-10-3473-2_4

Montesinos, E., & Bonaterra, A. (2009). *Microbial pesticides. Encyclopedia of microbiology* (3rd ed.). Elsevier Inc.

Muminah, B., Hazarin, S., & Fahruddin, D. B. (2015). Isolation and screening of exopolysaccharide producing bacterial (EPS) from potato rhizosphere for soil aggregation. *International Journal of Current Microbiology and Applied Sciences*, *4*(6), 341–349.

Murray, J. D. (2011). Invasion by invitation: Rhizobial infection in legumes. *Molecular Plant-Microbe Interactions*, *24*(6), 631–639. doi:10.1094/MPMI-08-10-0181 PMID:21542766

Naik, K., Mishra, S., Srichandan, H., Singh, P. K., & Sarangi, P. K. (2019). Plant growth promoting microbes: Potential link to sustainable agriculture and environment. *Biocatalysis and Agricultural Biotechnology*, *21*, 101326. doi:10.1016/j.bcab.2019.101326

Nie, P., Li, X., Wang, S., Guo, J., Zhao, H., & Niu, D. (2017). Induced systemic resistance against Botrytis cinerea by Bacillus cereus AR156 through a JA/ET-and NPR1-dependent signaling pathway and activates PAMP-triggered immunity in Arabidopsis. *Frontiers in Plant Science*, *8*, 238. doi:10.3389/fpls.2017.00238 PMID:28293243

Nion, Y. A., & Toyota, K. (2015). Recent trends in control methods for bacterial wilt diseases caused by Ralstonia solanacearum. *Microbes and Environments*. PMID:25762345

Nutaratat, P., Monprasit, A., Srisuk, N.(2017). High-yield production of indole-3-acetic acid by Enterobacter sp. DMKU-RP206, a rice phyllosphere bacterium that possesses plant growth-promoting traits. *Biotech, 7*(5), 305.

Ogut, M., Er, F., & Kandemir, N. (2010). Phosphate solubilization potentials of soil Acinetobacter strains. *Biology and Fertility of Soils*, *46*(7), 707–715. doi:10.100700374-010-0475-7

Okon, Y., & Labandera-Gonzalez, C. A. (1994). Agronomic applications of Azospirillum: An evaluation of 20 years worldwide field inoculation. *Soil Biology & Biochemistry*, *26*(12), 1591–1601. doi:10.1016/0038-0717(94)90311-5

Olanrewaju, O. S., Glick, B. R., & Babalola, O. O. (2017). Mechanisms of action of plant growth promoting bacteria. *World Journal of Microbiology & Biotechnology*, *33*(11), 197. doi:10.100711274-017-2364-9 PMID:28986676

Onofre-Lemus, J., Hernández-Lucas, I., Girard, L., & Caballero-Mellado, J. (2009). ACC (1-aminocyclopropane-1-carboxylate) deaminase activity, a widespread trait in Burkholderia species, and its growth-promoting effect on tomato plants. *Applied and Environmental Microbiology*, *75*(20), 6581–6590. doi:10.1128/AEM.01240-09 PMID:19700546

Percival, D. C., Abbey, J., Lu, H., & Harris, L. (2016). Use of biofungicides to address conventional Botrytis blight control challenges in wild blueberry production. In *XI International Vaccinium Symposium 1180* (pp. 241-248). Academic Press.

Pereg, L., & McMillan, M. (2015). Scoping the potential uses of beneficial microorganisms for increasing productivity in cotton cropping systems. *Soil Biology & Biochemistry*, *80*, 349–358. doi:10.1016/j.soilbio.2014.10.020

Pérez-Montaño, F., Alías-Villegas, C., Bellogín, R. A., Del Cerro, P., Espuny, M. R., Jiménez-Guerrero, I., & Cubo, T. (2014). Plant growth promotion in cereal and leguminous agricultural important plants: From microorganism capacities to crop production. *Microbiological Research*, *169*(5-6), 325–336. doi:10.1016/j.micres.2013.09.011 PMID:24144612

Petatán-Sagahón, I., Anducho-Reyes, M. A., Silva-Rojas, H. V., Arana-Cuenca, A., Tellez-Jurado, A., Cárdenas-Álvarez, I. O., & Mercado-Flores, Y. (2011). Isolation of bacteria with antifungal activity against the phytopathogenic fungi Stenocarpella maydis and Stenocarpella macrospora. *International Journal of Molecular Sciences*, *12*(9), 5522–5537. doi:10.3390/ijms12095522 PMID:22016606

Petrozza, A., Santaniello, A., Summerer, S., Di Tommaso, G., Di Tommaso, D., Paparelli, E., & Cellini, F. (2014). Physiological responses to Megafol® treatments in tomato plants under drought stress: A phenomic and molecular approach. *Scientia Horticulturae*, *174*, 185–192. doi:10.1016/j.scienta.2014.05.023

Pieterse, C. M., Zamioudis, C., Berendsen, R. L., Weller, D. M., Van Wees, S. C., & Bakker, P. A. (2014). Induced systemic resistance by beneficial microbes. *Annual Review of Phytopathology*, 52. PMID:24906124

Postma, J., Os, E. V., & Bonants, P. J. M. (2008). *Pathogen detection and management strategies in soilless plant growing systems*. Academic Press.

Raza, W., & Shen, Q. (2010). Growth, Fe^{3+} reductase activity, and siderophore production by Paenibacillus polymyxa SQR-21 under differential iron conditions. *Current Microbiology*, *61*(5), 390–395. doi:10.100700284-010-9624-3 PMID:20358373

Rex Consortium. (2013). Heterogeneity of selection and the evolution of resistance. *Trends in Ecology & Evolution*, *28*(2), 110–118. doi:10.1016/j.tree.2012.09.001 PMID:23040463

Riaz, U., Aziz, H., Anum, W., Mehdi, S. M., Murtaza, G., & Jamil, M. (2020a). Biofortification Technologies Used in Agriculture in Relation to Micronutrients. In *Plant Micronutrients* (pp. 225–239). Springer. doi:10.1007/978-3-030-49856-6_9

Riaz, U., Mehdi, S. M., Iqbal, S., Khalid, H. I., Qadir, A. A., Anum, W., & Murtaza, G. (2020b). Biofertilizers: Eco-Friendly Approach for Plant and Soil Environment. In *Bioremediation and Biotechnology* (pp. 189–213). Springer. doi:10.1007/978-3-030-35691-0_9

Rojas-Tapias, D., Moreno-Galván, A., Pardo-Díaz, S., Obando, M., Rivera, D., & Bonilla, R. (2012). Effect of inoculation with plant growth-promoting bacteria (PGPB) on amelioration of saline stress in maize (*Zea mays*). *Applied Soil Ecology*, *61*, 264–272. doi:10.1016/j.apsoil.2012.01.006

Romano, I., Ventorino, V., & Pepe, O. (2020). Effectiveness of plant beneficial microbes: Overview of the methodological approaches for the assessment of root colonization and persistence. *Frontiers in Plant Science*, *11*, 6. doi:10.3389/fpls.2020.00006 PMID:32076431

Romera, F. J., García, M. J., Lucena, C., Martínez-Medina, A., Aparicio, M. A., Ramos, J., Alcántara, E., Angulo, M., & Pérez-Vicente, R. (2019). Induced systemic resistance (ISR) and Fe deficiency responses in dicot plants. *Frontiers in Plant Science*, *10*, 287. doi:10.3389/fpls.2019.00287 PMID:30915094

Rouphael, Y., Cardarelli, M., & Colla, G. (2015a). Role of arbuscular mycorrhizal fungi in alleviating the adverse effects of acidity and aluminium toxicity in zucchini squash. *Scientia Horticulturae*, *188*, 97–105. doi:10.1016/j.scienta.2015.03.031

Rouphael, Y., Franken, P., Schneider, C., Schwarz, D., Giovannetti, M., Agnolucci, M., & Colla, G. (2015b). Arbuscular mycorrhizal fungi act as biostimulants in horticultural crops. *Scientia Horticulturae*, *196*, 91–108. doi:10.1016/j.scienta.2015.09.002

Saad, M. M., Kandil, M., & Mohammed, Y. M. (2020). Isolation and Identification of Plant Growth-Promoting Bacteria Highly Effective in Suppressing Root Rot in Fava Beans. *Current Microbiology*, *77*(9), 1–11. doi:10.100700284-020-02015-1 PMID:32372106

Sachdev, S., & Singh, R. P. (2018). Root colonization: Imperative mechanism for efficient plant protection and growth. *MOJ Ecology & Environmental Sciences*, *3*, 240–242.

Sandhya, V., & Ali, S. Z. (2015). The production of exopolysaccharide by Pseudomonas putida GAP-P45 under various abiotic stress conditions and its role in soil aggregation. *Microbiology*, *84*(4), 512–519. doi:10.1134/S0026261715040153

Sarkar, A., Ghosh, P. K., Pramanik, K., Mitra, S., Soren, T., Pandey, S., Mondal, M. H., & Maiti, T. K. (2018). A halotolerant Enterobacter sp. displaying ACC deaminase activity promotes rice seedling growth under salt stress. *Research in Microbiology*, *169*(1), 20–32. doi:10.1016/j.resmic.2017.08.005 PMID:28893659

Schmiedeknecht, G., Bochow, H., & Junge, H. (1998). Use of Bacillus subtilis as biocontrol agent. II. Biological control of potato diseases/Anwendung von *Bacillus subtilis* als Mittel für den biologischen Pflanzenschutz. II. Biologische Bekämpfung von Kartoffelkrankheiten. Zeitschrift für Pflanzenkrankheiten und Pflanzenschutz. *Journal of Plant Diseases and Protection*, 376–386.

Schoenborn, L., Yates, P. S., Grinton, B. E., Hugenholtz, P., & Janssen, P. H. (2004). Liquid serial dilution is inferior to solid media for isolation of cultures representative of the phylum-level diversity of soil bacteria. *Applied and Environmental Microbiology*, *70*(7), 4363–4366. doi:10.1128/AEM.70.7.4363-4366.2004 PMID:15240320

Schumpp, O., & Deakin, W. J. (2010). How inefficient rhizobia prolong their existence within nodules. *Trends in Plant Science, 15*(4), 189–195. doi:10.1016/j.tplants.2010.01.001 PMID:20117958

Schwarzott, D., & Walker, C. (2001). A new fungal phylum, the Glomermycota: Phylogeny and evolution. *Mycological Research, 105*(12), 1413–1421. doi:10.1017/S0953756201005196

Sharma, S., Kumar, V., & Tripathi, R. B. (2011). Isolation of phosphate solubilizing microorganism (PSMs) from soil. *Journal of Microbiology and Biotechnology Research, 1*(2), 90–95.

Shen, X., Hu, H., Peng, H., Wang, W., & Zhang, X. (2013). Comparative genomic analysis of four representative plant growth-promoting rhizobacteria in Pseudomonas. *BMC Genomics, 14*(1), 271. doi:10.1186/1471-2164-14-271 PMID:23607266

Singh, V. K., Singh, A. K., Singh, P. P., & Kumar, A. (2018). Interaction of plant growth promoting bacteria with tomato under abiotic stress: A review. *Agriculture, Ecosystems & Environment, 267*, 129–140. doi:10.1016/j.agee.2018.08.020

Sivasakthi, S., Usharani, G., & Saranraj, P. (2014). Biocontrol potentiality of plant growth promoting bacteria (PGPR)-Pseudomonas fluorescens and Bacillus subtilis: A review. *African Journal of Agricultural Research, 9*(16), 1265–1277.

Smith, S. E., & Read, D. J. (2010). *Mycorrhizal symbiosis*. Academic Press.

Souza, R. D., Ambrosini, A., & Passaglia, L. M. (2015). Plant growth-promoting bacteria as inoculants in agricultural soils. *Genetics and Molecular Biology, 38*(4), 401–419. doi:10.1590/S1415-475738420150053 PMID:26537605

Spaepen, S., & Vanderleyden, J. (2011). Auxin and plant-microbe interactions. *Cold Spring Harbor Perspectives in Biology, 3*(4), a001438. doi:10.1101/cshperspect.a001438 PMID:21084388

Spaepen, S., Vanderleyden, J., & Remans, R. (2007). Indole-3-acetic acid in microbial and microorganism-plant signaling. *FEMS Microbiology Reviews, 31*(4), 425–448. doi:10.1111/j.1574-6976.2007.00072.x PMID:17509086

Stein, T., Hayen-Schneg, N., & Fendrik, I. (1997). Contribution of BNF by Azoarcus sp. BH72 in Sorghum vulgare. *Soil Biology & Biochemistry, 29*(5-6), 969–971. doi:10.1016/S0038-0717(96)00211-8

Susilowati, D. N., Riyanti, E. I., Setyowati, M., & Mulya, K. (2018). Indole-3-acetic acid producing bacteria and its application on the growth of rice. In AIP Conference Proceedings: 2018. AIP Publishing LLC. doi:10.1063/1.5050112

Trabelsi, D., & Mhamdi, R. (2013). Microbial inoculants and their impact on soil microbial communities: A review. *BioMed Research International, 2013*, 1–11. doi:10.1155/2013/863240 PMID:23957006

Van Loon, L. C., Bakker, P. A. H. M., & Pieterse, C. M. J. (1998). Systemic resistance induced by rhizosphere bacteria. *Annual Review of Phytopathology, 36*(1), 453–483. doi:10.1146/annurev.phyto.36.1.453 PMID:15012509

Venturi, V., & Keel, C. (2016). Signaling in the rhizosphere. *Trends in Plant Science, 21*(3), 187–198. doi:10.1016/j.tplants.2016.01.005 PMID:26832945

Verbon, E. H., & Liberman, L. M. (2016). Beneficial microbes affect endogenous mechanisms controlling root development. *Trends in Plant Science*, *21*(3), 218–229. doi:10.1016/j.tplants.2016.01.013 PMID:26875056

Verma, S. K., & White, J. F. (2019). *Seed Endophytes*. Springer. doi:10.1007/978-3-030-10504-4

Vos, I. A., Pieterse, C. M., & Van Wees, S. C. (2013). Costs and benefits of hormone-regulated plant defences. *Plant Pathology*, *62*, 43–55. doi:10.1111/ppa.12105

Wallace, R. L., Hirkala, D. L., & Nelson, L. M. (2018). Mechanisms of action of three isolates of Pseudomonas fluorescens active against postharvest grey mold decay of apple during commercial storage. *Biological Control*, *117*, 13–20. doi:10.1016/j.biocontrol.2017.08.019

Woo, S. L., & Pepe, O. (2018). Microbial consortia: Promising probiotics as plant biostimulants for sustainable agriculture. *Frontiers in Plant Science*, *9*, 1801. doi:10.3389/fpls.2018.01801 PMID:30564264

Yu, S., Teng, C., Bai, X., Liang, J., Song, T., Dong, L., Jin, Y., & Qu, J. (2017). Optimization of siderophore production by Bacillus sp. PZ-1 and its potential enhancement of phytoextration of Pb from soil. *Journal of Microbiology and Biotechnology*, *27*(8), 1500–1512. doi:10.4014/jmb.1705.05021 PMID:28633518

Zhang, J., Howell, C. R., & Starr, J. L. (1996). Suppression of Fusarium colonization of cotton roots and Fusarium wilt by seed treatments with *Gliocladium virens* and *Bacillus subtilis. Biocontrol Science and Technology*, *6*(2), 175–188. doi:10.1080/09583159650039377

ADDITIONAL READING

Banerjee, M. R., Yesmin, L., & Vessey, J. K. (2006). *Plant-growth-promoting rhizobacteria as biofertilizers and biopesticides. Handbook of microbial biofertilizers*. Food Products Press.

Beattie, G. A. (2007). Plant-associated bacteria: survey, molecular phylogeny, genomics and recent advances. In *Plant-associated bacteria* (pp. 1–56). Springer.

Borriss, R. (2015). Bacillus, a plant-beneficial bacterium. In *Principles of plant-microbe interactions* (pp. 379–391). Springer.

Compant, S., Clément, C., & Sessitsch, A. (2010). Plant growth-promoting bacteria in the rhizo-and endosphere of plants: Their role, colonization, mechanisms involved and prospects for utilization. *Soil Biology & Biochemistry*, *42*(5), 669–678. doi:10.1016/j.soilbio.2009.11.024

Finkel, O. M., Castrillo, G., Paredes, S. H., González, I. S., & Dangl, J. L. (2017). Understanding and exploiting plant beneficial microbes. *Current Opinion in Plant Biology*, *38*, 155–163. doi:10.1016/j.pbi.2017.04.018 PMID:28622659

Higa, T., & Wididana, G. N. (1991). The concept and theories of effective microorganisms. In *Proceedings of the first international conference on Kyusei nature farming. US Department of Agriculture, Washington, DC, USA* (pp. 118-124).

Javaid, A. (2010). Beneficial microorganisms for sustainable agriculture. In *Genetic engineering, biofertilisation, soil quality and organic farming* (pp. 347–369). Springer. doi:10.1007/978-90-481-8741-6_12

Pliego, C., Kamilova, F., & Lugtenberg, B. (2011). Plant growth-promoting bacteria: fundamentals and exploitation. In *Bacteria in agrobiology: Crop ecosystems* (pp. 295–343). Springer. doi:10.1007/978-3-642-18357-7_11

KEY TERMS AND DEFINITIONS

Actinomycetes: Unicellular gram positive bacteria found in a variety of habitats.

Bio-Control Agent: It is the living organism/biological agent which is utilized for controlling insect pests. In other words they are natural enemies of crop pests.

Bio-Fertilizer: It is a type of fertilizer comprising of living organisms, which when applied to the plants, soil colonizes and promote plant growth wellbeing by providing them with primary nutrients.

Diazotrophs: Bacteria and archaea that are capable of converting atmospheric nitrogen gas into more usable form (ammonia). They are grown without external sources of fixed nitrogen.

Endophytic Bacteria: Type of bacteria that lives inside plants and improve the plant growth, health and development under normal as well as challenging conditions.

Pattern Recognition Receptors: They are the proteins used for recognition of molecules found frequently in pathogens or the molecules which are released by damaged cells.

Siderophore: It is an iron chelating compound produced/secreted by microorganisms and helps to transport iron across cell membranes.

Chapter 6
Soil Microbiome for Plant Growth and Bioremediation

Monika Yadav
Banasthali Vidyapeeth, Rajasthan, India

Sonu Kumari
Banasthali Vidyapeeth, Rajasthan, India

Junaid Ahmad Malik
https://orcid.org/0000-0003-4411-2015
Government Degree College, Bijbehara, India

Suphiya Khan
Banasthali Vidyapeeth, Rajasthan, India

ABSTRACT

Terrestrial soil is a complex part of the ecosystem hosting bacteria, fungi, protists, animals, and huge source of nutrients to plants. These soil-dwelling organisms exhibit an array of interactions with plants to span the full range of ecological possibilities. In the 19th century, many different bacterial strains were described as having plant growth favouring potential like Pseudomonas, Azospirillum, and even crop seeds were coated with bacterial cultures to improve growth and yield. The soil microbial community also recognized their considerable role to improve the soil health via energy transfer, catalyzing reactions, and nutrient mineralization. Thus, soil microorganisms and enzymatic process are generally regarded as rate-limiting steps in decomposition and nutrient cycling.

INTRODUCTION

The naturally prevalent sub-structure and shelter for plants, higher species, and diverse microorganisms such as bacteria, fungi, algae, annelids, and invertebrates is the soil ecosystem. The relationship between soil, plants and microorganisms is like an organization (Figure 1) that influences health and productivity rate of plants. The soil microbiome plays diverse role in transformation of nutrients via decomposition,

DOI: 10.4018/978-1-7998-7062-3.ch006

mineralization, and preservation of available nutrients. The phyto-microbiome community might be divided into the rhizosphere, phyllosphere (phyllo-microbiome) and endosphere. Rhizosphere is situated in the region of the roots of the plants (Kembel et al. 2014). The phyllosphere consist of the microorganisms which are present in the plant aerial parts (Lundberg et al. 2012) and the endosphere refers to the microorganisms reside within the plant (Berg et al. 2014). The microbial community structure depends upon the interactions existing among soil biosphere, plant and microorganism. These relations are facilitated by compounds which are confined by plants or microorganisms as exudates (East 2013; Hardoim et al., 2015).

Figure 1. Interaction between plant, microbiome and soil

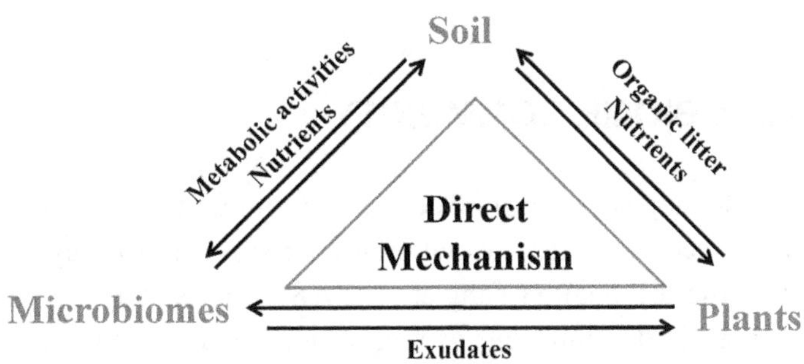

Nowadays, the agricultural crop is experiencing numerous constraints such as soil health, climate variations and demographic development. As microorganisms possess the strong potential as biofertilizers or biopesticides which impart a curiosity to incorporate them as alternative to fertilizers in agricultural soils (Mendes et al., 2013).

Like microbial activity, enzymatic activities are also acting as a critical index of soil fertility (Rohrbacher and St-Arnaud 2016). These soil enzymes synthesized by microorganisms play a key role in catalyzing various metabolic and biochemical processes such as nutrient recycling, soil quality improvement, soil decontamination and degradation of organic matters (Dong et al. 2015). The enzymatic process is generally regarded as a rate-limiting step in the decomposition, mineralization and recycling of nutrients (Bing et al., 2012). Analysis of soil enzymes helps to establish correlation with soil physico-chemical characters, soil productivity, microbial activity, biochemical cycling of nutrients in soil and to evaluate the succession stage of an ecosystem (Jackson et al., 2012). Soil enzyme activities are sensitive to the change in soil environment caused by natural and anthropogenic factors. Hence, enzyme activities can be considered as effective indicators and biomarkers to assess the nature and quality of soil environment (Dong et al., 2015). The soil enzymes play an elementary role to facilitate the development of plants via establishing various biogeochemical cycles. As the study of soil microbiome/ enzyme functional diversity has been strengthened to boost the understanding of the linkages among the resource availability, microbial community structure, function and ecosystem processes. Study of soil microorganisms and enzymes gives information about the resource availability, release of nutrients in soil by degradation of organic matter and functioning of ecosystem. Nearly all soil functional processes, both chemical and biological processes, depend on enzymatic catalysis (Maron *et al.*, 2007). Soil enzymes are mostly generated from the exudates of soil microorganisms, the decaying of plant and animal residues or even from dead cells (Peng, *et al.*, 2003). For both kingdoms, environmental variables appear to be more important in determining community composition and function (Bahram et al., 2018).

In soil rhizosphere various bacteria, fungi, and algae are allied with plant roots and stimulate plant growth as a bio-fertilizer. The plant growth-promoting rhizobacteria (PGPR) has a potential to protect plants from pests, detoxify toxic metals, degrade xenobiotic compounds and accelerate nutrients adsorption (Ahemad and Khan 2012). For bacteria, pH is the predominant environmental variable that drives community composition, along with carbon and oxygen quality/quantity, soil moisture, nitrogen (N) and phosphorus (P) availability (Fierer, 2017). Bacterial functional diversity is strongly associated with mean annual precipitation whereas for fungi, C/N ratio is the strongest predictor of community composition and function, which may reflect higher energy requirements and niche specialization compared with bacteria (Bahram et al., 2018).

TYPES OF SOIL MICROBIOME COMMUNITY

Soil ecosystem possess a wide range of microbial community inhabitants; various bacteria, fungi, algae, protozoa, actinomycetes and each one has special characteristic role to recycle nutrients, decompose organic matter and maintain the soil health (Yang et al. 2018). Soil bacteria, on the basis of their carbon and energy source are further divided into four groups which are as follows:

1. Photoautotrophs: These bacteria obtain energy from the sunlight and fix carbon as nutrient resource and play a major component to facilitate N-fixing e.g., Cyanobacteria (Aislabie and Deslippe 2013).
2. Photoheterotrophs: These bacteria assimilate carbon dioxide during photosynthesis in presence of an electron donor (Dijkhuizen and Harder 1984).
3. Chemoautotrophs: Enlist those bacteria which acquire carbon and energy by utilizing inorganic compounds and favour the nitrification activities e.g., *Nitrosomonas* and *Nitrobacteria* (Aislabie and Deslippe 2013).
4. Chemoheterotrophs**:** These bacteria nurture upon the pre-formed organic matter as a resource of carbon and energy (Boschker et al. 2014).

Mycorrhizal fungi play an important role in nutrient and carbon cycle by forming hyphae organization with plant roots and enhance the biomass and diversity of the soil e.g., ectomycorrhizal fungi (Smith et al. 2003). Lichens are also involved in nitrogen-carbon cycle and originate by the symbiotic relationship existing between the pigmented algae and fungus e.g., Cyanobacteria (Abdel-Raouf et al. 2012).

Commonly two main forms of soil enzymes are found in nature that play important role in catalyzing various microbial biochemical reactions and facilitate soil ecology (Mclaren 1975). The important application of numerous enzymes has been described in Table 1 (Das Varma, 2010).

1. Constitutive enzymes: These are present in a constant amount within microorganisms for their metabolic activity. Addition and elimination of any substrate does not affect the activity of these enzymes (Das and verma, 2011), for example, phosphatase and urease enzymes.
2. Inductive enzymes: These inductive enzymes are found in low amount but concentration may vary due to presence of substrate. Amidase and cellulose enzymes are examples of inductive enzymes (Kandeler, 2015).

Table 1. List of enzymes produced by soil microbiome as indicator

Soil Enzyme	Source	Indicator of Microbial Activity
Dehydrogenase	Microorganisms	C-cycle, microbial oxidative activity
β-glucosidase	Fungi, bacteria	C-cycle
α-Amylase	Plants, animals, microorganisms	C-cycle
Urease	Microorganisms, plants	N-cycle
Phosphatase	Microorganisms, plants	P-cycle
Protease	Microorganisms, plants	N-cycle
Aryl sulfatase	Plants, animals, microorganisms	S-cycle
Cellulase	Bacteria, fungi, termites, ruminant	C-cycle
Amidase	Prokaryotes and eukaryotes	N-cycle
Dioxygenase	Soil bacteria	C-cycle
Laccase	Bacteria, fungi, insects, plants	C- N-cycle
Lipase	Bacteria, Actinomycetes	C- N-cycle

HABITATS OF SOIL MICROBIOME COMMUNITY

Grassland Soil Microbiome

Grasslands are predominantly defined by community grasses (members of the family Poaceae) and other low-growing, non-woody plant species (Gibson, 2008). Grassland habitats are estimated to occupy ~26% of the global land area and store an estimated 20% of the total soil carbon stock (Ramankutty et al., 2008). As below-ground component of grasslands form a substantial sink of global C stocks, and modelling studies suggest that grassland ecosystems remain a highly reliable sink of C in future climate change scenarios (Hu et al., 2001) as they possess a greater ability to adapt to extreme weather events than other vegetation types (Vicente-Serrano et al., 2013). Many grasses and non-woody plant species occurring in grasslands form symbiotic plant–root associations with arbuscular mycorrhizal fungi (Smith & Read, 2008). Arbuscular mycorrhizal fungi possess the ability to enhance the performance of their plant–host symbiont by assisting the uptake of limiting nutrients from the soil (Smith & Read, 2008). This is achieved through the production of external mycelia by the arbuscular mycorrhizal fungi, which increase the effective absorptive surface area of the roots to facilitate the adsorption of limiting nutrients (e.g. inorganic N or P) that are distant from the plant roots or within soil pores that are too small for plant roots to access (Lambers et al., 2008; Smith & Gianinazzi-Pearson, 1988). In addition to nutrient uptake, arbuscular mycorrhizal fungi also benefit plants by enhancing their ability to acquire water, reducing their susceptibility to soilborne pathogens (Cameron et al., 2013) and enabling inter-plant signaling through a common mycorrhizal network that boosts plant defence responses against herbivory and diseases (Gorzelak et al., 2015). In return, the plant–host supplies the obligate biotrophic arbuscular mycorrhizal fungi symbionts with C in the form of plant derived photosynthates (Smith & Read, 2008). By enhancing growth of their plant symbionts, arbuscular mycorrhizal fungi have a crucial role in driving the primary productivity and C sequestration of grasslands through the increased capture and input of C by plants into the soil as organic C stocks. The drought, fire and extreme precipitation events in grassland result in changes in soil moisture that influence microbial community as well as plant growth in soil. Nitrogen is an essential nutrient for driving primary productivity in grasslands, but is continually lost through grazing activity, leaching and background volatilization (Fay et al., 2015). While dinitrogen (N_2) is highly abundant in the atmosphere in gaseous form unaccesible for the biota, so the gaseous N_2 must be 'fixed' into ammonia (NH_3) that can then be further converted to other more assimilable forms of N. In grasslands below-ground symbiotic and asymbiotic diazotrophic microorganisms are responsible for the biological N-fixation of atmospheric N via the nitrogenase enzyme (Ledgard & Steele, 1992).

The rate of decomposition and the immobilisation of nutrients into the soil solution may vary depending on the environment and land-use/management practices of the grassland (Anderson, 1991; Ochoa-Hueso et al., 2019). Decomposition rates in semi-arid alpine grasslands are expected to be comparatively lower than those of mesic grasslands of similar nutrient status occurring at a lower altitude because high temperature and soil moisture promotes microbial decomposition activity (Solly et al., 2014). Competitiveness and dominance of plant species in grassland communities have been related to the differences in degree of dependency on mycorrhizal fungi by plants for nutrient uptake (Klironomos, 2003). Mycorrhizal fungal networks facilitate the plant–plant interactions by assisting the transfer of resources among plants connected to the mycorrhizal network (Leake et al., 2004) which could promote species co-existence in grassland ecosystems with stressful environmental conditions (van der Heijden & Horton, 2009). Recent advancements in molecular techniques provide novel opportunities, such as better manipulation of the soil microbiota and plant communities to effectively restore degraded grasslands and enhance the resilience and resistance of grassland ecosystems to global change (Craven et al., 2016).

Forest Soil Microbiome

Forests cover four billion hectares globally and are often considered as one of the most important C sinks on Earth. Each year, approximately two billion tonnes of carbon dioxide (CO_2) are absorbed through the leaves of trees (Fay et al., 2015). While a large proportion of this C is allocated to the aboveground and below-ground biomass of a trees (~42%), a similar amount of this is exuded by the roots into the rhizosphere soil (~44%), where it can be mineralized by microbial communities (Pan et al., 2011). However, characterizing the complexity of the forest soil microbiome remains challenging considering the highly contrasting environmental conditions where different types of forests occur. At global scales, forest ecosystems are mainly shaped by climatic conditions and soil types, resulting in three extensive biomes: tropical, temperate and boreal (in order of distance from the equator) (Kirschbaum et al., 2000). Forest ecosystems might have possibilities to transform from sinks to sources of CO_2 in the future under influence of high temperature, drought and fire conditions. In boreal forests, low temperatures restrict microbial activities, organic matter decomposition and nutrient cycling, resulting in C accumulation in soil organic matter and N depletion. In contrast, the warm and wet conditions of tropical forests enhance microbial activities and encourage a rapid turnover of soil organic matter and nitrogen enrichment (Malhi et al., 1999; Reinsch et al., 2017). Both fungal and bacterial community responses towards climate changes may vary with their respective ecosystems. This is expected due to the variations in the type and quality of litter depending on the plant communities. Consequently, forest microbiomes are often characterized by high fungal/bacterial ratios, and consistently harbour a lower bacterial diversity than grasslands or agricultural soils (Roesch et al., 2007). The forest symbiotic interactions between ectomycorrhizal fungi and their hosts are relatively well studied (Finlay et al., 1990). The mycelia of ectomycorrhizal fungi can cover several square metres and are able to reach extended areas of soil where they can assist the acquisition of soluble nutrients (e.g. N, P) and water, mobilise nutrients via the release of extracellular enzymes (Hagerberg et al., 2016). The exchange of nutrients, C, and water between the ectomycorrhizal fungi and host tree occurs via the 'Hartig net', which is composed of the hyphae of the ectomycorrhizal fungi that enclose the host's root cells (Hobbie & Hogberg, 2012). In litter of a coniferous forest, *Betaproteobacteria*, *Bacteroidetes* and *Acidobacteria* incorporate more C than fungi (Stursova´ et al., 2012). Nutrients such as phosphorus, potassium, magnesium, calcium and iron often occur as insoluble minerals in forest soils. As such, mineral weathering, occurring through biotic or abiotic processes, increases soil nutrient availability and is a key process in biogeochemical cycles in forest ecosystems. Microbial communities are strongly involved in this process through the production of siderophores, protons or organic acids like citrate, gluconate, oxalate, succinate (Richardson & Simpson, 2011).

Dessert Soil Microbiome

Desert soils – comprising mostly of sand, generally represent an extreme environment for soil microorganisms. While deserts are present in many different parts of the world, they all share a combination of extreme temperatures and low water availability (Lugtenberg, 2015). As there is little plant litter in desert soils, soil microorganisms here inhabit a very different environment to those found in other habitats (Aguirre-Garrido et al., 2012). Extremophiles (bacteria and archaea) are the first colonisers in desert environments (Mapelli et al., 2012) due to their adaptations to survive in harsh physical and chemical conditions (Colica et al., 2014). The dessert soil multicellular organisms include lichens, mosses and fungi, which form living soil surface that play crucial roles in the ecological security and health of the desert region (Li et al., 2018). While the complexity of desert plants is related to environmental factors like moisture, pH, climate, lithology, temperature, and nutrient and organic matter content (Kaplan et al., 2013). The ability of desert plants to adapt to drought stress in desert environments has also been shown to be related to the bacterial composition of the soil microbiome (Shelef et al., 2013). Many rhizobacteria associated study with desert plants has identified *Bacillus* sp., *Enterobacter* sp. and *Pseudomonas* sp. inhibiting phytopathogenic fungi (Kumar et al., 2014).

Peatland Soil Microbiome

Peatlands are characterised by an accumulation of dead organic material on the soil surface due to water-saturated conditions that prevent the complete decomposition of plant material (Joosten & Clarke, 2002). Although covering only 2.84% of land, representing 4.23 million km2 (Xu et al., 2018), peatlands play a significant role in soil C storage and cycling (Page et al., 2011). Peatland microorganisms by controlling organic C turnover, nutrient uptake and mineralisation strongly influence the plant productivity and ecosystem functioning (Andersen et al., 2013). The peatlands have the potential to produce methane (CH_4) by sequestering carbon, but sensitive to pH, hydrologic regime, mineral element percentage and vegetation (Andersen et al., 2013). Moreover, it is unclear whether fungal (Golovchenko et al., 2007) or bacterial (Winsborough & Basiliko, 2010) biomass plays a more important structural and functional overall role in peatland ecosystems. In all peatland microbial communities and decomposition processes are influenced by carbon quality, depth and redox conditions (Morales et al., 2006).

Microbial activity and functional decomposition differ among shrub peatlands from forested and sedge peatlands (Fisk et al., 2003). Changes in the fungal community composition between peatland types are interlinked to changes in litter type (Andersen et al., 2013), and evidence suggests that litter type may be an even more important driver of below-ground fungal community structure as a whole than abiotic factors such as groundwater levels (Trinder et al., 2008).

Tundra Soil Microbiome

Tundra refers to all kinds of rock and soil containing ice that experience temperatures below 0°C for a significant portion of the year. According to the duration of freezing state, tundra can be generally divided into short-term freeze, seasonal tundra and permafrost. The tundra is a highly heterogeneous and dynamic landscape with unique hydrothermal characteristics. Permafrost is characterized by stable low temperatures, low nutrient inputs and continuous exposure to low levels of radiation (Gilichinsky, 2002). Microbial abundance is relatively low (100 to 400 cells per gram of dry weight) in the Arctic, Antarctic, Qinghai–Tibet Plateau and Siberian permafrost (Vishnivetskaya et al., 2006). Bacterial taxa found in these regions are *Cellulomonas, Arthrobacter, Planococcus, Pseudomonas,* and genera from *Acidobacteria, Firmicutes, Bacteroidetes, Proteobacteria* and *Gammaproteobacteria* (Ganzert et al., 2011).

Soil temperature is perhaps the most important factor in permafrost microbial communities, affecting soil respiration, decomposition of soil organic matter, nitrogen mineralisation, denitrification, plant productivity and nutrient uptake by vegetation (Callesen et al., 2003). It is generally believed that the critical point of microbial metabolic activity is −8 °C, and no microbial activity below −12 °C has been detected (Margesin, 2012). Bakermans et al. (2003) found that the growth of microorganisms in the Siberian permafrost was normal at 22 °C and slowed to negligible at −10 °C. The upper, or 'active', layer of permafrost undergoes an annual freeze–thaw process, which leads to changes in temperature, water, organic matter, and pH in the active layer or seasonal tundra soil (Gilichinsky, 2002). These changes affect the formation of microbial cell membranes, as well as the growth and metabolism of microorganisms. The freeze–thaw cycles decrease the metabolic activity of microorganisms and affect the expression of deaminase, that results in release of nitrous oxide, a greenhouse gas (Sharma et al., 2006). Furthermore, Wagner et al. (2007) showed that higher permafrost temperatures resulted in significant increases in methanogenesis that affects the microbial activity and temperature.

Soil total C is another primary regulator of the structure of soil bacterial community in some tundra soils which influences C mineralisation and thus plays an important role in soil microbial diversity (Jangid et al., 2008). The pH in tundra soil is mainly determined by the content of organic acids (Hobbie & Gough, 2004). The Arctic and Siberian permafrost soils are acidic, while Antarctic and Chinese Qinghai– Tibet Plateau permafrost soils are alkaline (Hobbie & Gough, 2004). Water is one of the most important limiting factors for bacterial growth in tundra soil. There is a significant positive correlation between the number of microorganisms in permafrost and soil moisture content, potentially as the liquid water film in permafrost is too thin to encapsulate microorganisms or enable microbial cells to migrate within it (Wang et al., 2011). The heterotrophic activity of microorganisms is limited when soil moisture content is low and long-term water shortage can also cause the death or dormancy of microorganisms in permafrost (Graham et al., 2012). Low water content in permafrost additionally affects the fluidity of protein and soil enzyme activities, and the low temperatures affect the fluidity of cell membranes, which together are essential for the survival of soil microorganisms.

ROLE OF MICROBIOME TO ENHANCE SOIL FERTILITY

Soil ecosystem being as complex ecosystem hosts various microorganisms like bacteria, fungi, protists etc. These microbes assist different activities to sustain soil health via providing physical support, filtering of pollutants, nutrient cycling, detoxification of waste and greenhouse gas regulation (Johns 2017). Carbon (C), phosphorous (P) and sulfur (S) released by microbes during nutrient cycling and organic matter degradation is further captivated by the plants for their growth contributing to bioremediation (Bing et al., 2012). The roles of microorganisms to maintain soil fertility are follows:

1. **Nitrogen Fixtation**: Rhizobium in symbiosis process fixes atmospheric nitrogen and make it available to legumes (Shridhar 2012).
2. **Enhancing phosphorous accessibility**: Fungi form a symbiotic relationship with plant root which enhance phosphorus uptake by the plant e.g., arbuscular mycorrhizal fungi (Richardson and Simpson 2011).
3. **Pathogen control**: Few soil protozoans e.g., Amoebae (*Acanthamoeba castellanii*) help in controlling certain plant disease by consuming disease-causing pathogenic microorganism (Saha et al. 2016).
4. **Soil quality improvement**: During breakdown of organic matter, some microorganisms release substances that incorporate with soil components and recover soil structure (Torsvik and Ovreas 2002).
5. **Pesticide degeneration**: Microorganisms in soil formed enzymes such as dehydogenases and hydrolytic enzymes that degrade agricultural toxic chemical (pesticides) in soil (Iqbal and Bartakke 2014).
6. **Recycling of nutrients**: Soil microorganisms breakdown huge quantity of plant/animal residue into functional organic componentss in soil (Shradha et al. 2011).

ROLE OF SOIL MICROBIOME AND ENZYMES TO ENHANCE PLANT GROWTH

Nutrient accessibility towards plants is the most considerable factor as nutrients like N, S and P are usually find to bound with organic molecules and minimally available to plants. In soil ecosystem, soil and plant biomass are directly dependent on soil microbes growth to mineralize the organic form of nutrients. The plant microbiome community comprises of beneficial, neutral or pathogenic microorganisms. The beneficial plant growth-promoting bacteria (PGPB) through modulating endogenous hormone levels enhance plant growth by producing phytohormones like auxin, cytokinin, and gibberellin. Diverse extracellular enzymes producing microbial community reside in soil ecosystem and being as catalyst trigger various biological processes like breakdown of inorganic waste to organic component, which stimulate plant development. For example, antioxidant enzymes like SOD (superoxide dismutases), CAT (chloramphenicol acetyltransferase) and POX (proline oxidase) reduce the level of superoxide and hydrogen peroxide in plants.

As a Source of Nutrients

Microorganisms produce beneficial effects on plant health by accelerating nutrient availability, assimilation and growth by suppressing disease caused by pathogens (Table 2). In soil ecosystem phosphorus (P), sulphur (S) and nitrogen (N) are organically bound and act as crucial elements to keeping a healthy nutritional life for plants (Ahemad and Kibret 2014). These elements (N, P, S) involves in the formation of major structure of nucleic acids, enzymes, and proteins. N is considered one of the most growth-limiting nutrients as in vast majority it is contained within N_2 molecules (Dalton and Krammer 2006). Consequently, atmospheric nitrogen (N_2) fixing and nutrient mineralization by microorganisms is necessary to for the healthy development of plants (Kim and Rees 1994). The plant root exudates release components like terpenoids or flavonoids that are responsible to form plant-microbe symbiotic relation (Reyes et al. 2002). These exudates play crucial role to promote mycorrhizal interaction via stimulating hyphal branching that boost up the nutrient uptake by plants. Many kinds of nitrogen fixing bacteria are recognized as non-symbiotic or free living bacteria which include *Cyanobacteria, Nostoc, Anabaena* and *Clostridium*. The second kind comprises the symbiotic bacteria that favour mutualistic interaction e.g., *Rhizobium, Frankia* and *Azospirillum* (Battacharrya and Jha 2012). In legumes, microorganism relations are also coupled with P-fixation along with N-fixation. In soil ecosystem very less P (only 0.1%) is usable by plants (Zhu et al. 2011). So, the problems associated with P availability and high cost of fertilizers drive attention towards the environmental alternative approach for improving crop yields without affecting environment. Natural occurred phosphorus-solubilizing microbes (PSMs) solubilize and mineralize P by transforming inorganic and organic soil P into available forms for plants e.g., fungi (*Penicillium, Aspergillus*), bacteria (*Rhodococus, Chryseobacterium, Phyllobacterium*) and actinomycetes (Sardar et al. 2007). *Pseudomonas* and *Bacillus* have been used in various researches to demonstrate that organic S mineralization triggers the growth-promoting phenotypes (Chung et al. 2005).

Table 2. List of plant growth promoting microbes (PGPMs)

Microorganism	Crop	Traits	References
Pseudomonas fluorescens	Arabidopsis	Plant growth increment	(Iavicoli et al., 2003)
Pseudomonas putida	Tomato and cucumber	Reduction in Pb and Cd uptake	(Rezzonoco et al., 2005)
Pseudomonas BA-8	Strawberry	Plant growth increment	(Pirlak and Kose, 2009)
Bacillus cereus	Legumes	Lowers the chromium toxicity to seedlings	(Vessey and Buss, 2002)
Bacillus pumilus	Wheat	Plant growth increment	(Hafeez et al., 2006)
Bacillus sp.	Banana, Raspberry	Elongate root and shoot height	(Orhan et al., 2006)
Agrobacterium amazonense	Rice	Nitrogen accumulation	(Rodrigues et al., 2008)

As the Indicators of Soil Health

Soil health mainly depends upon the three interrelated components: physical, chemical and biological fertility (Kibblewhite et al. 2008). The physical fertility component refers to the structure, texture and composition of soil whereas the chemical fertility component relies on pH conditions such as acidic, alkaline or saline conditions. The biological fertility component is most complex and diverse component as it involves the microorganisms and their interaction with soil. Microorganisms and their respective enzymes produce lots of gummy substances like polysaccharides and mucilage by them acts as a absolute factor in formative soil structure. Microbes provide indication to soil condition via responding quickly to any transform occurs in the physical and chemical property of the soil (Shonkor and Das 2011). These responses within microorganism originate due to the production of respective enzyme which exhibits useful sensing characteristic towards environment (Shonkor and Das 2011). Microorganisms like algae play important role to maintain soil fertility by adding organic matter to soil after their death. Microbiome also prevents the soil erosion by acting as a cementing agent to bind soil particles and also increase the water retention capacity for longer. Soil protozoans, the single celled organisms play major role in maintaining the microbial equilibrium in soil by feeding on bacteria. Even viruses also influence the soil ecology via controlling nutrient concentration and transferring gene from host to host (Kumar et al., 2013). Similarly, few enzymes like cellulase, amylase, dehydrogenase, and urease are catalysed various biochemical reactions and serve as energy source to microbes (Dic et al. 1996). Enzymes like oxidoreductases, oxygenases and hydrolytic enzymes have potential to detect and detoxify the toxic elements in soil (Karigar and Rao 2011). Dehydrogenase usually integrated with microbial respiration closely depends upon the air-water condition of soil ecosystem (Whiteley and Lee 2006).

Bioremediation of Soil

Bioremediation is an eco-friendly and cost-effective biological mechanism that recycles wastes into another useful form which is further reused by organisms. In this technology the soil microorganisms enzymatically decompose, eradicate or immobilize and transform the hazardous pollutants into less harmful products (Karigar and Rao 2011). Soil biosphere produce huge number of microbes and microbial enzymes which degrade the contaminants to drive their nutritional and energy requirements as listed in Table 3. Activity of bioremediator (fungi, bacteria) is highly controlled and optimized by various factors like pH, temperature, soil type and minerals availability. In situ and ex situ bioremediation methods by microbiome are the basic methods and are represented in Figure 2.

In situ type bioremediation process comprises decomposition of toxic contaminants on site via natural spontaneous process. The functioning of this method influenced by various factors like temperature, pH, oxygen, nutrient without altering soil structure and composition (Margesin and Schinner 2001, Mulligana and Yongb 2004). This technique ought to be less expensive than ex-situ. Some in-situ techniques might be superior like bioventing, biosparging. Bioventing involves the controlled stimulation of airflow by delivering oxygen to unsaturated zone and nutrient amendment to microbes to achieve harmless transformed state of pollutant. In bioventing also air is injected into soil surface to stimulate microbiome to promote pollutant remediation on site but phytoremediation technique relies on microbes-plant interaction in polluted site to reduce the toxic effect of pollutant.

Whereas, the ex situ bioremediation process involves the dig or pump out of contaminants from contaminated soil to the spot of the bioremediation treatment. Ex situ technique is majorly influenced by the types and concentration of pollutant, location, treatment cost and transformation method (Philp and Atlas 2005). Ex situ biopile mediated remediation process includes nutrient amendment, microbial activities and aeration to above ground excavated polluted soil to enhance bioremediation. It means that the inside raw materials of a bioreactor are transformed into particular products by undertaking different biological reactions in a bioreactor-based approach. Landfarming ex-situ bioremediation is the simplest technique in which polluted soils are usually tilled which brings aeration, nutrients addition and irrigation.

Figure 2. Different applied strategies of bioremediation to mitigate toxic effect of pollutant on soil and plant

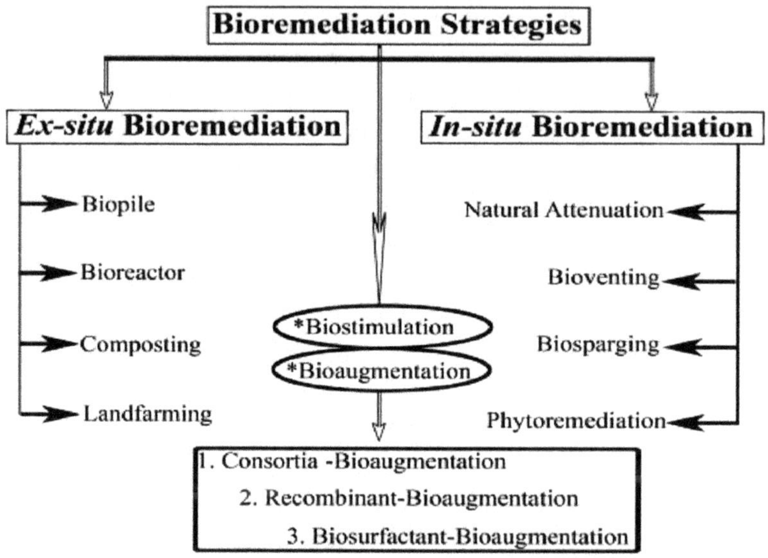

Table 3. List of important enzymes used for bioremediation

Type of Enzyme	Action Mode
Oxidoreductases	Detoxification and humification of toxic organic compounds through oxidative coupling reaction
Peroxidases	Catalyze oxidation of organic and inorganic compounds
Oxygenases	Detoxification or degradation of organic substrates
Laccases	Catalyze the oxidation of a broad range of phenolic and aromatic substrates
Hydrolytic enzymes	Hydrolysis of organic pollutant by breaking the bonds thus reduced the toxicity of the compound

(Fierer et al., 2017)

FACTOR AFFECTING THE ENZYMATIC ACTIVITIES

Soil enzymes are an enhancer or catalyst for several biochemical reactions, so suppression or lack of soil enzyme largely affects the soil fertility and agronomic productivity. Several physicochemical factors including temperature, pH, chemicals and pesticides, nature of soil or composition, soil texture, soil fertility, diversity of microbes and plant community etc. affect the activities of soil enzymes (Kheyrodin 2014) which are described in Table 4.

Table 4. Application and factors affecting the activities of enzymes produced by plants and microorganisms

Soil Enzyme	Applications	Factors Affecting Activity
Dehydrogenase	Food industry, textile industry,pharmaceuticals and biormediation	Water content, temperature, pesticides, pollution
β-glucosidase	Cellulose biodegradation, protein engineering, biofuel production	Water, pH, minerals, oxygen content, temperature and fungicides
α-Amylase	Food, textile and pharmaceuticals industry	Soil type, management practice, vegetation type
Urease	Blood urea analysis, alcoholic beverages, wastewater	Organic matter content, soil depth, management practices, temperature, pH
Phosphatase	ELISA application, enzyme immunoassay, precipitation of heavy metals	Organic matter content, management practices, temperature, pH, crop species
Protease	Protein engineering, fertilizer,biodegradable materials	C-N availability, humic acid
Aryl sulfatase	Analytical endocrinology	Heavy metals pollution, Organic matter content, management practices, temperature, pH,
Cellulase	Textile industry, paper industry, digestibility of animal feed	temperature, pH, enzyme-substrate concentration
Amidase	Food industry, analytical applications	C-N availability, humic acid concentration
Lipase	Control of oil spills, detergent production, paper and pulp industry	C-N availability, humic acid concentration

APPROACHES TO STUDY SOIL MICROBIOME

In any environmental ecology the soil microbial community is considered as the complex and heterogeneous community (Azubuike et al., 2016). With the advancement in sequences technique, the characterization of microbial components and their relation with ecosystem functioning becomes a wide spread application nowadays (Delmont et al. 2011). Over the last decades, metagenomics and metaproteomics (Maron et al. 2007; Becher et al. 2013) approaches provide a huge potential to explore the extensive range of soil biosphere microbiome.

Metagenomics

The term metagenomics was first reported by Handelsman to describe the functional potential of the soil microbial community. Soil metagenomics facilitate the exploration of the biosynthetic pathway and collective genome of soil microflora (Handelsman et al. 1998). It helps to assess untapped genetic diversity of uncultivated microbial species by providing the collective DNA information of native soil microbial community. This method involves the following steps:

- Soil DNA extraction
- Screening of genome
- Clone library construction
- Data sequencing and analysis

Soil metagenomics study provide prospect to confine new bioactive pharmaceutical products like antibiotic, biofuels, and enzymes (Ling et al. 2015). The new approach will bring light towards the myriad capabilities of the microbiome that facilitate the nutrient cycling and shape of the ecosystem. Metagenomics with the help of bioinformatics provide identification of key metabolites and how genes influence each other's activities in serving collective function (Jansson and Baker 2016). A schematic experimental work plan for metagenomics is illustrated in Figure 3.

Figure 3. A schematic workflow for the experimental designs of metagenomics

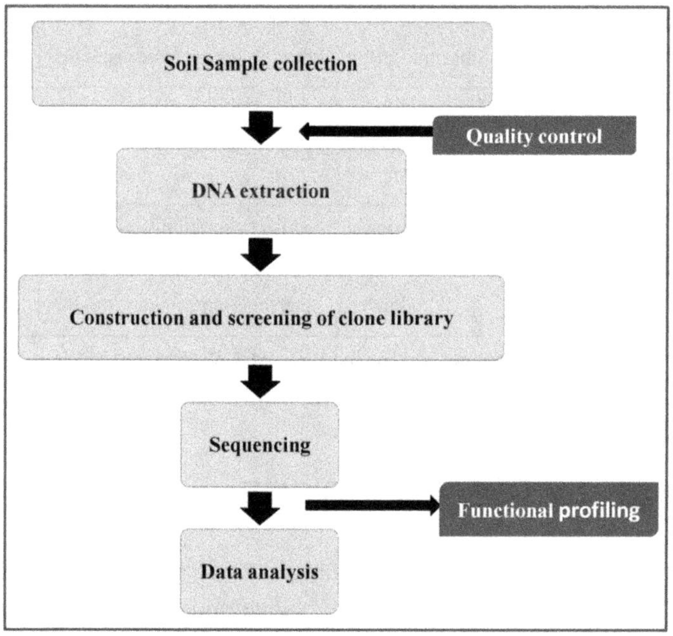

Metaproteomics

Metaproteomics or community proteogenomics involves the study of all protein samples recovered directly from environment so act as a balancing loom to metagenomics. Metaproteomics provides profiling and characterization of soil microbiome proteins at a wide scale. Metaproteomics in combination with top-down or bottom-up approach provides information about the functional characteristic, linking genetics and diversity of the microbial components at a given point in time (Maron et al. 2007). The quantity and quality of the thousands of protein components isolated from the microbial community are assessed with the help of mass spectrometry (MS) (Simon and Daniel 2011). Metaproteomics study is typically based upon three basic steps:

1. Isolation, purification and improvement of the quantity/quality of concerned protein.
2. Extraction and acquisition of both MS and MS/MS level protein/peptide.
3. Data analysis and functional characterization of microbiome community.

A schematic experimental work plan for metaproteomics approach is illustrated in Figure 4.

Figure 4. A schematic workflow for the experimental design of metaproteomics study

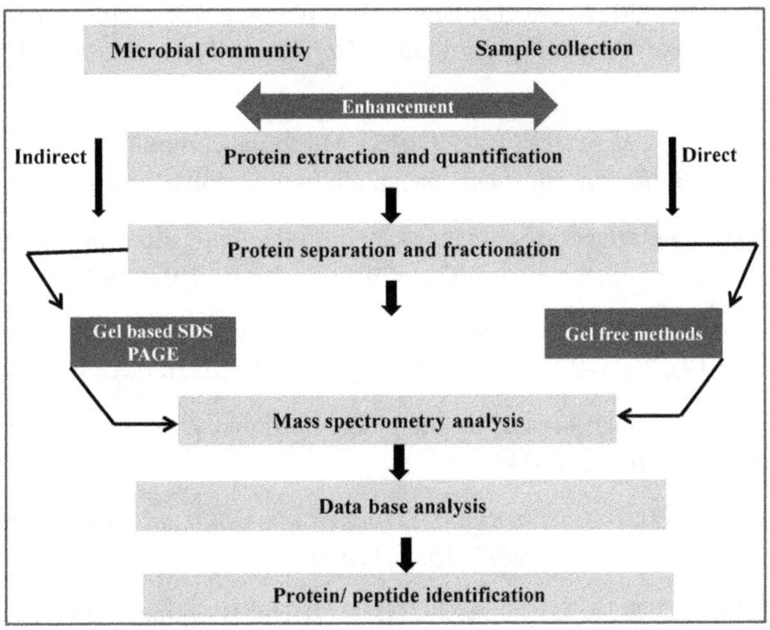

CONCLUSION AND FUTURE PROSPECTS

Enhancement in the soil productivity without affecting ecology via microorganisms and enzymes is a most challenging task in the present scenario. In this chapter we have discussed the co-relation and co-independency of microorganisms with soil to regulate or maintain the ecosystem. In addition plant microbiota and their interactions are highly diverse and microorganisms play major role to facilitate plant growth and production rate by assessing nutrients availability to plants. Further enzymes and microbes possess strong activity to detox the pollutants and hazardous chemicals reside in soil through bioremediation process. The advancement of new sequencing technology through different "omics approaches" have revealed an enormous amount of soil microbiome profiling data to evolve novel agroeconomical microbial pathways, metabolites and antibiotics which serves huge industrial and pharmaceutical potential to mankind. Modeling of plant and soil microbiome and their interactions are essential to envisage future ecosystem functions.

REFERENCES

Abdel-Raouf, N., Al-Homaidan, A. A., & Ibraheem, I. B. M. (2012). Agricultural importance of algae. *African Journal of Biotechnology*, *11*(54), 11648–11658. doi:10.5897/AJB11.3983

Ahemad, M., & Khan, M. S. (2012). Effect of fungicides on plant growth promoting activities of phosphate solubilizing *Pseudomonas putida* isolated from mustard (*Brassica compestris*) rhizosphere. *Chemosphere*, *86*(9), 945–950. doi:10.1016/j.chemosphere.2011.11.013 PMID:22133911

Ahemad, M., & Kibret, M. (2014). Mechanisms and applications of plant growth promoting rhizobacteria: Current perspective. *Journal of King Saud University. Science*, *26*(1), 1–20. doi:10.1016/j.jksus.2013.05.001

Andersen, R., Chapman, S. J., & Artz, R. R. E. (2013). Microbial communities in natural and disturbed peatlands: A review. *Soil Biology & Biochemistry*, *57*, 979–994. doi:10.1016/j.soilbio.2012.10.003

Anderson, J. M. (1991). The effects of climate change on decomposition processes in grassland and coniferous forests. *Ecological Applications*, *1*(3), 326–347. doi:10.2307/1941761 PMID:27755768

Armengaud, J., Marie Hartmann, E., & Bland, C. (2013). Proteogenomics for environmental microbiology. *Proteomics*, *13*(18-19), 2731–2742. doi:10.1002/pmic.201200576 PMID:23636904

Arrigo, K. R. (2005). Marine microorganisms and global nutrient cycles. *Nature*, *437*(7057), 349–355. doi:10.1038/nature04159 PMID:16163345

Atlas, R. M., & Philp, J. (2005). *Bioremediation: Applied microbial solutions for real-world environmental cleanup*. ASM Press. doi:10.1128/9781555817596

Azubuike, C. C., Chikere, C. B., & Okpokwasili, G. C. (2016). Bioremediation techniques–classification based on site of application: Principles, advantages, limitations and prospects. *World Journal of Microbiology & Biotechnology*, *32*(11), 180. doi:10.100711274-016-2137-x PMID:27638318

Bakermans, C., Tsapin, A. I., Souza-Egipsy, V., Gilichinsky, D. A., & Nealson, K. H. (2003). Reproduction and metabolism at– 10 C of bacteria isolated from Siberian permafrost. *Environmental Microbiology*, *5*(4), 321–326. doi:10.1046/j.1462-2920.2003.00419.x PMID:12662179

Becher, D., Bernhardt, J., Fuchs, S., & Riedel, K. (2013). Metaproteomics to unravel major microbial players in leaf litter and soil environments: C hallenges and perspectives. *Proteomics*, *13*(18-19), 2895–2909. doi:10.1002/pmic.201300095 PMID:23894095

Berg, G., Grube, M., Schloter, M., & Smalla, K. (2014). Unravelling the plant microbiome: Looking back and future perspectives. *Frontiers in Microbiology*, (5), 175–175.

Bhattacharyya, P. N., & Jha, D. K. (2012). Plant growth-promoting rhizobacteria (PGPR): Emergence in agriculture. *World Journal of Microbiology & Biotechnology*, *28*(4), 1327–1350. doi:10.100711274-011-0979-9 PMID:22805914

Bing-Cheng, Y. U. A. N., & Dong-Xia, Y. U. E. (2012). Soil microbial and enzymatic activities across a chronosequence of Chinese pine plantation development on the loess plateau of China. *Pedosphere*, *22*(1), 1–12. doi:10.1016/S1002-0160(11)60186-0

Boschker, H. T., Vasquez-Cardenas, D., Bolhuis, H., Moerdijk-Poortvliet, T. W., & Moodley, L. (2014). Chemoautotrophic carbon fixation rates and active bacterial communities in intertidal marine sediments. *PLoS One*, *9*(7), e101443. doi:10.1371/journal.pone.0101443 PMID:25003508

Callesen, I., Liski, J., Raulund-Rasmussen, K., Olsson, M. T., Tau-Strand, L., Vesterdal, L., & Westman, C. J. (2003). Soil carbon stores in Nordic well-drained forest soils—Relationships with climate and texture class. *Global Change Biology*, *9*(3), 358–370. doi:10.1046/j.1365-2486.2003.00587.x

Cameron, D. D., Neal, A. L., van Wees, S. C., & Ton, J. (2013). Mycorrhiza-induced resistance: More than the sum of its parts? *Trends in Plant Science*, *18*(10), 539–545. doi:10.1016/j.tplants.2013.06.004 PMID:23871659

Chung, H., Park, M., Madhaiyan, M., Seshadri, S., Song, J., Cho, H., & Sa, T. (2005). Isolation and characterization of phosphate solubilizing bacteria from the rhizosphere of crop plants of Korea. *Soil Biology & Biochemistry*, *37*(10), 1970–1974. doi:10.1016/j.soilbio.2005.02.025

Craven, D., Isbell, F., Manning, P., Connolly, J., Bruelheide, H., Ebeling, A., ... Beierkuhnlein, C. (2016). Plant diversity effects on grassland productivity are robust to both nutrient enrichment and drought. *Philosophical Transactions of the Royal Society B: Biological Sciences, 371*(1694), 20150277.

Dalton, D. A., & Kramer, S. (2007). Nitrogen-fixing bacteria in non-legumes. In *Plant-Associated Bacteria* (pp. 105–130). Springer.

Das, S. K., & Varma, A. (2010). Role of enzymes in maintaining soil health. In *Soil enzymology* (pp. 25–42). Springer. doi:10.1007/978-3-642-14225-3_2

Delmont, T. O., Robe, P., Clark, I., Simonet, P., & Vogel, T. M. (2011). Metagenomic comparison of direct and indirect soil DNA extraction approaches. *Journal of Microbiological Methods*, *86*(3), 397–400. doi:10.1016/j.mimet.2011.06.013 PMID:21723887

Dick, R. P., Breakwell, D. P., & Turco, R. F. (1997). Soil enzyme activities and biodiversity measurements as integrative microbiological indicators. *Methods for Assessing Soil Quality, 49*, 247-271.

Dijkhuizen, L., & Harder, W. (1984). Current views on the regulation of autotrophic carbon dioxide fixation via the Calvin cycle in bacteria. *Antonie van Leeuwenhoek*, *50*(5-6), 473–487. doi:10.1007/BF02386221 PMID:6099093

Dong, W. Y., Zhang, X. Y., Liu, X. Y., Fu, X. L., Chen, F. S., Wang, H. M., Sun, X. M., & Wen, X. F. (2015). Responses of soil microbial communities and enzyme activities to nitrogen and phosphorus additions in Chinese fir plantations of subtropical China. *Biogeosciences*, *12*(18), 5537–5546. doi:10.5194/bg-12-5537-2015

East, R. (2013). Soil science comes to life: Plants may be getting a little help with their tolerance of drought and heat. *Nature*, *501*, 18–19. doi:10.1038/501S18a

Erturk, Y., Ercisli, S., Haznedar, A., & Cakmakci, R. (2010). Effects of plant growth promoting rhizobacteria (PGPR) on rooting and root growth of kiwifruit (Actinidia deliciosa) stem cuttings. *Biological Research*, *43*(1), 91–98. doi:10.4067/S0716-97602010000100011 PMID:21157636

Fay, P. A., Prober, S. M., Harpole, W. S., Knops, J. M., Bakker, J. D., Borer, E. T., ... Adler, P. B. (2015). Grassland productivity limited by multiple nutrients. *Nature Plants*, *1*(7), 1–5. doi:10.1038/nplants.2015.80 PMID:27250253

Fierer, N. (2017). Embracing the unknown: Disentangling the complexities of the soil microbiome. *Nature Reviews. Microbiology*, *15*(10), 579–590. doi:10.1038/nrmicro.2017.87 PMID:28824177

Finlay, R. D., Ek, H., Odham, G., & Söderström, B. (1990). Mycelial uptake, translocation and assimilation of 15N-labelled nitrogen by ectomycorrhizal Pinus sylvestris plants. *Agriculture, Ecosystems & Environment*, *28*(1-4), 133–137. doi:10.1016/0167-8809(90)90028-C

Fisk, M. C., Ruether, K. F., & Yavitt, J. B. (2003). Microbial activity and functional composition among northern peatland ecosystems. *Soil Biology & Biochemistry*, *35*(4), 591–602. doi:10.1016/S0038-0717(03)00053-1

Ganzert, L., Lipski, A., Hubberten, H. W., & Wagner, D. (2011). The impact of different soil parameters on the community structure of dominant bacteria from nine different soils located on Livingston Island, South Shetland Archipelago, Antarctica. *FEMS Microbiology Ecology*, *76*(3), 476–491. doi:10.1111/j.1574-6941.2011.01068.x PMID:21314705

Gilichinsky, D. A. (2002). Permafrost model of extraterrestrial habitat. In *Astrobiology* (pp. 125–142). Springer. doi:10.1007/978-3-642-59381-9_9

Golovchenko, A. V., Tikhonova, E. Y., & Zvyagintsev, D. G. (2007). Abundance, biomass, structure, and activity of the microbial complexes of minerotrophic and ombrotrophic peatlands. *Microbiology*, *76*(5), 630–637. doi:10.1134/S0026261707050177 PMID:18069333

Gorzelak, M. A., Asay, A. K., Pickles, B. J., & Simard, S. W. (2015). Inter-plant communication through mycorrhizal networks mediates complex adaptive behaviour in plant communities. *AoB Plants*, *7*, 7. doi:10.1093/aobpla/plv050 PMID:25979966

Graham, D. E., Wallenstein, M. D., Vishnivetskaya, T. A., Waldrop, M. P., Phelps, T. J., Pfiffner, S. M., ... Elias, D. A. (2012). Microbes in thawing permafrost: The unknown variable in the climate change equation. *The ISME Journal*, *6*(4), 709–712. doi:10.1038/ismej.2011.163 PMID:22094350

Hafeez, F. Y., Yasmin, S., Ariani, D., Zafar, Y., & Malik, K. A. (2006). *Plant growth-promoting bacteria as biofertilizer*. Academic Press.

Hagerberg, D., Thelin, G., & Wallander, H. (2003). The production of ectomycorrhizal mycelium in forests: Relation between forest nutrient status and local mineral sources. *Plant and Soil*, *252*(2), 279–290. doi:10.1023/A:1024719607740

Handelsman, J., Rondon, M. R., Brady, S. F., Clardy, J., & Goodman, R. M. (1998). Molecular biological access to the chemistry of unknown soil microbes: A new frontier for natural products. *Chemistry & Biology*, *5*(10), R245–R249. doi:10.1016/S1074-5521(98)90108-9 PMID:9818143

Hardoim, P. R., Van Overbeek, L. S., Berg, G., Pirttilä, A. M., Compant, S., Campisano, A., Döring, M., & Sessitsch, A. (2015). The hidden world within plants: Ecological and evolutionary considerations for defining functioning of microbial endophytes. *Microbiology and Molecular Biology Reviews*, *79*(3), 293–320. doi:10.1128/MMBR.00050-14 PMID:26136581

Hobbie, E. A., & Högberg, P. (2012). Nitrogen isotopes link mycorrhizal fungi and plants to nitrogen dynamics. *The New Phytologist*, *196*(2), 367–382. doi:10.1111/j.1469-8137.2012.04300.x PMID:22963677

Hobbie, S. E., & Gough, L. (2004). Litter decomposition in moist acidic and non-acidic tundra with different glacial histories. *Oecologia*, *140*(1), 113–124. doi:10.100700442-004-1556-9 PMID:15164284

Hu, S., Chapin, F. S. III, Firestone, M. K., Field, C. B., & Chiariello, N. R. (2001). Nitrogen limitation of microbial decomposition in a grassland under elevated CO 2. *Nature*, *409*(6817), 188–191. doi:10.1038/35051576 PMID:11196641

Iavicoli, A., Boutet, E., Buchala, A., & Métraux, J. P. (2003). Induced systemic resistance in Arabidopsis thaliana in response to root inoculation with Pseudomonas fluorescens CHA0. *Molecular Plant-Microbe Interactions*, *16*(10), 851–858. doi:10.1094/MPMI.2003.16.10.851 PMID:14558686

Jackson, L. E., Bowles, T. M., Hodson, A. K., & Lazcano, C. (2012). Soil microbial-root and microbial-rhizosphere processes to increase nitrogen availability and retention in agroecosystems. *Current Opinion in Environmental Sustainability*, *4*(5), 517–522. doi:10.1016/j.cosust.2012.08.003

Jangid, K., Williams, M. A., Franzluebbers, A. J., Sanderlin, J. S., Reeves, J. H., Jenkins, M. B., Endale, D. M., Coleman, D. C., & Whitman, W. B. (2008). Relative impacts of land-use, management intensity and fertilization upon soil microbial community structure in agricultural systems. *Soil Biology & Biochemistry*, *40*(11), 2843–2853. doi:10.1016/j.soilbio.2008.07.030

Jansson, J. K., & Baker, E. S. (2016). A multi-omic future for microbiome studies. *Nature Microbiology*, *1*(5), 1–3. doi:10.1038/nmicrobiol.2016.49 PMID:27572648

Johns, C. (2017). Living soils: the role of microorganisms in soil health. *Fut Direct Intl*, 1-7.

Joosten, H., & Clarke, D. (2002). Wise use of mires and peatlands. International Mire Conservation Group and International Peat Society, 304.

Kandeler, E. (2007). Physiological and biochemical methods for studying soil biota and their function. In *Soil microbiology, ecology and biochemistry* (pp. 53–83). Academic Press. doi:10.1016/B978-0-08-047514-1.50007-X

Karigar, C. S., & Rao, S. S. (2011). Role of microbial enzymes in the bioremediation of pollutants: A review. *Enzyme Research*, *2011*, 2011. doi:10.4061/2011/805187 PMID:21912739

Kembel, S. W., O'Connor, T. K., Arnold, H. K., Hubbell, S. P., Wright, S. J., & Green, J. L. (2014). Relationships between phyllosphere bacterial communities and plant functional traits in a neotropical forest. *Proceedings of the National Academy of Sciences of the United States of America*, *111*(38), 13715–13720. doi:10.1073/pnas.1216057111 PMID:25225376

Khan, M. S., Zaidi, A., & Wani, P. A. (2007). Role of phosphate-solubilizing microorganisms in sustainable agriculture—A review. *Agronomy for Sustainable Development*, *27*(1), 29–43. doi:10.1051/agro:2006011

Kheyrodin, H. (2014). Methodology for measurement of enzyme activity in soil. *World J Biol Med Science*, *1*(1), 18–25.

Kibblewhite, M. G., Ritz, K., & Swift, M. J. (2008). Soil health in agricultural systems. *Philosophical Transactions of the Royal Society of London. Series B, Biological Sciences*, *363*(1492), 685–701. doi:10.1098/rstb.2007.2178 PMID:17785275

Kim, J., & Rees, D. C. (1994). Nitrogenase and biological nitrogen fixation. *Biochemistry*, *33*(2), 389–397. doi:10.1021/bi00168a001 PMID:8286368

Kirschbaum, M. U. (2000). Will changes in soil organic carbon act as a positive or negative feedback on global warming? *Biogeochemistry*, *48*(1), 21–51. doi:10.1023/A:1006238902976

Kishi, R. N. I., Júnior, R. F. G., Val-Moraes, S. P., & Kishi, L. T. (2017). Soil Microbiome and Their Effects on Nutrient Management for Plants. In Probiotics in Agroecosystem (pp. 117-143). Academic Press.

Klironomos, J. N. (2003). Variation in plant response to native and exotic arbuscular mycorrhizal fungi. *Ecology*, *84*(9), 2292–2301. doi:10.1890/02-0413

Kumar, S., Chaudhuri, S., & Maiti, S. K. (2013). Soil dehydrogenase enzyme activity in natural and mine soil-a review. *Middle East Journal of Scientific Research*, *13*(7), 898–906.

Lambers, H., Raven, J. A., Shaver, G. R., & Smith, S. E. (2008). Plant nutrient-acquisition strategies change with soil age. *Trends in Ecology & Evolution*, *23*(2), 95–103. doi:10.1016/j.tree.2007.10.008 PMID:18191280

Leake, J., Johnson, D., Donnelly, D., Muckle, G., Boddy, L., & Read, D. (2004). Networks of power and influence: The role of mycorrhizal mycelium in controlling plant communities and agroecosystem functioning. *Canadian Journal of Botany*, *82*(8), 1016–1045. doi:10.1139/b04-060

Ledgard, S. F., & Steele, K. W. (1992). Biological nitrogen fixation in mixed legume/grass pastures. *Plant and Soil*, *141*(1-2), 137–153. doi:10.1007/BF00011314

Ling, L. L., Schneider, T., Peoples, A. J., Spoering, A. L., Engels, I., Conlon, B. P., ... Jones, M. (2015). A new antibiotic kills pathogens without detectable resistance. *Nature*, *517*(7535), 455–459. doi:10.1038/nature14098 PMID:25561178

Lundberg, D. S., Lebeis, S. L., Paredes, S. H., Yourstone, S., Gehring, J., Malfatti, S., & Edgar, R. C. (2012). Defining the core Arabidopsis thaliana root microbiome. *Nature*, *488*(7409), 86–90. doi:10.1038/nature11237 PMID:22859206

Malhi, Y. A., Baldocchi, D. D., & Jarvis, P. G. (1999). The carbon balance of tropical, temperate and boreal forests. *Plant, Cell & Environment*, *22*(6), 715–740. doi:10.1046/j.1365-3040.1999.00453.x

Margesin, R. (2012). Psychrophilic microorganisms in alpine soils. In *Plants in Alpine Regions* (pp. 187–198). Springer. doi:10.1007/978-3-7091-0136-0_14

Margesin, R., & Schinner, F. (2001). Bioremediation (natural attenuation and biostimulation) of diesel-oil-contaminated soil in an alpine glacier skiing area. *Applied and Environmental Microbiology*, *67*(7), 3127–3133. doi:10.1128/AEM.67.7.3127-3133.2001 PMID:11425732

Maron, P. A., Ranjard, L., Mougel, C., & Lemanceau, P. (2007). Metaproteomics: A new approach for studying functional microbial ecology. *Microbial Ecology*, *53*(3), 486–493. doi:10.100700248-006-9196-8 PMID:17431707

Mendes, R., Garbeva, P., & Raaijmakers, J. M. (2013). The rhizosphere microbiome: Significance of plant beneficial, plant pathogenic, and human pathogenic microorganisms. *FEMS Microbiology Reviews*, *37*(5), 634–663. doi:10.1111/1574-6976.12028 PMID:23790204

Mohapatra, R. K., Srichandan, H., Mishra, S., & Parhi, P. K. (2019). Native Soil Bacteria: Potential Agent for Bioremediation. *Soil Microenvironment for Bioremediation and Polymer Production*, 17-34.

Morales, S. E., Mouser, P. J., Ward, N., Hudman, S. P., Gotelli, N. J., Ross, D. S., & Lewis, T. A. (2006). Comparison of bacterial communities in New England Sphagnum bogs using terminal restriction fragment length polymorphism (T-RFLP). *Microbial Ecology*, *52*(1), 34–44. doi:10.100700248-005-0264-2 PMID:16729225

Mulligana, C. N., & Yongb, R. N. (2004). Natural attenuation of contaminatedsoils. *Environment International*, *30*(4), 587–601. doi:10.1016/j.envint.2003.11.001 PMID:15031019

Ochoa-Hueso, R., Delgado-Baquerizo, M., King, P. T. A., Benham, M., Arca, V., & Power, S. A. (2019). Ecosystem type and resource quality are more important than global change drivers in regulating early stages of litter decomposition. *Soil Biology & Biochemistry*, *129*, 144–152. doi:10.1016/j.soilbio.2018.11.009

Orhan, E., Esitken, A., Ercisli, S., Turan, M., & Sahin, F. (2006). Effects of plant growth promoting rhizobacteria (PGPR) on yield, growth and nutrient contents in organically growing raspberry. *Scientia Horticulturae*, *111*(1), 38–43. doi:10.1016/j.scienta.2006.09.002

Page, S. E., Rieley, J. O., & Banks, C. J. (2011). Global and regional importance of the tropical peatland carbon pool. *Global Change Biology*, *17*(2), 798–818. doi:10.1111/j.1365-2486.2010.02279.x

Pırlak, L., & Köse, M. (2009). Effects of plant growth promoting rhizobacteria on yield and some fruit properties of strawberry. *Journal of Plant Nutrition*, *32*(7), 1173–1184. doi:10.1080/01904160902943197

Ramankutty, N., Evan, A. T., Monfreda, C., & Foley, J. A. (2008). Farming the planet: 1. Geographic distribution of global agricultural lands in the year 2000. *Global Biogeochemical Cycles*, *22*(1), n/a. doi:10.1029/2007GB002952

Rastogi, G., Sbodio, A., Tech, J. J., Suslow, T. V., Coaker, G. L., & Leveau, J. H. (2012). Leaf microbiota in an agroecosystem: Spatiotemporal variation in bacterial community composition on field-grown lettuce. *The ISME Journal*, *6*(10), 1812–1822. doi:10.1038/ismej.2012.32 PMID:22534606

Reinsch, S., Koller, E., Sowerby, A., De Dato, G., Estiarte, M., Guidolotti, G., ... Liberati, D. (2017). Shrubland primary production and soil respiration diverge along European climate gradient. *Scientific Reports*, *7*(1), 43952. doi:10.1038rep43952 PMID:28256623

Reyes, I., Bernier, L., & Antoun, H. (2002). Rock phosphate solubilization and colonization of maize rhizosphere by wild and genetically modified strains of Penicillium rugulosum. *Microbial Ecology*, *44*(1), 39–48. doi:10.100700248-002-1001-8 PMID:12019460

Rezzonico, F., Binder, C., Défago, G., & Moënne-Loccoz, Y. (2005). The type III secretion system of biocontrol Pseudomonas fluorescens KD targets the phytopathogenic Chromista Pythium ultimum and promotes cucumber protection. *Molecular Plant-Microbe Interactions*, *18*(9), 991–1001. doi:10.1094/MPMI-18-0991 PMID:16167769

Richardson, A. E., & Simpson, R. J. (2011). Soil Microorganisms Mediating Phosphorus Availability: Phosphorus Plant Physiology. *Plant physiology (Bethesda)*, *156*(3), 989–996. doi:10.1104/pp.111.175448

Rodrigues, E. P., Rodrigues, L. S., de Oliveira, A. L. M., Baldani, V. L. D., dos Santos Teixeira, K. R., Urquiaga, S., & Reis, V. M. (2008). Azospirillum amazonense inoculation: Effects on growth, yield and N 2 fixation of rice (Oryza sativa L.). *Plant and Soil*, *302*(1-2), 249–261. doi:10.100711104-007-9476-1

Roesch, L. F., Fulthorpe, R. R., Riva, A., Casella, G., Hadwin, A. K., Kent, A. D., Daroub, S. H., Camargo, F. A. O., Farmerie, W. G., & Triplett, E. W. (2007). Pyrosequencing enumerates and contrasts soil microbial diversity. *The ISME Journal*, *1*(4), 283–290. doi:10.1038/ismej.2007.53 PMID:18043639

Rohrbacher, F., & St-Arnaud, M. (2016). Root exudation: The ecological driver of hydrocarbon rhizoremediation. *Agronomy (Basel)*, *6*(1), 19. doi:10.3390/agronomy6010019

Saha, S., Loganathan, M., Rai, A. B., Singh, A., & Garg, R. (2016). Role of microbes in soil health improvement. *SATSA Mukhapatra Ann Tech*, (20), 53–62.

Sardar, K. H. A. N., Qing, C. A. O., Hesham, A. E. L., Yue, X., & He, J. Z. (2007). Soil enzymatic activities and microbial community structure with different application rates of Cd and Pb. *Journal of Environmental Sciences (China)*, *19*(7), 834–840. doi:10.1016/S1001-0742(07)60139-9 PMID:17966871

Sharma, S., Szele, Z., Schilling, R., Munch, J. C., & Schloter, M. (2006). Influence of freeze-thaw stress on the structure and function of microbial communities and denitrifying populations in soil. *Applied and Environmental Microbiology*, *72*(3), 2148–2154. doi:10.1128/AEM.72.3.2148-2154.2006 PMID:16517665

Sharma, S. B., Sayyed, R. Z., Trivedi, M. H., & Gobi, T. A. (2013). Phosphate solubilizing microbes: Sustainable approach for managing phosphorus deficiency in agricultural soils. *SpringerPlus*, *2*(1), 587. doi:10.1186/2193-1801-2-587 PMID:25674415

Shekher, R., Sehgal, S., Kamthania, M., & Kumar, A. (2011). Laccase: Microbial sources, production, purification, and potential biotechnological applications. *Enzyme Research*. PMID:21755038

Shridhar, B. S. (2012). nitrogen fixing microorganisms. *International Journal of Microbiology Research*, *3*(1), 46–52.

Simon, C., & Daniel, R. (2011). Metagenomic analyses: Past and future trends. *Applied and Environmental Microbiology*, *77*(4), 1153–1161. doi:10.1128/AEM.02345-10 PMID:21169428

Smith, S. E., & Gianinazzi-Pearson, V. (1988). Physiological interactions between symbionts in vesicular-arbuscular mycorrhizal plants. *Annual Review of Plant Physiology and Plant Molecular Biology*, *39*(1), 221–244. doi:10.1146/annurev.pp.39.060188.001253

Smith, S. E., & Read, D. J. (2008). *Mycorrhizal Symbiosis* (3rd ed.). Academic Press.

Smith, S. E., Smith, F. A., & Jakobsen, I. (2003). Mycorrhizal fungi can dominate phosphate supply to plants irrespective of growth responses. *Plant Physiology*, *133*(1), 16–20. doi:10.1104/pp.103.024380 PMID:12970469

Solly, E. F., Schöning, I., Boch, S., Kandeler, E., Marhan, S., Michalzik, B., Müller, J., Zscheischler, J., Trumbore, S. E., & Schrumpf, M. (2014). Factors controlling decomposition rates of fine root litter in temperate forests and grasslands. *Plant and Soil*, *382*(1-2), 203–218. doi:10.100711104-014-2151-4

Štursová, M., Žifčáková, L., Leigh, M. B., Burgess, R., & Baldrian, P. (2012). Cellulose utilization in forest litter and soil: Identification of bacterial and fungal decomposers. *FEMS Microbiology Ecology*, *80*(3), 735–746. doi:10.1111/j.1574-6941.2012.01343.x PMID:22379979

Torsvik, V., & Øvreås, L. (2002). Microbial diversity and function in soil: From genes to ecosystems. *Current Opinion in Microbiology*, *5*(3), 240–245. doi:10.1016/S1369-5274(02)00324-7 PMID:12057676

Trinder, C. J., Johnson, D., & Artz, R. R. (2008). Interactions among fungal community structure, litter decomposition and depth of water table in a cutover peatland. *FEMS Microbiology Ecology*, *64*(3), 433–448. doi:10.1111/j.1574-6941.2008.00487.x PMID:18430005

Van Der Heijden, M. G., & Horton, T. R. (2009). Socialism in soil? The importance of mycorrhizal fungal networks for facilitation in natural ecosystems. *Journal of Ecology*, *97*(6), 1139–1150. doi:10.1111/j.1365-2745.2009.01570.x

Vessey, J. K., & Buss, T. J. (2002). Bacillus cereus UW85 inoculation effects on growth, nodulation, and N accumulation in grain legumes: Controlled-environment studies. *Canadian Journal of Plant Science*, *82*(2), 282–290. doi:10.4141/P01-047

Vicente-Serrano, S. M., Gouveia, C., Camarero, J. J., Beguería, S., Trigo, R., López-Moreno, J. I., ... Morán-Tejeda, E. (2013). Response of vegetation to drought time-scales across global land biomes. *Proceedings of the National Academy of Sciences of the United States of America*, *110*(1), 52–57. doi:10.1073/pnas.1207068110 PMID:23248309

Vishnivetskaya, T. A., Petrova, M. A., Urbance, J., Ponder, M., Moyer, C. L., Gilichinsky, D. A., & Tiedje, J. M. (2006). Bacterial community in ancient Siberian permafrost as characterized by culture and culture-independent methods. *Astrobiology*, *6*(3), 400–414. doi:10.1089/ast.2006.6.400 PMID:16805696

Wagner, D., Gattinger, A., Embacher, A., Pfeiffer, E.-M., Schloter, M., & Lipski, A. (2007). Methanogenic activity and biomass in Holocene permafrost deposits of the Lena Delta, Siberian Arctic and its implication for the global methane budget. *Global Change Biology*, *13*(5), 1089–1099. doi:10.1111/j.1365-2486.2007.01331.x

Wang, L., Dong, X. P., Zhang, W., Zhang, G. S., Liu, G. X., & Feng, H. Y. (2011). Quantitative characters of microorganisms in permafrost at different depths and their relation to soil physicochemical properties. *Bingchuan Dongtu*, *33*, 436–441.

Whiteley, C. G., & Lee, D. J. (2006). Enzyme technology and biological remediation. *Enzyme and Microbial Technology*, *38*(3-4), 291–316. doi:10.1016/j.enzmictec.2005.10.010

Winsborough, C., & Basiliko, N. (2010). Fungal and bacterial activity in northern peatlands. *Geomicrobiology Journal, 27*(4), 315–320. doi:10.1080/01490450903424432

Xu, J., Morris, P. J., Liu, J., & Holden, J. (2018). PEATMAP: Refining estimates of global peatland distribution based on a meta-analysis. *Catena, 160*, 134–140. doi:10.1016/j.catena.2017.09.010

Yang, M., Yang, D., & Yu, X. (2018). Soil microbial communities and enzyme activities in sea-buckthorn (Hippophae rhamnoides) plantation at different ages. *PLoS One, 13*(1), e0190959. doi:10.1371/journal.pone.0190959 PMID:29324845

Zhu, F., Qu, L., Hong, X., & Sun, X. (2011). Isolation and characterization of a phosphate-solubilizing halophilic bacterium Kushneria sp. YCWA18 from Daqiao Saltern on the coast of Yellow Sea of China. *Evidence-Based Complementary and Alternative Medicine*. PMID:21716683

Chapter 7
Microbial Products and Their Role in Soil Health and Sustainable Agriculture

Amrendra Kumar
Indian Council of Agricultural Research, India

Swati Agarwal
Banasthali Vidyapith, India

ABSTRACT

Microbial products are being used from ages in known as well as unknown forms. Some common products harvested from microbes include proteins, amino acids, antibiotics, antibodies, secondary metabolites, organic acids, lipids, and so on. It also includes antivirals, polymers, surfactants, enzyme inhibitors, nutraceuticals, and many industrial and agricultural products. Moreover, sometimes the whole single celled microbes are harvested as a rich source of protein called single cell proteins. In a nutshell, all these products cover almost every economic sector like food, feed, agriculture, healthcare, fuel, textile, and pharmaceutical. Hence, these microbial products have serious socio-economic impressions and have unleashed enormous possibilities in terms of commercial production. However, only a small fraction of microbial products are exploited, and a larger chest remains to be achieved. In the chapter, the importance of microbes in the production of proteins, enzymes, and secondary metabolites are discussed in detail with special emphasis on sustainable agriculture.

INTRODUCTION

Microorganisms have been utilized since ancient human civilization. In early 6000 BC at Babylonians and Sumerians there is first evidence of commercial utilization of yeast for the production of alcoholic beverages from barley (Prajapati and Nair, 2008). The microbial products have gained recognition globally for their widespread applications in several industries like chemicals, food, biofuel-bioenergy, textile, leather, agriculture, pharmaceuticals etc. Microbial production of chemicals, enzymes and sec-

DOI: 10.4018/978-1-7998-7062-3.ch007

ondary metabolites (Figure 1) are rapidly gaining interest because of less time consuming fermentation reactions, low energy input, ecofriendly, required low energy and utilize cheap raw materials for their growth (Singh et al., 2016). Moreover, with the help of modern molecular biology techniques such as, recombinant DNA technology and protein engineering, microorganisms can be manipulated to produce large quantities of products with desired properties (Parket al., 2019). The role of microbial community in agriculture sector is also significant. It is expected to reach nearly $306 billion by the end of this year, 2020 (McWilliams, 2015).

Figure 1. Important microbial products

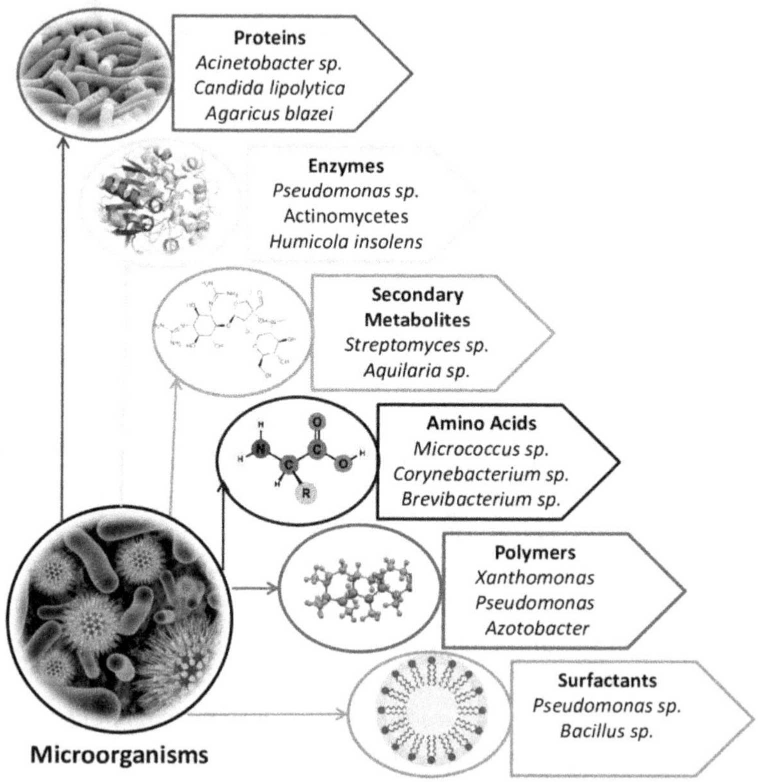

Traditionally, the farming and agricultural practices were entirely dependent upon the natural fertility of soil where soil microbiota played crucial role in productivity and maintaining the soil quality. With the advent of industrialization, the agricultural sector witnessed an unprecedented surge in the crop productivity due to the availability of chemical fertilizers, pest control agents and development of agricultural technology. Further commercialization of agriculture provoked the use of such chemicals at an alarming rate contaminating the soil, water, environment and other natural resources. It even destroyed the niche of beneficial microbes in the soil, decreasing the natural fertility and quality of soil. These impacts could be reversed and sustainability could be attained if such chemical practices are limited and microbes and microbial products are given a chance.

In the present chapter, the center of attention are microbes including bacteria, fungi, algae, yeast archaea, products derived from them along with their applications in vivid industrial sectors. The applications of different microbes in the agricultural sector are emphasized along with their role in attaining sustainable agriculture.

IMPORTANT MICROBIAL PRODUCTS

Proteins

Microbial proteins (MP) are the dried cells of microorganisms, which includes bacteria, algae, fungi and yeast (Matassaet al., 2016). Generally, MPs are used as protein-rich food and feed additives for animal consumption (Rangharet al., 2019). Since ancient time's different microorganism are used as a part of diet all over the world. Nowadays MPs are used as a replacement of animal or vegetable protein. Single-cell proteins (SCPs) are the substituted new word for microbial proteins since 60s (Nangul and Bhatia, 2020). Mostly they are the protein extract or microbial biomass, which are used as food sources or feed additives.

Conventional protein has several drawbacks over MPs, such as protein produced from conventional crop has shortage of land and some environmental disasters (drought, wind, flood etc.) (Ali et al., 2017). MPs are rich in essential amino acids, such as lysine, leucine, arginine, methionine, histidine etc. (Anderson and Jackson, 2000).Presently more than 25% of the world's population suffers from malnutrition and hunger (Behrman et al., 2004; Prosekov and Ivanova, 2018), therefore, MPs can be used as a good replacement of traditional protein sources (Matassa et al., 2016). MPs are also rich in fats, carbohydrates, vitamins, minerals and nucleic acids (Sumanet al., 2015).

Enzymes

Enzymes play an important role in catalyzing the biochemical reactions and used in all the stages of metabolism (Matson et al., 1994). Certain enzymes are widely used as organic catalysts in various processes, which require their mass production. Enzymes are produced with the help of various microbes known as microbial enzymes (ME). ME are known to be superior enzymes have wide applications in several industries. ME are far better than conventional enzymes due to its ease in availability, fast growth rate and great variations. Microbial cells can be easily manipulated using recombinant DNA technology based on the enzyme required, elevated production and scientific development (Singh et al., 2016).

Different microorganisms are used to produces several classes of enzymes such as, *Pseudomonas*, *Bacillus*, *Clostridium*, and some fungi for proteases; fungal species *Trichoderma*, *Penicillium* and *Aspergillus* for xylanases; bacteria such as *Cellulomonas, Cellvibrio* and *Pseudomonas* for cellulases etc (Table 1). MEs are more specific to perform specialized catalytic reactions in compare to conventional enzymes. With the helpof protein engineering, metagenomics and biochemical-reaction engineering different new enzymes have been designed (Liu and Kokare, 2017). Some modern molecular biology techniques have also been also used to manipulate the genetic makeup of microbes to improve the quality and performance of their enzymes for their wider applications in different industries (textile, cosmetics, detergent, paper, polymer etc.).

Table 1. Microbial enzymes and their applications

Industries	Enzymes Used	Applications	References
Pharmaceuticals	Penicillin acylase, peroxidase	Antibiotics (synthesis or semi-synthetic) AntimicrobialsAntibacterials	Sanchez and Demain, 2011
Paper and pulp	Amylase, ligninase, xylanase, cellulase, hemicellulase, esterase, lipase, protease	Starch degradation aiding, sizing, deinking etc.) Smoothening of fibers (cellulase and hemicellulase) Soften paper by remove lignin using lipases and ligninase	Lakshmidevi and Muthukumar, 2010
Textiles	Amylase, keratinase cellulase, protease, pectinase, peroxidase, catalase	Degumming of raw silk (biopolishing) Denims fabric finishing Wool treatment Cotton softening	Sarkar et al., 2017
Detergents	Amylase, mannanase, cellulase, protease, lipase	Remove insoluble starch and protein after staining, Cleaning agents Removingfats and oils	Hasan et al., 2006
Food, dairy and beverage	Protease, amylase, lipase, lactase, amyloglucosidase, pectinase, phospholipase, laccase	Degradation of starch and proteins into sugars Processing fruit juices and beer Production of cheese and glucose Stability and conditioning of dough	Mehta and Sehgal, 2019
Leather	Protease, lipase	Unhearing, bating, depicking	Choudhary et al., 2004
Ethanol production	Cellulase, ligninase, mannanase	Formation of ethanol	Martins et al., 2011
Molecular biology	DNA ligase, polymerase, restriction enzymes	Manipulate DNA in genetic engineering. DNA restriction and the polymerase chain reaction. Important in forensic science.	Adrio and Demain, 2014
Animal feed	Phytase, xylanase	Increase total phosphorus content for growth, Digestibility	Brandelli et al., 2015

Secondary Metabolites

Microbial secondary metabolites (SMs) are defined as a low molecular mass result of secondary metabolism. Microorganisms usually produce SMs during the idiophase (late growth phase). SMs are not essential for the growth and reproduction of microorganisms but it serves various other functions in nature (Mousa and Raizada, 2013). SMs are essentially used as anti-infective drugs. The anti-infective SMs had a market value of 55 billion dollars in year 2000 (Barber et al., 2004), but in a very short time period its market increased rapidly and reaches upto 66 billion dollars in year 2007 (Demain and Sanchez, 2009).

SMs are produced via several unique enzymatic pathways with the help of group of proteins or a specific protein and multifunctional polypeptides, e.g., polyketide synthases, peptide synthetases etc. (Demain, 2008). SMs include antibiotics, enzyme inhibitors, pigments, toxins, immuno-modulating agents, antiturnour agents, pheromones, receptor (antagonists and agonists), pesticides, animals/plants growth promoters, ecological effectors for symbiosis and other competitions. Most of the SMs are small (less than 1500 Da) in structure and produced by non-ribosomal systems of higher molecular weight (3000-4000 Da, 32-34 residues) known as lantobiotics (Alwendawi et al., 2019). Lantobiotics include subtilin produced by *Bacillus subtilis*, nisin by *Streptococcus lactis* and epidermin by *Staphylococcus epidermidis* (Demain, 2008).

Amino Acids

Amino acids have several application such as food raw material, cosmetics, pharmaceutical, biofertilizer, molecular biology and medical industries (Brandelli et al., 2010). Diversified applications of amino acids have played a significant role in boosting the research activities in this particular field. Conventional methods, such as amino acid extraction from natural sources (plant or animals) or through chemical synthesis, have largely been replaced by modern biotechnological approaches. Ongoing research on the production of amino acids through microorganisms started during the late 1940s. By the end of the 1950s, several microbial amino acids were produced (Yuan et al., 2017). The best example of microbial amino acid is l-glutamic acid produced by *Corynebacterium glutamicum* (Hirasawa and Shimizu, 2016). This progress had a great economic effect in the field of amino acid production and consequently, l-lysine was successfully produced and commercialized using a mutant strain of *C. glutamicum* (Kinoshita et al., 1958).

During the last three decades, a large number of mutant strains of different microorganisms have been constructed to produce several amino acids for various industrial purposes. Biotechnology has certainly played a major role in the fermentative production of various amino acids. Some examples of microbial amino acids are L-amino acids e.g. alanine (from pyruvate), leucine (from α-ketoisocaproic acid) and phenylalanine (from phenyl pyruvate) produced from dehydrogenases of *Bacillus megaterium* (Brautaset et al., 2007). In addition, the recent advances in genome analysis revolutionized the microbial strain improvement techniques. With these advancements, the amino acid production and consumption is expected to increase more in the near future (Ovaa, 2014).

Antibiotics

Microbial antibiotics are produced by both fungus and bacteria but more than 50% of them are solely obtained from *Streptomyces sp.* alone. *Streptosporangium, Streptoverticillium Actinoplanes, Micromonospora, Nocardia, Actinomadura*, and *Thermoactinomyces* are some important genera (Berdy, 2005).Some *Bacillus sp.* also produced different medically useful antibiotics (Kumar et al., 2020). Some important microbial antibiotics are listed in Table 2. In previous section we discussed about secondary metabolites produced by microorganism and their characteristics. SMs mostly have several valuable therapeutic characters and have been used as medical products such as, antibiotics, anti-tumor agents, and cholesterol-lowering compounds etc.

Variety of microorganisms (bacteria, fungi and actinomycetes) are producing antibiotics at large commercial scale. More than 8000 variety of antibiotics were solely isolated from bacterial cultures (gram + and -) and of fungi, but out of them only about 100 of these have been used commercially to treat diseases in humans, animals and plants (Kim et al., 2005). In addition, around 2500 antibiotic active substances have been reported in lichen, algae, higher animals and plants (Shrestha and Clair, 2013).

Table 2. List of commercially produced antibiotics

Antibiotics	Microorganisms	Activity	Chemical Nature	References
Amphotericin B	*Streptomyces nodosus*	Antifungal	Polyene	Haque et al., 2017
Bacitracin	*Bacillus subtilis*	Gram$^+$ bacteria	Peptide	Hassan et al., 2020
Cephalosporin C	*Ciphaloporium acrimonium*	Gram$^+$ and $^-$ bacteria	Peptide	Adinarayana et al., 2003
Cycloheximide	*Streptomyces griseus*	Antifungal	Dicarboximide	Sottorff et al., 2019
Fungimycin	*Streptomyces coelicolor*	Antifungal	Polyene	Mcdaniel et al.,1965
Gentamicin	*Micromonospora purpurea*	Gram$^+$ bacteria	Aminoglycoside	Meenavilli et al., 2008
Gramicidin	*Basillus brevis*	Gram$^+$ bacteria	Peptide	Berditsch et al., 2007
Griseofulvin	*Penicillium griseofulvum*	Gram$^+$ and $^-$ bacteria	Spirolactone	Saykhedkar and Singhal, 2004
Kanamycin	*Streptomyces kanamyceticus*	Gram$^+$ and $^-$ bacteria, mycobacteria	Aminoglycoside	Yanai et al., 2006
Neomycin	*Streptomyces fradiae*	Gram$^+$ and $^-$ bacteria	Aminoglycoside	Adinarayana et al.,2003
Pimaricin	*Streptomyces natalensis*	Antitumor	Polyene	Recio et al., 2004
Penicillin G	*Streptomyces chrysogenum*	Gram$^+$ bacteria	Peptide	Van Den Berg et al.,2008
Polymyxin B	*Bacillus polymyxa*	Antifungal	Peptide	Deng et al., 2011
Streptomycin	*Streptomyces griseus*	Gram$^+$ and $^-$ bacteria, mycobacteria	Aminoglycoside	Hong et al., 2007
Tetracycline	*Streptomyces spp.*	Gram$^+$ and $^-$ bacteria	Tetracyclin	El-Naggar et al., 2006
Trichomycin	*Streptomyces hachijoensis*	Antifungal	Polyene	Liu et al., 2012

Polymers

Polymers also called as macromolecules that are basically made up of small low molecular weight monomers (Gandini, 2008). This process in known as condensation reaction and it involves removal of water molecule. Polymers have several bonds such as, ester, amide, urethane, sulfide, and ether bonds (Ulery et al., 2011).Condensation polymerization involves diamines and dicarboxylic acids. Building blocks (monomers) for polymer formation must have carbon-carbon double bonds and microbially produced chemicals for polymer should satisfy these criteria. Diamines (cadaverine and putrescine), dicarboxylic acids (adipic, fumaric, glucaric, malic, and succinic acids), and diols (propanediols and butanediols) are commonly used monomers for polymerization reactions.

Various microbial polymers are already available in markets and they are generally produced via medium- to large-scale fermentations. The world's annual production of polymers is around 2,000 tons for polysaccharides dextran, 100,000 tons for xanthan and up to 100,000 tons for the polyesters (Jem et al., 2010). Microbial production of biopolymers has several advantages over other sources due to their unlimited renewable resources, biodegradable and biocompatible nature. Biopolymers are biodegradable, and this property makes them more attractive in comparison to oil based polymers.When exposed to the environment they are fully degradable in CO_2 and H_2O.

Biosurfactants

Biosurfactants are synthesized by microorganisms have diverse group of active molecules. These molecules have property to reduce the surface tensions of aqueous solutions. This particular property makes surfactant a potential candidate for de-emulsification processes and enhances oil recovery. Biosurfactants are more useful in comparison to chemically synthesized surfactants due to its lower toxicity, better environmental compatibility, higher biodegradability, specific activity at extreme conditions (temperature, pH, and salinity), higher foaming, high selectivity and using renewable sources as feedstock (Muthusamy et al., 2008).

Microorganisms utilize different types of organic compounds as carbon source for their growth. Insoluble hydrocarbon [C_xH_y]) when used as substrate for the microbial growth, biosurfactants are secreted by microbes help facilitate its diffusion inside the cell to be utilized as carbon source. Some microorganisms (bacteria and yeasts) excrete ionic chemical substances, which emulsifies the C_xH_y substrate in the growth medium, are also surfactants. Some examples of biosurfactants are rhamnolipids produced by *Pseudomonas sp.* (Haba et al., 2003), sophorolipids by *Torulopsis sp.* (Inoue and Ito, 1982), trehalose corynomycolatesby *Arthrobacter sp.* and *Mycobacterium sp.* (Tzvetkov et al., 2003), emulsan by *Acinetobacter sp.* (Fondi et al., 2012), surfactin and subtilisin by *Bacillus subtilis* (Deng et al., 2011) etc.

Antibodies

Nowadays, biopharmaceutical market gets significant attention in global economy. Biopharmaceutical production via microbes has several advantages over chemically synthesized pharmaceutical products (Sanchez and Demain, 2011). Microorganisms produce several compounds with high molecular weight such as proteins, which have ability to carry out highly enantio- and regioselective reactions. These kind of selective reactions are hard to achieve by chemically synthesized proteins. Repeated implementation of immobilized enzymes is achieved with the use of microorganisms ultimately resulting in the reduction of the overall production costs. Another advantage of microbially synthesized antibodies over chemically synthesized ones is that the microbes do not generate organic and inorganic pollutants, such as toluene and mercury (Gupta and Shukla, 2017).

More recently, microbial antibodies are not only used in therapeutic applications but also in immune-detection, purification, and bioseparation applications. Antibodies having antigen binding properties, can be easily manipulated and cultivated in microbes. Among the commercially available antibodies, *E. coli* is the most popular system for the production. Additionally, microbial derived antibodies don't require post translational modification for their biological activity (Terpe, 2006).

Chemicals

Nowadays, microbial production of chemicals has also gained a lot of attention due to its cost effective nature. Large amount of chemicals from renewable resources can be produced. Microbial production of chemicals can be produced from either anaerobic (high yield and productivity) or aerobic processes (less-efficient). Utilization of O_2 molecule plays an important role in product formation via energy generation and redox metabolism. Some is important chemicals which are produced by microorganisms are mainly organic solvents, including ethanol, acetone, butanol and citric acid.

Bioethanol is produced by microorganisms via fermentation of sugars in chemical industry (Singh et al., 2017). In anaerobic conditions, microbes produce ethanol and CO_2 by utilizing glucose. Some microorganisms such as *Kluyveromyces marxianus*, *Saccharomyces cerevisiae*, *Escherichia coli, Pichia kudriavzevii* and *S. diastaticus*are are used to produce ethanol from sugar substrates. Out of them *S. cerevisiae* is commonly used microorganism for bioethanol production via fermentation due to its high efficiency of ethanol conversion from sugars.

For acetone and butanol production, the most common microbial species used is *Clostridium acetobutylicum* via fermentation process. This bacterium, *Clostridium acetobutylicum* has ability to ferment huge amount of carbohydrates such as glucose, starch, sucrose, fructose, lactose, lignocelluloses and maltose. In chemical industry, acetone and butanol are widely used as solvents.

Likewise, citric acid is also a very important chemical in pharmaceutical and food industries. Different microbes (bacteria, fungi and yeast) such as *Arthrobacter paraffineus*, *Mucor piriformis*, *P. janthinellum*, *Ustulina vulgaris, P. purpurogenum, Saccharomycopsis lipolytica, Paecilomyces divaricatum, Penicillium citrinum*, *Botrytis* sp., *Trichoderma viride* etc. are used widely for the commercial production of citric acid. Among these microorganisms, *Aspergillus niger* is most commonly used species due to its ability to ferment variety of cheap raw materials, ease in handling and higher yields of citric acid production (Show et al., 2015).

MICROBES AND THEIR PRODUCTS

There are different types of microorganisms that are used for the large-scale production of industrial products and industrial microbiology is a field of microbiology, which studies the industrially important microorganism for large scale production. Such microorganism includes natural microbes, laboratory microbes or even genetically modified microbes. Here we are going to discuss some naturally occurring microorganisms. Natural microorganisms are mainly divided into 5 different class, 1) Archea, 2) Bacteria, 3) Algae, 4) Fungi and 5) Yeast. The following is a brief overview of these microorganisms and the industrial roles they play.

Archaea

Archaea are classified as prokaryotic microorganisms which have special abilities to survive in extreme conditions (Oren, 2014). Their nature to sustain in harsh environment makes them important for several commercialized processes. They can be manipulated and exploited for the commercial production of several products. An example is extremophile archaea. This group of archea got particular interest due to their biochemistry (production of some enzymes and biomolecules) which helps them to sustain their life in extreme conditions. They can survive in very high to very low temperature, extreme drought, acidic, alkaline or even in presence of radiations. An archaea named *Pyrococcus furiosus*is used to isolate specific enzyme DNA polymerases. DNA polymerases are one of the most important enzymes used in molecular biology. It is used in DNA polymerization reaction at very high temperature, but its unique thermostable property makes it suitable for such an application. Some other important enzymes are also isolated from *Pyrococcus* species, which includes specific types of amylases and galactosidases thatworks well in food processing and baking industry (Sundarram and Murthy, 2014).

Corynebacteria are another example of archaea. They have very diverse origin (Lombard and Moreira, 2011). They are used industrially for the production of numerous amino acids and nutritional factors. The most common amino acid produced is glutamic acid which is used as a very common food additive, also known as monosodium glutamate. *Corynebacterium* is also used in the steroid conversion reactions which includes the degradation of hydrocarbons (Donova, 2007). This particular reaction is very important in the development of several pharmaceutical products.

Bacteria

Different species of bacteria are exploited for several commercial product formations. They have capability to produce several metabolites which are used to manufacture antibiotics, food products, beverages, vitamins, chemicals, solvents, vaccines, drugs, enzymes, fuels etc. With the help of genetic engineering, bacteria such as *Lactococcus* and *Streptococcus* can be programmed to produce economically important products at large scale (Gaspar et al., 2013). Both the above-mentioned bacteria are highly exploited in food and dairy industry to produce yogurt, buttermilk, cheeses, including cottage cheese and cream cheese, sour cream, cultured butter etc. Some acetic acid and lactic acid bacteria also plays an important role in making pickles such as sauerkraut, olives and cucumber pickles. Other food products that are also processed via fermentation with the help of bacteria includes cocoa, teas, coffee, sausages, soy sauce and other amazing flavored foods from our everyday lives.

Other than food and dairy, economically important bacteria are also helpful in medicine and pharmaceutical industry. Drug and pharmaceutical products include vaccines, antibiotics and some medically-useful enzymes. Most of the antibiotics are produced by soil habitat bacteria. Most common types of antibiotics are rifamycin, streptomycin, tetracyclines, ivermectin, erythromycin (*Actinomycete* and *Streptomyces*) and bacitracin, polymyxin (*Bacillus* and *Paenibacillus*). Bacterial products are commonly used in preparation of several vaccines for immunizing agents against infectious disease. Examples of vaccines which are produced by bacteria are mainly against whooping cough, cholera, diphtheria, typhoid fever and tetanus (Germanier, 2012).

Algae

Algae are used to produce several foods and live feed for aquatic organisms. Most of the aquatic organisms are bivalve molluscs, abalone, crustaceans and some fishes. Therapeutic supplements and drugs are also produced with the help of algae such as astaxanthin, polyunsaturated fatty acid (DHA and EPA), polysaccharides (β-glucan dominate) and β-carotene(Pulz and Gross 2004; Spolaore et al., 2006). Some algae are exploited for bioenergy generation such as biohydrogen, biodiesel, biomethane etc. CO_2-mitigation is a process by which CO_2 is captured and sequestered, are also get possible with the help of micro-algae, this research is under process now (Li et al., 2007). Some important species of algae used in large scale production of useful products includes *Chaetoceros*, *Dunaliella*, *Isochrysis*, *Chlorella* and *Arthrospira*. Currently, the market of algal biomass gained a size of around 5,000 t y^{-1} of dry matter and generates more than US$ 1.25 billion per year of turnover (Prabhu et al., 2019).

Fungi

Fungi are used to produce several commercially important products such as enzymes, medicines, food items, textiles, leather, timber, rubber, plastic, etc. The substrates they utilized for product formation ranges from simple sugars to complex carbohydrates. They generally produce extracellular enzymes such as cellulases, xylanases and pectinases that help in releasing soluble components from insoluble materials. Hundreds of fungal species are tremendously important to man. In fact, our lives are intimately linked to those of the fungi (Alexopoulos et al., 1996).

More than 120,000 species of fungi have been discovered. They play an important role in medicine, agriculture (maintaining the fertility of the soil and causing crop and fruit diseases) and many food and textile industries (fabric, leather etc.). Some of the fungi are used experimental models in some important biological processes (Bennet, 1996; Benka-Coker and Olumagin, 1996). They also produce industrially important chemicals such as alcohols, acetone and enzymes as well as play crucial role in fermentation processes like in the production of alcoholic beverages, vinegar, cheese and bread dough.

Yeast

Yeast has wide variety of applications in different industries such as food, brewing, biofuel, enzyme, baking, molecular biology, etc. Application in food industry includes making wine, brewing, baking, distillation processes and in biomass production (single cell proteins [SCP]). In addition to brewing products, yeast enzymes are also used to produce baking products. It helps in making fermented dough. Baker's yeast (*Saccharomyces cerevisiae*) is commercially available in market in the form of dry cells (dehydrated cells) and instant cells (lyophilized) to produce bread and bakery products (Chavan and Chavan, 2011). Baker's yeast is mainly a brewery yeast which is produced via submerged fermentation process in the presence of oxygen molecules.

Recently, yeast is also exploited in the field of biofuel industry. Biofuel industry uses the metabolic process for conversion of carbohydrate compounds into carbon dioxide and ethyl alcohols in anaerobic condition. Industrially important yeast species are *Saccharomyces cerevisiae*, *Candida*, *Endomycopsis* and *Kluyveromyces* (Turker, 2014). Different approaches are used to manipulate genes of yeast to generate new variants according to specific needs of industry. Additionally the manipulated variants of yeast species are more tolerant to physiological stress such as heat, pH, salt and for high ethanol production.

COMMERCIAL APPLICATION OF DIFFERENT MICROBIAL PRODUCTS IN AGRICULTURE

Agricultural sector accounts for tremendously increasing microbes and their products to not only increase the crop production but also for crop protection and soil quality improvement as well as remediation. Microbes and their products can be used either as bioinoculants (microorganisms itself) or biostimulants. Biostimulants includes the microbial products like phytohormones, amino acids and vitamins that stimulate the growth and development of plants. Similarly, the bioinoculants helps in decomposition of organic matters in soil and further fixation, solubilization and mineralization of nutrients in soil. Apart from this, microbes have also played a crucial role in phytoremediation, removing toxic chemicals and metals ions from the soil and making land fertile and suitable for cultivation once again.

Enzymes Used in Feed and Fodder

Use of enzymes in agriculture is primarily in the feed sector. The enzymes are used in animal diet formulation to increases the availability of nutrients to animals by helping digesting the high amount of starch and degrading proteins into its constituent amino acids. Enzymes like, phytase, proteases, glucanases, α-amylases, α-galactosidases, xylanase and β-glucanase (Singh et al., 2016), primarily used in poultry feed and animal fodder. Phytases works on phytic acids present in cereal based foods to help utilize natural phosphorus present in phytic acids. Xylanase and β-glucanase help monogastric animals to digest fodder high in cellulose, hemicellulose and starch. Proteases help in easier digestion and uptake of supplement proteins provided to poultry feeds. These enzymes not only increases the nutritional quality of feed but also enhances the meat quality (Adrio and Demain, 2014).

Single Cell Proteins as Fodder and Feed Supplements

Microbial proteins or single cell proteins (SCPs) are the proteins produced either from bacteria, algae, fungi or protists. Advantages of SCPs are their high content of protein, ease of production and processing and economical manufacturing (Jones et al., 2020). Bacteria like *Methylococcus capsulatus, Cupravidus nectar, Methylophilus methylotropus, Areomonas hydrophylla* etc are utilized for SCP production (Bhalla et al., 2007; Nasseri et al., 2011; Ritala et al., 2017). Similarly, protists like *Schizochytrium limacinum*; microalgae like *Chlorella vulgaris, Desmodesmus* sp. while yeasts like *Saccharomyces cerevisiae* and *Candida utilis* are commercially utilized. The popularity of SCPs is majorly due to their high protein content. SCPs from bacteria, fungi and microalgae constitutes 50-80% of protein while the same is less for protists constituting only 10-20% (Jones et al., 2020). A fairly well summarized microbes in the production of SCPs can be found in reference (Ritala et al., 2017). These microbial proteins serve as a critical and economical source of protein to poultry feed as well as fodder.

Biopesticides and Biofertilizers

Biopesticide is any biological product obtained from plants, animals and microbes that can act against or help controlling the agricultural pest. Biopesticides of microbial origin includes microorganism like bacteria, fungi, viruses and protozoa (Sharma and Malik, 2012). The most common agents are *Trichoderma*, a biofungicide; *Phytopthora*, a bioherbicide and *Bacillus thuringiensis*, a bioinsecticide (Gupta and Dikshit, 2010). *Bacillus thuringiensis* produces a protein crystal, called Bt-δ-endotoxin during the bacterial spore formation which when consumed by certain categories of insect larve, causes lysis of gut cell (Gill et al., 1992). Similarly, *Trichodermaacts* against various soil borne diseases causing root rot and wilt among pulses like chickpea, gram and groundnut (Nargund et al., 2007). Among viruses, a special family called baculoviruses infects and kills their host pests by taking over the metabolic processes of host insects. These classes of viruses mainly infect the lepidopteran and sawfly pests. Use of biopesticides decreases the dependence upon similar competitive chemical agents, which causes much-anticipated environmental and agricultural hazard.

On the other hand, biofertilizers decreases the dependence upon chemical fertilizers which again poses serious concerns upon the soil productivity in long run as well as negative environmental consequences. Biofertilizers include the preparations of specific microorganisms that help and improves the accessibility and availability of nutrients in soil for an enhanced absorption by plants. These biological agents work by improving the soil quality in terms of availability of micro and macro-nutrients, minerals, bioactive compounds and decomposing organic matters (Bhardwajet al., 2014; Singh and Prasad, 2011; Youssef and Eissa, 2014). The most commonly utilized bifertilizers of bacterial origin are *Rhizobium*, *Azotobacter* and *Azospirillum.* These are well known as atmospheric nitrogen fixers and works basically at the roots of plants. *Azolla,* a blue green algae also falls under this category. Apart from nitrogen fixation, certain microbes help solubilizing phosphates (*Pseudomonas*, *Agrobacterium*, *Micrococcus*, *Flavobacterium*, etc), zinc (*Thiobacillus thiooxidans*, *Bacillus subtilis*, *Saccharomyces* sp.), potassium (*Aspergillus*, *Azotobacter*, *Clostridium*, *Rhizobium meliloti*, etc), aluminum silicates (*Bacillus globisporus* Q12) and so on. Some fungal species forms symbiotic association with plant roots forming mycorrhiza. This mutual association benefits the plant with the availability of nutrients like phosphate, calcium, zinc and copper while the fungal partner gets the sugar for its survival. *Glomus mosseae* is a fungi that forms symbiotic association in roots of corn resulting in increased plant growth and better harvest (Chen et al.,2017). Further details on biofertilizers and biopesticides are provided elsewhere (Abbey et al., 2019; Dhir, 2017).

Other Bioactive Compounds

Some bioactive compounds of microbial origin includes avermectin and milbemycins as insecticides(produced by *Strptomyces avermitilis* and *Streptomyces hygroscopicus*, respectively), phoslactomycin, dapiramicin and irumamycinas fungicides (isolated from *Streptomyces nigrescens*, *Micromonospora* sp. and *Streptomyces flavus* subsp. *irumaensis*, respectively) and anisomycin, hydantocidin, cornexistin, herboxidiene as herbicides (isolated from *Streptomyces* sp., *S. hygroscopicus* SANK 13584, *Paecilomyces variotti* SANK 21086, *Streptomyces chromofuscus*, respectively) (Tanaka and Omura, 1993).

Apart from this, microbes are also utilized for production of plant growth regulator, called phytohormones like gibberellins, cytokinins and auxins. Fungi like *Gibberella fujikuroi*, *Phaeospheria* sp. and *Sphaceloma* sp. are commercially utilized to extract gibberellins. Similarly, Gibberllins are also produced by bacteria like *Acetobacter diazotropicus*, *Bacillus pumilus*, *Azospirillum brasilense*, etc (MacMillan, 2001) Auxin, cytokinin and absicsic acid are also reported to be produced form bacterial species like *Proteus mirabilis*, *Bacillus megaterium*, *Bacillus cereus*, *Proteus vulgaris* etc.(Karadenizet al., 2006). Some fungi, *Funalia trogii* ATCC 200800 and *Terametes versicolor* ATCC 200801 thriving on waste water from oil mill and alcohol industry have also been reported to be potent in production of Gibberellic acids, cytokinin, abscisic acid and indole acetic acid (Yürekli et al.,1999).

BIOTECHNOLOGICAL INTERVENTIONS IN LARGE SCALE PRODUCTION OF MICROBIAL PRODUCTS

Commercially, every sector of economy is directly or indirectly consuming the microbial products that are used in food, agriculture, chemical, energy, medicine and numerous others. Its huge demand is satisfied by the large scale production which is made possible by the emergence of recombinant DNA technology, metabolic pathway engineering and strain improvement methodologies. These approaches make microbes suitable for enhanced production as well as large scale cultivation.

The goal of strain improvement is not only aimed at enhanced production but should also include lower fermentation periods, complete utilization of complex raw materials, decrease cell death, lower production of undesired metabolites and enhanced extracellular secretion of desired product/s (Saxena, 2015).This can be achieved by a number of methodologies like isolating superior microbes out of spontaneous or deliberate mutations caused by chemical (5-bromouracil, hydroxylamine, ethyl methane sulphonate, etc.) or physical (X-rays, UV rays) mutagens. For example, UV mutagenesis of *Penicillium chrysogenum*, *Gibberella fujikuroi*and *Acremonium chrysogenum* was carried out for enhanced production of penicillin, gibberellic acid and cephalosporin, respectively (Kardos and Demain, 2011; Lale et al., 2006; Saxena, 2015). Similar approaches has also been utilized to screen mutants resistant towards toxic metabolites and abiotic stress (Fiedurek et al., 2017).

Recombinant DNA technology has made it possible to manipulate the genetic constituents of microbe for improved industrial production of enzymes and amino acids (Gouka et al.,1997; Mahalik et al., 2014). This includes over expression of structural genes, genome engineering, ribosome engineering, precursor engineering and mutagenesis. For example, α-amylase gene from *Bacillus amyloliquefaciens* and benzyl penicillin acylase gene of *E. coli* were cloned on multi copy plasmid resulting in 45 fold increase in production of penicillin (Saxena, 2015) and 2,500 fold increase in α-amylase production (Palva 1982). Similar genetic improvements have been performed on countless microorganisms to cater the increasing consumption (Fiedurek et al., 2017; Paes and Almeida, 2014; Pham et al., 2019).

Increment in the production can also be achieved by targeting a desired alteration of certain metabolic pathways. This concept comes under the shed of metabolic engineering. It involves approaches like heterologous expression of entire gene cluster, redirecting metabolic pathway, engineering regulatory network, precoursor mediated stimulation, genetic knockout, gene insertion and deletion (Ko et al., 2020; Kumar and Prasad, 2011). For example, heterologous gene *crtEYIBZ* from *Pantoea ananatis* and *trCrBKT* gene from *Chlamydomonas reinhardtii* has been introduced in *E. coli* for the production of astaxanthin through astaxanthin pathway, which has vivid applications in pharmaceutical and cosmetic industry (Park et al., 2018). Similarly, gene encoding spidroid I protein (main component of spider silk protein) form *Nephila calvipes* was introduced in E. coli for hastle free expression of silk protein in large quantities (Xia et al., 2010). Recently, metabolic engineering has been used to enhance production of amino acids like lysine, valine, arginine, threonine, etc., industrially important chemicals like succinic acid, butanol, 1,4 butanediol, 1,3 propanediol, etc, and drugs like arteminism (Ko et al., 2020; Lee and Kim, 2015).

IMPACT OF COMMERCIALIZED AGRICULTURE ON SOIL MICROBIOTA

The ever growing demand of essential commodities by our ever expanding population has put our agricultural system to its extreme. To attain higher productivity, we relied on the usage of chemical fertilizers, pesticides, insecticides, herbicides and all other sorts of agricultural aids. No doubt their applications has increased the crop productivity and agricultural output but at the same time we have destroyed the microbial niches of beneficial soil microbes, decreased the overall fertility of soil, polluted the water bodies and aquatic lives, created resistant pathogenic microbes, and faced losses in terms of socio-economic factors.

Unprecedented use of chemical fertilizers affects the soil microbiota by altering the soil minerals as well as soil pH. Higher concentrations of N suppresses the population of *Azotobacter* population in soil while high P content negatively impacts the mycorrhizal formation (Bagyaraj and Revanna, 2016). The application of excessive herbicides has resulted in death of sensitive microbes like *Azotobacter* (Milošević and Govedarica, 2002). Herbicides like 2,4 D adversely affects the activities and population of *Rhizobium* sp., *Nitrobacter* sp. and other soil-friendly purple non-sulfur bacteria in soil (Chalam et al., 1997, Fabra et al., 1997; Fox et al., 2001). Another herbicide, glyphosate can cause reduction up to 40% of symbiotic mycorrhiza formation (Zaller et al., 2014). Similarly, Cu based fungicides are also toxic to *Rhizobium*, *Pseudomonas*, *Trichoderma* and *Asperzillus* sp. (Kyei-Boahen et al., 2001; Virág et al., 2007). Several pesticides like glyphosate, paraquat, atrazine and carbaryl are also reported to influence the soil enzymatic activities (Sannino and Gianfreda, 2001). Dinoseb, an insecticide inhibits the soil's nitrogenase activity while chlorpyrifos and its derivatives inhibit the beneficial biological activities of *Bacillus subtilis*, *Trichoderma harzianum*, *Mycobacterium phlei*, *Pseudomonas fluorescences* and *Fusarium oxysporum* (Niewiadomska, 2004; Virág et al., 2007). A comprehensive list of microbes affected by various herbicides, pesticides, fungicides and insecticides is available in a review by (Meena et al., 2020).

With these consequences, there should be controlled usage of chemical agents in agricultural sector so that the native beneficial microbes can thrive and support the soil quality, fertility and sustainability.

SUSTAINABLE AGRICULTURE: NEED OF THE HOUR

The population of world is ever increasing and it is expected to reach aroun 9 billion by 2050. In order to meet the continuous and uncompromised food supply, no stone was left unturned to increase the productivity and agricultural output. The use of chemical fertilizers, pesticides, insecticides, herbicides and other chemical agents were utilized since the time of industrial revolution to tackle the challenges of increased food and feed demands. Not only this, more and more forest cover was cleared for making more space for agricultural land. Nowadays, situation of agricultural land availability versus their net productivity has become an issue of debate. In the hoard to increase the agricultural produce and in the hassle to make it readily available, somewhere the sustainability was left far behind.

The concept of sustainable agriculture comes when we care about our current needs and demands without compromising it for the future generations. The microbial community in the soil depends significantly on the history of cultivation (Buckley and Schmidt, 2001). In the upper sections, we have discussed how the soil quality and soil microbiota has been affected by the application of chemical agents. The same can be restored by the proper soil management, organic farming, opting for crop rotation, substituting for chemical fertilizers and promoting biological agents for majority of agricultural practices. In the recent years, with much effort from the scientific community, the agricultural community has understood the importance of sustainable agriculture.

Recent developments in this regard are quite promising and encouraging as the soil microbiota and sustainable agriculture are interdependent (Glick, 2018; M Tahat et al., 2020; Mishra et al., 2016; Singh et al., 2011). The increased application of organic fertilizers have been reported to have great impact on soil properties and bacterial communities (Wu et al., 2020). Organic fertilizers help recruiting beneficial bacteria and enriches the microbial population (Lin et al., 2019) while substituting manure in place of chemical fertilizers improved the soil quality and biological functions (Luan et al.,2020). Now, the microbial community has been observed as promising probiotic as plant biostimulant for sustainable agriculture (Woo and Pepe, 2018) and the microbial resources can be utilized to even improve the fertilizers efficiency (Bargaz et al., 2018). Moreover, Actinobacteria has been found to better suit the warmer and drier soils (Araujo et al.,2020) that can cater the needs of warming planet. These are just a few examples to note that the sustainable agriculture has already settling in along with the conventional and classical methodology of agriculture. In spite of this, there is a long way to go to attain further sustainability along with continued productivity.

REFERENCES

Abbey, L., Abbey, J., Leke-Aladekoba, A., Iheshiulo, E. M. A., & Ijenyo, M. (2019). Biopesticides and Biofertilizers: Types, Production, Benefits, and Utilization. *Byproducts from Agriculture and Fisheries: Adding Value for Food, Feed, Pharma, and Fuels*, 479-500.

Adinarayana, K., Ellaiah, P., Srinivasulu, B., Devi, R. B., & Adinarayana, G. (2003). Response surface methodological approach to optimize the nutritional parameters for neomycin production by *Streptomyces marinensis* under solid-state fermentation. *Process Biochemistry*, *38*(11), 1565–1572. doi:10.1016/S0032-9592(03)00057-8

Adrio, J. L., & Demain, A. L. (2014). Microbial enzymes: Tools for biotechnological processes. *Biomolecules*, *4*(1), 117–139. doi:10.3390/biom4010117 PMID:24970208

Alexopoulos, C. J., Mims, C. W., & Blackwell, M. (1996). Introductory mycology (No. Ed. 4). John Wiley and Sons.

Alwendawi, S. A. (2019). In vitro assessment the potential antioxidant and antitumor activities of Bifido bacterium derived bacteriocins. *International Journal of Drug Delivery Technology*, *9*(02), 207–216. doi:10.25258/ijddt.9.2.15

Anderson, R. F., & Jackson, R. W. (1958). Essential amino acids in microbial proteins. *Applied Microbiology*, *6*(5), 369–373. doi:10.1128/AM.6.5.369-373.1958 PMID:13571982

Araujo, R., Gupta, V. V., Reith, F., Bissett, A., Mele, P., & Franco, C. M. (2020). Biogeography and emerging significance of *Actinobacteria* in Australia and Northern Antarctica soils. *Soil Biology & Biochemistry*, *146*, 107805. doi:10.1016/j.soilbio.2020.107805

Bagyaraj, D. J., & Revanna, A. (2016). Effect of chemical fertilizers on the beneficial soil microorganisms. *Fertilizers and Environment News*, *2*, 10–11.

Barber, M. S., Giesecke, U., Reichert, A., & Minas, W. (2004). Industrial enzymatic production of cephalosporin-based β-lactams. In *Molecular Biotechnolgy of Fungal beta-Lactam Antibiotics and Related Peptide Synthetases* (pp. 179–215). Springer. doi:10.1007/b99261

Bargaz, A., Lyamlouli, K., Chtouki, M., Zeroual, Y., & Dhiba, D. (2018). Soil microbial resources for improving fertilizers efficiency in an integrated plant nutrient management system. *Frontiers in Microbiology*, *9*, 1606. doi:10.3389/fmicb.2018.01606 PMID:30108553

Behrman, J., Alderman, H., & Hoddinott, J. (2004). Hunger and malnutrition. *Global crises. Global Solutions*, *363*, 420.

Benka-Coker, M. O., & Olumagin, A. (1996). Effects of waste drilling fluid on bacterial isolates from a mangrove swamp oilfield location in the Niger Delta of Nigeria. *Bioresource Technology*, *55*(3), 175–179. doi:10.1016/0960-8524(95)00165-4

Berditsch, M., Afonin, S., & Ulrich, A. S. (2007). The ability of *Aneurinibacillus migulanus* (*Bacillus brevis*) to produce the antibiotic gramicidin S is correlated with phenotype variation. *Applied and Environmental Microbiology*, *73*(20), 6620–6628. doi:10.1128/AEM.00881-07 PMID:17720841

Berdy, J. (2005). Bioactive microbial metabolites. *The Journal of Antibiotics*, *58*(1), 1–26. doi:10.1038/ja.2005.1 PMID:15813176

Bhalla, T. C., Sharma, N. N., & Sharma, M. (2007). *Production of metabolites, industrial enzymes, amino acid, organic acids, antibiotics, vitamins and single cell proteins*. Academic Press.

Bhardwaj, D., Ansari, M. W., Sahoo, R. K., & Tuteja, N. (2014). Biofertilizers function as key player in sustainable agriculture by improving soil fertility, plant tolerance and crop productivity. *Microbial Cell Factories*, *13*(1), 1–10. doi:10.1186/1475-2859-13-66 PMID:24885352

Brandelli, A., Daroit, D. J., & Riffel, A. (2010). Biochemical features of microbial keratinases and their production and applications. *Applied Microbiology and Biotechnology*, *85*(6), 1735–1750. doi:10.100700253-009-2398-5 PMID:20039036

Brandelli, A., Sala, L., & Kalil, S. J. (2015). Microbial enzymes for bioconversion of poultry waste into added-value products. *Food Research International*, *73*, 3–12. doi:10.1016/j.foodres.2015.01.015

Brautaset, T., Jakobsen, Ø. M., Josefsen, K. D., Flickinger, M. C., & Ellingsen, T. E. (2007). *Bacillus methanolicus*: A candidate for industrial production of amino acids from methanol at 50 C. *Applied Microbiology and Biotechnology*, *74*(1), 22–34. doi:10.100700253-006-0757-z PMID:17216461

Buckley, D. H., & Schmidt, T. M. (2001). The structure of microbial communities in soil and the lasting impact of cultivation. *Microbial Ecology*, *42*(1), 11–21. doi:10.1007002480000108 PMID:12035077

Chalam, A. V., Sasikala, C., Ramana, C. V., Uma, N. R., & Rao, P. R. (1997). Effect of pesticides on the diazotrophic growth and nitrogenase activity of purple nonsulfur bacteria. *Bulletin of Environmental Contamination and Toxicology*, *58*(3), 463–468. doi:10.1007001289900357 PMID:9008058

Chavan, R. S., & Chavan, S. R. (2011). Sourdough technology—a traditional way for wholesome foods: A review. *Comprehensive Reviews in Food Science and Food Safety*, *10*(3), 169–182. doi:10.1111/j.1541-4337.2011.00148.x

Chen, M., Yang, G., Sheng, Y., Li, P., Qiu, H., Zhou, X., Huang, L., & Chao, Z. (2017). Glomus mosseae inoculation improves the root system architecture, photosynthetic efficiency and flavonoids accumulation of liquorice under nutrient stress. *Frontiers in Plant Science*, *8*, 931. doi:10.3389/fpls.2017.00931 PMID:28638391

Choudhary, R. B., Jana, A. K., & Jha, M. K. (2004). *Enzyme technology applications in leather processing*. Academic Press.

Demain, A. L. (2008). A new opportunity for industry. *Secondary Metabolites: Their Function and Evolution*, *171*, 3. PMID:1302184

Demain, A. L., & Sanchez, S. (2009). Microbial drug discovery: 80 years of progress. *The Journal of Antibiotics*, *62*(1), 5–16. doi:10.1038/ja.2008.16 PMID:19132062

Deng, A., Wu, J., Zhang, G., & Wen, T. (2011). Molecular and structural characterization of a surfactant-stable high-alkaline protease AprB with a novel structural feature unique to subtilisin family. *Biochimie*, *93*(4), 783–791. doi:10.1016/j.biochi.2011.01.011 PMID:21281692

Dhir, B. (2017). Biofertilizers and biopesticides: eco-friendly biological agents. In *Advances in Environmental Biotechnology* (pp. 167–188). Springer. doi:10.1007/978-981-10-4041-2_10

Donova, M. V. (2007). Transformation of steroids by actinobacteria: A review. *Applied Biochemistry and Microbiology*, *43*(1), 1–14. doi:10.1134/S0003683807010012 PMID:17345852

El-Naggar, M. Y., El-Assar, S. A., & Abdul-Gawad, S. M. (2006). Meroparamycin production by newly isolated *Streptomyces sp.* strain MAR01: Taxonomy, fermentation, purification and structural elucidation. *Journal of Microbiology (Seoul, Korea)*, *44*(4), 432–438. PMID:16953179

Fabra, A., Duffard, R., & De Duffard, A. E. (1997). Toxicity of 2, 4-dichlorophenoxyacetic acid to *Rhizobium sp* in pure culture. *Bulletin of Environmental Contamination and Toxicology*, *59*(4), 645–652. doi:10.1007001289900528 PMID:9307432

Fiedurek, J., Trytek, M., & Szczodrak, J. (2017). Strain improvement of industrially important microorganisms based on resistance to toxic metabolites and abiotic stress. *Journal of Basic Microbiology*, *57*(6), 445–459. doi:10.1002/jobm.201600710 PMID:28370185

Fondi, M., Orlandini, V., Emiliani, G., Papaleo, M. C., Maida, I., Perrin, E., ... Fani, R. (2012). *Draft genome sequence of the hydrocarbon-degrading and emulsan-producing strain* Acinetobacter venetianus *RAG-1T*. Academic Press.

Fox, J. E., Starcevic, M., Kow, K. Y., Burow, M. E., & McLachlan, J. A. (2001). Endocrine disrupters and flavonoid signalling. *Nature*, *413*(6852), 128–129. doi:10.1038/35093163 PMID:11557969

Gandini, A. (2008). Polymers from renewable resources: A challenge for the future of macromolecular materials. *Macromolecules*, *41*(24), 9491–9504. doi:10.1021/ma801735u

Gaspar, P., Carvalho, A. L., Vinga, S., Santos, H., & Neves, A. R. (2013). From physiology to systems metabolic engineering for the production of biochemicals by lactic acid bacteria. *Biotechnology Advances*, *31*(6), 764–788. doi:10.1016/j.biotechadv.2013.03.011 PMID:23567148

Germanier, R. (Ed.). (2012). *Bacterial vaccines*. Academic Press.

Gill, S. S., Cowles, E. A., & Pietrantonio, P. V. (1992). The mode of action of *Bacillus thuringiensis* endotoxins. *Annual Review of Entomology*, *37*(1), 615–634. doi:10.1146/annurev.en.37.010192.003151 PMID:1311541

Glick, B. R. (2018). Soil microbes and sustainable agriculture. *Pedosphere*, *28*(2), 167–169. doi:10.1016/S1002-0160(18)60020-7

Gouka, R. J., Punt, P. J., & Van Den Hondel, C. A. M. J. J. (1997). Efficient production of secreted proteins by *Aspergillus*: Progress, limitations and prospects. *Applied Microbiology and Biotechnology*, *47*(1), 1–11. doi:10.1007002530050880 PMID:9035405

Gupta, S., & Dikshit, A. K. (2010). Biopesticides: An ecofriendly approach for pest control. *Journal of Biopesticides*, *3*, 186.

Gupta, S. K., & Shukla, P. (2017). Microbial platform technology for recombinant antibody fragment production: A review. *Critical Reviews in Microbiology*, *43*(1), 31–42. doi:10.3109/1040841X.2016.1150959 PMID:27387055

Haba, E., Pinazo, A., Jauregui, O., Espuny, M. J., Infante, M. R., & Manresa, A. (2003). Physicochemical characterization and antimicrobial properties of rhamnolipids produced by *Pseudomonas aeruginosa* 47T2 NCBIM 40044. *Biotechnology and Bioengineering*, *81*(3), 316–322. doi:10.1002/bit.10474 PMID:12474254

Haque, F., Sajid, M., Cameotra, S. S., & Battacharyya, M. S. (2017). Anti-biofilm activity of a sophorolipid-amphotericin B niosomal formulation against *Candida albicans*. *Biofouling*, *33*(9), 768–779. doi:10.1080/08927014.2017.1363191 PMID:28946803

Hasan, F., Shah, A. A., & Hameed, A. (2006). Industrial applications of microbial lipases. *Enzyme and Microbial Technology*, *39*(2), 235–251. doi:10.1016/j.enzmictec.2005.10.016

Hassan, M. U., Nayab, H., Rehman, T. U., Williamson, M. P., Haq, K. U., Shafi, N., & Shafique, F. (2020). Characterization of bacteriocins produced by *Lactobacillus spp.* isolated from the traditional Pakistani yoghurt and their antimicrobial activity against common foodborne pathogens. *BioMed Research International*.

Hirasawa, T., & Shimizu, H. (2016). Recent advances in amino acid production by microbial cells. *Current Opinion in Biotechnology*, *42*, 133–146. doi:10.1016/j.copbio.2016.04.017 PMID:27151315

Hong, B., Phornphisutthimas, S., Tilley, E., Baumberg, S., & McDowall, K. J. (2007). Streptomycin production by *Streptomyces griseus* can be modulated by a mechanism not associated with change in the adpA component of the A-factor cascade. *Biotechnology Letters*, *29*(1), 57–64. doi:10.100710529-006-9216-2 PMID:17120093

Inoue, S., & Ito, S. (1982). Sophorolipids from *Torulopsis bombicola* as microbial surfactants in alkane fermentations. *Biotechnology Letters*, *4*(1), 3–8. doi:10.1007/BF00139273

Jem, K. J., van der Pol, J. F., & de Vos, S. (2010). Microbial lactic acid, its polymer poly (lactic acid), and their industrial applications. In *Plastics from bacteria* (pp. 323–346). Springer. doi:10.1007/978-3-642-03287-5_13

Jones, S. W., Karpol, A., Friedman, S., Maru, B. T., & Tracy, B. P. (2020). Recent advances in single cell protein use as a feed ingredient in aquaculture. *Current Opinion in Biotechnology*, *61*, 189–197. doi:10.1016/j.copbio.2019.12.026 PMID:31991311

Karadeniz, A., Topcuoğlu, Ş. F., & Inan, S. (2006). Auxin, gibberellin, cytokinin and abscisic acid production in some bacteria. *World Journal of Microbiology & Biotechnology*, *22*(10), 1061–1064. doi:10.100711274-005-4561-1

Kardos, N., & Demain, A. L. (2011). Penicillin: The medicine with the greatest impact on therapeutic outcomes. *Applied Microbiology and Biotechnology*, *92*(4), 677–687. doi:10.100700253-011-3587-6 PMID:21964640

Kinoshita, S., Nakayama, K., & Kitada, S. (1958). Production of aspartic acid from fumaric acid by microorganism. *Hakko Kyokaishi*, *16*, 517–520.

Ko, Y. S., Kim, J. W., Lee, J. A., Han, T., Kim, G. B., Park, J. E., & Lee, S. Y. (2020). Tools and strategies of systems metabolic engineering for the development of microbial cell factories for chemical production. *Chemical Society Reviews*, *49*(14), 4615–4636. doi:10.1039/D0CS00155D PMID:32567619

Kumar, R. R., & Prasad, S. (2011). Metabolic engineering of bacteria. *Indian Journal of Microbiology*, *51*(3), 403–409. doi:10.100712088-011-0172-8 PMID:22754024

Kumar, S. N., Siji, J. V., Ramya, R., Nambisan, B., & Mohandas, C. (2020). Improvement of antimicrobial activity of compounds produced by *Bacillus sp.* associated with a *Rhabditid sp.* (entomopathogenic nematode) by changing carbon and nitrogen sources in fermentation media. *Journal of Microbiology, Biotechnology and Food Sciences*, *9*(4), 1424–1438.

Kyei-Boahen, S., Slinkard, A. E., & Walley, F. L. (2001). Rhizobial survival and nodulation of chickpea as influenced by fungicide seed treatment. *Canadian Journal of Microbiology*, *47*(6), 585–589. doi:10.1139/w01-038 PMID:11467735

Lakshmidevi, R., & Muthukumar, K. (2010). Enzymatic saccharification and fermentation of paper and pulp industry effluent for biohydrogen production. *International Journal of Hydrogen Energy*, *35*(8), 3389–3400. doi:10.1016/j.ijhydene.2009.12.165

Lale, G., Jogdand, V. V., & Gadre, R. V. (2006). Morphological mutants of *Gibberella fujikuroi* for enhanced production of gibberellic acid. *Journal of Applied Microbiology*, *100*(1), 65–72. doi:10.1111/j.1365-2672.2005.02754.x PMID:16405686

Lee, S. Y., & Kim, H. U. (2015). Systems strategies for developing industrial microbial strains. *Nature Biotechnology*, *33*(10), 1061–1072. doi:10.1038/nbt.3365 PMID:26448090

Li, T., Liu, J., Bai, R., Ohandja, D. G., & Wong, F. S. (2007). Biodegradation of organonitriles by adapted activated sludge consortium with acetonitrile-degrading microorganisms. *Water Research*, *41*(15), 3465–3473. doi:10.1016/j.watres.2007.04.033 PMID:17544472

Lin, W., Lin, M., Zhou, H., Wu, H., Li, Z., & Lin, W. (2019). The effects of chemical and organic fertilizer usage on rhizosphere soil in tea orchards. *PLoS One*, *14*(5), e0217018. doi:10.1371/journal.pone.0217018 PMID:31136614

Liu, Q., Liu, C., Yu, J., Yan, J., & Qi, X. (2012). Analysis of the ketosynthase genes in *Streptomyces* and its implications for preventing reinvestigation of polyketides with bioactivities. *The Journal of Agricultural Science*, *4*(7), 262–270.

Liu, X., & Kokare, C. (2017). Microbial enzymes of use in industry. In *Biotechnology of Microbial Enzymes* (pp. 267–298). Academic Press. doi:10.1016/B978-0-12-803725-6.00011-X

Luan, H., Gao, W., Huang, S., Tang, J., Li, M., Zhang, H., Chen, X., & Masiliūnas, D. (2020). Substitution of manure for chemical fertilizer affects soil microbial community diversity, structure and function in greenhouse vegetable production systems. *PLoS One*, *15*(2), e0214041. doi:10.1371/journal.pone.0214041 PMID:32084129

MacMillan, J. (2001). Occurrence of gibberellins in vascular plants, fungi, and bacteria. *Journal of Plant Growth Regulation*, *20*(4), 387–442. doi:10.1007003440010038 PMID:11986764

Mahalik, S., Sharma, A. K., & Mukherjee, K. J. (2014). Genome engineering for improved recombinant protein expression in *Escherichia coli*. *Microbial Cell Factories*, *13*(1), 177. doi:10.118612934-014-0177-1 PMID:25523647

Martins, D. A. B., Prado, H. F. A., Leite, R. S. R., Ferreira, H., Moretti, M. M. S., Silva, R., & Gomes, E. (2011). Agroindustrial wastes as substrates for microbial enzymes production and source of sugar for bioethanol production. *Integrated Waste Management*, *2*, 319–361.

Matassa, S., Boon, N., Pikaar, I., & Verstraete, W. (2016). Microbial protein: Future sustainable food supply route with low environmental footprint. *Microbial Biotechnology*, *9*(5), 568–575. doi:10.1111/1751-7915.12369 PMID:27389856

Matson, S. W., Bean, D. W., & George, J. W. (1994). DNA helicases: Enzymes with essential roles in all aspects of DNA metabolism. *BioEssays*, *16*(1), 13–22. doi:10.1002/bies.950160103 PMID:8141804

Mcdaniel, L. E., Schaffner, C. P., & Bailey, E. G. (1965). *U.S. Patent No. 3,182,004*. Washington, DC: U.S. Patent and Trademark Office.

McWilliams, A. (2015). *Microbial products: technologies, applications and global markets*. Academic Press.

Meena, R. S., Kumar, S., Datta, R., Lal, R., Vijayakumar, V., Brtnicky, M., ... Pathan, S. I. (2020). Impact of agrochemicals on soil microbiota and management: A review. *Land (Basel)*, *9*(2), 34. doi:10.3390/land9020034

Meenavilli, H., Potumarthi, R., & Jetty, A. (2008). Gentamicin production by *Micromonospora echinospora* (Me-22) in stirred tank reactor: Effect of various parameters. *Journal of Basic Microbiology*, *48*(1), 53–58. doi:10.1002/jobm.200700116 PMID:18247396

Mehta, P. K., & Sehgal, S. (2019). Microbial enzymes in food processing. In *Biocatalysis* (pp. 255–275). Springer. doi:10.1007/978-3-030-25023-2_13

Milošević, N. A., & Govedarica, M. M. (2002). Effect of herbicides on microbiological properties of soil. *Zbornik Matice Srpske za Prirodne Nauke*, (102), 5–21. doi:10.2298/ZMSPN0201005M

Mishra, J., Prakash, J., & Arora, N. K. (2016). Role of beneficial soil microbes in sustainable agriculture and environmental management. *Climate Change and Environmental Sustainability*, *4*(2), 137–149. doi:10.5958/2320-642X.2016.00015.6

Mousa, W. K., & Raizada, M. N. (2013). The diversity of anti-microbial secondary metabolites produced by fungal endophytes: An interdisciplinary perspective. *Frontiers in Microbiology*, *4*, 65. doi:10.3389/fmicb.2013.00065 PMID:23543048

Muthusamy, K., Gopalakrishnan, S., Ravi, T. K., & Sivachidambaram, P. (2008). Biosurfactants: Properties, commercial production and application. *Current Science*, 736–747.

Nangul, A., & Bhatia, R. (2020). Microorganisms: A marvelous source of single cell proteins. *Journal of Microbiology, Biotechnology and Food Sciences*, *9*(4), 15–18.

Nargund, V. B., Amaresh, Y. S., Sreenivas, A. G., & Nadagouda, S. (2007). *Trichoderma harzianum*–a potential bioagent for seed and soil borne diseases management in Upper Krishna project command area of Karnataka, India. *International Journal of Agricultural Sciences*, *3*, 158–160.

Niewiadomska, A. (2004). Effect of Carbendazim, Imazetapir and Thiram on nitrogenase activity, the number of microorganisms in soil and yield of Red Clover (*Trifolium pratense* L.). *Polish Journal of Environmental Studies*, *13*(4).

Oren, A. (2014). Halophilic archaea on Earth and in space: Growth and survival under extreme conditions. *Philosophical Transactions - Royal Society. Mathematical, Physical, and Engineering Sciences*, *372*(2030), 20140194. doi:10.1098/rsta.2014.0194 PMID:25368347

Ovaa, H. (2014). Unnatural amino acid incorporation in *E. coli*: Current and future applications in the design of therapeutic proteins. *Frontiers in Chemistry*, *2*, 15. PMID:24790983

Paes, B. G., & Almeida, J. R. (2014). Genetic improvement of microorganisms for applications in biorefineries. *Chemical and Biological Technologies in Agriculture*, *1*(1), 21. doi:10.118640538-014-0021-1

Palva, I. (1982). Molecular cloning of α-amylase gene from *Bacillus amyloliquefaciens* and its expression in *B. subtilis*. *Gene*, *19*(1), 81–87. doi:10.1016/0378-1119(82)90191-3 PMID:6183169

Park, S. R., Yoon, Y. J., Pham, J. V., Yilma, M. A., Feliz, A., Majid, M. T., ... Song, M. C. (2019). A review of the microbial production of bioactive natural products and biologics. *Frontiers in Microbiology*, *10*, 1404. doi:10.3389/fmicb.2019.01404 PMID:31281299

Park, S. R., Yoon, Y. J., Pham, J. V., Yilma, M. A., Feliz, A., Majid, M. T., ... Song, M. C. (2019). A review of the microbial production of bioactive natural products and biologics. *Frontiers in Microbiology*, *10*, 1404. doi:10.3389/fmicb.2019.01404 PMID:31281299

Park, S. Y., Binkley, R. M., Kim, W. J., Lee, M. H., & Lee, S. Y. (2018). Metabolic engineering of *Escherichia coli* for high-level astaxanthin production with high productivity. *Metabolic Engineering*, *49*, 105–115. doi:10.1016/j.ymben.2018.08.002 PMID:30096424

Prabhu, M., Chemodanov, A., Gottlieb, R., Kazir, M., Nahor, O., Gozin, M., Israel, A., Livney, Y. D., & Golberg, A. (2019). Starch from the sea: The green macroalga *Ulva ohnoi* as a potential source for sustainable starch production in the marine biorefinery. *Algal Research*, *37*, 215–227. doi:10.1016/j.algal.2018.11.007

Prajapati, J. B., & Nair, B. M. (2008). The history of fermented foods. Handbook of Fermented Functional Foods, 1-24.

Prosekov, A. Y., & Ivanova, S. A. (2018). Food security: The challenge of the present. *Geoforum*, *91*, 73–77. doi:10.1016/j.geoforum.2018.02.030

Pulz, O., & Gross, W. (2004). Valuable products from biotechnology of microalgae. *Applied Microbiology and Biotechnology*, *65*(6), 635–648. doi:10.100700253-004-1647-x PMID:15300417

Ranghar, S., Agrawal, S., & Agrawal, P. K. (2019). Microbial products: protein, enzyme, secondary metabolites and chemicals. In *Microbial Interventions in Agriculture and Environment* (pp. 347–384). Springer. doi:10.1007/978-981-32-9084-6_17

Recio, E., Colinas, Á., Rumbero, Á., Aparicio, J. F., & Martín, J. F. (2004). PI factor, a novel type quorum-sensing inducer elicits pimaricin production in *Streptomyces natalensis. The Journal of Biological Chemistry*, *279*(40), 41586–41593. doi:10.1074/jbc.M402340200 PMID:15231842

Sanchez, S., & Demain, A. L. (2011). Enzymes and bioconversions of industrial, pharmaceutical, and biotechnological significance. *Organic Process Research & Development*, *15*(1), 224–230. doi:10.1021/op100302x

Sannino, F., & Gianfreda, L. (2001). Pesticide influence on soil enzymatic activities. *Chemosphere*, *45*(4-5), 417–425. doi:10.1016/S0045-6535(01)00045-5 PMID:11680737

Sarkar, S., Banerjee, A., Halder, U., Biswas, R., & Bandopadhyay, R. (2017). Degradation of synthetic azo dyes of textile industry: A sustainable approach using microbial enzymes. *Water Conservation Science and Engineering*, *2*(4), 121–131. doi:10.100741101-017-0031-5

Saxena, S. (2015). Strategies of strain improvement of industrial microbes. In *Applied Microbiology* (pp. 155–171). Springer.

Saykhedkar, S. S., & Singhal, R. S. (2004). Solid-state fermentation for production of griseofulvin on rice bran using *Penicillium griseofulvum. Biotechnology Progress*, *20*(4), 1280–1284. doi:10.1021/bp0343662 PMID:15296463

Sharma, S., & Malik, P. (2012). Biopesticides: Types and applications. *International Journal of Advances in Pharmacy Biological Chemistry*, *1*(4), 2277–4688.

Show, P. L., Oladele, K. O., Siew, Q. Y., Aziz Zakry, F. A., Lan, J. C. W., & Ling, T. C. (2015). Overview of citric acid production from *Aspergillus niger. Frontiers in Life Science*, *8*(3), 271–283. doi:10.1080/21553769.2015.1033653

Shrestha, G., & Clair, L. L. S. (2013). Lichens: A promising source of antibiotic and anticancer drugs. *Phytochemistry Reviews*, *12*(1), 229–244. doi:10.100711101-013-9283-7

Singh, J. S., Pandey, V. C., & Singh, D. P. (2011). Efficient soil microorganisms: A new dimension for sustainable agriculture and environmental development. *Agriculture, Ecosystems & Environment*, *140*(3-4), 339–353. doi:10.1016/j.agee.2011.01.017

Singh, R., Kumar, M., Mittal, A., & Mehta, P. K. (2016). Microbial enzymes: industrial progress in 21st century. *Biotech, 6*(2), 174.

Singh, R. R., & Prasad, K. (2011). Effect of bio-fertilizers on growth and productivity of wheat (*Triticum aestivum*). *International Journal of Farm Sciences*, *1*(1), 1–8.

Singh, S., Adak, A., Saritha, M., Sharma, S., Tiwari, R., Rana, S., & Nain, L. (2017). Bioethanol production scenario in India: Potential and policy perspective. In *Sustainable Biofuels Development in India* (pp. 21–37). Springer. doi:10.1007/978-3-319-50219-9_2

Sottorff, I., Wiese, J., Lipfert, M., Preußke, N., Sönnichsen, F. D., & Imhoff, J. F. (2019). Different secondary metabolite profiles of phylogenetically almost identical *Streptomyces griseus* strains originating from geographically remote locations. *Microorganisms*, *7*(6), 166. doi:10.3390/microorganisms7060166 PMID:31174336

Spolaore, P., Joannis-Cassan, C., Duran, E., & Isambert, A. (2006). Optimization of *Nannochloropsis oculata* growth using the response surface method. *Journal of Chemical Technology & Biotechnology: International Research in Process, Environmental & Clean Technology*, *81*(6), 1049–1056.

Suman, G., Nupur, M., Anuradha, S., & Pradeep, B. (2015). Single cell protein production: A review. *International Journal of Current Microbiology and Applied Sciences*, *4*(9), 251–262.

Sundarram, A., & Murthy, T. P. K. (2014). α-amylase production and applications: A review. *Journal of Applied & Environmental Microbiology*, *2*(4), 166–175.

Tahat, M., Alananbeh, K., Othman, Y., & Leskovar, D. (2020). Soil Health and Sustainable Agriculture. *Sustainability*, *12*(12), 4859. doi:10.3390u12124859

Tanaka, Y., & Omura, S. (1993). Agroactive compounds of microbial origin. *Annual Review of Microbiology*, *47*(1), 57–87. doi:10.1146/annurev.mi.47.100193.000421 PMID:8257109

Terpe, K. (2006). Overview of bacterial expression systems for heterologous protein production: From molecular and biochemical fundamentals to commercial systems. *Applied Microbiology and Biotechnology*, *72*(2), 211–222. doi:10.100700253-006-0465-8 PMID:16791589

Türker, M. (2014). Yeast biotechnology: diversity and applications. In *27th VH yeast conference: advances in science and industrial production of baker's yeast, Istanbul* (pp. 14-15). Academic Press.

Tzvetkov, M., Klopprogge, C., Zelder, O., & Liebl, W. (2003). Genetic dissection of trehalose biosynthesis in *Corynebacterium glutamicum*: Inactivation of trehalose production leads to impaired growth and an altered cell wall lipid composition. *Microbiology*, *149*(7), 1659–1673. doi:10.1099/mic.0.26205-0 PMID:12855718

Ulery, B. D., Nair, L. S., & Laurencin, C. T. (2011). Biomedical applications of biodegradable polymers. *Journal of Polymer Science. Part B, Polymer Physics*, *49*(12), 832–864. doi:10.1002/polb.22259 PMID:21769165

Van Den Berg, M. A., Albang, R., Albermann, K., Badger, J. H., Daran, J. M., Driessen, A. J., ... Joardar, V. (2008). Genome sequencing and analysis of the filamentous fungus *Penicillium chrysogenum*. *Nature Biotechnology*, *26*(10), 1161–1168. doi:10.1038/nbt.1498 PMID:18820685

Virág, D., Naár, Z., & Kiss, A. (2007). Microbial toxicity of pesticide derivatives produced with UV-photodegradation. *Bulletin of Environmental Contamination and Toxicology*, *79*(3), 356–359. doi:10.100700128-007-9230-7 PMID:17639315

Woo, S. L., & Pepe, O. (2018). Microbial consortia: Promising probiotics as plant biostimulants for sustainable agriculture. *Frontiers in Plant Science*, *9*, 1801. doi:10.3389/fpls.2018.01801 PMID:30564264

Wu, L., Jiang, Y., Zhao, F., He, X., Liu, H., & Yu, K. (2020). Increased organic fertilizer application and reduced chemical fertilizer application affect the soil properties and bacterial communities of grape rhizosphere soil. *Scientific Reports*, *10*(1), 1–10. doi:10.103841598-020-66648-9 PMID:32533037

Xia, X. X., Qian, Z. G., Ki, C. S., Park, Y. H., Kaplan, D. L., & Lee, S. Y. (2010). Native-sized recombinant spider silk protein produced in metabolically engineered Escherichia coli results in a strong fiber. *Proceedings of the National Academy of Sciences of the United States of America*, *107*(32), 14059–14063. doi:10.1073/pnas.1003366107 PMID:20660779

Yanai, K., Murakami, T., & Bibb, M. (2006). Amplification of the entire kanamycin biosynthetic gene cluster during empirical strain improvement of *Streptomyces kanamyceticus. Proceedings of the National Academy of Sciences of the United States of America*, *103*(25), 9661–9666. doi:10.1073/pnas.0603251103 PMID:16766657

Youssef, M. M. A., & Eissa, M. F. M. (2014). Biofertilizers and their role in management of plant parasitic nematodes. A review. *Journal of Biotechnology and Pharmaceutical Research*, *5*(1), 1–6.

Yuan, Y., Liu, Z. Q., Jin, H., Sun, S., Liu, T. J., Wang, X., Fan, H.-J., Hou, S.-K., & Ding, H. (2017). Photodynamic antimicrobial chemotherapy with the novel amino acid-porphyrin conjugate 4I: In vitro and in vivo studies. *PLoS One*, *12*(5), e0176529. doi:10.1371/journal.pone.0176529 PMID:28493985

Yürekli, F., Yesilada, O., Yürekli, M., & Topcuoglu, S. F. (1999). Plant growth hormone production from olive oil mill and alcohol factory wastewaters by white rot fungi. *World Journal of Microbiology & Biotechnology*, *15*(4), 503–505. doi:10.1023/A:1008952732015

Zaller, J. G., Heigl, F., Ruess, L., & Grabmaier, A. (2014). Glyphosate herbicide affects belowground interactions between earthworms and symbiotic mycorrhizal fungi in a model ecosystem. *Scientific Reports*, *4*(1), 5634. doi:10.1038rep05634 PMID:25005713

Chapter 8
Multifaceted Potential of Plant Growth Promoting Rhizobacteria (PGPR): An Overview

Dwaipayan Sinha
https://orcid.org/0000-0001-7870-8998
Government General Degree College, Mohanpur, India

Suchetana Mukherjee
https://orcid.org/0000-0002-0417-1910
Department of Botany, Sripat Singh College, Jiaganj, India

Dakshayani Mahapatra
Department of Physiology, Government General Degree College, Mohanpur, India

ABSTRACT

Plant growth-promoting rhizobacteria (PGPR) is a unique group of bacteria that colonize the rhizosphere and roots of plants. They are involved in a plethora of interaction with the host plant and benefit the host plant from nutritional and pathological point of view. The beneficial role of PGPR extends from fixation of atmospheric nitrogen, solubilization of phosphates, siderophore production, synthesis of plant growth regulators, and conferring protection to plants through production of antibiotics and ultimately helping the plants in acquiring resistance. The microbes are also being used for bioremediation purposes and thus act as an eco-friendly cleansing agent. PGPR has gained immense interest in the scientific community and have emerged as a very reliable tool for eco-friendly and sustainable approach for crop production. PGPR is a potent candidate of bioprospection for sustainable use in agriculture and bioremediation process for the overall benefit of mankind.

DOI: 10.4018/978-1-7998-7062-3.ch008

INTRODUCTION

Plants have a very intricate relation with the soil as most of their advanced representatives are anchored to it through their root system. The soil acts as an important substrate and pool of minerals, nutrients and water for the plants. Out of the entire soil pool, the region which is in immediate vicinity of the plant is of utmost importance as it is in this region, all the interaction of the plant with the soil takes place. The importance of this zone was felt by the scientists long back and in 1904, German agronomist and plant physiologist Lorenz Hiltner first coined the term "rhizosphere" to illustrate the plant-root interface, a part of the word having its genesis from Greek word "rhiza", indicating root. According to Hiltner, 'rhizospehere' is the area around the root of a plant which is dwelled by a unique variety of microorganisms (McNear and David, 2013). The rhizosphere can further be divided into three zones viz. The innermost is the endorhizosphere which includes portions of cortex and endodermis which is generally inhabited by microbes and cations. The middle region forms the rhizoplane which is directly adjacent to the root and includes the epidermal region and mucilage. The outermost region is the ectorhizosphere which extends from rhizoplane into the soil (McNear and David, 2013; Chaparro et al., 2014). Soil contains various types of microorganisms of which bacteria are most predominant. The type and number of bacteria may vary due to abiotic factors like moisture, temperature, soil nutrient and the presence of other flora in the soil (Glick et al., 1999). The interaction of these bacteria with the plants may be beneficial, harmful or neutral (Lynch and Whipps, 1990). A plant under field condition is not a stand-alone member but often forms a community with microorganisms called the phytomicrobiomes. The rhizomicrobiome (association between microbe and root) among them is most extensively studied (Backer et al., 2018).

As mentioned earlier, the rhizosphere is inhabited by a wide range of bacteria and majority of them are associated with the plant roots. These root associated bacteria that often enters into a symbiotic relation with the plants are called rhizobacteria. The definition of rhizobacteria can be further refined based on their beneficial activity. This lead to the introduction of a new phrase namely 'plant growth promoting rhizobacteria (PGPR) which may be defined as the soil bacteria which inhabits on or around the root surface and directly or indirectly involved in promotion of plant growth and development through production of growth promoting substances or sequestering minerals or secreting a number of regulatory chemicals around the rhizosphere (Ahemad and Kibret, 2014). The term PGPR was first introduced by Kloepper and coworkers in the late 1970s (Tailor and Joshi, 2014) and from then onwards there has a continuous increase in interest on PGPR and its beneficial activities. In the 1990s, the original definition of PGPR was revised in order to accommodate a number of bacteria that have beneficial activity towards the plant but are present outside the rhizosphere (Bashan and Holguin, 1998; Goswami et al., 2016). Importance of PGPR was soon felt in the scientific world for the betterment of agricultural yield and sustainable development.

The hike in global population has resulted in an increased demand in food supply. In the 20th century Green Revolution resulted in the attainment of global food security mainly through chemical inputs and improved crop variety. But the extensive use of chemical fertilizers, pesticides and herbicides took a heavy toll on the environment. To fulfil this global demand of food, fibre and fuel with minimised environmental stress, taking climate change into consideration, put forward the concept of 'Fresh' Green revolution or Bio- Revolution which emphasises on the utilization of phytomicrobiome for biological inputs and crop improvements (Timmusk et al., 2017).

PGPR forms an important entity of phytomicrobiome. Classic example of a PGPR is the symbiotic bacterium *Rhizobium* which forms nodules in leguminous plants (Jones, 2007). Apart from *Rhizobium* members of *Bacillus* and *Pseudomonas* species are important member of PGPR (Podile and Kishore, 2006). *Rhizobium* symbiotically fix nitrogen to increase the growth and yield of plants while *Bacillus* and *Pseudomonas* effect plant growth by production of growth promoting secondary metabolites (Mahmud et al., 2020;Prieto et al., 2011;Shahid et al., 2017). They directly benefit the plant growth by supplying nutrient (which include nitrogen phosphorous and iron) through fixation and mobilization and hormone synthesis. They help the plant indirectly by decreasing susceptibility to disease (by the production of antibiotics and lytic enzymes, siderophores, by competition between pathogen and non-pathogen) and by triggering induced systemic resistance (ISR) (Glick, 2012; Prasad et al., 2019). PGPR is classified into two groups- ePGPR (extracellular PGPR), which inhabit the rhizosphere and spaces between cells of root cortex and iPGPR (intracellular PGPR), which inhabit the nodules (Gouda et al., 2018).

In addition to the global demand of food, environmental pollution is also creating havoc at present. Almost every component of the environment is polluted. The contamination of biosphere has been primarily due to the unplanned and haphazard deposition of the heavy metals from various sources like oil spillage, industries, mining, and effluents from agricultural, domestic and industrial origins and so on throughout the world (He et al., 2007; Souza et al., 2014). Exposure to the heavy metals, oils and various hydrocarbons for a prolonged period through contaminated food and water causes bio magnification of these metals in human system. Evidences have shown the carcinogenicity of these heavy metals through production of reactive oxygen species (ROS) and generation of tissue level oxidative stress (Tchounwou et al., 2012). Metals like cadmium inhibits cellular antioxidant system and form a metallothionein and cadmium complex causing conformational changes in the renal tubular cells and degradation of function (Andrews, 2000). It is associated with diseases of bone and kidneys (Dudley et al., 1985), endocrine disruption (Yang et al., 2015) and reproductive toxicity (Jahan et al., 2014). Whereas, arsenic inhibits DNA repair process through ROS generation and abnormal gene expression leads to cell damage (Shi et al., 2004; Hartwig et al., 2002). As has been known to cause cancer of liver, prostate, skin, etc (Goering et al., 1999). Chromium has been found to enter cells in ionized form via a specific membrane transport system (Eastmond et al., 2008). It causes damage of blood cells and degradation of live and kidney function. Various carcinogenicity of lung, skin and kidney has been associated with chromium and its compounds (Dartsch et al., 1998). Nickel and lead similarly generates ROS producing cellular damage leading to carcinogenesis (Kim et al., 2015). In addition to metal contamination, spillage of oils and contamination of oils in the soil is also a matter of concern (Iturbe et al., 2007). The major constituents of oil are the hydrocarbons. Contamination by hydrocarbons has a detrimental effect on the soil and alters the function and growth of microbial community. This inturn affects the biogeochemical networks and primary producers (Truskewycz et al., 2019). Plants growing in the soil polluted by hydrocarbons also adversely affected by direct toxicity, impaired ability to acquire nutrients, inhibition of photosynthesis and biomass accumulation (Nie et al., 2011). The petroleum hydrocarbons contamination also results in reduction of quality and productivity of the soil and makes it unsuitable for agriculture (Koshalf and Ball, 2017). In animal system petroleum hydrocarbons have been reported to cause oxidative stress which then becomes the precursor of a number of complicated human diseases (Khanna and Gharpure, 2017). There have been reports that petroleum hydrocarbons also results in occurrence of cancers (Stenehjem et al., 2017). Cancers are also caused by groups of pesticides and insecticides which are frequently applied in agricultural fields to ward off the insect pests (Alavanja and Bonner, 2018). In order to counteract these environmental problems there have been an increase in focus of use of PGPR for remediation of contaminated source (Sampaio et al., 2019; Patel et al., 2016). Thus the relevance of PGPR is gradually widening both in agricultural and environmental domains. The play an active role in both the spheres and benefit the both plants and animals directly or indirectly through their unique physiological mechanisms. In this chapter, various beneficial activities of the PGPR would be discussed. Efforts have been taken to describe their mechanism of action which bring about the beneficial activity to the plant along with future prospects related to research and practical applications of these beneficial microbes.

RELEVANCE OF PGPR

PGPR might have adapted themselves to the surrounding environment for sustaining their life processes but their physiological properties and behaviour towards plants have made them special. In the process of sustaining themselves, PGPR have improvised their physiological machinery through which they distinguish themselves from other microbes and provide a wide array of beneficial role for the plants. PGPR thus gained the potential to influence the plant through a wide array of beneficial activities. The influence of PGPR on the plants may be direct or indirect (Vacheron et al., 2013). Direct effect of PGPR is the promotion of plant growth which it does so through the production of plant growth regulators (Beneduzi et al., 2012). In addition to it, PGPR is also reported to fix atmospheric nitrogen (Vejan et al., 2016) and enhances the uptake of phosphorus (de Freitas et al., 1997), the two major elements required for plant growth. In addition to it, PGPR is also known to facilitate uptake of iron through production of siderophores (Patel et al., 2018). PGPR also indirectly influence the plant and consequently its growth through induction of systemic resistance (Serteyn et al., 2020). These properties of PGPR have been observed, investigated and scientifically validated by scientists ever since their discovery. The very unique feature of PGPR is that they are eco-friendly and have an intricate relationship with the environment and plants. Unlike chemical fertilizers and pesticides which negatively affects the ecosystem, these microbes have only positivity with respect to their effect towards environment. The present day world is severely affected by pollution of different types among which pollution by pesticides and hydrocarbons pose serious health threats (Nicolopoulou-Stamati et al., 2016; O'Callaghan-Gordo et al., 2016).PGPR have proven to be a very suitable candidate for remediation of sites contaminated with these pollutants (Hassen et al., 2018; Murray et al., 2019). Thus PGPR actively participates in a wide array of biotic activities in the soil ecosystem primarily by maintaining a dynamic nutrient turnover and sustainable for crop production (Gupta et al., 2015). Any plant for their optimal growth and production requires nutrients and protection from pathogens. Though plants are equipped nutrient quenching capacity and inbuilt resistance from pathogens but under unfavourable conditions their inbuilt capacity weakens. Under those circumstances, supplementation with nutrients or addition of external agents which confers protection from diseases becomes a necessity. In those cases, the role of PGPR becomes extremely relevant which by their inbuilt physiological machinery benefits the plants. To a scientist, the relevance of PGPR may be broadly classified into two parts namely academic relevance and practical relevance both of which are further interconnected to one another. The academic relevance relates to further investigation of these microbes in molecular and genetic level so that desirable traits related to plant growth promotion; disease resistance and remediation process can be engineered in their genetic stature. This might further be elaborated to the practical relevance domain to check their efficacy in the field condition. This fortifies their sustainable use especially in agricultural domain as they reduce the requirement of otherwise harmful chemical fertilizers and also decrease the overall production costs. Thus PGPR undoubtedly stands unique amongst all the microbes and stands as potent candidate in agricultural and environmental sustainability (Figure 1). In the subsequent sections various beneficial roles of PGPR will be discussed in details.

Figure 1. Schematic representation of various beneficial activity of Plant growth promoting rhizobacteria on plants

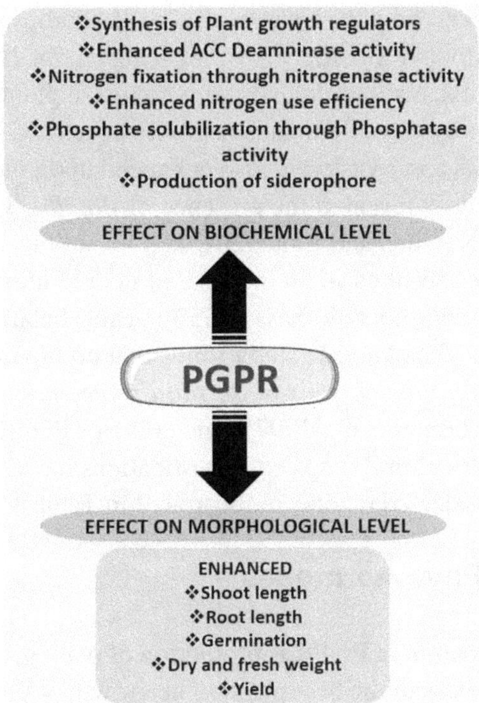

BENEFICIAL ACTIVITIES OF PGPR

PGPR for Growth of Plants

One of the most striking effects of PGPR is induction of plant growth. It is for this reason they are presently being used for betterment of growth and robustness of agricultural crops (Backer et al., 2018). Studies have been conducted on the growth promoting effect of PGPR or various agricultural crop species with a promising outcome. In one study it was shown that paddy plants subjected to salt stress when treated with PGPR namely *Bacillus pumilus* and *Pseudomonas pseudoalcaligenes* resulted in alleviation of stress response. In addition to it, the plants showed higher germination (16%), survival (8%), dry weight (27%), plant height (31%) and reduced cellular concentration of sodium (71%) and Calcium (36%). These results are all indicative of promotion of growth and alleviation of salt stress by the plants upon inoculation of PGPR (Jha and Subramanian, 2013). In a recent study from China it was shown that application of PGPR in the form of *Pseudomonas mosselii, Bacillus thuringeinsis* and Bacillus *sp* JBS-28 in paddy fields contaminated with arsenic resulted in a decrease in arsenic concentration and promoted rice growth as observed by increased grain yield in both green house and field condition

(Xiao et al., 2020). Investigations on the growth of wheat upon inoculation of PGPR have also been undertaken. In one study it was reported that inoculation of PGPR namely *Planomicrobium chinense, Bacillus cereus* and *Pseudomonas fluorescens* in wheat plants growing under draught stress resulted in significant enhancement of yield parameters such as plant height, spike length, grain yield and weight and also improved soil fertility. PGPR also enhanced the accumulation of macronutrients, total NO_3-N and P concentration and soil moisture content in the rhizosphere thus indicating its beneficial effect towards drought tolerance and overall growth (Khan et al., 2019). Studies show that inoculation of PGPR isolated from rhizosphere of halophytes namely *Atriplex leucoclada, Haloxylon salicornicum, Lespedeza bicolor, Suaeda fruticosa,* and *Salicornica virginica* to maize plants under induced drought stress resulted in accumulation of osmolytes and enhanced activity of antioxidant enzymes. In addition to it, length and dry mass of the shoots and roots were also enhanced upon inoculation of PGPR. 16S rRNA amplification and sequence analysis revealed the presence of *Bacillus sp* and *Arthrobacter pascens* in the rhizosphere of the halophytes (Ullah and Bano, 2015). A recent meta-analysis study showed that co inoculation of PGPR to soyabean resulted in an increase in nodule number (11.40%), nodule biomass (6.47%), root biomass (12.84%), and shoot biomass (6.53%) of soyabean crops thus indicating its growth promoting effects (Zeffa et al., 2020).Another study states that co inoculation of soybean with *Bradyrhizobium japonicum* and *Serratia liquefaciens* or *Serratia proteamaculans* resulted in an increase in grain yield and protein content (Dashti et al., 1997). The effect of PGPR on growth parameters are not only confined to the standard agricultural crops but investigations have also been done on several other plants. The effects of PGPR on selected plants are illustrated in Table 1.

PGPR and Synthesis of Phytohormones

One of the striking beneficial function of PGPR is promotion of plant growth and yield. They largely do so by modulating the growth process of the host plants. The growth of plant is largely governed by plant growth regulators. These plant growth regulators, also termed as phytohormones are the chemical messengers synthesized by the plants and control the overall growth and development. The site of generation of these phytohormones and site of action is quite distant from one another (Ahmed and Hasnain, 2014). According to classical concept, there are five different types of phytohormones present in the plant namely auxins, gibberellins, abscisic acid, cytokinins, and ethylene (Wang and Irvin, 2011). However there has been an increasing evidence that other molecules such as brassinosteroids (Cheon et al., 2013; Asami et al., 2005), jasmonates (Huang et al., 2017), salicylic acid (Ko et al., 2020), strigolactones (Zwanenburg and Blanco-Ania,2018) also possess plant growth promoting activities . These phytohormones have their own biosynthetic pathways in the plant body and involve a plethora of genes and enzymes. All of them have shown to influence the growth processes of plants directly or passively. The role of PGPR in plant growth process seems to be quite interesting. They either synthesize phytohormones directly (Patel and Saraf, 2017) which are then taken up by the plant or they can up regulate genes that are involved in the Phytohormone production (Tsukanova et al., 2017).The different activities of PGPR related to Phytohormone synthesis is tabulated in Table 2. The mechanism by which these bacteria induce growth promotion by involving phytohormones would be discussed in the next phase.

Table 1. Effects of PGPR on growth parameters of selected plants

S.No.	Name of the Plant	Name of the Inoculated PGPR	Effects on Growth Parameters	Reference
1.	*Saccharum sp.*	*Bacillus subtilis* and *Bacillus pumilus*	Enhancement of shoot and root growth.	Santos et al., 2018
2.	*Pennisetum glaucum*	*Acetobacter* strains UOM Ab9, Ab11 *Azospirillum* strain UOM Az3, *Pseudomonas* spp. UOM ISR 17	Promotion of growth as exemplified by increase in height, fresh weight, dry weight and leaf area.	Jogaiah et al., 2010
3.	*Pisum sativum*	*Bacillus paramycoides, Bacillus wiedmannii, Bacillus amyloliquefaciens*	Increase in shoot and root length and fresh matter.	Osman and Yin,2018
4.	*Brassica juncea*	Rhizobacterial isolates from the roots	Increase in shoot dry weight.	Sharma et al., 2018
5.	*Brassica campestris*	*Pseudomonas putida, Burkholderia cepacia, Burkholderia sp.*	Significant increase in yield of mustard.	Dutta et al., 2017
6.	*Brassica napus*	*Azotobacter chroococcum, Azospirillum brasilense, Paenibacillus polymyxa*	Increase in seed yield per plant and per hectare.	El-Howeity and Asfour, 2012
7.	*Brassica napus*	*Azotobacter, Pseudomonas*	Increase in number of pods per plant, seeds per pod and 1000- seed weight.	Naseri et al., 2013
8.	*Phaseolus vulgaris*	*Pseudomonas fluorescens P-93, Azospirillum lipoferum S-21, Rhizobium.*	Significant increase in pod per plant, number of seeds per pod, weight of 100 seed, weight of seeds per plant, weight of pods per plant, total dry matter.	Yadegari and Rahmani, 2008
9.	*Hordeum vulgare*	Nitrogen fixers: *Bacillus licheniformis RC02, Rhodobacter capsulatus RC04, Paenibacillus polymyxa RC05, Pseudomonas putida RC06, Bacillus OSU-142* *Phosphate solubilizers: Bacillus megaterium RC01, Bacillus M-13*	Increase in root and shoot weight.	ÇakmakÇi et al., 2007
10.	*Solanum lycopersicum*	*Pseudomonas putida, Pseudomonas fluorescens, Serratia marcescens, Bacillus amyloliquefaciens, Bacillus subtilis, Bacillus cereus*	Increase in shoot dry weight, plant height, number of fruits per plant, highest weight in comparison to negative control (nematode infected).	Almaghrabi et al., 2013
11.	*Ipomoea batatas*	*Bacillus cereus, Achromobacter xylosoxidans*	Increase in shoot length, root length, shoot fresh weight, shoot dry weight, root fresh weight, root dry weight.	Dawwam et al., 2013
12.	*Malus domestica cvs. 'Starkrimson' and 'Granny Smith'*	*Pseudomonas, Bacillus OSU-142*	Significant increase in yield per trunk cross-section area (13.3–118.5%), fruit weight (4.2–7.5%), shoot length (20.8–30.1%), and shoot diameter (9.0–19.8%) and in 'Starkrimson' and yield per Trunk Cross Sectional Area (TCSA; 14.9%) and fruit weight (6.5–8.7%) in 'Granny Smith' as compared to control.	Pirlak et al., 2007
13.	*Malus domestica*	*Bacillus M3, Bacillus OSU-142, Microbacterium FS01*	Significant increase in cumulative yield by 26%-88%, fruit weight by 13.9% to 25.5%, shoot length by 16.4% to 29.6%, shoot diameter by 15.9% to 18.4% in comparison to control.	Karlidag et al., 2007
14.	*Prunus avium*	*Pseudomonas BA-8, Bacillus OSU-142*	Significant increase in yield per trunk cross sectional area, fruit weight and shoot length as compared to control.	Esitken et al., 2006
15.	*Prunus cerasus*	*Bacillus mycoides T8, Bacillus subtilis OSU-142*	Significant increase in yield per tree, shoot length and leaf area.	Arikan and Pirlak, 2016
16.	*Fragari sp*	*Pseudomonas BA-8, Bacillus OSU-142, Bacillus M-3*	Increase in yield and fruit weight of the plant.	Pırlak and K˙ose, 2009
17.	*Rubus idaeus*	*Bacillus strains OSU-142, Bacillus M-3*	Significant increase in yield, cane length, number of cluster per cane, number of berry berries per cane.	Orhan et al., 2006
18.	*Mangifera indica*	*Burkholderia caribensis* XV, *Rhizobium sp.* XXV	Promotion of growth as exemplified by increase in biomass of root (89%),stem (34%),Leaves (51%), foliar area (53%), floral fate (100%) and flower buds (100%).	Santos-Villalobos et al., 2013

Table 2. Influence of PGPR on the Phytohormone synthetic machinery at various levels

Strain	Host Plant	Action With Respect to Phytohormone Modulation	Action on Morphology	Reference
Aeromonas punctata PNS-1	*Arabidopsis thaliana*	Auxin: Enhancement of endogenous production of Auxin. Auxin: Enhancement of auxin synthesis in-vitro by the bacteria with supplementation of tryptophan. 1-aminocyclopropane-1-carboxylate (ACC) deaminase: Presence of ACC deaminase activity.	Root: Significant increase in primary root length and lateral root density.	Iqbal and Hasnain, 2013
Azospirillum brasilense FP2	*Triticum aestivum*	Transcription Factor: Down regulation of ETTIN/ARF3 which mediates auxin dependent flowering and fruiting by binding on auxin receptor (AuxRec). Enzyme: Down regulation of aldoketose reductase. Protein: Upregulation of calmodulin-dependent auxin-induced protein SAUR (small auxin up RNA). Gene: Decreased Expression of ACO (*acc oxidase*) indicating decreased amount of ethylene production.	Mass: Improvement in root mass and plant mass.	Camilios-Neto et al.,2014
Azospirillum brasilense Sp245	Invitro batch cultures	Auxin: Synthesis of IAA in a standard minimal medium containing malate and tryptophan. Gene: Correlation of IAA synthesis and expression of ndole-3-pyruvate decarboxylase gene (*ipdC*). A key enzyme for biosynthesis of IAA.	n/a	Ona et al., 2005
Azospirillum brasilense Sp245,	*Arabidopsis thaliana*	Gene: Low expression of two nitrilases namely NIT1 and NIT2 which catalyses the conversion of indole-3-acetonitrile (IAN) to IAA. Upregulation of GH3.2, GH3.3, GH3.4, GH3.5 and GH3.12 which are family of genes that encode enzymes required to conjugate IAA with amino acids, an important function in auxin homeostasis.	Root: Increase in number of lateral roots and root hairs.	Spaepen et al.,2014
Burkholderia phytofirmans PsJN	*Arabidopsis thaliana*	Genes involved in auxin pathways: Upregulation of anthranilate synthase 1 (ASA1, At5G05730) which is involved in catalyzing the rate limiting step of biosynthesis of tryptophan, IAA induced gene (IAA1, AT4G14560), member of Aux/IAA transcription factor gene family, auxin responsive SAUR protein gene (SAUR68, At1G29510). Upregulation of AtGA3ox1 (Gibberellin 3-beta-dioxygenase, At1g15550) involved in catalyzing final step of gibberllin biosynthesis. Down regulation of as an auxin efflux carrier gene (At1G76520).		Poupin et al., 2013
Aneurinibacillus aneurinilyticus, Paenibacillus sp	*Phaseolus vulgaris*	Enzyme: ACC deaminase activity. IAA: Production of IAA.	Root and Shoot: Enhancement of root and shoot growth under salt stress.	Gupta and Pandey, 2019
Leclercia adecarboxylata MO1	*Solanum lycopersicum*	IAA: Production of IAA. Enzyme: Expression of acdS, the gene encoding for ACC deaminase.	Root and Shoot: Improvement of shoot length, root length, shoot fresh weight, root fresh weight and stem diameters under salt stress condition.	Kang et al., 2019
Sphingomonas sp. LK18, *Sphingomonas sp.* LK16, *Methylobacterium radiotolerans* LK17, *Bacillus subtilis* LK14, *Bacillus subtilis* LK15	*Solanum lycopersicum*	IAA: Production of IAA. Enzyme: ACC deaminase activity.	Root and shoot: Improvement of Shoot length, Shoot weight, Root length.	Latif Khan et al., 2016.
Leifsonia soli SE134	Cucumber, GA-deficient mutant rice cultivar Waito-C	Enzyme: Production of Gibberellins in culture.	Cucumber: Increase in biomass, hypocotyl, and root lengths . Waito-C rice: Increase in growth.	Kang et al., 2014
Pseudomonas fluorescens G20-18	Culture medium	Cytokinin: Production of isopentenyl adenosine (IPA), trans-zeatin ribose (ZR) and cytokinin dihydrozeatin riboside.		García de Salamone et al., 2001.

PGPR and Nitrogen Fixation

Nitrogen is one of the most essential macronutrient that is required by the plant for their growth. They are either absorbed by the plant directly from the soil through root hairs (Kiba and Krapp, 2016) or gets benefitted by the symbiotic bacteria which harbours themselves in the root nodules of leguminous plants and help in nitrogen fixation (Suzaki et al., 2015). PGPR can be an effective agent for absorption and fixation of nitrogen from the atmosphere and initiate plant growth. In this section the details of selected PGPR which possess the potential of nitrogen fixation would be discussed. The growth promoting and nitrogen fixing activity of a large number of PGPR on standard agricultural crops have been investigated. In a study conducted at University Agricultural Park, Universiti Putra Malaysia, it was observed that maize plants treated with PGPR strains *Klebsiella sp*. Br1, *Klebsiella pneumoniae* Fr1, *Bacillus pumilus* S1r1 and *Acinetobacter sp*. S3r2, and a reference strain *Bacillus subtilis* UPMB10 promoted a positive response in in-vitro tests. In pot experiments, it was found that maize plants treated with PGPR had an increased dry biomass prior to anthesis and ear harvest along with significant increase in nitrogen up-take. The nitrogen fixing capacity was also enhanced along with a delay in nitrogen remobilization and senescence indicating possibility of greater grain production (Kuan et al., 2016). Another interesting study from Chile reveals the presence of significant loads of bacteria in the rhizosphere of wheat plant. In addition to it, quantitative PCR analysis indicated the presence of 10^{12}~10^{13} copies per gram of 16S rRNA gene in the rhizosphere and 10^{7}~10^{8} copies per gram are present in the root endosphere while the *nif H* gene copy varied from 10^{5}~10^{6} and 10^{5} per gram of sample in rhizosphere and root endosphere respectively. The counts of putative nitrogen fixing bacteria were 10^{3} and 10^{2}~10^{3} CFU per gram of sample in rhizosphere and root endosphere. 16S rRNA sequencing revealed the presence of members belonging to Proteobacteria (*Bosea* and *Roseomonas*), Actinobacteria (*Georgenia, Mycobacterium, Microbacterium, Leifsonia*, and *Arthrobacter*), Bacteroidetes (*Chitinophaga*) and Firmicutes (*Bacillus* and *Psychrobacillus*). The study indicates the involvement of putative nitrogen fixing bacteria in the rhizosphere and root endosphere of wheat plant (Rilling et al., 2018). Another report from Afghanistan reveals that rhizobacteria isolated from rice fields showed positive responses on growth and physiological parameters of rice. It was found that among 98 bacteria isolated, 54% exhibited nitrogenase activity, 89% synthesized IAA while 40% produced siderophore and solubilized phosphates. Among the various rhizobacterial strains, *Pseudomonas resinovorans*, and *P. straminea* exhibited nitrogen fixing ability while *Rhizobium borbori* and *R. rosettiformans* exhibited interrelations with rice plants and fixation of nitrogen. *Enterobacter ludwigii* and *Pseudomonas putida* synthesized large quantities of IAA and also fixed nitrogen. Rice plants inoculated with bacterial strains in most cases exhibited a significant increase in dry weights of roots and shoots indicating the involvement of nitrogen fixation and phytohormones in the overall growth and development (Habibi et al., 2019). In a study performed on soybean, it was shown that inoculated by *Bacillus amyloliquefaciens* strain LL2012 along with natural symbiont *Bradyrhizobium japonicum* resulted in alteration of growth parameters and improved nodulation. Investigations also revealed that *Bacillus amyloliquefaciens* strain LL2012 was positive for nitrogen fixation, synthesized high quantities of auxin, gibberellins and salicylic acid in chemically defined medium and enhanced the capacity of *B. japonicum* to colonize the roots (Masciarelli et al., 2014). The effect of co-inoculation of PGPR (*Pseudomonas fluorescens* P-93 and *Azospirillum lipoferum* S-21) and *Rhizobium sp* on nodulation, nitrogen fixation, and yield *Phaseolus vulgaris* L. was also investigated in a study. It was observed that treatment with PGPR significantly increased nodule number, dry weight and quantity of nitrogen fixed in addition to increase in seed yield and protein content. On the other hand co-inoculation of PGPR with

Rhizobium sp resulted in an increase of nitrogen derived from the atmosphere (Yadegari et al., 2010). Population diversity of bacterial endophytes from *Corchorus capsularis* have been studied in one of the study. The results indicate that six isolates namely *Micrococcus sp.* strain MBL_B10, *Micrococcus sp.* strain MBL_B11, *Bacillus sp.* LTW29, *Pseudomonas psychrotolerans* strain MBL_B23, *Pseudomonas monteilii* strain MBL_B24, *Staphylococcus warneri* strain MBL_B25 were found to be positive for growth on nitrogen free solid media out of which *Bacillus pumilus* strain MBL_B12 was found to possess a 1200bp band for *nifH* gene whose sequencing showed similarity with sequence of *nifH* gene of NCBI database. In addition to it, acdS gene coding for ACC deaminase was also noted in some of the endophytic strains while some had the ability to solubilize phosphates. Increase in growth of the seedlings were also observed upon inoculation of the bacteria (Haidar et al., 2018) . Extracellular PGPR have been isolated and screened from the rhizosphere of tomato. Identification of the bacterial strains were done with the help of 16s rRNA analysis followed by BLAST analysis which finally revealed 38 isolates belonging to 9 different genera namely *Pseudomonas* (10), *Stenotrophomonas* (7), *Klebsiella* (5), *Chryseobacterium* and *Enterobacter* (4), *Sphingobacterium* and *Kosakonia* (3) and *Aeromonas, Delftia* and (1). Among these 38 screened isolates, 29 of them were found to be diazotrophs most of which belonged to the genus *Pseudomonas* (Guerrieri et al., 2020). The response of potato plant after inoculation with PGPR was also observed in a study conducted in Pakistan. Five bacterial strains were isolated from Rhizospheric soil samples of potato growing` areas and molecular characterization based on 16S rRNA gene sequence analysis revealed the presence of *Azospirillum sp, Agrobacterium sp, Pseudomonas sp., Enterobacter* sp while TN42 was a *Rhizobium sp.* Acetylene reduction assays confirmed that *Azospirillum sp, Pseudomonas sp*, and *Rhizobium sp* possessed nitrogenase activities. Out of three nitrogen fixing species, *nifH* gene could be amplified for *Azospirillum sp.* A 360 base pair fragment was obtained and sequenced which showed 99% similarity with an *Azospirillum brasilense* partial *nifH* gene, encoding a nitrogenase iron protein (Naqqash et al., 2016). The plant growth promoting activity and nitrogen fixing capacity of bacteria isolated from rhizosphere of sugarcane was also investigated. Molecular analysis revealed that the isolated strains belonged to the genus *Pseudomonas spp.* Nitrogen fixation capacity of the bacterial strain was determined by acytelene reduction assay. It was observed that all the strains were nitrogen fixers as evident from the nitrogenase activity among which *Pseudomonas putida* had a higher nitrogen fixing capacity. In addition to it, 10 strains were found to be positive for *nifH* gene amplification and produced an amplified fragment of about 360 bp all of which were related to partial *nifH* gene (Li et al., 2017). Thus it is evident that the bacterial population residing in the rhizosphere of a number of plants helps in fixation of nitrogen in addition to promoting growth. The mechanism of nitrogen fixation aided by the PGPR will be dealt later in the chapter.

PGPR and Solubilization of Minerals

Nutrition plays a pivotal role in the overall growth and development of a plant. The plants have evolved themselves in a very specialized way to absorb nutrients from the surrounding atmosphere. Apart from the autotrophic mode of nutrition in which the plant photosynthesizes their food in the form of carbohydrates (organic), they also rely on a sequestration mode for absorbing their nutrients from the soil. In the process, they largely absorb the elements which are required for their growth and normal physiology. Absorption of elements is also a highly sophisticated process and involves the action of a number of compounds that are secreted by the roots in order to degrade the complex minerals present in the soil (Canarini et al., 2019). However there are certain elements whose mineral form has a limited bioavailability. Classic example is that of phosphorus. Though phosphorus is an essential element, its bioavailability in the soil is largely finite due to number of reactions including adsorption, immobilization, or precipitation (Dixon et al., 2020). Another example is that of Zinc, which also have a limited bioavailability and decreases with increase in pH (Duffner et al., 2012). The beauty of PGPR is that it increases the bioavailability of a number of elements specially phosphate (Alori et al., 2017). There are also reports that PGPR also facilitates uptake of iron (Zhou et al., 2016) and zinc (Shakeel et al., 2015). Among various minerals that are solubilized by PGPR, its phosphate solubilizing activity has been most elaborately explored. The phosphate solubilizing activity of PGPR is thus elaborated in this section.

Phosphorus stands next to nitrogen in terms of growth and productivity determination of a plant (Mohidin et al., 2015). It is abundant in the soil in both organic and inorganic forms and its availability is restricted as it occurs mostly in insoluble form (Sharma et al., 2013). The average phosphorus content of the soil is about 0.05% (w/w) but only 0.1% of it is available making it a limiting factor for plant growth and development (Chen and Liu, 2019). In the soil, phosphorus is present either in the form of insoluble inorganic mineral such as apatite, strengite, and variscite (Shen et al.,2011) or as organic forms which include inositol phosphate (Turner et al., 2002), phosphomonoesters (Mc Laren et al.,2015), and phosphotriesters (Singh and Satyanarayana, 2011). Since most of the phosphorus in the soil is not bioavailable, to overcome the shortage, frequent fertilization of agricultural crops with the help of phosphate fertilizers is required. However, plants absorb a very little amount of these fertilizers and majority of them is rapidly converted into insoluble complex (Ahemad and Kibret, 2014). Thus relevance of alternative approaches to enhance the uptake of Phosphorus from the soil comes into picture. In this case, the PGPR or more appropriately the phosphate soluble microorganisms prove handy to enhance phosphorus availability and consequent uptake by the soil. Extensive research have been undertaken to find appropriate PGPR for solubilization of phosphates for agricultural and other crops. The details of PGPR related to phosphate solubilization is illustrated in Table 3.

Production of Siderophores and Chelation of Metals by PGPR

Siderophores are organic compounds having low molecular weight and are synthesized by microorganisms and plants growing under iron deficient condition (Ahmed and Holmström, 2014). Siderophores may be categorised into three main classes based on the chemical nature of the moieties which donates the oxygen ligands for coordinating with Fe (III). They are catecholates (catecholates and phenolates; also termed as "aryl caps"), hydroxamates, and (α-hydroxy-) carboxylates. In addition to it, there are siderophores whose chemical structures is a 'hybrid' of structures of two standard classes and thus shares features common to both groups. Such types of siderophores are called mixed siderophores (Miethke and Marahiel, 2007).The various types of siderophores are illustrated in Table 4. In general siderophores are often associated with virulence of the bacteria often resulting in occurrence of disease (Zawadzka et al., 2009; Su et al., 2016). However, siderophores secreted by the bacteria are also beneficial to the plants (Olanrewaju et al., 2017). Table 5 represents details of various siderophore producing PGPR and their beneficial effects on host plants.

Table 3. List of PGPR having phosphate solubilization efficacy

S.No.	PGPR Involved	Beneficiary Plant	Role in Plant Growth Promotion	Reference
1.	*Pseudomonas fluorescens* and *Pentoea ananatis*	*Pisum sativum*	Higher solubilization of phosphate when both bacteria are co cultured instead of single bacteria.	Anwar et al.,2019
2.	*Pseudomonas geniculata*	*Cicer arietinum*	Enhancement of available phosphorus in the rhizosphere.	Gopalakrishnan et al., 2015
3.	*Rhizobium cellulosilyticum, Rhizobium taibaishanense*	*Glycine max*	Solubilization of phosphate in Invitro assay.	Igiehon et al., 2019
4.	*Rhizobium panacihumi*	ginseng-cultivated soil	Solubilization of phosphate in Invitro assay.	Kang et al., 2019
5.	*Bacillus cereus YL6*	*Glycine max* *Triticum aestivum* *Brassica rapa*	Solubilization of phosphates in culture medium along with phosphorus concentration dependent activities of acid, alkaline and neutral phosphatases.	Ku et al.,2018
6.	*Mesorhizobium cicero,* *Mesorhizobium loti,* *M. huakuii CCBAU 2609*	*Cicer arietinum*	Solubilization of inorganic phosphate. Acid phosphatase activity.	Brigido et al., 2017
7.	*Azocarus sp* CIB	*Oryza sativa* *Nicotiana tabacum*	Solubilization of inorganic phosphate.	Fernández et al., 2014
8.	*Burkholderia contaminans* KNU17BI1	*Zea mays*	Solubilization of inorganic phosphate	Tagele et al., 2018
9	*Serratia marcescens* CDP-13	*Triticum aestivum*	Inorganic phosphate solubilization.	Singh and Jha, 2016
10.	*Streptomyces sp*	*Cicer arietinum*	Increased phosphorous availability in the rhizosphere.	Gopalakrishnan et al., 2015
11.	*Cellulosimicrobium funkei*	*Phaseolus vulgaris*	Solubilization of phosphates.	Karthik et al., 2016
20.	*Pantoea sp,* *Kosakonia sp,* *Bacillus sp*	*Lycopersicon esculentum*	Solubilization of phosphates.	Chakdar et al., 2018

Sequestration of Toxic Heavy Metals by PGPR

Heavy metals are naturally occurring elements that possess a high atomic weight with a density multiple times than that of water. Heavy metals also include metalloids such as arsenic and are intricately related to induction of toxicity. Other heavy metals including cadmium, chromium, lead and mercury also induce high degree of toxicity and are of grave concern with respect to public health (Tchounwou et al., 2012). Presently contamination by pollutants is posing a threat to the human population and also animal world (Jan et al., 2015; Cvjetkoet al., 2014). In addition to it, heavy metals also affect the plants in adverse way (Yousefi et al., 2011). Metal toxicity adversely affects yield of crops, soil biomass and fertility. It inhibits germination of seeds, elongation of roots, development of seedlings and also affects physiological parameters such as alteration of chloroplast, inhibition of electron transport chain, enzymes associated with Calvin cycle, impaired uptake of necessary elements and carbon dioxide deficiency due to stomatal closure (Sethy and Ghosh, 2013).Thus there has been a constant effort to sequester heavy metal contamination from the environment using various physical, chemical and biological processes (Sharma et al., 2018).One of the most innovative way to remove heavy metal from the environment is through the PGPR (Hassan et al.,2017;Pandey et al.,2013). The PGPR on one hand mediates the sequestering of heavy metals from the surrounding atmosphere especially from the soil and on the other hand helps the plant to alleviate the heavy metal stress. In this section we would discuss about the role of PGPR in removal of heavy metal contaminants and ameliorating heavy metal related stress in plants.

Table 4. Details of various categories of siderophores produced by bacteria.

S.No.	Type of Siderophores	Name	IUPAC Name	Source Microorganism	Reference
1.	Catecholate	Enterobactin	N-[(3S,7S,11S)-7,11-bis[(2,3-dihydroxybenzoyl)amino]-2,6,10-trioxo-1,5,9-trioxacyclododec-3-yl]-2,3-dihydroxybenzamide	*Streptomyces spp*	Fiedler et al., 2001
2.		Vibriobactin	(4R,5S)-N-[3-[3-[(2,3-dihydroxybenzoyl)amino]propyl-[(4S,5R)-2-(2,3-dihydroxyphenyl)-5-methyl-4,5-dihydro-1,3-oxazole-4-carbonyl]amino]propyl]-2-(2,3-dihydroxyphenyl)-5-methyl-4,5-dihydro-1,3-oxazole-4-carboxamide	*Vibrio cholerae*	Stoebner et al.,1992
3.	Phenolate	Yersiniabactin	(4S)-2-[(1S)-1-hydroxy-1-[(4R)-2-[(4R)-2-(2-hydroxyphenyl)-4,5-dihydro-1,3-thiazol-4-yl]-1,3-thiazolidin-4-yl]-2-methylpropan-2-yl]-4-methyl-5H-1,3-thiazole-4-carboxylic acid	*Yersinia pestis*	Bobrov et al.,2014
		Pyochelin	(4R)-2-[(4R)-2-(2-hydroxyphenyl)-4,5-dihydro-1,3-thiazol-4-yl]-3-methyl-1,3-thiazolidine-4-carboxylic acid	*Pseudomonas aeruginosa*	Braud et al.,2009
4.	Hydroxamate	Alcaligin	(8S,18S)-1,8,11,18-tetrahydroxy-1,6,11,16-tetrazacycloicosane-2,5,12,15-tetrone	*Bordetella pertussis, B. bronchiseptica, Alcaligenes denitrificans*	Moore et al.,1995
		Desferrioxamine B(3-)	N-[5-[[4-[5-[acetyl(oxido)amino]pentylamino]-4-oxobutanoyl]-oxidoamino]pentyl]-N'-(5-aminopentyl)-N'-oxidobutanediamide	*Streptomyces pilosus*	Chiani et al., 2010
5.	Carboxylate	Staphyloferrin A	2-[2-[[4-carboxy-4-[(3,4-dicarboxy-3-hydroxybutanoyl)amino] butyl]amino]-2-oxoethyl]-2-hydroxybutanedioic acid	*Staphylococcus aureus*	Laakso et al., 2016
		Achromobactin	1-[2-[(3R)-3-carboxy-5-[[(3S)-3-carboxy-3-(2-carboxy-2-hydroxy-5-oxopyrrolidin-1-yl)propyl]amino]-3-hydroxy-5-oxopentanoyl]oxyethyl]-2-hydroxy-5-oxopyrrolidine-2-carboxylic acid	*Erwinia chrysanthemi*	Douet et al., 2009
6.	Catecholate-hydroxamate	Heterobactin B	N-[(4R)-4-amino-5-[[2-[[(3S)-1-hydroxy-2-oxopiperidin-3-yl]amino]-2-oxoethyl]amino]-5-oxopentyl]-2,3-dihydroxybenzamide	*Rhodococcus erythropolis*	Carran et al.,2001
7.	Phenolate Hydroxamate	Mycobactin T	2-[(4R)-4-[[(2S)-6-[icosanoyl(oxido)amino]-1-[(2R)-4-[[(3S)-1-oxido-2-oxoazepan-3-yl]amino]-4-oxobutan-2-yl] oxy-1-oxohexan-2-yl]carbamoyl]-4,5-dihydro-1,3-oxazol-2-yl] phenolate;iron(6+)	*Mycobaterium tuberculosis*	Juárez-Hernández et al., 2012
8.	Citrate-catecholate	Petrobactin	4-[4-[3-[(3,4-dihydroxybenzoyl)amino]propylamino] butylamino]-2-[2-[4-[3-[(3,4-dihydroxybenzoyl)amino] propylamino]butylamino]-2-oxoethyl]-2-hydroxy-4-oxobutanoic acid	*Bacillus anthracis*	Wilson, 2010
9.	Citrate-Hydroxamate	Aerobactin	4-[[(1S)-5-[acetyl(hydroxy)amino]-1-carboxypentyl]amino]-2-[2-[[(1S)-5-[acetyl(hydroxy)amino]-1-carboxypentyl]amino]-2-oxoethyl]-2-hydroxy-4-oxobutanoic acid	*Escherichia coli*	Genuini et al., 2019

Table 5. Details of siderophore producing PGPR and their beneficial action on host plants

S.No.	Siderophore Producing PGPR	Beneficiary Plant	Name of Siderophore/Growth Promoting Effect	Genes Involved	Reference
1.	*Rhizobium* sp.	*Cyamopsis tetragonoloba*			Dhull and Gera,2018
2.	*Rhizobium* BICC 651	*Cicer arietinum* L		*sidC, sidE, sidB*, and *sidA*	Datta and Chakraborty,2014
3.	*Rhizobium leguminosarum*	*Vicia sp*	Name of Siderophore: vicibactin.	*vbsGSO, vbsADL, vbsC* and *vbsP*	Carter et al., 2002
4.	*Chryseobacterium spp.* C138 *Pseudomonas fluorescens* N21.4	*Lycopersicon esculentum*	Significant increase in plant yield, chlorophyll and iron content upon treatment with bacterial siderophores and bacteria and supplemented with iron.		Radzki et al., 2013
5.	*Pseudomonas* sp.	Wheat (*Triticum aestivum*), Chickpea (*Cicer arientinum*), Lathyrus (*Lathyrus sativus*), Greengram (*Vigna radiata*), Blackgram (*Vigna mungo*), Bottlegourd (*Lagenaria siceraria*) and Rice (*Oryza sativa*)	Invitro inhibitory effect against *Rhizoctonia solani, Sclerotium rolfsii.* Positive correlation between siderophore production and antagonistic effect. Positive effect on growth parameters such as root length, shoot length.		Priyanka et al., 2017
6.	*Pseudomonas japonica*	*Zea mays*	Production of siderophores and solubilization of zinc. Significant increase in plant height, fresh and dry weight.		Eshaghi et al.,2019
7.	*Serratia* sp.	*Zea mays*	Production of siderophores and solubilization of zinc, phosphate and potash. Synthesis of IAA, GA3 and positive for ACC deaminase. Significant increase of shoot length, root length and total chlorophyll content.		Kour et al.,2019
8.	*Arthrobacter globiformis*	*Zea mays*	Production of siderophores, chelation and dissolution of various iron complex. Enhancement of plant biomass, uptake of iron and phosphate, and protein and chlorophyll contents.		Sharma et al., 2016
9.	*Bacillus, Oceanobacillus, Halomonas* (halotolerant bacteria)	*Triticum turgidum* subsp. *durum*	Production of siderophore and solubilization of phosphates. Capacity of nitrogen fixation, ACC deaminase activity and auxin production. Antagonistic against *Fusarium culmorum.* Ability of host plant to withstand high salinity with increase of germination and seedling growth.		Albdaiwi et al., 2019
10.	Fluorsecent *Pseudomonas*	Cicer arietinum	Production of siderophores. Inhibitory activity against *Rhizoctonia solani* and *Sclerotium rolfsii.* Increase in yield and bundle weight.		Kotasthane et al.,2017
11.	*Pseudomonas* strains GRP3A and PRS9	Zea mays	Production of siderophores. Antagonistic against *Colletotrichum dematium, Rhizoctonia solani* and *Sclerotium rolfsii.* Significant increase in percentage of germination, shoot length and root length.		Sharma and Johri, 2003
12.	*Pseudomonas* sp.	*Lycopersicon esculentum*	Production of siderophores along with solubilization of phosphates. Production of IAA. Increase in seed germination, seedling height. Promotion of plant length, increase of collar diameter and number of leaves.		Qessaoui et al., 2019
13.	*Delftia tsuruhatensis* MTQ3		NRPS knock out mutants failed to synthesize siderophores.	nonribosomal peptide synthetase (NRPS)	Guo et al.,2016
14.	*Rhizobium oryzihabitans* sp. *nov*	*Oryza sativa*	Production of siderophores, ACC deaminase and IAA. Increase in e length of stem and fresh weight of seedlings inoculated with bacteria.		Zhao et al., 2020
15.	*Bacillus cereus*		Name of Siderophore: Petrobactin and Bacillibactin.	Petrobactin biosynthesis genes (*asbABCDEF*) Bacillibactin biosynthesis genes- dhb cluster (*dhbACBEF*)	Zeng et al., 2018
16.	*Enterobacteriaceae, Pseudomonaceae*		Name of Siderophore: Pyoverdine Achromobactin.	pvd, pyoverdine homologous genes, *fpvA,mbtH,fhu acrA* and *acrB*	Gupta et al., 2014

A study reported that *Lysinibacillus varians* strain KUBM17 and *Pseudomonas putida* strain KUBM18 isolated from the rhizospheric soil contaminated with industrial, sewage or agrochemical waste were tolerant to high concentrations of lead and cadmium. In addition to it, the bacterial strains exhibited remarkable lead and cadmium removal potential. The bacterial strains also exhibited promising plant growth promoting traits with respect to synthesis of IAA, solubilization of phosphates, fixation of nitrogen and siderophore production. In addition to it, the mentioned bacteria also improved the germination pattern and growth parameters of radish plants grown in presence of lead and calcium (Pal and Sengupta, 2019). In another study it was shown that inoculation of *Bacillus cereus* in *Oryza sativa* cultivars cv. Basmati grown under cadmium stress resulted in lesser accumulation of the metals within the plant as compared to the uninoculated control. In addition to it, inoculation of the bacteria resulted in the improvement of growth parameters and chlorophyll content which was previously reduced due to cadmium toxicity (Jan et al., 2019).Similar study on rice revealed that *Enterobacter* sp. exhibited a potential to remove cadmium from the growth medium. Additionally rice seedlings inoculated with the bacteria exhibited a significant decrease in cadmium uptake as compared to uninoculated plant treated with cadmium. The bacteria also enhanced growth and physiological parameters of rice seedlings (Mitra et al., 2018). A recent report states that co inoculation of mixed compost biochar and *Bacillus amyloliquefaciens* resulted in decrease of lead concentration (43%) in spinach root over the control (Zafar-ul-Hye et al., 2020). The potential of PGPR as a mediator of phytoremediation has also been explored. In a sturdy it was reported that *Acinetobacter sp* FQ-44 possessed the potential to absorb copper and zinc and thus can play a major role in limiting phytotoxicity. In addition to it, it was further reported that inoculation of *Acinetobacter sp* FQ-44 resulted in increase of root length, shoot length, fresh weight and percentage of germination in *Brassica napus*. The bacteria also increased uptake of copper by the plant hinting at its potent phytoextraction potential (Fang et al., 2016). In another recent study it was shown that coinnoculation of *Paenibacillus mucilaginosus* and the metal-resistant rhizobium *Sinorhizobium meliloti* in alfa alfa resulted in higher accumulation of copper when grown in copper contaminated soil again establishing itself as a potent candidate of phytoextraction (Ju et al., 2019).In another study it was shown that *Pseudomonas fluorescens* K23 was highly effective in absorbing lead on the cell surface (12.90 $\pm$ 0.85 mg lead g^{-1} Dry weight). In addition to it, highest value of intracellular accumulation of lead was observed in case of *Luteibacter sp*. K20 (2.34 $\pm$ 0.33 mg lead g^{-1} Dry weight). It was further reported that *Lathyrus sativus* plant inoculated with PGPR showed increased uptake of lead than the uninoculated plant. This result hints on the probable hyper accumulation potential of the plant when associated with PGPR (Abdelkrim et al., 2018).

Arsenic is another element which is found in a variety of forms in the environment, the most predominant form being inorganic arsenic, present in drinking water which is not only toxic but also carcinogenic and highly bioavailable (Anetor et al., 2007).Arsenic can cause a number of disorders namely skin lesions, respiratory and nervous disorders, and different types of cancers (Chung et al., 2014). There are a number of physicochemical-based conventional methodologies which removes arsenic. However these techniques are expensive and produce byproducts that create further toxicity. Thus there are constant search of sustainable and cost effective mode for removal of arsenic from surrounding environment (Bahar et al., 2013). Bacterial and fungal species have the potential of removing arsenic from contaminated soil and water and claims to be suitable candidates as a bioremeddiator and an effective cost effective alternative of physico chemical methods (Mitra et al., 2017). A number of processes such as oxidation, reduction, methylation and intracellular bioaccumulation are adopted by microorganisms for bioremediation of arsenic (Satyapal et al., 2018). Among various microorganisms, PGPR also play an important role in removal of arsenic. The role of selected PGPR in environmental management of arsenic contamination is tabulated in Table 6. Chromium is another heavy metal that is of wide industrial useage and is considered to be toxic. Presently chromium is a serious environmental pollutant and contaminates both soil and water. The toxicity of chromium depends upon the oxidation state ranging from low toxicity metal form to highly toxic hexavalent form (Tchounwou et al., 2012). The important toxic effects upon inhalation or ingestion of chromium in hexavalent state include dermatitis, skin reactions, ulcerations in skin and mucous membrane, nasal septum perforation, allergic asthmatic reactions, lung cancer, gastric problems, hepatocellular deficiency, and renal oligo anuric deficiency (Baruthio et al.,1992). In plants chromium alters the germination process, growth of roots, stems and leaves which further affect the dry matter production and yields (Shanker et al., 2005). PGPRs also possess the potential to modulate chromium stress and removal of chromium from the environment. Table 7 depicts the role of PGPR in modulation of chromium stress and contamination. The mechanistic aspects of remediation of heavy metals through PGPR will be discussed in the next section.

Table 6. Role of PGPR in management of arsenic contamination and stress

S.No.	Name of PGPR	Beneficiary Plant	Beneficial Effect	Reference
1.	*Bacillus aryabhattai* MCC3374	*Oryza sativa*	Production of arsenate reductase which transformed As(V) to As(III). Higher removal efficiency of As (V) as compared to As (III). Formation of complex of Arsenic and polarized group of cell surface of bacteria. Detection of pellets of arsenic. Improvement of morphological and biochemical parameters of rice along with increased activities of amylase and protease. Positive for PGP traits such as nitrogen fixation, phosphate solubilization, siderophore production, ACC deaminase activity and EPS production.	Ghosh et al.,2018
2.	*Sphingomonas paucimobilis*		Biosorption of arsenic	Titah et al.,2018
3.	*Agrobacterium radiobacter*	*Populus deltoides*	Enhancement of arsenic removal efficiency by *Populas deltoides* upon inoculation with bacteria. Increase in dry weight, plant height and root collar diameter of plant growing under arsenic stress upon inoculation with bacteria. Physiological parameters such sugar content, protein content, chlorophyll content, catalase activity, SOD activity increased upon inoculation of bacteria . Decrease in MDA upon inoculation of bacteria.	Wang et al.,2011
4.	*Bacillus flexus* and *Acinetobacter junii*		Removal of arsenic from the media. Detection for arsC gene	Marwa et al.,2018
5.	*Pseudomonas sp.* P1III2, *Delftia sp.* P2III5, *Variovorax sp.* P4III4, *Pseudoxanthomonas sp.* P4V6, and *Bacillus sp.* MPV12	*Pteris vittata*	Mixed inoculation resulted in enhanced arsenic accumulation in fronds and roots. Reduction in arsenic content in the soil. Significant increase in frond biomass, frond growth and dry weight.	Lampis et al., 2015
6.	*Pseudomonas, Actinomyces, Azotobacter, Azospirillum, Bacillus.*	*Eichhornia crassipes*	Higher phytoaccumulation of arsenic by the plant.	Kaur et al.,2018

Table 7. Role of PGPR in management of chromium contamination and stress

S.No.	Name of PGPR	Beneficiary Plant	Beneficial Effect	Reference
1.	*Agrobacterium fabrum* and *Leclercia adecarboxylata* + Iron	*Zea mays*	Significant increase in plant height, plant fresh weight and plant dry weight as compared to control indicating ameliorating action. Increase in nitrogen, phosphorus and potassium levels in roots and leaves as compared to control.	Danish et al.,2019
2.	*Cellulosimicrobium funkei* – homologue	*Phaseolus vulgaris*	Amelioration of inhibitory effect of chromium on seed germination, shoot length, root length, biomass and chlorophyll content. Decrease in proline and MDA content upon inoculation with bacteria. Reduction in activities of catalase, peroxidase and polyphenol oxidase. Significantly lower accumulation of chromium in roots and shoots.	Karthik et al., 2016
3.	*Bacillus subtilis* MNU16		Reduction of chromium (VI) to Chromium (III). Deposition of chromium particles in the cell. Positive for PGP traits such as production of IAA, siderophore, phosphate solubilization and ACC deaminase activity.	Upadhyay et al.,2017
4.	*Sinorhizobium sp.* SAR1		Adsorption of chromium and biotransformation from Chromium (VI) to chromium (III).	Jobby et al.,2019
5.	*Bacillus sp.* MH778713	*Prosopis laevigata*	Capacity to remove hexavalent chromium. Increase in dry weight, fresh weight, length of leaf, root and stem upon inoculation of bacteria.	Ramírez et al.,2019

Biodegradation of Pesticides by PGPR

Pesticides are compounds or a mixture of compounds that are largely used in agriculture or in public health protection programs for protecting plants from attack of pests, weeds, diseases and also humans from vector borne diseases such as malaria, schistosomiasis, and dengue (Nicolopoulou-Stamati et al., 2016). Pesticides are highly persistent in the environment and can be bio transformed into a number of products which further interacts with the living organism in a number of ways (Lushchak et al., 2018). The pesticide residues contribute to a large proportion of environmental pollution (Jayraj et al., 2017). It is also known that pesticides accumulate in living organism resulting in long term chronic effects (Damalas and Koutroubas, 2016). Thus removal of pesticide residues from the environment is a matter of great priority and a number of approaches are adopted for the purpose (Chen et al., 2019; Lozowicka et al., 2016). PGPR have also become a very effective tool in removal and degradation of pesticides. In this section the role of PGPR in removal and degradation of pesticides will be discussed. Table 8 illustrates the beneficial role of PGPR in degradation and removal of pesiticides.

Table 8. Role of PGPR in degradation and or removal of pesticides

S.No.	Name of Bacteria	Name of Pesticide/s	Beneficial Action	Reference
1.	*Bacillus subtilis* GB03, *Bacillus subtilis* FZB24, *Bacillus amyloliquefaciens* IN937a and *Bacillus pumilus* SE34	acibenzolar-S-methyl, metribuzin, napropamide, propamocarb hydrochloride and thiamethoxam	Degradation of all pesticides in the liquid medium. Significant reduction in half-life of acibenzolar-S-methyl in soil upon treatment with bacterium. Enhanced degradation of propamocarb hydrochloride and thiamethoxam in presence of *B. amyloliquefaciens* IN937a and *B. pumilus* SE34.	Myresiotis et al., 2011.
2.	*Pseudomonas rhizophila* S211	Pentachlorophenol	Production of rhamnolipid biosurfactant and ability and enhancement of solubility of pentachlorophenol. Genome sequence identified the presence of genes responsible for production of ACC deaminase, putative dioxygenases, auxin, pyroverdin, exopolysaccharide levan and rhamnolipid biosurfactant.	Hassen et al., 2018
3.	*Bacillus sp.* KF984414 *Bacillus sp.* LN849696	Endosulfan	Degradation of endosulfan both in broth and soil. Production of IAA, siderophores and solubilization of phosphates.	Rani and Kumar, 2017
4.	*Bacillus subtilis* strains DR-39, CS-126, TL-171, and TS-204	Profenofos	Significant reduction in half-life of profenofos as compared to uninoculated control indicating enhancement in degradation.	Salunkhe et al., 2013
5.	*Ochrobactrum sp.* CPD-03	Chlorpyrifos and 3,5,6-trichloro-2-pyridinol	Degradation of chlorpyrifos and its metabolite 3,5,6-trichloro-2-pyridinol. Exhibited growth promoting activity in rice seedling.	Nayak et al., 2019
6.	*Microbacterium sp.* P27	Lindane	High percentage of degradation of Lindane (82.7) along with positive results for PGP traits namely production of IAA and ACC deaminase activity	Singh and Singh, 2019
7.	*Chryseobacterium* sp. PYR2	α-HCH, β-HCH, γ-HCH, σ-HCH, o,p'-DDT, and p,p'-DDT	Degradation of HCHs and DDT by the bacteria in liquid medium. Degradation of DDT in soil.	Qu et al., 2015
8.	*Serratia marcescens* NCIM 2919	DDT	Degradation of DDT as evident of detection of four metabolites of DDT degradation pathway namely 2,2-bis (chlorophenyl)-1,1-dichloroethane (DDD), 2,2-bis (chlorophenyl)-1,1-dichloroethylene (DDE), 2,2-bis (chlorophenyl)-1-chloroethylene (DDMU), and 4-chlorobenzoic acid (4-CBA).	Neerja et al., 2016

Biodegradation of Hydrocarbons by PGPR

PGPRs are also used for bioremediation of hydrocarbons especially for removal of petroleum contamination in soil. In one study, it was shown that inoculation of two bacteria namely *Klebsiella* sp. D5A and *Pseudomonas* sp. SB along with seeds of *Festuca arundinacea* resulted in maximum hydrocarbon removal particularly the C21 – C34 aliphatic hydrocarbons and polycyclic aromatic hydrocarbons. In addition to it, inoculation of the bacteria also resulted in increase of shoot and root biomass of *Festuca* plant (Hou et al., 2015). In another study it was observed that endophytes isolated from *Lotus corniculatus* and *Oenothera biennis* growing on petroleum hydrocarbon polluted sites as the potential to grown on medium containing diesel. Phylogenetic analysis revealed that majority of the strain belongs to genera *Rhizobium, Pseudomonas, Stenotrophomonas*, and *Rhodococcus*. Most of the genera (90%) had the potentiality to grow on diesel oil while 20% of the isolates used n-hexadecane as the sole carbon source. PCR analysis revealed that 40% of the bacteria had P450 gene which encode cytochrome P450-type alkane hydroxylase (CYP153). In vitro tests revealed that the bacteria had PGP traits including synthesis of IAA, hydrogen cyanide, production of siderophores, and solubilization of phosphate. Gene encoding ACC deaminase was also present in 40% of the bacteria. The results were indicative of the fact that the strains have the dual capacity of growth promotion as well as removal of hydrocarbons (Pawlik et al., 2017). *Streptomyces sp* isolated from sandy soil contaminated with oil is also reported to thrive on petroleum as a sole carbon source with removal efficiency of as high as 98% after 7 days of incubation. The isolates had the potential to degrade n-alkanes (C6-C30), aromatic and polycyclic aromatic hydrocarbons and also possessed plant growth promoting features such as production of siderophores, solubilization of phosphates, nitrogen fixation etc. (Baoune et al., 2018). It was also reported that *Zea mays-Streptomyces* sp. Hlh1 was effective in removal of hydrocarbons (C8-C30) from contaminated soil. In addition to it, inoculation with bacteria also resulted in significant plant development along with increase in photosynthetic pigments (Baoune et al., 2019). In another study it was shown that co-inoculation of *Bacillus cereus* CPOU13 and *Bacillus subtilis* SPC14 resulted in effective degradation of phenanthrene, anthracene and pyrene in In-vitro condition (Rao et al., 2016). In addition to it, a study reports that mixed inoculation of *Bacillus thuringiensis* B3 and *Bacillus cereus* B6 along with two fungi namely *Geomyces pannorum* HR and *Geomyces* sp. strain HV resulted in removal of significant percentage (87.45%) of total petroleum hydrocarbon from soil samples treated with crude oil (Maddela et al., 2017). In a similar study it was also found that *B. thuringiensis* B3 and *B. cereus* B6, isolated from crude oil-contaminated sites had the potential to degrade n-alkane fractions (C8-C40) of utilized lubricating oil (Raju et al., 2017). Another interesting report states that a new actinomycetes strain by the name *Nocardiopsis* sp.*mrinalini* 9 isolated from *Hibiscus rosa-sinensis* was able to degrade diesel and plastics thereby claiming to be a potential candidate of degradation of not only diesel hydrocarbon but also hydrocarbons of non-degradable plastics (Singh and Sedhuraman, 2015). Thus it is quite evident that PGPRs can be very well used for remediation of petroleum hydrocarbons. In this section various beneficial activities of PGPR have been elaborated with appropriate examples. In the next section, the mechanistic aspect by which is responsible for their beneficial role would be discussed.

Production of Antibiotics by PGPR

Some PGPR are known to produce antibiotics which help in control of plant pathogens. This forms the basis of bio control of plant pathogens with the help of bacteria. The basis of antibiosis is secretion of molecules which is either fatal to the pathogen or reduce the growth and has been extensively investigated in last few decades (Whipps, 2001; Lugtenberg and Kamilova, 2009; Dowling and O'Gara, 1994). The antibiotics constitute a wide range of low molecular weight compounds that are detrimental to the growth and metabolic activities of microorganisms (Duffy et al., 2003). Six classes of antibiotic compounds are designated which are related to bio control of root diseases. They are phenazines, phloroglucinols, pyoluteorin, pyrrolnitrin, cyclic lipopeptides all of which are diffusible and the volatile hydrogen cyanide (Hans and Defago, 2005). In addition to it, lipopeptide biosurfactants which are produced by strains of *Pseudomonas* and *Bacillus* are also potent bio control agents due to their positive effect on competitive interactions with a wide range of pathogens (Beneduzi et al., 2012). Table 9 illustrates selected antibiotic producing strains of bacteria and their target organisms.

Table 9. Details of antibiotics produced by selected PGPRs

S.No.	Name of PGPR	Antibiotics Produced	Pathogen Inhibited	References
1	*Pseudomonas fluorescens* CHA0	2,4-diacetylphloroglucinol (DAPG)		Schinder-Keel et al., 2000
2.	*Pseudomonas sp.* LBUM223	Phenazine	*Streptomyces scabies*	Arseneault et al., 2013
3.	*Pseudomonas fluorescens* LBUM636	Phenazine-1-Carboxylic Acid	*Phytophthora infestans*	Morrison et al., 2017
4.	*Pseudomonas cepacia* B37w	Pyrrolnitrin	*Fusarium sambucinum*	Burkhead et al., 1994
5.	*Pseudomonas fluorescens* (BL915)	pyrrolnitrin	*Rhizoctonia solani*	Hill et al., 1994
6.	*Pseudomonas fluorescens* strain CHA0	Pyoluteorin	*Pythium ultimum*	Maurhofer et al.,1994
7.	*Rhizobacterium, Paenibacillus polymyxa* M-1	Polymyxin P	*Erwinia spp*	Niu et al., 2013
8.	*Bacillus subtilis*	fengycin, and iturin A	*Podosphaera fusca*	Romero et al.,2007
9.	*Bacillus subtilis*	iturin-like lipopeptides	*Xanthomonas campestris* pv. *cucurbitae*	Zeriouh et al., 2011
10.	*Bacillus cereus* UW85	zwittermicin A	*Phytophthora medicaginis*	Silo-Suh et al., 1994

MECHANISM OF ACTION OF PGPR

Growth of Plants and Production of Plant Growth Regulator

It is a well-established fact that PGPR is intricately associated with enhancement of growth of plant. This enhancement depends upon a number of factors including presence of optimal levels of nutrients, solubilization of phosphates, production of siderophores, and production of ACC deaminase for breakdown of ethylene and plant growth promoters (Ahmed and Hasnain, 2014). PGPRs are known to produce auxins and stimulate growth in various plants (Asari et al., 2017). It is estimated that 80% of the bacterial species residing in the rhizosphere are capable of synthesizing auxin or IAA (Talboys et al., 2014). The main precursor of auxin biosynthesis is tryptophan (Zhao, 2014). Based on the intermediates involved in tryptophan dependent IAA biosynthesis, five different pathways have been demarcated in the bacteria namely indole-3-acetamide (IAM), indole-3-pyruvic acid (IPA), indole-3-acetonitrile, tryptamine, and tryptophan side-chain oxidase pathways (Figure 2). In addition to it, tryptophan independent pathway is also reported to exist in bacteria though no enzymes involved in the pathway have been characterized (Li et al., 2018). Genes involved in IAA biosynthesis have been identified and characterized in bacteria. Selected genes involved in IAA biosynthesis in bacteria are tabulated in Table 10. It is reported that in plant pathogen *Pseudomonas savastanoi* pv. *savastanoi*, which synthesizes high quantities of IAA, the *iaaM* and *iaaH* genes are located on pIAA1 plasmid. Both genes are clustered on iaa operon and loss of pIAA1 plasmid results in loss of IAA synthesizing ability of the bacteria. Consequently bacteria lacking pIAA1 plasmid are incapable of forming galls in plants (Spaepen and Vanderleyden, 2011).

Figure 2. Various pathways leading to biosynthesis of IAA (Auxin) in PGPR
1: Indole-3-acetamide pathway; 2: Indole-3-pyruvic acid pathway; 3: Tryptamine pathway; 4: Indole-3-acetonitrile pathway; 5: Side-chain oxidase pathway
Enzymes involved: A= Nitrile Hydratase; B=Nitrilase; C=Indole acetamide hydrolase (IAH); D=AldA & AldB; E=indole pyruvate decarboxylase (IPDC); F=: Tryptophan 2-monooxygenase (TMO)

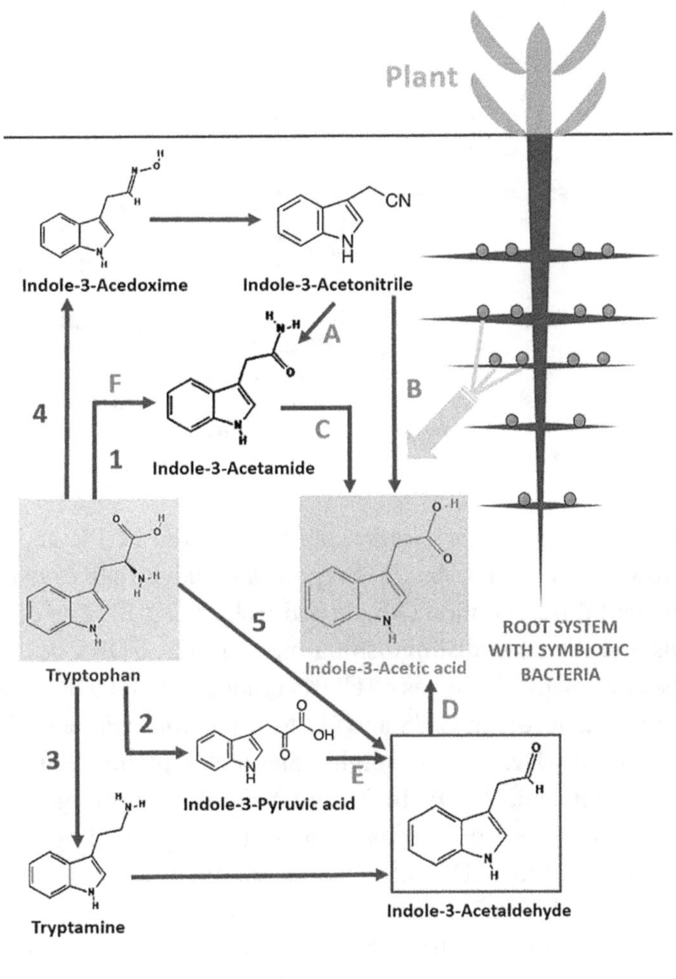

Figure 3. Schematic representation of wall loosening by the action of Indole-3-Acetic Acid (Auxin)

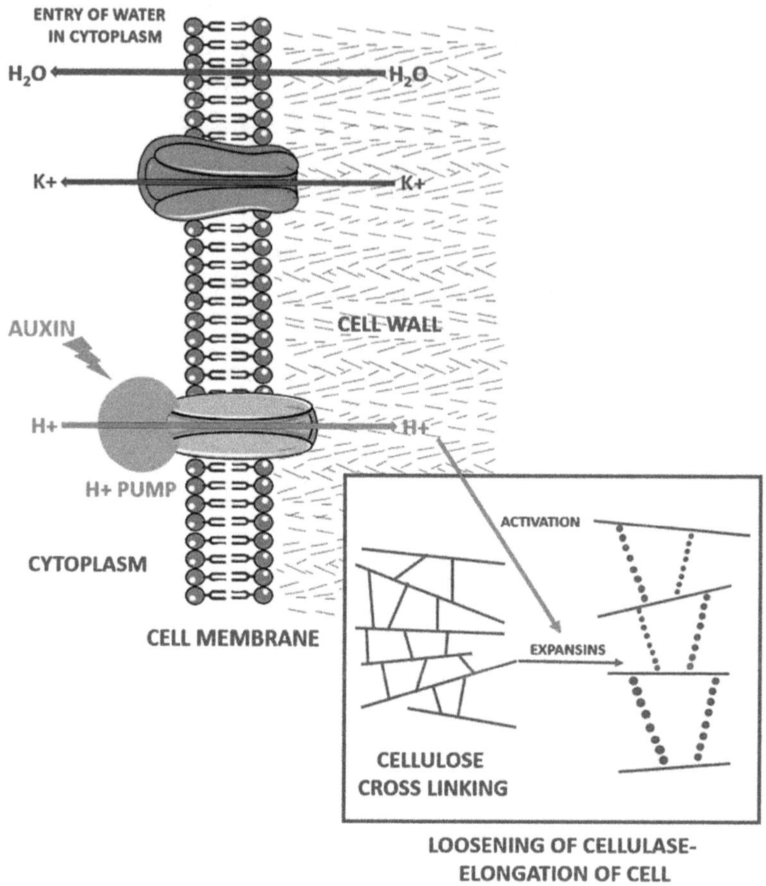

Plants are the direct beneficiary of the synthesized IAA by the PGPR and it is this hormone that primarily controls the growth of the plant. Auxin is reported to induce cell division or proliferation, cell elongation or growth and cell differentiation (Majda and Robert, 2018). The cell division is controlled by a plethora of proteins and the most distinguishing ones are the CDKs (Qi and Zhang, 2019). It is reported that plants possess as many as 6 types of CDKs among which CDKA represented by CDKA;1 in *Arabidopsis* plays an important role in G1/S and G2/M transitions (Francis, 2011).It is reported that auxin induces the expression of CDKA;1, the single homolog of mammalian Cdk1 and Cdk2 present in *Arabidopsis* (Perrot-Rechenmann, 2010). In *Arabidopsis*, the mitotic gene CYCD3; 1 acts as an important determinant for G1/S transition and its expression is regulated by auxin (Hu et al., 2003).In addition to it, expression of two of the CDK inhibitors namely KRP1 and KRP2 are down regulated in presence of auxin (Himanen et al., 2002; Sanz et al., 2011; Li et al., 2016). These indicate that auxin is responsible for triggering cell division through modulation of key molecules involved in divisional process. Auxin is also involved in cell wall loosening and expansion (Lewis et al., 2013). This involves structural and molecular modifications in the cell wall which results in relaxation of cell wall tension. Auxin is reported to activate H+ ATPase at the plasmamembrane resulting in extrusion of protons and consequent acidification, activation of expansins and ultimately loosening of the wall (Rober-Kleber et al., 2003; .Barbez et al., 2017) (Figure 3). Activation of plasma membrane H+ ATPase also results in hyperpolarization of membrane potential which in turn activates voltage dependent inward K+ channels (Okumura et al., 2016; Majda and Robert, 2018; Claussen et al., 1997).It is reported in maize that auxin induced expression of *Zmk1* and *Zmk2* which codes for inward potassium channels (Philippar et al., 1999). This results in uptake of potassium ions by the cell along with uptake of water molecules required for cell wall expansion (Hasanuzzaman et al., 2018).

Auxin activates proton pump which results in extrusion of proton in the cell wall (apoplast). This results in acidification of cell wall, activation of expansins and ultimately loosening of wall materials which initiates elongation.

Potassium enters inside the cytoplasm which then facilitates entry of water due to difference in osmotic potential. Entry of water increases turgor aiding in elongation process.

Table 10. Genes involved in biosynthesis of IAA in bacteria

Genes	Protein Encoded	Function	Source Bacteria	Reference
iaaM	tryptophan-2-monooxygenase	Converts tryptophan to IAM	*Agrobacterium vitis*	Oetiker et al.,1999
iaaH	Indole-3-Acetamide Hydrolase	conversion of IAM to IAA	*Alcaligenes faecalis* subsp. *parafaecalis*	Mishra et al., 2016
bam	*Bradyrhizobium amidehydrolase*	conversion of IAM to IAA	*Bradyrhizobium japonicum*	Sekine et al., 1989
ipdC	indole-3-pyruvate decarboxylase	Decarboxylation of IPA to indole-3-acetaldehyde	*Azospirillum brasilense*	Spaepen et al.,2007

Fixation of Nitrogen

The most abundant element present in the atmosphere is nitrogen and plays an important role for plant growth (Zerkle and Mikhail, 2017). Though atmosphere contains 78% of nitrogen but still it is not available to the plants (Mancinelli, 1996). Nitrogen requires to be converted to ammonia for assimilation in biological system (Behie and Bidochka, 2013). Nitrogen in the form of ammonia is a unique molecule and have both organic and inorganic entity and is assimilated by plants through the process of biological nitrogen fixation (BNF) (Mus et al., 2016). Biological nitrogen fixation are done by bacteria and are called diazotrophs (Che et al., 2018). Diazotrophs encode nitrogenase which converts gaseous nitrogen to ammonia (Burén et al., 2018). Nitrogenase is an enzyme complex and is highly conserved amongst the diazotrophs (Santi et al., 2013). Nitrogenase is a two component system and consists of a MoFe protein namely the dinitrogenase or component I which is responsible for reducing nitrogen to ammonia and an electron transfer Fe protein, the dinitrogenase reductase which constitutes component II (Figure 4). A reducing source in the form of MgATP is also required by nitrogenase. The reducing source undergoes hydrolysis along with association and dissociation of Fe protein and MoFe protein in a catalytic cycle which involves a single electron transfer. It was also reported that MoFe protein possess two metal clusters: the iron-molybdenum cofactor (Fe-Mo Co) which provides the active site substrate binding and P-cluster involved in electrons transfer from Fe protein to FeMo-Co (Hoffman et al., 2014) (Figure 4). The overall reaction catalyzed by nitrogenase is represented as follows (Hu and Ribbe, 2015):

$$N_2 + 8H^+ + 16MgATP + 8e^- \rightarrow 2NH_3 + H_2 + 16MgADP + 16Pi$$

Figure 4. Schematic representation of nitrogenase enzyme complex and biological nitrogen fixation

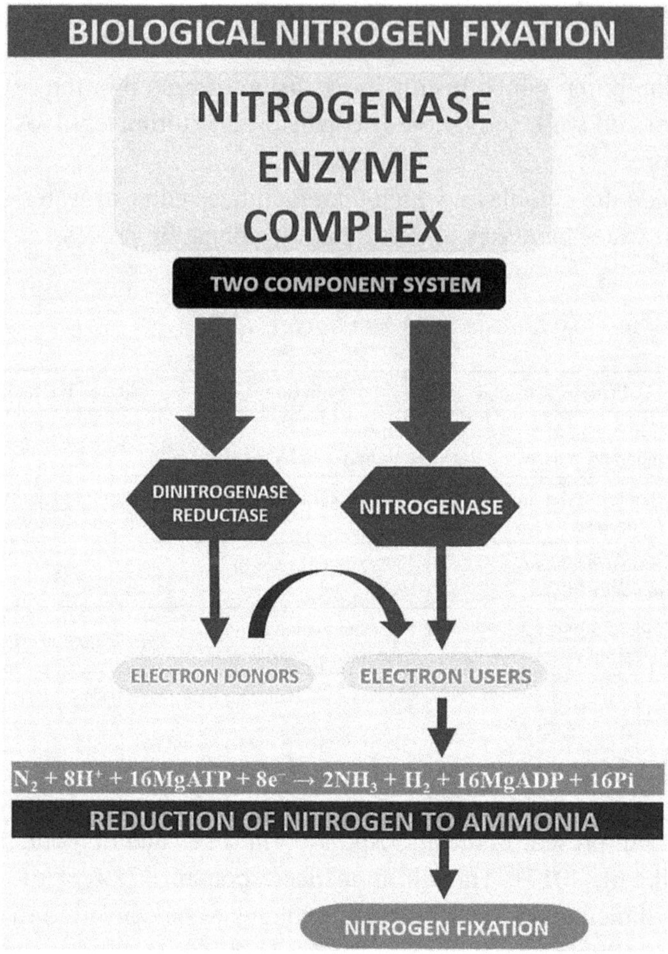

Nitrogen fixation involves *nif* genes which are present in both symbiotic and free living microorganisms. They consists of structural genes, involved in activation of iron protein, iron molybdenum cofactor biosynthesis, electron donation and regulatory genes that are required for synthesis and function of enzyme (Ahemad et al., 2013).Formation of root nodule is an important step for housing symbiotic bacterium into host plant. The symbiotic bacteria secrete the nodulation factors for effective formation of nodules which are acylated chitin oligomeric backbone with various substitutions at the (non)reducing-terminal and/or nonterminal residues. The nodulation factors are responsible for root hair formation and deformation alkalization of inside and outside of cells, depolarization of membrane potential, changes in ion fluxes, early expression of nodulin gene, and nodule primordial formation (D'Haeze et al.,2015).Ethylene acts as a negative regulator of nodule formation (Guinel et al., 2015) and one strains of *Rhizobium* increase the number of nodules in the host by synthesizing rhizobitoxine which inhibits 1-aminocyclopropane-1-carboxylate (ACC) synthase, key enzyme in ethylene biosynthesis (Sugawara et al.,2006). Some rhizobial strain ACC deaminase removes ACC, the precursor of ethylene. (Tittabutr et al., 2015). The symbiotic association of cyanobacteria with some bryophytes is guided by the hormogonia (Meeks and Elhai, 2002; Adams and Duggan, 2008). They are short filaments that are released from parental filaments. They differ from the vegetative filaments through their gliding capacity and small size. At the time of symbiotic association with the plants, the hormogonia function as infective units. Under nitrogen deprivation, hormogonium inducing factor (HIF) is produced by the host which leads to differentiation of vegetative filaments into hormogonia. It then chemotactically migrates towards the host cell infection sites. After infection, the host release hormogonia repressing factor which attenuates further proliferation of hormogonia which enables the cyanobacteria to shift its focus towards heterocyst formation for nitrogen fixation (Rai et al., 2000; Meeks and Elhai, 2002).

Phosphate Solubilization

Major proportion of phosphorous present in the soil exists in insoluble inorganic forms and hence is not available to the plants (Wan et al., 2020). The plants are able to absorb phosphates only in their monobasic and dibasic forms (Souza et al.,2015).In soil solubilization of inorganic phosphorus takes place by the action of low molecular weight organic acids such as gluconic acid (Alori et al., 2017), formic acid (Li et al.,2019), 2-ketogluconic acid (Hwangbo et al., 2003), citric acid, oxalic acid (Zeng et al., 2018), lactic acid (Gulati et al., 2010), isovaleric acid, malonic acid, fumaric acid, succinic acid, isobutyric acid, acetic acid (Kshetri et al.,2017), pyruvic acid and glycolic acid (Yang et al., 2018), which are synthesized by phosphate solubilizing bacteria. These organic acids solubilize mineral phosphate as a result of mineral exchange or chelation of cations bound to the phosphate by their free carboxyl and hydroxyl moiety (Saeid et al., 2018). This results in increased phosphorus availability with is ultimately absorbed by the plants (Oteino et al., 2015). Mineralization of phosphates by phosphatases is another way of making phosphorus available to plants (Cabugao et al., 2017).Phosphatases catalyzes the hydrolysis of wide variety of phosphomonoesters(Arai et al., 2014) .In addition to it, phosphatases catalyze trans phosphorylation reactions through transfer of phosphoryl group to alcohol in presence of phosphate acceptors (Wildberger et al.,2015). Acid phosphatases are located in bacterial cell wall and the surrounding extracellular polymeric substances (Behera et al., 2017) (Figure 5). A two-step reaction is involved in hydrolysis of phosphate esters via as shown below (Gandhi et al., 2012);

(i) $E + ROPO_3H^- \leftrightarrow E \bullet ROPO_3H^- \leftrightarrow ROH + (E\text{-}PO_3^-)$

(ii) $E\text{-}PO_3^- + H_2O \leftrightarrow E + Pi$

Figure 5. Schematic representation of procedure of phosphate solubilization by PGPR

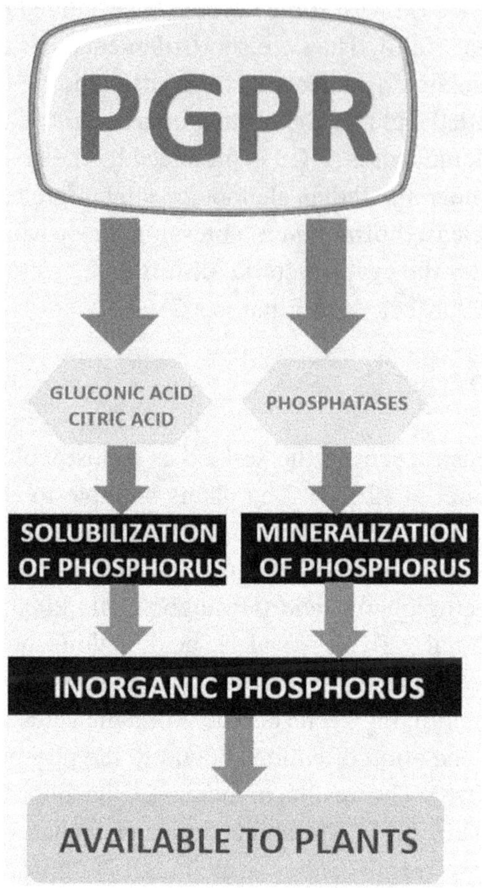

Siderophore Production and Chelation of Metals

Bacterial siderophore plays a major role in chelation of metals. In this section the mechanistic aspect of chelation is discussed with iron as a model metal. Iron is a vital element and plays an important role for growth of plant (Morrissey and Lou Guerinot, 2009). In aerobic condition, iron is available as Fe+3 form (Carpenter and Payne, 2014) and is more likely to form insoluble oxides and hydroxides thus making unavailable for plants and microbes (Ju et al., 2019). Two strategies are adopted by plants for absorbing iron. The first one requires the action of microbial ferric reductases which catalyzes the conversion of Fe^{+3} into Fe^{+2} with flavin mononucleotide, flavin adenine dinucleotide, and riboflavin as cofactor and NAD(P)H as hydrogen donor (Zhang et al., 2013) . In this process, weak Fe^{+2}-Chelate complex is formed which ultimately dissociates thereby liberating Fe^{+2} for transport or incorporation into the cell (Schröder et al., 2003).Secretion of siderophores which are low molecular weight iron chelators having strong affinity towards iron happens to be another strategy of iron chelation (Wilson et al.,2016).In gram negative bacteria, siderophores chelate iron molecules to form a complex which are too large to be transferred through porins (Andrews et al., 2003). They thus require outer membrane receptors for their uptake into periplasmic space (Wilson et al., 2016). Inside the periplasm, siderophores are taken over by periplasmic siderophore binding proteins which transport them to the cytoplasmic membrane transporters for further transport to the cytoplasm (Wilde et al., 2017). After binding with the periplasmic binding protein, the siderophores move towards cytoplasm of the cell and is aided by ABC transporter protein complex which is coupled to hydrolysis during the siderophore transportation process (Brillet et al., 2012; Bailey et al., 2018). The bacterial ABC transporters consists of four structural domains: two transmembrane domain forming a channel for passage of ferric siderophore and two nucleotide binding domain ATP hydrolysis (Zolnerciks et al., 2011; Krewulak et al., 2008).In gram positive bacteria, the siderophore binding protein is associated to a permease (Tonziello et al., 2019). The binding to a ferric siderophore results in conformational change in siderophore binding protein-permease complex which aids in transport of ferric siderophores through membranes into the cytoplasm (Wilson et al., 2016). Inside the bacterial cell, Fe^{+3} is reduced to Fe^{+2} along with loss of affinity for siderophores and reduced iron is finally assembled into iron containing molecules or stored in ferritins (Aznar et al., 2014).Plants have the capacity to take up iron from bacterial siderophores through direct uptake, chelate degradation, or ligand exchange reactions (Kurth et al.,2016). The schematic representation of Iron uptake by the bacteria is depicted in Figure 6.

Figure 6. Schematic representation of Uptake of Iron (III) with the help of siderophore and conversion to Iron (II) by PGPR

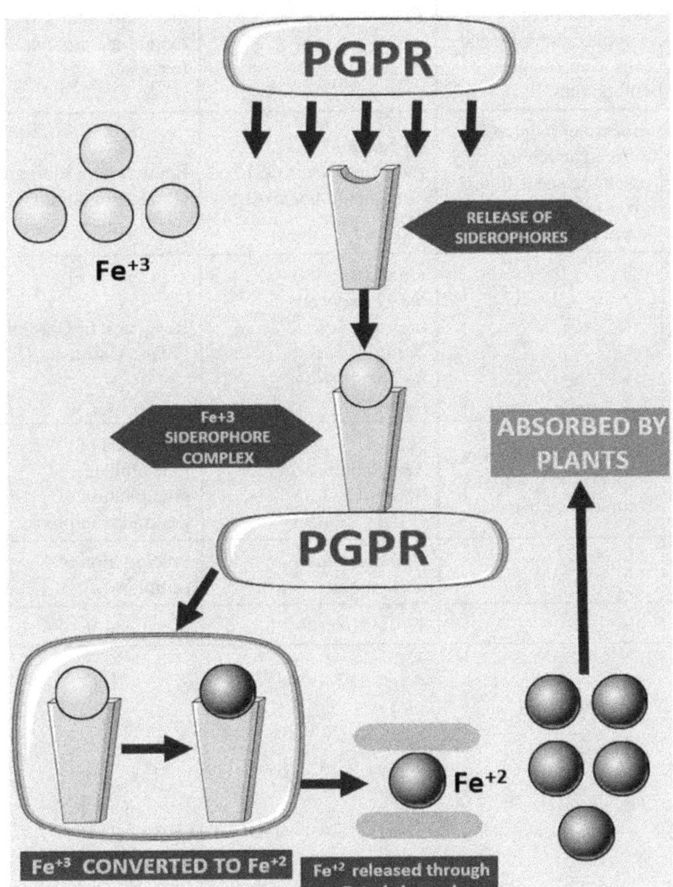

Table 11. Mechanisms involved in removal and/or absorption of heavy metals by bacteria

Name of the Process	Mechanism	Name of the Bacteria Involved	Metals Absorbed	References
Biosorption				
Metabolism independent – Biosorption	Passive uptake by the cell surface and complexation in the cell wall.	*Kocuria sp.* CRB15	Copper.	Hansda et al.,2017
Metabolism dependent-Bioaccumulation	Metabolism dependent uptake and related to physiological processes of bacteria.	*Cupriavidus necator* GX_5	Cadmium.	Li et al., 2018
Intracellular sequestration	Complexation of metals by molecules of cytoplasm.	*Pseudomonas putida*	Cadmium-bound to pseudothioneins.	Higham et al., 1986
		Rhizobium leguminosarum bv. *viciae*	Cadmium-GSH involved in sequestration.	Lima et al., 2006
		Escherischia coli	Mercury and arsenite-Involved in resistance.	Latinwo et al.,1998
Extracellular Sequestration	Accumulation of metal ions by cellular components in periplasm or complexation of metal ions as insoluble compounds.	*Pseudomonas syringae*	CopA, CopB (periplasmic proteins), and CopC (outer membrane proteins) bind to copper.	Cha and Cooksey, 1991
Metal reduction	Process of reducing more toxic higher oxidation states to less toxic lower oxidation states of a metal.	*Geobacter metallireducens* GS15, *Shewanella oneidensis* MR1	Reduction of Plutonium (IV) to Plutonium (III).	Boukhalfa et al., 2007
		Geobacter metallireducens, Desulfovibrio desulfuricans, Sulfurospirillum barnesii	Reduction of Chromium (VI) to Chromum (III).	Chovanec et al.,2012
Metal Precipitation	Generally reduction of sulphates to insoluble sulphides by bacteria.	*Klebsiella planticola* Strain (Cd-1)	Reduction of thiosulphate and precipitation of Cadmium sulphide.	Sharma et al., 2000
		Pseudomonas aeruginosa	Precipitation of cadmium.	Wang et al., 1997
		Vibrio harveyi	Precipitation of lead.	Mire et al., 2004

Table 12. Enzymes involved in detoxification process

Enzyme	Organism	Pesticide Degraded	Reference
glyphosate oxidoreductase	*Pseudomonas spp* GA07 and GC04	Glyphosate	Zhao et al., 2015
	Ochrobactrum sp	Glyphosate	Hadi et al., 2013
C-P lyase and glyphosate oxidoreductase	*Bacillus cereus* CB4	Glyphosate	Fan et al., 2012
Monooxygenases			
Ese	*Arthrobacter* sp	Endosulfan	Weir et al.,2006
Esd	*Mycobacterium* sp	β-Endosulfan	Sutherland et al.,2002
CYP101	*Pseudomonas putida*	polychlorinated benzenes	Jones et al.,2001
Phosphotriesterase			
OpdA	*Agrobacterium radiobacter* P230	phosmet and fenthion	Horne et al.,2002
Phosphotriesterase enzyme system similar to Opd	*Enterobacter Strain* B-14	Chlorpyrifos	Singh et al.,2004

Bacterial Extraction of Heavy Metals

As discussed before a large number of PGPR have the capacity to remove and sequester heavy metals. These are done by the bacteria through their own mechanisms. One of them involves production of organic compounds. It is reported that a number of bacteria produce organic compounds which can trap and chelate metals (Mishra et al., 2017). This often results in reduction of toxicity. In addition to it, bacteria also uptake metals on their cell surface (Mullen et al., 1989). Nonspecific binding of metals with the bacterial slime layer and extracellular polysaccharide are involved in the process (Gupta and Diwan, 2017; Nocelli et al., 2016). Another type of uptake is the metabolism dependent transport of heavy metals. This process involves the bioaccumulation and is dependent on a variety of physical, chemical and biological mechanisms (Igiri et al., 2018). The various mechanism of removal of heavy metals by bacteria is tabulated in Table 11.

Biodegradation of Pesticides and Hydrocarbons

Bacterial enzymes are the key factors for the degradation of pesticides. Metabolism of pesticides by bacteria occurs in three phases. In phase I metabolism involves largely oxidation, oxidation, reduction, and hydrolysis along with introduction of polar group in hydrophobic molecules i: e production of derivatives which contains -OH, -COOH, -NH_2, and -SH functional groups (Lushchak et al., 2018). The second phase involves adding of a pesticide or pesticide metabolite to a sugar moiety or an amino acid which results in increase of solubility and reduce toxicity (Koppel et al., 2018; Luschak et al.,2018). In the third phase, the phase II metabolites are converted into nontoxic secondary conjugates (Ortiz-Hernández et al.,2013). The enzyme system involved in detoxification and metabolism of pesticides are illustrated in Table 12.

Similar to the pesticide metabolism, there are also sets of enzymes responsible for degradation of hydrocarbons. Initially the bacteria are attracted chemotactically towards alkanes from a zone of low concentration to a zone of high concentration (Parales et al., 2000). It is reported that *Flavimonas oryzihabitans* showed chemotactic behaviour towards gas oil and hexadecane while *Pseudomonas aeruginosa* PAO1 showed chemotactic response towards hexadecane (Lanfranconi et al., 2003; Smits et al., 2003). It was further assumed that tlpS gene, located downstream of the alkane hydroxylase gene alkB1 in PAO1 genome, encode membrane-bound methyl-accepting chemotaxis proteins (MCP) which plays role in chemotaxis (Smits et al., 2003). Once the bacteria are in contact with the hydrocarbons, they are uptaken by the bacterial cell. Most bacteria are reported to secrete surfactants that help in emulsification of hydrocarbons (Ron and Rosenberg, 2002). Biosurfactants increase the surface area, reduce surface tension of oil/ water phase thus improving the access of bacteria to the oil phase (Fenibo et al., 2019; Santos et al., 2016). It is reported in *Pseudomonas putida* that alkL in the alk operon plays a vital role in transport of alkane into the cell (Wang and Shao, 2013). Similarly Transcriptome analysis of *Alcanivorax borkumensis* Sk2 revealed the presence of alkane-induced gene blc which encodes inter membrane lipoprotein Blc, presumed to be involved in uptake of alkane (Sabirova et al., 2011). The bacteria which degrade short chain alkanes (C2-C4) generally possess enzymes related to methane monooxygenases. These enzymes hydroxylate alkanes to their respective alcohols which are further oxidised to aldehydes by the respective dehydrogenase and finally to carboxylic acid and carbon dioxide by further action of dehydrogenases (Ji et al., 2013). The bacteria which degrades medium length alkanes (C5–C17) are equipped with cytochrome P450 (CYP4B1 and 4B2) and integral membrane non-heme iron monooxygenases, such as AlkB (Funhoff et al., 2007; Gregson et al., 2018). Long chain alkanes (>C18) are metabolized by alkane hydroxylases (Elumalai et al., 2017).Examples of such hydroxylase is AlmA which is a monoxygenase from *Acinetobacter* (Wang and Shao, 2012) and LadA, which is a thermophilic soluble long chain alkane monooxygenase from *Geobacillus* (Feng et al., 2007).

Production of Antibiotics

Antibiosis in PGPR plays a crucial role in plant disease management. Numerous antibiotics have been isolated from bacteria and they exert a number of action to inhibit the pathogen. One mechanism is inhibition of synthesis of pathogen cell wall (Suryadi et al., 2019).The antibiotics are also reported to influence membrane structure of the pathogen cells and inhibit the formation of initiation complexes on small subunit of ribosome (Maksimov et al., 2011). The biosynthetic genes responsible for antibiotic production have a well organised biosynthetic gene cluster. *PhlD* gene is responsible for 2,4-diacetylphloroglucinol (DAPG) biosynthesis (Fernando et al., 2015). DAPG is a phenolic polyketide compound. DAPG is one of the primary constituent which acts as biocontrol agent in PGPR (Beneduzi et al., 2012). Though uncertain the basis of disease control is the interaction between Phl-producing root associated microorganism and pathogen. Condensation of 3 molecules of acetyl CoA and 1 molecule of malonyl CoA produce mono acetyl phloroglucinol (MAPG), which is transacetylated to form DAPG (Kenawy et al., 2019). Strains producing Zwittermicin A possess zma R gene which give self-resistance against own antibiotic. Antibiotics produced by PGPR possess broad spectrum activity. ISR mediated plant defence is trigerred by bacterial antibiotics. It has been reported that pyochelis and pyocyanine act synergistically causing cell damage by producing active oxygen species whereas accumulation of DAGP in root act as a signal triggering ISR (Fernando et al., 2015).

Figure 7. Multifarious beneficial activities of PGPR

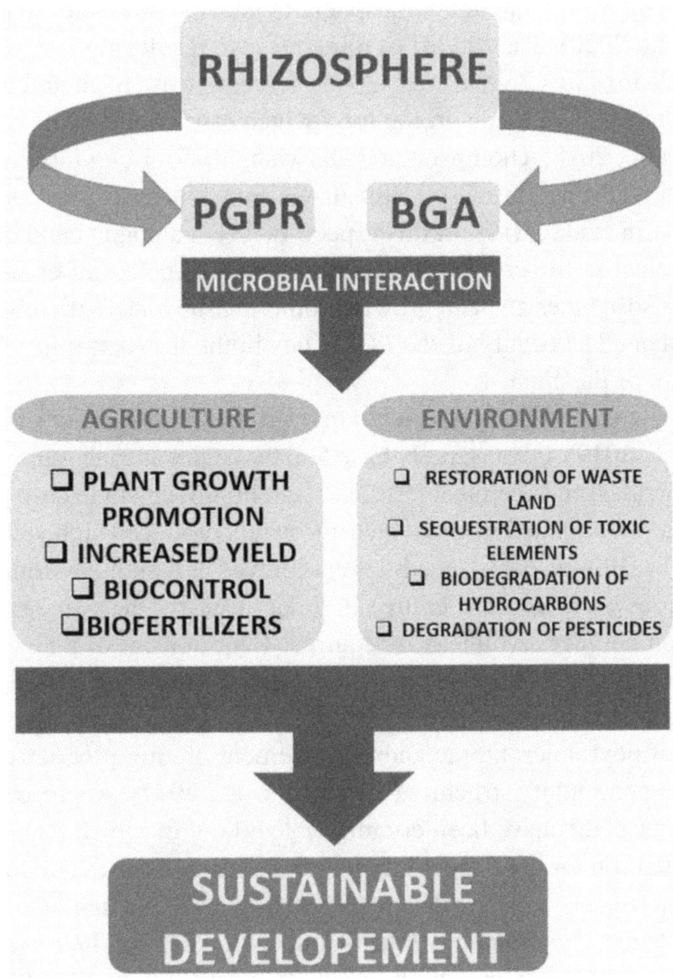

FUTURE PROSPECTS

PGPR plays an important role in sustainable agriculture and environmental system (Figure 7). These are beneficial bacteria that promote the growth and yield of plant both symbiotically and non-symbiotically. The increase in yield of the plants is largely due to the production of plant growth promoters by the bacteria and also due to effective supply of biologically available minerals by the microbes to the plants. The agriculture in 21st century is facing a tough challenge in terms of decline in productivity and degradation of agro-ecological sustainability (Prasad et al., 2019; Wang et al., 2018). The decrease in productivity is largely due to the climate change and unplanned anthropogenic activity which in turn perturbs the environmental balance resulting in drought, heavy rainfalls, temperature fluctuations, salinity, and insect pest attacks thereby increasing threats of starvation (Raza et al., 2019). In addition to it, long term use of chemical fertilizer also resulted in some negative effects to the environment. The productivity cost does not scale linearly and results in large-scale wastage of mineral resources. Moreover, millions of tons of synthetic compounds containing mineral nutrients are deposited in the soil which are ultimately not absorbed by the plants. It is also reported that 50% of nitrogen and 90% of phosphorus are run off from agricultural fields and escape into the atmosphere to generate greenhouse gases thereby polluting the environment (Ye et al., 2020). The run off of minerals also results in massive eutrophication which again results in large scale toxicity (Beman et al., 2005). Excessive use of chemical fertilizers also results in accumulation of a number of elements in the food which results in decline of quality and problems in safety issues (Ward et al., 2018; Thompson and Darwish, 2019). Long term application of chemical fertilizers also results in acidification of soil, imbalance in nutritional status and deterioration of the rhizosphere microflora (Lin et al., 2019). In this aspect, PGPR is a potent candidate for minimizing the negative effects of chemical fertilizers. PGPR has a tremendous potential of sequestering excess elements from the soil and also synthesize plant growth promoting substances thereby improving the overall growth process of the plant. The beauty of the PGPR lies in the fact that it is self replicatory and well sustain in the rhizosphere of the plant.

In the 20th century, the green revolution resulted in overwhelming gains with respect to crop productivity (Bailey-Serres et al., 2019). It was largely based on two main advances namely chemical inputs in terms of pesticides, herbicides, and chemical fertilizers and improvement in crop plants through targeted breeding and advanced genetic manipulations which ultimately produced high yielding varieties (Pingali, 2012; Backer et al., 2018). However this success was achieved at a high environmental cost and a deep negative impact on the ecosystem. In 21st Century, in order to satisfy the every growing demand of food for a booming population another revolution is required. Requirement of a new revolution is required which should be distinct from the previous conventional green revolution and should be environmental friendly. This green revolution should be more appropriately termed as a 'bio revolution' which should be based on utilization of phytomicrobiome and improvement of crop productivity through manipulation of phytomicrobiome community structure (Timmusk et al., 2017). At present a number of bacteria which promotes growth of plants have been commercialized out of which *Pseudomonas* and *Bacillus* have attracted special attention for their capability to reduce plant diseases (Reddy, 2012). Additionally *Azotobacter* (Chaudhary et al., 2013), *Azospirillum* (Zeffa et al., 2019), *Paenibacillus* (Khan et al., 2008), *Serratia* (George et al., 2013), *Burkholderia* (Parra-Cota et al., 2014) and *Herbaspirillum* (Dall'Asta et al., 2019), have shown promising outcome with respect to increase in crop yield.

The negative effect of climate change is also likely to impose more stress on crop plants and also risk of extinction of important agricultural species, decrease in productivity and quality (Abdallah et al., 2014; Zhao et al., 2017). As the climate change makes its negative stride across the globe, high quality agricultural lands are likely to be lost due to sea level rise and salinization (Gornall et al., 2010), erosion (Ozsoy and Aksoy, 2015), and desertification (Zabel et al., 2014). This implies that more productivity is required from comparatively lesser area of land leading to more stressful situation. The situation is in such an alarming condition that it is estimated by the year 2050 global agricultural production may need to be increased by 60%–110% to meet these increasing demands and to provide food security to nearly 870 million chronically undernourished population (Ray et al., 2013). Under such circumstances, the phytomicrobiome may play a very crucial role which can suffice nutrients to the plants without causing much damage to the environment. As a result of anthropogenic activity and imbalance in climatic condition, soil quality is highly perturbed and is often contaminated with high quantity of salts or metal ions. Under such conditions, PGPR proves to be a suitable candidate for not only increasing crop productivity but also and effective means of bioremediation of soil. It is also a promising component in integrated pest management promoting sustainable agriculture. It not only increase yield of plant significantly but also decrease the amount of pesticide and chemical fertilizer used (Antoun and Kloepper, 2001).

However there are also several challenges which requires to be resolved. Development of new PGPR inocula, based on laboratory screening considering some specific parameters namely nitrogen fixation, ACC deaminase activity, auxin synthesis and solubilization of phosphates also requires to be tested in field condition in order to check its final efficacy. Development of inocula with prolonged shelf life and high rhizosphere colonization rate also requires to be developed. In addition to it, standardization of dose of PGPR application along with suitable vectors or carriers should also be developed and introduced in order to affect successful colonization by minimizing the antagonism of other microbes (Backer et al., 2018).

Plant breeding has played a very important role in success of green revolution (Bhargava and Srivastava, 2019). However, this breeding programme is limited to conventional techniques involving plants for selection of desired traits. More efforts require to be given on microbiome-based plant breeding for production of heritable PGPR community which can enhance crop productivity (Trivedi et al., 2017). However green revolution resulted in introduction of inorganic fertilizers, herbicides and pesticides into the soil whose detrimental effects are clearly visible at present (Patra et al., 2016). Though initially use of these chemical agents were positively correlated with increase in yield and productivity but later on they posed major environmental threats as contaminants and run offs in water bodies and soil. Thus designing microbial inoculant for dual purpose of bioremediation and plant growth promotion should be taken into consideration (Baez-Rogelio et al., 2017). After the development of inoculant, its further commercialization is equally important. Prior to commercialization, evaluation of its safety and efficacy data is of utmost importance and proper guidelines requires to be framed which can be more or less applicable to all countries with minor modifications depending upon the requirement. Development of effective PGPR strain from isolation to commercialization is a daunting and time taking task. In this case, collaboration between various industries, academic institutions and government organizations are of utmost importance for successful development of products. Academic institutions in collaboration with industries should offer training programme so that the knowledge on PGPR is propagate in different levels of society starting from the scientific strata and ultimately reaching the common farmers who would play the final role in implementing the inoculant in the fields.

As mentioned earlier, contamination of soil is proving to be a threat to humans and animals and PGPR have proved to be an effective tool in remediation purpose. Though these microbes have their own machinery by which they have the capacity to break down complex organic molecules or sequester metal contaminants, however further investigation of their remediating potentials needs to be studied. Since there is continuous increase in the industrialization, the number and types of pollutants are also on the rise. Thus there is a need to genetically engineer the microbes with the potential to degrade most complex chemicals. Governmental policies also requires to be implemented in order to popularize the use of PGPR in both agricultural and remediation processes. This should be equipped with strict measures to crub haphazard disposal of industrial contaminants.

CONCLUSIVE REMARKS

PGPR is a ubiquitous member of the soil microbial community and is an excellent choice for bioremediation (Backer et al., 2018). PGPR takes part in phytoremediation through processes like phytoextraction, phytostabilization and reduces the bioavailability of these heavy metals (He et al., 2007). This in turn would cause reduced contamination of soil and water in the long term that would again help decrease the bio magnification of these carcinogenic entities in the human systems, finally leading to decline in ROS generation, cellular damage and cancer. Thus it becomes an urgent need to adapt to more environmental friendly sustainable means of restoring the natures balance with the help of microbes. PGPR not only requires to be extensively bioprospected in agricultural arena for increasing the crop productivity in order to fulfil the ever growing demand of food for an increasing global population but also requires to be applied for remediation purpose to cleanse the environmental contamination. These would ultimately result in a healthier environment on one hand and also benefit the human race by increasing crop productivity and satisfying the demand for food. Thus these PGPR needs to be bioprospected for overall benefit of mankind and environment.

REFERENCES

Abdallah, N. A., Moses, V., & Prakash, C. S. (2014). The impact of possible climate changes on developing countries: The needs for plants tolerant to abiotic stresses. *GM Crops and Food: Biotechnology in Agriculture and the Food Chain*, *5*(2), 77–80. doi:10.4161/gmcr.32208 PMID:25075960

Abdelkrim, S., Jebara, S. H., Saadani, O., Chiboub, M., Abid, G., & Jebara, M. (2018). Effect of Pb-resistant plant growth-promoting rhizobacteria inoculation on growth and lead uptake by Lathyrus sativus. *Journal of Basic Microbiology*, *58*(7), 579–589. doi:10.1002/jobm.201700626 PMID:29737549

Adams, D. G., & Duggan, P. S. (2008). Cyanobacteria–bryophyte symbioses. *Journal of Experimental Botany*, *59*(5), 1047–1058. doi:10.1093/jxb/ern005 PMID:18267939

Ahemad, M., & Kibret, M. (2014). Mechanisms and applications of plant growth promoting rhizobacteria: Current perspective. *Journal of King Saud University. Science*, *26*(1), 1–20. doi:10.1016/j.jksus.2013.05.001

Ahmed, A., & Hasnain, S. (2014). Auxins as one of the factors of plant growth improvement by plant growth promoting rhizobacteria. *Polish Journal of Microbiology*, *63*(3), 261–266. doi:10.33073/pjm-2014-035 PMID:25546935

Ahmed, E., & Holmström, S. J. (2014). Siderophores in environmental research: Roles and applications. *Microbial Biotechnology*, *7*(3), 196–208. doi:10.1111/1751-7915.12117 PMID:24576157

Alavanja, M. C., & Bonner, M. R. (2012). Occupational pesticide exposures and cancer risk: A review. *Journal of Toxicology and Environmental Health. Part B, Critical Reviews*, *15*(4), 238–263. doi:10.1080/10937404.2012.632358 PMID:22571220

Albdaiwi, R. N., Khaymi-Horani, H., Ayad, J. Y., Alananbeh, K. M., Kholoud, M., & Al-Sayaydeh, R. (2019). Isolation and characterization of halotolerant plant growth promoting rhizobacteria from durum wheat (Triticum turgidum subsp. durum) cultivated in saline areas of the dead sea region. *Frontiers in Microbiology*, *10*, 1639. doi:10.3389/fmicb.2019.01639 PMID:31396175

Almaghrabi, O. A., Massoud, S. I., & Abdelmoneim, T. S. (2013). Influence of inoculation with plant growth promoting rhizobacteria (PGPR) on tomato plant growth and nematode reproduction under greenhouse conditions. *Saudi Journal of Biological Sciences*, *20*(1), 57–61. doi:10.1016/j.sjbs.2012.10.004 PMID:23961220

Alori, E. T., Glick, B. R., & Babalola, O. O. (2017). Microbial phosphorus solubilization and its potential for use in sustainable agriculture. *Frontiers in Microbiology*, *8*, 971. doi:10.3389/fmicb.2017.00971 PMID:28626450

Andrews, G. K. (2000). Regulation of metallothionein gene expression by oxidative stress and metal ions. *Biochemical Pharmacology*, *59*(1), 95–104. doi:10.1016/S0006-2952(99)00301-9 PMID:10605938

Andrews, S. C., Robinson, A. K., & Rodríguez-Quiñones, F. (2003). Bacterial iron homeostasis. *FEMS Microbiology Reviews*, *27*(2-3), 215–237. doi:10.1016/S0168-6445(03)00055-X PMID:12829269

Anetor, J. I., Wanibuchi, H., & Fukushima, S. (2007). Arsenic exposure and its health effects and risk of cancer in developing countries: Micronutrients as host defence. *Asian Pacific Journal of Cancer Prevention*, *8*(1), 13. PMID:17477765

Antoun, H., & Kloepper, J. W. (2001). Plant growth promoting rhizobacteria. Encyclopedia of Genetics, 1477-1480.

Anwar, M. S., Paliwal, A., Firdous, N., Verma, A., Kumar, A., & Pande, V. (2018). Co-culture development and bioformulation efficacy of psychrotrophic PGPRs to promote growth and development of Pea (Pisum sativum) plant. *The Journal of General and Applied Microbiology*. PMID:30381611

Arai, S., Yonezawa, Y., Ishibashi, M., Matsumoto, F., Adachi, M., Tamada, T., Tokunaga, H., Blaber, M., Tokunaga, M., & Kuroki, R. (2014). Structural characteristics of alkaline phosphatase from the moderately halophilic bacterium *Halomonas sp.* 593. *Acta Crystallographica. Section D, Biological Crystallography*, *70*(Pt 3), 811–820. doi:10.1107/S1399004713033609 PMID:24598750

Arikan, Ş., & Pirlak, L. (2016). Effects of plant growth promoting rhizobacteria (PGPR) on growth, yield and fruit quality of sour cherry (*Prunus cerasus* L.). *Erwerbs-Obstbau*, *58*(4), 221–226. doi:10.100710341-016-0278-6

Arseneault, T., Goyer, C., & Filion, M. (2013). Phenazine production by *Pseudomonas sp.* LBUM223 contributes to the biological control of potato common scab. *Phytopathology*, *103*(10), 995–1000. doi:10.1094/PHYTO-01-13-0022-R PMID:23883153

Asami, T., Nakano, T., & Fujioka, S. (2005). Plant brassinosteroid hormones. *Vitamins and Hormones*, *72*, 479–504. doi:10.1016/S0083-6729(05)72014-8 PMID:16492480

Asari, S., Tarkowská, D., Rolčík, J., Novák, O., Palmero, D. V., Bejai, S., & Meijer, J. (2017). Analysis of plant growth-promoting properties of *Bacillus amyloliquefaciens* UCMB5113 using *Arabidopsis thaliana* as host plant. *Planta*, *245*(1), 15–30. doi:10.100700425-016-2580-9 PMID:27541497

Aznar, A., Chen, N. W., Rigault, M., Riache, N., Joseph, D., Desmaële, D., Mouille, G., Boutet, S., Soubigou-Taconnat, L., Renou, J. P., Thomine, S., Expert, D., & Dellagi, A. (2014). Scavenging iron: A novel mechanism of plant immunity activation by microbial siderophores. *Plant Physiology*, *164*(4), 2167–2183. doi:10.1104/pp.113.233585 PMID:24501001

Backer, R., Rokem, J. S., Ilangumaran, G., Lamont, J., Praslickova, D., Ricci, E., Subramanian, S., & Smith, D. L. (2018). Plant Growth-Promoting Rhizobacteria: Context, Mechanisms of Action, and Roadmap to Commercialization of Biostimulants for Sustainable Agriculture. *Frontiers in Plant Science*, *9*, 1473. doi:10.3389/fpls.2018.01473 PMID:30405652

Baez-Rogelio, A., Morales-García, Y. E., Quintero-Hernández, V., & Muñoz-Rojas, J. (2017). Next generation of microbial inoculants for agriculture and bioremediation. *Microbial Biotechnology*, *10*(1), 19–21. doi:10.1111/1751-7915.12448 PMID:27790851

Bahar, M. M., Megharaj, M., & Naidu, R. (2013). Bioremediation of arsenic-contaminated water: Recent advances and future prospects. *Water, Air, and Soil Pollution*, *224*(12), 1722. doi:10.100711270-013-1722-y

Bailey, D. C., Bohac, T. J., Shapiro, J. A., Giblin, D. E., Wencewicz, T. A., & Gulick, A. M. (2018). Crystal Structure of the Siderophore Binding Protein BauB Bound to an Unusual 2:1 Complex Between Acinetobactin and Ferric Iron. *Biochemistry*, *57*(48), 6653–6661. doi:10.1021/acs.biochem.8b00986 PMID:30406986

Bailey-Serres, J., Parker, J. E., Ainsworth, E. A., Oldroyd, G., & Schroeder, J. I. (2019). Genetic strategies for improving crop yields. *Nature*, *575*(7781), 109–118. doi:10.103841586-019-1679-0 PMID:31695205

Baoune, H., Aparicio, J. D., Acuña, A., El Hadj-khelil, A. O., Sanchez, L., Polti, M. A., & Alvarez, A. (2019). Effectiveness of the *Zea mays-Streptomyces* association for the phytoremediation of petroleum hydrocarbons impacted soils. *Ecotoxicology and Environmental Safety*, *184*, 109591. doi:10.1016/j.ecoenv.2019.109591 PMID:31514081

Baoune, H., El Hadj-Khelil, A. O., Pucci, G., Sineli, P., Loucif, L., & Polti, M. A. (2018). Petroleum degradation by endophytic *Streptomyces spp.* isolated from plants grown in contaminated soil of southern Algeria. *Ecotoxicology and Environmental Safety*, *147*, 602–609. doi:10.1016/j.ecoenv.2017.09.013 PMID:28923725

Barbez, E., Dünser, K., Gaidora, A., Lendl, T., & Busch, W. (2017). Auxin steers root cell expansion via apoplastic pH regulation in *Arabidopsis thaliana. Proceedings of the National Academy of Sciences of the United States of America*, *114*(24), E4884–E4893. doi:10.1073/pnas.1613499114 PMID:28559333

Baruthio, F. (1992). Toxic effects of chromium and its compounds. *Biological Trace Element Research*, *32*(1-3), 145–153. doi:10.1007/BF02784599 PMID:1375051

Bashan, Y., & Holguin, G. (1998). Proposal for the division of plant growth-promoting rhizobacteria into two classifications: biocontrol-PGPB (plant growth-promoting bacteria) and PGPB. *Soil Biology & Biochemistry*, *30*(8-9), 1225–1228. doi:10.1016/S0038-0717(97)00187-9

Behera, B. C., Yadav, H., Singh, S. K., Mishra, R. R., Sethi, B. K., Dutta, S. K., & Thatoi, H. N. (2017). Phosphate solubilization and acid phosphatase activity of *Serratia sp.* isolated from mangrove soil of Mahanadi river delta, Odisha, India. *Journal of Genetic Engineering and Biotechnology*, *15*(1), 169–178. doi:10.1016/j.jgeb.2017.01.003 PMID:30647653

Behie, S. W., & Bidochka, M. J. (2013). Insects as a nitrogen source for plants. *Insects*, *4*(3), 413–424. doi:10.3390/insects4030413 PMID:26462427

Beneduzi, A., Ambrosini, A., & Passaglia, L. M. (2012). Plant growth-promoting rhizobacteria (PGPR): Their potential as antagonists and biocontrol agents. *Genetics and Molecular Biology*, *35*(4, suppl), 1044–1051. doi:10.1590/S1415-47572012000600020 PMID:23411488

Bhargava, A., & Srivastava, S. (2019). *Participatory Plant Breeding: Concept and Applications*. Springer Singapore. doi:10.1007/978-981-13-7119-6

Bobrov, A. G., Kirillina, O., Fetherston, J. D., Miller, M. C., Burlison, J. A., & Perry, R. D. (2014). The *Yersinia pestis* siderophore, yersiniabactin, and the ZnuABC system both contribute to zinc acquisition and the development of lethal septicaemic plague in mice. *Molecular Microbiology*, *93*(4), 759–775. doi:10.1111/mmi.12693 PMID:24979062

Boukhalfa, H., Icopini, G. A., Reilly, S. D., & Neu, M. P. (2007). Plutonium (IV) reduction by the metal-reducing bacteria *Geobacter metallireducens* GS15 and Shewanella oneidensis MR1. *Applied and Environmental Microbiology*, *73*(18), 5897–5903. doi:10.1128/AEM.00747-07 PMID:17644643

Braud, A., Hannauer, M., Mislin, G. L., & Schalk, I. J. (2009). The Pseudomonas aeruginosa pyochelin-iron uptake pathway and its metal specificity. *Journal of Bacteriology*, *191*(11), 3517–3525. doi:10.1128/JB.00010-09 PMID:19329644

Brígido, C., Glick, B. R., & Oliveira, S. (2017). Survey of plant growth-promoting mechanisms in native Portuguese chickpea Mesorhizobium isolates. *Microbial Ecology*, *73*(4), 900–915. doi:10.100700248-016-0891-9 PMID:27904921

Brillet, K., Ruffenach, F., Adams, H., Journet, L., Gasser, V., Hoegy, F., Guillon, L., Hannauer, M., Page, A., & Schalk, I. J. (2012). An ABC transporter with two periplasmic binding proteins involved in iron acquisition in *Pseudomonas aeruginosa. ACS Chemical Biology*, *7*(12), 2036–2045. doi:10.1021/cb300330v PMID:23009327

Burén, S., & Rubio, L. M. (2018). State of the art in eukaryotic nitrogenase engineering. *FEMS Microbiology Letters*, *365*(2), fnx274. doi:10.1093/femsle/fnx274 PMID:29240940

Burkhead, K. D., Schisler, D. A., & Slininger, P. J. (1994). Pyrrolnitrin Production by Biological Control Agent *Pseudomonas cepacia* B37w in Culture and in Colonized Wounds of Potatoes. *Applied and Environmental Microbiology*, *60*(6), 2031–2039. doi:10.1128/AEM.60.6.2031-2039.1994 PMID:16349289

Cabugao, K. G., Timm, C. M., Carrell, A. A., Childs, J., Lu, T. S., Pelletier, D. A., Weston, D. J., & Norby, R. J. (2017). Root and Rhizosphere Bacterial Phosphatase Activity Varies with Tree Species and Soil Phosphorus Availability in Puerto Rico Tropical Forest. *Frontiers in Plant Science*, *8*, 1834. doi:10.3389/fpls.2017.01834 PMID:29163572

Cakmakci, R., Dönmez, M. F., & Erdoğan, Ü. (2007). The effect of plant growth promoting rhizobacteria on barley seedling growth, nutrient uptake, some soil properties, and bacterial counts. *Turkish Journal of Agriculture and Forestry*, *31*(3), 189–199.

Camilios-Neto, D., Bonato, P., Wassem, R., Tadra-Sfeir, M. Z., Brusamarello-Santos, L. C., Valdameri, G., Donatti, L., Faoro, H., Weiss, V. A., Chubatsu, L. S., Pedrosa, F. O., & Souza, E. M. (2014). Dual RNA-seq transcriptional analysis of wheat roots colonized by Azospirillum brasilense reveals up-regulation of nutrient acquisition and cell cycle genes. *BMC Genomics*, *15*(1), 378. doi:10.1186/1471-2164-15-378 PMID:24886190

Canarini, A., Kaiser, C., Merchant, A., Richter, A., & Wanek, W. (2019). Root exudation of primary metabolites: Mechanisms and their roles in plant responses to environmental stimuli. *Frontiers in Plant Science*, *10*, 157. doi:10.3389/fpls.2019.00157 PMID:30881364

Carpenter, C., & Payne, S. M. (2014). Regulation of iron transport systems in Enterobacteriaceae in response to oxygen and iron availability. *Journal of Inorganic Biochemistry*, *133*, 110–117. doi:10.1016/j.jinorgbio.2014.01.007 PMID:24485010

Carrano, C. J., Jordan, M., Drechsel, H., Schmid, D. G., & Winkelmann, G. (2001). Heterobactins: A new class of siderophores from *Rhodococcus erythropolis* IGTS8 containing both hydroxamate and catecholate donor groups. *Biometals*, *14*(2), 119–125. doi:10.1023/A:1016633529461 PMID:11508844

Carter, R. A., Worsley, P. S., Sawers, G., Challis, G. L., Dilworth, M. J., Carson, K. C., Lawrence, J. A., Wexler, M., Johnston, A. W., & Yeoman, K. H. (2002). The vbs genes that direct synthesis of the siderophore vicibactin in *Rhizobium leguminosarum*: Their expression in other genera requires ECF sigma factor RpoI. *Molecular Microbiology*, *44*(5), 1153–1166. doi:10.1046/j.1365-2958.2002.02951.x PMID:12028377

Cha, J. S., & Cooksey, D. A. (1991). Copper resistance in Pseudomonas syringae mediated by periplasmic and outer membrane proteins. *Proceedings of the National Academy of Sciences of the United States of America*, *88*(20), 8915–8919. doi:10.1073/pnas.88.20.8915 PMID:1924351

Chakdar, H., Dastager, S. G., Khire, J. M., Rane, D., & Dharne, M. S. (2018). Characterization of mineral phosphate solubilizing and plant growth promoting bacteria from termite soil of arid region. *Biotech*, *8*(11), 463.

Chaparro, J. M., Badri, D. V., & Vivanco, J. M. (2014). Rhizosphere microbiome assemblage is affected by plant development. *The ISME Journal*, *8*(4), 790–803. doi:10.1038/ismej.2013.196 PMID:24196324

Chaudhary, D., Narula, N., Sindhu, S. S., & Behl, R. K. (2013). Plant growth stimulation of wheat (*Triticum aestivum* L.) by inoculation of salinity tolerant Azotobacter strains. *Physiology and Molecular Biology of Plants: An International Journal of Functional Plant Biology*, *19*(4), 515–519.

Che, R., Deng, Y., Wang, F., Wang, W., Xu, Z., Hao, Y., Xue, K., Zhang, B., Tang, L., Zhou, H., & Cui, X. (2018). Autotrophic and symbiotic diazotrophs dominate nitrogen-fixing communities in Tibetan grassland soils. *The Science of the Total Environment*, *639*, 997–1006. doi:10.1016/j.scitotenv.2018.05.238 PMID:29929338

Chen, Q., & Liu, S. (2019). Identification and Characterization of the Phosphate-Solubilizing Bacterium Pantoea sp. S32 in Reclamation Soil in Shanxi, China. *Frontiers in Microbiology*, *10*, 2171. doi:10.3389/fmicb.2019.02171 PMID:31608027

Chen, X., Zhou, Q., Liu, F., Peng, Q., & Teng, P. (2019). Removal of nine pesticide residues from water and soil by biosorption coupled with degradation on biosorbent immobilized laccase. *Chemosphere*, *233*, 49–56. doi:10.1016/j.chemosphere.2019.05.144 PMID:31163308

Cheon, J., Fujioka, S., Dilkes, B. P., & Choe, S. (2013). Brassinosteroids regulate plant growth through distinct signaling pathways in *Selaginella* and *Arabidopsis*. *PLoS One*, *8*(12), e81938. doi:10.1371/journal.pone.0081938 PMID:24349155

Chiani, M., Akbarzadeh, A., Farhangi, A., Mazinani, M., Saffari, Z., Emadzadeh, K., & Mehrabi, M. R. (2010). Optimization of culture medium to increase the production of desferrioxamine B (Desferal) in *Streptomyces pilosus*. *Pakistan Journal of Biological Sciences*, *13*(11), 546–550. doi:10.3923/pjbs.2010.546.550 PMID:21848068

Chovanec, P., Sparacino-Watkins, C. E., Zhang, N., Basu, P., & Stolz, J. (2012). Microbial reduction of chromate in the presence of nitrate by three nitrate respiring organisms. *Frontiers in Microbiology*, *3*, 416. doi:10.3389/fmicb.2012.00416 PMID:23251135

Chung, J. Y., Yu, S. D., & Hong, Y. S. (2014). Environmental source of arsenic exposure. *Journal of Preventive Medicine and Public Health*, *47*(5), 253–257. doi:10.3961/jpmph.14.036 PMID:25284196

Claussen, M., Lüthe, H., Blatt, M., & Böttger, M. (1997). Auxin-induced growth and its linkage to potassium channels. *Planta*, *201*(2), 227–234. doi:10.1007/BF01007708

Cvjetko, P., Zovko, M., & Balen, B. (2014). Proteomics of heavy metal toxicity in plants. *Archives of Industrial Hygiene and Toxicology*, *65*(1), 1-18.

D'haeze, W., & Holsters, M. (2002). Nod factor structures, responses, and perception during initiation of nodule development. *Glycobiology*, *12*(6), 79R–105R. doi:10.1093/glycob/12.6.79R PMID:12107077

Dall'Asta, P., Velho, A. C., Pereira, T. P., Stadnik, M. J., & Arisi, A. (2019). *Herbaspirillum seropedicae* promotes maize growth but fails to control the maize leaf anthracnose. *Physiology and Molecular Biology of Plants: An International Journal of Functional Plant Biology*, *25*(1), 167–176.

Damalas, C. A., & Koutroubas, S. D. (2016). Farmers' Exposure to Pesticides: Toxicity Types and Ways of Prevention. *Toxics*, *4*(1), 1. doi:10.3390/toxics4010001 PMID:29051407

Danish, S., Kiran, S., Fahad, S., Ahmad, N., Ali, M. A., Tahir, F. A., Rasheed, M. K., Shahzad, K., Li, X., Wang, D., Mubeen, M., Abbas, S., Munir, T. M., Hashmi, M. Z., Adnan, M., Saeed, B., Saud, S., Khan, M. N., Ullah, A., & Nasim, W. (2019). Alleviation of chromium toxicity in maize by Fe fortification and chromium tolerant ACC deaminase producing plant growth promoting rhizobacteria. *Ecotoxicology and Environmental Safety*, *185*, 109706. doi:10.1016/j.ecoenv.2019.109706 PMID:31561073

Dartsch, P. C., Hildenbrand, S., Kimmel, R., & Schmahl, F. W. (1998). Investigations on the nephrotoxicity and hepatotoxicity of trivalent and hexavalent chromium compounds. *International Archives of Occupational and Environmental Health*, *71*, S40–S45. PMID:9827879

Dashti, N., Zhang, F., Hynes, R., & Smith, D. L. (1997). Application of plant growth-promoting rhizobacteria to soybean (*Glycine max* [L.] Merr.) increases protein and dry matter yield under short-season conditions. *Plant and Soil*, *188*(1), 33–41. doi:10.1023/A:1004295827311

Datta, B., & Chakrabartty, P. K. (2014). Siderophore biosynthesis genes of *Rhizobium sp*. isolated from *Cicer arietinum* L. *Biotech, 4*(4), 391-401.

Dawwam, G. E., Elbeltagy, A., Emara, H. M., Abbas, I. H., & Hassan, M. M. (2013). Beneficial effect of plant growth promoting bacteria isolated from the roots of potato plant. *Annals of Agricultural Science, 58*(2), 195–201. doi:10.1016/j.aoas.2013.07.007

De Freitas, J. R., Banerjee, M. R., & Germida, J. J. (1997). Phosphate-solubilizing rhizobacteria enhance the growth and yield but not phosphorus uptake of canola (Brassica napus L.). *Biology and Fertility of Soils, 24*(4), 358–364. doi:10.1007003740050258

de los Santos-Villalobos, S., de Folter, S., Délano-Frier, J. P., Gómez-Lim, M. A., Guzmán-Ortiz, D. A., & Pena-Cabriales, J. J. (2013). Growth promotion and flowering induction in mango (*Mangifera indica* L. cv "*Ataulfo*") trees by Burkholderia and Rhizobium Inoculation: Morphometric, biochemical, and molecular events. *Journal of Plant Growth Regulation, 32*(3), 615–627. doi:10.100700344-013-9329-5

Dhull, S., & Gera, R. (2018). Siderophore Production by Rhizobia Isolated from Cluster Bean [*Cyamopsis tetragonoloba* (L.) Taub.] Growing in Semi-Arid Regions of Haryana, India. *International Journal of Current Microbiology and Applied Sciences, 7*(3), 3187–3191. doi:10.20546/ijcmas.2018.703.368

Dixon, M., Simonne, E., Obreza, T., & Liu, G. (2020). Crop Response to Low Phosphorus Bioavailability with a Focus on Tomato. *Agronomy (Basel), 10*(5), 617. doi:10.3390/agronomy10050617

Douet, V., Expert, D., Barras, F., & Py, B. (2009). Erwinia chrysanthemi iron metabolism: The unexpected implication of the inner membrane platform within the type II secretion system. *Journal of Bacteriology, 191*(3), 795–804. doi:10.1128/JB.00845-08 PMID:18978048

Dowling, D. N., & O'Gara, F. (1994). Metabolites of Pseudomonas involved in the biocontrol of plant disease. *Trends in Biotechnology, 12*(4), 133–141. doi:10.1016/0167-7799(94)90091-4

Dudley, R. E., Gammal, L. M., & Klaassen, C. D. (1985). Cadmium-induced hepatic and renal injury in chronically exposed rats: Likely role of hepatic cadmium-metallothionein in nephrotoxicity. *Toxicology and Applied Pharmacology, 77*(3), 414–426. doi:10.1016/0041-008X(85)90181-4 PMID:3975909

Duffner, A., Hoffland, E., & Temminghoff, E. J. (2012). Bioavailability of zinc and phosphorus in calcareous soils as affected by citrate exudation. *Plant and Soil, 361*(1-2), 165–175. doi:10.100711104-012-1273-9

Duffy, B., Schouten, A., & Raaijmakers, J. M. (2003). Pathogen self-defense: Mechanisms to counteract microbial antagonism. *Annual Review of Phytopathology, 41*(1), 501–538. doi:10.1146/annurev.phyto.41.052002.095606 PMID:12730392

Dutta, S., Datta, J. K., & Mandal, N. C. (2017). Evaluation of indigenous rhizobacterial strains with reduced dose of chemical fertilizer towards growth and yield of mustard (Brassica campestris) under old alluvial soil zone of West Bengal, India. *Annals of Agrarian Science, 15*(4), 447–452. doi:10.1016/j.aasci.2017.02.015

Eastmond, D. A., MacGregor, J. T., & Slesinski, R. S. (2008). Trivalent chromium: Assessing the genotoxic risk of an essential trace element and widely used human and animal nutritional supplement. *Critical Reviews in Toxicology*, *38*(3), 173–190. doi:10.1080/10408440701845401 PMID:18324515

El-Howeity, M. A., & Asfour, M. M. (2012). Response of some varieties of canola plant (Brassica napus L.) cultivated in a newly reclaimed desert to plant growth promoting rhizobacteria and mineral nitrogen fertilizer. *Annals of Agricultural Science*, *57*(2), 129–136. doi:10.1016/j.aoas.2012.08.006

Elumalai, P., Parthipan, P., Karthikeyan, O. P., & Rajasekar, A. (2017). Enzyme-mediated biodegradation of long-chain n-alkanes (C 32 and C 40) by thermophilic bacteria. *Biotech, 7*(2), 1-10.

Eshaghi, E., Nosrati, R., Owlia, P., Malboobi, M. A., Ghaseminejad, P., & Ganjali, M. R. (2019). Zinc solubilization characteristics of efficient siderophore-producing soil bacteria. *Iranian Journal of Microbiology*, *11*(5), 419. doi:10.18502/ijm.v11i5.1961 PMID:32148673

Esitken, A., Pirlak, L., Turan, M., & Sahin, F. (2006). Effects of floral and foliar application of plant growth promoting rhizobacteria (PGPR) on yield, growth and nutrition of sweet cherry. *Scientia Horticulturae*, *110*(4), 324–327. doi:10.1016/j.scienta.2006.07.023

Fan, J., Yang, G., Zhao, H., Shi, G., Geng, Y., Hou, T., & Tao, K. (2012). Isolation, identification and characterization of a glyphosate-degrading bacterium, *Bacillus cereus* CB4, from soil. *The Journal of General and Applied Microbiology*, *58*(4), 263–271. doi:10.2323/jgam.58.263 PMID:22990486

Fang, Q., Fan, Z., Xie, Y., Wang, X., Li, K., & Liu, Y. (2016). Screening and Evaluation of the Bioremediation Potential of Cu/Zn-Resistant, Autochthonous Acinetobacter sp. FQ-44 from *Sonchus oleraceus* L. *Frontiers in Plant Science*, *7*, 1487. doi:10.3389/fpls.2016.01487 PMID:27746807

Feng, L., Wang, W., Cheng, J., Ren, Y., Zhao, G., Gao, C., Gao, C., Tang, Y., Liu, X., Han, W., Peng, X., & Liu, R. (2007). Genome and proteome of long-chain alkane degrading *Geobacillus thermodenitrificans* NG80-2 isolated from a deep-subsurface oil reservoir. *Proceedings of the National Academy of Sciences of the United States of America*, *104*(13), 5602–5607. doi:10.1073/pnas.0609650104 PMID:17372208

Fenibo, E. O., Ijoma, G. N., Selvarajan, R., & Chikere, C. B. (2019). Microbial Surfactants: The Next Generation Multifunctional Biomolecules for Applications in the Petroleum Industry and Its Associated Environmental Remediation. *Microorganisms*, *7*(11), 581. doi:10.3390/microorganisms7110581 PMID:31752381

Fernández, H., Prandoni, N., Fernández-Pascual, M., Fajardo, S., Morcillo, C., Díaz, E., & Carmona, M. (2014). Azoarcus sp. CIB, an anaerobic biodegrader of aromatic compounds shows an endophytic lifestyle. *PLoS One*, *9*(10), e110771. doi:10.1371/journal.pone.0110771 PMID:25340341

Fernando, W. D., Nakkeeran, S., & Zhang, Y. (2005). Biosynthesis of antibiotics by PGPR and its relation in biocontrol of plant diseases. In *PGPR: biocontrol and biofertilization* (pp. 67–109). Springer.

Fiedler, H. P., Krastel, P., Müller, J., Gebhardt, K., & Zeeck, A. (2001). Enterobactin: The characteristic catecholate siderophore of Enterobacteriaceae is produced by Streptomyces species. *FEMS Microbiology Letters*, *196*(2), 147–151. doi:10.1111/j.1574-6968.2001.tb10556.x PMID:11267771

Francis, D. (2011). A commentary on the G2/M transition of the plant cell cycle. *Annals of Botany*, *107*(7), 1065–1070. doi:10.1093/aob/mcr055 PMID:21558458

Funhoff, E. G., & Van Beilen, J. B. (2007). Alkane activation by P450 oxygenases. *Biocatalysis and Biotransformation*, *25*(2-4), 186–193. doi:10.1080/10242420701379254

Gandhi, N. U., & Chandra, S. B. (2012). A comparative analysis of three classes of bacterial non-specific acid phosphatases and archaeal phosphoesterases: Evolutionary perspective. *Acta Informatica Medica*, *20*(3), 167. doi:10.5455/aim.2012.20.167-173 PMID:23322973

García de Salamone, I. E., Hynes, R. K., & Nelson, L. M. (2001). Cytokinin production by plant growth promoting rhizobacteria and selected mutants. *Canadian Journal of Microbiology*, *47*(5), 404–411. doi:10.1139/w01-029 PMID:11400730

Genuini, M., Bidet, P., Benoist, J. F., Schlemmer, D., Lemaitre, C., Birgy, A., & Bonacorsi, S. (2019). ShiF acts as an auxiliary factor of aerobactin secretion in meningitis *Escherichia coli* strain S88. *BMC Microbiology*, *19*(1), 298. doi:10.118612866-019-1677-2 PMID:31847813

Ghosh, P. K., Maiti, T. K., Pramanik, K., Ghosh, S. K., Mitra, S., & De, T. K. (2018). The role of arsenic resistant Bacillus aryabhattai MCC3374 in promotion of rice seedlings growth and alleviation of arsenic phytotoxicity. *Chemosphere*, *211*, 407–419. doi:10.1016/j.chemosphere.2018.07.148 PMID:30077937

Glick, B., Patten, C., Holguin, G., & Penrose, D. (1999). Overview of plant growth-promoting bacteria. Biochemical and Genetic Mechanisms Used by Plant Growth Promoting Bacteria, 1-13.

Glick, B. R. (2012). Plant growth-promoting bacteria: Mechanisms and applications. *Scientifica*. PMID:24278762

Goering, P. L., Aposhian, H. V., Mass, M. J., Cebrián, M., Beck, B. D., & Waalkes, M. P. (1999). The enigma of arsenic carcinogenesis: role of metabolism. *Toxicological Sciences: An Official Journal of the Society of Toxicology*, *49*(1), 5-14.

Gopalakrishnan, S., Srinivas, V., Alekhya, G., & Prakash, B. (2015). Effect of plant growth-promoting *Streptomyces sp.* on growth promotion and grain yield in chickpea (Cicer arietinum L). *Biotech, 5*(5), 799-806.

Gopalakrishnan, S., Srinivas, V., Prakash, B., Sathya, A., & Vijayabharathi, R. (2015). Plant growth-promoting traits of Pseudomonas geniculata isolated from chickpea nodules. *Biotech, 5*(5), 653-661.

Gornall, J., Betts, R., Burke, E., Clark, R., Camp, J., Willett, K., & Wiltshire, A. (2010). Implications of climate change for agricultural productivity in the early twenty-first century. *Philosophical Transactions of the Royal Society of London. Series B, Biological Sciences*, *365*(1554), 2973–2989. doi:10.1098/rstb.2010.0158 PMID:20713397

Goswami, D., Thakker, J. N., & Dhandhukia, P. C. (2016). Portraying mechanics of plant growth promoting rhizobacteria (PGPR): A review. *Cogent Food & Agriculture*, *2*(1), 1127500. doi:10.1080/23311932.2015.1127500

Gouda, S., Kerry, R. G., Samal, D., Mahapatra, G. P., Das, G., & Patra, J. K. (2018). Application of Plant Growth Promoting Rhizobacteria in Agriculture. *Advances in Microbial Biotechnology,* 74-86.

Gregson, B. H., Metodieva, G., Metodiev, M. V., Golyshin, P. N., & McKew, B. A. (2018). Differential protein expression during growth on medium versus long-Chain alkanes in the obligate marine hydrocarbon-degrading bacterium *Thalassolituus oleivorans* MIL-1. *Frontiers in Microbiology*, *9*, 3130. doi:10.3389/fmicb.2018.03130 PMID:30619200

Guerrieri, M. C., Fanfoni, E., Fiorini, A., Trevisan, M., & Puglisi, E. (2020). Isolation and Screening of Extracellular PGPR from the Rhizosphere of Tomato Plants after Long-Term Reduced Tillage and Cover Crops. *Plants*, *9*(5), 668. doi:10.3390/plants9050668 PMID:32466288

Guinel, F. C. (2015). Ethylene, a hormone at the center-stage of nodulation. *Frontiers in Plant Science*, *6*, 1121. doi:10.3389/fpls.2015.01121 PMID:26834752

Gulati, A., Sharma, N., Vyas, P., Sood, S., Rahi, P., Pathania, V., & Prasad, R. (2010). Organic acid production and plant growth promotion as a function of phosphate solubilization by *Acinetobacter rhizosphaerae* strain BIHB 723 isolated from the cold deserts of the trans-Himalayas. *Archives of Microbiology*, *192*(11), 975–983. doi:10.100700203-010-0615-3 PMID:20821196

Guo, H., Yang, Y., Liu, K., Xu, W., Gao, J., Duan, H., Du, B., Ding, Y., & Wang, C. (2016). Comparative Genomic Analysis of *Delftia tsuruhatensis* MTQ3 and the Identification of Functional NRPS Genes for Siderophore Production. *BioMed Research International*, *2016*, 3687619. doi:10.1155/2016/3687619 PMID:27847812

Gupta, A., Gopal, M., Thomas, G. V., Manikandan, V., Gajewski, J., Thomas, G., Seshagiri, S., Schuster, S. C., Rajesh, P., & Gupta, R. (2014). Whole genome sequencing and analysis of plant growth promoting bacteria isolated from the rhizosphere of plantation crops coconut, cocoa and arecanut. *PLoS One*, *9*(8), e104259. doi:10.1371/journal.pone.0104259 PMID:25162593

Gupta, G., Parihar, S. S., Ahirwar, N. K., Snehi, S. K., & Singh, V. (2015). Plant growth promoting rhizobacteria (PGPR): current and future prospects for development of sustainable agriculture. *Journal of Microbial and Biochemical Technology, 7*(2), 96-102.

Gupta, P., & Diwan, B. (2017). Bacterial exopolysaccharide mediated heavy metal removal: A review on biosynthesis, mechanism and remediation strategies. *Biotechnology Reports (Amsterdam, Netherlands)*, *13*, 58–71. doi:10.1016/j.btre.2016.12.006 PMID:28352564

Gupta, S., & Pandey, S. (2019). ACC Deaminase Producing Bacteria With Multifarious Plant Growth Promoting Traits Alleviates Salinity Stress in French Bean (*Phaseolus vulgaris*) Plants. *Frontiers in Microbiology*, *10*, 1506. doi:10.3389/fmicb.2019.01506 PMID:31338077

Haas, D., & Défago, G. (2005). Biological control of soil-borne pathogens by fluorescent pseudomonads. *Nature Reviews. Microbiology*, *3*(4), 307–319. doi:10.1038/nrmicro1129 PMID:15759041

Habibi, S., Djedidi, S., Ohkama-Ohtsu, N., Sarhadi, W. A., Kojima, K., Rallos, R. V., Ramirez, M., Yamaya, H., Sekimoto, H., & Yokoyama, T. (2019). Isolation and Screening of Indigenous Plant Growth-promoting Rhizobacteria from Different Rice Cultivars in Afghanistan Soils. *Microbes and Environments*, *34*(4), 347–355. doi:10.1264/jsme2.ME18168 PMID:31527341

Hadi, F., Mousavi, A., Noghabi, K. A., Tabar, H. G., & Salmanian, A. H. (2013). New bacterial strain of the genus *Ochrobactrum* with glyphosate-degrading activity. *Journal of Environmental Science and Health. Part B, Pesticides, Food Contaminants, and Agricultural Wastes*, *48*(3), 208–213. doi:10.1080/03601234.2013.730319 PMID:23356342

Haidar, B., Ferdous, M., Fatema, B., Ferdous, A. S., Islam, M. R., & Khan, H. (2018). Population diversity of bacterial endophytes from jute (*Corchorus olitorius*) and evaluation of their potential role as bioinoculants. *Microbiological Research*, *208*, 43–53. doi:10.1016/j.micres.2018.01.008 PMID:29551211

Hansda, A., & Kumar, V. (2017). Cu-resistant Kocuria sp. CRB15: a potential PGPR isolated from the dry tailing of Rakha copper mine. *Biotech, 7*(2), 132.

Hartwig, A., & Schwerdtle, T. (2002). Interactions by carcinogenic metal compounds with DNA repair processes: Toxicological implications. *Toxicology Letters*, *127*(1-3), 47–54. doi:10.1016/S0378-4274(01)00482-9 PMID:12052640

Hasanuzzaman, M., Bhuyan, M. H. M., Nahar, K., Hossain, M., Mahmud, J. A., Hossen, M., Masud, A. A., & Fujita, M. (2018). Potassium: A vital regulator of plant responses and tolerance to abiotic stresses. *Agronomy (Basel)*, *8*(3), 31. doi:10.3390/agronomy8030031

Hassan, T. U., Bano, A., & Naz, I. (2017). Alleviation of heavy metals toxicity by the application of plant growth promoting rhizobacteria and effects on wheat grown in saline sodic field. *International Journal of Phytoremediation*, *19*(6), 522–529. doi:10.1080/15226514.2016.1267696 PMID:27936865

Hassen, W., Neifar, M., Cherif, H., Najjari, A., Chouchane, H., Driouich, R. C., Salah, A., Naili, F., Mosbah, A., Souissi, Y., Raddadi, N., Ouzari, H. I., Fava, F., & Cherif, A. (2018). *Pseudomonas rhizophila* S211, a New Plant Growth-Promoting Rhizobacterium with Potential in Pesticide-Bioremediation. *Frontiers in Microbiology*, *9*, 34. doi:10.3389/fmicb.2018.00034 PMID:29527191

He, Z. L., & Yang, X. E. (2007). Role of soil rhizobacteria in phytoremediation of heavy metal contaminated soils. *Journal of Zhejiang University. Science. B.*, *8*(3), 192–207. doi:10.1631/jzus.2007.B0192 PMID:17323432

Higham, D. P., Sadler, P. J., & Scawen, M. D. (1986). Cadmium-binding proteins in *Pseudomonas putida*: Pseudothioneins. *Environmental Health Perspectives*, *65*, 5–11. PMID:3709466

Hill, D. S., Stein, J. I., Torkewitz, N. R., Morse, A. M., Howell, C. R., Pachlatko, J. P., Becker, J. O., & Ligon, J. M. (1994). Cloning of Genes Involved in the Synthesis of Pyrrolnitrin from *Pseudomonas fluorescens* and Role of Pyrrolnitrin Synthesis in Biological Control of Plant Disease. *Applied and Environmental Microbiology*, *60*(1), 78–85. doi:10.1128/AEM.60.1.78-85.1994 PMID:16349167

Himanen, K., Boucheron, E., Vanneste, S., de Almeida Engler, J., Inzé, D., & Beeckman, T. (2002). Auxin-mediated cell cycle activation during early lateral root initiation. *The Plant Cell*, *14*(10), 2339–2351. doi:10.1105/tpc.004960 PMID:12368490

Hoffman, B. M., Lukoyanov, D., Yang, Z. Y., Dean, D. R., & Seefeldt, L. C. (2014). Mechanism of nitrogen fixation by nitrogenase: The next stage. *Chemical Reviews*, *114*(8), 4041–4062. doi:10.1021/cr400641x PMID:24467365

Horne, I., Sutherland, T. D., Harcourt, R. L., Russell, R. J., & Oakeshott, J. G. (2002). Identification of an opd (organophosphate degradation) gene in an Agrobacterium isolate. *Applied and Environmental Microbiology*, *68*(7), 3371–3376. doi:10.1128/AEM.68.7.3371-3376.2002 PMID:12089017

Hou, J., Liu, W., Wang, B., Wang, Q., Luo, Y., & Franks, A. E. (2015). PGPR enhanced phytoremediation of petroleum contaminated soil and rhizosphere microbial community response. *Chemosphere*, *138*, 592–598. doi:10.1016/j.chemosphere.2015.07.025 PMID:26210024

Hu, Y., & Ribbe, M. W. (2015). Nitrogenase and homologs. *JBIC Journal of Biological Inorganic Chemistry*, *20*(2), 435–445. doi:10.100700775-014-1225-3 PMID:25491285

Hu, Y., Xie, Q., & Chua, N. H. (2003). The Arabidopsis auxin-inducible gene ARGOS controls lateral organ size. *The Plant Cell*, *15*(9), 1951–1961. doi:10.1105/tpc.013557 PMID:12953103

Huang, H., Liu, B., Liu, L., & Song, S. (2017). Jasmonate action in plant growth and development. *Journal of Experimental Botany*, *68*(6), 1349–1359. doi:10.1093/jxb/erw495 PMID:28158849

Hwangbo, H., Park, R. D., Kim, Y. W., Rim, Y. S., Park, K. H., Kim, T. H., Sun, J.S., & Kim, K. Y. (2003). 2-Ketogluconic acid production and phosphate solubilization by *Enterobacter intermedium. Current Microbiology, 47*(2), 87-92.

Igiehon, N. O., Babalola, O. O., & Aremu, B. R. (2019). Genomic insights into plant growth promoting rhizobia capable of enhancing soybean germination under drought stress. *BMC Microbiology*, *19*(1), 159. doi:10.118612866-019-1536-1 PMID:31296165

Igiri, B. E., Okoduwa, S. I., Idoko, G. O., Akabuogu, E. P., Adeyi, A. O., & Ejiogu, I. K. (2018). Toxicity and bioremediation of heavy metals contaminated ecosystem from tannery wastewater: A review. *Journal of Toxicology*, *2018*, 2018. doi:10.1155/2018/2568038 PMID:30363677

Iqbal, A., & Hasnain, S. (2013). Aeromonas punctata PNS-1: A promising candidate to change the root morphogenesis of Arabidopsis thaliana in MS and sand system. *Acta Physiologiae Plantarum*, *35*(3), 657–665. doi:10.100711738-012-1106-8

Iturbe, R., Flores, C., Castro, A., & Torres, L. G. (2007). Sub-soil contamination due to oil spills in zones surrounding oil pipeline-pump stations and oil pipeline right-of-ways in Southwest-Mexico. *Environmental Monitoring and Assessment*, *133*(1-3), 387–398. doi:10.100710661-006-9593-y PMID:17286169

Jahan, S., Khan, M., Ahmed, S., & Ullah, H. (2014). Comparative analysis of antioxidants against cadmium induced reproductive toxicity in adult male rats. *Systems Biology in Reproductive Medicine*, *60*(1), 28–34. doi:10.3109/19396368.2013.843039 PMID:24156729

Jan, A. T., Azam, M., Siddiqui, K., Ali, A., Choi, I., & Haq, Q. M. (2015). Heavy metals and human health: Mechanistic insight into toxicity and counter defense system of antioxidants. *International Journal of Molecular Sciences*, *16*(12), 29592–29630. doi:10.3390/ijms161226183 PMID:26690422

Jan, M., Shah, G., Masood, S., Iqbal Shinwari, K., Hameed, R., Rha, E. S., & Jamil, M. (2019). Bacillus cereus enhanced phytoremediation ability of rice seedlings under cadmium toxicity. *BioMed Research International*, *2019*, 2019. doi:10.1155/2019/8134651 PMID:31428647

Jayaraj, R., Megha, P., & Sreedev, P. (2016). Organochlorine pesticides, their toxic effects on living organisms and their fate in the environment. *Interdisciplinary Toxicology*, *9*(3-4), 90–100. doi:10.1515/intox-2016-0012 PMID:28652852

Jha, Y., & Subramanian, R. B. (2013). Paddy plants inoculated with PGPR show better growth physiology and nutrient content under saline condition. *Chilean Journal of Agricultural Research*, *73*(3), 213–219. doi:10.4067/S0718-58392013000300002

Ji, Y., Mao, G., Wang, Y., & Bartlam, M. (2013). Structural insights into diversity and n-alkane biodegradation mechanisms of alkane hydroxylases. *Frontiers in Microbiology*, *4*, 58. doi:10.3389/fmicb.2013.00058 PMID:23519435

Jobby, R., Jha, P., Gupta, A., Gupte, A., & Desai, N. (2019). Biotransformation of chromium by root nodule bacteria *Sinorhizobium sp.* SAR1. *PLoS One*, *14*(7), e0219387. doi:10.1371/journal.pone.0219387 PMID:31361751

Jogaiah, S., Shivanna, R. K., Gnanaprakash, P. H., & Hunthrike, S. S. (2010). Evaluation of plant growth-promoting rhizobacteria for their efficiency to promote growth and induce systemic resistance in pearl millet against downy mildew disease. *Archiv für Phytopathologie und Pflanzenschutz*, *43*(4), 368–378. doi:10.1080/03235400701806377

Jones, J. P., O'Hare, E. J., & Wong, L. L. (2001). Oxidation of polychlorinated benzenes by genetically engineered CYP101 (cytochrome P450cam). *European Journal of Biochemistry*, *268*(5), 1460–1467. doi:10.1046/j.1432-1327.2001.02018.x PMID:11231299

Jones, K. M., Kobayashi, H., Davies, B. W., Taga, M. E., & Walker, G. C. (2007). How rhizobial symbionts invade plants: The Sinorhizobium–Medicago model. *Nature Reviews. Microbiology*, *5*(8), 619–633. doi:10.1038/nrmicro1705 PMID:17632573

Ju, M., Navarreto-Lugo, M., Wickramasinghe, S., Milbrandt, N. B., McWhorter, A., & Samia, A. C. S. (2019). Exploring the chelation-based plant strategy for iron oxide nanoparticle uptake in garden cress (*Lepidium sativum*) using magnetic particle spectrometry. *Nanoscale*, *11*(40), 18582–18594. doi:10.1039/C9NR05477D PMID:31528944

Ju, W., Liu, L., Fang, L., Cui, Y., Duan, C., & Wu, H. (2019). Impact of co-inoculation with plant-growth-promoting rhizobacteria and *Rhizobium* on the biochemical responses of alfalfa-soil system in copper contaminated soil. *Ecotoxicology and Environmental Safety*, *167*, 218–226. doi:10.1016/j.ecoenv.2018.10.016 PMID:30342354

Juárez-Hernández, R. E., Franzblau, S. G., & Miller, M. J. (2012). Syntheses of mycobactin analogs as potent and selective inhibitors of *Mycobacterium tuberculosis. Organic & Biomolecular Chemistry*, *10*(37), 7584–7593. doi:10.1039/c2ob26077h PMID:22895786

Kang, J. P., Huo, Y., Kim, Y. J., Ahn, J. C., Hurh, J., Yang, D. U., & Yang, D. C. (2019). *Rhizobium panacihumi* sp. *nov*., an isolate from ginseng-cultivated soil, as a potential plant growth promoting bacterium. *Archives of Microbiology*, *201*(1), 99–105. doi:10.100700203-018-1578-z PMID:30259064

Kang, S. M., Khan, A. L., You, Y. H., Kim, J. G., Kamran, M., & Lee, I. J. (2014). Gibberellin production by newly isolated strain Leifsonia soli SE134 and its potential to promote plant growth. *Journal of Microbiology and Biotechnology*, *24*(1), 106–112. doi:10.4014/jmb.1304.04015 PMID:24100624

Kang, S. M., Shahzad, R., Bilal, S., Khan, A. L., Park, Y. G., Lee, K. E., Asaf, S., Khan, M. A., & Lee, I. J. (2019). Indole-3-acetic-acid and ACC deaminase producing Leclercia adecarboxylata MO1 improves Solanum lycopersicum L. growth and salinity stress tolerance by endogenous secondary metabolites regulation. *BMC Microbiology*, *19*(1), 80. doi:10.118612866-019-1450-6 PMID:31023221

Karlidag, H., Esitken, A., Turan, M., & Sahin, F. (2007). Effects of root inoculation of plant growth promoting rhizobacteria (PGPR) on yield, growth and nutrient element contents of leaves of apple. *Scientia Horticulturae*, *114*(1), 16–20. doi:10.1016/j.scienta.2007.04.013

Karthik, C., Oves, M., Thangabalu, R., Sharma, R., Santhosh, S. B., & Arulselvi, P. I. (2016). *Cellulosimicrobium funkei*-like enhances the growth of *Phaseolus vulgaris* by modulating oxidative damage under Chromium (VI) toxicity. *Journal of Advanced Research*, *7*(6), 839–850. doi:10.1016/j.jare.2016.08.007 PMID:27668092

Kaur, P., Singh, S., Kumar, V., Singh, N., & Singh, J. (2018). Effect of rhizobacteria on arsenic uptake by macrophyte *Eichhornia crassipes* (Mart.) Solms. *International Journal of Phytoremediation*, *20*(2), 114–120. doi:10.1080/15226514.2017.1337071 PMID:28613914

Kenawy, A., Dailin, D. J., Abo-Zaid, G. A., Abd Malek, R., Ambehabati, K. K., Zakaria, K. H. N., Sayyed, R. Z., & El Enshasy, H. A. (2019). Biosynthesis of antibiotics by PGPR and their roles in biocontrol of plant diseases. In *Plant Growth Promoting Rhizobacteria for Sustainable Stress Management* (pp. 1–35). Springer. doi:10.1007/978-981-13-6986-5_1

Khan, N., Bano, A., & Babar, M. A. (2019). The stimulatory effects of plant growth promoting rhizobacteria and plant growth regulators on wheat physiology grown in sandy soil. *Archives of Microbiology*, *201*(6), 769–785. doi:10.100700203-019-01644-w PMID:30843087

Khan, Z., Kim, S. G., Jeon, Y. H., Khan, H. U., Son, S. H., & Kim, Y. H. (2008). A plant growth promoting rhizobacterium, *Paenibacillus polymyxa* strain GBR-1, suppresses root-knot nematode. *Bioresource Technology*, *99*(8), 3016–3023. doi:10.1016/j.biortech.2007.06.031 PMID:17706411

Khanna, S., & Gharpure, A. S. (2017). Petroleum Carcinogenicity and Aerodigestive Tract: In Context of Developing Nations. *Cureus*, *9*(4), e1202. doi:10.7759/cureus.1202 PMID:28573078

Kiba, T., & Krapp, A. (2016). Plant Nitrogen Acquisition Under Low Availability: Regulation of Uptake and Root Architecture. *Plant & Cell Physiology*, *57*(4), 707–714. doi:10.1093/pcp/pcw052 PMID:27025887

Kim, H. S., Kim, Y. J., & Seo, Y. R. (2015). An overview of carcinogenic heavy metal: Molecular toxicity mechanism and prevention. *Journal of Cancer Prevention*, *20*(4), 232–240. doi:10.15430/JCP.2015.20.4.232 PMID:26734585

Koo, Y. M., Heo, A. Y., & Choi, H. W. (2020). Salicylic acid as a safe plant protector and growth regulator. *The Plant Pathology Journal*, *36*(1), 1–10. doi:10.5423/PPJ.RW.12.2019.0295 PMID:32089657

Koppel, N., Rekdal, V. M., & Balskus, E. P. (2017). Chemical transformation of xenobiotics by the human gut microbiota. *Science*, *356*(6344), eaag2770. doi:10.1126cience.aag2770 PMID:28642381

Koshlaf, E., & Ball, A. S. (2017). Soil bioremediation approaches for petroleum hydrocarbon polluted environments. *AIMS Microbiology*, *3*(1), 25–49. doi:10.3934/microbiol.2017.1.25 PMID:31294147

Kotasthane, A. S., Agrawal, T., Zaidi, N. W., & Singh, U. S. (2017). Identification of siderophore producing and cynogenic fluorescent Pseudomonas and a simple confrontation assay to identify potential bio-control agent for collar rot of chickpea. *Biotech, 7*(2), 137.

Kour, R., Jain, D., Bhojiya, A. A., Sukhwal, A., Sanadhya, S., Saheewala, H., Jat, G., Singh, A., & Mohanty, S. R. (2019). Zinc biosorption, biochemical and molecular characterization of plant growth-promoting zinc-tolerant bacteria. *Biotech, 9*(11), 421.

Krewulak, K. D., & Vogel, H. J. (2008). Structural biology of bacterial iron uptake. *Biochimica et Biophysica Acta (BBA)- Biomembranes*, *1778*(9), 1781–1804. doi:10.1016/j.bbamem.2007.07.026

Kshetri, L., Pandey, P., & Sharma, G. D. (2018). Rhizosphere mediated nutrient management in *Allium hookeri* Thwaites by using phosphate solubilizing rhizobacteria and tricalcium phosphate amended soil. *Journal of Plant Interactions*, *13*(1), 256–269. doi:10.1080/17429145.2018.1472307

Ku, Y., Xu, G., Tian, X., Xie, H., Yang, X., & Cao, C. (2018). Root colonization and growth promotion of soybean, wheat and Chinese cabbage by *Bacillus cereus* YL6. *PLoS One*, *13*(11), e0200181. doi:10.1371/journal.pone.0200181 PMID:30462642

Kuan, K. B., Othman, R., Abdul Rahim, K., & Shamsuddin, Z. H. (2016). Plant growth-promoting rhizobacteria inoculation to enhance vegetative growth, nitrogen fixation and nitrogen remobilisation of maize under greenhouse conditions. *PLoS One*, *11*(3), e0152478. doi:10.1371/journal.pone.0152478 PMID:27011317

Kurth, C., Kage, H., & Nett, M. (2016). Siderophores as molecular tools in medical and environmental applications. *Organic & Biomolecular Chemistry*, *14*(35), 8212–8227. doi:10.1039/C6OB01400C PMID:27492756

Laakso, H. A., Marolda, C. L., Pinter, T. B., Stillman, M. J., & Heinrichs, D. E. (2016). A heme-responsive regulator controls synthesis of staphyloferrin B in *Staphylococcus aureus. The Journal of Biological Chemistry*, *291*(1), 29–40. doi:10.1074/jbc.M115.696625 PMID:26534960

Lampis, S., Santi, C., Ciurli, A., Andreolli, M., & Vallini, G. (2015). Promotion of arsenic phytoextraction efficiency in the fern Pteris vittata by the inoculation of As-resistant bacteria: A soil bioremediation perspective. *Frontiers in Plant Science*, *6*, 80. doi:10.3389/fpls.2015.00080 PMID:25741356

Lanfranconi, M. P., Alvarez, H. M., & Studdert, C. A. (2003). A strain isolated from gas oil-contaminated soil displays chemotaxis towards gas oil and hexadecane. *Environmental Microbiology*, *5*(10), 1002–1008. doi:10.1046/j.1462-2920.2003.00507.x PMID:14510854

Latif Khan, A., Ahmed Halo, B., Elyassi, A., Ali, S., Al-Hosni, K., Hussain, J., Al-Harrasi, A., & Lee, I. J. (2016). Indole acetic acid and ACC deaminase from endophytic bacteria improves the growth of Solarium lycopersicum. *Electronic Journal of Biotechnology*, *19*(3), 58–64. doi:10.1016/j.ejbt.2016.02.001

Latinwo, L. M., Donald, C., Ikediobi, C., & Silver, S. (1998). Effects of intracellular glutathione on sensitivity of *Escherichia coli* to mercury and arsenite. *Biochemical and Biophysical Research Communications*, *242*(1), 67–70. doi:10.1006/bbrc.1997.7911 PMID:9439611

Lewis, D. R., Olex, A. L., Lundy, S. R., Turkett, W. H., Fetrow, J. S., & Muday, G. K. (2013). A kinetic analysis of the auxin transcriptome reveals cell wall remodeling proteins that modulate lateral root development in *Arabidopsis*. *The Plant Cell*, *25*(9), 3329–3346. doi:10.1105/tpc.113.114868 PMID:24045021

Li, C., Li, Q., Wang, Z., Ji, G., Zhao, H., Gao, F., Su, M., Jiao, J., Li, H., & Li, H. (2019). Environmental fungi and bacteria facilitate lecithin decomposition and the transformation of phosphorus to apatite. *Scientific Reports*, *9*(1), 1–8. doi:10.103841598-019-51804-7 PMID:31653926

Li, G., Ma, J., Tan, M., Mao, J., An, N., Sha, G., Zhang, D., Zhao, C., & Han, M. (2016). Transcriptome analysis reveals the effects of sugar metabolism and auxin and cytokinin signaling pathways on root growth and development of grafted apple. *BMC Genomics*, *17*(1), 150. doi:10.118612864-016-2484-x PMID:26923909

Li, H. B., Singh, R. K., Singh, P., Song, Q. Q., Xing, Y. X., Yang, L. T., & Li, Y. R. (2017). Genetic diversity of nitrogen-fixing and plant growth promoting Pseudomonas species isolated from sugarcane rhizosphere. *Frontiers in Microbiology*, *8*, 1268. doi:10.3389/fmicb.2017.01268 PMID:28769881

Li, M., Guo, R., Yu, F., Chen, X., Zhao, H., Li, H., & Wu, J. (2018). Indole-3-acetic acid biosynthesis pathways in the plant-beneficial bacterium *Arthrobacter pascens* ZZ21. *International Journal of Molecular Sciences*, *19*(2), 443. doi:10.3390/ijms19020443 PMID:29389906

Li, X., Li, D., Yan, Z., & Ao, Y. (2018). Biosorption and bioaccumulation characteristics of cadmium by plant growth-promoting rhizobacteria. *RSC Advances*, *8*(54), 30902–30911. doi:10.1039/C8RA06270F

Lima, A. I. G., Corticeiro, S. C., & Figueira, E. M. D. A. P. (2006). Glutathione-mediated cadmium sequestration in *Rhizobium leguminosarum*. *Enzyme and Microbial Technology*, *39*(4), 763–769. doi:10.1016/j.enzmictec.2005.12.009

Lin, W., Lin, M., Zhou, H., Wu, H., Li, Z., & Lin, W. (2019). The effects of chemical and organic fertilizer usage on rhizosphere soil in tea orchards. *PLoS One*, *14*(5), e0217018. doi:10.1371/journal.pone.0217018 PMID:31136614

Lozowicka, B., Jankowska, M., Hrynko, I., & Kaczynski, P. (2016). Removal of 16 pesticide residues from strawberries by washing with tap and ozone water, ultrasonic cleaning and boiling. *Environmental Monitoring and Assessment*, *188*(1), 51. doi:10.100710661-015-4850-6 PMID:26694708

Lugtenberg, B., & Kamilova, F. (2009). Plant-growth-promoting rhizobacteria. *Annual Review of Microbiology*, *63*(1), 541–556. doi:10.1146/annurev.micro.62.081307.162918 PMID:19575558

Lushchak, V. I., Matviishyn, T. M., Husak, V. V., Storey, J. M., & Storey, K. B. (2018). Pesticide toxicity: A mechanistic approach. *EXCLI Journal*, *17*, 1101. PMID:30564086

Lynch, J. M., & Whipps, J. M. (1990). Substrate flow in the rhizosphere. *Plant and Soil*, *129*(1), 1–10. doi:10.1007/BF00011685

Maddela, N. R., Scalvenzi, L., & Venkateswarlu, K. (2017). Microbial degradation of total petroleum hydrocarbons in crude oil: A field-scale study at the low-land rainforest of Ecuador. *Environmental Technology*, *38*(20), 2543–2550. doi:10.1080/09593330.2016.1270356 PMID:27928937

Mahmud, K., Makaju, S., Ibrahim, R., & Missaoui, A. (2020). Current Progress in Nitrogen Fixing Plants and Microbiome Research. *Plants (Basel, Switzerland)*, *9*(1), 97. doi:10.3390/plants9010097 PMID:31940996

Majda, M., & Robert, S. (2018). The role of auxin in cell wall expansion. *International Journal of Molecular Sciences*, *19*(4), 951. doi:10.3390/ijms19040951 PMID:29565829

Maksimov, I. V., Abizgil'Dina, R. R., & Pusenkova, L. I. (2011). Plant growth promoting rhizobacteria as alternative to chemical crop protectors from pathogens. *Applied Biochemistry and Microbiology*, *47*(4), 333–345. doi:10.1134/S0003683811040090

Mancinelli, R. L. (1996). The nature of nitrogen: An overview. *Life Support & Biosphere Science*, *3*(1-2), 17–24. PMID:11539154

Marwa, N., Singh, N., Srivastava, S., Saxena, G., Pandey, V., & Singh, N. (2019). Characterizing the hypertolerance potential of two indigenous bacterial strains (*Bacillus flexus* and *Acinetobacter junii*) and their efficacy in arsenic bioremediation. *Journal of Applied Microbiology*, *126*(4), 1117–1127. doi:10.1111/jam.14179 PMID:30556924

Masciarelli, O., Llanes, A., & Luna, V. (2014). A new PGPR co-inoculated with Bradyrhizobium japonicum enhances soybean nodulation. *Microbiological Research*, *169*(7-8), 609–615. doi:10.1016/j.micres.2013.10.001 PMID:24280513

Maurhofer, M., Keel, C., Haas, D., & Défago, G. (1994). Pyoluteorin production by *Pseudomonas fluorescens* strain CHA0 is involved in the suppression of *Pythium* damping-off of cress but not of cucumber. *European Journal of Plant Pathology*, *100*(3-4), 221–232. doi:10.1007/BF01876237

McLaren, T. I., Smernik, R. J., McLaughlin, M. J., McBeath, T. M., Kirby, J. K., Simpson, R. J., Guppy, C. N., Doolette, A. L., & Richardson, A. E. (2015). Complex Forms of Soil Organic Phosphorus-A Major Component of Soil Phosphorus. *Environmental Science & Technology*, *49*(22), 13238–13245. doi:10.1021/acs.est.5b02948 PMID:26492192

McNear, D. H. Jr. (2013). The rhizosphere-roots, soil and everything in between. *Nature Education Knowledge*, *4*(3), 1.

Meeks, J. C., & Elhai, J. (2002). Regulation of cellular differentiation in filamentous cyanobacteria in free-living and plant-associated symbiotic growth states. *Microbiology and Molecular Biology Reviews*, *66*(1), 94–121. doi:10.1128/MMBR.66.1.94-121.2002 PMID:11875129

Michael Beman, J., Arrigo, K. R., & Matson, P. A. (2005). Agricultural runoff fuels large phytoplankton blooms in vulnerable areas of the ocean. *Nature*, *434*(7030), 211–214. doi:10.1038/nature03370 PMID:15758999

Miethke, M., & Marahiel, M. A. (2007). Siderophore-based iron acquisition and pathogen control. *Microbiology and Molecular Biology Reviews*, *71*(3), 413–451. doi:10.1128/MMBR.00012-07 PMID:17804665

Mire, C. E., Tourjee, J. A., O'Brien, W. F., Ramanujachary, K. V., & Hecht, G. B. (2004). Lead precipitation by *Vibrio harveyi*: Evidence for novel quorum-sensing interactions. *Applied and Environmental Microbiology*, *70*(2), 855–864. doi:10.1128/AEM.70.2.855-864.2004 PMID:14766565

Mishra, J., Singh, R., & Arora, N. K. (2017). Alleviation of heavy metal stress in plants and remediation of soil by rhizosphere microorganisms. *Frontiers in Microbiology*, *8*, 1706. doi:10.3389/fmicb.2017.01706 PMID:28932218

Mishra, P., Kaur, S., Sharma, A. N., & Jolly, R. S. (2016). Characterization of an Indole-3-Acetamide Hydrolase from *Alcaligenes faecalis* subsp. *parafaecalis* and Its Application in Efficient Preparation of Both Enantiomers of Chiral Building Block 2, 3-Dihydro-1, 4-Benzodioxin-2-Carboxylic Acid. *PLoS One*, *11*(7), e0159009. doi:10.1371/journal.pone.0159009 PMID:27391673

Mitra, A., Chatterjee, S., & Gupta, D. K. (2017). Potential role of microbes in bioremediation of arsenic. In *Arsenic Contamination in the Environment* (pp. 195–213). Springer. doi:10.1007/978-3-319-54356-7_10

Mitra, S., Pramanik, K., Sarkar, A., Ghosh, P. K., Soren, T., & Maiti, T. K. (2018). Bioaccumulation of cadmium by *Enterobacter sp.* and enhancement of rice seedling growth under cadmium stress. *Ecotoxicology and Environmental Safety*, *156*, 183–196. doi:10.1016/j.ecoenv.2018.03.001 PMID:29550436

Mohidin, H., Hanafi, M. M., Rafii, Y. M., Abdullah, S. N. A., Idris, A. S., Man, S., Idris, J., & Sahebi, M. (2015). Determination of optimum levels of nitrogen, phosphorus and potassium of oil palm seedlings in solution culture. *Bragantia*, *74*(3), 247–254. doi:10.1590/1678-4499.0408

Moore, C. H., Foster, L. A., Gerbig, D. G. Jr, Dyer, D. W., & Gibson, B. W. (1995). Identification of alcaligin as the siderophore produced by *Bordetella pertussis* and *B. bronchiseptica*. *Journal of Bacteriology*, *177*(4), 1116–1118. doi:10.1128/JB.177.4.1116-1118.1995 PMID:7860593

Morrison, C. K., Arseneault, T., Novinscak, A., & Filion, M. (2017). Phenazine-1-Carboxylic Acid Production by *Pseudomonas fluorescens* LBUM636 Alters *Phytophthora infestans* Growth and Late Blight Development. *Phytopathology*, *107*(3), 273–279. doi:10.1094/PHYTO-06-16-0247-R PMID:27827009

Morrissey, J., & Guerinot, M. L. (2009). Iron uptake and transport in plants: The good, the bad, and the ionome. *Chemical Reviews*, *109*(10), 4553–4567. doi:10.1021/cr900112r PMID:19754138

Mullen, M. D., Wolf, D. C., Ferris, F. G., Beveridge, T. J., Flemming, C. A., & Bailey, G. W. (1989). Bacterial sorption of heavy metals. *Applied and Environmental Microbiology*, *55*(12), 3143–3149. doi:10.1128/AEM.55.12.3143-3149.1989 PMID:2515800

Murray, E. W., Greenberg, B. M., Cryer, K., Poltorak, B., McKeown, J., Spies, J., & Gerwing, P. D. (2019). Kinetics of phytoremediation of petroleum hydrocarbon contaminated soil. *International Journal of Phytoremediation*, *21*(1), 27–33. doi:10.1080/15226514.2018.1523870 PMID:30701992

Mus, F., Crook, M. B., Garcia, K., Garcia Costas, A., Geddes, B. A., Kouri, E. D., Paramasivan, P., Ryu, M. H., Oldroyd, G., Poole, P. S., Udvardi, M. K., Voigt, C. A., Ané, J. M., & Peters, J. W. (2016). Symbiotic Nitrogen Fixation and the Challenges to Its Extension to Nonlegumes. *Applied and Environmental Microbiology*, *82*(13), 3698–3710. doi:10.1128/AEM.01055-16 PMID:27084023

Myresiotis, C. K., Vryzas, Z., & Papadopoulou-Mourkidou, E. (2012). Biodegradation of soil-applied pesticides by selected strains of plant growth-promoting rhizobacteria (PGPR) and their effects on bacterial growth. *Biodegradation*, *23*(2), 297–310. doi:10.100710532-011-9509-6 PMID:21870159

Naqqash, T., Hameed, S., Imran, A., Hanif, M. K., Majeed, A., & van Elsas, J. D. (2016). Differential response of potato toward inoculation with taxonomically diverse plant growth promoting rhizobacteria. *Frontiers in Plant Science*, *7*, 144. doi:10.3389/fpls.2016.00144 PMID:26925072

Naseri, R., Maleki, A., Naserirad, H., Shebibi, S., & Omidian, A. (2013). Effect of plant growth promoting rhizobacteria (PGPR) on reduction nitrogen fertilizer application in rapeseed (Brassica napus L.). *Middle East Journal of Scientific Research*, *14*(2), 213–220.

Nayak, T., Panda, A. N., Adhya, T. K., Das, B., & Raina, V. (2019). Biodegradation of Chlorpyrifos and 3, 5, 6-trichloro-2-pyridinol (TCP) by *Ochrobactrum* sp. CPD-03: Insights from genome analysis on organophosphorus pesticides degradation, chemotaxis and PGPR activity. *bioRxiv*. doi:10.1101/2019.12.12.866210

Neerja, G. (2016). Biodegradation of 1, 1, 1-trichloro-2, 2-bis (4-chlorophenyl) ethane (DDT) by using *Serratia marcescens* NCIM 2919. *Journal of Environmental Science and Health. Part B, Pesticides, Food Contaminants, and Agricultural Wastes*, *51*(12), 809–816. doi:10.1080/03601234.2016.1208455 PMID:27494385

Nicolopoulou-Stamati, P., Maipas, S., Kotampasi, C., Stamatis, P., & Hens, L. (2016). Chemical pesticides and human health: The urgent need for a new concept in agriculture. *Frontiers in Public Health*, *4*, 148. doi:10.3389/fpubh.2016.00148 PMID:27486573

Nie, M., Wang, Y., Yu, J., Xiao, M., Jiang, L., Yang, J., Fang, C., Chen, J., & Li, B. (2011). Understanding plant-microbe interactions for phytoremediation of petroleum-polluted soil. *PLoS One*, *6*(3), e17961. doi:10.1371/journal.pone.0017961 PMID:21437257

Niu, B., Vater, J., Rueckert, C., Blom, J., Lehmann, M., Ru, J. J., Chen, X. H., Wang, Q., & Borriss, R. (2013). Polymyxin P is the active principle in suppressing phytopathogenic *Erwinia spp.* by the biocontrol rhizobacterium *Paenibacillus polymyxa* M-1. *BMC Microbiology*, *13*(1), 137. doi:10.1186/1471-2180-13-137 PMID:23773687

Nocelli, N., Bogino, P. C., Banchio, E., & Giordano, W. (2016). Roles of extracellular polysaccharides and biofilm formation in heavy metal resistance of rhizobia. *Materials (Basel)*, *9*(6), 418. doi:10.3390/ma9060418 PMID:28773540

O'Callaghan-Gordo, C., Orta-Martínez, M., & Kogevinas, M. (2016). Health effects of non-occupational exposure to oil extraction. *Environmental Health: A Global Access Science Source*, *15*, 56.

Oetiker, J. H., Lee, D. H., & Kato, A. (1999). Molecular analysis of a tryptophan-2-monooxygenase gene (IaaM) of *Agrobacterium vitis. DNA Sequence*, *10*(4-5), 349–354. doi:10.3109/10425179909033963 PMID:10727091

Okumura, M., Inoue, S. I., Kuwata, K., & Kinoshita, T. (2016). Photosynthesis activates plasma membrane H+-ATPase via sugar accumulation. *Plant Physiology*, *171*(1), 580–589. doi:10.1104/pp.16.00355 PMID:27016447

Olanrewaju, O. S., Glick, B. R., & Babalola, O. O. (2017). Mechanisms of action of plant growth promoting bacteria. *World Journal of Microbiology & Biotechnology*, *33*(11), 197. doi:10.100711274-017-2364-9 PMID:28986676

Ona, O., Van Impe, J., Prinsen, E., & Vanderleyden, J. (2005). Growth and indole-3-acetic acid biosynthesis of Azospirillum brasilense Sp245 is environmentally controlled. *FEMS Microbiology Letters*, *246*(1), 125–132. doi:10.1016/j.femsle.2005.03.048 PMID:15869971

Orhan, E., Esitken, A., Ercisli, S., Turan, M., & Sahin, F. (2006). Effects of plant growth promoting rhizobacteria (PGPR) on yield, growth and nutrient contents in organically growing raspberry. *Scientia Horticulturae*, *111*(1), 38–43. doi:10.1016/j.scienta.2006.09.002

Ortiz-Hernández, M. L., Sánchez-Salinas, E., Dantán-González, E., & Castrejón-Godínez, M. L. (2013). Pesticide biodegradation: mechanisms, genetics and strategies to enhance the process. *Biodegradation-Life of Science*, 251-287.

Osman, N. I., & Yin, S. (2018). Isolation and characterization of pea plant (Pisum sativum L.) growth-promoting Rhizobacteria. *African Journal of Microbiological Research*, *12*(34), 820–828. doi:10.5897/AJMR2018.8859

Otieno, N., Lally, R. D., Kiwanuka, S., Lloyd, A., Ryan, D., Germaine, K. J., & Dowling, D. N. (2015). Plant growth promotion induced by phosphate solubilizing endophytic *Pseudomonas* isolates. *Frontiers in Microbiology*, *6*, 745. doi:10.3389/fmicb.2015.00745 PMID:26257721

Ozsoy, G., & Aksoy, E. (2015). Estimation of soil erosion risk within an important agricultural sub-watershed in Bursa, Turkey, in relation to rapid urbanization. *Environmental Monitoring and Assessment*, *187*(7), 419. doi:10.100710661-015-4653-9 PMID:26059559

Pal, A. K., & Sengupta, C. (2019). Isolation of Cadmium and Lead Tolerant Plant Growth Promoting Rhizobacteria: *Lysinibacillus varians* and *Pseudomonas putida* from Indian Agricultural Soil. *Soil and Sediment Contamination: An International Journal*, *28*(7), 601–629. doi:10.1080/15320383.2019.1637398

Pandey, S., Ghosh, P. K., Ghosh, S., De, T. K., & Maiti, T. K. (2013). Role of heavy metal resistant *Ochrobactrum sp.* and *Bacillus spp.* strains in bioremediation of a rice cultivar and their PGPR like activities. *Journal of Microbiology (Seoul, Korea)*, *51*(1), 11–17. doi:10.100712275-013-2330-7 PMID:23456706

Parales, R. E., Ditty, J. L., & Harwood, C. S. (2000). Toluene-degrading bacteria are chemotactic towards the environmental pollutants benzene, toluene, and trichloroethylene. *Applied and Environmental Microbiology*, *66*(9), 4098–4104. doi:10.1128/AEM.66.9.4098-4104.2000 PMID:10966434

Patel, P., Trivedi, G., & Saraf, M. (2018). Iron biofortification in mungbean using siderophore producing plant growth promoting bacteria. *Environmental Sustainability*, *1*(4), 357–365. doi:10.100742398-018-00031-3

Patel, P. R., Shaikh, S. S., & Sayyed, R. Z. (2016). Dynamism of PGPR in bioremediation and plant growth promotion in heavy metal contaminated soil. *Indian Journal of Experimental Biology*, *54*(4), 286–290. PMID:27295926

Patel, T., & Saraf, M. (2017). Biosynthesis of phytohormones from novel rhizobacterial isolates and their in vitro plant growth-promoting efficacy. *Journal of Plant Interactions*, *12*(1), 480–487. doi:10.1080/17429145.2017.1392625

Patra, S., Mishra, P., Mahapatra, S. C., & Mithun, S. K. (2016). Modelling impacts of chemical fertilizer on agricultural production: A case study on Hooghly district, West Bengal, India. *Modeling Earth Systems and Environment*, *2*(4), 1–11. doi:10.100740808-016-0223-6

Pawlik, M., Cania, B., Thijs, S., Vangronsveld, J., & Piotrowska-Seget, Z. (2017). Hydrocarbon degradation potential and plant growth-promoting activity of culturable endophytic bacteria of *Lotus corniculatus* and *Oenothera biennis* from a long-term polluted site. *Environmental Science and Pollution Research International*, *24*(24), 19640–19652. doi:10.100711356-017-9496-1 PMID:28681302

Perrot-Rechenmann, C. (2010). Cellular responses to auxin: Division versus expansion. *Cold Spring Harbor Perspectives in Biology*, *2*(5), a001446. doi:10.1101/cshperspect.a001446 PMID:20452959

Philippar, K., Fuchs, I., Luthen, H., Hoth, S., Bauer, C. S., Haga, K., Thiel, G., Ljung, K., Sandberg, G., Bottger, M., Becker, D., & Hedrich, R. (1999). Auxin-induced K+ channel expression represents an essential step in coleoptile growth and gravitropism. *Proceedings of the National Academy of Sciences of the United States of America*, *96*(21), 12186–12191. doi:10.1073/pnas.96.21.12186 PMID:10518597

Pingali, P. L. (2012). Green revolution: Impacts, limits, and the path ahead. *Proceedings of the National Academy of Sciences of the United States of America*, *109*(31), 12302–12308. doi:10.1073/pnas.0912953109 PMID:22826253

Pırlak, L., & Köse, M. (2009). Effects of plant growth promoting rhizobacteria on yield and some fruit properties of strawberry. *Journal of Plant Nutrition*, *32*(7), 1173–1184. doi:10.1080/01904160902943197

Pirlak, L., Turan, M., Sahin, F., & Esitken, A. (2007). Floral and foliar application of plant growth promoting rhizobacteria (PGPR) to apples increases yield, growth, and nutrient element contents of leaves. *Journal of Sustainable Agriculture*, *30*(4), 145–155. doi:10.1300/J064v30n04_11

Podile, A. R., & Kishore, G. K. (2007). Plant growth-promoting rhizobacteria. In *Plant-associated bacteria* (pp. 195–230). Springer.

Poupin, M. J., Timmermann, T., Vega, A., Zuñiga, A., & González, B. (2013). Effects of the plant growth-promoting bacterium Burkholderia phytofirmans PsJN throughout the life cycle of *Arabidopsis thaliana*. *PLoS One*, *8*(7), e69435. doi:10.1371/journal.pone.0069435 PMID:23869243

Prasad, M., Srinivasan, R., Chaudhary, M., Choudhary, M., & Jat, L. K. (2019). Plant growth promoting rhizobacteria (PGPR) for sustainable agriculture: perspectives and challenges. In PGPR Amelioration in Sustainable Agriculture (pp. 129-157). Woodhead Publishing.

Prieto, P., Schilirò, E., Maldonado-González, M. M., Valderrama, R., Barroso-Albarracín, J. B., & Mercado-Blanco, J. (2011). Root hairs play a key role in the endophytic colonization of olive roots by Pseudomonas spp. with biocontrol activity. *Microbial Ecology*, *62*(2), 435–445. doi:10.100700248-011-9827-6 PMID:21347721

Priyanka, T. A., Kotasthane, A. S., Kosharia, A., Kushwah, R., Zaidi, N. W., & Singh, U. S. (2017). Crop specific plant growth promoting effects of ACCd enzyme and siderophore producing and cynogenic fluorescent *Pseudomonas. Biotech, 7*(1).

Qessaoui, R., Bouharroud, R., Furze, J. N., El Aalaoui, M., Akroud, H., Amarraque, A., Vaerenbergh, J. V., Tahzima, R., Mayad, E. H., & Chebli, B. (2019). Applications of New Rhizobacteria Pseudomonas Isolates in Agroecology via Fundamental Processes Complementing Plant Growth. *Scientific Reports, 9*(1), 12832. doi:10.103841598-019-49216-8 PMID:31492898

Qi, F., & Zhang, F. (2020). Cell Cycle Regulation in the Plant Response to Stress. *Frontiers in Plant Science, 10*, 1765. doi:10.3389/fpls.2019.01765 PMID:32082337

Qu, J., Xu, Y., Ai, G. M., Liu, Y., & Liu, Z. P. (2015). Novel Chryseobacterium sp. PYR2 degrades various organochlorine pesticides (OCPs) and achieves enhancing removal and complete degradation of DDT in highly contaminated soil. *Journal of Environmental Management, 161*, 350–357. doi:10.1016/j.jenvman.2015.07.025 PMID:26203874

Radzki, W., Gutierrez Mañero, F. J., Algar, E., Lucas García, J. A., García-Villaraco, A., & Ramos Solano, B. (2013). Bacterial siderophores efficiently provide iron to iron-starved tomato plants in hydroponics culture. *Antonie van Leeuwenhoek, 104*(3), 321–330. doi:10.100710482-013-9954-9 PMID:23812968

Rai, A. N., Söderbäck, E., & Bergman, B. (2000). Cyanobacterium-plant symbioses. *The New Phytologist, 147*(3), 449–481. doi:10.1046/j.1469-8137.2000.00720.x

Raju, M. N., Leo, R., Herminia, S. S., Morán, R. E. B., Venkateswarlu, K., & Laura, S. (2017). Biodegradation of Diesel, Crude Oil and Spent Lubricating Oil by Soil Isolates of *Bacillus spp. Bulletin of Environmental Contamination and Toxicology, 98*(5), 698–705. doi:10.100700128-017-2039-0 PMID:28210752

Ramírez, V., Baez, A., López, P., Bustillos, M. D. R., Villalobos, M. A., Carreño, R., Contreras, J. L., Muñoz-Rojas, J., Fuentes-Ramírez, L. E., Martínez, J., & Munive, J. A. (2019). Chromium hyper-tolerant *Bacillus sp.* MH778713 assists phytoremediation of heavy metals by mesquite trees (*Prosopis laevigata*). *Frontiers in Microbiology, 10*, 1833. doi:10.3389/fmicb.2019.01833 PMID:31456770

Rani, R., & Kumar, V. (2017). Endosulfan degradation by selected strains of plant growth promoting rhizobacteria. *Bulletin of Environmental Contamination and Toxicology, 99*(1), 138–145. doi:10.100700128-017-2102-x PMID:28484804

Rao, P. M., Anitha, Y., & Satyaprasad, K. (2016). Combined effect of Bacillus cereus CPOU13 and *B. subtilis* SPC14 on polycyclic aromatic hydrocarbons degradation in vitro. *International Journal of Bioassays, 5*(4), 505–4511.

Ray, D. K., Mueller, N. D., West, P. C., & Foley, J. A. (2013). Yield Trends Are Insufficient to Double Global Crop Production by 2050. *PLoS One, 8*(6), e66428. doi:10.1371/journal.pone.0066428 PMID:23840465

Raza, A., Razzaq, A., Mehmood, S. S., Zou, X., Zhang, X., Lv, Y., & Xu, J. (2019). Impact of Climate Change on Crops Adaptation and Strategies to Tackle Its Outcome: A Review. *Plants (Basel, Switzerland), 8*(2), 34. doi:10.3390/plants8020034 PMID:30704089

Reddy, P. P. (2012). Plant growth-promoting rhizobacteria (PGPR). In *Recent advances in crop protection* (pp. 131–158). Springer. doi:10.1007/978-81-322-0723-8_10

Rilling, J. I., Acuña, J. J., Sadowsky, M. J., & Jorquera, M. A. (2018). Putative Nitrogen-Fixing Bacteria Associated With the Rhizosphere and Root Endosphere of Wheat Plants Grown in an Andisol From Southern Chile. *Frontiers in Microbiology*, *9*, 2710. doi:10.3389/fmicb.2018.02710 PMID:30524385

Rober-Kleber, N., Albrechtová, J. T., Fleig, S., Huck, N., Michalke, W., Wagner, E., Speth, V., Neuhaus, G., & Fischer-Iglesias, C. (2003). Plasma membrane H+-ATPase is involved in auxin-mediated cell elongation during wheat embryo development. *Plant Physiology*, *131*(3), 1302–1312. doi:10.1104/pp.013466 PMID:12644680

Romero, D., de Vicente, A., Rakotoaly, R. H., Dufour, S. E., Veening, J. W., Arrebola, E., Cazorla, F. M., Kuipers, O. P., Paquot, M., & Pérez-García, A. (2007). The iturin and fengycin families of lipopeptides are key factors in antagonism of *Bacillus subtilis* toward *Podosphaera fusca. Molecular plant-microbe interactions. Molecular Plant-Microbe Interactions*, *20*(4), 430–440. doi:10.1094/MPMI-20-4-0430 PMID:17427813

Ron, E. Z., & Rosenberg, E. (2002). Biosurfactants and oil bioremediation. *Current Opinion in Biotechnology*, *13*(3), 249–252. doi:10.1016/S0958-1669(02)00316-6 PMID:12180101

Sabirova, J. S., Becker, A., Lünsdorf, H., Nicaud, J. M., Timmis, K. N., & Golyshin, P. N. (2011). Transcriptional profiling of the marine oil-degrading bacterium *Alcanivorax borkumensis* during growth on n-alkanes. *FEMS Microbiology Letters*, *319*(2), 160–168. doi:10.1111/j.1574-6968.2011.02279.x PMID:21470299

Saeid, A., Prochownik, E., & Dobrowolska-Iwanek, J. (2018). Phosphorus solubilization by *Bacillus* species. *Molecules (Basel, Switzerland)*, *23*(11), 2897. doi:10.3390/molecules23112897 PMID:30404208

Salunkhe, V. P., Sawant, I. S., Banerjee, K., Rajguru, Y. R., Wadkar, P. N., Oulkar, D. P., Naik, D. G., & Sawant, S. D. (2013). Biodegradation of profenofos by *Bacillus subtilis* isolated from grapevines (*Vitis vinifera*). *Journal of Agricultural and Food Chemistry*, *61*(30), 7195–7202. doi:10.1021/jf400528d PMID:23806113

Sampaio, C., de Souza, J., Damião, A. O., Bahiense, T. C., & Roque, M. (2019). Biodegradation of polycyclic aromatic hydrocarbons (PAHs) in a diesel oil-contaminated mangrove by plant growth-promoting rhizobacteria. *Biotech, 9*(4), 155.

Santi, C., Bogusz, D., & Franche, C. (2013). Biological nitrogen fixation in non-legume plants. *Annals of Botany*, *111*(5), 743–767. doi:10.1093/aob/mct048 PMID:23478942

Santos, D. K. F., Rufino, R. D., Luna, J. M., Santos, V. A., & Sarubbo, L. A. (2016). Biosurfactants: Multifunctional biomolecules of the 21st century. *International Journal of Molecular Sciences*, *17*(3), 401. doi:10.3390/ijms17030401 PMID:26999123

Santos, R. M., Kandasamy, S., & Rigobelo, E. C. (2018). Sugarcane growth and nutrition levels are differentially affected by the application of PGPR and cane waste. *MicrobiologyOpen*, *7*(6), e00617. doi:10.1002/mbo3.617 PMID:29653035

Sanz, L., Dewitte, W., Forzani, C., Patell, F., Nieuwland, J., Wen, B., Quelhas, P., De Jager, S., Titmus, C., Campilho, A., Ren, H., Estelle, M., Wang, H., & Murray, J. A. (2011). The *Arabidopsis* D-type cyclin CYCD2; 1 and the inhibitor ICK2/KRP2 modulate auxin-induced lateral root formation. *The Plant Cell*, *23*(2), 641–660. doi:10.1105/tpc.110.080002 PMID:21357490

Satyapal, G. K., Mishra, S. K., Srivastava, A., Ranjan, R. K., Prakash, K., Haque, R., & Kumar, N. (2018). Possible bioremediation of arsenic toxicity by isolating indigenous bacteria from the middle Gangetic plain of Bihar, India. *Biotechnology Reports (Amsterdam, Netherlands)*, *17*, 117–125. doi:10.1016/j.btre.2018.02.002 PMID:29541605

Schnider-Keel, U., Seematter, A., Maurhofer, M., Blumer, C., Duffy, B., Gigot-Bonnefoy, C., Reimmann, C., Notz, R., Défago, G., Haas, D., & Keel, C. (2000). Autoinduction of 2, 4-diacetylphloroglucinol biosynthesis in the biocontrol agent *Pseudomonas fluorescens* CHA0 and repression by the bacterial metabolites salicylate and pyoluteorin. *Journal of Bacteriology*, *182*(5), 1215–1225. doi:10.1128/JB.182.5.1215-1225.2000 PMID:10671440

Schröder, I., Johnson, E., & De Vries, S. (2003). Microbial ferric iron reductases. *FEMS Microbiology Reviews*, *27*(2-3), 427–447. doi:10.1016/S0168-6445(03)00043-3 PMID:12829278

Sekine, M., Watanabe, K., & Syono, K. (1989). Molecular cloning of a gene for indole-3-acetamide hydrolase from *Bradyrhizobium japonicum*. *Journal of Bacteriology*, *171*(3), 1718–1724. doi:10.1128/JB.171.3.1718-1724.1989 PMID:2646294

Serteyn, L., Quaghebeur, C., Ongena, M., Cabrera, N., Barrera, A., Molina-Montenegro, M. A., Francis, F., & Ramírez, C. C. (2020). Induced Systemic Resistance by a Plant Growth-Promoting Rhizobacterium Impacts Development and Feeding Behavior of Aphids. *Insects*, *11*(4), 234. doi:10.3390/insects11040234 PMID:32276327

Sethy, S. K., & Ghosh, S. (2013). Effect of heavy metals on germination of seeds. *Journal of Natural Science, Biology, and Medicine*, *4*(2), 272. doi:10.4103/0976-9668.116964 PMID:24082715

Shahid, I., Rizwan, M., Baig, D. N., Saleem, R. S., Malik, K. A., & Mehnaz, S. (2017). Secondary Metabolites Production and Plant Growth Promotion by *Pseudomonas chlororaphis* and *P. aurantiaca* Strains Isolated from Cactus, Cotton, and Para Grass. *Journal of Microbiology and Biotechnology*, *27*(3), 480–491. doi:10.4014/jmb.1601.01021 PMID:27974729

Shakeel, M., Rais, A., Hassan, M. N., & Hafeez, F. Y. (2015). Root Associated Bacillus sp. Improves Growth, Yield and Zinc Translocation for Basmati Rice (Oryza sativa) Varieties. *Frontiers in Microbiology*, *6*, 1286. doi:10.3389/fmicb.2015.01286 PMID:26635754

Shanker, A. K., Cervantes, C., Loza-Tavera, H., & Avudainayagam, S. (2005). Chromium toxicity in plants. *Environment International*, *31*(5), 739–753. doi:10.1016/j.envint.2005.02.003 PMID:15878200

Sharma, A., & Johri, B. N. (2003). Growth promoting influence of siderophore-producing Pseudomonas strains GRP3A and PRS9 in maize (Zea mays L.) under iron limiting conditions. *Microbiological Research*, *158*(3), 243–248. doi:10.1078/0944-5013-00197 PMID:14521234

Sharma, M., Mishra, V., Rau, N., & Sharma, R. S. (2016). Increased iron-stress resilience of maize through inoculation of siderophore-producing *Arthrobacter globiformis* from mine. *Journal of Basic Microbiology*, *56*(7), 719–735. doi:10.1002/jobm.201500450 PMID:26632776

Sharma, P. K., Balkwill, D. L., Frenkel, A., & Vairavamurthy, M. A. (2000). A new Klebsiella planticola strain (Cd-1) grows anaerobically at high cadmium concentrations and precipitates cadmium sulfide. *Applied and Environmental Microbiology*, *66*(7), 3083–3087. doi:10.1128/AEM.66.7.3083-3087.2000 PMID:10877810

Sharma, R., Sindhu, S., & Sindhu, S. S. (2018). Bioinoculation of mustard (*Brassica juncea* L.) with beneficial rhizobacteria: A sustainable alternative to improve crop growth. *International Journal of Current Microbiology and Applied Sciences*, *7*(5), 1375–1386. doi:10.20546/ijcmas.2018.705.163

Sharma, S., Tiwari, S., Hasan, A., Saxena, V., & Pandey, L. M. (2018). Recent advances in conventional and contemporary methods for remediation of heavy metal-contaminated soils. *Biotech, 8*(4), 216.

Sharma, S. B., Sayyed, R. Z., Trivedi, M. H., & Gobi, T. A. (2013). Phosphate solubilizing microbes: Sustainable approach for managing phosphorus deficiency in agricultural soils. *SpringerPlus*, *2*(1), 587. doi:10.1186/2193-1801-2-587 PMID:25674415

Shen, J., Yuan, L., Zhang, J., Li, H., Bai, Z., Chen, X., Zhang, W., & Zhang, F. (2011). Phosphorus dynamics: From soil to plant. *Plant Physiology*, *156*(3), 997–1005. doi:10.1104/pp.111.175232 PMID:21571668

Shi, H., Shi, X., & Liu, K. J. (2004). Oxidative mechanism of arsenic toxicity and carcinogenesis. *Molecular and Cellular Biochemistry*, *255*(1-2), 67–78. doi:10.1023/B:MCBI.0000007262.26044.e8 PMID:14971647

Silo-Suh, L. A., Lethbridge, B. J., Raffel, S. J., He, H., Clardy, J., & Handelsman, J. (1994). Biological activities of two fungistatic antibiotics produced by *Bacillus cereus* UW85. *Applied and Environmental Microbiology*, *60*(6), 2023–2030. doi:10.1128/AEM.60.6.2023-2030.1994 PMID:8031096

Singh, B., & Satyanarayana, T. (2011). Microbial phytases in phosphorus acquisition and plant growth promotion. *Physiology and Molecular Biology of Plants*, *17*(2), 93–103. doi:10.100712298-011-0062-x PMID:23572999

Singh, B. K., Walker, A., Morgan, J. A. W., & Wright, D. J. (2004). Biodegradation of chlorpyrifos by *Enterobacter* strain B-14 and its use in bioremediation of contaminated soils. *Applied and Environmental Microbiology*, *70*(8), 4855–4863. doi:10.1128/AEM.70.8.4855-4863.2004 PMID:15294824

Singh, M. J., & Sedhuraman, P. (2015). Biosurfactant, polythene, plastic, and diesel biodegradation activity of endophytic *Nocardiopsis sp.* mrinalini9 isolated from *Hibiscus rosasinensis* leaves. *Bioresources and Bioprocessing*, *2*(1), 2. doi:10.118640643-014-0034-4

Singh, R. P., & Jha, P. N. (2016). The multifarious PGPR *Serratia marcescens* CDP-13 augments induced systemic resistance and enhanced salinity tolerance of wheat (Triticum aestivum L.). *PLoS One*, *11*(6), e0155026. doi:10.1371/journal.pone.0155026 PMID:27322827

Singh, T., & Singh, D. K. (2019). Rhizospheric Microbacterium sp. P27 showing potential of lindane degradation and plant growth promoting traits. *Current Microbiology*, *76*(7), 888–895. doi:10.100700284-019-01703-x PMID:31093691

Smits, T. H., Witholt, B., & van Beilen, J. B. (2003). Functional characterization of genes involved in alkane oxidation by *Pseudomonas aeruginosa*. *Antonie van Leeuwenhoek*, *84*(3), 193–200. doi:10.1023/A:1026000622765 PMID:14574114

Souza, E. C., Vessoni-Penna, T. C., & de Souza Oliveira, R. P. (2014). Biosurfactant-enhanced hydrocarbon bioremediation: An overview. *International Biodeterioration & Biodegradation*, *89*, 88–94. doi:10.1016/j.ibiod.2014.01.007

Souza, R. D., Ambrosini, A., & Passaglia, L. M. (2015). Plant growth-promoting bacteria as inoculants in agricultural soils. *Genetics and Molecular Biology*, *38*(4), 401–419. doi:10.1590/S1415-475738420150053 PMID:26537605

Spaepen, S., Bossuyt, S., Engelen, K., Marchal, K., & Vanderleyden, J. (2014). Phenotypical and molecular responses of *Arabidopsis thaliana* roots as a result of inoculation with the auxin-producing bacterium *Azospirillum brasilense*. *The New Phytologist*, *201*(3), 850–861. doi:10.1111/nph.12590 PMID:24219779

Spaepen, S., & Vanderleyden, J. (2011). Auxin and plant-microbe interactions. *Cold Spring Harbor Perspectives in Biology*, *3*(4), a001438. doi:10.1101/cshperspect.a001438 PMID:21084388

Spaepen, S., Versées, W., Gocke, D., Pohl, M., Steyaert, J., & Vanderleyden, J. (2007). Characterization of phenylpyruvate decarboxylase, involved in auxin production of *Azospirillum brasilense*. *Journal of Bacteriology*, *189*(21), 7626–7633. doi:10.1128/JB.00830-07 PMID:17766418

Stenehjem, J. S., Robsahm, T. E., Bråtveit, M., Samuelsen, S. O., Kirkeleit, J., & Grimsrud, T. K. (2017). Aromatic hydrocarbons and risk of skin cancer by anatomical site in 25 000 male offshore petroleum workers. *American Journal of Industrial Medicine*, *60*(8), 679–688. doi:10.1002/ajim.22741 PMID:28692192

Stoebner, J. A., Butterton, J. R., Calderwood, S. B., & Payne, S. M. (1992). Identification of the vibriobactin receptor of *Vibrio cholerae*. *Journal of Bacteriology*, *174*(10), 3270–3274. doi:10.1128/JB.174.10.3270-3274.1992 PMID:1315733

Su, Q., Guan, T., & Lv, H. (2016). Siderophore biosynthesis coordinately modulated the virulence-associated interactive metabolome of uropathogenic Escherichia coli and human urine. *Scientific Reports*, *6*(1), 1–11. doi:10.1038rep24099 PMID:27076285

Sugawara, M., Okazaki, S., Nukui, N., Ezura, H., Mitsui, H., & Minamisawa, K. (2006). Rhizobitoxine modulates plant–microbe interactions by ethylene inhibition. *Biotechnology Advances*, *24*(4), 382–388. doi:10.1016/j.biotechadv.2006.01.004 PMID:16516430

Suryadi, Y., Susilowati, D. N., & Fauziah, F. (2019). Management of Plant Diseases by PGPR-Mediated Induced Resistance with Special Reference to Tea and Rice Crops. In *Plant Growth Promoting Rhizobacteria for Sustainable Stress Management* (pp. 65–110). Springer. doi:10.1007/978-981-13-6986-5_4

Sutherland, T. D., Horne, I., Harcourt, R. L., Russell, R. J., & Oakeshott, J. G. (2002). Isolation and characterization of a *Mycobacterium* strain that metabolizes the insecticide endosulfan. *Journal of Applied Microbiology, 93*(3), 380–389. doi:10.1046/j.1365-2672.2002.01728.x PMID:12174035

Suzaki, T., Yoro, E., & Kawaguchi, M. (2015). Leguminous plants: Inventors of root nodules to accommodate symbiotic bacteria. *International Review of Cell and Molecular Biology, 316*, 111–158. doi:10.1016/bs.ircmb.2015.01.004 PMID:25805123

Tagele, S. B., Kim, S. W., Lee, H. G., Kim, H. S., & Lee, Y. S. (2018). Effectiveness of multi-trait Burkholderia contaminans KNU17BI1 in growth promotion and management of banded leaf and sheath blight in maize seedling. *Microbiological Research, 214*, 8–18. doi:10.1016/j.micres.2018.05.004 PMID:30031484

Tailor, A. J., & Joshi, B. H. (2014). Harnessing plant growth promoting rhizobacteria beyond nature: A review. *Journal of Plant Nutrition, 37*(9), 1534–1571. doi:10.1080/01904167.2014.911319

Talboys, P. J., Owen, D. W., Healey, J. R., Withers, P. J., & Jones, D. L. (2014). Auxin secretion by *Bacillus amyloliquefaciens* FZB42 both stimulates root exudation and limits phosphorus uptake in *Triticum aestivum. BMC Plant Biology, 14*(1), 51. doi:10.1186/1471-2229-14-51 PMID:24558978

Tchounwou, P. B., Yedjou, C. G., Patlolla, A. K., & Sutton, D. J. (2012). Heavy metal toxicity and the environment. In *Molecular, clinical and environmental toxicology* (pp. 133–164). Springer. doi:10.1007/978-3-7643-8340-4_6

Thompson, L. A., & Darwish, W. S. (2019). Environmental Chemical Contaminants in Food: Review of a Global Problem. *Journal of Toxicology, 2019*, 2345283. doi:10.1155/2019/2345283 PMID:30693025

Timmusk, S., Behers, L., Muthoni, J., Muraya, A., & Aronsson, A. C. (2017). Perspectives and challenges of microbial application for crop improvement. *Frontiers in Plant Science, 8*, 49. doi:10.3389/fpls.2017.00049 PMID:28232839

Titah, H. S., Abdullah, S., Idris, M., Anuar, N., Basri, H., Mukhlisin, M., Tangahu, B. V., Purwanti, I. F., & Kurniawan, S. B. (2018). Arsenic Resistance and Biosorption by Isolated Rhizobacteria from the Roots of *Ludwigia octovalvis. International Journal of Microbiology, 2018*, 3101498. doi:10.1155/2018/3101498 PMID:30723505

Tittabutr, P., Sripakdi, S., Boonkerd, N., Tanthanuch, W., Minamisawa, K., & Teaumroong, N. (2015). Possible role of 1-aminocyclopropane-1-carboxylate (ACC) deaminase activity of *Sinorhizobium sp.* BL3 on symbiosis with mung bean and determinate nodule senescence. *Microbes and Environments, 30*(4), ME15120. doi:10.1264/jsme2.ME15120 PMID:26657304

Tonziello, G., Caraffa, E., Pinchera, B., Granata, G., & Petrosillo, N. (2019). Present and future of siderophore-based therapeutic and diagnostic approaches in infectious diseases. *Infectious Disease Reports, 11*(2). Advance online publication. doi:10.4081/idr.2019.8208 PMID:31649808

Trivedi, P., Schenk, P. M., Wallenstein, M. D., & Singh, B. K. (2017). Tiny microbes, big yields: Enhancing food crop production with biological solutions. *Microbial Biotechnology, 10*(5), 999–1003. doi:10.1111/1751-7915.12804 PMID:28840959

Truskewycz, A., Gundry, T. D., Khudur, L. S., Kolobaric, A., Taha, M., Aburto-Medina, A., Ball, A. S., & Shahsavari, E. (2019). Petroleum Hydrocarbon Contamination in Terrestrial Ecosystems-Fate and Microbial Responses. *Molecules (Basel, Switzerland)*, *24*(18), 3400. doi:10.3390/molecules24183400 PMID:31546774

Tsukanova, K. A., Meyer, J. J. M., & Bibikova, T. N. (2017). Effect of plant growth-promoting Rhizobacteria on plant hormone homeostasis. *South African Journal of Botany*, *113*, 91–102. doi:10.1016/j.sajb.2017.07.007

Turner, B. L., Papházy, M. J., Haygarth, P. M., & McKelvie, I. D. (2002). Inositol phosphates in the environment. *Philosophical Transactions of the Royal Society of London. Series B, Biological Sciences*, *357*(1420), 449–469. doi:10.1098/rstb.2001.0837 PMID:12028785

Ullah, S., & Bano, A. (2015). Isolation of plant-growth-promoting rhizobacteria from rhizospheric soil of halophytes and their impact on maize (*Zea mays* L.) under induced soil salinity. *Canadian Journal of Microbiology*, *61*(4), 307–313. doi:10.1139/cjm-2014-0668 PMID:25776270

Upadhyay, N., Vishwakarma, K., Singh, J., Mishra, M., Kumar, V., Rani, R., Mishra, R. K., Chauhan, D. K., Tripathi, D. K., & Sharma, S. (2017). Tolerance and Reduction of Chromium (VI) by *Bacillus* sp. MNU16 Isolated from Contaminated Coal Mining Soil. *Frontiers in Plant Science*, *8*, 778. doi:10.3389/fpls.2017.00778 PMID:28588589

Vacheron, J., Desbrosses, G., Bouffaud, M. L., Touraine, B., Moënne-Loccoz, Y., Muller, D., Legendre, L., Wisniewski-Dyé, F., & Prigent-Combaret, C. (2013). Plant growth-promoting rhizobacteria and root system functioning. *Frontiers in Plant Science*, *4*, 356. doi:10.3389/fpls.2013.00356 PMID:24062756

Vejan, P., Abdullah, R., Khadiran, T., Ismail, S., & Nasrulhaq Boyce, A. (2016). Role of Plant Growth Promoting Rhizobacteria in Agricultural Sustainability-A Review. *Molecules (Basel, Switzerland)*, *21*(5), 573. doi:10.3390/molecules21050573 PMID:27136521

Wan, W., Qin, Y., Wu, H., Zuo, W., He, H., Tan, J., Wang, Y., & He, D. (2020). Isolation and Characterization of Phosphorus Solubilizing Bacteria With Multiple Phosphorus Sources Utilizing Capability and Their Potential for Lead Immobilization in Soil. *Frontiers in Microbiology*, *11*, 752. doi:10.3389/fmicb.2020.00752 PMID:32390988

Wang, B., Liu, L., O'Leary, G. J., Asseng, S., Macadam, I., Lines-Kelly, R., Yang, X., Clark, A., Crean, J., Sides, T., Xing, H., Mi, C., & Yu, Q. (2018). Australian wheat production expected to decrease by the late 21st century. *Global Change Biology*, *24*(6), 2403–2415. doi:10.1111/gcb.14034 PMID:29284201

Wang, C. L., Michels, P. C., Dawson, S. C., Kitisakkul, S., Baross, J. A., Keasling, J. D., & Clark, D. S. (1997). Cadmium removal by a new strain of *Pseudomonas aeruginosa* in aerobic culture. *Applied and Environmental Microbiology*, *63*(10), 4075–4078. doi:10.1128/AEM.63.10.4075-4078.1997 PMID:9327571

Wang, Q., Xiong, D., Zhao, P., Yu, X., Tu, B., & Wang, G. (2011). Effect of applying an arsenic-resistant and plant growth–promoting rhizobacterium to enhance soil arsenic phytoremediation by Populus deltoides LH05-17. *Journal of Applied Microbiology*, *111*(5), 1065–1074. doi:10.1111/j.1365-2672.2011.05142.x PMID:21895895

Wang, W., & Shao, Z. (2012). Diversity of flavin-binding monooxygenase genes (almA) in marine bacteria capable of degradation long-chain alkanes. *FEMS Microbiology Ecology*, *80*(3), 523–533. doi:10.1111/j.1574-6941.2012.01322.x PMID:22304419

Wang, W., & Shao, Z. (2013). Enzymes and genes involved in aerobic alkane degradation. *Frontiers in Microbiology*, *4*, 116. doi:10.3389/fmicb.2013.00116 PMID:23755043

Wang, Y. H., & Irving, H. R. (2011). Developing a model of plant hormone interactions. *Plant Signaling & Behavior*, *6*(4), 494–500. doi:10.4161/psb.6.4.14558 PMID:21406974

Ward, M. H., Jones, R. R., Brender, J. D., de Kok, T. M., Weyer, P. J., Nolan, B. T., Villanueva, C. M., & van Breda, S. G. (2018). Drinking Water Nitrate and Human Health: An Updated Review. *International Journal of Environmental Research and Public Health*, *15*(7), 1557. doi:10.3390/ijerph15071557 PMID:30041450

Weir, K. M., Sutherland, T. D., Horne, I., Russell, R. J., & Oakeshott, J. G. (2006). A single monooxygenase, ese, is involved in the metabolism of the organochlorides endosulfan and endosulfate in an *Arthrobacter sp. Applied and Environmental Microbiology*, *72*(5), 3524–3530. doi:10.1128/AEM.72.5.3524-3530.2006 PMID:16672499

Whipps, J. M. (2001). Microbial interactions and biocontrol in the rhizosphere. *Journal of Experimental Botany*, *52*(suppl_1), 487–511. doi:10.1093/jxb/52.suppl_1.487 PMID:11326055

Wildberger, P., Pfeiffer, M., Brecker, L., Rechberger, G. N., Birner-Gruenberger, R., & Nidetzky, B. (2015). Phosphoryl transfer from α-d-glucose 1-phosphate catalyzed by *Escherichia coli* sugar-phosphate phosphatases of two protein superfamily types. *Applied and Environmental Microbiology*, *81*(5), 1559–1572. doi:10.1128/AEM.03314-14 PMID:25527541

Wilde, E. J., Hughes, A., Blagova, E. V., Moroz, O. V., Thomas, R. P., Turkenburg, J. P., Raines, D. J., Duhme-Klair, A. K., & Wilson, K. S. (2017). Interactions of the periplasmic binding protein CeuE with Fe(III) n-LICAM^{4-} siderophore analogues of varied linker length. *Scientific Reports*, *7*(1), 45941. doi:10.1038rep45941 PMID:28383577

Wilson, B. R., Bogdan, A. R., Miyazawa, M., Hashimoto, K., & Tsuji, Y. (2016). Siderophores in iron metabolism: From mechanism to therapy potential. *Trends in Molecular Medicine*, *22*(12), 1077–1090. doi:10.1016/j.molmed.2016.10.005 PMID:27825668

Wilson, M. K., Abergel, R. J., Arceneaux, J. E., Raymond, K. N., & Byers, B. R. (2010). Temporal production of the two *Bacillus anthracis* siderophores, petrobactin and bacillibactin. *Biometals*, *23*(1), 129–134. doi:10.100710534-009-9272-x PMID:19816776

Xiao, A. W., Li, Z., Li, W. C., & Ye, Z. H. (2020). The effect of plant growth-promoting rhizobacteria (PGPR) on arsenic accumulation and the growth of rice plants (*Oryza sativa* L.). *Chemosphere*, *242*, 125136. doi:10.1016/j.chemosphere.2019.125136 PMID:31654806

Yadegari, M., & Rahmani, H. A. (2010). Evaluation of bean (Phaseolus vulgaris) seeds inoculation with *Rhizobium phaseoli* and plant growth promoting Rhizobacteria (PGPR) on yield and yield components. *African Journal of Agricultural Research*, *5*(9), 792–799.

Yadegari, M., Rahmani, H. A., Noormohammadi, G., & Ayneband, A. (2010). Plant growth promoting rhizobacteria increase growth, yield and nitrogen fixation in Phaseolus vulgaris. *Journal of Plant Nutrition*, *33*(12), 1733–1743. doi:10.1080/01904167.2010.503776

Yang, O., Kim, H. L., Weon, J. I., & Seo, Y. R. (2015). Endocrine-disrupting chemicals: Review of toxicological mechanisms using molecular pathway analysis. *Journal of Cancer Prevention*, *20*(1), 12–24. doi:10.15430/JCP.2015.20.1.12 PMID:25853100

Yang, P., Zhou, X. F., Wang, L. L., Li, Q. S., Zhou, T., Chen, Y. K., Zhao, Z. Y., & He, B. Y. (2018). Effect of Phosphate-Solubilizing Bacteria on the Mobility of Insoluble Cadmium and Metabolic Analysis. *International Journal of Environmental Research and Public Health*, *15*(7), 1330. doi:10.3390/ijerph15071330 PMID:29941813

Ye, L., Zhao, X., Bao, E., Li, J., Zou, Z., & Cao, K. (2020). Bio-organic fertilizer with reduced rates of chemical fertilization improves soil fertility and enhances tomato yield and quality. *Scientific Reports*, *10*(1), 177. doi:10.103841598-019-56954-2 PMID:31932626

Yousefi, N., Chehregani, A., Malayeri, B., Lorestani, B., & Cheraghi, M. (2011). Investigating the effect of heavy metals on developmental stages of anther and pollen in *Chenopodium botrys* L. (Chenopodiaceae). *Biological Trace Element Research*, *140*(3), 368–376. doi:10.100712011-010-8701-6 PMID:20499206

Zabel, F., Putzenlechner, B., & Mauser, W. (2014). Global agricultural land resources—A high resolution suitability evaluation and its perspectives until 2100 under climate change conditions. *PLoS One*, *9*(9), e107522. doi:10.1371/journal.pone.0107522 PMID:25229634

Zafar-Ul-Hye, M., Tahzeeb-Ul-Hassan, M., Abid, M., Fahad, S., Brtnicky, M., Dokulilova, T., Datta, R., & Danish, S. (2020). Potential role of compost mixed biochar with rhizobacteria in mitigating lead toxicity in spinach. *Scientific Reports*, *10*(1), 12159. doi:10.103841598-020-69183-9 PMID:32699323

Zawadzka, A. M., Abergel, R. J., Nichiporuk, R., Andersen, U. N., & Raymond, K. N. (2009). Siderophore-mediated iron acquisition systems in *Bacillus cereus*: Identification of receptors for anthrax virulence-associated petrobactin. *Biochemistry*, *48*(16), 3645–3657. doi:10.1021/bi8018674 PMID:19254027

Zeffa, D. M., Fantin, L. H., Koltun, A., de Oliveira, A. L., Nunes, M. P., Canteri, M. G., & Gonçalves, L. S. (2020). Effects of plant growth-promoting rhizobacteria on co-inoculation with Bradyrhizobium in soybean crop: A meta-analysis of studies from 1987 to 2018. *PeerJ*, *8*, e7905. doi:10.7717/peerj.7905 PMID:31942248

Zeffa, D. M., Perini, L. J., Silva, M. B., de Sousa, N. V., Scapim, C. A., Oliveira, A., Amaral Júnior, A., & Azeredo Gonçalves, L. S. (2019). Azospirillum brasilense promotes increases in growth and nitrogen use efficiency of maize genotypes. *PLoS One*, *14*(4). doi:10.1371/journal.pone.0215332 PMID:30998695

Zeng, Q., Xie, J., Li, Y., Gao, T., Xu, C., & Wang, Q. (2018). Comparative genomic and functional analyses of four sequenced Bacillus cereus genomes reveal conservation of genes relevant to plant-growth-promoting traits. *Scientific Reports*, *8*(1), 1–10. doi:10.103841598-018-35300-y PMID:30451927

Zeriouh, H., Romero, D., Garcia-Gutierrez, L., Cazorla, F. M., de Vicente, A., & Perez-Garcia, A. (2011). The iturin-like lipopeptides are essential components in the biological control arsenal of *Bacillus subtilis* against bacterial diseases of cucurbits. *Molecular plant-microbe interactions. Molecular Plant-Microbe Interactions*, *24*(12), 1540–1552. doi:10.1094/MPMI-06-11-0162 PMID:22066902

Zerkle, A. L., & Mikhail, S. (2017). The geobiological nitrogen cycle: From microbes to the mantle. *Geobiology*, *15*(3), 343–352. doi:10.1111/gbi.12228 PMID:28158920

Zhang, C., Meng, X., Li, N., Wang, W., Sun, Y., Jiang, W., Guan, G., & Li, Y. (2013). Two bifunctional enzymes with ferric reduction ability play complementary roles during magnetosome synthesis in *Magnetospirillum gryphiswaldense* MSR-1. *Journal of Bacteriology*, *195*(4), 876–885. doi:10.1128/JB.01750-12 PMID:23243303

Zhao, C., Liu, B., Piao, S., Wang, X., Lobell, D. B., Huang, Y., Huang, M., Yao, Y., Bassu, S., Ciais, P., Durand, J. L., Elliott, J., Ewert, F., Janssens, I. A., Li, T., Lin, E., Liu, Q., Martre, P., Müller, C., ... Asseng, S. (2017). Temperature increase reduces global yields of major crops in four independent estimates. *Proceedings of the National Academy of Sciences of the United States of America*, *114*(35), 9326–9331. doi:10.1073/pnas.1701762114 PMID:28811375

Zhao, H., Tao, K., Zhu, J., Liu, S., Gao, H., & Zhou, X. (2015). Bioremediation potential of glyphosate-degrading *Pseudomonas spp*. strains isolated from contaminated soil. *The Journal of General and Applied Microbiology*, *61*(5), 165–170. doi:10.2323/jgam.61.165 PMID:26582285

Zhao, J., Zhao, X., Wang, J., Gong, Q., Zhang, X., & Zhang, G. (2020). Isolation, Identification and Characterization of Endophytic Bacterium *Rhizobium oryzihabitans* sp. *nov*., from Rice Root with Biotechnological Potential in Agriculture. *Microorganisms*, *8*(4), 608. doi:10.3390/microorganisms8040608 PMID:32331293

Zhao, Y. (2014). Auxin biosynthesis. *The Arabidopsis Book / American Society of Plant Biologists*, *12*, e0173. doi:10.1199/tab.0173 PMID:24955076

Zheng, B. X., Ibrahim, M., Zhang, D. P., Bi, Q. F., Li, H. Z., Zhou, G. W., Ding, K., Peñuelas, J., Zhu, Y. G., & Yang, X. R. (2018). Identification and characterization of inorganic-phosphate-solubilizing bacteria from agricultural fields with a rapid isolation method. *AMB Express*, *8*(1), 47. doi:10.118613568-018-0575-6 PMID:29589217

Zhou, C., Guo, J., Zhu, L., Xiao, X., Xie, Y., Zhu, J., Ma, Z., & Wang, J. (2016). *Paenibacillus polymyxa* BFKC01 enhances plant iron absorption via improved root systems and activated iron acquisition mechanisms. *Plant physiology and biochemistry. PPB*, *105*, 162–173. PMID:27105423

Zolnerciks, J. K., Andress, E. J., Nicolaou, M., & Linton, K. J. (2011). Structure of ABC transporters. *Essays in Biochemistry*, *50*(1), 43. PMID:21967051

Zwanenburg, B., & Blanco-Ania, D. (2018). Strigolactones: New plant hormones in the spotlight. *Journal of Experimental Botany*, *69*(9), 2205–2218. doi:10.1093/jxb/erx487 PMID:29385517

Chapter 9
Role of Bacillus spp. in Agriculture:
A Biofertilization and Bioremediation Perspective

Mridul Umesh
CHRIST University (Deemed), India

Ann Mary Sebastian
CHRIST University (Deemed), India

Adhithya Sankar S.
CHRIST University (Deemed), India

Allwyn Vyas G.
SRM Institute of Science and Technology, India

Thazeem Basheer
Vellalar Institutions, Maruthi Nagar, India

Kumaresan Priyanka
Bharathiar University, India

ABSTRACT

The advent of the industrial revolution and intensified agricultural practices have posed irreversible impairment in the soil by accumulating various xenobiotic compounds. Soil, being a core constituent of Earth, not only supports plant growth but also acts as a water filter, buffering pollutants and conserving myriad microorganisms. Untreated industrial effluents, dumping of plastics, and overuse of pesticides are some of the major contaminants enrooted for soil pollution causing severe threats to living beings and the biosphere. Bioremediation using microbes has been recommended as a safe and viable method for the soil fertility restoration due to their adaptive nature modulated by the environment. Among the microbes, Bacillus sp is considered as an effective bioremediating agent as they are the warehouse of copious enzymes, eco-friendly products, and plant growth-promoting metabolites that play a key role in agriculture, textile, food, leather, and beverage industries and thereby ensure soil sustainability.

BACKGROUND

Human activities over the last few decades led to a high pollution status over the exploitation of natural resources and its reprehensible wastes disposal (Figure 1). Rapid increase in global population coupled

DOI: 10.4018/978-1-7998-7062-3.ch009

with accelerated level of Industrialization leads to exponential increase in accumulation of noxious waste in environment. As the quality of life in the biosphere is directly related to the quality of environment, the accumulation of these obnoxious pollutants has a direct correlation to human health. The frequent use of conventional methods to decontaminate the polluted soil leads to unintended alteration of the physicochemical and biological characteristics of that soil. These traditional methods, although widely applied, often fails to prove as ecofriendly and sustainable strategies for pollution management and for restoring soil fertility. As a result, multiple measures were put forward to determine the most useful strategies to deal with polluted areas. Soil microbes take part in degradation and transformation of contamination in soil as they are major contributors in carbon, nitrogen, phosphorus, oxygen, sulfur and heavy metal cycles (Chandra et al., 2019; Teng & Chen, 2019). Microbial remediation offers a promising potential to reinstate contaminated soil in an ecofriendly manner thereby emphasizing a sustainable waste management strategy. The multifunctional microbial enzyme system clearly makes them important candidates for restoring the physicochemical properties of contaminated soil by wide array of process for removing or mitigating environmental contaminants. Exploring the mechanisms that control the growth and activity of microbial enzymes in the contaminated areas can open new windows towards their widespread application in bioremediation.

INTRODUCTION

Environment materialized with non-renewable resources like air, land and water is acclaimed not only for aesthetic appearance but also to sustain the vibrant living of corporeal creatures. Their intact correlation contributes to the sustenance of humans in concert with other living entities. But the advent of science and technological progression to ease the lifestyle of an overgrown population has derogated the holistic function and intrinsic value of indispensable reserves (Kalavathy, 2004). Blooming of industries and rapid urbanization poses a significant challenge in resources management in the past few decades. Pollution is defined as the undesirable alterations occurring by physical, chemical or biological means which adversely affects the wellbeing of humans and environment (Wong, 2012). Contamination of natural resources occurs in a number of ways, among which soil pollution is of paramount concern as it acts as a universal sink for various pollutants (Kirpichtchikova et al., 2006).

Soil characterized by organic and inorganic layers forms the basis for agriculture. It nourishes plants and microbes, maintains biodiversity for the habitual regulation of biogeochemical cycles and a balanced ecosystem (Dixit et al., 2015). Despite their intrinsic values, soils get contaminated naturally or by anthropomorphic sources. Soil contamination arises from several activities like discharge of untreated industrial effluents, leaching of solid waste, overuse of pesticides and herbicides and runoff from storage tanks resulting in the accumulation of xenobiotic compounds at high concentrations (Khan et al., 2008; Wuana & Okieimen, 2011). Particles that depreciate soil fecundity and reduce biotic balance are reflected as soil pollutants among which heavy metals and industrial effluents leave a significant challenge in the disposal site. Almost every wastewater constitutes a substantial amount of heavy metals besides micronutrients and non-degradable organics. Their deleterious effects are influenced by the nature of contaminants and the concentration. The emanation of these untreated or partially treated effluents poses an irreparable damage on agricultural soil and water bodies (Okereke et al., 2016). Thus, there arises an urgent need to imply an effective technology to eliminate these toxins which ensures eco-friendly matrices. Keeping this in view, this chapter summarizes the key concepts of the role of bacteria in soil bioremediation with specific reference to *Bacillus sp*. The chapter addresses the role of *Bacillus sp*., in bioremediation heavy metals, industrial effluents and solid waste management. It further discusses the importance of *Bacillus sp*., in sustainable agriculture, enzyme production for toxic waste management and production of ecofriendly biopolymers. All the key points presented in this work directly or indirectly correlate with the human quest for finding promising methods for ensuring soil sustainability by employing microbial biotechnological principles and practices.

Figure 1. Sources of soil contamination

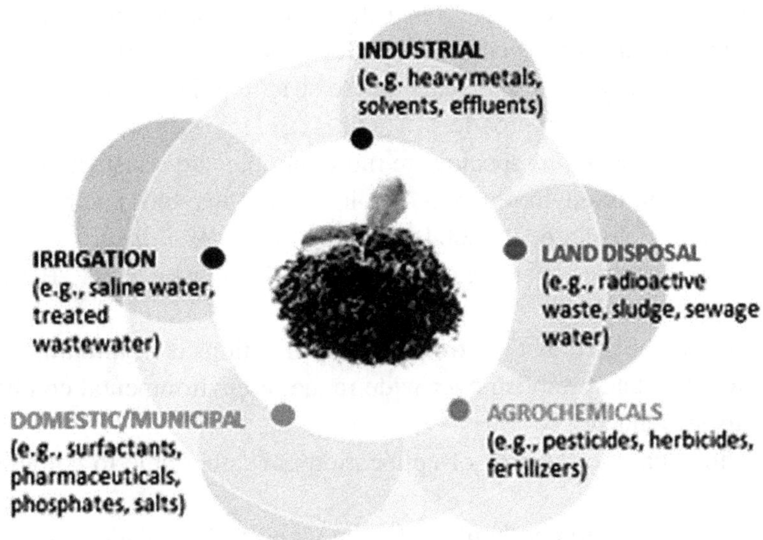

BACILLUS SP. IN AGRICULTURE

Many *Bacillus* sp., have been known to induce plant growth promotion (hence known as Plant Growth Promoting Bacteria) and have found a wide application in microbial inoculant formulations. This is mainly attributed to the spore-forming ability of PGPB that makes them more resistant and survives under the field conditions (Radhakrishnan et al., 2017). It is this property that led to the formulation of Alinit in 1897 the first bacteriological fertilizer which was inoculated onto cereals. An increase in crop yield as high as 40% had been reported by using Alinit (Kilian et al., 2000).

The major strategies used by Plant Growth Promoting Bacteria (PGPBs) which in turn results in significant increase of nutrient uptake are nitrogen fixation and phosphate solubilization. The uptake of essential compounds, mediated by the interaction between bacteria and roots help to prevent toxic compound accumulation (Arora, 2015). In addition to these, they can produce various siderophores, plant hormones, lytic enzymes such as protease, cellulase and cyanides. These compounds are considered to exhibit a phytostimulant effect on the plants and can function as rhizomediators and biopesticides (Stamenković et al., 2018). PGPBs like *Bacillus* are generally indigenous to the plant rhizosphere and the soil ecosystem, where they suppress a broad spectrum of bacterial, fungal and nematode diseases as well as provide protection against viral diseases. The use of beneficial microorganisms is speculated to be an environmentally sound choice to increase crop yields and reduce disease incidence (Calvo et al., 2010)

. Enzymes produced by various bacteria including biocontrol *Bacillus* are implicated in indirect plant growth promotion because of their inhibitory effects on various factors affecting plant growth (Glick, 2012; Ahemad & Kibret, 2014). Generally, competition for nutrients, exclusion of niche, and induction of systemic resistance and production of antifungal metabolites are the chief modes of biocontrol.

Considering the present scenario of pollution beyond the limits and urgency to curb it in all possible ways, the replacement of broad-spectrum synthetic chemicals which is extensively being used as pesticides, insecticides, herbicides, and fertilizers in the agricultural sector using rhizobacterial inoculants has proved to be an effective and reliable alternative that drives more research in this field. That could possibly explain the fact that, though the practice of using microbial inoculant formulations have started in the 20th century, a surge is still witnessed in investing on formulated microbial inoculants by many agricultural biotechnology companies even recently (Kaminsky et al., 2019). Moreover, the value of chemo pesticides is seen to decline over the years due to strict regulations imposed and the preference for environment-friendly and safe agro products (Borriss, 2011).

Compared to other species, *Bacillus* species are more durable and resistant hence can be subjected to extreme chemical, physical, or environmental conditions that may not favor the formulation of less resistant microbial biomass (Schisler et al., 2004). In order to use *Bacillus* or any other such species as a microbial inoculant, ideally there are several requirements (Kaminsky et al., 2019);

- It must be genetically stable so that the frequency of mutations is minimum.
- Should be able to adapt and establish over wide range of environmental conditions such that the applicability is also extended.
- Must not be confined only to the area of application but rather able to colonize extended rhizospheric area.
- Performs the predicted function without losing the viability for an adequate time period that ensures the maintenance of required population density to infect plant roots.
- Leaves no direct impact on human health like the synthetic fertilizers by leaving back toxic chemicals.
- Do not dominate over the indigenous taxa and suppress their growth.

BACILLUS SP. BASED FORMULATIONS AND BIOFERTILIZERS

A microbial formulation is defined as a mixture of the biomass of the useful microorganism(s) along with the substances necessary for its survival and effective establishment. The process of formulation includes several steps from cultivation of the *Bacillus* spp. under optimised conditions, its purification by centrifugation and incorporation with suitable carriers that support their growth. Carriers for preparing inocula are designed to provide an adequate microenvironment that ensures its viability and improves shelf life of the inoculant formulation. The selection of a desirable carrier, ideally depends on the ease of availability, stability, economical and eco-friendly aspects, readiness of application and appreciably good moisture retentivity and pH buffering capacity (Malusá et al., 2012). Finally the shelf life of the formulation is ensured so that it remains viable for at least 6 months so as to retain the commercial standards (Stamenković-Stojanović et al., 2019).

Formulations can be of various forms like liquids i.e., suspensions in water, oil or emulsions; dry products like wettable powders, dust and granules; or as micro encapsulations wherein the microbial biomass is encapsulated in a protective layer that is usually rendered inert (Schisler et al., 2004). The major drawback associated with liquids, though easily formulated, is that they have comparatively lesser shelf life and are also difficult for proper packing and storage (Chumthong et al., 2008). To tackle this problem, the endospore forming *Bacillus* sp. are a solution, as they could be formulated into compact solid forms with amendments like carriers, stabilizers, protectants and other supplements (Stamenkovic-Stojanovic et al., 2019). *Bacillus*-based bio-fertilizers have proved to be highly effective in comparison with *Pseudomonas*-based fertilizers on a commercial scale due to the production of active metabolites and spore-forming character of *Bacillus* spp., which in turn extends the viability of cells in the formulated products (Haas and Defago, 2005).

BACILLUS SP. IN BIOPLASTIC PRODUCTION FOR SOIL SUSTAINABILITY

Plastic has become an unavoidable menace in the world today. Its non-biodegradable nature can drastically reduce the soil fertility and water quality (Ojumu et al., 2004). Biopolymers are ecofriendly polymers which possess the properties of plastic and can be easily degraded by microorganism into basic biomolecules which do not have any harmful effect on nature. Some common biopolymers are cellulose, polyhydroxyalkanoates (PHA), chitosan etc. PHAs are hydroxyalkanoates polyesters which are produced and accumulated as storage granules in many prokaryotes (Preethi & Vineetha, 2015). These microbes produce PHA when there is a limit on any one of the essential nutrients with excess carbon source (Kourmentza et al., 2017). An important fact is that PHA can be produced by growing organisms in media supplemented with agro-waste or food waste; this helps us to solve the issue of waste accumulation in nature. PHA degradation happens via both intracellular and extracellular pathways (Hiraishi & Taguchi, 2013). The degradation starts when the cells face stress due to limited carbon source and leads to formation of acetyl-CoA, which enters the Kreb's cycle for energy production (Lemes et al., 2015).

Previous literature reveals that there are about 150 different PHA molecules (Chen, 2009; Ojumu et al., 2004). 80% of the different varieties of PHAs are present in different bacteria (Lee, 1996a), and the majority of them are found in *Bacillus spp.* The PHA biosynthesis starts right after glycolysis of glucose to pyruvate. The pyruvates are then converted to acetyl CoA by pyruvate dehydrogenase enzyme. Two acetyl CoA molecules are condensed by the enzyme β-ketothiolase to form acetoacetyl CoA, which is further reduced to 3-hydroxybutyryl-CoA by the action of acetoacetyl-CoA dehydrogenase. The 3-hydroxybutyryl-CoA molecules are polymerized Polyhydroxybutyrate (PHB) by the enzyme P(3HB) polymerase (Aldor & Keasling, 2003; Lee, 1996b; Mohapatra et al., 2017; Rehm, 2003). PHA polymers have a high extension to break, low tensile strength, and low melting temperature; these properties make them an ideal candidate to replace many synthetic polymers. PHAs can be used in food packaging (Khosravi-Darani & Bucci, 2015; Koller, 2014), drug delivery and other biomedical applications (Chee et al., 2019; Umesh et al., 2018; Umesh & Thazeem, 2019) due to their biocompatibility and wide range of properties. Some important biomedical uses of PHA include their use in degradable sutures, cardiological stents, orthopedic tools, tissue engineering scaffolds and nerve guides (Bonartsev et al., 2019; Umesh & Preethi, 2017). In industries they are used to make synthetic papers, thermoformed articles and binders. PHAs are also used in agriculture for encapsulation of fertilizers and seeds, controlled release of insecticides and nitrogen fixations (Philip et al., 2007). A list of PHA producing *Bacillus sp.,* is presented in Table 1.

Table 1. List of some Bacillus sp., producing PHA

Bacillus sp.	Substrate	PHA Yield	Reference
Bacillus thuringiensis IAM 12077	Jackfruit seed powder	3.93 g/L	(Gowda & Shivakumar, 2014)
Bacillus cereus SPV	Molasses (sugarcane)	61.07% of cell dry weight	(Akaraonye et al., 2012)
Bacillus aryabhattai PKV01	Sweet sorghum juice	4.36 g/L	(Tanamool et al., 2013)
Bacillus mycoides RLJ B-017.	Sucrose	69±4% dry cell weight	(Borah et al., 2002)
Bacillus subtilis NCDC0671	Orange peel hydrolysate, Carica papaya waste	5.09 g/L 4.2 g/L	(Umesh et al., 2018; Umesh et al., 2017)
Bacillus drentensis BP17	Pineapple peel solution	5.55 g/L	(Penkhrue et al., 2020)
Bacillus megaterium TISTR 1814	Cantaloupe waste extract	1.1 g/L	(Rehman et al., 2020)
Bacillus sphaericus NCIM 5149	Jackfruit seed hydrolysate	2.2g/L	(Ramadas et al., 2010)
Bacillus endophyticus MTCC 9021	Distillery waste	6.45 ± 0.07 g/L	(Priyanka et al., 2020)

BACILLUS SP. IN HEAVY METALS BIOREMEDIATION

Elements having high atomic number greater than 40 and high density above 5g/cm^3 are termed as heavy metals or metalloids (Masindi & Muedi, 2018). Heavy metals are insecure by three features: persistent, toxicity and bioaccumulation. Bioavailability and their dosage determine the degree of pollution in the receiving site. Worldwide more than 5 million sites covering 20 million hectares of land was heavily polluted with metals(loids) beyond the restricted limits (Liu et al., 2018). Metals like nickel, zinc, manganese, and iron are required by plants in trace amounts to sustain the biological functions are essential elements. Conversely metals like arsenic, chromium, mercury and lead intervene in metabolic activities even at lower concentration are non-essential elements. Though some metals are essential for biochemical regulation their presence beyond threshold limit leads to harmful impact in the living system (Sidhu, 2016). As heavy metals are resilient to chemical or microbial degradation their bioaccumulation impairs physiological function by increasing reactive oxygen species and inhibiting biomolecule synthesis leading to various disorders (Järup, 2003).

Both natural and anthropogenic activities upsurge the intrusion of heavy metal in the ecosystem. Dissemination of heavy metals from earth crust occurs through pedagogical processes like, weathering of rocks, volcanic eruptions and surface winds. Mining, smelting, application of fertilizers, burning of fossil fuels and farming agriculture land with industrial effluent are major contributors of anthropogenic share for bio-magnification of heavy metals. It alters physical and chemical properties of soil which eventually modifies the chemical composition of grown crops and poses health risk to the food chain, (Lin & Aarts, 2012; Onakpa et al., 2018).The aforementioned activities have directed the academicians to devise effective and eco-friendly techniques in order to revitalize polluted soil and increase food security (Hu et al., 2017). Contemplating the site of pollution, remedial procedures are categorized as in-situ and ex-situ. Criteria like toxicity, degree of contamination, ecological conditions, expense and legislative strategies determines the bioremediation methods. Numerous approaches have been practised to alleviate heavy metals from contaminated sources using chemical, biological and physical strategies. Excavation and thermal treatment are generally followed in field remediation. Membrane filtration, chemical precipitation, adsorption, vapour exaction and electrochemical treatment are followed for detoxification of heavy metals (Gomes et al., 2013; C. Li et al., 2019). Nevertheless, these conventional procedures are expensive, laborious, ineffectual, and at times liberate toxic sludge which are unsafe for constant monitoring. Hence bioremediation has emerged as state-of-the art techniques to mitigate the hazardous components and their ill-effects through biological activities. The innate feature of plants or microorganisms serves as an effective tool to render them innocuous and promulgate toxin free environments (Ayangbenro & Babalola, 2017). The outcome of microbial remediation depends on the ionic nature and the composition of microbial cell walls and their interactions are influenced by pH, temperature and certain organic acid which might alter the bioavailability of contaminants. Microbial clean-up of heavy metals constitutes different mechanisms like biosorption, bioaccumulation, biotransformation, bio precipitation, bioleaching and bioaugmentation (Igiri et al., 2018; Fang et al., 2010; Mishra & Malik, 2013; Verma & Sharma, 2017; Medfu et al., 2020; Verma & Kuila, 2019).

Various mechanisms of microbial remediation have been depicted in Figure 2.

Figure 2. Mechanisms of microbial remediation

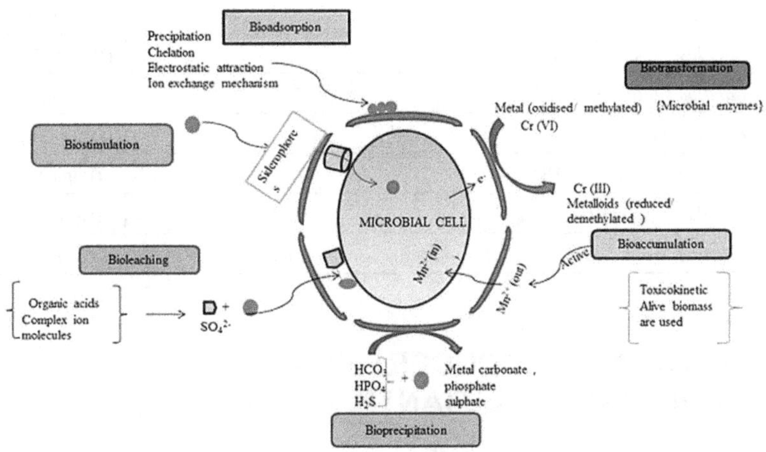

Numerous microorganisms like bacteria, fungi, algae and yeast have been identified for detoxification. Use of indigenous or autochthonous microbes improves the efficacy rather than extrinsic microbes either in consortium or single entity. *Bacillus, Micrococcus, Flavobacterium, Rhodococcus, Desulfovibrio, Methylobacterium,* are the active members of detoxification. Moreover their chemical moiety (teichoic acid) and enlarged surface area intensifies their binding affinity towards heavy metal (Mosa et al., 2016).

Among the resisting bacteria, *Bacillus* and related genera gained unique importance due to rapid replication, spore formation and tolerance to adverse conditions. These gram positive bacteria are ubiquitous and produce copious enzymes which enable their multifaceted role in the industrial and agricultural arena in a cost benefit manner. Their genetic stability and exposition of different mechanisms in an eco-friendly approach either in association or single entities enroutes effective detoxification. *Bacillus* enriches soil in several ways: promoting plant growth, heavy metals removal and as biopesticide for antagonistic pathogen.(Shafi et al., 2017) Myriad of *Bacillus* have been investigated to combat heavy metal contamination from soil has been depicted in Table 2.

Table 2. List of Bacillus sp., involved in heavy metal bioremediation

Metal	Microbes	Reference
Arsenic (As)	*Bacillus aryabhattai* AS6	(Ghosh et al., 2018)
	Bacillus cereus W2	(Miyatake & Hayashi, 2011)
	Bacillus firmus L-148	(Bagade et al., 2020)
Chromium (Cr)	*Bacillus subtilis* MNU16	(Upadhyay et al., 2017)
	Bacillus circulans MN1	(Chaturvedi, 2011)
	Bacillus cereus 332	(Li et al., 2020)
Lead (Pb)	*Bacillus toyonensis* SCE1	(Mathew et al., 2019)
	Bacillus species AS2	(Cephidian et al., 2016)
Mercury (Hg)	*Bacillus thuringiensis* CASKS3	(Saranya et al., 2019)
	Bacillus cereus BW 03 (pPW-05)	(Dash & Das, 2015)
Cadmium (Cd)	*Bacillus circulans* EB1	(Yilmaz & Ensari, 2005)
	Bacillus megaterium BM18-2	(Wu et al., 2019)
Nickle (Ni)	*Bacillus subtilis* SJ-101	(Zaidi et al., 2006)
	Bacillus thuringiensis KUNi1	(Das et al., 2014)
Copper (Cu)	*Bacillus cereus* KTSMBNL 81	(Pugazhendhi et al., 2018)
	Bacillus sp.,505Y11	(Esertaş et al., 2020)
Zinc (Zn)	*Bacillus cereus, Bacillus subtilis and Bacillus sphaericus*	(Costa & Duta, 2001)

BACILLUS SP., AS ENZYME PRODUCERS FOR BIOREMEDIATION OF SOIL POLLUTANTS

Amidst other biologically active substances, enzymes can effectively act upon the substrates constituting the pollutants transforming it to a detoxified state. They may be considered as a great substitute to hurdle the limitations of microbes as they demand low energy and have minimum effect on the environment (Piotrowska-Długosz, 2017). Enzymes were first suggested for waste management in the 1930s but enzymes were not used until the 1970s to prey on specific contaminants in waste. Hydrolases, dehalogenases, transferases and oxidoreductases are the important enzyme classes that are being used in remediation of contaminated and polluted environments. Their key producers are bacteria, fungi, (predominantly white-red fungi), and plants. The conversion of various xenobiotic substances has been evaluated predominantly under experimental conditions for several of these enzymes (Whiteley & Lee, 2006).

The enzymatic mediated degradation of pollutants transforming into a non toxic form greatly relies on microorganisms. The growth and functioning of the microorganisms depend upon environmental conditions for a successful bioremediation. Therefore, this implementation requires modification of environmental conditions so that growth of microbes and degradation can continue rapidly (Karigar & Rao, 2011). An extensive range of extracellular enzymes are produced by different strains of *Bacillus* sp. The extracellular enzymes which are generally used in industrial purposes are extracted from various species of *Bacillus* (Raddadi *et al., 2012).* The involvement of *Bacillus* enzymes and its mechanism in bioremediation of various contaminants are shown in (Table 3).

Microbial Enzymatic Bioremediation of Heavy Metals - Chromium (Cr^{6+})

Microbial oxidoreductases, via their oxidation mechanism, are capable of degrading natural and artificial contaminants, reversing toxicity induced by xenobiotics, and reducing heavy metals (Okino-Delgado et al., 2019; Sharma et al., 2018; Singh & Geetanjali, 2013). Microorganisms harvest energy from the reaction with the help of enzymes and break chemical bonds to aid in movement of electrons from a reduced donor to another acceptor. Eventually the contaminants get oxidized and are harmless as a result of this redox reaction. The phenolic byproducts from lignin degradation in soil are humified by oxidoreductases, therefore used in decolorization of azo dyes and transformation of heavy metals ((Okino-Delgado et al., 2019; Sharma et al., 2018; Singh & Geetanjali, 2013). For humans and animals, trace amounts of heavy metals such as Cu, Fe, Co, Mo, Zn, Ni, Mn, and V are essential, they are hazardous beyond permissible limits. Other heavy metals don't have any benefits for living beings. Consequently, they are significantly life threatening. However, the widespread use of heavy metals (Figure 3) will alter the geochemical cycles and biochemical equilibrium for human purposes. Therefore, the excess amounts of harmful heavy metals are released directly into the soil and water bodies. Indiscriminate assimilation of heavy metal concentrations can pose a threat to humans and aquatic species (Senthil Kumar & Gunasundari, 2018).

In the presence of metal ions, microbes in the immediate environment execute various mechanisms like reduction, bioaccumulation, biosorption and efflux to utilize those ions for metabolic activities in trace amounts, to resist the toxicity, and detoxification of excess metal ions. Chromium in its hexavalent state is portable and extremely hazardous. Chromate resistant organisms are found to have chromate reductases, and are capable of reducing toxic hexavalent chromium to its trivalent state catalyzed through class I ("tight") and class II ("semi-tight") mechanisms by the transfer of electron donors to Cr (VI) where, Reactive Oxygen Species (ROS) are produced simultaneously. They have recently gained special attention for their possible use in the bioremediation method. ChrR, YieF, NemA and LpDH are various chromate reductases located either in cytoplasm or membrane bound from different bacterial sources (Baldiris et al., 2018; Mala et al., 2015; Thatoi et al., 2014). Some chromate reductases that were produced from cytoplasm of *Pseudomonas putida* have reduced hexavalent chromium extracellularly (Priester et al., 2006). The extracellular chromate reductases are soluble in nature and are exudated into the media for the reduction of hexavalent chromium (e.g. nitrate reductases, ferrireductases, flavin reductases and flavin proteins). Most of these enzymes are produced when chromium ions are present in their environment, thus they are intensively regulated (Cheung & Gu, 2007). On the other hand, the intracellular mechanisms are carried out by reduction of hexavalent chromium in cytosol by cytoplasmic reductases. In this process electron donors play an important role as intermediate reductants (NADH and NADPH) (Puzon et al., 2005).

Table 3. Potential Bacillus sp., and its enzymes and mechanism involved in bioremediation of Chromium and synthetic dyes

Enzyme	Organism	Contaminant	Mechanism	Reference
Chromate reductase	*Bacillus* sp. RE	Chromium	Chromate tolerant Extracellular reduction	(Elangovan et al., 2006)
Chromate reductase	*Bacillus methylotrophicus*	Chromium	Extracellular reduction	(Mala et al., 2015)
Chromate reductase	*Bacillus* sp.	Chromium	Chromate tolerant Extracellular reduction	(Prusty et al., 2019)
Chromate reductase	*Bacillus* sp. MNU16	Chromium	Chromate tolerant Rhizoremediation	(Upadhyay et al., 2017)
Chromate reductase	*Bacillus* sp. Strain FM1	Chromium	Heavy metal Resistant	(Masood & Malik, 2011)
Chromate reductase	*Bacillus cereus* G1DM20 *Bacillus fusiformis* G1DM22 *Bacillus sphaericus* G1DM64	Chromium	NADH enhanced Extracellular reduction	(Desai et al., 2008)
Chromate reductase	*Bacillus subtilis*	Chromium	Membrane bound reduction	(Mangaiyarkarasi et al., 2011)
Azoreductase	*Bacillus badius*	Amaranth Dye	NADH and NADPH dependent	(Misal et al., 2011)
Azoreductase	*Bacillus cereus*	Indigoid Compounds	NADH dependent	(Pricelius et al., 2007)
Laccases	*Bacillus pumilus*	Acetosyringone, Indigo Carmine	2,20-Azino-bis (3-ethylbenzthiazoline-6-sulphonic acid), 2,6-dimethoxyphenol	(Reiss et al., 2011)
NADH–DCIP reductase	*Bacillus* sp ADR	Reactive Orange 16	NADH dependent	(Telke et al., 2009)
Phenol oxidase	*Bacillus* sp ADR	Reactive Orange 16	Extracellular	(Telke et al., 2009)
Laccases	*Bacillus* sp ADR	Many Dyes	Extracellular	(Telke et al., 2011)
Cot A Laccase	*Bacillus subtilis*	Sudan Orange G	Direct degradation	(Pereira et al., 2009)
Laccases	*Bacillus subtilis* WD23	Many Dyes	Spore laccase	(Wang et al., 2010)
Azoreductase	*Bacillus velezensis*	Direct red 28	NADH dependent	(Bafana et al., 2008)

Microbial Enzymatic Bioremediation of Synthetic Dyes - Azo Dyes

Colors as an important aspect in the human world have an influence on choices of food, clothes and even in everyday choices. Two percent of basic dyes to fifty percent of reactivate dyes are lost in the wastewater produced by the dyeing industries. Each year textile industries discharge about 28,000 tons of dyes around the world (Jin et al., 2007; Pande et al., 2019). In contrast with organic compounds that are of natural sources (degrade readily in the environment), degradation of synthetic dyes are significantly difficult for the endogenous microorganisms (Ali, 2010; Shedbalkar et al., 2008). Dyes are generally made of two electron systems, chromophores and autochromes. Chromophores help in absorption of light in dye molecules, the electron system's overall energy is transformed by auxochromes to concentrate the colors of chromophores. Around 70 percent of dyes produced globally are azo dyes making them the most popular type of synthetic dyes in textile industries (R. P. Singh et al., 2014). As compared to the natural dyes, azo dyes are manufactured greatly due to their ease of use, cost efficient synthesis, durability, and availability of color range (Saratale et al., 2011). These dyes can be degraded by enzymes such as lignin peroxidase (LiP), manganese peroxidase, (MnP), and laccase. These enzymes can degrade the dyes by the direct method and indirect method mediated by reduction-oxidation compounds (Figure 4). Some bacterial strains are also capable of degrading dyes aerobically which are catalyzed by azoreductases. Azoreductases have been isolated from *Bacillus, Enterococcus, Staphylococcus aureus, E.coli, Pseudomonas aeruginosa, Shigella flexneri, R. sphaeroides, Xenophilus azovorans, and Pigmentiphaga kullae* (Bafana et al., 2011).

Figure 3. Heavy metals Lead, Chromium, Mercury, Arsenic, Cadmium are extremely hazardous to living species, however trace amounts of zinc are required

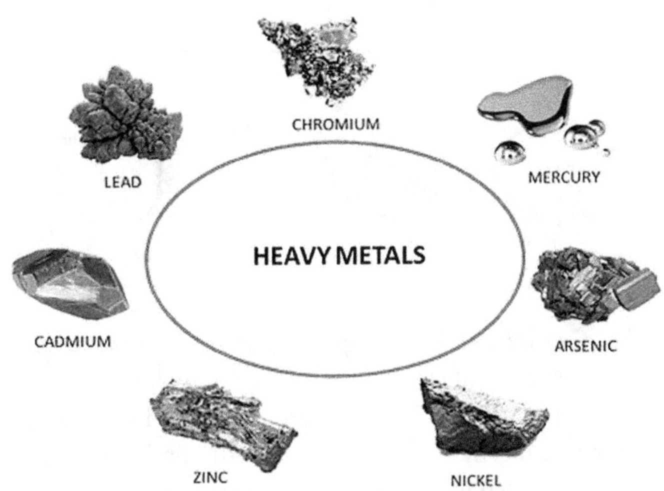

The dominant groups of enzymes expressed in azo dye degrading microorganisms are azoreductases for decolorization/degradation of these dyes. Azo dyes can be decolorized into their respective colorless products, aromatic amines, by reductive azo bond cleavage. These enzymes are active only in the presence of intermediary reductants NADH, NADPH and $FADH_2$ acting as electron donors inside the cell or extracellularly. The activity of intracellular azoreductases were highly suspected in recent years, as sulfonate groups of higher molecular weight are constituted in these dyes making the dyes difficult to get transported into the cells. Therefore, the reduction of dyes are not dependent on absorption by the microorganisms. The breakdown of azo bonds processed by the azoreductase is then transformed into resulting amines by aerobic degradation. However, intracellular degradations are reported in some microorganisms in which low molecular weight mediators transfer electrons between outer membrane bound NADH facilitated azoreductases and azo dyes. It is suggested that these mediators are the resultant products of bacterial metabolism of specific substrates or additives that are supplemented externally (Aranganathan et al., 2013; Singh et al., 2015). Kudlich et al., (1997) proposed a two-way enzyme system that reduces azo dyes anaerobically (Figure 5). However, the fate of azo dye reduction is carried out by redox reactions and these mediators are dependent on contribution of electrons by the cytosolic reducing enzymes. They also suggested the possibility of redox reactions involving a dehydrogenase enzyme synthesized in cytoplasm exudated extracellularly without accumulating inside the cell.

Figure 4. Various methods of degradation of azo dyes (RM- Redox mediator; ED- Electron donor)

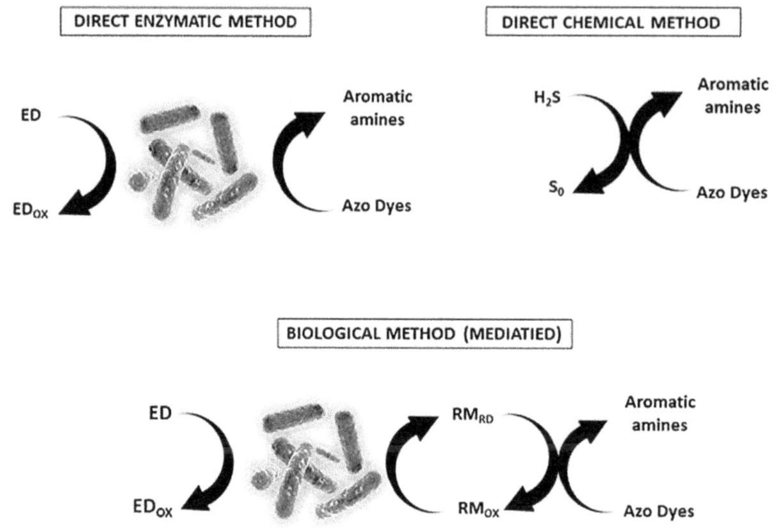

Multicopper oxidases (MCOs) are an all-encompassing group of enzymes that oxidize a diverse array of aromatic phenolic and non-phenolic substrates. Laccases and metalloxidase are two functional classes of multicopper oxidases (Guan et al., 2018). Laccases are catalyzed by mono-electronic oxidation of the substrate reducing it to water making them more eco- friendly and can be produced from the fungus, bacteria, higher plants, insects and also in lichen, etc. They engage in the polymerization of monomers involved in the breakdown of a wide range of industrial pollutants. Considering that laccases are one of the earliest enzymes ever identified and significant to the decomposition of xenobiotics; pulp, paper, textile as well as food industrial waste bioremediation processes are achieved more effectively by using this enzyme over the last few years (Chandra & Chowdhary, 2015; Guan et al., 2018; Guauque-Torres & Bustos, 2019). Laccase catalyzes aromatic, organometallics and non-phenolic compounds. The diffusible electron carriers act as an intermediary for the catalysis of non-phenolic compounds with higher redox potential than laccases alone, called mediators, are organic compounds of low molecular weight that constitute Laccase Mediator System (LMS) (Morozova et al., 2007). Laccases are highly effective in decolorization/degradation of azo dyes due to their higher specificity towards substrates, expanded capabilities in reactions and in some cases; they are not dependent on any mediators. During some dye decolorization process laccases oxidize the compound directly without directly breaking down the azo bonds through the interaction of highly nonspecific radical processes (Kalme et al., 2009).

Figure 5. Proposed mechanism for degradation of azo dyes by azoreductase

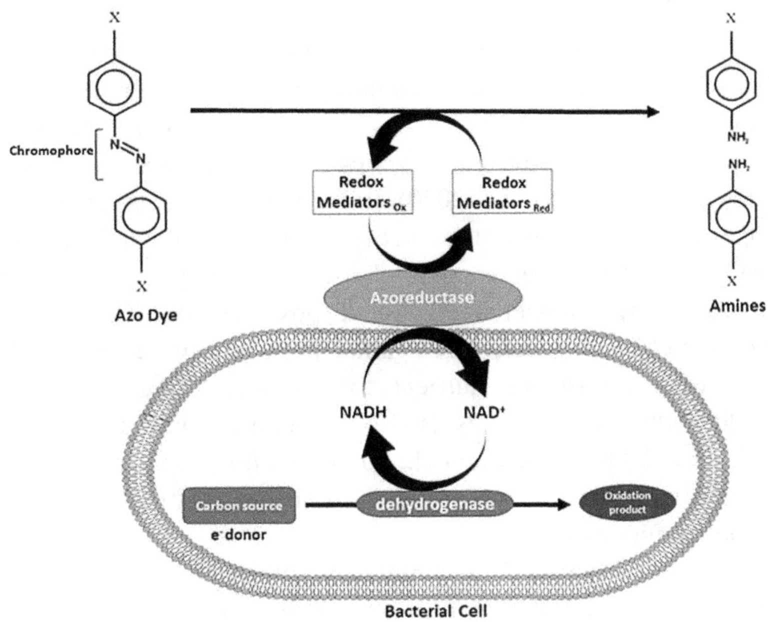

Microbial Enzymatic Bioremediation of Hydrocarbons

Hydrocarbon contamination from petrochemical industries arises as a major pollutant threat as well as accidental spillages being a specific distress to the environment. Burial and incineration being a conventional disposal method of hydrocarbons will not be feasible when the disposal becomes extremely expensive and the contamination is in higher quantity. Hydrocarbon utilization of microorganisms can naturally metabolize the threat of contamination and can clean the polluted sites. The amount of the hydrocarbons accumulated in the soil and nature of the hydrocarbons affect the degradation by microorganism as it is a dynamic process. The non-availability of microorganisms plays a crucial limitation of biodegradation of the pollutants. Petroleum hydrocarbons which are bound to the soil are very staineous to be degraded or eliminated. Under aerobic conditions the degradation of hydrocarbons is rapid and an absolute degradation can be achieved (Das & Chandran, 2011; Varjani, 2017).

Degradation of aromatic and aliphatic hydrocarbons can be carried out aerobically and anaerobically. Aerobic degradation by microorganisms employs oxygenase enzymes Monooxygenases catalyzes by mono-electronic oxidation of the substrate reducing it to water. The accurate and effective influx of single oxygen into organic substrates is challenging in a non-enzymatic process, whereas dioxygenases add two atoms of oxygen and both degrade the aromatic ring in the chemical structures (Singh & Geetanjali, 2013). The aromatic hydrocarbons are aerobically degraded by iron-containing catechol dioxygenases. They cleave the aromatic rings by transferring oxygen molecules to catechol as well as its derivatives and the resulting compounds enter the Citric Acid cycle to be oxidized into carbon dioxide (Peixoto et al., 2011).

BACILLUS SP. IN SOLID WASTE MANAGEMENT

Rapid increase in the population rate and economy has led to an enormous generation of waste in recent years, amongst which solid waste generation, accumulation and unhealthy disposal play a significant part. Contemporary solid waste management practices such as stabilization of the material before landfill, possible energy recovery, source reduction and recycling require strict governmental approvals and may vary among different regions (Rastogi et al., 2020). In India, the predominant waste disposal practices include landfilling, open dumping, incineration and composting that contaminate soil, groundwater and air drastically. Decrease in soil fertility in terms of loss of productivity is the ultimate effect of environmental contamination caused by improper solid waste disposal methods (Pan et al., 2012).

On the other hand, microbial intervention has gained popularity in recent decades as microbes are capable of transforming organic waste into nutrients for plants and can also reduce carbon-to-nitrogen ratio to advocate soil fecundity. Microorganisms also maintain the nutrient flow, reducing ecological imbalance (Novinscak et al., 2008; Umsakul et al., 2010). *Bacillus spp.* stands out for its vibrant solid waste composting ability, enabling soil fertility which has been supported by ample reports. They are mesophilic and cellulolytic microbes.

Available unscientific disposal methods for Municipal Solid Waste (MSW) cause serious problems because of the obnoxious secondary pollutants released into the soil and water. Shifting the organic waste from unsafe landfills to economically viable composting has proved to improve soil properties (porosity, texture, organic matter and NPK content) promoting sustainable agricultural practices and applications (Kjerstadius et al., 2016). Awasthi et al., (2016) co-composted MSW along with sludge, using mixed microbial culture - *Candida rugopelliculosa, Bacillus casei,* Trichoderma, *Lactobacillus buchneri* and white-rot fungi which resulted in the enhancement of the mineralization rate and reduction of nitrogen loss. Composting of MSW using *Bacillus* isolates - *B. subtilis, B. tequilensis, B. venezuelans* and *B. amyloliquefaciens* reduced composting time and finest quality compost was achieved (Voběrková et al., 2017). Furthermore, good quality compost with surplus humic acid was achieved by Ding et al., (2016) when a bacterial consortium consisting of *Bacillus, Lactobacillus, Pseudomonas* and others were used during the starting stage of waste composting. A study consisting of seven organic waste substrates (fruit wastes, leaves, vegetable wastes, hay, wheat straw, newspaper, and rice husks) was reported by (Pan et al., 2012) where *Bacillus* isolates served as hydrolyzers and *Pseudomonas* isolates as nitrogen fixers. This consortium paved the way for organic fertilizer production.

Bioremediation is one of the present-day technologies appreciated to recycle industrial wastes, as it is a cleaner technology which substitutes improper disposal methods. Globally, leather is one of the most broadly retailed products. The substantial growth of tanneries is combined with issues in discharging and dumping of the effluents and solid wastes respectively. Major problem of the industry is that wastes remain unutilized or underutilized, releasing secondary environmental pollutants into soil, water and air (Basheer & Umesh, 2018). Alternatively, Zerdani et al., (2004) isolated eight strains of *Bacillus* sp. from Morocco compost soil, among which *Bacillus subtilis* and *Bacillus licheniformis* hydrolyzed tannery solid waste more potently into protein hydrolysates for its use as soil fertilizers. Extracellular alkaline protease producing *Bacillus subtilis* was isolated from tannery waste which efficiently degraded leather shavings, trimmings and splittings into finer hydrolysates (Aftab et al., 2006). With the help of alkaline protease producing *Bacillus cereus* 1173900, (Ravindran et al., 2011) were successfully able to liquefy tannery fleshing waste, as the alkaline pH of the fleshing waste spoils the characteristics of soil and eventually affects plant growth when buried in landfills. It's a matter of fact that fertilizer from fleshing waste acted similarly to that of organic compost. Slow nitrogen release concept has been applied due to which fleshing waste fertilizer is added to the soil before planting (Sundar et al., 2011).

In order to reduce soil deterioration due to the unsafe landfill burial of tannery solid waste, valorization of the same into a proteinaceous feed ingredient via lactic acid fermentation using *Lactobacillus plantarum* was attempted by Thazeem et al., (2015); Thazeem et al., (2020). Utilizable products with multifunctionalities were obtained which could possibly serve as a protein source for livestock and as a nitrogen source for soil as well. Lipolytic and proteolytic *Bacillus cereus* and *Bacillus subtilis* respectively liquefied tannery fleshing waste completely, enabling bio-methanization process (Sundar et al., 2011).

Likewise, keratinous wastes' degradation is a troublesome issue in the poultry farms (due to chicken feathers) and in leather industries (during de-hairing process) that notably harm the soil's fertility. Poultry feather waste is a well-known solid waste whose current disposal methods are not economically and ecologically valued (Joardar & Rahman, 2018). On the flip side, tannery effluents are the repositories of microorganisms. A beneficiary attempt was made where keratinolytic microbes were isolated from tannery aeration tank effluent (6A - *Bacillus mycoides*, 8A - *Bacillus cereus*, 11A - *Bacillus vallismortis*, 12A - *Bacillus mojavensis*); Isolate 8A showed maximum degree of feather degradation and isolates 8A, 11A and 12A exhibited vibrant de-hairing activity (Preethi et al., 2015). This could eventually replace the traditional burning/burying methods, favoring sustainable soil properties. An analogous study performed by (Thazeem et al., 2016 a) revealed the presence of proteolytic *Bacillus subtilis, Bacillus flexus and Bacillus endophyticus* in tannery lime effluent which actively solubilized poultry feathers, substituting conventional feather processing systems. Chemical de-hairing of raw leather hides is a major contributor of chemical load that adversely affects soil and water bodies. Thus, vigorous hair removal was observed when hides were immersed in the culture supernatant of *Bacillus pumilus* isolated from tannery lime effluent, thereby promoting chemical-free soil (Thazeem et al., 2016 b). A homologous study exploring proteolytic *Bacillus* strains' (isolated from tannery effluent) biodegradative, de-hairing and de-staining activities proved noteworthy (Thazeem et al., 2017). Maximum degradation of chicken feathers by *Bacillus licheniformis*; nutritious hydrolyzed feather powder; significant de-hairing activity of all the *Bacillus* isolates via dip method and complete de-staining of stained clothes further reinforced the proteolytic efficiency of the strains used. Liquefied tannery solid waste, poultry feather waste and de-haired keratinous waste could be beneficially used as bio-fertilizers in agricultural sectors.

Paul et al., (2018) investigated a novel technique of applying degraded feather waste as a bioactive nitrogen input for wheat crops, which may enhance soil texture and agro-system. Disposal and management of Fish Solid Waste (FSW) is one of the biggest challenges being faced by the environmental scientists because of its devastating effects on soil and water quality. S. Mohapatra et al., (2017) efficiently converted this FSW into a valuable substrate for polyhydroxybutyrate (PHB) production using *Bacillus subtilis* (KP172548).

Amount of agro waste increases with increase in the demand for food and their inappropriate disposal pose serious soil and groundwater damage. As a choice, microbes are vital entities that degrade and decompose agricultural wastes into useful form. Various agricultural wastes such as groundnut oil cake, coconut oil cake, linseed oil cake, cottonseed meal, soybean meal and wheat bran were advantageously utilized as substrates by Elumalai et al., (2020) to produce protease under submerged fermentation using *Bacillus subtilis* B22. According to Sadh et al., (2018), *Bacillus licheniformis* MTCC 1483 was the bacterium used to beneficially treat wheat straw, sugarcane baggase, maize straw and paddy straw through solid state fermentation. Aromatic plant waste composts were investigated by Zaccardelli et al., (2020) for potential isolates that exert bio-control against soil-borne diseases. Researchers were able to isolate spore forming strains of *Bacillus amyloliquefaciens* and *Bacillus subtilis* from waste, which could act as soil fertility promoters and aid in plant diseases management. Nayak & Mukherjee, (2015) discussed the role of *Bacillus pumilus* (marine bacterium) in degrading numerous agro wastes within five weeks; the role of *Bacillus pumilus* and *Bacillus atrophaeus* in degradation and assimilation of lignin; the effectiveness of several *Bacillus spp* isolated from soil in the degradation of xylans/mannan and also the potential of *Bacillus licheniformis* and *Bacillus cereus* (soil bacteria) to exhibit higher pectinolytic activity.

Not only the above mentioned wastes, but also *Bacillus spp* mark a remarkable contribution in biomedical waste management. Generally biomedical wastes are incinerated, whose ash leads to heavy metals migration problem that eventually affects the soil texture and harms plant growth. Microorganisms serve as a great tool for reducing the toxicity of the incineration ash. In a study conducted by Heera et al., (2014), metal tolerant *Bacillus sp.* KGMDI reduced the alkalinity, hardness and heavy metals level of the biomedical waste incineration ash, which positively influenced the soil nature, groundwater and surface water. Thus, the role of *Bacillus spp.* in environment-friendly modern biotechnological approaches is exemplary, as they are adaptable and fast growing strains that produce essential metabolites. Their importance in solid waste management is flourishing with multiple benefits to the environment, especially soil.

FUTURE RESEARCH DIRECTIONS

The versatile nature of *Bacillus sp.,* makes them an important candidate for bioprocess and bioremediation. Although extensive research has been carried out to explore the potential of these bacteria in maintaining the sustainability of soil, still in-depth molecular studies is required to understand the relationship between various species of *Bacillus* in promoting plant growth and mitigation of environmental pollution. Proteomics, metabolomics and optimization studies are required to understand and enhance the potential of *Bacillus sp.,* to serve as the most promising candidate for soil bioremediation.

CONCLUSION

Soil serves as the backbone for sustainable development of every economy as it supports various sectors ranging from agriculture to therapeutics with its diverse microflora. Rapid urbanization and industrialization have clearly accelerated a stress on the physicochemical properties of soil resulting in various types of pollution. Bacterial bioremediation method employing *Bacillus sp.,* was found to have a promising role in maintaining the sustainability of soil. Bacterial mechanism of bioremediation has a direct and indirect role in maintaining the integrity of soil through production of soil friendly compounds like biopolymers or through enzyme production for mitigating toxic products. Future works and genetic modification of *Bacillus sp.,* can revolutionize their application in improving the quality of soil thereby driving towards sustainable development.

REFERENCES

Aftab, M. N., Hameed, A., & Ikram-ul-Haq, C. (2006). Biodegradation of leather waste by enzymatic treatment. *The Chinese Journal of Process Engineering*, *6*(3), 462–465.

Ahemad, M., & Kibret, M. (2014). Mechanisms and applications of plant growth promoting rhizobacteria: Current perspective. *Journal of King Saud University. Science*, *26*(1), 1–20. doi:10.1016/j.jksus.2013.05.001

Akaraonye, E., Moreno, C., Knowles, J. C., Keshavarz, T., & Roy, I. (2012). Poly(3-hydroxybutyrate) production by *Bacillus cereus* SPV using sugarcane molasses as the main carbon source. *Biotechnology Journal*, *7*(2), 293–303. doi:10.1002/biot.201100122 PMID:22147642

Aldor, I. S., & Keasling, J. D. (2003). Process design for microbial plastic factories: Metabolic engineering of polyhydroxyalkanoates. *Current Opinion in Biotechnology*, *14*(5), 475–483. doi:10.1016/j.copbio.2003.09.002 PMID:14580576

Ali, H. (2010). Biodegradation of Synthetic Dyes—A Review. *Water, Air, and Soil Pollution*, *213*(1-4), 251–273. doi:10.100711270-010-0382-4

Aranganathan, V., Kanimozhi, A. M., & Palvannan, T. (2013). Statistical optimization of synthetic azo dye (orange II) degradation by azoreductase from *Pseudomonas oleovorans* PAMD_1. *Preparative Biochemistry & Biotechnology*, *43*(7), 649–667. doi:10.1080/10826068.2013.772063 PMID:23768111

Arora, N. K. (Ed.). (2015). *Plant Microbes Symbiosis: Applied Facets*. Springer.

Awasthi, M. K., Wang, Q., Ren, X., Zhao, J., Huang, H., Awasthi, S. K., Lahori, A. H., Li, R., Zhou, L., & Zhang, Z. (2016). Role of biochar amendment in mitigation of nitrogen loss and greenhouse gas emission during sewage sludge composting. *Bioresource Technology*, *219*, 270–280. doi:10.1016/j.biortech.2016.07.128 PMID:27497088

Ayangbenro, A. S., & Babalola, O. O. (2017). A New Strategy for Heavy Metal Polluted Environments: A Review of Microbial Biosorbents. *International Journal of Environmental Research and Public Health*, *14*(1), 94. Advance online publication. doi:10.3390/ijerph14010094 PMID:28106848

Bafana, A., Chakrabarti, T., & Devi, S. S. (2008). Azoreductase and dye detoxification activities of *Bacillus velezensis* strain AB. *Applied Microbiology and Biotechnology*, *77*(5), 1139–1144. doi:10.100700253-007-1212-5 PMID:18034237

Bafana, A., Devi, S. S., & Chakrabarti, T. (2011). Azo dyes: past, present and the future. *Environmental Review*, *19*, 350–371.

Bagade, A., Nandre, V., Paul, D., Patil, Y., Sharma, N., Giri, A., & Kodam, K. (2020). Characterisation of hyper tolerant *Bacillus firmus* L-148 for arsenic oxidation. *Environmental Pollution*, *261*, 1–13. doi:10.1016/j.envpol.2020.114124 PMID:32078878

Baldiris, R., Acosta-Tapia, N., Montes, A., Hernández, J., & Vivas-Reyes, R. (2018). Reduction of Hexavalent Chromium and Detection of Chromate Reductase (ChrR) in *Stenotrophomonas maltophilia*. *Molecules (Basel, Switzerland)*, *23*(2), 1–20. doi:10.3390/molecules23020406 PMID:29438314

Basheer, T., & Umesh, M. (2018). Valorization of Tannery Solid Waste Materials Using Microbial Techniques: Microbes in Tannery Solid Waste Management. In *Handbook of Research on Microbial Tools for Environmental Waste Management* (pp. 127–145). IGI Global. doi:10.4018/978-1-5225-3540-9.ch007

Bonartsev, A. P., Bonartseva, G. A., Reshetov, I. V., Kirpichnikov, M. P., & Shaitan, K. V. (2019). Application of Polyhydroxyalkanoates in Medicine and the Biological Activity of Natural Poly(3-Hydroxybutyrate). *Acta Naturae*, *11*(2), 4–16. doi:10.32607/20758251-2019-11-2-4-16 PMID:31413875

Borah, B., Thakur, P. S., & Nigam, J. N. (2002). The influence of nutritional and environmental conditions on the accumulation of poly-beta-hydroxybutyrate in *Bacillus mycoides* RLJ B-017. *Journal of Applied Microbiology*, *92*(4), 776–783. doi:10.1046/j.1365-2672.2002.01590.x PMID:11966920

Borriss, R. (2011). Use of Plant-Associated Bacillus Strains as Biofertilizers and Biocontrol Agents in Agriculture. In D. K. Maheshwari (Ed.), *Bacteria in Agrobiology: Plant Growth Responses* (pp. 41–76). Springer Berlin Heidelberg. doi:10.1007/978-3-642-20332-9_3

Calvo, P., Ormeño-Orrillo, E., Martínez-Romero, E., & Zúñiga, D. (2010). Characterization of *Bacillus* isolates of potato rhizosphere from andean soils of Peru and their potential PGPR characteristics. *Brazilian Journal of Microbiology*, *41*(4), 899–906. doi:10.1590/S1517-83822010000400008 PMID:24031569

Cephidian, A., Makhdoumi, A., Mashreghi, M., & Mahmudy Gharaie, M. H. (2016). Removal of anthropogenic lead pollutions by a potent *Bacillus* species AS2 isolated from geogenic contaminated site. *International Journal of Environmental Science and Technology*, *13*(9), 2135–2142. doi:10.100713762-016-1023-2

Chandra, D., & General, T. Nisha, & Chandra, S. (2019). Microorganisms: an asset for decontamination of soil. In Smart Bioremediation Technologies (pp. 319–345). Elsevier.

Chandra, R., & Chowdhary, P. (2015). Properties of bacterial laccases and their application in bioremediation of industrial wastes. *Environmental Science. Processes & Impacts*, *17*(2), 326–342. doi:10.1039/C4EM00627E PMID:25590782

Chaturvedi, M. K. (2011). Studies on chromate removal by chromium-resistant *Bacillus sp.* isolated from tannery effluent. *The Journal of Electronic Publishing: JEP*, *2*(1), 76–82.

Chee, J. Y., Lakshmanan, M., Jeepery, I. F., Hairudin, N. H. M., & Sudesh, K. (2019). The potential application of *Cupriavidus necator* as polyhydroxyalkanoates producer and single cell protein: A review on scientific, cultural and religious perspectives. *Applied Food Biotechnology*, *6*(1), 19–34.

Chen, G.-Q. (2009). A microbial polyhydroxyalkanoates (PHA) based bio- and materials industry. *Chemical Society Reviews*, *38*(8), 2434–2446. doi:10.1039/b812677c PMID:19623359

Cheung, K. H., & Gu, J.-D. (2007). Mechanism of hexavalent chromium detoxification by microorganisms and bioremediation application potential: A review. *International Biodeterioration & Biodegradation*, *59*(1), 8–15. doi:10.1016/j.ibiod.2006.05.002

Chumthong, A., Kanjanamaneesathian, M., Pengnoo, A., & Wiwattanapatapee, R. (2008). Water-soluble granules containing *Bacillus megaterium* for biological control of rice sheath blight: Formulation, bacterial viability and efficacy testing. *World Journal of Microbiology & Biotechnology*, *24*(11), 2499–2507. doi:10.100711274-008-9774-7

da Costa, A. C. A., & Duta, F. P. (2001). Bioaccumulation of copper, zinc, cadmium and lead by *Bacillus sp., Bacillus cereus, Bacillus sphaericus* and *Bacillus subtilis. Brazilian Journal of Microbiology*, *32*(1), 1–5. doi:10.1590/S1517-83822001000100001 PMID:30637653

Das, N., & Chandran, P. (2011). Microbial degradation of petroleum hydrocarbon contaminants: An overview. *Biotechnology Research International*, *2011*, 1–14. doi:10.4061/2011/941810 PMID:21350672

Das, P., Sinha, S., & Mukherjee, S. K. (2014). Nickel Bioremediation Potential of *Bacillus thuringiensis* KUNi1 and Some Environmental Factors in Nickel Removal. *Bioremediation Journal*, *18*(2), 169–177. doi:10.1080/10889868.2014.889071

Dash, H. R., & Das, S. (2015). Bioremediation of inorganic mercury through volatilization and biosorption by transgenic *Bacillus cereus* BW-03(pPW-05). *International Biodeterioration & Biodegradation*, *103*, 179–185. doi:10.1016/j.ibiod.2015.04.022

Desai, C., Jain, K., & Madamwar, D. (2008). Evaluation of in vitro Cr(VI) reduction potential in cytosolic extracts of three indigenous *Bacillus sp.* isolated from Cr(VI) polluted industrial landfill. *Bioresource Technology*, *99*(14), 6059–6069. doi:10.1016/j.biortech.2007.12.046 PMID:18255287

Ding, L., Cheng, J., Xia, A., Jacob, A., Voelklein, M., & Murphy, J. D. (2016). Co-generation of biohydrogen and biomethane through two-stage batch co-fermentation of macro- and micro-algal biomass. *Bioresource Technology*, *218*, 224–231. doi:10.1016/j.biortech.2016.06.092 PMID:27371795

Dixit, R., Wasiullah, Malaviya, D., Pandiyan, K., Singh, U., Sahu, A., Shukla, R., Singh, B., Rai, J., Sharma, P., Lade, H., & Paul, D. (2015). Bioremediation of Heavy Metals from Soil and Aquatic Environment: An Overview of Principles and Criteria of Fundamental Processes. *Sustainability: Science Practice and Policy*, *7*(2), 2189–2212. doi:10.3390u7022189

Elangovan, R., Abhipsa, S., Rohit, B., Ligy, P., & Chandraraj, K. (2006). Reduction of Cr(VI) by a *Bacillus sp. Biotechnology Letters*, *28*(4), 247–252. doi:10.100710529-005-5526-z PMID:16555008

Elumalai, P., Lim, J.-M., Park, Y.-J., Cho, M., Shea, P. J., & Oh, B.-T. (2020). Agricultural waste materials enhance protease production by *Bacillus subtilis* B22 in submerged fermentation under blue light-emitting diodes. *Bioprocess and Biosystems Engineering*, *43*(5), 821–830. doi:10.100700449-019-02277-5 PMID:31919603

Fang, L., Huang, Q., Wei, X., Liang, W., Rong, X., Chen, W., & Cai, P. (2010). Microcalorimetric and potentiometric titration studies on the adsorption of copper by extracellular polymeric substances (EPS), minerals and their composites. *Bioresource Technology*, *101*(15), 5774–5779. doi:10.1016/j.biortech.2010.02.075 PMID:20227874

Ghosh, P. K., Maiti, T. K., Pramanik, K., Ghosh, S. K., Mitra, S., & De, T. K. (2018). The role of arsenic resistant *Bacillus aryabhattai* MCC3374 in promotion of rice seedlings growth and alleviation of arsenic phytotoxicity. *Chemosphere*, *211*, 407–419. doi:10.1016/j.chemosphere.2018.07.148 PMID:30077937

Glick, B. R. (2012). Plant growth-promoting bacteria: Mechanisms and applications. *Scientifica*, *2012*, 1–16. doi:10.6064/2012/963401 PMID:24278762

Gomes, H. I., Dias-Ferreira, C., & Ribeiro, A. B. (2013). Overview of in situ and ex situ remediation technologies for PCB-contaminated soils and sediments and obstacles for full-scale application. *The Science of the Total Environment*, *445-446*, 237–260. doi:10.1016/j.scitotenv.2012.11.098 PMID:23334318

Gowda, V., & Shivakumar, S. (2014). Agrowaste-based Polyhydroxyalkanoate (PHA) production using hydrolytic potential of *Bacillus thuringiensis* IAM 12077. *Brazilian Archives of Biology and Technology*, *57*(1), 55–61. doi:10.1590/S1516-89132014000100009

Guan, Z.-B., Luo, Q., Wang, H.-R., Chen, Y., & Liao, X.-R. (2018). Bacterial laccases: Promising biological green tools for industrial applications. *Cellular and Molecular Life Sciences: CMLS*, *75*(19), 3569–3592. doi:10.100700018-018-2883-z PMID:30046841

Guauque-Torres, M. P., & Bustos, A. Y. (2019). Laccases for Soil Bioremediation. In A. Kumar & S. Sharma (Eds.), *Microbes and Enzymes in Soil Health and Bioremediation* (Vol. 16, pp. 165–209). Springer Singapore. doi:10.1007/978-981-13-9117-0_8

Heera, S., Kunal, & Rajor, A. (2014). Bacterial treatment and metal characterization of biomedical waste ash. *Journal of Waste Management*, *2014*, 1–7. doi:10.1155/2014/956316

Hiraishi, T., & Taguchi, S. (2013). Protein Engineering of Enzymes Involved in Bioplastic Metabolism. In T. Ogawa (Ed.), *Protein Engineering - Technology and Application* (pp. 133–165). InTech. doi:10.5772/55552

Hu, B., Chen, S., Hu, J., Xia, F., Xu, J., Li, Y., & Shi, Z. (2017). Application of portable XRF and VNIR sensors for rapid assessment of soil heavy metal pollution. *PLoS One*, *12*(2), 1–13. doi:10.1371/journal.pone.0172438 PMID:28234944

Igiri, B. E., Okoduwa, S. I. R., Idoko, G. O., Akabuogu, E. P., Adeyi, A. O., & Ejiogu, I. K. (2018). Toxicity and Bioremediation of Heavy Metals Contaminated Ecosystem from Tannery Wastewater: A Review. *Journal of Toxicology*, *2018*, 1–17. doi:10.1155/2018/2568038 PMID:30363677

Järup, L. (2003). Hazards of heavy metal contamination. *British Medical Bulletin*, *68*(1), 167–182. doi:10.1093/bmb/ldg032 PMID:14757716

Jin, X.-C., Liu, G.-Q., Xu, Z.-H., & Tao, W.-Y. (2007). Decolorization of a dye industry effluent by *Aspergillus fumigatus* XC6. *Applied Microbiology and Biotechnology*, *74*(1), 239–243. doi:10.100700253-006-0658-1 PMID:17086413

Joardar, J. C., & Rahman, M. M. (2018). Poultry feather waste management and effects on plant growth. *International Journal of Recycling of Organic Waste in Agriculture*, *7*(3), 183–188. doi:10.100740093-018-0204-z

John Sundar, V., Gnanamani, A., Muralidharan, C., Chandrababu, N. K., & Mandal, A. B. (2011). Recovery and utilization of proteinous wastes of leather making: A review. *Reviews in Environmental Science and Biotechnology*, *10*(2), 151–163. doi:10.100711157-010-9223-6

Kalavathy, S. (2004). The Multidisciplinary nature of environmental studies. *Environmental Studies, 1.*

Kalme, S., Jadhav, S., Jadhav, M., & Govindwar, S. (2009). Textile dye degrading laccase from *Pseudomonas desmolyticum* NCIM 2112. *Enzyme and Microbial Technology*, *44*(2), 65–71. doi:10.1016/j.enzmictec.2008.10.005

Kaminsky, L. M., Trexler, R. V., Malik, R. J., Hockett, K. L., & Bell, T. H. (2019). The Inherent Conflicts in Developing Soil Microbial Inoculants. In Trends in Biotechnology (Vol. 37, Issue 2, pp. 140–151). doi:10.1016/j.tibtech.2018.11.011

Karigar, C. S., & Rao, S. S. (2011). Role of microbial enzymes in the bioremediation of pollutants: A review. *Enzyme Research*, *2011*, 805187. doi:10.4061/2011/805187 PMID:21912739

Khan, S., Cao, Q., Zheng, Y. M., Huang, Y. Z., & Zhu, Y. G. (2008). Health risks of heavy metals in contaminated soils and food crops irrigated with wastewater in Beijing, China. *Environmental Pollution*, *152*(3), 686–692. doi:10.1016/j.envpol.2007.06.056 PMID:17720286

Khosravi-Darani, K., & Bucci, D. Z. (2015). Application of poly (hydroxyalkanoate) in food packaging: Improvements by nanotechnology. *Chemical and Biochemical Engineering Quarterly*, *29*(2), 275–285. doi:10.15255/CABEQ.2014.2260

Kirpichtchikova, T. A., Manceau, A., Spadini, L., Panfili, F., Marcus, M. A., & Jacquet, T. (2006). Speciation and solubility of heavy metals in contaminated soil using X-ray microfluorescence, EXAFS spectroscopy, chemical extraction, and thermodynamic modeling. *Geochimica et Cosmochimica Acta*, *70*(9), 2163–2190. doi:10.1016/j.gca.2006.02.006

Kjerstadius, H., Saraiva, A. B., & Spångberg, J. (2016). Can source separation increase sustainability of sanitation management. *Impact on Nutrient Recovery, Climate Change and Eutrophication of Two Sanitation Systems for a Hypothetical Urban Area in Southern Sweden Using Life Cycle Assessment (opublicerat).* https://va-tekniksodra.se/2016/10/can-source-separation-increase-sustainability-of-sanitation-management/

Koller, M. (2014). Poly (hydroxyalkanoates) for food packaging: Application and attempts towards implementation. *Applied Food Biotechnology, 1*(1), 3-15.

Kourmentza, C., Plácido, J., Venetsaneas, N., Burniol-Figols, A., Varrone, C., Gavala, H. N., & Reis, M. A. M. (2017). Recent Advances and Challenges towards Sustainable Polyhydroxyalkanoate (PHA) Production. *Bioengineering (Basel, Switzerland)*, *4*(2), 55. Advance online publication. doi:10.3390/bioengineering4020055 PMID:28952534

Kudlich, M., Keck, A., Klein, J., & Stolz, A. (1997). Localization of the enzyme system involved in anaerobic reduction of azo dyes by *Sphingomonas* sp. strain BN6 and effect of artificial redox mediators on the rate of azo dye reduction. *Applied and Environmental Microbiology*, *63*(9), 3691–3694. doi:10.1128/AEM.63.9.3691-3694.1997 PMID:16535698

Lee, S. Y. (1996a). Bacterial polyhydroxyalkanoates. *Biotechnology and Bioengineering*, *49*(1), 1–14. doi:10.1002/(SICI)1097-0290(19960105)49:1<1::AID-BIT1>3.0.CO;2-P PMID:18623547

Lee, S. Y. (1996b). Plastic bacteria? Progress and prospects for polyhydroxyalkanoate production in bacteria. *Trends in Biotechnology*, *14*(11), 431–438. doi:10.1016/0167-7799(96)10061-5

Lemes, A. P., Montanheiro, T. L. A., Passador, F. R., & Durán, N. (2015). Nanocomposites of polyhydroxyalkanoates reinforced with carbon nanotubes: chemical and biological properties. In V. K. Thakur & M. K. Thakur (Eds.), *Eco-friendly Polymer Nanocomposites: Processing and Properties* (pp. 79–108). Springer India. doi:10.1007/978-81-322-2470-9_3

Li, C., Zhou, K., Qin, W., Tian, C., Qi, M., Yan, X., & Han, W. (2019). A review on heavy metals contamination in soil: Effects, sources, and remediation techniques. *Soil and Sediment Contamination: An International Journal*, *28*(4), 380–394. doi:10.1080/15320383.2019.1592108

Li, M.-H., Gao, X.-Y., Li, C., Yang, C.-L., Fu, C.-A., Liu, J., Wang, R., Chen, L.-X., Lin, J.-Q., Liu, X.-M., Lin, J.-Q., & Pang, X. (2020). Isolation and identification of chromium reducing *Bacillus Cereus* species from chromium-contaminated soil for the biological detoxification of chromium. *International Journal of Environmental Research and Public Health*, *17*(6), 1–13. doi:10.3390/ijerph17062118 PMID:32209989

Lin, Y.-F., & Aarts, M. G. M. (2012). The molecular mechanism of zinc and cadmium stress response in plants. *Cellular and Molecular Life Sciences: CMLS*, *69*(19), 3187–3206. doi:10.100700018-012-1089-z PMID:22903262

Liu, L., Li, W., Song, W., & Guo, M. (2018). Remediation techniques for heavy metal-contaminated soils: Principles and applicability. *The Science of the Total Environment*, *633*, 206–219. doi:10.1016/j.scitotenv.2018.03.161 PMID:29573687

Malusá, E., Sas-Paszt, L., & Ciesielska, J. (2012). Technologies for beneficial microorganisms inocula used as biofertilizers. *TheScientificWorldJournal*, *2012*, 1–13. doi:10.1100/2012/491206 PMID:22547984

Mary Mangaiyarkarasi, M. S., Vincent, S., Janarthanan, S., Subba Rao, T., & Tata, B. V. R. (2011). Bioreduction of Cr(VI) by alkaliphilic *Bacillus subtilis* and interaction of the membrane groups. *Saudi Journal of Biological Sciences*, *18*(2), 157–167. doi:10.1016/j.sjbs.2010.12.003 PMID:23961119

Masindi, V., & Muedi, K. L. (2018). Environmental Contamination by Heavy Metals. In H. E.-D. M. Saleh & R. F. Aglan (Eds.), *Heavy Metals*. InTech. doi:10.5772/intechopen.76082

Masood, F., & Malik, A. (2011). Hexavalent chromium reduction by *Bacillus sp.* strain FM1 isolated from heavy-metal contaminated soil. *Bulletin of Environmental Contamination and Toxicology*, *86*(1), 114–119. doi:10.100700128-010-0181-z PMID:21181113

Mathew, B. B., Biju, V. G., & Nideghatta Beeregowda, K. (2019). Accumulation of lead (Pb II) metal ions by *Bacillus toyonensis* SCE1 species, innate to industrial-area ground water and nanoparticle synthesis. *Applied Nanoscience*, *9*(1), 49–66. doi:10.100713204-018-0892-8

Medfu Tarekegn, M., Zewdu Salilih, F., & Ishetu, A. I. (2020). Microbes used as a tool for bioremediation of heavy metal from the environment. *Cogent Food & Agriculture*, *6*(1), 74. doi:10.1080/23311932.2020.1783174

Misal, S. A., Lingojwar, D. P., Shinde, R. M., & Gawai, K. R. (2011). Purification and characterization of azoreductase from alkaliphilic strain *Bacillus badius*. *Process Biochemistry*, *46*(6), 1264–1269. doi:10.1016/j.procbio.2011.02.013

Mishra, A., & Malik, A. (2013). Recent Advances in Microbial Metal Bioaccumulation. *Critical Reviews in Environmental Science and Technology, 43*(11), 1162–1222. doi:10.1080/10934529.2011.627044

Miyatake, M., & Hayashi, S. (2011). Characteristics of Arsenic Removal by Bacillus cereus Strain W2. Resources Processing, 58(3), 101–107. doi:10.4144/rpsj.58.101

Mogharabi, M., & Faramarzi, M. A. (2014). Laccase and laccase-mediated systems in the synthesis of organic compounds. *Advanced Synthesis & Catalysis, 356*(5), 897–927. doi:10.1002/adsc.201300960

Mohapatra, S., Maity, S., Dash, H. R., Das, S., Pattnaik, S., Rath, C. C., & Samantaray, D. (2017). Bacillus and biopolymer: Prospects and challenges. In Biochemistry and Biophysics Reports (Vol. 12, pp. 206–213). doi:10.1016/j.bbrep.2017.10.001

Mohapatra, S., Sarkar, B., Samantaray, D. P., Daware, A., Maity, S., Pattnaik, S., & Bhattacharjee, S. (2017). Bioconversion of fish solid waste into PHB using *Bacillus subtilis* based submerged fermentation process. *Environmental Technology, 38*(24), 3201–3208. doi:10.1080/09593330.2017.1291759 PMID:28162048

Morozova, O. V., Shumakovich, G. P., Shleev, S. V., & Yaropolov, Y. I. (2007). Laccase-mediator systems and their applications: A review. *Applied Biochemistry and Microbiology, 43*(5), 523–535. doi:10.1134/S0003683807050055 PMID:18038679

Mosa, K. A., Saadoun, I., Kumar, K., Helmy, M., & Dhankher, O. P. (2016). Potential Biotechnological Strategies for the Cleanup of Heavy Metals and Metalloids. *Frontiers in Plant Science, 7*, 1–14. doi:10.3389/fpls.2016.00303 PMID:27014323

Nayak, S., & Mukherjee, A. K. (2015). Management of Agricultural Wastes Using Microbial Agents. In Singh & Sarkar (Eds.), Waste Management: Challenges, Threats and Opportunities (pp. 65-91). Nova Scientific.

Novinscak, A., Surette, C., Allain, C., & Filion, M. (2008). Application of molecular technologies to monitor the microbial content of biosolids and composted biosolids. *Water Science and Technology: A Journal of the International Association on Water Pollution Research, 57*(4), 471–477.

Ojumu, T. V., Yu, J., & Solomon, B. O. (2004). Production of Polyhydroxyalkanoates, a bacterial biodegradable polymer. In African Journal of Biotechnology (Vol. 3, Issue 1, pp. 18–24). doi:10.5897/AJB2004.000-2004

Okereke, J. N., Ogidi, O. I., & Obasi, K. O. (2016). Environmental and health impact of industrial wastewater effluents in Nigeria-A Review. *International Journal of Advanced Research in Biological Sciences, 3*(6), 55-67.

Okino-Delgado, C. H., Zanutto-Elgui, M. R., do Prado, D. Z., Pereira, M. S., & Fleuri, L. F. (2019). Enzymatic Bioremediation: Current Status, Challenges of Obtaining Process, and Applications. In P. Arora (Ed.), *Microbial Metabolism of Xenobiotic Compounds. Microorganisms for Sustainability* (pp. 79–101). Springer. doi:10.1007/978-981-13-7462-3_4

Onakpa, M. M., Njan, A. A., & Kalu, O. C. (2018). A Review of Heavy Metal Contamination of Food Crops in Nigeria. *Annals of Global Health, 84*(3), 488–494. doi:10.29024/aogh.2314 PMID:30835390

Pan, I., Dam, B., & Sen, S. K. (2012). Composting of common organic wastes using microbial inoculants. *Biotech, 2*(2), 127–134.

Pande, V., Pandey, S. C., Joshi, T., Sati, D., Gangola, S., Kumar, S., & Samant, M. (2019). Biodegradation of toxic dyes: a comparative study of enzyme action in a microbial system. In *Smart Bioremediation Technologies* (pp. 255–287). Elsevier. doi:10.1016/B978-0-12-818307-6.00014-7

Paul, T., Mandal, A., & Mondal, K. C. (2018). *Waste to value aided fertilizer: an alternative cleaning technique for poultry feathers waste disposal.* Ann Microbiol Immunol.

Peixoto, R. S., Vermelho, A. B., & Rosado, A. S. (2011). Petroleum-degrading enzymes: Bioremediation and new prospects. *Enzyme Research, 2011*, 475193. doi:10.4061/2011/475193 PMID:21811673

Penkhrue, W., Jendrossek, D., Khanongnuch, C., Pathom-Aree, W., Aizawa, T., Behrens, R. L., & Lumyong, S. (2020). Response surface method for polyhydroxybutyrate (PHB) bioplastic accumulation in *Bacillus drentensis* BP17 using pineapple peel. *PLoS One*, *15*(3), 1–21. doi:10.1371/journal.pone.0230443 PMID:32191752

Pereira, L., Coelho, A. V., Viegas, C. A., Santos, M. M. C. D., Robalo, M. P., & Martins, L. O. (2009). Enzymatic biotransformation of the azo dye Sudan Orange G with bacterial CotA-laccase. *Journal of Biotechnology*, *139*(1), 68–77. doi:10.1016/j.jbiotec.2008.09.001 PMID:18938200

Philip, S., Keshavarz, T., & Roy, I. (2007). Polyhydroxyalkanoates: Biodegradable polymers with a range of applications. *Journal of Chemical Technology and Biotechnology (Oxford, Oxfordshire)*, *82*(3), 233–247. doi:10.1002/jctb.1667

Piotrowska-Długosz, A. (2017). The use of enzymes in bioremediation of soil xenobiotics. In M. Z. Hashmi, V. Kumar, & A. Varma (Eds.), *Xenobiotics in the Soil Environment* (Vol. 49, pp. 243–265). Springer International Publishing. doi:10.1007/978-3-319-47744-2_17

Preethi, K., Anand, M., & Thazeem, B. (2015). Isolation and identification of keratinolytic bacteria from tannery effluent: A study on their biodegradative and dehairing activity. *International Journal of Multidisciplinary Research and Development*, *2*(10), 227–234.

Preethi, K., & Vineetha, U. M. (2015). Water hyacinth: A potential substrate for bioplastic (PHA) production using *Pseudomonas aeruginosa. International Journal of Applied Research in Veterinary Medicine*, *1*(11), 349–354.

Pricelius, S., Held, C., Murkovic, M., Bozic, M., Kokol, V., Cavaco-Paulo, A., & Guebitz, G. M. (2007). Enzymatic reduction of azo and indigoid compounds. *Applied Microbiology and Biotechnology*, *77*(2), 321–327. doi:10.100700253-007-1165-8 PMID:17891390

Priester, J. H., Olson, S. G., Webb, S. M., Neu, M. P., Hersman, L. E., & Holden, P. A. (2006). Enhanced exopolymer production and chromium stabilization in Pseudomonas putida unsaturated biofilms. *Applied and Environmental Microbiology*, *72*(3), 1988–1996. doi:10.1128/AEM.72.3.1988-1996.2006 PMID:16517647

Priyanka, K., Umesh, M., Thazeem, B., & Preethi, K. (2020). Polyhydroxyalkanoate biosynthesis and characterization from optimized medium utilizing distillery effluent using *Bacillus endophyticus* MTCC 9021: A statistical approach. *Biocatalysis and Biotransformation*, 1–13. doi:10.1080/10242422.2020.1789112

Prusty, J. S., Rath, B. P., & Thatoi, H. (2019). Production optimization and application of extracellular chromate reductase from *Bacillus sp.* for bioremediation of hexavalent Chromium. In R. Kundu, R. Narula, R. Paul, & S. Mukherjee (Eds.), *Environmental Biotechnology For Soil and Wastewater Implications on Ecosystems* (Vol. 36, pp. 103–108). Springer Singapore. doi:10.1007/978-981-13-6846-2_13

Pugazhendhi, A., Ranganathan, K., & Kaliannan, T. (2018). biosorptive removal of copper(ii) by *Bacillus cereus* isolated from contaminated soil of electroplating industry in India. *Water, Air, and Soil Pollution*, *229*(3), 1780. doi:10.100711270-018-3734-0

Puzon, G. J., Roberts, A. G., Kramer, D. M., & Xun, L. (2005). Formation of soluble organo-chromium(III) complexes after chromate reduction in the presence of cellular organics. *Environmental Science & Technology*, *39*(8), 2811–2817. doi:10.1021/es048967g PMID:15884380

Raddadi, N., Crotti, E., Rolli, E., Marasco, R., Fava, F., & Daffonchio, D. (2012). The most important Bacillus species in biotechnology. In Bacillus thuringiensis Biotechnology (pp. 329–345). doi:10.1007/978-94-007-3021-2_17

Radhakrishnan, R., Hashem, A., & Abd Allah, E. F. (2017). Bacillus: A biological tool for crop improvement through bio-molecular changes in adverse environments. *Frontiers in Physiology*, *8*, 1–14. doi:10.3389/fphys.2017.00667 PMID:28932199

Ramadas, N. V., Soccol, C. R., & Pandey, A. (2010). A statistical approach for optimization of polyhydroxybutyrate production by *Bacillus sphaericus* NCIM 5149 under submerged fermentation using central composite design. *Applied Biochemistry and Biotechnology*, *162*(4), 996–1007. doi:10.100712010-009-8807-5 PMID:19812909

Rastogi, M., Nandal, M., & Khosla, B. (2020). Microbes as vital additives for solid waste composting. *Heliyon*, *6*(2), 1–11. doi:10.1016/j.heliyon.2020.e03343 PMID:32095647

Ravindran, B., Kumar, A. G., Bhavani, P. S. A., & Sekaran, G. (2011). Solid-state fermentation for the production of alkaline protease by *Bacillus cereus* 1173900 using proteinaceous tannery solid waste. *Current Science*, *100*(5), 726–730.

Rehm, B. H. A. (2003). Polyester synthases: Natural catalysts for plastics. *The Biochemical Journal*, *376*(1), 15–33. doi:10.1042/bj20031254 PMID:12954080

Rehman, Z. U., Ali, H. H., & Akbar, A. (2020). Production of polyhydroxybutyrate (PHB) from soil bacterium (*Bacillus megaterium* TISTR 1814) with Cantaloupe waste extract as potential carbon source. *Pak-Euro Journal of*. https://readersinsight.net/PJMLS/article/view/1293

Reiss, R., Ihssen, J., & Thöny-Meyer, L. (2011). *Bacillus pumilus* laccase: A heat stable enzyme with a wide substrate spectrum. *BMC Biotechnology*, *11*(1), 1–11. doi:10.1186/1472-6750-11-9 PMID:21266052

Sadh, P. K., Duhan, S., & Duhan, J. S. (2018). Agro-industrial wastes and their utilization using solid state fermentation: A review. *Bioresources and Bioprocessing*, *5*(1), 1–15. doi:10.118640643-017-0187-z

Sandana Mala, J. G., Sujatha, D., & Rose, C. (2015). Inducible chromate reductase exhibiting extracellular activity in *Bacillus methylotrophicus* for chromium bioremediation. *Microbiological Research*, *170*, 235–241. doi:10.1016/j.micres.2014.06.001 PMID:24985094

Saranya, K., Sundaramanickam, A., Shekhar, S., & Swaminathan, S. (2019). Biosorption of mercury by *Bacillus thuringiensis* (CASKS3) isolated from mangrove sediments of southeast coast India. *Indian Journal of Geo-Marine Sciences*, *48*(2), 143–150.

Saratale, R. G., Saratale, G. D., Chang, J. S., & Govindwar, S. P. (2011). Bacterial decolorization and degradation of azo dyes: A review. *Journal of the Taiwan Institute of Chemical Engineers*, *42*(1), 138–157. doi:10.1016/j.jtice.2010.06.006

Schisler, D. A., Slininger, P. J., Behle, R. W., & Jackson, M. A. (2004). Formulation of *Bacillus spp.* for biological control of plant diseases. *Phytopathology*, *94*(11), 1267–1271. doi:10.1094/PHYTO.2004.94.11.1267 PMID:18944465

Senthil Kumar, P., & Gunasundari, E. (2018). Bioremediation of Heavy Metals. In S. J. Varjani, A. K. Agarwal, E. Gnansounou, & B. Gurunathan (Eds.), *Bioremediation: Applications for Environmental Protection and Management* (Vol. 201–202, pp. 165–195). Springer Singapore. doi:10.1007/978-981-10-7485-1_9

Shafi, J., Tian, H., & Ji, M. (2017). *Bacillus* species as versatile weapons for plant pathogens: A review. *Biotechnology, Biotechnological Equipment*, *31*(3), 446–459. doi:10.1080/13102818.2017.1286950

Sharma, B., Dangi, A. K., & Shukla, P. (2018). Contemporary enzyme based technologies for bioremediation: A review. *Journal of Environmental Management*, *210*, 10–22. doi:10.1016/j.jenvman.2017.12.075 PMID:29329004

Shedbalkar, U., Dhanve, R., & Jadhav, J. (2008). Biodegradation of triphenylmethane dye cotton blue by *Penicillium ochrochloron* MTCC 517. *Journal of Hazardous Materials*, *157*(2-3), 472–479. doi:10.1016/j.jhazmat.2008.01.023 PMID:18282658

Sidhu, G. P. S. (2016). Heavy metal toxicity in soils: Sources, remediation technologies and challenges. *Advances in Plants & Agriculture Research*, *5*(1), 445–446.

Singh, R. (2013). Exploring Flavin as Catalyst for the Remediation of Halogenated Compounds. In *New and Future Developments in Catalysis*. doi:10.1016/B978-0-444-53870-3.00015-0

Singh, R. L., Singh, P. K., & Singh, R. P. (2015). Enzymatic decolorization and degradation of azo dyes – A review. *International Biodeterioration & Biodegradation*, *104*, 21–31. doi:10.1016/j.ibiod.2015.04.027

Singh, R. P., Singh, P. K., & Singh, R. L. (2014). Bacterial decolorization of textile azo dye acid orange by *Staphylococcus hominis* RMLRT03. *Toxicology International*, *21*(2), 160–166. doi:10.4103/0971-6580.139797 PMID:25253925

Stamenković, S., Beškoski, V., Karabegović, I., Lazić, M., & Nikolić, N. (2018). Microbial fertilizers: A comprehensive review of current findings and future perspectives. *Spanish Journal of Agricultural Research, 16*(1), 1–18. doi:10.5424jar/2018161-12117

Stamenković-Stojanović, S., Karabegović, I., Beškoski, V., Nikolić, N., & Lazić, M. (2019). Bacillus based microbial formulations: Optimization of the production process. *Hemijska Industrija, 73*(3), 169–182. doi:10.2298/HEMIND190214014S

Stamenkovic-Stojanovic, S., Karabegovic, I., Beskoski, V., Nikolic, N., & Lazic, M. (2019). *Bacillus* based microbial formulations: Optimization of the production process. *Chemistry & Industry, 73*(3), 169–182. doi:10.2298/HEMIND190214014S

Tanamool, V., Imai, T., Danvirutai, P., & Kaewkannetra, P. (2013). An alternative approach to the fermentation of sweet sorghum juice into biopolymer of poly-β-hydroxyalkanoates (PHAs) by newly isolated, *Bacillus aryabhattai* PKV01. *Biotechnology and Bioprocess Engineering; BBE, 18*(1), 65–74. doi:10.100712257-012-0315-8

Telke, A. A., Ghodake, G. S., Kalyani, D. C., Dhanve, R. S., & Govindwar, S. P. (2011). Biochemical characteristics of a textile dye degrading extracellular laccase from a *Bacillus* sp. ADR. *Bioresource Technology, 102*(2), 1752–1756. doi:10.1016/j.biortech.2010.08.086 PMID:20855194

Telke, A. A., Kalyani, D. C., Dawkar, V. V., & Govindwar, S. P. (2009). Influence of organic and inorganic compounds on oxidoreductive decolorization of sulfonated azo dye C.I. Reactive Orange 16. *Journal of Hazardous Materials, 172*(1), 298–309. doi:10.1016/j.jhazmat.2009.07.008 PMID:19640646

Teng, Y., & Chen, W. (2019). Soil Microbiomes—a Promising Strategy for Contaminated Soil Remediation: A Review. *Pedosphere, 29*(3), 283–297. doi:10.1016/S1002-0160(18)60061-X

Thatoi, H., Das, S., Mishra, J., Rath, B. P., & Das, N. (2014). Bacterial chromate reductase, a potential enzyme for bioremediation of hexavalent chromium: A review. *Journal of Environmental Management, 146*, 383–399. doi:10.1016/j.jenvman.2014.07.014 PMID:25199606

Thazeem, B., Beryl, G. P., & Umesh, M. (2017). A comparative study on alkaline protease production from *Bacillus spp.* and their biodegradative, dehairing and destaining activity. *International Journal of Academic Research and Development, 2*(2), 74–79.

Thazeem, B., Preethi, K., & Umesh, M. (2015). Characterization and fermentative utilization of tannery fleshings using *Lactobacillus plantarum. International Journal of Recent Scientific Research, 6*(3), 3037–3041.

Thazeem, B., Preethi, K., Umesh, M., & Radhakrishnan, S. (2018). Nutritive characterization of delimed bovine tannery fleshings for their possible use as a proteinaceous aqua feed ingredient. *Waste and Biomass Valorization, 9*(8), 1289–1301. doi:10.100712649-017-9922-0

Thazeem, B., Umesh, M., Mani, V. M., Beryl, G. P., & Preethi, K. (2020). Biotransformation of bovine tannery fleshing into utilizable product with multifunctionalities. *Biocatalysis and Biotransformation*, 1–19. doi:10.1080/10242422.2020.1786071

Thazeem, B., Umesh, M., & Vikas, O. V. (2016a). Bioconversion of poultry feather into feather meal using proteolytic *Bacillus* Species – A comparative study. *International Journal of Advances in Scientific Research, 1*(1), 14–16.

Thazeem, B., Umesh, M., & Vikas, O. V. (2016b). Isolation and identification of proteolytic *Bacillus pumilus* from tannery lime effluent and its dehairing activity. *International Journal of Academic Research and Development, 1*(2), 5–7.

Umesh, M., Mani, V. M., Thazeem, B., & Preethi, K. (2018). Statistical optimization of process parameters for bioplastic (PHA) Production by *Bacillus subtilis* NCDC0671 using orange peel-based medium. *Iranian Journal of Science and Technology. Transaction A, Science, 42*(4), 1947–1955. doi:10.100740995-017-0457-9

Umesh, M., & Preethi, K. (2017). Fabrication of antibacterial bioplastic sheet using orange peel medium and its antagonistic effect against common clinical pathogens. *Research Journal of Biotechnology, 12*(7), 67–74.

Umesh, M., Priyanka, K., Thazeem, B., & Preethi, K. (2017). Production of Single Cell Protein and Polyhydroxyalkanoate from *Carica papaya* Waste. *Arabian Journal for Science and Engineering, 42*(6), 2361–2369. doi:10.100713369-017-2519-x

Umesh, M., Priyanka, K., Thazeem, B., & Preethi, K. (2018). Biogenic PHA nanoparticle synthesis and characterization from Bacillus subtilis NCDC0671 using orange peel medium. *International Journal of Polymeric Materials and Polymeric Biomaterials, 67*(17), 996–1004. doi:10.1080/00914037.2017.1417284

Umesh, M., & Thazeem, B. (2019). Biodegradation Studies of Polyhydroxyalkanoates extracted from *Bacillus Subtilis* NCDC 0671. *Research Journal of Chemistry and Environment, 23*(6), 107–114.

Umsakul, K., Dissara, Y., & Srimuang, N. (2010). Chemical, physical and microbiological changes during composting of the water hyacinth. *Pakistan Journal of Biological Sciences, 13*(20), 985–992. doi:10.3923/pjbs.2010.985.992 PMID:21319457

Upadhyay, N., Vishwakarma, K., Singh, J., Mishra, M., Kumar, V., Rani, R., Mishra, R. K., Chauhan, D. K., Tripathi, D. K., & Sharma, S. (2017). Tolerance and Reduction of Chromium(VI) by *Bacillus sp.* MNU16 Isolated from Contaminated Coal Mining Soil. *Frontiers in Plant Science, 8*, 778. doi:10.3389/fpls.2017.00778 PMID:28588589

Üreyen Esertaş, Ü. Z., Uzunalioğlu, E., Güzel, Ş., Bozdeveci, A., & Alpay Karaoğlu, Ş. (2020). Determination of bioremediation properties of soil-borne *Bacillus sp.* 5O5Y11 and its effect on the development of *Zea mays* in the presence of copper. *Archives of Microbiology, 202*(7), 1817–1829. doi:10.100700203-020-01900-4 PMID:32440759

Varjani, S. J. (2017). Microbial degradation of petroleum hydrocarbons. *Bioresource Technology, 223*, 277–286. doi:10.1016/j.biortech.2016.10.037 PMID:27789112

Verma, N., & Sharma, R. (2017). Bioremediation of Toxic Heavy Metals: A Patent Review. *Recent Patents on Biotechnology, 11*(3), 171–187. doi:10.2174/1872208311666170111111631 PMID:28078980

Verma, S., & Kuila, A. (2019). Bioremediation of heavy metals by microbial process. *Environmental Technology & Innovation*, *14*, 100369. doi:10.1016/j.eti.2019.100369

Voběrková, S., Vaverková, M. D., Burešová, A., Adamcová, D., Vršanská, M., Kynický, J., Brtnický, M., & Adam, V. (2017). Effect of inoculation with white-rot fungi and fungal consortium on the composting efficiency of municipal solid waste. *Waste Management (New York, N.Y.)*, *61*, 157–164. doi:10.1016/j.wasman.2016.12.039 PMID:28065548

Wang, C. L., Zhao, M., Wei, X. D., Li, T. L., & Lu, L. (2010). Characteristics of spore-bound laccase from *Bacillus subtilis* WD23 and its use in dye decolorization. *Advanced Materials Research*, *113–116*, 226–230. . doi:10.4028/www.scientific.net/AMR.113-116.226

Whiteley, C. G., & Lee, D.-J. (2006). Enzyme technology and biological remediation. *Enzyme and Microbial Technology*, *38*(3-4), 291–316. doi:10.1016/j.enzmictec.2005.10.010

Wong, M. H. (2012). *Environmental Contamination: Health Risks and Ecological Restoration*. CRC Press. doi:10.1201/b12531

Wu, J., Kamal, N., Hao, H., Qian, C., Liu, Z., Shao, Y., Zhong, X., & Xu, B. (2019). Endophytic *Bacillus megaterium* BM18-2 mutated for cadmium accumulation and improving plant growth in Hybrid *Pennisetum*. *Biotechnology Reports (Amsterdam, Netherlands)*, *24*, 1–9. doi:10.1016/j.btre.2019.e00374 PMID:31763195

Wuana, R. A., & Okieimen, F. E. (2011). Heavy Metals in Contaminated Soils: A Review of Sources, Chemistry, Risks and Best Available Strategies for Remediation. *International Scholarly Research Network Ecology*, *2011*, 1–20. doi:10.5402/2011/402647

Yilmaz, E. I., & Ensari, N. Y. (2005). Cadmium biosorption by *Bacillus circulans* strain EB1. *World Journal of Microbiology & Biotechnology*, *21*(5), 777–779. doi:10.100711274-004-7258-y

Zaccardelli, M., Sorrentino, R., Caputo, M., Scotti, R., De Falco, E., & Pane, C. (2020). Stepwise-selected *Bacillus amyloliquefaciens* and *B. subtilis* strains from composted aromatic plant waste able to control soil-borne diseases. *Collection FAO: Agriculture*, *10*(2), 1–15. doi:10.3390/agriculture10020030

Zaidi, S., Usmani, S., Singh, B. R., & Musarrat, J. (2006). Significance of *Bacillus subtilis* strain SJ-101 as a bioinoculant for concurrent plant growth promotion and nickel accumulation in *Brassica juncea*. *Chemosphere*, *64*(6), 991–997. doi:10.1016/j.chemosphere.2005.12.057 PMID:16487570

Zerdani, I., Faid, M., & Malki, A. (2004). Digestion of solid tannery wastes by strains of *Bacillus sp.* isolated from compost in morocco. *International Journal of Agriculture and Biology*, *6*(5), 758–761.

KEY TERMS AND DEFINITIONS

Bio Augmentation: Introduction of site-specific or exogenic microbes in the polluted site to expedite the bio-degradation rate of contaminants.

Bio Precipitation: It takes place either in cell surface or inside the microbes through enzymes or secondary metabolites making it sparingly soluble for bioremediation.

Bio-Sorption: Biosorption is the passive mode of decontaminating pollutants wherein live or dead microbial matrix was used as sorbents to adsorb heavy metals on the surface of biomass and linking with exo-polysaccharide.

Bioaccumulation: Pollutants accumulate not only on the surface of the cell but also inside the cell as they are metabolically active and provide many binding sites. Rate of accumulation depends on the sensitivity of microbes as this mechanism is toxicokinetic beyond defined limits.

Bioleaching: The solubility of sulfides forms of heavy metals were enhanced by secretion of organic acids or complex substances to flexibly eliminate metals from polluted sites.

Biotransformation: The heavy metal toxicity is lowered via catalytic reactions like oxidation or reduction to facilitate the availability.

Chapter 10
Potassium Solubilizing Bacteria: An Insight

Anandkumar Naorem
https://orcid.org/0000-0002-4632-0662
ICAR-Central Arid Zone Research Institute, India

Shiva Kumar Udayana
Krishi Vigyan Kendra, India

Sachin Patel
ICAR-Central Arid Zone Research Institute, India

ABSTRACT

Potassium (K) is one of the essential nutrients required for plants. Although the total pool of K in the soil is generally large, the bioavailable portion is meager. There are several mechanisms through which the insoluble K can be made available through soil microbes called "potassium solubilizing bacteria" or KSB. They play an important role in increasing the solubility of K for proper crop establishment under potassium deficient soils through the production of organic and inorganic acids, acidolysis, polysaccharides, complexolysis, chelation, and exchange reactions. Moreover, they also produce specific exopolysaccharides and biofilm that enhances the weathering of the K-rich minerals and increase the K concentration in the soil solution. Hence, the production and management of biological fertilizers containing KSB can be an effective alternative to chemical fertilizers. This chapter presents the underlying mechanisms and their role in providing sufficient K to the crops.

INTRODUCTION

There has been an increasing human population projected to reach 8.9 billion by 2050 (Wood, 2001). Therefore, to feed the burgeoning population, it is a significant challenge to increase agricultural productivity from gradually shrinking cultivable land. Moreover, climate change, desertification, and environmental pollution add to the present concerns in agricultural production. On the other hand,

DOI: 10.4018/978-1-7998-7062-3.ch010

researchers are seeking eco-friendly methods or techniques that can sustain agricultural productivity without jeopardizing the environment. Soil-plant-microbe interaction is one of the most complex yet essential relationships explored for sustainable agricultural production. After nitrogen and phosphorus, potassium (K) is the third most important major essential element having a pivotal role in the metabolism, growth, and development of plants (Zia-ul-hassan and Arshad, 2010). K^+ activates some critical enzymes that are involved in plant metabolism. K^+ plays a crucial role in several metabolic processes, including photosynthesis, accumulation of sugars, and plant growth (Zhao et al., 2001; Wang et al., 2012). Sufficient levels of K in plants improve the shelf life and quality of the crop harvest. Low levels of plant K restrict root growth, affecting the uptake of water and nutrients from the soil (Mengel and Kirkby, 2001; White, 2003). K is also vital for proper seed development and maintaining the quality of fiber crops (Akhtar et al., 2003; White and Karley, 2010). It has a massive role in increasing the water use efficiency of the crops, which is highly essential in dryland areas with frequent water scarcity. It is involved in the metabolism of organic acids, fats, carbohydrates, protein synthesis, and increasing resistance to abiotic stress, including drought and frost (Rehm and Schmitt, 2002). K deficiency in plants limits crop growth and development, leading to lesser crop yield and low-quality harvest (White and Karley, 2010). Insufficient K levels in plants increase the plant's susceptibility to several diseases and insect attacks (Armengaud et al., 2010). Therefore, it is essential to study the K pool in the soil and the availability of each pool to the plants and soil microbes. This chapter will emphasize the role of soil microorganisms in K solubilization in the soil and its associated mechanisms of action, prospects and concerns of using potassium solubilizing bacteria (KSB) in sustainable agriculture.

POTASSIUM DEFICIENCY: THE FORGOTTEN ELEMENT

Although K is the eighth-most abundant essential element on earth, a small portion of total K is present in soluble forms. Due to the introduction of high yielding varieties and intensive cropping system, K deficiency in soil is increasing worldwide. In intensive conventional agricultural practice, K fertilizers' application is the common practice of supplying K to the plants through the soil system. Potassic fertilizers are relatively costlier than other inorganic fertilizers that increase production costs (Kumar et al., 2015). Most of the farmers either neglect using K fertilizer or apply K fertilizers in significant amounts, leading to a massive loss of inputs (Mohammadi and Sohrabi, 2012). As K deficiency symptoms are not as conspicuous as other major essential elements, the supplementation of lost K from the soil is mostly ignored by the farmers (Panday et al., 2018). As a result, most commercial crops are reported with K deficiency (Xiao et al., 2017). K deficiency has been increasingly reported in Southeast, Oceania and Africa due to limited K-rich mineral resources. The cost of K fertilizers has been increasing every year, and the heavy use of K fertilizers can degrade the soil quality and pollute the environment.

FACTORS AFFECTING POTASSIUM DYNAMICS IN THE SOIL

K is the most abundant cation in plant cells (White, 2003) and constitutes approximately 2.6% of the earth's crust. Naturally, igneous and sedimentary rocks are rich in K. Mineral soil K concentration could range from 0.04-3% in the lithosphere (Syers, 2003). The surface soil can contain 3000 to 1000000 kg K per hectare of soil. Around 98% of the total K is found in non-exchangeable form, and the remaining 2%

is soluble and is available for plant uptake (Figure 1). The non-exchangeable form of K is firmly bound in negatively charged interlayers of the mineral, and it is moderately to sparsely available for plant uptake governed by the existing soil conditions. As there is an equilibrium between available and unavailable pools of K in the soil system, these trapped K^+ is released when the K level in the soil solution decreases through plant uptake or leaching or erosion related losses. However, the release of the exchangeable and non-exchangeable K from the minerals depends on several factors (Jackson, 1964):

1. Weathering of the K-rich minerals such as feldspars and mica;
2. K levels in other pools;
3. Physical and chemical characteristics of K-bearing minerals;
4. Physicochemical properties of soil system such as soil reaction, electrical conductivity, soil aeration, temperature etc.

The increasing order of pools that generates higher bioavailable K is mineral K < fixed K < exchangeable K < solution K (Sparks, 2000). The K in soil solution is the only pool from which the plant and the microorganisms can take up K. In addition to it, solution K is susceptible to several losses such as leaching, runoff, erosion and fixation.

Figure 1. The potassium cycle in soil depicting the equilibrium maintained between the pools. The underlying processes are shown in italics.

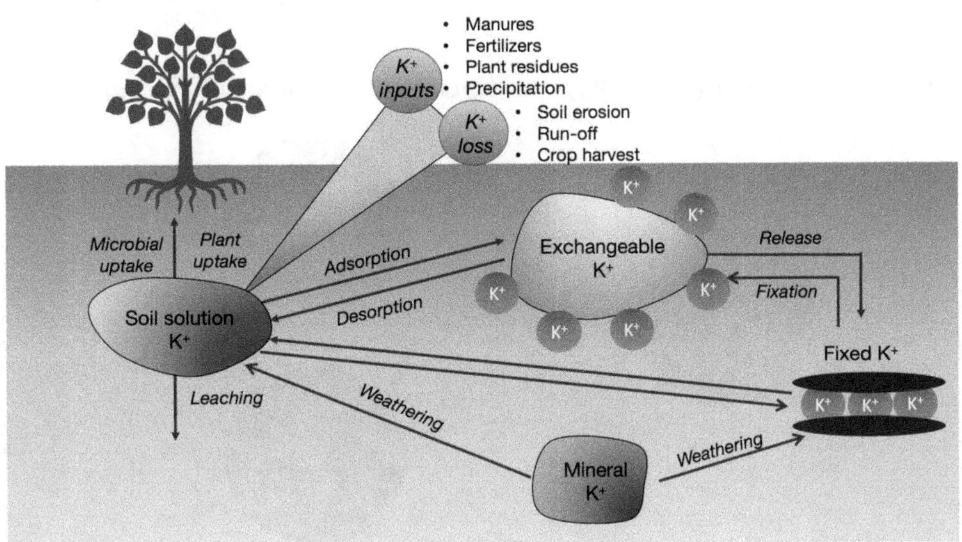

THE IMPORTANCE OF POTASSIUM SOLUBILIZING BACTERIA

Soil rhizosphere is teeming with enormous, diverse microbial groups of several classes that are developed through functional relationships. However, establishing a specific microbial group in the rhizosphere depends upon soil, plant and environmental factors (Verma et al., 2014). Several microbes that are capable of performing plant growth-promoting activities colonize the rhizosphere. This beneficial group of bacteria is commonly known as "Plant growth-promoting rhizobacteria", out of which K solubilizers are one of them (Khan et al., 2007). There are certain beneficial bacteria in the soil system that can help solubilize the K from the minerals and are referred to as "Potassium solubilizing bacteria (KSB)" (Maurya et al., 2014). These beneficial soil bacteria can convert the insoluble form of K to the available K in the soil (Requena et al., 1997; Khan et al., 2007; Zeng et al., 2012; Abhilash et al., 2013). KSB also plays a crucial role in K cycling in the soil system (Diep and Hieu, 2013). Among the soil microbiota, bacteria is one of the active members in K solubilization. There is a corpus of scientific evidence on soil bacteria's plant growth-promoting abilities (Parmar and Sindhu, 2013). Some examples of KSBs are *Bacillus, Burkholderia, Acidithiobacillus, Paenibacillus, Pseudomonas*, etc. (Bennett et al., 1998; Rajawat et al., 2012; Zeng et al., 2012; Syed and Patel, 2014), among which *Bacillus mucilaginosus* and *B. circulanscan* have been often reported as effective K solubilizers (Lian et al., 2002; Meena et al., 2016). *B. mucilaginosus* could release 4.29 mg K per liter in media enriched with muscovite mica (Sugumaran and Janarthanam, 2007). Therefore, KSBs are isolated from different types of soils and are studied *in-vitro* for their K-solubilizing abilities (Mirminachi et al., 2002; Sheng et al., 2008; Basak and Biswas, 2012). The identification of effective KSB strains is the first step in understanding the K-solubilization *via* soil microbes. There is a substantial KSB population in the soil rhizosphere, comprising both aerobic and non-aerobic strains of KSB. Due to more favourable microclimate and easy availability of food and energy sources, KSBs are found in larger numbers in the rhizosphere than in the non-rhizosphere zone (Padma and Sukumar, 2015). Generally, KSBs are isolated by serial dilution plate method, and they are grown in a suitable growth medium, keeping the pH around neutral (pH=7). The K-solubilizing capacity is identified by developing halo zones around the bacterial colonies (Rajawat et al., 2016) (Figure 2).

Figure 2. A conventional method of detecting potassium solubilizing capacity of a culturable soil bacteria

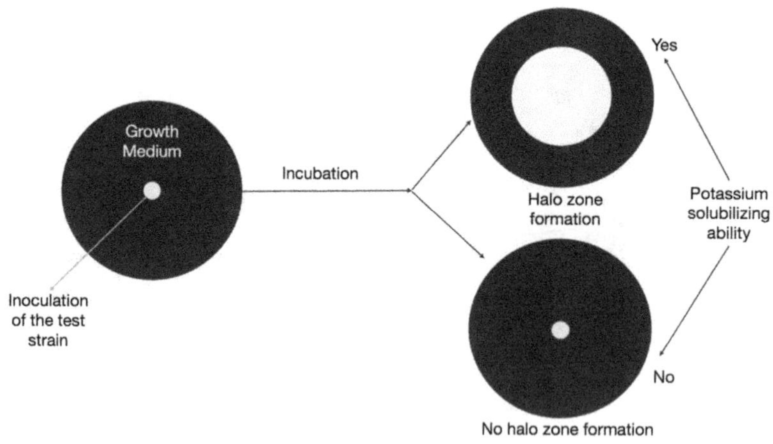

MECHANISMS OF POTASSIUM SOLUBILIZATION IN THE SOILS

Although there is a myriad of literature on K solubilizing through beneficial soil microorganisms, very little information is available on the mechanism of microbial induced K solubilization. However, the microbes involved in K solubilization utilizes unique mechanisms, including redox reactions and organic acids' production, to enhance the weathering of minerals (Uroz et al., 2009). There are several mechanisms through which KSB solubilizes K from K-bearing minerals. Although fungi and bacteria are widely studied for K solubilizing capacity, bacteria take the central role in K solubilization in many cases. Based on the mechanisms followed by these bacteria, there are four main ways through which mineral K can be solubilized microbially such as (Lian et al., 2008; Parmar and Sindhu, 2013; Meena et al., 2015) (Figure 3):

1. Direct process of K solubilization;
2. Indirect process of K solubilization;
3. Secretion of exopolysaccharides and
4. Production of biofilm on mineral surfaces

Figure 3. The underlying mechanisms of potassium solubilization through beneficial soil microbes

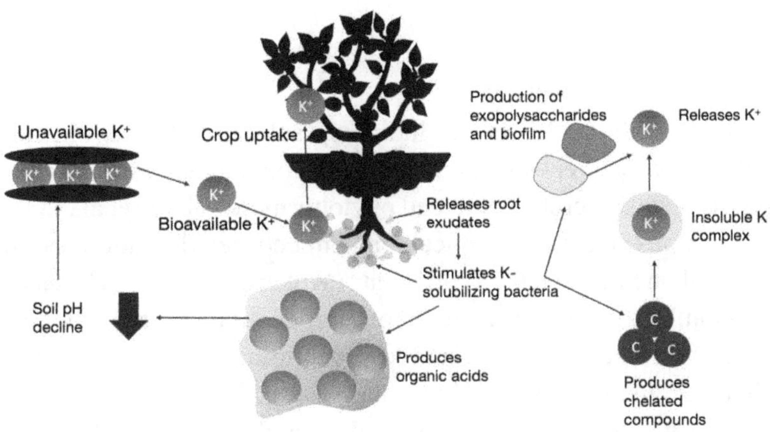

Direct Mechanism: Production of Organic Acids

Among all these mechanisms mentioned above, the most prominent is the production of organic acids. Under the direct method of K solubilization, the K solubilizers produce organic acids (Goldsetin, 1994; Zarjani et al., 2013), enhance rhizospheric acidolysis (Basak et al., 2016) and carbonic acid-mediated weathering (Han et al., 2006). The production of organic acids, coupled with H^+ ions, lowers down the soil pH of the surrounding area (Basak et al., 2016). Lower soil pH increases the bioavailability of several other nutrients such as Fe, Mg, and K, etc. Additionally, the microbial respiration and microbial decomposition of organic substrates release carbonic acid and enhance the chemical weathering of the minerals *via* the proton-promoted dissolution process (Meena et al., 2014). Tartaric acid is one such organic acid that can release from minerals (Zarjani et al., 2013). The possible ways of organic acid mediated chemical weathering are (Lian et al., 2008):

1. Organic acids adhere to the surfaces of K-bearing mineral and extract K from the mineral particles through electron transfer;
2. Organic acids destroy the oxygen links in the minerals;
3. Organic acids also chelate ions in the soil solution.

Silicate-solubilizing bacteria produce several types of organic acids and release K from the insoluble K complex. Some of the organic acids produced by KSBs are oxalic acids, α-ketogluconic acid, lactic acid, malonic acid, citric acid, propionic acid, fumaric acid, succinic acid (Saiyad et al., 2015). The organic acids lower the soil pH, resulting in K discharge into the soil solution (Goldsetin, 1994). The release of acids by KSBs also increases the chelation of K and acidolysis of rhizospheric minerals. Low rhizospheric pH increases protonation and further enhances the acidolysis process (Zarjani et al., 2013). KSB can also act on phlogopite K-bearing minerals and improve the weathering rate of these minerals. KSB helps in aluminum chelation and solubilization of phlogopite minerals' crystal framework by producing organic acids (Abou-el-Seoud and Abdel-Megeed, 2012). One such example of KSB capable of enhanced weathering of phlogopite minerals is *B. altitudinis* (Huang et al., 2013). The H^+ ions produced by KSBs displace K^+ and other essential ions from the minerals, such as Ca^{2+} and Fe^{2+} (Huang et al., 2013).

Indirect Methods

Chelation

In addition to these direct methods, KSBs can indirectly solubilize K-rich minerals by chelating the cations bound to K silicate or by direct attachment of KSBs on the mineral surfaces or producing metal complexing ligands or releasing microbial-induced phytohormones (Uroz et al., 2009). The chelation of ions in the soil solution creates an electrochemical gradient between the cation and the anion concentrations, thereby indirectly enhancing the K dissolution rate (Welch et al., 2002). *B. mucilaginosus* have been reported as an active K-solubilizer through triple actions: production of carboxylic acids, polysaccharides (Malinovskaya et al., 1990; Lin et al., 2002), and low weight molecular ligands that could improve K solubilization from feldspar and muscovite (Sheng and Huang, 2002).

The weathering of phlogopite minerals is increased through acid dissolution or by the Al chelation process of the crystal structure. The KSBs react with Al and Si of the K-minerals and form organic-metal complexes, due to which the trapped K is released into the soil solution (Romheld and Kirkby, 2010). This chelation mechanism of KSBs is similar to that of EDTA that releases trapped K (Prakash and Verma, 2016). In addition to organic acids, KSB also produces high molecular weight organic ligands and polymers that form complexes with the mineral ions and weaken the metal-oxygen bonding (Basak et al., 2016). The ligand-ion complex formation affects the saturation state of the solution. KSB can produce slime layers and certain enzymes around the mineral surface and enhance the ion diffusion from the mineral surface. These microbially produced exopolysaccharides contain –COO- groups that can efficiently form a complex with mineral ions, alter the solution's saturation state and increase K solubilization. These organic compounds further manipulate the surrounding pH and increase K solubilization rate (Welch et al., 2002).

Production of Extracellular Polymers

KSB also produces extracellular polymers that can increase K-solubilization from the K-bearing minerals (Sheng and He, 2006). These polymers help to attach KSBs in the mineral structure (Welch and Vandevivere, 1994). Polymers such as exopolysaccharides can form complexes with the mineral framework ions and increase feldspars dissolution rate. KSB also produces biofilms that create a favorable microenvironment around the microbial cells and enhance the weathering of K-bearing minerals. Biofilm formation around the aluminosilicate minerals increases the residence time of water and improves the weathering process. Biofilms not only accelerates the weathering process but also regulates the denudation losses. It also promotes the corrosion of K-bearing shales and releases K (Man et al., 2014).

Exopolysaccharide Production

Organic acids readily adsorb the capsular exopolysaccharides produced by KSBs. Therefore, when these polymers are attached to the mineral surface of the K-bearing minerals, it also delivers a significant amount of organic acids that further increase the K-solubilization (Liu et al., 2012). Some examples of KSBs that can exude capsular exopolysaccharides are *Thiobacillus, Bacillus* and *Clostridium*. The inoculation of these KSBs in the soil showed higher microbial-mediated solubilization of K from feldspars and illite (Sheng and He, 2006). The chemical composition of the extracellular polysaccharides governs the K-solubilization. Higher protein content in the microbial exopolysaccharides results in increased formation of the microbe-mineral complex, through which the KSBS releases organic acids and lowers the pH of the surrounding soil, ultimately increasing K-solubilization.

Moreover, the exopolysaccharides bind solution K^+ and induce the K equilibrium to supply more K from the exchangeable and non-exchangeable form. On the other hand, the inoculation of KSBs in K-deficient soils increases the release of K into the soil solution (Meena et al., 2016). This is because the beneficial bacteria, in search of energy sources, act on K-minerals, and the microbes' interaction with the minerals lead to K-solubilization.

Biofilm Formation

KSBs also produce biofilms, but they are lesser recognized than other mechanisms (Sangeeth et al., 2012). The plant-microbe interaction in biofilm-induced K-solubilization is relatively simple. The bacterial cells of KSBs are attached to the mineral surfaces with the biofilm produced by KSBs. The biofilm is composed of proteins, DNA, and extracellular polysaccharides. The biofilm favors the growth of other beneficial bacteria and thereby increases the microbial population of KSBs around the mineral surfaces of the K-rich minerals (Nagaraju et al., 2017). Similar to other mechanisms, biofilm formation enhances the release of organic acids and lowers soil rhizospheric pH that further helps in K dissolution (Balogh-Brunstad et al., 2008). Simply, biofilm creates a suitable environment for the growth and development of a diverse population of soil microbes that help in K-solubilization as a microbial consortium. The microbes present in this zone are protected by the production of exopolysaccharides (Gadd, 2007).

BENEFICIAL SOIL FUNGI IN POTASSIUM SOLUBILIZATION

K-solubilization is also carried out by a group of soil fungi such as *Aspergillus* and *Fusarium*. Like bacteria, fungi also produce organic acids and help in K- solubilization (Vassileva et al., 2000). Some examples of the organic acids released by soil fungi are gluconic acid, oxalic acid, and citric acid. The K-solubilizing capacity of these organic acids works differently on different types of K-bearing minerals. For example, oxalic acids could significantly release trapped K from feldspars, whereas tartaric acids in illite and gluconate (Argelis et al., 1993). Fungi also chelate the mineral elements and secrete polymers that can significantly enhance K-solubilization (Lian et al., 2008). With the help of hyphae, fungi exert bio-physical forces and increase the weathering of K-minerals (Xia et al., 2008). Fungi release acids in the rhizosphere and help in the dissolution of silicate rocks (Leyval and Berthelin, 1989). Some reports on certain ectomycorrhizal fungi showed that they could exude unique organic compounds and form microscopic holes in the K-bearing minerals (Van Scholl et al., 2008). This process increases the susceptibility of the minerals to weathering and ultimately releases mineral K into the soil solution.

BENEFITS OF CROP INOCULATION WITH KSBs

Co-inoculation of *B. mucilaginosus* and *Azotobacter chroococcum* in Sudan grass (*Sorghum vulgar*) enhances the N and K availability, resulting in increased plant biomass and nutrient acquisition (Basak and Biswas, 2012). Inclusion of a KSB (*B. circulans*) in a microbial consortium of *Azotobacter, Azospirillum*, and *B. megaterium* improved the tuber size, total chlorophyll content, starch content, leaf area, and uptake of macronutrients in potato (*Solanum tuberosum*) (Abdel-Salam and Shams, 2012). Under the hydroponic system, K assimilation of maize (*Zea mays*) and wheat (*Triticum aestivum*) can be substantially improved with the inoculation of *B. mucilaginosus* (Singh et al., 2010). *Pseudomonas putida* was reported to enhance the quality of tea parameters such as flavour indices, caffeine content, colour, and theaflavin content (Bagyalakshmi et al., 2012). Under the field experiment, inoculation of crops with KSBs has shown significantly higher biomass yield and improved crop harvest quality (Girgis, 2006; Supanjani et al., 2006).

BIOTECHNOLOGICAL IMPLICATIONS FOR SUSTAINABLE AGRICULTURE

KSBs not only help in solubilizing of K and increasing the K availability to plants but also actively involve in stress mitigation. Crop yield fluctuates due to several biotic and abiotic factors. The incidence of pests and diseases and extreme temperature, salinity and droughts often lessen the agricultural productivity. The K nutrition can significantly influence the plant health by reducing the impact of these biotic and abiotic stresses. In most of the crops, proper K nutrition provides better protection of the crops from diseases and pests attack than the crops grown in K poor soils. Effects of K nutrition on disease incidence can vary and be affected by the type and amount of pathogens, the type of crop and environmental conditions. KSBs adjust the uptake and ratio of nutrients by the host plant and lessen the effect of the stress on the plant. It improves the nutritional status of the plant for proper growth. Inoculation of certain KSBs such as Arbuscular mycorrhizal symbiosis can regulate the uptake of sodium and chloride ions by the plants and decrease the effects of salinity on the plant health while improving the absorption of other macro and micro-nutrients needed for plant growth and disease resistance.

KSBs AS BIOFERTILIZERS

To maintain optimum plant growth and development, it is important to maintain soil fertility. The conventional intensive farming depletes the soil available nutrients and thus lead to decreased agricultural productivity. Therefore, soil fertility is enhanced through integrated use of chemical as well as organic products. The beneficial soil microorganisms such as KSBs can be used to reduce the intensive use of chemicals in agricultural production. Therefore, biofertilizers are often recommended as a part of fertilization regime in farming. KSB biofertilizers are becoming an important component of integrated nutrient management systems in sustainable form of agriculture. These biofertilizers are environmentally safer and cost-effective than the inorganic chemicals. KSBs, when applied as biofertilizers, not only solubilize K and increase the K availability in the soil, but also improve the resistance of the plant from abiotic and biotic stress through certain mechanisms. Some of the common KSBs widely used as biofertilizers are *B. mucilaginosus, Paenibacillus* spp., E. *hormaechei, B. circulans* etc. Han et al. (2006) reported that the sole application of rock phosphate and potassium did not enhance the soil phosphorus and potassium content, but the application of phosphate and K solubilizing bacteria increase P and K availability in the soil and improved the growth of the test crops (pepper and cucumber). Several other research trials also demonstrated the improved growth of the crops with the application of KSBs, for example application of *B. mucilaginosus* and *B. subtilis* as biofertilizers enhance maize growth (Abou-el-Seoud and Abdel-Megeed, 2012). KSBs can be inoculated with other beneficial microbes, subjected to the positive compatibility between the strains. The co-inoculation of KSBs with other PGPRs can perform several beneficial actions to improve plant growth and development. Application of KSB biofertilizers with other organic amendments can be useful in increasing the crop yield.

CONCLUSION AND FUTURE PERSPECTIVES

The K rich minerals are found in significant amount in the lithosphere that can be explored for fertilization for crops. The poor application of potassic fertilizers and the high cost of these fertilizers have restricted the increase in crop yield and an increasing fear of global potassium deficiency. The low K availability of these K-bearing minerals can be enhanced by using potassium solubilizing bacteria. It is an eco-friendly method in which certain beneficial bacteria could convert the unavailable form of K to bioavailable form of K. These KSBs have great potential in sustainable agriculture and need to be explored further to understand its underlying mechanisms of potassium solubilization in the soil. To date, most of the KSBs are isolated and studied using the growth culture medium. However, 1% of the total soil microbiota can be grown in-vitro. Therefore, the remaining 98% must be identified and studied for its K-solubilizing capacity. Culture-independent methods are being developed and employed in isolating and in-depth study of unculturable KSBs. In addition to it, several environmental factors influence the K-solubilization process that includes both soil and plant factors. These factors are rather complex and affect each other. Therefore, it is crucial to conduct experiments on the effect of these environmental, soil, and plant factors on the growth and development of KSBs. In addition to these, there is a need to check the compatibility of KSBs with other soil beneficial microbes to understand their relationships and increase its effectiveness. The most important is the gaps in the demonstration and transfer of technology that need to be filled in order to increase the use of KSB biofertilizers and reduce intensive use of chemicals in the soil.

FUNDING

This research received no specific grant from any funding agency in the public, commercial, or not-for-profit sectors.

REFERENCES

Abdel-Salam, M. A., & Shams, A. S. (2012). Feldspar-K fertilization of potato (*Solanum tuberosum* L.) augmented by biofertilizer. *Journal of Agriculture and Environmental Sciences*, *12*, 694–699.

Abhilash, P. C., Dubey, R. K., Tripathi, V., Srivastava, P., Verma, J. P., & Singh, H. B. (2013). Remediation and management of POPs-contaminated soils in a warming climate: Challenges and perspectives. *Environmental Science and Pollution Research International*, *20*(8), 5879–5885. doi:10.100711356-013-1808-5 PMID:23677754

Abou-el-Seoud, I. I., & Abdel-Megeed, A. (2012). Impact of rock materials and biofertilizations on P and K availability for maize (*Zea mays*) under calcareous soil conditions. *Saudi Journal of Biological Sciences*, *19*(1), 55–63. doi:10.1016/j.sjbs.2011.09.001 PMID:23961162

Akhtar, M. E., Sardar, A., Ashraf, M., Akhtar, M., & Khan, M. Z. (2003). Effect of potash application on seed cotton yield and yield components of selected cotton varieties–I. *Asian Journal of Plant Sciences*, *2*, 602–604. doi:10.3923/ajps.2003.602.604

Argelis, D. T., Gonzala, D. A., Vizcaino, C., & Gartia, M. T. (1993). Biochemical mechanism of stone alteration carried out by filamentous fungi living in monuments. *Biogeochemistry*, *19*, 129–147.

Armengaud, P., Breitling, R., & Amtmann, A. (2010). Coronatine-intensive 1 (COI1) mediates transcriptional responses of *Arabidopsis thaliana* to external potassium supply. *Molecular Plant*, *3*(2), 390–405. doi:10.1093/mpsq012 PMID:20339157

Bagyalakshmi, B., Ponmurugan, P., & Balamurugan, A. (2012). Impact of different temperature, carbon and nitrogen sources on solubilization efficiency of native potassium solubilizing bacteria from tea (*Camellia sinensis). Journal of Biological Research (Thessaloniki)*, *3*, 36–42.

Balogh-Brunstad, Z., Keller, C. K., Gill, R. A., Bormann, B. T., & Li, C. Y. (2008). The effect of bacteria and fungi on chemical weathering and chemical denudation fluxes in pine growth experiments. *Biogeochemistry*, *88*(2), 153–167. doi:10.100710533-008-9202-y

Basak, B., & Biswas, D. (2012). *Modification of waste mica for alternative source of potassium: evaluation of potassium release in soil from waste mica treated with potassium solubilizing bacteria (KSB)*. Lap Lambert Academic Publishing.

Basak, B. B., Sarkar, B., Biswas, D. R., Sarkar, S., Sanderson, P., & Naidu, R. (2016). Bio-intervention of naturally occurring silicate minerals for alternative source of potassium: Challenges and opportunities. *Advances in Agronomy*, *141*, 115–145. doi:10.1016/bs.agron.2016.10.016

Bennett, P. C., Choi, W. J., & Rogera, J. R. (1998). Microbial destruction of feldspars. *Mineral Management*, *8*(1), 149–150. doi:10.1180/minmag.1998.62A.1.79

Diep, C. N., & Hieu, T. N. (2013). Phosphate and potassium solubilizing bacteria from weathered materials of denatured rock mountain, Ha Tien, Kie^n Giang province Vietnam. *American Journal of Life Sciences*, *1*(3), 88–92. doi:10.11648/j.ajls.20130103.12

Gadd, G. M. (2007). Geomycology: Biogeochemical transformations of rocks, minerals, metals and radionuclides by fungi, bioweathering and bioremediation. *Mycological Research, 11*(1), 3–49. doi:10.1016/j.mycres.2006.12.001 PMID:17307120

Girgis, M. G. Z. (2006). Response of wheat to inoculation with phosphate and potassium mobilizers and organic amendment. *Annals of Agricultural Sciences Shams Univ Cairo, 51*(1), 85–100.

Goldstein, A. H. (1994). Involvement of the quinoprotein glucose dehydrogenase in the solubilization of exogenous mineral phosphates by gram-negative bacteria. In A. Torriani-Gorni, E. Yagil, & S. Silver (Eds.), *Phosphate in microorganisms: Cellular and molecular biology* (pp. 197–203). ASM Press.

Han, H. S., Supanjani, & Lee, K. D. (2006). Effect of co-inoculation with phosphate and potassium solubilizing bacteria on mineral uptake and growth of pepper and cucumber. *Plant, Soil and Environment, 52*(3), 130–136. doi:10.17221/3356-PSE

Huang, Z., He, L., Sheng, X., & He, Z. (2013). Weathering of potash feldspar by *Bacillus sp.* L11. Wei sheng wu xue bao. *Acta Microbiologica Sinica, 53*, 1172–1178. PMID:24617258

Jackson, M. L. (1964). Chemical composition of soils. In Chemistry of the Soil (pp. 71-141). Reinhold.

Khan, M. S., Zaidi, A., & Wani, P. A. (2007). Role of phosphate-solubilizing microorganisms in sustainable agriculture – a review. *Agronomy for Sustainable Development, 27*(1), 29–43. doi:10.1051/agro:2006011

Kumar, A., Bahadur, I., Maurya, B. R., Raghuwanshi, R., Meena, V. S., Singh, D. K., & Dixit, J. (2015). Does a plant growth promoting rhizobacteria enhance agricultural sustainability? *Journal of Pure & Applied Microbiology, 9*, 715–724.

Leyval, C., & Berthelin, J. (1989). Interactions between *Laccaria laccata, Agrobacterium radiobacter* and beech roots: Influence on P, K, Mg, and Fe mobilization from minerals and plant growth. *Plant and Soil, 117*(1), 103–110. doi:10.1007/BF02206262

Lian, B., Fu, P. Q., Mo, D. M., & Liu, C. Q. (2002). A comprehensive review of the mechanism of potassium release by silicate bacteria. *Acta Mineralogica Sinica, 22*, 179–182.

Lian, B., Wang, B., Pan, M., Liu, C., & Teng, H. H. (2008). Microbial release of potassium from K-bearing minerals by thermophilic fungus *Aspergillus fumigatus. Geochimica et Cosmochimica Acta, 72*(1), 87–98. doi:10.1016/j.gca.2007.10.005

Lin, Q., Rao, Z., Sun, Y., Yao, J., & Xing, L. (2002). Identification and practical application of silicate-dissolving bacteria. *Agricultural Sciences in China, 1*, 81–85.

Liu, D., Lian, B., & Dong, H. (2012). Isolation of *Paenibacillus sp.* and assessment of its potential for enhancing mineral weathering. *Geomicrobiology Journal, 29*(5), 413–421. doi:10.1080/01490451.2011.576602

Malinovskaya, I. M., Kosenko, L. V., Votselko, S. K., & Podgorskii, V. S. (1990). Role of *Bacillus mucilaginosus* polysaccharide in degradation of silicate minerals. *Microbiology, 59*, 49–55.

Man, L. Y., Cao, X. Y., & Sun, D. S. (2014). Effect of potassium-solubilizing bacteria-mineral contact mode on decomposition behavior of potassium-rich shale. *Zhongguo Youse Jinshu Xuebao, 24*, 48–52.

Maurya, B. R., Meena, V. S., & Meena, O. P. (2014). Influence of Inceptisol and Alfisol's potassium solubilizing bacteria (KSB) isolates on release of K from waste mica. *Vegetos*, *27*(1), 181–187. doi:10.5958/j.2229-4473.27.1.028

Meena, V. S., Maurya, B. R., & Bahadur, I. (2014). Potassium solubilization by bacterial strain in waste mica. *Bangladesh Journal of Botany*, *43*(2), 235–237. doi:10.3329/bjb.v43i2.21680

Meena, V. S., Maurya, B. R., Verma, J. P., Aeron, A., Kumar, A., Kim, K., & Bajpai, V. K. (2015). Potassium solubilizing rhizobacteria (KSR): Isolation, identification, and K-release dynamics from waste mica. *Ecological Engineering*, *81*, 340–347. doi:10.1016/j.ecoleng.2015.04.065

Meena, V. S., Maurya, B. R., Verma, J. P., & Meena, R. S. (2016). *Potassium solubilizing microorganisms for sustainable agriculture*. Springer India., doi:10.1007/978-81-322-2776-2

Mengel, K., & Kirkby, E. A. (2001). *Principles of plant nutrition* (5th ed.). Kluwer Acad Publishers. doi:10.1007/978-94-010-1009-2

Mirminachi, F., Zhang, A., & Roehr, M. (2002). Citric acid fermentation and heavy metal ions. *Acta Biotechnologica*, *22*(3-4), 363–373. doi:10.1002/1521-3846(200207)22:3/4<363::AID-ABIO363>3.0.CO;2-A

Mohammadi, K., & Sohrabi, Y. (2012). Bacterial biofertilizers for sustainable crop production: A review. *American Journal of Agricultural and Biological Sciences*, *7*, 307–316.

Nagaraju, Y., Triveni, S., Subhashreddy, R., & Jhansi, P. (2017). Biofilm formation of zinc solubilizing, potassium releasing bacteria on the surface of fungi. *International Journal of Current Microbiology and Applied Sciences*, *6*(4), 2037–2047. doi:10.20546/ijcmas.2017.604.241

Padma, S. D., & Sukumar, J. (2015). Response of mulber- ry to inoculation of potash mobilizing bacterial isolate and other bio-inoculants. *Global Journal of Bio-Science and BioTechnology*, *4*, 50–53.

Panday, S. C., Choudhary, M., Singh, S., Meena, V. S., Mahanta, D., Yadav, R. P., Pattanayak, A., & Bisht, J. K. (2018). Increasing farmer's income and water use efficiency as affected by long-term fertilization under a rainfed and supplementary irrigation in a soybean-wheat cropping system of Indian mid-Himalaya. *Field Crops Research*, *219*, 214–221. doi:10.1016/j.fcr.2018.02.004

Parmar, P., & Sindhu, S. S. (2013). Potassium solubilization by rhizosphere bacteria: Influence of nutritional and environmental conditions. *Journal of Microbiology Research (Rosemead, Calif.)*, *3*, 25–31.

Prakash, S., & Verma, J. P. (2016). Global perspective of potash for fertilizer production. In V. S. Meena, B. R. Maurya, J. P. Verma, & R. S. Meena (Eds.), *Potassium Solubilizing Microorganisms for Sustainable Agriculture* (pp. 327–331). Springer. doi:10.1007/978-81-322-2776-2_23

Rajawat, M. V. S., Singh, S., Singh, G., & Saxena, A. K. (2012). Isolation and characterization of K-solubilizing bacteria isolated from different rhizospheric soil. In *Proceeding of 53rd annual conference of association of microbiologists of India* (p. 124). Academic Press.

Rajawat, M. V. S., Singh, S., Tyagi, S. P., & Saxena, A. K. (2016). A modified plate assay for rapid screening of potassium-solubilizing bacteria. *Pedosphere*, *26*(5), 768–773. doi:10.1016/S1002-0160(15)60080-7

Rehm, G., & Schmitt, M. (2002). *Potassium for Crop Production*. University of Minnesota.

Requena, B. N., Jimenez, I., Toro, M., & Barea, J. M. (1997). Interactions between plant growth promoting rhizobacteria (PGPR), arbuscular mycorrhizal fungi and *Rhizobium spp.*in the rhizosphere of *Anthyllis cytiisoides*, a model legume for revegetation in Mediterranean semi-arid ecosystem. *The New Phytologist*, *136*(4), 667–677. doi:10.1046/j.1469-8137.1997.00786.x

Romheld, V., & Kirkby, E. A. (2010). Research on potassium in agriculture: Needs and prospects. *Plant and Soil*, *335*(1-2), 155–180. doi:10.100711104-010-0520-1

Saiyad, S. A., Jhala, Y. K., & Vyas, R. V. (2015). Comparative efficiency of five potash and phosphate solubilizing bacteria and their key enzymes useful for enhancing and improvement of soil fertility. *International Journal of Scientific Research*, *5*, 1–6.

Sangeeth, K. P., Bhai, R. S., & Srinivasan, V. (2012). *Paenibacillus glucanolyticus*, a promising potassium solubilizing bacterium isolated from black pepper (Piper nigrum L.) rhizosphere. *Journal of Spices and Aromatic Crops*, *21*(2), 118–124.

Sheng, X. F., & He, L. Y. (2006). Solubilization of potassium-bearing minerals by a wild-type strain of *Bacillus edaphicus* and its mutants and increased potassium uptake by wheat. *Canadian Journal of Microbiology*, *52*(1), 66–72. doi:10.1139/w05-117 PMID:16541160

Sheng, X. F., & Huang, W. Y. (2002). Mechanism of potassium release from feldspar affected by the strain NBT of silicate bacterium. *Turang Xuebao*, *39*, 863–871.

Sheng, X. F., Zhao, F., He, H., Qiu, G., & Chen, L. (2008). Isolation, characterization of silicate mineral solubilizing *Bacillus globisporus* Q12 from the surface of weathered feldspar. *Canadian Journal of Microbiology*, *54*(12), 1064–1068. doi:10.1139/W08-089 PMID:19096461

Singh, G., Biswas, D. R., & Marwaha, T. S. (2010). Mobilization of potassium from waste mica by plant growth promoting rhizobacteria and its assimilation by maize (*Zea mays*) and wheat (*Triticum aestivum* L.): A hydroponics study under phytotron growth chamber. *Journal of Plant Nutrition*, *33*(8), 1236–1251. doi:10.1080/01904161003765760

Sparks, D. L. (2000). Bioavailability of soil potassium. In M. E. Sumner (Ed.), *Handbook of soil science*. CRC Press.

Sugumaran, P., & Janarthanam, B. (2007). Solubilization of potassium containing minerals by bacteria and their effect on plant growth. *World Journal of Agricultural Sciences*, *3*, 350–355.

Supanjani, H. H. S., Jung, S. J., & Lee, K. D. (2006). Rock phosphate and potassium rock solubilizing bacteria as alternative sustainable fertilizers. *Agronomy for Sustainable Development*, *26*(4), 233–340. doi:10.1051/agro:2006020

Syed, B. A., & Patel, B. (2014). Investigation and correlation of soil biotic and abiotic factors affecting agricultural productivity in semi-arid regions of north Gujarat, India. *International Journal of Research Studies in Biosciences*, *2*, 18–29.

Syers, J. K. (2003). Potassium in soils: Current concepts. In *Proceedings of the IPI Golden Jubilee Congress 1952–2002 on Feed the soil to feed the people: The role of potash in sustainable agriculture*. (pp. 301–10). International Potash Institute.

Uroz, S., Calvaruso, C., Turpault, M. P., & Frey-Klett, P. (2009). Mineral weathering by bacteria: Ecology, actors and mechanisms. *Trends in Microbiology*, *17*(8), 378–387. doi:10.1016/j.tim.2009.05.004 PMID:19660952

Van Schöll, L., Kuyper, T. W., Smits, M. M., Landeweert, R., Hoffland, E., & Van Breemen, N. (2008). Rock-eating mycorrhizas: Their role in plant nutrition and biogeochemical cycles. *Plant and Soil*, *303*(1-2), 35–47. doi:10.100711104-007-9513-0

Vassileva, M., Azcon, R., Barea, J., & Vassilev, N. (2000). Rock phosphate solubilization by free and encapsulated cells of *Yarrowia lipolytica. Process Biochemistry*, *35*(7), 693–697. doi:10.1016/S0032-9592(99)00132-6

Verma, J. P., Yadav, J., Tiwari, K. N., & Jaiswal, D. K. (2014). Evaluation of plant growth promoting activities of microbial strains and their effect on growth and yield of chickpea (*Cicer arietinum* L.) in India. *Soil Biology & Biochemistry*, *70*, 33–37. doi:10.1016/j.soilbio.2013.12.001

Wang, N., Hua, H., Eneji, A. E., Li, Z., Duan, L., & Tian, X. (2012). Genotypic variation in photosynthetic and physiological adjustment to potassium deficiency in cotton (*Gossypium hirsutum*). *Journal of Photochemistry and Photobiology*, *110*, 1–8. doi:10.1016/j.jphotobiol.2012.02.002 PMID:22387141

Welch, S. A., Taunton, A. E., & Banfield, J. F. (2002). Effect of microorganisms and microbial metabolites on apatite dissolution. *Geophysical Journal of the Royal Astronomical Society*, *19*, 343–367.

Welch, S. A., & Vandevivere, P. (1994). Effect of microbial and other naturally occurring polymers on mineral dissolution. *Geomicrobiology Journal*, *12*(4), 227–238. doi:10.1080/01490459409377991

White, A. F. (2003). Natural weathering rates of silicate minerals. *Treatise on Geochemistry*, *5*, 133–168. doi:10.1016/B0-08-043751-6/05076-3

White, P. J., & Karley, A. J. (2010). "Potassium," in Plant Cell Monographs 17. In R. Hell & R. R. Mendel (Eds.), *Cell Biology of Metals and Nutrients* (pp. 199–224). Springer-Verlag. doi:10.1007/978-3-642-10613-2_9

Wood, N. (2001). Nodulation by numbers: The role of ethylene in symbiotic nitrogen fixation. *Trends in Plant Science*, *6*(11), 501–502. doi:10.1016/S1360-1385(01)02128-8 PMID:11701355

Xiao, B., Lian, B., Sun, L., & Shao, W. (2012). Gene transcription response to weathering of K-bearing minerals by *Aspergillus fumigatus. Chemical Geology*, *306-307*, 1–9. doi:10.1016/j.chemgeo.2012.02.014

Xiao, Y., Wang, X., Chen, W., & Huang, Q. (2017). Isolation and identification of three potassium solubilizing bacteria from rape rhizospheric soil and their effects on ryegrass. *Geomicrobiology Journal*, *34*(10), 873–880. doi:10.1080/01490451.2017.1286416

Zarjani, J. K., Aliasgharzad, N., Oustan, S., Emadi, M., & Ahmadi, A. (2013). Isolation and characterization of potassium solubilizing bacteria in some Iranian soils. *Archives of Agronomy and Soil Science*, *59*(12), 1713–1723. doi:10.1080/03650340.2012.756977

Zeng, X., Liu, X., Tang, J., Hu, S., Jiang, P., Li, W., & Xu, L. (2012). Characterization and potassium-solubilizing ability of *Bacillus Circulans* Z 1–3. *Advanced Science Letters*, *10*(1), 173–176. doi:10.1166/asl.2012.3726

Zhao, D., Oosterhuis, D. M., & Bednarz, C. W. (2001). Influence of potassium deficiency on photosynthesis, chlorophyll content, and chloroplast ultrastructure of cotton plants. *Photosynthetica*, *39*(1), 103–109. doi:10.1023/A:1012404204910

Zia-ul-Hassan & Arshad, M. (2010). *Cotton growth under potassium deficiency stress is influenced by photos*. Academic Press.

KEY TERMS AND DEFINITIONS

Potassium Fixation: The conversion of bioavailable potassium into any form of potassium that are not available to plants and soil microbes.

Potassium Pools: Different types of potassium in the soil that vary in sizes and its bioavailabilty to plants and soil microbiota.

Potassium Solubilization: The conversion of unavailable form of potassium to bioavailable potassium through microbial or chemical induced processes.

Chapter 11
Bacterial Siderophores for Enhanced Plant Growth

Himanshi Verma
University of Delhi, India

Meghna Jindal
University of Delhi, India

Shabir A. Rather
Northwest A&F University, Yangling, China

ABSTRACT

The soil is a repository of microorganisms such as bacteria, fungi, algae, and protozoa. Among these, more bacteria are found, most of which are located in the rhizosphere region of the soil. The rhizosphere, under the direct control of plant root secretions, is the complex, narrow area of the soil. It is densely populated with microorganisms (mostly bacteria) that interact with the plants. These interactions influence the growth of the plant directly or indirectly. Plant growth-promoting rhizobacteria (PGPR) inhabiting the rhizosphere colonizes the plant roots and increases plant growth via different mechanisms. Iron is an essential micronutrient required by almost all life forms including plants. Oxidation of Fe2+ (soluble) to Fe3+ (insoluble) due to the soil's aerobic conditions limits its bioavailability. Siderophores are selective low molecular weight ferric ion chelators secreted by bacteria to acquire iron from the surrounding. They bind to iron (Fe3+) with high specificity as well as high affinity. By helping the insolubilisation of iron, it promotes the growth and yield.

INTRODUCTION

Soil is a storehouse of microorganisms like bacteria, fungi, protozoa, and algae. Among these, bacteria are found in more amount, which is mostly concentrated in the rhizosphere region of the soil. The rhizosphere is the dynamic, narrow region of the soil where plant roots are easily accessible and are densely populated with microorganisms (mostly bacteria). Maximum interactions between plant roots and the

DOI: 10.4018/978-1-7998-7062-3.ch011

fauna occur in this region (Pahari et al., 2017). These interactions influence the growth of the plant directly or indirectly. Plant Growth Promoting Bacteria (PGPB) inhabiting the rhizosphere or colonizing the plant tissues and organs increases plant growth via different mechanisms (Pahari et al., 2017). PGPB fulcrum the plant growth either directly (by facilitating nutrient uptake from the surrounding via nitrogen fixation, siderophore production, production of growth phytohormones, phosphate solubilization) or indirectly (by providing defence against harmful pathogens via lytic enzymes production, the release of antibiotics, siderophore production or induced systemic response) (Glick, 2012).

Plants require nutrients for their proper growth and development. Iron is one of the essential micronutrients and is indispensable for plant growth (Lurthy et al., 2020). It is required for chlorophyll development, for the catalytic activity of proteins involved in essential cellular metabolic processes like respiration, photosynthesis, DNA synthesis, and defence against ROS (Rout et al., 2015). The total amount of iron in the soil is more than that required by plants. However, the aerobic and alkaline conditions of soil cause oxidation of Fe^{2+}(soluble) to Fe^{3+} (insoluble) which form insoluble complexes thereby limiting the availability of the usable form of iron to plants (Lurthy et al., 2020). Hence plants need to have mechanisms for the acquisition of iron.

Siderophores are selective low molecular weight ferric ion chelators secreted by bacteria to acquire iron from the surrounding (Aznar et al., 2013). These siderophores are synthesised in response to iron deficiency. By helping the insolubilisation of iron, it promotes the growth and yield of the plant since they can also utilise these siderophores (Schmidt, 1999). Not only this, siderophores can chelate various other metals apart from the ferric ion. As a result, they also help promote plant growth by inducing uninhibited auxin synthesis in phytohormone-producing bacteria by chelating metals in conditions where specific metal ions are present in the medium which may hinder auxin synthesis thus, increasing the phytoremediation potential of plants (Dimkpa et al., 2008). Siderophores have shown potential use in alleviating metal toxicity leading to bioremediation and promoting plant growth (Khuong et al., 2020). Siderophore-producing PGPBs like *Pseudomonas spp., Azospirillum spp.* etc. are used as biocontrol agents and as biofertilizers in agriculture inoculants for application on certain plant parts. The inoculants comprise antibiotics and siderophores which help in the biocontrol of diseases caused by various phytopathogens. Siderophores produced from one microbe can also be used by other microbes that do not produce siderophores (Scavino and Pedraza, 2013).

Hence bacterial siderophores have an extensive range of applications, thereby promoting sustainable agriculture and reducing chemical-based products. This chapter highlights the role of iron as an essential and limiting micro-nutrient for living organisms, siderophore production by bacteria particularly PGPB, siderophore mechanism of action, types of siderophores and their role in promoting sustainable agriculture.

IRON: AN ESSENTIAL MICROELEMENT

Iron is the fourth most abundant element in the lithosphere and the third most limiting nutrient for plant growth. Iron is an essential micronutrient that is indispensable for plant growth (Tripathi et al., 2018). Plants require it for chlorophyll development and stability, for the catalytic activity of proteins involved in essential cellular metabolic processes like respiration, photosynthesis, DNA synthesis, and defence against ROS (Rout et al., 2015). Despite being present in high quantity, it is not readily available for use by plants growing in aerobic and neutral to alkaline pH soil. This is so because the aerobic and alkaline conditions of soil cause oxidation of Fe^{2+}(soluble) to Fe^{3+} (insoluble) which form insoluble complexes thereby limiting the availability of the usable form of iron to plants (Rout et al., 2015). For example: in aerobic soil, Fe^{3+} is present majorly as a constituent of highly insoluble oxyhydroxide polymers (Briat et al., 2007). Plants growing in such soils, therefore, have adopted specific strategies to acquire iron in a usable form that is essential for their growth (Rout et al., 2015).

Role of Iron in Plant Growth

Plants require iron in optimum quantity for their growth and development. Too low or too high levels of iron can be toxic to them. Approximately 80-85% of iron is present in photosynthetic tissues of the plants where it is used for the synthesis and stability of chlorophyll molecules, for the biosynthesis of heme proteins like cytochromes and non-heme proteins like iron Sulphur proteins (Fe-S), which forms an essential component of photosynthetic and mitochondrial electron transport chain (ETC) (Hell and Stephen, 2003). Therefore, iron is involved in the respiration process, in photosynthesis which determines plant growth and productivity.

Too low quantities of iron can lead to chlorosis as it will hamper chlorophyll production, stunted root growth, and lead to the overall low development of the plant (Rout et al., 2015). Above optimum iron levels may lead to necrotic pitting, speckling, and bronzing, producing harmful reactive oxygen species (ROS) which causes peroxidation of membrane lipids, damage to the cellular structure and may eventually cause cell death (Crichton et al., 2002). The iron thus is required in optimum quantity to carry out and form part of essential processes involved in proper growth and development of plants (Tripathi et al., 2018).

Role of Iron in Plant Metabolism and Enzyme Activities

Iron forms the primary and most critical constituent of nearly all redox systems in plants. Due to its redox properties, affinity towards metalloprotein sites, ability to form complexes with different kinds of ligands, it forms part of many enzymes; electron carriers have a structural role in the prosthetic group of enzymes involved in essential biological processes (Grotz and Guerinot, 2006). For example, it helps in the development of the porphyrin structure of chlorophyll as it controls the rate of delta-aminolevulinic acid (ALA) formation (which is the precursor of porphyrins); it acts as a cofactor of many enzymes that are important for plant hormone synthesis, such as ethylene, abscisic acid, forms a prosthetic group of cytochromes, involved in Fe-S clusters formation, in DNA synthesis via the action of the ribonucleotide reductase, it also interacts with non-heme proteins like ferredoxin, superoxide dismutase, etc. (Rout et al., 2015). Hence iron is necessary for plant metabolism.

Deficiency of Iron in Plants

When the amount of iron in plants is below its optimum level, it leads to its deficiency. The very first symptom of iron in plants is interveinal chlorosis of young leaves where veins remain green. However, the rest of the portion of the leaf turns yellow as the chlorophyll synthesis in plants is hampered and also due to changes in the expression and assembly of different photosynthetic components (Tripathi et al., 2018). Other effects include stunted root growth, poor nutritional quality, low yield and productivity, etc. iron deficiency decreases its interactions with other micronutrients like Zn and Mn. It has been observed that Iron deficiency causes a decrease in sulfur concentrations in the roots and shoots of plants, and this is probably a consequence of the iron requirements for sulfur assimilation (Hell and Stephan, 2003). Since iron forms part of enzymes involved in various life-sustaining processes of the plant (like respiration, DNA synthesis, photosynthesis, etc.) so its deficiency disturbs the normal physiology and metabolism of plants which severely affects its growth and yield (Rout et al., 2015). Hence iron is an essential micronutrient for plant growth and development.

IRON ACQUISITION BY HIGHER PLANTS

Iron, the fourth-most abundant micronutrient, is present in the soil in quantity more than that required by plants. However, due to its limited bioavailability plants have adopted different ways to obtain iron from the soil in its usable form (Hell and Stephan, 2003). For the significant acquisition of iron plants have two species-dependent strategies (as proposed by Rumheld and Marschner, 1986) under different soil conditions:

Strategy I

This strategy to obtain iron is used by dicots and non-gramineous monocots (Kobayashi and Nishizawa, 2012). In this case, plants take up iron by following three ways: i) solubilisation of Fe^{3+}, which is present as part of insoluble complexes by releasing protons (H^+) through P-type ATPase present in the plasma membrane. The released protons acidify the surrounding soil and cause reduction of Fe^{3+} into Fe^{2+} (soluble form) which can quickly be taken up by plants (Kobayashi and Nishizawa, 2012). ii) Preferential transport of Fe^{2+} through plasma membrane Fe- transporters after reducing Fe^{3+} at the root surface (Tripathi et al., 2018). iii) reduction by ferric (Fe^{3+}) chelate reductase, a membrane-bound integral membrane protein that enhances the mobilization of iron. The reduced soluble form of iron (Fe^{2+}) thus generated via this strategy is taken up and transported from the root through divalent cation transporter types IRT1 and is used by the plants for different purposes (Rout et al., 2015). This strategy is quite effective under iron-deficient conditions.

Strategy II (Phytosiderophores)

Only gramineous monocots, i.e., grasses, use this strategy. It involves releasing specific ferric ion chelators by plants in the rhizosphere under iron-deficient conditions (Takagi et al., 1984). These Fe^{3+} chelating compounds are known as Phytosiderophores. Chemically these are mugineic acids (MAs) or their modified derivatives which are synthesized via a series of reactions involving condensation of three S-adenosyl-methionine molecules that form the precursor nicotianamine (NA) (Kobayashi et al., 2001). The mugineic acid released solubilize Fe^{3+} resulting in the formation of Fe^{3+}-MA complexes that are taken up by specific transporters (Takagi et al., 1984). In maize, the complex is taken up via H^+/ Fe (III)-PS complex symporter is known as yellow stripe 1 (YS1). The yellow stripe 1 is named after the phenotype, yellow striped leaves, in maize plants that had a mutation in genes coding for this transporter (Rout et al., 2015). The iron (Fe^{3+}) complex thus taken up is reduced to Fe^{2+} and this reduced iron is transferred to nicotianamine which is translocated to shoot as well as other plant parts and is used to carry out life-sustaining processes (Rout et al., 2015).

PLANT GROWTH PROMOTING BACTERIA (PGPB)

Among the different types of micro-organisms found in the soil, bacteria are the most dominating one (Glick, 2012). The type of vegetation characterises the abundance and the diversity of bacteria found in the soil found there and the types of soil conditions like temperature, pH, moisture, and the type of chemicals present in the soil (Glick, 2012). Among these bacteria, those that help in enhancing the growth of the plant are known as Plant Growth Promoting Bacteria (PGPB). These PGPB can be free-living or symbiotic inhabiting the rhizosphere or colonize inner plant tissues or organs. PGPB includes the following genera: *Pseudomonas, Bacillus, Azotobacter, Enterobacter, Rhizobium, Azoarcus, Herbaspirillum, Azospirillum,* etc. (Scavino and Pedraza, 2013). Those colonizing the rhizospheric region of the soil are known as Plant Growth Promoting Rhizobacteria (PGPR) that can be grouped into two classes based on their place of colonization on the roots: extracellular PGPR (ePGPR) and intracellular PGPR (iPGPR). ePGPR, majorly colonizing the rhizoplane or intercellular spaces between the root cortical cells, includes *Azotobacter, Azospirillum, Agrobacterium, Bacillus, Flavobacterium, Pseudomonas, Arthrobacter*, etc. (Pahari et al., 2017). iPGPR, inhabiting specialized nodular structure, includes *Rhizobium, Bradyrhizobium, Mesorhizobium, Allorhizobium* (Pahari et al., 2017). PGPB had evolved different mechanisms to promote plant growth. Direct mode of plant growth includes: i) facilitating nitrogen fixation either symbiotically or symbiotically, ii) help in combating environmental stress by production of hormones like auxin, cytokinin and gibberellins, iii) promotes plant growth by providing them resources via solubilisation of nutrients like phosphorous, iv) reduces ethylene production for better plant development during early growth stages, v) aids in iron acquisition by producing siderophores (Scavino and Pedraza, 2013). The indirect mode includes: i) synthesis of antibiotics and antitoxin to protect against phytopathogens, ii) by providing induced systemic resistance, iii) starving the pathogens of iron as well as other beneficial metals by itself utilising it the most via their high-affinity siderophore production, and iv) by synthesizing lytic enzymes to kill the harmful microbes (Glick, 2012).

Bacterial siderophores, produced by different bacteria belonging to the genera *Rhizobium, Pseudomonas, Bacillus, Pseudomonasaeruginosa,* enterobacteria, etc., have proved to be beneficial for plant growth as they enhance their iron utilisation efficiency. It has been observed that among siderophores produced by bacteria belonging to different genera, those belonging to genera *Pseudomonas* produces siderophores with relatively low Km i.e. they have a high affinity for ferric ions and even showed promising results on plant growth (Beneduzi et al., 2012).

SIDEROPHORES

Siderophores are selective low molecular weight (less than 2000 Da), water-soluble, organic molecules that bind to ferric ions (Fe^{3+}) with high affinity and specificity. Nearly all aerobic bacteria and fungi secrete them under iron-deficient conditions, where these molecules act as high-affinity ferric ion chelators (Beneduzi et al., 2012). Siderophores are present in high concentration in the rhizosphere region of the soil than in the bulk soil as siderophore-producing bacteria are found to colonize and interact with plant roots in the rhizosphere (Beneduzi et al., 2012). Over more than 500 different siderophores have been discovered to date (Glick, 2012). They help in modifying the interactions that occur among the different types of organisms inhabiting the rhizosphere. This is mediated either by competing with another microorganism for iron available in the rhizosphere and depriving them of the same or helping other microbes that do not produce siderophore or have low competence (Scavino and Pedraza, 2013). For example, bacterial siderophores with a high affinity for iron deprive the pathogens, like pathogenic fungi, whose siderophores have a low affinity for iron (Scavino and Pedraza, 2013). While on the other hand, it may even help other microorganisms that cannot produce siderophore by allowing them to utilize their iron (Fe^{3+})- siderophore complex via their receptors for the same and meet their iron requirements. Ferrochrome, a common fungal siderophore, is utilized by many rhizospheric microorganisms like *Yersinia, Erwinia* that has receptors for the same (Scavino and Pedraza, 2013). Not only these microorganisms but even plants have great potential to utilize bacterial siderophores for iron uptake from the limited rhizospheric iron pool of the soil. By helping insolubilisation and extraction of iron from mineral or organic complexes, it alleviates growth and yield of a plant (Pahari et al., 2017).

Mechanism of Action of Bacterial Siderophore

Many Gram-positive and Gram-negative bacteria can synthesize ferric ion chelators, known as siderophores, to acquire iron from iron-deficient soil (Beneduzi et al., 2012). These bacterial siderophores are of diverse types with varying affinity. The successful uptake of iron by these bacteria via ferric ion siderophores involves an extensive transport system that consists of siderophore synthesis and degrading enzymes, siderophore receptors, membrane transporters, membrane reductases, and regulatory proteins (Crowley et al., 1991). In general, these components of the transport system like membrane proteins, receptors, etc. for different siderophore types might differ in terms of their sequence but are almost similar in structure and function (Krewulak and Voge, 2008). The most predominant and most extensively studied is enterobactin, a siderophore produced by enterobacteria like *Escherichia coli* (Scavino and Pedraza, 2013). The components involved in the uptake of ferric ion-enterobactin complex and the function of each is discussed below:

Outer Membrane Receptor: Fep A

Fep A serves as the receptor for the ferric-enterobactin complex present on the outer membrane of the *E. coli* cell (Krewulak and Voge, 2008). Fep A is composed of two major components: i) β- barrel domain and ii) N- terminal cork or plug domain (Krewulak and Voge., 2008).

β- Barrel Domain

This domain is made up of 3 components: i) 22 anti-parallel -β stranded β - barrel, ii) 10- short periplasmic loops that vary in length from 2- 10 residues, and iii) 11 extracellular loops that can be 2 - 37 residues long that extends 30-40 Å above the outer bilayer. β- the barrel is 70 Å in height and extends above the lipid bilayer membrane (Krewulak and Voge, 2008). The whole barrel is stabilized by interstrand hydrogen bonds and salt bridges formed between strand 1 and 22. Approximately 40-50% of the β-barrel is comprised of those 11 extracellular loops that perform two essential functions: one to interact with the ferric-enterobactin complex and secondly, it can prohibit the entry of non-desirable solutes by occluding the opening of the β- barrel (Krewulak and Voge, 2008).

N- terminal Cork or Plug Domain

The N- terminal of the receptor is globular and is also known as the cork/plug/hatch domain since it occludes the β-barrel by forming two salt bridges between two conserved arginines (R) residues in the globular domain and two conserved glutamic acids (E) residues in the β- barrel domain and around 40-70 hydrogen bonds (Krewulak and Voge, 2008). The cork domain comprises mainly of i) a central mixed 4 stranded β-sheet with surrounding loops and helices, ii) 3 apices, namely A, B, and C that along with extracellular loops, is involved in siderophore binding and iii) a TonB box near the N- terminus, that interacts with the TonB protein spanning the periplasm. Binding of the ferric-siderophore complex to the receptor causes maximum conformational changes in the cork domain (Krewulak and Voge, 2008). The interaction of the TonB box with TonB protein facilitates the transport of siderophore the outer membrane receptor, and this is mediated by coupling of proton motive force of cytoplasmic membrane with that of the outer membrane (Celia et al., 2019).

The C terminal portion of all outer membrane receptors contains a conserved phenylalanine or tryptophan residue that is important for correct folding and insertion of the receptor in the outer membrane (Krewulak and Voge, 2008).

Periplasmic Siderophore Binding Proteins (PSBPs)

Different types of PSBPs found in the bacterial cell's periplasm share structural similarity despite their low sequence similarity (often less than 10%). PSBPs consist of domains that are connected either via two or three β - strands or via a long α-helix and the loops of each domain consist of a mixed α/β structure (Krewulak and Voge, 2008). The function of these PSBPs is to escort the ferric-siderophore complex to the cytoplasmic membrane transporters so that it can further be transported into the bacterial cell cytoplasm (Krewulak and Voge, 2008).

TonB–ExbB–ExbD (Ton Complex)

Transport of ferric-siderophore complex into the periplasm is an active process; i.e., it requires energy. Three proteins, namely i) ExbD, ii) ExbB and iii) TonB couples the cytoplasmic membrane proton motive force to the outer membrane and hence facilitate the transport (Celia et al., 2019).

TonB is a 26-kDa cytoplasmic or inner membrane protein that interacts with the other two proteins ExbB and ExbD, to form an energy transducing complex (Krewulak and Voge, 2008). It consists of three domains: I) N-terminal domain made up of a 32-residues long membrane-spanning helix and a small cytoplasmic region, II) a central domain that extends from amino acid residue 33 to 102 which lies in the periplasm. The domain is characterized by the presence of proline-rich repeats of Pro-Glu and Pro-Lys present from residues 66-102 (Celia et al., 2019). The primary function of this domain is to increase the energy transduction efficiency under a specific condition, III) Carboxy terminal domain that extends from amino acid residue 103 to 239. It is this region of the TonB protein that interacts with the TonB box of the cork domain of the outer membrane receptor protein (Celia et al., 2019).

ExbD is a 17-kDa dimeric inner membrane protein that is around 141 amino acid residues long. It is similar in topology to the TonB protein, i.e., it too consists of N terminal cytoplasmic region followed by a single membrane-spanning helix, a periplasmic central domain and a C- terminal domain (Krewulak and Voge, 2008).

ExbB is a 244 amino acid long pentameric cytoplasmic-membrane protein that comprises of a: I) small periplasmic N terminal domain, ii) three trans-membrane helix (TM) namely TM1, TM2, and TM3, with a small periplasmic loop between TM 2 and TM 3, iii) a large C-terminal domain (Pramanik et al., 2010).

ExbB and ExbD together form a proton channel that translocates protons across the cytoplasmic membrane to form a proton gradient. The PMF generated or energy derived from this proton gradient is coupled to the TonB protein that interacts with the TonB box of the outer membrane receptor protein (Celia et al., 2019).

ATP-Binding Cassette Transporters

The ferric-siderophore complex, once inside the periplasm, interacts, and binds to the PSBP and is translocated across the inner membrane into the cytoplasm via an ABC transporter protein (Krewulak and Voge, 2008). ABC transporter translocates the complex across the membrane by utilising the energy obtained from ATP hydrolysis coupled to the transport of the complex. The ABC transporter in bacteria is made up of four structural domains, i.e., two membrane-spanning domains that form the channel for the passage of ferric-siderophore complex and two nucleotide-binding domains that hydrolyze ATP (Crowley et al., 1991).

Once inside the cytoplasm, the cytoplasmic reductase reduces Fe^{3+} to $Fe^{2+,}$ leading to the dissociation of siderophore from the complex as its affinity for Fe^{2+} is very low (Crowley et al., 1991). The dissociated or deferrisiderophore is then either destroyed as in the case of enterobactin or recycled back to the environment example aerobactin (Crowley et al., 1991).

Bacterial Siderophores for Uptake of Iron by Plants

Plants growing in the soil that inhabits siderophore-producing bacteria in its rhizosphere are advantageous in utilizing iron more efficiently as they can exploit these ferric ions chelating bacterial siderophores (Schmidt, 1999). The ability to utilize bacterial siderophores is majorly the function of high capacity and high-affinity receptors and reductases located in the plasma membrane of the rhizodermal cells (Schmidt, 1999). These reductases are capable of externally reducing the ferric ion to ferrous that can be quickly taken up by plants and utilized for carrying out processes that help in growth and development.

TYPES OF SIDEROPHORES

More than 500 types of siderophores have been discovered. Siderophores are classified into four major types depending upon the oxygen ligand used for complex formation with ferric ion, namely hydroxymate, catecholate, carboxylate, and salicylate. Apart from these, various other kinds of siderophores have also been reported and classified depending upon their characteristic coordination structure (Sheng et al., 2020). Many bacteria are capable of producing more than one kind of siderophore, which helps them in colonizing various kinds of environments more efficiently (CESA-LUNA, Catherine et al., 2020).

Figure 1. Mechanism of uptake of ferric-enterobactin complex by Gram-negative bacteria Escherichia coli: 1) binding of Fe^{3+}- Enterobactin complex to Fep A receptor and its uptake into the periplasm, 2) binding of Fe^{3+}- Enterobactin complex to Periplasm Siderophore Binding Protein (PSBP), 3) transport of Fe^{3+}- Enterobactin complex into the cytoplasm via ABC transporter, 4) reduction of Fe^{3+} to Fe^{2+} by cytoplasmic reductase and release of enterobactin, 5) degradation of enterobactin.

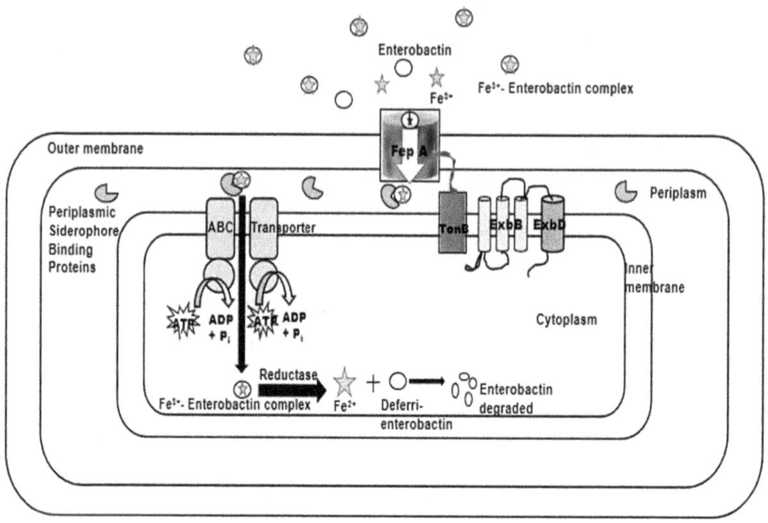

Hydroxymate Type

Hydroxymate type of siderophores are hydrophilic and produced by both fungi and bacteria (Baakza et al., 2004). The majority of siderophores produced by fungi are of hydroxymate type except for zygomycetes which produce carboxylate type of siderophore (Silva-Baila˜o et al., 2014). Bacterial hydroxymates comprise acylated and hydroxylated alkylamines, whereas fungal hydroxymates are based on acylated and hydroxylated ornithine (Baakza et al., 2004). The ferric ion chelating complex in hydroxymate siderophores comprises a carboxyl group bound to adjacent nitrogen. Ferrichrome; a fungal siderophore produced by *Ustilago sphaerogena* is an example of this type (Paul and Dubey, 2015).

Catecholate Type

Catecholate type of siderophores are hydrophobic and present only in bacteria and comprise only of hydroxyl and catecholate groups (Baakza et al., 2004). Some of the unique properties of catecholate include complex stability, lipophilicity and resistance to high environmental pH. Catecholates contain a dihydroxybenzoic acid bound to an amino acid (Winklemnan, 2002). Enterobactin produced by E. coli is one of the best-studied examples of this type (Paul and Dubey, 2015).

Carboxylate Type

Carboxylate type of siderophores is present in fungi belonging to Zygomycota (Mucorales). They are also found in a few bacteria (*Staphylococcus hyicus and Rhizobium meliloti*). They comprise carboxy and hydroxy groups which interact with the ferric ion (Baakza et al., 2004). Microbes like fungi living in the acidic environment produce carboxylate type of siderophores. However, they have less affinity for the ferric ion at physiological pH than catecholates, which have a stronger affinity for ferric ions at the same pH. All these ligands are arranged and organized in different types of structures, which include peptides, aminnoalkanes and citric acid-based ligands. The siderophore organization and ligand type determine the ferric ion-siderophore complex (Scavino and Pedraza, 2013).

A miscellaneous mixed type of siderophore containing both hydroxymate and catecholate groups as in heterobactin produced by *Rhosococcus erythropolis* has also been identified (Paul and Dubey, 2015).

Salicylate Type

A fourth type of siderophore called salicylate type (SA) also exists. Certain plant growth-promoting bacteria belonging to the genus *Azospirillum* produce salicylic acid which has siderophore activity and also acts as a precursor for the synthesis of various catechole types of siderophores. Salicylic acid also helps in plant defense by activating plant defense mechanisms like localised and systemic acquired resistance (Scavino and Pedraza, 2013).

Table 1. Types of siderophores with examples of organisms producing different types of siderophores

Type	Examples	References
Hydroxymate	Ferrichrome from *Ustilago sphaerogena*	Paul and Dubey, 2015
Catecholate	Enterobactin from *Escherichia coli*	Paul and Dubey, 2015
Carboxylate	Zygomycota(Mucrorales),*Staphylococcus hyicus,* and *Rhizobium meliloti*	Baazka et al., 2004
Salicylate	*Azospirillum* spp.	Scavino and Pedraza, 2013
Mixed type (containing both hydroxymate and catecholate groups)	Heterobactin from *Rhosococcus erythropolis*	Paul and Dubey, 2015

ROLES OF SIDEROPHORES

Chemical-based products like fertilizers, pesticides, and weedicides are extensively used in agriculture to provide nutrients to plant, remove pests and control weeds respectively. This ultimately helps in achieving better crop quality, higher yields and lesser losses. PGPBs majority of which produce siderophores are an environmentally friendly and cost-effective alternative to the harmful, environmentally polluting, and expensive chemical-based products which dominate the agricultural market in today's day and age.

Biocontrol

Biocontrol or biological control is an environmentally friendly way in which individual organisms are used to control or prevent the growth of pests like insects, mites, pathogenic microbes etc. (Flint and Dreistadt, 1998). Bacteria producing siderophores play an essential role in controlling or preventing the growth of certain pathogenic microbes. *Brevibacillus brevis* GZDF3 is a PGPR isolated from the rhizosphere of *Pinellia ternata* which is an essential herb used in traditional Chinese medicine. It produces siderophores which were shown to help in biocontrol against fungal pathogen *Candida albicans*. It also shows biocontrol against other pathogenic microbes. This bacterial strain produces large amounts of siderophores and shows strong antagonistic activity making it a promising biocontrol agent (Sheng et al., 2020). Soil-borne fluorescent pseudomonads produce multiple kinds of siderophores which suppress the disease by competing for iron. They produce several types of siderophores like salicylic acid, pyoverdine, azotobactin, pyochelin and pseudomonine (Scavino and Pedraza, 2013). Black pepper in Malaysia is prone to many controlled diseases using chemical-based products that are hazardous, have harmful health effects and affect crop quality in the long run. A biological approach was adopted to replace these chemical-based treatments.

Seven indigenous rhizobacteria (*Bacillus subtilis, Bacillus siamensis, Brevibacillus gelatini, Pseudomonas geniculata, Pseudomonas beteli, Burkholderia ubonensis, and Burkholderia territorii*) were antagonistic to *Fusarium solani*, a soil-borne disease-causing fungus of black pepper. All these rhizobacteria produced antifungal siderophores. On the application of these bacteria on the plant, increased plant growth along with enhanced root development via IAA secretion was observed. As a result of siderophore production, these bacteria were able to colonize at the plant roots' rhizosphere and assisted the plant growth by providing them with iron nutrition and competitively inhibiting the growth of phytopathogens by competing for iron. Hence, they acted as both biocontrol agents by suppressing fungal growth and biofertilizer by promoting plant growth (Lau et al., 2020). Siderophore producing bacteria which are endophytic, colonize the plant cells, thereby hogging the ecological niche of other microorganisms due to the production of siderophores. A notable example of this phenomenon is bacterial strains of the genus *Burkhholderia* which colonize rice plants and could be of great importance in preventing a pathogen attack in young plants (Loaces et al., 2011). The genus Azospirillum consists of certain bacteria that promote plant growth by the production of salicylic acid. Salicylic acid is also a type of siderophore along with its plant defense activating properties. It can activate defense mechanisms like localised and systemic acquired resistance in plants against pathogens. It also acts as a precursor in the synthesis of some catechole types of siderophores which include yersiniabactin, pyoverdine and pyochelin. The role of salicylic acid produced by bacteria in plant induced systemic resistance is still under debate (Scavino and Pedraza, 2013).

Biofertilizer

Biofertilizer is a substance comprising of micro-organisms (living) which, when applied on plant parts or soil, provides nutrition to the plants and promotes their growth. It is an environmentally sound way of providing nutrients to the plant instead of harmful chemical-based formulations (Vessey, 2003). Siderophores have been shown to act as biofertilizers by making certain nutrients available to the plant.

Iron

Iron is an essential element for the growth and development of the plant. The deficiency of iron can lead to chlorosis in plants. Iron provided in the form of chemical-based fertilizers is harmful to the environment. Moreover, it is highly stable in soil and can enter drinking water. Siderophores are iron chelators and useful in supplementing iron to the plants. Siderophores from bacterial strain *Chryseobacterium* C138 can provide iron to the iron-starved tomato plants via its roots, thus acting as a potent organic biofertilizer. Siderophores were equally effective in their function irrespective of the bacterial presence or absence. Iron-starved tomato plants treated with the siderophores from this strain showed increased yield, iron content and chlorophyll over the positive controls. Hence, strain C138 could be used as an economically feasible and effective organic iron chelator or biofertilizer (Radzki et al., 2013). Siderophores from *Pseudomonas* strain GRP3 showed enhanced chlorophyll production and helped in the reversal of iron chlorosis in mung beans (Sharma et al., 2003). Siderophores are claimed to have restored chlorosis caused due to iron deficiency in peanuts by *Paenibacillus illinoisensis* and *Bacillus sp.* growing in calcareous soil (Liu et al., 2017).

Nitrogen

Pseudomonas aeruginosa producing hydroxymate type of siderophores helped to enhance nodulation and nitrogen fixation in *Vigna radiata* (mung bean plant). The degree of nodulation and nitrogen fixation was more in plants infected with *Pseudomonas aeruginosa* than plants infected with just *Bradyrhizobium* strain. The ecological advantages posed by these bacterial siderophores encourages their use in the form of inoculants along with root nodule bacteria (Mahmoud and Abd-Alla, 2001). Four strains of bacteria belonging to *Enterobacter, Pseudomonas, Ochrobactrum* and C*ellulosimicrobium*; when inoculated, were shown to have plant growth promoting attributes in tomato plantlets. All four strains produced siderophores which helped in providing the plant with iron nutrition and competed with pathogens for iron, thereby inhibiting their growth. Nitrogen fixation, IAA production and phosphate solubilisation were among other properties shown by these strains. These attributes help weak plant growth and development of the root system. This further enhances nutrient uptake and subsequently plant growth (Pérez-Rodriguez et al., 2020). Siderophores from *Azotobacter vinelandii* have also been shown to help in nitrogen fixation by binding to Molybdenum and vanadium, two crucial metals needed in the nitrogen fixation process under conditions present in a limited amount in diazotrophic cultures (Kraepiel et al., 2009).

Bioremediation

Bioremediation involves micro-organisms (living) or plants for the reduction or degradation of environmental pollutants into non-toxic or less toxic forms (Zouboulis and Moussas, 2011). Siderophores have also been shown to help in bioremediation. Siderophores help in promoting plant growth in a variety of both direct and indirect ways (Scavino and Pedraza, 2013). Siderophores can chelate various other metals like molybdenum, manganese, cobalt and nickel apart from the ferric ion. Non-essential metals like plutonium, americium, thorium and uranium have also been shown to bind to siderophores. A drop in the level of siderophore production has been observed when exposed to toxic levels of trace elements (Edberg et al., 2010). Certain phytohormone, like auxin-producing bacteria, is also capable of promoting plant growth. Dimkpa et al., 2008 studied *Streptomyces* strains that could produce both siderophores and auxin simultaneously which put a light on their potential as future plant growth promoters and in phytoremediation soil contaminated with metals. It was shown that siderophores helped promote plant growth by inducing uninhibited auxin synthesis in phytohormone producing bacteria by chelating specific metal ions (Al^{3+}, Cd^{2+}, Cu^{2+}, and Ni^{2+}) which, if present in the medium, may hinder auxin synthesis. This, in turn, enhanced the phytoremediation potential of plants.

Siderophores from acid and manganese resistant purple non-sulfur bacteria or PNSB (*Rhodopseudomonas palustris* strains TLS12, VNS19, VNS32, VNS62, and VNW95, and *Rhodopseudomonas har-*

woodiae strain TLW42) have shown potential use in alleviating metal toxicity by binding to Manganese ions resulting in the formation of immobilized siderophore–manganese complexes in acidic conditions leading to bioremediation. The PNSBs also showed bioremediation by adsorption of manganese ions to exopolymeric substances (EPS) which was the primary mechanism by which the PNSBs reduced manganese toxicity bioaccumulation and siderophore production. They showed more adsorption of Manganese ions by releasing exopolymeric substances than their biomass (bioaccumulation). PNSBs also released nutrients like NH^{4+} by nitrogen fixation and PO_4^{3-} by phosphate solubilisation. The bacteria also released plant growth-promoting substances (PGPS) like indole-3-acetic acid (IAA), 5-aminolevulinic acid (ALA) and siderophores. The plant nutrients and PGPS helped in increasing the pH and promoting plant growth. Hence, the PNSBs showed potential in bioremediation, nutrient release and plant growth promotion leading to better fertility and cultivation in acid sulfate soil conditions. Manganese ion resistance mechanisms result in the release of plant nutrients and PGPS which can be isolated and used in bioremediation and as biofertilizers on acid sulfate soils thereby leading to plant growth promotion (Khuong et al., 2020).

Pseudomonas fluorescens produces fluorescent pyoverdine siderophore, which could mobilise or leach iron, nickel and cobalt from mine waste (acid-leached ore) of a former uranium mine (Edberg et al., 2010). Siderophores from *Agrobacterium radiobacter* have been shown to remove about 54 percent of the pollutants from a soil contaminated with metals (Pahari et al., 2017). Aesbestos containing products are carcinogenic and their use has been banned in several countries. However, large amounts of asbestos are used in buildings leading to a high asbestos waste generation which needs to be removed in an environment-friendly manner. Biodegradation by using bacterial siderophore pyoverdine which releases iron from asbestos, is one such mechanism. Example of such bacteria includes *Pseudomonas aeruginosa* and *Pseudomonas mandelli.* Iron present in asbestos confers its carcinogenic properties. It catalyzes many reactions including lipid peroxidation, precursors for tumor development, DNA damage, oxygen consumption and formation of ROS. Hence, removing iron from asbestos using pyoverdines is an eco-friendly approach for asbestos-containing waste treatment (David et al., 2020).

CONCLUSION

Iron is an essential limiting micronutrient needed by all living organisms. It is essential for the growth and development of plants. However, the bioavailability of iron is limited as it remains in oxidised form (Fe^{3+}) in soil due to aerobic conditions and remains unavailable to the plants for utlilisation. Siderophores are ferric ion chelating molecules secreted by bacteria and other microbes, which solubilize iron and make it available for plants. Various types of bacterial siderophores exist which have numerous other roles apart from providing iron nutrition to the plants. Siderophore-producing bacteria promote plant growth in various ways, both direct and indirect ranging from biocontrol against pathogens to biofertlizers and bioremediation. Thus, it can be concluded that siderophores are an environmentally friendly alternative to expensive chemicals which cause environmental pollution and health problems. Hence, they pave the way for a more sustainable agriculture method and hold a great future potential to replace artificial methods with natural ones. There is an increasing awareness among the masses and more people are adopting organic products and shifting towards organic farming. Hence, siderophore formulated products would be up-and-coming in shifting towards organic farming.

Table 2. Roles of bacterial siderophores in plant growth promotion with examples

Role	Siderophore Producing Bacteria	Function	References
Biocontrol	*Brevibacillus brevis* GZDF3	Biocontrol against fungal pathogen *Candida albicans*	Sheng et al., 2020
	Soil-borne fluorescent pseudomonads	Produce multiple kinds of siderophores which suppress the disease by competing for iron	Scavino and Pedraza, 2013
	Bacillus subtilis, Bacillus siamensis, Brevibacillus gelatini, Pseudomonas geniculata, Pseudomonas beteli, Burkholderia ubonensis and Burkholderia territorii	Produce antifungal siderophores against *Fusarium solani*, increased plant growth along with enhanced root development	Lau et al., 2020
	Bacterial strains of the genus *Burkhholderia*	Prevent pathogen attack by endophytically colonizing plant roots of rice and hogging the ecological niche of microbes by the production of siderophores	Loaces et al., 2011
	Azospirillum spp.	Produces salicylic acid which activates plant defense mechanisms like localised and systemic acquired resistance against plant pathogens	Scavino and Pedraza, 2013
Biofertilizer	IRON *Chryseobacterium* C138	Provide iron to the iron-starved tomato plants	Radzki et al., 2013
	Pseudomonas strain GRP3	Chlorosis reversal in mung beans	Sharma et al., 2003
	Paenibacillus illinoisensis and *Bacillus* sp.	Chlorosis reversal in peanut	Liu et al., 2017
	NITROGEN *Pseudomonas aeruginosa, Bradyrhizobium*	Enhance nodulation and nitrogen fixation in *Vigna radiata* (mung bean plant)	Mahmoud and Abd-Alla, 2001
	Enterobacter, Pseudomonas, Ochrobactrum and *Cellulosimicrobium*	Plant growth promotion in tomato plantlets, provide iron nutrition and biocontrol against pathogens	Pérez-Rodriguez et al., 2020
	Azotobacter vinelandii	Siderophores produced help in nitrogen fixation by binding to Molybdenum and vanadium	Kraepiel et al., 2009
Bioremediation	*Streptomyces* strains	Produce both siderophores and auxin simultaneously, promotion of plant growth and phytoremediation of metal contaminated soil	Dimkpa et al., 2008
	Purple non-sulfur bacteria (Rhodopseudomonas palustris strains TLS12, VNS19, VNS32, VNS62, and VNW95, and *Rhodopseudomonas harwoodiae* strain TLW42)	Siderophores produced helped in reducing manganese ion toxicity	Khuong et al., 2020
	Pseudomonas fluorescens	Pyoverdine siderophore produced was able to mobilise or leach iron, nickel and cobalt from mine waste (acid-leached ore) of a former uranium mine	Edberg et al., 2010

REFERENCES

Aznar, A., Chen, N. W., Rigault, M., Riache, N., Joseph, D., Desmaële, D., ... Thomine, S. (2014). Scavenging iron: A novel mechanism of plant immunity activation by microbial siderophores. *Plant Physiology*, *164*(4), 2167–2183. doi:10.1104/pp.113.233585 PMID:24501001

Baakza, A., Vala, A. K., Dave, B. P., & Dube, H. C. (2004). A comparative study of siderophore production by fungi from marine and terrestrial habitats. *Journal of Experimental Marine Biology and Ecology*, *311*(1), 1–9. doi:10.1016/j.jembe.2003.12.028

Beneduzi, A., Ambrosini, A., & Passaglia, L. M. (2012). Plant growth-promoting rhizobacteria (PGPR): Their potential as antagonists and biocontrol agents. *Genetics and Molecular Biology*, *35*(4), 1044–1051. doi:10.1590/S1415-47572012000600020 PMID:23411488

Briat, J. F., Curie, C., & Gaymard, F. (2007). Iron utilization and metabolism in plants. *Current Opinion in Plant Biology*, *10*(3), 276–282. doi:10.1016/j.pbi.2007.04.003 PMID:17434791

Celia, H., Botos, I., Ni, X., Fox, T., De Val, N., Lloubes, R., Jiang, J., & Buchanan, S. K. (2019). Cryo-EM structure of the bacterial Ton motor subcomplex ExbB–ExbD provides information on structure and stoichiometry. *Communications Biology*, *2*(1), 1–6. doi:10.103842003-019-0604-2 PMID:31602407

Cesa-Luna, C., Baez, A., Quintero-Hernández, V., De la Cruz-Enríquez, J., Castañeda-Antonio, M. D., & Muñoz-Rojas, J. (2020). The importance of antimicrobial compounds produced by beneficial bacteria on the biocontrol of phytopathogens. *Acta Biologica Colombiana*, *25*(1), 140–154. doi:10.15446/abc.v25n1.76867

Crichton, R. R., Wilmet, S., Legssyer, R., & Ward, R. J. (2002). Molecular and cellular mechanisms of iron homeostasis and toxicity in mammalian cells. *Journal of Inorganic Biochemistry*, *91*(1), 9–18. doi:10.1016/S0162-0134(02)00461-0 PMID:12121757

Crowley, D. E., Wang, Y. C., Reid, C. P. P., & Szaniszlo, P. J. (1991). Mechanisms of iron acquisition from siderophores by microorganisms and plants. In *Iron Nutrition and Interactions in Plants* (pp. 213–232). Springer. doi:10.1007/978-94-011-3294-7_27

David, S. R., Ihiawakrim, D., Regis, R., & Geoffroy, V. A. (2020). Efficiency of pyoverdines in iron removal from flocking asbestos waste: An innovative bacterial bioremediation strategy. *Journal of Hazardous Materials*, *122532*. Advance online publication. doi:10.1016/j.jhazmat.2020.122532 PMID:32200235

Dimkpa, C. O., Svatos, A., Dabrowska, P., Schmidt, A., Boland, W., & Kothe, E. (2008). Involvement of siderophores in the reduction of metal-induced inhibition of auxin synthesis in *Streptomyces* spp. *Chemosphere*, *74*(1), 19–25. doi:10.1016/j.chemosphere.2008.09.079 PMID:18986679

Edberg, F., Kalinowski, B. E., Holmström, S. J., & Holm, K. (2010). Mobilization of metals from uranium mine waste: The role of pyoverdines produced by *Pseudomonas fluorescens*. *Geobiology*, *8*(4), 278–292. doi:10.1111/j.1472-4669.2010.00241.x PMID:20456501

Flint, M. L., & Dreistadt, S. H. (1998). *Natural Enemies Handbook: The Illustrated Guide to Biological Pest Control* (J. K. Clark, Ed.). University of California Press.

Glick, B. R. (2012). Plant growth-promoting bacteria: Mechanisms and applications. *Scientifica*. PMID:24278762

Grotz, N., & Guerinot, M. L. (2006). Molecular aspects of Cu, Fe and Zn homeostasis in plants. *Biochimica et Biophysica Acta (BBA)-. Molecular Cell Research*, *1763*(7), 595–608.

Hell, R., & Stephan, U. W. (2003). Iron uptake, trafficking and homeostasis in plants. *Planta*, *216*(4), 541–551. doi:10.100700425-002-0920-4 PMID:12569395

Khuong, N. Q., Kantachote, D., Nookongbut, P., Xuan, L. N. T., Nhan, T. C., Xuan, N. T. T., & Tantirungkij, M. (2020). Potential of Mn 2+-Resistant Purple Nonsulfur Bacteria Isolated from Acid Sulfate Soils to Act as Bioremediators and Plant Growth Promoters via Mechanisms of Resistance. *Journal of Soil Science and Plant Nutrition*, *20*(4), 1–15. doi:10.100742729-020-00303-0

Kobayashi, T., & Nishizawa, N. K. (2012). Iron uptake, translocation, and regulation in higher plants. *Annual Review of Plant Biology*, *63*(1), 131–152. doi:10.1146/annurev-arplant-042811-105522 PMID:22404471

Kobayashi, T., Ogo, Y., Aung, M. S., Nozoye, T., Itai, R. N., Nakanishi, H., Yamakawa, T., & Nishizawa, N. K. (2010). The spatial expression and regulation of transcription factors IDEF1 and IDEF2. *Annals of Botany*, *105*(7), 1109–1117. doi:10.1093/aob/mcq002 PMID:20197292

Kraepiel, A. M., Bellenger, J. P., Wichard, T., & Morel, F. M. (2009). Multiple roles of siderophores in free-living nitrogen-fixing bacteria. *Biometals: An International Journal on the Role of Metal Ions in Biology, Biochemistry, and Medicine, 22*(4), 573–581. doi:10.100710534-009-9222-7

Krewulak, K. D., & Vogel, H. J. (2008). Structural biology of bacterial iron uptake. *Biochimica et Biophysica Acta (BBA)- Biomembranes*, *1778*(9), 1781–1804. doi:10.1016/j.bbamem.2007.07.026

Lau, E. T., Tani, A., Khew, C. Y., Chua, Y. Q., & San Hwang, S. (2020). Plant growth-promoting bacteria as potential bio-inoculants and biocontrol agents to promote black pepper plant cultivation. *Microbiological Research*, *240*, 126549. doi:10.1016/j.micres.2020.126549 PMID:32688172

Liu, D., Yang, Q., Ge, K., Hu, X., Qi, G., Du, B., Liu, K., & Ding, Y. (2017). Promotion of iron nutrition and growth on peanut by Paenibacillus illinoisensis and Bacillus sp. strains in calcareous soil. *Brazilian Journal of Microbiology, 48*(4), 656-670. doi:10.1016/j.bjm.2017.02.006

Loaces, I., Ferrando, L., & Scavino, A. F. (2011). Dynamics, diversity and function of endophytic siderophore-producing bacteria in rice. *Microbial Ecology*, *61*(3), 606–618. doi:10.100700248-010-9780-9 PMID:21128071

Lurthy, T., Cantat, C., Jeudy, C., Declerck, P., Gallardo, K., Barraud, C., Leroy, F., Ourry, A., Lemanceau, P., Salon, C., & Mazurier, S. (2020). Impact of Bacterial Siderophores on Iron Status and Ionome in Pea. *Frontiers in Plant Science*, *11*, 730. doi:10.3389/fpls.2020.00730 PMID:32595663

Mahmoud, A., & Abd-Alla, M. (2001). Siderophore production by some microorganisms and their effect on *Bradyrhizobium*-Mung Bean symbiosis. *International Journal of Agriculture and Biology*, *3*(2), 157–162.

Pahari, A., Pradhan, A., Nayak, S. K., & Mishra, B. B. (2017). Bacterial siderophore as a plant growth promoter. In *Microbial Biotechnology* (pp. 163–180). Springer. doi:10.1007/978-981-10-6847-8_7

Paul, A., & Dubey, R. (2014). Characterization of Protein involved in Nitrogen Fixation and Estimation of Co-Factor. *International Journal of Advanced Biotechnology and Research*, *5*(4), 582–597.

Pérez-Rodriguez, M. M., Piccoli, P., Anzuay, M. S., Baraldi, R., Neri, L., Taurian, T., Lobato Ureche, M. A., Segura, D. M., & Cohen, A. C. (2020). Native bacteria isolated from roots and rhizosphere of *Solanum lycopersicum* L. increase tomato seedling growth under a reduced fertilization regime. *Scientific Reports*, *10*(1), 15642. doi:10.103841598-020-72507-4 PMID:32973225

Pramanik, A., Zhang, F., Schwarz, H., Schreiber, F., & Braun, V. (2010). ExbB protein in the cytoplasmic membrane of Escherichia coli forms a stable oligomer. *Biochemistry*, *49*(40), 8721–8728. doi:10.1021/bi101143y PMID:20799747

Radzki, W., Gutierrez Mañero, F. J., Algar, E., Lucas García, J. A., García-Villaraco, A., & Ramos Solano, B. (2013). Bacterial siderophores efficiently provide iron to iron-starved tomato plants in hydroponics culture. *Antonie van Leeuwenhoek*, *104*(3), 321–330. doi:10.100710482-013-9954-9 PMID:23812968

Rout, G. R., & Sahoo, S. (2015). Role of iron in plant growth and metabolism. *Reviews in Agricultural Science*, *3*(0), 1–24. doi:10.7831/ras.3.1

Scavino, A. F., & Pedraza, R. O. (2013). The role of siderophores in plant growth-promoting bacteria. In *Bacteria in agrobiology: crop productivity* (pp. 265–285). Springer. doi:10.1007/978-3-642-37241-4_11

Schmidt, W. (1999). Mechanisms and regulation of reduction-based iron uptake in plants. *The New Phytologist*, *141*(1), 1–26. doi:10.1046/j.1469-8137.1999.00331.x

Sharma, A., Johri, B. N., Sharma, A. K., & Glick, B. R. (2003). Plant growth-promoting bacterium Pseudomonas sp. strain GRP3 influences iron acquisition in mung bean (Vigna radiata L. Wilzeck). *Soil Biology & Biochemistry*, *35*(7), 887–894. doi:10.1016/S0038-0717(03)00119-6

Sheng, M. M., Jia, H. K., Zhang, G. Y., Zeng, L. N., Zhang, T. T., Long, Y. H., Lan, J., Hu, Z. Q., Zeng, Z., Wang, B., & Liu, H. M. (2020). Siderophore Production by Rhizosphere Biological Control Bacteria *Brevibacillus brevis* GZDF3 of *Pinellia ternata* and Its Antifungal Effects on *Candida albicans*. *Journal of Microbiology and Biotechnology*, *30*(5), 689–699. doi:10.4014/jmb.1910.10066 PMID:32482934

Silva-Bailão, M. G., Bailão, E. F. L. C., Lechner, B. E., Gauthier, G. M., Lindner, H., Bailão, A. M., Haas, H., & de Almeida Soares, C. M. (2014). Hydroxamate production as a high affinity iron acquisition mechanism in *Paracoccidioides* spp. *PLoS One*, *9*(8), e105805. doi:10.1371/journal.pone.0105805 PMID:25157575

Takagi, S. I., Nomoto, K., & Takemoto, T. (1984). Physiological aspect of mugineic acid, a possible phytosiderophore of graminaceous plants. *Journal of Plant Nutrition*, *7*(1-5), 469–477. doi:10.1080/01904168409363213

Tripathi, D. K., Singh, S., Gaur, S., Singh, S., Yadav, V., Liu, S., ... Dubey, N. K. (2018). Acquisition and homeostasis of iron in higher plants and their probable role in abiotic stress tolerance. *Frontiers in Environmental Science*, *5*, 86. doi:10.3389/fenvs.2017.00086

Vessey, J. K. (2003). Plant growth promoting rhizobacteria as biofertilizers. *Plant and Soil*, *255*(2), 571–586. doi:10.1023/A:1026037216893

Winkelmann, G. (2002). Microbial siderophore-mediated transport. *Biochemical Society Transactions*, *30*(4), 691–696. doi:10.1042/bst0300691 PMID:12196166

Zouboulis, A. I., & Moussas, P. A. (2011). Groundwater and Soil Pollution: Bioremediation. Encyclopedia of Environmental Health, 1037–1044. doi:10.1016/B978-0-444-52272-6.00035-0

KEY TERMS AND DEFINITIONS

Biocontrol: Biocontrol or biological control is an environmentally friendly way in which certain organisms are used to control or prevent the growth of pests like insects, mites, pathogenic microbes, etc.

Biofertilizer: Biofertilizer is a substance comprising of micro-organisms (living) which, when applied on plant parts or soil provides nutrition to the plants and promotes their growth. It is an environmentally sound way of providing nutrients to the plant instead of harmful chemical-based formulations.

Bioremediation: Bioremediation involves the use of micro-organisms (living) or plants for reduction or degradation of environmental pollutants into non-toxic or less toxic forms.

Phytosiderophores: Fe^{3+} chelating compounds (mugineic acids (MAs) or its modified derivatives) released by plants in the rhizosphere under iron deficient conditions.

Plant Growth Promoting Bacteria (PGPB): Bacteria those that help in enhancing the growth of the plant are known as Plant Growth Promoting Bacteria (PGPB). These can be free living or symbiotic inhabiting the rhizosphere or colonize inner plant tissues or organs.

Rhizosphere: A dynamic, narrow region of the soil where plant roots are easily accessible and are densely populated with microorganisms (especially bacteria). It is the region where maximum interactions between plant roots and the fauna take place.

Siderophores: Selective low molecular weight (less than 2000 Da), water soluble, organic molecules that binds to ferric ions (Fe^{3+}) with high affinity and specificity. They are secreted by nearly all aerobic bacteria and fungi under iron deficient conditions.

Chapter 12
Plant Growth-Promoting Rhizobacteria (PGPR): A Unique Strategy for Sustainable Agriculture

Podduturi Vanamala
Telangana Social Welfare Residential Degree College for Women, Kamareddy, India

Uzma Sultana
Telangana Social Welfare Residential Degree College for Women, Kamareddy, India

Podduturi Sindhura
Telangana Social Welfare Residential Degree College for Women, Bhupalpally, India

Mir Zahoor Gul
https://orcid.org/0000-0002-1088-4255
Department of Biochemistry, University College of Science, Osmania University, Hyderabad, India

ABSTRACT

With a substantial decline in the use of synthetic chemicals, the growing demand for agricultural production is a critical concern in today's world. The use of plant growth-promoting rhizobacteria (PGPR) has been found to be an environmentally sound way of increasing agricultural productivity by promoting plant growth either through a direct or indirect mechanism. PGPRs are commonly occurring soil microbes that colonize the root system, which is an ideal location for interactions with plant microbes. PGPRs can provide an enticing way of reducing the use of toxic chemicals and can affect plant growth and development, either through releasing plant growth regulators or other bioactive stimulants and by taking up nutrients through fixation and mobilization, minimizing adverse effects of microbial pathogens on crops by using numerous mechanisms. In addition, they also play a significant role in soil fertility. This chapter aims to explore the diversified plant growth mechanisms that promote rhizobacteria in fostering crop yields and promoting sustainable agriculture.

DOI: 10.4018/978-1-7998-7062-3.ch012

INTRODUCTION

Non-pathogenic strains of soil bacteria that multiply in the rhizosphere and grow in and around the root tissues, by stimulating host plant growth by different biological mechanisms are defined as plant growth-promoting rhizobacteria (PGPR). Rhizospheric zones surrounding roots are the hotspots for all the microbial interactions due to the presence of various organic, biochemical metabolites secreted by the roots. These root secretions majorly include water-soluble sugars, vitamins, organic acids, amino acids, sugar-phosphate esters, phenolics, and amino compounds (Uren, 2000). Root exudates serve as the major source of nourishment for the microorganisms and they attract huge number of microbial populations when compared to the non-rhizospheric soils. These root secretions govern the plant-microbial interactions via the chemotaxis mediated response. Microbial interactions in response to the exudates secreted by the plant play a significant role in successful root colonization. The success of plant growth-promoting rhizobacteria depends upon the colonization of roots, as the colonization of roots is the first crucial stage during the interaction between the PGPR and the host plant. After successful root colonization PGPR facilitates host plant growth (Ahemad et al., 2014; Goswami et al., 2016), through direct or indirect mechanisms (Ortíz-Castro et al., 2009).

Direct mechanisms include nitrogen fixation, potassium, phosphorus, zinc solubilization, production of siderophores, and phytohormones. Direct mechanisms help in enhancing soil fertility. Indirect mechanisms comprised of production of antibiotics, exopolysaccharides hydrolytic enzymes, and cyanide compound. Indirect mechanisms do not affect soil fertility directly but help in maintaining soil health by repressing pathogenic soil microorganisms, mentioned in Figure 1. PGPR repress the phytopathogens by; **1)** exhibiting direct antagonistic activities against the pathogens (Beneduzi et al., 2012), **2)** competing for space and nutrients (Kumari and Srivastava 1999), and **3)** triggering induced systemic resistance (ISR) in plants (Egamberdieva et al., 2017). Induced systemic resistance elevates the defense capacity of host plants against the phytopathogens and pests to overcome the biotic stresses.

Biotic and abiotic stress is the major cause of yield loss in agriculturally important crops. Rainfed crops mainly suffer from abiotic stresses like nutrient deficiencies and environmental factors like high temperature, drought, salinity, and pH of soils. Crop loss due to physiological modulation in plants is observed against abiotic stresses. These stresses cause a 50-82% significant decrease in agricultural productivity. Among biotic stresses pathogenic microorganisms, pests and weeds cause enormous damage to the crop. The pathogenic microflora damages the root hairs, lateral roots, release toxins and destroys the plants (Singh et al., 2014; Mishra et al., 2015). Nearly 7-15% of the crops are damaged by various soil-borne, bacteria, fungi oomycetes, and nematodes.

Microorganisms that colonize the roots and possess the ability for salt-tolerant, nutrient uptake and produce compatible solutes play a significant role in abiotic stress management. Plants are affected by salt stress in three different ways viz., ionic toxicity, osmotic imbalance, and decrease in nutrient uptake (Selvakumar et al., 2014). Proline is vital compatible solute for both bacteria and plants to respond against the osmotic imbalance and ion toxicity. Proline can influence cell proliferation and apoptosis and regulates specific gene expression to reduce salt stress (Ahmad et al., 2016). Thus, the above facts illustrate that the rhizobacteria with plant growth-promoting abilities can be used as a suitable bio inoculant to promote plant growth and enhance productivity through different mechanisms in addition to the accumulation of proline as osmo-regulators.

Figure 1. Direct and indirect mechanisms of PGPR in promoting plant growth

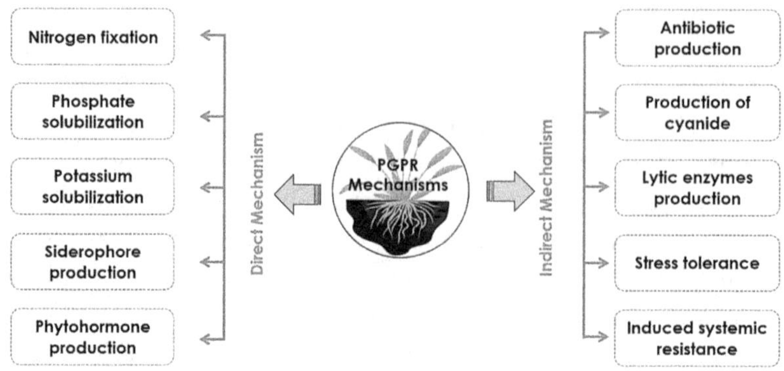

Other environmental abiotic stresses like heavy metal accumulation are a major concern that poses severe adverse effects upon the plant, animal, and human health. A normal concentration of metals is required for the proper physiological functioning of plants. Concentrations directly above the required levels cause toxic effects and limit plant growth. Besides limiting plant growth, mineral toxicity also has negative effects on the crop yield of plants by accelerating the synthesis of ethylene (Safronova et al., 2006). Hence soils polluted with heavy metals are restricted for agriculture. Many strategies have been developed to counteract heavy metal accumulation for the reclamation of agricultural lands. The application of plant growth-promoting rhizobacteria has been a promising approach in reduction strategies of heavy metal tolerance. Therefore, the use of PGPR in the form of biofertilizer improves fertility and increases agricultural productivity. Thus, plant growth-promoting rhizobacteria represent an essential component of biofertilizer technology to replace or reduce the use of chemical inorganic fertilizers.

PGPR AS BIOFERTILIZERS

Soil rich in nutrients provides nutrients for optimum plant growth and enhances agricultural production (Ney et al., 2019). Although, crop productivity is often limited by available soil nutrients, especially nitrogen (Vitousek and Howarth, 1991). Nitrogen content in the atmosphere is highest and constitutes about 78% of all atmospheric gases. Despite its abundance in the atmosphere nitrogen is present in inadequate amounts in soil and is not directly taken up by the plants (Hedin et al., 2009). Current soil management strategies are mostly based on synthetic nitrogen fertilizers which cause soil contamination and a serious threat to the environment and human health. Therefore, reducing dependence on nitrogenous fertilizers in agriculture in the developed world and developing countries may lead to potential gains in the plant, soil, and human health. Biological nitrogen fixation has drawn attention to achieve sustainable agricultural goals in economically important food and forage crops (Sulieman and Tran, 2016). It has been estimated that worldwide, biological nitrogen fixation contributes to the average production of 200 million tons of nitrogen annually (Graham, 1992; Peoples et al., 2009). The utilization of beneficial microbes as biofertilizers has gained significant interest in the agricultural industry for their primary importance in food safety and sustainable crop production. The application of plant growth-promoting rhizobacteria is the most promising approach in enhancing plant nutrition and has been proven to be an environmentally sound way of increasing crop yields without posing environmental contamination (Calvo et al., 2014). Depending upon their interactions with plants, PGPR can be categorized into symbiotic bacteria, where they live in the intercellular spaces of the host and free-living rhizobacteria which live outside plant cells (Gray and Smith, 2005).

Nitrogen-Fixing Bacteria

Nitrogen is a major essential nutrient and the main element of nucleic acids and protein and other organic nitrogenous compounds. Although its concentration in air is high, the Nitrogen concentration in soil, seawater, and rocks is relatively less and its availability is often a limiting factor for plant growth and crop production. Chemical production of nitrogenous fertilizers like urea requires an enormous amount of energy that releases a 10-fold or even greater amount of CO_2 equivalent (Zhang et al., 2013). Moreover, only 30-40% of the chemical fertilizers applied in fields are used by the plants (Prasad, 2009), while the rest contaminates and cause severe environmental complications. The pollution triggered by chemical nitrogenous fertilizers has been expected to cost the European Union a huge amount that could be anywhere between EUR 70 and 320 billion/year (Sutton et al., 2011).

Biological nitrogen fixation appears as an alternate strategy and can be exploited as an alternate approach to decrease the input of nitrogen fertilizers in agronomy and their undesirable environmental impacts. Biological nitrogen fixation is a natural process of transforming atmospheric nitrogen (N_2) into a simple soluble nontoxic form (NH_4^+) which can be utilized by the plant cells for the synthesis of several biomolecules. Nitrogen fixation is one of the major sources of nitrogen for plants and a crucial process in distributing this nutrient in the ecosystem. Biological Nitrogen fixation is carried out enormously by prokaryotes: bacteria and archaea (Graham, 1992; Peoples et al., 2009).

Diazotrophic bacteria are present in several phyla (Boyd et al., 2013) and the representative members of this phylum are found to engage in nitrogen-fixing symbiosis with plants (Hardoim et al., 2015). The nitrogen-fixing trait of plant growth-promoting rhizobacteria has been identified among the bacilli and specifically among the proteobacteria (Schmid and Hartmann, 2007). Nitrogen fixation is a dynamic and energetically expensive ATP molecules are utilized because 16 molecules of ATP are required to breakdown a nitrogen molecule and additionally, 12 ATP molecules are utilized for ammonium assimilation and transport. In a symbiotic association, nodulating plants must provide 12 g of glucose to their bacterial partners to benefit 1 g Nitrogen in part (Buscot et al., 2005). BNF enables the plants to use an inert form of nitrogen from the atmosphere after biotransformation of nitrates and nitrites into Ammonia mediated by nitrogen-fixing organisms (Bohlool et al., 1992).

Diazotrophs can fix atmospheric nitrogen and can convert atmospheric nitrogen into more utilizable compound ammonia through the action of the nitrogenase enzyme. Biological nitrogen fixation is mediated by an enzyme called Nitrogenase. The nitrogenase enzyme is a metalloenzyme complex comprised of an iron-protein homodimer and an iron-molybdenum protein heterodimer and is encoded by *nif*HDK genes (Peters et al., 2011; Rubio and Ludden, 2008).

Rhizobia, a group of soil bacteria that includes the genera *Rhizobium, Bradyrhizobium, Sinorhizobium, Azorhizobium,* and *Mesorhizobium*) are gram-negative bacteria that inhabit as nitrogen-fixing endosymbionts in the stem and root nodules formed on host plants (Graham and Vance, 2003). The common microbial species that associate endosymbiotically with legumes of the Fabaceae (Papilionaceae) family is the gram-negative alpha-proteobacteria (Schultze, 1998; Oldroyd and Downie, 2008; Desbrosses, 2011). *Rhizobium* species live in a mutualistic relationship with the roots of leguminous plants that has the potential to reduce atmospheric nitrogen to ammonia for their host through the formation of nodules (a new differentiated special organ). Rhizobia form a symbiotic association with plant species such as *Arachis. hypogaea, Acacia sp., Cajanus cajan, Cicer arietinum, Cercis Canadensis, Glycine max, Lotus corniculatus, Lens culinaris, Medicago sativa, Pisum sativum, Phaseolus vulgaris,* and *Trifolium sp.* (Verma et al., 2010). *Parasponia andersonii* the only non-legume of family cannabaceae exhibit unique nitrogen-fixing symbiosis and nodulation by rhizobia (Sytsma, 2002). Rhizobia proliferate and fix nitrogen within infectious threads and are not released into cells in symbiosomes. The degree of specificity between rhizobia and legumes varies. For instance, Nod factors secreted by *Rhizobium loti* and *Rhizobium etli* are identical, but they have different hosts (*Lotus spp.* and *Phaseolus spp.* respectively). Additionally, nod factors secreted by different rhizobia that nodulate the same plant varies. *Rhizobium etli* and *Rhizobium tropici* produce two different nod factors (acetylfucosylated and sulfated nod factors) respectively but can nodulate the same plant *Phaseolus vulgaris* common bean (Perret et al., 2000).

In addition to fixing nitrogen and promoting plant growth, *Rhizobium japonicum* employs antagonistic activity against phytopathogens *Fusarium solani* and *Macrophomina phaseolina,* causative agents of soybean root rot. Therefore, *Rhizobium japonicum* is considered an important bacterium in the management of root rot diseases. According to previous studies, rhizobia reported a significant increase in seed germination and improvement of crop yields and plant health management through reducing the attack of soil-borne pathogens (Sheikh et al., 2006; Mazen et al., 2008).

Numerous diazotrophic strains (*Azotobacter, Azospirillum, Rhizobia, Bradyrhizobium, Ensifer, Pseudomonas, Klebsiella*) have been reported to amplify the plant growth and grain yield of chickpea, wheat, rice, bean, and pea. These strains produce phyto-stimulators and secondary metabolites (Gopalakrishnan et al., 2017). Gopalakrishnan et al. (2018) demonstrated that rhizobia also act as PGP by producing phytohormones (IAA, GA_3) organic acids, and siderophores (Iron binding compounds) that have resulted in stimulation of root and stem growth of chickpea (*Cicer arietinum*). Some of the *Bradyrhizobial* strains recovered from rice rhizosphere and *Azotobacter caulinodans* associated with Sesbania rostrate have the potential to fix N_2 in a free-living state under low oxygen concentrations (Yanni et al., 1997). Moreover, Gopalakrishnan et al. (2015) and Das et al. (2017) described that rhizobia can act as biocontrol agents against phytopathogenic fungi (*Rhizoctania solani, Fusarium oxysporium, Macrophomina phaseolina,* and *Sclerotium rolfsii*) through the production of hydrocyanic acid, antimycotic enzymes/antimicrobial agents. Rhizobia in the rhizosphere exhibited biocontrol ability against these pathogens showed high efficiency under greenhouse and field conditions (Nelson, 2004; Siddiqui, 2006; Akhtar and Siddiqui, 2009).

Azotobacter

Azotobacter has been used as a biofertilizer and was first described in 1901 by Martinus Beijerinck. Biofertilizers are environmental friendly and often counteract plant pathogens. *Azotobacter* is ubiquitous, aerobic, free-living, gram-negative, soil-borne bacteria commonly found in soil, water, and sediments. *Azotobacter* belongs to kingdom bacteria and is included under phyla *Proteobacteria,* class *gamma proteobacteria*, order *Pseudomonadales* and family Azotobacteraceae (Kennedy et al., 2005). *Azotobacter* is chemoorganotrophs and capable of utilizing sugars, alcohols, and organic acid salts for their growth. *Azotobacter* is a common free-living diazotroph found in agricultural soils without any symbiotic association with plants. *Azotobacter* plays different beneficial roles like the production of plant growth hormones (Indole-3-acetic acid [IAA] and gibberellins), biofilms, Exopolysaccharides, hydrolytic enzymes, antifungal substances, vitamins (riboflavin), and siderophores and notably, they can fix atmospheric nitrogen (Myresiotis, 2012). For its plant growth-promoting activity for sustainable agriculture (Chennappa et al., 2014; Jimenez et al., 2011; Aquilanti et al., 2004), several *Azotobacter* species *A. agilis, A. brasilense, A. paspali, A. insignis, A. tropicalis, A. salinestris, A. vinelandii* are extensively studied. *A. agilis* and *A. chrococcum* were first isolated and cultured by Beijerinck. *A. vinelandii* is greatly studied and its genome has been sequenced (Setubal et al., 2009).

Azotobacter can solubilize insoluble phosphates in the soil. *Azotobacter* mutants capable of releasing 1.5-1.7 µg phosphorous/mL from the supernatant of insoluble tricalcium phosphate were isolated by Kumar et al. (2001). The majority of nitrogen-fixing organisms fix nitrogen under anaerobic conditions as the nitrogenase enzyme that plays a very important role in nitrogen fixation is inactivated by the presence of oxygen. On contrary, *Azotobacter* has unique potential as they can carry out nitrogen fixation aerobically because of having maximum respiratory quotient among all biological systems studied (Haddock and Jones, 1977). This property of *Azotobacter* is exploited in using this organism as a biofertilizer. Seed inoculation with wild-type *Azotobacter* has improved the yield of cereals like corn, oat, rice, barley, wheat, pearl millet, and sorghum. The enhanced yield of oil seeds like sunflower and mustards has also been reported. Seed inoculation of vegetable crops like carrot, onion, potato, chilies, sugar beets, tomato, and beans with *Azotobacter* enriched the crop yield (Mrkovacki and Milic, 2001). Many *Azotobacter* species are found to produce IAA in the range of 2.09-33.28 µg/mL (Chenneppa et al., 2013). *Azotobacter* produces gibberellins that promote cell division, flowering, and seed growth and reverse the dormancy induced by Abscisic acid (ABA). IAA is responsible for cell division, cellular differentiation of plant tissue, and has a role in stimulating root elongation. *A. vinelandii* strain ATCC 12837 and *Azotobacter chrococcum* ATCC H23 (CECT- 4435) produced niacin, riboflavin, biotin, and pantothenic acid (Revillas et al., 2000). Ahmad et al. (2005) reported that *Azotobacter* has the potential to produce a high amount of IAA (7.3-32.8 mg/mL) in agriculture. *Azotobacter* needs a high amount of organic carbon for their growth and is less active in soils deficit in poor organic content (Bhosale et al., 2013; Barrera and Soto, 2010).

Inoculation of seeds with engineered *A. chroococcum* HKD 15 (Bageshwar et al., 2017) enhanced the corn yield of wheat by 60%. On the other hand, seed inoculation with wild type *A. chroococcum* CBD15 resulted in only a 10% enhancement of yield. Recent research studies on *A. vinelandii* revealed that siderophores besides iron mobilization were used in the uptake of molybdenum (Mo), vanadium (V), and nitrogenase cofactors. Extensive studies of siderophore production in *A. vinelandii* showed that siderophore production was increased in the presence of limited iron, whereas under Mo limitation, there was increased production of catechol type of siderophore (McRose et al., 2017).

Phosphate Solubilizing Bacteria

Phosphorus is the second key element after nitrogen as a mineral nutrient in terms of plant requirements. Phosphorus has a major role in N-fixation in legumes, root development, flower, and seed formation and to impart resistance to plant diseases. It also accounts for the early maturation of crops like cereal, and legumes. Phosphorus accounts for about 0.2 - 0.8% of the plant's dry weight. Although phosphorus being rich in soils it is one of the major plant growth-limiting nutrients. Anions of phosphate are highly reactive and get immobilized through precipitation with cations such as Ca^{2+}, Mg^{2+}, Fe^{3+}, and Al^{3+} (Gyaneshwar et al., 2002). In these forms, phosphorus is highly insoluble and unavailable to plants due to which phosphorus deficiency is seen in most of the soils worldwide. Several studies have shown that phosphate solubilizing microorganisms (PSMs) solubilize the fixed 'P' in the soil resulting in higher crop yields (Mahanta et al., 2014). Many different bacterial species can solubilize insoluble inorganic phosphate compounds like dicalcium phosphate, tricalcium phosphate, and hydroxyapatite, and rock phosphate. Phosphate solubilizing microorganisms (PSM's) are common in the rhizosphere of many crops. PSMs differ from one soil to another, phosphate solubilizing bacteria constitute 1-50% total microbial population. Phosphate-solubilizing bacteria outnumber phosphate solubilizing fungi by 2-150 times.

The primary mechanisms for mineral phosphate solubilization are through the production of acid phosphatases and organic acids like gluconic, citric, maleic, succinic, glyoxalic, and fumaric acids (Aeron et al., 2011). Non- specific acid phosphatases produced by bacteria are extracellular molecules of enzymes that catalyze the hydrolysis of a wide variety of phosphomonoesters and transphosphorylation reactions. Among the different classes of phosphatase enzymes released by PSM, phosphomonoesterases are the most abundant (Nannipieri et al., 2011). Depending upon the pH optima enzymes are classified into acid and alkaline phosphatases. Acid phosphatases predominate in acid soils, and alkaline phosphatases are more abundant in alkaline and neutral soils. Organic acids produced by PSM's solubilize insoluble phosphate by lowering of pH and chelation of cations. Among the various organic acids, solubilization by gluconic acid seems to be the major mechanism followed by Gram-negative bacteria (Kim et al., 1997). Gluconic acid is produced by the oxidative metabolism of glucose-by-glucose dehydrogenase (GDH), which requires pyrroloquinoline quinine (PQQ) as a cofactor. Besides enzymes and organic acids, other mechanisms of phosphorus solubilization include inorganic acids produced by chemoautotrophic bacteria (Khan et al., 2014). Inorganic acids like hydrochloric acid can solubilize phosphate but are less effective than organic acids at the same pH. In *in vitro* conditions, phosphorus solubilization generally related to the degree of acidification of the media as measured by a fall in pH. The most important bacterial genera of mineral phosphate solubilizers include *Bacillus* and *Pseudomonas* while *Aspergillus* and *Penicillium* form fungal genera (Saritha and Tollamadugu, 2019). Soil *Bacillus* mineralizes fixed organic phosphates through the release of extracellular enzymes like phosphoesterases, phosphodiesterases, phytases, and phospholipases. Mixed cultures of PSMs (*Bacillus*, *Streptomyces,* and *Pseudomonas*) are most effective in mineralizing organic phosphate (Walpola and Yoon, 2012).

PGPR AS BIOCONTROL AGENTS

Plant microbial interactions can be beneficial or deleterious, and in a few instances, they are neutral too. The beneficial rhizobacteria are generally referred to as plant growth-promoting rhizobacteria (Kloepper et al., 1989). PGPR, accumulate in the root region, and they protect plants from pathogens by antagonistic mechanisms. To control Plant diseases chemical pesticides are used worldwide but the continuous exercise of using pesticides reduces the quality of soil and poses severe environmental issues (Guo et al., 2013; Dun-chun et al., 2016). Hence biocontrol of phytopathogens is contemplated as an efficient alternative for eco-friendly agriculture (Compant et al., 2005). The application of beneficent rhizobacteria to the soil is advantageous over conventional pest control practices, as it is non-toxic, stimulates plant growth, and prevent the devastation that occurred because of the presence of different pathogens (Olanrewaju et al., 2017). There is an abundant number of bacterial forms that interact with the plant roots. With innovations in the field of soil microbiology, there is the development of several ways they operate to increase the efficacy of biocontrol agents (BCAs). PGPR has widespread use in sustainable irrigation because of its metabolic versatility and the ability to produce antifungal agents and excellent root colonization capability.

Production of Antibiotics

PGPR strains found in association with cereal crops produce several antifungal agents capable of controlling fungal diseases (Ongena et al., 1999; Bloemberg and Lugtenberg 2001; Antibiosis is another highly effective way for controlling soil-borne pathogens associated with several crops (Handelsman and Stab, 1996). The PGPR strains produce antibiotics, which control the accumulation of fungal root pathogens in the rhizosphere region (Haas and Défago, 2005). A wide variety of antibiotics produced by the *Bacillus spp.*, *Stenotrophomonas spp,* and *actinomycetes* members, include Oligomycin A, Kanosamine, Xanthobaccin. In addition to these few bacteria also produce numerous antimicrobial compounds which include 2,4-diacetylphloroglucinol (DAPG), hydrogen cyanide, Oomycin, phenazine, amphisin, which possess phytotoxic, antioxidant, and anti-tumor properties. The antibiotics produced by rhizobacteria cause the cellular damage of pathogenic forms such as *Pythium spp.* and block the formation of zoospores (deSouza et al., 2003).

The phylogenetic reviews on biocontrol strains of rhizobacteria reported that there is a positive correspondence on control of plant diseases and production of antibiotics (Vincent et al., 1991). The most common bacteria found in this rhizosphere region are *Pseudomonas and Bacillus spp.* (Dutta and Podile, 2010). A special focus is there on soil-borne pathogen *fluorescent pseudomonads* (Walsh et al., 2001). The plant-microbe interactions are significant for plant growth development and for controlling plant pathogens. Another aspect of PGPR is formulating a live microorganism as BCA for integrated disease control management. The rhizobacteria occupy the root region of crop plants and are the most effective means of biocontrol, as they suppress soil-borne pathogens. They became an important alternative to the use of chemical antimicrobial agents, because of enhanced consciousness of detrimental effects of chemical products on health and environment as mentioned in various studies (Raupach and Kloepper, 1998; Walsh et al., 2001; Kobayashi et al., 2002). PGPR with biocontrol efficiency provides indelible shielding from soil-borne pathogens. Rhizobacteria are used as inoculants in soil, they rapidly colonize the rhizosphere region. Several PGPR forms are widely used as soil inoculants on cereals, but most of them are applied as organic biofertilizers, not as biocontrol agents (Ryder et al., 1999).

Seed Treatment and Rhizosphere Competence

The potential of PGPR as a biocontrol agent was reported in several cereal crops including maize, rice, sorghum, and chickpea. Root rot caused by *F. verticilloidees*, in maize crop has been subsided by incorporating *B. amyloliquifacience* in the treatment of the seeds (Pereira et al., 2009). The functional principle of a biological control agent is a determining factor to develop efficacious disease control methods. It is often believed that rhizobacteria must subjugate the root surface of a plant to be an efficient biocontrol agent. *Bacillus* and *pseudomonas spp.*, are the most important bacteria, associated with several crops. They secrete various metabolites capable of suppressing the accumulation of most of the bacterial and fungal pathogens in the rhizosphere (Rangarajan et al., 2003). Most of the rhizobacteria that accumulate in the rhizosphere region exert antifungal properties, protecting plants from a wide range of fungal pathogens in the soil. To be a consistent performer as a BCA, bacteria must accumulate sufficiently in the crop soil (Bloemberg and Lugtenberg, 2001).

Siderophore Production

Iron Sequestering Bacteria

Iron is an inevitable nutrient for the metabolism of plants and associated microorganisms. Iron plays an important role in various physiochemical pathways in plants. It is a key component of many vital enzymes, cytochromes of the electron transport chain, involved in the synthesis of chlorophyll, and it is crucial for the maintenance of the structure and function of the chloroplast. Iron is required as a cofactor to control enzymatic reactions in all microorganisms. In iron limiting conditions there will be great competition for the uptake of iron in the rhizosphere. Microorganisms use active strategies for the uptake of iron from the soil by different processes like acidification, reduction, and chelation by secretion of iron-chelating molecules. Under iron starvation plant growth-promoting bacteria secrete, specialize iron-chelating molecules known as siderophores. In 1973, the term siderophore was first coined by Lankford to describe low molecular weight compounds that have an exceptionally high affinity towards ferric ions. Different iron-chelating ligands of siderophores are hydroxamates, oxazoline, phenolates, carboxylate, α-hydroxy carboxylate, and ketohydroxyl bidentate. Many plant growth-promoting bacteria like *Azotobacter, Rhizobium, Pseudomonas, Bacillus, Azospirillum, and Serratia* secrete various types of siderophores. These iron-binding molecules with high redox potential will form a complex with an iron later the ferrous form of iron is transferred to plants through the apoplastic pathway in roots (Crowley, 2006). Plant growth-promoting microbes produce high-affinity siderophores when compared to phytopathogens which produce low-affinity siderophores. Microorganisms producing high-affinity siderophores colonize efficiently in the rhizosphere.

Biocontrol agents exert their antifungal or antibacterial activity against various pathogens affecting plants by secreting siderophores, which preferentially chelate iron (Fe^{+3}) to meet the requirements of the cell (Neilands 1995; Wandersman and Delepelaire, 2004). Siderophore reduces the levels of available iron in the rhizosphere region, which in turn prevents the proliferation of bacterial and fungal pathogens (Olanrewaju et al., 2017; O'sullivan, and O'gara, 1992). The growth of the fungal pathogens is inhibited due to the siderophores, as they require iron for their metabolism and sporulation. The fluorescent *pseudomonads* are efficient in iron chelation, found in the rhizosphere region of various crops. They secrete two major types of siderophores, pseudobactin and pyochlins. Pseudobactin is a fluorescent pigmented pyoverdins. In several studies, it is reported that *Pseudomonas* strain B324 secrete pyoverdins, which plays important role in controlling Phythium root rot disease of wheat (Loper and Henkels, 1999). Thus, the siderophore production by rhizosphere organisms is an important mechanism to protect the plant against root pathogens.

Cell Wall Degrading Enzymes

The cell wall degrading enzymes secreted by various PGPR surrounding rhizosphere region, contribute to antibiosis and antifungal properties to suppress the fungal pathogens (Chet et al., 1990; Kobayashi et al., 2002). Cell wall degrading enzymes includes cellulase, chitinase, protease, and β-1,3-glucanase synthesized by PGPR strains exert an efficient inhibitory effect on the proliferation of fungal growth. The enzymes chitinase and beta-glucanases produced by *Pseudomonas flourescens* LPK2, *Sinorhizobium fredii* KCC5 strains, degrade the structural components of the fungal cell wall and suppress the wilt caused by *Fusarium udum*. *Paenibacillus* and *Streptomyces spp.* reported to exert biocontrol activity against root pathogen, *Fusarium oxysporium.* Similarly, *Bacillus cepacia* produces β-1,3-glucanase which degrades the cell walls of fungal pathogens, *R. solani, P. ultimatum, and S. rolfsi* (Compant et al., 2005). *Pseudomonas, Bacillus, Alcaligens,* and *aeromonas* produce hydrogen cyanide that possesses antifungal activity (Compant et al., 2005; Guo et al., 2013; Olanrewaju et al., 2017). The other species of *bacillus* include *B. licheniformis, B. thuringiensis, B. cereus,* and *B. circulans* secrete cell wall degrading enzyme, chitinase (Sadfi et al., 2001). Chitinolytic activities are also found in gram-negative bacteria, *Pseudomonas aeruginosa, Serratia marcescens, Enterobacter agglomerans,* and *P. fluorescens* (Nelson and Sorenson, 1999).

Induction of Systemic Resistance

The colonization of diverse rhizobacteria may induce systemic resistance in the plant, which gives non-specific protection against most of the soilborne pathogens (van Loon et al., 1998; Silva et al., 2004). The Induced Systemic Resistance (ISR) is a key mechanism, specific for the recognition of pathogens by plant receptors (Pieterse et al., 2014). The mechanism of disease bio-control by nonpathogenic rhizosphere bacteria is different from systemic acquired resistance (SAR), which is induced by pathogen by producing salicylic acid (Bakker et al., 2014). PGPR exert ISR in most of the plants by enhancing the physiological and mechanical strength of the host, by inducing the production of antagonistic chemicals such as Chitinase, peroxidase, and the proteins related to pathogenesis (Ramamoorthy et al., 2001; Nandakumar et al., 2001; Silva et al., 2004). *P. fluorescens* strains showed an adverse effect on vegetative growth of *R. solani* by inducing resistance in the rice plant (Radjacommare et al., 2004). Resistance induced by PGPR by activating genes encode for cell wall degrading enzymes and few other biocontrol agents. *S. marcescens* B2 strain inhibits the growth of common root pathogens including *F. oxysporum,* due to the induced systemic resistance (Someya et al., 2000). Several strains of Bacillus including *B. subtilis* AF1 usually found in the rhizosphere, also capable of inducing systemic resistance against various root pathogens.

Quorum Sensing

Quorum sensing (QS) is a community regulation process where bacteria sense their population density through a cell to the cellular communication system. Quorum sensing signaling is mediated by auto-inducer molecules that control the microbiological functions of agricultural, medical, and industrial importance. The N-acyl homoserine lactones (AHLs) are reported in most of the signaling systems. In few gram-negative bacteria some other molecules, including diketopiperazines, furanosyl borate diester, and c-butyrolactone were reported in density-dependent signaling (Holden et al., 1999; Chen et al., 2002; Yamada and Nihira, 1998). Pathogenic bacteria use quorum sensing as a key mechanism to control the expression of virulence factors which includes biofilm formation, secretion of toxins, and hydrolytic enzymes. Interruption of the regulation system could be a valuable tool to control the activities of plant pathogens. Various signal interference mechanisms, enzymatic and non-enzymatic quench the QS and blocks biofilm formation (Zhang and Dong 2004; Ren et al., 2001). Several studies focused on the quorum sensing capacity of the pathogen by impairing the signaling system required to produce virulence factors (Olanrewaju et al., 2017). Few strains of rhizobacteria can detoxify the virulence factors, produced by *Xanthomonas albilineans, Fusarium,* and other phytopathogens which are also considered as a mechanism of biocontrol (Compant et al., 2005).

PGPR IN DROUGHT TOLERANCE

Drought is the major constraints in the field of Agriculture. Drought impairs normal growth, reduces the crop yield and this problem will be more severe in the future. Inclusive research is being carried out for discovering innovative approaches to increase the stress tolerance in plants which involves the adaption of traditional water-saving irrigation and the production of drought-tolerant plants by genetic engineering. One most convenient, alternate approach in this connection is the use of beneficial rhizobacteria, which is comparatively less expensive in practice. PGPR can facilitate plant growth by enhancing tolerance against biotic as well as abiotic stress (Bashan and Holguin, 1998; Cassan et al., 2009).

Production of Phytohormones

The most common procedures used by rhizobacteria to vitalize the growth of the plant include nitrogen fixation in the rhizosphere, phosphate solubilization, secretion of iron sequestering siderophores and Phytohormones production such as abscisic acid (ABA), gibberellic acid, cytokinins, and IAA (Glick et al., 1999). The Possible explanation for drought tolerance by PGPR includes induced systemic tolerance by bacterial exopolysaccharides secreted in the rhizosphere. They are also linked to catabolism of molecules, such as bacterial ACC deaminase (1-aminocyclopropane-1-carboxylate deaminase), by reducing plant ethylene levels in roots (Mayak et al., 2004; Arshad et al., 2008). The rhizobacteria must be able to survive and overshadow with native microflora, in the root region for a successful application, especially in drought-affected soils (Bashan, 1998). The drought-tolerant bacteria thus are advantageous over other bacteria in promoting plant growth to overcome the adverse conditions. Such rhizobacteria are naturally adapted to drought, found in association with various crops. The microbial species, with stress alleviating potential are used for sustainable agriculture and are reported to possess an important role in helping plants to cope with drought.

ACC Deaminase Activity

ACC (1-amio cyclopropane-1-carboxylic acid) is the essential precursor to produce plant hormone ethylene. Ethylene is a key phytohormone for normal growth in plants but at high concentrations, it induces defoliation and leads to reduced crop production. PGPR found in rhizosphere and rhizoplane, was ecofriendly, and has an excellent effect in augmenting the plant growth and stress tolerance (Shrivastava and Kumar, 2015; Turan et al., 2017; Gouda et al., 2018; Grobelak et al., 2018; Nagargade et al., 2018). The bacteria were found to reduce the stress by hydrolyzing ACC to ammonia and α-ketobutyrate, by the enzyme ACC deaminase. Thus, reduce the level of ethylene to minimize the abiotic stress and equip the plant with enhanced resistance to drought stress and salinity stress (Pourbabaee et al., 2016; Ravanbakhsh et al., 2017; Ghosh et al., 2018; Saikia et al., 2018). ACC deaminase activity can modify the pathway of ethylene biosynthesis and produce IAA, which strengthens the root system of the arabidopsis plant (Desbrosses et al., 2009). The primary mechanism of ACC deaminase function includes the destruction of ethylene and diminishing the accumulation of ethylene levels and construct a healthy root system needed to cope with salinity stress. Several bacteria including *Achromobacter, Azospirillum, Pseudomonas,* and *Rhizobium* reported having ACC deaminase activity as stated in several research reviews (Ghosh et al., 2003; Govindasamy et al., 2008; Duan et al., 2009). PGPR strains expressing the enzyme ACC deaminase can stimulate plant growth and development especially under environmental stress like salinity, waterlogging, and drought (Ghosh et al., 2018). They are also helpful in increasing dry matter of root and aerial parts in canola (*Brassica napus*) if the seeds are inoculated with ACC deaminase producing gene. Modified genes of rhizobacteria that express ACC deaminase producing genes have been found to be useful in biological control of plant diseases. In canola, it is observed that the PGPR inoculated strain significantly improved the saline resistance and lowered the levels of ethylene by interfering with the salt-induced pathway of ethylene synthesis (Cheng et al., 2007).

Production of Volatiles

Various bacterial strains produce volatile organic compounds which constitutes an important mechanism in promoting plant growth, enhancing plant biomass, and drought tolerance. As per certain reviews few members of PGPR, including *Bacillus subtilis* GB03, *B. amyloliquefacience* IN937a and members of Enterobacteriaceae produce volatile organic compounds that promoted the abiotic stress tolerance of Arabidopsis thaliana (Ryu et al., 2003). Enzymes responsible for the production of Acetoin are identified in a few plants like carrot, maize, rice, and tobacco (Forlani et al., 1999). The volatiles produced by PGPR accumulate in sufficient concentration and trigger the signaling system to mediate plant-microbe interactions. In addition to aceetoin, few other volatile compounds like terpenes, jasmonates, and components of green leaves are also identified to be potent signal molecules (Farmer, 2001). Still, the actual role of volatiles in the signaling system of the plant is not established clearly. There is a scope in this area to understand the mechanism involved in the plant rhizobacteria signaling system.

Induction of volatile compounds observed in plants subjected to a multitude of stresses (Loreto and Schnitzler, 2010; Holopainen and Gershenzon, 2010). This volatility induces systemic response as they serve as signals to activate the defense system in plants (Heil and Silva Bueno, 2007; Choudhary et al., 2008; Niinemets, 2010). The role of stress-induced volatiles explained with inoculation of *B. thuringiensis* AZP2 in wheat seedlings under drought stress, which assisted in higher survival of plants, enhanced rate of photosynthesis, and reduced the liberation of volatiles (Timmusk et al., 2014). Microbial volatile 2R, 3R-butanediol produced by *Pseudomonas chlororaphis* 06, stop the water loss by stomatal closure. Root colonization with *Azospirillum brasilense* prevents the alterations in root morphology under drought stress. Rhizobacteria reduce the membrane potential in wheat seedlings and phospholipid content by modifying the proton efflux activities across the cell membrane (Bashan et al., 1992; 2004). Changes in the flexibility of root cell membrane induced by PGPR enhance the drought tolerance in plants (Dimkpa et al., 2009).

Antioxidant Defense System

In drought-affected plants, the reactive oxygen species (ROS) namely, superoxide ion (O^{-}_{2}), Hydroxyl (^{-}OH), and hydrogen peroxide (H_2O_2), are generated which react with biomolecules and cause oxidative stress that impairs the normal metabolism of a plant cell. To resist this antioxidant system activates in stress conditions. Inoculation with PGPR ameliorates the strength of plant cells and activates the antioxidant defense system that prevents ROS accumulation and alleviates the adverse effects occurred during stress (Gusain et al., 2015; Miller et al., 2010). This defense system comprises enzymatic components include catalase, ascorbate peroxidase, superoxide dismutase, glutathione peroxidase, and glutathione reductase. Certain non-enzymatic components also involved in controlling the oxidative damage, contain cysteine, ascorbic acid, and glutathione. The phytohormones especially IAA and ethylene play a significant role in activating signal transduction pathways in stress caused by salinity (Wani et al., 2016).

Co-inoculation of PGPR

Compared to the root inoculation with a single strain, the combination of mycorrhizal fungi and bacteria elicit higher drought tolerance in plants. Many reviews state that the application of *Rhizobium tropici* with different species of pseudomonas resulted in enhanced crop yield than with rhizobium as a single inoculant. The co-inoculation improves the root nodulation in drought-stressed plants. When PGPR strain *Pseudomonas mendocina palleroni* co-inoculated with the arbuscular mycorrhizal fungus was reported to increase the root phosphatase activity (Kohler et al., 2008).

HEAVY METAL TOLERANCE BY PGPR

Heavy metals accumulation has become a major obstacle to the worldwide agroecosystems (Shahid et al., 2015). Heavy metals reduce plant growth by influencing the physiological, biochemical, and molecular mechanisms of the plant. Earlier many research studies have reported the pessimistic impact of heavy metals on food crop production and human health (Shahid et al., 2015). Therefore, remediation of heavy metals from the soils is very much essential for the reclamation of contaminated soils. Numerous physic-biochemical methods have been adopted for the nullification of heavy metals from the agricultural soils. Among these methods, biological remediation is considered the most effective method for the removal of toxic metals. One such technology is the application of PGPR. Several studies have been reported where PGPR acts as a potential agent to promote the abiotic stress tolerance including heavy metal tolerance (Dary et al., 2010; Tiwari et al., 2016, 2017). These rhizobacteria have developed many mechanisms to tolerate heavy metal stress viz; i) transport of metals across cytoplasmic membrane ii) bioaccumulation to the cell walls iii)entrapment of metals in the extracellular capsules iv) precipitation of heavy metals and v) metal detoxification via oxidation-reduction reactions (Zubair et al., 2016). Few species of bacteria remove carbon, nitrogen, and phosphorus compounds while others remove toxic metals, chemical pesticides, herbicides, and non-biodegradable compounds in multistep processes.

CONCLUSION

PGPR is highly diverse and promotes plant growth by various mechanisms. A special focus is required in this field for a further understanding of mechanisms imparting positive effects in crop production. They offer an attractive environment-friendly biological control of plant disease. New formulations are designed to utilize all the beneficial factors by employing resistance-inducing bacteria and antagonistic bacteria. Siderophore-producing bacteria are found in the rhizosphere contributing to antibiosis and elicit an effective pathogen control. In addition to this, rhizobacteria can produce various antimicrobial compounds used in defense strategies. The efficacy and durability of biocontrol agents can be improved through the application of new formulations, screening procedures, and innovative integrated disease management practices. The application of PGPR in sustainable agriculture becomes the key requirement for the world, because of its deleterious effects on the environment. PGPR greatly reduced the chemical inputs in soil and pollution. The use of rhizobacteria demonstrated in various studies found effective in promoting plant growth and as biocontrol agents (Kumar et al., 2017a, b; Singh et al., 2017 a, b, c). PGPR is very effective at promoting plant growth development, but in few instances, some of the bacterial species are reported to have inhibitory effects under certain specific conditions, hence the selection of specific PGPR is of prime importance in achieving maximum crop yields concerning plant growth and development.

Drought stress is the most affecting environmental constraint of the food industry. Abiotic stress is one of the most destructive stress, with increased adverse effects over the past decades in sustainable agriculture. Drought stress reduces the availability of soil nutrients carried to the roots through the water. Because of this stress, the diffusion of the nutrients in the rhizosphere decreases. Furthermore, the transport of water-soluble nutrients such as Mg, Ca, nitrate, sulfate, and Silicon also diminishes drastically (Barber, 1995; Selvakumar et al., 2012). PGPR has an important role in inducing tolerance for heavy metals and drought tolerance to resolve the issues related to agricultural productivity in plants. The bacteria colonizing the roots have an impact on the plant and modify the soil properties. They even show the impact on osmotic response and induce the activation of novel genes which play a prime role in plant growth management under drought stress. The development of drought-tolerant, heavy metal tolerant plants by the application of genetic engineering is a practicable approach but it is cost-intensive, whereas the root inoculation with potential PGPR to alleviate drought stress in dryland agriculture provides the most feasible alternate mechanism to deal with future food security issues.

ACKNOWLEDGMENT

The authors are thankful to Prof. Sashidhar Rao Beedu (Department of Biochemistry, UCS, Osmania University, Hyderabad) for his constant motivation, encouragement, and support.

REFERENCES

Aeron, A., Kumar, S., Pandey, P., & Maheshwari, D. K. (2011). Emerging Role of Plant Growth Promoting Rhizobacteria in Agrobiology. In D. Maheshwari (Ed.), *Bacteria in Agrobiology: Crop Ecosystems*. Springer. doi:10.1007/978-3-642-18357-7_1

Ahemad, M., & Kibret, M. (2014). Mechanisms and applications of plant growth promoting rhizobacteria: Current perspective. *Journal of King Saud University. Science*, *26*(1), 1–20. doi:10.1016/j.jksus.2013.05.001

Ahmad, F., Ahmad, I., & Khan, M. S. (2005). Indole acetic acid production by the indigenous isolates of *Azotobacter* and fluorescent *Pseudomonas* in the presence and absence of tryptophan. *Turkish Journal of Biology*, *29*, 29–34.

Ahmad, P., Rasool, S., Gul, A., Sheikh, S. A., Akram, N. A., Ashraf, M., Kazi, A. M., & Gucel, S. (2016). Jasmonates: Multifunctional roles in stress tolerance. *Frontiers in Plant Science*, *7*, 813. doi:10.3389/fpls.2016.00813 PMID:27379115

Akhtar, M. S., & Siddiqui, Z. A. (2009). Use of plant growth promoting rhizobacteria for the biocontrol of root-rot disease complex of chickpea. *Australasian Plant Pathology*, *38*(1), 44–50. doi:10.1071/AP08075

Aquilanti, L., Favilli, F., & Clementi, F. (2004). Comparison of different strategies for isolation and preliminary identification of *Azotobacter* from soil samples. *Soil Biology & Biochemistry*, *36*(9), 1475–1483. doi:10.1016/j.soilbio.2004.04.024

Arshad, M., Shaharoona, B., & Mahmood, T. (2008). Inoculation with *Pseudomonas* spp. containing ACC-deaminase partially eliminates the effects of drought stress on growth, yield, and ripening of pea (*Pisum sativum* L.). *Pedosphere*, *18*(5), 611–620. doi:10.1016/S1002-0160(08)60055-7

Bageshwar, U. K., Srivastava, M., Pardha-Saradhi, P., Paul, S., Gothandapani, S., Jaat, R. S., Shankar, P., Yadav, R., Biswas, D. R., Kumar, P. A., Padaria, J. C., Mandal, P. K., Annapurna, K., & Dasa, H. K. (2017). An environmentally friendly engineered *Azotobacter* strain that replaces a substantial amount of urea fertilizer while sustaining the same wheat yield. *Applied and Environmental Microbiology*, *83*(15), e00590–e17. doi:10.1128/AEM.00590-17 PMID:28550063

Bakker, P. A. H. M., Ran, L., & Mercado-Blanco, J. (2014). Rhizobacterial salicylate production provokes headaches. *Plant and Soil*, *382*(1-2), 1–16. doi:10.100711104-014-2102-0

Barber, S. A. (1995). *Soil Nutrient Bioavailability A Mechanistic Approach* (2nd ed.). Wiley.

Barrera, A. D., & Soto, E. (2010). Biotechnological uses of *Azotobacter vinelandii*: Current state limits and prospects. *African Journal of Biotechnology*, *9*(33), 5240–5250.

Bashan, Y. (1998). Inoculants of plant growth-promoting bacteria for use in agriculture. *Biotechnology Advances*, *16*(4), 729–770. doi:10.1016/S0734-9750(98)00003-2

Bashan, Y., de Bashan, L. E., Prabhu, S. R., & Hernandez, J. P. (2014). Advances in plant growthpromoting bacterial inoculant technology: Formulations and practical perspectives (1998- 2013). *Plant and Soil*, *378*(1-2), 1–33. doi:10.100711104-013-1956-x

Bashan, Y., & Holguin, G. (1998). Proposal for the division of plant growth-promoting rhizobacteria into two classifications: biocontrol-PGPB (plant growth-promoting bacteria) and PGPB. *Soil Biology & Biochemistry*, *30*(8-9), 1225–1228. doi:10.1016/S0038-0717(97)00187-9

Bashan, Y., Menéndez, A., & Toledo, G. (1992). Responses of soybean and cowpea root membranes to inoculation with Azospirillum brasilense. *Symbiosis*, *13*, 217–228.

Beneduzi, A., Ambrosini, A., & Passaglia, L. M. P. (2012). Plant growth-promoting rhizobacteria. (PGPR): Their potential as antagonists and biocontrol agents. *Genetics and Molecular Biology*, *35*(4), 1044–1051. doi:10.1590/S1415-47572012000600020 PMID:23411488

Bhosale, H. J., Kadam, T. A., & Bobade, A. R. (2013). Identification and production of *Azotobacter vinelandii* and its antifungal activity against *Fusarium oxysporum. Journal of Environmental Biology*, *34*(2), 177–182. PMID:24620576

Bloemberg, G. V., & Lugtenberg, B. J. J. (2001). Molecular basis of plant growth promotion and biocontrol by rhizobacteria. *Current Opinion in Plant Biology*, *4*(4), 343–350. doi:10.1016/S1369-5266(00)00183-7 PMID:11418345

Bohlool, B., Ladha, J., Garrity, D., & George, T. (1992). Biological nitrogen fixation for sustainable agriculture: A perspective. *Plant and Soil*, *141*(1-2), 1–11. doi:10.1007/BF00011307

Boyd, E. S., & Peters, J. W. (2013). New insights into the evolutionary history of biological nitrogen fixation. *Frontiers in Microbiology*, *4*, 201. doi:10.3389/fmicb.2013.00201 PMID:23935594

Buscot, F., & Varma, A. (2005). *Microorganisms in Soils: Roles in Genesis and Functions*. Springer.

Calvo, P., Nelson, L. M., & Kloeppe, J. W. (2014). Agricultural uses of plant biostimulants. *Plant and Soil*, *383*(1-2), 3–41. doi:10.100711104-014-2131-8

Cassan, F., Maiale, S., Masciarelli, O., Vidal, A., Luna, V., & Ruiz, O. (2009). Cadaverine production by *Azospirillum brasilense* and its possible role in plant growth promotion and osmotic stress mitigation. *European Journal of Soil Biology*, *45*(1), 12–19. doi:10.1016/j.ejsobi.2008.08.003

Chen, X., Schauder, S., Potier, N., Van Dorsselaer, A., Pelczer, I., Bassler, B. L., & Hughson, F. M. (2002). Structural identification of a bacterial quorum-sensing signal containing boron. *Nature*, *415*(6871), 545–549. doi:10.1038/415545a PMID:11823863

Cheng, Z., Park, E., & Glick, B. R. (2007). 1-Aminocyclopropane-1-carboxylate deaminase from *Pseudomonas putida*UW4 facilitates the growth of canola in the presence of salt. *Canadian Journal of Microbiology*, *53*(7), 912–918. doi:10.1139/W07-050 PMID:17898846

Chennappa, G., Adkar-Purushothama, C. R., Suraj, U., Tamilvendan, K., & Sreenivasa, M. Y. (2014). Pesticide tolerant *Azotobacter* isolates from paddy growing areas of northern Karnataka, India. *World Journal of Microbiology & Biotechnology*, *30*(1), 1–7. doi:10.100711274-013-1412-3 PMID:23813305

Chet, I., Ordentilich, A., Shapira, R., & Oppenheim, A. (1990). Mechanisms of biocontrol of soil borne plant pathogens by rhizobacteria. *Plant and Soil*, *129*(1), 85–92. doi:10.1007/BF00011694

Choudhary, D. K., Johri, B. N., & Prakash, A. (2008). Volatiles as priming agents that initiate plant growth and defence responses. *Current Science*, *94*(5), 595–604.

Compant, S., Duffy, B., Nowak, J., Clément, C., & Barka, E. A. (2005). Use of plant growth-promoting bacteria for biocontrol of plant diseases: Principles, mechanisms of action, and future prospects. *Applied and Environmental Microbiology*, *71*(9), 4951–4959. doi:10.1128/AEM.71.9.4951-4959.2005 PMID:16151072

Crowley, D. E. (2006). Microbial Siderophores in the Plant Rhizosphere. In L. L. Barton & J. Abadia (Eds.), *Iron Nutrition in Plants and Rhizospheric Microorganisms*. Springer. doi:10.1007/1-4020-4743-6_8

Dary, M., Chamber-Pérez, M. A., Palomares, A. J., & Pajuelo, E. (2010). *In situ* phyto stabilisation of heavy metal polluted soils using *Lupinus luteus* inoculated with metal resistant plant-growth promoting rhizobacteria. *Journal of Hazardous Materials*, *177*(1-3), 323–330. doi:10.1016/j.jhazmat.2009.12.035 PMID:20056325

Das, K., Prasanna, R., & Saxena, A. K. (2017). Rhizobia: A potential biocontrol agent for soilborne fungal pathogens. *Folia Microbiologica*, *62*(5), 425–435. doi:10.100712223-017-0513-z PMID:28285373 de Souza, J. T., Arnould, C., Deulvot, C., Lemanceau, P., Gianinazzi-Pearson, V., & Raaijmakers, J. M. (2003). Effect of 2,4-diacetylphloroglucinol on pythium: Cellular responses and variation in sensitivity among propagules and species. *Phytopathology*, *93*(8), 966–975. doi:10.1094/PHYTO.2003.93.8.966 PMID:18943863

Desbrosses, G., Contesto, C., Varoquaux, F., Galland, M., & Touraine, B. (2009). PGPR-Arabidopsis interactions is a useful system to study signaling pathways involved in plant developmental control. *Plant Signaling & Behavior*, *4*(4), 321–323. doi:10.4161/psb.4.4.8106 PMID:19794852

Desbrosses, G. J., & Stougaard, J. (2011). Root nodulation: A paradigm for how plant-microbe symbiosis influences host developmental pathways. *Cell Host & Microbe*, *10*(4), 348–358. doi:10.1016/j.chom.2011.09.005 PMID:22018235

Dimkpa, C., Weinand, T., & Asch, F. (2009). Plant-rhizobacteria interactions alleviate abiotic stress conditions. *Plant, Cell & Environment*, *32*(12), 1682–1694. doi:10.1111/j.1365-3040.2009.02028.x PMID:19671096

Duan, J., Muller, K. M., Charles, T. C., Vesely, S., & Glick, B. R. (2009). 1-Aminocyclopropane-1-carboxylate (ACC) deaminase genes in rhizobia from Southern Saskatchewan. *Microbial Ecology*, *57*(3), 423–436. doi:10.100700248-008-9407-6 PMID:18548183

Dun-chun, H. E., Jia-sui, Z. H. A. N., & Lian-hui, X. I. E. (2016). Problems, challenges and future of plant disease management: From an ecological point of view. *Journal of Integrative Agriculture*, *154*(4), 705–715.

Dutta, S., & Podile, A. R. (2010). Plant growth promoting rhizobacteria (PGPR): The bugs to debug the root zone. *Critical Reviews in Microbiology*, *36*(3), 232–244. doi:10.3109/10408411003766806 PMID:20635858

Egamberdieva, D., Wirth, S. J., Shurigin, V. V., Hashem, A., & Abd-Allah, E. F. (2017). Endophytic bacteria improve plant growth, symbiotic performance of chickpea. (*Cicer arietinum* L.). and induce suppression of root rot caused by Fusarium under salt stress. *Frontiers in Microbiology*, *8*, 1887. doi:10.3389/fmicb.2017.01887 PMID:29033922

Farmer, E. E. (2001). Surface-to-air signals. *Nature*, *411*(6839), 854–856. doi:10.1038/35081189 PMID:11459069

Forlani, G. M., Mantelli, M., & Nielsen, E. (1999). Biochemical evidence for multiple acetoin-forming enzymes in cultured plant cells. *Phytochemistry*, *50*(2), 255–262. doi:10.1016/S0031-9422(98)00550-0

Ghosh, P. K., De, T. K., & Maiti, T. K. (2018). Role of ACC Deaminase as a Stress Ameliorating Enzyme of Plant Growth-Promoting Rhizobacteria Useful in Stress Agriculture: A Review. In V. Meena (Ed.), *Role of Rhizospheric Microbes in Soil*. Springer., doi:10.1007/978-981-10-8402-7_3

Ghosh, S. J. N., Penterman, R. D., Little, R., Chavez, R., & Glick, B. R. (2003). Three newly isolated plant growth-promoting bacilli facilitate the seedling growth of canola, Brassica campestris. *Plant Physiology and Biochemistry*, *41*(3), 277–281. doi:10.1016/S0981-9428(03)00019-6

Glick, B. R., Patten, C. L., Holguin, G., & Penrose, D. M. (1999). *Biochemical and Genetic Mechanisms Used by Plant Growth Promoting Bacteria*. London Imperial College Press. doi:10.1142/p130

Gopalakrishnan, S., Sathya, A., Vijayabharathi, R., Varshney, R.K., Gowda, C.L.L., & Krishnamurthy, L. (2015). Plant growth promoting rhizobia Challenges and opportunities. *Biotech, 5*(4), 355-377.

Gopalakrishnan, S., Srinivas, V., & Samineni, S. (2017). Nitrogen fixation, plant growth and yield enhancements by diazotrophic growth-promoting bacteria in two cultivars of chickpea. (Cicero arietinum L.). *Biocatalysis and Agricultural Biotechnology*, *11*, 116–123. doi:10.1016/j.bcab.2017.06.012

Gopalakrishnan, S. V., Srinivas, V., Vemula, A. K., Samineni, S., & Rathore, A. (2018). Influence of diazotrophic bacteria on nodulation, nitrogen fixation, growth promotion and yield traits in five cultivars of chickpea. *Biocatalysis and Agricultural Biotechnology*, *15*, 35–42. doi:10.1016/j.bcab.2018.05.006

Goswami, D., Thakker, J. N., Dhandhukia, P. C., & Moral, M. T. (2016). Portraying mechanics of plant growth promoting rhizobacteria. (PGPR). A review. *Cogent Food & Agriculture*, *2*(1), 1. doi:10.1080/23311932.2015.1127500

Gouda, S., Kerry, R. G., Das, G., Paramithiotis, S., Shin, H. S., & Patra, J. K. (2018). Revitalization of plant growth promoting rhizobacteria for sustainable development in agriculture. *Microbiological Research*, *206*, 131–140. doi:10.1016/j.micres.2017.08.016 PMID:29146250

Govindasamy, V., Senthilkumar, M., Gaikwad, K., & Annapurna, K. (2008). Isolation and characterization of ACC deaminase gene from two plant growth-promoting rhizobacteria. *Current Microbiology*, *57*(4), 312–317. doi:10.100700284-008-9195-8 PMID:18654819

Graham, P. H. (1992). Stress tolerance in Rhizobium and Bradyrhizobium, and nodulation under adverse soil conditions. *Canadian Journal of Microbiology*, *38*(6), 475–484. doi:10.1139/m92-079

Graham, P. H., & Vance, C. P. (2003). Legumes: Importance and constraints to greater use. *Plant Physiology*, *131*(3), 872–877. doi:10.1104/pp.017004 PMID:12644639

Gray, E. J., & Smith, D. L. (2005). Intracellular and extracellular PGPR: Commonalities and distinctions in the plant-bacterium signaling processes. *Soil Biology & Biochemistry*, *37*(3), 395–412. doi:10.1016/j.soilbio.2004.08.030

Grobelak, A., Kokot, P., Hutchison, D., Grosser, A., & Kacprzak, M. (2018). Plant growth-promoting rhizobacteria as an alternative to mineral fertilizers in assisted bioremediation-sustainable land and waste management. *Journal of Environmental Management*, *227*, 1–9. doi:10.1016/j.jenvman.2018.08.075 PMID:30170232

Guo, R. F., Yuan, G. F., & Wang, Q. M. (2013). Effect of NaCl treatments on glucosinolate metabolism in broccoli sprouts. *Journal of Zhejiang University. Science. B.*, *14*(2), 124–131. doi:10.1631/jzus.B1200096 PMID:23365011

Gusain, Y. S., Singh, U. S., & Sharma, A. K. (2015). Bacterial mediated amelioration of drought stress in drought tolerant and susceptible cultivars of rice (*Oryza sativa* L.). *African Journal of Biotechnology*, *14*(9), 764–773. doi:10.5897/AJB2015.14405

Gyaneshwar, P., Naresh Kumar, G., Parekh, L. J., & Poole, P. S. (2002). Role of soil microorganisms in improving P nutrition of plants. *Plant and Soil*, *245*(1), 83–93. doi:10.1023/A:1020663916259

Haas, D., & Défago, G. (2005). Biological control of soil-borne pathogens by fluorescent pseudomonads. *Nature Reviews. Microbiology*, *3*(4), 307–319. doi:10.1038/nrmicro1129 PMID:15759041

Haddock, B. A., & Jones, C. W. (1977). Bacterial respiration. *Bacteriological Reviews*, *41*(1), 47–99. doi:10.1128/BR.41.1.47-99.1977 PMID:140652

Handelsman, J., & Stabb, E. V. (1996). Bio-control of soil borne plant pathogens. *The Plant Cell*, *8*(10), 1855–1869. doi:10.2307/3870235 PMID:12239367

Hardoim, P. R., van Overbeek, L. S., Berg, G., Pirttilä, A. M., Compant, S., Campisano, A., Döring, M., & Sessitsch, A. (2015). The hidden world within plants: Ecological and evolutionary considerations for defining functioning of microbial endophytes. *Microbiology and Molecular Biology Reviews*, *79*(3), 293–320. doi:10.1128/MMBR.00050-14 PMID:26136581

Hedin, L. O., Brookshire, E. J., Menge, D. N., & Barron, A. R. (2009). The nitrogen paradox in tropical forest ecosystems. *Annual Review of Ecology, Evolution, and Systematics*, *40*(1), 613–635. doi:10.1146/annurev.ecolsys.37.091305.110246

Heil, M., & Silva, B. J. C. (2007). Within-plant signaling by volatiles leads to induction and priming of an indirect plant defense in nature. *Proceedings of the National Academy of Sciences of the United States of America*, *104*(103), 5467–5472. doi:10.1073/pnas.0610266104 PMID:17360371

Holden, M. T., Ram Chhabra, S., de Nys, R., Stead, P., Bainton, N. J., Hill, P. J., Manefield, M., Kumar, N., Labatte, M., England, D., Rice, S., Givskov, M., Salmond, G. P., Stewart, G. S., Bycroft, B. W., Kjelleberg, S., & Williams, P. (1999). Quorum-sensing cross talk: Isolation and chemical characterization of cyclic dipeptides from Pseudomonas aeruginosa and other gram-negative bacteria. *Molecular Microbiology*, *33*(6), 1254–1266. doi:10.1046/j.1365-2958.1999.01577.x PMID:10510239

Holopainen, J. K., & Gershenzon, J. (2010). Multiple stress factors and the emission of plant VOCs. *Trends in Plant Science*, *15*(3), 176–184. doi:10.1016/j.tplants.2010.01.006 PMID:20144557

Jiménez, D. J., Montaña, J. S., & Martínez, M. M. (2011). Characterization of free nitrogen fixing bacteria of the genus Azotobacter in organic vegetable-grown Colombian soils. *Brazilian Journal of Microbiology*, *42*(3), 846–858. doi:10.1590/S1517-83822011000300003 PMID:24031700

Kennedy, C., Rudnick, P., MacDonald, T., & Melton, T. (2005). Genus Azotobacter. In G. M. Garirity. Bergey's Manual of Systematic Bacteriology, 2, 384–401.

Khan, M. S., Zaidi, A., & Ahmad, E. (2014). Mechanism of Phosphate Solubilization and Physiological Functions of Phosphate-Solubilizing Microorganisms. In M. Khan, A. Zaidi, & J. Musarrat (Eds.), *Phosphate Solubilizing Microorganisms*. Springer., doi:10.1007/978-3-319-08216-5_2

Kim, K., Jordan, D., & McDonald, G. (1997). Effect of phosphate-solubilizing bacteria and vesicular-arbuscular mycorrhizae on tomato growth and soil microbial activity. *Biology and Fertility of Soils*, *26*(2), 79–87. doi:10.1007003740050347

Kloepper, J. W., Lifshitz, R., & Zablotowicz, R. M. (1989). Free-living bacterial inocula for enhancing crop productivity. *Trends in Biotechnology*, *7*(2), 39–43. doi:10.1016/0167-7799(89)90057-7

Kobayashi, D. Y., Reedy, R. M., Bick, J., & Oudemans, P. V. (2002). Characterization of a chitinase gene from Stenotrophomonas maltophilia strain 34S1 and its involvement in biological control. *Applied and Environmental Microbiology*, *68*(3), 1047–1054. doi:10.1128/AEM.68.3.1047-1054.2002 PMID:11872449

Kohler, J., Hernaındez, J. A., Caravaca, F., & Roldaın, A. (2008). Plant-growth promoting rhizobacteria and arbuscular mycorrhizal fungi modify alleviation biochemical mechanisms in water stressed plants. *Functional Plant Biology*, *35*(2), 141–151. doi:10.1071/FP07218 PMID:32688765

Kumar, A., Singh, A. K., Kaushik, M. S., Mishra, S. K., Raj, P., Singh, P. K., & Pandey, K. D. (2017a). Interaction of turmeric (*Curcuma longa* L.) with beneficial microbes: a review. *Biotech, 7*(6), 357. doi:10.100713205-017-0971-7

Kumar, A., Verma, H., Singh, V. K., Singh, P. P., Singh, S. K., & Ansari, W. A. (2017b). Role of *Pseudomonas* sp. in Sustainable Agriculture and Disease Management. In V. Meena, P. Mishra, J. Bisht, & A. Pattanayak (Eds.), *Agriculturally Important Microbes for Sustainable Agriculture*. Springer. doi:10.1007/978-981-10-5343-6_7

Kumar, V., Behl, R. K., & Narula, N. (2001). Effect of phosphate solubilizing strains of *Azotobacter chroococcum* on yield traits and their survival in the rhizosphere of wheat genotypes under field conditions. *Acta Agronomica Hungarica*, *49*(2), 141–149. doi:10.1556/AAgr.49.2001.2.4

Kumari, V., & Srivastava, J. (1999). Molecular and biochemical aspects of rhizobacterial ecology with emphasis on biological control. *World Journal of Microbiology & Biotechnology*, *15*(5), 535–543. doi:10.1023/A:1008958912647

Loper, J. E., & Henkels, M. D. (1999). Utilization of heterologous siderophores enhances levels of iron available to *Pseudomonas putida* in the rhizosphere. *Applied and Environmental Microbiology*, *65*(12), 5357–5363. doi:10.1128/AEM.65.12.5357-5363.1999 PMID:10583989

Loreto, F., & Schnitzler, J. P. (2010). Abiotic stresses and induced BVOCs. *Trends in Plant Science*, *15*(3), 154–166. doi:10.1016/j.tplants.2009.12.006 PMID:20133178

Mahanta, D., Rai, R. K., Mishra, S. D., Raja, A. J., & Varghese, E. (2014). Influence of phosphorus and biofertilizers on soybean and wheat root growth and properties. *Field Crops Research*, *166*, 1–9. doi:10.1016/j.fcr.2014.06.016

Mayak, S., Tirosh, T., & Glick, B. R. (2004). Plant growth-promoting bacteria that confer resistance to water stress in tomatoes and peppers. *Plant Science*, *166*(2), 525–530. doi:10.1016/j.plantsci.2003.10.025

Mazen, M. M., El-Batanony, N. H., Abd El-Monium, M. M., & Massoud, O. N. (2008). Cultural filtrate of Rhizobium spp. and arbuscular mycorrhiza are potential biological control agents against root rot fungal diseases of faba bean. *Global Journal of Biotechnology and Biochemistry*, *3*, 32–41.

McRose, D. L., Baars, O., Morel, F. M. M., & Kraepiel, A. M. L. (2017). Siderophore production in *Azotobacter vinelandii* in response to Fe-, Mo- and V-limitation. *Environmental Microbiology, 19*(9), 3595–3605. doi:10.1111/1462-2920.13857 PMID:28703469

Miller, G., Suzuki, N., Ciftci-Yilmaz, S., & Mittler, R. (2010). Reactive oxygen species homeostasis and signalling during drought and salinity stresses. *Plant, Cell & Environment, 33*(4), 453–467. doi:10.1111/j.1365-3040.2009.02041.x PMID:19712065

Mishra, S., Singh, A., Keswani, C., Saxena, A., Sarma, B. K., & Singh, H. B. (2015). Harnessing Plant-Microbe Interactions for Enhanced Protection Against Phytopathogens. In N. Arora (Ed.), *Plant Microbes Symbiosis: Applied Facets*. Springer. doi:10.1007/978-81-322-2068-8_5

Mrkovacki, N., & Milic, V. (2001). Use of Azotobacter chroococcum as potentially useful in agricultural application. *Annals of Microbiology, 51*, 145–158.

Myresiotis, C. K., Vryzas, Z., & Papadopoulou-Mourkidou, E. (2012). Biodegradation of soil-applied pesticides by selected strains of plant growth-promoting rhizobacteria (PGPR) and their effects on bacterial growth. *Biodegradation, 23*(2), 297–310. doi:10.100710532-011-9509-6 PMID:21870159

Nagargade, M., Tyagi, V., & Singh, M. K. (2018). Plant Growth-Promoting Rhizobacteria: A Biological Approach Toward the Production of Sustainable Agriculture. In V. Meena (Ed.), *Role of Rhizospheric Microbes in Soil*. Springer. doi:10.1007/978-981-10-8402-7_8

Nandakumar, R., Babu, S., Viswanathan, R., Raguchander, Y., & Samiyappan, R. (2001). Induction of systemic resistance in rice against sheath blight disease by *Pseudomonas fluorescens. Soil Biology & Biochemistry, 33*(4-5), 603–612. doi:10.1016/S0038-0717(00)00202-9

Nannipieri, P., Giagnoni, L., Landi, L., & Renella, G. (2011). Role of Phosphatase Enzymes in Soil. In E. Bünemann, A. Oberson, & E. Frossard (Eds.), *Phosphorus in Action. Soil Biology* (Vol. 26). Springer. doi:10.1007/978-3-642-15271-9_9

Neilands, J. B. (1995). Siderophores: Structure and function of microbial Iron transport compounds. *The Journal of Biological Chemistry, 270*(45), 26723–26726. doi:10.1074/jbc.270.45.26723 PMID:7592901

Nelson, L. M. (2004). Plant Growth Promoting Rhizobacteria (PGPR): Prospects for New Inoculants. *Crop Management, 3*(1), 1–7. doi:10.1094/CM-2004-0301-05-RV

Nelson, M. N., & Sorenson, J. (1999). Chitinolytic activity of *Pseudomonas fluorescens* isolates from barley and sugar beet rhizosphere. *FEMS Microbiology Ecology, 30*(3), 217–227. doi:10.1111/j.1574-6941.1999.tb00650.x PMID:10525178

Ney, L., Franklin, D., Mahmud, K., Cabrera, M., Hancock, D., Habteselassie, M., Newcomer, Q., & Fatzinger, B. (2019). Rebuilding Soil Ecosystems for Improved Productivity in Biosolarized Soils. *International Journal of Agronomy, 2019*, 5827585. Advance online publication. doi:10.1155/2019/5827585

Niinemets, Ü. (2010). Mild versus severe stress and BVOCs: Thresholds, priming and consequences. *Trends in Plant Science, 15*(3), 145–153. doi:10.1016/j.tplants.2009.11.008 PMID:20006534

O'sullivan, D. J., & O'gara, F. (1992). Traits of fluorescent *Pseudomonas* spp. involved in suppression of plant root pathogens. *Microbiological Reviews*, *56*(4), 662–676. doi:10.1128/MR.56.4.662-676.1992 PMID:1480114

Olanrewaju, O. S., Glick, B. R., & Babalola, O. O. (2017). Mechanisms of action of plant growth promoting bacteria. *World Journal of Microbiology & Biotechnology*, *33*(11), 197. doi:10.100711274-017-2364-9 PMID:28986676

Oldroyd, G. E., & Downie, J. A. (2008). Coordinating nodule morphogenesis with rhizobial infection in legumes. *Annual Review of Plant Biology*, *59*(1), 519–546. doi:10.1146/annurev.arplant.59.032607.092839 PMID:18444906

Ongena, M., Daayf, F., Jacques, P., Thonart, P., Benhamou, N., Paulitz, T.C., Cornélis, P., Koedam, N., & Bélanger, R.R, (1999). Protection of cucumber against Pythium root rot by fluorescent pseudomonads: predominant role of induced resistance over siderophores and antibiosis. *Plant Pathology, 48*, 66-76. doi:10.1046/j.1365-3059.1999.00315.x

Ortíz-Castro, R., Contreras-Cornejo, H. A., Macías-Rodríguez, L., & López-Bucio, J. (2009). The role of microbial signals in plant growth and development. *Plant Signaling & Behavior*, *4*(8), 701–712. doi:10.4161/psb.4.8.9047 PMID:19820333

Peoples, M. B., Hauggaard-Nielsen, H., & Jensen, E. S. (2009). The Potential Environmental Benefits and Risks Derived from Legumes in Rotations. In D. W. Emerich & H. B. Krishnan (Eds.), *Nitrogen Fixation in Crop Production*. John Wiley & Sons. doi:10.2134/agronmonogr52.c13

Perret, X., Staehelin, C., & Broughton, W. J. (2000). Molecular basis of symbiotic promiscuity. *Microbiology and Molecular Biology Reviews: MMBR*, *64*(1), 180–201. doi:10.1128/MMBR.64.1.180-201.2000 PMID:10704479

Peters, J. W., Boyd, E. S., Hamilton, T., & Rubio, L. M. (2011). Biochemistry of Mo-nitrogenase. In J. W. B. Moir (Ed.), *Nitrogen cycling in bacteria: Molecular Analysis*. Caister Academic Press.

Pieterse, C. M., Zamioudis, C., Does, D. V. der & Van Wees, S.C. (2014). Signalling Networks Involved in Induced Resistance. In Induced Resistance for Plant Defense. John Wiley & Sons. doi:10.1002/9781118371848.ch4

Pourbabaee, A., Bahmani, E., Alikhani, H., & Emami, S. (2016). Promotion of wheat growth under salt stress by halo tolerant bacteria containing ACC deaminase. *Journal of Agricultural Science and Technology*, *18*, 855–864.

Prasad, R. (2009). Efficient fertilizer use: The key to food security and better environment. *Journal of Tropical Agriculture*, *47*(1-2), 1–17.

Radjacommare, R., Kandan, A., Nandakumar, R., & Samiyappan, R. (2004). Association of the Hydrolytic Enzyme Chitinase against *Rhizoctonia solani* in Rhizobacteria-treated Rice Plants. *Journal of Phytopathology*, *152*(6), 365–370. doi:10.1111/j.1439-0434.2004.00857.x

Ramamoorthy, V., Viswanathan, R., Raguchander, T., Prakasam, V., & Samayapan, R. (2001). Induction of systemic resistance by plant growth promoting rhizobacteria in crop plants against pests and diseases. *Crop Protection (Guildford, Surrey)*, *20*(1), 1–11. doi:10.1016/S0261-2194(00)00056-9

Raupach, G. S., & Kloepper, J. W. (1998). Mixtures of plant growth-promoting rhizobacteria enhance biological control of multiple cucumber pathogens. *Phytopathology*, *88*(11), 1158–1164. doi:10.1094/PHYTO.1998.88.11.1158 PMID:18944848

Ravanbakhsh, M., Sasidharan, R., Voesenek, L. A. C. J., Kowalchuk, G. A., & Jousset, A. (2017). ACC deaminase-producing rhizosphere bacteria modulate plant responses to flooding. *Journal of Ecology*, *105*(4), 979–986. doi:10.1111/1365-2745.12721

Ren, D., Sims, J. J., & Wood, T. K. (2001). Inhibition of biofilm formation and swarming of *Escherichia coli* by (*5Z*) -4-bromo-5-(bromomethylene)-3-butyl-2(5H)-furanone. *Environmental Microbiology*, *3*(11), 731–736. doi:10.1046/j.1462-2920.2001.00249.x PMID:11846763

Revillas, J., Rodelas, B., Pozo, C., Martínez-Toledo, M., & González-López, J. (2000). Production of B-group vitamins by two *Azotobacter* strains with phenolic compounds as sole carbon source under diazotrophic and adiazotrophic conditions. *Journal of Applied Microbiology*, *89*(3), 486–493. doi:10.1046/j.1365-2672.2000.01139.x PMID:11021581

Rubio, L. M., & Ludden, P. W. (2008). Biosynthesis of the iron-molybdenum cofactor of nitrogenase. *Annual Review of Microbiology*, *62*(1), 93–111. doi:10.1146/annurev.micro.62.081307.162737 PMID:18429691

Ryder, M. H., Yan, Z., Terrace, T. E., Rovira, A. D., Tang, W., & Correll, R. L. (1999). Use of Bacillus isolated in China to suppress take-all and *rhizoctonia* root rot, and promote seedling growth of glass house grown wheat in Australian soils. *Soil Biology & Biochemistry*, *31*(1), 19–29. doi:10.1016/S0038-0717(98)00095-9

Ryu, C. M., Farag, M. A., Hu, C. H., Reddy, M. S., Wei, H. X., Paré, P. W., & Kloepper, J. W. (2003). Bacterial volatiles promote growth in Arabidopsis. *Proceedings of the National Academy of Sciences of the United States of America*, *100*(8), 4927–4932. doi:10.1073/pnas.0730845100 PMID:12684534

Sadfi, N., Cherif, M., Fliss, I., Boudabbous, A., & Antoun, H. (2001). Evaluation of bacterial isolates from salty soils and *Bacillus thuringiensis* strains for the biocontrol of *Fusarium* dry rot of potato tubers. *Journal of Plant Pathology*, *83*, 101–118.

Safronova, V. I., Stepanok, V. V., Engqvist, G. L., Alekseyev, Y. V., & Belimov, A. A. (2006). Root-associated bacteria containing 1- minocyclopropane- 1-carboxylate deaminase improve growth and nutrient uptake by pea genotypes cultivated in cadmium supplemented soil. *Biology and Fertility of Soils*, *42*(3), 267–272. doi:10.100700374-005-0024-y

Saikia, J., Sarma, R. K., Dhandia, R., Yadav, A., Bharali, R., Gupta, V. K., & Saikia, R. (2018). Alleviation of drought stress in pulse crops with ACC deaminase producing rhizobacteria isolated from acidic soil of Northeast India. *Scientific Reports*, *8*(1), 3560. doi:10.103841598-018-21921-w PMID:29476114

Saritha, M., & Tollamadugu, N. V. K. V. P. (2019). The Status of Research and Application of Biofertilizers and Biopesticides: Global Scenario. In V. Buddolla (Ed.), *Recent Developments in Applied Microbiology and Biochemistry*. Academic Press., doi:10.1016/B978-0-12-816328-3.00015-5

Schmid, M., & HartMann, A. (2007). Molecular Phylogeny and Ecology of Root Associated Diazotrophic α- and β-Proteobacteria. In *Associative and Endophytic Nitrogen-fixing Bacteria and Cyanobacterial Associations. Nitrogen Fixation: Origins, Applications, and Research Progress* (Vol. 5). Springer. doi:10.1007/1-4020-3546-2_2

Schultze, M., & Kondorosi, A. (1998). Regulation of symbiotic root nodule development. *Annual Review of Genetics*, *32*(1), 33–57. doi:10.1146/annurev.genet.32.1.33 PMID:9928474

Schultze, M., & Kondorosi, A. (1998). Regulation of symbiotic root nodule development. *Annual Review of Genetics*, *32*(1), 33–57. doi:10.1146/annurev.genet.32.1.33 PMID:9928474

Selvakumar, G., Panneerselvam, P., & Ganeshamurthy, A. N. (2012). Bacterial Mediated Alleviation of Abiotic Stress in Crops. In D. Maheshwari (Ed.), *Bacteria in Agrobiology: Stress Management*. Springer. doi:10.1007/978-3-662-45795-5_10

Selvakumar, G., Panneerselvam, P., & Ganeshamurthy, A. N. (2014). Biosafety of novel bioinoculants. *Journal of Biofertilizers & Biopesticides*, *5*(2), 145.

Setubal, J. C., dos Santos, P., Goldman, B. S., Ertesvåg, H., Espin, G., Rubio, L. M., Valla, S., Almeida, N. F., Balasubramanian, D., Cromes, L., Curatti, L., Du, Z., Godsy, E., Goodner, B., Hellner-Burris, K., Hernandez, J. A., Houmiel, K., Imperial, J., Kennedy, C., ... Wood, D. (2009). Genome sequence of Azotobacter vinelandii, an obligate aerobe specialized to support diverse anaerobic metabolic processes. *Journal of Bacteriology*, *191*(14), 4534–4545. doi:10.1128/JB.00504-09 PMID:19429624

Shahid, M., Khalid, S., Abbas, G., Shahid, N., Nadeem, M., & Sabir, M. (2015). Heavy Metal Stress and Crop Productivity. In K. Hakeem (Ed.), *Crop Production and Global Environmental Issues*. Springer. doi:10.1007/978-3-319-23162-4_1

Sheikh, L. I., Dawar, S., Zaki, M. J., & Ghaffar, A. (2006). Efficacy of *Bacillus thuringiensis* and *Rhizobium meliloti* with nursery fertilizers in the control of root infecting fungi on mung bean and okra plants. *Pakistan Journal of Botany*, *38*, 465–473.

Shrivastava, P., & Kumar, R. (2015). Soil salinity: A serious environmental issue and plant growth promoting bacteria as one of the tools for its alleviation. *Saudi Journal of Biological Sciences*, *22*(2), 123–131. doi:10.1016/j.sjbs.2014.12.001 PMID:25737642

Siddiqui, Z. A. (2005). PGPR: Prospective Biocontrol Agents of Plant Pathogens. In Z. A. Siddiqui (Ed.), *PGPR: Biocontrol and Biofertilization*. Springer. doi:10.1007/1-4020-4152-7_4

Silva, H. S. A., Romeiro, R., Macagnan, D., Halfeld-Vieira, B. A., Pereira, M. C. B., & Mounteer, A. (2004). Rhizobacterial induction of systemic resistance in tomato plants: Non-specific protection and increase in enzyme activities. *Biological Control*, *29*(2), 288–295. doi:10.1016/S1049-9644(03)00163-4

Singh, A., Jain, A., Sarma, B. K., Upadhyay, R. S., & Singh, H. B. (2014). Rhizosphere competent microbial consortium mediates rapid changes in phenolic profiles in chickpea during *Sclerotium rolfsii* infection. *Microbiological Research*, *169*(5-6), 353–360. doi:10.1016/j.micres.2013.09.014 PMID:24168925

Singh, M., Kumar, A., Singh, R., & Pandey, K. D. (2017). Endophytic bacteria: a new source of bioactive compounds. *Biotech, 7*(5), 315. doi:10.100713205-017-0942-z

Singh, R., Pandey, D. K., Kumar, A., & Singh, M. (2017b). PGPR isolates from the rhizosphere of vegetable crop *Momordica charantia*: Characterization and application as biofertilizer. *International Journal of Current Microbiology and Applied Sciences*, *6*(3), 1789–1802. doi:10.20546/ijcmas.2017.603.205

Singh, V.K., Singh, A.K. & Kumar, A. (2017c). Disease management of tomato through PGPB: current trends and future perspective. *Biotech, 7*, 255. doi:10.100713205-017-0896-1

Someya, N., Kataoka, N., Komagata, T., Hirayae, K., Hibi, T., & Akutsu, K. (2000). Biological Control of Cyclamen Soilborne Diseases by *Serratia marcescens* Strain B2. *Plant Disease*, *84*(3), 334–340. doi:10.1094/PDIS.2000.84.3.334 PMID:30841252

Sulieman, S., & Tran, L. P. (2016). *Legume Nitrogen Fixation in a Changing Environment- Achievements and Challenges*. Springer Nature Switzerland AG.

Sutton, M. A., Oenema, O., Erisman, J. W., Leip, A., van Grinsven, H., & Winiwarter, W. (2011). Too much of a good thing. *Nature*, *472*(7342), 159–161. doi:10.1038/472159a PMID:21478874

Sytsma, K. J., Morawetz, J., Pires, J. C., Nepokroeff, M., Conti, E., Zjhra, M., Hall, J. C., & Chase, M. W. (2002). *Urticalean rosids*: Circumscription, rosid ancestry, and phylogenetics based on rbcL, trnL-F, and ndhF sequences. *American Journal of Botany*, *89*(9), 1531–1546. doi:10.3732/ajb.89.9.1531 PMID:21665755

Timmusk, S., Islam, A., Abd El, D., Lucian, C., Tanilas, T., & Kannaste, A. (2014). Drought-tolerance of wheat improved by rhizosphere bacteria from harsh environments: Enhanced biomass production and reduced emissions of stress volatiles. *PLoS One*, *9*(5), 1–13. doi:10.1371/journal.pone.0096086 PMID:24811199

Tiwari, S., Lata, C., Chauhan, P. S., & Nautiyal, C. S. (2016). *Pseudomonas putida* attunes morphophysiological, biochemical and molecular responses in *Cicer arietinum* L. during drought stress and recovery. *Plant Physiology and Biochemistry*, *99*, 108–117. doi:10.1016/j.plaphy.2015.11.001 PMID:26744996

Tiwari, S., Lata, C., Chauhan, P. S., Prasad, V., & Prasad, M. (2017). A Functional Genomic Perspective on Drought Signalling and its Crosstalk with Phytohormone-mediated Signalling Pathways in Plants. *Current Genomics*, *18*(6), 469–482. doi:10.2174/1389202918666170605083319 PMID:29204077

Turan, M., Yildirim, E., Kitir, N., Unek, C., Nikerel, E., & Ozdemir, B. S. (2017). Beneficial Role of Plant Growth-Promoting Bacteria in Vegetable Production under Abiotic Stress. In A. Zaidi & M. Khan (Eds.), *Microbial Strategies for Vegetable Production*. Springer. doi:10.1007/978-3-319-54401-4_7

Uren, N. C. (2000). Types, amounts, and possible functions of compounds released into the rhizosphere by soil-grown plants. In *The Rhizosphere - Biochemistry and Organic Substances at the Soil-Plant Interface* (2nd ed.). CRC Press.

van Loon, L. C., Bakker, P. A., & Pieterse, C. M. (1998). Systemic resistance induced by rhizosphere bacteria. *Annual Review of Phytopathology*, *36*(1), 453–483. doi:10.1146/annurev.phyto.36.1.453 PMID:15012509

Verma, J. P., Yadav, J., Tiwari, K. N., Lavakush, & Singh, V. (2010). Impact of plant growth promoting rhizobacteria on crop production. *International Journal of Agricultural Research*, *5*(11), 954–983. doi:10.3923/ijar.2010.954.983

Vitousek, P. M., & Howarth, R. W. (1991). Nitrogen limitation on land and in the sea: How can it occur? *Biogeochemistry*, *13*(2), 87–115. doi:10.1007/BF00002772

Walpola, B., & Yoon, M. (2012). Prospectus of phosphate solubilizing microorganisms and phosphorus availability in agricultural soils. A review. *African Journal of Microbiological Research*, *6*(37), 6600–6605.

Walsh, U. F., Morrissey, J. P., & O'Gara, F. (2001). *Pseudomonas* for biocontrol of phytopathogens: From functional genomics to commercial exploitation. *Current Opinion in Biotechnology*, *12*(3), 289–295. doi:10.1016/S0958-1669(00)00212-3 PMID:11404107

Wandersman, C., & Delepelaire, P. (2004). Bacterial iron sources: From siderophores to hemophores. *Annual Review of Microbiology*, *58*(1), 611–647. doi:10.1146/annurev.micro.58.030603.123811 PMID:15487950

Wani, S. H., Kumar, V., Shriram, V., & Sah, S. K. (2016). Phytohormones and their metabolic engineering for abiotic stress tolerance in crop plants. *The Crop Journal*, *4*(3), 162–176. doi:10.1016/j.cj.2016.01.010

Yamada, Y., & Nihira, T. (1998). Microbial hormones and microbial chemical ecology. In D. H. R. Barton & K. Nakanishi (Eds.), *Comprehensive natural products chemistry*. Elsevier Sciences.

Yanni, Y. G., Rizk, R. Y., Corich, V., Squartini, A., Ninke, K., Philip-Hollingsworth, S., Orgambide, G., de Bruijn, F., Stoltzfus, J., Buckley, D., Schmidt, T. M., Mateos, P. F., Ladha, J. K., & Dazzo, F. B. (1997). Natural endophytic association between *Rhizobium leguminosarum* bv. trifolii and rice roots and assessment of its potential to promote rice growth. *Plant and Soil*, *194*(1/2), 99–114. doi:10.1023/A:1004269902246

Zhang, L. H., & Dong, Y. H. (2004). Quorum sensing and signal interference: Diverse implications. *Molecular Microbiology*, *53*(6), 1563–1571. doi:10.1111/j.1365-2958.2004.04234.x PMID:15341639

Zhang, W. F., Dou, Z.-X., He, P., Ju, X.-T., Powlson, D., Chadwick, D., Norse, D., Lu, Y.-L., Zhang, Y., Wu, L., Chen, X.-P., Cassman, K. G., & Zhang, F.-S. (2013). New technologies reduce greenhouse gas emissions from nitrogenous fertilizer in China. *Proceedings of the National Academy of Sciences of the United States of America*, *110*(21), 8375–8380. doi:10.1073/pnas.1210447110 PMID:23671096

Zubair, M., Shakir, M., Ali, Q., Rani, N., Fatima, N., Farooq, S., Shafiq, S., Kanwal, N., Ali, F., & Nasir, I. A. (2016). Rhizobacteria and phytoremediation of heavy metals. *Environmental Technology Reviews*, *5*(1), 112–119. doi:10.1080/21622515.2016.1259358

Chapter 13
An Account on Mycoviruses and Their Applications

Md. Idrish Raja Khan
College of Fisheries, Central Agricultural University, Tripura, India

ABSTRACT

Mycoviruses are obligate parasites of fungi and can infect the majority of the fungal groups. They remain mysterious to various communities throughout the globe. Mycoviruses are responsible for certain changes in fungal hyphae, which could be asymptomatic and may cause a reduction or elimination of the virulence capacity of fungal hosts by the process called hypovirulence. Such fungal-virus system could be valuable for the development of novel biocontrol approaches against fungal pathogens for the development of a sustainable environment. There are adequate reports where mycovirus has been employed as a biocontrol approach against the pathogenic fungi in the fields of agriculture and other allied sciences. The prime focus of this review is to emphasize naturally available mycoviruses and strategies to adopt the mycovirus therapy which could serve as an excellent alternative strategy against chemical prophylactic and therapeutic approaches.

INTRODUCTION

Presently the global scenario is shifting from food security to nutritional security and in the present context; agriculture and other allied production sectors have emerged as a potential avenue to achieve the ever-increasing nutritional requirement of population and can play an instrumental role in accomplishing the sustainable development goals (SDGs) of the United Nation by 2030. To achieve these targets, there is a genuine necessity for good husbandry practices with minimal detrimental impact on the environment. However, in the present era of technological advancement and intensification, the production sectors are laden with additional vulnerability to different stressors particularly to pathogenic organisms. It is the irresponsible and indiscriminate use of available resources that are underpinning the environmental and health hazards, ultimately leading to catastrophic events; for example, the wonder drug of the 20th century *i.e.* antibiotics and other chemotherapeutics are under the risk of running out of power and may lead to the development of antimicrobial resistance (AMR) or superbugs. The gravity of situation can be

DOI: 10.4018/978-1-7998-7062-3.ch013

recognized by the report from CDC (Center for Disease Control, USA), which reveals that the humans are exposed to more than 300 different kinds of environmental chemicals or their metabolites, which are associated with various health hazards (CDC, 2017).

So it is high time to think about some alternatives approaches those are environment-friendly and at the same time with preventive/ therapeutic potential (Khan et al., 2018; Khan et al., 2019; Reverter et al., 2014). As an answer to present concern, the biocontrol approach can be encouraging alternative. The word biological control coined by H. S. Smith (1919) to indicate the potential of naturally available enemies of invasive organisms or opportunistic pathogens and defined as "the action of any organism in maintaining the invasive organism's population density" (De Bach, 1964). The biocontrol strategy takes the advantage of fundamental ecological interactions between organisms which includes parasitism, predation, competition, pathogenicity *etc.* Since the method involves application of natural or biological approach the possibilities of adverse effect or developing any resistance is nullified.

The flourishing biomedical and agriculture sector has been presently accompanied by the emergence or re-emergence of several infectious diseases. Given that the sector usually serves as an invitation to various etiological agents, thereby causing severe economic loss and intense mortality. The gravity of the present circumstance and backdrop information, urge researchers to explore the alternative eco-friendly approach to ascertain the sustainability of environment such as mycovirus. The present review is an account on the current state of knowledge about the mycoviruses and their application against an array of fungal host.

MYCOVIRUS

Even after almost six decades from its first report from nature, mycoviruses are still very unique and new to humankind. Sometime they are also termed as mycophage. Like other viruses or bacteriophages, mycoviruses haven't got much attention maybe because most mycoviruses cause asymptomatic infection. The first-ever report of mycovirus came in the year 1948 from a mushroom (*Agaricus bisporus*) farm owned of Pennsylvania, USA and they named the symptoms as La France disease, later based on the symptoms people also started to referring as watery stripe disease/ X disease/ dieback/ brown disease (Hollings, 1962). Soon after the first report, similar kind of infections was also reported from Japan, Australia and Europe (Ghabrial et al., 2015). But it was the year 1962, when Hollings reported the first-ever exact association of virus-like particles with diseased mushroom sporophores which ultimately gives rise to another branch of science *i.e.* mycovirology (Ghabrial & Suzuki, 2009). Later, Sanderlin and Ghabrial (1978) concluded that "the mycoviruses are infectious particle and cause infection to filamentous fungal pathogen". Since the first report of mycovirus, *Cryphonectria parasitica* hypovirus 1 (CHV1) is the best-known mycovirus in the history of mycovirology and the rest remains underexplored.

Like viruses, mycovirus do not possess any extracellular phase in their life cycle. For their fundamental needs of energy and genetic activities, they depend upon their host cell, that's why sometimes they referred as "genetic parasites" (Kaya et al., 2015). Usually, mycovirus is having a diameter of 25±5.0 nm with naked isometric particulate nature (Xie and Jiang, 2014). Because of their minute and asymptomatic existence, it becomes difficult to isolate them, which led the foundation for extraction of dsRNA from fungus mycelium through molecular methods to confirm the existence of a mycoviral infection (Kaya et al., 2015). However, with the recent findings, the perception for all mycovirus posses dsRNA as genetic material has substantially changed, as they also possess other forms of the genetic construct. While sharing some common attributes with viruses, mycovirus also includes some unique characteristics like, most involves an extracellular course for infection, intercellular transmission through sporulation, cell division, cell fusion *etc.* (Son et al., 2015).

GENOMIC COMPOSITION OF MYCOVIRUS

The recent 10th ICTV (International Committee on Taxonomy of Viruses, London, UK) report on virus classification has listed more than 250 mycoviruses widely distributed over 16 families, 17 taxa including 22 genera; however, about 20% of isolated mycoviruses remained un-assigned to any genus or family (Table 1) (ICTV, 2014; King et al., 2011; Kotta-Loizou & Coutts, 2017; Maimaiti et al., 2020; NCBI, 2014; Xie & Jiang, 2014). The morphology of mycoviruses varies from flexuous rods, rigid rods, enveloped bacilliform particles, club-shaped particles, Herpesvirus-like viruses and isometric; among all, the isometric form is dominant morphology (Varga et al., 2003). Slightly more than 250 mycoviral sequences have been submitted to NCBI (National Center for Biotechnology Information, Maryland, USA) database where most of the mycoviruses belong to RNA virus family; however, *Rhizidiomyces* virus is only one to belong DNA virus (Ozkan-Kotiloglu & Coutts, 2018). According to latest ICTV report, mycoviruses are accommodated into several groups with different genomic composition, represented by seven dsRNA linear genomes (*Endornaviridae, Totiviridae, Chrysoviridae, Partitiviridae, Megabirnaviridae, Reoviridae* and *Quadriviridae*), six positive sense ssRNA linear genomes (*Narnaviridae, Alphaflexiviridae, Gammaflexiviridae, Barnaviridae Hypoviridae* and *Deltaflexiviridae*), reverse transcribing linear ssRNA virus families (*Metaviridae* and *Pseudoviridae*), negative-sense ssRNA linear genome (*Mymonaviridae*), ssDNA circular genomes and rarely dsDNA (Ghabrial et al., 2015; Kotta-Loizou & Coutts, 2017; Lefkowitz et al., 2018; Li et al., 2019; Wang & Jin, 2017).

The diverse genetic makeup of mycovirus considered to be very significant for their identification purposes. The genetic material may be segmented or non-segmented where genome size varies from 3.7 to 4.9 kbp with single ORF (RnQV1-Rosellinia necatrix quadrivirus 1) to 5.31 kbp with triple ORF (Chalara elegans RNA Virus 1- CeRV1) (Chiba et al., 2009; Park, James & Punja, 2005). But there are several reports where, the range is bit different, from 1.4 to 2.4 kbp with a single ORF (*Partitiviridae*) to 9 to 13 kbp with two overlapping ORFs (*Hypoviridae*) (Flores et al., 2015). In an experiment of Flores et al. (2015), reported satellite element of small RNA (0.9 to 1.4 kbp) with the main genome (3.7 to 5 kbp) in Basydiomicetous yeast, *Xanthophyllomyces dendrorhous*. Among them the most frequent one is dsRNA (Xie & Jiang, 2014); however, Pearson (2009) observed that about 30% of mycoviruses have positive-sense ssRNA as their genome. Some of the mycoviruses (especially dsDNA) does not fit the narrative of mycoviruses which are frequently considered as virus-like particles (VLP) (Goker et al., 2014).

Table 1. Classification of mycovirus by International Commission on Taxation of Viruses (ICTV)

Genome Type	Family	Genus	Genomic Feature and Morphology	Examples
I. Positive sense ssRNA	1. *Narnaviridae* (Naked RNA)	A. Narnavirus (Narnavirus infect fungi, oomycetes and protista and are localized in the cytoplasm)	Single linear, segmented, nucleoprotein complex without a cap, genome size ~ 2.3 to 3.6 kb, non-enveloped	-*Saccharomyces* 20S RNA narnavirus -*Saccharomyces* 23S RNA narnavirus
		B. Mitovirus (most commonly found, exclusively infect fungi only)	Virion complex composed of nucleoprotein, non-enveloped	-*Gremmeniella mitovirus S1* -*Ophiostoma mitovirus* 3a (OMV3a) -*Cryphonectria mitovirus 1*
	2. *Barnaviridae* (Bacilliform RNA)	A. Barnavirus	Single linear, genome ~ 4.0 kb in size, 19 x 50 nm, non-enveloped, bacilliform	-Mushroom bacilliform virus
	3. *Gammaflexiviridae*	A. Mycoflexivirus	Encapsidated in filamentous virions, ~ 600 to 720 nm flexuous, single linear, genome 6.8 kb in length	-Botrytis virus *Y* -Botrytis virus F
	4. *Alphaflexiviridae*	A. Botrexvirus	Two genomic segments encapsidated in spherical virions ~ 600 to 720 nm flexuous, single linear molecule, genome length 7.0 kb	-Botrytis virus X
		B. Sclerodarnavirus	Not enveloped, single linear molecule, genome length ~ 5.4 kb	-*Sclerotinia sclerotiorum* debilitation-associated RNA virus (SsDRV)
	5. *Hypoviridae*	A. Hypovirus	Monopartite, linear, genome length ~ 9 to 13 kb, pleomorphic vesicles, 50-80 nm, unconventionally encapsulated genomes and accommodated in lipid pleomorphic vesicles of host	-*Cryphonectria hypovirus* 1, 2, 3 and 4 (CHV-1, CHV-2, CHV-3, CHV-4) *Valsaceratosperma* hypovirus 1 -*Sclerotinia sclerotiorum hypovirus* 1 and 2 -*Phomopsis longicolla hypovirus* 1 -*Fusariumgr aminearum hypovirus* 1 (FgHV1)
	6. *Deltaflexiviridae*		linear, positive-sense ssRNA, length of genetic material is 8,246 nucleotides	-*Fusarium graminearum* deltaflexivirus 1 (FgDFV1)
II. Negative sense ssRNA	1.*Mycomononegaviridae*		Single linear (some time circular also), genome length is ~ 10 kb	-*Sclerotinia sclerotiorum* negative-stranded RNA virus 1
III. Positive sense dsRNA	1. *Totoviridae* dsRNA	A. Totivirus (Infects protozoans)	4-43 nm size, mono-segmented, bicistronic, unsealed, size of genome ~ 4.6 to 7.0 kbp and icosahedral virions	-*Saccharomyces cereviceae* LA(L1) -*Saccharomyces cereviceae* L-BC (BA) -*Aspergillus niger virus* S -*Aspergillus foetidus virus* S
		B. Victorivirus (Exclusively infect fungi*)*	Non-enveloped, non segmented genomes and icosahedral virions	-*Agaricus bisporus virus* 1 -*Magnaporthe oryzae virus* 1 (MoV1) -*Epichloë festucae virus* 1 (EfV1) -*Chalara elegans* RNA virus 1 (CeRV1) -*Helminthosporium victoriae virus* 190 S
		C. Satellite dsRNAs	Sub-viral nucleic acid molecules require helper viruses	-Satellites of *Saccharomyces cerevisiae* L-A virus -Satellite of *Ustilago maydis* killer M virus
	2. *Chrysoviridae*	A. Chrysovirus	30-35 nm, segmented, non-enveloped, multi-component with four monocistronic genome segments,enome length ~2.4 to 3.6 kbp and separately encapsulated in virus particles	-*Agaricus bisporus virus* 1 -*Penicillum brevicopactus virus* -*Penicillum chrysogenum virus* -*Penicillum cyaneo-fulvum virus*
	3. *Reoviridae*	A. Mycoreovirus	Linear and multi-segmented genome (10 to 12 segments are encapsulated within a large icosahedral virion) with a length of 0.7 to 4.1 kbp, isometric ~80 nm and non enveloped	-Mycoreovirus 1 [*Cryphonectria parasitica mycoreovirus*-1 (9B21)] -Mycoreovirus 2 -Mycoreovirus 3 (*Rosellinia necatrix* W370 *virus*)
	4.*Partitiviridae* (Partitivirus)	A. Partitivirus -Gammapartitivirus (Exclusively infect fungi) -Alphapartitivirus -Betapartitivirus, infect both fungi and plants -Deltapartitivirus exclusively infect both protozoa and plants	Isometric 30 to 35 nm, bi-segmented (separately each one is encapsulated in distinct icosahedral virions), multi-component, genome ~1.4–2.4 kbp in length, non-enveloped	-*Aspergillus ochreseous virus* -*Fusarium solani virus* 1 -*Fusarium poae virus* -*Penicillum stolineferum virus* S -*Penicillum stolineferum virus* F
	5. *Endornaviridae*	A. Endornavirus	Linear, ~14 to 17 kbp long genome, not enveloped, encapsidated amd non-segmented genomes accommodated in lipid vesicles of host origin	-*Phytophthora endornavirus* 1 (PEV1)
	6. *Megabirnaviridae*	A. Megabirnavirus	Bisegmented, separately encapasulated in 50 nm of isometric particles	-*Rosellinia necatrix* megabirna *virus* 1
	7. *Quadriviridae*	A. Quadrivirus	45 nm isometric virus particles are separately packed with four segments of mono-cistronic genome, having a length of 3.7 to 4.9 kbp	-*Rosellinia necatrix* quadrivirus 1
	8. Botybirnavirus	-	Two linear dsRNA segments, 6.2- 5.8 kbp in size	-*Sclerotinia sclerotiorum* mycotymovirus 1 (SsMTV1/SZ-150) -*Sclerotinia sclerotiorum* botybirnavirus 3 (SsBV3/SZ-150)

continued on following page

Table 1. Continued

Genome Type	Family	Genus	Genomic Feature and Morphology	Examples
IV. Positive sense ssRNA RT (Reverse transcription)	1. *Pseudoviridae*	A. Hemivirus	50 nm diameter, un-segmented, unsealed, isometric to quasi-isometric	-*Saccharomyces paradoxus* Ty5 *virus* -*Candida albicans* Tca5 *virus* -*Candida albicans* Tca2 *virus*
		B. Pseudovirus	30 to 40 nm in diameter, non-segmented, unsealed, isometric to quasi-isometric	-*Saccharomyces cereviceae* Ty1 *virus*, Ty2 *virus* and Ty4 *virus*
	2. *Metaviridae*	A. Metavirus	50 nm diameter, un-segmented, unsealed, irregular, ovoid, irregular, enveloped nucleoprotein complex	-*Fusarium oxysporum skippy virus* -*Saccharomyces cereviceae* Ty3 *virus* -*Cladosporium fulvum* T-1 *virus*
V. dsDNA	1. *Adenoviridae*	A. *Rhizidiomyces*	Not enveloped, 60 nm diameter,	-*Rhizidiomyces virus* (RhiV)

(Source: Fauquet et al., 2005; Ghabrial and Suzuki, 2009; Ghabrial et al., 2015; ICTVdB 2002; King et al., 2011; Kotta-Loizou and Coutts, 2017; Lefkowitz et al., 2018; Li et al., 2019 Nibert et al., 2014)

MYCOVIRUS INDUCED HYPOVIRULENCE

Hypovirulence is a development in which, mycovirus renders the capacity of any fungal pathogen to cause any infection (Li et al., 2019; Nuss, 2005). The effects caused by mycoviruses are influenced by the host and ecological conditions (Hyder et al., 2013). Commonly, mycovirus infection cause change from asymptomatic (cryptic symptoms) to the alteration in virulence through significant morphological and physiological variations includes cytological alteration, changes in colony morphology, sluggish mycelia growth, variations of sporulation, slow invasion, toxin secretion *etc* (Ghabrial & Suzuki, 2009; Milgroom & Hillman, 2011; Nuss, 2005). During 1905 there was first report of chestnut blight disease in North America which hampered the growth of stem and root, however by the 1950s the situation turned around and the infected trees were not killed and the lesions were also healed. This phenomenon eventually led to the development of biocontrol aspect through hypovirulent strains of mycoviruses to kill harmful fungal pathogens (Longkumer and Ahmad, 2020).

The effectiveness of hypovirulence depends on the natural spread, influenced by the environmental factors and the triple interaction between mycovirus, fungal pathogen and host. Mycoviruses can transmit through both vertical and horizontal mode via spores and anastomosis (cytoplasmic exchange between two different fungal hyphae following fusion) respectively; however, *Sclerotinia gemycircular virus 1* (formerly named *Sclerotinia sclerotiorum* hypovirulence-associated DNA virus 1) infect extracellularly to their fungal host (Yu et al., 2013). The mode of action of hypovirulence can be broadly categorized into two categories, in the first extracellular pathway, the selected mycoviral agents are allowed to infect host fungi or transferred on mantle protrusions of the target host. The very same concept is applied in a member from totividae family, where infectious myonecrosis virus (IMNV) binds to surface fungi of the shrimps which are responsible for fouling over exoskeleton (Keceli, 2017). In the second process, the desired gene sequence is identified which is responsible for producing some antifungal compounds, is incorporated into the mycovirus genome with the help of modern genetic engineering approach. Through this method a broad spectrum yeast killer toxin-producing gene is transferred to mycovirus which works against the fungal host, as for example Keceli (2017) worked on *Candida albicans,* and infected host with specific mycovirus to eliminate its pathogenicity. Some yeasts, including *Zygosaccharomyces*, *Ustilago maydis*, *Saccharomyces* and *Hanseniaspora* can also lead to the production of lethal toxin (Bruenn, 2005). These toxin-secreting "killer yeasts" may kill the susceptible yeasts strains to get additional nourishment (Park et al., 1996).

Hypovirulent or debilitated strains of pathogenic fungi, carrying transmissible mycoviruses, are galvanizing researchers because of their potential capability as a biocontrol agents, additionally specific probes can be designed for deciphering the mechanisms of fungal pathogenesis (Sharma et al., 2018). Isolation of double-stranded RNA is a vital non-specific indicator of the mycovirus presence, which can be used as a precise diagnostic probe for the mycovirus identification. Fungal pathogens cause catastrophic outbreak of diseases among in all major crops and ultimately affected human kind. The optimization of mycovirus based biocontrol therapy might serves an avenue as an efficient, environment-friendly alternative method to counter the great concerns as a consequence to the abuse of the chemical therapeutics. Mycovirus treatment may also help in eliminating the probability of rousing anti-microbial resistance issue (Sharma et al., 2018).

STRATEGIES FOR MYCOVIRUS BASED ANTIFUNGAL THERAPY

The selection of potential mycovirus isolates from nature requires consideration of certain systematic approaches followed by *in vitro* and *in vivo* characterization of potential isolate.

The ideal characteristics of a potential mycovirus candidate:

- Non-infectious to host
- Should not be immunogenic to host
- Should have an extracellular mode of transmission
- Should not have the capacity to integrate into the host genome
- Should have a wide range of antimycotic activity with 100% killing ability
- Should have the capacity to proliferate within pathogenic fungi
- Should be proficient to be acquired with high purity and large quantities for direct application
- Should have a genome that facilitates genetic manipulation for future advancement

The above-mentioned criteria provide a broad line strategy for the selection of appropriate natural isolate of mycovirus. Since mycovirus are accountable for acting against a specific fungal pathogen, so the selection of the pathogenic condition or agent is the primary step in therapy. Additionally, several other chronological steps should be followed for mycovirus therapy:

- Isolation of mycovirus from natural resources
- *In-vitro* propagation of the isolated mycovirus
- Characterization (both phenotypic and genotypic) of isolated mycovirus
- Mycovirus typing
- Standardization of probable application strategies

Table 2. Application of mycoviruses

S. N.	Fungus	Main Host or Disease	Mycoviruses	Family	References
1.	*Diplodia pinea*	*Pinus* spp.	*Sphaeropsis sapinea* RNA *virus* 1 and 2	*Totiviridae*	Preisig et al. (1998)
2.	*Helminthosporium victoriae* (Ascomycete)	Victoria blight of oats	-	-*Totiviridae* -*Chrysoviridae*	Ghabrial et al. (2002)
3.	*Heterobasidion annosum* complex	Various	-Heterobasidion partitivirus -Heterobasidion partitivirus –P -Heterobasidion partitivirus (2-pa1 7-pa1,1,2,3,4,5,6,7 and 8)	-*Partitiviridae*	Ihrmark et al. (2002); Vainio et al. (2014)
4.	*Diaporthe perjuncta* (Ascomycete)	Diaporthe diseases of stone fruits	-	Unclassified ssRNA virus related to tombus viruses	Chu et al. (2002)
5.	*Rosellinia necatrix* (Ascomycete)	White root rot	-	*Reoviridae*	Wei et al. (2003)
6.	*Gremmeniella abietina*	-*Abies* spp. -*Pinus* spp. -*Larix* spp. - *Picea* spp.	-*Gremmeniella abietina* RNA virus (L1 and MS 1) -*Gremmeniella abietina* mitocondrial RNA virus -S1	-*Narnaviridae* -*Partitiviridae* -*Totiviridae*	Tuomivirta and Hantula (2003 a; b)
7.	-*Homoeocarpa* (Ascomycete) -*Homoeocarpa* (Ascomycete)	Dollar spot disease of Turfgrass	-	*Narnaviridae* (genus *Mitovirus*)	Deng et al. (2003)
8.	*Cryphonectria parasitica*	*Castanea* spp. (Chestnut blight)	-*Cryphonectria hypovirus* (1, 2, 3 and 4) -*Cryphonectria mitovirus* 1 (CHV1) -*Cryphonectria mitovirus* (1 and 2)	-*Reoviridae* -*Hypoviridae* -*Narnaviridae* (genus *mitovirus*)	Hillman and Suzuki (2004); Suzuki et al. (2004)
9.	*Ophiostoma novo-ulmi*	*Ulmus* spp.	-*Ophiostoma novo-ulmi mitoviruses* (2,3a, 3b, 1a, 1b and 1c) -*Ophiostoma novo-ulmi mitoviruses* (4-Ld, 5-Ld, 6-Ld and 7 Ld)	-*Narnaviridae*	Doherty et al. (2006); Hintz et al. (2013)
10.	*Rosellinia necatrix*	Various	-*Rosellinia necatrix megabirnavirus* 1 and 2	-*Megabirnaviridae* -*Partitiviridae*	Chiba et al. (2009)
11.	*Botrytis cinerea*	Various	-*Botrytis cinerea mitovirus* 1	-*Narnaviridae*	Wu et al. (2010)
12.	*Verticillium dahliae*	Various	-*Verticillium dahliae chrysovirus* 1	-*Chrysoviridae*	Cao et al. (2011)
13.	*Diplodia scrobiculata*	*Pinus* spp.	-*Diplodia scrobiculata* RNA virus 1	-*Chrysoviridaerelated*	De Wet et al. (2011)
14.	*Fusarium graminearum*	*Fusarium* spp.	*Fusarium graminearum Hypovirus* 2 (FgHV2 and 1)	-*Hypoviridae*	Li et al. (2019); Wang et al. (2013)
15.	*F. circinatum*	-*Pseudotsuga menziesii* -*Pinus* spp.	-*Fusarium circinatum mitovirus* (1, 2-1 and 2-2)	-*Narnaviridae*	Martinez-Alvarez et al. (2014)
16.	*Botryosphaeria dothiea*	-*Pyrus* spp. -*Malus* spp. -*Eucalyptus* spp.	-*Botryosphaeria dothidea partitivirus* 1 -*Botryosphaeria dothidea chrysovirus* 1	-*Chrysoviridae* -*Partitiviridae*	Wang et al. (2014)
17.	*Hymenoscyphus fraxineus*	*Fraxinus* spp.	-*Hymenoscyphus fraxineus mitovirus* 1	-*Narnaviridae*	Schoebel et al. (2014)
18.	*Verticillium albo-atrum*	Various	-*Verticillium albo-atrum partitivirus* 1	-*Partitiviridae*	Canizares et al. (2014)
19.	*Aspergillus fumigatus*	*Aspergillus* spp.	-*Aspergillus fumigatus tetramycovirus*-1 (AfuTmV-1)	-	Kanhayuwa et al. (2015)
20.	*Penicillium digitatum*	-	-*Penicillium digitatum* Narna-like virus 1 (PdNLV1) -*Penicillium digitatum polymycovirus* 1 (PdPmV1)	-Polymycoviruses	Niu et al. (2018)
21.	*Rhizoctonia solani*	*Zoysia japonica*	-*Rhizoctonia solani* dsRNA virus 2	-Alphapartitivirus	Picarelli et al. (2019)
22.	*A. fumigatus*	*Aspergillus* spp.	-*Aspergillus fumigatus chrysovirus* (AfuCV41362)	-*Chrysoviridae*	Takahashi-Nakaguchi et al. (2020)
23.	*Alternaria dianthicola*	-	-*Alternaria dianthicola* dsRNA virus 1" (AdRV1).	-*Partitiviridae*	Hu et al. (2020)
24.	*Botryosphaeria dothidea*	-	-*Botryosphaeria dothidea botourmiavirus* 1 (BdBOV-1)	-*Botourmiaviridae*	Yang et al. (2020)
25.	*Leptosphaeria biglobosa*	*Brassica napus*	Novel double-stranded RNA quadrivirus	-	Shah et al. (2020)
26.	*Corynespora cassiicola*	Rubber leaf fall disease	-*Corynespora cassiicola bipartite mycovirus* 1 (CcBV1)	Unassigned dsRNA mycoviruses	Wang et al. (2020)
27.	*Podosphaera xanthii*	Melon powdery mildew	Red clover powdery mildew-associated totivirus 1 and 2, red clover powdery mildew associated totivirus 2–2, YP_009182176	-*Totiviridae*	Maimaiti et al. (2020)

- Assessment of therapeutic efficacy against experimental infection through *in-vitro* and *in-vivo* trials
- Complete genome sequencing for the identification of any possible virulent or toxic genes in the mycovirus before the field based trials
- Therapy must withstand the different agro-climatic culture condition
- Therapy should have long term sustainability with appropriate storage medium to facilitate maximum shelf life
- Therapy should have regulatory consent from public and scientific agencies *etc.*

The mycovirology is gaining momentum among researchers to explore their biology, mode of action and possible action in the field of biomedical and agri-allied sectors. The current level of knowledge lays the foundation for the application of mycovirus against an array of fungal pathogens. The experimental trials have revealed the broad host range and capability to induce hypovirulence in experimental infection to heterogeneous hosts. Till date, there are very few successful examples in human biomedical science; however, in agriculture and other allied sciences the use is going on successfully over the past decades (Table 2).

ADVANTAGES OF MYCOVIRUS THERAPY

There are some advantages of mycovirus therapy over chemotherapy. These advantages are:

- Since mycovirus are naturally occurring so attaining regulatory approval would be easy compared to chemical therapeutics
- Since mycovirus are specific to their host, so no harm to normal non-pathogenic microbial flora (non targeted fungi)
- Since mycovirus and host fungi are found in the same environmental condition which signifies their ability to survive in the different physico-chemical condition
- Like chemotherapy there won't be any magnification or residual or anti-microbial resistance problem *etc.*

LIMITATIONS OF MYCOVIRUS THERAPY

In addition to the advantages of mycovirus application, there are few constrain related to the therapy:

- The application requires the exact recognition of virulent fungal species
- Since the scarcity of available information about mycovirus, optimization (isolation and characterization) of them from new environment such as from the aquatic ecosystem would be difficult
- Proper regulatory approvals may have problems because of lack of awareness and scanty of scientific validation.

CONCLUSION

The studies have shown the potential of mycoviruses as biological control and its capability in regulating the application of chemical remedial agents. However, there is an urgent need for the researches to understand the biology of mycoviruses and to conduct *in-vitro* and *in-vivo* trials to ensure its efficacy against fungal pathogens. The modern approaches such as genomics, transcriptomics, and proteomics can be coupled with bioinformatic analysis which may enable researchers to identify the potential species and possibly help in revealing the exact mechanism of the host-virus interactions at the cellular and molecular level. The application of mycovirus in various sectors such as application in biomedical science, agriculture, fisheries and aquaculture, *etc.* can open new avenues as sustainable alternative in the near future. However, genuine challenge for the successful application of mycoviruses lies in the development of field-based implementation with consideration of significant factors such as the source of inoculum, mode of infection, form of colonization *etc.* The possibilities are immense with the application of mycovirus but lack of scientific information and corroborative data is a major drawback in its application which gives endless possibilities which need to be investigated.

REFERENCES

Bruenn, J. (2005). The Ustilago maydis killer toxins. In M. Schmitt, R. Schaffrath, & M. J. Schmitt (Eds.), *Microbial Protein Toxins* (pp. 157–174). Springer.

Canizares, M. C., Perez Artes, E., & Garcia-Pedrajas, M. D. (2014). The complete nucleotide sequence of a novel partitivirus isolated from the plant pathogenic fungus *Verticillium albo-atrum. Archives of Virology*, *159*(11), 3141–3144. doi:10.100700705-014-2156-6 PMID:24986717

Cao, Y. F., Zhu, X. W., Xiang, Y., Li, D. Q., Yang, J. R., Mao, Q. Z., & Chen, J. S. (2011). Genomic characterization of a novel dsRNA virus detected in the phytopathogenic fungus *Verticillium dahliae* Kleb. *Virus Research*, *159*(1), 73–78. doi:10.1016/j.virusres.2011.04.029 PMID:21571013

Center for Disease Control. (2017). *Biomonitoring Program Activities: Environmental Chemicals.* National Biomonitoring Program. https://www.cdc.gov/biomonitoring/environmental_chemicals.html

Chiba, S., Salaipeth, L., Lin, Y. H., Sasaki, A., Kanematsu, S., & Suzuki, N. (2009). A novel bipartite double-stranded RNA Mycovirus from the white root rot Fungus *Rosellinia necatrix*: Molecular and biological characterization, taxonomic considerations, and potential for biological control. *Journal of Virology*, *83*(24), 12801–12812. doi:10.1128/JVI.01830-09 PMID:19828620

Chu, Y. M., Jeon, J. J., Yea, S. J., Kim, Y. H., Yun, S. H., Lee, Y. W., & Kim, K. H. (2002). Double-stranded RNA mycovirus from *Fusarium graminearum. Applied and Environmental Microbiology*, *68*(5), 2529–2534. doi:10.1128/AEM.68.5.2529-2534.2002 PMID:11976130

De Bach, P. (1964). *Biological control of insect pests and weeds.* Reihold.

De Wet, J., Bihon, W., Preisig, O., Wingfield, B. D., & Wingfield, M. J. (2011). Characterization of a novel dsRNA element in the pine endophytic fungus *Diplodia scrobiculata. Archives of Virology*, *156*(7), 1199–1208. doi:10.100700705-011-0978-z PMID:21442227

Deng, F., Xu, R., & Boland, G. J. (2003). Hypovirulence-Associated Double-Stranded RNA from *Sclerotinia homoeocarpa* Is Conspecific with *Ophiostoma novoulmi* Mitovirus 3a-Ld. *Phytopathology, 93*(11), 1407–1414. doi:10.1094/PHYTO.2003.93.11.1407 PMID:18944069

Doherty, M., Coutts, R. H. A., Brasier, C. M., & Buck, K. W. (2006). Sequence of RNA-dependent RNA polymerase genes provides evidence for three more distinct mitoviruses in *Ophiostoma novo-ulmi* isolate Ld. *Virus Genes, 33*(1), 41–44. doi:10.100711262-005-0029-5 PMID:16791417

Fauquet, C. M., & Fargette, D. (2005). International Committee on Taxonomy of Viruses and the unassigned species. *Virology Journal, 2*(1), 64. doi:10.1186/1743-422X-2-64 PMID:16105179

Flores, O., Alcaíno, J., Fernandez-Lobato, M., Cifuentes, V., & Baeza, M. (2015). Characterization of virus-like particles and identification of capsid proteins in *Xanthophyllomyces dendrorhous. Virus Genes, 50*(2), 253–259. doi:10.100711262-015-1171-3 PMID:25663143

Ghabrial, S., & Suzuki, N. (2009). Viruses of plant pathogenic fungi. *Annual Review of Phytopathology, 47*(1), 353–384. doi:10.1146/annurev-phyto-080508-081932 PMID:19400634

Ghabrial, S. A., Caston, J. R., Jiang, D., Nibert, M. L., & Suzuki, N. (2015). 50-plus years of fungal viruses. *Virology, 479*, 356–368. doi:10.1016/j.virol.2015.02.034 PMID:25771805

Ghabrial, S. A., Soldevila, A. I., & Havens, W. M. (2002). Molecular genetics of the viruses infecting the plant pathogenic fungus *Helminthosporium victoriae*. In S. Tavantzis (Ed.), *Molecular Biology of Double-Stranded RNA: Concepts and Applications in Agriculture* (pp. 213–236). Forestry and Medicine.

Goker, M., Scheuner, C., Klenk, H. P., Stielow, J. B., & Menzel, W. (2014). Codivergence of mycoviruses with their hosts. *PLoS One, 6*(7), 22252. doi:10.1371/journal.pone.0022252 PMID:21829452

Hillman, B. I., & Suzuki, N. (2004). Viruses of the Chestnut bligth fungus, *Cryphonectria parasitica. Advances in Virus Research, 63*, 423–472. doi:10.1016/S0065-3527(04)63007-7 PMID:15530566

Hintz, W. E., Carneiro, J. S., Kassatenko, I., Varga, A., & James, D. (2013). Two novel mitoviruses from a Canadian isolate of the Dutch elm pathogen *Ophiostoma novoulmi* (93-1224). *Virology Journal, 10*(1), 252. doi:10.1186/1743-422X-10-252 PMID:23924036

Hollings, M. (1962). Viruses associated with a die-back disease of cultivated mushroom. *Nature, 196*(4858), 962–965. doi:10.1038/196962a0

Hu, Z., Guo, J., Da Gao, B., & Zhong, J. (2020). A novel mycovirus isolated from the plant-pathogenic fungus *Alternaria dianthicola. Archives of Virology, 165*(9), 2105–2109. doi:10.100700705-020-04700-9 PMID:32556598

Hyder, R., Pennanen, T., Hamberg, L., Vainio, E. J., Piri, T., & Hantula, J. (2013). Two viruses of *Heterobasidion* confer beneficial, cryptic or detrimental effects to their hosts in different situations. *Fungal Ecology, 6*(5), 387–396. doi:10.1016/j.funeco.2013.05.005

ICTV. (2014). *International Committee on Taxonomy of Viruses.* http://ictvonline.org/index.asp

ICTVdB. (2002). *The Universal Virus Database, version 4.* https://www.ncbi.nlm.nih.gov/ICTVdb/ICTVdB/

Ihrmark, K., Johannesson, H., Stenstrom, E., & Stenlid, J. (2002). Transmission of double-stranded RNA in *Heterobasidion annosum. Fungal Genetics and Biology*, *36*(2), 147–154. doi:10.1016/S1087-1845(02)00011-7 PMID:12081468

Kanhayuwa, L., Kotta-Loizou, I., Ozkan, S., Gunning, A. P., & Coutts, R. H. (2015). A novel mycovirus from *Aspergillus fumigatus* contains four unique dsRNAs as its genome and is infectious as dsRNA. *Proceedings of the National Academy of Sciences of the United States of America*, *112*(29), 9100–9105. doi:10.1073/pnas.1419225112 PMID:26139522

Kaya, A. G. A., Dogmus Lehtijarvi, H. T., & Lehtijarvi, A. (2015). The usage of mycoviruses in biological control against tree pathogenic fungi. *Istanbul Üniversitesi Orman Fakültesi Dergisi*, *1*, 60–71.

Keceli, S. A. (2017). Mycoviruses and importance in mycology. *Mikrobiyoloji Bulteni*, *51*(4), 404–412. PMID:29153071

Khan, M. I. R., Saha, R. K., & Saha, H. (2018). Muli bamboo (*Melocanna baccifera*) leaves ethanolic extract a non-toxic phyto-prophylactic against low pH stress and saprolegniasis in *Labeo rohita* fingerlings. *Fish & Shellfish Immunology*, *74*, 609–619. doi:10.1016/j.fsi.2017.11.047 PMID:29183812

Khan, M. I. R., Singh, M., & Monsang, S. J. (2019). Prophylactic potential of ethanolic leaves extract of Muli bamboo (*Melocanna baccifera*) against *Aeromonas hydrophila* infection in *Labeo rohita. Journal of Entomology and Zoology Studies*, *7*(2), 288–291.

King, A. M. Q., Lefkowitz, E. M., Adams, J., & Carstens, E. B. (2011). *Virus Taxonomy: Ninth Report of the International Committee on Taxonomy of Viruses (ICTV)*. Elsevier Academic.

Kotta-Loizou, I., & Coutts, R. H. A. (2017). Mycoviruses in Aspergilli: A comprehensive review. *Frontiers in Microbiology*, *8*, 1699. doi:10.3389/fmicb.2017.01699 PMID:28932216

Lefkowitz, E. J., Dempsey, D. M., Hendrickson, R. C., Orton, R. J., Siddell, S. G., & Smith, D. B. (2018). Virus taxonomy: The database of the Internatinal Committee on Taxonomy of Viruses (ICTV). *Nucleic Acids Research*, *46*(D1), D708–D717. doi:10.1093/nar/gkx932 PMID:29040670

Li, P., Bhattacharjee, P., Wang, S., Zhang, L., Ahmed, I., & Guo, L. (2019). Mycoviruses in *Fusarium* Species: An Update. *Frontiers in Cellular and Infection Microbiology*, *9*, 257. doi:10.3389/fcimb.2019.00257 PMID:31380300

Longkumer, I. Y., & Ahmad, M. (2020). Potential of Mycovirus in the Biological Control of Fungal Plant Pathogens: A Review. *Agricultural Reviews (Karnal)*, *41*(3). Advance online publication. doi:10.18805/ag.R-1896

Maimaiti, Y., Ding, L., Chai, M., Jing, X., Yang, D., Han, S., Sun, L., & Chen, W. (2020). Identification and analysis of new mycoviruses from melon powdery mildew. *Journal of Plant Pathology*, 1–6.

Martinez-Alvarez, P., Pando, V., & Diez, J. J. (2014). Alternative species to replace Monterey pine plantations affected by pitch canker caused by *Fusarium circinatum* in northern Spain. *Plant Pathology*, *63*(5), 1086–1094. doi:10.1111/ppa.12187

Milgroom, M. G., & Hillman, B. I. (2011). The ecology and evolution of fungal viruses. In *Studies in viral ecology: microbial and botanical host systems* (pp. 217–253). John Wiley and Sons. doi:10.1002/9781118025666.ch9

NCBI. (2014). *National Center for Biotechnology Information*. https://www.ncbi.nlm.nih.gov

Nibert, M. L., Ghabrial, S. A., Maiss, E., Lesker, T., Vainio, E. J., Jiang, D., & Suzuki, N. (2014). Taxonomic reorganization of family Partitiviridae and other recent progress in partitivirus research. *Virus Research, 188*, 128–141. doi:10.1016/j.virusres.2014.04.007 PMID:24768846

Niu, Y., Yuan, Y., Mao, J., Yang, Z., Cao, Q., Zhang, T., Wang, S., & Liu, D. (2018). Characterization of two novel mycoviruses from *Penicillium digitatum* and the related fungicide resistance analysis. *Scientific Reports, 8*(1), 1–12. doi:10.103841598-018-23807-3 PMID:29615698

Nuss, D. L. (2005). Hypovirulence: Mycoviruses at the fungal-plant interface. *Nature Reviews. Microbiology, 3*(8), 632–642. doi:10.1038/nrmicro1206 PMID:16064055

Ozkan-Kotiloglu, S., & Coutts, R. H. A. (2018). Multiplex detection of *Aspergillus fumigatus* Mycoviruses. *Viruses, 10*(5), 247. doi:10.3390/v10050247 PMID:29738445

Park, C. M., Banerjee, N., Koltin, Y., & Bruenn, J. A. (1996). The *Ustilago maydis* virally encoded KP1 killer toxin. *Molecular Microbiology, 20*(5), 957–963. doi:10.1111/j.1365-2958.1996.tb02537.x PMID:8809749

Park, Y., James, D., & Punja, Z. K. (2005). Co-infection by two distinct totivirus-like double-stranded RNA elements in *Chalara elegans* (*Thielaviopsis basicola*). *Virus Research, 109*(1), 71–85. doi:10.1016/j.virusres.2004.10.011 PMID:15826915

Pearson, M. N., Beever, R. E., Boine, B., & Arthur, K. (2009). Mycoviruses of filamentous fungi and their relevance to plant pathology. *Molecular Plant Pathology, 10*(1), 115–128. doi:10.1111/j.1364-3703.2008.00503.x PMID:19161358

Picarelli, M. A. S., Forgia, M., Rivas, E. B., Nerva, L., Chiapello, M., Turina, M., & Colariccio, A. (2019). Extreme diversity of mycoviruses present in isolates of *Rhizoctonia solani* AG2-2 LP from *Zoysia japonica* from Brazil. *Frontiers in Cellular and Infection Microbiology, 9*, 244. doi:10.3389/fcimb.2019.00244 PMID:31355150

Preisig, O., Wingfield, B. D., & Wingfield, M. J. (1998). Coinfection of a fungal pathogen by two distinct double-stranded RNA viruses. *Virology, 252*(2), 399–406. doi:10.1006/viro.1998.9480 PMID:9878619

Reverter, M., Bontemps, N., Lecchini, D., Banaigs, B., & Sasal, P. (2014). Use of plant extracts in fish aquaculture as an alternative to chemotherapy: Current status and future perspectives. *Aquaculture (Amsterdam, Netherlands), 433*, 50–61. doi:10.1016/j.aquaculture.2014.05.048

Sanderlin, R. S., & Ghabrial, S. A. (1978). Physicochemical properties of two distinct types of virus-like particles from *Helminthosporium victoriae*. *Virology, 87*(1), 142–151. doi:10.1016/0042-6822(78)90166-6 PMID:664249

Schoebel, C. N., Zoller, S., & Rigling, D. (2014). Detection and genetic characterisation of a novel mycovirus in *Hymenoscyphus fraxineus*, the causal agent of ash dieback. *Infection, Genetics and Evolution*, *28*, 78–86. doi:10.1016/j.meegid.2014.09.001 PMID:25219345

Shah, U. A., Kotta-Loizou, I., Fitt, B. D., & Coutts, R. H. (2020). Mycovirus-induced hypervirulence of *Leptosphaeria biglobosa* enhances systemic acquired resistance to *Leptosphaeria maculans* in *Brassica napus*. *Molecular Plant-Microbe Interactions*, *33*(1), 98–107. doi:10.1094/MPMI-09-19-0254-R PMID:31652089

Sharma, M., Guleria, S., Singh, K., Chauhan, A., & Kulshrestha, S. (2018). Mycovirus associated hypovirulence, a potential method for biological control of Fusarium species. *Virusdisease*, *29*(2), 134–140. doi:10.100713337-018-0438-4 PMID:29911145

Son, M., Yu, J., & Kim, K. H. (2015). Five Questions about Mycoviruses. *PLoS Pathogens*, *11*(11), e1005172. doi:10.1371/journal.ppat.1005172 PMID:26539725

Suzuki, N., Supyani, S., Maruyama, K., & Hillman, B. I. (2004). Complete genome sequence of Mycoreovirus-1/Cp9B21., a member of a novel genus within the family *Reoviridae,* isolated from the chestnut blight fungus *Cryphonectria parasitica*. *The Journal of General Virology*, *85*(11), 3437–3448. doi:10.1099/vir.0.80293-0 PMID:15483262

Takahashi-Nakaguchi, A., Shishido, E., Yahara, M., Urayama, S. I., Sakai, K., Chibana, H., Kamei, K., Moriyama, H., & Gonoi, T. (2020). Analysis of an intrinsic mycovirus associated with reduced virulence of the human pathogenic fungus *Aspergillus fumigatus*. *Frontiers in Microbiology*, *10*, 3045. doi:10.3389/fmicb.2019.03045 PMID:32010101

Tuomivirta, T. T., & Hantula, J. (2003a). *Gremmeniella abietina* mitochondrial RNA virus S1 is phylogenetically related to the members of the genus *Mitovirus*. *Archives of Virology*, *148*(12), 2429–2436. doi:10.100700705-003-0195-5 PMID:14648296

Tuomivirta, T. T., & Hantula, J. (2003b). Two unrelated double-stranded RNA molecule patterns in *Gremmeniella abietina* type A code for putative viruses of the families *Totiviridae* and *Partitiviridae*. *Archives of Virology*, *148*(12), 2293–2305. doi:10.100700705-003-0194-6 PMID:14648287

Vainio, E. J., Müller, M. M., Korhonen, K., Piri, T., & Hantula, J. (2014). Viruses accumulate in aging infection centers of a fungal forest pathogen. *The ISME Journal*, *9*(2), 497–507. doi:10.1038/ismej.2014.145 PMID:25126757

Varga, J., Rigo, K., Molnar, J., Toth, B., Szencz, S., & Teren, J. (2003). Mycotoxin production and evolutionary relationships among species of *Aspergillus* section Clavati. *Antonie van Leeuwenhoek*, *83*(2), 191–200. doi:10.1023/A:1023355707646 PMID:12785313

Wang, L., Jiang, J., Wang, Y., Hong, N., Zhang, F., Xu, W., & Wang, G. (2014). Hypovirulence of the phytopathogenic fungus *Botryosphaeria dothidea*: Association with a coinfecting chrysovirus and a partitivirus. *Journal of Virology*, *88*(13), 7517–7527. doi:10.1128/JVI.00538-14 PMID:24760881

Wang, M., & Jin, H. (2017). Spray-induced gene silencing: A powerful innovative strategy for crop protection. *Trends in Microbiology*, *25*(1), 4–6. doi:10.1016/j.tim.2016.11.011 PMID:27923542

Wang, S., Kondo, H., Liu, L., Guo, L., & Qiu, D. (2013). A novel virus in the family Hypoviridae from the plant pathogenic fungus *Fusarium graminearum. Virus Research, 174*(1-2), 69–77. doi:10.1016/j.virusres.2013.03.002 PMID:23499998

Wang, Y., Zhao, H., Xue, C., Xu, C., Geng, Y., & Zang, R., Guo, Y., Wu, H., & Zhang, M. (2020). Complete genome sequence of a novel mycovirus isolated from the phytopathogenic fungus *Corynespora cassiicola* in China. *Archives of Virology*, 1–4. PMID:32757057

Wei, C. Z., Osaki, H., Iwanami, T., Matsumoto, N., & Ohtsu, Y. (2003). Molecular characterization of dsRNA segments 2 and 5 and electron microscopy of a novel reovirus from a hypovirulent isolate of the plant pathogen *Rosellina necatrix. The Journal of General Virology, 84*(9), 2431–2437. doi:10.1099/vir.0.19098-0 PMID:12917464

Wu, M., Zhang, L., Li, G., Jiang, D., & Ghabrial, S. A. (2010). Genome characterization of a debilitation-associated mitovirus infecting the phytopathogenic fungus *Botrytis cinerea. Virology, 406*(1), 117–126. doi:10.1016/j.virol.2010.07.010 PMID:20674953

Xie, J., & Jiang, D. (2014). New insights into mycoviruses and exploration for the biological Control of Crop Fungal Diseases. *Annual Review of Phytopathology, 52*, 3.1–3.24.

Yang, M., Zhou, X., Zhai, L., Xiao, F., Hong, N., & Wang, G. (2020). Molecular characterization of a novel mycovirus infecting the phytopathogenic fungus *Botryosphaeria dothidea. Archives of Virology, 165*(7), 1667–1670. doi:10.100700705-020-04629-z PMID:32328855

Yu, X., Li, B., Fu, Y., Xie, J., Cheng, J., Ghabrial, S. A., Li, G., Yi, X., & Jiang, D. (2013). Extracellular transmission of a DNA mycovirus and its use as a natural fungicide. *Proceedings of the National Academy of Sciences of the United States of America, 110*(4), 1452–1457. doi:10.1073/pnas.1213755110 PMID:23297222

KEY TERMS AND DEFINITIONS

Anti-Microbial Resistance: AMR is the ability of pathogenic organisms to develop resistance to the effects of medication.

Anti-Mycotic: Any agent that is fungicidal or fungistatic and capable enough in eliminating their growth and further proliferation.

Asymptomatic: Exhibiting no symptoms of a particular illness or abnormality.

Chemotherapy: The treatment of any disease through chemical substances that binds to specific cells or specifically kills microbes.

Hypovirulence: A kind of biological phenomenon in which the virulence capacity or pathogenicity of a pathogen is eliminated or reduced by being infected with a virus.

ICTV: The International Committee on Taxonomy of Viruses authorizes and organizes the taxonomic classification and nomenclatures for viruses. Headquarter is located in London, UK.

Prophylactic: A medication or a treatment designed and used preventing the spread or occurrence of disease or infection.

Therapeutic: The branch of medicine concerned with the treatment of disease.

Section 3

Microbes and Site Remediation

Chapter 14
Application of Dehalogenase Enzymes in Bioremediation of Halogenated Pollutants:
A Short Review

Raghunath Satpathy
https://orcid.org/0000-0001-5296-8492
Gangadhar Meher University, Samabalpur, India

ABSTRACT

The halogenated hydrocarbons have been widely used by human beings. They are xenobiotic and toxic. The microbes having a specific group of hydrolase enzymes, known as dehalogenases, that actually break the carbon-halogen bonds of the halogenated substances and subsequently convert them into their non-toxic forms. In this chapter, the categories of dehalogenase enzymes possessed by microorganisms are narrated. The overall source, mechanism of catalysis, and structural aspects of the haloalkane dehalogenase enzymes have been discussed with special focus to the bioremediation of 1, 2 dichloroethane.

INTRODUCTION

The halogenated compounds (both aliphatic and aromatic) are xenobiotic in nature and are being used widely from many years. Therefore, the progressive accumulation of these substances in the environment created a global threat to human health in recent times (Atashgahi et al., 2018; Zhu et al., 2017). The substances like dichlorodiphenyltrichloroethane (DDT), polychlorinated biphenyl (PCB), dioxins harm human health as they have been reported to be persistent (Besis & Samara, 2012; Chaudhry & Chapalamadugu, 1991; Fetzner, 2002). A basic step in the bio-degradation of organohalides is the cleavage of the carbon-halogen bond known as *dehalogenation* that is shown by a diverse group of microorganisms as described by many researchers. There are some key hydrolase group of enzymes from different microbial sources which are responsible for dehalogenation of these substances by catalysing the cleavage of carbon–halogen bonds of these molecules (Janssen, 2004; Satpathy, 2019; Satpathy et

DOI: 10.4018/978-1-7998-7062-3.ch014

al., 2015). The bioremediation process of these substances can be carried out successfully by using microbe based dehalogenation. Hence, it is essential to obtain a deeper understanding of the process at the molecular level. In addition to this, a thorough study is also required about the toxicity of organohalide compounds and how they are directly related to the bioaccumulation in the food chain, food web and subsequently causes environmental contamination (Millow et al., 2012; Byun et al., 2013). The microbe based dehalogenation reactions are considered as very important because of their potential application in biotechnological line in the bioremediation of such environmental pollutants (Satpathy et al., 2016; Byun et al., 2013). Several dehalogenase enzymes are available in microorganisms that exhibit diversity in the catalyzation of the halogenated pollutants and depend on the exposure condition of the microbes. The dehalogenase enzymes are classified in many types such as haloalkane dehalogenases, halohydrin dehalogenases, haloacetate dehalogenases, etc. based on the types of the substrates they catalyse. Further, eight types of dehalogenase enzymes and the reaction mechanism of the substrate has been described by Janssen *et al.* in 1994. In current literature, many attempts are made by the researchers to discover suitable novel microorganisms that are frequently used to replace toxic halogenated substances thereby playing a major role in the recycling of these compounds (Kurihara & Esaki, 2008; Janssen et al., 2001; Janssen, 2007). Several methods have been developed by researchers to replace these toxic halogenated compounds to their non-toxic form by using the microorganisms. Therefore, microbial resources play a major role in the recycling of halogenated compounds. This also opens the door to thorough analysis and understanding the specific diverse microbial community which can be used for the biodegradation and biotransformation processes of these compounds. In addition to this, the biodegradation study of the halogenated substances in the environment will lead to understanding the carbon–halogen bond cleaving process as well as to quantify the metabolic potential of the microbes (Satpathy et al., 2017; Erable et al., 2006).

In this chapter, the mechanism, factors of haloalkane dehalogenase enzymes and their discovery from the microbial sources as well as their impact on the environmental cleaning process has been elaborated.

CATEGORIES OF DEHALOGENATION ENZYMES AND MECHANISMS

The microorganisms possess several types of dehalogenase enzyme systems which are involved for catalyzation of different types of halogenated substances to which they are exposed. Hamid et al. (2013) classified the dehalogenase enzymes as haloalkane dehalogenases, halohydrin dehalogenases, haloacetate dehalogenases, dichloromethane dehalogenases and D- and L-haloalkanoic acid dehalogenases (Hamid et al., 2013; Janssen et al., 1994). Eight categories of dehalogenase enzymes and their reaction mechanisms for the substrates as described by Janssen *et al.* in 1994 has been shown in Table 1.

IDENTIFICATION OF DEHALOGENASE ENZYME-PRODUCING BACTERIA

In nature, the several bacterial species constitute a major group that cause dehalogenation of halogenated pollutants by the help of dehalogenase enzymes. However, the identification and study of the dehalogenation pattern of the different haloalkane pollutants are interesting and challenging too. The common method that is followed for the identification and characterization of potential bacterial species for the dehalogenation reaction has been shown in Figure 1.

Table 1. Types of halogenation reactions and mechanisms

Types	Name of Reaction	General Mechanism	Examples of the Enzymes
A	Hydrolytic dehalogenation	$R\text{-}X + H_2O \rightarrow R\text{-}OH + H^+ + X^-$	Haloalkane dehalogenase, 2-haloacid dehalogenase
B	Glutathione substitution	$R\text{-}X + GSH \rightarrow [GS\text{-}R\text{-}X] + HOH \rightarrow GSH + R\text{-}COH + X$	Dichloromethane dehalogenase, Tetrachloro-p- hydroquinone dehalogenase
C	Hydration	$C2X2HCOOH + H_2O \rightarrow CH2(CHO)COOH$	4-chlorobenzozte dehalogenase
D	Intramolecular substitution	$R\text{-}X^1 + XH \rightarrow R\text{-}X + H^+ + X^{1-}$	3-chlorobenzoate hydroxylase
E	Dehydrohalogenation	$R\text{-}CH_2\text{-}CHX\text{-}R \rightarrow R\text{-}CH{=}CH\text{-}R + H^+ + X$	g-HCH dehydrochlorinase
F	Reduction	$R\text{-}X + 2\,[H] \rightarrow R\text{-}H + H^+ + X^-$	Chlorobenzene Dioxygenase
G	Oxygenation	$R\text{-}CH_2\text{-}X + O_2 + 2[H^+] \rightarrow R\text{-}CHO + H^+ + X^-$	Pentachlorophenol 4 monoxygenase

Figure 1. Common method for identification of bacterial strains containing dehalogenases enzymes

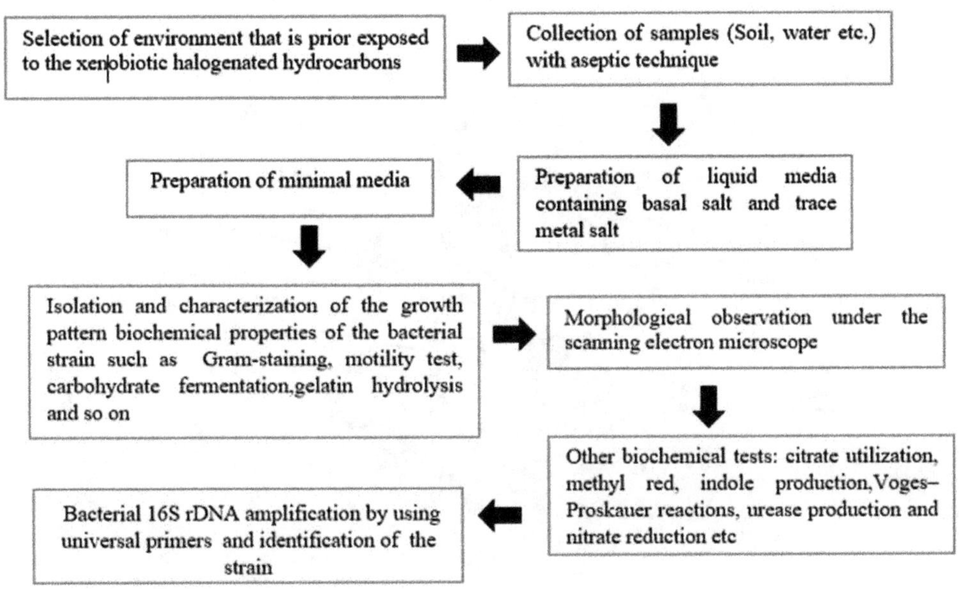

DEHALOGENATION MECHANISM OF HALOALKANE DEHALOGENASE ENZYMES

The haloalkane dehalogenase enzymes from the microbial sources play a crucial role to remediate the toxic haloalkane compounds, therefore play an important role in biotechnological applications. Haloalkane dehalogenase having EC No. 3.8.1.5 is an important enzyme in general that catalyzes the following chemical reaction;

Haloalkane + H_2O → Primary alcohol + Halide

This enzyme has the systematic name as 1-haloalkane halidohydrolase and belongs to the hydrolases family that specifically act on the carbon-halogen bonds in the haloalkane compounds (Satpathy et al., 2015). Haloalkane dehalogenases are found in many bacteria and also participate in the metabolic pathways of 1, 2-dichloroethane, chloro-butane, hexachlorocyclohexane, 1, 3-dichloropropene and so on. Based on the sequence of phylogenetic analysis, the haloalkane dehalogenases (HLDs) can be divided into three subfamilies such as HLD-I, HLD-II and HLD-III based on difference in the catalytic residues, substrate selectivity and the architecture of the domain (Chovancová et al., 2007). Based on the enzyme activity and substrate specificity, Koudelakova et al. (2011) analysed the substrate specificity of 30 haloalkane compounds and determined the substrate specificity profiles of HLDs and accordingly classified HLDs into four specificity subfamilies that are different from the phylogenetic superfamilies. However, the specificity subfamilies do not agree with the phylogenetic subfamilies as described above, and it was suggested that the architecture of the active site is important for classification rather than sequence homology.

Figure 2. Structure of haloalkane dehalogenase PDB ID 2BFN, with substrate 1, 2-dichloropropane with active site residues (blue colour) and presence of chloride ion (hydrolase domains are shown in yellow) (Monincová et al., 2007)

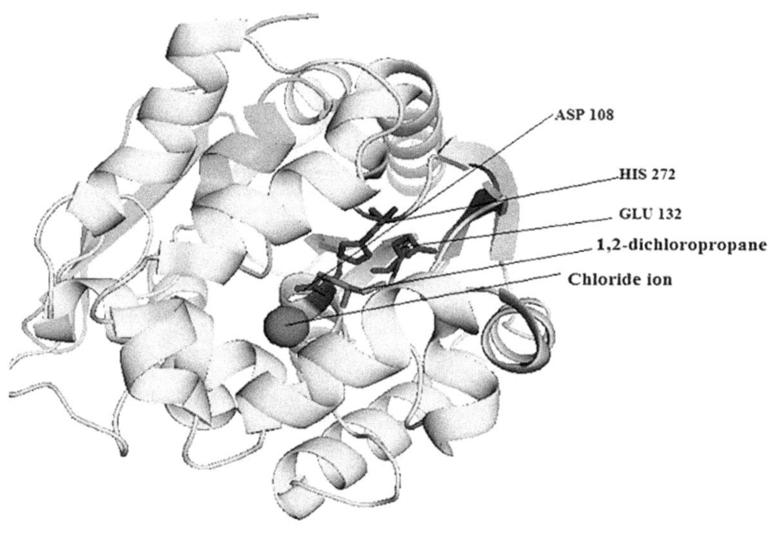

Figure 2 shows the crystal structure of haloalkane dehalogenase enzyme complex with 1, 2, 3-trichloropropane, available as 2BFN code in PDB (Protein Data bank). In the crystal structure, α/β-hydrolase domain is the important feature of the enzyme involved in the dehalogenation of haloalkane compounds. Also, this domain is the most common structure found in the case of hydrolase proteins irrespective of their divergence in their amino acid sequence (Otyepka & Damborský, 2012; Marek et al., 2000). The three conserved amino acid residues (catalytic triad) have been observed that are the key highly conserved residues for α/β hydrolases activity referred to as catalytic triad. In general, the catalytic triad consists of an aspartic acid (nucleophilic), a histidine residue (basic) and an aspartic or glutamic acid (acidic). The first two residues such as aspartic acid and the histidine show rigid nature in the enzyme structure responsible for catalytic activity. The conserved aspartic acid residues act as nucleophile and displace a halide ion from the halogenated substance. All the catalytic triad residues act simultaneously and play a prominent role in dehalogenation reaction such as halide ion binding, the formation of hydrogen bond as well as maintain proper orientation of the nucleophile (Hesseler et al., 2011; Pavlova et al., 2009; Li & Shao, 2014). The haloalkane compounds are characterized, as the best substrates for the haloalkane dehalogenase enzymes. Therefore, the catalytic activity of the purified enzymes has been analysed with the substrates to identify the best substrate upon which the enzyme can best act. Also, quantification of the optimum enzyme-substrate activity can be performed at specific environmental conditions such as at different temperature and pH. The common methods that are followed by the researchers for quantification of these enzymes have been shown in Figure 3.

Figure 3. In-vitro methods for Biochemical characterization of the haloalkane dehalogenase enzymes

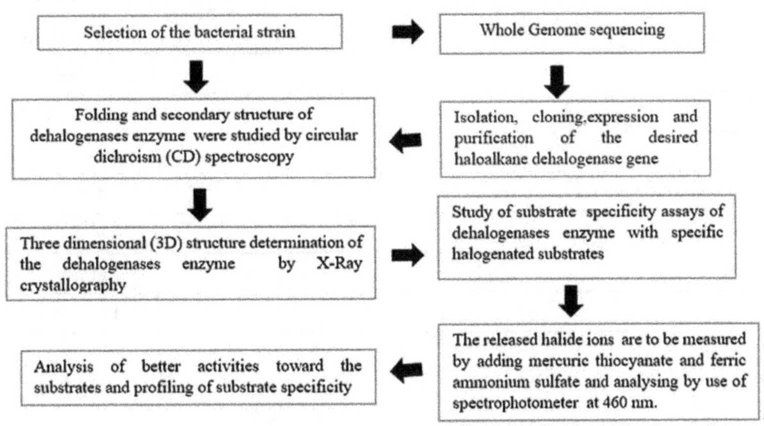

CASE STUDY: MICROBIAL DEHALOGENATION OF 1, 2 DICHLOROETHANE BY HALOALKANE DEHALOGENASE

Chemically, compound 1,2-dichloroethane is a chlorinated ethane and most commonly known as ethylene dichloride (EDC). 1, 2-Dichloroethane is synthetic hence is a xenobiotic chemical and non-biodegradable in its activity. Considering the physical property of the compound it is a colourless, liquid in nature, and having chloroform like odour. 1, 2-dichloroethane is mostly used for the production of vinyl chloride, which is the precursor molecule for the synthesis of polyvinyl chloride (PVC) used for furniture, pipe and automobile parts and so on (Manfred et al., 2006). (Table 2).

Other properties of the compound have been confirmed as flammable, genotoxic as well as carcinogenic in nature (Gwinn et al., 2011; Doucette et al., 2010). Since large quantities of chlorinated aliphatic hydrocarbons such as 1, 2 dichloroethane are produced for industrial and commercial uses, so they also constitute common contaminants of soil as well as groundwater. Also, it has been observed that the non-biotic degradation mode such as hydrolysis and photolysis are not much effective in terms of time as well as cost in comparison to the biodegradation by microbes in case of the perennial pollutants like 1, 2-Dichloroethane (Ware, 1988; Zok et al., 1998). As per a report, in case of 1, 2-Dichloroethane the abiotic degradation occurs at 15°C, pH 7, and in the presence of 1 mM total sulphide content (as that of groundwater condition) and the half-life was computed as 23 years (WHO, 1995). So biotic mode of degradation would be a desirable model that can be followed for the remediation purpose of this compound (Barbash & Reinhard, 1989). The metabolic pattern of the microbial dehalogenation of 1,2 dichloroethane is shown in Figure 4.

Table 2. Structural and physical properties of 1,2 dichloroethane

S. No.	Property	Remark
1	Structure	
2	IUPAC name	1,2-dichloroethane
3	SMILE notation	C(CCl)Cl
4	Chemical formula	C2H4 Cl2
5	Molecular weight	98.96 g/mol
6	Odour	Chloroform like
7	Density	1,253 g/cm3
8	Boiling point	84°C
9	Melting point	-35°C
10	Water solubility	0.87g/100ml
11	Dipole moment	1.80D
12	Hazardous	Toxic and carcinogenic

Figure 4. Degradation pathway of 1, 2 dichloroethane biodegradation
(Obtained from the University of Minnesota Biocatalysis/Biodegradation Database, http://eawag-bbd.ethz.ch/dce/dce_image_map.html)

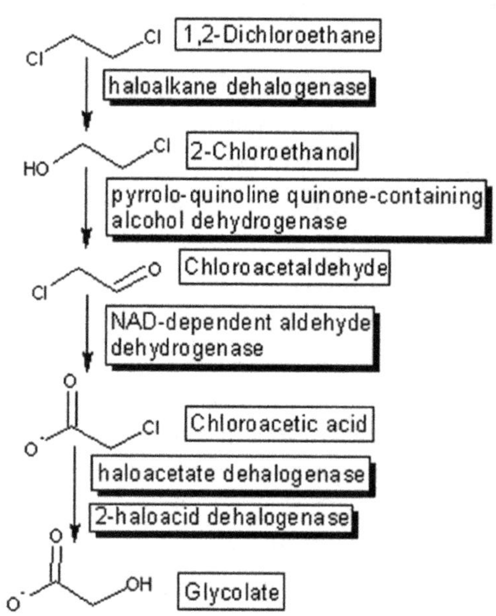

Olaniran et al. (2007) identified about three bacterial genera *Paenibacillus*, *Bacillus*, and *Microbacterium* from the pulp mill wastewater effluent which can degrade 1,2-dichloroethane. After isolation, the degradation rate constant of the substance was calculated and the removal of the pollutant molecules was quantified by using gas chromatographic study. Finally, the bacterial genus was identified by the 16S rRNA gene sequencing methods.

The dicchloroethane degrading bacteria have been utilized to purify polluted groundwater. For the purpose, the low concentrations of 1, 2-dichloroethane was inoculated in two different bioreactors (having glass bead type and granular activated carbon as carrier type). By applying two pure cultures such as with controlled oxygen supply in the two different types of bioreactors shows the 80% disappearance of dichloroethane in 3 weeks was effective by the consumption of oxygen and decrease in pH as well as the formation of chloride ions in case of glass bead type of bioreactors (Stucki et al., 1992). Van der Zaan et al. (2009) in their research studied the metabolism, pathways of biodegradation and the effect of external factors that are responsible for degradation of the pollutant 1,2-dichloroethane. Biodegradation study showed, in case of reductive dechlorination resulted in chloroethane, or ethene, respectively, as part of the major dechlorination products. Different reductively dehalogenating species of microorganisms such as *Dehalococcoides* spp., *Dehalobacter* spp., *Desulfitobacterium* spp. and *Sulfurospirillum* spp. were identified by 16S ribosomal RNA gene-targeted PCR and sequence analysis. Also, the anaerobic oxidation of 1,2-dichloroethane was obtained under denitrifying or iron-reducing conditions. The degradation study of 1,2-dichloroethane contaminated with the groundwater was performed. The parameters of the study like yields, maximum specific growth rates, and half-saturation coefficients were identified by applying Dehalococcoides culture. Yu et al. (2013) isolated a potential *Pseudomonas* sp. strain DCA1, from a 1,2-dichloroethane degrading biofilm, that utilizes 1, 2-dichloroethane as the sole carbon and energy source without using the as vitamins, for optimal growth. Oxygen and NAD(P)H are required for this initial step. The bacterial strain produces a monooxygenase is responsible for the degradation in strain DCA1 (Hage et al., 1999). Janssen et al (1994) described the aerobic biodegradation of the 1,2-dichloroethane by using the *Pseudomonas* sp. strain DE2 and *Xanthobacter autotrophicus* GJ10.The detail pathways of the dehalogenation process was thoroughly studied. 1,2-dichloroethane converts to the 2-chloroethanol and further converted to the chloroacetaldehyde by alcohol dehydrogenase. Chloroacetaldehyde is further dehydrogenated to form chloroacetic acid by the chloroacetate dehalogenases enzyme (Janssen et al., 1985). The *X. autotrophicus* GJ10 strain has been used in several studies for the biological treatment aiming in removal of 1, 2-dichloroethane contaminated in synthetic waste water (dos Santos & Livingston, 1993; dos Santos & Livingston, 1994; dos Santos & Livingston, 1995). A comparative study about the affinity of 1, 2 dichloroethane was carried out by van den Wijngaard et al. (1992, 1993). The isolated bacterial strain *Ancylobacter aquaticus* AD25 having higher degradation affinity towards the substrate was obtained in comparison to the strain *X. autotrophicus* GJ10. Also, another, the strain *Pseudomonas* sp. strain DCA1, has been confirmed of having high affinity for dichloroethane and possess a novel pathway for dichloroethane degradation (Stucki et al., 1983). Degradation of pathway of 1,2-DCE proceeds via 2-chloroethanol, chloroacetaldehyde and chloroacetate to glycolate was established by Janssen et al. (1994). The chromosomal genes responsible for the conversion process were identified as alcohol dehydrogenase and the haloacid dehalogenases, in addition to the plasmid encoded genes such as haloalkane dehalogenase and aldehyde dehydrogenase. The alternative pathway of the 1,2-dichloroethane was studied by using CO_2 under aerobic conditions by Dinglasan-Panlilio et al. (2006). Some of the organisms such as Betaproteobacteria belonging to the genus *Thauera* have also been isolated that can reduce 1,2-dichloroethane to ethene via dihaloclimination under anaerobic, fermentative conditions. A methanogenic sludge was prepared by De Wildeman et al. (2002) to study the dechlorination process of 1,2-dichloroethane to ethane in reactors. In the study, ethanol was supplied as the sole source of carbon and energy. 16S rRNA sequencing analysis of the isolated bacterial strain showed the bacterium was closely related to *Acetobacterium wieringae*. Some perspectives of anaerobic in situ bioremediation of groundwater polluted with chloroethanes are presented. Popova-Kroumova et al. (2015) described a method for degradation of the 1,2-dichloroethane by using a constant electric field utilizing the strain *Xanthobacter autotrophicus* GJ10 Subsequent mathematical modelling showed that, effect of the electric field on microbial specific growth rate and the rate of microbial decay was estimated quantitatively in terms of the corresponding rate constants by considering the microbial growth and field influence on biomass growth and the effect of substrate and product inhibition (Popova-Kroumova et al., 2015). The major technical and scientific issues are illustrated by comparing two examples, that of 1,2-dichloroethane where the successful full-scale application of pump-and-treat bio-treatment processes has been achieved, and 1,2,3-trichloropropane, for which protein and genetic engineering yielded effective bacterial cultures that still await application (Janssen & Stucki, 2020). The anaerobic mode of the dehalogenation by halorespiring bacteria *Vulcanibacillus* spp., that degrades the 1, 2 dichloroethane have been reported. Soils contaminated with 1,2-dichloroethane were used as the starting material and degradation was analysed by using the gas chromatography and detoxification potential of the bacteria was analysed (Ngivprom et al., 2020). The bacterial dehalogenases enzymes of the bacteria have been cloned in the plants and the phytodehalogenation potential has been evaluated. Mena-Benitez et al. (2008) described the halogenated aliphatic compounds that include 1,2-dichloroethane which can be degraded by the plant *Nicotiana tabacum*, that lack the enzymatic activity. Making the genetically engineered plant by inserting the two enzymes haloalkane dehalogenase (DhlA) and haloacid dehalogenase (DhlB) from the bacterium *Xanthobacter autotrophicus* GJ10, show the potential of the transgenic tobacco to degrade the 1, 2 dichloroethane by converting it to 2-chloroethanol (Mena-Benitez et al., 2008).

CONCLUSION

Biodegradation by a microorganism of halogenated organic compounds is the major process involved for the removal of these toxic compounds from the environment. Research on microbial biodegradation pathways revealed that these xenobiotic compounds are used as the sole carbon and energy sources by the microbes. The microbes possess dehalogenases enzymes that are responsible for catalyzation of the carbon-halogen bonds in the compound. This has raised interest in the identification of novel microorganisms that produce the dehalogenases enzymes with improved catabolic activities. Especially, the 1,2-dichloroethane is widely studied haloalkane compound that is toxic, non-biodegradable, the major pollutant of soil and ground water as well as having a possible carcinogenic effect. Several microorganisms have been identified that are potentially involved in the degradation process of the pollutant in many environments such as aerobic, anaerobic mode. Bioreactors have been developed to purify the 1,2-dichloroethane contaminated water by using specific microbial strains. Interestingly, the bacterial haloalkane dehalogenase gene is utilized to generate genetically recombinant plant that successfully degrades 1,2-dichloroethane. Recently, many categories of the hydrolytic, reductive, and oxygenolytic dehalogenases have been identified and characterized; however, detailed structural information is not available in protein data bank (PDB). Also, the exact mechanism of the dehalogenation process is to be studied yet in all types of microbes adapting to the environment. So, there is a great scope to conduct further research on the functional aspects of these enzymes.

REFERENCES

Atashgahi, S., Shetty, S. A., Smidt, H., & de Vos, W. M. (2018). Flux, impact, and fate of halogenated xenobiotic compounds in the gut. *Frontiers in Physiology*, *9*, 888. doi:10.3389/fphys.2018.00888 PMID:30042695

Barbash, J. E., & Reinhard, M. (1989). Abiotic dehalogenation of 1, 2-dichloroethane and 1, 2-dibromoethane in aqueous solution containing hydrogen sulfide. *Environmental Science & Technology*, *23*(11), 1349–1358. doi:10.1021/es00069a004

Besis, A., & Samara, C. (2012). Polybrominated diphenyl ethers (PBDEs) in the indoor and outdoor environments–a review on occurrence and human exposure. *Environmental Pollution*, *169*, 217–229. doi:10.1016/j.envpol.2012.04.009 PMID:22578798

Byun, G. H., Moon, H. B., Choi, J. H., Hwang, J., & Kang, C. K. (2013). Biomagnification of persistent chlorinated and brominated contaminants in food web components of the Yellow Sea. *Marine Pollution Bulletin*, *73*(1), 210–219. doi:10.1016/j.marpolbul.2013.05.017 PMID:23768977

Chaudhry, G. R., & Chapalamadugu, S. (1991). Biodegradation of halogenated organic compounds. *Microbiology and Molecular Biology Reviews*, *55*(1), 59–79. PMID:2030673

Chovancová, E., Kosinski, J., Bujnicki, J. M., & Damborský, J. (2007). Phylogenetic analysis of haloalkane dehalogenases. *Proteins*, *67*(2), 305–316. doi:10.1002/prot.21313 PMID:17295320

De Wildeman, S., Nollet, H., Van Langenhove, H., Diekert, G., & Verstraete, W. (2002). Reductive biodegradation of 1, 2-dichloroethane by methanogenic granular sludge: Perspectives for in situ remediation. *Water Science and Technology*, *45*(10), 43–48. doi:10.2166/wst.2002.0284 PMID:12188575

Dinglasan-Panlilio, M. J., Dworatzek, S., Mabury, S., & Edwards, E. (2006). Microbial oxidation of 1, 2-dichloroethane under anoxic conditions with nitrate as electron acceptor in mixed and pure cultures. *FEMS Microbiology Ecology*, *56*(3), 355–364. doi:10.1111/j.1574-6941.2006.00077.x PMID:16689868

dos Santos, L. F., & Livingston, A. G. (1993). A novel bioreactor system for the destruction of volatile organic compounds. *Applied Microbiology and Biotechnology*, *40*(1), 151–157. doi:10.1007/BF00170444

dos Santos, L. F., & Livingston, A. G. (1994). Extraction and biodegradation of a toxic volatile organic compound (1, 2-dichloroethane) from waste-water in a membrane bioreactor. *Applied Microbiology and Biotechnology*, *42*(2-3), 421–431. doi:10.1007002530050273 PMID:7765782

dos Santos, L. F., & Livingston, A. G. (1995). Novel membrane bioreactor for detoxification of VOC wastewaters: Biodegradation of 1, 2-dichloroethane. *Water Research*, *29*(1), 179–194. doi:10.1016/0043-1354(94)00137-V

Doucette, W. J., Hall, A. J., & Gorder, K. A. (2010). Emissions of 1, 2-dichloroethane from holiday decorations as a source of indoor air contamination. *Ground Water Monitoring and Remediation*, *30*(1), 67–73. doi:10.1111/j.1745-6592.2009.01267.x

Erable, B., Goubet, I., Lamare, S., Legoy, M. D., & Maugard, T. (2006). Bioremediation of halogenated compounds: Comparison of dehalogenating bacteria and improvement of catalyst stability. *Chemosphere*, *65*(7), 1146–1152. doi:10.1016/j.chemosphere.2006.04.007 PMID:16723151

Fetzner, S. (2002). Biodegradation of xenobiotics. *Biotechnology*, *10*, 215–246.

Gwinn, M. R., Johns, D. O., Bateson, T. F., & Guyton, K. Z. (2011). A review of the genotoxicity of 1, 2-dichloroethane (EDC). *Mutation Research/Reviews in Mutation Research*, *727*(1-2), 42–53. doi:10.1016/j.mrrev.2011.01.001 PMID:21255676

Hage, J. C., & Hartmans, S. (1999). Monooxygenase-mediated 1, 2-dichloroethane degradation by Pseudomonas sp. strain DCA1. *Applied and Environmental Microbiology*, *65*(6), 2466–2470. doi:10.1128/AEM.65.6.2466-2470.1999 PMID:10347028

Hamid, A. A. A., Wong, E. L., Joyce-Tan, K. H., Shamsir, M. S., Hamid, T. H. T. A., & Huyop, F. (2013). Molecular modelling and functional studies of the non-stereospecific α-haloalkanoic acid Dehalogenase (DehE) from Rhizobium sp. RC1 and its association with 3-chloropropionic acid (β-chlorinated aliphatic acid). *Biotechnology, Biotechnological Equipment*, *27*(2), 3725–3736. doi:10.5504/BBEQ.2012.0142

Hesseler, M., Bogdanović, X., Hidalgo, A., Berenguer, J., Palm, G. J., Hinrichs, W., & Bornscheuer, U. T. (2011). Cloning, functional expression, biochemical characterization, and structural analysis of a haloalkane dehalogenase from Plesiocystis pacifica SIR-1. *Applied Microbiology and Biotechnology*, *91*(4), 1049–1060. doi:10.100700253-011-3328-x PMID:21603934

Janssen, D. B. (2004). Evolving haloalkane dehalogenases. *Current Opinion in Chemical Biology*, *8*(2), 150–159. doi:10.1016/j.cbpa.2004.02.012 PMID:15062775

Janssen, D. B. (2007). Biocatalysis by dehalogenating enzymes. *Advances in Applied Microbiology*, *61*, 233–252. doi:10.1016/S0065-2164(06)61006-X PMID:17448791

Janssen, D. B., Oppentocht, J. E., & Poelarends, G. J. (2001). Microbial dehalogenation. *Current Opinion in Biotechnology*, *12*(3), 254–258. doi:10.1016/S0958-1669(00)00208-1 PMID:11404103

Janssen, D. B., Pries, F., & van der Ploeg, J. R. (1994). Genetics and biochemistry of dehalogenating enzymes. *Annual Review of Microbiology*, *48*(1), 163–191. doi:10.1146/annurev.mi.48.100194.001115 PMID:7826004

Janssen, D. B., Scheper, A., Dijkhuizen, L., & Witholt, B. (1985). Degradation of halogenated aliphatic compounds by Xanthobacter autotrophicus GJ10. *Applied and Environmental Microbiology*, *49*(3), 673–677. doi:10.1128/AEM.49.3.673-677.1985 PMID:3994371

Janssen, D. B., Scheper, A., & Witholt, B. (1984). Biodegradation of 2-chloroethanol and 1, 2-dichloroethane by pure bacterial cultures. *Progress in Industrial Microbiology*, *20*, 169–178.

Janssen, D. B., & Stucki, G. (2020). Perspectives of genetically engineered microbes for groundwater bioremediation. *Environmental Science. Processes & Impacts*, *22*(3), 487–499. doi:10.1039/C9EM00601J PMID:32095798

Janssen, D. B., van der Ploeg, J. R., & Pries, F. (1994). Genetics and biochemistry of 1, 2-dichloroethane degradation. *Biodegradation*, *5*(3-4), 249–257. doi:10.1007/BF00696463 PMID:7765836

Koudelakova, T., Chovancova, E., Brezovsky, J., Monincova, M., Fortova, A., Jarkovsky, J., & Damborsky, J. (2011). Substrate specificity of haloalkane dehalogenases. *The Biochemical Journal*, *435*(2), 345–354. doi:10.1042/BJ20101405 PMID:21294712

Kurihara, T., & Esaki, N. (2008). Bacterial hydrolytic dehalogenases and related enzymes: Occurrences, reaction mechanisms, and applications. *Chemical Record (New York, N.Y.)*, *8*(2), 67–74. doi:10.1002/tcr.20141 PMID:18366103

Li, A., & Shao, Z. (2014). Biochemical characterization of a haloalkane dehalogenase DadB from Alcanivorax dieselolei B-5. *PLoS One*, *9*(2), e89144. doi:10.1371/journal.pone.0089144 PMID:24586552

Manfred, R., Wilhelm, L., Gerhard, P., Adolf, T. L., & Ernest, L. (2006). Chlorinated hydrocarbons Ullmann's Encyclopedia of Industrial Chemistry.

Marek, J., Vévodová, J., Smatanová, I. K., Nagata, Y., Svensson, L. A., Newman, J., Takagi, M., & Damborský, J. (2000). Crystal structure of the haloalkane dehalogenase from Sphingomonas paucimobilis UT26. *Biochemistry*, *39*(46), 14082–14086. doi:10.1021/bi001539c PMID:11087355

Mena-Benitez, G. L., Gandia-Herrero, F., Graham, S., Larson, T. R., McQueen-Mason, S. J., French, C. E., Rylott, E. L., & Bruce, N. C. (2008). Engineering a catabolic pathway in plants for the degradation of 1, 2-dichloroethane. *Plant Physiology*, *147*(3), 1192–1198. doi:10.1104/pp.108.119008 PMID:18467461

Millow, C. J., Mackintosh, S. A., Lewison, R. L., Dodder, N. G., & Hoh, E. (2015). Identifying bioaccumulative halogenated organic compounds using a nontargeted analytical approach: Seabirds as sentinels. *PLoS One*, *10*(5), e0127205. doi:10.1371/journal.pone.0127205 PMID:26020245

Monincová, M., Prokop, Z., Vévodová, J., Nagata, Y., & Damborský, J. (2007). Weak activity of haloalkane dehalogenase LinB with 1, 2, 3-trichloropropane revealed by X-ray crystallography and microcalorimetry. *Applied and Environmental Microbiology*, *73*(6), 2005–2008. doi:10.1128/AEM.02416-06 PMID:17259360

Ngivprom, U., Milintawisamai, N., & Reungsang, A. (2020). *Reductive dechlorination of 1, 2-dichloroethane to ethylene by anaerobic enrichment culture containing Vulcanibacillus spp. Walailak Journal of Science and Technology.*

Olaniran, A. O., Naidoo, S., Masango, M. G., & Pillay, B. (2007). Aerobic biodegradation of 1, 2-dichloroethane and 1, 3-dichloropropene by bacteria isolated from a pulp mill wastewater effluent in South Africa. *Biotechnology and Bioprocess Engineering; BBE*, *12*(3), 276–281. doi:10.1007/BF02931104

Otyepka, M., & Damborský, J. (2002). Functionally relevant motions of haloalkane dehalogenases occur in the specificity-modulating cap domains. *Protein Science*, *11*(5), 1206–1217. doi:10.1110/ps.ps3830102 PMID:11967377

Pavlova, M., Klvana, M., Prokop, Z., Chaloupkova, R., Banas, P., Otyepka, M., & Damborsky, J. (2009). Redesigning dehalogenase access tunnels as a strategy for degrading an anthropogenic substrate. *Nature Chemical Biology*, *5*(10), 727–733. doi:10.1038/nchembio.205 PMID:19701186

Popova-Kroumova, P., Vasileva, E., & Beschkov, V. (2015). Modelling of 1, 2-dichloroethane Biodegradation, Stimulated by Constant Electric Field. *Biomath Communications, 2*(1).

Satpathy, R. (2019). Computational Tools and Techniques to Predict Aquatic Toxicity of Some Halogenated Pollutants. In *Handbook of Research on the Adverse Effects of Pesticide Pollution in Aquatic Ecosystems* (pp. 318–337). IGI Global. doi:10.4018/978-1-5225-6111-8.ch018

Satpathy, R., Konkimalla, V. B., & Ratha, J. (2015). Application of bioinformatics tools and databases in microbial dehalogenation research: A review. *Applied Biochemistry and Microbiology*, *51*(1), 11–20. doi:10.1134/S0003683815010147 PMID:25842899

Satpathy, R., Konkimalla, V. B., & Ratha, J. (2016). In silico phylogenetic analysis and molecular modelling study of 2-haloalkanoic acid dehalogenase enzymes from bacterial and fungal origin. *Advances in Bioinformatics*, *2016*, 2016. doi:10.1155/2016/8701201 PMID:26880911

Satpathy, R., Konkimalla, V. B., & Ratha, J. (2017). Microbial dehalogenation: 3-chloropropanoic acid (3-CPA) degradation as a case study. *Microbiology*, *86*(1), 32–41. doi:10.1134/S0026261716060175

Satpathy, R., Konkimalla, V. S. B., & Ratha, J. (2015). Dehalobase: A database of dehalogenase and other allied enzymes. *International Journal of Applied Research on Information Technology and Computing*, *6*(1), 33–37. doi:10.5958/0975-8089.2015.00004.4

Stucki, G., Krebser, U., & Leisinger, T. (1983). Bacterial growth on 1, 2-dichloroethane. *Experientia*, *39*(11), 1271–1273. doi:10.1007/BF01990365 PMID:6641901

Stucki, G., Thüer, M., & Bentz, R. (1992). Biological degradation of 1, 2-dichloroethane under groundwater conditions. *Water Research*, *26*(3), 273–278. doi:10.1016/0043-1354(92)90023-W

Van den Wijngaard, A. J., Van der Kamp, K. W., Van der Ploeg, J., Pries, F., Kazemier, B., & Janssen, D. B. (1992). Degradation of 1, 2-dichloroethane by Ancylobacter aquaticus and other facultative methylotrophs. *Applied and Environmental Microbiology*, *58*(3), 976–983. doi:10.1128/AEM.58.3.976-983.1992 PMID:1575500

van den Wijngaard, A. J., Wind, R. D., & Janssen, D. B. (1993). Kinetics of bacterial growth on chlorinated aliphatic compounds. *Applied and Environmental Microbiology*, *59*(7), 2041–2048. doi:10.1128/AEM.59.7.2041-2048.1993 PMID:16348981

van der Zaan, B., de Weert, J., Rijnaarts, H., de Vos, W. M., Smidt, H., & Gerritse, J. (2009). Degradation of 1, 2-dichloroethane by microbial communities from river sediment at various redox conditions. *Water Research*, *43*(13), 3207–3216. doi:10.1016/j.watres.2009.04.042 PMID:19501382

Ware, G. W. (1988). 1, 2-Dichloroethane. In *Reviews of Environmental Contamination and Toxicology* (pp. 69–79). Springer. doi:10.1007/978-1-4612-3922-2_6

WHO. (1995). *1,2-Dichloroethane* (2nd ed.). World Health Organization. https://apps.who.int/iris/handle/10665/37243

Yu, R., Peethambaram, H. S., Falta, R. W., Verce, M. F., Henderson, J. K., Bagwell, C. E., Brigmon, R. L., & Freedman, D. L. (2013). Kinetics of 1, 2-dichloroethane and 1, 2-dibromoethane biodegradation in anaerobic enrichment cultures. *Applied and Environmental Microbiology*, *79*(4), 1359–1367. doi:10.1128/AEM.02163-12 PMID:23263950

Zhu, Y., Boye, A., Body-Malapel, M., & Herkovits, J. (2017). *The toxic effects of xenobiotics on the health of humans and animals*. Academic Press.

Zok, S., Boutonnet, J. C., De Rooij, C., Garny, V., Lecloux, A., Papp, R., Thompson, R. S., & Van Wijk, D. (1998). Euro Chlor risk assessment for the marine environment OSPARCOM region: North sea-Chloroform. *Environmental Monitoring and Assessment*, *53*(3), 401–424. doi:10.1023/A:1006010515371

Chapter 15
Microbe-Assisted Phytoremediation of Petroleum Hydrocarbons

Haruna Yahaya Ismail
https://orcid.org/0000-0003-0190-4338
University of Maiduguri, Nigeria

Ahmad Ali Farouq
Usmanu Danfodiyo University, Sokoto, Nigeria

Abdullahi Bako Rabah
Usmanu Danfodiyo University, Sokoto, Nigeria

Aminu Bayawa Muhammad
Usmanu Danfodiyo University, Sokoto, Nigeria

Ibrahim Alkali Allamin
University of Maiduguri, Nigeria

Umar Balarabe Ibrahim
Usmanu Danfodiyo University, Sokoto, Nigeria

Usman Ali Bukar
University of Maiduguri, Nigeria

ABSTRACT

Petroleum is an important source of hydrocarbons, which are one of the major environmental contaminants that disturb ecosystem functioning and stability. In the past few decades, a number of approaches employed in the remediation of polluted soil, water, and aquifers have experienced setbacks. Recently, phytoremediation is gaining more attention due to its numerous benefits. Different mechanisms are used in phytoremediation; however, the integration of microorganisms and plant species to achieve remediation has been alluring. Phytoremediation provides a solution to one of the dreadful problems of pollution in situ, devoid of secondary contamination. Phytoremediation addresses pressing environmental pollution problems, and it also provides other important ecosystem services. In this review, a concise discussion of phytoremediation in synergy with microbes will be provided.

INTRODUCTION

The word "phytoremediation" was first described by Ilga Raskin ("*phyto*" Gr. Plant; and "*remediation*" L. able to cure) in the early 1990s. It is a general term that describes the process of using plants to decrease

DOI: 10.4018/978-1-7998-7062-3.ch015

the quantity, mobility, or toxicity of contaminants in contaminated media like soil or groundwater (Van Epps, 2006). It is also defined as a form of bioremediation that involves a plants-microbial synergism to detoxify contaminants. Phytoremediation is a green process in which vegetative plants remove or degrade contaminants from an environment (Cameselle et al., 2013).

As an *in-situ* bioremediation technique, phytoremediation employ the inherent abilities of living plants. Over the years, interactions among plants, microorganisms water, and soil have been demonstrated to play a significant role in manipulating environmental components and it is on this principle the concepts of phytoremediation are established (Ahalya and Ramachandra, 2006). General information on phytoremediation has been developed from a number of laboratory and field studies such as in constructed wetlands, oil spills, and accumulation of heavy metals by agricultural plants (Sumiahadi and Acar, 2018). The technology is solar-energy driven and operates based on the principles of using nature to cleanse nature which showcase its eco-friendliness (Nwaaichi et al., 2015). Currently, phytoremediation as a promising technology solving the problem of different pollutions faced by mankind. Phytoremediation in addition to addressing environmental pollution problems, it also provides several ecosystem services (Chakravarty et al., 2017).

Phytoremediation is applied in terrestrial and aquatic environments as a beginning or finishing treatment option after initial clean-up processes (Ahalya and Ramachandra, 2006). Presently, phytoremediation is the only known most-passive cleanup technology in which growing, and in some cases harvesting the plants on a contaminated site renders it safe; especially where the levels of the contaminants are low or moderate. It is used to clean up heavy metals, organic pollutants (e.g. pesticides, petroleum hydrocarbons, and solvents), explosives, radioactive contaminants, and landfill leachates. In essence, phytoremediation is the most economical cleanup technology for different organic and inorganic pollutants (Pilon-Smits, 2005). One of the successful areas of phytoremediation application is in the treatment of hydrocarbon polluted media. A number of investigations have reported encouraging findings in the decontamination of soil and water using phytoremediation (Frick et al., 1999). Although hydrocarbon degradation can spontaneously proceed due to microbial activity, a number of studies have shown that the presence of plant species increases the rate of disappearance of the hydrocarbon contaminants (McIntosh et al., 2017; Rodriguez- Campos et al., 2018; Riskuwa-Shehu and Ismail, 2018). In addition, researchers have established that plants' role is majorly indirectly because the fundamental mechanism for the cleanup of hydrocarbon contaminants is rhizodegradation (Frick et al., 1999, Germinda et al., 2002; Hall et al., 2011; Lu et al., 2019).

In rhizodegradation, microorganisms and plants are involved both individually and in synergy for the degradation or transformation of petroleum hydrocarbons into products that are environmentally less harmful and less persistent than the parent compounds (Germinda et al., 2002; Kotoky et al., 2018). The interaction between plants and microorganisms in the rhizosphere is the primary mechanism by which petroleum hydrocarbons are degraded in soils. In the rhizosphere, plants increase significantly the microbial activity by nutrient supplementation through root exudation (Rohrbacher and St-Arnaud, 2016). The plants make oxygen available either by excreting oxygen or by creating void spaces in the subsurface that allows for greater oxygen diffusion from the atmosphere (Tsao, 2003; Van Epps, 2006). Microbial populations benefit plants through recycling and solubilization of mineral nutrients as well as by supplying vitamins, amino acids, auxins, cytokinins, and gibberellins, which stimulate plant growth (Vaziri et al., 2013).

A number of empirical evidence on successful remediation of hydrocarbons using plants are documented (Frick et al., 1999; Zand et al., 2009). Anyasi and Atagana (2017) screened 28 plant species of

Nigerian origin for phytoremediation of hydrocarbon contaminants; taking into cognizance the phytotoxic effects of the contaminants and their uptake. Among the studied plant species, *Aspilla Africana*, *Chromolaena odorata* and *Uvaria chamae* were chosen to be best candidates for phytoremediation of polycyclic aromatic hydrocarbons (PAHs). In a study to investigate the mechanisms that promote hydrocarbon degradation in soil during phytoremediation (Siciliano et al., 2003), it was established that the system increase the catabolic potential of contaminated rhizosphere soil by altering the functional composition of the microbial community, in contrast to a bulk control soil under field conditions. Zand et al. (2009) observed a decrease in TPH by 96.3% in a phytoremediation study using tall fescue. Studies by Fatima et al. (2018) reported an effective crude-oil remediation of contaminated soil at an oil exploration and production company using plants-bacterial synergism. High rate of oil disappearance (80%) was observed in a synergistic setup compared to using a plant or microbe alone. Ubogu et al. (2019) observed comparatively the effectiveness of biostimulation, bioaugmentation, and phytoremediation of hydrocarbon contaminated mangrove swamp soil using *Phragmites australis* and *Eichhornia crassipes*. The study concluded that rhizoremediation was more effective than biostimulation and bioaugmentation techniques. Van Epps (2006) have reported a number of successful field trials and pilot studies of hydrocarbon phytoremediation.

PETROLEUM HYDROCARBONS

Since the beginning of the last century, crude oil and gas had become indispensable resources for modern life as fuels and raw materials. Petroleum refining yield over 2500 products, including the common ones like LPG, gasoline, kerosene, aviation fuel, diesel fuel, fuel oils, lubricating oils, and raw materials for petrochemical industry (United States Environmental Protection Agency [USEPA], 2011). The abundance and multipurpose nature of oil and gas facilitated the unprecedented economic growth around the world and improvement in human health (Allison and Mandler, 2018).

The majority of the compounds present in crude oils are hydrocarbons which exist as gases, liquids, and solids. Hydrocarbons present in crude petroleum could reach up to 97% by weight (e.g., in the lighter paraffinic crude oils) or ≤ 50% by weight as in heavy asphaltic crude oils (Speight, 2006). However, crude oils containing as little as 50% of hydrocarbon components still retain most of the essential characteristics of the hydrocarbons. The hydrocarbon present in crude oil is grouped into saturated hydrocarbons, unsaturated hydrocarbons, and aromatics (Weisman, 1998). In general, crude oils contain the classes of hydrocarbons shown in Table 1.

Other organic compounds containing Sulphur (hydrogen sulfide, mercaptans, etc.), Nitrogen (quinotine, pyridine, pyrrole, indole, carbazole), and Oxygen (naphthenic acids, phenols, some other organic acids) are found in varying proportions among petroleum from different sources. Their presence in most instances is undesirable due to problems associated with refining, storage, and consumption of the products. For example, compounds of Sulphur, Nitrogen, and Oxygen cause foul odor, color alteration of refined products, and corrosion of oil facility respectively (Speight, 2006). Trace petroleum constituents are metallic derivatives and porphyrins.

Table 1. Hydrocarbon compounds present in petroleum mixture

Group	Hydrocarbon Family	Distinguishing Characteristics	Major Hydrocarbons	Remarks
Saturated	Paraffins (Alkanes)	They have straight carbon chain	Methane, ethane, propane, butane, pentane, hexane	General formula C_nH_{2n+2} Boiling point increases as the number of carbon atom increases. With number of carbon 25-40, paraffin becomes waxy.
	Isoparaffins (Iso alkanes)	Straight carbon chains with branches	Isobutane, Isopentane, Neopentane, Isooctane	The number of possible isomers increases in geometric progression as the number of carbon atoms increases.
	Naphthenes	5 or 6 carbon atoms in ring	Cyclopentane, Methyl cyclopentane, Dimethyl cyclopentane,	General formula $C_nH_{2n+2-2Rn}$ Rn is number of naphthenic ring The average crude oil contains about 50% by weight naphthenes. Naphthenes are modestly good.
Unsaturated	Olefins (Alkenes)	One pair of carbon atoms	Ethylene, Propylene	General formula C_nH_{2n} Olefins are not present in crude oil, but are formed during process. Undesirable in the finished product because of their high reactivity. Low molecular weight olefins have good antiknock properties.
Aromatics	Aromatics	6 carbon atom in ring with three around linkage.	Benzene, Toluene, Xylene, Ethyl Benzene, Cumene, Naphthaline	Aromatics are not desirable in kerosene and lubricating oil. Benzene is carcinogenic and hence undesirable part of gasoline.

(Source: Mall, 2007 with modifications)

Polycyclic aromatic hydrocarbons are among the most considered compounds as a result of their toxicity, carcinogenicity, and mutagenicity (Harvey et al., 2002). The major source of PAHs is crude petroleum, however; they are predominantly introduced to the environment through natural and anthropogenic combustion processes (Speight, 2006). The release of PAHs from natural sources is as a result of spontaneous fires from forests and grassland and also volcanic emissions. On the other hand, the anthropogenic sources are diverse ranging from simple processes of incineration of wood for cooking and heating to complex industrial activities such as refining of crude petroleum, manufacturing of chemicals, and vehicle emissions (D'Souza et al., 2015). Soil and sediments are the main sinks for all the PAHs derived from pyrogenic, petrogenic, and biological activities in the environment (Abdel-Shafy and Mansour, 2015).

PETROLEUM HYDROCARBON CONTAMINATION

Petroleum hydrocarbon contamination is the overflow of hydrocarbon compounds from petrochemical process into pristine environments. Environmental contamination with petroleum and its derived products are frequent events although sometimes in small scales (Fingas et al., 2001). Along the oil and gas value chain, activities such as refining, storage, transportation, sales, equipment maintenance, bunkering and sabotage result in emission and overflow of petroleum hydrocarbons to immediate environment

(Wang et al., 2017). Transportation of petroleum has contributed to majority of oil spills in the world. For marketing and consumption of petroleum and its refined products, its transportation from oil fields to its target destination is necessary. Presently, petroleum products, and essentially all natural gas, are conveyed through tankers and pipelines, laid over million miles away (International Tanker Owners Pollution Federation [ITOPF], 2019). During the course of transportation, spillage of the products occurs. Although the spill may amount to less than 0.001% of the quantity transported, its recurrences may add up to millions of gallons spilled per annum (ITOPF, 2019).

Petroleum hydrocarbon contamination of soil and groundwater may comparatively be low (15% of all pollution) in developed countries like United Kingdom (Stroud et al., 2007) but is much more pronounced in developing countries like Nigeria which ranked the 8th country with proven petroleum reserves (Organization of the Petroleum Exporting Countries [OPEC], 2019). Alarming pollution incidences due to oil spillage are often reported and earlier data have reported that over 13 million tons of hydrocarbons were spilled in the Niger Delta region in the last six decades (Sam et al., 2017) which caused much damage to the environment (Kaddafa, 2012; Ite et al., 2013). Theoretically, 240 thousand barrels of crude oil is spilled in the region per annum on the average (Ordinioha and Brisibe, 2013) making it one of the most polluted oil field in the world. In other parts of the world, spillage of both crude and refined petroleum products is occasionally reported. Substantial quantity of engine oil are discharged into the environment during engine oil replacement and disposed into the vicinities as commonly practiced by industries, motor mechanics and generator users (Odjegba and Sadiq, 2002). Rising public awareness and concerns about the effects of petroleum hydrocarbon pollution in the environment has led to the evolution of various treatment technologies which can serve to prepare the hydrocarbon pollutants for recycling, or for final disposal in a manner safer than disposal without treatment (Sylvia, 2019).

Methods of Petroleum Hydrocarbon Remediation

Since the time when the world's early major oil spills occurred, (Michel and Fingas, 2016), enormous resources have been dedicated towards oil recovery and environmental cleanup (Sebastián et al., 2014). The world has seen the evolution of different cleanup technologies in the last five decades (Atlas, 1981; Ram et al., 1993; Streche et al., 2018; Maceiras, 2020). Popular among the treatment methods are physical, chemical, thermal and biological (Watson, 1996, de Souza et al., 2013; Wang et al., 2017; Xuezhi et al., 2020). The goal of the remediation techniques is to meet any or all the following:

1. elimination or alteration of contaminants,
2. extraction or separation from an environment, and/or
3. Immobilization of the contaminants.

Before selection of appropriate technology for hydrocarbon clean up, feasibility study focusing on the cost implication, environmental suitability and time frame is recommended. Biological method has always been described as eco-friendly and less costly than the other techniques. In the sketch below (Figure 1), various techniques under the remediation options are outlined.

Figure 1. Remediation techniques of petroleum hydrocarbons

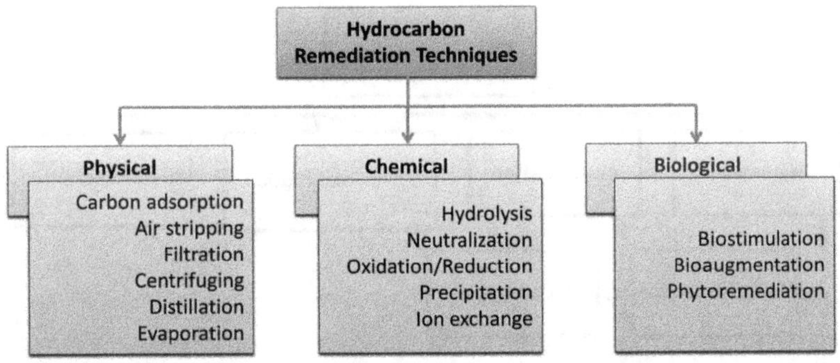

PHYTOREMEDIATION OF PETROLEUM HYDROCARBONS

Phytoremediation is a remediation technique that exploits plants and microorganisms to decontaminate a polluted environment. Phillips (2008) considers phytoremediation as the use of plants and the microbial communities associated with them to sequester, degrade, or prevent the mobility of xenobiotic contaminants. In natural ecosystems, plants remove and utilize substances generated by nature. Since the inception of phytoremediation, a remarkable body of knowledge on the use of plants to remediate a wide variety of both inorganic and organic compounds has been produced (Phillips, 2008). This may be due to the fact that the success of microbial degradation has been limited with petroleum-based constituents rather than residual organic and metal pollutants. Vegetation-based remediation however shows potential for accumulating, immobilizing and transforming complex compounds into low level of persistent contaminants (Sutar et al., 2012).

The fundamental principles governing phytoremediation include (Kathi and Khan, 2011);

1. Absorption of organic compounds from the root zone.
2. Processing and deposition of these chemicals via lignification, volatilization, metabolization, or mineralization.
3. Enzymatic degradation of complex organic molecules into simpler molecules (ultimately carbon dioxide and water).
4. Enrichment of the root zone with nutrients, carbon and oxygen which promotes microbial activity.

Mechanisms in Phytoremediation

There are different mechanisms of phytoremediation based on the contaminants treated and the environment. Nature of contaminants and impacted environment determines the phytoremediation approach that is suitable for the distinct condition to be treated. Defining this process is essential to understanding the role to be played by a chosen plant species. The fate of a particular contaminant and effective phytoremediation protocols needed must be understood to avoid undesirable consequences. Various mechanism employed in phytoremediation are outlined in Figure 2.

Figure 2. Mechanisms in phytoremediation

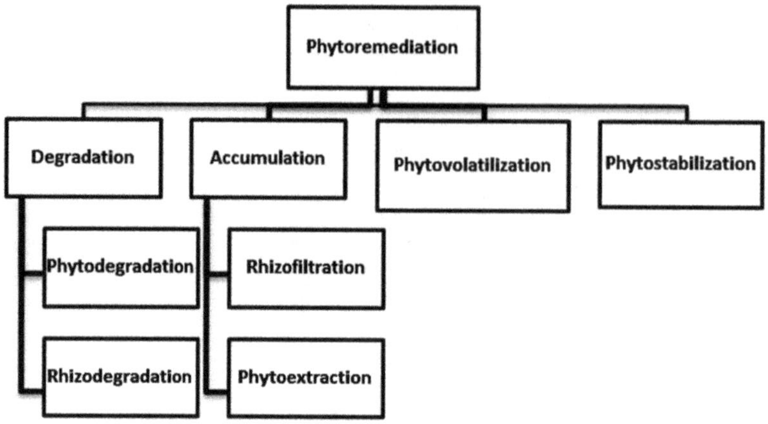

Plant and microbial synergism remediate petroleum hydrocarbons through three fundamental mechanisms in soil and groundwater. These mechanisms include degradation, containment, as well as transfer of the hydrocarbons to atmosphere (Cunningham et al., 1996). In containment process, plants reduce or eliminate the bioavailable contaminants from the environment. Plants contain petroleum hydrocarbons by accumulation within the plants, adsorption on the root surface and as organic pumps that allows its isolation within the root zone, thus limiting its spread. Indirectly, humification – a process that bind contaminants into soil organic matter as a result of enzymatic activities is exercised. Humification is enhanced by increasing soil organic matter content (Cunningham et al., 1996). In the case of hydrocarbon transfer from soil to atmosphere, plants absorb and translocate the compounds and then get liberated into atmosphere by transpiration (Frick et al., 1999). However, the process may lead to subsequent contamination of the atmosphere which results to breach of air quality regulatory standards. In degradation however, plants and microorganisms play a direct or indirect role in the breakdown of hydrocarbons into simpler products that are generally considered less toxic and less recalcitrant the parent compounds. There are speculations on the effectiveness of direct hydrocarbon degradation process by plants and Frick et al. (1999) suggested the degradation pathway as follows;

n-alkane → primary alcohol → fatty acids → acetyl-CoA → various compounds

Conversely, the indirect role of plants in degradation hydrocarbons is well established and considerable body of information is available (Gunther et al., 1996). The plants employ three mechanisms to accomplish degradation. These include alteration of soil's physical and chemical conditions by plants and their root systems, enhancement of rhizosphere effect through root exudation and release of root-associated enzymes capable of transforming organic contaminants through co-metabolism (Frick et al., 1999). Researchers reported variation in hydrocarbon degradation from as little as 5% to greater than 50% using different plant species (Phillips, 2008). Degradation refers to breakdown or transformation of complex or toxic substances to simpler and less toxic ones and is believed to be the major mechanism for organic contaminants cleanup. Depending on the type of contaminant and plant species, two types of degradation have been identified in hydrocarbon phytoremediation: phytodegradation and rhizodegradation.

Phytodegradation

Phytodegradation is a process that involves uptake, metabolism, and degradation of contaminants inside plant system (Wenzel 2009). Typically, it is a contaminant destruction process which is also known as phytotransformation. The degradation of organic pollutants is driven by metabolic processes of plants (Prasad and Freitas 2003; Sharma and Juwarkar, 2015). Contaminants such as chlorinated solvents, herbicides, pesticides, and other organic contaminants may be eliminated or reduced through phytodegradation process. In this process, plant metabolism contributes to the reduction of contaminant by means of different metabolic processes such as transformation, breakdown, or volatilization, and the simpler breakdown products then are incorporated into plant tissues. Certain enzymes in plants such as dehalogenases, oxygenases, and reductases are responsible for the breakdown of contaminants into simpler forms (Ghosh and Singh 2005; Sharma and Juwarkar, 2015). The process may also take place in soils, sediments, sludge, surface water, or groundwater with the help of plant's extracellular enzymes.

Rhizodegradation

Rhizodegradation is the breakdown of contaminants as a result of stimulation of microbial communities by plant roots which eventually cause the destruction of contaminants. Microorganisms including bacteria, fungi, and actinomycetes breakdown organic contaminants into less toxic products or sometimes completely mineralize them to carbon dioxide and water. Though plants and microorganisms can degrade petroleum hydrocarbons independent of one another, the literature suggests that rhizosphere effect is the primary mechanism responsible for the degradation process in phytoremediation efforts (Frick et al., 1999). Plant roots enhance the abundance and diversity of microbial populations in the rhizosphere which ultimately leads to improvement in contaminant biodegradation (Rani and Juwarkar 2012; Sharma and Juwarkar, 2015). Different compounds like sugars, amino acids, organic acids, fatty acids, sterols, growth factors, nucleotides, flavanones, enzymes, etc., are released from plant roots as exudates which increase the activity of the rhizospheric microflora and thus increased biodegradation. Rhizodegradation primarily occurs in soil and the nature of released exudates determines the soil physicochemical conditions which dictates the rate of contaminants' bioavailability (Pivetz 2001; Sharma and Juwarkar, 2015). The plant roots positively affect the soil properties which create more favorable conditions for soil microflora.

Metabolism of Hydrocarbons by Plants

Following exposure to petroleum hydrocarbons, plants withstand their effect by lowering, transforming, and degrading the harmful contaminants in specialized cells adapted for detoxification process (Sandermann, 1994; Sun et al., 2015). Once in the plants' rhizosphere, they drift to the roots but some lipophilic compounds limit their uptake or cause their accumulation in the partly suberized cortex of the root. Hydrocarbon lipophilicity and its adsorption capacity to soil particles limit its uptake by plants. Therefore, the most likely compounds taken up by plants are those with moderate octanol-water partition coefficient (log K_{ow} 0.5 – 3.0). Hydrocarbons with lower log K_{ow} are water soluble and not firmly attached to roots and passively transported through plant membranes; whereas, those with higher log K_{ow} (> 3.0) can only adsorb to the surface of the roots with high proportion of lipids – uptake and translocation is restricted (Schnoor et al., 1995; Siciliano and Germinda, 1998; Farrel and Germinda, 2002).

However, there are divergent views on the ability of plants to uptake hydrocarbons where plants' inability to uptake hydrocarbon is an approved standard by the Canadian Council of Ministers of the Environment; CCME, 2008). A number of findings are in support of this standard (Nwaaichi et al., 2011; Lu et al., 2010). Some other studies by different researchers however, are of the opinion that plant uptake hydrocarbon at different capacity depending on their physiology (Radwan et al., 2000; Palmroth et al., 2002; Wild et al., 2005; Basumatary et al., (2012); Naidoo and Naidoo, 2016; Patowary et al., 2017; Anyasi and Atagana 2018). Despite these findings, Hunt et al. (2018) described them as numerically inconsequential and generally lack reliable data to back their conclusions; because majority of the investigations were not aimed at determining hydrocarbon uptake and/or its distribution but focused on determining the rates of phytoremediation. Furthermore, methodological inconsistencies, inadequate description of environmental conditions and analytical procedures and irreconcilable measurements marred the findings (Doucette et al., 2018; Hunt et al. 2018).

Where hydrocarbon uptake is believed to have taken place, the compounds are prevented from detoxification and metabolism but transferred into symplast to avoid the suberized casparian strips barrier in the root endodermis. They are later translocated by the transpiration stream along the xylem into other tissues of the plant (root and shoot) (Kathi and Khan, 2011). The metabolism is enzyme catalyzed and occurs in three phases. Phase-I is catalyzed by P-450 enzymes complex responsible for transformation reactions like hydroxylation, N and O-alkyl group removal and Sulphyl group oxidation. In phase-II of the metabolism, conjugation of the earlier transformed compounds with polar molecules of plants origin occurs (Kvesitadze et al., 2009; Pandey and Bajpai, 2019). This stage is central in hydrocarbon detoxification by plants and it is facilitated by the activities of transferases (Aken et al., 2010). If the formed conjugates are soluble, they can totally disintegrate into CO_2 and H_2O for the plant's benefit, but if they are insoluble, they are transferred by exocytosis to the apoplast and become part of the cell wall (Komossa et al., 1995; Kathi and Khan, 2011; Schwitzgue´bel, 2017). This describes the Phase-III or last stage of hydrocarbon metabolism in plants.

Phytoxicity

Petroleum hydrocarbons induce toxic effects on different plant species during germination and growth (Agbogidi, 2010) especially in heavily contaminated environment (Chaineau et al., 1997). Plants seeds are seriously damaged due to the fact that some oil fractions have the capacity to wet and strongly penetrate into seed coat and embryo, which result to destruction and loss of seed viability (Kathi and Khan, 2011; Ismail et al., 2019). There are reports that show that phytotoxic effects on seeds is correlated with hydrophobic properties of oils that prevent and/or reduce exchange of water and gases which disrupts the metabolism or cause acute toxicity that destroys the embryo (Amadi et al., 1992). After emergence, hydrocarbons are known to reduce growth and yield of crops even at low concentrations (Ali, 2019). Individual hydrocarbon fractions are ideal for testing hydrocarbon toxicity and as such, it is obscure to figure out the toxicity of petroleum mixture without knowing the parent constituents. The amount of TPH observed depends on the nature of solvent used in hydrocarbon extraction in which volatile compounds are lost during solvent concentration, which cause wrong estimate of plant hydrocarbon contents (ATSDR, 1999). However, Chaineau et al. (1997) have shown that light aromatics and naphtha to be more phytotoxic in seven different plant species. Van Epps (2006) reported that TPH reaching up to 810 mg/kg consisting of compounds between 5 and 28 carbons atoms in a soil sampled from a gas station can elicit phytotoxic effects in willow and poplar trees. Studies by Somtrakoon and Chouychai (2013) have shown the toxicity of different PAHs on the germination and growth of sweet corn, waxy corn, and rice in which both single and mixed PAH treatment delayed germination and growth. Petroleum hydrocarbons impede plant growth by reducing the growth rate, soil fertility and plants resistance to pests and diseases (Wang et al., 2017).

Plants Used in Hydrocarbon Phytoremediation

One of the major focuses in phytoremediation is to identify a plant species that is resistant or tolerant to a particular contaminant with a view to maximizing its potential for remediation. Plants growing on soils with underlying contaminants or on the boundary of polluted sites are commonly resistant or tolerant (Vaziri et al., 2013). There are some plant species with better remediation properties than other species; therefore, more efficient species should be selected for phytoremediation of hydrocarbons (Rodriguez et al., 2005).

For a sustainable phytoremediation process, the use of plant species that are economically and ecologically valuable has been suggested (Pandey et al., 2015). Additionally, some of the desirable qualities include being indigenous, ability to propagate easily and rapidly, fast growing, high biomass production, abundant root system, ability to concentrate pollutants, withstand harsh conditions, inedible, perennial and ecologically stable (Pandey and Bajpai, 2019). More so, it is advantageous that the selected species or its product could be valorized and should also be valuable to society in terms of energy and environmental services (Pandey and Bajpai, 2019).

Currently, more than hundred plant species that have some desirable qualities and the potentials for soil and water remediation have been identified (Yaqoob et al., 2019). This includes a broad range of plants such as trees (e.g. poplar trees), edible plants (e.g. rice), aquatic weeds (e.g. duckweed) and terrestrial grasses (Chakravarty et al., 2017). Trees, legumes and grasses are frequently used in hydrocarbon remediation, with trees majorly selected for remediation of BTEX as against grasses which are more commonly used for remediation of PAHs and TPH.

Frequently, leguminous plants and grasses are considered most promising in hydrocarbon phytoremediation (Aprill and Sims, 1990; Qiu et al., 1997; Van Epps, 2006; Ruley et al., 2019). This is because; grass have the largest root surface area, penetrate deep into soil, genetically diverse and easily grow under unfavorable soil conditions (Aprill and Sims, 1990). Legumes however fix nitrogen; thus, limiting microbial completion for nitrogen which become limited in oil-contaminated sites (Aprill and Sims, 1990; Frick et al., 1999). They are also diverse with variety of propagation methods and able to grow in almost all forms of terrestrial environments due to enhanced defense and nutrient acquisition (Hall et al., 2011). Like grass, legumes provide oxygen in soil environment which stimulate microbial activities and subsequent promotion of hydrocarbon biodegradation (Peer et al. 2006).

Comparatively, studies have shown that legumes can do better than grasses in hydrocarbon phytoremediation which might reflect its better nutrient supplementation capacity (Yateem et al., 2000; Diab, 2008). However, little differences were observed between grasses and trees with regards to successful reduction of hydrocarbon concentration within the same time frame (Cook and Hesterberg, 2013). Herbs and shrubs are also important in treating hydrocarbon pollutants depending on the plant species, nature of the pollutants as well as prevailing environmental conditions (Frick et al. 1999). Available information shows that most of the successful phytoremediation projects were carried out using trees which show their ability for enduring extended period of application as compared to other plant species (Yan, 2012). Table 2 shows a list of some often reported plant species with hydrocarbon phytoremediation potentials.

Table 2. Plants with phytoremediation ability

Plant Species	Hydrocarbons	Comment	Reference
Medicago sativa L. and *Medicago falcata* L. (Leguminosae)	Oil sludge	Stimulate microbial growth and decrease major oil fractions	Panchenko et al. (2017)
Festuca arundinacea Schreb. (Poaceae)	PAHs	The abundance of PAH degrading bacteria in the rhizosphere was substantially increased; most of 4-ring PAHs were degraded	Huang et al., 2004; Parrish et al., 2005). Sun et al. (2011)
Trifolium repens L. (Leguminosae) *Lolium perenne* L. (Poaceae)	PHC	Significantly reduced the hydrocarbon concentration to undetectable limits	Germaine et al. (2015).
Trifolium repens Trifolium pretense (Fabaceae)	Diesel, PAHs	Enhanced degradation of diesel; root exudates facilitated PAHs bioavailability and increased biodegradation rate	Ying et al. (2018), Davin et al. (2019)
Sorghum bicolor L. *Hordeum vulgare* L. (Poaceae)	PHC	significant reduction in the concentration of petroleum hydrocarbons	Asiabadi et al. (2018)
Cynodon dactylon L. (Poaceae)	PHC	About 50% reduction in PHC concentration, with amendment using organic fertilizer	Basumatary and Bordoloi (2016)
Prairie grass	PHC	Significant reduction in TPH	April and Sims (1990)
Heliamthus annus (Asteraceae)	PHC and heavy metal co-contamination	58% reduction in TPH and reduction in heavy metal concentration was observed	Vitor et al. (2018)
Salix smithiana L. *Salix viminalis* L. (Salicaceae)	PAHs	PAHs were removed by 50.9% after three years of soil In synergy with white rot fungi, caused the highest PAH removal rate.	Košnář et al. (2020) Ma et al. (2020)
Triticum aestivum L. (Poaceae)	PHC	Fertilizer application enhanced the degradation	Masu et al. (2013)
Jatropha carcus L. (Euphorbiaceae)	PHC	caused 78.8% reduction of TPH with compost amendments	Bertrand, (2020)

PLANT-MICROBE SYNERGISM FOR HYDROCARBON DEGRADATION

It has been known for long that plants like animals have microbiota which is present in their endosphere, rhizosphere and phyllosphere. This include plants' normal flora which consists of a few dominant species called the core microbiome which are constantly associated with a given plant irrespective of environmental influence; the major microbiome which determine plant fitness and few other microbes in the endosphere whose roles are not clearly understood (Nataraja et al., 2019). Plant-microbe interactions can be beneficial, harmful or neutral based on the effects to the host (Imam et al., 2016). Different types of interactions are known to exist including mutualism, pathogenesis, and parasitism (Hu¨ckelhoven, 2007; Phillips, 2008; Singh et al., 2019). Earlier studies by Paungfoo-Lonhienne et al. (2010) have demonstrated a predatory relationship in which microbes enter root cells and are later digested to release nitrogen for growth. There are enough evidences that plants-microbes association dramatically influence each other's lifestyles and health trajectories (Zhang et al., 2014; O'Banion et al., 2019).

Recently, the interaction between plants and microbes has been exploited to remove environmental contaminants from soil which offers a cheaper, safer, and eco-friendly alternative to available methods (Singh et al., 2019). Although soil contamination with pollutants affects biological functions, synergistic plant-microbe interaction plays a crucial role in improving soil quality and plant performance (Radwan et al., 1995; Siciliano and Germida, 1998; Velmourougane et al., 2017). In this process, microorganisms degrade organic contaminants or make inorganic pollutants bioavailable for uptake by plants (Chaudhry et al., 2005). For the microbes to grow, multiply and subsequently degrade contaminants, they require essential nutrients from plants, while plants benefit from the detoxification of pollutants by the microbes (Siciliano and Germida 1998; Manoharachary and Mukerji, 2006).

The key roles played by microorganisms in microbe-assisted phytoremediation of petroleum hydrocarbons are plant growth promotion. This is made through minimizing phytotoxicity, promoting extensive root system, improving pumping capacity, providing mobilization due to surfactant production, enhancing stabilization due to secretion of chelators and detoxification as a result of sequestration on cell walls (Thijs et al., 2016; Vangronsveld et al., 2019). Studies by Montalbán et al (2017) have shown that endophytic bacteria significantly decreased contaminant-induced stress and increased contaminant uptake into the plants. Conversely, plants offer the microbes a micro-environment, nutrients, electron acceptors, growth factors and water for growth (Ashraf et al., 2013). There is increasing efforts to further elucidate plant-microbe interaction and how the networks operate in the environment (Sahu et al., 2020). Recently, focus has been shifted to using genetic and bioinformatics approaches to give clear understanding of the interconnectivity between plants and microbes in remediation processes (Agarwal et al., 2020; Sharma et al., 2020).

Interaction in the Rhizosphere

Rhizosphere is the term used to describe the portion of soil surrounding plant root system and under its influence (Shukla et al., 2011; Correa-Garcia et al., 2018). It is indefinite soil zone with a varying microbial abundance and diversity in which substantial microbial alteration in the soil is pronounced adjacent to roots and subside as it far away. Rhizoplane is the external surface of plant root together with any closely adhering particles of soil or debris (Manoharachary and Mukerji, 2006). To obtain rhizoplane soil, plant roots are gently removed from soil and transferred to a fresh sterile solution and shaken vigorously (Bolton et al., 1992). The size of this zone is determined by the soil type, plant type and soil conditions (Manoharachary and Mukerji, 2006).

Microorganisms are found in three distinct sites of the rhizosphere: (1) the endosphere; (2) the rhizoplane usually as biofilm and (3) the soil (ectorhizosphere) influenced by the plant roots (Rohrbacher and St-Arnaud, 2016). Although the rhizosphere covers some distance around the root in soils, its size and shape is difficult to assess despite the fact that recent understandings showed that it is quasi-stationary (Kuzyakov and Razavi, 2019). The rhizosphere associated with peanut and soybean roots was estimated to reach about 0.2 mm thick using electron microbeam analysis and scanning electron microscope (Bolton et al., 1992). Rhizosphere microorganisms are either harmful or beneficial. The beneficial effects occur in either of the following ways (Bais et al., 2006);

1. The first hypothesis suggests that there is aggressive colonization of roots by beneficial microbes which displaces the harmful ones and consequently leads to promotion of plant growth.

2. The second hypothesis believed that the beneficial microbes directly attack and kill the harmful ones. Beneficial microorganisms may produce hormones such as auxins and kinetins that bring about plant growth promotion.

The stimulatory effect on microorganisms in the rhizosphere by plants is called the rhizosphere effect (Manoharachary and Mukerji, 2006). The magnitude of the rhizosphere effect is estimated using a ratio known as rhizosphere effect (R/S) ratio (Nie et al., 2010). The R/S ratio is calculated by dividing the number of microorganisms (or the rate of a biochemical process) in rhizosphere soil by the number of microorganisms in control (bulk) soil (Diab, 2008). The rhizosphere effect may vary greatly from as much as 100% and decrease steadily as it moves away from the root zone. As such, it is up to 100 times richer in microbial biomass but poorer in diversity than bulk soil (Hrynkiewicz et al., 2009; Rohrbacher and St-Arnaud, 2016). Bolton et al. (1992) has listed a number of factors that affect rhizosphere effect including plant root exudates, plant root cell lyses, nutrients and water availability.

The predominant microbial species that inhabit the rhizosphere are fungi and bacteria (Bais et al., 2006). Rhizosphere effect may increase fungal and bacterial abundance by 2 - 20 times greater than in the bulk soil (Phillips, 2008). Due the large number of microbes in the rhizosphere, the available nutrients become limited and as a result, there is high competition for nutrients. Therefore, different microbial species have evolved special adaptations for survival ranging from antagonism to synergism, both among themselves and with the plant. Due to the wide microbial diversity, several kind of interactions within the microbial community and between the host plants is possible. The understanding of fundamentals of these interactions is essential for their use in plant growth promotion and remediation of contaminated soils (Hrynkiewicz et al., 2009).

There are reports on successful remediation of contaminated environments using rhizosphere effects. Walton et al. (1994) have opined that the first function of microorganisms in favor of plants during remediation is preventing phytotoxic effects to plants due to the contaminants. This is achieved when microorganisms degrade contaminants in the rhizosphere prior to plant uptake. The works of Rasolomanana and Balandreau (1987) and Radwan et al. (1995) using *Oryza sativa* and *Senecio glaucus* respectively are in support of this hypothesis. In return, essential nutrients for microbial growth are exuded following plants' successful establishment. Kotoky and Pandey (2019) reported that benzo (a) pyrene was reduced by more than 85% compared with 68.22% in bulk soil; using the synergistic effect of *Bacillus flexus* S1I26, *Paenibacillus* sp. S1I8 and plant *Melia azadirachta* respectively. In a field treatment of a highly contaminated site over a two years period, a decrease in hydrocarbon concentration by 30% was observed, which was double that of bulk soils (Siciliano et al., 2003). Similarly, in another field study conducted on crude oil contaminated site, 42% and 50% PHC was removed using ryegrass (*Lolium annual*) and St. Augustine grass (*Stenotaphrum secundatum*) respectively during 21 months treatment period (Nedunuri et al., 2000, Gerhardt et al., 2009). More so, in a study involving PAHs degradation using *Cajanus cajan* and *Lablab purpureus*, only 12.18% and 25.40% respectively of the initial PAHs concentration was recovered with complete disappearance or significant depletion of Naphthalene, pyrene, fluorene, fluoranthene and indeno (1,2,3-c, d) pyrene (Riskuwa-Shehu and Ismail, 2018).

Interaction in the Endosphere

Endophytic microbes are microorganisms residing inside plant tissues including root cortex and xylem during all or part of their life cycle. Plant are colonized through vascular or apoplast system and dead and hollow hyaline cells. Endophytes are endogenous, evolving from internal organelles such as mitochondria or chloroplast (Hardoim et al., 2008; Anyasi et al., 2019). It is also opined that, endophytes originate from different niches including rhizosphere, caulosphere, laimosphere, phyllosphere, anthosphere, carposphere and spermosphere (Compant et al., 2012, Verma et al., 2017). Despite having similar colonization behaviors with pathogens (Kumar et al., 2014), endophytes remain associated with plant devoid of disease causation (Sessitsch et al., 2002) but rather promote plants' growth (Strobel and Daisy, 2003). Endophytes promote growth by increasing plant's biomass (Płociniczak et al., 2017), induce disease resistance and enhance tolerance to ecological stress (Anyasi et al., 2019). Beside improving plant fitness, they are also involved in recycling nutrients (Sturz et al. 2000) and serve as important source of metabolites which are of health and environmental significance (Strobel and Daisy 2003).

Endophytic microbes are cryptic and they commonly manifest in an ecosystem during decomposition of the host, because they initiate microbial colonization of dead plant tissues (Oses et al. 2008, Sudha et al., 2016). Covertly, they are capable of enhancing phosphate solubilization and uptake, fix atmospheric nitrogen, produce siderophores, and hormones such as auxin, abscisins, ethylene, gibberellins, and indole acetic acid (IAA) which are important in regulating plant growth (Firakova et al., 2007; Puente et al., 2009; Weyens et al., 2009; Sudha et al., 2016).

Many studies have elucidated the crucial role played by endophytes in hydrocarbon phytoremediation (Germaine et al., 2009; Ma et al., 2011; Shehzadi et al., 2016; Marín et al., 2018). They play both direct and indirect roles. Indirectly, an endophyte partake in remediation through enhancing the growth of plants with phytoremediation ability (Glick 2010; Płociniczak et al., 2017) while directly, through degradation of pollutants by itself (Sessitsch et al., 2012; Mitter et al., 2013; Wu et al., 2019). The work of Płociniczak et al. (2017) has demonstrated the indirect role of bacterial endophyte *Rhodococcus erythropolis* CD106 strain during phytoremediation of an aged hydrocarbon-polluted soil. In the study, the bacterial endophyte significantly increased the biomass of ryegrass (*Lolium perenne*) which results to decrease in hydrocarbon concentration by 31.2% against the control after 210 days.

In some few years back, there were deliberate attempts to design plant-endophyte consortium for successful phytoremediation (Beckers et al., 2016). There are promising results from such attempts due to the fact microbes in form of inoculants easily colonize selected plants species (Barac et al., 2004; Weyens et al., 2010). The microbes not only degrade contaminants but also facilitate transfer of catabolic genes to non-degrader communities present in different plants tissue (Taghavi et al., 2005). Kaneez et al. (2018) reported that, when *Brachiaria mutica* and *Leptochloa fusca* were inoculated with bacterial endophytes, there was increase in abundance and expression of *alkB* gene in the rhizosphere as well as in the endosphere of the plants than in un-vegetated soil. Improvement in plant performance, hydrocarbon degradation as well as overall soil health was observed. Significant plant development, enhanced photosynthetic pigments and significant higher degradation rates were also observed in phytoremediation of petroleum hydrocarbons impacted soils using *Zea mays* and endophytic *Streptomyces* sp. Hlh1 inoculant (Baoune et al., 2019). Correspondingly, Mitter et al. (2019) have shown that inoculation of bacterial consortium to *Melilotus officinalis* reduced the phytotoxicity, increased biomass and enhanced the remediation of diesel polluted soil. There is relative calmness within plant tissues compared to soil environment and this might allow better endomicrobial performance. In some instances, inoculation of the endophytes can be done repeatedly to ensure continuous presence of the inoculated organisms (Vangronsveld et al., 2019).

Role of Root Exudates

Traditionally, plant root system is known to offer support and conduction of nutrients and water to the aerial parts, however, studies have shown that plants also release substantial amount of organic molecules to soil through discharge from roots or exudation (Rohrbacher and St-Arnaud, 2016). Plant roots exudation can be active or passive (Hoang et al., 2021), and may amount to 40% of a plant's total photosynthate (Gerhardt et al., 2009). Different types of complex (organic acids, sugars, phenolic compounds, polysaccharides, and humic compounds) and simple (amino acids, monosaccharides etc.) organic molecules, are secreted through plant roots and they are collectively referred to as root exudates (Rohrbacher and St-Arnaud, 2016; Hoang et al., 2021). These exudates provide nutrient source for the growing microbes at the rhizospheric regions and help in effective colonization (Singh et al., 2019). In addition to mucilage secreted from roots, worn out cells from root caps, decayed roots and starvation of the root cells also serves as source of nutrients for the microbes (Gupta et al., 2020).

There is a great diversity in the type and abundance of plant exudates which is a function of plant species, its age, health status and external biotic and abiotic influences (Liu et al., 2019). The quantity of exudates in the rhizosphere varies and are more concentrated at the root tips and lateral branching (Shukla et al., 2011). Root exudates can be grouped into four based on the way they are produced. There are passive exudates, secondary plant metabolites, lysates and mucilage (Martin et al., 2014; Gupta at al., 2020). Different plant species secret specific exudates and the primary constituents in the exudates dictate the rhizosphere community structure (Zhang et al., 2014; Mhlongo et al., 2018). Zhang et al. (2017) has observed the selective effect of rice-rice rotations on soil community over 31 years period of cultivation. Root exudates stimulate microbial community shift in contaminated soils through two different ways: alteration of microbial catabolic genes expression and specific selection of microbial strains (Siciliano et al., 2003; Gupta et al., 2020).

Rhizospheric microorganisms significantly relay on exudates as carbon and energy sources. Since most of the exudates are readily available sources of nutrients, microbial species become easily attracted through chemotaxis, leading to colonization and increased biomass (Hoang et al., 2021). Plant roots serve as attachment sites for microbes and provide oxygen for metabolic activities including contaminant degradation (Martin et al., 2014). As a result, beneficial rhizosphere microbiome may be selectively attracted towards roots thereby leading to increased metabolic activities (Correa-García et al., 2018). There is evidence that certain exudates specifically trigger enzymatic pathways for degradation of particular hydrocarbon compounds. They may also act as analogues to particular contaminants especially if they have related chemical structures (Singer et al., 2003).

Likewise, root exudates actively modulate the composition, diversity, and microbial activities in the rhizosphere. The availability of organic contaminants for microbial metabolism is equally enhanced by the exudates (Correa-García et al., 2018). Some of the root exudates (e.g. phenolics and flavonoids) act as inducers of genes for degradation pathways by rhizosphere microorganisms due to their resemblance with contaminants and as a result, catabolic genes for contaminants are boosted within the rhizosphere (Hoang et al., 2021). Studies by Shukla et al. (2011) revealed that degradation of PAHs and their derivatives in *Sorghum* sp. rhizosphere might be linked to enzymatic activity of oxidoreductases released from the roots as exudates.

However, root exudates offer special benefits to their host plant in addition to that of the microbes. Research findings have shown that the growth of competing plant species close to the host is inhibited through root exudation (Schandry and Becker, 2019). They also use exudates to attract beneficial microbes and regulate rhizosphere microbial community composition (Vieira et al., 2020). Flavonoids present in root exudates of legumes activate the *Rhizobium meliloti* genes coding for the nodulation process (Becard et al., 1995). The root cells are protected by defense proteins like phytoalexins and other unknown chemicals from pathogenic bacteria (Flores et al., 1999). In some cases, the plants and microbially produced compounds are further degraded to yield allelopathic or other toxic compounds, which are inhibitory to pathogenic microbes (Velmourougane et al., 2017).

CONCLUSION

In the last two decades, phytoremediation technology has evolved into a more promising cleanup technology towards achieving safer or cleaner environments. Among the prominent in its astounding qualities is being a solar-driven naturally occurring system which reflects its high public acceptance. There is minimal site destruction, low environmental impact and aesthetically pleasing. However, phytoremediation is slower than other cleanup technologies and not suitable where the target contaminants present an immediate danger to human and environment health. The technology is only applicable to low and moderately contaminated sites but not in heavily polluted sites; except where other conventional technologies could not meet an exhaustive cleanup. Process optimization including tilling, aeration, nutrient supplementation and microbial inoculation are essential in enhancing plant performance. In the years ahead, concerted efforts need to focus on developing more efficient plant and microbial synergy through transgenesis. Problems associated with hydrocarbon bioavailability and migrations need to be tackled through special supplement formulations. Proper understanding of rhizosphere metagenome and abiotic factors in the microbial milieu is essential towards achieving successful remediation; especially now that hydrocarbon exploration in more remote and fragile environments is ongoing; and portends a risk of more oil spills accidents in tumultuous environments.

REFERENCES

Abdel-Shafy, H. I., & Mansour, M. S. (2016). A review on polycyclic aromatic hydrocarbons: source, environmental impact, effect on human health and remediation. *Egyptian Journal of Petroleum, 25*(1), 107-123. . doi:10.1016/j.ejpe.2015.03.011

Agarwal, P., Giri, B. S. & Rani, R. (2020). Unraveling the role of rhizospheric plant-microbe synergy in phytoremediation: A genomic perspective. *Current Genentics*, 21.

Agbogidi, O. M., Dolor, E. D., & Okechukwu, E. M. (2007). Evaluation of *Tectona grandis* (Linn.) and *Gmelina arborea* (Roxb.) for phytoremediation in crude oil contaminated soils. *ACS. Agriculturae Conspectus Scientificus*, *72*(2), 149–152.

Agency for Toxic Substances and Disease Registry. (1999). *Toxicological profile for total petroleum hydrocarbons (TPH). Toxicol. profile Total Pet. Hydrocarb.* TPH.

Ahalya, N., & Ramachandra, T. Y. (2006). Phvtoremediation: Processes and mechanisms. *Journal of Ecobiology*, *18*(1), 33–38.

Aken, B. V., Correa, P. A., & Schnoor, J. L. (2010). Phytoremediation of polychlorinated biphenyls: New trends and promises. *Environmental Science & Technology*, *44*(8), 276–2776. doi:10.1021/es902514d PMID:20384372

Ali, W. A. (2019). Biodegradation and phytotoxicity of crude oil hydrocarbons in an agricultural soil. *Chilean Journal of Agricultural Research*, *79*(2), 266–277. doi:10.4067/S0718-58392019000200266

Allison, E., & Mandler, B. (2018). *Petroleum and the environment*. American Geoscience Institute.

Amadi, A., Dickson, A. A., & Maate, G. O. (1993). Remediation of oil polluted soils: Effects of organic and inorganic nutrient supplements on the performance of maize (*Zea mays* L.). *Water, Air, and Soil Pollution*, *66*(1-2), 59–76. doi:10.1007/BF00477060

Anyasi, R. O., & Atagana, H. I. (2017). Assessment of plants at petroleum contaminated site for phytoremediation. *Proceedings of the international conference of recent trends in environmental science and engineering (RTESE'17)*. 10.11159/rtese17.105

Anyasi, R. O. & Atagana, H. I. (2018). Profiling of plants at petroleum contaminated site for 820 phytoremediation. *Int. J. Phytoremediation, 20*, 352–361. . doi:10.1080/15226514.2017.1393386

Anyasi, R. O., Atagana, H. I., & Sutherland, R. (2019). Identification and characterization of PAH-degrading endophytes isolated from plants growing around a sludge dam. *International Journal of Phytoremediation*, *21*(7), 672–682. doi:10.1080/15226514.2018.1556585 PMID:30942084

Aprill, W., & Sims, R. C. (1990). Evaluation of the use of prairie grasses for stimulating polycyclic aromatic hydrocarbon treatment in soil. *Chemosphere*, *20*(1-2), 253–265. doi:10.1016/0045-6535(90)90100-8

Ashraf, M. A., Asif, M., Zaheer, A., Malik, A., Ali, Q., & Rasool, M. (2013). Plant growth promoting rhizobacteria and sustainable agriculture: A review. *African Journal of Microbiological Research*, *7*(9), 704–709. doi:10.5897/AJMR12.936

Asiabadi, F. I., Mirbagheri, S. A., Najafi, P., & Moatar, F. (2018). Concentrations of petroleum hydrocarbons at different depths of soil following phytoremediation. *Environmental Engineering and Management Journal*, *17*(9), 2129–2135. doi:10.30638/eemj.2018.211

Atlas, R. M. (1981). Microbial degradation of petroleum hydrocarbons: An environmental perspective. *Microbiological Reviews*, *45*(1), 180–209. doi:10.1128/MR.45.1.180-209.1981 PMID:7012571

Bais, H. P., Weir, T. L., Perry, L. G., Gilroy, S., & Vivanco, J. M. (2006). The role of root exudates in rhizosphere interactions with plants and other organisms. *Annual Review of Plant Biology*, *57*(1), 233–266. doi:10.1146/annurev.arplant.57.032905.105159 PMID:16669762

Baoune, H., Aparicio, J. D., Acuña, A., El Hadj-khelil, A. O., Sanchez, L., Polti, M. A., & Alvarez, A. (2019). Effectiveness of the *Zea mays-Streptomyces* association for the phytoremediation of petroleum hydrocarbons impacted soils. *Ecotoxicology and Environmental Safety*, *184*, 109591. doi:10.1016/j.ecoenv.2019.109591 PMID:31514081

Barac, T., Taghavi, S., Borremans, B., Provoost, A., Oeyen, L., Colpaert, J. V., Vangronsveld, J., & van der Lelie, D. (2004). Engineered endophytic bacteria improve phytoremediation of water-soluble, volatile, organic pollutants. *Nature Biotechnology*, *22*(5), 583–588. doi:10.1038/nbt960 PMID:15077119

Basumatary, B., & Bordoloi, S. (2016). Phytoremediation of crude oil-contaminated soil using *Cynodon dactylon* (L.) Pers. In A. Ansari, S. Gill, R. Gill, G. Lanza, & L. Newman (Eds.), *Phytoremediation*. Springer. doi:10.1007/978-3-319-41811-7_3

Basumatary, B., Saikia, R., Bordoloi, S., Das, H. C., & Sarma, H. P. (2012). Assessment of potential plant species for phytoremediation of hydrocarbon-contaminated areas of upper Assam, India. *Journal of Chemical Technology and Biotechnology (Oxford, Oxfordshire)*, *87*(9), 1329–1334. doi:10.1002/jctb.3773

Becard, G., Taylor, L. P., & Douds, D. D. (1995). Flavonoids are not necessary plant signal compounds in arbuscular mycorrhizal symbiosis. *Molecular Plant-Microbe Interactions*, *8*(2), 252–258. doi:10.1094/MPMI-8-0252

Beckers, B., Op De Beeck, M., Thijs, S., Truyens, S., Weyens, N., Boerjan, W., & Vangronsveld, J. (2016). Performance of 16s rDNA primer pairs in the study of rhizosphere and endosphere bacterial microbiomes in metabarcoding studies. *Frontiers in Microbiology*, *7*, 650. doi:10.3389/fmicb.2016.00650 PMID:27242686

Bertrand, F. N. (2020). Phytoremediation of petroleum hydrocarbon-contaminated soils with two plant species: *Jatropha curcas* and *Vetiveria zizanioides* at Ghana Manganese Company Ltd. *International Journal of Phytoremediation*, 1–10. doi:10.1080/15226514.2020.1803204 PMID:32805144

Bolton, H., Fredrickson, J. K., & Elliot, L. E. (1992). Microbial ecology of the rhizosphere. Marcel Dekker.

Cameselle, C., Chirakkara, R. A., & Reddy, K. R. (2013). Electrokinetic-enhanced phytoremediation of soils: Status and opportunities. *Chemosphere*, *93*(4), 626–636. doi:10.1016/j.chemosphere.2013.06.029 PMID:23835413

Canadian Council of Ministers of the Environment (CCME). (2008). *Canada-wide standard for petroleum hydrocarbons (PHC) in soil: Scientific rationale*. Supporting technical document. PN 1399, 1–383.

Chaîneau, C. H., Morel, J. L., & Oudot, J. (1997). Phytotoxicity and plant uptake of fuel oil hydrocarbons. *Journal of Environmental Quality*, *26*(6), 1478–1483. doi:10.2134/jeq1997.00472425002600060005x

Chakravarty, P., Bauddh, K., & Kumar, M. (2017). Phytoremediation: A multidimensional and ecologically viable practice for the cleanup of environmental contaminants. In *Phytoremediation potential of bioenergy plants* (pp. 1–46)., doi:10.1007/978-981-10-3084-0_1

Chaudhry, Q., Blom-Zandstra, M., Gupta, S., & Joner, E. (2005). Utilising the synergy between plants and rhizosphere microorganisms to enhance breakdown of organic pollutants in the environment. *Environmental Science and Pollution Research International*, *12*(1), 34–48. doi:10.1065/espr2004.08.213 PMID:15768739

Compant, S., Sessitsch, A., & Mathieu, F. (2012). The 125th anniversary of the first postulation of the soil origin of endophytic bacteria – a tribute to M.L.V. Galippe. *Plant and Soil*, *356*(1-2), 299–301. doi:10.100711104-012-1204-9

Cook, R. L., & Hesterberg, D. (2013). Comparison of trees and grasses for rhizoremediation of petroleum hydrocarbons. *International Journal of Phytoremediation*, *15*(9), 844–860. doi:10.1080/15226514.2012.760518 PMID:23819280

Correa-Garcia, S., Armand, P. S., St-Arnaud, M., & Yergeau, E. (2018). Rhizoremediation of petroleum hydrocarbons: A model system for plant microbiome manipulation. *Microbial Biotechnology*, *11*(5), 819–832. doi:10.1111/1751-7915.13303 PMID:30066464

Cunningham, S. D., Anderson, T. A., Schwab, P. A., & Hsu, F. C. (1996). Phytoremediation of soils contaminated with organic pollutants. *Advances in Agronomy*, *56*, 55–114. doi:10.1016/S0065-2113(08)60179-0

D'Souza, R., Varun, M., Lakhani, A., Singla, V., & Paul, M. S. (2015). PAH Contamination of urban soils and phytoremediation. In *Phytoremediation: Management of environmental contaminants* (Vol. 1). doi:10.1007/978-3-319-10395-2_15

Davin, M., Starren, A., Marit, E., Lefébure, K., Fauconnier, M.-L., & Colinet, G. (2019). Investigating the effect of *Medicago sativa* L. and *Trifolium pratense* L. root exudates on PAHs bioremediation in an aged-contaminated soil. *Water, Air, and Soil Pollution*, *230*(12), 296. doi:10.100711270-019-4341-4

de Souza, R. B., Maziviero, T. G., Christofoletti, C. A., & Fontanetti, C. S. (2013). Soil contamination with heavy metals and petroleum derivatives: Impact on edaphic fauna and remediation strategies. In M. C. H. Soriano (Ed.), *Soil processes and current trend in quality assessment*. doi:10.5772/52868

Diab, E. (2008). Phytoremediation of oil contaminated desert soil using the rhizosphere effects. *Global Journal of Environmental Research*, *2*(2), 66–73. doi:10.1016/j.envpol.2020.114787

Doucette, W. J., Shunthirasingham, C., Dettenmaier, E. M., Zaleski, R. T., Fantke, P., & Arnot, J. A. (2018). A review of measured bioaccumulation data on terrestrial plants for organic chemicals: Metrics, variability, and the need for standardized measurement protocols. *Environmental Toxicology and Chemistry*, *37*(1), 21–33. doi:10.1002/etc.3992 PMID:28976607

Farrell, R. E., & Germida, J. J. (2002). *Phytotechnologies: Plant-based Systems for remediation of oil impacted soils*. Available from https://www.esaa.org/wp-content/uploads/2015/06/02-09FarrellPaper

Fatima, K., Imran, A., Amin, I., Khan, Q. M., & Afzal, M. (2018). Successful phytoremediation of crude-oil contaminated soil at an oil exploration and production company by plants-bacterial synergism. *International Journal of Phytoremediation*, *20*(7), 675–681. doi:10.1080/15226514.2017.1413331 PMID:29723052

Fingas, M. F., Fieldhouse, B., Sigouin, L., Wang, Z., & Mullin, J. V. (2001). Dispersant effectiveness testing: laboratory studies of fresh and weathered oils. *Proceedings of the 24th Arctic and Marine Oil spill Program Technical Seminar*, 551-566.

Firáková, S., Šturdíková, M., & Múčková, M. (2007). Bioactive secondary metabolites produced by microorganisms associated with plants. *Biologia.*, *62*(3), 251–257. doi:10.247811756-007-0044-1

Flores, H. E., Vivanco, J. M., & Loyola-Vargas, V. M. (1999). Radicle biochemistry: The biology of root-specific metabolism. *Trends in Plant Science*, *4*(6), 220–226. doi:10.1016/S1360-1385(99)01411-9 PMID:10366878

Frick, C. M., Farrell, R. E., & Germida, J. J. (1999). *Assessment of phytoremediation as an in-situ technique for cleaning oil-contaminated sites. Petroleum Technology Alliance of Canada*. PTAC.

Gerhardt, K. E., Huang, X. D., Glick, B. R., Bruce, M., & Greenberg, B. M. (2009). Phytoremediation and rhizoremediation of organic soil contaminants: Potential and challenges. *Plant Science*, *176*(1), 20–30. doi:10.1016/j.plantsci.2008.09.014

Germaine, K. J., Byrne, J., Liu, X., Keohane, J., Culhane, J., Lally, R. D., Kiwanuka, S., Ryan, D., & Dowling, D. N. (2015). Ecopiling: A combined phytoremediation and passive biopiling system for remediating hydrocarbon impacted soils at field scale. *Front. Plant Sci.*, *5*, 756. doi:10.3389/fpls.2014.00756 PMID:25601875

Germaine, K. J., Keogh, E., Ryan, D., & Dowling, D. N. (2009). Bacterial endophyte mediated naphthalene phytoprotection and phytoremediation. *Fems. Microbiol.*, *296*(2), 226–234. doi:10.1111/j.1574-6968.2009.01637.x PMID:19459954

Germinda, J. J., Frick, C. M., & Farrell, R. E. (2002). Phytoremediation of oil-contaminated soils. *Developments. Soil Science*, *28B*, 169–186.

Ghosh, M., & Singh, S. P. (2005). A review on phytoremediation of heavy metals and utilization of It's by products. *Aust J Energy Environ*, *6*(4), 214–231.

Glick, B. R. (2010). Using soil bacteria to facilitate phytoremediation. *Biotechnology Advances*, *28*(3), 367–374. doi:10.1016/j.biotechadv.2010.02.001 PMID:20149857

Gunther, T., Dornberger, U., & Fritsche, W. (1996). Effects of ryegrasss on biodegradation of hydrocarbons in soil. *Chemosphere*, *33*(2), 203–215. doi:10.1016/0045-6535(96)00164-6 PMID:8696773

Gupta, A., Patel, A. K., Gupta, D., Singh, G., & Mishra, V. K. (2020). Rhizospheric remediation of organic pollutants from the soil; a green and sustainable technology for soil clean up. In *Abatement of Environmental Pollutants* (pp. 263–286). Elsevier. doi:10.1016/B978-0-12-818095-2.00013-8

Hall, J., Soole, K., & Bentham, R. (2011). Hydrocarbon phytoremediation in the family fabacea—A review. *International Journal of Phytoremediation*, *13*(4), 317–332. doi:10.1080/15226514.2010.495143 PMID:21598795

Hardoim, P. R., van Overbeek, L. S., & Elsas, J. D. (2008). Properties of bacterial endophytes and their proposed role in plant growth. *Trends in Microbiology*, *16*(10), 463–471. doi:10.1016/j.tim.2008.07.008 PMID:18789693

Harvey, P. J., Campanella, B. F., Castro, P. M. L., Harms, H., Lichtfouse, E., Schäffner, A. R., Smrcek, S., & Werck-Reichhart, D. (2002). Phytoremediation of polyaromatic hydrocarbons, anilines and phenols. *Environmental Science and Pollution Research International*, *9*(1), 29–47. doi:10.1007/BF02987315 PMID:11885416

Hoanga, S. A., Lamba, D., Seshadria, B., Sarkarb, B., Choppalaa, G., Kirkhamc, M. B., & Bolana, N. S. (2021). Rhizoremediation as a green technology for the remediation of petroleum hydrocarbon-contaminated soils. *Journal of Hazardous Materials*, *401*, 123282. doi:10.1016/j.jhazmat.2020.123282 PMID:32634659

Hrynkiewicz, K., Baum, C., Niedojadło, J., & Dahm, H. (2009). Promotion of mycorrhiza formation and growth of willows by the bacterial strain *Sphingomonas* sp. 23L on Fly Ash. *Biology and Fertility of Soils*, *45*(4), 385–394. doi:10.100700374-008-0346-7

Huang, X. D., El-Alawi, Y., Penrose, D. M., Glick, B. R., & Greenberg, B. M. (2004). A multi-process phytoremediation system for removal of polycyclic aromatic hydrocarbons from contaminated soils. *Environmental Pollution*, *130*(3), 465–476. doi:10.1016/j.envpol.2003.09.031 PMID:15182977

Hu¨ckelhoven, R. (2007). Transport and secretion in plant_microbe interactions. *Current Opinion in Plant Biology*, *10*(6), 573–579. doi:10.1016/j.pbi.2007.08.002 PMID:17875397

Hunt, L. J., Duca, D., Dan, T., & Knopper, L. D. (2018). Petroleum hydrocarbon (PHC) uptake in plants: A literature review. *Environmental Pollution*, *245*, 472–484. doi:10.1016/j.envpol.2018.11.012 PMID:30458377

Imam, J., Singh, P. K., & Shukla, P. (2016). Plant microbe interactions in post genomic era: Perspectives and applications. *Frontiers in Microbiology*, *7*, 1488. doi:10.3389/fmicb.2016.01488 PMID:27725809

International Tanker Owners Pollution Federation (ITOPF). (2019). *Oil Tanker Spill Statistics*. Available at http://www.itopf.org

Ismail, H. Y., Riskuwa-Shehu, M. L., Allamin, I. A., Farouq, A. A., & Abakwak, C. S. (2019). Biostimulation Potentials of Vigna Species (L.) in Hydrocarbon Impacted Soil. *American Journal of Bioscience and Bioengineering*, *7*(1), 22–27. doi:10.11648/j.bio.20190701.15

Ite, A. E., Ibok, U. J., Ite, M. U., & Petters, S. W. (2013). Petroleum exploration and production: Past and present environmental issues in the Nigeria's Niger Delta. *Nature*, *1*, 78–90.

Kaddafa, A. A. (2012). Oil exploration and spillage in the Niger Delta of Nigeria. *Civ. Environ. Res.*, *2*, 38–51.

Kaneez, F., Asma, I., Imran, A., Qaiser, M. K., & Muhammad, A. (2018). Successful phytoremediation of crude-oil contaminated soil at an oil exploration and production company by plants-bacterial synergism. *International Journal of Phytoremediation*, *20*(7), 675–681. doi:10.1080/15226514.2017.1413331 PMID:29723052

Kathi, S., & Khan, A. B. (2011). Phytoremediation approaches to PAH contaminated soil. *Indian Journal of Science and Technology*, *4*(1), 56–63. doi:10.17485/ijst/2011/v4i1.15

Komossa, D., Langebartels, C., & Sandermann, H. Jr. (1995). Metabolic processes for organic chemicals in plants. In *Plant contamination: Modeling and simulation of organic chemical processes* (pp. 60–103). CRC Press.

Košnář, Z., Mercl, F., & Tlustoš, P. (2020). Long-term willows phytoremediation treatment of soil contaminated by fly ash polycyclic aromatic hydrocarbons from straw combustion. *Environmental Pollution*, *264*, 114787. doi:10.1016/j.envpol.2020.114787 PMID:32559881

Kotoky, R., & Pandey, P. (2019). Rhizosphere mediated biodegradation of benzo (A) pyrene by surfactin producing soil bacilli applied through *Melia azadirachta* rhizosphere. *International Journal of Phytoremediation*, •••, 1–10. PMID:31522524

Kotoky, R., Rajkumari, J., & Pandey, P. (2018). The rhizosphere microbiome: Significance in rhizoremediation of polyaromatic hydrocarbon contaminated soil. *Journal of Environmental Management, 217*, 858–870. doi:10.1016/j.jenvman.2018.04.022 PMID:29660711

Kumar, A., Singh, R., Yadav, A., Giri, D. D., Singh, P. K., & Pandey, K. D. (2016). Isolation and characterization of bacterial endophytes of *Curcuma longa* L. *Biotech, 6*(1), 60. PMID:28330130

Kuzyakov, Y., & Razavi, B. S. (2019). Rhizosphere size and shape: Temporal dynamics and spatial stationarity. *Soil Biology & Biochemistry, 135*, 343–360. doi:10.1016/j.soilbio.2019.05.011

Kvesitadze, E., Sadunishvili, T., & Kvesitadze, G. (2009). Mechanisms of organic contaminants uptake and degradation in plants. *World Academy of Science, Engineering and Technology, 55*, 458–468.

Liu, C., Li, C., Zheng, F., Zhang, H. & Yu, H. (2019). Composition identification and allelopathic effect of root exudates of ginseng in different continuous cropping years. *Acta Microsc, 28.*

Lu, H., Wanga, W., Li, F., & Zhu, L. (2019). Mixed-surfactant-enhanced phytoremediation of PAHs in soil: Bioavailability of PAHs and responses of microbial community structure. *The Science of the Total Environment, 653*, 658–666. doi:10.1016/j.scitotenv.2018.10.385 PMID:30759591

Lu, M., Zhang, Z., Sun, S., Wei, X., Wang, Q., & Su, Y. (2010). The use of goosegrass (*Eleusine indica*) to remediate soil contaminated with petroleum. *Water, Air, and Soil Pollution, 209*(1-4), 181–189. doi:10.100711270-009-0190-x

Ma, X., Li, X., Liu, J., Cheng, Y., Zhai, F., Sun, Z., & Han, L. (2020). Enhancing Salix viminalis L.–mediated phytoremediation of polycyclic aromatic hydrocarbon–contaminated soil by inoculation with Crucibulum laeve (white-rot fungus). *Environmental Science and Pollution Research International, 27*(33), 41326–41341. doi:10.100711356-020-10125-3 PMID:32681334

Ma, Y., Rajkumar, M., Luo, Y., & Freitas, H. (2011). Inoculation of endophytic bacteria on host and non-host plants—Effects on plant growth and Ni uptake. *Journal of Hazardous Materials, 195*, 230–237. doi:10.1016/j.jhazmat.2011.08.034 PMID:21872991

Maceiras, R. (2016). Emerging technologies for soil remediation of hydrocarbons. *Pharm Anal Chem, 2*(01), 102. doi:10.4172/2471-2698.1000e102

Mall, I. D. (2007). *Petrochemical process technology*. Sanat Pronters.

Manoharachary, C., & Mukerji, K. G. (2006). Rhizosphere biology – an Overview. In K. G. Mukerji, C. Manoharachary, & J. Singh (Eds.), *Microbial activity in the rhizosphere soil biology*. Springer-Verlag Berlin Heidelberg. doi:10.1007/3-540-29420-1_1

Martin, B. C., George, S. J., Price, C. A., Ryan, M. H., & Tibbett, M. (2014). The role of root exuded low molecular weight organic anions in facilitating petroleum hydrocarbon degradation: Current knowledge and future directions. *The Science of the Total Environment, 472*, 642–653. doi:10.1016/j.scitotenv.2013.11.050 PMID:24317170

Masu, S., Cojocariu, L., Horablaga, N. M., Bordean, D. M., & Borozan, A. B. (2013). The Effects of Triticum aestivum species for the phytoremediation of petroleum-contaminated soil. International Multidisciplinary Scientific GeoConference, 1, 963-970.

McIntosh, P., Schulthess, C. P., Kuzovkina, Y. A., & Guillard, K. (2017). Bioremediation and phytoremediation of total petroleum hydrocarbons (TPH) under various conditions. *International Journal of Phytoremediation, 19*(8), 755–764. doi:10.1080/15226514.2017.1284753 PMID:28165761

Mhlongo, M. I., Piater, L. A., Madala, N. E., Labuschagne, N., & Dubery, I. A. (2018). The chemistry of plant–microbe interactions in the rhizosphere and the potential for metabolomics to reveal signaling related to defense priming and induced systemic resistance. *Front. Plant Sci.*, *9*, 112. doi:10.3389/fpls.2018.00112 PMID:29479360

Michel, J., & Fingas, M. (2016). Oil spills: Causes, consequences, prevention, and countermeasures. Fossil Fuels, 159-201.

Mitter, B., Brader, G., Afzal, M., Compant, S., Naveed, M., Trognitz, F., & Sessitsch, A. (2013). Advances in elucidating beneficial interactions between plants, soil and bacteria. In *Advances in Agronomy* (Vol. 121, pp. 381–445). Academic Press.

Mitter, E. K., Kataoka, R., Renato de Freitas, J., & Germida, J. J. (2019). Potential use of endophytic root bacteria and host plants to degrade hydrocarbons. *International Journal of Phytoremediation, 21*(9), 928–938. doi:10.1080/15226514.2019.1583637 PMID:30907105

Montalbán, B., Thijs, S., Lobo, M. C., Weyens, N., Ameloot, M., Vangronsveld, J., & Pérez-Sanz, A. (2017). Cultivar and metal-specific effects of endophytic bacteria in Helianthus tuberosus exposed to Cd and Zn. *International Journal of Molecular Sciences*, *18*(10), 20–26. doi:10.3390/ijms18102026 PMID:28934107

Naidoo, G., & Naidoo, K. (2016). Uptake of polycyclic aromatic hydrocarbons and their cellular effects inthe mangrove *Bruguiera gymnorrhiza. Marine Pollution Bulletin, 113*(1-2), 193–199. doi:10.1016/j.marpolbul.2016.09.012 PMID:27634737

Nataraja, K. N., Suryanarayanan, T. S., Shaanker, R. U., Senthil-Kumar, M., & Oelmuller, R. (2019). Plant–microbe interaction: Prospects for crop improvement and management. *Plant Physiology Reproduction*, *24*(4), 461–462. doi:10.100740502-019-00494-4

Nedunuri, K., Lowell, C., Meade, W., Vonderheide, A., & Shann, J. (2009). Management practices and phytoremediation by native grasses. *International Journal of Phytoremediation*, *12*(2), 200–214. doi:10.1080/15226510903213928 PMID:20734616

Nie, M., Yang, Q., Jiang, L. F., Fang, C. M., Chen, J. K., & Li, B. (2010). Do plants modulate biomass allocation in response to petroleum pollution? *Biology Letters*, *6*(6), 811–814. doi:10.1098/rsbl.2010.0261 PMID:20484231

Nwaaichi, E. O., Frac, M., Nwoha, P. A., & Eragbor, P. (2015). Enhanced phytoremediation of crude oil-polluted soil by four plant species: Effect of inorganic and organic bioaugumentation. *International Journal of Phytoremediation*, *17*(12), 1253–1261. doi:10.1080/15226514.2015.1058324 PMID:26090948

O'Banion, B. S., O'Neal, L., Alexander, G., & Lebeis, S. L. (2020). Bridging the gap between single-strain and community-level plant-microbe chemical interaction. *Molecular Plant-Microbe Interactions*, *3*(2), 124–134. doi:10.1094/MPMI-04-19-0115-CR PMID:31687914

Odjegba, V. J., & Sadiq, A. O. (2002). Effects of spent engine oil on growth parameters, chlorophyll and protein level of *Amaranthus hybrious*. *The Environmentalist*, *22*(1), 23–28. doi:10.1023/A:1014515924037

Ordinioha, B., & Brisibe, S. (2013). The human health implications of crude oil spills in the Niger Delta, Nigeria: An interpretation of published studies. *Nigerian Medical Journal*, *54*(1), 10–16. doi:10.4103/0300-1652.108887 PMID:23661893

Organization of the Petroleum Exporting Countries (OPEC). (2019). OPEC Share of World Crude Oil Reserves. *OPEC Annual Statistical Bulletin, 17*, 52. Available at: https://www.opec.org/opec_web/en/76.html

Oses, R., Valenzuela, S., Freer, J., Sanfuentes, E., & Rodriguez, J. (2008). Fungal endophytes in xylem of healthy chilean trees and their possible role in early wood decay. *Fungal Diversity*, *33*, 77–86.

Palmroth, M. R. T., Pichtel, J., & Puhakka, J. A. (2002). Phytoremediation of subarctic soil contaminated with diesel fuel. *Bioresource Technology*, *84*(3), 221–228. doi:10.1016/S0960-8524(02)00055-X PMID:12118697

Panchenko, L., Muratova, A., & Turkovskaya, O. (2017). Comparison of the phytoremediation potentials of Medicago falcata L. And Medicago sativa L. in aged oil-sludge-contaminated soil. *Environmental Science and Pollution Research International*, *24*(3), 3117–3130. doi:10.100711356-016-8025-y PMID:27858273

Pandey, V. C., & Bajpai, O. (2019). Phytoremediation: From theory toward practice. In Phytomanagement of polluted sites (pp. 1-49). Elsevier. doi:10.1016/B978-0-12-813912-7.00001-6

Pandey, V. C., Pandey, D. N., & Singh, N. (2015). Sustainable phytoremediation based on naturally colonizing and economically valuable plants. *Journal of Cleaner Production*, *86*, 37–39. doi:10.1016/j.jclepro.2014.08.030

Parrish, Z. D., Banks, M. K., & Schwab, A. P. (2005). Effect of root death and decay on dissipation of polycyclic aromatic hydrocarbons in the rhizosphere of yellow sweet clover and tall fescue. *Journal of Environmental Quality*, *34*(1), 207–216. doi:10.2134/jeq2005.0207 PMID:15647551

Patowary, R., Patowary, K., Devi, A., Kalita, M. C., & Deka, S. (2017). Uptake of total petroleum hydrocarbon (TPH) and polycyclic aromatic hydrocarbons (PAHs) by *Oryza sativa* L. grown in soil contaminated with crude oil. *Bulletin of Environmental Contamination and Toxicology*, *98*(1), 120–126. doi:10.100700128-016-1990-5 PMID:27896384

Paungfoo-Lonhienne, C., Rentsch, D., Robatzek, S., Webb, R. I., Sagulenko, E., Näsholm, T., Schmidt, S., & Lonhienne, T. G. A. (2010). Turning the table: Plants consume microbes as a source of nutrients. *PLoS One*, *5*(7), e11915. doi:10.1371/journal.pone.0011915 PMID:20689833

Peer, W. A., Baxter, I. R., Richards, E. L., Freeman, J. L., & Murphy, A. S. (2006). Phytoremediation and hyperaccumulator plants. In *Molecular Biology of Metal Homeostasis and Detoxification* (pp. 299–340). Springer.

Phillips, A. L. (2008). *The Relationship between plants and their root-associated microbial communities in hydrocarbon phytoremediation systems* (Ph.D. Thesis). College of Graduate Studies and Research University of Saskatchewan, Saskatoon, Canada.

Pilon-Smits, E. (2005). Phytoremediation. *Annual Review of Plant Biology*, *56*(1), 15–39. doi:10.1146/annurev.arplant.56.032604.144214 PMID:15862088

Pivetz, B. E. (2001). *Ground water issue: phytoremediation of contaminated soil and ground water at hazardous waste sites*. National Risk Management Research Lab ADA OK.

Płociniczak, T., Fic, E., Pacwa-Płociniczak, M., Pawlik, M., & Piotrowska-Seget, Z. (2017). Improvement of phytoremediation of an aged petroleum hydrocarbon-contaminated soil by *Rhodococcus erythropolis* CD 106 strain. *International Journal of Phytoremediation*, *19*(7), 614–620. doi:10.1080/15226514.2016.1278420 PMID:28103078

Prasad, M. N. V., & Freitas, H. M. O. (2003). Metal hyperaccumulation in plants biodiversity prospecting for phytoremediation technology. *Electronic Journal of Biotechnology*, *6*(3), 285–321. doi:10.2225/vol6-issue3-fulltext-6

Puente, M. E., Li, C. Y., & Bashan, Y. (2009). Endophytic bacteria in cacti seeds can improve the development of cactus seedlings. *Environmental and Experimental Botany*, *66*(3), 402–408. doi:10.1016/j.envexpbot.2009.04.007

Qiu, X., Leland, T. W., Shah, S. I., Sorensen, D. L., & Kendall, E. W. (1997). Field study: grass remediation for clay soil contaminated with polycyclic aromatic hydrocarbons. In *Phytoremediation of soil and water contaminants*. American Chemcial Society. 10.1021/bk-1997-0664.ch014

Radwan, S., Sorkhoh, N. A., & El-Nemr, I. (1995). Oil biodegradation around roots. *Nature*, *376*(6538), 302. doi:10.1038/376302a0 PMID:7630395

Radwan, S. S., Al-Awadhi, H., & El-Nemr, I. M. (2000). Cropping as a phytoremediation practice for oily desert soil with reference to crop safety as food. *International Journal of Phytoremediation*, *2*(4), 383–396. doi:10.1080/15226510008500046

Ram, N. M., David, H. B., Robert, F., & Maureen, L. (1993). A decision framework for selecting remediation technologies at hydrocarbon- contaminated sites. *Journal of Soil Contamination*, *2*(2), 1–24. doi:10.1080/15320389309383436

Rani, R., & Juwarkar, A. (2012). Biodegradation of phorate in soil and rhizosphere of *Brassica juncea* L. (Indian Mustard) by a microbial consortium. *International Biodeterioration & Biodegradation*, *71*, 36–42. doi:10.1016/j.ibiod.2012.04.004

Rasolomanana, J. L., & Balandreau, J. (1987). Role de la rhizosphere dans la biodegradation decomposes recalcitrants: Cas d'une riziere polluee par des residus petroliers. *Revue D'Ecologie et de Biologie du Sol.*, *24*(3), 443–457.

Riskuwa-Shehu, M. L., & Ismail, H. Y. (2018). Isolation of endophytic bacteria and phytoremediation of soil contaminated with polycyclic aromatic hydrocarbons using *Cajanus cajan* and *Lablab purpereus. Bioremediation Science and Technology Research,* *6*(1), 26-30. https://journal.hibiscuspublisher.com/index.php/BSTR/issue/view/42

Rodriguez, L., Lopez-Bellido, F. J., Carnicer, A., Recreo, F., Tallos, A., & Monteagudo, J. M. (2005). Mercury recovery from soils by phytoremediation. In *Book of environmental chemistry* (pp. 197–204). Springer. doi:10.1007/3-540-26531-7_18

Rodriguez-Campos, J., Perales-Garcia, A., Hernandez-Carballo, J., Martinez-Rabelo, F., Hernández-Castellanos, B., Barois, I., & Contreras-Ramos, S. M. (2018). Bioremediation of soil contaminated by hydrocarbons with the combination of three technologies: Bioaugmentation, phytoremediation, and vermiremediation. *Journal of Soils and Sediments*, *19*(4), 1981–1994. doi:10.100711368-018-2213-y

Rohrbacher, F., & St-Arnaud, M. (2016). Root exudation: The ecological driver of hydrocarbon rhizoremediation. *Agronomy (Basel)*, *6*(19), 1–27. doi:10.3390/agronomy6010019

Ruley, J. A., Tumuhairwe, J. B., Amoding, A., Opolot, E., Oryem-Origa, H., & Basamba, T. (2019). Assessment of plants for phytoremediation of hydrocarbon-contaminated soils in the Sudd Wetland of South Sudan. *Plant, Soil and Environment*, *65*(9), 463–469. doi:10.17221/322/2019-PSE

Sahu, J., Vaishnav, A., & Singh, H. B. (2020). Insights in plant-microbe interaction through genomics approach (Part 1). *Current Genetics*, *21*(3), 155–156. PMID:33071608

Sam, K., Coulon, F., & Prpich, G. (2017). Management of petroleum hydrocarbon contaminated sites in Nigeria: Current challenges and future direction. *Land Use Policy*, *64*, 133–144. doi:10.1016/j.landusepol.2017.01.051

Sandermann, H. (1994). Higher plant metabolism of xenobiotics: the "green liver" concept. *Pharmacogenetics, 4*, 225-241.

Schandry, N., & Becker, C. (2019). Allelopathic plants: Models for studying plant–interkingdom interactions. *Trends in Plant Science*, *6*, 234–256. PMID:31837955

Schnoor, J. L., Licht, L. A., McCutcheon, S. C., Wolfe, N. L., & Carreira, L. H. (1995). Phytoremediation of organic and nutrient contaminants. *Environmental Science & Technology*, *29*(7), 318–323. doi:10.1021/es00007a747 PMID:22667744

Schwitzgue'bel, J. P. (2017). Phytoremediation of soils contaminated by organic compounds: Hype, hope and facts. *Journal of Soils and Sediments*, *17*(5), 1492–1502. doi:10.100711368-015-1253-9

Sebastián, F. A., Méndez, V., Aguila, P., & Seeger, M. (2014). Bioremediation of petroleum hydrocarbons: Catabolic genes, microbial communities, and applications. *Applied Microbiology and Biotechnology*, *98*(11), 4781–4794. Advance online publication. doi:10.100700253-014-5684-9 PMID:24691868

Sessitsch, A., Hardoim, P., Döring, J., Weilharter, A., Krause, A., Woyke, T., & Reinhold-Hurek, B. (2012). Functional characteristics of an endophyte community colonizing rice roots as revealed by metagenomic analysis. *Molecular Plant-Microbe Interactions*, *25*(1), 28–36. doi:10.1094/MPMI-08-11-0204 PMID:21970692

Sessitsch, A., Reiter, B., Pfeifer, U., & Wilhelm, E. (2002). Cultivation-independent population analysis of bacterial endophytes in three potato varieties based on eubacterial and Actinomycetes-specific PCR of 16S rRNA genes. *FEMS Microbiology Ecology*, *39*(1), 23–32. doi:10.1111/j.1574-6941.2002.tb00903.x PMID:19709181

Shahzadi, A., Saddiqui, S., & Bano, A. (2016). The response of maize (*Zea mays* L.) plant assisted with bacterial consortium and fertilizer under oily sludge. *International Journal of Phytoremediation*, *18*(5), 521–526. doi:10.1080/15226514.2015.1115964 PMID:26587972

Sharma, J. K., & Juwarkar, A. A. (2015). Phytoremediation: General account and its application. In B. Bahadur (Ed.), *Plant Biology and Biotechnology: Plant Genomics and Biotechnology* (Vol. 2, pp. 469–497). Springer India. doi:10.1007/978-81-322-2283-5_34

Sharma, M., Sudheer, S., Usmani, Z., Rani, R. & Gupta, P. (2020). Deciphering the omics of plant- microbe interaction: perspectives and new insights. *Curr. Gen.*, 21.

Shukla, K. P., Sharma, S., Singh, N. K., Singh, V., Tiwari, K., & Singh, A. S. (2011). Nature and role of root exudates: Efficacy in bioremediation. *African Journal of Biotechnology*, *10*, 9717–9724.

Siciliano, S. D., & Germida, J. J. (1998). Mechanisms of phytoremediation: Biochemical and ecological interactions between plants and bacteria. *Environmental Reviews*, *6*(1), 65–79. doi:10.1139/a98-005

Siciliano, S. D., Germida, J. J., Banks, K., & Greer, C. W. (2003). Changes in microbial community composition and function during a polyaromatic hydrocarbon phytoremediation field trial. *Applied and Environmental Microbiology*, *69*(1), 483–489. doi:10.1128/AEM.69.1.483-489.2003 PMID:12514031

Singer, A. C., Crowley, D. E., & Thompson, I. P. (2003). Secondary plant metabolites in phytoremediation and biotransformation. *Trends in Biotechnology*, *21*(3), 123–130. doi:10.1016/S0167-7799(02)00041-0 PMID:12628369

Singh, P. P., Kujur, A., Yadav, A., Kumar, A., Singh, S. K., & Prakash, B. (2019). Mechanisms of Plant-Microbe Interactions and its Significance for Sustainable Agriculture. In PGPR Amelioration in Sustainable Agriculture. doi:10.1016/B978-0-12-815879-1.00002-1

Somtrakoona, K., & Chouychai, W. (2013). Phytotoxicity of single and combined polycyclic aromatic hydrocarbons toward economic crops. *Russian Journal of Plant Physiology: a Comprehensive Russian Journal on Modern Phytophysiology*, *60*(1), 139–148. doi:10.1134/S1021443712060155

Speight, J. G. (2006). *The Chemistry and Technology of Petroleum*. CRC Press. doi:10.1201/9781420008388

Streche, C., Cocârţă, D. M., Istrate, I., & Badea, A. A. (2018). Decontamination of petroleum-contaminated soils using the electrochemical technique: Remediation degree and energy consumption. *Scientific Reports*, *8*(1), 3272. doi:10.103841598-018-21606-4 PMID:29459642

Strobel, G., & Daisy, B. (2003). Bioprospecting for microbial endophytes and their natural products. *Microbiology and Molecular Biology Reviews*, *67*(4), 491–502. doi:10.1128/MMBR.67.4.491-502.2003 PMID:14665674

Stroud, J., Paton, G., & Semple, K. T. (2007). Microbe-aliphatic hydrocarbon interactions in soil: Implications for biodegradation and bioremediation. *Journal of Applied Microbiology*, *102*(5), 1239–1253. doi:10.1111/j.1365-2672.2007.03401.x PMID:17448159

Sturz, A. V., & Nowak, J. (2000). Endophytic communities of rhizobacteria and the strategies required to create yield enhancing associations with crops. *Applied Soil Ecology*, *15*(2), 183–190. doi:10.1016/S0929-1393(00)00094-9

Sudha, V., Govindaraj, R., Baskar, K., Al-Dhabi, N. A., & Duraipandiyan, V. (2016). Biological properties of endophytic fungi. *Brazilian Archives of Biology and Technology*, *59*(0), e16150436. doi:10.1590/1678-4324-2016150436

Sumiahadi, A., & Acar, R. (2018). A review of phytoremediation technology: Heavy metals uptake by plants. *IOP Conference Series. Earth and Environmental Science*, *142*, 12–23. doi:10.1088/1755-1315/142/1/012023

Sun, J., Wu, X., & Gan, J. (2015). Uptake and metabolism of phthalate esters by edible plants. *Environmental Science & Technology*, *49*(14), 8471–8478. doi:10.1021/acs.est.5b01233 PMID:26090545

Sun, M., Fu, D., Teng, Y., Shen, Y., Luo, Y., Li, Z., & Christie, P. (2011). In situ phytoremediation of PAH-contaminated soil by intercropping alfalfa (*Medicago sativa* L.) with tall fescue (*Festuca arundinacea* Schreb.) and associated soil microbial activity. *Journal of Soils and Sediments*, *11*(6), 980–989. doi:10.100711368-011-0382-z

Sutar, H., & Das, C. K. (2012). A review on: Bioremediation. *International Journal of Research in Chemistry and Environment*, *2*(1), 3–21.

Sylvia, A. (2019). Introduction of petroleum hydrocarbons contaminants and its human effects. *Journal of Environmental Science and Public Health, 3*, 1-9.

Taghavi, S., Barac, T., Greenberg, B., Borremans, B., Vangronsveld, J., & van der Lelie, D. (2005). Horizontal gene transfer to endogenous endophytic bacteria from poplar improves phytoremediation of toluene. *Applied and Environmental Microbiology*, *71*(12), 8500–8505. doi:10.1128/AEM.71.12.8500-8505.2005 PMID:16332840

Thijs, S., Sillen, W., Rineau, F., Weyens, N., & Vangronsveld, J. (2016). Towards an enhanced understanding of plant-microbiome interactions to improve phytoremediation: Engineering the metaorganism. *Frontiers in Microbiology*, *7*, 341. doi:10.3389/fmicb.2016.00341 PMID:27014254

Tsao, D. T. (2003). Overview of phytotechnologies. In T. Scheper & D. T. Tsao (Eds.), *Advances in Biochemical Engineering Biotechnology* (Vol. 78, pp. 1–50). Springer.

Ubogu, M., Odokuma, L. O., & Akponah, E. (2019). Enhanced rhizoremediation of crude oil–contaminated mangrove swamp soil using two wetland plants (Phragmites australis and Eichhornia crassipes). *Brazilian Journal of Microbiology*, *50*(3), 715–728. doi:10.100742770-019-00077-3 PMID:30993597

USEPA. (2011). *Petroleum refining*. OAR, Office of Air Quality Planning and Standards (OAQPS). Available on: https://www3.epa.gov/ttnchie1/ap42/ch05/final/c05s01.pdf

Van Epps, A. (2006). *Phytoremediation of Petroleum Hydrocarbons*. U.S. Environmental Protection Agency.

Vangronsveld, J., Weyens, N., Thijs, S., Dubin, D., Clemmens, M., Van Geert, K., van den Eeckhaut, M., van den Bossche, P., van Gestel, G., Bruneel, N., Crauwels, L. & Lemmens, C. (2019). Phytoremediation – Code of Good Practice. *Ovam*, 1-132.

Varma, P. K., Uppala, S., Pavuluri, K., Chandra, K. J., Chapala, M. M., & Kumar, K. V. K. (2017). Endophytes: Role and functions in crop health. In *Plant-Microbe Interactions in Agro-Ecological Perspectives*. doi:10.1007/978-981-10-5813-4_15

Vaziri, A., Panahpour, E., & Beni, M. H. M. (2013). Phytoremediation, a method for treatment of petroleum hydrocarbon contaminated soils. *International Journal of Farm and Allied Sciences*, *2*(21), 909–913.

Velmourougane, K., Saxena, G., & Prasanna, R. (2017). Plant-microbe interactions in the rhizosphere: Mechanisms and their ecological benefits. In *Plant-microbe interactions in agro-ecological perspectives*. doi:10.1007/978-981-10-6593-4_7

Verma, S. K., & Gond, S. K. (2017). Fungal endophytes representing diverse habitats and their role in plant protection. *Developments in Fungal Biology and Applied Mycology*, 135-157.

Vieira, S., Sikorski, J., Dietz, S., Herz, K., Schrumpf, M., Bruelheide, H., Scheel, D., Friedrich, M. W., & Overmann, J. (2020). Drivers of the composition of active rhizosphere bacterial communities in temperate grasslands. *The ISME Journal*, *14*(2), 463–475. doi:10.103841396-019-0543-4 PMID:31659233

Vitor, S. L., Eliana, F. C. S., & Fernando, J. S. O. (2018). Biosurfactant- assisted phytoremediation of multi-contaminated industrial soil using sunflower (*Helianthus annuus* L.). *Journal of Environmental Science and Health. Part A, Toxic/Hazardous Substances & Environmental Engineering*, *53*(7), 609–616. doi:10.1080/10934529.2018.1429726 PMID:29388890

Walton, B. T., Hoylman, A. M., Perez, M. M., Anderson, T. A., Johnson, T. R., Guthrie, E. A., & Christman, R. F. (1994). Rhizosphere microbial communitites as a plant defense against toxic substances in soils. In T. A. Anderson & J. R. Coats (Eds.), *Bioremediation Through Rhizosphere Technology* (pp. 82–92). American Chemical Society. doi:10.1021/bk-1994-0563.ch007

Wang, S., Xu, Y., Lin, Z., Zhang, J., Norbu, N., & Liu, W. (2017). The harm of petroleum-polluted soil and its remediation research. *AIP Conference Proceedings*, *1864*, 020222. doi:10.1063/1.4993039

Watson, J. G. (1996). Physical/chemical treatment of organically contaminated soils and sediments. *Journal of the Air & Waste Management Association*, *46*(10), 993–1003. doi:10.1080/10473289.1996.10467536 PMID:28065141

Weisman, W. (1998). Total Petroleum Hydrocarbon Criteria Working Group Series: Vol. 1. *Analysis of petroleum hydrocarbons in environmental media*. Amherst Scientific Publishers.

Wenzel, N., van der Lelie, D., Taghavi, S., & Vangronsveld, J. (2009). Phytoremediation: Plant-endophyte partnerships take the challenge. *Current Opinion in Biotechnology*, *20*(2), 248–254. doi:10.1016/j.copbio.2009.02.012 PMID:19327979

Weyens, N., Croes, S., Dupae, J., Newman, L., van der Lelie, D., Carleer, R., & Vangronsveld, J. (2010). Endophytic bacteria improve phytoremediation of Ni and TCE co- contamination. *Environmental Pollution*, *158*(7), 2422–2427. doi:10.1016/j.envpol.2010.04.004 PMID:20462680

Weyens, N., van der Lelie, D., Taghavi, S., & Vangronsveld, J. (2009). Phytoremediation: Plant-endophyte partnerships take the challenge. *Current Opinion in Biotechnology*, *20*(2), 248–254. doi:10.1016/j.copbio.2009.02.012 PMID:19327979

Wild, E., Dent, J., Thomas, G. O., & Jones, K. C. (2005). Direct observation of organic contaminant uptake, storage, and metabolism within plant roots. *Environmental Science & Technology*, *39*(10), 3695–3702. doi:10.1021/es048136a PMID:15952374

Wu, K., Dumat, C., Li, H., Xia, H., Li, Z., & Wu, J. (2019). Responses of soil microbial community and enzymes during plant-assisted biodegradation of di-(2-ethylhexyl) phthalate and pyrene. *International Journal of Phytoremediation*, *21*(7), 683–692. doi:10.1080/15226514.2018.1556586 PMID:30924369

Xuezhi, D., Anum, A. A., Ishaq, M., Tariq, S., & Qudratullah, K. (2020). Remediation methods of crude oil contaminated soil. *World J Agri & Soil Sci.*, *4*(3), 34–46. doi:10.33552/WJASS.2020.04.000595

Yan, L. (2012). *The use of plants, including trees, to remediate oil contaminated soils: a review and empirical study* (MSc thesis). Department of Forestry, University of Helsinki. Available on: https://core.ac.uk/download/pdf/14926191.pdf

Yaqoob, A., Nasim, F. H., Sumreen, A., & Munawar, N. (2019). Current scenario of phytoremediation: Progresses and limitations. *International Journal of Biosciences*, *14*(3), 191–206. doi:10.12692/ijb/14.3.191-206

Yateem, A., Balba, M. T., El-Nawawy, A. S. N., & AlAwadhi, N. (2000). Plants-associated microflora and the remediation of oil contaminated soil. *International Journal of Phytoremediation*, *2*(3), 183–191. doi:10.1080/15226510009359031

Ying, X., Yizhi, S., David, M. J., Meng, L., Huigang, L., & Yingping, H. (2018). Se enhanced phytoremediation of diesel in soil by *Trifolium repens. Ecotoxicology and Environmental Safety, 154*, 137-144. doi:10.1016/j.ecoenv.2018.01.061

Zand, A. D., Bidhendi, G. N., & Mehrdadi, M. (2009). Phytoremediation of total petroleum hydrocarbons (TPHs) using plant species in Iran. *Turkish Journal of Agriculture and Forestry*, *34*, 429–438. doi:10.3906/tar-0903-2

Zhang, N., Wang, D., Liu, Y., Li, S., Shen, Q., & Zhang, R. (2014). Effects of different plant root exudates and their organic acid components on chemotaxis, biofilm formation and colonization by beneficial rhizosphere-associated bacterial strains. *Plant and Soil*, *374*(1-2), 689–700. doi:10.100711104-013-1915-6

Zhang, X., Zhang, R., Gao, J., Wang, X., Fan, F., Ma, X., Yin, H., Zhang, C., Feng, K., & Deng, Y. (2017). Thirtyone years of rice-rice-green manure rotations shape the rhizosphere microbial community and enrich beneficial bacteria. *Soil Biology & Biochemistry*, *104*, 208–217. doi:10.1016/j.soilbio.2016.10.023

KEY TERMS AND DEFINITIONS

Biodegradation: The conversion of complex toxic molecules into smaller less harmful ones using biological agents like bacteria, fungi, plants.

Contaminant: Is any potentially undesirable substance (physical, chemical, or biological) usually in high concentration that has potential danger to plants, animals, and environment.

Endophytes: Are microorganisms (bacteria and fungi) that predominantly live in plant tissues as symbiont or commensals.

Endosphere: Is an internal region in a plant inhabited by microorganisms.

Exudates: Are secretions from plant roots containing a range of organic compounds.

PAHs: Is a group of compounds comprised of two or more condensed aromatic hydrocarbon rings.

Rhizosphere: Is a thin region of a medium (soil or water) around and under the influence of plant roots.

TPH: Is a parameter for quantifying environmental contamination originating from various hydrocarbon products.

Chapter 16
Microbial Bioremediation of Heavy Metals: A Genetic and Omics Approach

Asha Laxman Giriyan
The Energy and Resources Institute (TERI), India

Vikrant B. Berde
Arts, Commerce, and Science College, Lanja, India

Elroy J. Pereira
The Energy and Resources Institute (TERI), India

Chanda Vikrant Parulekar-Berde
Goa University, Goa, India

ABSTRACT

Heavy metals are found naturally. Anthropogenic activities and rapid industrialization have led to their unprecedented release into the environment. Being non-biodegradable in nature, they persist in the environment. Prolonged exposure and accumulation of these metals poses a serious threat to the ecosystem. Conventional treatment of contaminated material whether soil or water involves expensive chemical or physical methods which are arduous, energy demanding, and carry the risk of secondary contamination. It is thus necessary to adopt a sustainable remediation process to mitigate this problem. Biological remediation processes are preferable as they are environmentally safe, techno-economically feasible, and do not generate toxic byproducts. Microbial bioremediation is particularly attractive as it allows remediation processes by tapping naturally occurring catabolic capacities to transform, accumulate, and adsorb metals for detoxification. It is a comparatively low-cost technology. Therefore, microbial bioremediation is promising as an alternative to physico-chemical methods.

DOI: 10.4018/978-1-7998-7062-3.ch016

INTRODUCTION

Heavy metals comprise of very heterogenous elements that differ widely in their chemical properties as well as biological functions. Heavy metals are naturally occurring elements that comprises of metals and metalloids with high density of more the 5 gr cm^{-3}, high atomic weights in the range of 63.5–200.6 g mol^{-1} and toxicity at very low concentrations (Srivastava and Majumdar, 2008). Some of the examples are Pb, As, Hg, Cd, Zn, Ag, Cu, Fe, Cr, Ni, Pd, Tl and Pt. These are further grouped as essential such as copper, iron, nickel, zinc and non-essential such as cadmium, mercury.

Presence of heavy metals such as lead, mercury, cadmium, can be a threat to human health and environment at very low concentrations (Bagal-Kestwal et al., 2008). Though some essential heavy metals are required in trace amounts as cofactors of enzymes in metabolic pathways to carry out the biochemical and physiological functions (WHO/FAO/IAEA, 1996). Examples of these metals are copper, selenium, zinc, manganese, iron etc. However, even these heavy metals in higher concentration (Chang et al., 1996; ATSDR, 2002; Tchounwou et al., 2008) or in different transitions state (ATSDR, 2002; Harvey and Mc Ardle, 2008; Tchounwou et al., 2008; Stern, 2010), are toxic. Similar damaging consequences are observed in other essential elements. Cadmium at concentrations as low as 0.003 mgL^{-1} or less may be allowed while for mercury, the allowed values being 0.006 mgL^{-1} (Ali and Khan, 2018). The concentration of the heavy metal determines the beneficial or toxic effect (Chang et al., 1996; Tchounwou et al., 2008). Remediation of toxic heavy metals from the environment is thus mandatory and a challenging job protecting the environment from hazardous effects of heavy metals. There are a variety of strategies for the removal or reduction of heavy metals and these are classified as biological, chemical and physical approaches (Lim et al., 2014).

HEAVY METAL TOXICITY

Heavy metals are natural components; however they cannot be biodegraded, tend to persist in nature. Because of this property, the heavy metals are recognized as a major environmental concern. Entry of heavy metals in the environment is either via natural sources and/ or anthropogenic sources that include industrial discharge, automobiles exhaust, and mining (Fergusson, 1990; Bradl, 2002; He et al., 2005). Some human activities such as agriculture practices also contribute to increased heavy metal concentrations, ultimately threatening ecological and human health. Various activities contribute to heavy metal poisoning such as contaminated air, industrial exposure, medicines, etc. Heavy metals in environment are also due to emissions occurring naturally as in the case of volcanic eruptions, forest fires, sprays of sea salt, weathering of rocks, soil erosion, etc. Hence under certain environmental conditions natural weathering will lead to release of metals in the biosphere in bound forms as oxides, sulphates, phosphates, others. Heavy metals along with other salts and minerals comprises the inorganic pollutants (Wong, 2012) and are mostly added due to human indulgence (Shallari et al., 1998; Herawati et al., 2000; Goyer, 2001; He et al., 2005), becoming toxic due to bioaccumulation in the food chains (Salomons et al., 1995). All spheres of environment are affected by toxic nature of the heavy metals, resulting in disturbances in the food chains as well as severe health problems.

Environmental pollution due to industrial, agricultural and household emissions as well as natural emissions including mine tailings, use of paints, fertilisers, pesticides, irrigation using waste water, coal combustion, petrochemical usage and spillage, soil erosion, heavy metal leaching, weathering of soil or rocks, volcanic eruptions, etc, are sources of heavy metal pollution (Nriagu, 1989; Fergusson, 1990; Shallari et al., 1998; Bradl, 2002; He et al., 2005; Zhuang et al., 2013; Ventura et al., 2017; Afzal et al., 2018; Soleimani et al., 2018).

Agricultural runoff containing excess of chemical fertilisers and pesticides as well as runoff from mining areas, add significant quantities of heavy metals to the runoff water that enters the water bodies. The heavy metals may be in trace amounts, but may be toxic to the plants and life in the water bodies. The metals sink and get concentrated in the sediment of the water bodies (Musilova et al., 2016).

The metals enter the food chains and food webs eventually getting bio-accumulated and becoming more toxic, leading to severe health issues (Lee et al., 2002). Presence of heavy metals in the water bodies such as estuaries, which support aquatic life are highly affected, and have consequences on life cycles, breeding, spawning, accumulation in tissues, etc. significant histopathological alterations in tissues have been reported by number of workers (Akif et al., 2002; Ahmed et al., 2014; Rezania et al., 2016). Pollution of water bodies with heavy metals is a worldwide problem especially due to the toxicity of heavy metals and their tendency to persist in environment, bioaccumulate and bio-magnify in the food chains (Rajaei et al., 2012).

Heavy metals being non-biodegradable tend to persist in the environment (Merian, 1984; Atkins and Jones, 1997). Heavy metals and metalloids enter soil from anthropogenic sources as well as lithogenic sources (Alloway, 2013). In urban regions, contamination of soil with heavy metals is mostly due to emissions from traffic on roads while excessive use of fertilisers, pesticides and other chemicals for agricultural practices, are source of contamination in soils in rural areas (Semu and Singh, 1996). Soil is the major sink for the heavy metals as even the heavy metals in the water source tends to sink and reach the sediment. Humans, animals, plants and ecosystems are at a risk of exposure and the different ways apart from absorption by plants and food chains; also involve direct ingestion, consumption of heavy metal contaminated water (Musilova et al., 2016).

MICROBIAL REMEDIATION OF HEAVY METALS

Bioremediation of toxic metals is a convenient method, as it is inexpensive, environment friendly (Kang et al., 2016) and can be coupled with other physicochemical treatment technologies. Moreover, it is a non-invasive technique that does not have a large footprint and leaves the ecosystem intact (Vidali, 2001). Microorganisms tolerate the toxicity of metals and survive in harsh environmental conditions by adopting various ingenious resistance and detoxification mechanisms, which include biotransformation by oxidation/methylation, extrusion by metal efflux pumps, use of enzymes, intracellular and extracellular metal sequestration, production of exopolysaccharide (EPS) and synthesis of chelators like metallothioneins and bio surfactants (Dixit et al., 2015). Microorganisms can also nullify/reduce the toxic effects of metal by chemical modifications such as valence conversion or volatilization. Further, the anionic nature of their cell surface due to presence of hydroxyl, phosphoryl, amine, carboxyl, ester, sulfydryl, thioether, and thiol groups enables them to bind metal cation (Ramasamy et al., 2006). They also possess metabolic pathways which use the toxic pollutants as an energy source for their growth through respiration, fermentation and co-metabolism.

Heavy metals like Cr, As, Hg and Fe undergo oxidation and reduction cycles. Microbial reduction enhances the solubility of ions like Fe (III) and As (V) by reducing them to Fe(II) and As(III), respectively. The process facilitates leaching of metals from soil (Yin et al., 2019). Microorganisms play major role in the transformation of metals in nature. They possess enzymes that can degrade or detoxify the heavy metals in order to adapt themselves to toxic metals in the ecosystem and thus gain resistance. They exhibit the mechanisms of bioaccumulation, biomineralization, biosorption, and biotransformation. Microbial metabolites such as carboxylic acids and amino acids bring about metal ion chelation. These mechanisms are exploited for *in situ* or *ex situ* remediation purposes.

REMEDIATION MECHANISMS

Microorganisms immobilize metals acting as sinks for metals by implementing different mechanisms of bioremediation.

Biosorption

Microbes accumulate heavy metals by two main mechanism of either adsorption or absorption. The process of adsorption unlike absorption occurs when the heavy metal is in the liquid state, whereas adsorption is a surface phenomenon involving cell surface complexation of heavy metals which can then be absorbed into the cell (Lloyd, 2002). Biosorption mechanisms can be (i) metabolism dependent or (ii) non-metabolic dependent; and based on the site of biosorption can be either: (i) extra cellular precipitation/ accumulation, (ii) cell surface sorption or (iii) intracellular accumulation.

In non -metabolism dependent mechanism, the structural makeup of the microbial cell surface can trap heavy metal ions (Malik, 2004) where the heavy metal uptake is as a result of physicochemical interaction between the functional groups present on the cellular surface and the metal. The amount ions trapped contingent on the kinetic equilibrium and concentration of metal on the cell surface and involves numerous processes including ion exchange, electrostatic interaction, redox potential surface complexation and precipitation. Biosorption involves passive metal uptake and can thus be carried out by either dead or living biomass of the microorganism (Fomina, 2014).

The metabolism dependent mechanism is generally linked with the microbial defense system wherein it is stimulates in the presence of toxic metal (Ahalya et al., 2003).The heavy metals are transported across the cell membranes through interaction with ions present on the surface of cell such as magnesium, sodium, potassium, phosphorus which can be complexed with metal ions. In addition, negatively charged anions like phosphoric acid and carboxyl anionic groups on the cell surface interact with heavy metal surfaces which mostly carry a cationic group allowing the metal ions to bind or transport through the cell membrane. As a result, microbes adsorb heavy metal ions rapidly and reduce the concentration in the environment. Biosorption by microorganism is inexpensive and can accomplish quick removal of multiple heavy metals such as Zn, Cu, Pb, Cd, Cr and As due to the variety of cellular assemblies such as peptidoglycans like N-acetylmuramic acid and poly-N acetylglucosamine (Rasmussen, 2007).

Bioaccumulation

Bioaccumulation is another method in which heavy metals are taken up either through cellular metabolic activity or by active uptake. It is a metabolically-active process where heavy metals enter the intracellular space using importer complexes through the lipid bilayer. Once inside, heavy metal ions can be sequestered by peptide ligands systems and proteins (Malik, 2004; Mishra and Malik, 2013). Immobilization of metal ions can also be performed by microbial biofilms, primarily due to the binding capacity of the exopolysaccharides (EPS) (Zhang et al., 2010).

Heavy metals attach on EPS present on the cell wall by interactions with proton or micro-precipitation of metals (Comte et al., 2008). Presence of amino, carboxyl, phosphoryl, and sulfo groups create negative charge to the cell surface of the biomass making them potential ion exchange sites and metal sinks.EPS can efficiently bind various heavy metals and influence the distribution of heavy metals in microorganisms. Therefore, EPS can resist microorganisms against heavy metal toxicity and moreover be used as an important bioremediation tool.

EPS is secreted by the encased microbes and can be produced by bacteria, cyanobacteria, microalgae (Boonchai et al., 2014; Parikh and Madamwar, 2006), yeasts (Pavlova and Grigorova, 1999), fungi (Elisashvili et al., 2009), and protists (Lee Chang et al., 2014). Chemically, EPS comprise biosynthetic polymers such as polysaccharides, enzymes, structural proteins, nucleic acids, lipids, and other compounds such as humic acids (Flemming and Wingender, 2010). EPS has many roles and in addition to being involved in the formation of biofilm, it has functions in cellular growth and incorporates water channels allowing nutrient and oxygen transport (Shukla et al., 2014; Singh et al., 2006). The net anionic allows the biopolymer to efficiently sequester positively-charged heavy metal ions (Shukla et al., 2014, 2017). Several researchers have reported the biosorption of heavy metals such as Cd(II), Cr(VI), Pb(II), Hg(II), Ni(II), Zn(II), Cu(II), and Mn(II) using bacterial biomass.

Bioleaching

Bioleaching is a bioprocess involving the utilization of heavy metals from insoluble ores though microbial mediation wherein microbes generate low molecular organic acids which dissolve heavy metals and promote its leaching into the soil (Chanmugathas et al., 2016). Microorganisms actively contribute in redox reactions and change in heavy metal valency, thereby altering heavy metal activity, and their subsequent mobility and/or toxicity.

BIOREMEDIATION OF HEAVY METALS BY ALGAE

Among the brown, green and red algal groups, brown algae have demonstrated adequate capacity for biosorption (phycoremediation). Metal ion biosorption depends on type and structure of the algal biomass, and the valency of the heavy metal ion. Presence of hydroxyl, sulphate, phosphate groups etc., in algal proteins act as potential sites for binding metal ions forming complexes during its remediation (Abbas et al., 2014; Romera et al., 2007). Cations present in the cell wall get replaced by heavy metal ions via ion exchange. Microalgae also consume other heavy metals such as boron, cobalt, iron, copper and manganese as trace elements for enzyme functioning and cellular metabolism (Sun et al., 2015). Cyanobacteria species such as *Phormidium, Oscillatoria* and *Spirogyra* showing tolerance towards heavy metals have with active reactive binding sites which form complexes with heavy metals, leading to flocculation and reduction in their environmental concentration (Balaji et al., 2016).

Algae like microbes also synthesize antioxidant enzymes such as ascorbate peroxidase, catalase, superoxide dismutase and peroxidase and non- enzymatic antioxidants (such as carotenoids, proline and glutathione) (Upadhay et al., 2016) both of which are involved in the reduction of free radicals and reactive oxygen species generated by intracellular interactions with heavy metal ions. Remediation of heavy metals by micro algae is achieved by two main mechanisms: (i) extracellular adsorption or biosorption and (ii) intracellular diffusion and bioaccumulation of heavy metal ions. In addition to polymeric and exopolymeric substances, the cell wall also comprises of cellulose and alginate, lipids and organic protiens capable of binding heavy metals. Adsorption of heavy metals on the surface of microalgae also involves ionic exchange and the formation of covalent bonds between cations (such as uronic acid in exopolysaccaride) present in the cell wall and heavy metal ions.

BIOREMEDIATION OF HEAVY METALS BY FUNGI

Fungi cell wall is a complex and flexible structure composed of chitin, polysaccharide, polyphosphates, peptidoglycan and proteins rich in metal-binding ligands which help them to eliminate metal toxicity. Among the metal binding groups, amines are more active in metal uptake as they bind to both anionic and cationic metal species by electrostatic interaction and surface complexation respectively. Extracellular and intracellular precipitation, valence conversion and energetic uptake mechanisms are employed by fungi to minimize metal toxicity.

Fungi, due to their high capacity for metal uptake and recovery are extensively studied for biosorbtion for toxic metal with both active and dead fungal biomass which are known to play a significant role in removal of inorganic chemicals. Srivastava and Thakur (2006) reported efficient chromium removal up to 65% from synthetic medium by *Aspergillus sp*. Similarly, *Coprinopsis atramentaria* was also found to be an as effective accumulator of heavy metal ions effectively bioccumulating cadmium and lead. The process of bioaccumulation is dictated by the initial metal ion concentration as well as the pH of the medium. Yeast strains (such as *Saccharomyces cerevisiae, Rhodotorula pilimanae* and *Pichia guilliermondii*) have also been employed for bioconversion of toxic Cr^{6+} to less harmful Cr^{3+}. For similar biotransformation, researchers have also reportedly used dead fungal biomass of *Aspergillus niger, Rhizopus oryzae, S. cerevisiae,* and *Penicillium chrysogenum.*

BIOREMEDIATION OF HEAVY METALS BY BACTERIA

Bacteria are ubiquitous and the most abundant microorganism thriving in a wide range of environmental conditions. Bacteria have been extensively used to bioremediate heavy metal pollutants due to several advantages such as microscopic size, ease of cultivation and quick growth rate with several metal remediation methods established based on bacteria such as *Escherichia, Pseudomonas, Bacillus* and *Micrococcus*. Among them, *Escherichia coli* K-12 was found to absorb almost 30 kinds of metal ions-the widest variety of metal ions from the soil (Dasola, 2014). Bacterial cells generally adsorb heavy metals which get accumulated on the cell surface through the polysaccharide biofilm which comprise functional groups Adsorption capacities of heavy metals by bacteria from 1 mg/g to 500 mg/g. *Pseudomonas aeruginosa*, a mercury-resistant strain, has shown a maximum mercury uptake capacity of nearly 180 mg/g (Yin.et al. 2018).

The biomass of *Arthrobacter viscosus* can adsorb Cr^{6+} with high adsorption capacity. Both living and dead cells can reduce Cr^{6+} to Cr^{3+}in aqueous solution. When the initial concentrations of Cr^{6+} is lower than 100 mg/L, all the Cr^{6+} can be removed under acidic pH ranging 1 - 2. Hlihor et al. (2017) reported the removal of Cr^{6+} through biofilms of *Staphylococcus epidermidis*. Dave et al. (2010) reported the

Eichhornia spp. biomass can achieve 85% copper removal from 100 ppm Cu^{2+} containing solution. The Zn^{2+} can be absorbed by *Rhodobacter capsulatus* with a maximal uptake capacity of 164 mg/g, Bacterial biomass of *Bacillus cereus* RC-1 have the biosorption capacity for Cd^{2+}, with about 24.01 mg/g and 31.95 mg/g for living and dead cells, respectively. *Bacillus cereus* has been reported to remediate Cd^{2+} and Arsenic, similarly *Ochrobactrum* sp. for Cd^{2+} and *Bacillus arsenicus* for arsenic.

Microbes also produce iron-chelating substances called siderophores, which enhance mobility and reduce bioavailability. Sulphate-reducing bacteria such as *Desulfovibriode sulfuricans* have the ability to convert sulphate to hydrogen sulphate which then reacts with heavy metals such as Cd and Zn to insoluble forms of these metal sulphides. Bacteria can remediate heavy metals through functional groups, such as carboxyl, aldehydes and ketones present in their cell walls.

FACTORS INFLUENCING BIOREMEDIATION

The growth and performance of bioremediation by microbial cells are influenced by various biotic and abiotic factors and are subjected multiphasic heterogeneous surroundings influencing the bioremediation process. Insufficient information about the influencing factors often reduces the efficiency of the bioremediation process. Three major factors that need to be considered are;

1. Physico-chemical features or abiotic factors of the environment,
2. Biotic or biological factors, and
3. Physico-chemical and climatic conditions are the major factors influencing the metabolic rates in microorganism.

The extent of metal toxicity and bioremediation also depends on the concentration of metal ions, its chemical form and factors such as redox potential which depend on environmental factors like pH, temperature, and presence of low organic acids and humic acids. These factors alter the ionic state of heavy metals thereby affecting transformation, transportation, and its bioavailability. For example, at acidic pH the adsorbent surface is more positively charged thus reducing the attraction between adsorbent and metal cations leading to increased toxicity. Similarly, higher temperature increases the rate of adsorption across the cell wall and leads to higher bioavailability of the heavy metals due to increased solubility. However, rise in temperature also enhances enzyme activity and microbial metabolism resulting in accelerated bioremediation. Hence efficiency of bioremediation depends both on environmental factors as well as intrinsic properties of microorganisms.

GENETIC BASIS OF HEAVY METAL RESISTANCE

Microorganisms have developed three methods of resistance to survive heavy metal pollution: (i) efflux of the toxic metal extracellularly by membrane transporters, (ii) transformation into inert or non-toxic forms and (iii) biosorption. Although metal efflux forms the first line of defense, they are often coupled as is in the case of enzymatic conversion of absorbed metal which then precipitates as salt (Ranaweera et al., 2015; Das et al., 2016) (Figure 1). Several metal-resistant genes towards metals such as arsenic, cadmium, chromium, copper, mercury, nickel and lead have been expressed (Table 1).

*Figure 1. Simplified illustration of the genetic mechanisms involved in bacterial resistance to toxic metals. Mechanisms of resistance include efflux systems, biotransformation by intracellular and/or extracellular enzymes (***enzyme***), sequestration by metallothioneins (precipitates-* **ppts***) and metal ion (***M+***) removal (bioaccumulation/precipitation/biotranformation/immobilization) by extracellular polymeric substances (***EPS***). Furthermore, metabolic enzymes and the DNA repair system also protect from metal induced oxidative stress and DNA damage (not shown in figure).*

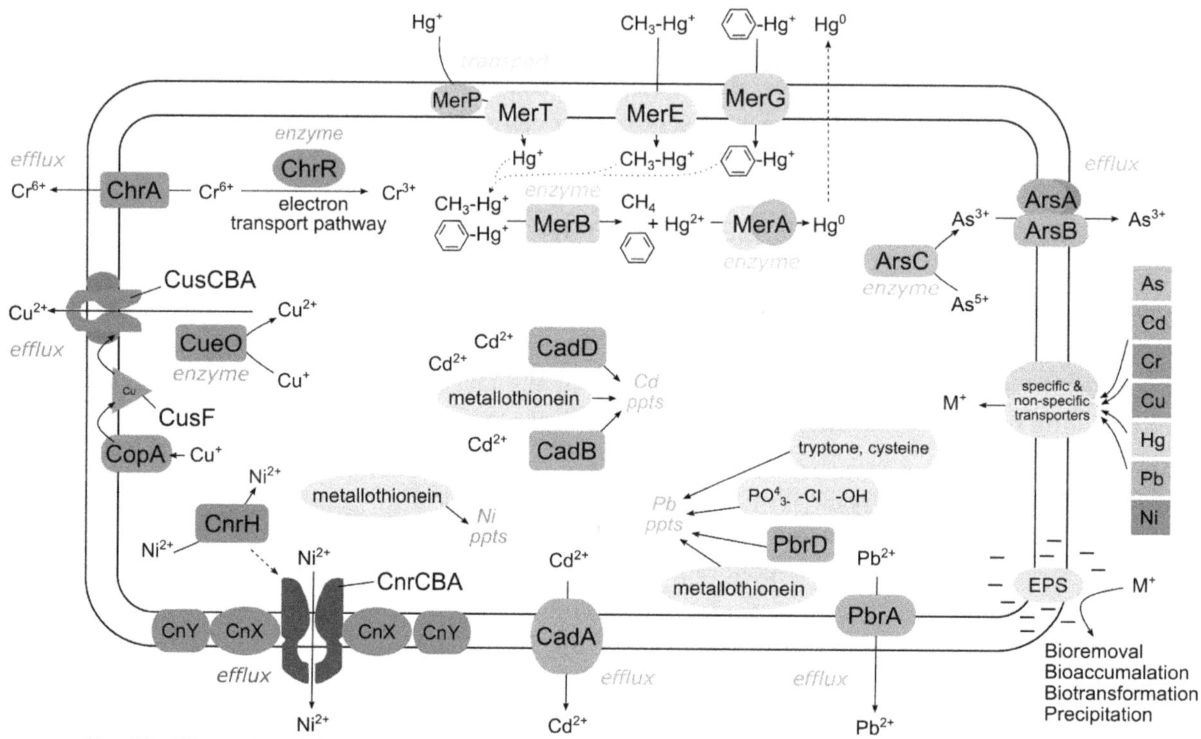

Simplified illustration of the genetic mechanisms involved in bacterial resistance to toxic metals

Mechanisms of resistance include efflux systems, biotransformation by intracellular and/or extracellular enzymes (**enzyme**), sequestration by metallothioneins (precipitates- **ppts**) and metal ion (**M+**) removal (bioaccumulation/precipitation/biotranformation/immobilization) by extracellular polymeric substances (**EPS**). Furthermore, protective metabolic enzymes/proteins (such as *superoxide dismutase*, *catalase*, etc.) and the DNA repair system also protect from metal induced oxidative stress and DNA damage (not shown in figure).

MOLECULAR APPROACHES TO STUDY MICROBIAL HEAVY METAL REMEDIATION

Knowledge on abiotic and biotic microbial interactions is useful not only in the field of heavy metal bioremediation, but in various other fields as well. To implement bioremediation, biological contribution and impact on the ecosystem needs to be evaluated. This can be done by analysis of microbial communities that partake in in situ bioremediation. At present, most studies on or related to metal bioremediation processes are based on the "treatability study" where samples from sites contaminated with metal are incubated under controlled conditions and the metal immobilization, transformation rates, etc. are documented. Although these studies generally evaluate potential microbial metabolic activities, they provide little information on the bioremediating microbes. There are some studies that focus on isolating and characterizing theses microbes relying on culture-dependent techniques. This is problematic as more

Table 1. Bacterial genes conferring heavy metal resistance/detoxification

Heavy Metal	Gene	Protein/ Enzyme Encoded	Reported Function	Reference
Arsenic	*arsC*	Arsenate reductase	Reduction of arsenate [As^{5+}] to arsenite [As^{3+}] in response to arsenic-containing substances	Carlin et al. (1995) Ji & Silver (1992) Oden et al. (1994)
	arsB	Arsenic efflux pump protein	Arsenite transmembrane transporter activity	
	arsA	Arsenic ABC transporter ATPase	Binds with ArsB to efflux As^{3+} and enables detoxification of arsenic-containing substances	
Cadmium	*cadA*	Cadmium transporting ATPase	Couples the hydrolysis of ATP with the export of cadmium and is involved in cadmium resistance	Hsieh et al. (2010)
	cadB	Cadmium resistance protein B	Hypothesized to enable binding of cadmium at the cell membrane and protects the bacterial cell	Crupper et al. (1999) Smith & Novick (1972)
	cadD	Cadmium resistance protein		
Chromium	*chrA*	Chromate transport protein	Chromate transmembrane transporter activity which reduces chromate accumulation and is essential for chromate resistance	Díaz-Magaña et al., 2009)
	chrR	Chromate reductase	Oxidoreductase activity transforming Cr^{6+} to less toxic Cr^{3+}	Gonzalez et al., 2005 Cervantes & Campos-García (2007)
Copper	*cusF*	Cation efflux system protein CusF	Part of cation efflux systems that mediates copper resistance (binds one copper per polypeptide), cellular copper ion homeostasis and detoxification	Yu et al. (2014)
	copA/copB	Copper-exporting P-type ATPase	Exports Cu^{+} from the cytoplasm to the periplasm. Plays a role in cellular response to copper ion and its detoxification	Lee et al. (2002) Munson et al. (2000)
	cusCBA	Cation efflux system protein complex	Provides copper resistance through cation efflux	Yu et al. (2014)
	cueO	Copper oxidase	Plays a role copper efflux and is involved in periplasmic detoxification of copper	Djoko et al. (2010)
Lead	*pbrA*	P-type Pb^{2+} efflux ATPase	Plays a role in lead export from the cytoplasm and providing lead resistance	Borremans et al. (2001) Hynninen et al. (2009)
	pbrB	phosphatase PbrB		
	pbrD	Pb^{2+}-binding protein	Putative protein that reduces lead toxicity by binding to it intracellularly	
Mercury	*merP*	Mercuric transport protein periplasmic component	MerP acts as a mercury scavenger that specifically binds to a mercuric ion in the periplasm passing it to the cytoplasmic mercuric reductase MerA via the mercuric transport protein MerT. MerA is responsible for Hg^{2+} reduction and volatilizing mercury as Hg^0	Hamlett et al. (1992) Moore et al. (1990) Dash & Das, (2012)
	merT	Mercuric transport protein MerT		
	merA	Mercuric reductase		
	Mere	Broad-spectrum mercury transporter MerE	Broad mercury transporter that mediates the transport of both methyl-mercury (CH_3-Hg^+) and inorganic mercury (Hg^{2+}) across the membrane	Barkay et al. (2003) Kiyono & Pan-Hou (1999) Schneiker et al. (2001)
	merG		Phenylmercury resistance	
	merB	Organmercurial lyase	Cleaves the carbon-mercury bond of organomercurials to form Hg^{2+}, which is subsequently detoxified by mercuric reductase	Dash & Das, (2012) Schaefer et al. (2011
Nickel	*cnrCBA*	Nickel and cobalt resistance protein CnrB	Membrane-bound protein complex that catalyzes the energy-dependent efflux of Ni^{2+} and Co^{2+}.	Grass et al. (2000)
	Nre	Nickel/cobalt efflux system	Functions similarly like the *cnr* operon	Grass et al., (2005)
Multi-metal (Pb, Cd, Ni)	bmtA, smtAB	Metallothionein	Binds metal ions under high metal stress conditions	Abdel-Monem et al., 2010 Liu et al., 2021 Naik et al., 2012a Naik et al., 2012b Robinson et al. (1990)

than 99% of indigenous microbes that are uncultivable due to their complex culture conditions requirements. These limitations can be overcome by a number of molecular techniques which are invaluable in investigating diversity and structure of microbial communities applicable on both culturable as well non-culturable microbes (Tomotada and Masao, 2001; Head et al., 2003; Bursle and Robson, 2016).

Traditional DNA-Based Molecular Tools

Traditional DNA-based molecular tools are accurate, reproducible and utilize genes such as ITS, 18S rRNA and 16S rRNA as biomarkers for microbial identification and characterization. Cultivation dependent techniques require gene amplification and sequencing of extracts from microorganisms in culture (Reller et al., 2007). Cultivation independent techniques useful in the field of microbial ecology can outline complex microbial diversity by extracting nucleic acids from the environmental sources thus helping in understanding the community structure and diversity without biases characteristic to culture analysis (Fakruddin and Mannan, 2013). This is done by a combination of several molecular tools such as genetic fingerprinting, quantitative PCR, fluorescence in situ hybridization (FISH) and DNA microarray.

Genetic fingerprinting techniques provide a of the community diversity pattern/profile and include temperature/denaturing gradient gel electrophoresis (TGGE/DGGE), random amplified polymeric DNA (RAPD), single-stranded conformation polymorphism (SSCP), amplified ribosomal DNA restriction analysis (ARDRA), terminal restriction fragment length polymorphism (T-RFLP/RFLP), and length heterogeneity PCR (LH-PCR) as some of the most notable genetic fingerprinting techniques. These methods allow for concurrent visualization and analysis of multiple samples allowing the comparison of diverse microbial communities from different locations, different time periods or, before and after the advent of heavy metal contamination (Muyzer, 1999). Quantitative PCR is used to determine the presence and assortment of operative and taxonomic gene markers associated with microbial communities in environmental samples (Dymond, 2013). In FISH, fluorescent dyes or probes are used in conjugation with an oligonucleotide probe to measure accumulated amplicons in each cycle of PCR. Probes of 16S rRNA are conventionally used to detect presence and evaluate different microbial communities. DNA from environmental samples can also be used to generate fluorescently labeled amplified PCR products that hybridise with designed molecular probes immobilized on the surface of DNA microarray chip, providing rapid, accurate and reliable characterization of environmental samples (Ranjbar et al., 2017).

Advanced Omic Tools

Microbial bioremediation makes use of indigenous microbial communities to detoxify environmental contamination. This rate of detoxification is influenced by the nature and scale of heavy metal contamination, environmental conditions and composition of native microbial communities (Chakraborty et al., 2012). Besides research interest and a few commercial applications, the success of microbial bioremediation remains limited and is yet to be actualized. This is due to lack of information on microbial flora inhabiting contaminated sites (since most microbes are not readily culturable), poor knowledge of their metabolic capabilities and lack of insight on how indigenous communities respond to environmental conditions such as metal contamination. Optimization of the bioremediation process will involve integration of complex variables such as community structure, metabolic function, fate of heavy metals in the environment and interaction (Lovley, 2003).

DNA-based molecular techniques are inadequate for this purpose as specific information on gene expression under *in situ* conditions, is unavailable. Hence, high throughput molecular approaches such as metagenomics 1, metatranscriptomics 2, metaproteomics 3 and metabolomics 4 are gaining importance in the field of bioremediation as they provide a link between genetic and functional properties within diverse microbial communities indigenous to heavy metal contaminated environments. This involves understanding microbial communities from contaminated sites, from the functional aspects as well as the discovering novel microbes not accessible through traditional culturing methods (Moran, 2009; Miller, 2007; Riesenfeld et al., 2004; Wilmes & Bond, 2006).

Potential Application of Omic Tools in Heavy Metal Bioremediation

So far, bioremediation has been largely attributed to culturable microbes. However, most catabolic potential in nature is unexplored due to the difficulty in replicating specific environmental conditions. In order to gain greater access to this catabolic potential, "omics" technology can be exploited to yield information prior to their cultivation, with the intent of studying enzymatic activities related to bio-transformation (Malla et al., 2018). Metagenomics in particular, has increased understanding of how microbes develop abilities to precipitate or immobilize metals. It allows for differentiation of sites with metal contamination into areas where indigenous microflora is capable of remediation through intensive *ex situ* treatment or by *in situ* bioaugmentation. It has helped categorize key microbial processes and community components that could greatly complement the cleanup process, especially when the process is carried out by a consortia and cross talk is necessary. It can also be used to furnish metagenomics databases that provide good collection of relevant genes for novel strain development with targeted bioremediation efforts (Thomas et al., 2012). However, it is limited in interpreting gene expression and activity. Meta-transcriptomics and metaproteomics have made it possible to ascertain the gene activity within an environment.

In addition to identification of novel genes and proteins, protein folding, conformational stability of the proteins and physiological changes that microbes undergo can be predicted and analyzed (Beauvais-Flück et al., 2017; Izrael- Zivkoví et al., 2018). Metabolomics allows analysis of microbial cellular metabolites and to understand the dynamic and functional operations of indigenous microbial communities. Microbes subjected to environmental stressors release numerous low molecular weight metabolites which can be mapped by metabolome analysis for their functional roles (Tanaka et al., 2007; Booth et al., 2011a).

Table 2 presents the integration and implementation of various omics tools for (i) uncovering cellular functions in response to toxic heavy metals, (ii) studying the stress response, (iii) discovering novel genes/proteins involved in the biotransformation of heavy metals and, (iv) isolating heavy metal-resistant microbes. This can thus provide additional insight into the development of new remediation techniques, improving pre-existing ones or a combination of each- such as favoring/designing metal resistant strains within contaminated environments and possibly combining the natural with engineered strains (Tomotada and Masao, 2001; Valls and de Lorenzo, 2002; Zhao and Poh, 2008; Altimira et al., 2012).

Table 2. Application of omic tools in heavy metal bioremediation

Heavy Metal Contaminant	Omic Tool	Application	Reference
Copper	Genomics	DGGE analysis to study microbial diversity in Cu and non-Cu polluted soil	(Altimira et al., 2012)
Multi-metal	Genomics	Composition of cultivable multi-metal resistant bacterial communities 16S rDNA gene sequencing	(Domingues et al., 2020)
Multi-metal	Genomics	Genomic sequence determination of multi-metal resistant *Pseudomonas putida* ATH-43	(Rodríguez-Rojas et al., 2016)
Multi-metal	Metagenomics	Change in microbial diversity of AM fungi found in contaminated soil	(Hassan et al., 2011)
Cobalt (Co) and Lead (Pb)	Proteomics	Differentially expressed proteins in *Klebsiella pneumoniae*	(Bar et al., 2007)
Cadmium (Cd)	Proteomics	Differentially expressed proteins in *Pseudomonas* spp.	(Izrael- Zivkoví et al., 2018; Jain & Bhatt, 2013)
Mercury (Hg)	Transcriptomics	Analysis of physiological and tolerance response of microalga to Hg	(Beauvais-Flück et al., 2017)
Cadmium (Cd)	Metabolomics	Metabolomic analysis of Cd stress resistance in yeast cells	(Tanaka et al., 2007)
Copper (Cu)	Metabolomics	Metabolomic analysis of *Pseudomonas* spp.in tolerance and resistance response	(Booth et al., 2011b)

GENETIC ENGINEERING OF MICROBES FOR ENHANCED BIOREMEDIATION

Molecular biology can also be used to maximize metabolic capacities of microbes. This is done be doing modifications to pathways using genetic engineering tools (Tahri et al., 2013). As outlined in Section (Genetic Basis of Heavy Metal Resistance) numerous genes responsible for metal tolerance, biotransformation etc., have been identified especially from microorganisms exposed to metal contamination. Hence, microbial strains have the potential to remove metal contamination through genetic manipulation or modification.

Desired or potential genes are first screened using genetic screens with the arrival of omic tools greatly speeding the process. Candidate genes can be either isolated from microbial cells and then copied, or artificially synthesized. Once isolated, the gene is ligated into a plasmid with other genetic elements and inserted into the target microorganism (Beardmore and Porter, 2003). Modification of the target/host microorganism as a result can be done through inhibition/promotion/addition/removal of an enzyme to modify a competitive pathway or toxic byproduct, amplification of or engineering a single gene or gene cluster to improve synthesis of existing products, providing a pathway for degradation of toxic metabolites (pathway switching), etc. (Das et al., 2016; Yang et al., 1998).

The key to developing effectual bioremediation technologies of heavy metal contaminated environments is to understand the microbial physiology of such niches. Genetic engineering can thus generate microorganisms having favorable catalytic potential, which are capable in removing any environmental pollutant. For example: (i) improved nickel binding capability in recombinant *Staphylococcus xylosus* and *Staphylococcus carnosus* strains with the introduction of H1 or H2 peptides in their surface proteins (Samuelson et al., 2000), and (ii) *Ralstonia eutropha* was modified with DNA sequences encoding mouse metallothionein I and was found to have enhanced ability to immobilize Cd2+ ions (Valls & de Lorenzo, 2002).

CONCLUSION

Although genetic engineering has produced strains capable of reducing toxicity of heavy metals in laboratory settings under controlled conditions, the translation of these abilities into actual bioremediation strategies has been forthcoming. This could be due to:

1. The microbial strain or strains in question acquired through traditional enrichment procedures not performing the bulk of the remedial work and being far less significant ecologically in natural conditions
2. Use of fast growers which inevitably leads to buildup of unnecessary biomass
3. General uncertainty with the release of foreign GEM into the environment
4. Unfavorable field conditions for engineered microbes due to limited knowledge of engineered microbes.

Molecular applications have been restricted to a few bacteria (*Escherichia coli, Pseudomonas putida, Bacillus subtilis*, etc.). Moreover, the adverse effect of the deliberate release of GEM for bioremediation on indigenous microflora in not evident. For practical biotechnological applications, it is necessary for the development of a greater and more diverse number of GEM. Additionally, their performance in terms of survival, amenability to HGT, catalytic ability to cell mass ratio (maximum catalytic ability to minimum cell mass would be optimal) and effect of indigenous microflora in a real-world environmental situation must be tested. Frequently microbes are designed for use under laboratory conditions overlooking field requirements for bio-remedial processes. This needs to be addressed through further research.

REFERENCES

Abdel-Monem, M. O., Al-Zubeiry, A. H. S., & Al-Gheethi, A. A. S. (2010). Biosorption of nickel by *Pseudomonas cepacia* 120S and *Bacillus subtilis* 117S. *Water Science and Technology*, *61*(12), 2994–3007. doi:10.2166/wst.2010.198 PMID:20555195

Afzal, M. S., Ashraf, A., & Nabeel, M. (2018). Characterization of industrial effluents and groundwater of Hattar industrial estate, Haripur. *Advances in Agriculture and Environmental Science: Open Access*, *1*(2), 70–77.

Agency for Toxic Substances and Disease Registry (ATSDR). (2002). *Toxicological Profile for Copper*. Centers for Disease Control.

Ahalya, N., Ramachandra, T. V., & Kanamadi, R. D. (2003). Biosorption of heavy metals. *Research Journal of Chemistry and Environment*, *7*(4), 71–78.

Ahmed, M. K. E., Parvin, M. M., Islam, M. S., Akter, S., Khan, S., & Al-Mamun, M. H. (2014). Lead- and cadmium-induced histopathological changes in gill, kidney and liver tissue of freshwater climbing perch *Anabas testudineus* (Bloch, 1792). *Chemistry and Ecology*, *30*(6), 532–540. doi:10.1080/02757540.2014.889123

Akif, M., Khan, A. R., & Sok, K., Min, Hussain, Z., Maal-Abrar, Khan, M., & Yousafjai, A. M. (2002). Textile effluents and their contribution towards aquatic pollution in the Kabul River (Pakistan). *Journal of the Chemical Society of Pakistan*, *24*, 106–111.

Ali, H., & Khan, E. (2018). What are heavy metals? Long-standing controversy over the scientific use of the term' heavy metals'- proposal of a comprehensive definition. *Toxicological and Environmental Chemistry*, *100*(1), 6–19. doi:10.1080/02772248.2017.1413652

Alloway, B. J. (2013). Sources of heavy metals and metalloids in soils. In *Heavy Metals in Soils: Trace Metals and Metalloids in Soils and their Bioavailability* (pp. 11–50). Springer. doi:10.1007/978-94-007-4470-7_2

Altimira, F., Yãez, C., Bravo, G., González, M., Rojas, L. A., & Seeger, M. (2012). Characterization of copper-resistant bacteria and bacterial communities from copper-polluted agricultural soils of central Chile. *BMC Microbiology*, *12*(1), 193. doi:10.1186/1471-2180-12-193 PMID:22950448

Atkins, P., & Jones, L. (1997). Chemistry—Molecules, Matter and Change. W. H. Freeman.

Bagal-Kestwal, D., Karve, M., Kakade, B., & Pillai, V. (2008). Invertase inhibition based electrochemical sensor for the detection of heavy metal ions in aqueous system: Application of ultra-microelectrode to enhance sucrose biosensor's sensitivity. *Biosensors & Bioelectronics*, *24*(4), 657–664. doi:10.1016/j.bios.2008.06.027 PMID:18667298

Balaji, S., Kalaivani, T., Sushma, B., Pillai, C. V., Shalini, M., & Rajasekaran, C. (2016). Characterization of sorption sites and different stress response of microalgae isolates against application towards phytoremediation. *International Journal of Phytoremediation*, *18*(8), 747–753. doi:10.1080/15226514.2015.1115960 PMID:26587690

Bar, C., Patil, R., Doshi, J., Kulkarni, M. J., & Gade, W. N. (2007). Characterization of the proteins of bacterial strain isolated from contaminated site involved in heavy metal resistance-A proteomic approach. *Journal of Biotechnology*, *128*(3), 444–451. doi:10.1016/j.jbiotec.2006.11.010 PMID:17210198

Barkay, T., Miller, S. M., & Summers, A. O. (2003). Bacterial mercury resistance from atoms to ecosystems. *FEMS Microbiology Reviews*, *27*(2-3), 355–384. doi:10.1016/S0168-6445(03)00046-9 PMID:12829275

Beardmore, J. A., & Porter, J. S. (2003). *Genetically modified organisms and aquaculture*. Food and Agriculture Organization of the United Nations.

Beauvais-Flück, R., Slaveykova, V. I., & Cosio, C. (2017). Cellular toxicity pathways of inorganic and methyl mercury in the green microalga *Chlamydomonas reinhardtii*. *Scientific Reports*, *7*(1), 1–12. doi:10.103841598-017-08515-8 PMID:28808314

Boonchai, R., Kaewsuk, J., & Seo, G. (2014). Effect of nutrient starvation on nutrient uptake and extracellular polymeric substance for microalgae cultivation and separation. *Desalination and Water Treatment*, *55*(2), 360–367. doi:10.1080/19443994.2014.939501

Booth, S. C., Workentine, M. L., Weljie, A. M., & Turner, R. J. (2011a). Metabolomics and its application to studying metal toxicity. *Metallomics*, *3*(11), 1142–1152. doi:10.1039/c1mt00070e PMID:21922109

Booth, S. C., Workentine, M. L., Wen, J., Shaykhutdinov, R., Vogel, H. J., Ceri, H., Turner, R. J., & Weljie, A. M. (2011b). Differences in metabolism between the biofilm and planktonic response to metal stress. *Journal of Proteome Research*, *10*(7), 3190–3199. doi:10.1021/pr2002353 PMID:21561166

Borremans, B., Hobman, J. L., Provoost, A., Brown, N. L., & Van Der Lelie, D. (2001). Cloning and functional analysis of the *pbr* lead resistance determinant of *Ralstonia metallidurans* CH34. *Journal of Bacteriology*, *183*(19), 5651–5658. doi:10.1128/JB.183.19.5651-5658.2001 PMID:11544228

Bradl, H. (2002). *Heavy Metals in the Environment: Origin, Interaction and Remediation* (Vol. 6). Academic Press.

Bursle, E., & Robson, J. (2016). Non-culture methods for detecting infection. *Australian Prescriber*, *39*(5), 171–175. doi:10.18773/austprescr.2016.059 PMID:27789929

Carlin, A., Shi, W., Dey, S., & Rosen, B. P. (1995). The *ars* operon of *Escherichia coli* confers arsenical and antimonial resistance. *Journal of Bacteriology*, *177*(4), 177. doi:10.1128/JB.177.4.981-986.1995 PMID:7860609

Cervantes, C., & Campos-García, J. (2007). Reduction and efflux of chromate by bacteria. Molecular Microbiology of Heavy Metals, 407–419. doi:10.1007/7171_2006_087

Chakraborty, R., Wu, C. H., & Hazen, T. C. (2012). Systems biology approach to bioremediation. *Current Opinion in Biotechnology*, *23*(3), 483–490. doi:10.1016/j.copbio.2012.01.015 PMID:22342400

Chang, L. W., Magos, L., & Suzuki, T. (Eds.). (1996). *Toxicology of Metals*. CRC Press.

Chanmugathas, P., & Bollag, J. M. (1988). A column study of the biological mobilization and speciation of cadmium in soil. *Archives of Environmental Contamination and Toxicology*, *17*(2), 229–237. doi:10.1007/BF01056029

Das, S., Dash, H. R., & Chakraborty, J. (2016). Genetic basis and importance of metal resistant genes in bacteria for bioremediation of contaminated environments with toxic metal pollutants. *Applied Microbiology and Biotechnology*, *100*(7), 2967–2984. doi:10.100700253-016-7364-4 PMID:26860944

Dash, H. R., & Das, S. (2012). Bioremediation of mercury and the importance of bacterial *mer* genes. *International Biodeterioration & Biodegradation*, *75*, 207–213. doi:10.1016/j.ibiod.2012.07.023

Dasola, A. M., Adeyemi, A. L., Tunbosun, L. A., Abidemi, O. O., & Razaq, S. O. (2014). Kinetic and equilibrium studies of the heavy metal remediation potential of *Helix pomentia*. *African Journal of Pure and Applied Chemistry*, *8*, 123–133.

Dave, S., Maitri, D., & Tipre, D. (2010). Copper remediation by *Eichhornia* spp. and sulphate-reducing bacteria. *Journal of Hazardous Materials*, *173*(1-3), 231–235. doi:10.1016/j.jhazmat.2009.08.073 PMID:19747776

Dermont, G., Bergeron, M., Mercier, G., & Richerlaflèche, M. (2008). Soil washing for metal removal: A review of physical/chemical technologies and field applications. *Journal of Hazardous Materials*, *152*(1), 1–31. doi:10.1016/j.jhazmat.2007.10.043 PMID:18036735

Díaz-Magaña, A., Aguilar-Barajas, E., Moreno-Sánchez, R., Ramírez-Díaz, M. I., Riveros-Rosas, H., Vargas, E., & Cervantes, C. (2009). Short-chain chromate ion transporter proteins from *Bacillus subtilis* confer chromate resistance in *Escherichia coli*. *Journal of Bacteriology*, *191*(17), 5441–5445. doi:10.1128/JB.00625-09 PMID:19581367

Dixit, R., Wasiullah, Malaviya, D., Pandiyan, K., Singh, U., Sahu, A., Shukla, R., Singh, B., Rai, J., Sharma, P., Lade, H., & Paul, D. (2015). Bioremediation of heavy metals from soil and aquatic environment: An overview of principles and criteria of fundamental processes. *Sustainability*, *7*(2), 2189–2212. doi:10.3390u7022189

Djoko, K. Y., Chong, L. X., Wedd, A. G., & Xiao, Z. (2010). Reaction mechanisms of the multicopper oxidase CuO from *Escherichia coli* support its functional role as a cuprous oxidase. *Journal of the American Chemical Society*, *132*(6), 2005–2015. doi:10.1021/ja9091903 PMID:20088522

Domingues, V. S., de Souza Monteiro, A., Júlio, A. D. L., Queiroz, A. L. L., & dos Santos, V. L. (2020). Diversity of metal-resistant and tensoactive-producing culturable heterotrophic bacteria isolated from a copper mine in Brazilian Amazonia. *Scientific Reports*, *10*(1), 1–12. doi:10.103841598-020-62780-8 PMID:32277075

Dymond, J. S. (2013). Explanatory chapter: Quantitative PCR. *Methods in Enzymology*, *529*, 279–289. doi:10.1016/B978-0-12-418687-3.00023-9 PMID:24011054

Fakruddin, M., & Mannan, K. S. (2013). Methods for analyzing diversity of microbial communities in natural environments. *Ceylon Journal of Science. Biological Sciences*, *42*(1), 19. doi:10.4038/cjsbs.v42i1.5896

Fergusson, J. E. (1990). *The Heavy Elements: Chemistry, Environmental Impact and Health Effects*. Pergamon Press.

Flemming, H.-C., Wingender, J., Szewzyk, U., Steinberg, P., Rice, S. A., & Kjelleberg, S. (2016). Biofilms: An emergent form of bacterial life. *Nature Reviews. Microbiology*, *14*(9), 563–575. doi:10.1038/nrmicro.2016.94 PMID:27510863

Fomina, M., & Gadd, G. M. (2014). Biosorption: Current perspectives on concept, definition and application. *Bioresource Technology*, *160*, 3–14. doi:10.1016/j.biortech.2013.12.102 PMID:24468322

Gonzalez, C. F., Aekerley, D. F., Lynch, S. V., & Matin, A. (2005). ChrR, a soluble quinone reductase of *Pseudomonas putida* that defends against H_2O_2. *The Journal of Biological Chemistry*, *280*(24), 22590–22595. doi:10.1074/jbc.M501654200 PMID:15840577

Goyer, R. A. (2001). Toxic effects of metals. In C. D. Klaassen (Ed.), *Cassarett and Doull's Toxicology: The Basic Science of Poisons* (pp. 811–867). McGraw-Hill Publisher.

Grass, G., Fricke, B., & Nies, D. H. (2005). Control of expression of a periplasmic nickel efflux pump by periplasmic nickel concentrations. *Biometals*, *18*(4), 437–448. doi:10.100710534-005-3718-6 PMID:16158236

Grass, G., Große, C., & Nies, D. H. (2000). Regulation of the cnr cobalt and nickel resistance determinant from *Ralstonia* sp. strain CH34. *Journal of Bacteriology*, *182*(5), 1390–1398. doi:10.1128/JB.182.5.1390-1398.2000 PMID:10671463

Hamlett, N. V., Landale, E. C., Davis, B. H., & Summers, A. O. (1992). Roles of the Tn21 merT, merP, and merC gene products in mercury resistance and mercury binding. *Journal of Bacteriology*, *174*(20), 6377–6385. doi:10.1128/JB.174.20.6377-6385.1992 PMID:1328156

Harvey, L. J., & McArdle, H. J. (2008). Biomarkers of copper status: A brief update. *British Journal of Nutrition*, *99*(S3), S10–S13. doi:10.1017/S0007114508006806 PMID:18598583

Hassan, S. E. D., Boon, E., St-Arnaud, M., & Hijri, M. (2011). Molecular biodiversity of arbuscular mycorrhizal fungi in trace metal-polluted soils. *Molecular Ecology*, *20*(16), 3469–3483. doi:10.1111/j.1365-294X.2011.05142.x PMID:21668808

He, Z. L., Yang, X. E., & Stoffella, P. J. (2005). Trace elements in agro-ecosystems and impacts on the environment. *Journal of Trace Elements in Medicine and Biology*, *19*(2–3), 125–140. doi:10.1016/j.jtemb.2005.02.010 PMID:16325528

Head, I., Singleton, I., & Milner, M. (2003). *Bioremediation; a critical review*. Caister Academic Press.

Herawati, N., Suzuki, S., Hayashi, K., Rivai, I. F., & Koyoma, H. (2000). Cadmium, copper and zinc levels in rice and soil of Japan, Indonesia and China by soil type. *Bulletin of Environmental Contamination and Toxicology*, *64*, 33–39. doi:10.1007001289910006 PMID:10606690

Hlihor, R. M., Figueiredo, H., Tavares, T., & Gavrilescu, M. (2017). Biosorption potential of dead and living *Arthrobacter viscosus* biomass in the removal of Cr (VI): Batch and column studies. *Process Safety and Environmental Protection*, *108*, 44–56. doi:10.1016/j.psep.2016.06.016

Hsieh, P., Lin, H., Lin, T., & Wang, J. (2010). CadC regulates *cad* and *tdc* operons in response to gastrointestinal stresses and enhances intestinal colonization of *Klebsiella pneumoniae*. *The Journal of Infectious Diseases*, *202*(1), 52–64. doi:10.1086/653079 PMID:20497056

Hynninen, A., Touzé, T., Pitkänen, L., Mengin-Lecreulx, D., & Virta, M. (2009). An efflux transporter PbrA and a phosphatase PbrB cooperate in a lead-resistance mechanism in bacteria. *Molecular Microbiology*, *74*(2), 384–394. doi:10.1111/j.1365-2958.2009.06868.x PMID:19737357

Izrael- Zivkoví, L., Rikaloví, M., Gojgí Cvijoví, G., Kazazí, S., Vrví, M., Bř ceski, I., Bě skoski, V., Loň Careví, B., GopčevíGopˇGopčeví, K., & KaradžíKaradží, I. (2018). Cadmium specific proteomic responses of a highly resistant Pseudomonas aeruginosa san ai. *RSC Advances*, *8*, 10549–10560.

Jain, S., & Bhatt, A. (2013). Proteomic analysis of diversified extremophilic strains of *Pseudomonas* in the presence of cadmium. *Agricultural Research*, *2*(4), 354–359.

Ji, G., & Silver, S. (1992). Reduction of arsenate to arsenite by the ArsC protein of the arsenic resistance operon of *Staphylococcus aureus* plasmid pI258. *Proceedings of the National Academy of Sciences of the United States of America*, *89*(20), 9474–9478.

Kang, C.-H., Kwon, Y.-J., & So, J.-S. (2016). Bioremediation of heavy metals by using bacterial mixtures. *Ecological Engineering*, *89*, 64–69.

Kiyono, M., & Pan-Hou, H. (1999). The merG gene product is involved in phenylmercury resistance in Pseudomonas strain K-62. *Journal of Bacteriology*, *181*(3), 726–730.

Lee, G., Bigham, J. M., & Faure, G. (2002). Removal of trace metals by coprecipitation with Fe, Al and Mn from natural waters contaminated with acid mine drainage in the Ducktown Mining District, Tennessee. *Applied Geochemistry*, *17*(5), 569–581.

Lee, S. M., Grass, G., Rensing, C., Barrett, S. R., Yates, C. J. D., Stoyanov, J. V., & Brown, N. L. (2002). The Pco proteins are involved in periplasmic copper handling in *Escherichia coli. Biochemical and Biophysical Research Communications*, *295*(3), 616–620.

Lee Chang, K. J., Nichols, C. M., Blackburn, S. I., Dunstan, G. A., Koutoulis, A., & Nichols, P. D. (2014). Comparison of thraustochytrids *Aurantiochytrium* sp., *Schizochytrium* sp., *Thraustochytrium* sp., and *Ulkenia* sp. for production of biodiesel, long-chain omega-3 oils, and exopolysaccharide. *Marine Biotechnology (New York, N.Y.)*, *16*, 396–411.

Lim, J. L., Wilhelmus, M. M., deVries, H. E., Drukarch, B., Hoozemans, J. J., & van Horssen, J. (2014). Antioxidative defense mechanisms controlled by Nrf2: State of the art and clinical perspectives in neurodegenerative diseases. *Archives of Toxicology*, *88*, 1773–1786.

Liu, Y., Serrano, A., Wyman, V., Marcellin, E., Southam, G., Vaughan, J., & Villa-Gomez, D. (2020). Nickel complexation as an innovative approach for nickel-cobalt selective recovery using sulfate-reducing bacteria. *Journal of Hazardous Materials*, *402*, 123506.

Lloyd, J. R. (2002). Bioremediation of metals; the application of micro-organisms that make and break minerals. *Microbiology Today*, *29*, 67–69.

Lovley, D. R. (2003). Cleaning up with genomics: Applying molecular biology to bioremediation. *Nature Reviews. Microbiology*, *1*(1), 35–44.

Malik, A. (2004). Metal bioremediation through growing cells. *Environment International*, *30*, 261–278.

Malla, M. A., Dubey, A., Yadav, S., Kumar, A., Hashem, A., & Abd-Allah, E. F. (2018). Understanding and designing the strategies for the microbe-mediated remediation of environmental contaminants using omics approaches. *Frontiers in Microbiology*, *9*, 1132.

Mani, D., & Kumar, C. (2014). Biotechnological advances in bioremediation of heavy metals contaminated ecosystems: An overview with special reference to phytoremediation. *International Journal of Environmental Science and Technology*, *11*, 843–872.

Menn, F.-M., Easter, J. P., & Sayler, G. S. (2008). Genetically engineered microorganisms and bioremediation. *Biotechnology (Faisalabad)*, *11*, 441–463.

Merian, E. (1984). Introduction on environmental chemistry and global cycles of chromium, nickel, cobalt beryllium, arsenic, cadmium and selenium, and their derivatives. *Toxicological and Environmental Chemistry*, *8*, 9–38.

Miller, M. G. (2007). Environmental metabolomics: A SWOT analysis (strengths, weaknesses, opportunities, and threats). *Journal of Proteome Research*, *6*, 540–545.

Mishra, A., & Malik, A. (2013). Recent advances in microbial metal bioaccumulation. *Critical Reviews in Environmental Science and Technology*, *43*, 1162–1222.

Mohapatra, R. K., Parhi, P. K., Patra, J. K., Panda, C. R., & Thatoi, H. N. (2017). Biodetoxification of toxic heavy metals by marine metal resistant bacteria - a novel approach for bioremediation of the polluted saline environment. In J. Patra, C. Vishnuprasad, & G. Das (Eds.), *Microbial Biotechnology* (pp. 343–376). Springer.

Monchy, S., Benotmane, M. A., Janssen, P., Vallaeys, T., Taghavi, S., Van Der Lelie, D., & Mergeay, M. (2007). Plasmids pMOL28 and pMOL30 of *Cupriavidus metallidurans* are specialized in the maximal viable response to heavy metals. *Journal of Bacteriology*, *189*(20), 7417–7425.

Moore, M. J., Distefano, M. D., Zydowsky, L. D., Cummings, R. T., & Walsh, C. T. (1990). Organomercurial lyase and mercuric ion reductase: Nature's mercury detoxification catalysts. *Accounts of Chemical Research*, *23*(9), 301–308.

Moran, M. A. (2009). Metatranscriptomics: Eavesdropping on complex microbial communities-large-scale sequencing of mRNAs retrieved from natural communities provides insights into microbial activities and how they are regulated. *Microbe*, *4*(7), 329.

Munson, G. P., Lam, D. L., Outten, F. W., & O'Halloran, T. V. (2000). Identification of a copper-responsive two-component system on the chromosome of *Escherichia coli* K-12. *Journal of Bacteriology*, *182*(20), 5864–5871.

Musilova, J., Arvay, J., Vollmannova, A., Toth, T., & Tomas, J. (2016). Environmental contamination by heavy metals in region with previous mining activity. *Bulletin of Environmental Contamination and Toxicology*, *97*, 569–575.

Muyzer, G. (1999). Genetic fingerprinting of microbial communities-present status and future perspectives. Methods of microbial community analysis. In *Proceedings of 8th International symposium on microbial ecology*. Atlantic Canada Society for Microbial Ecology.

Naik, M. M., Pandey, A., & Dubey, S. K. (2012a). *Pseudomonas aeruginosa* strain WI-1 from Mandovi estuary possesses metallothionein to alleviate lead toxicity and promotes plant growth. *Ecotoxicology and Environmental Safety*, *79*, 129–133.

Naik, M. M., Shamim, K., & Dubey, S. K. (2012b). Biological characterization of lead-resistant bacteria to explore role of bacterial metallothionein in lead resistance. *Current Science*, *103*(4), 426–429.

Nriagu, J. O. (1989). A global assessment of natural sources of atmospheric trace metals. *Nature*, *338*, 47–49.

Oden, K. L., Gladysheva, T. B., & Rosen, B. P. (1994). Arsenate reduction mediated by the plasmid-encoded ArsC protein is coupled to glutathione. *Molecular Microbiology*, *12*(2), 301–306.

Parikh, A., & Madamwar, D. (2006). Partial characterization of extracellular polysaccharides from cyanobacteria. *Bioresource Technology*, *97*, 1822–1827.

Pearce, C. I., Lloyd, J. R., & Guthrie, J. T. (2003). The removal of color from textile wastewater using whole bacterial cells: A review. *Dyes and Pigments*, *58*(3), 179–196.

Rajaei, G., Mansouri, B., Jahantigh, H., & Hamidian, A. H. (2012). Metal concentrations in the water of Chah nimeh reservoirs in Zabol, Iran. *Bulletin of Environmental Contamination and Toxicology*, *89*(3), 495–500.

Ramasamy, K., Kamaludeen, S., & Parwin, B. (2006). Bioremediation of metals microbial processes and techniques. In S. N. Singh & R. D. Tripathi (Eds.), *Environmental Bioremediation Technologies* (pp. 173–187). Springer Publication.

Ranaweera, I., Shrestha, U., Ranjana, K. C., Kakarla, P., Willmon, T. M., Hernandez, A. J., Mukherjee, M. M., Barr, S. R., & Varela, M. F. (2015). Structural comparison of bacterial multidrug efflux pumps of the major facilitator superfamily. *Trends in Cell & Molecular Biology*, *10*, 131–140.

Ranjbar, R., Behzadi, P., Najafi, A., & Roudi, R. (2017). DNA microarray for rapid detection and identification of food and water borne bacteria: From dry to wet lab. *The Open Microbiology Journal*, *11*(1), 330–338.

Rasmussen, L. D., Sørensen, S. J., Turner, R. R., & Barkay, T. (2000). Application of a mer-lux biosensor for estimating bioavailable mercury in soil. *Soil Biology & Biochemistry*, *32*, 639–646.

Reller, L., Weinstein, M., & Petti, C. (2007). Detection and identification of microorganisms by gene amplification and sequencing. *Clinical Infectious Diseases*, *44*, 1108–1114.

Rezania, S., Taib, S. M., Md Din, M. F., Dahalan, F. A., & Kamyab, H. (2016). Comprehensive review on phytotechnology: Heavy metals removal by diverse aquatic plants species from wastewater. *Journal of Hazardous Materials*, *318*, 587–599.

Riesenfeld, C. S., Schloss, P. D., & Handelsman, J. (2004). Metagenomics: Genomic analysis of microbial communities. *Annual Review of Genetics*, *38*, 525–552.

Robinson, N. J., Gupta, A., Fordham-Skelton, A. P., Croy, R. R. D., Whitton, B. A., & Huckle, J. W. (1990). Prokaryotic metallothionein gene characterization and expression: Chromosome crawling by ligation-mediated PCR. *Proceedings of the Royal Society of London. Series B, Biological Sciences*, *242*(1305), 241–247.

Rodríguez-Rojas, F., Tapia, P., Castro-Nallar, E., Undabarrena, A., Muñoz-Díaz, P., Arenas-Salinas, M., Díaz-Vásquez, W., Valdés, J., & Vásquez, C. (2016). Draft genome sequence of a multi-metal resistant bacterium *Pseudomonas putida* ATH-43 isolated from Greenwich Island, Antarctica. *Frontiers in Microbiology, 7*, 1777.

Roy, M., & McDonald, L. M. (2015). Metal uptake in plants and health risk assessments in metal-contaminated smelter soils. *Land Degradation & Development, 26,* 785-792.

Salomons, W., Forstner, U., & Mader, P. (1995). *Heavy Metals: Problems and Solutions*. Springer-Verlag.

Samuelson, P., Wernérus, H., Svedberg, M., & Ståhl, S. (2000). Staphylococcal surface display of metal-binding polyhistidyl peptides. *Applied and Environmental Microbiology, 66*(3), 1243–1248.

Schaefer, J. K., Rocks, S. S., Zheng, W., Liang, L., Gu, B., & Morel, F. M. M. (2011). Active transport, substrate specificity, and methylation of Hg(II) in anaerobic bacteria. *Proceedings of the National Academy of Sciences of the United States of America, 108*(21), 8714–8719.

Schneiker, S., Keller, M., Dröge, M., Lanka, E., Pühler, A., & Selbitschka, W. (2001). The genetic organization and evolution of the broad host range mercury resistance plasmid pSB102 isolated from a microbial population residing in the rhizosphere of alfalfa. *Nucleic Acids Research, 29*(24), 5169–5181.

Semu, E., & Singh, B. R. (1996). Accumulation of heavy metals in soils and plants after long-term use of fertilizers and fungicides in Tanzania. *Fertilizer Research, 44*(3), 241–248.

Shallari, S., Schwartz, C., Hasko, A., & Morel, J. L. (1998). Heavy metals in soils and plants of serpentine and industrial sites of Albania. *The Science of the Total Environment, 19209*, 133–142.

Shukla, S. K., Mangwani, N., Karley, D., & Rao, T. S. (2017). Bacterial biofilms and genetic regulation for metal detoxification. Handbook of Metal-Microbe Interactions and Bioremediation, 317.

Shukla, S. K., Mangwani, N., Rao, T. S., & Das, S. (2014). Biofilm-mediated bioremediation of polycyclic aromatic hydrocarbons. Microbial Biodegradation and Bioremediation, 203–232.

Singh, P., & Cameotra, S. S. (2004). Enhancement of metal bioremediation by use of microbial surfactants. *Biochemical and Biophysical Research Communications, 319*(2), 291–297.

Smith, K., & Novick, R. P. (1972). Genetic studies on plasmid-linked cadmium resistance in *Staphylococcus aureus. Journal of Bacteriology, 112*(2), 761–772.

Soleimani, M., Amini, N., Sadeghian, B., Wang, D., & Fang, L. (2018). Heavy metals and their source identification in particulate matter (PM 2.5) in Isfahan City, Iran. *Journal of Environmental Sciences (China)*, 72.

Srivastava, N. K., & Majumder, C. B. (2008). Novel biofiltration methods for the treatment of heavy metals from industrial waste water. *Journal of Hazardous Materials, 151*, 18.

Srivastava, S., & Thakur, I. S. (2006). Isolation and process parameter optimization of *Aspergillus* sp. for removal of chromium from tannery effluent. *Bioresource Technology, 97*(10), 1167–1173.

Stern, B. R. (2010). Essentiality and toxicity in copper health risk assessment: Overview, update and regulatory considerations. *Toxicology Environment and Health, 73*(2), 114–127.

Tahri, N., Bahafid, W., Sayel, H., & El Ghachtouli, N. (2013). Biodegradation: involved microorganisms and genetically engineered microorganisms. In Biodegradation - Life of Science. IntechOpen.

Tanaka, Y., Higashi, T., Rakwal, R., Wakida, S., & Iwahashi, H. (2007). Quantitative analysis of sulfur-related metabolites during cadmium stress response in yeast by capillary electrophoresis–mass spectrometry. *Journal of Pharmaceutical and Biomedical Analysis*, *44*(2), 608–613.

Tchounwou, P., Newsome, C., Williams, J., & Glass, K. (2008). Copper-induced cytotoxicity and transcriptional activation of stress genes in human liver carcinoma cells. *Metal Ions in Biology and Medicine*, *10*, 285–290.

Thomas, T., Gilbert, J., & Meyer, F. (2012). Metagenomics - a guide from sampling to data analysis. *Microbial Informatics and Experimentation*, *2*(1), 3.

Tomotada, I., & Masao, N. (2001). Current Bioremediation Practice and Perspective. Journal of Bioscience and Bioengineering, 92.

Trace Elements in Human Nutrition and Health. (1996). WHO/FAO/IAEA. World Health Organization.

Valls, M., & de Lorenzo, V. (2002). Exploiting the genetic and biochemical capacities of bacteria for the remediation of heavy metal pollution. *FEMS Microbiology Reviews*, *26*(4), 327–338.

Ventura, L. M. B., Mateus, V. L., de Almeida, A. C. S. L., Wanderley, K. B., Taira, F. T., Saint'Pierre, T. D., & Gioda, A. (2017). Chemical composition of fine particles (PM2.5): Water-soluble organic fraction and trace metals. *Air Quality, Atmosphere & Health*, *10*, 845–852.

Verma, S., & Kuila, A. (2019). Bioremediation of heavy metals by microbial process. *Environmental Technology and Innovation*, *14*, 100369.

Vidali, M. (2001). Bioremediation. An overview. *Pure and Applied Chemistry*, *73*(7), 1163–1172.

Wang, Y., Guo, J., & Liu, R. (2001). Biosorption of heavy metals by bacteria isolated from activated sludge. *Applied Biochemistry and Biotechnology*, *91–93*, 171–184.

Wilmes, P., & Bond, P. L. (2006). Metaproteomics: Studying functional gene expression in microbial ecosystems. *Trends in Microbiology*, *14*, 92–97.

Wong, M. H. (2012). *Environmental Contamination: Health Risks and Ecological Restoration*. Taylor & Francis Group.

Yang, Y. T., Bennett, G. N., & San, K. Y. (1998). Genetic and metabolic engineering. *Electronic Journal of Biotechnology*, *1*, 49–60.

Yin, K., Wang, Q., Lu, M., & Chen, L. (2019). Microorganism remediation strategies towards heavy metals. *Chemical Engineering Journal*, 360.

Yu, P., Yuan, J., Deng, X., Ma, M., & Zhang, H. (2014). Subcellular targeting of bacterial CusF enhances Cu accumulation and alters root to shoot Cu translocation in *Arabidopsis*. *Plant & Cell Physiology*, *55*(9), 1568–1581.

Zhao, B., & Poh, C. L. (2008). Insights into environmental bioremediation by microorganisms through functional genomics and proteomics. *Proteomics*, *8*, 874–881.

Zhuang, P., Li, Z.-a., McBride, M. B., Zou, B., & Wang, G. (2013). Health risk assessment for consumption of fish originating from ponds near Dabaoshan mine, South China. *Environmental Science and Pollution Research International*, *20*(8), 5844–5854.

Chapter 17
Potential of Thallophytes in Degradation of Dyes

Sumira Malik
https://orcid.org/0000-0001-5077-1493
Amity Institute of Biotechnology, Amity University, Ranchi, India

Shilpa Prasad
https://orcid.org/0000-0002-9547-1782
Amity Institute of Biotechnology, Amity University, Ranchi, India

Shreya Ghoshal
Amity Institute of Biotechnology, Amity University, Ranchi, India

Shashank Shekhar
Shivalik Institute of Professional Studies, Dehradun, India

Tanvi Kumari
Amity Institute of Biotechnology, Amity University, Ranchi, India

Ankita Agrawal
Amity University, Ranchi, India

Bijaya Samal
Amity University, Ranchi, India

ABSTRACT

Synthetic dyes cause hazardous health-related problems in humans and affect the biological system underwater. They also have a negative impact on the nutritive value of soils and thereby on crops. Until now there is no effective method to remove the harmful component of dyes from the environment. However, the integrated treatment using bio agents with implication of physical and chemical processes can be effective in the treatment of dye effluents. From the complex azo dyes to their dissociation via thallophytes is a new scope for sustenance. Various studies have supported that laccases have the capability to degrade synthetic dyes that have different chemical structures. Thallophytes have been used to degrade the complex dyes with varying ranges of temperature and pH. Thallophytes have recently been used to treat the textile effluents with effective higher temperature and alkaline pH with decreasing BOD and thus cleaning them from environment in an eco-friendly and cost-efficient manner.

DOI: 10.4018/978-1-7998-7062-3.ch017

INTRODUCTION

Thallophytes belong to polyphyletic group (organisms that evolved and grouped together but don't share an immediate common ancestor) that are classified based on similarity of characteristics. They were formerly categorized as a sub-kingdom of kingdom Plantae, which include fungus, lichens, algae, bacteria, slime moulds and bryophytes. They are commonly characterized as "Thalloid plants". On the other hand, dye is essentially a coloured substance that, when added to the material, exposes the chemical bond to the substrate to give different colours to the materials it forms. Dyes are basically organic compounds that are produced naturally or chemically and mainly characterized as natural dyes and synthetic dyes. As the name suggest, the natural dyes are those which have been taken from natural sources i.e., plants, minerals etc. On the other hand, the synthetic dyes are manufactured by using complex aromatic molecular structures that are very tough to break down. This stiffness of synthetic dye causes a serious concern of pollution in the environment and ecology that includes aquatics, soil properties, humans and many more. Thus, to overcome this issue some major methods and mechanisms has been adopted that could help in the degradation of dyes. Among these mechanisms different activities related to thallophytes is also involved.

Synthetic dyes are commonly used in textiles, food entities, paper, furniture, and cosmetic industries (Silveira et al, 2009). Textile industries are the main platform that uses synthetic dyes as well as generation free wastewater with dye (Hassaan & Nemr, 2017). The released wastewater that also contains synthetic dye contaminates water and soil leading to environmental pollution. Most of the dyes that are used in textile industries are azo dyes which contain diazotized amine associated with amine or phenol groups and also one or more azo groups. They are mainly cost-efficient and also quite easy to use which makes them the widely used synthetic dyes.

DYES: CLASSIFICATION AND ENVIRONMENTAL IMPACT

Dyes get absorbed into the pores that are present on material because the shape of the dyes are like narrow strips of papers having length and breadth but very less thickness which assist them to move and acquire the place into polymer system. The size of the dye molecules is smaller as compared to the size of the pores present in fibres and there is an affinity between the fibres and dyes due to force of attraction. It is important to recognise that dyes have a preference for vegetables, animals or humans to pick the right dye because different materials have different chemical structure due to which they require different dyes (Mathur et al, 2006).

Classification of Dyes and their Molecular Structure

Natural Dyes

Natural dyes are the dyes which are obtained from different natural resources like mineral or insects, vegetable matter or can be manufactured from petrochemical feedstock in the factory (Mathur et al, 2006). Indigo is the most popular natural dye that is well known for its blue color which is prepared by processing the leaves of plant *Indigofera tinctora* through fermentation. The red color natural dye is extracted from a resinous protective secretion called Lac of a tiny insect. Due to complex production of natural dyes they are less used than synthetic dyes (Figure 1).

Figure 1. Natural dyes

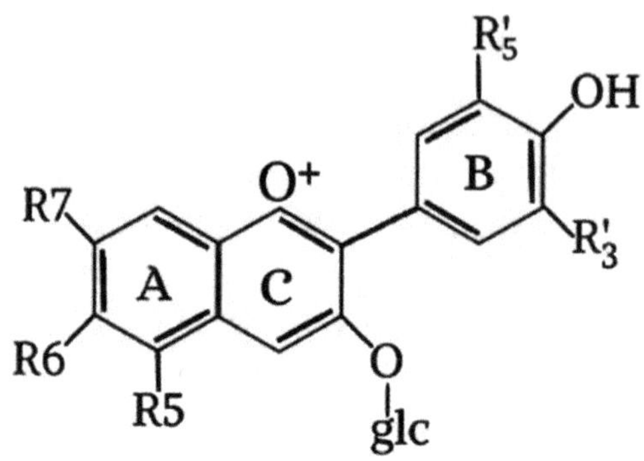

Synthetic Dyes

Synthetic dyes are the dyes which are not obtained from any natural resources. The first synthetic dye mauviene was made from coal tar. Dyes originating from sources like iron oxide and minerals give a brown color (Figure 2). Buff derived from ferrous sulphate is also used for coloring fibres (Keharia & Madamwar, 2003).

Figure 2. Synthetic dyes

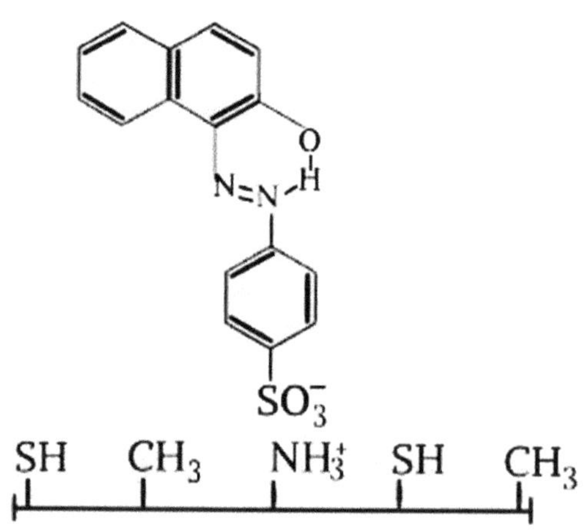

Direct Dyes

Direct dyes are cheapest and easiest to apply but have poor color fastness, also known as salt dyes or cotton colors (Figure 3). They are used to viscose rayon, dye cotton and other vegetable fibres. Direct dyes are readily soluble in water but the process of dying cotton fabrics with direct dyes is not so fast.

To accelerate the process Sodium Chloride is added as an electrolyte. To make them fast in fabric Sodium Carbonate is used for warm color and copper sulphate for cool color (Burkinshaw et al, 1995; Burkinshaw & Salihu, 2017).

Figure 3. Direct dyes

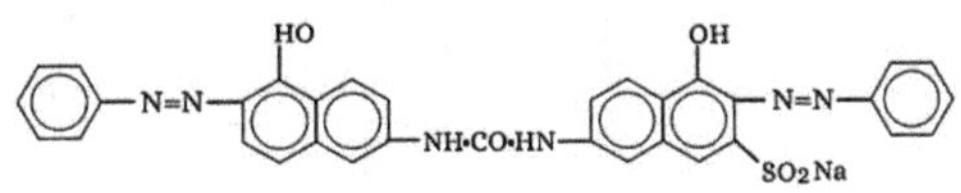

Acid Dyes

Acid dyes are applied under acidic conditions (Figure 4). They are soluble in water and are mostly used in wool and silk (Nunn et al, 1979). Quantity of H_2SO_4 decides the amount of dye that will get absorbed. Acid dyes are inexpensive and very economical.

Figure 4. Acid dyes

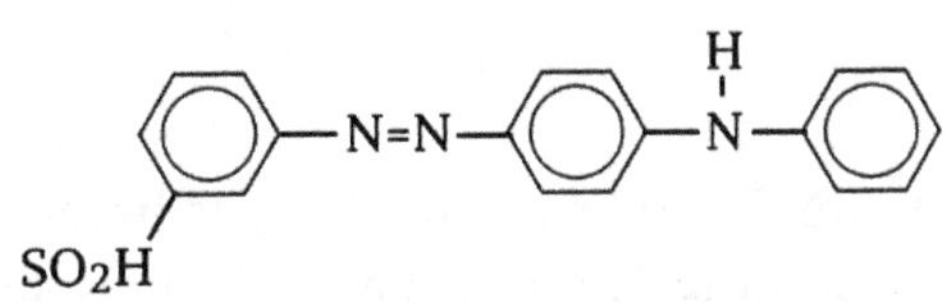

Vat Dyes

Vat dyes are expensive dyes due to their initial cost and method of application. Vat dyes are good for cotton, rayon and linen but can also be applied to wool, nylon and polyester (Burkinshaw et al, 2013; Figure 5). They are hot water dyes but are insoluble in water. This problem could be solved by using strong reducing agent like Sodium Hydrosulphite that is dissolved in Sodium Hydroxide. Vat dyes are available in liquid form and powder form (Burkinshaw et al, 2013).

Azoic Dyes

Azoic dyes are applied on cotton through two stages (Figure 6). First one includes the treatment with naphthol and in second stage the naphtholated material is treated with diazotozed base and diazotozed salt. Due to the coupling reaction that occurs between naphthol and diazo component the development of color takes place in-situ. Azoic dyes have poor to excellent fastness to light. They are also known as ice dyes as ice is frequently used for bringing the dyes to low temperature. Azoic colors give more bright and high intensity color than any other dye (Rajaguru et al, 2002).

Figure 5. Vat dyes

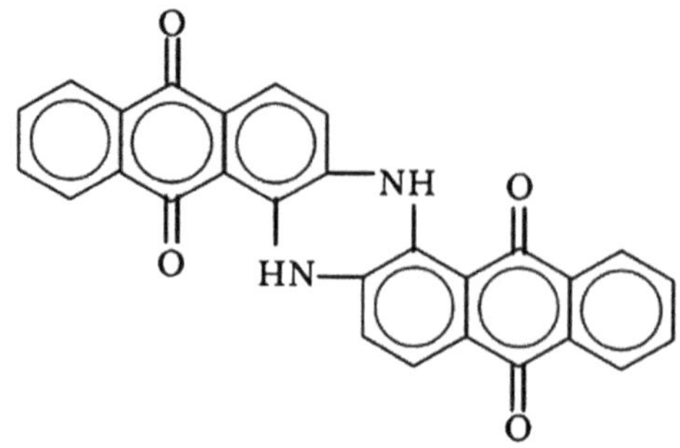

Figure 6. Azoic dyes

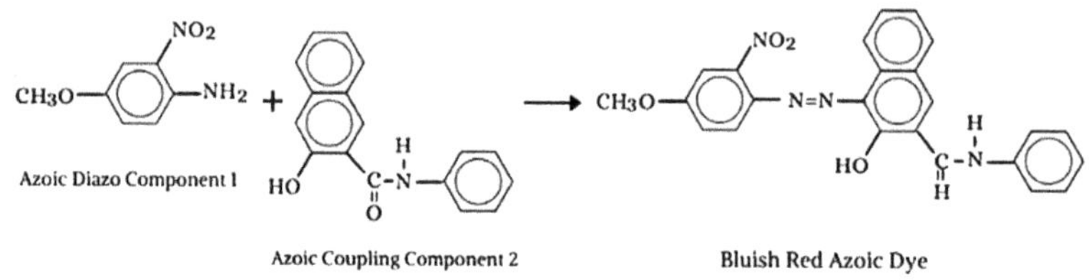

Basic Dyes

Basic dye is applied to cotton, silk, wool, acrylic and modacrylic fibres. The initial coal tar dye was the basic dye but because of the lack of specific dye sites in acrylic fibres they were difficult to dye (Figure 7). Therefore, basic dyes were introduced into the fibres which gave brilliant color to them (Taylor, 2000).

Figure 7. Basic dyes

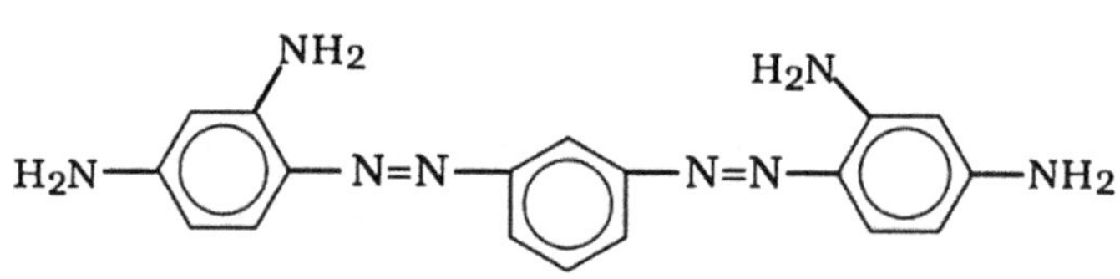

Reactive Dyes

Reactive dyes were first developed by I.C.I., U.K. in 1956 (Allen, 1971). The fastness properties of reactive dyes are very excellent as it works due to the chemical reaction between dye and the fibre. Natural fibres, man-made cellulosic fibres, polyamide fibres and natural protein fibres are mostly colored with reactive dyes (Taylor, 2000) (Figure 8).

Figure 8. Reactive dyes

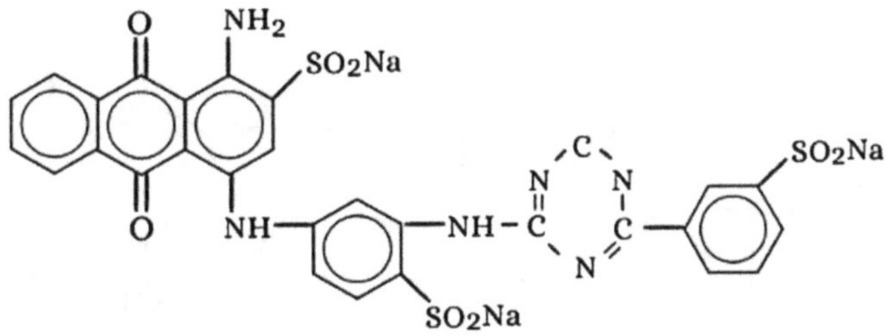

Disperse Dyes

Cellulose diacetate, cellulose triacetate and polyster fibres are mostly dyed with disperse dyes (Figure 9). Acrylic and nylon fibres are dyed to a lesser extent. To achieve satisfactory dyeing the help of high pressure and high temperature is taken as polyster fibres are hydrophobic and have significant crystalline content (Clark, 2011).

Figure 9. Disperse dyes

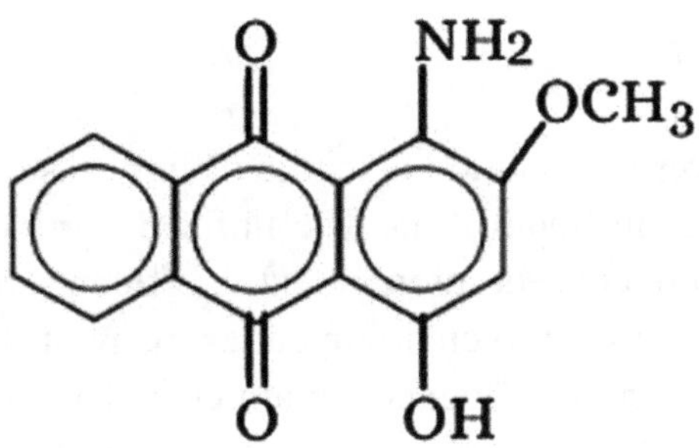

Sulphur Dyes

Mostly natural and man-made cellulosic fibres are dyed with sulphur dyes (Figure 10). Reduction of dye by using sodium sulphide or sodium hydrosulphite results in the production of water-soluble dye (Holme, 2002). To obtain satisfactory rate of dye the dye liquor is heated. The reduced sulphur dye is then converted into its original insoluble form by oxidation due to an oxidizing agent like sodium perborate once the dye is in the fibre (Shore, 1995).

Figure 10. Sulphur dyes

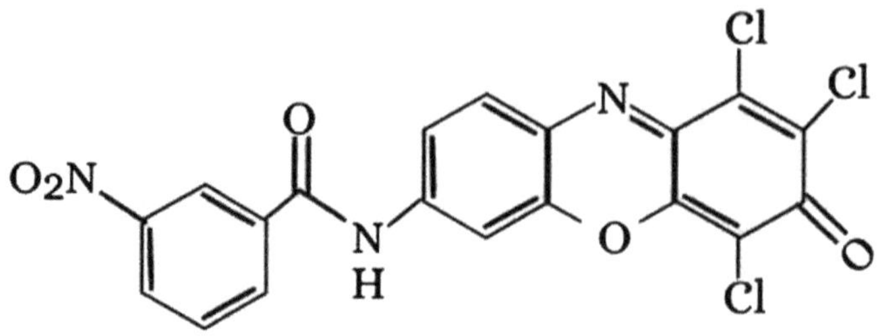

Effect of Dyes on Ecosystem and Mankind

The industries manufacturing dyes relatively represents a small part of the overall chemical industries while on the other hand incomplete exhaustion of dyes during the dyeing process has become a major concern in recent years (Ananthashankar, 2012; Hassaan, 2016). All dyes and chemicals that are introduced in environment must be treated with care as with long term and accidental over exposure they can be hazardous for human health. Exposure to chemicals acting as irritants during the dyeing process gives rise to most predominant health problems (Elliott et al, 1954).

Effect on Aquatic Ecosystem

One of the most important sources of pollution is textile dyeing wastewater. India is one of the major contributors of textile wastewater in South Asia. The unfixed portions of dye on fabrics which get washed out with waters are high in concentration in textile effluents. These effluents are rich in dyes and chemicals which pose a major threat to environment as many of them are non-biodegradable and carcinogenic (Kdasi et al, 2004). These effluents contain chemicals used in various processes which reduce the nutritive values like protein, carbohydrate and lipids in freshwater female crabs. The presence of azo dyes in water affects the aesthetic merit, transparency and water-gas solubility. Due to the reduction of penetrating light through water the photosynthetic activity under the water decreases which cause oxygen deficiency and de-regulation of the biological cycle of aquatic biota (Apostol et al, 2012). It affects the biota of aquatic system by accumulating the sediments in fishes and other aquatic life forms, decomposing the pollutants which results in carcinogenesis and mutagenesis. Owing to the high thermal and photo stability that biodegradation dyes can resist, they can persist in the environment for a long period of time. Azo dyes can also damage the DNA of aquatic organisms after getting ingested and metabolized by the intestinal microorganisms. Dyes can also result in the reduction of RBCs in some fishes. Protein content, pigment content and nutritive content of many algae are also affected by the dyes.

Effects on Soil Properties

Effluents from dye industries contain harmful chemicals or poisonous toxins which when enter into the soil affects the germination rate of different plants consequently decreases the soil fertility (Manu et al, 2003). This can result into low nutritive property of the crops (Savin & Butnaru, 2008). These dyes have the capability to alter the chemical and physical properties of soil.

Effects on Humans

Many respiratory problems can arise due to the inhalation of dye particles and can also affect the person's immune system which has symptoms like itching, watery eyes and sneezing (Hassaan, 2016). Exposure to chemicals present in the dyes can cause skin irritation, blocked nose, sneezing and sore eyes due to the presence of formaldehyde-based resins, acetic acid, ammonia etc. (Nese et al, 2007). They become more dangerous where they are metabolized by liver enzymes. Absorption is faster than ingestion so more dyes can be absorbed at a small timeframe which is very much dangerous for the human health.

METHODS OF TREATMENT

The first commercial synthetic dye was discovered by William Henry Perkin in 1856 (Holme, 2006). Such dyes are soluble in water, effortlessly absorbed, and relatively quick in coloration when compared to the biological dye (Pandey et al, 2016). By the end of 19th century, about ten thousand synthetic dyes were already produced and were used in manufacturing (Robinson et al, 2001). Also, the advancement of worldwide textile industries during the years since then had shown a sudden increase in the usage of the synthetic dyes, led to the sudden rise in pollution due to the wastewater polluted with the dyestuff (Banat et al, 1996; Jin et al, 2007; Doble & Kumar, 2007). Therefore, treatment of effluents comprising dyes and their metabolites is vital before they are discharged to the environment (Figure 11).

Figure 11. Methods of treatment of dyes

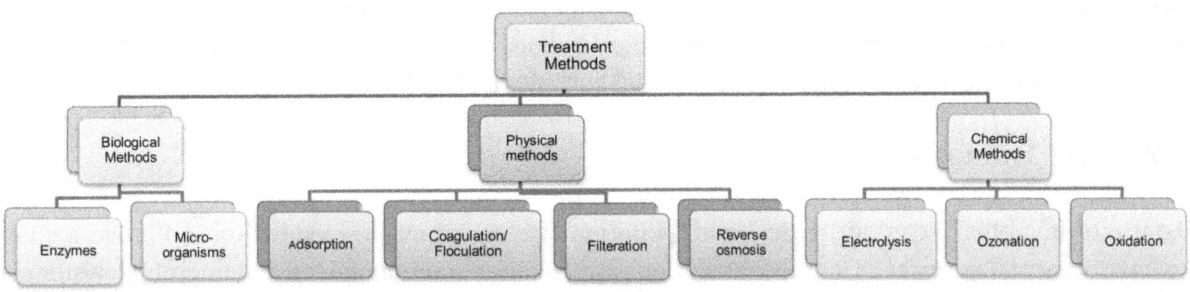

Physical Treatments

Physical methods are based on coagulation-flocculation of dyes and are effective for the removal of mainly sulphur based dyes and disperse dyes. In case of acid, direct reactive and vat dyes they show very low coagulation-flocculation capacity. Although, the application of these techniques has been limited by the low color removal efficiency and also due to large amount of sludge that is produced (Vandevivere et al, 1998). Due to their higher efficiency, adsorption methods have been attracting considerable interest. High affinity properties together with capacity for target compounds and the possibility of adsorbent regeneration are the characteristics to select an adsorbent. Activated carbon (AC) has been very effective adsorbent for different types of dyes (Robinson et al, 2001). Low-cost adsorbent materials such as peat, fly ash, bentonite clay, ion exchangers, polymeric resins and many different biological materials such as maize cobs and stalks as well as wheat straw are used for the removal of dye from wastewater (Ramakrishna & Viraraghavan, 1997). However, regeneration and disposal, high sludge production issues, low effectiveness with regard to wide range of dyes and also high cost are the related problems that has been limiting this practical application of adsorbents (Anjaneyulu et al, 2005; Karcher et al, 2001; Dos Santos et al, 2007). However, some significant drawbacks also include high cost, potential membrane fouling and production of some secondary waste streams which will need further treatment (Robinson et al, 2001).

Chemical Treatments

The decomposition of dye molecules has been enabled by chemical oxidation methods, and such approaches use different oxidizing agents, such as ozone (O_3), hydrogen peroxide (H_2O_2) and permanganate (MnO_4).The dye molecules became susceptible to degradation due to modification in the chemical composition of a compound that takes place in the presence of oxidizing agents (Metcalf, 2003). As it have high reactivity with many azo dyes and lack of alteration of the reaction volume due to its gaseous state, and the good color removal efficiencies, ozonation has been found to be efficient (Alaton et al, 2002). However, its practical application is limited by its short lifetime ineffectiveness towards capacity, and high cost of ozone (Anjaneyulu et al, 2005). The majority of color removal techniques work by concentrating the color into sludge and also by the complete destruction of the colored molecules. To treat color containing wastewater, an aerobic granular sludge method can also be used (Adav et al, 2009). There are some drawbacks that are associated with this method like being economically unefficient; unable to completely remove the recalcitrant azo dyes or their organic metabolites because of the color fastness, stability and resistance of azo dyes towards degradation (Anjaneyulu et al, 2005); generating a good amount of sludge that may lead to secondary pollution; thereby increasing the cost of treatment methods; and also involves tough procedures (Eichlerova et al, 2007; Forgacs et al, 2004).

Biological Treatments

The use of microbial technique to deal with pollution also called as bioremediation is a broad area in the environmental sciences. The biodegradation of recalcitrant compounds in the microbial system is mainly based on the action of the biotransformation enzymes (Saratale et al, 2007). Studies suggests the degradation of organic substances rendered by enzymatic methods, like those which are associated with laccase (Hatvani & Mecs, 2001), lignin peroxidases (Duran & Esposito, 2000), NADH-DCIP reductase (Bhosale et al, 2006), tyrosinase (Zhang & Flurkey, 1997), hexane oxidase (Saratale et al, 2007) and aminopyrine N-demethylase (Salokhe& Govindwar, 1999).

THALLOPHYTES AND THEIR ROLE IN DEGRADATION OF DYES

Degradation of Dyes by Fungi

Filamentous fungi are omnipresent in the environment, colonizing ecological niches like soil, plants and living materials. The potential of fungi to quickly transform their metabolism to differing carbon together with nitrogen sources is important for survival. This metabolic action is accomplished through the generation of a huge set of intra and extracellular enzymes that are proficient to reduce several complex organic pollutants which include organic manure, polyaromatic hydrocarbons, dye effluents and steroids compounds (Saratale et al, 2007). Comprehensive studies have been performed on white-rot fungi that are used to develop different bioprocesses for the mineralization of synthetic dye. It has already reported that production of laccase by *Phanerochaete chrysosporium* and *Neurospora crassa* can also be used for removal of pigments and also phenol from liquid waste. On the other hand, *Trametes versicolor, Aspergillus ochraceus, Bjerkandera adusta*, species of *Pleurotus and Phlebia*, etc. has also attained much attention regarding this (Saratale et al, 2011; Singh & Singh, 2015). Although, the long hydraulic retention time that is required for decolorization to get done completely also limits the performance of the fungal decolorization system (Banat et al, 1996; Chang et al, 2004), as well as the preservation of fungi in bioreactors is also a matter to care (Stolz, 2001).

Degradation of Dyes by Yeast

Biological decolorization of dyes by yeast is mediated by azoreductases present in yeast which catalyze separation of azo groups (–N=N–). Some ascomycetes yeast species like *Candida oleophila, Debaryomyces polymorphus,* and *Candida zeylanoides* involve in the reductive cleavage of azo groups. Decolorization of these strains is because of azo bond reduction that forms the corresponding amines. Recently, a study based on the enzymes responsible for the decolorization of Methyl Red and Malachite Green by involvement of *Saccharomyces cerevisiae* MTCC 463 revealed about different levels of the activities of lignin peroxidase, laccase, NADH-DCIP reductase, azoreductase, tyrosinase, and aminopyrine N-demethylase. It is also suggested by the studies that these products can be further degraded into aliphatic amines that might be done by using oxidative enzymes such as lignin peroxidase and laccase (Jafaria et al, 2014). *Saccharomyces cerevisiae* cells have shown bioaccumulation of certain reactive textile dye like Remazol Blue, Remazol Black B, and Remazol Red RB during the process of growth in molasses (Aksu & Donmez, 2003). Studies suggested that some species of yeast can act as a efficient dye adsorbent and can uptake higher dye concentration (Safarikova et al, 2005), as well as some ascomycetes yeast species, such as *Debaryomyces polymorphus*, *Candida tropicalis* and *Issatchenkia accidentalis* decolorize azo dyes(Saratale et al, 2011). *Galactomyces geotrichum* MTCC 1360 can effectively decolorize triphenylmethane, azo and reactive high exhaust textile dyes (Jadhav et al, 2008). *Trichosporon beigelii* NCIM-3326 has shown to the capability to decolorize Navy Blue HER, with the involved enzymatic mechanisms and toxicity of degradation products (Saratale et al, 2009a).

Degradation of Dyes by Algae

Photosynthetic organisms such as algae and cyanobacteria have ubiquitous distribution. They are found in different habitats worldwide and are involved in wastewater decolorization also. Studies suggest that algae and cyanobacteria can degrade azo dyes by an induced form of an azoreductase (Vijayaraghavan & Yun, 2007a). Colour removal of algae is due to three naturally occurring different mechanisms of assimilative utilization of chromophores that can be utilized for the assembly of algal biomass and CO_2 and H_2O transformation of coloured molecules to non-coloured ones, and also adsorption of chromophores on algal biomass. Several species of *Chlorella* and *Oscillatoria* have also been reported to degrade azo dyes into aromatic amines, and further metabolize them to simpler organic compounds (Acuner & Dilek, 2004). Reports showed that more than 30 azo compounds can be biodegraded by the different species of

Chlorella pyrenodasa, *Chlorella vulgaris* and *Oscillatoria tenuis*into simpler aromatic amines (Yan & Pan, 2004). Algae play an important role in the method of removal of azo dyes and aromatic amines in stabilization ponds. This biosorption process could be adopted as a cost-effective and efficient approach for the decolorization of effluents, and it can be a viable alternative to more costly materials (Banat et al, 1996; Daeshwar et al, 2007).

Degradation of Dyes by Bacteria

The method of decolorization and degradation of azo dyes occurs naturally under conventional anaerobic, facultative anaerobic and aerobic conditions by different groups of bacteria. The microbial degradation of azo dyes include the process of reductive separation of azo bonds by using azoreductase enzymes under anaerobic conditions that result in the formation of colorless solutions and also potentially harmful aromatic amines (Van der Zee & Villaverde, 2005). Intermediate metabolites are also formed like aromatic amines which are further degraded either aerobically or anaerobically (Joshi et al, 2008). Studies have shown about the utilization of microbial biocatalysts to degrade dyes from the effluent (Chang et al, 2004; Hu, 2001). The bacterial decolorization and degradation of certain dyes has been of interest due to high degree of biodegradation and mineralization and it is applicable to a wide variety of azo dyes, inexpensive and environmentally-friendly process, and also produces less sludge (Khehra et al, 2006; Rai et al, 2005; Saratale et al, 2009c; Verma & Madamwar, 2003).

METHODOLOGY AND MECHANISM

Mycoremediation: Degradation by Fungi

Global industrialization is leaving a hazardous impact on living beings. Taking the account of textile industries, the dye effluents discharged by them pollute water bodies and soil to a large extent. It alters chemical oxygen demand (COD), total organic carbon (TOC), pH, colour, biological oxygen demand (BOD) (Banat et al, 1996; Akan et al, 2009). Fungi can degrade a wide range of die effluents. Several fungal organisms have been employed in degradation of dye molecules. Various genera of fungi such as *Aspergillus niger* and white-rot fungi *Phanerochatee chrysosporium* and other fungal organisms involved in mycoremediation approach showed a great impact in the process of detoxification of dye toxicants.

Fungi come out to play an important role in degradation of dyes with their enzymatic actions and adsorption, accumulation and absorption mechanism. Mycoremediation of dyes using fungi is cheap, environment friendly and effective method for treatment of dye effluents.

Fungi and Dye Degradation

The dye effluents released by chemical/dye/textile industries may accommodate some heavy toxic metals like Hg, Cd, Fe, Co, Mn, Mg, Cr and Ni which are mutagenic and carcinogenic and can affect humans directly or indirectly.

Microorganisms have the ability to degrade these toxicants with environment friendly mechanisms. Fungi like *Phlebia radiate, Bjerkandera adusta, Pleurotus* spp., *Phanerochaete chrysosporium* and *Prametes versicolor* have the ability to produce enzymes like laccase which can degrade lignin and abroad structural spectrum of dyes (Hammel et al, 1992; Heinfling et al, 1997; Paszczynski & Crawford, 1991; Podgornik et al, 1999; Singh et al, 2006). A report by Heinfling et al, in 1997 reported that 95% of HRB 8 dye have been decolourised within four days by the mechanism of *Tarmetes versicolor* and *Bjerkandera adusta.*

Dye Degradation Mechanism

Methods of remediation comprises of at least three steps- minor change among organic molecules leaving their molecular structure still and unharmed; complex organic molecules undergoes fragmentation in a way that their fragments can be restructured as they were previously; and then the conversion or organic molecules to minerals i.e. called complete mineralization (Akan et al, 2009; Singh & Singh, 2010).

Absorption of dye molecules by microbes has been observed as an important mechanism of dye degradation (Knapp et al, 1995). Singh & Singh (2017) has reported that hydrophobic-hydrophillic interplay between the dyes and fungus might be effective in absorption of dye molecules as fungal hyphae was observed to absorb dye molecules and the dye decolourized upon increased in concentration of enzymes or cell mass. The enzymatic action and adsorption by cell masses observed to be effective integrated phenomenon.

Adsorption of indigo, acid violet 7 and acid green 27 by *Tarmetes versicolor* on their living and dead mycelia was reported by Wong et al, in 1999. Adsorption of dye molecules has been observed as a primary mechanism of dye degradation. Further, the whole mechanism of dye decolourization by fungi can be categorised into biodegradation, bioaccumulation and biosorption (Husain et al, 2006; Knapp et al, 1995; Singh, 2003).

Biodegradation is a process in which complex molecules are simplified by certain enzymatic action. The process is dependent on metabolic activities and is very energy intensive. Similarly, bioaccumulation is also a metabolically dependent process in which growing cells collect the toxicants in their cytoplasm. Unlike, biodegradation and bioaccumulation, biosorption is metabolically independent and process involves the absorption of toxicants on cellular surface of fungi, thus the dye degradation can be occurred by both living and dead fungi (Srinivasan & Viraraghavan, 2010). Further, to enhance the biosorption and decolourisation capacity of fungi various treatments like heat treatment or soda/acidic treatment can also be applied (Yin et al, 1999, Yan & Viraraghavan, 2000). Kapoor et al, in 1999 have reported that soda treated *Aspergillus niger* removed Pb^{2+}, Cu^{2+}, Cd^{2+} more efficiently by the mechanism of biosorption.

Role of Fungal Enzymes

Fungal enzymes like manganese peroxidases and other extracellular peroxidases have a great role to play in decolourisation of dye (Gold et al, 1988). An edible mushroom *Lentinus edodes* has been absolved to produce manganese peroxidases (Ollikka et al, 1998, Bumpus et al, 1985). Vyas et al, has reported in 1999 that mycelium growth of extracellular ligninolytic enzyme on solid medium produced by *Lentinus edodes* have decolourised various dyes like remazol brilliant blue (RBBR) and poly-478 and synthetic dyes has also been reported to degraded by several enzymes involved in lignin degradation such as laccase, manganese dependent peroxidase and lignin peroxidase (Vyas & Molitoris, 1995).

White rot fungi are also known for its enzyme production efficient for decolourisation of dye (Wesenberg et al, 2003). White rot fungi of Phelephora sp. was reported for dye degradation by their mechanism of ligninolytic activity. *Trichoderma harzianum* and *aspergillus flavus* has also been reported for degradation of several dyes like direct green, bromophenol blue, congo red and acid red (Singh & Singh, 2015).

Mechanism of Aspergillus niger and Phanerochaete chrysosporium

Methodology and Mechanisms

The process of degradation of dye effluents comprises of various steps (Figure 12). The very first step towards examining the decolourisation of selected dye molecules is selection of fungal dye strains (Park et al, 2007; Raniet al, 2014).

Figure 12. Flowchart of treatment of dyes

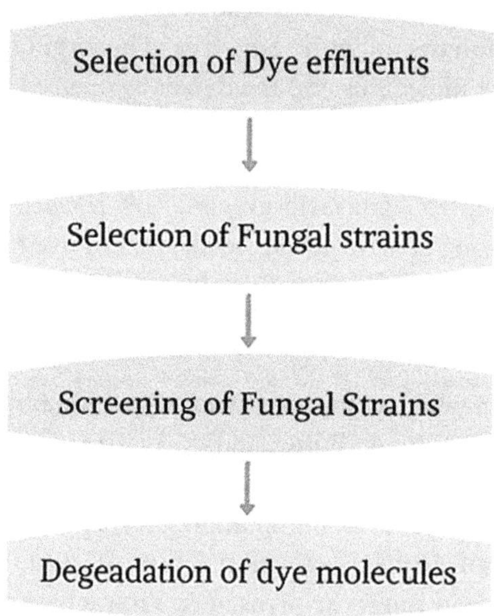

Dyes

The dyes which are used for the purpose mainly are nigrosine, basic fuchsin and malachite green. The structure of these dyes is depicted in Figure 13, 14 and 15 respectively.

Figure 13. Nigrosine

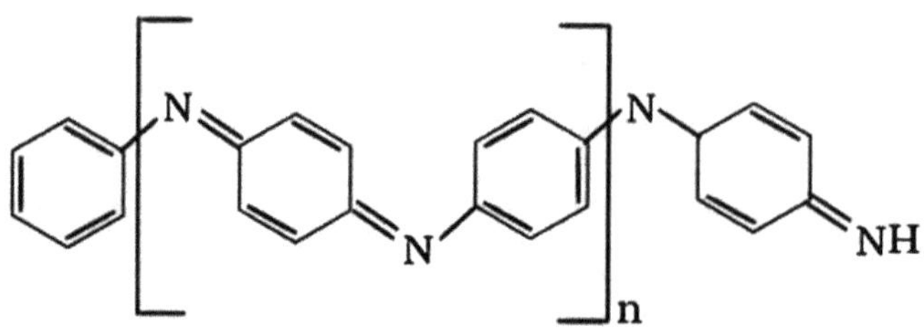

Figure 14. Malachite green

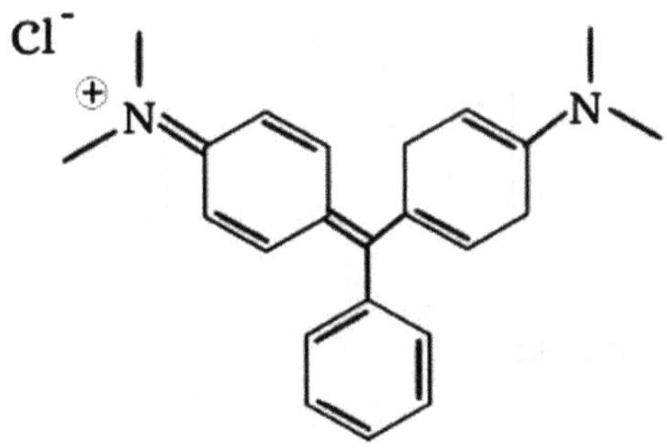

Figure 15. Basic fuchsin

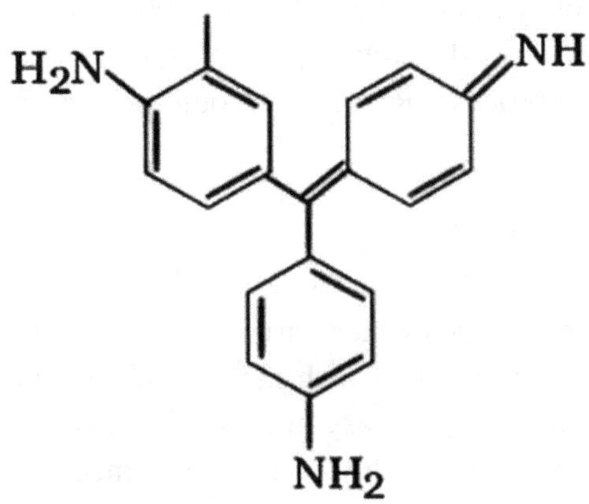

Selection of Sources for Fungal Strains

The very crucial step is the selection of site for the collection of fungal strains. The identification of fungal strains is done based on their morphological characteristics. The soil isolated strains of fungus such as *Phanerochaete chrysosporium* and *Aspergillus niger* are examined for further process of dye decolourisation (Rani et al, 2014) (Table 1 & Table 2).

Table 1. The table shows the compared degradation degree of dyes in liquid medium using Stationery and Shaking method

Fungi	Malachite Green		Nigrosine		Basic Fuchsin		Dye Mixture	
	Stationery	Shaking	Stationery	Shaking	Stationery	Shaking	Stationery	Shaking
Aspergillus niger	Less	Less	Moderate	Moderate	Highest	Highest	Least	Least
Phanerochaete chrysosporiun	Moderate	Less	Highest	Highest	Less	Moderate	Least	Least

Table 2. The table shows the degradation degree of dyes in solid medium under Tube overlay method

Fungi	Malachite Green	Nigrosine	Basic Fuchsin	Dye Mixture
Aspergillus niger	Moderate	Highest	Low	Least
Phanerochaete chrysosporiun	Highest	Low	Moderate	Least

Degradation of Dyes by Laccases

Laccases are the multifaceted enzymes that can catalyze oxidation reactions coupled to four-electron reduction of molecular oxygen to water. They possess multicopper and are widely distributed in higher plants and fungi especially in many white-rot fungi. Laccases provide a useful tool for the biodegradation of different chemical structures of synthetic dyes (Wong, 1999). Phenol oxidase is a range of enzymes produced by some fungi and among its different types, the specific one classified as laccase performs particular reaction in remediation process (Rodriguez Couto & Herrera, 2006; Bibi et al, 2011; Zhou & Xiang, 2013).

Mechanisms of Azo Dye Degradation

Azo dyes are a part of aromatic compounds which carry one or more than one azo bonds (-N=N-). Azo dyes are replaced with naphthalene or benzene which carries different functional groups (-Cl, -NO_2, -OH, -CO, -NH_2 and CH_3). Azo dyes can be cleaved asymmetrically or symmetrically in enzymatic degradation pathway (Telkeet al, 2009a) by using highly non-specific mechanism i.e. free radical mechanism which produces phenolic products.

Degradation of azo dye by laccases begins with asymmetric cleavage of the azo bond along with oxidative cleavage, deamination, dihydroxylation, demethylation and desulfonation, depending on the dye structure (Adnanet al, 2015; Telkeet al, 2011; Telkeet al, 2009b; Zhenget al, 2016; Yang et al, 2015). The first step is carbocation for the decolorization of mono azo dye by laccases, that results in the formation of an electron-deficient reaction centre which forms the intermediates that are highly reactive. This can be liable to the nucleophilic attack by –OH, –SO_3 or halogen nucleophiles, results in cleavage of azo bond asymmetrically (Telke et al, 2010). p-N, p-hydroxybenzene sulfonic acid and N'-dimethylamine phenyldiazine are the degradation products formed from Methyl Orange by laccase. Albeit, there compounds are marked as toxic (Wanget al, 2008).

Bis azo dyes degradation is a more intricate process. These dyes are cleaved in the same way as azo dyes are cleaved, that is asymmetrically by laccases (Siet al, 2013; Adnanet al, 2015; Zheng et al, 2016) however electrons are required for reduction in this reaction (Nam & Renganathan, 2000). As in the catalytic centre, electrons are moved to azo dye, laccases carry four histidine-rich copper binding domains (Zhenget al, 2016). There is a formation of naphthalene amine when biodegradation of the bis azo dye Congo Red takes place by high-redox potential laccases from *Trametes pubescens* (Shleevet al, 2007). It has also been observed that after phytotoxic test azo dye Congo Red is detoxified by purified laccase, which concludes that degradation of Congo Red does not complete in this phase but it probably continues forming other non-toxic degradation products (Si et al, 2013) (Figure 16).

Figure 16. Mechanism of degradation of congo red with the help of purified laccases fromTrametes pubescens

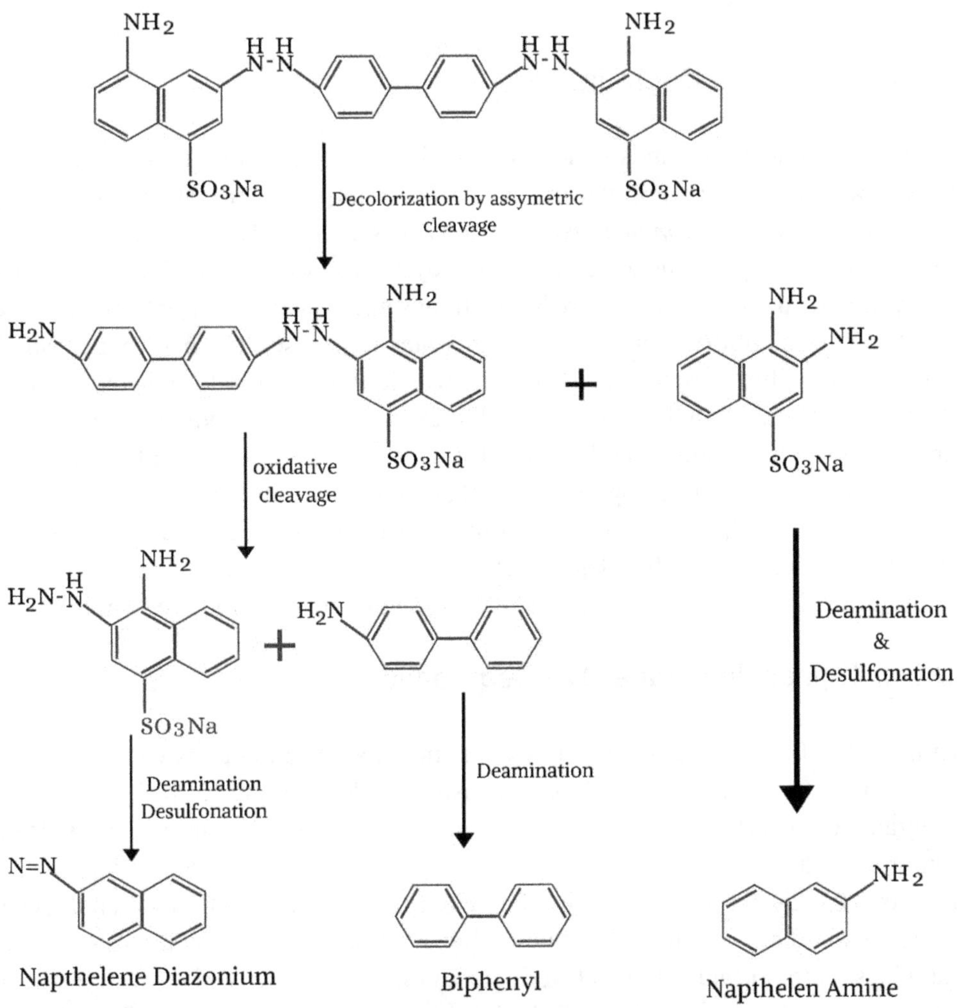

Comparably, cleavage of bis azo bond initiates the bis azo dye Reactive Black 5 biodegradation by laccase from *Trichoderma artroviride* F03 which is followed by hydroxylation, sulphonationand deamination. The mechanism of biodegradation carried on with the aromatic ring fission of naphthalene-1,2,8-triol, where its oxygenated the ring at C_1 and C_2 position was cleaved to 2-(2-carboxy-ethyl)-6-hydroxy-benzoic acid via 8-hydroxy-[1,2]-naphthoquinine. It has been reported that by two possible pathways, 2-(2-carboxyethyl)-6-hydroxy-benzoic acid can be degenerated;

1. It goes through methylation and decarboxylation to form 2,4-ditertbutylphenol, and
2. By decarboxylation mechanism, it is converted to benzoic acid (Adnanet al, 2015).

Furthermore, toxic aromatic amines did not be generated by laccase from *T. Artroviride* D03 (Adnanet al, 2015). It has also been found that the biodegradation of Acid Black 172 by laccase does not form toxic amines (Zhenget al, 2016).

Mechanisms of Indigo Dye Degradation

Indigo dyes are the organic compounds which are mostly distinguished by their peculiar blue colour. Earlier, indigo dye was extracted naturally from leaves of various plants such as *Indigofera suffructicosa* Mill., *Indigoferatinctoria* L., *Polygonum tinctorium* and *Isatis tinctoria* L. At present time large amounts of Indigo dyes are produced by chemical synthesis by Baeyer-Drewson reaction known as synthetic dyes. These dyes carry intra- and inter-molecular hydrogen bonds due to which they not soluble in water, ether and alcohol that is why, solubilisation of this dye in water with a sodium hydroxide (NaOH/base) is needed to induce its solubility in water. Indigo Carmine, a derivative of Indigo mainly used in different sectors like textile, food, and medicine for its color but these dyes are extremely toxic.

To start the degradation of Indigo and its derivatives, at first, electrochemical oxidation to dehydro indigo is required, succeeded by nucleophile attack that causes the integration of O-atoms into degradation products (Campos et al, 2001). Indigo and its derivatives are degraded by laccases via isatine (indol-2,3-dion) formation, which is then degraded to anthranilic acid (2-aminobenzoic acid) involuntary by isatic acid decarboxylation, which is intermediate formed hydrolytically after isatine degradation.

Mechanisms of Triphenylmethane Dye Degradation

Triphenylmethane (TPM) dyes are considered as synthetic organic compounds which have acute colour and are used in various sectors such as food, paper, cosmetics, leather and textile industries, and also in medicines. Degradation of TPM dye requires more time as these are resistant to enzymatic decolorization (Forootanfar et al, 2012).

Degradation of TPM dye by Laccases is by the oxidation of methyl carbon which attached in TPM dye structure, results in the formation of stable products that are affected by p-substituted phenyl. N-demethylation plays an important role in TPM degradation (Bibi et al, 2011). Laccases degrade TPM dyes but not able to degrade non-substituted TPM dyes completely (Casas et al, 2009). Malachite Green (MG) is a common type of TPM dye mostly used in farmed fish for managing of fungal and protozoan infection and there are two parallel pathways proposed for fast and efficient degradation of MG by laccase from Cerrena sp. (Yang et al, 2015) likely depended on laccase type and reaction state.

In the first pathway, demethylation of MG occurs and continues by polymerization or degradation of MG for the destruction of chromophore. The second pathway begins with hydroxylation of MG to its carbinol form (Fischer et al, 2011) which simply degenerates.

Mechanisms of Anthraquinone Dye Degradation

Anthraquinone dyes are known to be most abundant group of dyes and are second most important class of textile dyes (Baughman &Weber, 1994). These dyes provide variety of colours such as violet, green and blue which have long-term colour stability (Menget al, 2003). Remazol Brilliant Blue R (RBBR) and Acid Blue 129 are the anthraquinone dyes mainly used in process of textile production. In production of polymeric dye, anthraquinone dyes were used as reactive dyes and these dyes are often toxic and act as pollutants.

Anthraquinone dyes can get decolorized by laccases more effectively than any other classes of dyes (Zeng et al, 2011). Laccases catalyse anthraquinone dyes that decolorize azo dye by acting as oxidation-reduction (redox) mediators (Zeng et al, 2011). Laccase degrade the anthraquinone dyes by breaking down the chromophore of dye and forms smaller molecules which have lower toxicity levels. Laccase from *T. pubescens* biodegrade RBBR by various reaction pathways such as hydroxylation, reduction, oxidation and deamination (Osma et al, 2010) (Figure 17).

Figure 17. Pathway for degradation of Remazol Brilliant Blue R (RBBR) with the help of laccase of Trametes pubescens

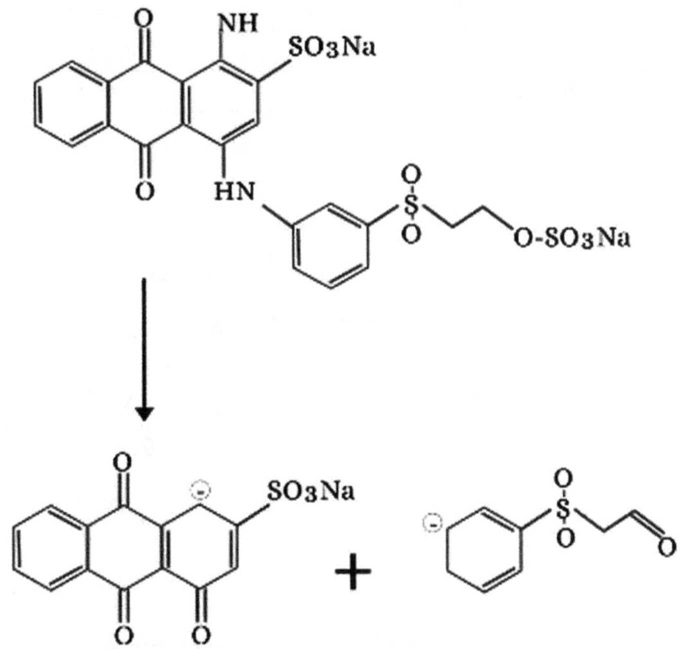

Laccase are able to oxidise various organic compounds like di- and mono- phenols or their derivatives along with carboxy-, hydroxyl-, amino-, sulpho- or methoxy- functional groups by radical mechanism. Phenols have low redox potential due to which known as typical substrates for laccases that permits electron substraction by the help of Cu T1. Hence, the capacity of laccase enzymes to oxidise molecules is measured by redox potential of Cu T1 (E^0 Cu T1). E^0 can be measured by the use of potentiometric titrations for different laccases (Table 3).

Table 3. Several values of redox potential E^0 (Cu T1) for laccases that are extracted from different organisms

Organism	MW (kDa)	pH	E^0 (Cu T1) (mV vs. NHE)	References
Coprinus cinereus	58 kDa	5.5 pH	0.55	Schneideret al, 1999
Rhus vernicifera	NOT FOUND	5.5-8.5 pH	0.410	Johnsonet al, 2003
Pycnoporus sanguineus	67 kDa	4.5 pH	0.747	Zimbardi et al, 2016
Trametes ochracea	64±2 kDa	3.7- 4.9 pH	0.79±0.1	Shleevet al, 2005
Cerrena maxima	67±4 kDa	4.0-6.0 pH	0.75±0.05	Shleevet al, 2005
Pleurotus ostreatus	NOT FOUND	3.0 pH	0.588	Dai et al, 2016
Trametes hirsuta	70±2 kDa	3.5-4.5 pH	0.78±0.1	Shleevet al, 2005
Ganoderma sp.	62 kDa	3.0-5.0 pH	0.63	Sharmaet al, 2013
Melanocarpus alomyces	NOT FOUND	8.0 pH	0.46±0.01	Kruuset al, 2002
Coriolopsis fulvocinerea	65±2 kDa	3.9-5.2 pH	0.78±0.1	Shleevet al, 2005
Marasmius guercophilus C30	65 kDa	5.7 pH	0.73	Klonowskaet al, 2002

Figure 18. Orange II

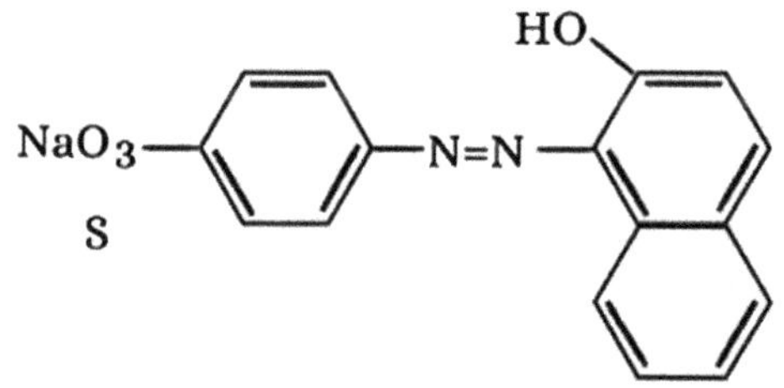

Figure 19. Methyl Red

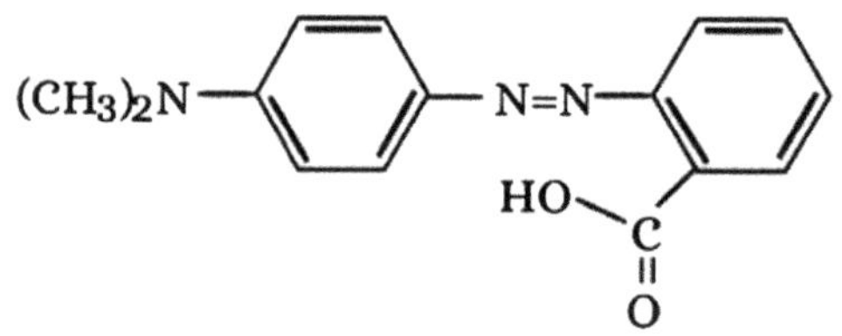

Figure 20. Methyl Orange

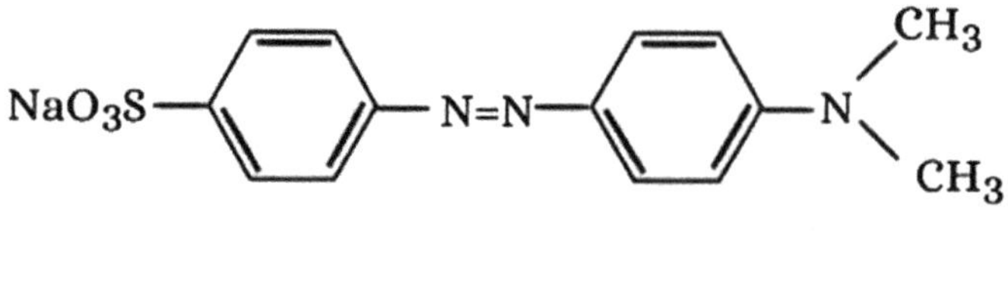

Degradation of Dyes by Yeasts

In earlier sections, we have discussed the discharge of dye effluents from industries and their hazardous effects. Among the dye constituents, the largest consuming and contributing group to the humans and dye effluents respectively are azo dyes.

Yeasts are capable to resist unfavourable conditions like intense organic wastewater or salt concentration as in case of industrial discharge, low pH, etc. These characteristics stand to be great advantage in bioremediation of dyes (Ramalho et al, 2002).

Azo Dyes

Azo dyes have a great diversity among them. There are thousands of azo dye variants. Some examples of azo dyes are reactive blue 171, reactive red 141, methyl orange, reactive yellow 84, reactive black 5, orange II, methyl red, etc. (Figure 18, 19 and 20).

Biodegradation Process

Previously in this chapter the enzymatic action of several fungi like *Thelelophera* sp., *Tricoderma, Harzianum* and *Aspergillus flavus* have been discussed that explains about the way they decolourise various dyes like direct green, bromophenol glue, congo red and acid red. Similarly, several species of yeasts like *Candido albicans*, *Saccharomycus*, *Cerevisiae*, *Candida tropicalis*, *Candida oleophila*, *Issatchenkia occidentalis*, *Candida zeylanoides* and *Debaryomyces polymorphus* plays a great role in bioremediation/ biodegradation of azo dyes with their vital enzymatic action (Martins et al, 1999; Ramalho et al, 2002; Yang et al, 2003; Ramalho et al, 2005; Ramalho et al, 2004; Lucas et al, 2006; Vitor & Corso, 2008).

The enzymatic action of yeasts in the azo dye biodegradation can be categorized as the mechanism of two reactions- oxidative reactions and reductive actions. The oxidative breakdown of azo dye molecules can be performed by the manganese dependent peroxidase, laccase (ligninolytic enzymes) and lignin peroxidase. Generally, the oxidative breakdown of azo dyes by ligninolytic enzymatic action found to be the leading cause behind the formation of diazene and benzoquinane derivativedue to the nucleophilic water attack and carbonium ion formation. Further, diazene is oxidized to breakdown of nitrogen molecules which leads the production of hydrperoxide derivative (Chivukula et al, 1995).

Non-Biodegradation Process

Yeasts possess a major non-enzymatic mechanism for the decolourisation of azo dye molecules. The whole mechanism can be categorised into two types- bioaccumulation and biosorption. Bioaccumulation is the process in which uptake of toxicants occurs by the action of living microbes. In case of bioaccumulation by yeasts, yeast metabolism may further lead to biotransformation of dye by redox reactions. Unlike biosorption, bio accumulation does not lead to the formation of any additional sludge hence no further treatment is required. Although, the employment of living yeasts in the process also have certain limitations like it cannot deal with highly with toxic effluents.

Biosorption is a process that involves absorption of toxicants on cellular surface of living or dead yeasts with the help of physical-chemical interactions. There is a lot of advantage using the mechanism of dead yeasts in biosorption. The dead yeasts biomass can perform better in extreme conditions of pH and temperature without accelerating the growth of nutrients (Meehan et al, 2000). By adopting this phenomenon, the waste is biomass produced by the beer industries can also be utilised. It has some drawbacks, however, as the biosorption mechanism by yeast azo dye molecules leads to the creation of sludge that requires further treatment.

Degradation of Dyes by Algae

Azo Dye Degradation by Chlorella vulgaris

Chlorella vulgaris is recognized as the microalgae widely used in synthetic azo dye degradation. Biomass based on *Chlorella* is also used as biosorbent which basically functions for the removal of malachite green. For the removal of colour from dyes and wastewater, both living and non-living algae can be used for the purpose. *Caulerpa scalpelliformis* and *Caulerpa lentillifera* are living biomass of macroalgae that are used to remove basic dyes by the help of biosorption. From the mono-azo dye tectilon yellow 2G, 63-69% colour was removed by *Chlorella vulgaris* by transforming it into aniline (Aravindhan et al, 2007). Waste biomass of *Corynebacterium glutamicumas* can be utilized as a biosorbent for the eviction of reactive black 5 from the aqueous solution. At pH 1 and initial concentration of dye i.e., 500 mh/l, biosorption role of *Corynebacterium glutamicum* for reactive black 5 was highest (Vijayaraghavan & Yun, 2007b).

Biosorption of ramazol red RR, ramazol golden yellow RNL and ramazol black B can be done by using dried *Chlorella vulgaris* (Aksu & Donmez, 2003). The biomass of algae shows the maximum dye uptake limit for all the dyes at initial pH of 2.0, where, Ramazol black B was the dye which adsorbed more efficiently by using biosorbent. Variety of species of *Chlorella pyrenoidosa, Oscillatoria tenuis, Chlorella vulgaris* and *Spirogyra* can degrade large number of azo dyes to certain limit, suggesting reduction are linked to the molecular structure of dyes and species of algae used.

Degradation of Dyes by Bacteria

Like other microorganisms, bacteria also play a vital role in degradation of dye molecules. It has a major role to play in biodegradation of triphenyl methane and azo dyes. There are several advantages of using bacteria in the bioremediation of dyes. Bacterial biodegradation of dyes leaves non-toxic complete mineralised end products and are inexpensive at the same time. Several bacteria like *Protus vulgaris* NCIM-202, *Pseudomonas* sp., *Citrobacter* sp., *Aeromonas hydrophila, Acinetobacter calcoaceticus, Bacillus* sp., *Exiguobacterium* sp., *Pseudomonas aeruginosa strain* BCH and *Pseudomonas* sp. The dye degradation process of bacteria comprises of various mechanisms such as the mechanism of enzymatic and non-enzymatic action under aerobic and anaerobic conditions (Wu et al, 2005).

Immobilized Bacteria

There are several reports suggesting that microorganism immobilization is effective in wastewater treatment (Chibata et al, 1981; Hyde et al, 1991; Zeroual et al, 2001; Isaka et al, 2007). There are several methods which can be opted in immobilization of bacteria. The four main methods involved in the bacterial cell in immobilization are covalent binding, microencapsulation, matrix entrapment and adsorption (Monsan et al, 1982). Among all methods, entrapment in Polyvinyl alcohol (PVA) in gel beads has been observed as the best method because of their higher operational and stability advantage and other benefits like less toxicity to system, low cost and easy use. Some precautions are also taken to preserve mechanical strength of bacterial cells during immobilization for the degradation process of dyes. Based on the abilities of immobilized bacterial cells in degradation of azo dyes, they have been subjected to aerobic and anaerobic condition (Georgiou et al, 2005).

Aerobic Condition

Most dyes are non-biodegradable under aerobic condition. Many studies have been done in last decades to find a suitable method for degradation of dye in aerobic condition. Bacterial strain of *Enterobacter agglonerans* with fluidized bed reactor (FBR) during different support systems was incubated with methyl red dye under aerobic condition. A complete decolourisation of methyl red was found after six hours of incubation (Zeroual et al, 2001).

Anaerobic Condition

The concept of anaerobic decolourisation in dye degradation has been practised since 1970s. Anaerobic bacteria were immobilized using reticulated sintered glass for the azo dye degradation process. Full degradation of dye molecules was found among azo dyes in less than four hours HRT (Hydraulic Retention Times). Methane and biogas production were also observed in the process (Khan et al, 2013).

FACTORS INFLUENCING DECOLOURIZATION

The biodegradation of synthetic dyes and other chemicals in textile effluent depends on the chemical, physical, and biological processes, likewise some environmental factors. The major factors influencing the decolourization of dyes include

1. Dye structure
2. Dye concentration
3. Carbon and nitrogen sources in dyes
4. Temperature and pH
5. Dissolved oxygen

Dye Structure

Dyes having simpler structures and low molecular weight shows colour removal at higher rates. The character of substituents on the aromatic ring has been shown to own an impact on oxidation reaction. Studies demonstrated that methyl and methoxy substituents donates electron and enhances the enzymatic degradation of azo phenols, while electron withdrawing chloro-, fluoro-, and nitro- substituents inhibit oxidation (Singh & Singh, 2015).

Dye Concentration

Studies show that, increasing the dye concentration gradually decreases the decolorization rate. It can be because of dyes, which is toxic on the microorganisms. Various other possible reasons may be the inadequate cell to dye ratio or may be blockage of active sites of azo reductase by dye molecules with different structures (Jadhav et al, 2008; Sani et al, 1999; Saratale et al, 2009a; Tony et al, 2009a; Tony et al, 2009b). It has been reported that removal of dyes using species of algae is dependent on the concentration of dye and the biomass of algae (Venkata Mohan et al., 2002). However, Yang et al. (2011) found that there was a negative relation between the dye removal and the initial dye concentration while using *Shewanellao neidensis* to decolorize the acid yellow 199. Concentration of dyes were selected based on the highest decolorization percentage and are used to optimize the pH and temperature for effective decolorization of dyes. LG1 for RB5 and DR1 was taken in 200 mg concentration and of LG1 for DB71 was 300 mg and pH of the growth media is optimized for decolorization.

In the lichen *Permelia perlata*, the study effect for the dye concentration SR24 shown that upon increasing the concentration the decolorization showed 99, 80, 76, 75 and 72% decolorization for 50, 100, 150, 200 and 250 mgL^{-1} concentrations at 24 hours. Fungus *Aspergillus ochraceus* showed maximum decolorization at concentration 50 mg/L at 24 hours (Kadam et al, 2014). However, in case of microalgae *Cosmarium* species, the decolorization was at 10 mgL^{-1} (Daeshwar et al, 2007). Symbiosis of algae and fungus is the product lichen; thus, it might be more resistant to the increasing dye concentration. Significant decolorization activity of *P. perlata* was shown at higher dye concentration ranges at 250 mgL^{-1}.

Carbon and Nitrogen Sources

Most of the microorganisms generally cannot utilize the carbon and/or a nitrogen source present in the dyes for their growth. The types of bacterial cultures that cannot utilize the carbon or nitrogen source present requires carbohydrate sources, complex organic sources like yeast extract, peptone, or a combination of both the organic source (Khan et al, 2013).

Temperature and pH

Yang and his colleagues have reported that *S. oneidensis* decolorize acid yellow199 (Yang et al., 2011). Rate of decolorization decreases at more acidic or alkaline pH and increases at optimal pH. Textile industrial processes take place mainly under the alkaline conditions thus, the high pH tolerance is important. The optimal often being between 6.0 and 10.0 (Saratale et al, 2011). It is observed that with the increase in the range of temperature (optimum) the decolorization rate increases but increasing the temperature further drastically decreases the rate. At very high temperature the azo reductase can be denatured.

It has been reported that the effect of temperature and pH on decolorization of SR24 (Solvent red 24) was via lichen *P. perlata* (Kulkarni et al, 2014). Temperature of 27 °C is optimum for the degradation of true-blue dye via *Aspergillus niger* which is a fungi and also showed 48, 50, 85, 99 and 79% decolorization at pH 2, 4, 6, 8 and 10, respectively (Ponraj et al, 2011). Alkaline pH i.e., pH 8 is suitable for the decolorization of SR24 dye. Similarly, pH 9 (again alkaline) was found to be suitable in case of degradation via microalgae *Cosmarium* sp. (Daeshwar et al, 2007). The algal and fungal systems mostly prefer mesophilic temperature and pH conditions for the biodegradation activity, but lichens usually operate at alkaline pH which is slightly more than the mesophilic range hence these can be applied to the textile effluent treatment processes with higher temperature and alkaline pH (Kulkarni et al, 2014).

Dissolved Oxygen

Different groups of bacteria which are used can decolorize the dyes under conditions like aerobic, anaerobic, and facultative anaerobic conditions. The reductive enzyme activities are higher in anoxic conditions that break down the synthetic dye's structure. For azo dye reduction processes the dissolved oxygen (DO) is taken as inhibitor for both the molecules that act as electron acceptors and oxygen is a much stronger oxidant (Jafaria et al, 2014). Upon treatment of red dye effluent from *Spirogyra* species there was a decrease in BOD reported. Similarly, the BOD in blue dye effluent treated from *Spirogyra* and *Oscillatoria* was found to get decreased significantly (Brahmbhatt et al, 2016).

CONCLUSION

Synthetic dyes have detrimental impact on the environment. Some of the dyes cause hazardous health related problems in human. The harmful chemicals present in dyes affects the biological system under water. They also have bad impact on the nutritive value of soils and thereby on crops. There is no effective method for extracting the dangerous dye portion from the ecosystem so far. Integrated therapy of bioagents involving physical and chemical agents can, however, be effective in the treatment of dye effluents. The composition and structure of the dye, appropriate parameters such as temperature and pH, and the dissolved oxygen are the factors affecting the decolorisation. From the complex azo dyes to their dissociation via the thallophytes is a new scope for sustenance.

The literature indicates that algae, bacteria, and fungi have been well used for the treatment of dyes in aerobic and anaerobic environments. The comprehensive method of remediation, along with the extraction of enzymes, lipids and biofuels, as subsequent intervention, appears to be the best technique for economic agriculture. Various studies have supported that laccases have the capability to degrade synthetic dyes that have different chemical structures. Types of laccase-producing organism affects the efficiency of biodegradation that determines the rates of dye biodegradation and redox potential of laccases. Nonetheless, numerous plant and bacterial species produce low-redox potential laccases, and filamentous fungi produce medium to high-redox potential laccases. Thallophytes have been used to degrade complex dyes across different temperature and pH ranges. Recently, thallophytes such as algae and lichens have been used to treat textile effluents with efficient higher temperature and alkaline pH with diminishing BOD and therefore in a convenient and cost-effective way to remove them from the setting.

The oxidation of toxic dyes can be tested by a combined consortium of the strongest types of algae, bacteria, and fungi. Toxicity assessments specifically demonstrate which strain is better for recycled water for future applications.

REFERENCES

Acuner, E., & Dilek, F. B. (2004). Treatment of Tectilon Yellow 2G by Chlorella vulgaris. *Process Biochemistry*, *39*(5), 623–631. doi:10.1016/S0032-9592(03)00138-9

Adav, S. S., Lee, D. J., & Lai, J. Y. (2009). Treating chemical industries influent using aerobic granular sludge: Recent development. *Journal of the Taiwan Institute of Chemical Engineers*, *40*(3), 333–336. doi:10.1016/j.jtice.2009.02.002

Adnan, L. A., Sathishkumar, P., Mohd Yusoff, A. R., & Hadibarata, T. (2015). Metabolites characterisation of laccase mediated Reactive Black 5 biodegradation by fast growing ascomycete fungus *Trichoderma atroviride* F03. *International Biodeterioration & Biodegradation*, *104*, 274–282. doi:10.1016/j.ibiod.2015.05.019

Akan, J. C., Abdulrahman, F. I., Ayodele, J. T., & Ogugbuaja, V. O. (2009). Impact of tannery and textile effluent on the chemical characteristics of Challawa River, Kano State, Nigeria. *Australian Journal of Basic and Applied Sciences*, *3*(3), 1933–1947.

Aksu, Z., & Donmez, G. (2003). A Comparative Study on the Biosorption Characteristics of some Yeasts for Remazol Blue Reactive Dye. *Chemosphere*, *50*(8), 1075–1083. doi:10.1016/S0045-6535(02)00623-9 PMID:12531715

Alaton, A., Balcioglu, I. A., & Bahnemann, D. W. (2002). Advanced Oxidation of a Reactive Dyebath Effluent: Comparison of O3, H2O2/UV-C and TiO2/UV-A Processes. *Water Research*, *36*(5), 1143–1154. doi:10.1016/S0043-1354(01)00335-9 PMID:11902771

Allen, R. L. M. (1971). Reactive dyes. In *Colour Chemistry. Studies in Modern Chemistry*. Springer. doi:10.1007/978-1-4615-6663-2_13

Ananthashankar, R. (2012). *Treatment of textile effluent containing reactive red 120 dye using advanced oxidation* (M.Sc. Thesis). Dalhousie University, Halifax, Nova Scotia, Canada.

Apostol, L. C., Pereira, L., Pereira, R., Gavrilescu, M., & Alves, M. M. (2012). Biological decolorizartion of Xanthene dyes by anaerobic granular biomass. *Biodegradation*, *23*(5), 725–737. doi:10.100710532-012-9548-7 PMID:22437968

Anjaneyulu, Y., Sreedhara Chary, N., & Raj, S. S. D. (2005). Decolourization of Industrial Effluents-Available Methods and Emerging Technologies—A Review. *Reviews in Environmental Science and Biotechnology*, *4*(4), 245–273. doi:10.100711157-005-1246-z

Aravindhan, R., Rao, J. R., & Nair, B. U. (2007). Removal of basic yellow dye from aqueous solution by sorption on green alga Caulerpa scalpelliformis. *Journal of Hazardous Materials*, *142*(1-2), 68–76. doi:10.1016/j.jhazmat.2006.07.058 PMID:16938392

Banat, I. M., Nigam, P., Singh, D., & Marchant, R. (1996). Microbial decolorization of textile-dyecontaining effluents: A review. *Bioresource Technology*, *58*(3), 217–227. doi:10.1016/S0960-8524(96)00113-7

Baughman, G. L., & Weber, E. J. (1994). Transformation of dyes and related compounds in anoxic sediment: Kinetics and products. *Environmental Science & Technology*, *28*(2), 267–276. doi:10.1021/es00051a013 PMID:22176172

Bhosale, S., Saratale, G., & Govindwar, S. (2006). Biotransformation enzymes in *Cunninghamella blakesleeana* (NCIM-687). *Journal of Basic Microbiology*, *46*(6), 444–448. doi:10.1002/jobm.200510117 PMID:17139609

Bibi, I., Bhatti, H. N., & Asgher, M. (2011). Comparative study of natural and synthetic phenolic compounds as efficient laccase mediators for the transformation of cationic dye. *Biochemical Engineering Journal*, *56*(3), 225–231. doi:10.1016/j.bej.2011.07.002

Brahmbhatt, N. H., & Jasrai, R. T. (2016). The Role of Algae in Bioremediation of Textile Effluent. *International Journal of Engineering Research and General Science*, *4*(1), 443–451.

Bumpus, J., Tien, M., Wright, D., & Aust, S. (1985). Oxidation of persistent environmental pollutants by a white rot fungus. *Science*, *228*(4706), 1434–1436. doi:10.1126cience.3925550 PMID:3925550

Burkinshaw, S. M. (1995). *Chemical Principles of Synthetic Fibre Dyeing*. Springer Science & Business Media. doi:10.1007/978-94-011-0593-4

Burkinshaw, S. M., & Salihu, G. (2019). The role of auxiliaries in the immersion dyeing of textile fibres part 2: Analysis of conventional models that describe the manner by which inorganic electrolytes promote direct dye uptake on cellulosic fibres. *Dyes and Pigments*, *161*, 531–545. doi:10.1016/j.dyepig.2017.08.034

Burkinshaw, S. M., Jeong, D. S., & Chun, T. I. (2013). The coloration of poly (lactic acid) fibres with indigoid dyes: Part 2: Wash fastness. *Dyes and Pigments*, *97*(2), 374–387. doi:10.1016/j.dyepig.2012.12.026

Campos, R., Kandelbauer, A., Robra, K., Cavaco-Paulo, A., & Gübitz, G. (2001). Indigo degradation with purified laccases from Trametes hirsuta and Sclerotium rolfsii. *Journal of Biotechnology*, *89*(2-3), 131–139. doi:10.1016/S0168-1656(01)00303-0 PMID:11500206

Casas, N., Parella, T., Vicent, T., Caminal, G., & Sarrà, M. (2009). Metabolites from the biodegradation of triphenylmethane dyes by *Trametes versicolor* or laccase. *Chemosphere*, *75*(10), 1344–1349. doi:10.1016/j.chemosphere.2009.02.029 PMID:19298999

Chang, J. S., Chen, B. Y., & Lin, Y. S. (2004). Stimulation of bacterial decolorization of an azo dye by extracellular metabolites from Escherichia coli strain NO3. *Bioresource Technology*, *91*(3), 243–248. doi:10.1016/S0960-8524(03)00196-2 PMID:14607483

Chibata, I., & Tosa, T. (1981). Use of immobilized cells. *Annual Review of Biophysics and Bioengineering*, *10*(1), 197–216. doi:10.1146/annurev.bb.10.060181.001213 PMID:7020575

Chivukula, M., & Renganathan, V. (1995). Phenolic azo dye oxidation by laccase from *Pyricularia oryzae*. *Applied and Environmental Microbiology*, *61*(12), 4374–4377. doi:10.1128/AEM.61.12.4374-4377.1995 PMID:16535191

Clark, M. (2011). *Handbook of Textile and Industrial Dyeing: Principles, Processes and Types of Dyes* (Vol. 1). Elsevier.

Daeshwar, N., Ayazloo, M., Khataee, A. R., & Pourhassan, M. (2007). Biological Decolorization of Dye Solution Containing Malachite Green by Microalgae *Cosmarium* sp. *Bioresource Technology*, *98*(6), 1176–1182. doi:10.1016/j.biortech.2006.05.025 PMID:16844368

Doble, M., & Kumar, A. (2007). *Biotreatment of Industrial Effluents. Resource Conservation Recycling*. Elsevier.

Dos Santos, A. B., Cervantes, F. J., & van Lier, J. B. (2007). Review Paper on Current Technologies for Decolourisation of Textile Wastewaters: Perspectives for Anaerobic Biotechnology. *Bioresource Technology*, *98*(12), 2369–2385. doi:10.1016/j.biortech.2006.11.013 PMID:17204423

Duran, N., & Esposito, E. (2000). Potential Applications of Oxidative Enzymes and Phenol oxidase-like Compounds in Wastewater and Soil Treatment. A Review. *Applied Catalysis B: Environmental*, *28*(2), 83–99. doi:10.1016/S0926-3373(00)00168-5

Eichlerova, I., Homolka, L., Benada, O., Kofroňová, O., Hubálek, T., & Nerud, F. (2007). Decolorization of Orange G and Remazol Brilliant Blue R by the white rot fungus *Dichomitus squalens*: Toxicological evaluation and morphological study. *Chemosphere*, *69*(5), 795–802. doi:10.1016/j.chemosphere.2007.04.083 PMID:17604080

Elliott, A., Hanby, W., & Malcolm, B. (1954). The near infra-red absorption spectra of natural and synthetic fibres. *British Journal of Applied Physics*, *5*(11), 377–381. doi:10.1088/0508-3443/5/11/301

Fischer, A. R., Werner, P., & Goss, K.-U. (2011). Photodegradation of malachite green and malachite green carbinol under irradiation with different wavelength ranges. *Chemosphere*, *82*(2), 210–214. doi:10.1016/j.chemosphere.2010.10.019 PMID:21035831

Forgacs, E., Cserháti, T., & Oros, G. (2004). Removal of synthetic dyes from wastewaters: A review. *Environment International*, *30*(7), 953–971. doi:10.1016/j.envint.2004.02.001 PMID:15196844

Forootanfar, H., Moezzi, A., Aghaie-Khozani, M., Mahmoudjanlou, Y., Ameri, A., Niknejad, F., & Faramarzi, M. A. (2012). Synthetic dye decolorization by three sources of fungal laccase. *Iranian Journal of Environmental Health Sciences & Engineering*, *9*(1), 27. doi:10.1186/1735-2746-9-27 PMID:23369690

Georgiou, D., Hatiras, J., & Aivasidis, A. (2005). Microbial immobilization in a two-stage fixed-bed-reactor pilot plant for on-site anaerobic decolorization of textile wastewater. *Enzyme and Microbial Technology*, *37*(6), 597–605. doi:10.1016/j.enzmictec.2005.03.019

Gold, M. H., Glenn, J. K., & Alic, M. (1988). Use of polymeric dyes in lignin biodegradation assays. *Biomass Part B: Lignin, Pectin, and Chitin*, *181*, 74–78. doi:10.1016/0076-6879(88)61011-1

Hammel, K. E., Gai, W. Z., Green, B., & Moen, M. A. (1992). Oxidative degradation of phenanthrene by the ligninolytic fungus *Phanerochaete chrysosporium. Applied and Environmental Microbiology*, *58*(6), 1832–1838. doi:10.1128/AEM.58.6.1832-1838.1992 PMID:1622259

Hassaan, M. A. (2016). *Advanced oxidation processes of some organic pollutants in fresh and seawater* (PhD Thesis). Faculty of Science, Port Said University.

Hassaan, M. A., & Nemr, A. E. (2017). Health and Environmental Impacts of Dyes: Mini Review. *American Journal of Environmental Science and Engineering*, *1*(3), 64–67.

Hatvani, N., & Mecs, I. (2001). Production of Laccase and Manganese Peroxidase by *Lentinus edodes* on Malt Containing by Product of the Brewing Process. *Process Biochemistry*, *37*(5), 491–496. doi:10.1016/S0032-9592(01)00236-9

Heinfling, A., Bergbauer, M., & Szewzyk, U. (1997). Biodegradation of azo and phthalocyanine dyes by *Trametes versicolour* and *Bjerkandera adusta. Applied Microbiology and Biotechnology, 48*(2), 261–266. doi:10.1007002530051048

Holme, I. (2002). Recent developments in colorants for textile applications. *Surface Coatings International. Part B, Coatings Transactions, 85*(4), 243–264. doi:10.1007/BF02699548

Holme, I. (2006). Sir William Henry Perkin: A review of his life, work and legacy. *Coloration Technology, 122*(5), 235–251. doi:10.1111/j.1478-4408.2006.00041.x

Hu, T. L. (2001). Kinetics of Azoreductase and Assessment of Toxicity of Metabolic Products from Azo Dyes by *Pseudomonas luteola. Water Science and Technology, 43*(2), 261–269. doi:10.2166/wst.2001.0098 PMID:11380189

Husain, Q. (2006). Potential applications of the oxidoreductive enzymes in the decolorization and detoxification of textile and other synthetic dyes from polluted water: A review. *Critical Reviews in Biotechnology, 26*(4), 201–221. doi:10.1080/07388550600969936 PMID:17095432

Hyde, F. W., Hunt, G. R., & Errede, L. A. (1991). Immobilization of bacteria and *Saccharomyces cerevisiae* in poly (tetrafluoroethylene) membranes. *Applied and Environmental Microbiology, 57*(1), 219–222. doi:10.1128/AEM.57.1.219-222.1991 PMID:2036008

Isaka, K., Yoshie, S., Sumino, T., Inamori, Y., & Tsuneda, S. (2007). Nitrification of landfill leachate using immobilized nitrifying bacteria at low temperatures. *Biochemical Engineering Journal, 37*(1), 49–55. doi:10.1016/j.bej.2007.03.008

Jadhav, S. U., Jadhav, M. U., Kagalkar, A. N., & Govindwar, S. P. (2008). Decolorization of Brilliant Blue G dye mediated by degradation of the microbial consortium of *Galactomyces geotrichum* and *Bacillus* sp. *Journal of the Chinese Institute of Chemical Engineers, 39*(6), 563–570. doi:10.1016/j.jcice.2008.06.003

Jafaria, N., Soudib, M. R., & Kasra-Kermanshahi, R. (2014). Biodegradation Perspectives of Azo Dyes by Yeasts. *Microbiology, 83*(5), 484–497. doi:10.1134/S0026261714050130

Jin, X., Liu, G., Xu, Z., & Tao, W. (2007). Decolorization of a Dye Industry Effluent by *Aspergillus fumigatus* XC6. *Applied Microbiology and Biotechnology, 74*(1), 239–243. doi:10.100700253-006-0658-1 PMID:17086413

Johnson, D. L., Thompson, J. L., Brinkmann, S. M., Schuller, K. A., & Martin, L. L. (2003). Electrochemical Characterization of Purified *Rhus vernicifera* Laccase: Voltammetric Evidence for a Sequential Four-Electron Transfer. *Biochemistry, 42*(34), 10229–10237. doi:10.1021/bi034268p PMID:12939151

Joshi, T., Iyengar, L., Singh, K., & Garg, S. (2008). Isolation, identification and application of novel bacterial consortium TJ-1 for the decolourization of structurally different azo dyes. *Bioresource Technology, 99*(15), 7115–7121. doi:10.1016/j.biortech.2007.12.074 PMID:18289845

Kadam, A. A., Kulkarni, A. N., Lade, H. S., & Govindwar, S. P. (2014). Exploiting the potential of plant growth promoting bacteria in decolorization of dye Disperse Red 73 adsorbed on milled sugarcane bagasse under solid state fermentation. *International Biodeterioration & Biodegradation, 86*, 364–371. doi:10.1016/j.ibiod.2013.10.012

Kapoor, A., Viraraghavan, T., & Cullimore, D. R. (1999). Removal of heavy metals using the fungus *Aspergillus niger. Bioresource Technology, 70*(1), 95–104. doi:10.1016/S0960-8524(98)00192-8

Karcher, S., Kornmüller, A., & Jekel, M. (2001). Screening of commercial sorbents for the removal of reactive dyes. *Dyes and Pigments, 51*(2-3), 111–125. doi:10.1016/S0143-7208(01)00066-3

Kdasi, A., Idris, A., Saed, K., & Guan, C. (2004). Treatment of textile wastewater by advanced oxidation processes: A review. *Global Nest: The International Journal, 6*(3), 222–230.

Keharia, H., & Madamwar, D. (2003). Bioremediation concepts for treatment of dye containing waste water: A review. *Indian Journal of Experimental Biology, 41*(9), 1068–1075. PMID:15242298

Khan, R., Bhawana, P., & Fulekar, M. H. (2013). Microbial decolorization and degradation of synthetic dyes: A review. *Reviews in Environmental Science and Biotechnology, 12*(1), 75–97. doi:10.100711157-012-9287-6

Khehra, M. S., Saini, H. S., Sharma, D. K., Chadha, B. S., & Chimni, S. S. (2006). Biodegradation of azo dye C.I. Acid Red 88 by an anoxic–aerobic sequential bioreactor. *Dyes and Pigments, 70*(1), 1–7. doi:10.1016/j.dyepig.2004.12.021

Klonowska, A., Gaudin, C., Fournel, A., Asso, M., Le Petit, J., Giorgi, M., & Tron, T. (2002). Characterization of a low redox potential laccase from the basidiomycete C30. *European Journal of Biochemistry, 269*(24), 6119–6125. doi:10.1046/j.1432-1033.2002.03324.x PMID:12473107

Knapp, J. S., Newby, P. S., & Reece, L. P. (1995). Decolorization of dyes by wood-rotting basidiomycete fungi. *Enzyme and Microbial Technology, 17*(7), 664–668. doi:10.1016/0141-0229(94)00112-5

Kruus, K., Kiiskinen, L. L., Saloheimo, M., Haklinen, N., Rouvinen, J., Paananen, A., Linder, M., & Viikari, L. (2002). A novel laccase from the ascomycete Melanocarpus albomyces. *Applied Microbiology and Biotechnology, 59*, 198–204. doi:10.100700253-002-1012-x PMID:12111146

Kulkarni, A. N., Kadam, A. A., Kachole, M. S., & Govindwar, S. P. (2014). Lichen *Permelia perlata*: A novel system for biodegradation and detoxification of disperse dye Solvent Red 24. *Journal of Hazardous Materials, 276*, 461–468. doi:10.1016/j.jhazmat.2014.05.055 PMID:24929306

Lucas, M. S., Amaral, C., Sampaio, A., Peres, J. A., & Dias, A. A. (2006). Biodegradation of the diazo dye Reactive Black 5 by a wild isolate of *Candida oleophila. Enzyme and Microbial Technology, 39*(1), 51–55. doi:10.1016/j.enzmictec.2005.09.004

Manu, B. (2003). *Decolourization of indigo and azo dye in semicontinous reactors with long hydraulic retention time* (PhD Thesis). IIT Bombay.

Martins, M. A. M., Cardoso, M. H., Queiroz, M. J., Ramalho, M. T., & Campus, A. M. O. (1999). Biodegradation of azo dyes by the yeast *Candida zeylanoides* in batch aerated cultures. *Chemosphere, 38*(11), 2455–2460. doi:10.1016/S0045-6535(98)00448-2 PMID:10204232

Mathur, N., Bhatnagar, P., & Bakre, P. (2006). Assessing mutagenicity of textile dyes from Pali (Rajasthan) using Ames bioassay. *Applied Ecology and Environmental Research, 4*(1), 111–118. doi:10.15666/aeer/0401_111118

Meehan, C., Banat, I., McMullan, G., Nigam, P., Smyth, F., & Marchant, R. (2000). Decolorization of Remazol Black-B using a thermotolerant yeast, *Kluyveromyces marxianus* IMB3. *Environment International*, *26*(1-2), 75–79. doi:10.1016/S0160-4120(00)00084-2 PMID:11345742

Meng, Q., Yan, W., Yu, M., & Huang, D. (2003). A study of third-order nonlinear optical properties for anthraquinone derivatives. *Dyes and Pigments*, *56*(2), 145–149. doi:10.1016/S0143-7208(02)00123-7

Metcalf, E. (2003). *Waste water Engineering: Treatment and Reuse* (4th ed.). McGraw-Hill.

Monsan, P. (1982). Les methodes immobilisation enzymes. In G. Durand & P. Monsan (Eds.), *Les enzymes, productions utilizations industrielles* (pp. 81–118). Gauthier-Villards.

Nam, S., & Renganathan, V. (2000). Non-enzymatic reduction of azo dyes by NADH. *Chemosphere*, *40*(4), 351–357. doi:10.1016/S0045-6535(99)00226-X PMID:10665399

Nese, T., Sivri, N., & Toroz, I. (2007). Pollutants of textile industry wastewater an assessment of its discharge limits by water quality standards. *Turkish Journal*, *7*, 97–103.

Nunn, D. M. (1979). *The Dyeing of Synthetic-Polymer and Acetate Fibres*. Dyers Co. Publications Trust.

Ollikka, P., Harjunpää, T., Palmu, K., Mäntsälä, P., & Suominen, I. (1998). Oxidation of crocein orange G by lignin peroxidase isoenzymes. *Applied Biochemistry and Biotechnology*, *75*(2-3), 307–321. doi:10.1007/BF02787783 PMID:10230025

Osma, J. F., Toca-Herrera, J. L., & Rodriguez-Couto, S. (2010). Transformation pathway of Remazol Brilliant Blue R by immobilised laccase. *Bioresource Technology*, *101*(22), 8509–8514. doi:10.1016/j.biortech.2010.06.074 PMID:20609582

Pandey, A. K., Sarada, D. V., & Kumar, A. (2016). Microbial Decolorization and Degradation of Reactive Red 198. *Proceedings of the National Academy of Sciences*, 805–815.

Park, C., Lee, M., Lee, B., Kim, S.-W., Chase, H. A., Lee, J., & Kim, S. (2007). Biodegradation and biosorption for decolorization of synthetic dyes by *Funalia trogii*. *Biochemical Engineering Journal*, *36*(1), 59–65. doi:10.1016/j.bej.2006.06.007

Paszczynski, A., & Crawford, R. L. (1991). Degradation of azo compounds by ligninase from *Phanerochaete chrysosporium*: Involvement of veratryl alcohol. *Biochemical and Biophysical Research Communications*, *178*(3), 1056–1063. doi:10.1016/0006-291X(91)90999-N PMID:1872828

Podgornik, H., Grgić, I., & Perdih, A. (1999). Decolorization rate of dyes using lignin peroxidases of *Phanerochaete chrysosporium*. *Chemosphere*, *38*(6), 1353–1359. doi:10.1016/S0045-6535(98)00537-2

Ponraj, M., Jamunarani, P., & Zambare, V. (2011). Isolation and optimization of culture condition for decolourization of true blue using dye decolorizing fungi. *Asian Journal of Experimental Biological Sciences*, *2*, 270–277.

Rai, H., Bhattacharya, M., Singh, J., Bansal, T. K., Vats, P., & Banerjee, U. C. (2005). Removal of Dyes from the Effluent of Textile and Dyestuff Manufacturing Industry: A Review of Emerging Techniques with Reference to Biological Treatment. *Critical Reviews in Environmental Science and Technology*, *35*(3), 219–238. doi:10.1080/10643380590917932

Rajaguru, P., Vidya, L., Baskarasethupathi, B., Kumar, P. A., Palanivel, M., & Kalaiselvi, K. (2002). Genotoxicity evaluation of polluted ground water in human peripheral blood lymphocytes using the comet assay. *Mutation Research/Genetic Toxicology and Environmental Mutagenesis*, *517*(1-2), 29–37. doi:10.1016/S1383-5718(02)00025-6 PMID:12034306

Ramakrishna, K. R., & Viraraghavan, T. (1997). Dye Removal Using Low Cost Adsorbents. *Water Science and Technology*, *36*(2-3), 189–196. doi:10.2166/wst.1997.0516

Ramalho, P. A., Cardoso, M. H., Cavaco-Paulo, A., & Ramalho, M. T. (2004). Characterization of Azo Reduction Activity in a Novel Ascomycete Yeast Strain. *Applied and Environmental Microbiology*, *70*(4), 2279–2288. doi:10.1128/AEM.70.4.2279-2288.2004 PMID:15066823

Ramalho, P. A., Paiva, S., Cavaco-Paulo, A., Casal, M., Cardoso, M. H., & Ramalho, M. T. (2005). Azo Reductase Activity of Intact Saccharomyces cerevisiae Cells Is Dependent on the Fre1p Component of Plasma Membrane Ferric Reductase. *Applied and Environmental Microbiology*, *71*(7), 3882–3888. doi:10.1128/AEM.71.7.3882-3888.2005 PMID:16000801

Ramalho, P. A., Scholze, H., Cardoso, M. H., Ramalho, M. T., & Oliveira-Campos, A. (2002). Improved conditions for the aerobic reductive decolourisation of azo dyes by *Candida zeylanoides*. *Enzyme and Microbial Technology*, *31*(6), 848–854. doi:10.1016/S0141-0229(02)00189-8

Rani, B., Kumar, V., Singh, J., Bisht, S., Teotia, P., Sharma, S., & Kela, R. (2014). Bioremediation of dyes by fungi isolated from contaminated dye effluent sites for bio-usability. *Brazilian Journal of Microbiology*, *45*(3), 1055–1063. doi:10.1590/S1517-83822014000300039 PMID:25477943

Robinson, T., McMullan, G., Marchant, R., & Nigam, P. (2001). Remediation of dyes in textile effluent: A critical review on current treatment technologies with a proposed alternative. *Bioresource Technology*, *77*(3), 247–255. doi:10.1016/S0960-8524(00)00080-8 PMID:11272011

Rodriguez Couto, S., & Toca Herrera, J. L. (2006). Industrial and biotechnological applications of laccases: A review. *Biotechnology Advances*, *24*(5), 500–513. doi:10.1016/j.biotechadv.2006.04.003 PMID:16716556

Safarikova, M., Ptackova, L., Kibrikova, I., & Safarik, I. (2005). Biosorption of water-soluble dyes on magnetically modified *Saccharomyces cerevisiae* subsp. uvarum cells. *Chemosphere*, *59*(6), 831–835. doi:10.1016/j.chemosphere.2004.10.062 PMID:15811411

Salokhe, M. D., & Govindwar, S. P. (1999). Effect of carbon source on the biotransformation enzymes in Serratia marcescens. *World Journal of Microbiology & Biotechnology*, *15*(2), 229–232. doi:10.1023/A:1008875404889

Sani, R. K., & Banerjee, U. C. (1999). Decolorization of triphenylmethane dyes and textile and dyestuff effluent by Kurthia sp. *Enzyme and Microbial Technology*, *24*(7), 433–437. doi:10.1016/S0141-0229(98)00159-8

Saratale, G. D., Humnabadkar, R. P., & Govindwar, S. P. (2007). Study of mixed function oxidase system in Aspergillus ochraceus (NCIM 1146). *Indian Journal of Microbiology*, *47*(4), 304–309. doi:10.100712088-007-0056-0 PMID:23100682

Saratale, R. G., Saratale, G. D., Chang, J. S., & Govindwar, S. P. (2011). Bacterial decolorization and degradation of azo dyes: A review. *Journal of the Taiwan Institute of Chemical Engineers*, *42*(1), 138–157. doi:10.1016/j.jtice.2010.06.006

Saratale, R. G., Saratale, G. D., Chang, J. S., & Govindwar, S. P. (2009c). Ecofriendly degradation of sulfonated diazo dye C.I. Reactive Green 19A using *Micrococcus glutamicus* NCIM-2168. *Bioresource Technology*, *100*(17), 3897–3905. doi:10.1016/j.biortech.2009.03.051 PMID:19375909

Saratale, R. G., Saratale, G. D., Chang, J. S., & Govindwar, S. P. (2009a). Decolorization and biodegradation of textile dye Navy blue HER by Trichosporon beigelii NCIM-3326. *Journal of Hazardous Materials*, *166*(2-3), 1421–1428. doi:10.1016/j.jhazmat.2008.12.068 PMID:19157708

Saratale, R. G., Saratale, G. D., Kalyani, D. C., Chang, J. S., & Govindwar, S. P. (2009b). Enhanced decolorization and biodegradation of textile azo dye Scarlet R by using developed microbial consortium-GR. *Bioresource Technology*, *100*(9), 2493–2500. doi:10.1016/j.biortech.2008.12.013 PMID:19157864

Savin, I., & Butnaru, R. (2008). Wastewater characteristics in textile finishing mills. *Environmental Engineering and Management Journal*, *7*(6), 859–864. doi:10.30638/eemj.2008.113

Schneider, P., Caspersen, M. B., Mondorf, K., Halkier, T., Skov, L. K., Østergaard, P. R., Brown, K. M., Brown, S. H., & Xu, F. (1999). Characterization of a Coprinus cinereus laccase. *Enzyme and Microbial Technology*, *25*(6), 502–508. doi:10.1016/S0141-0229(99)00085-X

Sharma, K. K., Shrivastava, B., Sastry, V. R. B., Sehgal, N., & Kuhad, R. C. (2013). Middle-redox potential laccase from *Ganoderma* sp.: Its application in improvement of feed for monogastric animals. *Scientific Reports*, *3*(1), 1299. doi:10.1038rep01299 PMID:23416696

Shleev, S., Jarosz-Wilkolazka, A., Khalunina, A., Morozova, O., Yaropolov, A., Ruzgas, T., & Gorton, L. (2005). Direct electron transfer reactions of laccases from different origins on carbon electrodes. *Bioelectrochemistry (Amsterdam, Netherlands)*, *67*(1), 115–124. doi:10.1016/j.bioelechem.2005.02.004 PMID:15941673

Shleev, S., Nikitina, O., Christenson, A., Reimann, C. T., Yaropolov, A. I., Ruzgas, T., & Gorton, L. (2007). Characterization of two new multiforms of *Trametes pubescens* laccase. *Bioorganic Chemistry*, *35*(1), 35–49. doi:10.1016/j.bioorg.2006.08.001 PMID:16989887

Shore, J. (1995). Dyeing with reactive dyes. *Cellulosics Dyeing*, 189-245.

Si, J., Peng, F., & Cui, B. (2013). Purification, biochemical characterization and dye decolorization capacity of an alkali-resistant and metal-tolerant laccase from Trametes pubescens. *Bioresource Technology*, *128*, 49–57. doi:10.1016/j.biortech.2012.10.085 PMID:23196221

Silveira, E., Marques, P. P., Silva, S. S., Lima-Filho, J. L., Porto, A. L. F., & Tambourgi, E. B. (2009). Selection of Pseudomonas for industrial textile dyes decolourization. *International Biodeterioration & Biodegradation*, *63*(2), 230–235. doi:10.1016/j.ibiod.2008.09.007

Singh, L. (2003). *Microbial degradation of hazardous dyes: A fungal approach*. University of Delhi.

Singh, L., & Singh, V. P. (2010). Biodegradation of textiles dyes, Bromophenol blue and congo red by fungus-*Aspergillus flavus. Environment & We: An International Journal of Science and Technology, 5*(4), 235–242.

Singh, L., & Singh, V. P. (2017). Decolourization of azo (acid red) and anthraquinonic (basic blue) dyes by the fungus *Aspergillus flavus. International Journal of Biomedical Engineering and Clinical Science, 3*(1), 1. doi:10.11648/j.ijbecs.20160301.11

Singh, L., & Singh, V. P. (2015). Textile dyes degradation: A microbial approach for biodegradation of pollutants. In Microbial degradation of synthetic dyes in wastewaters (pp. 187-204). Springer. doi:10.1007/978-3-319-10942-8_9

Singh, V. P., Singh, L., Singh, I., & Kumar, R. (2006). Microbial Degradation of Hazardous Dyes. In *Current Concepts in Botany* (pp. 273–285). IK International Publishing House Pvt Ltd.

Srinivasan, A., & Viraraghavan, T. (2010). Decolorization of dye wastewaters by biosorbents: A review. *Journal of Environmental Management, 91*(10), 1915–1929. doi:10.1016/j.jenvman.2010.05.003 PMID:20627542

Stolz, A. (2001). Basic and applied aspects in the microbial degradation of azo dyes. *Applied Microbiology and Biotechnology, 56*(1-2), 69–80. doi:10.1007002530100686 PMID:11499949

Taylor, J. A. (2000). Recent developments in reactive dyes. *Review of Progress in Coloration and Related Topics, 30*(1), 93–108. doi:10.1111/j.1478-4408.2000.tb03785.x

Telke, A. A., Ghodake, G. S., Kalyani, D. C., Dhanve, R. S., & Govindwar, S. P. (2011). Biochemical characteristics of a textile dye degrading extracellular laccase from a *Bacillus* sp. *ADR. Bioresource Technology, 102*(2), 1752–1756. doi:10.1016/j.biortech.2010.08.086 PMID:20855194

Telke, A. A., Kadam, A. A., Jagtap, S. S., Jadhav, J. P., & Govindwar, S. P. (2010). Biochemical characterization and potential for textile dye degradation of blue laccase from *Aspergillus ochraceus* NCIM-1146. *Biotechnology and Bioprocess Engineering; BBE, 15*(4), 696–703. doi:10.100712257-009-3126-9

Telke, A. A., Kalyani, D. C., Dawkar, V. V., & Govindwar, S. P. (2009a). Influence of organic and inorganic compounds on oxidoreductive decolorization of sulfonated azo dye C.I. Reactive Orange 16. *Journal of Hazardous Materials, 172*(1), 298–309. doi:10.1016/j.jhazmat.2009.07.008 PMID:19640646

Tony, B. D., Goyal, D., & Khanna, S. (2009a). Decolorization of textile azo dyes by aerobic bacterial consortium. *International Biodeterioration & Biodegradation, 63*(4), 462–469. doi:10.1016/j.ibiod.2009.01.003

Tony, B. D., Goyal, D., & Khanna, S. (2009b). Decolorization of Direct Red 28 by mixed bacterial culture in an up-flow immobilized bioreactor. *Journal of Industrial Microbiology & Biotechnology, 36*(7), 955–960. doi:10.100710295-009-0574-3 PMID:19390882

Van der Zee, F. P., & Villaverde, S. (2005). Combined anaerobic–aerobic treatment of azo dyes—A short review of bioreactor studies. *Water Research, 39*(8), 1425–1440. doi:10.1016/j.watres.2005.03.007 PMID:15878014

Vandevivere, P. C., Bianchi, R., & Verstraete, W. (1998). Treatment and Reuse of Wastewater from the Textile Wet-processing Industry: Review of Emerging Technologies. *Journal of Industrial Microbiology & Biotechnology*, *72*, 289.

Venkata Mohan, S., Chandrasekhar Rao, N., Krishna Prasad, K., & Karthikeyan, J. (2002). Treatment of simulated Reactive Yellow 22 (Azo) dye effluents using *Spirogyra* species. *Waste Management (New York, N.Y.)*, *22*(6), 575–582. doi:10.1016/S0956-053X(02)00030-2 PMID:12214968

Verma, P., & Madamwar, D. (2003). Decolourization of synthetic dyes by a newly isolated strain of Serratia marcescens. *World Journal of Microbiology & Biotechnology*, *19*(6), 615–618. doi:10.1023/A:1025115801331

Vijayaraghavan, K., & Yun, Y. S. (2007a). Chemical Modification and Immobilization of *Corynebacterium glutamicum* for Biosorption of Reactive Black 5 from Aqueous Solution. *Industrial & Engineering Chemistry Research*, *46*(2), 608–617. doi:10.1021/ie061158g

Vijayaraghavan, K., & Yun, Y. S. (2007b). Utilization of fermentation waste (*Corynebacterium glutamicum*) for biosorption of Reactive Black 5 from aqueous solution. *Journal of Hazardous Materials*, *141*(1), 45–52. doi:10.1016/j.jhazmat.2006.06.081 PMID:16879915

Vitor, V., & Corso, C. R. (2008). Decolorization of textile dye by Candida albicans isolated from industrial effluents. *Journal of Industrial Microbiology & Biotechnology*, *35*(11), 1353–1357. doi:10.100710295-008-0435-5 PMID:18712543

Vyas, B. R., & Molitoris, H. P. (1995). Involvement of an extracellular H_2O_2-dependent ligninolytic activity of the white-rot fungus *Pleurotus ostrus* in the decolorization of remazol brilliant blue R. *Applied and Environmental Microbiology*, *61*(11), 3919–3927. doi:10.1128/AEM.61.11.3919-3927.1995 PMID:8526504

Wang, X., Cheng, X., Sun, D., & Qi, H. (2008). Biodecolorization and partial mineralization of Reactive Black 5 by a strain of *Rhodopseudomonas palustris. Journal of Environmental Sciences (China)*, *20*(10), 1218–1225. doi:10.1016/S1001-0742(08)62212-3 PMID:19143346

Wesenberg, D., Kyriakides, I., & Agathos, S. N. (2003). White-rot fungi and their enzymes for the treatment of industrial dye effluents. *Biotechnology Advances*, *22*(1-2), 161–187. doi:10.1016/j.biotechadv.2003.08.011 PMID:14623049

Wong, Y. (1999). Laccase-catalyzed decolorization of synthetic dyes. *Water Research*, *33*(16), 3512–3520. doi:10.1016/S0043-1354(99)00066-4

Wu, J. Y., Hwang, S. C. J., Chen, C. T., & Chen, K. C. (2005). Decolorization of azo dye in a FBR reactor using immobilized bacteria. *Enzyme and Microbial Technology*, *37*(1), 102–112. doi:10.1016/j.enzmictec.2005.02.012

Yan, G. Y., & Viraraghavan, T. (2000). Effect of pretreatment on the biosorption of heavy metals on *Mucor rouxii. Water S.A.*, *26*, 119–123.

Yan, H., & Pan, G. (2004). Increase in Biodegradation of Dimethyl Phthalate by *Closterium lunula* Using Inorganic Carbon. *Chemosphere*, *55*(9), 1281–1285. doi:10.1016/j.chemosphere.2003.12.019 PMID:15081769

Yang, J., Yang, X., Lin, Y., Ng, T. B., Lin, J., & Ye, X. (2015). Laccase-Catalyzed Decolorization of Malachite Green: Performance Optimization and Degradation Mechanism. *PLoS One*, *10*(5), e0127714. doi:10.1371/journal.pone.0127714 PMID:26020270

Yang, Q., Yang, M., Pritsch, K., Yediler, A., Hagn, A., Schloter, M., & Kettrup, A. (2003). Decolorization of synthetic dyes and production of manganese-dependent peroxidase by new fungal isolates. *Biotechnology Letters*, *25*(9), 709–713. doi:10.1023/A:1023454513952 PMID:12882171

Yang, Y. Y., Du, L. N., Wang, G., Jia, X. M., & Zhao, Y. H. (2011). The decolorisation capacity and mechanism of *Shewanella oneidensis* MR-1 for Methyl Orange and Acid Yellow 199 under microaerophilic conditions. *Water Science and Technology*, *63*(5), 956–963. doi:10.2166/wst.2011.275 PMID:21411946

Yin, P., Yu, Q., Jin, B., & Ling, Z. (1999). Biosorption removal of cadmium from aqueous solution by using pretreated fungal biomass cultured from starch wastewater. *Water Research*, *33*(8), 1960–1963. doi:10.1016/S0043-1354(98)00400-X

Zeng, X., Cai, Y., Liao, X., Zeng, X., Li, W., & Zhang, D. (2011). Decolorization of synthetic dyes by crude laccase from a newly isolated *Trametes trogii* strain cultivated on solid agro-industrial residue. *Journal of Hazardous Materials*, *187*(1-3), 517–525. doi:10.1016/j.jhazmat.2011.01.068 PMID:21315513

Zeroual, Y., Moutaouakkil, A., & Blaghen, M. (2001). Volatilization of mercury by immobilized bacteria (Klebsiella pneumoniae) in different support by using fluidized bed bioreactor. *Current Microbiology*, *43*(5), 322–327. doi:10.1007002840010310 PMID:11688795

Zhang, X., & Flurkey, W. (1997). Phenol oxidases in Portabella Mushrooms. *Journal of Food Science*, *62*(1), 97–100. doi:10.1111/j.1365-2621.1997.tb04376.x

Zheng, F., Cui, B.-K., Wu, X.-J., Meng, G., Liu, H.-X., & Si, J. (2016). Immobilization of laccase onto chitosan beads to enhance its capability to degrade synthetic dyes. *International Biodeterioration & Biodegradation*, *110*, 69–78. doi:10.1016/j.ibiod.2016.03.004

Zhou, X., & Xiang, X. (2013). Effect of different plants on azo-dye wastewater biodecolorization. *Procedia Environmental Sciences*, *18*, 540–546. doi:10.1016/j.proenv.2013.04.073

Zimbardi, A., Camargo, P., Carli, S., Aquino Neto, S., Meleiro, L., Rosa, J., De Andrade, A. R., Jorge, J. A., & Furriel, R. (2016). A High Redox Potential Laccase from Pycnoporus sanguineus RP15: Potential Application for Dye Decolorization. *International Journal of Molecular Sciences*, *17*(5), 672. doi:10.3390/ijms17050672 PMID:27164083

Chapter 18
Remediation of Bauxite Residue Through Integrated Approach of Microbes and Plantation: A Case Study

Kumud Dubey
https://orcid.org/0000-0001-6367-4617
Forestry Research Centre for Eco-Rehabilitation, Prayagraj, India

K. P. Dubey
Environment, Forest, and Climate Change Department, Lucknow, India

ABSTRACT

Bauxite residue (red mud) is an industrial waste bye product of Alumina industry. It is toxic and highly alkaline in nature having heavy metals. Its disposal is the paramount environmental issue in Alumina industry. In the present study, bioremediation of red mud was carried out through cyanobacteria amendments and plantation. Two cyanobacterial species (viz. Phormidium and Oscillatoria) were found promising after studying their effect on physico-chemical characteristics of red mud. Seeds of selected tree species (viz. Dalbergia sissoo, Prosopis juliflora, Acacia auriculiformis, Pithecellobium dulce, Cassia siamia) were procured, and a nursery of these tree species was raised. Performances of two cyanobacteria (viz. Phormidium and Oscillatoria sps.) in combinations with PSB and VAM on red mud are very encouraging and hold considerable promise for bioremediation and revegetation of red mud. Inoculated seedlings of P. juliflora, P. dulce, A. auriculiformis, and C. siamia performed well for red mud revegetation.

INTRODUCTION

To meet the high requirement of materials, the natural wealth is being exploited to its utmost level. In consequence of which there is exhaustion of these precious resources as well as accrual of diverse types of waste products. Red mud is an industrial waste material which is produced during alumina extraction

DOI: 10.4018/978-1-7998-7062-3.ch018

from bauxite ore using concentrated NaOH at increased temperature during Bayer's process. The volume depends on composition of raw material (Das and Thakur 1995; Thakur and das, 2003; Sutar *et. al.*, 2014).

Aluminium is an element existing in plenty on earth and it is considered as third most plenteous element (8%) in the earth's crust, next only to oxygen and silicon. It is light and tough metal with excellent thermal and electrical conductivity. It is easy to fabricate consumer goods and is non-magnetic in nature having high resistance to several chemicals and resultant corrosion. Non toxicity makes it a very useful metal. Bauxite ($Al_2O_3.xH_2O$) is economic ore of Aluminium, consisting of high concentration of Aluminium compounds in association with silica; iron oxide; titanium dioxide and few others minor and traces of impurities (Krishna, 2003). For extracting Aluminium from bauxite, it is first treated with caustic soda which produces refined Aluminium oxide. Red mud (Bauxite residue) is a byproduct of this refining process. Due to the large amount of iron compounds, it is red in colour. After refinement, red mud is separated and dumped off. Red mud is highly alkaline in nature and contains oxides and salts of six major oxides of Fe, Al, Ti, Si, Na, Ca, and several minor trace elements. Around 1.5–2.5 tons of red mud is produced per ton of alumina extracted. The main ecological risks linked with red mud dumping, is due to its high pH and presence of traces of heavy metals & radionuclides. Establishment of vegetation on these red mud dumping sites is very important for reducing the ecological risk (Wehr *et. al.*, 2006). Its alkalinity; fine nature, high level of oxides of iron and other metals, nutrients deficiency and devoid of beneficial microbes make red mud dumping sites a poor substrate to re-vegetate. The present study was carried out to make this red mud suitable for re-vegetation through integrated approach of bioremediation technology.

Bioremediation is a low cost and eco-friendly technique through utilization of plants and microbes to clean up moderately contaminated areas by absorbing/ adsorbing the toxic material or by converting the toxic molecules to lesser toxic form and reducing its bioavailability. It is a low cost, effective alternative to the conventional methods of remediation procedures and may be considered as promising technology for remediation of contaminated sites.

As the red mud is disposed and dumped in nearby area of the refinery, these dumping sites occupy large area and are problematic due to ecological risks associated with these sites. To leave these dumping sites as such is not prudent. Recent development in bio/phyto-remediation technology has enabled us to remediate and reclaim such degraded ecosystems. Therefore, the bio-remediation of red mud seems to be the rationally practical and expedient step for the ecologically safe disposal of this highly toxic industrial residue and for sustainable reclamation of these red mud dumping sites.

Therefore, the red mud was first remediated to the less toxic form and made it appropriate for plantation. Beneficial microbes, cyanobacteria, phosphate solubilizing bacteria (PSB) and vesicular arbuscular mycorrhiza (VAM) were used for biological remediation of red mud. Suitability of plant species for the plantation was also studied and species suitable for vegetation of such sites have been screened and recommended for the reclamation of such sites.

METHODOLOGY

Site Survey and Sample Collection

M/s Hindalco Industries Ltd. (HINDALCo), UP, India, was chosen for study as it is the only Aluminium production industry functioning in Uttar Pradesh, India. The company is situated at Renukoot, Sonbhadra

District of Uttar Pradesh. HINDALCo was visited for surveying Red Mud production site with consent of HINDALCo authorities. HINDALCo dumps red mud after carrying out a drying process, called 'Dry Stacking of Red Mud' (Dubey and Dubey, 2011; Dubey, 2012).

Red Mud Sample Collection and Analysis

Three samples of red mud were collected randomly from red mud dumping site and sample for analysis was prepared by carefully mixing of all three samples (Dubey and Dubey, 2011; Dubey, 2012). The sample was analyzed for pH, conductivity, available nitrogen and organic matter as the methods described by Piper (1944).

Culturing of Cyanobacteria and Mass Propagation

Four cyanobacteria viz. *Oscillatoria* sp., *Lyngbya* sp., *Phormadium* sp. and *Microcystis* sp. were chosen for the study and cultured and propagated on liquid BG11 medium and utilized as source inoculums for propagation in bulk (Rippka et al., 1979).

Bulk Propagation of Cyanobacteria

Bulk production of cyanobacteria was done in tanks in outdoor situation. For outdoor bulk cultivation of cyanobacterial biofertilizers, the specific strain cultures had been used as starter inoculum. Mixture of two acclimatized strains of cyanobacteria, e.g. species of *Oscilittoria* and *Lymgya* was also used as starter inoculums for bulk production. Mass cultivation of cyanobacteria was done in tanks by using following steps:

1. About 10 Kg of farm soil (collected from open place for 1.0 m^2 area of the tank) and 100 g of superphosphate (SSP) added.
2. The tank was watered up to about 10 cm height.
3. The pH of the tank was adjusted to 7.0 by adding lime.
4. The insecticide e.g. malathion (2.0 ml) was added to protect the culture from mosquitoes and other flies.
5. The mixture was stired well and allowed to settle down the soil.
6. When water layer became clear, 100 g of starter inoculums was strewn on the surface water.
7. The temperature was maintained between 35-40 °C (requisite range for summer) for achieving optimum growth of cyanobacteria.
8. The water level to about 10 cm during this period had been maintained.
9. After drying, the algal mat got separated from the soil and formed flakes. Production varies according to ambience and species. These were collected, dried and used for the experimental trials.

The algal flakes may also be used as starter inoculums for further propagation (Dubey, 2012).

To Study the Promising Cyanobacteria on Red Mud Amendments With Different Treatments in Nursery Conditions

Experimentation was laid out in trays for studying the effect of cyanobacteria (blue green algae) viz., *Lyngbya, Phormidium, Oscillatoria* and *Microcystis* sps. on red mud amendments with different treatments in nursery conditions for selection of promising cyanobacteria for further studies. Effect on red mud was studied with different bio-amendments. Cyanobacterial growth was observed on red mud with different amendments. Treatments were Control 1 (Red mud), Control 2 (Red mud mixed with normal soil in 1:1 amended with 10g Bone Meal, a bio-source of phosphorus), Red mud mixed with normal soil and 10.0g of Bone Meal inoculated with 5.0g *Phormidium,* Red mud mixed with normal soil and Bone Meal inoculated with 5.0g *Oscillatoria,* Red mud mixed with normal soil and Bone Meal inoculated with 5.0g *Lyngbya,* Red mud mixed with normal soil and Bone Meal inoculated with 5.0g *Microcystis.* Soil pH, EC, Organic matter and nitrogen was noted down as indicators of bioremediation (Dubey, 2012).

Effect of Promising Blue Green Algae/ Bio-Inoculants Combinations on the Growth Performance of Selected Plant Species in Nursery

The effect of promising Blue Green Algae (BGA) / bioinoculants combinations on the growth performance of chosen plant species was studied. A nursery pot experimentation was laid out for studying the combinations of Blue Green Algae (BGA) with bio-amendments on the growth performance of chosen plant species viz.: *Dalbergia sissoo, Prosopis juliflora, Acacia auriculiformis, Pithecellobium dulce* and *Cassia siamia* on red mud. These selected species are well-known species of the Sonbhadra district, Uttar Pradesh, India, where this HINDALCo industry situated. Combination of potential BGA species i.e. *Phormidium* and *Oscillatoria* was used for cyanobacterial inoculation in 1:1 proportion. Treatments carried out were: Control 1(Red mud: Normal Soil), Control 2 comprised of equal amount of Normal Soil, Sand and Farm Yard manure (FYM) without red mud, T1 (Red mud: Normal Soil amended with 10g of each bone meal, FYM and Cyanobacteria), T2 (Red mud: Normal Soil amended with 10g of each bone meal, FYM and Cyanobacteria and PSB), T3 (Red mud: Normal Soil amended with 10g of each bone meal, FYM and Cyanobacteria and VAM). Control 2 was treated as positive control for the trial. Seeds of selected plant species *Dalbergia sissoo, Prosopis juliflora, Acacia auriculiformis, Pithecellobium dulce* and *Cassia siamia* were sown and growth was observed.

RESULTS

The composition of Red Mud sample, provided by the HINDALCo Industries, Renukoot, Sonebhadra, Uttar Pradesh, India, has been depicted in (Figure 1).The pH and EC, organic matter and nitrogen were depicted in Figure 2 and Figure 3, respectively.

Figure 1. The composition of Red Mud sample

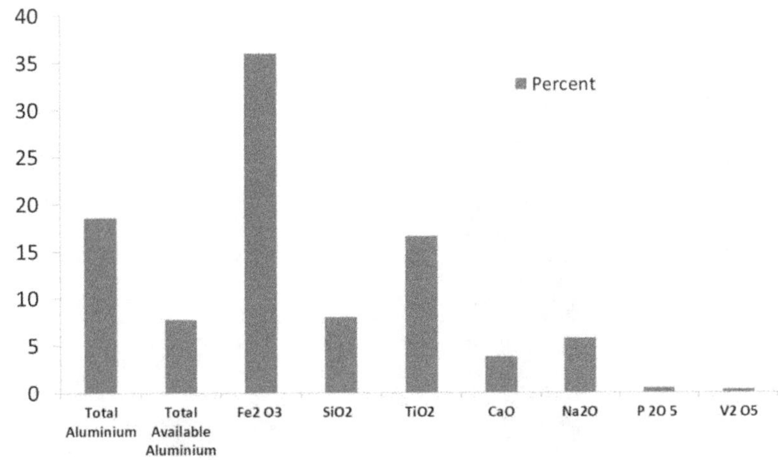

Figure 2. pH and EC of Red Mud

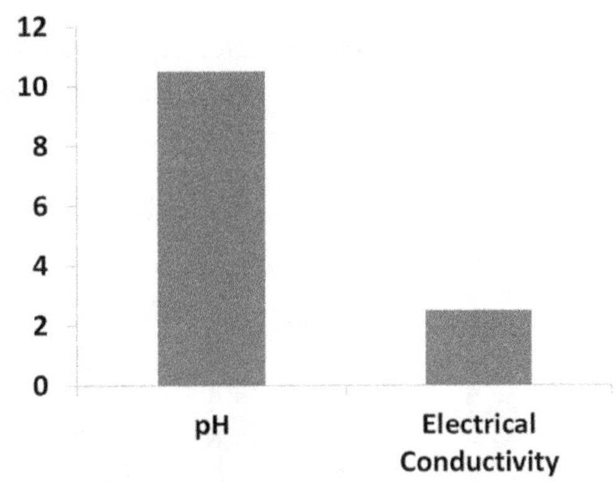

Promising Cyanobacteria on Red Mud Amendments with Different Treatments in Nursery Conditions

For selection of promising Cyanobacteria for further studies, experimentation was conducted in trays in nursery conditions. Effect on red mud was studied with different bio-amendments. Cyanobacterial growth was observed on red mud with different amendments. Soil pH, EC, Organic carbon and nitrogen was observed after 45 days and depicted in Figure 4a, 4b, 4c and 4d, respectively.

Based on these findings, the physical growth of these Cyanobacteria in Red mud amended medium and their effect on physico-chemical characteristics of red mud, the promising cyanobacterial species were selected for bioremediation. Two cyanobacterial species viz. *Phormidium* and *Oscillatoria* were found to be the most promising for bioremediation of red mud. Both cyanobacteria (Blue green algae) were used in further experimentations.

Figure 3. Organic Carbon and Nitrogen of Red Mud

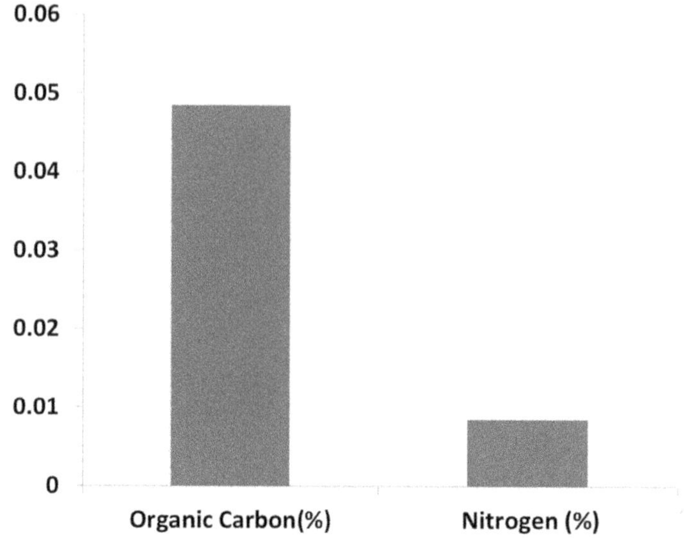

Figure 4. Effect of Cyanobacterial inoculation on characteristics of Red mud with different treatments (a) pH, (b) Electrical Conductivity, (c) Organic Carbon, and (d) Nitrogen

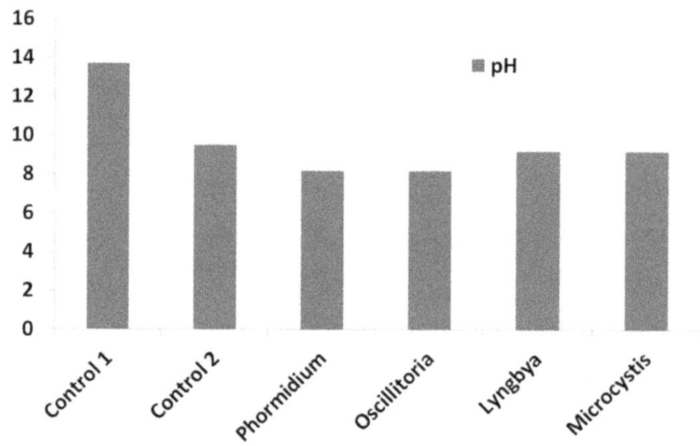

Figure 5. Growth performance (height in meter) of the selected tree species with different bio treatments: (a) Prosopis juliflora, (b) Pithecellobium dulce (c) Cassia siamia (d) Acacia auriculiformis, (e) Dalbergia sissoo

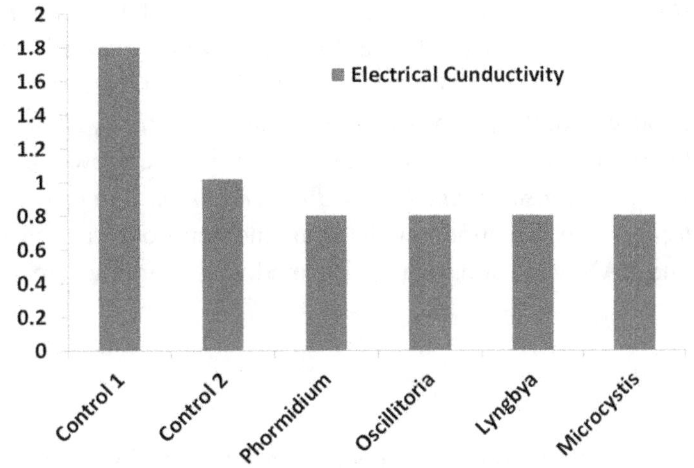

Figure 6. Comparative growth performance (height in meter) of the selected plant species Prosopis juliflora, Acacia ariculiformis, Dalbergia sissoo, Cassia siamia and Pithecellobium dulce with different bio treatments

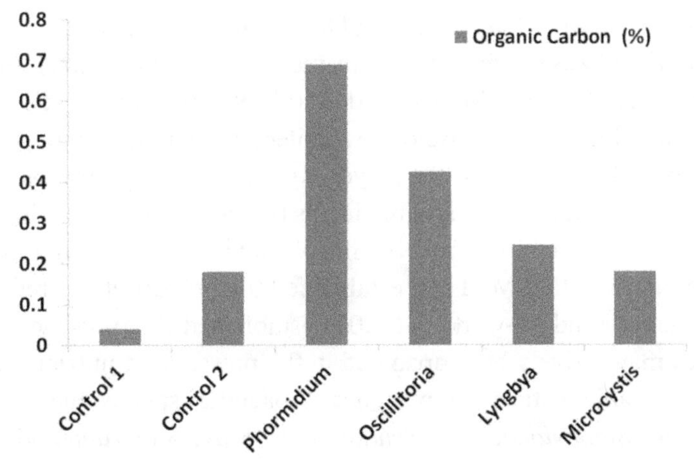

Effect of Promising Blue Green Algae/ Bioinoculants Combinations on the Growth Performance of Selected Plant Species in Nursery

The effect of promising Blue Green Algae/ bioinoculants combinations on the growth performance of selected plant species was studied in nursery. A nursery pot experiment was laid out for studying the combinations of Cyanobacteria on the growth performance of selected plant species viz.: *Dalbergia sissoo, Prosopis juliflora, Acacia auriculiformis, Pithecellobium dulce* and *Cassia siamia* on red mud with different treatments. Combination of promising *Phormidium* and *Oscillatoria* was used for Cyanobacterial inoculation. The growth data (height in meter) after six months of each species was depicted in

Figure 5 (a) *Prosopis juliflora,* (b) *Pithecellobium dulce* (c) *Cassia siamia* (d) *Acacia auriculiformis,* (e) *Dalbergia sissoo*. As far as growth was concerned, the effect of cyanobacterial and bio amendments was found to have positive effects in comparison to control 1 and it was significantly at par with the control 2 in case of *Prosopis juliflora, Pithecellobium dulce, Cassia siamia* and *Acacia auriculiformis*. Control 2 was considered as positive control without red mud. In case of *Dalbergia sissoo*, the growth with treatments was significantly less in comparison to control 2. As far as comparative growth performance of all selected species was concerned, the performance of *Prosopis juliflora, Pithecellobium dulce* on red mud was found better in comparison to other species (Figure 6). The growth performance of selected species in descending order is: *Prosopis juliflora* > *Pithecellobium dulce* > *Casia siamia* > *Acacia auriculiformis* > *Dalbergia sissoo* . Treatment with the amendments of Cyanobacteria *Phormidium* and *Oscillatoria* with PSB and VAM were at par and performed well over other treatments.

DISCUSSION

From the above findings, it was interpreted that *Phormidium* and *Oscillatoria* performed better in comparison to *Lyngbya* and *Microcystis* species on Red Mud amended medium. Since *Phormidium* and *Oscillatoria* species can flourish well in alkaline conditions (Vijayakumar et. al., 2005, 2007,Dubey et. al., 2011; Vijayakumar,2012 ; Vijayakumar and Manoharan, 2012; Abdulsada, 2014; Amin et. al., 2013; EL-Sheekh and Mahmoud, 2017; EL-Sheekh et.al.,2016; Kaur et. al., 2019; Shabana, 2019), they can also develop further resistance towards alkalinity of Red Mud. Hence, these two Cyanobacteria may be used for bioremediation of Red Mud (Dubey, 2012; Dubey and Dubey, 2011).

Bioremediation of Red Mud was previously tried by means of a fungus *Aspergillus tubingensis* and the role of this fungus in reducing pH of Red Mud amended soils was examined by Krishna et al. (2005). The reduction in pH was mainly due to the release of low molecular weight organic acid during the process of phosphate solubilization through which their hydroxyl and carboxyl groups chelate cations which bound to phosphate, thus converting it into soluble forms (Gizaw *et. al.*, 2017; Xiao *et. al.*, 2018). Elias *et. al.* (2016) also reported the reduction in pH by this phosphate solubilizing fungi. Valerie Ee (1999) investigated the microbiology of Red Mud and established the probability of reduction in alkalinity of Red Mud by bacteria. Hamdy and co-workers (2001) established that presence of damaged bacterial cells in Bauxite residues may actively be propagated in the presence of nutrients and/or mulch of Bermuda grass as source of organic matter. The presence of bacterial species like *Bacillus, Lactobacillus, Leuconostoc, Micrococcus, Staphylococcus, Pseudomonas, Flavobacterium* and *Enterobacter* etc. had been observed in treated Bauxite residues. Chemicals like Gypsum were also found helpful in reducing the pH of Red Mud and to make it suitable for plantation (Courtney and Timpson, 2005; Courtney *et. al.*, 2009; Wong and Ho, 1993; Babu and Reddy, 2011). Menzies and co-workers (2004) re-vegetated Red Mud dumping sites by neutralizing it with Seawater.

The role of Cyanobacteria for biological remediation of Red Mud has not been considered. Cyanobacteria are an important group of micro-organisms capable to carry out both photosynthesis as well as nitrogen fixation non-symbiotically. Therefore, it has the additional advantage of being photo-synthesizers. Through photosynthesis it may supply organic carbon at a fast rate and proliferating itself in favourable conditions. It may also perform as a significant means for carbon fixation concurrently (Hall *et. al.*, 1995). It synthesizes a carpet or coating like layer on soil surface. This layer may act as resistant coating to metals and metalloids and also adsorb these toxic elements from the surroundings. These Cyanobacterial coatings are an perfect arrangement for bioremediation of toxic mine dumps which are dangerous to health of human population, residing in nearby area (Subbarao, 2017 4th ed.). It was recommended by several people that cyanobacterial coating could have huge possibility to bio-remediate degraded soils and mined out areas (Eldridge, 1996; Doudle and Williams, 2010; Dubey and Dubey, 2011). It was projected that the formation of cyanobacterial coatings may act as early pioneer of re-development of the soil. It may enhance the soil organic carbon and may also have the prospective to ameliorate the soil situation by capturing/ adsorbing the heavy metals, thereby enhancing the restoration. BGA have also been accounted as means for bio-remediation of alkaline/ sodic soil due to their capability to exude carbon acidic compounds and fix sodium in its biomass (Jeganathan, 2006; Kaushik, 1989; Kaushik, and Krishnamurti, 1981; Kaushik, and Subhashini, 1985; Subhashini and Kaushik, 1981). Improvement in the quality of Fly-Ash by some species of BGA was considered by Rai and co-workers (2000). Significance of biological Nitrogen sequestration by BGA in forestation, thus improving the soil characteristics, was illustrated by Umali and Krishnapillay (2002). BGA has also been observed to take out toxic matters from the surroundings through their bio-sorption/ bio-adsorption to exterior poly-saccharides and their intracellular accrual utilizing metal sequestering metallothionin proteins. *Spirulina platensis*, a BGA, was found to contain noticeable levels of Mercury and Lead, after culturing on polluted sites (Slotton *et. al*, 1989). It showed the involvement of BGA in absorbing the harmful metals from surroundings. BGA both adsorbs and absorb metals. The role of *Phormidium* in adsorption of heavy trace metals has been observed (Sadettin and Donmez, 2007; Wang *et. al.*, 1998). The *Phormidium* and *Oscillatoria* cyanobacterial morphotypes had also been found effective for bio-remediating the sites polluted with petroleum products (Cohen, 2002). BGA has also been utilized efficiently for bio-remediating difficult sites (Kumar and Dubey, 2009, Vardhan and Dubey, 2009). The BGA layer may help in surface soil stabilization, escalating water infiltration and dropping wind and water erosion. BGA crust consequently, proffer the role of soil shield, as well as initiate biological functions that stay alive under inconsiderate circumstances (Doudle and Williams, 2010).These conclusions strongly support the statement that Cyanobacteria may participate a main function in detoxifying Red Mud by fixing heavy metals, formation of resistant layer on Red Mud dumps by this means dropping the ecological menace due to water and wind erosion and elevating its nutrient status in an ecologically sustainable way.

In above study, *Phormidium* and *Oscillatoria* performed better. Inoculation of these cyanobacteria augmented the carbon matter and available nitrogen. However pH and Electrical Conductivity of the soil diminished. *Lyngbya* and *Microcystis* had no noteworthy effect on organic matter, available nitrogen, pH and EC. *Phormidium* and *Oscillatoria* selected for their effect on growth performance on chosen tree species.

The effect of potential Blue green algae/ bioinoculants combinations on the growth performance of selected plant species was studied in nursery. Seedling growth in all selected tree species in case of treatment T2 (Red mud: Normal Soil amended with bone meal+FYM inoculated with Cyanobacteria and Phosphobacteria PSB) and T3 (Red mud: Normal Soil amended with bone meal+ FYM inoculated with Cyanobacteria and VAM) performed well in comparison of other treatments and were at par with the seedlings grown in control 2 (Normal Soil: Sand: FYM). Therefore, the combination of BGA with PSB and VAM may be used for inoculation. The performance of seedlings of chosen tree species inoculated with these microbes had been studied on red mud and it was fond that *Prosopis juliflora, Pithecellobium dulce*, *Acacia auriculiformis* and *Cassia siamia* performed well. These plant species may be used for the re-vegetation of the red mud dumping site, after treating it with Cyanobacteria, PSB and VAM.

CONCLUSION

In current study, performances of two of the four studied Cyanobacteria *viz. Phormidium* and *Oscillatoria* species on Red Mud are very encouraging and hold considerable promise for bioremediation of Red Mud. From these results, it may be inferred that inoculation of these Cyanobacteria in combination with PSB and VAM, would decrease pH, minimize toxicity by immobilizing heavy metals and improve its fertility by increasing organic matter and nitrogen content of Red Mud. The establishment of cyanobacterial crust shall also be helpful in reducing environmental risks due to water and wind erosion. Therefore, it may be concluded that synergetic functioning of these beneficial microbes play an important role in bioremediating Red Mud in an environmentally safe and sustainable manner. This derived prospect to bioremediate Red Mud may further be explored by using these microbes in consortia.

The performance of selected tree species viz. *Dalbergia sissoo, Prosopis juliflora, Acacia auriculiformis, Pithecellobium dulce* and *Cassia siamia* inoculated with cyanobacteria with phosphor-bacteria and VAM inoculation was studied and It was observed that *Prosopis juliflora* and *Pithecellobium dulce* performed well over other species. *Acacia auriculiformis* and *Cassia siamia* also performed well. These species viz. *Dalbergia sissoo, Prosopis juliflora, Acacia auriculiformis, Pithecellobium dulce* and *Cassia siamia* are leguminous and fix nitrogen. These tree species are also reported for phytoremediation of contaminated Soil (Kumar *et. al.*, 2002; Hemlata *et. al.*, 2009; Kumar *et. al.*, 2013; Manikandan *et. al.*, 2016; Prasad and Tewari, 2016). For instances, Kalam *et. al.* (2019) studied the heavy metal extraction capacity of D. sissoo and its ability to compart- mentalize heavy metals in tissues of three aboveground plant organs. They observed that *Dalbergia sissoo* has ability to phytoremediate the tannery effluent area near Ganga River in Kanpur, India. They found that a significant amount of heavy metals may be locked into the wood of the plant, where the metals can be retained for long time without reentering the soil through decomposition. In India, the timber of this tree is very commonly used in making furniture, doors and other building material, and thus, the significant amounts of Cr, Cu, and Ni localized in the annual wood rings can become locked in for a long period of time. These features, in addition to the wide distribution and high biomass, make *D. sissoo* an ideal candidate for the phytoremediation of Cr and Ni (Kalam *et. al.*, 2019).

Therefore from the above study, it may be concluded that theses four species with Cyanobacteria viz. *Phormidium* and *Oscillatoria* species in combinations with PSB and VAM inoculation may be used for re-vegetating the red mud dumping site.

Recently Rui *et al.* (2020) studied ecological succession of red mud disposal site through natural activity of microbial to fix essential minerals and hoarding it in soil, thereby ameliorating the soil quality. It is slow process and result of long-term natural weathering process. This process may be stimulated by the artificial inoculation of above studied beneficial microbial consortia before vegetation. It may increases in nutrient cycling, at faster rate, through microbial metabolic process. The resultant microbial biomass may also be used as a biological indicator to evaluate red mud quality and a better medium for revegetation. Liao *et. al.* (2018) had isolated *Penicillium oxalicum*, an alkali-resistant acid-producing fungus screened from red mud disposal sites. They found that it could effectively grow and release the organic acids under extreme alkaline and saline conditions and significantly reduce the pH of the red mud by producing organic acids oxalic acid, formic acid and acetic acid (Zhang et. al., 2020). Therefore, the combined application of this fungus and above studied microorganism could reduce the alkalinity of red mud more effectively, which may provide theoretical basis and practical reference for bioremediation in the red mud disposal sites.

Development of flora on these Red Mud dumping sites is necessary for reducing the ensuing ecological threat. In the present study, performances of two Cyanobacteria viz. *Phormidium* and *Oscillatoria* sps in combinations with Phosphobacteria and VAM on Red Mud are very encouraging and hold considerable promise for bioremediation of Red Mud. Inoculated seedlings of *Prosopis juliflora, Pithecellobium dulce, Acacia auriculiformis* and *Cassia siamia* performed well on red mud. These plant species with above discussed microbes may be used for re-vegetating the red mud site. Organic acid producing phosphate solubilizing fungi like *Penicillium oxalicum* and *Aspergillus tubingensis* may also be used in combination with above discussed microbes i.e. cyanobacteria, PSB and VAM, before plantation to remediate the red mud pond conditions and make it suitable for vegetation. It was expected that the findings of the proposed study will provide suitable, eco-friendly and cost-effective bioremediation methods for Bauxite residue.

ACKNOWLEDGMENT

Authors acknowledge ICFRE, Dehradun for financial support and Divisional Forest Officers of Renukoot Territorial Forest Division, Uttar Pradesh, India and their associated forest officials for providing the local logistics, staff, infrastructural and field survey support.

REFERENCES

Abdulsada, Z. K. (2014). *Evaluation of microalgae for secondary and tertiary wastewater treatment* (Master Thesis). Ottawa-Carleton Institute for Environmental Engineering Ottawa, Ontario, Canada.

Amin, A., Naik, A. T. R., Azhar, M., & Nayak, H. (2013). Bioremediation of different waste waters—A review. *Canadian Journal of Fisheries and Aquatic Sciences*, *7*(2), 7–17.

Babu, G., & Reddy, M. S. (2011). *Aspergillus tubingensis* Improves the Growth and Native Mycorrhizal Colonization of Bermudagrass in Bauxite Residue. *Bioremediation Journal*, *15*(3), 157–164. doi:10.1080/10889868.2011.598486

Cohen, Y. (2002). Bioremediation of oil by marine microbial mats. *International Microbiology*, *5*(4), 189–193. doi:10.100710123-002-0089-5 PMID:12497184

Courtney, R., Mullen, G., & Harrington, T. (2009). An Evaluation of Revegetation Success on Bauxite Residue. *Restoration Ecology*, *17*(3), 350–358. doi:10.1111/j.1526-100X.2008.00375.x

Courtney, R. G., & Timpson, J. P. (2005). Reclamation of Fine Fraction Bauxite Processing Residue (Red Mud) Amended with Coarse Fraction Residue and Gypsum. *Water, Air, and Soil Pollution*, *164*(1-4), 91–102. doi:10.100711270-005-2251-0

Das, S. N., & Thakur, R. S. (1995). International series on environment, Red mud analysis and utilization. Publication and Information Directorate.

Doudle, S., & Williams, W. (2010). Can we kick-start mining rehabilitation with cyanobacterial crusts? In *Proceedings of the 16th Biennial Conference of the Australian Rangeland Society*. Australian Rangeland Society.

Dubey, K. (2012). Bio-remediation of Bauxite residue (red mud) generated from Aluminium industry by using blue green algae / bio-inoculants. ICFRE Project Report.

Dubey, K., & Dubey, K. P. (2011). A Study of the Effect of Red Mud Amendments on the Growth of Cyanobacterial Species. *Bioremediation Journal*, *15*(3), 133–139. doi:10.1080/10889868.2011.598483

Dubey, S. K., Dubey, J., Mehra, S., Tiwari, P., & Bishwas, A. J. (2011). Potential use of cyanobacterial species in bioremediation of industrial effluents. *African Journal of Biotechnology*, *10*(7), 1125–1132.

El-Sheekh, M. M. & Mahmoud, Ya-G. (2017). Technological approach of bioremediation using microbial tools: bacteria, fungi, and algae. In Handbook of research on inventive bioremediation techniques. IGI Global.

El-Sheekh, M. M., Farghl, A. A., Galal, H. R., & Bayoumi, H. S. (2016). Bioremediation of different types of polluted water using microalgae. *Rend. Fis. Acc. Lincei.*, *27*(2), 401–410. doi:10.100712210-015-0495-1

Eldridge, D. J. (1996). Cryptogamic soil crusts: fixers of the desert. In *Proc 1996 Workshop on rehabilitation of arid and semi-arid lands*. Goldfields Land Rehabilitation Group.

Elias, F., Woyessa, D., & Muleta, D. (2016). Phosphate Solubilization Potential of Rhizosphere Fungi Isolated from Plants in Jimma Zone, Southwest Ethiopia. *International Journal of Microbiology*, *5472601*, 1–11. Advance online publication. doi:10.1155/2016/5472601 PMID:27688771

Gizaw, B., Tsegay, Z., Tefera, G., Aynalem, E., Wassie, M., & Abatneh, E. (2017). Phosphate Solubilizing Fungi Isolated and Characterized from Teff Rhizosphere Soil Collected from North Showa and Gojam, Ethiopia. *J Fertil Pestic*, *8*(02), 180. doi:10.4172/2471-2728.1000180

Hall, D. O., Markov, S. A., Watanabe, Y., & Rao, K. K. (1995). The potential applications of cyanobacterial photosynthesis for clean technology. *Photosynthesis Research*, *46*(1-2), 159–167. doi:10.1007/BF00020426 PMID:24301578

Hemlata, P., Jambhulkar, P., & Juwarkar, A. (2009). Assessment of bioaccumulation of heavy metals by different plant species grown on fly ash dump. *Ecotoxicology and Environmental Safety*, *72*(4), 1122–1128. doi:10.1016/j.ecoenv.2008.11.002 PMID:19171381

Jeganathan, K. (2006). *Bioremediation studies on oil refinery industry effluent using* Oscillatoria earli Gartner (M.Phil dissertation). Bharathidasan University, Tiruchirapalli, India.

Kalam, S. U., Naushin, F., Khan, F. A., & Rajakaruna, N. (2019). Long-term phytoremediating abilities of *Dalbergia sissoo* Roxb. (Fabaceae). *SN Applied Sciences*, *1*(501), 1–8. doi:10.100742452-019-0510-8

Kaur, H., Rajor, A., & Singh Kaleka, A. S. (2019). Role of Phycoremediation to Remove Heavy Metals from Sewage Water: Review Article. *Journal of Environmental Science and Technology*, *12*(1), 1–9. doi:10.3923/jest.2019.1.9

Kaushik, B. D. (1989). Reclamative potential of cyanobacteria in salt-affected soils. *Phykos (Algiers)*, *28*, 101–109.

Kaushik, B. D., & Krishnamurti, G. S. R. (1981). Effect of blue-green algae and gypsum application on physico-chemical properties of alkali soils. *Phykos (Algiers), 20*, 91–94.

Kaushik, B. D., & Subhashini, D. (1985). Amelioration of salt-affected soils with blue-green algae. II Improvement in soil properties. *Proc Indian Natl Sci Acad Part B, 51*, 386–389.

Krishna, P. (2003). *Bioremediation of bauxite residue (red mud) using microbes* (Dissertation). Thapar Institute of Engineering and Technology, Patiala, Punjab, India

Krishna, P., Reddy, M. S., & Patnaik, S. K. (2005). *Aspergillus Tubingensis* Reduces the pH of the Bauxite Residue (Red Mud) Amended Soils. *Water, Air, and Soil Pollution, 167*(1-4), 201–209. doi:10.100711270-005-0242-9

Kumar, A., Tripathi, R. D., Singh, N., Rai, U. N., & Singh, S. N. (2002). Biochemical Responses of *Cassia siamea* Lamk. Grown on Coal Combustion Residue Fly-ash. *Bulletin of Environmental Contamination and Toxicology, 68*(5), 675–683. doi:10.1007001280307 PMID:12068933

Kumar, K., & Dubey, K. (2009). Microbial-amelioration of Sodic/Alkaline Soil. In *National Seminar on Frontiers in Biotechnology* (NSFB-2009). Department of Biotechnology, Bharathiar University.

Kumar, S. R., Arumugam, T., Anandakumar, C. R., Balakrishnan, S., & Rajave, D. S. (2013). Use of Plant Species in Controlling Environmental Pollution- A Review. *Bull. Env. Pharmacol. Life Sci., 2*(2), 52–63.

Liao, J., Jiang, J., Xue, S., Qingyu, C., Wu, H., Manikandan, R., Hartley, W., & Huang, L. (2018). A novel acid-producing fungus isolated from bauxite residue: The potential to reduce the alkalinity. *Geomicrobiology Journal, 35*(10), 840–847. doi:10.1080/01490451.2018.1479807

Manikandan, M., Kannan, V., Mahalingam, K., Vimala, A., & Chun, S. (2016). Phytoremediation potential of chromium containing tannery effluent-contaminated soil by native Indian timber-yielding tree species. *Preparative Biochemistry & Biotechnology, 46*(1), 100–108. doi:10.1080/10826068.2015.1045607 PMID:26177918

Menzies, N. W., Fulton, I. M., & Morrell, W. J. (2004). Seawater Neutralization of Alkaline Bauxite Residue and Implications for Revegetation. *Journal of Environmental Quality, 33*(5), 1877–1884. doi:10.2134/jeq2004.1877 PMID:15356249

Piper, C. S. (1944). Soil and Plant Analysis. Interscience Publishers.

Prasad, M. N. V., & Tewari, J. C. (2016). Prosopis juliflora (Sw) DC: Potential for Bioremediation and Bioeconomy. In Bioremediation and Bioeconomy. Elsevier. doi:10.1016/B978-0-12-802830-8.00003-4

Rai, U., Tripathi, R., Singh, N., Kumar, A., Ali, M. B., Pal, A., & Singh, S. N. (2000). Amelioration of Fly-Ash by Selected Nitrogen Fixing Blue Green Algae. *Bulletin of Environmental Contamination and Toxicology, 64*(2), 294–301. doi:10.1007001289910043 PMID:10656898

Rippka, R., Deruelles, J., Waterbery, J. B., Herdman, M., & Stainer, R. Y. (1979). Generic assignments, strain histories and properties, pure cultures of cyanobacteria. *Journal of General Microbiology, 111*, 1–61.

Rui, Q. C., Hao, X., & William, W. (2020). Ecological Stoichiometry of Microbial Biomass Carbon, Nitrogen and Phosphorus on Bauxite Residue Disposal Areas. *Geomicrobiology Journal*, *37*(5), 467–474. doi:10.1080/01490451.2020.1722768

Sadettin, S., & Donmez, G. (2006). Simultaneous bioaccumulation of reactive dye and chromium (VI) by using thermophilic *Phormidium* sp. *Enzyme and Microbial Technology*, *41*(1-2), 175–180. doi:10.1016/j.enzmictec.2006.12.015

Shabana, E. F., Senousy, H. H., & Khourshid, E. B. (2019). Pharmaceutical wastewater treatment using free and immobilized Cyanobacteria. *Egyptian Journal of Phycology*, *20*(1), 123–154. doi:10.21608/egyjs.2019.116025

Slotton, D. G., Goldman, C. R., & Frank, A. (1989). Commercially Grown *Spirulina* Found to Contain Low Levels of Mercury and Lead. *Nutrition Reports International*, *40*(2), 1165–1172.

Subba Rao, N. S. (2017). *Bio-fertilizers in Agriculture and Forestry* (4th ed.). Oxford & IBH Publishing Co. Pvt. Ltd.

Subhashini, D., & Kaushik, B. D. (1981). Amelioration of sodic soils with blue green algae. *Australian Journal of Soil Research*, *19*(3), 361–367. doi:10.1071/SR9810361

Sutar, H., Mishra, S. C., Sahoo, S. K., Chakraverty, A. P., & Mahrana, H. S. (2014). Progress of Red Mud Utilization: An Overview. *American Chemical Science Journal*, *4*(3), 255–279. doi:10.9734/ACSJ/2014/7258

Thakur, R. S., & Das, S. N. (2003). Red mud Analysis and utilisation of metal values. Publication and Information Directorate (CSIR) and Willy Eastern Ltd.

Umali, G. M., & Krishnapillay, B. (2002). Biological nitrogen fixation in tree species. *Basic principles of biotechnology and their application in forestry*, 145-148.

Valarie, E. (1999). Bioremediation of bauxite residue using indigenous bacteria. Minerals council of Australia Environmental Workshop, 311.

Vardhan, V., & Dubey, K. (2009). Bioremediation of Soil Contaminated with Petroleum Crude Oil. In *National Seminar on Frontiers in Biotechnology (NSFB-2009)*. Department of Biotechnology, Bharathiar University.

Vijayakumar, S. (2012). Potential Applications of Cyanobacteria in Industrial Effluents-A Review. *Journal of Bioremediation & Biodegradation*, *3*(154), 1–4. doi:10.4172/2155-6199.1000154

Vijayakumar, S., & Manoharan, C. (2012). Treatment of dye industry effluent using free and immobilized cyanobacteria. *Bioremed. Biodeg.*, *3*(165), 1–6. doi:10.4172/2155-6199.1000165

Vijayakumar, S., Thajuddin, N., & Manoharan, C. (2005). Role of cyanobacteria in the treatment of dye industry effluent. *Pollution Research*, *24*(1), 69–74.

Vijayakumar, S., Thajuddin, N., & Manoharan, C. (2007). Biodiversity of cyanobacteria in industrial effluents. *Acta Botanica Malacitana*, *32*, 27–34. doi:10.24310/abm.v32i0.7026

Wang, T., Weissman, J., Ramesh, G., Varadarajan, R., & Benemann, J. R. (1998). Heavy Metal Binding and Removal by Phormidium. *Bulletin of Environmental Contamination and Toxicology*, *60*(5), 739–744. doi:10.1007001289900688 PMID:9595189

Wehr, J. B., Fulton, I., & Menzies, N. W. (2006). Revegetation strategies for bauxite refinery residue: A case study of Alcan Gove in Northern Territory, Australia. *Environmental Management*, *37*(3), 297–306. doi:10.100700267-004-0385-2 PMID:16456629

Wong, W. C., & Ho, G. E. (1993). Use of Waste Gypsum in the Revegetation on Red Mud Deposits: A Greenhouse Study. *Waste Management & Research*, *11*(3), 249–256. doi:10.1177/0734242X9301100306

Xiao, C., Liu, T., Guang, X., & Ruan, C. (2018). Characteristics and Mechanisms of Biosolubilization of Rock Phosphate by *Aspergillus japonicas. Brazilian Archives of Biology and Technology*, *60*(0), 1–21. doi:10.1590/1678-4324-2017160541

Zhang, Y., Xue, R., He, X., Cheng, Q., Hartley, W., & Xue, S. (2020). Effect of acid production by Penicillium oxalicum on physicochemical properties of bauxite residue. *Geomicrobiology Journal*, *37*(10), 929–936. doi:10.1080/01490451.2020.1801907

Section 4

Microbial Bioremediation: Tools and Technologies

Chapter 19
The Use of Micro-Algal Technologies for Soil and Agronomic Improvements

Joan Nyika
https://orcid.org/0000-0001-8300-6990
Technical University of Kenya, Kenya

ABSTRACT

Microalgae are promising tools in improving soil fertility and agricultural production in the era of increased population and the need for food security, which is mostly hindered by climate change. The microbes have the ability to sequester atmospheric carbon dioxide, produce metabolites with many applications in addition surviving and growing in harsh environmental conditions. In this chapter, microalgae species of the cyanobacteria and green algae groups are established as good soil biofertilizers and conditioners which are crucial in nutrient cycling, improved soil structure, and increased soil microbial activity. These are requirements for better crop production. Microalgae are also crucial biocontrol agents that suppress and kill plant pathogens and pests, regulate the production of phytohormones, and in bio-remediation of polluted soils. Their use is therefore a road map to sustainable agriculture and food security. To ensure their optimal use, extensive research is necessary to understand the mechanisms of action behind the benefits.

INTRODUCTION

Modern day faces a challenge in meeting the food demands through sustainable agricultural activities despite the growing global population and other agricultural issues. While the need to increase food and biomass productivity is urgent, climate change is also predominant and therefore, innovative technologies and products that facilitate increased crop productivity, yield and quality and at the same time reducing the resultant carbon footprint from agricultural activities must be sought (Ronga et al., 2019). Traditional agricultural activities are heavily influenced by non-renewable inputs such as pesticides and fertilizers (Chiaiese et al., 2018). The introduction of these agrochemicals however poses environmental

DOI: 10.4018/978-1-7998-7062-3.ch019

and human health threats in addition to extra costs considering the increased nutrient mining during intensified agricultural activities (Costa et al., 2019). Furthermore, there is mounting public concern to regulate the use of these agrochemicals using stringent legal frameworks and hence their application is limited in optimizing agricultural production (Renuka et al., 2018).

Microalgae have been identified as potential alternatives to conventional agrochemical inputs. They are categorized based on their cell structure, life cycle and pigmentation. Existence literature has estimated the number of microalgae species to be approximately 800, 000 while only about 50, 000 have been described (Suganya et al., 2016). These microbes have diversified uses and it is possible to choose varied strains that have specific biochemical composition and the capacity to grow under different environmental conditions. Generally, algae are categorized into 1) multicellular, 2) filamentous, 3) colonial and 4) unicellular algae according to biologists (Nabti, Jha, & Hartmann, 2017). From the four categories, algae can be micro or macro based on size. The latter are macroscopic with a maximum length of at least 60 m and are multicellular while microalgae are microscopic and have a size ranging from approximately 1 to 900 µm (Ronga et al., 2019). Microalgae grow in fresh and marine water and are photosynthetic in nature. They can also be grown in wastewater, which reduces their production costs. The dominating microalgae species that are available commercially include *Dunaliella* spp., *Arthrospira* spp., *Chlorella* spp., *Chaetoceros* spp. and *Isochrysis* spp (Priyadarshani & Rath, 2012).

Microalgae consist of a variety of components including carbohydrates, proteins, pigments and lipids in addition to biomass that make them suitable for use in crop, pharmaceutical, animal feed, food and fuel production. Some of the components of the microbes are as shown in Figure 1. The current and emerging applications of microalgae are also shown in the representation. For purposes of agricultural production, microalgae comprises of micro- and macro-nutrients essential for plant growth and production. It is for this reason that microalgae have been used as biofertilisers and biostimulants (Garcia-Gonzalez & Sommerfeld, 2016; Khan et al., 2009; Shaaban, 2001; Shaaban, 2001). These microbes are gaining significant attention towards sustainable agriculture (Renuka et al., 2018). This book chapter explores the uses of microalgae for improved soil and agricultural production towards food security and greener ecosystems.

TECHNOLOGIES FOR MICROALGAE PRODUCTION

Microalgae are eukaryotic and prokaryotic microorganisms that have the capacity to produce lipids, proteins and carbohydrates through photosynthesis. They are fast maturing and can survive in harsh terrestrial and aquatic environs owing to their simple multicellular and unicellular structure. Some of the common examples include diatoms (Bacillariophyta), green algae (Chlorophyta), golden algae (Chrysophyceae) and cyanobacteria or blue-green algae (Cyanophyceae) (Mostafa, 2013). In agricultural applications cyanobacteria and blue-green algae species are commonly applied. Arable land, nutrients, water and sunlight are the growth requirements for algae and the organisms can fix CO_2 ten times better than terrestrial plants. Apart from growing them, microalgae species can be produced due to their commercial and economic applications from preserving water, recovering nutrients and wastewater (Ronga et al., 2019). The conventional approach uses open or raceway ponds to produce microalgae while the modern approach uses closed photobioreactors or hybrid systems (Khan et al., 2009).

Open ponds are of different sizes and shapes but the raceway design is the commonest. It occurs as a closed loop, rectangular and fitted with a recirculation channel. The ponds work at 15 to 20 cm depth to obtain productivities and concentrations of 60-100 mg/L/day and 1 g dry weight/L, respectively (Pulz,

2001). The pond system has a paddlewheel that circulates and mixes algal biomass where flow is regulated around baffles and bends within the channel. The system built on compacted earth can be of different sizes. Culture input occurs at the paddlewheel where flow begins during the day and at completion of the circulation loop, broth is harvested. The system however loses water via evaporation and is susceptible to contamination by foreign species since it is exposed to the environment (Khan et al., 2009). They can however tolerate high dissolved oxygen levels, resist predation, are cost effective compared to alternatives and grow rapidly. Microalgae species such as *Pleurochrysis* spp., *Phaeodactylum* spp., *Anabaena* spp., *Dunaliella* spp. and *Arthrospira* spp. have been produced using this system (Ronga et al., 2019).

Figure 1. Composition of microalgae and its current and emerging applications
(Ronga et al., 2019)

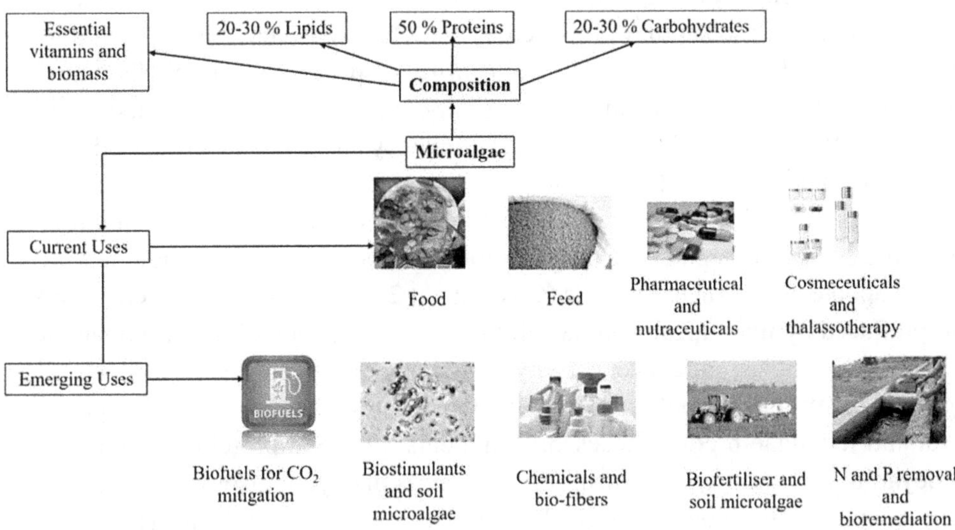

Closed bioreactors have lower footprint and higher productivity dependent on yield basis and reactor volume, respectively and they save chemicals, energy and water. The system emits a single microalgae species for longer periods and they are designed as bubble column, semi-hallow spheres, plate or tubular reactors though the latter is the commonest (Sato et al., 2006). Tubular photobioreactors have an array of parallel transparent glass or plastic tubes. Their solar collector tubes are of 0.1 m diameter to allow light to penetrate to the culture broth and ensure high biomass productivity. The broth is input from a reservoir to a solar collector and back to the reservoir for the process to be continuous (Khan et al., 2009; Sato et al., 2006). Closed photobioreactor systems were used to produce *Tetraselmis* spp., *Haematococcus* spp., *Chlorella* spp., *Nannochloropsis* spp., *Arthrospira* spp., *Phaeodactylum* spp. and *Porphyridium* spp. (Ronga et al., 2019).

Hybrid systems of both open ponds and closed bioreactors are combined to optimize on their advantages. Although open ponds are lucrative and proficient in microalgal growth their vulnerability to contamination makes the process expensive unless combined with a closed system. In hybrid systems, open systems are inoculated with the target microalgae strain that was previously grown in a bioreactor. Inoculums used are large to grow in open systems faster prior to contamination. Additionally, the ponds used as batch cultures are flushed and cleaned regularly to minimize contamination (Schenk et al., 2008). Schenk et al. (2008) used the hybrid system to cultivate *Haematococcus pluvialis* and produce astaxanthin.

MICROALGAE BIOACTIVE COMPOUNDS

Microalgae are essential sources of bioactive metabolites that are categorized into two: 1) primary and 2) secondary metabolites. Primary metabolites include carotenoids, β-carotene, lycopene, astaxanthin and phycobiliproteins while secondary metabolites include sterols, polyunsaturated fatty acids (PUFA), biofuels, vitamins, proteins and enzymes (Suganya et al., 2016). Carotenoids occur during the light photosynthetic phase as pigment for photo-protection of the photosynthetic system from free radicals and scavenging reactive oxygen species. Carotenoids are used as antioxidants and colorants in many foods. Examples of carotenoids is astaxanthin produced by the Chlorophyceae family of microalgae such as *Heamatococcus, Dunaliela, Chlamydomonas* and *Chlorella* spp. and xanthophylls whose formation is regulated by the presence of salts, metal ions, light intensity, oxidation, temperature and nitrogen-limitation (Trentacoste et al., 2015). Astaxanthin has anti-inflammatory and anti-cancerous activity since it slows down the growth of cancerous cells. β-carotene, a provitamin is also produced by microalgae prior to its transformation to retinol that has anti-carcinogenic properties, enhances the differentiation of regulatory proteins during the cell cycle, regulates growth factors that look like insulin and condenses the growth while promoting apoptosis of cancerous cells (Costa et al., 2019). Lycopene, which is a non-provitamin A carotenoid produced by microalgae and has biological activity including antioxidant activity via the scavenging of peroxyl radicals, anticancer activity by slowing down growth of cancer cells, prevention of DNA damage oxidatively and promoting the breakdown of carcinogenic enzymes (Chiaiese et al., 2018). Cyanobacteria and red algae produce accessory pigments, phycobiliproteins such as allophycocyanin (APC), phycocyanin (PC) and phycoerythrin (PE) organices in phycobilisome complexes. The pigments are used as nutraceuticals, for biotechnological applications and as natural dyes (Priyadarshani & Rath, 2012). Examples of microalgae species, the products obtained from the microbes and their various applications are summarized in Table 1 (Ashokkumar et al., 2013).

Apart from the primary metabolites, secondary microalgae products have a role to play in various applications of these microbes as shown in Table 1. Secondary metabolites are a variety and include lipids, proteins, carbohydrates, nucleic acids among other macromolecules and peptides, polyketides and isoprenoids among other molecules (Trentacoste et al., 2015). They indicate the unique adaptations of microalgae to their diverse environs. It is from these secondary metabolites that a variety of products used in the pharmaceutical, cosmetic, food and nutrition sectors as well as agricultural sector are produced. This book chapter explores the role of microalgae in improving soil fertility and crop productivity as a whole for sustainable agricultural practices.

Table 1. Applications of various products from named microalgae species

Microalgae Species	Products	Applications
Ulkenia spp	Decosahexanoic acid	Nutrition and pharmaceuticals
Spirulina platensis	Biomass protein, phycocyanin, γ-linolenic acid	Cosmetics and healthy foods
Schizochytrium sp.	Decosahexanoic acid	Nutrition and pharmaceuticals
Scenedesmus almeriensis	B-carotene, lutein	Cosmetics, nutrition and pharmaceuticals
Porphyridium cruentum	Polysaccharides and arachidonic acid	Nutrition, cosmetics and pharmaceuticals
Phaedactylum tricorntum	Fatty acids, lipids and eicosapentanoic acid	Biofuel and nutrition
Odontella aurita *Nannochloropsis gaditana*	Fatty acids	Baby food, cosmetics and pharmaceuticals
Nannochloropsis gaditana	Eicosapentanoic acid	Nutrition and pharmaceuticals
Muriellopsis sp.	Lutein	Nutrition and pharmaceuticals
Lyngbya majuscule	Immune modulators	
Isochrysis galbana	Flucoxanthin, carotenoids and fatty acids	Animal nutrition, cosmetics and pharmaceuticals
Haematococcus pluvialis	Lutein, cantaxanthin, astaxanthin and carotenoids	Healthy foods, feed additives and pharmaceuticals
Galdiera suphuraria	Phycocyanin	Nutrition and pharmaceuticals
Dunaliella salina	β-carotene, carotenoids	Food and feed supplements
Diacronema vlkianum	Fatty acids	Nutrition and pharmaceuticals
Crythecodinium conhi	Decosahexanoic acid	
Chlorella vulgaris	Pigments, biomass	Food supplements and healthy foods
Chlorella spp., Chlorella elipsodea, Coccomyxa acidophila	β-carotene and lutein	Nutrition and pharmaceuticals
Scenedesmus, Botryococcus	Fuel molecules	Energy production
Red algae	Phycoerythrin	Biotechnology
Teraselmis sp. Isochrysis sp.	Aquaculture feed	Animal feed

(Ashokkumar & Rengasamy, 2012; Coates et al., 2013; Trentacoste et al., 2015)

APPLICATIONS OF MICROALGAE IN SOIL AND AGRONOMIC IMPROVEMENTS

Microalgae particularly eukaryotic green algae and cyanobacteria are useful in production of secondary bioactive metabolites and in the mobilization and mineralization of the metabolites. These phenomena have a role to play in improved fertility of soils and productivity of crops (Gayathri et al., 2015). The microbes play a crucial role in enhancing productivity of aquatic and terrestrial ecosystem via nitrogen (N) fixation and photosynthesis to improve nutrient transformation and their cycling (Prasanna et al., 2016). Algalization that commonly refers to N-fixation by cyanobacteria not only improves soil fertility for optimal agricultural productivity but also serves as an alternative to chemical fertilization (Etesami & Alikhani, 2016). The capacity of microalgae to grow in harsh environs such as wastelands, salty areas and metal contaminated land facilitate their use in reclamation of such habitats (Prasanna et al., 2016).

Cyanobacteria among other microalgae are used as biocontrol agents that fight against plant pathogens including nematodes, fungi and bacteria through a mechanism of action involving production of biocidal chemicals including majusculonic acid and benzoic acid as well as hydrolytic enzymes (Gupta et al., 2013). The microalgal chemicals fight against the pathogens by invading their cytoplasmic membrane and preventing the metabolism of proteins (Gayathri et al., 2015). In some cases, the microalgae colonize some plant organelles and persist in the rhizosphere, which is an antagonistic activity towards pathogens as the microbes release enzymes and metabolites to inhibit the thriving of the pathogens (Gupta et al., 2013). These lines of defence are established to result to enhance crop yields and plant immunity against pathogens (Swain, Paidesetty, & Padhy, 2017). Agricultural use of cyanobacteria to enhance crop yields, growth and productivity and in the modulation of soil fertility through enhanced nutrient supplementation and microbial activity is widely researched and reported as summarized in Table 2 (Renuka et al., 2018).

Table 2. Examples of microalgae species with the potential to fight against pathogens and pests of various plants

Microalgae Species	Crop	Target Plant Pathogen
Anabaena sp. and *Bacillus sp.*	cotton	*Rhizoctonia solani*
Scytonema MKU 106	Cotton	*Styleptaderogate, Heliothis larvae, Helicoverpa armigera*
Microcoleus vaginatus	Tomato	*Meloidogyne incognita*
Oscillatoria chlorina	Tomato	*M. arenaria*
Aulosira fertilissima	Vegetable, wheat and paddy crops	*M. triticoryzae*
Calothris spp., Anabaena sp., Bacillus subtilis, A. oscillarioides	Tomato	*R. solani, Pythium aphanidermatum P. debaryanum, Fusarium oxysporum*
Calothrix, Nostoc, Nodularia, Anabaena and Oscillatoria sp.	Rice	*Alternaria alternate*
Oscillatoriatenuis FK 109, *Nostocommune* FK-103	Rice	*Phytophthora capsici*
Nostoc commune FA-103	Tomato	*F. oxysporum f. sp lycopersici, R. solani, F. moniliforme and P. debaryanum*
Anabaena sp.		

(Renuka et al., 2018)

Microalgae are also being used in bio-fertilisation to improve grain yield, nutritional and quality levels of fruits, faster plant growth and soil fertility (Gupta et al., 2013). They are also good soils conditioners and a promising tool to sustainable agriculture (Garcia-Gonzalez & Sommerfeld, 2016). Although many studies have focused on the use of microalgae for N fertilisation, recent studies also confirmed cyanobacteria inoculation increases the availability of macronutrients such as potassium (K), phosphorous (P) and carbon (C) as well as micronutrients such as iron (Fe), copper (Cu) and zinc (Zn) to soils for plant uptake (Prasanna et al., 2016, 2017). The research too has expanded from exclusively rice to the growth of other crops such as grains while the exclusive use of heterocystous microalgae strains has been expanded to non-heterocystous. Other inoculants of cyanobacteria introduced to soils as dippings of biofertilizer slurry, seed priming, field broadcasts, foliar additives and seed dressings have positive effects on yields, plant growth and germination rates of vegetables, horticultural crops and cereals (Coppens et al., 2016; Prasanna et al., 2016; Renuka et al., 2018; Swain et al., 2017). More details on these applications are presented in the following sections and summarized in Figure 2.

Figure 2. Various applications of microalgae in the improvement of soil fertility and agricultural production

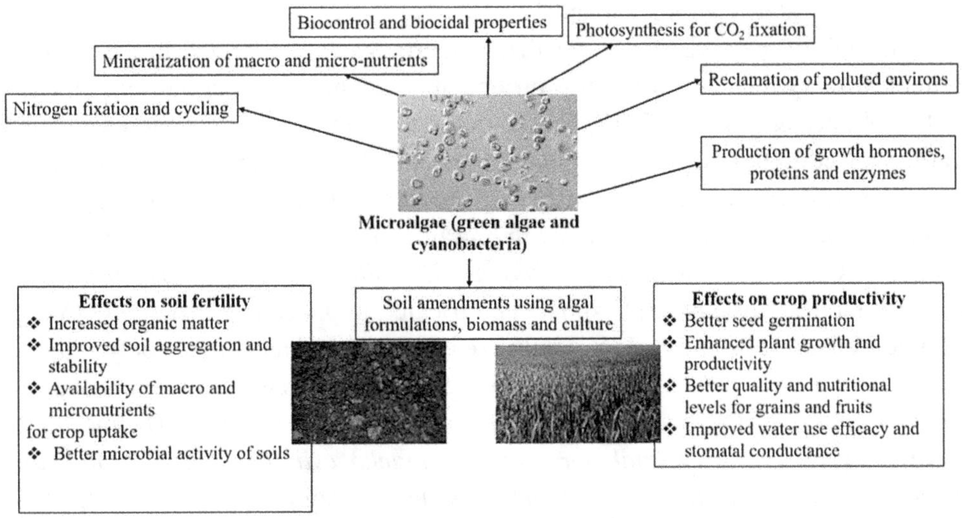

The Role of Microalgae in Nitrogen Fixation and Cycling

Cyanobacteria group of microalgae have heterocyst cells that have the capacity to fix atmospheric N, which is important to plants and various macro and micro flora and fauna. In this case, they do not compete for N with crops and enhance soil fertility (Renuka et al., 2018). Inoculation of cyanobacteria and their associated consortia showed improved soil fertility (Renuka et al., 2016), and saved on the use of chemical fertilizers by 25 to 40% (Prasanna et al., 2017). Some of the microalgae species used for this purpose include *Anabaena sp., N. entophytum* and *O. augustissima* (Swarnalakshmi et al., 2013). Other species used for this purpose include *Scytonema, Aulosira, Westiellop, Nannochloropsis, Klebsormidium* and *Ulothrix* species (Ronga et al., 2019). In this biofertilisation process, positive effects were also reported on the yields of certain crops including rice (Prasanna et al., 2012). In the use of microalgae for nitrogen fixation, the risk of nitrogen leaching is present although, it is lower compared to the use of chemical fertilisers (Mager & Thomas, 2011). Therefore, extensive research should be conducted to ensure that nitrogen fixation is optimized with minimal N-leaching possibilities during agricultural activities.

Microalgae in Soil Fertility Improvement

The use of heavy machinery for frequent and excessive tillage is associated with alterations of the soil structure and stability and consequently, capacity to mobilize nutrients and during soil-water infiltration. High organic carbon content of soils and a quality structure is key in sustainable crop production. Microalgae particularly green algae and cyanobacteria are essential sources of organic matter for agro-ecosystems through their ability to assimilate atmospheric carbon dioxide during photosynthesis in the microbes' biomass. Therefore, they increase organic matter pool through exopolysaccharides (EPS) excretion into soil, which makes the environment suitable for growth of microflora and fauna (Renuka et al., 2017, 2018). A number of studies reported increased microbial biomass, organic matter and microbial activity following the inoculation with microalgae (Nisha et al 2007; Uysal et al., 2015; Renuka et al., 2016). Similarly, Yilmaz and Sonmez (2017) conducted pot experiments in greenhouse conditions and established that microalgae acted as biofertilisers that increased organic carbon in soils improving its aeration, structure and water retention capacity.

Apart from the ability to work as a biofertilizer, microalgae are involved in the solubilisation and mineralization of primary micro- and macro-nutrients in soils that are essential for plant growth (Coppens et al., 2016; Yilmaz & Sonmez, 2017). Through the production of siderophores and organic acids, microalgae are involved in biomineralization (Renuka et al., 2018). Examples of organic acids include humic acid, which plays a role in mineral weathering. According to Bai et al. (2016), cyanobacteria such as *Microcystis aeruginosa* secretes EPS that are important in phenanthrene bio-absorption equivalent to biological pumping. Cyanobacteria such as *Anabaena variabilis* and *Westiellopsis prolifica* were noted to solubilize insoluble Mussorie rock phosphate and tricalcium phosphate efficiently (Yandigeri et al., 2010). In alkaline wetland soils, microalgae solubilized magnesium carbonates commonly referred to us dypingite according to Power et al. (2007). Siderophores serve as organic compounds that chelate ferric iron in environs it is deficient to make it bioavailable to plants and microbes (Ahmed & Holmstrom, 2014). Microalgal species such as *Anabaena spp.*, *Anabaena cylindrical* and *Anabaena flos aquae* from the cyanobacteria family chelate micronutrients such as copper and iron using their produced siderophores (Goldman et al., 1983). *Chorella spp.* and *Scenedesmus incrassatulus* among other green algae also produce siderophores involved in chelation of iron (Benderliev et al., 2003). Enrichment of plant parts particularly grains with nutrients such as zinc, copper, manganese and iron using bacteria consortia complexed with green algae for bio-fortification has been successfully reported in several studies (Prasanna et al., 2015; Manjunath et al., 2016; Renuka et al., 2017). However, the translocation mechanisms used during bio-fortification from soil to roots and up the plant is less understood. The gap necessitates extensive research to gain more insight on the involved mechanisms.

Microalgae in Improving and Modulating Soil Microbial Activity

The use of cyanobacteria inoculants is shown to influence microbiomes of the rhizosphere, which has ultimate effects on the quantity and structure of microbes important in solubilisation and mineralization of nutrients (Priya et al., 2015; Manjunath et al., 2016; Ranjan et al., 2016). A case example is the use of *Calothrix elenkinii* that has been shown to offer advantages to rhizosphere microbiome and the plant at large (Priya et al., 2015; Ranjan et al., 2016). Improvement and modulation of microbial activity in this context occurs via the production of EPS after cyanobacteria inoculants are introduced in soils. The EPS are sources of organic carbon for the growth of microbes and also facilitate the formation of biofilms and bioflocs in vegetal rhizosphere (Xiao & Zheng, 2016). Biofilms usually have a top layer of oxygenic phototrophic green algae, cyanobacteria and diatoms followed by heterotrophic fungi and bacteria (Bharti et al., 2017). EPS produced by heterotrophic and phototrophic partners result to a hydrated mix that aggregates the organisms and cells to improve their mechanical stability during micro-colony formation to enhance nutrient flow and cushion against predators and grazers (Xiao & Zheng, 2016). Additionally, EPS has inorganic components such as silica and carbonate as well as organic compounds such as nucleic acids, proteins and extracellular polysaccharides that are essential soil organic carbon sources (Bondoc et al., 2016). EPS enable light transmission to deeper regions of the biofilm hence, providing oxygen and organic carbon during N fixation and photosynthesis. This has been established in green algae and cyanobacteria (Renuka et al., 2016, 2017, 2018).

Microalgae Role in Colonization of Plant Tissues

Microalgae are reported to establish adequately in soils and colonize some plant organelles and some cyanobacteria species have symbiotic relationships with some of their non-vascular and vascular plants (Krings et al., 2009). Some of the examples are shown in Table 3. Cyanobacteria has symbiotic associations with vascular plants, pteridophytes, gymnosperms, fungi and microalgae as Santi et al. (2013) noted. With these relationships, the microalgae can infuse in different plant parts including parenchyma cells, intracellular spaces, sub-stomatal chambers, stomata and mycorrhizal arbuscule-regions to form intracellular coils and loops (Krings et al., 2009). These colonization tendencies by microalgae have been reported in rice and wheat (Bidyarani et al., 2015) and enhance nitrogen fixation while in chickpea; they improve the microbial communities of associated soils and in the nodules and rhizosphere of the plants (Prasanna et al., 2017; Ramakrishnan et al., 2017). Overall, the effect is improved soil fertility and bettered agricultural production.

Table 3. Microalgae species, the plants they colonize and the kinds of associations involved

Microalgae Species	Plant Species	Kind of Association
Leptolyngbya sp., *Tolypothris* sp. and *Nostoc* spp.	*Cycas, Bowenia, Lepidozamia and maccrozamia* (Cycads)	Endosymbiont
Nostoc sp. strain 2S9B	*Triticum aestivum* L. (Wheat)	Root colonization
Nostoc sp. strain 2S9B	*T. vulgare* (Wheat)	
Nostoc strains	*Oryza sativa* (rice)	Root and intracellular surfaces colonization
Anabaena biofilms with *Azobacter* and *Trichoderma*	*Gossyoium* spp. (cotton)	Root colonisation
Calothrix sp. and *Anabaena laxa*	*Oryza sativa* (rice) and *T. aestivum* L. (wheat)	Root and stem colonization

(Renuka et al., 2018)

Role of Microalgae in Production of Growth Hormones

Hormones serve crucial roles in plant growth and development and their external natural or synthetic supplementation is key in agricultural improvement particularly enhanced crop yields and productivity as well as weed elimination (Vats, 2015; Epp et al., 2016). Microalgae usually produce growth hormones such as jasmonic acid, cytokinins, abscisic acid, gibberellins, ethylene and auxins that are useful agricultural biostimulants. Table 4 summarizes some of the microalgae strains of the chlorophyte and cyanophyta groups and the hormones they produce useful as biostimulants. Most of the microbe strains being green algae and cyanobacteria have either inherent hormones or some excrete them while growing in suitable environs including growth medium (Lu & Xu, 2015; Romanenko et al., 2015). Studies such as those by Plaza et al. (2018) established that microalgae species such as *Arthrospira* spp. and *Scenedesmus* spp.

have high concentrations of abscisic acid, auxins, gibberellins and cytokinins. Similarly, Karthikeyan et al. (2007) demonstrated that cyanobacteria have the capacity to excrete growth-promoting hormones particularly indole acetic acid that enhance soil microbial activity for enhanced fertility and crop production.

Table 4. Groups and species of microalgae involved in production of biostimulants

Microalgae Species	Cyanobacterial/ Green Microalgae Group	Hormone Produced
Scenedesmus obliquus	Chlorophyta	Indole-3-acetic acid
Chlorella pyrenoidosa, S. armatus	Chlorophyta	
Scenedesmus, Chlorella and *Protococcus* sp.	Chlorophyta	Topolin and zeatin conjugates
Chlamydomonas, Tetracystis, Chlorosarcina, Coenochloris, Chlorella and *Anabaena* species	Chlorophyta	Cytokinin-kinetin, Indole-butyric acid and auxin
Nostoc calicola and *A. vaginicola*	Cyanophyta	Indole-3-propionic acid, Indole butyric acid, Indole-3-acetic acid
Aphanothece sp. MBDU 515	Cyanophyta	Indole-3-acetic acid
Nostoc spp.		
Phormidium animale and *Calothrix* spp.		Cytokinin-kinetin, Indole-butyric acid and auxin

(Renuka et al., 2018)

Role of Microalgae in Promoting Plant Protection Mechanisms

Cyanobacteria group of microalgae regulate the defence and protection mechanisms of plants through their ability to stimulate the pathogenesis and antioxidant effects of plants. This is evident through production and increased activity of phenylalanine ammonia lyase, polyphenol oxidase, peroxidase, catalase, chitinase and β-1, 3 endoglucanase (Priya et al., 2015). This has been demonstrated in wheat using *Anabaena* and *Calothrix* sp. (Babu et al., 2015), seed spice crop using cyanobacterial strains (Kumar et al., 2013) and in rice using *Calothrix elenkinii* (Priya et al., 2015). In the three studies, various plant organelles were found to have high levels of antioxidants and defence enzymes, which is synergistic to the building of individual plant immunity. Additionally, it was established that green algae inoculation enhances RNA activity and the production of nutrient assimilating enzymes as lines of defence for plants (Grzesik et al., 2017).

Role of Microalgae in Agricultural Pest and Disease Management

The use of agrochemicals in pest and pathogen control is associated with extensive pollution of agroecosystems and hence is not sustainable. It is from this precognition that alternative pathogen control methods have been exploited as viable alternatives. Of these alternatives, biological options are preferred as they are cost effective and friendly environmentally (Renuka et al., 2016, 2017). Common organisms used in biocontrol include fungi, bacteria and most recently cyanobacteria (Hernandez-Carlos & Gamboa-Angulo, 2011). By enhancing nutrition accretion for better plant immunity, cyanobacteria significantly reduce the use of harmful agrochemicals to prevent crop diseases and pests while being environmentally sensitive (Chaudhary et al., 2012; Prasanna et al., 2016). Antimicrobial effects against disease causing fungi and bacteria in plants and antibiotic properties have also been associated with green microalgae according to Mashjoor et al. (2016) and Navvaro et al. (2017).

Pathogen Control and Disease Suppression by Microalgae

Microalgae particularly cyanobacteria produce compounds such as carbamidocyclophane A, ambigol A, majusculonic acid, benzoic acid with anti-bacterial properties that have the capacity to kill and suppress disease causing nematodes, fungi and bacteria. Additionally, they produce biocidal metabolites and hydrolytic enzymes that weaken plant pathogens. The antimicrobial compounds work through mechanisms involving functional and structural modifications, inhibition of protein metabolism, inhibition of enzyme activity and disruption of cytoplasmic membrane for the target organisms (Swain et al., 2017). Microalgae also have bioactive compounds such as tocopherols and polyphenols in addition to pigments, oils, proteins and carbohydrates whose antimicrobial features are helpful in fighting soil borne diseases (Michalak & Chojnacka, 2015). This occurs in microalgal cultures in the form of seed treatments, foliar sprays, dried and fresh biomass, which are applied on the plants and soils. Examples of microalgae compounds with insecticidal, biocontrol and pesticidal features include majusculamide-like chemicals from *Anabaena laxa*, chlorine-based antibiotics from *Scytonema* sp. and benzoic acid from *Calothrix* sp. (Natarajan et al., 2012; Singh, 2014). Seventy *Anabaena* strains were found to have fungicidal properties due to their ability to produce hydrolytic enzymes (Prasanna et al., 2008). Similarly, *Calothris sp.* and *Anabaena spp.* were found to produce endoglucanases and chitosanase enzymes with anti- bacterial and anti-fungal characteristics (Gupta et al., 2013; Natarajan et al., 2012). Manjunath et al. (2010) established that *Calothrix spp.* were effective for suppressing the dumping off disease in vegetables while *Anabaena oscillarioides* was effective in suppressing fungal species such as *Rhizoctonia solani, Pythium aphanidermatum, P. debaryanum* and *Fusarium oxysporum* that are pathogenic in tomatoes (Dukare et al., 2011). El-Sheekh et al. (2006) found that *Nostoc muscorum* extracts possessed antimicrobial activity against gram negative and positive bacteria. Prasanna et al. (2016) and Babu et al. (2015) established that cyanobacteria formulations were effective in controlling rot disease in cotton roots. *Anabaena* sp. were found to be effective in reducing the damping off disease in tomatoes while at the same time acting as a biofertilizer (Chaudhary et al., 2012).

Pest Management Using Microalgae

The population of plant pests such as nematodes have been reduced using microalgae, which is synergistic to crop production. Cyanobacteria for instance produce nematicidal compounds and peptide toxins that suppress the thriving of pests. *Microcoleus vaginatus* cyanobacteria decreased the levels of *Meloidogyne incognita* nematode after inoculation in soils used for tomato growth (Khan & Park, 1999). Similarly, *M. incognita* populations reduced following the introduction of *Oscillatoria chlorine* that had nematicidal activity (Khan et al., 2007). The hatching of *Meloidogyne triticoryzae* also known as the root-knot nematode was inhibited following the introduction of *Aulosira fertilissima* in infected soils (Chandel, 2009). *Nostoc* cyanobacterial strains ATCC 53789 had nematicidal activity against *Caenorhabditis,* caused cytotoxicity on *Artemia salina* and antifungal activity on *Verticillium albo-atrum, Sclerotinis sclerotiorum, Rosellinia* sp., *Armillaria* sp. and *Phytophthora cambivora* (Biondi et al., 2004).

Role of Microalgae in Bioremediation and Soil Conditioning

Cyanobacteria and green microalgae have a ubiquitous nature and can withstand harsh environmental conditions (Qiao et al., 2015; Subramaniyam et al., 2016). Their survival in metal, oil and salt contaminated areas as well as drought-inflicted areas fits their use in reclamation of such wastelands (Monteiro et al., 2009; Trejo et al., 2012). Green algae species such as *Azospirillum brasilense* and *Chlorella sorokiniana* have been used to reclaim desert-eroded soils and results showed positive increases in organic carbon, organic matter and microbial biomass (Trejo et al., 2012). Similarly, *Scytonema, Nostoc* and *Oscillatoria* sp. were involved in revitalization of soil structure and microbial community in soils heated above 350°C (Acea, 2003). Both cyanobacteria and green algae species were found to produce polysaccharides that assist in restoring and stabilizing desert soils (Park et al., 2017; Rossi et al., 2017). In a study examining the role of *Phormidium tenue* produced polysaccharides on the shrub *Caragana korshinskii* that grows in desert soils, it was established that the cyanobacteria improved its nutritional features in addition to enhanced seed germination and plant growth (Xu et al., 2013).

Microalgae have a role to play in the degrading petroleum and oil in addition to supporting the growth of heterotrophic bacteria that degrade oil (Abed, 2010). Five species of cyanobacteria including *Oscillatoria, Halothece, Synechocystis, Dactylococcopsis salina* and *Aphanothece halophyletic*a were found to degrade n-alkane found in petroleum products according to Abed (2010). This activity is related and more synergistic in the presence of oil degrading bacteria where cyanobacteria help in EPS formation and with mutual association with bacteria to form biofilms. The degrading of oil and petroleum products has positive effects in restoring soil fertility at contaminated sites (Renuka et al., 2018). Both cyanobacteria and blue green algae trap excess sodium (Na^+) amounts from saline soils in their EPS matrix and prevent excess uptake by plants (Roeselers et al., 2008). As such, the microbes are essential in ameliorating sodic soils for agricultural use. In vitro experiments have confirmed that microalgae can remove heavy metal from contaminated regions (Monteiro et al., 2009; Subramaniyam et al., 2016; Hamed et al., 2017). Cyanobacteria was also established to bioaccumulate fly ash- sourced heavy metals following their inoculation in polluted soils in addition to improving their nitrogen and phosphorus content (Rai et al., 2000). Therefore, the microbes act as biofertilisers and effective inoculants in metal contaminated sites.

CONCLUSION

This book chapter explores the various uses of microalgae in soil and agricultural improvements for enhanced crop productivity. The use of both cyanobacteria and green algae as a biofertilizer, soil conditioners, biostimulants and in bioremediation was highlighted. Additionally, heterocystous cyanobacteria were identifies as essential in nutrient mineralization and cycling (N-fixation) which contributed to better crop yields and sustained soil fertility. The microalgae also promoted organic carbon enrichment leading to high levels of soil organic matter correspondent to fertile and productive soils. Although the benefits of microalgae in improving soil fertility and agricultural production have been established, the mechanisms behind these benefits are less understood. This trend necessitates further research to characterize and understand these microbes better and produce species with many facets of agricultural importance rather than using different species for each given purpose. Conclusively, microalgae are a promising bio-option to sustainable agricultural production if their uses are optimized.

REFERENCES

Abed, R. (2010). Interaction between cyanobacteria and aerobic heterotrophic bacteria in the degradation of hydrocarbons. *International Biodeterioration & Biodegradation*, *64*(1), 58–64. doi:10.1016/j.ibiod.2009.10.008

Acea, M. (2003). Cyanobacterial inoculation of heated soils: Effect on microorganisms of C and N cycles and on chemical composition in soil surface. *Soil Biology & Biochemistry*, *35*(4), 513–524. doi:10.1016/S0038-0717(03)00005-1

Ahmed, E., & Holmstrom, S. J. (2014). Siderophores in environmental research: Roles and applications. *Microbial Biotechnology*, *7*(3), 196–208. doi:10.1111/1751-7915.12117 PMID:24576157

Ashokkumar, V., & Rengasamy, R. (2012). Mass culture of *Botryococcus braunii Kutz.* under open raceway pond for biofuel production. *Bioresource Technology*, *104*, 394–399. doi:10.1016/j.biortech.2011.10.093 PMID:22115530

Babu, S., Bidyarani, N., Chopra, P., Monga, D., Kumar, R., Radha, P., Kranthi, S., & .... (2015). Evaluating microbe-plant interactions and varietal differences for enhancing biocontrol efficacy in root rot challenged cotton crop. *European Journal of Plant Pathology*, *142*(2), 345–362. doi:10.100710658-015-0619-6

Bai, L., Xu, H., Wang, C., Deng, J., & Jiang, H. (2016). Extracellular polymeric substances facilitate the biosorption of phenanthrene on cyanobacteria *Microcystis aeruginosa. Chemosphere*, *162*, 172–180. doi:10.1016/j.chemosphere.2016.07.063 PMID:27497347

Benderliev, K. M., Ivanova, N. I., & Pilarski, P. S. (2003). Singlet oxygen and other reactive oxygen species are involved in regulation of release of iron-binding chelators from *Scenedesmus* cells. *Biologia Plantarum*, *47*(4), 523–526. doi:10.1023/B:BIOP.0000041056.07819.df

Bharti, A., Velmourougane, K., & Prasanna, R. (2017). Phototrophic biofilms: Diversity, ecology and applications. *Journal of Applied Phycology*, 1–16.

Bidyarani, N., Prasanna, R., Chawla, G., Babu, S., & Singh, R. (2015). Deciphering the factors associated with the colonization of rice plants by cyanobacteria. *Journal of Basic Microbiology*, *55*(4), 407–419. doi:10.1002/jobm.201400591 PMID:25515189

Biondi, N., Piccardi, R., Margheri, M. C., Rodolfi, L., Smith, G. D., & Tredici, M. R. (2004). Evaluation of *Nostoc* strain ATCC 53789 as a potential source of natural pesticides. *Applied and Environmental Microbiology*, *70*(6), 3313–3320. doi:10.1128/AEM.70.6.3313-3320.2004 PMID:15184126

Bondoc, K. G., Heuschele, J., Gillard, J., Vyverman, W., & Pohnert, G. (2016). Selective silicate-directed motility in diatoms. *Nature Communications*, *7*(1), 10540. doi:10.1038/ncomms10540 PMID:26842428

Chandel, S. T. (2009). Nematicidal activity of the Cyanobacterium, *Aulosira fertilissima* on the hatch of *Meloidogyne triticoryzae* and *Meloidogyne incognita. Archiv für Phytopathologie und Pflanzenschutz*, *42*(1), 32–38. doi:10.1080/03235400600914363

Chaudhary, V., Prasanna, R., Nain, L., Dubey, S. C., Gupta, V., Singh, R., Jaggi, S., & Bhatnagar, A. K. (2012). Bioefficacy of novel cyanobacteria-amended formulations in suppressing damping off disease in tomato seedlings. *World Journal of Microbiology & Biotechnology*, *28*(12), 3301–3310. doi:10.100711274-012-1141-z PMID:22869418

Chiaiese, P., Corrado, G., Colla, G., Kyriacou, M. C., & Rouphael, Y. (2018). Renewable sources of plant biostimulation: Microalgae as a sustainable means to improve crop performance. *Frontiers in Plant Science*, *871*(12), 1–6. doi:10.3389/fpls.2018.01782 PMID:30581447

Coates, R., Trentacoste, E., & Gerwick, W. (2013). Bioactive and novel chemicals from microalgae. In A. Richmond & Q. Hu (Eds.), *Handbook of microalgal culture: applied phycology and biotechnology* (2nd ed., pp. 504–531). Wiley. doi:10.1002/9781118567166.ch26

Coppens, J., Grunert, O., Van Den Hende, S., Vanhoutte, I., Boon, N., Haesaert, G., & De Gelder, L. (2016). The use of microalgae as a high-value organic slow-release fertilizer results in tomatoes with increased carotenoid and sugar levels. *Journal of Applied Phycology*, *28*(4), 2367–2377. doi:10.100710811-015-0775-2

Costa, J. A. V., Freitas, B. C. B., Cruz, C. G., Silveira, J., & Morais, M. G. (2019). Potential of microalgae as biopesticides to contribute to sustainable agriculture and environmental development. *Journal of Environmental Science and Health. Part B, Pesticides, Food Contaminants, and Agricultural Wastes*, *54*(5), 366–375. doi:10.1080/03601234.2019.1571366 PMID:30729858

Dukare, A. S., Prasanna, R., Chandra Dubey, S., Nain, L., Chaudhary, V., Singh, R., & Saxena, A. K. (2011). Evaluating novel microbe amended composts as biocontrol agents in tomato. *Crop Protection (Guildford, Surrey)*, *30*(4), 436–442. doi:10.1016/j.cropro.2010.12.017

El-Sheekh, M. M., Osman, M. E., Dyab, M. A., & Amer, M. S. (2006). Production and characterization of antimicrobial active substance from the cyanobacterium *Nostoc muscorum*. *Environmental Toxicology and Pharmacology*, *21*(1), 42–50. doi:10.1016/j.etap.2005.06.006 PMID:21783637

Etesami, H., & Alikhani, H. A. (2016). Co-inoculation with endophytic and rhizosphere bacteria allows reduced application rates of N-fertilizer for rice plantquery id="q1">. *Rhizosphere*, *2*, 5–12. doi:10.1016/j.rhisph.2016.09.003

Garcia-Gonzalez, J., & Sommerfeld, M. (2016). Biofertilizer and biostimulant properties of the microalga Acutodesmus dimorphus. *Journal of Applied Phycology*, *28*(2), 1051–1061. doi:10.100710811-015-0625-2 PMID:27057088

Gayathri, M., Kumar, P. S., Prabha, A. M. L., & Muralitharan, G. (2015). In vitro regeneration of *Arachis hypogaea L.* and *Moringa oleifera Lam.* using extracellular phytohormones from *Aphanothece* sp. MBDU 515. *Algal Research*, *7*, 100–105. doi:10.1016/j.algal.2014.12.009

Goldman, S., Lammers, P., Berman, M., & Sanders-Loehr, J. (1983). Siderophore-mediated iron uptake in different strains of *Anabaena* sp. *Journal of Bacteriology*, *156*(3), 1144–1150. doi:10.1128/JB.156.3.1144-1150.1983 PMID:6227608

Grzesik, M., Romanowska-Duda, Z., & Kalaji, H. M. (2017). Effectiveness of cyanobacteria and green algae in enhancing the photosynthetic performance and growth of willow (*Salix viminalis L.*) plants under limited synthetic fertilizers application. *Photosynthetica, 55*(3), 1–12. doi:10.100711099-017-0716-1

Gupta, V., Ratha, S. K., Sood, A., Chaudhary, V., & Prasanna, R. (2013). New insights into the biodiversity and applications of cyanobacteria (blue-green algae)-Prospects and challenges. *Algal Research, 2*(2), 79–97. doi:10.1016/j.algal.2013.01.006

Hamed, S. M., Selim, S., Klöck, G., & AbdElgawad, H. (2017). Sensitivity of two green microalgae to copper stress: Growth, oxidative and antioxidants analyses. *Ecotoxicology and Environmental Safety, 144*, 19–25. doi:10.1016/j.ecoenv.2017.05.048 PMID:28599127

Hernandez-Carlos, B., & Gamboa-Angulo, M. M. (2011). Metabolites from freshwater aquatic microalgae and fungi as potential natural pesticides. *Phytochemistry Reviews, 10*(2), 261–286. doi:10.100711101-010-9192-y

Karthikeyan, S., Balasubramanian, R., & Iyer, C. (2007). Evaluation of the marine algae *Ulva fasciata* and *Sargassum* sp. for the biosorption of Cu (II) from aqueous solutions. *Bioresource Technology, 98*(2), 452–455. doi:10.1016/j.biortech.2006.01.010 PMID:16530408

Khan, S. A., Rashmi, Hussain, M. Z., Prasad, S., & Banerjee, U. C. (2009). Prospects of biodiesel production from microalgae in India. *Renewable & Sustainable Energy Reviews, 13*(9), 2361–2372. doi:10.1016/j.rser.2009.04.005

Khan, Z., Kim, Y. H., Kim, S. G., & Kim, H. W. (2007). Observations on the suppression of root knot nematode (*Meloidogyne arenaria*) on tomato by incorporation of cyanobacterial powder (*Oscillatoria chlorina*) into potting field soil. *Bioresource Technology, 98*(1), 69–73. doi:10.1016/j.biortech.2005.11.029 PMID:16458501

Khan, Z., & Park, S. D. (1999). Effects of inoculum level and time of *Microcoleus vaginatus* on control of *Meloidogyne incognita* on tomato. *Journal of Asia-Pacific Entomology, 2*(2), 93–96. doi:10.1016/S1226-8615(08)60036-9

Krings, M., Hass, H., Kerp, H., Taylor, T. N., Agerer, R., & Dotzler, N. (2009). Endophytic cyanobacteria in a 400-million-yr-old land plant: A scenario for the origin of a symbiosis? *Review of Palaeobotany and Palynology, 153*(1-2), 62–69. doi:10.1016/j.revpalbo.2008.06.006

Kumar, M., Prasanna, R., Bidyarani, N., Babu, S., Mishra, B. K., Kumar, A., Adak, A., Jauhari, S., Yadav, K., Singh, R., & Saxena, A. K. (2013). Evaluating the plant growth promoting ability of thermotolerant bacteria and cyanobacteria and their interactions with seed spice crops. *Scientia Horticulturae, 164*, 94–101. doi:10.1016/j.scienta.2013.09.014

Lu, Y., & Xu, J. (2015). Phytohormones in microalgae: A new opportunity for microalgal biotechnology? *Trends in Plant Science, 20*(5), 273–282. doi:10.1016/j.tplants.2015.01.006 PMID:25697753

Mager, D. M., & Thomas, A. D. (2011). Extracellular polysaccharides from cyanobacterial soil crusts: A review of their role in dryland soil processes. *Journal of Arid Environments, 75*(2), 91–97. doi:10.1016/j.jaridenv.2010.10.001

Mahmoud, S. (2001). Nutritional Status and Growth of Maize Plants as Affected by Green Microalgae as Soil Additives. *The Journal of Biological Sciences*, *1*(6), 475–479. doi:10.3923/jbs.2001.475.479

Manjunath, M., Kanchan, A., Ranjan, K., Venkatachalam, S., Prasanna, R., Ramakrishnan, B., Hossain, F., Nain, L., Shivay, Y. S., Rai, A. B., & Singh, B. (2016). Beneficial cyanobacteria and eubacteria synergistically enhance bioavailability of soil nutrients and yield of okra. *Heliyon*, *2*(2), e00066. doi:10.1016/j.heliyon.2016.e00066 PMID:27441245

Manjunath, M., Prasanna, R., Nain, L., Dureja, P., Singh, R., Kumar, A., Jaggi, S., & Kaushik, B. D. (2010). Biocontrol potential of cyanobacterial metabolites against damping off disease caused by *Pythium aphanidermatum* in solanaceous vegetables. *Archiv für Phytopathologie und Pflanzenschutz*, *43*(7), 666–677. doi:10.1080/03235400802075815

Mashjoor, S., Yousefzadi, M., Esmaeili, M. A., & Rafiee, R. (2016). Cytotoxicity and antimicrobial activity of marine macro algae (*Dictyotaceae* and *Ulvaceae*) from the Persian Gulf. *Cytotechnology*, *68*(5), 1717–1726. doi:10.100710616-015-9921-6 PMID:26507649

Michalak, I., & Chojnacka, K. (2015). Algae as production systems of bioactive compounds. *Engineering in Life Sciences*, *15*(2), 160–176. doi:10.1002/elsc.201400191

Monteiro, C. M., Marques, A. P., Castro, P. M., & Xavier Malcata, F. (2009). Characterization of *Desmodesmuspleiomorphus* isolated from a heavy metal-contaminated site: Biosorption of zinc. *Biodegradation*, *20*(5), 629–641. doi:10.100710532-009-9250-6 PMID:19225897

Mostafa, S. (2013). Microalgal biotechnology: Prospects and applications. *Plant Science*.

Nabti, E., Jha, B., & Hartmann, A. (2017). Impact of seaweeds on agricultural crop production as biofertilizer. *International Journal of Environmental Science and Technology*, *14*(5), 1119–1134. doi:10.100713762-016-1202-1

Natarajan, C., Prasanna, R., Gupta, V., Dureja, P., & Nain, L. (2012). Characterization of the fungicidal activity of *Calothrix elenkinii* using chemical methods and microscopy. *Applied Biochemistry and Microbiology*, *48*(1), 51–57. doi:10.1134/S0003683812010115 PMID:22567886

Navarro, F., Forján, E., Vázquez, M., Toimil, A., Montero, Z., Ruiz-Domínguez, M., Garbayo, I., Castaño, M. Á., Vílchez, C., & Vega, J. M. (2017). Antimicrobial activity of the acidophilic eukaryotic microalga *Coccomyxaonubensis*. *Phycological Research*, *65*(1), 38–43. doi:10.1111/pre.12158

Nisha, R., Kaushik, A., & Kaushik, C. P. (2007). Effect of indigenous cyanobacterial application on structural stability and productivity of an organically poor semi-arid soil. *Geoderma*, *138*(1–2), 49–56. doi:10.1016/j.geoderma.2006.10.007

Park, C., Li, X. R., Zhao, Y., Jia, R. L., & Hur, J. (2017). Rapid development of cyanobacterial crust in the field for combating desertification. *PLoS One*, *12*(6), e0179903. doi:10.1371/journal.pone.0179903 PMID:28644849

Plaza, B. M., Gómez-Serrano, C., Acién-Fernández, F. G., & Jimenez-Becker, S. (2018). Effects of microalgae hydrolysate foliar application (*Arthrospira platensis* and *Scenedesmus* sp.) on Petunia x hybrida growth. *Journal of Applied Phycology*, *30*(4), 2359–2365. doi:10.100710811-018-1427-0

Power, I. M., Wilson, S. A., Thom, J. M., Dipple, G. M., & Southam, G. (2007). Biologically induced mineralization of dypingite by cyanobacteria from an alkaline wetland near Atlin, British Columbia, Canada. *Geochemical Transactions*, *8*(1), 13. doi:10.1186/1467-4866-8-13 PMID:18053262

Prasanna, R., Bidyarani, N., Babu, S., Hossain, F., Shivay, Y. S., Nain, L., & Moral, M. T. (2015). Cyanobacterial inoculation elicits plant defense response and enhanced Zn mobilization in maize hybrids. *Cogent Food & Agriculture*, *1*(1), 998507. doi:10.1080/23311932.2014.998507

Prasanna, R., Joshi, M., Rana, A., Shivay, Y. S., & Nain, L. (2012). Influence of co-inoculation of bacteria-cyanobacteria on crop yield and C–N sequestration in soil under rice crop. *World Journal of Microbiology & Biotechnology*, *28*(3), 1223–1235. doi:10.100711274-011-0926-9 PMID:22805842

Prasanna, R., Kanchan, A., Kaur, S., Ramakrishnan, B., Ranjan, K., Singh, M. C., Hasan, M., Saxena, A. K., & Shivay, Y. S. (2016). *Chrysanthemum* Growth Gains from Beneficial Microbial Interactions and Fertility Improvements in Soil Under Protected Cultivation. *Horticultural Plant Journal*, *2*(4), 229–239. doi:10.1016/j.hpj.2016.08.008

Prasanna, R., Nain, L., Tripathi, R., Gupta, V., Chaudhary, V., Middha, S., Joshi, M., Ancha, R., & Kaushik, B. D. (2008). Evaluation of fungicidal activity of extracellular filtrates of cyanobacteria – possible role of hydrolytic enzymes. *Journal of Basic Microbiology*, *48*(3), 186–194. doi:10.1002/jobm.200700199 PMID:18506903

Prasanna, R., Ramakrishnan, B., Simranjit, K., Ranjan, K., Kanchan, A., Hossain, F., & Nain, L. (2017). Cyanobacterial and rhizobial inoculation modulates the plant physiological attributes and nodule microbial communities of chickpea. *Archives of Microbiology*, *199*(9), 1311–1323. doi:10.100700203-017-1405-y PMID:28669069

Priya, H., Prasanna, R., Ramakrishnan, B., Bidyarani, N., Babu, S., Thapa, S., & Renuka, N. (2015). Influence of cyanobacterial inoculation on the culturable microbiome and growth of rice. *Microbiological Research*, *171*, 78–89. doi:10.1016/j.micres.2014.12.011 PMID:25644956

Priyadarshani, I., & Rath, B. (2012). Commercial and industrial applications of micro algae –. *RE:view*, *3*(4), 89–100.

Pulz, O. (2001). Photobioreactors: Production systems for phototrophic microorganisms. *Applied Microbiology and Biotechnology*, *57*(3), 287–293. doi:10.1007002530100702 PMID:11759675

Qiao, K., Takano, T., & Liu, S. (2015). Discovery of two novel highly tolerant $NaHCO_3$ Trebouxiophytes: Identification and characterization of microalgae from extreme saline–alkali soil. *Algal Research*, *9*, 245–253. doi:10.1016/j.algal.2015.03.023

Rai, U. N., Tripathi, R. D., Singh, N., Kumar, A., Ali, M. B., Pal, A., & Singh, S. N. (2000). Amelioration of fly-ash by selected nitrogen fixing blue green algae. *Bulletin of Environmental Contamination and Toxicology*, *64*(2), 294–301. doi:10.1007001289910043 PMID:10656898

Ramakrishnan, B., Kaur, S., Prasanna, R., Ranjan, K., Kanchan, A., Hossain, F., Shivay, Y. S., & Nain, L. (2017). Microbial inoculation of seeds characteristically shapes the rhizosphere microbiome in desi and kabuli chickpea types. *Journal of Soils and Sediments*, *17*(8), 2040–2053. doi:10.100711368-017-1685-5

Ranjan, K., Priya, H., Ramakrishnan, B., Prasanna, R., Venkatachalam, S., Thapa, S., Tiwari, R., Nain, L., Singh, R., & Shivay, Y. S. (2016). Cyanobacterial inoculation modifies the rhizosphere microbiome of rice planted to a tropical alluvial soil. *Applied Soil Ecology*, *108*, 195–203. doi:10.1016/j.apsoil.2016.08.010

Renuka, N., Guldhe, A., Prasanna, R., Singh, P., & Bux, F. (2018). Microalgae as multi-functional options in modern agriculture: Current trends, prospects and challenges. *Biotechnology Advances*, *36*(4), 1255–1273. doi:10.1016/j.biotechadv.2018.04.004 PMID:29673972

Renuka, N., Prasanna, R., Sood, A., Ahluwalia, A. S., Bansal, R., Babu, S., Singh, R., Shivay, Y. S., & Nain, L. (2016). Exploring the efficacy of wastewater-grown microalgal biomass as a biofertilizer for wheat. *Environmental Science and Pollution Research International*, *23*(7), 6608–6620. doi:10.100711356-015-5884-6 PMID:26638970

Renuka, N., Prasanna, R., Sood, A., Bansal, R., Bidyarani, N., Singh, R., Shivay, Y. S., Nain, L., & Ahluwalia, A. S. (2017). Wastewater grown microalgal biomass as inoculants for improving micronutrient availability in wheat. *Rhizosphere*, *1*, 150–159. doi:10.1016/j.rhisph.2017.04.005

Roeselers, G., Van Loosdrecht, M., & Muyzer, G. (2008). Phototrophic biofilms and their potential applications. *Journal of Applied Phycology*, *20*(3), 227–235. doi:10.100710811-007-9223-2 PMID:19396356

Romanenko, E. A., Kosakovskaya, I. V., & Romanenko, P. A. (2015). Phytohormones of microalgae: Biological role and involvement in the regulation of physiological processes. Pt I. auxins, abscisic acid, ethylene. *International Journal on Algae*, *17*(3), 275–289. doi:10.1615/InterJAlgae.v17.i3.80

Ronga, D., Biazzi, E., Parati, K., Carminati, D., Carminati, E., & Tava, A. (2019). Microalgal biostimulants and biofertilisers in crop productions. *Agronomy (Basel)*, *9*(4), 1–22. doi:10.3390/agronomy9040192

Rossi, F., Li, H., Liu, Y., & De Philippis, R. (2017). Cyanobacterial inoculation (cyanobacterisation): Perspectives for the development of a standardized multifunctional technology for soil fertilization and desertification reversal. *Earth-Science Reviews*, *171*, 28–43. doi:10.1016/j.earscirev.2017.05.006

Santi, C., Bogusz, D., & Franche, C. (2013). Biological nitrogen fixation in non-legume plants. *Annals of Botany*, *111*(5), 743–767. doi:10.1093/aob/mct048 PMID:23478942

Sato, T., Usui, S., Tsuchiya, Y., & Kondo, Y. (2006). Invention of outdoor closed type photobioreactor for microalgae. *Energy Conversion and Management*, *47*(6), 791–799. doi:10.1016/j.enconman.2005.06.010

Schenk, P. M., Thomas-Hall, S. R., Stephens, E., Marx, U. C., Mussgnug, J. H., Posten, C., Kruse, O., & Hankamer, B. (2008). Second Generation Biofuels: High-Efficiency Microalgae for Biodiesel Production. *BioEnergy Research*, *1*(1), 20–43. doi:10.100712155-008-9008-8

Shaaban, M. (2001). Green Microalgae Water Extract as Foliar Feeding to Wheat Plants. *Pakistan Journal of Biological Sciences*, *4*(6), 628–632. doi:10.3923/pjbs.2001.628.632

Singh, S. (2014). A review on possible elicitor molecules of cyanobacteria: Their role in improving plant growth and providing tolerance against biotic or abiotic stress. *Journal of Applied Microbiology*, *117*(5), 1221–1244. doi:10.1111/jam.12612 PMID:25069397

Subramaniyam, V., Subashchandrabose, S. R., Thavamani, P., Chen, Z., Krishnamurti, G. S. R., Naidu, R., & Megharaj, M. (2016). Toxicity and bioaccumulation of iron in soil microalgae. *Journal of Applied Phycology*, *28*(5), 2767–2776. doi:10.100710811-016-0837-0

Suganya, T., Varman, M., Masjuki, H. H., & Renganathan, S. (2016). Macroalgae and microalgae as a potential source for commercial applications along with biofuels production: A biorefinery approach. *Renewable & Sustainable Energy Reviews*, *55*, 909–941. doi:10.1016/j.rser.2015.11.026

Swain, S. S., Paidesetty, S. K., & Padhy, R. N. (2017). Antibacterial, antifungal and antimycobacterial compounds from cyanobacteria. *Biomedicine and Pharmacotherapy*, *90*, 760–776. doi:10.1016/j.biopha.2017.04.030 PMID:28419973

Swarnalakshmi, K., Prasanna, R., Kumar, A., Pattnaik, S., Chakravarty, K., Shivay, Y. S., Singh, R., & Saxena, A. K. (2013). Evaluating the influence of novel cyanobacterial biofilmed biofertilizers on soil fertility and plant nutrition in wheat. *European Journal of Soil Biology*, *55*, 107–116. doi:10.1016/j.ejsobi.2012.12.008

Trejo, A., De-Bashan, L. E., Hartmann, A., Hernandez, J.-P., Rothballer, M., Schmid, M., & Bashan, Y. (2012). Recycling waste debris of immobilized microalgae and plant growth-promoting bacteria from wastewater treatment as a resource to improve fertility of eroded desert soil. *Environmental and Experimental Botany*, *75*, 65–73. doi:10.1016/j.envexpbot.2011.08.007

Trentacoste, E. M., Martinez, A. M., & Zenk, T. (2015). The place of algae in agriculture: Policies for algal biomass production. *Photosynthesis Research*, *123*(3), 305–315. doi:10.100711120-014-9985-8 PMID:24599393

Uysal, O., Uysal, F. O., & Ekinci, K. (2015). Evaluation of microalgae as microbial fertilizer. *European Journal of Sustainable Development*, *4*(2), 77. doi:10.14207/ejsd.2015.v4n2p77

Xiao, R., & Zheng, Y. (2016). Overview of microalgal extracellular polymeric substances (EPS) and their applications. *Biotechnology Advances*, *34*(7), 1225–1244. doi:10.1016/j.biotechadv.2016.08.004 PMID:27576096

Xu, Y., Rossi, F., Colica, G., Deng, S., De Philippis, R., & Chen, L. (2013). Use of cyanobacterial polysaccharides to promote shrub performances in desert soils: A potential approach for the restoration of desertified areas. *Biology and Fertility of Soils*, *49*(2), 143–152. doi:10.100700374-012-0707-0

Yandigeri, M. S., Yadav, A. K., Meena, K. K., & Pabbi, S. (2010). Effect of mineral phosphates on growth and nitrogen fixation of diazotrophic cyanobacteria Anabaena variabilis and Westiellopsis prolifica. *Antonie van Leeuwenhoek*, *97*(3), 297–306. doi:10.100710482-009-9411-y PMID:20069361

Yilmaz, E., & Sonmez, M. (2017). The role of organic/bio–fertilizer amendment on aggregate stability and organic carbon content in different aggregate scales. *Soil & Tillage Research*, *168*, 118–124. doi:10.1016/j.still.2017.01.003

Chapter 20
Nano-Bioremediation Technologies for Potential Application in Soil Reclamation

Arunima Nayak
Graphic Era University, Dehradun, India

Brij Bhushan
https://orcid.org/0000-0002-9360-3083
Graphic Era University, Dehradun, India

ABSTRACT

Rapid industrialization, urbanization, and use of modern agricultural practices have resulted in the rise in pollutant levels in soil. In this context, nano-bioremediation has emerged as a new tool for controlling soil pollution by the application of nanomaterials with subsequent use of bioremediation. Due to its cost-effectiveness, eco-friendliness, and sustainability, the use of bioremediation in soil reclamation has rapidly gained prominence. Nanomaterials have helped in remediating toxic soil environments, thereby improving microbial activity and bioremediation efficiency. The overall time as well as costs are greatly reduced. The major limitation of this technology is its longer treatment time and its ineffectiveness for a wide range of pollutants. The chapter has an aim to present an overview of the recent advances and applications in the field of nano-bioremediation of various polluted areas of the environment. Different classes of nanomaterials along with their properties as well as application towards removal of soil pollutants will be addressed.

INTRODUCTION

Increased anthropogenic activities e.g., industrialization and man-made activities have resulted in unprecedented rise in pollutant levels in the terrestrial environment. Soil has the capacity to degrade the pollutants only up to a specific limit. Excessive levels of pollutants are to some extent stored in the soil and are further transmitted to water bodies or to the biological food chain via the growing plants (Cec-

DOI: 10.4018/978-1-7998-7062-3.ch020

chinet al., 2017). The pollutants not only degrade the soil but also the water bodies which in turn has adverse implication to humans and animals. Pollution in soil is mainly caused by indiscriminate use of fertilizers, pesticides, herbicides, insecticides, dumping of organic wastes etc. The major pollutants identified in soil and which have major health implications are inorganic toxic substances like metal ions (Hg^{2+}, Cd^{2+}, Pb^{2+} etc.), organic wastes like pesticides, polycyclic aromatic hydrocarbons (PAHs), polychlorinated biphenyls (PCBs), dichlorodiphenyltrichloroethane (DDT), total petroleum hydrocarbon (TPH), nuclear wastes, plastics and sewage etc. Reclamation of soil facilitates recovery of ecosystem, helps to minimize adverse environmental impacts due to soil pollution, creates additional lands for agricultural or forestry uses, and enhances the carbon sequestration.

The different remediation methods adopted for soil reclamation is physico-chemical, thermal remediation, bioremediation, micro-remediation and vermi-remediation. Soil pollutants are converted to less toxic forms via the use of living organisms like plants, fungi, yeast or bacteria (bioremediation), via the use of soil microrganisms (micro-remediation) or via the use microbes like earthworm (vermi-remediation). Different plant species like *Brassica nigra* (Singh et al., 2015), *Helianthus annuus, Tithonia diversifolia* (Adesodun et al., 2010) and *Trifolium alexandrinum* (Bhatti et al., 2016) were used for removal of several toxic metal ions from contaminated soil. Some of the various microrganisms used for decontamination of polluted soil by Mani and Kumar (2014) were identified as *Pseudomonas aeruginosa, Chlorella vulgaris, Phormidiumvalderium, Stereumhirsutum, Citrobacter* sp.*, Chlorellavulgaris, Ganoderma applanatum, Volvariellavolvacea, Daedaleaquercina* etc. Vermi-remediation technique was used to remove heavy metals and polycyclic aromatic hydrocarbons (Rorat et al., 2017). Petroleum hydrocarbons (Njoku et al., 2017), fly ash (Saxena et al., 1998) and human excreta (Bajsa et al., 2004) was also removed from contaminated soil by vermi-remediation.

Bioremediation, micro-remediation and vermi-remediation are known to be cost-effective and environmentally friendly options for reclamation of soil (Lees and Senior 1995). Other advantages exhibited over physicochemical methods are their high selectivity, specificity, energy efficiency, minimal equipment requirement, etc. As per Azubuike et al. (2016), bioremediation of soil has restricted application in sites which are contaminated with highly toxic and hazardous pollutants. Bioremediation/ micro-remediation or vermi-remediation of a contaminated site operates in either of the two ways. In the first process, indigenous microrganisms (microbes inhabiting the site) play a major role in the clean-up process (Shankar et al., 2014); whereas, in the second process, exogenous microbes are added to the site to assist in the degradation of soil pollutants (Mukherjee and Bordoloi, 2011). In both processes, appropriate temperature, pH, nutrients etc. help in the growth of the microbes and thereby enhance the rate of pollutant degradation. The application of bioremediation can be in situ (within the contaminated site). Ex-situ applications require lesser treatment times and are used to treat diverse pollutant types and can be applied to different soil types (Dott et al., 1995). However, such techniques have their own limitation that is, they require longer treatment time for degradation of a toxic pollutant, typically in the range of several months to over a year.

More recently, nano-bioremediation has emerged as more effective, low cost and clean technology for soil reclamation (Otto et al., 2008). Nano-bioremediation technique has fast gained in popularity in the last decade mainly on account of its superiority over bioremediation methods of soil decontamination. It involves the use of reactive nanomaterials for enhancing the bio-catalytic activity and increasing the efficiency of microrganisms during the bioremediation of contaminated soil. The nanomaterials have the potential to transform via chemical reduction or via catalysis so as to minimize or eliminate the toxic pollutants (Otto et al., 2008). The high reactivity of the nanomaterials enables more efficient remediation of diverse pollutants as well as faster kinetics. The other advantages of the use of nanotechnology are its potential to remediate the contaminated site in situ and involve lesser dependencies on chemicals during the clean-up operation. Nano-bioremediation has the potential advantage of two remediation techniques viz. nanotechnology and bioremediation in the mitigation of soil pollutants. A review of literature reveals scarce literature on the application of nano-bioremediation for soil reclamation. Although the technology has immense potential for safe and clean reclamation of contaminated soil, yet more practical applications are required for full scale implementation. The present chapter has thus provided a detailed insight into the fundamentals and advantages of nanotechnology during the process of soil bioremediation. Different nanomaterials used till date for successful and eco-friendly remediation of soil pollutants are discussed and subsequently, the associated drawbacks of nanotechnology are highlighted. Finally, the immense potential of nano-bioremediation technology is brought to the knowledge of the readers as evident from laboratory scale studies conducted till date. It is recommended that more research is required for evaluating the effect of various environmental conditions on the efficacy of the technology. Future practical aspects of the technology for large scale implementation are required.

NANO-REMEDIATION FOR SOIL RECLAMATION

Use of nanomaterials during the bioremediation of soil has demonstrated higher decontamination efficiencies and faster kinetics as evident from literature. The reasons cited for their better performance are their high reactivity, better penetrabilityetc. Other properties exhibited during their application in contaminated site remediation are their easy application at the affected site, flexibility for in situ bioremediation, better mobility etc. Such properties have been cited due to their nano-scale particle size, high surface area to volume ratio and favorable surface coatings.

Literature study has revealed the application of various nanomaterials for bioremediation of soil for e.g., nanoscale zero valent iron (nZVI), metal oxide nanoparticles, carbon nanotubes, bimetallic nanoparticles, polymer-based nanomaterials etc.

Nano Zero Valent Iron (nZVI)

The nano zero valent iron (nZVI) has particle dimension in the nano range (<100 nm). It has a core shell structure in which the core consists of zero valent iron and the shell consists of hydroxides/oxides of iron formed as a result of the oxidation of the core zero valent iron. While the shell is responsible for effective chemisorption, the core zero valent iron helps in effective reduction of the contaminants via electron donation (Stumm and Morgan, 1996). Also, because of the nano-dimension, nZVI has large particle surface area and high surface reactivity (Zhang, 2003). Because of such advantageous characteristics, nZVI has demonstrated as a good agent for environmental remediation.

The last decade has seen immense work on nZVI for degradation of various organic contaminants like chlorinated organic solvents, organochlorine pesticides, polychlorinated biphenyls (PCBs), and organic dyes (Elliott & Zhang, 2006).) nZVI was synthesized under laboratory conditions via sodium borohydride reduction method and the same demonstrated a core shell structure with a surface area of 30 m^2/g (Liu et al., 2005). The particles helped in effective dechlorination of trichloroethylene (TCE) (80% of which was converted to ethane and 20% to C3-C6 coupling products). nZVI was applied in a contaminated site belonging to a metal fabrication industrial area of the Czech Republic. The study revealed significant reduction of chlorinated ethylene within a month of the injection of nZVI and approximately 50% removal was achieved within 5–6 months (Lacina et al., 2015). In a study by Ahn et al., application of 30 kg of nZVI helped in the removal of 95.7% of trichloroethylene within 60 days of its injection into the contaminated site. The study also reported that the lifetime of nZVI was more than 5 months and hence had the potential to be reused several times (Ahn et al., 2016). Positive results achieved from a pilot scale field study conducted at the Naval Air Station in Florida revealed the success of nano-remediation technology using nZVI. The study demonstrated that 300 pounds of nZVI could degrade abiotically 65-99% of volatile organic compounds (VOCs) within 5 weeks of injection into the site (Henn and Waddill, 2006). The test further revealed that nZVI brought about abiotic degradation for the initial 6-9 months followed by biological degradation as the primary degradation pathway. Various pilot tests were undertaken by Golder Associates between 2003 and 2005 in North America (United States and Canada) and Europe. Results revealed that in all such tests there was significant decrease in the concentration of chlorinated solvents over a short time frame since the injection of nZVI. Besides the chlorinated solvents, there was dramatic decrease in the sulphates and nitrates in some contaminated sites. The tests also revealed that application of nZVI for bioremediation caused minor change in microbial community structure (Mace et al., 2006). In a study conducted by Wang and Zhang, nZVI particles were synthesized and used for the degradation of halogenated organic compounds (HOCs) (Wang and Zhang, 1997). The particles showed a BET surface area of 33.5 m^2/g and the same showed great efficiency in dechlorinating several chlorinated aliphatic compounds at low metal-solution ratio of 2-5 g/100 mL. In one study, nZVI brought about 99% reduction of trichloroethylene within a few days of injection (Zhang, 2003). Similar successful results were achieved as a result of application of nZVI for remediation of several inorganic based contaminants. In a test by Cao et al., nZVI injection brought about degradation of perchlorate to chloride without generating any intermediate products (Cao et al., 2005). In a study conducted by Kanel et al., reduction of As(V) to As(III) was ensured within a short time frame by nZVI injection into the contaminated soil. Higher dosage of nZVI was required for complete elimination of As(V) from contaminated soil (Kanel et al., 2006). nZVI was used for soil washing as a pretreatment for reclamation of soil contaminated by Cu, Pb and Sb (Boente et al., 2018). High efficiency was observed for selective removal of the contaminants by nZVI. In a study by Sohn et al. (2006), the stability of ZVI was assessed with respect to time. It was demonstrated that approximately 5 nm of iron oxide coating was developed on the shell as a result of the high reactivity of nZVI in presence of air; the stability thus acquired decreased the overall reactivity of nitrate reduction by 50% as compared to that in pure nZVI. But the reactivity was still higher as compared to commercial grade iron powder. Ponder et al., (2000) synthesized supported nZVI from borohydride reduction of an aqueous iron salt in the presence of a support material and applied for remediation of Cr(VI) and Pb(II). The reduction rates of Cr(VI) to Cr(III) and Pb(II) to Pb were spontaneous and was approximately 30% higher as compared to iron powder. After 2 months, because of saturation of active sites, reactivity of nZVI decreased; demonstrating 4.8 times higher reactivity as compared to the iron powder. Laboratory synthesized nano scale iron powder (specific surface area of 35 m^2/g and having dimension of <100nm) was used for stabilization of chromium ore processing residue (COPR). The COPR had high concentration of Cr(VI). The study demonstrated that 1g of nZVI had the capacity to reduce or stabilize 65-110 mg of Cr(VI) in the COPR. In addition, the rate of Cr(VI) reduction was 25 times higher than that of commercial iron powders (Cao and Zhang, 2006). Both reduction as well as surface complexation was identified as the mechanism for Ni(II) remediation via the use of nZVI. The Ni(II) removal capacity was experimentallydetermined to be 0.13 g Ni(II)/g of Fe; which was found to be approximately 100% higher as compared to the commercial zeolites (Li and Zhang, 2006).

Although success rates for application of nZVI for soil remediation are very high, yet their stability and hence mobility in the soil environment has been less studied. In a study, it was demonstrated that the mobility of nZVI in a column soil bed setup was restricted due to the colloidal nature of iron particles. Field tests showed that mobility of nano sized iron particles is restricted to a few inches from the point of injection and the affecting factors have been identified as soil composition, nano particle dimension etc. (Sun, 2006). Promising results have been achieved for improving the mobility of nZVI via the use of delivery vehicles as has been denoted by Malloukand group at Penn State. The delivery vehicles are supports on which nZVI are synthesized. Such supports not only help in stabilizing the nZVI but also, they help in promoting their mobility in the soil environment. In order to improve deliverability of nZVI in soil, Schrick et al., (2004) used delivery vehicles like anionic hydrophilic carbon (C) and poly(acrylic acid)(PAA), both of which showed high binding efficiency and also created highly negative surface charges on nZVI. Development of surface negative charges helped in reducing the nano-particle aggregation (thereby enhancing stability in nZVI) and also reduced the filtration efficiency by aquifer materials. Soil column tests conducted in laboratory on Fe/C, Fe/PAA and nZVI (unsupported) revealed that use of support on nZVI promoted mobility through the soil bed. Similar results of improved mobility was demonstrated by Sun (2006) via using poly(vinyl alcohol-co-vinyl acetate-co-itaconic acid) (PV3A) as the driving vehicle or support for nZVI. Carboxy methyl cellulose (CMC) stabilized nZVI, as synthesized by He and Zhao (2007) demonstrated better soil deliverability and also promoted better Cr (VI) immobilization in both soil and water samples (Xu and Zhao, 2007). 0.08 g/L of CMC stabilized nZVI helped to reduce the leachability of Cr (VI) by 50%. During fixed bed studies, carried out with 5.7 bed volumes of 0.06 g/L of CMC stabilized nZVI, all Cr (VI) was efficiently reduced to Cr (III). Also, toxicity characteristic leaching procedure tests (TCLP) conducted on the stabilized nZVI revealed a toxicity reduction of 90%. In another test conducted by Liu et al. (2013), starch stabilized nZVI showed highly effective reduction of radionuclides in contaminated soils. Batch tests revealed that approximately 96% of perrhenate (ReO_4^-) reduction was achieved in 8hrs via the application of 560 mg/L of starch stabilized nZVI. The reduced perrhenate obtained as per the study was ReO_2. The stabilized nanoparticles were transferable through the soil packed column. Approximately 56% of perrhenate was reduced to ReO_2 when 14 pore volumes of 560 mg/L of nanoparticles were passed through the column packed with soil.

Metallic/Bi-Metallic/Metal Oxide Nanoparticles

Besides the iron nanoparticles, other metal oxide nano particles have also demonstrated good environmental remediation potential (Nutt et al., 2005; Lien and Zhang, 2005; Xu et al., 2005). But bare nanoparticles tend to agglomerate and also exhibit oxidation on exposure to atmosphere. This is because of their high surface reactivity. Hence, the surface of nanoparticles is coated with suitable support materials to enhance their stability. Literature has revealed that stabilized metal oxide nanoparticles have been efficiently used for soil remediation.

Carboxy methyl cellulose (CMC) stabilized MnO_2 nano particles were synthesized in the laboratory and used for in-situ remediation of soil contaminated by estradiol (Han et al., 2015). The characteristics of the synthesized nano particles were their stability and hence the CMC-MnO_2 particles were well dispersed in soil medium for several months; thereby retaining its efficiency. A comparative study revealed that in comparison to non-stabilized MnO_2, CMC stabilized MnO_2 nano particles brought about enhanced oxidation of 17β-estradiol in contaminated soil at a pH range of 6-7. Results further revealed that CMC-stabilized nanoparticles were transportable and deliverable in a sandy loamy soil under optimum injection pressure. Iron oxide nano particles were used for remediation of soil contaminated with As(V) and As(III) (Shipley et al., 2015).

In another study, iron oxide (Fe_3O_4) nano particles stabilized with polyacrylic acid was used for removal of Cd(II) ions from contaminated soils. In a study by Liang et al. (2014), starch stabilized magnetite nanoparticles demonstrated both soil deliverability as well as was found to be effective in immobilization of As(V) in a model sandy soil. Column tests revealed that when the soil was treated with 34 pore volumes of 0.1 g stabilized nanoparticles, As(V) was reduced by 93%. TCLP (toxicity characteristic leaching procedure) tests revealed a reduction of 83%. Various studies have demonstrated the application of Cerium oxide nano particles in the removal of various pollutants from soil contaminated with wastes from textiles, agricultural pesticides, pharmaceuticals and leather tanneries (Pradhan and Parida,2010; Dahle and Arai, 2015; Peng et al., 2015; Vivekananthan et al., 2014). In a study by He and Zhao (2005), synthesized palladized iron (Fe-Pd) stabilized with food grade starch exhibited a mean particle size of 14.1 nm and a surface area of 55 m^2/g. The stabilized nanoparticles (dosage of 1g/L) demonstrated enhanced reactivity by destroying 98% of trichloroethylene (TCE) and 80% of polychlorobromine (PCB) as compared to non-stabilized Fe-Pd nanoparticles. In a study conducted by Liu and Zhao (2007), a new class of CMC stabilized iron phosphate nanoparticles were synthesized for in-situ immobilization of Pb^{2+} in three model contaminated soil (calcareous, neutral and acidic). When the soils were treated with 0.61-3.0 mg (nanoparticles)/g(soil) for a continuous period of 56 days, the TCLP (toxicity characteristic leaching procedure) of Pb^{2+} was reduced by 85-95% whereas, PBET (physiologically based extraction test) bio-accessibility of $Pb^{2+}$was lowered by 31-47%.

Carbon-Based Nanoparticles

Among the different carbon-based nanomaterials used for environmental remediation, carbon nanotubes and graphene oxides have gained in importance among the scientific community because of their unique surface properties which enable them to retain pollutants via the process of adsorption (Gupta et al., 2016; Ouni et al., 2019; Fiyadh et al., 2019; Wang et al., 2019). Also, because of their large surface area, maximum pollutants are removed from application of lesser dosages of such nanomaterials. In a study by Kabbashi et al (2009), approximately 99% of Cu^{2+} was removed from aqueous solution via the application of a small dosage of CNTs (carbon nanotubes). Based on the good results obtained from CNTs as adsorbent in aqueous phase application, experiments were conducted to assess their performance for immobilization of metal ion pollutants from contaminated soil (Matos, 2016). Like other nanomaterials tending to aggregate in environmental samples, stability and mobility of CNTs were ensured via functionalization using various organic, inorganic moieties. Matos et al. (2017) used ultrasonication for obtaining dispersed suspension of MWCNTs (multi-walled carbon nanotube) with a non-ionic surfactant (Pluronic F-127). During application of stabilized MWCNTs in a model contaminated soil, it was found that the order of affinity for the heavy metal ions was: $Pb^{2+}>Cu^{2+}>Ni^{2+}>Zn^{2+}$. The permeability tests revealed the beneficial effect of MWCNTs for immobilization of heavy metals like Ni^{2+} and Zn^{2+} from contaminated soil. The study further revealed that CNTs have the capacity to remediate contaminated soil, helping in immobilization of toxic metal ions and finally minimizing the adverse impact of such toxic soil to environment and human health. Reduced graphene oxide (nGO) nanoparticles have demonstrated better reduction potential for various metal pollutants from soil as compared to nZVI (Baragaño et al.,

2020). Study revealed that nGO effectively immobilized Cu, Pb and Cd, but mobilized As and P (even at lower doses), while nZVI promoted significant immobilization for As and Pb, a poorer result for Cd, and an increased availability for Cu.

Use of Polymeric Nanoparticles

Various hydrophobic pollutants like polycyclic aromatic hydrocarbons (PAH) originating from petroleum by-products have demonstrated greater binding affinity to soil particles; thereby exhibiting minimum degradation by soil microrganisms. Such pollutants are classified as hazardous by Environment Protection Agency (EPA). The PAH exhibit minimum solubilization and mobility in soil. Also, because of their high binding to soil, bioavailability of PAH is negligible. Amphiphilic polyurethane (APU) nanoparticles were synthesized by Kim et al (2000) from polyurethane-acrylate-anionomer (UAA) precursor chains. Application of APU nanoparticles revealed exceptional binding of phenanthrene from soil sample with high recovery rate. In a similar study by Tungittiplakorn et al., (2004), APU nanoparticles were synthesized from a mixture of poly(ethylene glycol) modified polyurethane acrylate(PMUA) and polyurethane acrylate precursor chains and was further used for solubilizing a model PAH or phenanthrene from contaminated soil. Characterization studies carried out on the engineered nanoparticle reveal that APU nanoparticles had an amphiphilic polyethylene glycol (PEG) chain protruding outside and a hydrophobic polyurethane acrylate core. The hydrophobic core resulted in greater affinity for phenanthrene and hence causing greater desorption from aquifer sand soil while the hydrophilic exterior of the engineered nanoparticles caused greater particle mobility in soil. The study further revealed that the surface properties of the nanoparticles could be controlled depending on the contaminant type and soil conditions. The affinity for the contaminant can be controlled by altering the size of the hydrophobic core of the APU nanoparticle. On the other hand, the mobility of the APU suspension can be controlled by changing the size of the hydrophilic chains protruding from the surface of the nanoparticle. In another study conducted by Tungittiplakorn et al., (2005), the usefulness of the engineered APU nanoparticles was tested on enhancing the bioavailability of phenanthrene. The engineered polymeric nanoparticles demonstrated increased mineralization of the model phenanthrene under three conditions: phenanthrene dissolved in water, phenanthrene sorbed onto aquifer and phenanthrene dissolved in hexadecane (model non aqueous phase liquid) in the presence of aquifer material. The study further revealed that the properties of the engineered nanoparticles were found to be stable in the presence of a bacterial population which proved the reusability of nanoparticles after the biodegradation of phenanthrene by the bacteria. Thus, the application of polymeric engineered nanoparticles can help in enhancing the in-situ biodegradation rate during bioremediation of soil.

NANO-BIOREMEDIATION FOR SOIL RECLAMATION

Although both nanotechnology and bioremediation are effective in remediation of various toxic pollutants present in soil, yet each has its own disadvantages. Some of the major problems encountered during the use of nanomaterials for soil remediation are loss of reactivity with time, in-situ mobility or transport and inherent toxicity. Because of their high reactivity, bare nanoparticles tend to aggregate with each other and with soil particles. Thereby, the overall reactivity gets decreased with time. Clustering tendency with soil particles causes difficulty in transport to the contaminated site. Bare nanoparticles are also toxic to microbial community. Bioremediation has its own disadvantages like requirement of longer treatment times and sensitivity of microbes to environmental factors. Thus, the use of a single technology may be expensive and may not be effective or sustainable. The use of nanotechnology followed by use of bioremediation in a sequential manner is nano-bioremediation that ensures reclamation of contaminated soil in less time, with more efficiency and is environmentally friendly as compared to individual remediation technologies. Work on the techno-feasibility of nano-bioremediation by Cecchin et al. is being carried out on Brazilian residual clays contaminated with chlorinated organic contaminants (Cecchin et al., 2017). Work carried out thus far reveals the efficacy and sustainability of this technology as revealed from literature studies and the findings have been summarized in Table-1.

Table 1. Application of nano-bioremediation for reclamation of contaminated soil

Nanoparticle Used	Microbial Species	Contaminant in Soil	Reference
nZVI	organochlorine respiring bacteria	chlorinated aliphatic hydrocarbon	Koenig et al., 2016
nZVI	*Paracoccus* sp.	Nitrate	Liu et al., 2014
nZVI	*Sphingomonas*sp. PH07	polybrominated diethyl ether-PBDE	Kim et al., 2011
nZVI	*White rot fungi*	hexahydro-1,3,5-trinitro-1,3,5-triazine	Oh et al., 2001
nZVI	*Dehaococcoides*spp	trichloroethylene	Xiu et al., 2010
Pd-nZVI	*Tramates versicolor*	Triclosan	Bokare et al., 2010
Pd-nZVI	*Spingomonaswittichii RW1*	2,3,7,8 tetrachlorodibenzo p-dioxin	Bokare et al., 2012
Pd-nZVI	*Burkholderiaxenovorans* LB400	Polychlorinated biphenyl	Le et al., 2015
CMC stabilized Pd-nZVI	*Sphingomonas* sp. strain NM05	γ-HCH	Singh et al., 2013
Pd nanoparticles	*C. pasteurianum*	Cr(VI)	Chidambaran et al. 2010
Pd nanoparticles	*Shewanellaoneidensis*	Polychlorobiphenyl (PCB)	Windt et al., 2005
nZVI-C-A-beads	*Bacilius subtilis, E.Coli, A.junii*	Cr(VI)	Ravikumar et al. 2016
Fe_3O_4 nanoparticles	*Sphingomonas* sp.	Carbazole	Li et al., 2013
Fe_3O_4 nanoparticles/ gellan gum gel	*Sphingomonas* sp.	Carbazole	Wang et al., 2007
Fe_3O_4 magnetic nanoparticles	*Pseudomonas delafieldii*	Dibenzothiophene	Shan et al., 2014
Carbon nanotubes	*Shewanellaoneidensis*	Cr(VI)	Yan et al., 2013
polyvinyl alcohol (PVA), sodium alginate immobilized on MWCNTs	*P. aeruginosa*	Cr(VI)	Pang et al., 2011
nZVI and polyaspartate coated nZVI	*NA*	Trichloroethylene (TCE)	Kirschling et al. 2010
nZVI	*Sphingomonas sp. PH-07*	polybrominated diphenyl ether) (PBDE)	Kim et al., 2012

Koenig et al. (2016) combined nanotechnology involving nZVI and bioremediation using organochlorine respiring bacteria (ORB) to stabilize soil contaminated with chlorinated aliphatic hydrocarbon (CAH). CAH are recalcitrant organic compounds and are not removed completely by nZVI or ORB. The study involved a mixture of two CAH (1,2-dichloroethane which is degradable by ORB but not by nZVI, and 1,1,2-trichloroethane which is degraded by both). Results of the study demonstrated that when nZVIwas applied at a dosage of 0.5g/L, 1,1,2-trichloroethane was dechlorinated. But nZVI had a lethal effect on ORB at 0.5g/L; for which the activity of ORB was inhibited and was unable to dechlorinate 1,2dichloroethane. Results also have suggested that using both nZVI at a dosage of 0.1g/L and ORB have the potential to detoxify a wider range of CAHs as compared to individual remedy. Kim et al., (2011) used a combination of nZVI and aerobic bacterium (*Sphingomonas*sp. PH07) for the degradation of a persistent organic pollutant (polybrominated diethyl ether-PBDE). Incorporation of nZVI helped to break down the PBDE to lower molecular weight BDE (brominated diethyl ether) via debromination reaction. The reaction medium was aerobically treated for 4 days with the diphenyl ether degrading bacterium (*Sphingomonas*sp. PH07). As a result of the bacterial treatment, the low BDEs were bio-degraded to bromophenols. This hybrid method of nano-bioremediation has thus paved a treatment technology for sites contaminated with toxic halogenated environmental pollutants. In a study by Ravikumar et al. (2016), Cr(VI) removal efficiency was investigated using a fixed-bed column packed individually with nZVI-immobilized calcium alginate beads (nZVI–C–A beads) and a biofilm formed on nZVI–C–A beads. Results revealed that removal efficiency of Cr(VI) was 91.35% for column packed with nZVI–C–A beads while the same for biofilm-coated nZVI–C–A beads showed 97.84% removal efficiency. nZVI and white rot fungi were applied simultaneously for testing the degradation rate of RDX (hexahydro-1,3,5-trinitro-1,3,5-triazine) in a study conducted by Oh et al., (2001). Results revealed that nano-bioremediation caused increase in RDX degradation as compared to separate application of nZVI and fungi.

Singh et al. (2013) evaluated the combined effect of carboxy methyl cellulose (CMC) stabilized Pd-nZVI bimetallic nanoparticles and a *Sphingomonas* sp. strain NM05, on the degradation of γ-HCH in soil. Degradation efficiency for γ-HCH was approximately 1.7-2.1 times greater by the application of nano-bioremediation as compared to single technology application using either NM05 or CMC-Pd-nZVI. Results also showed that NM05 had better compatibility with the nanoparticles and showed better growth in the presence of CMC-Pd-nZVI as compared to the control systems. Nano-bioremediation has shown potential efficiency as an effective alternative remedial tool for γ-HCH contaminated soil.

Problems related to potential toxicity of nanoparticles on the microbial growth and activity have been addressed by various workers by either manipulation of appropriate dosage (Koenig et al., 2016) or by encapsulation of the nanoparticles. Li et al. (2010) used bare nZVI nanoparticles (100mg/L) for studying their bactericidal properties to gram-negative bacterium, *Escherichia coli*. Results revealed exposure of bare nanoparticles caused a 2.2-log inactivation after 10 min and a 5.2-log inactivation after 60 min in *E coli*. But the same nZVI encapsulated with poly(styrene sulfonate) (PSS), poly(aspartate) (PAP), or NOM (natural organic matter) resulted in decreased toxicity, when exposed to the bacterium causing less than 0.2-log inactivation after 60 min. Similar observation of reduced toxicity was obtained in a study by Ahn et al. (2010) during exposure of denitrifying bacteria to chitosan/sodium oleate modified-iron nanoparticles. Both nZVI and *Dehalococcoides* spp are known to dechlorinate trichloroethylene (TCE). Xiu et al., (2010) studied the effect of bare and encapsulated nZVI on the expression of gene coding for reductive dechlorination (tceA and vcrA) of TCE in *Dehalococcoides*spp. Bare nanoparticles (1g/L) caused a down regulation of tceA and vcrA genes (97- and 137-fold, respectively) with respect to time t=0 conditions after 72-h exposure to chlorinated ethenes. But nZVI encapsulated with maleic acid caused a significant up-regulation of tceA and vcrA by 3.0- and 3.5-fold, respectively after 48-h exposure. The study thus revealed that encapsulation of nanoparticles can reduce inhibition and enhance the expression of dechlorinating genes in *Dehalococcoides* spp.; thereby promoting the TCE degradation efficiency in the nano-bioremediation process. Kirschling et al. (2010) studied the effect of nZVI injection on the microbial community in three different aquifer materials from TCE (trichloroethylene) contaminated sites in Alameda Point, CA, Mancelona, MI, and Parris Island, SC. The study conducted over a 250days period showed that both nZVI and biodegradable polyaspartate coated nZVI exhibited no toxic effects on the native microorganisms. The coated nZVI brought about a bio-stimulation of the microbial community; thereby bringing about enhanced degradation of TCE. In a study conducted by Xiu et al., (2010), the effect of nZVI on anaerobic dechlorinating microorganisms was investigated for the degradation of TCE. Results of the study showed that the activity of the methanogens was initially inhibited due to the presence of nZVI; but after a lag phase, dechlorination activity and ethane production recovered. Although the laboratory study showed positive effects of nZVI on both the microbial activity and TCE reduction; yet pilot scale studies may help to identify potential limitations associated with scale-up operations. Various other studies have been conducted to test the efficiency of nZVI on bioremediation of sites contaminated with TCE (Liu and Lowry, 2006; Liu et al., 2007), chlorinated ethanes (Song and Carraway,2005), other chlorinated solvents (Comba et al, 2011), polychlorinated bromine (Lowry and Johnson, 2004).A more detailed description of such studies have been compiled in a review by Zhang (2003). The application of nZVI is known to induce production of cathodic H_2 (Fajardo et al., 2012; Liu et al., 2005; Němeček et al., 2015) and shift in redox potential which in turn is highly favorable for growth of microorganisms and thus has the potential to enhance bioremediation of various organic soil contaminants (Aulenta et al., 2006). Because of the possibility of aggregation of nZVI which in turn may affect its stability, Kim et al., 2010 doped nZVI with Pd and further immobilized in alginate bead and subsequently studied their degradation efficiency for TCE (trichloroethylene). Approximately 99.8% of TCE was degraded by application of 50g/L of the immobilized nZVI. A study was conducted to evaluate the effect of a nano-bioremediation technique along with a subsequent aerobic treatment for degradation of PBDEs (polybrominated diphenyl ether) (Kim et al., 2012). The nano-bioremediation was carried out with nZVI followed by the activity of diphenyl ether degrading bacteria *Sphingomonas sp.* PH-07. Results showed that the bacterial tolerance limit was 5g/L of nZVI in which the strain showed healthy growth by using non-brominated diphenyl ether as growth substrate. The debromination efficiency recorded was 67% during a 20day treatment period. The aerobic treatment for an additional 4 days treatment helped to mineralize the low brominated diphenyl ether. Simultaneous injection of nZVI and whey helped in nano-bioremediation of both Cr(VI) and chlorinated ethenes from contaminated site (Němeček et al., 2016). Treatment with nZVI alone was highly efficient in Cr(VI) reduction to Cr(III) but a low degradation was observed for chlorinated ethenes. Polycyclic aromatic hydrocarbons (PAHs) in crude oil have a serious adverse effect on living organisms. A study was conducted by Oyewole et al. (2019) to evaluate the synergistic effect of nano-bioremediation with *Alcaligenes faecalis* ADY25 and iron oxide nanoparticles on the biodegradation of PAHs in crude oil. The selected bacterial strain has the ability to degrade petrochemical products. With supply of iron oxide nanoparticles, it was found that the degradation of PAH increased. The study confirmed that the nanoparticles helped in reducing the lag phase of the microbes while increasing the duration of stationary and exponential phase. Highest bacterial count was observed on the 18th day during the addition of 200 mg of nanoparticles into the culture medium. Another study showed that the fungal isolates along with silver nanoparticles helped in the bioremediation of crude oil hydrocarbons (Al-Zaban et al., 2020). Various fungal species have shown promising biodegradation rates for crude oil hydrocarbons *(Aspergillus, Alternaria, Cladosporium, Eupenicillium, Fusarium, and Trichoderma spp)* (Zhang et al., 2015). The two fungal isolates selected for the study (*A. flavus AF15 and T. harzianum TH07*) demonstrated rapid crude oil degradation but when used in consortium with silver nanoparticles showed enhanced biodegradation.

A sequential application of bimetallic Pd-nZVI nanoparticles and *Burkholderiaxenovorans* LB400 helped in total transformation of polychlorinated biphenyl (PCB) to fewer toxic compounds like benzoic acid (Le et al., 2015). The nanoparticles resulted in dechlorination of tri-, tetra-, penta-, and hexachlorinated biphenyls to the tune of 99%, 92%, 84%, and 28% respectively. The subsequently biodegradation resulted in 90% elimination of the biphenyls produced after the dechlorination of PCB by bacterial metabolism. The biodegradation products and residual PCBs revealed low cytotoxicity toward *Escherichia coli*. Also, nZVI showed no toxic effects towards the microbial community. In a study by He et al., (2009), Pd coated Fe (Pd/Fe) was used to catalytically reduce pentachlobiphenyl (PCB) followed by a subsequent treatment with an aerobic bacteria for ensuring the biodegradation of the chemical by-products. The results of the study confirmed the application of an integrated Pd/Fe catalytic reduction-aerobic biodegradation as a feasible treatment option for PCB contaminated soil.

Nano-bioremediation technology was used for Cr(VI) reduction, in a study conducted by Yan et al., (2013). Carbon nanotubes (CNTs) were impregnated with Ca-alginate beads and were further used to immobilize *Shewanellaoneidensis* MR-1 for bringing about enhanced Cr(VI) reduction. As compared to the microbial cells and the alginate beads, the AL/CNT/cell beads demonstrated approximately 4 times higher reduction rates. Similar results of enhanced Cr(VI) degradation was achieved by Pang et al., (2011) who used *P. aeruginosa*immobilized inpolyvinyl alcohol (PVA), sodium alginate, and MWCNTs (multiwalled carbon nanotube) matrix.

CONCLUSION

Nano-bioremediation has all the potential benefits for safe and clean remediation of contaminated soil. Stabilized nanomaterials synthesized via effective coating and functionalization provides safe and effective reduction of the toxic contaminants to by-products that are conducive to biodegradation. Subsequent treatment with microbial community promotes biodegradation of the by-products to risk free levels. Thus, the technology has immense potential and is a sustainable approach for reclamation of many contaminated sites. Further studies are required to study the effect of various environmental parameters like pH, temperature, ionic strength, presence of inhibitory substances etc. on the efficacy of the nano-bioremediation technology. More applications are needed for full scale implementation of this technology.

REFERENCES

Adesodun, J., Atayese, M., Agbaje, T., Osadiaye, B., Mafe, O., & Soretire, A. (2010). Phytoremediation potentials of sunflowers (*Tithonia diversifolia* and *Helianthus annuus*) for metals in soils contaminated with zinc and lead nitrates. *Water, Air, and Soil Pollution, 207*(1-4), 195–201. doi:10.100711270-009-0128-3

Al-Zaban, M. I., Mahmoud, M. A., Al Harbi, M. A., & Bahatheq, A. M. (2020). Bioremediation of Crude Oil by Rhizosphere Fungal Isolates in the Presence of Silver Nanoparticles. *International Journal of Environmental Research and Public Health*, *17*(18), 6564. doi:10.3390/ijerph17186564 PMID:32916946

An, Y., Li, T., Jin, Z., Dong, M., Xia, H., & Wang, X. (2010). Effect of bimetallic and polymer coated Fe nanoparticles on biological denitrification. *Bioresource Technology*, *101*(24), 9825–9828. doi:10.1016/j.biortech.2010.07.110 PMID:20727742

Anjum, M., Miandad, R., Waqas, M., Gehany, F., & Barakat, M. A. (2016). Remediation of wastewater using various nanomaterials. *Arabian Journal of Chemistry*, *12*(8), 4897–4919. doi:10.1016/j.arabjc.2016.10.004

Aulenta, F., Di Tomassi, C., Cupo, C., Papini, M. P., & Majone, M. (2006). Influence of hydrogen on the reductive dechlorination of tetrachloroethene (PCE) to ethane in a methanogenic biofilm reactor: Role of mass transport phenomena. *Journal of Chemical Technology and Biotechnology (Oxford, Oxfordshire)*, *81*(9), 1520–1529. doi:10.1002/jctb.1562

Azubuike, C. C., Chikere, C. B., & Okpokwasili, G. C. (2016). Bioremediation techniques–classification based on site of application: Principles, advantages, limitations and prospects. *World Journal of Microbiology & Biotechnology*, *32*(11), 1–18. doi:10.100711274-016-2137-x PMID:27638318

Bajsa, O., Nair, J., Mathew, K., & Ho, G. E. (2004). *Pathogen Die-Off in Vermicomposting Process*. Paper presented at the International Conference on "Small Water and Wastewater Treatment Systems", Perth, Australia.

Baragaño, D., Forján, R., Welte, L., & Gallego, J. L. R. (2020). Nanoremediation of As and metals polluted soils by means of graphene oxide nanoparticles. *Scientific Reports*, *10*(1), 1896. doi:10.103841598-020-58852-4 PMID:32024880

Bhatia, M., Girdhar, A., Chandrakar, B., & Tiwari, A. (2013). Implicating nanoparticles as potential biodegradation enhancers: A review. *Journal of Nanomedicine & Nanotechnology*, *4*(175), 2. doi:10.4172/2157-7439.1000175

Bhatti, S. S., Sambyal, V., & Nagpal, A. K. (2016). Heavy metals bioaccumulation in Berseem (*Trifolium alexandrinum*) cultivated in areas under intensive agriculture, Punjab, India. *SpringerPlus*, *5*(1), 173. doi:10.118640064-016-1777-5 PMID:27026870

Boente, C., Sierra, C., Martínez-Blanco, D., Menéndez-Aguado, J. M., & Gallego, J. R. (2018). Nanoscalezero-valent iron-assisted soil washing for the removal of potentially toxic elements. *Journal of Hazardous Materials*, *350*, 55–65. doi:10.1016/j.jhazmat.2018.02.016 PMID:29448214

Bokare, V., Murugesan, K., Kim, J. H., Kim, E. J., & Chang, Y. S. (2012). Integrated hybrid treatment for theremediation of 2, 3, 7, 8-tetrachlorodibenzo-p-dioxin. *The Science of the Total Environment*, *435*, 563–566. doi:10.1016/j.scitotenv.2012.07.079 PMID:22909785

Bokare, V., Murugesan, K., Kim, Y. M., Jeon, J. R., Kim, E. J., & Chang, Y. S. (2010). Degradation of triclosan byan integrated nano-bio redox process. *Bioresource Technology*, *101*(16), 6354–6360. doi:10.1016/j.biortech.2010.03.062 PMID:20381343

Cao, J., Elliott, D., & Zhang, W.-X. (2005). Perchlorate reduction by nanoscale iron particles. *Journal of Nanoparticle Research*, *7*(4-5), 499–506. doi:10.100711051-005-4412-x

Cao, J., & Zhang, W. (2006). Stabilization of chromium ore processing residue (COPR) with nanoscale iron particles. *Journal of Hazardous Materials*, *132*(2-3), 213–219. doi:10.1016/j.jhazmat.2005.09.008 PMID:16621279

Cecchin, I., Reddy, K. R., Thomé, A., Tessaro, E. F., & Schnaid, F. (2017). Nano-bioremediation: Integration of nanoparticles and bioremediation for sustainable remediation of chlorinated organic contaminants in soils. *International Biodeterioration & Biodegradation*, *119*, 419–428. doi:10.1016/j.ibiod.2016.09.027

Chidambaram, D., Hennebel, T., Taghavi, S., Mast, J., Boon, N., Verstraete, W., van der Lelie, D., & Fitts, J. P. (2010). Concomitant microbial generation of palladium nanoparticles and hydrogen to immobilize chromate. *Environmental Science & Technology*, *44*(19), 7635–7640. doi:10.1021/es101559r PMID:20822130

Comba, S., Di Molfetta, A., & Sethi, R. (2011). A Comparison Between Field Applications of Nano-, Micro-, and Millimetric Zero-Valent Iron for the Remediation of Contaminated Aquifers. *Water, Air, and Soil Pollution*, *215*(1-4), 595–607. doi:10.100711270-010-0502-1

Dott, W., Feidieker, D., Steiof, M., Becker, P. M., & Kämpfer, P. (1995). Comparison of ex situ and in situ techniques for bioremediation of hydrocarbon-polluted soils. *International Biodeterioration & Biodegradation*, *35*, 301–316. doi:10.1007/978-94-011-1312-0_4

Fajardo, C., Ortiz, L. T., Rodriguez-Membibre, M. L., Nande, M., Lobo, M. C., & Martin, M. (2012). Assessing the impact of zero-valent iron (ZVI) nanotechnology on soil microbial structure and functionality: A molecular approach. *Chemosphere*, *86*(8), 802–808. doi:10.1016/j.chemosphere.2011.11.041 PMID:22169206

Fiyadh, S. S., AlSaadi, M. A., Jaafar, W. Z., AlOmar, M. K., Fayaed, S. S., Mohd, N. S., Hin, L. S., & El-Shafie, A. (2019). Review on heavy metal adsorption processes by carbon nanotubes. *Journal of Cleaner Production*, *230*, 783–793. doi:10.1016/j.jclepro.2019.05.154

Galdames, A., Mendoza, A., Orueta, M., de Soto García, I. S., Sánchez, M., Virto, I., & Vilas, J. L. (2017). Development of new remediation technologies for contaminated soils based on the applicationof zero-valent iron nanoparticles and bioremediation with compost. *Resour Effic Technol.*, *3*(2), 166–176. doi:10.1016/j.reffit.2017.03.008

Gupta, V., Moradi, O., Tyagi, I., Agarwal, S., Sadegh, H., Shahryari-Ghoshekandi, R., Makhlouf, A., Goodarzi, M., & Garshasbi, A. (2016). Study on the removal of heavy metal ions from industry waste by carbon nanotubes: effect of the surface modification: A review. *Critical Reviews in Environmental Science and Technology*, *46*(2), 93–118. doi:10.1080/10643389.2015.1061874

Han, B., Zhang, M., Zhao, D., & Feng, Y. (2015). Degradation of aqueous and soil-sorbed estradiol using a new class of stabilized manganese oxide nanoparticles. *Water Research*, *70*, 288–299. doi:10.1016/j.watres.2014.12.017 PMID:25543239

He, F., & Zhao, D. (2005). Preparation and characterization of a new class of starch-stabilized bimetallic nanoparticles for degradation of chlorinated hydrocarbons in water. *Environmental Science & Technology, 39*(9), 3314–3320. doi:10.1021/es048743y PMID:15926584

He, F., & Zhao, D. (2007). Manipulating the size and dispersibility of zerovalent iron nanoparticles by use of carboxymethyl cellulose stabilizers. *Environmental Science & Technology, 41*(17), 6216–6221. doi:10.1021/es0705543 PMID:17937305

He, N., Li, P., Zhou, Y., Fan, S., & Ren, W. (2006). Degradation of pentachlorobiphenyl by a sequential treatment using Pd coating iron and aerobic bacterium (H1). *Chemosphere, 76*(11), 1491–1497. doi:10.1016/j.chemosphere.2009.06.046 PMID:19596135

Henn, K. W., & Waddill, D. W. (2006). Utilization of nanoscale zerovalent iron for source remediation—A case study. *Remediation, 16*(2), 57–77. doi:10.1002/rem.20081

Kabbashi, N. A., Atiehc, M. A., Al-Mamuna, A., Mirghamia, M. E. S., Alama, M. D. Z., & Yahyaa, N. (2009). Kinetic adsorption of application ofcarbon nanotubes for Pb(II) removal from aqueous solution. *Journal ofEnvironmental Sciences (China), 21*(4), 539–544. doi:10.1016/S1001-0742(08)62305-0 PMID:19634432

Kanel, S. R., Greeneche, J. M., & Choi, H. (2006). Arsenic (V) removal from groundwater using nano scale zerovalent iron as a colloidal reactive barrier material. *Environmental Science & Technology, 40*(6), 2045–2050. doi:10.1021/es0520924 PMID:16570634

Kim, H., Hong, H.-J., Jung, J., Kim, S.-H., & Yang, J.-W. (2010). Degradation of trichloroethylene (TCE) by nanoscale zero-valent iron (nZVI) immobilized in alginate bead. *Journal of Hazardous Materials, 176*(1-3), 1038–1043. doi:10.1016/j.jhazmat.2009.11.145 PMID:20042289

Kim, J.-Y., Cohen, C., Shuler, M. L., & Lion, L. W. (2000). Use of Amphiphilic Polymer Particles for In Situ Extraction of Sorbed Phenanthrene from a Contaminated Aquifer Material. *Environmental Science & Technology, 34*(19), 4133–4139. doi:10.1021/es001021w

Kim, Y. M., Murugesan, K., Chang, Y. Y., Kim, E. J., & Chang, Y. S. (2011). Degradation of polybrominated diphenyl ethers by a sequential treatment with nanoscale zero valent iron and aerobic biodegradation. *Journal of Chemical Technology and Biotechnology (Oxford, Oxfordshire), 87*(2), 216–224. doi:10.1002/jctb.2699

Kim, Y. M., Murugesan, K., Chang, Y. Y., Kim, E. J., & Chang, Y. S. (2012). Degradation of polybrominated diphenyl ethers by a sequential treatment with nanoscale zero valent iron and aerobic biodegradation. *Journal of Chemical Technology and Biotechnology (Oxford, Oxfordshire), 87*(2), 216–224. doi:10.1002/jctb.2699

Kirschling, T. L., Gregory, K. B., Minkley, E. G. Jr, Lowry, G. V., & Tilton, R. D. (2010). Impact of Nanoscale Zero Valent Iron on Geochemistry and Microbial Populations in Trichloroethylene Contaminated Aquifer Materials. *Environmental Science & Technology, 44*(9), 3474–3480. doi:10.1021/es903744f PMID:20350000

Koenig, J. C., Boparai, H. K., Lee, M. J., O'Carroll, D. M., Barnes, R. J., & Manefield, M. J. (2016). Particles and enzymes: Combining nanoscale zero valent iron and organochlorine respiring bacteria for the detoxification of chloroethane mixtures. *Journal of Hazardous Materials*, *308*, 106–112. doi:10.1016/j.jhazmat.2015.12.036 PMID:26808236

Lacina, P., Dvorak, V., Vodickova, E., Barson, P., Kalivoda, J., & Goold, S. (2015). The application of nanosizedzero-valent iron for in situ remediation of chlorinated ethylenes in groundwater: A field case study. *Water Environment Research*, *87*(4), 326–333. doi:10.2175/106143015X14212658613596 PMID:26462077

Le, T. T., Nguyen, K. H., Jeon, J. R., Francis, A. J., & Chang, Y. S. (2015). Nano/bio treatment of polychlorinatedbiphenyls with evaluation of comparative toxicity. *Journal of Hazardous Materials*, *287*, 335–341. doi:10.1016/j.jhazmat.2015.02.001 PMID:25679799

Lees, Z. M., & Senior, E. (1995). Bioremediation: a practical solution to land pollution. In *Clean technology and the environment*. Springer.

Li, X., Elliott, D. W., & Zhang, W. (2006). Zero-Valent Iron Nanoparticles for Abatement of Environmental Pollutants: Materials and Engineering Aspects. *Critical Reviews in Solid State and Material Sciences*, *31*(4), 111–122. doi:10.1080/10408430601057611

Li, X. Q., & Zhang, W. (2006). Iron nanoparticles: The core-shell structure and unique properties for Ni(II) sequestration. *Langmuir*, *22*(10), 4638–4642. doi:10.1021/la060057k PMID:16649775

Li, Y., Du, X., Wu, C., Liu, X., Wang, X., & Xu, P. (2013). An efficient magnetically modified microbial cell biocomposite for carbazole biodegradation. *Nanoscale Research Letters*, *8*(1), 522. doi:10.1186/1556-276X-8-522 PMID:24330511

Li, Z., Greden, K., Alvarez, P. J. J., Gregory, K. B., & Lowry, G. V. (2010). Adsorbed polymer and NOM limits adhesion and toxicity of nano scale zerovalent iron to *E. coli*. *Environmental Science & Technology*, *44*(9), 3462–3467. doi:10.1021/es9031198 PMID:20355703

Liang, Q., & Zhao, D. (2014). Immobilization of arsenate in a sandy loam soil using starch-stabilized magnetite nanoparticles. *Journal of Hazardous Materials*, *271*, 16–23. doi:10.1016/j.jhazmat.2014.01.055 PMID:24584068

Lien, H. L., & Zhang, W.-X. (2001). Nanoscale iron particles for complete reduction of chlorinated ethenes. *Colloids and Surfaces. A, Physicochemical and Engineering Aspects*, *191*(1-2), 97–105. doi:10.1016/S0927-7757(01)00767-1

Liu, H. F., Qian, T. W., & Zhao, D. Y. (2013). Reductive immobilization of perrhenate in soil and groundwater using starch stabilized ZVI nanoparticles. *Chinese Science Bulletin*, *58*(2), 275–281. doi:10.100711434-012-5425-3

Liu, R., & Zhao, D. (2007). Reducing leachability and bio-accessibility of lead in soils using a new class of stabilized iron phosphate nanoparticles. *Water Research*, *41*(12), 2491–2502. doi:10.1016/j.watres.2007.03.026 PMID:17482234

Liu, Y., Li, S., Chen, Z., Megharaj, M., & Naidu, R. (2014). Influence of zero-valent iron nanoparticles on nitrate removal by *Paracoccus* sp. *Chemosphere*, *108*, 426–432. doi:10.1016/j.chemosphere.2014.02.045 PMID:24630453

Liu, Y., & Lowry, G. V. (2006). Effect of particle age (Fe0 content) and solution pH on NZVI reactivity: H_2 evolution and TCE dechlorination. *Environmental Science & Technology*, *40*(19), 6085–6090. doi:10.1021/es060685o PMID:17051804

Liu, Y., Phenrat, T., & Lowry, G. V. (2007). Effect of TCE concentration and dissolved groundwater solutes on NZVI-promoted TCE dechlorination and H_2 evolution. *Environmental Science & Technology*, *41*(22), 7881–7887. doi:10.1021/es0711967 PMID:18075103

Liu, Y. Q., Majetich, S. A., Tilton, R. D., Sholl, D. S., & Lowry, G. V. (2005). TCE dechlorination rates, pathways, and efficiency of nanoscale iron particles with different properties. *Environmental Science & Technology*, *39*(5), 1338–1345. doi:10.1021/es049195r PMID:15787375

Lowry, G. V., & Johnson, K. M. (2004). Congener-specific dechlorination of dissolved PCBs by microscale and nanoscale zerovalent iron in a water/methanol solution. *Environmental Science & Technology*, *38*(19), 5208–5216. doi:10.1021/es049835q PMID:15506219

Macé, C., Desrocher, S., Gheorghiu, F., Kane, A., Pupeza, M., Cernik, M., Kvapil, P., Venkatakrishnan, R., & Zhang, W.-X. (2006). Nanotechnology and groundwater remediation: A step forward in technology understanding. *Remediation*, *16*(2), 23–33. doi:10.1002/rem.20079

Mani, D., & Kumar, C. (2014). Biotechnological advances in bioremediation of heavy metals contaminated ecosystems: An overview with special reference to phytoremediation. *International Journal of Environmental Science and Technology*, *11*(3), 843–872. doi:10.100713762-013-0299-8

Matos, M. (2016). *Soil decontamination using nanomaterials* (MSc Thesis). University of Coimbra, Coimbra, Portugal.

Matos, M. P. S. R., Correia, A. A. S., & Rasteiro, M. G. (2017). Application of carbon nanotubes to immobilize heavy metals in contaminated soils. *Journal of Nanoparticle Research*, *19*(4), 126. doi:10.100711051-017-3830-x

Mukherjee, A. K., & Bordoloi, N. K. (2011). Bioremediation and reclamation of soil contaminated withpetroleum oil hydrocarbons by exogenously seeded bacterial consortium: A pilot-scale study. *Environmental Science and Pollution Research International*, *18*(3), 471–478. doi:10.100711356-010-0391-2 PMID:20835890

Nemecek, J., Pokorný, P., Lhotský, O., Knytl, V., Najmanov, P., Steinov, J., Cerník, M., Filipov, A., Filip, J., & Cajthaml, T. (2016). Combined nano-biotechnology for in-situ remediation of mixed contamination of groundwater by hexavalent chromium and chlorinated solvents. *The Science of the Total Environment*, *563-564*, 822–834. doi:10.1016/j.scitotenv.2016.01.019 PMID:26850861

Něměček, J., Pokorný, T. P., Lacinová, L., Černík, M., Masopustová, Z., Lhotský, O., Filipová, A., & Cajthaml, T. (2015). Combined abiotic and biotic in-situ reduction of hexavalent chromium in groundwater using nZVI and whey: A remedial pilot test. *Journal of Hazardous Materials*, *300*, 670–679. doi:10.1016/j.jhazmat.2015.07.056 PMID:26292054

Njoku, K. L., Nomba, E. U., & Olatunde, A. M. (2017). Vermiremediation of Crude Oil Contaminated Soil Using *Eudrillus euginae* and *Lumbricus terrestris. Journal of Biological and Environmental Sciences, 11*(31), 43–50.

Nutt, M. O., Hughes, J. B., & Wong, M. S. (2005). Designing Pd-on-Au Bimetallic Nanoparticle Catalysts for Trichloroethene Hydrodechlorination. *Environmental Science & Technology, 39*(5), 1346–1353. doi:10.1021/es048560b PMID:15787376

O'Hara, S., Krug, T., Quinn, J., Clausen, C., & Geiger, C. (2006). Field andlaboratory evaluation of the treatment of DNAPL sourcezones using emulsified zero-valent iron. *Remediation, 16*(2), 35–56. doi:10.1002/rem.20080

Oh, B. T., Just, C. L., & Alvarez, P. J. (2001). Hexahydro-1, 3, 5-trinitro-1, 3, 5-triazine mineralization byzerovalent iron and mixed anaerobic cultures. *Environmental Science & Technology, 35*(21), 4341–4346. doi:10.1021/es010852e PMID:11718353

Otto, M., Floyd, M., & Bajpai, S. (2008). Nanotechnology for site remediation. *Remediation, 19*(1), 99–108. doi:10.1002/rem.20194

Ouni, L., Ramazani, A., & Taghavi Fardood, S. (2019). An overview of carbon nanotubes role in heavy metals removal from wastewater. *Frontiers of Chemical Science and Engineering, 13*(2), 274–295. doi:10.100711705-018-1765-0

Oyewole, O. A., Raji, R. O., Musa, I. O., Enemanna, C. E., Abdulsalam, O. N., & Yakubu, J. G. (2019). Enhanced degradation of Crude Oil with *Alcaligenes faecalis* ADY25 and Iron oxide Nanoparticle. *International Journal of Applied Biological Research, 10*(2), 62–72.

Pang, Y., Zeng, G. M., Tang, L., Zhang, Y., Liu, Y. Y., Lei, X. X., Wu, M. S., Li, Z., & Liu, C. (2011). Cr(VI) reduction by Pseudomonas aeruginosa immobilized in a polyvinyl alcohol/sodium alginatematrix containing multi-walled carbon nanotubes. *Bioresource Technology, 102*(22), 10733–10736. doi:10.1016/j.biortech.2011.08.078 PMID:21937224

Peng, Y., Joseph, J. P., Minori, U., & Jason, C. W. (2015). Hetero aggregation of cerium oxide nanoparticles and nanoparticles of pyrolyzed biomass. *Environmental Science & Technology, 49*(22), 13294–13303. doi:10.1021/acs.est.5b03541 PMID:26461459

Ponder, S. M., Darab, J. G., & Mallouk, T. E. (2000). Remediation of Cr(VI) and Pb(II) aqueous solutions using supported, nanoscale zero-valent iron. *Environmental Science & Technology, 34*(12), 2564–2569. doi:10.1021/es9911420

Pradhan, G. K., & Parida, K. M. (2010). Fabrication of iron-cerium mixed oxide: An efficient photo catalyst for dye degradation. *International Journal of Engineering Science and Technology, 2*, 9.

Ravikumar, K. V. G., Kumar, D., Kumar, G., Mrudula, P., Natarajan, C., & Mukherjee, A. (2016). Enhanced Cr(VI) removal by nano zerovalent iron-immobilized alginate beads in the presence of a biofilm ina continuous-flow reactor. *Industrial & Engineering Chemistry Research, 55*(20), 5973–5982. doi:10.1021/acs.iecr.6b01006

Rorat, A., Wloka, D., Grobelak, A., Grosser, A., Sosnecka, A., Milczarek, M., Jelonek, P., Vandenbulcke, F., & Kacprzak, M. (2017). Vermi remediation of polycyclic aromatic hydrocarbons and heavy metals in sewage sludge composting process. *Journal of Environmental Management*, *187*, 347–353. doi:10.1016/j.jenvman.2016.10.062 PMID:27836561

Saxena, M., Chauhan, A., & Asokan, P. (1998). Flyash vermicompost from non-eco-friendly organic wastes. *Pollution Research*, *17*, 5–11.

Schrick, B., Hydutsky, B. W., Blough, J. L., & Mallouk, T. E. (2004). Delivery vehicles for zerovalent metal nanoparticles in soil and groundwater. *Chemistry of Materials*, *16*(11), 2187–2193. doi:10.1021/cm0218108

Shan, G., Xing, J., Zhang, H., & Liu, H. (2005). Bio-desulfurization of dibenzothiophene by microbial cells coated with magnetite nanoparticles. *Applied and Environmental Microbiology*, *71*(8), 4497–4502. doi:10.1128/AEM.71.8.4497-4502.2005 PMID:16085841

Shankar, S., Kansrajh, C., Dinesh, M. G., Satyan, R. S., Kiruthika, S., & Tharanipriya, A. (2014). Application of indigenous microbial consortiain bioremediation of oil-contaminated soils. *International Journal of Environmental Science and Technology*, *11*(2), 367–376. doi:10.100713762-013-0366-1

Shipley, H. J., Engates, K. E., & Guettner, A. M. (2011). Study of iron oxide nanoparticles in soil for remediation of arsenic. *Journal of Nanoparticle Research*, *13*(6), 2387–2397. doi:10.100711051-010-9999-x

Singh, R., Manickam, N., Mudiam, M. K. R., Murthy, R. C., & Misra, V. (2013). An integrated (nano-bio)technique for degradation of γ-HCH contaminated soil. *Journal of Hazardous Materials*, *258*, 35–41. doi:10.1016/j.jhazmat.2013.04.016 PMID:23692681

Singh, U. V., Abhishek, A., Bhaskar, M., Tandan, N., Ansari, N. G., & Singh, N. P. (2015). Phyto-extraction of heavy metals and biochemical changes with Brassica nigra L. grown in rayon grade paper mill effluent irrigated soil. *Bioinformation*, *11*(3), 138–144. doi:10.6026/97320630011138 PMID:25914448

Sohn, K., Kang, S. W., Ahn, S., Woo, M., & Yang, S. K. (2006). Fe(0) nanoparticles for nitrate reduction: Stability, reactivity, and transformation. *Environmental Science & Technology*, *40*(17), 5514–5519. doi:10.1021/es0525758 PMID:16999133

Song, H., & Carraway, E. R. (2005). Reduction of chlorinated ethanes by nanosized zerovalent iron: Kinetics, pathways, and effects of reaction conditions. *Environmental Science & Technology*, *39*(16), 6237–6245. doi:10.1021/es048262e PMID:16173587

Stumm, W., & Morgan, J. J. (1996). *Aquatic Chemistry: Chemical Equilibria and Rates in Natural Waters* (3rd ed.). JohnWiley & Sons, Inc.

Sun, Y. P. (2006). *Dispersion of nanoscale iron particles* (Doctoral dissertation). Department of Civil and Environmental Engineering, Lehigh University, Bethlehem, PA.

Tungittiplakorn, W., Cohen, C., & Lion, L. W. (2005). Engineeredpolymeric nanoparticles for bioremediation of hydrophobic contaminants. *Environmental Science & Technology*, *39*(5), 1354–1358. doi:10.1021/es049031a PMID:15787377

Tungittiplakorn, W., Lion, L. W., Cohen, C., & Kim, J. Y. (2004). Engineered polymeric nanoparticles for soil remediation. *Environmental Science & Technology*, *38*(5), 1605–1610. doi:10.1021/es0348997 PMID:15046367

Vivekananthan, V., Selvapriya, A., Janani, D., & Narendhar, C. (2014). Synthesis of mixed oxides of cerium-iron nanostructures for effective removal of heavy metals from wastewater. *Research Journal of Recent Sciences*, *3*, 212–217.

Wang, C. B., & Zhang, W. X. (1997). Synthesizing nanoscale iron particles for rapid and complete dechlorination of TCE and PCBs. *Environmental Science & Technology*, *31*(7), 2154–2156. doi:10.1021/es970039c

Wang, X., Gai, Z., Yu, B., Feng, J., Xu, C., Yuan, Y., Lin, Z., & Xu, P. (2007). Degradation of carbazole bymicrobial cells immobilized in magnetic gellan gum gel beads. *Applied and Environmental Microbiology*, *73*(20), 6421–6428. doi:10.1128/AEM.01051-07 PMID:17827304

Wang, Y., Pan, C., Chu, W., Vipin, A. K., & Sun, L. (2019). Environmental Remediation Applications of Carbon Nanotubes and Graphene Oxide: Adsorption and Catalysis. *Nanomaterials (Basel, Switzerland)*, *9*(3), 439. doi:10.3390/nano9030439 PMID:30875970

Windt, W. D., Aelterman, P., & Verstraete, W. (2005). Bioreductive deposition of palladium(0) nanoparticles on *Shewanella oneidensis* with catalytic activity towards reductive dechlorination of polychlorinated biphenyls. *Environmental Microbiology*, *7*(3), 314–325. doi:10.1111/j.1462-2920.2005.00696.x PMID:15683392

Xiu, Z., Jin, Z., Li, T., Mahendra, S., Lowry, G. V., & Alvarez, P. J. J. (2010). Effects of nano-scale zero-valent iron particles on a mixed culture dechlorinating trichloroethylene. *Bioresource Technology*, *101*(4), 1141–1146. doi:10.1016/j.biortech.2009.09.057 PMID:19819128

Xiu, Z. M., Gregory, K. B., Lowry, G. V., & Alvarez, P. J. (2010). Effect of bare and coated nanoscale zerovalent iron on tceA and vcrA gene expression in *Dehalococcoides* spp. *Environmental Science & Technology*, *44*(19), 7647–7651. doi:10.1021/es101786y PMID:20804135

Xu, J., Dozier, A., & Bhattacharyya, D. (2005). Synthesis of Nanoscale Bimetallic Particles in Polyelectrolyte Membrane Matrix for Reductive Transformation of Halogenated Organic Compounds. *Journal of Nanoparticle Research*, *7*(4-5), 449–467. doi:10.100711051-005-4273-3

Xu, Y., & Zhao, D. (2007). Reductive immobilization of chromate in water and soil using stabilized iron nanoparticles. *Water Research*, *41*(10), 2101–2108. doi:10.1016/j.watres.2007.02.037 PMID:17412389

Yan, F. F., Wu, C., Cheng, Y. Y., He, Y. R., Li, W. W., & Yu, H. Q. (2013). Carbon nanotubes promote Cr(VI) reduction by alginate-immobilized *Shewanella oneidensis* MR-1. *Biochemical Engineering Journal*, *77*, 183–189. doi:10.1016/j.bej.2013.06.009

Zhang, M., Liu, G. H., Song, K., Wang, Z., Zhao, Q., Li, S., & Ye, Z. (2015). Biological treatment of 2,4,6-trinitrotoluene (TNT) red water by immobilized anaerobic and aerobic microbial filters. *Chemical Engineering Journal*, *259*, 876–884. doi:10.1016/j.cej.2014.08.041

Zhang, W.-X. (2003). Nanoscale iron particles for environmental remediation: An overview. *Journal of Nanoparticle Research*, *5*(3/4), 323–332. doi:10.1023/A:1025520116015

Zhang, W. X., & Elliott, D. W. (2006). Applications of iron nanoparticlesfor groundwater remediation. *Remediation*, *16*(2), 7–21. doi:10.1002/rem.20078

Chapter 21
Nanomaterials for Soil Reclamation

Avni Jain
Banasthali Vidyapith, Niwai, India

Neha Singh
Banasthali Vidyapith, Niwai, India

Suphiya Khan
Banasthali Vidyapith, Niwai, India

ABSTRACT

The demand for the development of eco-friendly, sustainable, and adaptable technologies for the disinfection of the environmental contaminants is increasing nowadays. Nano-bioremediation is one such technique that has made possible the use of biosynthetic nanoparticles for soil pollution remediation. It is an effective, efficient, and feasible method for revitalizing soil potential and rendering it pollution free. Pollutants present in soil are a great threat to soil biota, environment, and in fact human health. Nanomaterials exhibit the unique chemical and physical properties because of which they have always received attention in the growing era of bioremediation. Use of nanotechnology for bioremediation is one such technology as it focuses mainly on the interaction between the contaminants, the microorganisms, and the nanomaterials being used for both the positive (i.e., stimulating) and negative or toxic environmental effects. Thus, this chapter focuses on the need to recover the polluted soil and application of nano-remediation technology for restoring soil's cultivation capacity.

INTRODUCTION

Environmental resilience is characterised as an important environmental interaction that prevents natural resources from declining or deteriorating, thus improving the quality of the environment for a longer period of time. However, the world's general definition of durability or sustainability is continuous development resulting in degeneration of the environment. The rapidly increasing polluted sites have

DOI: 10.4018/978-1-7998-7062-3.ch021

led to a wide range increment in the demand for the development of new ways and techniques and for quicker cleaning or purification of polluted sites and also to reduce the costs of technologies being used (Cecchin et al., 2016). The level of wastes and toxic materials in the environment is growing quickly with the on-going industrial development. Science and technology directly or indirectly add to increase the toxicity in environment. Due to the technological innovations across the world in different processes and products with no proper diligence to the environment, the disposal of materials or wastes into the environment is excessive and that too without proper management (Cecchin et al., 2016). Thus, the research to develop technologies to accelerate the decontamination of these sites as well as reduce the cost of these contaminant removal processes is increasingly promoted (Menendez-Vega et al., 2007). Various restoration strategies have been made to use for the conservation and betterment of environment and one such technique is Bioremediation, which involves the use of micro-organisms or also the use of nanoparticles (Zhang et al., 2020). Several technologies are available for deployment, both in-situ and ex-situ technologies. In in-situ treatment, there is no need to excavate the soils while in ex-situ, it involves polluted soil removal and off-site treatment under suitable maintained conditions (Cecchin et al., 2016). Ex-situ remediation involves the use of prefabricated bed and bioreactor. Over the years, in-situ bioremediation techniques are being used for remediation of various hydrocarbon-polluted sites specifically. In situ treatment is a very established and profitable method and it reduces expensive excavation process and emission (Menendez-Vega et al., 2007) Nanoparticles, on the other hand, have exclusive capabilities as depicted in Fig. 1, to sterilize or sanitize the environments from such harmful toxicants. They tend to provide an active base for microbial activities and thereby, triggering the cleaning process. Nano-bioremediation is the term preferred when the nanoparticles are used for pollutants removal and leading to growth in microbial activities. Nano-bioremediation is one of such kind of methods which received a lot of attention in the past few years. It aims at reducing the contaminant concentrations to risk-based levels, alleviating the additional environmental impacts simultaneously.

Figure 1. Illustration of nanomaterials in bioremediation

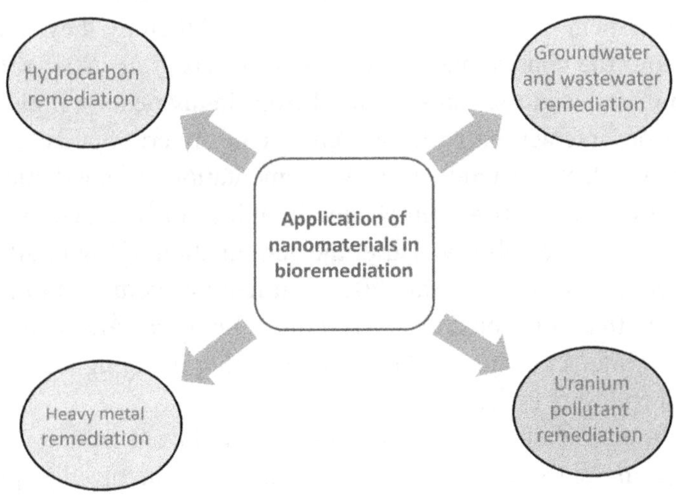

This method brings the benefits of both nanotechnology and bioremediation together to achieve a remediation that is more efficient, less time taking, and environment friendly than the individual processes (Singh et al., 2020) Nanoparticles (NPs), which are used in Nano-bioremediation can be either metallic or non-metallic and of differently shapes. Nanoparticles or Nanocomplexes which are used in purification processes are of the following types- Single metal NPs, Bimetallic NPs, Carbon-based NPs, and Modified NPs etc. There are various advantages of bioremediation processes over other conventional methods as it much is cost effective, has a very high competence, involves minimal use of chemical and biological sludge, they are generally selective to specific metals. Also, there is no use of supplementary nutrient requirements, it also implicates the bio-sorbent regeneration, and greater is the possibility of metal recovery (Davis et al., 2017). Nowadays, sustainable remediation is given a lot of importance as it aims in reduction of quantities to risk-based levels as well as to minimize the environmental impacts such as greenhouse gas emissions, waste generation and natural resource consumption, among others (Cecchin et al., 2016).

Polynuclear aromatic hydrocarbons (PAHs), are the organic hydrophobic soil contaminants which are difficult to remove. Different kind of Nanoparticles known as Amphiphilic polyurethane (APU) were synthesized for remediation PAHs contaminated soils (Rizwan et al., 2014) Due to the unique optical, thermal, electrical, chemical and physical properties, nanoparticles do have wide range applications (Davis et al., 2017) Use of nanoparticles or nanomaterials for contaminated soil site remediation has received a great importance and is continuously gaining a lot attention (Cecchin et al., 2016).

CONCEPT OF NANO-BIOREMEDIATION TECHNOLOGIES

Ecosystem is highly polluted by the pollutants and it becomes necessary to remove these contaminants. Nano-bioremediations is one such hybrid technology used for transforming and detoxifying the environment. For large scale clean-up, with less cost and lesser harmful by-products, both In-situ and Ex-situ methods are applicable. In past years, Biological processes are made to integrate with Nanomaterials or Nanoparticles for rapid increasing and promoting the environmental contaminants removal (Vazquez-Nunez et al., 2020) Large surface area per unit mass and quantum effects are the best suitable properties of nanoparticles that make them more reactive and useful., and thereby they show plasmon resonance and can also penetrate into the contaminated sites easily (Davis et al., 2017) Physical and Chemical properties of the contaminants and toxicants released from industries are highly variable, along with their cytotoxicity and various interactions with environment, i.e., microorganisms, plants, animals, water, minerals, organic matter, wind, etc., Therefore, the implementation of remediation technologies is quiet difficult (Vazquez-Nunez et al., 2020) A step-change in ability to remediate with lesser intermediate processes and quick results has been observed after the combination of Nanoparticles or Nanomaterials and Biotechnology is being used (Davis et al., 2017). In nano-bioremediation, the most important is sorption process. In this, both adsorption and absorption are involved. Adsorption is a process in which the pollutants and sorbent interact at surface level. On the other hand, in absorption the pollutants get inside the sorbent into deep layers forming a solution.

A further distinction is also available as Chemisorption and Physisorption. Chemisorption involves the chemical reactions taking place while only physical forces are involved in physisorption. With the different type of nanomaterials, photocatalytic processes are also used for degradation of contaminants. Biotransformation of the products of photocatalytic degradation is required so that its concentration in media is reduced. Some varieties of contaminants are degraded by the enzymes which are released in environment by living organisms. Nano-bioremediation technologies are also said to extend their field of applications. Thus, nano-bioremediation technique is advantageous over other conventional remediation techniques. However, protocols for nanoparticles toxicity measurement in soil and water, their interactions with biotic and abiotic components in environment and appliable frames for different materials are needed to be studied (Vazquez-Nunez et al., 2020).

Figure 2. Schematic diagram of Zero valent Iron

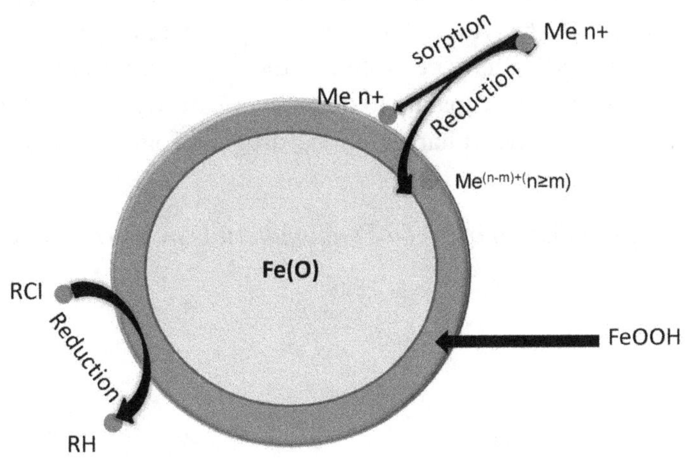

NANOMATERIALS AND NANOPARTICLES USED IN BIOREMEDIATION

For the bioremediations of contaminated sites or soil and removal of pollutants under variable environmental conditions, various nanomaterials have been successfully used. Several different types of nanomaterials have been tested to understand their ability to reduce contaminants with the help of living organisms. Many criteria like nano iron and its derivatives, dendrimers, carbon-based NMs, single-enzyme NPs, engineered polymeric and biogenic uraninite and metals other than iron are used for understanding.

Selecting the type of nanomaterial to be used depends on the type of contaminant present. For example, Iron nanoparticles are used for removal of heavy metals from soil, because of its magnetic properties. On the other hand, Carbon based nanomaterials are much applicable in trapping organic pollutants or heavy metals both from soil (Tripathi et al., 2018) Elemental or zero-valent metals, such as iron, nickel and palladium, have shown much positive results in highly toxic contaminated soils in nanoscale form. Also, they are of great use in stabilizing transitional metals such as chromium and arsenic and persistent organic compound's dehalogenation. Nanoparticles can also be applied for the remediation of tetrachloroethylene (TCE), and Polyvinyl biphenyls (PCB) and other such recalcitrant organic compounds. This method has proven to be much effective and quick as compared to conventional slow microbial degradation processes (Contreras et al., 2015).

Amongst all the nanomaterials being used, nanoscale zero-valent iron, known as nano-iron (nZVI), has resulted to be one of the most effective and practically injectable into subsurface environment at the contaminated sites. This is because of its low cost of production and lesser toxicity (Cecchin et al., 2016; Li et al. 2006; Zhang et al. 2006).

Developments in Nanobiotechnology have ensured the emergence of environmentally benign nanoparticles. Silver nanoparticles (AgNP's) have varying applications in non-linear optics, or intercalation in electrical batteries, optical receptors, as a catalyst and as an antibacterial. Although, the first nanoparticle which was used for cleaning of environment is Iron (NP) (Pandey, 2018). Iron-based technologies for remediation of contaminated land or groundwater remediation can be applied in two ways. On the basis of chemistry: technologies which use iron as a sorbent (adsorptive/immobilization technologies) or as an electron donor to break down or to convert contaminants into a less toxic or mobile form (reductive technologies) (Cecchin et al., 2016) However, it has been proved that many techniques involve both these methods. Zinc nanoparticles (ZnNPs), have been studied thoroughly all over the world as a semiconductor photo catalyst as it has an enormous capacity to degrade organic dyes. ZnNPs as a photocatalyst, can completely degrade a large variety of dyes, pharmaceutical drugs and other compounds (Davis et al., 2017).

Figure 3. Diagrammatic representation of several nanomaterials in bioremediation

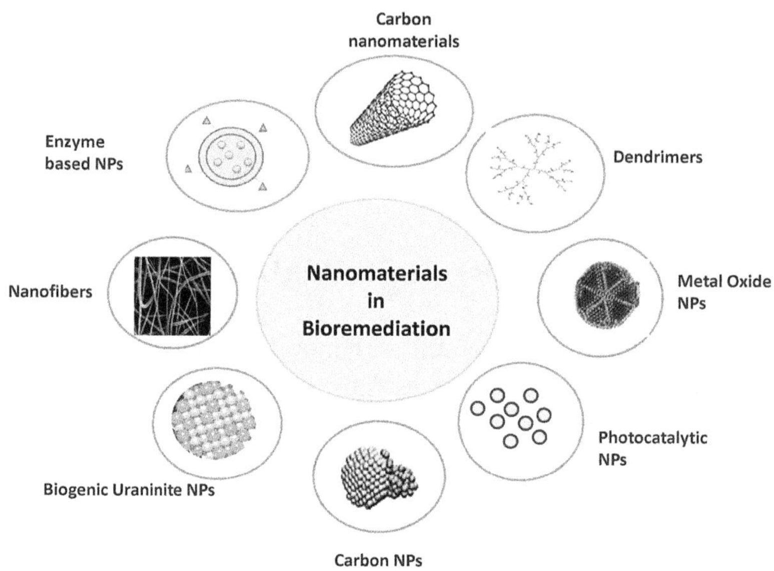

Noble nanoparticles like Gold and Silver also have enormous applications in various areas. Researchers have studied the potential applications of gold (Au) and Silver (Ag) nanoparticles in removal of organic dyes. Copper (Cu) nanoparticles have also shown great results when used for organic dye removal. Although, in general, nZVI have been considered as the most effective technique when nZVI mobility in the porous medium can be induced sufficiently and effectively so that it is distributed accurately in contaminated zones (Boente et al., 2017). In recent years, nZVI has been used in various areas and its application is increasing significantly, making it one of the most used in-situ remediation technique for toxic compound contaminated soil (Cecchin et al., 2016). It is said that nZVI has a much higher reactivity in soil as compared to other NPs but it is not the only property or feature which makes it this successful in field applications. They also show particulate aggregation control, mobility in porous environments, reactivity and longevity in the subsurface environment which make it much efficient technique for remediation of contaminated sites with nZVI under field conditions (Davis et al., 2017) Studies have shown that bacteria and plants have quiet high capability to immobilize metals and transform all the organic as well as inorganic contaminants. Although, it is important to be kept in mind that the type of nanomaterial being used and type of contaminant present is important as microbes and other living organisms respond to each NM and contaminant in a different way (Tripathi et al. 2018).

At global level, higher persistence of toxic compound like polychlorinated by-phenils (PCBs) and their long-range atmospheric transport, less or slow degradation and bioaccumulation is a major problem. For the nano-bioremediation of PCB-polluted soils, several methods like NPs catalyzed Fenton or Fenton-like, and persulfate activation has been proven effective. Nevertheless, for removal or degradation of PCBs, various techniques using carbon nanotubes (CNT) and Arthrobacter sp. are being developed (Vazquez-Nunez et al., 2020). In general, direct or indirect integration of nanoparticles have led to emerge as promising technique for controlling pollution and its harmful effects (Table 1). As an example, with the direct use of nZVI, organic pollutants like herbicides (*i.e.* atrazine, molinate) and pesticides (*i.e.* chlorpyrifos) are removed. Whereas, nanoparticles incorporated with phyto-remediation degrades organic pollutants effectively. (Tripathi et al., 2018) Nanotechnological applications in ecological remediation, is rapidly increasing from pilot scale to full-scale achievement in the treatment of chlorinated sites. Nanoscale Titanium oxide (TiO2), Carbon Nanotubes (CNTs), dendrimers, swellable organically-modified silica (SOMS) and metallo-porphyrinogens are some nanoproducts exclusively used for remediation of pollutants in both ex-situ or in-situ methods. Titanium oxide NPs (TiO_2) nanoparticles have proven to remove a wide range of chemical fertilizers, herbicides, insecticides and pesticides via photocatalysis through ex-situ management way of resources. Iron, Copper and Nickel nanoparticles have been synthesized biologically in combination with a metal-catalyst like gold, platinum, palladium, and nickel so as to increase the redox-reaction rates (Pandey, 2018).

However, in Bioaugmentation, Nanomaterials have shown no benefit, as in polluted environments they inhibit the microbial population (Vazquez-Nunez et al., 2020). Also, high carbon nano tubes concentrations have lessened the biological degradation rate by resisting the growth of bacterial species and also inhibiting the activity or microbes, whereas lower carbon nanotubes concentrations improve the rate of biodegradation by enhancing growth of bacteria and expression of degradation genes (Rajesha et al., 2017). Because of the large surface area in nanoparticles, their application in bioaugmentation could mitigate the limitations related to immobilization and entrapment of microorganisms (Vazquez-Nunez et al., 2020).

NANO-BIOREMEDIATION IN SOIL RECLAMATION

Various harmful inorganic compounds, organic pollutants, heavy metals and several other complex compounds in surface contaminated soil and ground water system has been introduced via industrial boom and population growth (Tosco et al., 2014; Santornchot et al., 2010). These contaminate needs to be eliminated as of the environment. Nanotechnology has been reported to play important role in addressing different effective and innovative solutions to many of the diverse environmental challenges (Reddy et al., 2014).

Table 1. List of contaminants and the nanoparticles used to remove them

Contaminants	Nanoparticles Used	Results	References
Arsenite	Al_2O_3 nanoparticles	They absorbed Arsenite at normal temperature and pH conditions.	Prabhakar and Samadder, (2018)
Arsenic(As)	CuO nanoparticles	Considerable amount of As was removed with CuO nanoparticles.	Reddy et al., (2013)
Benzophenone-3 (BP-3)	Zinc oxide nanoparticles	High level degradation of Benzophenone observed.	Rajesha et al., (2017)
Carbamazepine	Hematite nanoparticles	Approximately, 90% carbamazepine was absorbed with Hematite nanoparticles.	Rajendran and Sen, (2018)
Cu, Pb, Sb	Zero-valent Iron (nZVI)	Soil washing and metal removal efficiency increased.	Boente et al., 2018)
Cadmium (II), lead (II), and chromium (VI) ions	Cerium oxide nanoparticles	Effective heavy metal removal at pH 5 and & was observed.	Contreras et al., (2015)
EDCs (diclofenac, metoprolol, estrone, and chloramphenicol)	TiO_2 nanoparticles	With the help of photocatalytic activity of TiO_2 particles, EDCs were efficiently removed.	Czech and Rubinowska, (2013)
PAHs and metal contaminants	Magnetic nanoparticle adsorbents, (Mag-PCMA-T)	More than 85% PAHs and metals were removed.	Huang et al., (2016)
Pentachlorobiphenyl	Palladium nanoparticles	Stabilized Pd particles with supercritical fluid CO_2 removed almost all PCBs from soil at all the temperature ranges and at 200 atm pressure.	Wang and Chiu, (2009)
Phenol, bisphenol A, and atrazine	Reduced graphene oxide silver nanoparticles (rGO-Ag)	Photocatalytic degradation of these organic contaminants was observed in visible light range.	Bhunia and Jana, (2014)
17β-estradiol	Manganese oxide nanoparticles	Almost 88% estrogen removal from soil was observed with MnO2 nanoparticles. It was also stated that with increase in nanoparticles concentration, the degradation level also increased.	Han et al., (2017)
Tetracycline (TC)	Fe/Ni bimetallic nanoparticles	Due to ageing in nanoparticles of Fe and Ni, the TC removal efficiency reduced.	Dong et al., (2018)

Nano-remediation is an innovative technology that's help in reducing the pollutants level from the environment (Reddy et al., 2014). The technology comprises of many application that help in lowering down the level of toxicant such as purification and remediation of pollutants (Pyae et al., 2019), prevention of pollution/contaminates (Pandey, 2018), and detection of pollution. It served as an alternative to traditional treatments due to cost effective, environmental friendly, high efficiency, large surface area and nano size particles. Several metals like Silver, Palladium, Iron, Gold show promising result in the treatment of contaminated sites infected with numerous toxic materials due to their high surface area behaviour and size (Pandey, 2018).

Nano-bioremediation is a combination of both nanotechnology and bioremediation and forms an integrated technology that efficiently helps to remove the pollutants from contaminated sites as depicted in Fig. 5. It's an integrated approach that consumes less time, eco-friendly and more effective than other existing technology in remediation of toxic substances and could overcome the disadvantages of individual technologies to provide the better results. For example: nZVI integrated with microbial strains helps in remediation of pollutants in more efficient and effective way. Several hydrocarbons such as Chlorinated aliphatic hydrocarbons (CAH) are recalcitrant compounds those are not fully removed via neither nZVI nor organochlorine respiring bacteria (ORB). Koenig et al. (2016) combined both the technologies i.e. nZVI and organochlorine respiring bacteria (ORB) that shows the effective results for removal of CAHs at appropriate dosage, and by this a wide range of CAHs can be removed. The nZVI spent during the remediation process can be regenerated that remains in the existing bacterial environment via various minerals such as like cysteine and vitamins. For degradation of polybrominated diphenyl ethers (PBDEs) in aqueous solution, reductive-oxidative strategy consisting of nZVI and an aerobic bacterium (Sphingomonas sp. PH-07) prove to be highly efficient. In environmental substrates, microorganisms carrying out biotransformation of contaminants considered hydrogen as highly favorable electron donor. Many researchers explore the possibility of utilizing cathodic hydrogen (produced during corrosion of nZVI under anaerobic conditions) as an electron donor for contaminant-degrading microbes (Weathers et al. 1997; Liu et al. 2005). Carboxymethyl cellulose (CMC) stabilized bimetallic nanoparticles (CMC-Pd/nFe0) was integrated with Sphingomonas sp. strain NM05 for studying degradation of γ-HCH, synergistic effect on γ-HCH degradation was reported in other studies (Singh et al. 2013). It was further reported in case of integrated system, which further indicate that stabilized nanoparticles have some kind of biostimulatory effect on cell growth (Singh et al. 2013). In remediation of other contaminates, such as Multi-walled carbon nanotubes (CNTs) along with bioremediation approach can be successfully utilized. In calcium alginate beads, Shewanella oneidensis MR-1, a facultative Gram-negative bacterium comprising carbon nanotubes was immobilized to reduce Cr (VI) to Cr (III). In integrated system, Pd nanoparticles also had shown efficient property. Chidambaram et al. (2010) reported in situ synthesis of Pd nanoparticles using C. pasteurianum BC1 cells, wherein C. pasteurianum reduced the Pd (II) ions to Pd nanoparticles which were retained in the cell wall and cytoplasm of the cells in the form of bio-Pd.

ENGINEERED POLYMERIC NANOPARTICLES FOR SOIL

Several hydrophobic organic contaminants, such as polynuclear aromatic hydrocarbons (PAHs), sorb strongly to soils and are difficult to remove. Amphiphilic polyurethane (APU) NP has been synthesized for use in remediation of soil contaminated with PAHs. APU particles have the ability to enhance PAH desorption and transport in a manner comparable to that of surfactant micelles, but unlike the surface-active components of micelles, the individual cross-linked precursor chains in APU particles are not free to sorb to the soil surface. Thus, the APU particles are stable, independent of their concentration in the aqueous phase.

Figure 4. Schematic representation of nano-bioremediation of contaminated soil using incorporation of nZVI

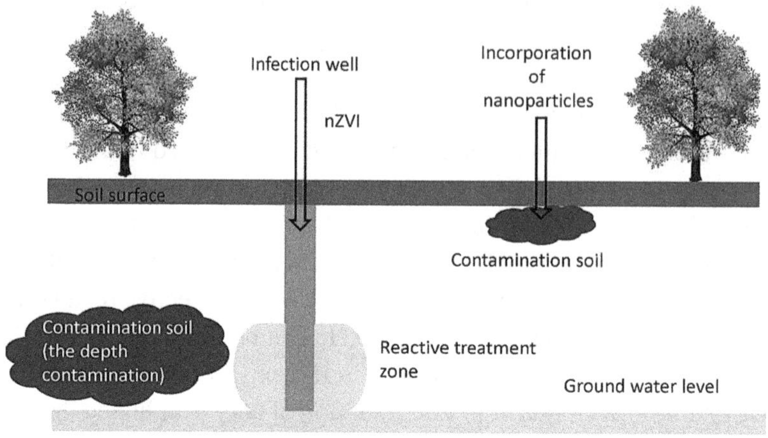

CONCLUSION AND FUTURE PROSPECTS

Nanotechnology is revolutionizing the way we live. Nano-bioremediation is a technique that refers to the use of nanomaterials to remove or lower the concentration of contamination to conducive level via biodegradation and also promote biodegradation of the contaminants to reach the risk assessment levels. Environment-friendly method of nanoparticles coupled with biological remediation enhances the sustainability to a great level. Through biosynthesis, the use of harmful or toxic chemicals or solvents is greatly reduced and solvents and it is simple, cost effective and time saving method. Many researchers have synthesized Zinc (Zn), Silver (Ag), Gold (Au), Iron (Fe) and Copper (Cu) nanoparticles biologically. However, during a bioremediation process, the synergetic effect of biotechnology in combination with nanoparticles and how they respond in diverse nature still lacks. Nano-bioremediation is highly cheap when compared to other technologies, it provides wide range of applications, can be coupled with biological treatments and give highly effective contaminant degradation and thus it contributes enormously to environment sustainability and enhances possibilities to face new challenges.

Overall, it is important to understand the nanoparticles, their effects on microbes in different soil conditions is very important to make a remediation technique to get better results in degrading pollutants from soil without negatively impacting the micro-organisms. For improving the quality of environment in developed and developing countries, nanotechnology remediation techniques have proven one of the best methods till date. However, to understand the mechanism of decontamination and bioremediation, significant amount of research has been performed. Furthermore, nanotechnology and bioremediation world markets are expected to continue growing and developing new niches to improve not only environmental aspects but also human lifestyle.

ACKNOWLEDGMENT

This study was funded by DBT BIG-BIRAC Ref No. BIRAC/CCAMP0723/BIG-13/18 and DST major project entitled "Low Cost-Renewable Energy Driven (LC-RED) Water Treatment Solution Centre" Ref No. DST/TM/WTI/WIC/2K17/124. We acknowledge Prof. Aditya Shastri for providing research amenities and bioinformatics centre for providing computational facilities.

REFERENCES

Bhunia, S. K., & Jana, N. R. (2014). Reduced graphene oxide-silver nanoparticle composite as visible light photocatalyst for degradation of colorless endocrine disruptors. *ACS Applied Materials & Interfaces*, *6*(22), 20085–20092. doi:10.1021/am505677x PMID:25296393

Boente, C., Sierra, C., Martínez-Blanco, D., Menéndez-Aguado, J. M., & Gallego, J. R. (2018). Nanoscale zero-valent iron-assisted soil washing for the removal of potentially toxic elements. *Journal of Hazardous Materials*, *350*, 55–65. doi:10.1016/j.jhazmat.2018.02.016 PMID:29448214

Cecchin, I., Reddy, K. R., Thomé, A., Tessaro, E. F., & Schnaid, F. (2017). Nanobioremediation: Integration of nanoparticles and bioremediation for sustainable remediation of chlorinated organic contaminants in soils. *International Biodeterioration & Biodegradation*, *119*, 419–428. doi:10.1016/j.ibiod.2016.09.027

Chidambaram, D., Hennebel, T., Taghavi, S., Mast, J., Boon, N., Verstraete, W., van der Lelie, D., & Fitts, J. P. (2010). Concomitant microbial generation of palladium nanoparticles and hydrogen to immobilize chromate. *Environmental Science & Technology*, *44*(19), 7635–7640. doi:10.1021/es101559r PMID:20822130

Contreras, A. R., Casals, E., Puntes, V., Komilis, D., Sánchez, A., & Font, X. (2015). Use of cerium oxide (CeO2) nanoparticles for the adsorption of dissolved cadmium (II), lead (II) and chromium (VI) at two different pHs in single and multi-component systems. *Global NEST Journal*, *17*(3), 536–543. doi:10.30955/gnj.001687

Czech, B., & Rubinowska, K. (2013). TiO_2-assisted photocatalytic degradation of diclofenac, metoprolol, estrone and chloramphenicol as endocrine disruptors in water. *Adsorption*, *19*(2-4), 619–630. doi:10.100710450-013-9485-8

Davis, A. S., Prakash, P., & Thamaraiselvi, K. (2017). Nanobioremediation technologies for sustainable environment. In *Bioremediation and Sustainable Technologies for Cleaner Environment* (pp. 13–33). Springer. doi:10.1007/978-3-319-48439-6_2

Dong, H., Jiang, Z., Deng, J., Zhang, C., Cheng, Y., Hou, K., Zhang, L., Tang, L., & Zeng, G. (2018). Physicochemical transformation of Fe/Ni bimetallic nanoparticles during aging in simulated groundwater and the consequent effect on contaminant removal. *Water Research*, *129*, 51–57. doi:10.1016/j.watres.2017.11.002 PMID:29128681

Huang, Y., Fulton, A. N., & Keller, A. A. (2016). Simultaneous removal of PAHs and metal contaminants from water using magnetic nanoparticle adsorbents. *The Science of the Total Environment*, *571*, 1029–1036. doi:10.1016/j.scitotenv.2016.07.093 PMID:27450251

Koenig, J. C., Boparai, H. K., Lee, M. J., O'Carroll, D. M., Barnes, R. J., & Manefield, M. J. (2016). Particles and enzymes: Combining nanoscale zero valent iron and organochlorine respiring bacteria for the detoxification of chloroethane mixtures. *Journal of Hazardous Materials*, *308*, 106–112. doi:10.1016/j.jhazmat.2015.12.036 PMID:26808236

Li, X. Q., Elliott, D. W., & Zhang, W. X. (2006). Zero-valent iron nanoparticles for abatement of environmental pollutants: Materials and engineering aspects. *Critical Reviews in Solid State and Material Sciences*, *31*(4), 111–122. doi:10.1080/10408430601057611

Liu, Y., Majetich, S. A., Tilton, R. D., Sholl, D. S., & Lowry, G. V. (2005). TCE dechlorination rates, pathways, and efficiency of nanoscale iron particles with different properties. *Environmental Science & Technology*, *39*(5), 1338–1345. doi:10.1021/es049195r PMID:15787375

Menendez-Vega, D., Gallego, J. L. R., Pelaez, A. I., de Cordoba, G. F., Moreno, J., Munoz, D., & Sanchez, J. (2007). Engineered in situ bioremediation of soil and groundwater polluted with weathered hydrocarbons. *European Journal of Soil Biology*, *43*(5-6), 310–321. doi:10.1016/j.ejsobi.2007.03.005

Pandey, G. (2018). Prospects of Nanobioremediation in Environmental cleanup. *Oriental Journal of Chemistry*, *34*(6), 2838–2850. doi:10.13005/ojc/340622

Pandey, G. (2018). Prospects of Nanobioremediation in Environmental cleanup. *Oriental Journal of Chemistry*, *34*(6), 2838–2850. doi:10.13005/ojc/340622

Pyae, H. A., Win, W. A., Yossapol, C., & Dararatana, S. (2019). Micro-particle ZVI inhibition threshold in cassava pulp bio-methanation. *Environment Asia*, *12*(Special Issue), 64–73.

Rajendran, K., & Sen, S. (2018). Adsorptive removal of carbamazepine using biosynthesized hematite nanoparticles. *Environmental Nanotechnology, Monitoring & Management*, *9*, 122–127. doi:10.1016/j.enmm.2018.01.001

Rajesha, J. B., Ramasami, A. K., Nagaraju, G., & Balakrishna, G. R. (2017). Photochemical elimination of Endocrine Disrupting Chemical (EDC) by ZnO nanoparticles, synthesized by gel combustion. *Water Environment Research*, *89*(5), 396–405. doi:10.2175/106143016X14733681696086 PMID:27779923

Reddy, K. J., McDonald, K. J., & King, H. (2013). A novel arsenic removal process for water using cupric oxide nanoparticles. *Journal of Colloid and Interface Science*, *397*, 96–102. doi:10.1016/j.jcis.2013.01.041 PMID:23452518

Reddy, K. R., Darnault, C. J., & Darko-Kagya, K. (2014). Transport of lactate-modified nanoscale iron particles in porous media. *Journal of Geotechnical and Geoenvironmental Engineering*, *140*(2), 04013013. doi:10.1061/(ASCE)GT.1943-5606.0001015

Reddy, K. R., Khodadoust, A. P., & Darko-Kagya, K. (2014). Transport and reactivity of lactate-modified nanoscale iron particles for remediation of DNT in subsurface soils. *Journal of Environmental Engineering*, *140*(12), 04014042. doi:10.1061/(ASCE)EE.1943-7870.0000870

Rizwan, M., Singh, M., Mitra, C. K., & Morve, R. K. (2014). Ecofriendly application of nanomaterials: nanobioremediation. *Journal of Nanoparticles*.

Santornchot, P., Satapanajaru, T., & Comfort, S. D. (2010). Application of nano-zero valent iron for treating metolachlor in aqueous solution. *World Academy of Science, Engineering and Technology*, *48*, 625–628.

Singh, R., Behera, M., & Kumar, S. (2020). Nano-bioremediation: An innovative remediation technology for treatment and management of contaminated sites. In *Bioremediation of Industrial Waste for Environmental Safety* (pp. 165–182). Springer. doi:10.1007/978-981-13-3426-9_7

Singh, R., Manickam, N., Mudiam, M. K. R., Murthy, R. C., & Misra, V. (2013). An integrated (nano-bio) technique for degradation of γ-HCH contaminated soil. *Journal of Hazardous Materials*, *258*, 35–41. doi:10.1016/j.jhazmat.2013.04.016 PMID:23692681

Tosco, T., Papini, M. P., Viggi, C. C., & Sethi, R. (2014). Nanoscale zerovalent iron particles for groundwater remediation: A review. *Journal of Cleaner Production*, *77*, 10–21. doi:10.1016/j.jclepro.2013.12.026

Tripathi, S., Sanjeevi, R., Anuradha, J., Chauhan, D. S., & Rathoure, A. K. (2018). Nano-bioremediation: nanotechnology and bioremediation. In Biostimulation Remediation Technologies for Groundwater Contaminants (pp. 202-219). IGI Global. doi:10.4018/978-1-5225-4162-2.ch012

Vázquez-Núñez, E., Molina-Guerrero, C. E., Peña-Castro, J. M., Fernández-Luqueño, F., & de la Rosa-Álvarez, M. (2020). Use of Nanotechnology for the Bioremediation of Contaminants: A Review. *Processes (Basel, Switzerland)*, *8*(7), 826. doi:10.3390/pr8070826

Wang, J. S., & Chiu, K. (2009). Destruction of pentachlorobiphenyl in soil by supercritical CO2 extraction coupled with polymer-stabilized palladium nanoparticles. *Chemosphere*, *75*(5), 629–633. doi:10.1016/j.chemosphere.2009.01.018 PMID:19211124

Weathers, L. J., Parkin, G. F., & Alvarez, P. J. (1997). Utilization of cathodic hydrogen as electron donor for chloroform cometabolism by a mixed, methanogenic culture. *Environmental Science & Technology*, *31*(3), 880–885. doi:10.1021/es960582d

Zhang, C., Wu, D., & Ren, H. (2020). Bioremediation of oil contaminated soil using agricultural wastes via microbial consortium. *Scientific Reports*, *10*(1), 1–8. PMID:32513982

Chapter 22
Quorum Quenching for Sustainable Environment

Sumira Malik
https://orcid.org/0000-0001-5077-1493
Amity Institute of Biotechnology, Amity University, Ranchi, India

Shilpa Prasad
https://orcid.org/0000-0002-9547-1782
Amity Institute of Biotechnology, Amity University, Ranchi, India

Tanvi Kumari
Amity Institute of Biotechnology, Amity University, Ranchi, India

Shreya Ghoshal
Amity Institute of Biotechnology, Amity University, Ranchi, India

Ankita Agrawal
Amity University, Ranchi, India

Shashank Shekhar
Shivalik Institute of Professional Studies, Dehradun, India

Bijaya Samal
Amity Institute of Biotechnology, Amity University, Ranchi, India

ABSTRACT

Quorum quenching is the process that prevents quorum sensing through the disruption of signalling cascade and bacterial communication among themselves mediated by the degradation of the signalling molecules. Therefore, quorum quenching has a considerable contribution in the negative regulation of threatening diseases and eventually increasing soil reclamation through different mechanism mediated by microorganisms in reclamation of soil. Quorum sensing has a significant contribution in enhancement of soil quality through microbial-based enzymes and mechanism in the versatile fields which are a component of the environment. The current chapter discusses the details of various direct and indirect mechanisms mediated by microbial systems that have a significant role in soil reclamation for the sustenance of the environment.

DOI: 10.4018/978-1-7998-7062-3.ch022

INTRODUCTION

Quorum Quenching (QQ) is a method used to prevent bacterial infections by interfering the QS system between bacterial cells and preventing the expression of QS- dependent genes. In recent years, it has been found that bacteria can develop biofilm, a kind of membrane complex, to defend themselves. Bacteria produce a large amount of extracellular muco polysaccharides due to the formation of biofilms with the help of QS mechanism. Bacteria can protect themselves from the activity of antibiotics and terminate the host immune functions by establishing biofilms. The mechanism of QQ is also very effective in sustaining environment and enhancing agriculture sector by playing its role in aquaculture and crop culture (Grandclément et al., 2016).

As now the increasing population is a global concern, the demand of food supply is also increasing enormously. Many experts have stated that in coming days there would be shortage of agricultural land for the production of grains and other food essentials. Quorum quenching comes out be an effective method for treatment by restricting regulations of antibiotics (Grandclément et al., 2016). Quorum signals by microbial organisms are also disrupting the cultivation of several crops and plant pathogens and hence affecting the net production. However, in case of crop culture, QS can be degraded either by plants itself or by the use of microbial bio control agents.

QUORUM QUENCHING AND QUORUM SENSING INHIBITORS

Several studies showed that there are a number of bacterial communities which are regulating and altering the gene expressions through the mechanism of Quorum sensing. They coordinate and regulate these processes with the help of QS signals either by host-micro interaction or microbe-microbe interaction (Grandclément et al., 2016).

The Quorum sensing mechanisms leads to the emergence of Quorum Quenching. Quorum Quenching is the environmental phenomenon of clearing or recycling of QS associated organisms by different mode of actions such as competitive inhibition, QS signal cleavage, etc. using different QQ chemical compounds or enzymes. The enzymes involved in the mechanisms of QQ are called as QQ enzymes and chemicals involved are called as QQ inhibitors (QSIs). Basically, all the mechanisms involved in the disruption of Quorum sensing are called as Quorum Quenching (Dong et al., 2001). Physical parameters also have a role to play in disturbance of QS signal pathways by altering pH and temperature (Byers et al. 2002; Delalande et al. 2005).

QQ Enzymes

Many studies have showed the degradation of N-acyl-homoserine lactones (AHL) by several enzymes. AHL is the QS molecule associated with gram-negative bacteria. The very first case of degradation in AHL was by the action of soil bacterial isolates of *Bacillus* and *Variovorax* genera (Dong et al. 2000; Greenburg et al. 2004). Mainly there are four catalytic classes involved in the degradation of AHL-cytochrome oxidases which are involved in the acyl chain oxidation (Chowdhary et al., 2007). Amidohydrolases which are involved in the breakdown of AHLs' amide bond and releasing of homoserine lactone and fatty acid (Y.H. Lin 2003); reductases which are involved in the conversion of 3-oxo-substituted

AHL to 3-hydroxyl-substituted AHL (Bijtenhoorn, P. et al., 2011) and lactonases which are involved in the opening of homoserine lactone ring (Zhang et al. 2002; Uroz et al. 2008).

In case of AHL degradation there are different architecture and amino acid sequence of enzymes involved in the process. Especially, in the case of lactonases, there are four families involved in the process varying in their structure and mechanisms. They are known as α/β – hydrolase fold lactonases, phosphotriesterase – like lactonases, paraoxonases and metallo – β – lactomase-like lactonases.

Quorum Sensing Inhibitor (QSI)

The prokaryotes' or eukaryotes' molecules involved in the disrupting of Quorum signals (QS) are known as Quorum sensing inhibitors (QSIs).

QSI Identification

There are numerous approaches for the identification of Quorum sensing inhibitors (QSIs). Screening of organism is mostly opted for the identification of QSI. Screening of various organisms associated with medicinal plants, tissues, cells and chemicals are done using several bacterial QS signal biosensors (Rai et al., 2015). Quorum sensing inhibitors (QSI) also play a great role in agriculture. Dulla & Lindow, 2009 reported the reduction of pathogen influenced infection among plants by introduction of Epiphytic bacteria like *Pseudomonas*. *Pseudomonas* bacterial species were found in influencing Quorum signal pathogens and playing its role in Quorum quenching. The contribution and mechanism of microorganisms acting as Quorum sensing inhibitors to influence Quorum signals in quorum quenching is explained in Figure 1.

QUORUM QUENCHING IN MICROBIAL SYSTEM AND THEIR USE IN GREEN SUSTENANCE

Quorum quenching (QQ) processes involves the disturbance of QS by degrading the AHL molecules. The main molecules for QQ are diverse like enzymes and chemical compounds; mode of action includes QS-signal cleavage, competitive inhibition, and targets (Byers et al. 2002; Yates et al. 2002; Delalande et al. 2005). Quorum quenching strategy is used to resist plant diseases, improving the fertility of the soil, crop improvement for green sustenance. There are a range of microorganisms that act as bio fertilizers and can be used to fight bacteria. A range of microorganisms are being studied particularly plant growth-promoting bacteria (PGPB) also known as plant growth-promoting rhizobacteria (PGPR) for the utilization in sustainable agriculture. They usually act by counteracting other bacteria by inducing stress tolerance, disease resistance and are effective in sustainable and environment-friendly agriculture (Table 1).

Pseudomonas segetis and Revolution in Biocontrol

Quorum quenching (QQ) targets the attenuation of virulence and reduces infection. Quorum sensing (QS) disruption is AHL degradation that can take place by various enzymes like lactonases, acylases and oxidoreductases.

Figure 1. Mechanism of microorganisms acting as Quorum sensing inhibitors to influence Quorum Signals

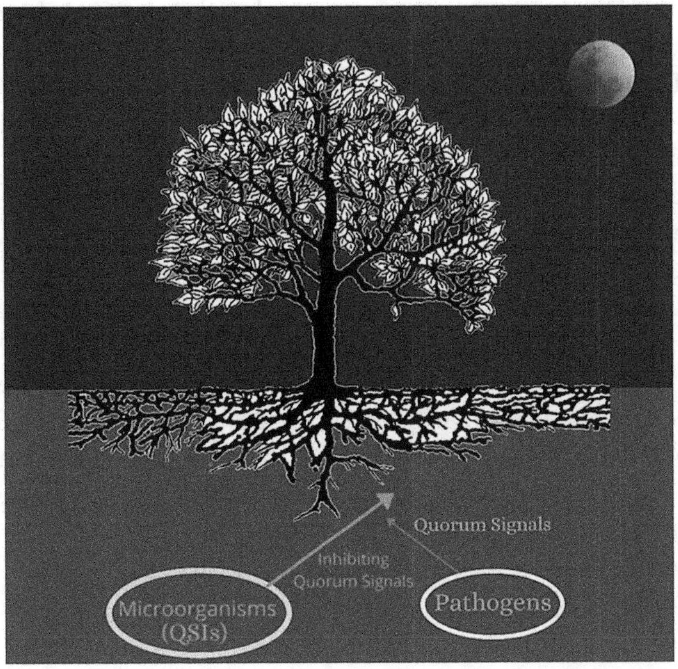

P. segetis strain P6 isolated from *Salicornia europaea* rhizosphere by Rodriguez and his colleagues in 2020 for the PGP and QQ activities of this species as a beneficial strategy to promote plant growth and to control bacterial infections. The experiments were carried out in tomato plant. Earlier no strain was studied which has both plant growth promotion and AHL degradation in combine, silencing of virulence of a bacteria phytopathogen via plant growth promoting bacterium to degrade the QS-signaling molecules commonly AHLs. This was also for the first time that this species was isolated from saline environment for studying biocontrol. This bacterium falls under PGPB which has high salt tolerance (halotolerant) and can grow in high NaCl concentrations. Due to these many reasons they are future of biotechnological applications in agriculture and possibly inducing the features to tolerate climatic stress, this salt tolerating PGPB can be used.

Rodrigues and his colleagues evaluated the feature like plant growth promotion using this strain P6 using biopriming techniques and in vivo experiments with tomato plants under sterile conditions. The results exhibited that strain P6 treated tomato plant showed increase in plant length and vigor index observed in the seeds and also it increased the weight of plants treated by P6 strain with respect to the plants not treated by them (negative control). This positive result was shown by this bacterium because of their PGP properties like acid/alkaline phosphatase and siderophore production and nitrogen fixation. Biopriming of tomato seed was done by *P. aeruginosa* and with several fluorescent *Pseudomonas* spp. and found 100% increase in the root length and 138-177% of increase in vigor index (Vaikuntapu et al. 2014; Conde et al. 2018). *P. geniculate* was added in tomato plant and observed 7% enhanced aerial weight and 9% enhanced root weight in tomato plant, on the other hand *P. fluorescens* increased 4.7% aerial dry weight (Gopalakrishnan et al. 2015; Siddiqui et al. 2001). *P. segetis* showed degradation of both long and short chain AHLs totally or partially. Rodriguez and his colleagues also did HPLC-

MRM to evaluate whether this AHL degradation in *P.segetis* is from intercellular systems, but found that this was not the case. However, they found a receptor LuxR that was enabling the bacteria to sense neighbouring exogenous/ endogenous signals and response to it. They also tested the capacity of this species to degrade AHLs produced by pathogens such as *Dickeya solani, Pectobacterium atrosepticum, P. carotovorum* subsp. *carotovorum* and *P. syringae* pv. tomato. For that they prepared co-culture of plant bacterial pathogen with P6. The silencing of bacterial virulence however could not support in all of them as P6 itself produces some degrading lactonases in the culture, but they observed affects in *D. solani* species where inhibition of caseinase, gelatinase and motility was observed and lipase was reduced by 70% (approx). Moreover, studies related to carrots and potato with P6 showed reduction in soft-rot maceration caused by *D. solani, P. atrosepticum* and *P. carotovorum* subsp. *carotovorum*. This reduced virulence is due to the activity of strain P6 quorum quenching and didn't inhibit the growth of bacteria but only silences their virulence. When an indoor greenhouse experimented was performed in tomato infected with *P. syringae* pv., tomato strain DC3000, *P. segetis* P6's QQ activities and PGP activities were evaluated and found that a significant reduction in the dead leaves and symptoms of necrosis and chlorosis in the plants and also enhanced the chlorophyll content. *Pseudomonas segetis* strain P6 is a plant growth-promoting quorum-quenching bacterium and shows a promising future and great potential as biocontrol-agent in agricultural sector (Rodriquez et al., 2020).

QQ in *Rhodococcus* Population

Cirou et al. (2011) explained that QS-degrading bacteria like *Rhodococcus* can degrade the phytopathogen like *Pectobacterium* on potato plant. They treated *Rhodococcus* with GCL (Gamma-caprolactone) in PHS that stimulated the growth of *Rhodococcus* and that is used to produce tubers in greenhouses. After this treatment, the population of NAHL-degraders reached 70% of the total cultivable bacteria and most of them were belonging to *R. erythropolis* species and it exhibited strong bio stimulation for the native QQ population. The isolates from PHS are recovered and most of the *Rhodococcus* bacteria were able to protect these potato tubers from the *Petrobacterium*. It means that the GCL treated QQ bio populations can protect potato from infection. Genes known for NAHL degradation in the population treated with this GCL are dad in *R. erythropolis* and att M in *A. tumifaciens*. GCL is found in different plants and are used in cosmetics, fragrance, and perfumes. The study found that these genes can be used as markers for detecting GCL-induced modification in rhizospheric populations and their structures (Cirou et al. 2011). GCL when introduced in hydroponic cultures from batch and PHS showed a positive impact in tissue of potato tubers. This was also a remarkable highlight from their work that after decline of the concentration of GCL after several weeks, the GCL was still impactful and showed metabolism. GCL metabolism and identification of its by-products were investigated that possibly supports the growth of *Rhodococcus* bacterial population. GCL treatment in this QQ bacterium stimulates the rapid degradation of GCL in hydroponic cultures and also stimulates the growth of bacterial population. This makes the use of *Rhodococcus* in controlling blackleg and soft rot diseases in potato plants and tubers. This biocontrol strategy can be used against the plant pathogen as studied by them and the virulence is controlled by QS (Cirou et al. 2011). They also paved a way for possible future studies related to QQ bacteria in ornamental plants, food crop plants like tomatoes and rice plants. According to the findings, gene involved in the degradation of N-AHL was identified in the screening of *Rhodococcus erythrolpolis* and was found to be the gene qsdA for quorum-sensing signal degradation. Gene qsdA encodes N-AHSL lactonase that was unrelated to aiiA and aiiD but this gene is involved in phosphotriesterases. This change the acyl-ring from C_6 to C_{14} thus degrades the AHL signals by disrupting their structure. They demonstrated that qsdA can act as a possible tool for quorum quenching procedures (Uroz & Heinonsalo, 2008).

Bacillus thuringiensis in AHL Degradation

Bacillus thuringiensis, a well-known gram-positive bacterium produces two AHL-Lactonases namely aiiA and AiiB. It was shown that *B. thuringiensis* produces lactonases that can interfere and suppress the QS signaling of plant bacterial pathogen *Erwinia carotovora* (*pectobacterium*). The aiiA expression gene in *E. carotovora* was shown to reduce virulence through signal interference, control, and prevention of infectious diseases in plants (Raddadi et al., 2007). *B. thuringiensis* produces siderophores and so is useful for biocontrol of phytopathogenic fungi due to competition for iron and provides plant with iron. Siderophore production is a common phenomenon for *B. cereus* and *B. anthracis.* It has been investigated that *B. thuringiensis* could act as biofertilizer and biostimulator that can promote plant growth. For that they tested 16 strains of *B. thuringiensis* and carried out phenotypic and PCR tests and evaluated different activities including solubilization of inorganic phosphates, phytohormones and siderophore production (Raddadi et al., 2008).

Biocontrol Potential of *B. thuringiensis* and Pathogenicity of *S. marcescens*

Various experiments were conducted on *P. heterophylla* in pots with purified isolate of *S. marcescens* that causes root rot and wilt disease on the seedlings. *S. marcescens* can be isolated from the leaves of infected heterophylla plant for the verification and is sequenced by 16S rDNA sequencing methods. This was due to the QS signals from the strains of *S. marcescens* which invaded the roots to cause root rot and wilt diseases in this plant. They also grew this plant under *B. thuringiensis* and *E. coli* DH5α to check whether these strains are pathogenic in the consecutive monoculture, as a result they did not show pathogenicity on seedlings of *heterophylla.* They evaluated effect of the isolated QQ strain *B. thuringiensis* on the pathogenicity of *S. marcescens. P. heterophylla* seedlings were treated by mixed strains of *S. marcescens* and *B. thuringiensis* in the ratio 7:3, 4:1 and 9:1 that soon developed the withered disease and died. By changing their ratios to 2:3 they showed no disease in the entire experiment in the monoculture system. This showed that *B. thuringiensis* can alleviate the disease.

Further *S. marcescens* with overexpression of aiiA gene was developed using the expression vector pBbudK. Biosensor strain CV026 produced violacein in the presence of wild *S. marcescens,* whereas excess aiiA in *S. marcescens* inhibited violacein production by CV026 implying that AHLs of the QS system were degraded in the recombinant strain of bacteria. This over expression of aiiA is used to treat *P. heterophylla* seedlings in the pots with sterilized soil. Further the seedlings of this plant grew well and showed no disease during the cultivation period (entire). The toxic effect was dependent on the SwrIR QS system of *S. marcescens* and *B. thuringiensis* secrets lactonases that degrades the QS signals and overexpression of aiiA resist the plant disease like wilt and rot (Zhang et al., 2016).

Pectobacterium atrosepticum with GCL in *Solanum tuberosum*

The use of NAHL-degrading strains for biocontrol strategy against *P. atrosepticum* causes potato blackleg. From N-acylhomoserine lactone (NAHL), gammacaprolactone (GCL), 6-caprolactone (6CL) and 4-heptanolide (HTN), 17 molecules were found that can degrade AHL which is the legend in QS-signalling. It has been found that the NAHL degrading bacteria recovered the rhizospheric soils in bulk. The consortia treated with 6CL, GCL and HTN has NHL-degrading bacteria in abundant in comparison to that of mannitol treated consortia (control). Additionally, the GCL and HTN consortia showed elevated biocontrol activity against the *P. atrosepticum* in soft rot assays as the bacteria causes maceration in potato. When GCL was provided to the culture (hydroponic) of *S. tuberosum* for several independent experiments, it was found that there was an elevation in the increase ratio of NAHL degrading bacteria among total cultivable bacteria (belonging to genera *Rhodococcus* and *Delftia*). Their work basically highlights the possibility of NAHL degrading bacteria to treat complex environment like rhizosphere. They used biosensor *C. violaceum* CV026 to detect 6CL, GCL and HTN. HPLC-MS was run and decrease in C6-HSL signal was observed. They observed that due to C6-HSL the GBL-ring was opening as it is showing a lactonase kind of activity in the consortia by disrupting the QS signals. Biocontrol activity of 6CL, HSL and HTN were detected against *P. Atrosepticum* CFBP 6276 where the virulence is regulated by legend Oc8-HSL. The consortia of 6CL, HSL and HTN were positively degrading or inactivating the legend signal (OC8-HSL) for the virulence of the *atrosepticum* bacteria and were reducing the symptom of the disease. Among these GCL and HTN were more suitable to improve the biocontrol activity of soil (Cirou et al., 2007). The treatment of *P. atrosepticum* with AHL degrading consortia positively decreases soft rot disease and maceration activity in *Solanum tuberosum*. This shows prospectus for AHL-degrading bacteria to be used for quenching the QS-signals in different phytopathogens like *atrosepticum* and degrade the virulence caused by them and also improve the soil rhizosphere.

QQ in Agrobacterium Clearing QS Signaling

Agrobacterium strains C58 and R10 have unique LuxI-LuxR systems termed as TraI-TraR which are regulated by Ti plasmid controlled by QS-signal OC8-HSL. These signals control amplification in terms of copy number increase, conjugation, virulence (Fuqua et al. 1995; Hwang et al.1994; Pappas et al. 2003; Lang et al. 2013). The opine regions in both the strains C58 and R10 i.e., agrocinopines and octopines respectively stimulated TraR gene transcription whose product dimerizes with two OC8-HSL molecules. TraR-OC8-HSL complex regulates QS genes including TraI. QS are associated in opine-rich environment such as tumor region and contribute to spreading of Ti plasmid. There are two cytoplasmic AHL-lactonases called aiiB and BlcC formerly AttM which is studied in *Agrobacterium* system (Zhang et al. 2002; Carlier et al. 2003). It has been reported that BlcC lactonase expressed by blcR mutant shows a strong delay in the Ti plasmid transfer (Zhang et al. 2002; Khan et al. 2009). All these findings suggest that the lactonases-mediated control of Ti plasmid transfer is non-permanent and lactonases only controls the acquisition of QS signals at early plant tumor development. Equally TraM anti-activator contributes to delay QS signal in the infection (Fuqua et al. 1995; Qin et al. 2007).

Several mechanisms have been proposed that were non-exclusive for answering why *A. tumefaciens* prevent QS-regulated transfer of Ti plasmid at the early stage of the infection. Kinetics of OC8-HSL-synthesis overcomes the OC8-HSL-degradation mediated by lactonase (Khan et al. 2009; Haudecoeur et al. 2009). Possible affinity of OC8-HSL for the TraR sensor would be much higher than for the lactonases (Zhu et al. 2001; Liu et al. 2007). The OC8-HSL molecule bound to TraR would be possibly protected from the activity of lactonases (Khan, S. R., & Farrand, S. K. 2009). Lactonase-encoding gene expression is controlled tightly by the plant tumor-derived compound like proline (free) that causes GABA-induced expression of BlcC-encoding gene (Haudecoeur et al., 2009). The aiiB lactonase displays high selectivity for AHLs (Liu et al., 2007), BlcC lactonase hydrolyzes other lactones like GBL and further converted to GHB. The other two genes of blc ABC operon encode SSA from GHB and SSA to succinic acid. In plant tumor GABA is gathered in high level (Chevrot et al. 2006; Deeken et al. 2006), while GBL or GHB presence in plant tumor remains unknown (Grandclément et al., 2016). GABA also secrets inducers so are naturally present in infection. It is most likely that Blc C has been selected for the degradation of plant-released compounds such as GBL and it has activities resembling aiiA for AHL degradation and affecting the tumorogenicity in plants. It does a competitive inhibition for QS and thus QQ in *Agrobacterium* clears the QS signal and helps in disease resistance.

P. aeruginosa and Infection Inhibition

Bacterial biofilms are pathogenic in nature and caused as a result when bacteria adhere to the plant surface. Bacterial biofilms are the main virulence factors dependent on quorum sensing, so it needs to be degraded for a healthy disease resistant plant from agricultural aspects for green sustenance. The quorum quenching activity in cell-free lysates was applied and studied *in vitro* for inhibition of biofilm formation in *P. aeruginosa* PAO1. The virulence factors in bacteria include this biofilm formation due to which drug resistance is stimulated and that causes several associated biofilm infections (Vallet et al., 2004). QQ enzymes as well as QSI have the potential to inhibit biofilm formation in *P. aeruginosa* species. For measuring the extent of QQ cell-free lysate in inhibition of biofilm, Rai et al. (2015) took PAO1 and PAO1-JP2 as model organisms. In results the wild type PAO1 exhibited >80% inhibition during cell-free lysate treatment. Similarly during cell-free lysate treatment of PAO1-JP2 in presence of C4-HSL exhibited greater than 50% inhibition of biofilm.

Additionally, recent reports directed that small cationic peptides (synthetic) can inhibit the biofilm formation in *P. aeruginosa* by reducing the motility of the bacteria. They prevented more than 50% biofilm formation by reducing swimming and swarming of bacterial groups (de la Fuente-Nunez et al., 2012). Furanone is capable of penetrating deeper into the biofilm matrix and infringe quorum sensing gene expression due to which inhibition of biofilm maturation results. The QQ strategy suggests the blocking of cell to cell communication and reducing the biofilm formation. This strategy is useful in the treatment of infection by drug resistant *P. aeruginosa*.

MECHANISM OF QUORUM QUENCHING

In theory, various mechanisms takes part in quorum sensing and are used in quenching quorum sensing and also in the prevention of microbial infections, some examples are mentioned in Table 2 and 3. Till today, several groups of enzymes and potent quorum-quenching chemicals has been identified which includes the halogenated furanone compounds that are produced by seaweed *Delisea pulchra* and also the derivatives (synthetic) that mainly aims to target R proteins (Givskov et al., 1996; Hentzer et al., 2003), the AIP analogues and the synthetic AHL (N-acyl homoserine lactone) that may take part with the corresponding quorum-sensing signals (Lyon et al. 2000; Smith et al., 2003) and also with the quorum-quenching enzymes which includes AHL-acylase, AHL-lactonase and paraoxonases (PONs), that degenerate AHL signals (Dong et al. 2000; Lin et al. 2003; Draganov et al. 2005; Ozer et al. 2005; Yang et al. 2005).

Table 1. Bacteria exhibiting quorum quenching activity and their contribution in green sustenance

Microbe	QQ (AHLase Detected)	Virulence Silencing	Features for Green Sustenance	References
Pseudomonas segetis	Acylase	Against *D. solani, P. atrosepticum,* biocontrol agent.	Plant growth promoter; increases plant length, root length and vigor; Siderophore production; nitrogen fixation; acid/alkaline phosphatase activity.	Rodríguez, 2000
Rhodococcus erythropolis	Lactonase, Acylase, Reductase	Against *Pectobacterium,* biocontrol agent.	Inhibits blackleg and soft rot disease in potato, positive plant growth promoter.	Cirou et al., 2011; Uroz et al., 2005; Uroz et al., 2008
Bacillus thuringiensis	Lactonase	Biocontrol againstS. *marcescens.*	Positive growth promoter; act as biofertilizer; soil improvement; solubilize inorganic phosphatase; secrets phytohormones; produces siderophores; inhibits root rot and wilt disease of potato tuber.	Raddadi et al., 2008; Zhang et al., 2016
Pectobacterium atrosepticum	Lactonase	Biocontrol against soft rot and blackleg disease when treated with GCL.	Positive plant growth promoter improves soil rhizosphere, controls potato skin maceration after treatment with GCL.	Cirou, 2007
Agrobacterium tumefaciens	Lactonase	Anti-activator and biocontrol for tumor in plants.	Tumor inactivator, anti-activator and delay QS.	Haudecoeur et al., 2009; Khan and Farrand, 2009; Zhang et al., 2002
Pseudomonas aeruginosa	Acylase	Reduces motility of bacteria including swimming, swarming; biofilm inhibition.	Infection resistance in plants.	Hentzer et al., 2003; Vallet et al., 2004

Table 2. Various key components and prospective of Quorum-Quenching strategy

Quorum-sensing Process	Mechanism	Key Component	Prospective of Quorum-Quenching Strategy
High-Population Density	(1) Signal reception (2)Activation and auto-induction of quorum-sensing regulon (3) Signal decay	LuxR-type (R) transcription factor Transcription factors dependent on quorum-sensing; I and R proteins associated in boosted production of AHL signal. AHL degradation enzyme and also its regulatory mechanisms.	R protein inhibitor Inhibitors for R and I proteins; Enzymes that degrade AHL signal Chemical persuading early expression of AHL degradation enzyme.
Low-Population Density	(4) Basal signal generation (5) Signal accumulation	LuxI-type (I) protein; Enzymes and Proteins associated in biosynthesis of S-adenosylmethionine (SAM) and acyl chain. Proteins associated in long-chain signal active efflux.	SAM biosynthesis inhibitor; fatty-acid biosynthesis inhibitor; I protein inhibitor. Active efflux inhibitor; AHL signal degradation enzyme

Mechanism Involved in AHL Signal Degradation

The AHLs molecules are found to be sensitive to alkaline pH and temperature (Yates et al., 2002; Byers et al., 2002; Delalande et al., 2005). Lactonolysis mechanism is involved which means the opening of the lactone ring, due to which AHLs are converted into QS-inactive and acyl homoserine derivatives. Degradation happens faster when the acyl side chain is shorter (Yates et al., 2002). This mechanism plays an important role in the regulation of their aggregation in plant or microbial environment. For example, in case of *Pectobacterium carotovorum* pv. *carotovorum* (Pcc), QS regulates the appearance of bacterial pathogenicity functions with the help of signal OHHL (N-3-oxo-hexanoyl homoserine lactone). During stationary phase, in the culture supernatant of the plant pathogen, this signal degradation has been registered (Byers et al., 2002).

It corresponds with the alkalinisation of medium which leads in microbial metabolism. Specially, plant defence mechanism involved the proton pump activation which induced on Pcc infection; it results in the alkalinisation of strong medium at the site of infection (Nachin et al., 2000). Degradation of pH mediated AHLs also takes place in different complex environments which are known as biofilm covering marine stromatolites (Decho, A., 2009) (Table 3). A Wide range of organisms are capable of degrading AHL molecules. At the first, in bacteria, degradation of AHLs was found (Cirou, A., 2007). Next, Basidiomycota and Ascomycota divisions of fungi were also found to degrade AHLs, whose feature is homologous to eukaryotic organisms. AHLs can also be degraded by various plants from the clade of legume such as *Trifolium pretense, Lotus corniculatus, Pachyrhizuserosus* (Delalande et al., 2005; Gotz et al., 2007; Uroz et al., 2008), as well as by porcine kidney (Xu et al., 2003; Yang et al., 2005). Although, organisms that are able to degrade AHLs should be generally in contact with the large communities of AHL-producing bacteria.

The stronger activity of AHL degradation is found in the root system of legume plants as compared to its aerial parts. Certainly, the aerial parts hold population of bacteria about one thousand times less dense as compared to root system. In degradation of AHL, various enzymatic activities are involved which are together called as AHLases; comes under three classes such as acylases (amidohydrolases), lactone hydrolases, and reductases or oxidases.

Mechanism of AHL Lactonase

AHL-lactonases have been identified from bacterial species that hydrolyse the homoserine lactone ring of AHL signals (Dong et al., 2005; Zhang et al., 2007). This enzyme carries a 'His104-X-His106-X-Asp108-His109' design due to which is considered as a member of metallo-hydrolase superfamily (Dong, Y., 2000). The 'His106-X-Asp108-His109K59X-His169-21X-Asp191' motif has been formulated by the site-directed mutagenesis based on sequence alignment of the aiiA homologues; this is required for the activity of AHL-lactonase enzyme (Dong et al. 2000).

Active site of AHL-lactonase contains two zinc ions (Kim et al. 2005; Liu et al. 2005) and by biochemical analysis, it has been proved that AHL-lactonase is a metalloprotein (Thomas, P., 2005). A number of ligands are coordinated to two zinc ions that include His104, His106, His169, His235 and Asp108, also single oxygen of a bridging carboxylate from Asp191 and a hydroxide ion/bridging water. All residues are totally maintained in AHL-lactonases.

Table 3. Organisms involved in AHL degradation

Belonging Class	Belonging Genus and Species	Detected AHLase	Genetic Determinant	References
Alphaproteobacteria	*Agrobacterium radiobacter*	Lactonase	Not defined	Uroz & Heinonsalo, 2008
	Sphingopyxis sp.	Not defined	Not defined	D'Angelo-Picard, 2005
	Bosea sp.	Not defined	Not defined	D'Angelo-Picard, 2005
	Sphingomonas sp.	Not defined	Not defined	D'Angelo-Picard, 2005
	Agrobacterium tumefaciens	Lactonase	*att*M, *aii*B	Carlier et al., 2003; Zhang et al., 2002
	Ochrobactrum sp.	Not defined	Not defined	Jafra, 2006
Firmicute	*Bacillus megaterium*	Oxidase	Not defined	Chowdhary, 2007
	Arthrobacter sp	Lactonase	*ahl*D	Park, 2003
	Bacillus spp.	Lactonase	*aiiA*	Dong, 2002
Gammaproteobacteria	*Pseudomonas aeruginosa*	Acylase	*qui*P	Huang, 2006
	Unknown (soil metagenome)	Lactonase	Not defined	Schipper, 2009
	Acinetobacter	Not defined	Not defined	Kang, 2004
	Shewanella sp.	Acylase	*aac*	Morohoshi, 2008
	Klebsiella pneumoniae	Lactonase	*ahl*K	Park, 2003
	Pseudomonas sp.	Acylase	Not defined	Huang, 2003
Acidobacteria	Not defined	Lactonase	*qlc*A	Riaz, 2008
Betaproteobacteria	*Variovoraxpardoxus*	Acylase	Not defined	Leadbetter et al., 2000; Chun et al., 2004
	Comamonas sp.	Acylase	Not defined	Uroz, 2007
	Ralstonia sp.	Acylase	*aii*D	Lin, 2003
	Delftiaacidovorans	Not defined	Not defined	Jafra, 2006
Actinobacteria	*Rhodococcus* spp.	Lactonase	*qsd*A	Park et al.,2006; Uroz et al., 2008
	Rhodococcus erythropolis	Reductase; Acylase	Not defined	Uroz et al., 2005
	Streptomyces sp.	Acylase	*ahl*M	Park, 2006

The substitution of His104 with serine is non-essential for the activity of AiiA240B1 (Dong, Y. et al, 2000), however on the basis of structural analysis it was found that the substitution with alanine is critical for aiiABTK (Kim, et al., 2005). Now, it has been observed that the replacement of His104 with alanine in aiiA240B1 ends the activity of enzyme. So, it is unlike that, aiiA240B1 could not be a metalloprotein (Wang et al., 2004), and the enzyme might also contain zinc ions. aiiA240B1 and aiiABTK shares high 90% amino acid identity (Dong et al., 2000; Kim et al., 2005).

By the crystal structure analysis of AHL-lactonase, it has been revealed that there is an ab/ba sandwich-fold in overall structure which contains two zinc ions in its active sites (Kim et al., 2005; Liu et al., 2005). These structural features are little bit similar to RNase Z proteins and glyoxalase II these are the members of metallo-b-lactamase superfamily (Cameron et al., 1999).

A catalytic mechanism of AHL-lactonase has been introduced on the basis of 3-D structures (three-dimensional structure) of AHL-lactonase, can or cannot be L-homoserine lactone included, and the reaction mechanism of RNase Z (Kim et al., 2005; Li de la Sierra-Gallay, I. et al., 2005) and binuclear metal-binding glyoxalase II (Cameron et al., 1999). A nucleophilic hydroxide ion /bridging water attack the substrate's carbonyl carbon. A carbonyl oxygen and lactone ring of AHL interacts with Zn2 and Zn1 ion, respectively, results in increasing the polarization of carbonyl bond that makes it more susceptible to a nucleophilic attack. After nucleophilic attack, there is a formation of a negatively charged intermediate which can be primarily stabilized by the interactions with Zn1 ion. Then, to form the ring-opened product, C-O bond of the lactone ring of AHL breaks itself. Here, Tyr194 behave as a normal acid for protonation of leaving group. Inverse of Paraoxonase (PON) enzymes and AHL-acylase that have changing substrate spectra, AHL-lactonase is the most specific AHL-degradation enzyme. Both short and long-chain hydrolyses by AHL-lactonase with the same efficiency and it shows little or no residue activity to other chemicals (Wang et al., 2004).

Mechanism of AHL Acylase

Various bacterial species such as *Variovorax paradoxus, Streptomyces* sp. and many more has been found that encodes AHL-acylase for degradation of AHL signals (Leadbetter &Greenberg. 2000; Huang et al., 2003; Lin et al., 2003; Park et al., 2005) (Table 4). There are mainly three identified AHL-acylases that is, AhlM from *Streptomyces* sp., PvdQ from *P. aeruginosa* PAO1, and aiiD from *Ralstonia* sp. XJ12B which shares several characteristics of Ntn hydrolases, including a signal peptide followed by an alpha subunit, spacer sequence and beta subunit (Park at al., 2005; Huang at al., 2003; Lin et al., 2003; Hewitt et al., 2000). Although, some differences also found in the substrate specificities among AHL-acylases. aiiD successfully degrades short-chain AHLs and also long-chain AHLs but with less efficiency (Lin at al., 2003).

PvdQ is not able to degrade AHLs whose acyl chains are shorter than eight carbons (Huang at al., 2003). Like this only, in degrading AHLs which are shorter than eight carbons, AhlM shows residue activity (Park at al., 2005). Moreover, aiiD is not able to degrade ampicillin and penicillin G, albeit AhlM catalyses the hydrolysis of penicillin G, proposing a wider substrate specificity (Lin et al. 2003). All these mentioned AHL-acylases shares similar structure with the cephalosporin acylase (CAD) from Pseudomonas diminuta (Lin et al. 2003; Park et al. 2005).

Amazingly, these acylases have non-identical residues in two corresponding positions shown by the sequence alignment of the three AHL-acylases with CAD (Leu50 and Asp57 in PvdQ, Leu50 and Ser57 in AhlM and Ile50 and Ser57 in aiiD). In addition, crystal structure analysis and mutagenesis of these AHL-acylase would be very difficult for explaining the molecular mechanism involving in substrate specificity and catalysis.

Table 4. Some examples of quorum-quenching molecules against microbial infections

Molecules of Quorum-quenching	Host	Effects	Reference
AHL-acylase *aiiD*	*P. aeruginosa*	decreases ability of swarming, attenuates nematode paralysation, and production of pyocyanin and elastase.	Lin et al., 2003
3-oxo-C12-(2-aminocyclohexanone)	*P. aeruginosa*	Reduction in production of biofilm formation and virulence factors.	Smith et al., 2003
DSF	*Candida albicans*	Inhibition of fungal dimorphic transition that is involved with virulence	Wang et al., 2004
Furanone	mouse	attenuates the virulence of *P. aeruginosa* in mouse models	Hentzer et al., 2003
Synthetic AIP-II	mouse	resistance to *S. aureus* infection is shown by treated mice.	Mayville, et al., 1999
AHL-lactonase *aii*A, *att*M, *aii*B	*Erwinia carotovora*	attenuates soft rot symptom on inoculated plants, and decreases extracellular pectolytic enzyme activities.	Dong et al., 2000
	Pseudomonas aeruginosa	decreases production of rhamnolipids, elastase, pyocyanin and hydrogen cyanide, and also inhibits bacterial swarming.	Reimmann et al., 2002
	Burkholderia thailandensis	prevents the b-haemolysis of sheep erythrocytes, and reduces the bacterial twitching motility and swarming.	Ulrich, 2004
	Escherichia coli	attenuates the pathogenicity of E. carotovora when co-inoculated	Lee et al., 2002
	Erwinia amylovora	tolerance to hydrogen peroxide and impairs extracellular polysaccharide production, and reduces the fire blight symptom on apple leaves	Molina et al., 2005
	Bacillus thuringiensis	On AHL-lactonase, the efficiency of biocontrol against *E. carotovora* infection is dependent	Dong et al., 2004
	Erwinia carotovora subsp. *atroseptica*	decreases maceration in potato tubers	Carlier at al ., 2003

Mechanism of QQ-Bacteria

There are various techniques which can be applied to increase the percentage of QQ-bacteria in the soil by the method of in situ method which helps in increasing the crop protection very first technique is by introducing selected signal bacteria (degrading bacteria) in soil; second, by introducing biodegradable compounds which promotes the community of QQ-bacteria for its growth residing in soils; and third, last but not least, by amalgam of both (Faure & Dessaux. 2007).

It has been observed that, biodegradable compounds are helpful in promoting the growth of QQ-bacteria in the rhizosphere of *Solanumtuberosum* when they are grown in hydroponic conditions (Cirou et al., 2012). Gamma-heptanolactone (GHL) and gamma-caprolactone (GCL or gamma-hexanolactone) are two investigated compound shows a short aliphatic carbon chain and a gamma-butyrolactone ring. When tried under hydroponic culture of *S. tuberosum*, they promote the growth of QQ-bacteria, specially, population of *Delftia* and *Rhodococcus* that can use GHL (Gamma-heptanolactone) and GCL (Gamma-caprolactone or gamma-hexanolactone) as a carbon source. So, use of GHL and GCL are referred as environmental friendly as they can be easily biodegraded by the community of bacteria they promote. These molecules can also be used in food industry as a flavouring agent as they are considered as very low or nontoxic compounds. These features build them acceptable compounds to expand sustainable, ecological and disease control procedures in the field.

QUORUM QUENCHING ENZYMES

Quorum quenching can be approached either by using small molecules for the inhibition, production, transportation and detection of quorum quenching signals or by the enzymes for the degradation of quorum sensing signalling molecules (Whitehead et al., 2001). Enzymes involved in the degradation of quorum quenching signalling molecules are known as quorum quenching enzymes. Quorum quenching enzymes can be used as potential antimicrobials for targeting pathogenic bacteria. Two types of quorum quenching enzymes which are being studied widely are lactonases and acylases. These enzymes target acyl-homoserine lactonase (AHLs). AHLs are produced by Gram-negative bacteria which are a predominant class of quorum sensing signals. The variation in the length of Acyl side chain and substituents of AHLs dedicate the specificity of the signal. Lactonases inactivate both short and long chain AHLs by the hydrolization of the ester bond of the lactone ring to yield acyl-homoserine whereas acylase is effective against AHLs with side chains longer than 10 carbon atoms. Both types of enzymes inactivate the AHL signalling molecule, however, acylase reaction is irreversible only. For the production of virulence factor both plants and human bacterial pathogens such as *Erwinia sterwatii, Pseudomonas aeruginosa* etc. depends on AHL quorum sensing signal. A strategy could be developed for the control of bacterial infection by the degradation of AHL signals produced by bacterial pathogen by concentrating an AHL signal which is a key factor in mediating the virulence gene expression. AHL lactonase isolated from a soil bacteria belonging to a Gram-positive *Bacillus* species is the first quorum quenching enzyme encoded by gene *aiiA* (Dong et al., 2000). Leadbetter and Greenberg reported a strain of *Variovorax paradoxus* (VAI-C) which was capable of using AHL molecule as the sole source of energy and nitrogen. The presence of homoserine lactone in the AHL metabolic mixture of V.*paradoxus* VAI-C suggest that the gene encoding for AHL-acylase should remain cloned and characterized for the production of AHL-acylase. In at least 10 bacterial species the quorum quenching activity has been demonstrated and documented and in most of the cases the corresponding genes encoding the AHL-degradation enzymes have been cloned and demonstrated. Taxonomically these organisms belong to 3 phyla of bacteria kingdom which are Proteobacteria, Actinobacteria and Firmicutes. This diverse distribution suggests the conservation of genes encoding AHL-degradation enzymes among many prokaryotic organisms. The sequence variation of AHL-degradation enzymes produced by the bacterial species is also mirrored by the taxonomical diversity of these bacterial species (Table 5). There are two cluster of prokaryotic AHL-lactonase which includes aiiA cluster and attM cluster (Dong et al., 2000; Dong et al., 2002; Lee et al., 2002; Reimmann et al., 2002; Ulrich, 2004). AHL-lactonases from *Bacillus* species that share more than 90% of peptide sequence identities comes under cluster aiiA whereas enzymes from *A.tumefaciens, Klebsiella pneumonia* and *Arthrobacter sp.* which share 39-58% homology in their peptide sequences comes under cluster Att M (Zhang et al., 2002).

Table 5. AHL degradation enzymes in some prokaryotes

Species	Gene	Enzymes	Reference
Bacillus sp. 240BI	*aiiA* gene	AHL lactonase	Dong et al., 2000
B. anthracis	*aiiA*homologues	AHL lactonase	Ulrich, 2004
B. cereus	*aiiA*homologues	AHL lactonase	Dong et al., 2002; Reimmann et al., 2002
B. thuringiensis	*aiiA*homologues	AHL lactonase	Dong et al., 2002; Lee et al., 2002
B. mycoides	*aiiA homologues*	AHL lactonase	Dong et al., 2002
*Ralstonia*strain XJI2B	*aiiD*	AHL acylase	Lin et al., 2003; Hu et al., 2003

BIOTECHNOLOGICAL APPLICATIONS OF QS INHIBITORS IN SOIL RECLAMATION

Byers et al. (2002) successfully terminated the bacterial QS system and further impeded the formation of biofilm by intervening with AI signal molecules, which provided a breakthrough for biological control of disease induced by biofilm formation. QQ can show its effect by interfering at different stages of the QS pathway, which generally includes four mechanisms: 1) By inhibiting the synthesis of signal molecules 2) By inhibiting the transportation of signal molecules 3) Degrading signal molecules by chemical or Biological methods 4) Inhibiting the combination of signal molecules and receptor. QQ is considered a promising biological management strategy and is predicted to become a new approach for the antibacterial drug treatment and biological management. QQ does not kill pathogens and only minimize the production of drug resistance (Von Bodman et al., 2008). Therefore, QQ is considered as a promising biological control strategy, and is expected to become a modern approach for treatment of antibacterial and biological control. Generally, QQ activity can be classified into two categories: small molecule QS inhibitors (QSIs) and macromolecule QQ substances both can affect the bacterial QS system independently.

With the sudden growth of the population, the worldwide demand for food and agricultural products is increasing at a rapid rate. Though, plant pathogens inflict enormous economic expenses to agriculture every year. Traditionally, antibiotics are identified as powerful agents to regulate bacterial pathogens. However, the extensive use of antibiotics has developed a sequence of difficulties, such as environmental deterioration, ecological equilibrium devastation, and drug resistance. Therefore, more and more attentions have been paid to biological control of plant diseases. QQ can effectively control plant diseases by regulating the expression of genes related to plant pathogens to enhance the efficiency of agricultural production. For these advantages, QQ is considered to be a possible opportunity or interrelated strategy for antibiotics. Associations between plant and bacteria are well known and prove helpful to each other. Bacteria living as epiphytes – *Pseudomonas*, *Pantoea* and *Erwinia* seem to help the plant by manipulating the QS behaviour of plant pathogens. Premature induction of QS can allow the host to activate its defence mechanisms. Epiphytic bacteria having an inherent genetic make-up to produce 10-fold higher quantities of QS signal (acyl homoserine lactones [AHLs]) – 3OC6-HSL – caused *Pseudomonas syringae* to prematurely induce its quorum sensing system (QSS). It allows tobacco plants to become resistant to pathogenic attack of P. syringae (Quinones et al., 2004; Dulla G.F. & Lindow.2009). Dong et al., 2000 transferred the plasmid carrying aiiA gene into *Erwinia carotovora* strain SCG1 and found that the expression of aiiA could interfere with QS system and inhibit the production of virulence factors. Several plants including Chinese cabbage, eggplant and potatoes were infected with the recombinant pathogens without getting soft rot symptoms. This is the first application of QQ in the area of biological disease control. *Pseudomonas aureofaciens* is a symbiotic bacterium that can regulate the production of phenazine antibiotics by AHLs-mediated QS system. At the same time, it can protect wheat against *Gaeumannomyces graminis* var. tritici and improve the resistance of wheat to fungal infection. The exchange of signals between bacteria allows them to coordinate many different physiological activities. In legume rhizobia, the establishment and regulation of the symbiotic interactions between nitrogen-fixing bacteria and plant hosts are closely related to the QS system. This symbiotic interaction can enhance the nitrogen fixation by stimulating the QS system in these bacteria and reducing the demand for fertilizer and financial investment for crop hosts (Table 6). It can protect the environment and maintain the ecological balance (Cao et al., 2009).

Table 6. Summary of QQ application in the field of agriculture

Application	Quencher	Reference
Reduction in pathogenicity of *Pseudomonas Syringae* on tobacco plant	Erwinia, Pantoea and Pseudomonas (epiphytic bacteria)	Quinones et al., 2004; Dulla & Lindow, 2009
Reduce maceration in potato	AHL- Lactonase genes: att M, aii B	Carlier et al., 2003
Reduced infection on plants: Cabbage and tobacco	Recombinant *Erwinia carotovora*	Dong et al., 2000
Transgenic lines of tobacco and potato	Aii Alactonase from *Bacillus* sp. Att Mlactonase from *A. tumefaciens*	Dong et al., 2001 D'Angelo-Picard et al., 2011

Bacterial plant microbes depend on smart regulatory networks to synchronize the infection method and induce specific virulence factors once involved with the host plant. Besides the perception of plant signals or nutrient accessibility, QS plays an important role within the initiation of the pathogenic cycle. Thus, QQ methods are currently thought of as great alternatives or complementary methods to the employment of pesticides (Mole et al., 2007). Different QS signalling molecule is generated, depending on the bacterial microbes such as AHL's, Al-2, 3-hydroxy palmitate methyl ester (3-OH-PAME) and diffusible signal factors. Most of these signals can be degraded by QQ enzymes (Table 7).

Table 7. Summary of QS signalling molecules produced by microbes and QQ enzymes involved in the degradation of signals

QS Signaling	Microbes	QQ Enzymes Involved in the Degradation of Signal Produced by Soil Microbes	Reference
AHL'S	*Agrobacterium tumefaciens, Dickeya* spp., *Erwinia* spp., *Pantoea* spp., *Pectobacterium* spp. and *P. syringae*	Lactonases or acylases produced by soil bacteria such as *A. tumefaciens* or *Bacillus* sp.	Shinohara et al., 2007; Newman et al., 2008
Al-2	*Erwinia* spp., *Pantoea* spp., *Pectobacterium* spp.	Lactonases	Shinohara et al., 2007; Newman et al., 2008
3-hydroxy palmitate methyl ester (3-OH- PAME)	*R. solanacearum*	Esterase produced by the soil bacterium *Ideonella* sp.	Shinohara et al., 2007; Newman et al., 2008
DSF (Diffusible signal factors) Family	*Xanthomonas* spp., *Xylellafastidiosa*	CarAB(a Carbamoyl phosphate synthetase produced by several *Pseudomonas* spp.	Shinohara et al., 2007; Newman et al., 2008
Aac	*Shewanella* sp. MIBO15	Acylases	Morohoshi et al., 2008
aiiC	*Anabaeba* sp. PCC7120	Acylases	Romero et al., 2008

Some soil bacteria such as A. tumefaciens or Bacillus sp. naturally produce lactonases to degrade AHL's signals (Dong et al., 2000; Carlier et al., 2003; Zhang et al., 2002). *Bacillus thuringiensis* was shown to produce lactonase enzyme called as aiiA, which degrades the QS signal, AHL's produced by *Pectobacterium carotovorum*, thereby reducing its pathogenicity on potatoes slices (Dong et al., 2004). *Bacillus sonorensis* isolated from the fermentation brineof Chinese soy sauce has the capability to destroy AHL signal (Yin et al., 2012). There are many Bacillus species which have the capacity to produce lactonase enzyme similar to that of *Bacillus marcorestinctum* (Han et al., 2010) and B. licheniformis (Mani et al., 2012).

In order to boost the potency of the *Bacillus thuringiensis* lactonase aiiA, a fusion with a secretive protein was developed to expand the dispersion of the lactonase within the surrounding, leading to show increased tolerance to *Pectobacterium carotovorum*on potato (Zhang et al., 2007). Another QQ method was also tested against bacterial plant pathogens: some plants were genetically modified using bacterial genes from *Bacillus* spp. or *A. tumefaciens* to produce lactonases. The first transgenic lines were reported in 2001, transforming tobacco and potato lines with the aiiA gene from *Bacillus*. The resulting transgenic lines showed an increased tolerance to *P. carotovorum* with symptoms only appearing after inoculation with very high bacterial concentrations (Dong et al., 2001). These results showed that QQ has been used as a successful approach to protect plants from bacterial pathogens in laboratory conditions. However, this demonstration was only achieved using plant GMO producing lactonases. QQ enzymes that may be used to treat and protect plants from bacterial infections is an attractive alternative to genetically modified plants but is however impaired by the poor stability of enzymes. To circumvent this issue, the development of environmentally stable and chemical-resistant enzymes is crucial. There is some drawback in the use of QQ methods for pest control could be the impact on symbiotic bacteria that are naturally present in the environment. The ecological impact of tobacco lines expressing the lactonase AttM from *A. tumefaciens* was shown to be minimal, as no major difference was recorded between the root microbiota of transgenic and WT tobacco lines (D'Angelo-Picard et al., 2011). However, if the bacterial populations were not impacted, some functions of bacteria using AHL mediated QS might have been altered. Nitrogen fixing plants like *Medicago truncatula* and *Pisum sativum* are known to exude chemical compounds which act as QS mimics in response to bacterial infections. These compounds give protection to plants against microbes by altering their QS-regulated expression of genes responsible for their virulence nature (Teplitski et al., 2000; Cirou et al., 2012; LaSarre et al., 2013). For agricultural purposes, these genetic tools can be used to control bacterial infection and, consequently, help in achieving higher crop yield.

SUMMARY

Quorum sensing is used by different microbial systems as discussed in *E. coli*, *A. fischeri*, *P. aeruginosa*, *A. baumannii*, *P. syringae* to communicate intracellularly and intercellularly. In the acyl-homoserine lactone (AHL)-dependent quorum sensing systems, the quorum sensing signal is detected by a transcription factor which is cytosolic, whereas the quorum-sensing signal auto inducing peptide (AIP) is detected by two component response regulatory system which is typically membrane associated. By quorum, the bacteria exhibit different virulence factors like biofilm formation and motility. Different bacteria have different pathway to control the gene regulation. There is an emergent need to develop the inhibition pathways for quorum sensing so that the virulence can be eliminated from either the bacteria or from the specific host like plants and animals to induce disease resistance.

The inhibition of quorum sensing signal mainly AHLs is termed as Quorum quenching. Quorum quenching is the disruption in the signal molecules AHL or AI-2. The main legends used for this disruption are quorum quenching enzymes that possible alters the gene which produces the signals in different bacteria. Bacteria like *P. segetis, B. thuringiensis, P. atrosepticum, A. tumefaciens, P. aeruginosa* are mainly used to degrade the AHL signals either by possible induced monocultures or by directly degrading their QS-signalling molecules. QQ is used in different domains to develop antibacterial and anti-disease strategies targeting pathogens. Development in QS disruption has applications in sustenance like crop treatment, soil fertility, improving rhizosphere and disease resistance for revolution in green sustenance.

CONCLUSION

This chapter deals with Quorum quenching; its mechanisms; quorum sensing and the various microbial systems that uses quorum sensing for major virulence factors, pathogenicity, tumours, bio-fouling etc.; possible enzymes for quorum quenching mechanism help the sustenance of different fields like crop improvement and vice versa in enhancement of soil reclamation process through quorum quenching.

REFERENCES

Bijtenhoorn, P., Schipper, C., Hornung, C., Quitschau, M., Grond, S., Weiland, N., & Streit, W. R. (2011). BpiB05: A novel metagenome-derived hydrolase acting on N-acylhomoserine lactones. *Journal of Biotechnology*, *155*(1), 86–94. doi:10.1016/j.jbiotec.2010.12.016 PMID:21215778

Byers, J. T., Lucas, C., Salmond, G. P., & Welch, M. (2002). Nonenzymatic turnover of an Erwinia carotovora quorum-sensing signaling molecule. *Journal of Bacteriology, 184*(4), 1163–1171. doi:10.1128/jb.184.4.1163-1171.2002 PMID:11807077

Cameron, A. D., Ridderström, M., Olin, B., & Mannervik, B. (1999). Crystal structure of human glyoxalase II and its complex with a glutathione thiolester substrate analogue. *Structure, 7*(9), 1067–1078. . doi:10.10160969-2126(99)80174-9

Carlier, A., Uroz, S., Smadja, B., Fray, R., Latour, X., Dessaux, Y., & Faure, D. (2003). The Ti plasmid of Agrobacterium tumefaciens harbors an attM-paralogous gene, aiiB, also encoding N-Acyl homoserine lactonase activity. *Applied and Environmental Microbiology, 69*(8), 4989–4993. doi:10.1128/AEM.69.8.4989-4993.2003 PMID:12902298

Chevrot, R., Rosen, R., Haudecoeur, E., Cirou, A., Shelp, B. J., Ron, E., & Faure, D. (2006). GABA controls the level of quorum-sensing signal in Agrobacterium tumefaciens. *Proceedings of the National Academy of Sciences of the United States of America, 103*(19), 7460–7464. doi:10.1073/pnas.0600313103 PMID:16645034

Chowdhary, P. K., Keshavan, N., Nguyen, H. Q., Peterson, J. A., González, J. E., & Haines, D. C. (2007). Bacillus megaterium CYP102A1 oxidation of acyl homoserine lactones and acyl homoserines. *Biochemistry, 46*(50), 14429–14437. doi:10.1021/bi701945j PMID:18020460

Chun, C. K., Ozer, E. A., Welsh, M. J., Zabner, J., & Greenberg, E. P. (2004). Inactivation of a Pseudomonas aeruginosa quorum-sensing signal by human airway epithelia. *Proceedings of the National Academy of Sciences of the United States of America, 101*(10), 3587–3590. doi:10.1073/pnas.0308750101 PMID:14970327

Cirou, A., Diallo, S., Kurt, C., Latour, X., & Faure, D. (2007). Growth promotion of quorum-quenching bacteria in the rhizosphere of Solanum tuberosum. *Environmental Microbiology, 9*(6), 1511–1522. doi:10.1111/j.1462-2920.2007.01270.x PMID:17504488

Cirou, A., Diallo, S., Kurt, C., Latour, X., & Faure, D. (2007). Growth promotion of quorum-quenching bacteria in the rhizosphere of Solanum tuberosum. *Environmental Microbiology, 9*(6), 1511–1522. doi:10.1111/j.1462-2920.2007.01270.x PMID:17504488

Cirou, A., Mondy, S., An, S., Charrier, A., Sarrazin, A., Thoison, O., DuBow, M., & Faure, D. (2012). Efficient bio stimulation of native and introduced quorum-quenching Rhodococcus erythropolis populations is revealed by a combination of analytical chemistry, microbiology, and pyrosequencing. *Applied and Environmental Microbiology, 78*(2), 481–492. doi:10.1128/AEM.06159-11 PMID:22081576

Cirou, A., Raffoux, A., Diallo, S., Latour, X., Dessaux, Y., & Faure, D. (2011). Gamma- caprolactone stimulates growth of quorum-quenching Rhodococcus populations in a large-scale hydroponic system for culturing Solanum tuberosum. *Research in Microbiology, 162*(9), 945–950. doi:10.1016/j.resmic.2011.01.010 PMID:21288487

Conde, M. I. R., Ocampo, S. A., Castañeda, G. C., Hernández, A. R. G., & Aguilar, G. M. B. (2018). Efect of fuorescent Pseudomonas on tomato seed germination and seedling vigor. *Revista Chapingo Serie Horticultura, 24*, 121–131. doi:10.5154/r.rchsh.2017.06.023

D'Angelo-Picard, C., Chapelle, E., Ratet, P., Faure, D., & Dessaux, Y. (2011). Transgenic plants expressing the quorum quenching lactonase AttM do not significantly alter root-associated bacterial populations. *Research in Microbiology*, *162*(9), 951–958. doi:10.1016/j.resmic.2011.01.009 PMID:21315818

D'Angelo-Picard, C., Faure, D., Penot, I., & Dessaux, Y. (2005). Diversity of N-acyl homoserine lactone-producing and -degrading bacteria in soil and tobacco rhizosphere. *Environmental Microbiology*, *7*(11), 1796–1808. doi:10.1111/j.1462-2920.2005.00886.x PMID:16232294

de la Fuente-Núñez, C., Korolik, V., Bains, M., Nguyen, U., Breidenstein, E. B., Horsman, S., Lewenza, S., Burrows, L., & Hancock, R. E. (2012). Inhibition of bacterial biofilm formation and swarming motility by a small synthetic cationic peptide. *Antimicrobial Agents and Chemotherapy*, *56*(5), 2696–2704. doi:10.1128/AAC.00064-12 PMID:22354291

Decho, A. W., Visscher, P. T., Ferry, J., Kawaguchi, T., He, L., Przekop, K. M., Norman, R. S., & Reid, R. P. (2009). Autoinducers extracted from microbial mats reveal a surprising diversity of N-acylhomoserine lactones (AHLs) and abundance changes that may relate to diel pH. *Environmental Microbiology*, *11*(2), 409–420. doi:10.1111/j.1462-2920.2008.01780.x PMID:19196272

Deeken, R., Engelmann, J. C., Efetova, M., Czirjak, T., Müller, T., Kaiser, W. M., Tietz, O., Krischke, M., Mueller, M. J., Palme, K., Dandekar, T., & Hedrich, R. (2006). An integrated view of gene expression and solute profiles of Arabidopsis tumors: A genome-wide approach. *The Plant Cell*, *18*(12), 3617–3634. doi:10.1105/tpc.106.044743 PMID:17172353

Delalande, L., Faure, D., Raffoux, A., Uroz, S., D'Angelo-Picard, C., Elasri, M., Carlier, A., Berruyer, R., Petit, A., Williams, P., & Dessaux, Y. (2005). N-hexanoyl-L-homoserine lactone, a mediator of bacterial quorum-sensing regulation, exhibits plant-dependent stability and may be inactivated by germinating Lotus corniculatus seedlings. *FEMS Microbiology Ecology*, *52*(1), 13–20. doi:10.1016/j.femsec.2004.10.005 PMID:16329888

Dong, Y. H., Gusti, A. R., Zhang, Q., Xu, J. L., & Zhang, L. H. (2002). Identification of quorum-quenching N-acyl homoserine lactonases from Bacillus species. *Applied and Environmental Microbiology*, *68*(4), 1754–1759. doi:10.1128/AEM.68.4.1754-1759.2002 PMID:11916693

Dong, Y. H., Wang, L. H., Xu, J. L., Zhang, H. B., Zhang, X. F., & Zhang, L. H. (2001). Quenching quorum-sensing-dependent bacterial infection by an N-acyl homoserine lactonase. *Nature*, *411*(6839), 813–817. doi:10.1038/35081101 PMID:11459062

Dong, Y. H., Xu, J. L., Li, X. Z., & Zhang, L. H. (2000). aiiA, an enzyme that inactivates the acylhomoserine lactone quorum-sensing signal and attenuates the virulence of Erwinia carotovora. *Proceedings of the National Academy of Sciences of the United States of America*, *97*(7), 3526–3531. doi:10.1073/pnas.97.7.3526 PMID:10716724

Dong, Y. H., & Zhang, L. H. (2005). Quorum sensing and quorum-quenching enzymes. *Journal of Microbiology (Seoul, Korea)*, *43*(Spec No), 101–109. PMID:15765063

Dong, Y. H., Zhang, X. F., Xu, J. L., & Zhang, L. H. (2004). Insecticidal Bacillus thuringiensis silences Erwinia carotovora virulence by a new form of microbial antagonism, signal interference. *Applied and Environmental Microbiology*, *70*(2), 954–960. doi:10.1128/AEM.70.2.954-960.2004 PMID:14766576

Draganov, D. I., Teiber, J. F., Speelman, A., Osawa, Y., Sunahara, R., & La Du, B. N. (2005). Human paraoxonases (PON1, PON2, and PON3) are lactonases with overlapping and distinct substrate specificities. *Journal of Lipid Research*, *46*(6), 1239–1247. doi:10.1194/jlr.M400511-JLR200 PMID:15772423

Dulla, G. F., & Lindow, S. E. (2009). Acyl-homoserine lactone-mediated cross talk among epiphytic bacteria modulates behavior of Pseudomonas syringae on leaves. *The ISME Journal*, *3*(7), 825–834. doi:10.1038/ismej.2009.30 PMID:19340082

Faure, D., & Dessaux, Y. (2007). Quorum sensing as a target for developing control strategies for the plant pathogen *Pectobacterium. European Journal of Plant Pathology*, *119*(3), 353–365. doi:10.100710658-007-9149-1

Fuqua, C., Burbea, M., & Winans, S. C. (1995). Activity of the Agrobacterium Ti plasmid conjugal transfer regulator TraR is inhibited by the product of the traM gene. *Journal of Bacteriology*, *177*(5), 1367–1373. doi:10.1128/JB.177.5.1367-1373.1995 PMID:7868612

Givskov, M., de Nys, R., Manefield, M., Gram, L., Maximilien, R., Eberl, L., Molin, S., Steinberg, P. D., & Kjelleberg, S. (1996). Eukaryotic interference with homoserine lactone-mediated prokaryotic signalling. *Journal of Bacteriology*, *178*(22), 6618–6622. doi:10.1128/JB.178.22.6618-6622.1996 PMID:8932319

Gopalakrishnan, S., Srinivas, V., Prakash, B., Sathya, A., & Vijayabharathi, R. (2015). Plant growth-promoting traits of Pseudomonas geniculata isolated from chickpea nodules. *Biotech, 5*(5), 653–661. . doi:10.100713205-014-0263-4

Götz, C., Fekete, A., Gebefuegi, I., Forczek, S. T., Fuksová, K., Li, X., Englmann, M., Gryndler, M., Hartmann, A., Matucha, M., Schmitt-Kopplin, P., & Schröder, P. (2007). Uptake, degradation and chiral discrimination of N-acyl-D/L-homoserine lactones by barley (Hordeum vulgare) and yam bean (Pachyrhizuserosus) plants. *Analytical and Bioanalytical Chemistry*, *389*(5), 1447–1457. doi:10.100700216-007-1579-2 PMID:17899036

Grandclément, C., Tannières, M., Moréra, S., Dessaux, Y., & Faure, D. (2016). Quorum quenching: Role in nature and applied developments. *FEMS Microbiology Reviews*, *40*(1), 86–116. doi:10.1093/femsre/fuv038 PMID:26432822

Haudecoeur, E., Planamente, S., Cirou, A., Tannières, M., Shelp, B. J., Moréra, S., & Faure, D. (2009). Proline antagonizes GABA-induced quenching of quorum-sensing in Agrobacterium tumefaciens. *Proceedings of the National Academy of Sciences of the United States of America*, *106*(34), 14587–14592. doi:10.1073/pnas.0808005106 PMID:19706545

Hentzer, M., Wu, H., Andersen, J. B., Riedel, K., Rasmussen, T. B., Bagge, N., Kumar, N., Schembri, M. A., Song, Z., Kristoffersen, P., Manefield, M., Costerton, J. W., Molin, S., Eberl, L., Steinberg, P., Kjelleberg, S., Høiby, N., & Givskov, M. (2003). Attenuation of Pseudomonas aeruginosa virulence by quorum sensing inhibitors. *The EMBO Journal*, *22*(15), 3803–3815. doi:10.1093/emboj/cdg366 PMID:12881415

Hewitt, L., Kasche, V., Lummer, K., Lewis, R. J., Murshudov, G. N., Verma, C. S., Dodson, G. G., & Wilson, K. S. (2000). Structure of a slow processing precursor penicillin acylase from Escherichia coli reveals the linker peptide blocking the active-site cleft. *Journal of Molecular Biology*, *302*(4), 887–898. doi:10.1006/jmbi.2000.4105 PMID:10993730

Hu, J. Y., Fan, Y., Lin, Y. H., Zhang, H. B., Ong, S. L., Dong, N., Xu, J. L., Ng, W. J., & Zhang, L. H. (2003). Microbial diversity and prevalence of virulent pathogens in biofilms developed in a water reclamation system. *Research in Microbiology*, *154*(9), 623–629. doi:10.1016/j.resmic.2003.09.004 PMID:14596899

Huang, J. J., Han, J. I., Zhang, L. H., & Leadbetter, J. R. (2003). Utilization of acyl-homoserine lactone quorum signals for growth by a soil pseudomonad and Pseudomonas aeruginosa PAO1. *Applied and Environmental Microbiology*, *69*(10), 5941–5949. doi:10.1128/AEM.69.10.5941-5949.2003 PMID:14532048

Huang, J. J., Petersen, A., Whiteley, M., & Leadbetter, J. R. (2006). Identification of QuiP, the product of gene PA1032, as the second acyl-homoserine lactone acylase of Pseudomonas aeruginosa PAO1. *Applied and Environmental Microbiology*, *72*(2), 1190–1197. doi:10.1128/AEM.72.2.1190-1197.2006 PMID:16461666

Hwang, I., Li, P. L., Zhang, L., Piper, K. R., Cook, D. M., Tate, M. E., & Farrand, S. K. (1994). TraI, a LuxI homologue, is responsible for production of conjugation factor, the Ti plasmid N-acylhomoserine lactone autoinducer. *Proceedings of the National Academy of Sciences of the United States of America*, *91*(11), 4639–4643. doi:10.1073/pnas.91.11.4639 PMID:8197112

Jafra, S., Przysowa, J., Czajkowski, R., Michta, A., Garbeva, P., & van der Wolf, J. M. (2006). Detection and characterization of bacteria from the potato rhizosphere degrading N-acyl-homoserine lactone. *Canadian Journal of Microbiology*, *52*(10), 1006–1015. doi:10.1139/w06-062 PMID:17110970

Kang, B. R., Lee, J. H., Ko, S. J., Lee, Y. H., Cha, J. S., Cho, B. H., & Kim, Y. C. (2004). Degradation of acyl-homoserine lactone molecules by Acinetobacter sp. strain C1010. *Canadian Journal of Microbiology*, *50*(11), 935–941. doi:10.1139/w04-083 PMID:15644910

Khan, S. R., & Farrand, S. K. (2009). The BlcC (AttM) lactonase of Agrobacterium tumefaciens does not quench the quorum-sensing system that regulates Ti plasmid conjugative transfer. *Journal of Bacteriology*, *191*(4), 1320–1329. doi:10.1128/JB.01304-08 PMID:19011037

Kim, M. H., Choi, W. C., Kang, H. O., Lee, J. S., Kang, B. S., Kim, K. J., Derewenda, Z. S., Oh, T. K., Lee, C. H., & Lee, J. K. (2005). The molecular structure and catalytic mechanism of a quorum-quenching N-acyl-L-homoserine lactone hydrolase. *Proceedings of the National Academy of Sciences of the United States of America*, *102*(49), 17606–17611. doi:10.1073/pnas.0504996102 PMID:16314577

Lang, J., Planamente, S., Mondy, S., Dessaux, Y., Moréra, S., & Faure, D. (2013). Concerted transfer of the virulence Ti plasmid and companion at plasmid in the Agrobacterium tumefaciens-induced plant tumour. *Molecular Microbiology*, *90*(6), 1178–1189. doi:10.1111/mmi.12423 PMID:24118167

LaSarre, B., & Federle, M. J. (2013). Exploiting quorum sensing to confuse bacterial pathogens. *Microbiology and molecular biology reviews. Microbiology and Molecular Biology Reviews*, *77*(1), 73–111. doi:10.1128/MMBR.00046-12 PMID:23471618

Leadbetter, J. R., & Greenberg, E. P. (2000). Metabolism of acyl-homoserine lactone quorum-sensing signals by Variovorax paradoxus. *Journal of Bacteriology, 182*(24), 6921–6926. doi:10.1128/JB.182.24.6921-6926.2000 PMID:11092851

Lee, S. J., Park, S. Y., Lee, J. J., Yum, D. Y., Koo, B. T., & Lee, J. K. (2002). Genes encoding the N-acyl homoserine lactone-degrading enzyme are widespread in many subspecies of Bacillus thuringiensis. *Applied and Environmental Microbiology, 68*(8), 3919–3924. doi:10.1128/AEM.68.8.3919-3924.2002 PMID:12147491

Li de la Sierra-Gallay, I., Pellegrini, O., & Condon, C. (2005). Structural basis for substrate binding, cleavage and allostery in the tRNA maturase RNase Z. *Nature, 433*(7026), 657–661. doi:10.1038/nature03284 PMID:15654328

Lin, Y. H., Xu, J. L., Hu, J., Wang, L. H., Ong, S. L., Leadbetter, J. R., & Zhang, L. H. (2003). Acyl-homoserine lactone acylase from Ralstonia strain XJ12B represents a novel and potent class of quorum-quenching enzymes. *Molecular Microbiology, 47*(3), 849–860. doi:10.1046/j.1365-2958.2003.03351.x PMID:12535081

Liu, D., Lepore, B. W., Petsko, G. A., Thomas, P. W., Stone, E. M., Fast, W., & Ringe, D. (2005). Three-dimensional structure of the quorum-quenching N-acyl homoserine lactone hydrolase from Bacillus thuringiensis. *Proceedings of the National Academy of Sciences of the United States of America, 102*(33), 11882–11887. doi:10.1073/pnas.0505255102 PMID:16087890

Liu, D., Thomas, P. W., Momb, J., Hoang, Q. Q., Petsko, G. A., Ringe, D., & Fast, W. (2007). Structure and specificity of a quorum-quenching lactonase (AiiB) from Agrobacterium tumefaciens. *Biochemistry, 46*(42), 11789–11799. doi:10.1021/bi7012849 PMID:17900178

Lyon, G. J., Mayville, P., Muir, T. W., & Novick, R. P. (2000). Rational design of a global inhibitor of the virulence response in Staphylococcus aureus, based in part on localization of the site of inhibition to the receptor-histidine kinase, Agr C. *Proceedings of the National Academy of Sciences of the United States of America, 97*(24), 13330–13335. doi:10.1073/pnas.97.24.13330 PMID:11087872

Mani, A., Hameed, S. S., Ramalingam, S., & Narayanan, M. (2012). Assessment of Quorum Quenching Activity of Bacillus Species Against Pseudomonas aeruginosa MTCC 2297. *Global Journal of Pharmacology, 6*(2), 118–125.

Mayville, P., Ji, G., Beavis, R., Yang, H., Goger, M., Novick, R. P., & Muir, T. W. (1999). Structure-activity analysis of synthetic autoinducing thiolactone peptides from Staphylococcus aureus responsible for virulence. *Proceedings of the National Academy of Sciences of the United States of America, 96*(4), 1218–1223. doi:10.1073/pnas.96.4.1218 PMID:9990004

Mole, B. M., Baltrus, D. A., Dangl, J. L., & Grant, S. R. (2007). Global virulence regulation networks in phytopathogenic bacteria. *Trends in Microbiology, 15*(8), 363–371. doi:10.1016/j.tim.2007.06.005 PMID:17627825

Molina, L., Rezzonico, F., Défago, G., & Duffy, B. (2005). Autoinduction in Erwinia amylovora: Evidence of an acyl-homoserine lactone signal in the fire blight pathogen. *Journal of Bacteriology, 187*(9), 3206–3213. doi:10.1128/JB.187.9.3206-3213.2005 PMID:15838048

Morohoshi, T., Nakazawa, S., Ebata, A., Kato, N., & Ikeda, T. (2008). Identification and characterization of N-acylhomoserine lactone-acylase from the fish intestinal Shewanella sp. strain MIB015. *Bioscience, Biotechnology, and Biochemistry*, *72*(7), 1887–1893. doi:10.1271/bbb.80139 PMID:18603799

Newman, K. L., Chatterjee, S., Ho, K. A., & Lindow, S. E. (2008). Virulence of plant pathogenic bacteria attenuated by degradation of fatty acid cell-to-cell signaling factors. *Molecular plant-microbe interactions. Molecular Plant-Microbe Interactions*, *21*(3), 326–334. doi:10.1094/MPMI-21-3-0326 PMID:18257682

Ozer, E. A., Pezzulo, A., Shih, D. M., Chun, C., Furlong, C., Lusis, A. J., Greenberg, E. P., & Zabner, J. (2005). Human and murine paraoxonase 1 are host modulators of Pseudomonas aeruginosa quorum-sensing. *FEMS Microbiology Letters*, *253*(1), 29–37. doi:10.1016/j.femsle.2005.09.023 PMID:16260097

Pappas, K. M., & Winans, S. C. (2003). A LuxR-type regulator from Agrobacterium tumefaciens elevates Ti plasmid copy number by activating transcription of plasmid replication genes. *Molecular Microbiology*, *48*(4), 1059–1073. doi:10.1046/j.1365-2958.2003.03488.x PMID:12753196

Park, S. Y., Hwang, B. J., Shin, M. H., Kim, J. A., Kim, H. K., & Lee, J. K. (2006). N-acylhomoserine lactonase producing Rhodococcus spp. with different AHL-degrading activities. *FEMS Microbiology Letters*, *261*(1), 102–108. doi:10.1111/j.1574-6968.2006.00336.x PMID:16842366

Park, S. Y., Kang, H. O., Jang, H. S., Lee, J. K., Koo, B. T., & Yum, D. Y. (2005). Identification of extracellular N-acylhomoserine lactone acylase from a Streptomyces sp. and its application to quorum quenching. *Applied and Environmental Microbiology*, *71*(5), 2632–2641. doi:10.1128/AEM.71.5.2632-2641.2005 PMID:15870355

Park, S. Y., Lee, S. J., Oh, T. K., Oh, J. W., Koo, B. T., Yum, D. Y., & Lee, J. K. (2003). AhlD, an N-acylhomoserine lactonase in Arthrobacter sp., and predicted homologues in other bacteria. *Microbiology (Reading, England)*, *149*(Pt 6), 1541–1550. doi:10.1099/mic.0.26269-0 PMID:12777494

Qin, Y., Su, S., & Farrand, S. K. (2007). Molecular basis of transcriptional antiactivation. TraM disrupts the TraR-DNA complex through stepwise interactions. *The Journal of Biological Chemistry*, *282*(27), 19979–19991. doi:10.1074/jbc.M703332200 PMID:17475619

Quinones, B., Pujol, C. J., & Lindow, S. E. (2004). Regulation of AHL production and its contribution to epiphytic fitness in Pseudomonassyringae. *Molecular Plant-Microbe Interactions*, *17*(5), 521–531. doi:10.1094/MPMI.2004.17.5.521 PMID:15141956

Raddadi, N., Cherif, A., Boudabous, A., & Daffonchio, D. (2008). Screening of plant growth promoting traits of Bacillus thuringiensis. *Annals of Microbiology*, *58*(1), 47–52. doi:10.1007/BF03179444

Raddadi, N., Cherif, A., Ouzari, H., Marzorati, M., Brusetti, L., Boudabous, A., & Daffonchio, D. (2007). Bacillus thuringiensis beyond insect biocontrol: Plant growth promotion and biosafety of polyvalent strains. *Annals of Microbiology*, *57*(4), 481–494. doi:10.1007/BF03175344

Rai, N., Rai, R., & Venkatesh, K. V. (2015). Quorum sensing biosensors. In V. C. Kalia (Ed.), *Quorum Sensing vs Quorum quenching: A Battle with No End in Sight* (pp. 173–183). Springer India.

Reimmann, C., Ginet, N., Michel, L., Keel, C., Michaux, P., Krishnapillai, V., Zala, M., Heurlier, K., Triandafillu, K., Harms, H., Défago, G., & Haas, D. (2002). Genetically programmed autoinducer destruction reduces virulence gene expression and swarming motility in *Pseudomonas aeruginosa* PAO1. *Microbiology (Reading, England)*, *148*(Pt 4), 923–932. doi:10.1099/00221287-148-4-923 PMID:11932439

Riaz, K., Elmerich, C., Moreira, D., Raffoux, A., Dessaux, Y., & Faure, D. (2008). A metagenomic analysis of rhizospheric bacteria extends the diversity of quorum-quenching lactonases. *Environmental Microbiology*, *10*(3), 560–570. doi:10.1111/j.1462-2920.2007.01475.x PMID:18201196

Rodríguez, M., Torres, M., Blanco, L., Béjar, V., Sampedro, I., & Llamas, I. (2020). Plant growth-promoting activity and quorum quenching-mediated biocontrol of bacterial phytopathogens by Pseudomonas segetis strain P6. *Scientific Reports*, *10*(1), 1–2. doi:10.103841598-020-61084-1 PMID:32139754

Romero, M., Diggle, S. P., Heeb, S., Cámara, M., & Otero, A. (2008). Quorum quenching activity in Anabaena sp. PCC 7120: Identification of AiiC, a novel AHL-acylase. *FEMS Microbiology Letters*, *280*(1), 73–80. doi:10.1111/j.1574-6968.2007.01046.x PMID:18194337

Schipper, C., Hornung, C., Bijtenhoorn, P., Quitschau, M., Grond, S., & Streit, W. R. (2009). Metagenome derived clones encoding two novel lactonase family proteins involved in biofilm inhibition in *Pseudomonas aeruginosa*. *Applied and Environmental Microbiology*, *75*(1), 224–233. doi:10.1128/AEM.01389-08 PMID:18997026

Shinohara, M., Nakajima, N., & Uehara, Y. (2007). Purification and characterization of a novel esterase (beta-hydroxypalmitate methyl ester hydrolase) and prevention of the expression of virulence by Ralstonia solanacearum. *Journal of Applied Microbiology*, *103*(1), 152–. doi:10.1111/j.1365-2672.2006.03222.x

Siddiqui, Z. A., Iqbal, A., & Mahmood, I. (2001). Efects of Pseudomonas fuorescens and fertilizers on the reproduction of Meloidogyne incognita and growth of tomato. *Applied Soil Ecology*, *16*(2), 179–185. doi:10.1016/S0929-1393(00)00083-4

Smith, K. M., Bu, Y., & Suga, H. (2003). Induction and inhibition of Pseudomonas aeruginosa quorum sensing by synthetic autoinducer analogs. *Chemistry & Biology*, *10*(1), 81–89. doi:10.1016/S1074-5521(03)00002-4 PMID:12573701

Teplitski, M., Robinson, J. B., & Bauer, W. D. (2000). Plants secrete substances that mimic bacterial N-acyl homoserine lactone signal activities and affect population density-dependent behaviors in associated bacteria. *Molecular Plant-Microbe Interactions*, *13*(6), 637–648. doi:10.1094/MPMI.2000.13.6.637 PMID:10830263

Thomas, P. W., Stone, E. M., Costello, A. L., Tierney, D. L., & Fast, W. (2005). The quorum-quenching lactonase from *Bacillus thuringiensis* is a metalloprotein. *Biochemistry*, *44*(20), 7559–7569. doi:10.1021/bi050050m PMID:15895999

Ulrich, R. L. (2004). Quorum quenching: Enzymatic disruption of N-acylhomoserine lactone-mediated bacterial communication in *Burkholderia thailandensis*. *Applied and Environmental Microbiology*, *70*(10), 6173–6180. doi:10.1128/AEM.70.10.6173-6180.2004 PMID:15466564

Uroz, S., Chhabra, S. R., Camara, M., Williams, P., Oger, P., & Dessaux, Y. (2005). N-Acylhomoserine lactone quorum-sensing molecules are modified and degraded by *Rhodococcus erythropolis* W2 by both amidolytic and novel oxidoreductase activities. *Microbiology*, *151*(10), 3313–3322. doi:10.1099/mic.0.27961-0 PMID:16207914

Uroz, S., & Heinonsalo, J. (2008). Degradation of N-acyl homoserine lactone quorum sensing signal molecules by forest root-associated fungi. *FEMS Microbiology Ecology*, *65*(2), 271–278. doi:10.1111/j.1574-6941.2008.00477.x PMID:18400006

Uroz, S., Oger, P., Chhabra, S. R., Cámara, M., Williams, P., & Dessaux, Y. (2007). N-acyl homoserine lactones are degraded via an amidolytic activity in Comamonas sp. strain D1. *Archives of Microbiology*, *187*(3), 249–256. doi:10.100700203-006-0186-5 PMID:17136382

Uroz, S., Oger, P. M., Chapelle, E., Adeline, M. T., Faure, D., & Dessaux, Y. (2008). A *Rhodococcus* qsdA-encoded enzyme defines a novel class of large-spectrum quorum-quenching lactonases. *Applied and Environmental Microbiology*, *74*(5), 1357–1366. doi:10.1128/AEM.02014-07 PMID:18192419

Vaikuntapu, P. R., Dutta, S., Samudrala, R. B., Rao, V. R. V. N., Kalam, S., & Podile, A. R. (2014). Preferential Promotion of *Lycopersicon esculentum* (Tomato) Growth by Plant Growth Promoting Bacteria Associated with Tomato. *Indian Journal of Microbiology*, *54*(4), 403–412. doi:10.100712088-014-0470-z PMID:25320438

Vallet, I., Diggle, S. P., Stacey, R. E., Cámara, M., Ventre, I., Lory, S., Lazdunski, A., Williams, P., & Filloux, A. (2004). Biofilm formation in Pseudomonas aeruginosa: Fimbrial cup gene clusters are controlled by the transcriptional regulator MvaT. *Journal of Bacteriology*, *186*(9), 2880–2890. doi:10.1128/JB.186.9.2880-2890.2004 PMID:15090530

Wang, L. H., Weng, L. X., Dong, Y. H., & Zhang, L. H. (2004). Specificity and enzyme kinetics of the quorum-quenching N-Acyl homoserine lactone lactonase (AHL-lactonase). *The Journal of Biological Chemistry*, *279*(14), 13645–13651. doi:10.1074/jbc.M311194200 PMID:14734559

Whitehead, N. A., Barnard, A. M., Slater, H., Simpson, N. J., & Salmond, G. P. (2001). Quorum-sensing in Gram-negative bacteria. *FEMS Microbiology Reviews*, *25*(4), 365–404. doi:10.1111/j.1574-6976.2001.tb00583.x PMID:11524130

Xu, F., Byun, T., Deussen, H. J., & Duke, K. R. (2003). Degradation of N-acylhomoserine lactones, the bacterial quorum-sensing molecules, by acylase. *Journal of Biotechnology*, *101*(1), 89–96. doi:10.1016/S0168-1656(02)00305-X PMID:12523973

Yang, F., Wang, L. H., Wang, J., Dong, Y. H., Hu, J. Y., & Zhang, L. H. (2005). Quorum quenching enzyme activity is widely conserved in the sera of mammalian species. *FEBS Letters*, *579*(17), 3713–3717. doi:10.1016/j.febslet.2005.05.060 PMID:15963993

Yates, E. A., Philipp, B., Buckley, C., Atkinson, S., Chhabra, S. R., Sockett, R. E., Goldner, M., Dessaux, Y., Cámara, M., Smith, H., & Williams, P. (2002). N-acylhomoserine lactones undergo lactonolysis in a pH-, temperature-, and acyl chain length-dependent manner during growth of *Yersinia pseudotuberculosis* and *Pseudomonas aeruginosa*. *Infection and Immunity*, *70*(10), 5635–5646. doi:10.1128/IAI.70.10.5635-5646.2002 PMID:12228292

Yin, W. F., Tung, H. J., Sam, C. K., Koh, C. L., & Chan, K. G. (2012). Quorum quenching *Bacillus sonorensis* isolated from soya sauce fermentation brine. *Sensors (Basel), 12*(4), 4065–4073. doi:10.3390120404065 PMID:22666018

Zhang, H. B., Wang, L. H., & Zhang, L. H. (2002). Genetic control of quorum-sensing signal turnover in Agrobacterium tumefaciens. *Proceedings of the National Academy of Sciences of the United States of America, 99*(7), 4638–4643. doi:10.1073/pnas.022056699 PMID:11930013

Zhang, L., Guo, Z., Gao, H., Peng, X., Li, Y., Sun, S., Lee, J. K., & Lin, W. (2016). Interaction of *Pseudostellaria heterophylla* with Quorum Sensing and Quorum Quenching Bacteria Mediated by Root Exudates in a Consecutive Monoculture System. *Journal of Microbiology and Biotechnology, 26*(12), 2159–2170. doi:10.4014/jmb.1607.07073 PMID:27666992

Zhang, L., Ruan, L., Hu, C., Wu, H., Chen, S., Yu, Z., & Sun, M. (2007). Fusion of the genes for AHL-lactonase and S-layer protein in *Bacillus thuringiensis* increases its ability to inhibit soft rot caused by Erwinia carotovora. *Applied Microbiology and Biotechnology, 74*(3), 667–675. doi:10.100700253-006-0696-8 PMID:17216466

Zhu, J., & Winans, S. C. (2001). The quorum-sensing transcriptional regulator TraR requires its cognate signaling ligand for protein folding, protease resistance, and dimerization. *Proceedings of the National Academy of Sciences of the United States of America, 98*(4), 1507–1512. doi:10.1073/pnas.98.4.1507 PMID:11171981

Chapter 23
Laccases for Soil Bioremediation: An Introduction

Rajalakshmi Sridharan
https://orcid.org/0000-0002-7194-3669
Stella Maris College (Autonomous), University of Madras, Chennai, India

Veena Gayathri Krishnaswamy
https://orcid.org/0000-0002-3012-8561
Stella Maris College (Autonomous), University of Madras, Chennai, India

ABSTRACT

Industrialization led to an increase in chemicals in the environment. The soil absorbs these chemicals and holds them for years until treated. The action of bacteria, fungi, and algae utilize the pollutants and generate energy. The bioremediation contains a diverse treatment process, but the effectiveness of the bioremediation increases by the enzymatic action. Laccase, a copper-containing enzyme, is versatile and oxidizes complex organic compounds without generating reactive oxygen species (ROS). This process is carried by laccase-mediated systems (LCMs) controlled by low redox potential. The presence of redox mediators oxidizes the chemical compounds at the higher rate, making laccase degradation of the pollutants effectively. The chapter provides a glimpse of soil bioremediation by bacteria and fungi as individual species and symbiotic species, the production of laccase enzyme by bacteria and fungi, methods adopted to enhance the enzyme activity, and degradation of pollutants in soil.

INTRODUCTION – SOIL POLLUTION

The environmental transition that impacts its physical, chemical, and biological features is referred to as pollution. The foreign substances in the environment sourced from different sites – household, industry, mining, automobile wastes, and radioactive wastes – cause undesirable environmental changes. Soil, mentioned as "Universal Sink" contains all types of pollutants (Doran et al., 1996; Havugimana et al., 2018).

DOI: 10.4018/978-1-7998-7062-3.ch023

The contamination affects the structure of the soil and is rendered unhealthy or abandoned if it crosses its threshold. Based on the nature of the pollutants, they are biodegradable and non-biodegradable. The heavy metals are considered biodegradable and persistent in soil. In turn, the biodegradable pollutants produce intermediates or products of toxic nature, increasing the soil toxicity (Sims & Cupples, 1999; Havugimana et al., 2018). The Persistent Organic Pollutants (POPs), plastics, heavy metals and xenobiotics have inert characteristics making degradation difficult. The POPs generate from sources such as industries, automobiles, waste incineration. The Stockholm Convention comprising 152 countries reported 12 highly toxic POPs and called them a "Dirty Dozen". It contains - dieldrin, aldrin, dioxins, chlordane, furans, mirex, DDT, endrin, heptachlor, hexachlorobenzene, PCBs, and toxaphene - (UNEP 2009). During 2017, the list was extended further by the addition of 16 more POPs - α-hexachlorocyclohexane, chlordecone, β-hexachlorocyclohexane, decabromo diphenyl ether, hexabromobiphenyl, hexabromodiphenyl ether/heptabromodiphenyl ether, hexachlorobutadiene, hexabromocyclododecane, lindane, pentachlorobenzene, perfluorooctane sulfonic acid, pentachlorophenol and its salts and esters, perfluorooctane sulfonyl fluoride, polychlorinated naphthalenes, short-chain chlorinated paraffin, endosulfan and its related isomers, tetrabromodiphenyl ether, and pentabromodiphenyl ether (Bull et al., 2014, Araki et al., 2014; Jarosiewicz et al., 2017). The ingestion of polluted soil by the animals enters the food chain, reaching the higher trophic levels leading to the accumulation of POPs. The diverse source of pollutants (Figure 1) has impeccable effects on plants, animals and humans.

Figure 1. Source of soil pollution

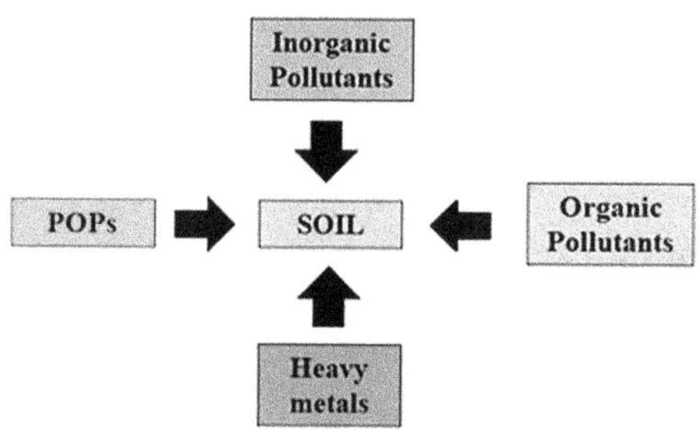

The agrochemical pollution in the soil contributes as large as synthetic fertilizers made of hydrocarbon. The applied pesticide passes through the soil horizons and gets absorbed by the soil particles. Thus, it remains in the soil for a prolonged period causing ill effects in living organisms. Based on the living systems' level of adsorption, the pesticides are classified into three categories (Mirsal 2008);

1. **Contact pesticides** – pesticides are remaining on the surface of the plants/soil.
2. **Quasi Systemic pesticides** – pesticides transported to the leaves cuticle and epidermis of the animals.
3. **Systemic pesticides** – pesticides in the internal systems of plants and animals.

The heavy metal pollution in the soil based on its quantity, the toxicity towards plants, animals, and human beings vary. The micronutrients - Fe, Mn, Cu, Zn, Mo – in higher concentrations affect the plant's developmental process while metals – Arsenic, Mercury, Lead, Cadmium – causes toxic effects on humans and animals even at mild exposure (Zwolak et al., 2019). The heavy metals in soil occur either as a combination or as free metal ions. The metal ions also adhere to the organic compounds or the silicates in the soil. The metals associated with silicates are less toxic than metals available in Free State (Marques et al., 2009; Ramos et al., 1994; Chibuike & Obiora, 2014). The heavy metal concentration in soil is controlled by the soil colloids' surface area naturally (Marques et al., 2009). The heavy metal contamination alters the biological and biochemical characteristics of the soil. As the toxicity of the metal depends on the pH, temperature, salinity, organic and inorganic materials in the soil, it drastically changes the diversity, composition, and role of the soil microbes (Friedlova, 2010; Nannipieri et al., 1997; Baath et al., 1989; Giller et al., 1998). The heavy metal tends to accumulate in living organisms' tissues, causing ill effects (Herawati et al., 2000; He et al., 2005). The sewage further increases the pollution from industries and municipal areas, treated and used for irrigation. It has led to the accumulation of heavy metals by plants (Zwolak et al., 2019). The heavy metal leaching is reported to occur likely at alkaline pH (Bielecka et al., 2009). The accumulation in plants could be prevented by adding phosphates, organic and inorganic materials in the soil (Paltseva et al., 2018; Zwolak et al., 2019). Apart from the addition of materials, the total concentration of the heavy metals in the soil determines plants' uptake (McBride et al., 2015).

The soil also contains radioactive elements released by the major source – nuclear power stations. The availability of these pollutants is also high in water and air. The elements such as 238U, 232Th, 222Ra, 87Rb, 137Cs, 239Pu, 241Am, 90Sr, 91Y, 14C, and 3H exist nature (ICRP, 2007). The industries contribute a greater part to the pollution by releasing dyes into the environment. Around 100,000 varieties of textile dyes are available of which 7×103 tons are synthetic dyes (Aksu, 2005; Fu & Veraraghavan, 2001, Tehrani & Holmberg, 2013). The chromophoric group produces the dye's color called the Azo group (-N=N-). The azo bond on cleavage by reduction produces carcinogenic amines (Savin & Butnaru, 2008; Puvaneshwari et al., 2006). This chapter explains soil's bioremediation contaminated with recalcitrant compounds and heavy metals by using laccase enzymes from bacteria and fungi.

BIOREMEDIATION METHODS

The elimination of the pollutants from the environment by the native microorganisms is termed bioremediation (Timmis & Pieper, 1999; Ang et al., 2005). Bioremediation is often characterised as the transformation or usage of toxins by microorganisms to obtain energy (Tangy et al., 2007; Abatenh et al., 2017). Bacteria, archaea, algae, and fungi are known as prime biological agents (Strong & Burgess, 2008). Bioremediation is a redox reaction that terminates the contaminated environment's chemical modification and microbiology (Yeung, 2009; Tandon & Singh, 2016; Ojuederie & Babalola, 2017). Bioremediation efficiency depends on pollutants' nature, concentration, physic-chemical characteristics, bioavailability, pH, soil, temperature, nutrients, and electron acceptors (Fantroussi & Agathos, 2005). The degradation or remediation of the pollutants occurs in a series of chemical reactions by the microbe-producing enzymes (Ahuja et al., 2004, Pereira & de Freitas, 2012; Okino-Delgado et al., 2019). Application of enzymes rather than the microorganisms serves as an advantage due to the enzymes' faster and homogeneous reaction. The enzymatic process uses certain concentrations of enzymes involved in the metabolism of inert pollutants under a controlled environment and substrate concentration (Brown et al., 2017; Okino-Delgado et al., 2019).

Microbial Degradation of Pollutants

Bioremediation is broadly classified as in-situ and ex-situ based on the site of action of microorganisms. In-situ bioremediation involves an onsite treatment process by enhancing the native or wild microorganisms using nutrient supplements. It is also enhanced by the bioaugmentation technique to speed up the remediation of contaminants. The ex-situ method involves transportation of the contaminated samples to the treatment plants for remediation (Rayu et al., 2012; Mani & Kumar, 2014; Azubuike et al., 2016; Ojuederie & Babalola, 2017). Degradation of pollutants occurs either by the aerobic or anaerobic reaction. The oxidation is an initial reaction occurring intracellularly, which on activation triggers the enzymatic reactions. Simultaneously, the degradation by peripheral pathway reaction occurs in a series of steps forming intermediates. The outline of aerobic degradation of pollutants is illustrated in Figure 2 (Das & Chandran, 2011, Mbachu et al., 2020).

Figure 2. Outline of aerobic degradation of pollutants
(Das and Chandran 2011, Mbachu et al., 2020)

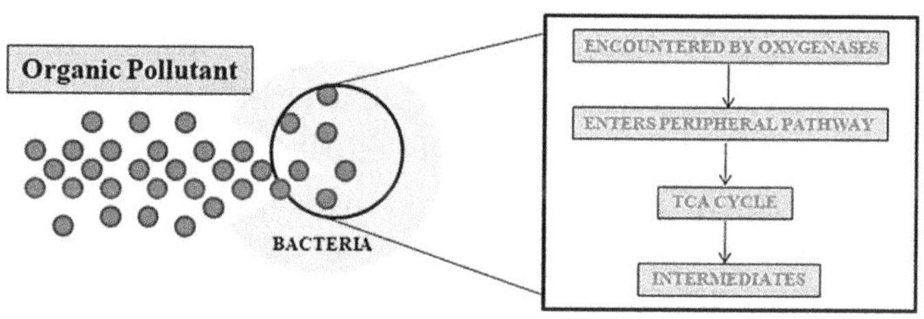

The microbial treatment of pollutants is made further effective by genetic engineering methods, which might reduce remediation duration. The bioremediation of soil using microbes prevents topsoil quality, maintains the biogeochemical cycle, enriches the soil-dwelling organisms, and mainly produces minimal wastes on treatment (Das & Adholeya 2011; Yap et al., 2019). Bioremediation is mitigated using individual native microorganisms or as co-culture or consortium, which has higher chances of increasing the degradation rate (Yap et al., 2019). The remediation of heavy metals by microorganisms either precipitates or changes the oxidation state of the metals. The process occurs step by step as follows (Jan et al., 2014, Ojuederie and Babalola, 2017);

1. Sequestration of heavy metals by metal-binding proteins and siderophores produced by bacteria or fungi.
2. Modification of biochemical pathways to hinder the metal uptake.
3. Enzymatic conversion of metals.
4. Metal reduction by efflux systems.

The microorganisms possess alternate plans to remove heavy metals from the soil. The uptake of heavy metals by the biomass with high pollutant degrading capacity involves accumulation (Intracellular or extracellular), precipitation, and cell surface adsorption, which might depend on cell metabolism or has its independent metabolism (Alburquerque et al., 2011; Beiyuan et al., 2017). The biosorption of heavy metals are affected by the negatively charged functional groups such as -OH, P, and C=O groups

(Dixit et al., 2015). The ion exchange of the heavy metals is aided by the uronic acid and sulfate groups present in the bacterial cell wall. The presence of alanine, glutamine amino acids, teichoic acid, lipoproteins, glycoproteins, phospholipids, and lipopolysaccharides acts as a ligand. These ligands bind to the heavy metals and mineralize them (Fomina & Gadd, 2014; Coelho et al., 2015; Gupta et al., 2015; Ayangbenro & Babalola, 2017). The algae have high biosorption capacity than other microbes as they form large amounts of biomass (Ayangbenro & Babalola, 2017). The degradation of pesticides by the bacteria occurs either by mineralization or the co-metabolism process (Ye et al., 2018). Hydrocarbon pollution is one of the problems treated by bioremediation. It occurs in a series, as given in Figure 3 (Wentzel et al., 2007; Varjani, 2017).

Figure 3. Bioremediation of Hydrocarbons
(Wentzel et al. 2007, Varjani 2017)

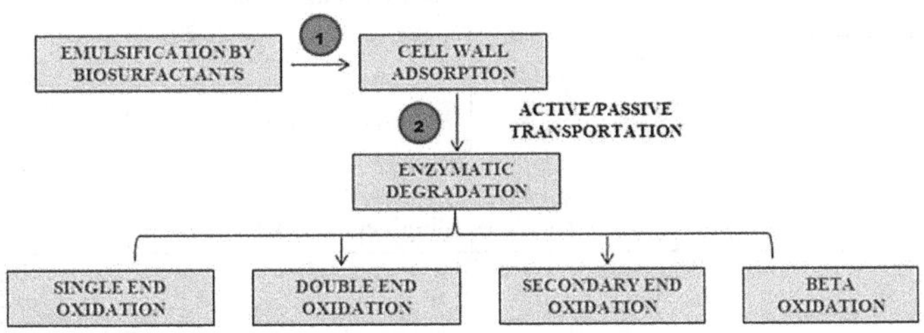

Enzymatic Degradation of Pollutants

The enzymatic degradation of pollutants in the environment nullifies the disadvantages of using microbes. Enzymes are specific (broad or narrow) to substrates, which involve transforming pollutants into simpler compounds and less toxic (Gianfreda & Bollag, 2003). The enzymes are produced either intracellularly or extracellularly by the microbes. The classes transferases, hydrolases, and oxidoreductases are involved in remediating pollutants (Whiteley & Lee, 2006). The enzymes cleave substrate bonds and favours electron transfer for microbes to yield energy (Singh et al., 2010). The oxidoreductases are involved in remediating phenolic compounds, azo dyes, metals, radioactive metals, and chlorinated phenolic compounds. The process of radioactive metal remediation involves the uptake of electrons from the organic substances where the radioactive metal acts as an electron acceptor (Leung, 2004; Vidali, 2001; Husain, 2006; Park et al., 2006). Oxygenase enzymes cleave the aromatic groups by incorporating oxygen molecules into them (Arora et al., 2009). These groups of enzymes also remediate pesticides and halogenated compounds. Monooxygenases are enzymes with the high region and stereoselectivity and independent of the cofactor (Arora et al., 2010). Peroxidases can be both haem and non-haem proteins involved in hormone regulation, cell wall, and lignin formation in plants (Hiner et al., 2002; Koua et al., 2009). Hydrolases are enzymes involved in the degradation of oil, organophosphate insecticides (Williams, 1977). Hydrolases include DNases, proteases, lipases, amylases, hemicellulases, and cellulases (Sánchez-Porro et al., 2003; Schmidt, 2006).

Table 1. Laccase producing bacterial strains

Name of Bacterial Strains	References
Bacillus subtilis MTCC 2414	Narayanan et al. 2015
Bacillus tequilensis SN4 (SN4LAC)	Sondhi et al.2014
Bacillus licheniformis LS04	Lu et al. 2012
Bacillus subtilis	Wang et al. 2011
Azospirillum lipoferum	Diamantidis et al. 2000
Aquisalibacillus elongatus	Rezaei et al. 2017
Pseudomonas extremorientalis	Neifar et al. 2016
Arthrographis sp. *Enterobacter cloacae*	Devasia and Nair 2016
Pseudomonas lurida *Lysinibacillus sphaericus*	Dhiman and Shirkot 2015
Rhodococcus sp., *Enterobacter* sp., *Staphylococcus saprophyticus* *Delftia tsuruhatensis*	Mongkolthanaruk et al., 2012
m γ-proteobacterium JB	Singh et al., 2010

Laccase - An Oxidoreductase Enzyme

Laccases contain four atoms of copper; type 1 paramagnetic copper is responsible for blue color and substrate oxidation. Type 2 copper and two 3rd type copper reduces the oxygen molecule to two water molecules. Phenolic compounds are oxidized by laccases and form nontoxic aromatic amines (Montaya et al., 2015). Laccase possesses low-substrate specificity, which explains its wide range of applications in the environment (Pant & Adholeya, 2009). Recent studies on discovery of novel laccase has been performed using metagenomic analysis (Datta et al., 2020). However, the catalytic degradation efficiency affects the mass transfer ability of laccase and pollutants which limits its application in soil bioremediation process (Sharma et al., 2018; Wang et al., 2020).

Microbial Sources of Laccase

Laccases are highly reported in fungi than in bacteria. White rot fungi produce laccase along with ligninolytic enzymes (Wong, 2009; Arora & Sharma, 2010). Most commonly reported laccase producing fungi are *Pleurotus ostreatus, Trametes Versicolor, Agaricus bisporus, P. eryngii, P. florida, P. pulmonarius, P. sajor-caju, T. hirsuta, T. pubescens, T. trogii,* and *T. villosa* (Baldrian, 2006; Arora & Sharma, 2010; Yang et al., 2017). The laccases produced by bacteria are less explored compared to fungi. Table 1 summarizes the laccase producing bacterial strains.

Role of Laccase in Bioremediation

Laccase has a wide range of applications in the field of environmental cleanup. The enzyme is reported to remediate pollutants such as PAHs, dyes, POPs, and phenolic compounds. Phenolic compounds are oxidized by laccases and form nontoxic aromatic amines (Zouari et al., 2006; Montaya et al., 2015).

The azo dye-containing phenolic group is oxidized by laccase enzyme, which generates phenoxy radical and oxidizes to carbonium ion (Camarero et al., 2005). Bacteria that produce Laccase enzymes are *Pseudomonas desmolyticum* (Kalme et al., 2009), *Bacillus* sps. (Dawkar et al., 2008), *Coriolus versicolor, Paraconiothyrium variabile, Tremetes versicolor* (Asadgol et al., 2014). The PAHs are degraded by laccase into quinines, which further forms carbon-di-oxide (Madhavi and Lele 2009). Apart from its use in pollutants degradation, laccase are used exclusively in textile azo dye degradation at varying pH, temperature, and nutrients (Pereira et al., 2009), (Masindi & Muedi, 2018). Ren et al., (2020) studied the degradation of 2,4-DCP contaminated soil by immobilizing laccase enzyme in an organic biofertilizer resulted in 58.6% degradation in 5 days.

CONCLUSION

Bioremediation is an environment-friendly, cost-effective, and highly explored method used to treat environmental pollutants. The microbes; fungi, bacteria, yeast, and algae are used in the process of remediation. The process aims to exploit natural methods available in the environment to bring its original characteristics. The enzymes produced by the microbes provide an advantage of better remediation and reusability compared to the microbes themselves. The application of enzymes at its optimum conditions results in better removal of toxic pollutants. Thus, this chapter provides insight into the effectiveness of bioremediation, especially by enzymes (mainly laccase), in removing organic and inorganic pollutants in the environment.

ACKNOWLEDGMENT

We extend our gratitude to The Management of Stella Maris College (Autonomous), affiliated to University of Madras, Chennai, Tamil Nadu, India.

REFERENCES

Abatenh, E., Gizaw, B., Tsegaye, Z., & Wassie, M. (2017). The role of microorganisms in bioremediation-A review. *Open Journal of Environmental Biology, 2*(1), 30-46.

Ahuja, S. K., Ferreira, G. M., & Moreira, A. R. (2004). Utilization of enzymes for environmental applications. *Critical Reviews in Biotechnology, 24*(2-3), 125–154. doi:10.1080/07388550490493726 PMID:15493529

Aksu, Z. (2005). Application of biosorption for the removal of organic pollutants: A review. *Process Biochemistry, 40*(3-4), 997–1026. doi:10.1016/j.procbio.2004.04.008

Alburquerque, J. A., De La Fuente, C., & Bernal, M. P. (2011). Improvement of soil quality after "alperujo" compost application to two contaminated soils characterised by differing heavy metal solubility. *Journal of Environmental Management*, *92*(3), 733–741. doi:10.1016/j.jenvman.2010.10.018 PMID:21035939

Araki, A., Saito, I., Kanazawa, A., Morimoto, K., Nakayama, K., Shibata, E., ... Saijo, Y. (2014). Phosphorus flame retardants in indoor dust and their relation to asthma and allergies of inhabitants. *Indoor Air*, *24*(1), 3–15. doi:10.1111/ina.12054 PMID:23724807

Arora, D. S., & Sharma, R. K. (2010). Ligninolytic fungal laccases and their biotechnological applications. *Applied Biochemistry and Biotechnology*, *160*(6), 1760–1788. doi:10.100712010-009-8676-y PMID:19513857

Arora, P. K., Kumar, M., Chauhan, A., Raghava, G. P., & Jain, R. K. (2009). OxDBase: A database of oxygenases involved in biodegradation. *BMC Research Notes*, *2*(1), 67. doi:10.1186/1756-0500-2-67 PMID:19405962

Arora, P. K., Srivastava, A., & Singh, V. P. (2010). Application of Monooxygenases in Dehalogenation, Desulphurization, Denitrification and Hydroxylation of Aromatic Compounds. *Journal of Bioremediation & Biodegradation*, *1*(03), 112. doi:10.4172/2155-6199.1000112

Asadgol, Z., Forootanfar, H., Rezaei, S., Mahvi, A. H., & Faramarzi, M. A. (2014). Removal of phenol and bisphenol-A catalyzed by laccase in aqueous solution. *Journal of Environmental Health Science & Engineering*, *12*(1), 93. doi:10.1186/2052-336X-12-93 PMID:25031840

Ayangbenro, A. S., & Babalola, O. O. (2017). A new strategy for heavy metal polluted environments: A review of microbial biosorbents. *International Journal of Environmental Research and Public Health*, *14*(1), 94. doi:10.3390/ijerph14010094 PMID:28106848

Azubuike, C. C., Chikere, C. B., & Okpokwasili, G. C. (2016). Bioremediation techniques–classification based on site of application: Principles, advantages, limitations and prospects. *World Journal of Microbiology & Biotechnology*, *32*(11), 180. doi:10.100711274-016-2137-x PMID:27638318

Bååth, E. (1989). Effects of heavy metals in soil on microbial processes and populations (a review). *Water, Air, and Soil Pollution*, *47*(3-4), 335–379. doi:10.1007/BF00279331

Baldrian, P. (2006). Fungal laccases–occurrence and properties. *FEMS Microbiology Reviews*, *30*(2), 215–242. doi:10.1111/j.1574-4976.2005.00010.x PMID:16472305

Beiyuan, J., Awad, Y. M., Beckers, F., Tsang, D. C., Ok, Y. S., & Rinklebe, J. (2017). Mobility and phytoavailability of As and Pb in a contaminated soil using pine sawdust biochar under systematic change of redox conditions. *Chemosphere*, *178*, 110–118. doi:10.1016/j.chemosphere.2017.03.022 PMID:28319738

Bielicka, A., Ryłko, E., & Bojanowska, I. (2009). Zawartość pierwiastków metalicznych w glebach i warzywach z ogrodów działkowych Gdańska i okolic. *Ochrona Środowiska i Zasobów Naturalnych*, (40).

Brown, L. D., Cologgi, D. L., Gee, K. F., & Ulrich, A. C. (2017). Bioremediation of oil spills on land. In *Oil Spill Science and Technology* (2nd ed., pp. 699–729). Gulf Professional Publishing. doi:10.1016/B978-0-12-809413-6.00012-6

Bull, S., Burnett, K., Vassaux, K., Ashdown, L., Brown, T., & Rushton, L. (2014). Extensive literature search and provision of summaries of studies related to the oral toxicity of perfluoroalkylated substances (PFASs), their precursors and potential replacements in experimental animals and humans. Area 1: Data on toxicokinetics (absorption, distribution, metabolism, excretion) in in vitro studies, experimental animals and humans. Area 2: Data on toxicity in experimental animals. Area 3: Data on observations in humans. *EFSA Supporting Publications*, *11*(4), 572E. doi:10.2903p.efsa.2014.EN-572

Camarero, S., Ibarra, D., Martínez, M. J., & Martínez, Á. T. (2005). Lignin-derived compounds as efficient laccase mediators for decolorization of different types of recalcitrant dyes. *Applied and Environmental Microbiology*, *71*(4), 1775–1784. doi:10.1128/AEM.71.4.1775-1784.2005 PMID:15812000

Chibuike, G. U., & Obiora, S. C. (2014). Heavy metal polluted soils: Effect on plants and bioremediation methods. *Applied and Environmental Soil Science*, *2014*, 2014. doi:10.1155/2014/752708

Coelho, L. M., Rezende, H. C., Coelho, L. M., de Sousa, P. A., Melo, D. F., & Coelho, N. M. (2015). Bioremediation of polluted waters using microorganisms. In N. Shiomi (Ed.), *Advances in Bioremediation of Wastewater and Polluted Soil*. InTech. doi:10.5772/60770

Das, M., & Adholeya, A. (2012). Role of microorganisms in remediation of contaminated soil. In *Microorganisms in environmental management* (pp. 81–111). Springer. doi:10.1007/978-94-007-2229-3_4

Das, N., & Chandran, P. (2011). Microbial degradation of petroleum hydrocarbon contaminants: An overview. *Biotechnology Research International*, *2011*, 1–13. doi:10.4061/2011/941810 PMID:21350672

Datta, S., Rajnish, K. N., Samuel, M. S., Pugazlendhi, A., & Selvarajan, E. (2020). Metagenomic applications in microbial diversity, bioremediation, pollution monitoring, enzyme and drug discovery. A review. *Environmental Chemistry Letters*, *18*(4), 1229–1241. doi:10.100710311-020-01010-z

Devasia, S., & Nair, A. J. (2016). Screening of potent laccase producing organisms based on the oxidation pattern of different phenolic substrates. *International Journal of Current Microbiology and Applied Sciences*, *5*(5), 127–137. doi:10.20546/ijcmas.2016.505.014

Dhiman, K., & Shirkot, P. (2015). Bioprospecting and molecular characterization of laccase producing bacteria from paper mills of Himachal Pradesh. *Proceedings of the National Academy of Sciences. India. Section B, Biological Sciences*, *85*(4), 1095–1103. doi:10.100740011-015-0541-x

Diamantidis, G., Effosse, A., Potier, P., & Bally, R. (2000). Purification and characterization of the first bacterial laccase in the rhizospheric bacterium *Azospirillum lipoferum*. *Soil Biology & Biochemistry*, *32*(7), 919–927. doi:10.1016/S0038-0717(99)00221-7

Dixit, R., Malaviya, D., Pandiyan, K., Singh, U. B., Sahu, A., Shukla, R., Singh, B. P., Rai, J. P., Sharma, P. K., & Lade, H. (2015). Bioremediation of heavy metals from soil and aquatic environment: An overview of principles and criteria of fundamental processes. *Sustainability*, *7*(2), 2189–2212. doi:10.3390u7022189

Doran, J. W., Sarrantonio, M., & Liebig, M. (1996). Soil health and sustainability. In D. L. Sparks (Ed.), *Advances in Agronomy* (Vol. 56). Academic Press.

El Fantroussi, S., & Agathos, S. N. (2005). Is bioaugmentation a feasible strategy for pollutant removal and site remediation? *Current Opinion in Microbiology*, *8*(3), 268–275. doi:10.1016/j.mib.2005.04.011 PMID:15939349

Fomina, M., & Gadd, G. M. (2014). Biosorption: Current perspectives on concept, definition and application. *Bioresource Technology*, *160*, 3–14. doi:10.1016/j.biortech.2013.12.102 PMID:24468322

Friedlova, M. (2010). The influence of heavy metals on soil biological and chemical properties. *Soil and Water Research*, *5*(1), 21–27. doi:10.17221/11/2009-SWR

Fu, Y., & Viraraghavan, T. (2001). Fungal decolorization of dye wastewaters: A review. *Bioresource Technology*, *79*(3), 251–262. doi:10.1016/S0960-8524(01)00028-1 PMID:11499579

Gianfreda, L., Sannino, F., Rao, M. A., & Bollag, J. M. (2003). Oxidative transformation of phenols in aqueous mixtures. *Water Research*, *37*(13), 3205–3215.

Giller, K. E., Witter, E., & Mcgrath, S. P. (1998). Toxicity of heavy metals to microorganisms and microbial processes in agricultural soils: A review. *Soil Biology & Biochemistry*, *30*(10-11), 1389–1414. doi:10.1016/S0038-0717(97)00270-8

Gupta, V. K., Nayak, A., & Agarwal, S. (2015). Bioadsorbents for remediation of heavy metals: Current status and their future prospects. *Environmental Engineering Research*, *20*(1), 1–18. doi:10.4491/eer.2015.018

Havugimana, E., Bhople, B. S., Kumar, A., Byiringiro, E., Mugabo, J. P., & Kumar, A. (2015). Soil pollution–major sources and types of soil pollutants. *Environmental Science and Engineering, 11*, 53-86.

He, Z. L., Yang, X. E., & Stoffella, P. J. (2005). Trace elements in agroecosystems and impacts on the environment. *Journal of Trace Elements in Medicine and Biology*, *19*(2-3), 125–140. doi:10.1016/j.jtemb.2005.02.010 PMID:16325528

Herawati, N., Suzuki, S., Hayashi, K., Rivai, I. F., & Koyama, H. (2000). Cadmium, copper, and zinc levels in rice and soil of Japan, Indonesia, and China by soil type. *Bulletin of Environmental Contamination and Toxicology*, *64*(1), 33–39. doi:10.1007001289910006 PMID:10606690

Hiner, A. N., Ruiz, J. H., López, J. N. R., Cánovas, F. G., Brisset, N. C., Smith, A. T., Arnao, M. B., & Acosta, M. (2002). Reactions of the class II peroxidases, lignin peroxidase andArthromyces ramosus peroxidase, with hydrogen peroxide catalase-like activity, compound III formation, and enzyme inactivation. *The Journal of Biological Chemistry*, *277*(30), 26879–26885. doi:10.1074/jbc.M200002200 PMID:11983689

Husain, Q. (2006). Potential applications of the oxidoreductive enzymes in the decolorization and detoxification of textile and other synthetic dyes from polluted water: A review. *Critical Reviews in Biotechnology*, *26*(4), 201–221. doi:10.1080/07388550600969936 PMID:17095432

ICRP Recommendations of the International Commission on Radiological Protection (Users Edition). (2007). ICRP Publication 103 (Users Edition). *Annals of the ICRP*, *37*(2-4), 1–332. PMID:18082557

Jan, A. T., Azam, M., Ali, A., & Haq, Q. M. R. (2014). Prospects for exploiting bacteria for bioremediation of metal pollution. *Critical Reviews in Environmental Science and Technology*, *44*(5), 519–560. doi:10.1080/10643389.2012.728811

Jarosiewicz, M., Duchnowicz, P., Włuka, A., & Bukowska, B. (2017). Evaluation of the effect of brominated flame retardants on hemoglobin oxidation and hemolysis in human erythrocytes. *Food and Chemical Toxicology*, *109*, 264–271. doi:10.1016/j.fct.2017.09.016 PMID:28893619

Kalme, S., Jadhav, S., Jadhav, M., & Govindwar, S. (2009). Textile dye degrading laccase from Pseudomonas desmolyticum NCIM 2112. *Enzyme and Microbial Technology*, *44*(2), 65–71. doi:10.1016/j.enzmictec.2008.10.005

Koua, D., Cerutti, L., Falquet, L., Sigrist, C. J., Theiler, G., Hulo, N., & Dunand, C. (2009). PeroxiBase: A database with new tools for peroxidase family classification. *Nucleic Acids Research*, *37*(Database, suppl_1), D261–D266. doi:10.1093/nar/gkn680 PMID:18948296

Leung, M. (2004). Bioremediation: Techniques for cleaning up a mess. *BioTeach Journal*, *2*, 18–22.

Lu, L., Zhao, M., Wang, T. N., Zhao, L. Y., Du, M. H., Li, T. L., & Li, D. B. (2012). Characterization and dye decolorization ability of an alkaline resistant and organic solvents tolerant laccase from Bacillus licheniformis LS04. *Bioresource Technology*, *115*, 35–40. doi:10.1016/j.biortech.2011.07.111 PMID:21868217

Madhavi, V., & Lele, S. S. (2009). Laccase: Properties and applications. *BioResources*, *4*(4), 1694–1717.

Mani, D., & Kumar, C. (2014). Biotechnological advances in bioremediation of heavy metals contaminated ecosystems: An overview with special reference to phytoremediation. *International Journal of Environmental Science and Technology*, *11*(3), 843–872. doi:10.100713762-013-0299-8

Marques, A. P., Rangel, A. O., & Castro, P. M. (2009). Remediation of heavy metal contaminated soils: Phytoremediation as a potentially promising clean-up technology. *Critical Reviews in Environmental Science and Technology*, *39*(8), 622–654. doi:10.1080/10643380701798272

Mbachu, A. E., Chukwura, E. I., & Mbachu, N. A. (2020). Role of Microorganisms in the Degradation of Organic Pollutants: A Review. *Energy and Environmental Engineering*, *7*(1), 1–11. doi:10.13189/eee.2020.070101

McBride, M. B., Shayler, H. A., Russell-Anelli, J. M., Spliethoff, H. M., & Marquez-Bravo, L. G. (2015). Arsenic and lead uptake by vegetable crops grown on an old orchard site amended with compost. *Water, Air, and Soil Pollution*, *226*(8), 265–279. doi:10.100711270-015-2529-9 PMID:26900187

Mirsal, I. A. (2008). Sources of soil pollution. In *Soil Pollution* (pp. 137–173). Springer. doi:10.1007/978-3-540-70777-6_7

Mongkolthanaruk, W., Tongbopit, S., & Bhoonobtong, A. (2012). Independent behavior of bacterial laccases to inducers and metal ions during production and activity. *African Journal of Biotechnology*, *11*(39), 9391–9398. doi:10.5897/AJB11.3042

Montaya. (2015). decolourisation of dyes with different molecular properties using free and immobilized laccases from *Trametes versicolor. Journal of Molecular Liquids*, *212*, 30–37. doi:10.1016/j.molliq.2015.08.040

Nannipieri, P., Badalucco, L., Landi, L., & Pietramellara, G. (1997). Measurement in assessing the risk of chemicals to the soil ecosystem. *Ecotoxicology: Responses, biomarkers and risk assessment. SOS Publications. Fair Haven, NJ, 7704*, 507–534.

Narayanan, M. P., Murugan, S., Eva, A. S., Devina, S. U., & Kalidass, S. (2015). Application of immobilized laccase from Bacillus subtilis MTCC 2414 on decolourization of synthetic dyes. *Res J Microbiol*, *10*(9), 421–432. doi:10.3923/jm.2015.421.432

Neifar, M., Chouchane, H., Mahjoubi, M., Jaouani, A., & Cherif, A. (2016). Pseudomonasextremorientalis BU118: a new salt-tolerant laccase-secreting bacterium with biotechnological potential in textile azo dye decolourization. *Biotech, 6*(1), 107.

Ojuederie, O. B., & Babalola, O. O. (2017). Microbial and plant-assisted bioremediation of heavy metal polluted environments: A review. *International Journal of Environmental Research and Public Health*, *14*(12), 1504. doi:10.3390/ijerph14121504 PMID:29207531

Okino-Delgado, C. H., Zanutto-Elgui, M. R., do Prado, D. Z., Pereira, M. S., & Fleuri, L. F. (2019). Enzymatic bioremediation: current status, challenges of obtaining process, and applications. In *Microbial Metabolism of Xenobiotic Compounds* (pp. 79–101). Springer. doi:10.1007/978-981-13-7462-3_4

Paltseva, A., Cheng, Z., Deeb, M., Groffman, P. M., Shaw, R. K., & Maddaloni, M. (2018). Accumulation of arsenic and lead in garden-grown vegetables: Factors and mitigation strategies. *The Science of the Total Environment*, *640-641*, 640–641. doi:10.1016/j.scitotenv.2018.05.296 PMID:29859443

Pant, D., & Adholeya, A. (2009). Concentration of fungal ligninolytic enzymes by ultrafiltration and their use in distillery effluent decolorization. *World Journal of Microbiology & Biotechnology*, *25*(10), 1793–1800. doi:10.100711274-009-0079-2

Park, J. W., Park, B. K., & Kim, J. E. (2006). Remediation of soil contaminated with 2, 4-dichlorophenol by treatment of minced shepherd's purse roots. *Archives of Environmental Contamination and Toxicology*, *50*(2), 191–195. doi:10.100700244-004-0119-8 PMID:16392021

Pereira, A. R. B., & de Freitas, D. A. F. (2012). Uso de micro-organismos para a biorremediação de ambientes impactados. *Revista Eletrônica em Gestão. Educação e Tecnologia Ambiental*, *6*(6), 995–1006.

Pereira, L., Coelho, A. V., Viegas, C. A., dos Santos, M. M. C., Robalo, M. P., & Martins, L. O. (2009). Enzymatic biotransformation of the azo dye Sudan Orange G with bacterial CotA-laccase. *Journal of Biotechnology*, *139*(1), 68–77. doi:10.1016/j.jbiotec.2008.09.001 PMID:18938200

Puvaneswari, N., Muthukrishnan, J., & Gunasekaran, P. (2006). *Toxicity assessment and microbial degradation of azo dyes*. Academic Press.

Ramos, L., Hernandez, L. M., & Gonzalez, M. J. (1994). Sequential fractionation of copper, lead, cadmium and zinc in soils from or near Donana National Park. *Journal of Environmental Quality*, *23*(1), 50–57. doi:10.2134/jeq1994.00472425002300010009x

Rayu, S., Karpouzas, D. G., & Singh, B. K. (2012). Emerging technologies in bioremediation: Constraints and opportunities. *Biodegradation*, *23*(6), 917–926. doi:10.100710532-012-9576-3 PMID:22836784

Ren, D., Cheng, Y., Huang, C., Wang, Z., Zhang, S., Zhang, X., & Gong, X. (2020). Study on Remediation-improvement of 2,4-Dichlorophenol Contaminated Soil by Organic Fertilizer Immobilized Laccase. *Soil and Sediment Contamination: An International Journal*, *30*(2), 201–215. doi:10.1080/15320383.2020.1828266

Rezaei, S., Shahverdi, A. R., & Faramarzi, M. A. (2017). Isolation, one-step affinity purification, and characterization of a polyextremotolerant laccase from the halophilic bacterium Aquisalibacillus elongatus and its application in the delignification of sugar beet pulp. *Bioresource Technology*, *230*, 67–75. doi:10.1016/j.biortech.2017.01.036 PMID:28161622

Sánchez-Porro, C., Martin, S., Mellado, E., & Ventosa, A. (2003). Diversity of moderately halophilic bacteria producing extracellular hydrolytic enzymes. *Journal of Applied Microbiology*, *94*(2), 295–300. doi:10.1046/j.1365-2672.2003.01834.x PMID:12534822

Savin, I. I., & Butnaru, R. (2008). Wastewater characteristics in textile finishing mills. [EEMJ]. *Environmental Engineering and Management Journal*, *7*(6).

Schmidt, O. (2006). *Wood and tree fungi*. Springer-Verlag.

Sharma, B., Dangi, A. K., & Shukla, P. (2018). Contemporary enzyme based technologies for bioremediation: A review. *Journal of Environmental Management*, *210*, 10–22. doi:10.1016/j.jenvman.2017.12.075 PMID:29329004

Sims, G. K., & Cupples, A. M. (1999). Factors controlling degradation of pesticides in soil. *Pesticide Science*, *55*(5), 598–601. doi:10.1002/(SICI)1096-9063(199905)55:5<598::AID-PS962>3.0.CO;2-N

Singh, G., Bhalla, A., Capalash, N., & Sharma, P. (2010). Characterization of immobilized laccase from γ-proteobacterium JB: Approach towards the development of biosensor for the detection of phenolic compounds. *Indian Journal of Science and Technology*, *2*(1), 48–53. doi:10.17485/ijst/2010/v3i1.8

Sondhi, S., Sharma, P., Saini, S., Puri, N., & Gupta, N. (2014). Purification and characterization of an extracellular, thermo-alkali-stable, metal tolerant laccase from Bacillus tequilensis SN4. *PLoS One*, *9*(5), e96951. doi:10.1371/journal.pone.0096951 PMID:24871763

Strong, P. J., & Burgess, J. E. (2008). Treatment methods for wine-related ad distillery wastewaters: A review. *Bioremediation Journal*, *12*(2), 70–87. doi:10.1080/10889860802060063

Tandon, P. K., & Singh, S. B. (2016). Redox processes in water remediation. *Environmental Chemistry Letters*, *14*(1), 15–25. doi:10.100710311-015-0540-4

Tang, C. Y., Kwon, Y. N., & Leckie, J. O. (2007). Fouling of reverse osmosis and nanofiltration membranes by humic acid—Effects of solution composition and hydrodynamic conditions. *Journal of Membrane Science*, *290*(1-2), 86–94. doi:10.1016/j.memsci.2006.12.017

Tehrani-Bagha, A. R., & Holmberg, K. (2013). Solubilization of Hydrophobic Dyes in Surfactant Solutions. *Materials (Basel)*, *6*(2), 580–608. doi:10.3390/ma6020580 PMID:28809328

UNEP. (2009). *Stockholm Convention on Persistent Organic Pollutants (POPs)*. http://irptc.unep.ch/pops

Varjani, S. J. (2017). Microbial degradation of petroleum hydrocarbons. *Bioresource Technology, 223*, 277–286. doi:10.1016/j.biortech.2016.10.037 PMID:27789112

Vidali, M. (2001). Bioremediation. an overview. *Pure and Applied Chemistry, 73*(7), 1163–1172. doi:10.1351/pac200173071163

Wang, C., Zhao, M., Lu, L., Wei, X., & Li, T. (2011). Characterization of spore laccase from Bacillus subtilis WD23 and its use in dye decolorization. *African Journal of Biotechnology, 10*(11), 2186–2192.

Wang, Z., Ren, D., Kang, C., Zhang, S., Zhang, X., Deng, Z., Huang, C., & Guo, H. (2020). Migration of heavy metals and migration-degradation of phenanthrene in soil using electro kinetic-laccase combined remediation system. *Journal of Environmental Science and Health. Part B, Pesticides, Food Contaminants, and Agricultural Wastes, 55*(8), 704–711. doi:10.1080/03601234.2020.1773719 PMID:32500809

Wentzel, A., Ellingsen, T. E., Kotlar, H., Zotchev, S. B., & Throne-Holst, M. (2007). Bacterial metabolism of long-chain n-alkanes. *Applied Microbiology and Biotechnology, 76*(6), 1209–1221. doi:10.100700253-007-1119-1 PMID:17673997

Whiteley, C. G., & Lee, D. J. (2006). Enzyme technology and biological remediation. *Enzyme and Microbial Technology, 38*(3-4), 291–316.

Williams, P. P. (1977). Metabolism of synthetic organic pesticides by anaerobic microorganisms. In *Residue reviews* (pp. 63–135). Springer. doi:10.1007/978-1-4612-6352-4_3

Yang, J., Li, W., Ng, T. B., Deng, X., Lin, J., & Ye, X. (2017). Laccases: Production, expression regulation, and applications in pharmaceutical biodegradation. *Frontiers in Microbiology, 8*, 832. doi:10.3389/fmicb.2017.00832 PMID:28559880

Yap, C. K., & Penge, S. H. T. (2019). Cleaning Up of Contaminated Soils by Using Microbial Remediation: A Review and Challenges to the Weaknesses. *American Journal of Biomedical Science & Research, 2*(3), 126–128.

Ye, X., Dong, F., & Lei, X. (2018). Microbial resources and ecology-microbial degradation of pesticides. *Natural Resources Conservation and Research, 1*(1).

Yeung, A. T. (2009). Remediation technologies for contaminated sites. In Y. Chen, X. Tang, & L. Zhan (Eds.), *Advances in Environmental Geotechnics* (pp. 328–369). Springer.

Zwolak, A., Sarzyńska, M., Szpyrka, E., & Stawarczyk, K. (2019). Sources of soil pollution by heavy metals and their accumulation in vegetables: A review. *Water, Air, and Soil Pollution, 230*(7), 164. doi:10.100711270-019-4221-y

Chapter 24
Rhizosphere Engineering and Soil Sustainability:
An Introduction

Samreen Nazeer
Nigde Omer Halisdemir University, Nigde, Turkey

Muhammad Zubair Akram
https://orcid.org/0000-0001-6797-6444
Nigde Omer Halisdemir University, Nigde Turkey

Madad Ali
University of Agriculture, Faisalabad, Pakistan

ABSTRACT

Soils are a vital part of agricultural production. Soil health plays a significant role in the best crop production. Nowadays, our lands are under immense pressure. This pressure may be in the form of climatic changes that affect crop productivity or may be due to population increment that forces our current food system to produce more food to meet consumer needs. Climatic changes affect soil sustainability in the wrong way. Salinity, drought, and heavy metals disturb land structure badly. As the population increases, it dramatically impacts the current production system to fulfill the present needs. In all these situations, agricultural soil sustainability is a challenging factor for soil scientists to make our agriculture sustainable because agricultural sustainability couldn't be possible without maintaining soil health. Many approaches are available to improve soil structure and health. Among these, plant growth-promoting rhizobacterium is a good option. It not only improves soil structure but also helps the plants under abiotic stress conditions.

DOI: 10.4018/978-1-7998-7062-3.ch024

SUSTAINABLE AGRICULTURE AND SOIL SUSTAINABILITY: AN INTRODUCTION

As the population is increasing tremendously, our agriculture is under enormous pressure to produce an adequate amount of food for this increasing population. So, our land and agriculture must be sustained accordingly and this can only be possible by the intervention of environmentally friendly techniques. In recent times, our agriculture is facing two types of challenges: firstly, to produce adequate food from the current disturbed system of cultivation and, secondly, sustainable livestock resources. Soil is a vital renewable resource and its sustainability is a more significant challenge for present researchers. Production from soils can be increased by many approaches like utilization of chemicals, agronomic techniques, and overuse of water for irrigation needs that lead to soil degradation. Ultimately, the soil cannot maintain its structure. On the other hand, soil health can also be affected by reducing soil organic matter contents and loss of living organisms in soils. All these things result in adverse effects on ecosystems that diminish the sustainability of agriculture (Meena et al., 2015).

Soil health is the main key factor in fulfilling both challenges of sustainable agriculture and soils (Meena et al., 2016). Excessive use of chemicals in pesticides, insecticides and fertilizer harms soil health. The usage of these chemicals now becomes an essential part of the modern agriculture system, but by using new generation techniques, it is possible to reduce the use of these chemicals without affecting the crop (Mia and Shamsuddin, 2010). Recent studies give a precise interaction mechanism between the roots of the rhizosphere of plants. As the seed starts its germination and the root develops from radicle, the organic matter present in the soil promotes root growth and microbial population in the soil, particularly in the root zone. That is called the "Rhizosphere effect" (Meena et al., 2018).

Rhizobacteria are the key elements that play an essential role in the sustainability of soils. The soils rich in microorganisms have a well-organized soil structure and these soil productions are higher than the disturbed soils. Plant growth-promoting rhizobacteria are currently used worldwide because of their low cost, effectiveness and eco-friendly characteristics. Different factors affecting soil activity include high or low soil temperature, deficit soil moisture, and low pH (less than 5.5) (Aliyu et al., 2013). Besides this, PGPR are beneficial to maintain soil structure and soil health.

By growing human population, food consumption also increases which is observed as a great challenge for agriculture. Besides, climatic change, land degradation and reduced resources like freshwater also significantly affect agricultural production. It has also been observed that by the next ten years, crop yield will be growing at a rate less than 1% and expansion of arable land will be very limited (Alexandratos and Bruinsma, 2012). Overall, food demand is increasing with declining crop production. Sustainable agriculture must meet two challenges, i.e., food security and human health (Godfray et al., 2010). For this purpose, different approaches can be used among which the genetic engineering approach is most important. Thus, the development of stress-resistant crops suitable for the current environmental conditions is significant (Shinozaki et al., 2015). In last two decades, considerable work has been done on the model plant *Arabidopsis* for improving response to current climatic conditions (Zhu, 2016). Now, genetically enhanced plants can be used as main crops in the areas where they are suitable.

Different Approaches to Soil Sustainability

There are many approaches and management methods that have been practiced worldwide for the soil sustainability. Here, we will discuss different strategies that ensure their efficiency, security, protection, vitality and acceptability. These approaches should have more financial returns by increasing efficiency day by day and supporting minimum security risks. Resources of water and soil should be protected for the future. A specific approach should be accepted socially and economically; if not, the system will fail in the future (Tuğrul, 2019). These approaches are the following (FESLM, 1993; ACSEDU, 2020; Altieri, 2017):

Optimizing Soil Nutrients Use

For agriculture crops, application methods, time and types of the nutrient are vital for beneficiary yields. It is certified that through different ways, 50% of applied nitrogen and 90% applied phosphorus is not available for the plant. High nitrogen losses lead to environmental pollution. Fertilizer use can be decreased by minimizing soil nutrient losses (caused by leaching, denitrification, evaporation and surface flow). The mulching approach can reduce the fertilizer amount, but it is expensive. A significant setback is the reduction of resources. Accordingly, if a cultivator utilizes lesser resources (e.g., chemicals, fertilizer, petroleum, cash and human resources), farm expenses will be lessened. There is less possibility of harm due to the wastage of residuals or overburden. The land and earth will face the minimum possible to run out of the necessary resources for soil sustainability (Rware, 2016).

Restorable Farming

This farming seeks to develop a farming system that can restore itself automatically or with a small number of resources. Practices such as vermicompost, green manures and recycled products can be useful for soil sustainability after harvesting each crop. Nowadays, permaculture maybe the only farming approach that can be restorable. Permaculture is a set of proposed postulates focused on entire practices such as studying, assuming, or precisely using the models and recover characteristics discovered in natural ecosystems, for example, the landscape may be either small (home garden) or large (agricultural farm). It may be harvested to supply products like woods, eggs, fruits, herbs and vegetables without damaging the environmental equilibrium. But it demands tiny input and once it is established, it persists to generate yields and stay sustainable.

Biodynamic Systems

This approach is focused on mobilizing biological systems. Living organisms in the soil, such as bacteria (actinobacteria) and worms (earthworms), convert organic matter into the usable plant form that ensures nutrients available to pastures and crops. Nature would help dispose of wastages (e.g., animal residuals) and persuade predators to get rid of pests and weeds under appropriate circumstances.

Organic Systems

Conventional organic systems involve naturally occurring inputs for fertilizers and insect/pest management and approach like crop rotation and composts. There are several soil associations such as the Biological Farmers Association (BFA) in Australia and the National Association for Sustainable Agriculture (NASAA) in the UK monitoring the development of organic systems.

Conservation Farming

This approach is established on the ideas of agricultural land resources that are previously present on the land. It requires different actions such as recognizing and preserving the level and quality of soil, water channels, water stream/slakes, nature strips, slopes, etc. Severe conventional agriculture practices can cause soil deprivation physically and chemically, such as a decline in organic matters limited biological exercises in the soil, which reduced crop yields. On the opposite, the procedure of sustainable agriculture reflects a sustainable and profitable farming approach. Conservation farming settled on three fundamental rules; 1) crop rotation, 2) soil-free agriculture, 3) Continuous soil surface covered with plants or detritus of plants (Giller, 2015).

Cover Crop and Rotation

This approach protects soil for extensive use for certain types of nutrients. Each type of crop utilized particular nutrients for growth. If we grow one crop season after season, soil fertility will decrease and weeds and pest problems will also appear. So, if we rest soil for one season by any covering crop (grasses) or practicing crop rotation techniques (legumes for nitrogen fixation), it will help soil to restore its quality.

Water Sustainability

In agriculture, water plays a vital role in crop production. Water management is essential for soil sustainability because plants use water for obtaining nutrients through the soil. Water sustainability should be maintained by ensuring water quality and calculating the required quantity for irrigation. To reduce waterborne erosion, it has to be assured that the water is infiltrated to the soil and irrigation design on the path of affection.

Soil Organisms

The soil gives residence to a vast proportion of biodiversity. The bond between soil organisms and soil capacities are perceived to be complicated. It is understood that soil living organisms tear down organic matter, making nutrients obtainable for plants and organisms. Soil organisms store soil nutrients and prevent any depletion by leaching. Soil organisms serve to sustain soil structure and earthworms play a vital role in bioturbation. Despite this, we don't know the critical features of how these organisms function and interact with the soil. The identification of glomalin in 1995 symbolizes that we still need to require more information to precisely answer some of the usual fundamental puzzles about the biogeochemical cycle within soils.

In well-balanced soil, plants flourish in well-effective and consistent conditions. The organic-matters content and its magnificent structure are crucial for soil richness and the presence of life in soil empowers soil cycles and enhances soil fertility. Organism's actions and organic matters in the soil would expand and put debris on the soil surface because that provides food supplies to plants. The main soil biota is reflected in Table 1.

Table 1. Soil biota and its characteristics

Soil Biota	Size Range	Examples
Megafauna	20 mm upward	Moles, rabbits and rodents
Macrofauna	2 to 20 mm	Woodlice, earthworms, beetles, centipedes, sludges, snails, ants and harvestmen
Mesofauna	100 mm to 2 mm	tardigrades, mites and springtails
Microfauna and Microflora	1 to 100 mm	yeasts, bacteria (commonly actinobacteria), fungi, protozoa, roundworms and rotifers

Source: Sachidanand, 2019

Bacteria and fungi participate as pivotal characters in sustaining better soil health. They serve as decomposers that tear down organic matter to generate available beneficiary nutrients in the soil. Soil detritivores such as earthworms ingest residues and decompose it. Saprotrophs, adequately exposed by fungi and bacteria, extract solvable nutrients from detritus. The ants (macrofauna) assist by tearing down the orgainic matter and enhancing the soil quality. Also, rodents, wood-eaters encourage the soil to be extra absorbent.

Soil Quality

Soil quality is the capability of the soil to operate for particular land use within different ecosystemic borders. This capability is an inborn feature of soil and differs from soil-to-soil. Some factors include organic matter content, accessible nutrients, tilth, compactness and rooting depth, which determines and influences the soil quality and health. Organic matter content, soil organism movement, acidic and saline nature of soil are associated with the soil's capability to accumulate a whole cycle of nutrients in favor of plant growth. Soil compactness, tilth, and accessible water utilizing capacity imitate a soil's ability to control and division the water-flow. Texture (e.g., loam, silt, clay) is a significant soil asset and a foundation for soil construction. For decades, soil quality improvements have been determined by increasing the organic matter content in agricultural soil. Quality gives an excellent contribution to soil sustainability as it is necessary to enhance and maintain soil capability to perform as humane desire. On the other hand, if a soil's ability to carry out advantageous functions has weakened, the soil quality will be ruined and will become challenging to achieve soil sustainability (Johnson, 1997; Cherubin, 2017).

GENETIC ENGINEERING FOR SUSTAINABLE SOILS

Genetic Engineering and Rhizosphere

The process by which pieces of DNA are transferred from one organism to another is called genetic engineering. The process involves creating recombinant DNA molecules by manipulating a DNA sequence. The DNA produced is then inserted in the desired host organism. Cloning is also an example of genetic engineering. A vector is needed to transfer a gene into a host cell. Vectors are either plasmids or viruses. We can brief the process by following steps (Koh *et al.*, 2015; Nicholl, 2008; Michels, 2002; Rocha-Martins et al., 2015);

1. DNA carrying a gene of interest is taken from a cell (include choosing target gene, extraction from the cell, gene isolation and modification).
2. The gene is inserted into the DNA of another cell (Transformation, Transfection, Transduction).
3. Gene targeting (Meganucleases and Zinc finger nucleases, TALEN and CRISPR).
4. The host cell now contains recombinant DNA (Regeneration and Confirmation).
5. The host cell multiplies.
6. The desired protein is produced.

All components of the rhizosphere (plants and microbes) can be engineered and soil can be amended to promote plant health and growth, from the field to the landscape scale. Plant engineering has led to valuable results in terms of resistance to high metal concentration in soil and resistance to pathogens. Aside from plant growth-promoting rhizobacteria living at the root surface, endophytic bacteria receive renewed attention. They have proved to be of interest, particularly in the context of tolerance to pollutants. A novel aspect of microbial engineering involves population engineering rather than single strain engineering. What is observed in the animal world, the plant's apparition and its associated microbial cortege are changing; they are not separate elements rather constituents of a superorganism, the holobiont (Simon et al., 2019).

All components of the rhizosphere can be engineered to promote plant health and growth; these two features strongly depend upon the interactions of living organisms with their environment. More generally, the plants (and the associated microbes) are no longer seen as 'individual' rather as a holobiont, in other words, a unit of selection in evolution. This concept holds great promise for future plant breeding programs.

Rhizosphere Engineering for Sustainable Soils

Underneath the soil, the rhizosphere is the narrow zone of plant root and soil interaction. Too much communication has been observed in this area. The roots, soils and the food web are also affected by this zone (Dessaux et al., 2016). The rhizosphere includes three zones: endorhizosphere, rhizoplane and ectorhizosphere (Walker, 2011). Endorhizosphere is a region of the root cortex and endodermis. At that region, microbes and other minerals, ions reside in apoplastic space between the cells. Rhizoplane is considered the middle area following root epidermal cells and mucilage.

In the same way, the ectorhizosphere is the outer zone which banquets from rhizoplane out into bulk soil (McNear Jr., 2013). One important thing to keep in mind is that the rhizosphere area is not of definite size and shape. However, it is the gradient zone for physical, biological and chemical properties along with the roots (McNear Jr., 2013).

The Rhizospheric zone is strongly influenced by metabolism by delivering CO_2 from plants. It is also influenced by photosynthate secreted to assemble root exudates (Estabrook and Yoder, 1998). Root exudates mainly include phytohormones that aid in enhancing rhizospheric interaction. It provides energy for microorganism's actions. These root exudates also act as chemical repellents and attractants in the rhizospheric zone (Bais et al., 2001). They also provide communication molecules to start physiological and biological interaction between plant roots and microorganisms. They enhance soil's chemical and physical properties by promoting microbial community, e.g., promoting nitrogen-fixing bacteria, inhibiting competing species of plants, and reducing fungal, insect and pest attacks (Nardi et al., 2000).

Global changing climate, rising temperature, water scarcity conditions and rise in atmospheric CO_2 significantly affects rhizospheric ecology. It ultimately affects communication between plant roots and soil due to which crops are affected, e.g., it has been estimated that with an increase of temperature from 1981 to 2005, a massive reduction in major cereals has been observed that costs five billion dollars per year (Lobell and Field, 2007). Due to global climatic changes, abiotic stresses increase day by day, including temperature, drought, and salinity. Drought stress severely affects photosynthesis and roots in the rhizospheric zone (Verslues, 2017). Due to salinity, ion toxicity occurs, leading to reduced plant growth and development (Negrão et al., 2017). Combining both stresses increases ethylene's level that is an inhibitor of root growth and affects many physiological pathways, e.g., hormonal imbalance and increasing susceptibility to pest and fungal attacks (Sun et al., 2016). For a plant's survival under such climatic conditions, extensive physiological adaptations are needed that mainly include hormonal balance that enhances root growth and development under such conditions. Rhizospheric engineering must be required under these abiotic stresses for plant survival.

Different approaches are existing to make the plant tolerant against abiotic stress conditions. Transcriptome engineering is an essential approach in which continuous overexpression of a gene producing proteins and pathways scavenges ROS and has tolerance against such adverse conditions. Still, this approach is limited due to more than one path in plant tolerance and pleiotropic growth (Reguera et al., 2012). The agrochemical approach also exists, but it is also limited due to its long-term usage and the cost and environmental contamination threats. An exciting system to cope with all these adverse conditions is beneficial mutualistic plant microorganisms. They enhance root growth and uptake of nutrients and increases biomass production (Mirshad and Puthur, 2017). This approach has many advantages, like it can be applied to more than one stress and it can be helpful to evade a wide variety of plant hosts (Vurukonda et al., 2016a).

Role of Rhizosphere Engineering under Abiotic Stress

Abiotic stresses such as drought, salinity and temperature have severe effects on plant growth. These abiotic stresses reduce plant growth, which leads to low yielding. Rhizosphere organisms act as plant growth promoters to help the plant to flourish. We should manipulate the genetics of these rhizosphere microorganisms according to our desire to be helpful under abiotic stresses. We discuss the following three abiotic stresses; drought, salinity and temperature and the role of rhizosphere organisms bellow:

Drought

Drought mainly considers the most important single threat to plant productivity. It affects the growth and development of a plant more than any other abiotic condition (Anjum et al., 2011). With the increment of temperature, drought threats to biomass production increase (Quinn et al., 2015). A sustainable rhizosphere and microbiota play a significant role in plant response to drought. Plant growth-promoting bacteria confers drought stress in many ways. They colonize the rhizosphere of plants and are producing many chemicals. They make exopolysaccharides, ACC, volatile organic compounds and many other phytohormones, i.e., Abscisic acid, GA and IAA. They also enhance the accumulation of antioxidant enzymes and increases the production of osmolytes. They alter root morphology and also regulate stress-responsive genes (Vurukonda et al., 2016b). *Azospirillum* spp. improves root morphology by IAA production that enhance its growth and lateral root formation and produces tolerance against drought in wheat (Arzanesh et al., 2011). In the same way, soybean plants with inoculation of gibberellin producing *Pseudomonas putida*, increase shoot length and fresh weight under water deficit conditions. These inoculated plants seem at a higher level of SA (salicylic acid), ABA and chlorophylls contents (Kang et al. 2014).

Salinity

Salinity is a major limiting factor for crop productivity. It enhances ion toxicity that ultimately increases the ionic concentration in the rhizosphere which disrupts metabolic balances. It also produces a water deficit by hyperosmotic character. Plants cope with this stress by producing different osmolytes and antioxidant enzymes that scavenge ROS (Upadhyay et al., 2011). In recent studies, inoculation of carotenoid-producing halotolerant bacterium *Dietzia natronolimnaea* in wheat reports an increased tolerance against salinity. Tolerance increases due to a higher level of proline and other antioxidant enzymes (Bharti et al., 2016). SOS activity, abscisic acid signaling and even iron transport seem to be improved in inoculated plants. Similarly, peanuts under salinity conditions inoculated with five bacterial isolates from *Klebsiella*, *Pseudomonas*, *Agrobacterium* and *Ochrobacterium* genera show an enhanced salinity tolerance by ionic homeostasis and less ROS production that promotes growth (Sharma et al., 2016).

Temperature

Heat shock is also a significant environmental stress that limits plant productivity. It affects both plants and microbial population growth and its homeostasis. Less work has been done to find potential PGPRs that allow the plants to alleviate temperature conditions. Experiments with inoculations of PGPR promoting plant growth under high temperature are significantly fewer. Till now, research has been focused on identifying the bacterium species that can survive at a high temperature of 60 ^{0}C (Rodriguez et al., 2008). A few studies have shown that, somehow, up to a certain extent, microbiota decrease the effect of high temperature (Barka et al., 2006; Dimkpa et al., 2009). Grapevine plants inoculated with *Bacillus phytofirmans* show a higher level of photosynthesis, carbohydrates, proline and phenolic contents (Barka et al., 2006).

Microbial Genetic Engineering

Microbes are the most important organisms that can contribute to soil sustainability. Over the past 50 years, different microorganisms are used to advance technologies in various sectors like medice, human and animal health, food processing, genetic engineering, environmental protection and agricultural biotechnology (Kavino et al., 2007). Although there have been many successive stories regarding microbial inoculation in agricultural biotechnologies, it didn't take too much of the scientist's attention because it is not easy to get consistent results from these populations. The microbes only perform well when provided with optimum temperature, pH and oxygen (Harish et al., 2009a). Nowadays, microbes are an excellent alternative to chemical fertilizers and pesticides that are widely applied in agriculture currently.

Plant growth-promoting rhizobacteria and cyanobacteria are present in the rhizosphere called microbes of soil. They produce many bioactive compounds in the soil that enhance soil structure, plant growth, and protection against pathogens (Harish et al., 2009b). Only efficient and microbial biota is suitable for sustainable agriculture. Microbial inoculation is a different approach to optimize soil and plant management practices, includes crop rotations, soil fertility restoration, crop husbandry recycling, soil quality sustainability and biocontrol of plant diseases. The utilization of PGPR enhances ACC deaminase activity for plant growth promotion under normal and stress conditions and genetic transformation of those genes that produce such enzymes (Sergeeva et al., 2006).

A study in 1995 revealed that *Anabaena* sp. and *Nostoc ellipsosporum* (two cyanobacteria) could break down lindane (a highly chlorinated aliphatic pesticide). The investigation showed that the genetic engineering approach could enhance this ability. It also provided qualitative evidence about these two strains which aid in the degradation of chlorinated pollutant 4-chlorobenzoate (Kuritz and Wolk, 1995). The first-ever study announces that cyanobacteria can be genetically modified for the degradation of organic pollutants. There are some studies present that also show that cyanobacteria can remediate heavy metals from contaminated soils. *Limnothrtix planctontca*, *Synechococcus leopoldtensts* and *Phormidium limnetica* strains can be used against Hg (Lefebvre et al., 2007). *Lyngbya* and *Gloeocapsa* are effective against Cr accumulation (Kiran et al., 2008).

The availability of micronutrients in the soil depends on crop species, root interaction with rhizobacterial microorganisms, and soil surrounding the plant's roots. Zinc is essential for a plant's normal growth and functioning, but its deficiency shows Zn deficiency symptoms. Zn may be deficient in plants not because of soil deficit, but because of the less bioavailability (Baruah, 2018).

Efficient soil microbes present in the soil are maybe in the form of plant growth-promoting rhizobacteria or cyanobacteria. Microbial interaction is present in the soil; because of this, the growth of plants is promoted. They also act as biological fertilizers and biocontrol agents in the soil. Some genetically modified strains can be used to remediate polluted chemicals from the soil, like heavy metals. These microbes also enhance the stability and productivity of desert soils. Soil structures improve by the employment of microbes. Ultimately all these amendments enhance soil and plants (Singh et al., 2011).

ADVANTAGES

Rhizosphere engineering is an essential system to postulate many advantages related to genetic engineering approaches for soil sustainability. It is already reported that plants interact with their rhizosphere by secreting carbon and other metabolites in the soil (Badri et al., 2012) and there exist genetic variations in these traits (Rovira, 1969). These variations lead to contribute to variations in the microbial community in soils (Bouffaud et al., 2014). A well-known fact is that microbes play an essential role in the survival of the plant in adverse conditions by aiding in host nutrient acquisition (Marschner et al., 1986) and changing growth pattern related conditions (Vacheron., 2013).

Genetic engineering approaches can highly achieve soil sustainability. The best example of this is the phytoremediation of heavy metals (using plants to remove trace elements from soil to make soils heavy metal-free). The plants used for this approach must have the ability to adapt to an area outside of its collection. They must have an extensive root system with a fast growth rate and a high capacity to store trace elements in their areal parts (Pilon-Smits, 2005). Such types of plants can be developed through a traditional breeding system. Such plants can be easily cultivated by genetic engineering that is rapid and provides novel genes which can be efficiently inserted into phytoremediatiing plants (Jagtap and Bapat et al., 2015).

Besides heavy metals, the genetic engineering approach can also develop such plants that are salt tolerant. Such transgenic plants can accumulate and tolerate higher concentrations of salts. In transgenic

Arabidopsis plants, the insertion of OsAP21 and SbAP37 results in higher growth and the plant shows a better tolerance against salt/drought and temperature conditions (Parveda et al., 2017). Similarly, the MYB transcription factor family is the most important transcription factor family in abiotic stress response. Overexpression of the OsMYB2 transcription factor in *Arabidopsis* makes the plant more tolerant to salinity than wild-type plants.

PGPRs also play an essential role in the sustainability of soils by affecting many genes. PGPRs are vital organisms for the economy. It has been estimated that Brazil saves up to 7 billion dollars per year by replacing N fertilizers with PGPRs (Hungria et al., 2013). PGPRs form a symbiotic relationship with plants. Previously, Beijerinck (1888) observed the symbiotic effect of PGPRs. Then most of the studies postulated that these microorganisms only belong to *Rhizobium* genera. Simultaneously, it was said that only a small number of organisms belong from this genera and the rest are from *Bradyrhizobium* (Meena et al., 2017). For a functional symbiotic relationship, two genes are most important and necessary; one is for N fixation and the other is involved in nodulation (Sammauria and Kumawat, 2018). Genes responsible for nodulation are named *nodABC* and those for N fixation called as *nifHDK*. *Rhizobium etli* overexpressing trehalose-6-phosphate synthase genes in *Phaseolus vulgaris* results in the upregulation of genes that are involved in stress tolerance. Besides this, they also overexpress those genes involved in carbon and nitrogen metabolism (Suárez et al., 2008). Similarly, *Bacillus* isolate 23-B with *Mesorhizobium ciceri* in chickpea results in a higher concentration of proline which improves germination, root and shoots growth due to which biomass increases and the plant shows a better growth (Sharma et al., 2013 a, b).

CHALLENGES

The following challenges are constrained to soil sustainability and genetic engineering:

Soil Biota Threats

The soil is the repository for most seeds of different weeds, insects/pests, nematodes, and pathogens ahead to plant disease, carrying advantageous living microbes. The sustainable control of soil resources is fundamental to the reduction of biotic plant threats. Despite, lack of knowledge of how agronomic exercises should be implemented, also exist; such as tillage and applying organic supplements and crop residuals influence plant biotic threats tenacity and transmission. This deficiency of information limits our capacity to produce reliable data on how soil management concludes both crop health and productivity. Direction needs to be elaborated on how to handle soil features that will improve soil capacity to overcome crop diseases and pests by sustainable soil management.

Availability of Cover Crop

Future studies are required to improve soil's crop-protection properties, crucial for flexible and sustainable crop health. Cover crops help to sustain soil quality but the availability of cover crops is problematic. Cover crops such as buckwheat, hairy vetch, grasses, clover and rye are not readily available. Significant management provocations can be discussed with the efficient application of cover and associate crops, optimizing accuracy tillage for seedbed development, covering crops to bio-remediate soil composition, classification of quality traits/crop flexibility markers and evaluation of germplasm which challenges the soil conditions.

Erosion

Over the previous 150 years, almost 50% of topsoil on the earth has been misplaced. In the extension of erosion, soil quality is influenced by different perspectives of agriculture. These perspectives contain compactness, destruction of soil structure, nutrient deterioration, and enhancement of salinity. These are quite practical and also critical issues. The consequences of soil erosion are somehow out of control, which leads to the loss of fertile soil. Soil erosion has been directed to enhanced contamination and sedimentation in streams/lakes and rivers, obstructing rivers and lakes resulting in farmness in fishes and other different species. And erosion affected lands are less capable of capturing water, which makes flooding more damaging. Sustainable land management may help reduce soil erosion, degradation, and loss of well-intentioned land, leading to desertification by minimizing the influence of agricultural practices and livestock activities.

Salinization

Soil salinity is an obstacle to crop production throughout the globe. According to accessible records, crops cultivated in saline soils are subjected to osmotic-stress, weak physical soil structure, imbalance in nutrients, and decreased crop yields. Limiting crop damages because of stress caused by salinity is an influential field of interest (Etesami and Noori, 2019).

Carbon Sequestration

Forest, grassland and savannas under human influence lose almost 30-40% of its soil organic carbon (SOC) content (Poeplau, 2015). If we change land-use methods, SOC will either increase or decrease unless it reaches the equilibrium level, affecting climate change. By increasing the input of organic carbon, we can improve the SOC. For that purpose, different approaches such as harvest residuals being left on harvested land, application of manure instead of chemical fertilizers and crop rotation (with perennial crops) can be used (Goglio et al., 2015).

Acidification

The acidity of soil can provoke disturbance to plants and soil living organisms; in plants, soil acidity results in shorter, short-lasting roots. Acidification of soils sometimes destroys the root tips, lessening growth (Haling et al., 2011). Plant height is undermined, and seed germination reduces too, which leads to lesser plant density. Some plants stunt their growth due to the acidification of soil. The diversity of macrofaunal and microorganisms has also been lessened by acidification (Horne et al., 1996; Slattery and Hollier, 2002).

Risks of Genetic Engineering

Risks of genetic engineering are the following (Dabrock, 2009; Ghareeb, 2009):

- Transference of the chosen gene into different species can advantage one individual but can damage another.
- The majority of people declared that use of genetic engineering means altering nature is an unethical act.
- Genetic engineering is costly and requires highly skillful scientists to use it. So, developing countries may not access and afford such technologies.
- Genetic engineering may have adverse effects on human and animal lives; for example, cancer in humans and toxicity in plant pollens kill beneficiary insects.

CONCLUSION

In conclusion, soil sustainability is an essential phenomenon to achieve the best products from a specific crop. Nowadays, soils are much more polluted by different types of pollutions among which water pollution is most important. When herbicides and pesticides are sprayed on crops, these chemicals mix with water and move in groundwater and pollute it. Similarly, factories and heavy industries also secrete trace elements to the land that are very bad for humans and plant health. As a result of which, heavy metals accumulate in the soil and pollute the environment. Due to these metals, soil structure is disturbed and soil losses its fertility. There are many approaches to reclaim such soil types, but genetic engineering approaches are mostly used nowadays. Transgenic plants formed by these techniques are very efficient in the reclamation of such soils. Many halophytes are edited by using genetic engineering techniques to make them more efficient. These plants improve the soil structure and fertility of the soil and ultimately enhance plant health.

SUMMARY

Population increment is the main issue nowadays by which food demand is increasing day by day. Agricultural lands are going under urbanization. Our lands are decreasing and the demand for food is rising. For this purpose, there is a need to increase our food from currently available resources. Another main reason for food shortage is the current climatic conditions. Due to global warming, our global temperature is rising. This increment of temperature severely affects our food production. Almost 15 to 20% of the land is under salinization. Salinity is the main issue in the recent era that is affecting the crops badly.

On the other hand, most countries are facing water shortage issues. Water deficiency conditions severely affect our land structure. The crop is affecting by drought and they are not able to grow naturally. Agricultural sustainability with agricultural soil sustainability is the main challenge for researchers.

Our lands are under enormous pressure to meet the food demand of the current population. It becomes a massive threat to global food security. Urbanization affects our land in a threatening manner. The water coming out from industries have a more significant impact on agricultural land. This water contains a vast amount of trace elements that cause pollution in land and water as well. Our underground water is contaminating because of this factory's wastes. Excessive use of chemicals in the form of pesticides, fungicides, insecticides and chemical fertilizer also pollutes our underground water because after applying these chemicals, they leach down in the soil and mix with groundwater. Because of toxic chemicals present in our soil, its structure becomes too poor. This poor structure is not able to fulfill the demand for food. Crops that grow in these soils are not able to get nutrients from such soils. As a result, the crops cannot grow well and at last end up significantly less productive.

There are many approaches to cope with all of the situations mentioned above. Many agronomic practices can be implemented to make the soil structure better. Organic matter can be added to the soil to make the soil fertility level at a certain level. Besides this, plant growth-promoting bacteria is the best choice to make soil health good. They colonize the soil and make all the nutrients available to the plants. They make the soil health in excellent conditions. There are many genetic approaches to detect such strains that are better for specific soils. Many studies are available which claim to make the soil best by the inoculation of certain organisms and are now used globally because of their eco-friendly behavior.

Genetic engineering also helps to make the soils in better conditions to make the production profitable. In the soils under salinity and heavy metal stress, by using genetic engineering techniques, transgenic plants are formed that extract these chemicals from soils and improve soil structure. Heavy metals are stores in the upper part of the plant and soil remains clean from these metals. In this way, soil structure can be improved and can be made useful and able to meet consumers' demand. Agricultural sustainability is highly dependent on modern genetic engineering techniques.

REFERENCES

ACSEDU. (2020). *Different Approaches to Sustainability*. Retrieved from https://www.acsedu.co.uk/Info/Agriculture/Sustainable-Agriculture/Different-Approaches-to-Sustainability.aspx

Alexandratos, N., & Bruinsma, J. (2012). *World agriculture towards 2030/2050: 2012 revision*. ESA.

Altieri, M. A., Nicholls, C. I., & Montalba, R. (2017). Technological approaches to sustainable agriculture at a crossroads: An agroecological perspective. *Sustainability*, *9*(3), 349. doi:10.3390u9030349

Anjum, S. A., Xie, X. Y., Wang, L. C., Saleem, M. F., Man, C., & Lei, W. (2011). Morphological, physiological and biochemical responses of plants to drought stress. *African Journal of Agricultural Research*, *6*(9), 2026–2032.

Arzanesh, M. H., Alikhani, H. A., Khavazi, K., Rahimian, H. A., & Miransari, M. (2011). Wheat (Triticum aestivum L.) growth enhancement by Azospirillum sp. under drought stress. *World Journal of Microbiology & Biotechnology*, *27*(2), 197–205. doi:10.100711274-010-0444-1

Badri, D. V., Chaparro, J. M., Manter, D. K., Martinoia, E., & Vivanco, J. M. (2012). Influence of ATP-binding cassette transporters in root exudation of phytoalexins, signals, and in disease resistance. *Frontiers in Plant Science*, *3*, 149. doi:10.3389/fpls.2012.00149 PMID:22783269

Bais, H. P., Loyola-Vargas, V. M., Flores, H. E., & Vivanco, J. M. (2001). Root-specific metabolism: The biology and biochemistry of underground organs. *In Vitro Cellular & Developmental Biology. Plant*, *37*(6), 730–741. doi:10.100711627-001-0122-y

Barka, E. A., Nowak, J., & Clément, C. (2006). Enhancement of chilling resistance of inoculated grapevine plantlets with a plant growth-promoting rhizobacterium, Burkholderiaphytofirmans strain PsJN. *Applied and Environmental Microbiology*, *72*(11), 7246–7252. doi:10.1128/AEM.01047-06 PMID:16980419

Baruah, R. (2018). Towards the Bioavailability of Zinc in Agricultural Soils. In *Role of Rhizospheric Microbes in Soil* (pp. 99–136). Springer. doi:10.1007/978-981-13-0044-8_4

Beijerinck, M. W. (1888). Die bacterien der papilionaceenknöllchen. *Botanische Zeitung*, *46*(46), 725–735.

Bharti, N., Pandey, S. S., Barnawal, D., Patel, V. K., & Kalra, A. (2016). Plant growth promoting rhizobacteria *Dietzianatronolimnaea* modulates the expression of stress responsive genes providing protection of wheat from salinity stress. *Scientific Reports*, *6*(1), 1–16. doi:10.1038rep34768 PMID:28442746

Bouffaud, M. L., Poirier, M. A., Muller, D., & Moënne-Loccoz, Y. (2014). Root microbiome relates to plant host evolution in maize and other Poaceae. *Environmental Microbiology*, *16*(9), 2804–2814. doi:10.1111/1462-2920.12442 PMID:24588973

Cherubin, M. R., Tormena, C. A., & Karlen, D. L. (2017). Soil quality evaluation using the soil management assessment framework (SMAF) in Brazilian oxisols with contrasting texture. *Revista Brasileira de Ciência do Solo*, *41*(0), 41. doi:10.1590/18069657rbcs20160148

Dabrock, P. (2009). Playing God? Synthetic biology as a theological and ethical challenge. *Systems and Synthetic Biology*, *3*(1-4), 47–54. doi:10.100711693-009-9028-5 PMID:19816799

Dessaux, Y., Grandclément, C., & Faure, D. (2016). Engineering the rhizosphere. *Trends in Plant Science*, *21*(3), 266–278. doi:10.1016/j.tplants.2016.01.002 PMID:26818718

Dimkpa, C., Weinand, T., & Asch, F. (2009). Plant–rhizobacteria interactions alleviate abiotic stress conditions. *Plant, Cell & Environment*, *32*(12), 1682–1694. doi:10.1111/j.1365-3040.2009.02028.x PMID:19671096

Estabrook, E. M., & Yoder, J. I. (1998). Plant-plant communications: Rhizosphere signaling between parasitic angiosperms and their hosts. *Plant Physiology*, *116*(1), 1–7. doi:10.1104/pp.116.1.1

Etesami, H., & Noori, F. (2019). Soil salinity as a challenge for sustainable agriculture and bacterial-mediated alleviation of salinity stress in crop plants. In *Saline Soil-based Agriculture by Halotolerant Microorganisms* (pp. 1–22). Springer. doi:10.1007/978-981-13-8335-9_1

FESLM. (1993). *An International Framework for Evaluating Sustainable Land Management Report 73*. Retrieved june 05, 2019, from http://www.fao.org/3/T1079E/t1079e04.htm#nature%20of%20sustainable%20land%20management%20(slm)

Ghareeb, B. (2009). Genetic engineering and transgenesis: advantages, assessment of risks and ethics. *Al-Quds Open University Research Journal*.

Giller, K. E., Andersson, J. A., Corbeels, M., Kirkegaard, J., Mortensen, D., Erenstein, O., & Vanlauwe, B. (2015). Beyond conservation agriculture. *Frontiers in Plant Science*, *6*, 870. doi:10.3389/fpls.2015.00870 PMID:26579139

Godfray, H. C. J., Beddington, J. R., Crute, I. R., Haddad, L., Lawrence, D., Muir, J. F., & Toulmin, C. (2010). Food security: The challenge of feeding 9 billion people. *Science*, *327*(5967), 812–818. doi:10.1126cience.1185383 PMID:20110467

Goglio, P., Smith, W. N., Grant, B. B., Desjardins, R. L., McConkey, B. G., Campbell, C. A., & Nemecek, T. (2015). Accounting for soil carbon changes in agricultural life cycle assessment (LCA): A review. *Journal of Cleaner Production*, *104*, 23–39. doi:10.1016/j.jclepro.2015.05.040

Haling, R. E., Simpson, R. J., Culvenor, R. A., Lambers, H., & Richardson, A. E. (2011). Effect of soil acidity, soil strength and macropores on root growth and morphology of perennial grass species differing in acid-soil resistance. *Plant, Cell & Environment*, *34*(3), 444–456. doi:10.1111/j.1365-3040.2010.02254.x PMID:21062319

Harish, S., Kavino, M., Kumar, N., Balasubramanian, P., & Samiyappan, R. (2009a). Induction of defense-related proteins by mixtures of plant growth promoting endophytic bacteria against Banana bunchy top virus. *Biological Control*, *51*(1), 16–25. doi:10.1016/j.biocontrol.2009.06.002

Harish, S., Kavino, M., Kumar, N., & Samiyappan, R. (2009b). Differential expression of pathogenesis-related proteins and defense enzymes in banana: Interaction between endophytic bacteria, Banana bunchy top virus and Pentalonianigronervosa. *Biocontrol Science and Technology*, *19*(8), 843–857. doi:10.1080/09583150903145000

Horne, J. E., Kalevitch, A. E., & Filimonova, M. V. (1996). Soil acidity effect on initial wheat growth and development. *Journal of Sustainable Agriculture*, *7*(2-3), 5–13. doi:10.1300/J064v07n02_03

Hungria, M., Nogueira, M. A., & Araujo, R. S. (2013). Co-inoculation of soybeans and common beans with rhizobia and azospirilla: Strategies to improve sustainability. *Biology and Fertility of Soils*, *49*(7), 791–801. doi:10.100700374-012-0771-5

Jagtap, U. B., & Bapat, V. A. (2015). Genetic engineering of plants for heavy metal removal from soil. In *Heavy Metal Contamination of Soils* (pp. 433–470). Springer. doi:10.1007/978-3-319-14526-6_22

Johnson, D. L., Ambrose, S. H., Bassett, T. J., Bowen, M. L., Crummey, D. E., Isaacson, J. S., & Winter-Nelson, A. E. (1997). Meanings of environmental terms. *Journal of Environmental Quality*, *26*(3), 581–589. doi:10.2134/jeq1997.00472425002600030002x

Kang, S. M., Radhakrishnan, R., Khan, A. L., Kim, M. J., Park, J. M., Kim, B. R., Shin, D.-H., & Lee, I. J. (2014). gibberellin secreting rhizobacterium, Pseudomonas putida H-2-3 modulates the hormonal and stress physiology of soybean to improve the plant growth under saline and drought conditions. *Plant Physiology and Biochemistry*, *84*, 115–124. doi:10.1016/j.plaphy.2014.09.001 PMID:25270162

Kavino, M., Harish, S., Kumar, N., Saravanakumar, D., Damodaran, T., Soorianathasundaram, K., & Samiyappan, R. (2007). Rhizosphere and endophytic bacteria for induction of systemic resistance of banana plantlets against bunchy top virus. *Soil Biology & Biochemistry*, *39*(5), 1087–1098. doi:10.1016/j.soilbio.2006.11.020

Kiran, B., Kaushik, A., & Kaushik, C. P. (2008). Metal-salt co-tolerance and metal removal by indigenous cyanobacterial strains. *Process Biochemistry*, *43*(6), 598–604. doi:10.1016/j.procbio.2008.01.019

Koh, H., Kwon, S., & Thomson, M. (2015). *Current Technologies in Plant Molecular Breeding: A Guide Book of Plant Molecular Breeding for Researchers*. Springer. doi:10.1007/978-94-017-9996-6

Kuritz, T., & Wolk, C. P. (1995). Use of filamentous cyanobacteria for biodegradation of organic pollutants. *Applied and Environmental Microbiology*, *61*(1), 234–238. doi:10.1128/AEM.61.1.234-238.1995 PMID:7534052

Lefebvre, D. D., Kelly, D., & Budd, K. (2007). Biotransformation of Hg (II) by cyanobacteria. *Applied and Environmental Microbiology*, *73*(1), 243–249. doi:10.1128/AEM.01794-06 PMID:17071784

Lobell, D. B., & Field, C. B. (2007). Global scale climate–crop yield relationships and the impacts of recent warming. *Environmental Research Letters*, *2*(1), 014002. doi:10.1088/1748-9326/2/1/014002

Marschner, H., Römheld, V., Horst, W. J., & Martin, P. (1986). Root-induced changes in the rhizosphere: Importance for the mineral nutrition of plants. *ZeitschriftfürPflanzenernährung undBodenkunde*, *149*(4), 441–456. doi:10.1002/jpln.19861490408

McNear, D. H. Jr. (2013). The rhizosphere-roots, soil and everything in between. *Nature Education Knowledge*, *4*(3), 1.

Meena, H., Meena, R. S., Rajput, B. S., & Kumar, S. (2016). Response of bio-regulators to morphology and yield of clusterbean [*Cyamopsistetragonoloba* (L.) Taub.] under different sowing environments. *Journal of Applied and Natural Science*, *8*(2), 715–718. doi:10.31018/jans.v8i2.863

Meena, R. S., Meena, P. D., Yadav, G. S., & Yadav, S. S. (2017). Phosphate solubilizing microorganisms, principles and application of microphos technology. *Journal of Cleaner Production*, *145*, 157–158. doi:10.1016/j.jclepro.2017.01.024

Meena, R. S., Vijayakumar, V., Yadav, G. S., & Mitran, T. (2018). Response and interaction of *Bradyrhizobium japonicum* and arbuscular mycorrhizal fungi in the soybean rhizosphere. *Plant Growth Regulation*, *84*(2), 207–223. doi:10.100710725-017-0334-8

Meena, V. S., Maurya, B. R., & Meena, R. S. (2015). Residual impact of wellgrow formulation and NPK on growth and yield of wheat (*Triticum aestivum*L.). *Bangladesh Journal of Botany*, *44*(1), 143–146. doi:10.3329/bjb.v44i1.22738

Mia, M. B., & Shamsuddin, Z. H. (2010). Rhizobium as a crop enhancer and biofertilizer for increased cereal production. *African Journal of Biotechnology*, *9*(37), 6001–6009.

Michels, C. A. (2002). Genetic Techniques for Biological Research: A Case Study Approach. John Wiley & Sons.

Mirshad, P. P., & Puthur, J. T. (2017). Drought tolerance of bioenergy grass Saccharum spontaneum L. enhanced by arbuscular mycorrhizae. *Rhizosphere*, *3*, 1–8. doi:10.1016/j.rhisph.2016.09.004

Nardi, S., Concheri, G., Pizzeghello, D., Sturaro, A., Rella, R., & Parvoli, G. (2000). Soil organic matter mobilization by root exudates. *Chemosphere*, *41*(5), 653–658. doi:10.1016/S0045-6535(99)00488-9 PMID:10834364

Negráo, S., Schmöckel, S. M., & Tester, M. (2017). Evaluating physiological responses of plants to salinity stress. *Annals of Botany*, *119*(1), 1–11. doi:10.1093/aob/mcw191 PMID:27707746

Nicholl, S. T. (2008). *An Introduction to Genetic Engineering*. Cambridge University Press. doi:10.1017/CBO9780511800986

Parveda, M., Kiran, B., Punita, D. L., & Kishor, P. K. (2017). Overexpression of SbAP37 in rice alleviates concurrent imposition of combination stresses and modulates different sets of leaf protein profiles. *Plant Cell Reports*, *36*(5), 773–786. doi:10.100700299-017-2134-z PMID:28393269

Pilon-Smits, E. (2005). Phytoremediation. *Annual Review of Plant Biology*, *56*(1), 15–39. doi:10.1146/annurev.arplant.56.032604.144214 PMID:15862088

Poeplau, C., & Don, A. (2015). Carbon sequestration in agricultural soils via cultivation of cover crops–A meta-analysis. *Agriculture, Ecosystems & Environment*, *200*, 33–41. doi:10.1016/j.agee.2014.10.024

Quinn, L. D., Straker, K. C., Guo, J., Kim, S., Thapa, S., Kling, G., & Voigt, T. B. (2015). Stress-tolerant feedstocks for sustainable bioenergy production on marginal land. *BioEnergy Research*, *8*(3), 1081–1100. doi:10.100712155-014-9557-y

Reguera, M., Peleg, Z., & Blumwald, E. (2012). Targeting metabolic pathways for genetic engineering abiotic stress-tolerance in crops. *Biochimica et Biophysica Acta (BBA)-. Gene Regulatory Mechanisms*, *1819*(2), 186–194. PMID:21867784

Rocha-Martins, M., Cavalheiro, G. R., Matos-Rodrigues, G. E., & Martins, R. P. (2015). From Gene Targeting to Genome Editing: Transgenic animals applications and beyond. *Anais da Academia Brasileira de Ciências*, *87*(2), 1323–1348. doi:10.1590/0001-3765201520140710 PMID:26397828

Rodriguez, R. J., Henson, J., Van Volkenburgh, E., Hoy, M., Wright, L., Beckwith, F., & Redman, R. S. (2008). Stress tolerance in plants via habitat-adapted symbiosis. *The ISME Journal*, *2*(4), 404–416. doi:10.1038/ismej.2007.106 PMID:18256707

Rovira, A. D. (1969). Plant root exudates. *Botanical Review*, *35*(1), 35–57. doi:10.1007/BF02859887

Rware, H., Kayuki, C., Macharia, M., & Oduor, G. (2016). Fertilizer use optimization approach: An innovation to increase agricultural profitability for African farmers. *African Journal of Agricultural Research*, *11*(38), 3587–3597. doi:10.5897/AJAR2016.11408

Sachidanand, B., Mitra, N., Kumar, V., Roy, R., & Mishra, B. (2019). Soil as a huge laboratory for microorganisms. *Agricultural Research & Technology: Open Access Journal, 22*(4).

Sammauria, R., & Kumawat, S. (2018). Legume Plant Growth-Promoting Rhizobacteria (PGPRs): Role in Soil Sustainability. In *Legumes for Soil Health and Sustainable Management* (pp. 409–443). Springer. doi:10.1007/978-981-13-0253-4_13

Sergeeva, E., Shah, S., & Glick, B. R. (2006). Growth of transgenic canola (Brassica napus cv. Westar) expressing a bacterial 1-aminocyclopropane-1-carboxylate (ACC) deaminase gene on high concentrations of salt. *World Journal of Microbiology & Biotechnology*, *22*(3), 277–282. doi:10.100711274-005-9032-1

Sharma, P., Khanna, V., & Kumari, P. (2013a). Efficacy of aminocyclopropane-1-carboxylic acid (ACC)-deaminase-producing rhizobacteria in ameliorating water stress in chickpea under axenic conditions. *African Journal of Microbiological Research*, *7*(50), 5749–5757. doi:10.5897/AJMR2013.5918

Sharma, S., Kulkarni, J., & Jha, B. (2016). Halotolerant rhizobacteria promote growth and enhance salinity tolerance in peanut. *Frontiers in Microbiology*, *7*, 1600. doi:10.3389/fmicb.2016.01600 PMID:27790198

Sharma, S. B., Sayyed, R. Z., Trivedi, M. H., & Gobi, T. A. (2013b). Phosphate solubilizing microbes: Sustainable approach for managing phosphorus deficiency in agricultural soils. *SpringerPlus*, *2*(1), 587. doi:10.1186/2193-1801-2-587 PMID:25674415

Shinozaki, K., Uemura, M., Bailey-Serres, J., Bray, E. A., & Weretilnyk, E. (2015). *Biochemistry and Molecular Biology of Plants* (B. B. Buchanan, W. Gruissem, & R. L. Jones, Eds.). Wiley.

Simon, J. C., Marchesi, J. R., Mougel, C., & Selosse, M. A. (2019). Host-microbiota interactions: From holobiont theory to analysis. *Microbiome*, *7*(1), 1–5. doi:10.118640168-019-0619-4 PMID:30635058

Singh, J. S., Pandey, V. C., & Singh, D. P. (2011). Efficient soil microorganisms: A new dimension for sustainable agriculture and environmental development. *Agriculture, Ecosystems & Environment*, *140*(3-4), 339–353. doi:10.1016/j.agee.2011.01.017

Slattery, B., & Hollier, C. (2002). *The impact of acid soils in Victoria*. Agriculture Victoria.

Suárez, R., Wong, A., Ramírez, M., Barraza, A., Orozco, M. D. C., Cevallos, M. A., Lara, M., Hernández, G., & Iturriaga, G. (2008). Improvement of drought tolerance and grain yield in common bean by overexpressing trehalose-6-phosphate synthase in rhizobia. *Molecular Plant-Microbe Interactions*, *21*(7), 958–966. doi:10.1094/MPMI-21-7-0958 PMID:18533836

Sun, H., Tao, J., Gu, P., Xu, G., & Zhang, Y. (2016). The role of strigolactones in root development. *Plant Signaling & Behavior*, *11*(1), e1110662. doi:10.1080/15592324.2015.1110662 PMID:26515106

Tuğrul, K. M. (2019). Soil Management in Sustainable Agriculture. In *Soil Management and Plant Nutrition for Sustainable Crop Production*. IntechOpen.

Upadhyay, S. K., Singh, J. S., & Singh, D. P. (2011). Exopolysaccharide-producing plant growth-promoting rhizobacteria under salinity condition. *Pedosphere*, *21*(2), 214–222. doi:10.1016/S1002-0160(11)60120-3

Vacheron, J., Desbrosses, G., Bouffaud, M. L., Touraine, B., Moënne-Loccoz, Y., Muller, D., & Prigent-Combaret, C. (2013). Plant growth-promoting rhizobacteria and root system functioning. *Frontiers in Plant Science*, *4*, 356. doi:10.3389/fpls.2013.00356 PMID:24062756

Verslues, P. E. (2017). Time to grow: Factors that control plant growth during mild to moderate drought stress. *Plant, Cell & Environment*, *40*(2), 177–179. doi:10.1111/pce.12827 PMID:27588960

Vurukonda, S. S. K. P., Vardharajula, S., Shrivastava, M., & SkZ, A. (2016a). Multifunctional Pseudomonas putida strain *FBKV2* from arid rhizosphere soil and its growth promotional effects on maize under drought stress. *Rhizosphere*, *1*, 4–13. doi:10.1016/j.rhisph.2016.07.005

Vurukonda, S. S. K. P., Vardharajula, S., Shrivastava, M., & SkZ, A. (2016b). Enhancement of drought stress tolerance in crops by plant growth promoting rhizobacteria. *Microbiological Research*, *184*, 13–24. doi:10.1016/j.micres.2015.12.003 PMID:26856449

Walker, V., Bertrand, C., Bellvert, F., Moënne-Loccoz, Y., Bally, R., & Comte, G. (2011). Host plant secondary metabolite profiling shows a complex, strain-dependent response of maize to plant growth-promoting rhizobacteria of the genus Azospirillum. *The New Phytologist*, *189*(2), 494–506. doi:10.1111/j.1469-8137.2010.03484.x PMID:20946131

Zhu, J. K. (2016). Abiotic stress signalling and responses in plants. *Cell*, *167*(2), 313–324. doi:10.1016/j.cell.2016.08.029 PMID:27716505

Compilation of References

Aarkrog, A. (1994). Source terms and inventories of anthropogenic radionuclides. In E. Holm (Ed.), *Radioecology. Lectures in Environmental Radioactivity* (pp. 21–38). World Scientific Publishing.

Aarkrog, A. (2003). Input of anthropogenic radionuclides into the World Ocean. *Deep-sea Research. Part II, Topical Studies in Oceanography*, *50*(17-21), 2597–2606. doi:10.1016/S0967-0645(03)00137-1

Abatenh, E., Gizaw, B., Tsegaye, Z., & Wassie, M. (2017). The role of microorganisms in bioremediation-A review. *Open Journal of Environmental Biology*, *2*(1), 30-46.

Abbas, S. H., Ismail, I. M., Mostafa, T. M., & Sulaymon, A. H. (2014). Biosorption of heavy metals: A review. *Journal of Chemical Science and Technology*, *3*, 74–102.

Abbes, C., Mansouri, A., Werfelli, N., & Landoulsi, A. (2018). Aerobic Biodegradation of DDT by *Advenella kashmirensis* and Its Potential Use in Soil Bioremediation. *Soil and Sediment Contamination: An International Journal*, *27*(6), 455–468. doi:10.1080/15320383.2018.1485629

Abbey, L., Abbey, J., Leke-Aladekoba, A., Iheshiulo, E. M. A., & Ijenyo, M. (2019). Biopesticides and Biofertilizers: Types, Production, Benefits, and Utilization. *Byproducts from Agriculture and Fisheries: Adding Value for Food, Feed, Pharma, and Fuels*, 479-500.

Abdallah, N. A., Moses, V., & Prakash, C. S. (2014). The impact of possible climate changes on developing countries: The needs for plants tolerant to abiotic stresses. *GM Crops and Food: Biotechnology in Agriculture and the Food Chain*, *5*(2), 77–80. doi:10.4161/gmcr.32208 PMID:25075960

Abd-Alrahman, S. H., & Salem-Bekhit, M. M. (2013). Microbial biodegradation of butachlor pollution (obsolete pesticide Machete 60% EC). *Asian Journal of Multidimensional Research*, *7*(4), 330–335.

Abdel-Aty, A. M., Ammar, N. S., Abdel Ghafar, H. H., & Ali, R. K. (2013). Biosorption of cadmium and lead from aqueous solution by fresh water alga *Anabaena sphaerica* biomass. *Journal of Advanced Research*, *4*(4), 367–374. doi:10.1016/j.jare.2012.07.004 PMID:25685442

Abdelkrim, S., Jebara, S. H., Saadani, O., Chiboub, M., Abid, G., & Jebara, M. (2018). Effect of Pb-resistant plant growth-promoting rhizobacteria inoculation on growth and lead uptake by Lathyrus sativus. *Journal of Basic Microbiology*, *58*(7), 579–589. doi:10.1002/jobm.201700626 PMID:29737549

Abdel-Monem, M. O., Al-Zubeiry, A. H. S., & Al-Gheethi, A. A. S. (2010). Biosorption of nickel by *Pseudomonas cepacia* 120S and *Bacillus subtilis* 117S. *Water Science and Technology*, *61*(12), 2994–3007. doi:10.2166/wst.2010.198 PMID:20555195

Abdel-Raouf, N., Al-Homaidan, A. A., & Ibraheem, I. B. M. (2012). Agricultural importance of algae. *African Journal of Biotechnology*, *11*(54), 11648–11658. doi:10.5897/AJB11.3983

Abdel-Salam, M. A., & Shams, A. S. (2012). Feldspar-K fertilization of potato (*Solanum tuberosum* L.) augmented by biofertilizer. *Journal of Agriculture and Environmental Sciences*, *12*, 694–699.

Abdel-Shafy, H. I., & Mansour, M. S. (2016). A review on polycyclic aromatic hydrocarbons: source, environmental impact, effect on human health and remediation. *Egyptian Journal of Petroleum*, *25*(1), 107-123. . doi:10.1016/j.ejpe.2015.03.011

Abdel-Shafy, H. I., & Mansour, M. S. M. (2018). Solid waste issue: Sources, composition, disposal, recycling, and valorization. *Egyptian Journal of Petroleum*, *27*(4), 1275–1290. doi:10.1016/j.ejpe.2018.07.003

Abdulsada, Z. K. (2014). *Evaluation of microalgae for secondary and tertiary wastewater treatment* (Master Thesis). Ottawa-Carleton Institute for Environmental Engineering Ottawa, Ontario, Canada.

Abedin, M. J., Cotter-Howells, J., & Meharg, A. A. (2002). Arsenic uptake and accumulation in rice (*Oryza sativa* L.) irrigated with contaminated water. *Plant and Soil*, *240*(2), 311–319. doi:10.1023/A:1015792723288

Abed, R. (2010). Interaction between cyanobacteria and aerobic heterotrophic bacteria in the degradation of hydrocarbons. *International Biodeterioration & Biodegradation*, *64*(1), 58–64. doi:10.1016/j.ibiod.2009.10.008

Abhilash, P. C., Dubey, R. K., Tripathi, V., Srivastava, P., Verma, J. P., & Singh, H. B. (2013). Remediation and management of POPs-contaminated soils in a warming climate: Challenges and perspectives. *Environmental Science and Pollution Research International*, *20*(8), 5879–5885. doi:10.100711356-013-1808-5 PMID:23677754

Abhilash, P. C., Srivastava, S., & Singh, N. (2011). Comparative bioremediation potential of four rhizospheric microbial species against lindane. *Chemosphere*, *82*(1), 56–63. doi:10.1016/j.chemosphere.2010.10.009 PMID:21044795

Abou-el-Seoud, I. I., & Abdel-Megeed, A. (2012). Impact of rock materials and biofertilizations on P and K availability for maize (*Zea mays*) under calcareous soil conditions. *Saudi Journal of Biological Sciences*, *19*(1), 55–63. doi:10.1016/j.sjbs.2011.09.001 PMID:23961162

Acea, M. (2003). Cyanobacterial inoculation of heated soils: Effect on microorganisms of C and N cycles and on chemical composition in soil surface. *Soil Biology & Biochemistry*, *35*(4), 513–524. doi:10.1016/S0038-0717(03)00005-1

Aceves-Diez, A. E., Estrada-Castaneda, K. J., & Castaneda-Sandoval, L. M. (2015). Use of Bacillus thuringiensis supernatant from a fermentation process to improve bioremediation of chlorpyrifos in contaminated soils. *Journal of Environmental Management*, *157*, 213–219. doi:10.1016/j.jenvman.2015.04.026 PMID:25910975

Ackerley, D. F., Gonzalez, C. F., Keyhan, M., Blake, R., & Matin, A. (2004). Mechanism of chromate reduction by the Escherichia coli protein, NfsA, and the role of different chromate reductases in minimizing oxidative stress during chromate reduction. *Environmental Microbiology*, *6*(8), 851–860. doi:10.1111/j.1462-2920.2004.00639.x PMID:15250887

ACSEDU. (2020). *Different Approaches to Sustainability*. Retrieved from https://www.acsedu.co.uk/Info/Agriculture/Sustainable-Agriculture/Different-Approaches-to-Sustainability.aspx

Acuner, E., & Dilek, F. B. (2004). Treatment of Tectilon Yellow 2G by Chlorella vulgaris. *Process Biochemistry*, *39*(5), 623–631. doi:10.1016/S0032-9592(03)00138-9

Adams, D. G., & Duggan, P. S. (2008). Cyanobacteria–bryophyte symbioses. *Journal of Experimental Botany*, *59*(5), 1047–1058. doi:10.1093/jxb/ern005 PMID:18267939

Adav, S. S., Lee, D. J., & Lai, J. Y. (2009). Treating chemical industries influent using aerobic granular sludge: Recent development. *Journal of the Taiwan Institute of Chemical Engineers*, *40*(3), 333–336. doi:10.1016/j.jtice.2009.02.002

Adesodun, J., Atayese, M., Agbaje, T., Osadiaye, B., Mafe, O., & Soretire, A. (2010). Phytoremediation potentials of sunflowers (*Tithonia diversifolia* and *Helianthus annuus*) for metals in soils contaminated with zinc and lead nitrates. *Water, Air, and Soil Pollution*, *207*(1-4), 195–201. doi:10.100711270-009-0128-3

Adinarayana, K., Ellaiah, P., Srinivasulu, B., Devi, R. B., & Adinarayana, G. (2003). Response surface methodological approach to optimize the nutritional parameters for neomycin production by *Streptomyces marinensis* under solid-state fermentation. *Process Biochemistry*, *38*(11), 1565–1572. doi:10.1016/S0032-9592(03)00057-8

Adki, V. S., Jadhav, J. P., & Bapat, V. A. (2012). Exploring the phytoremediation potential of cactus (Nopalea cochenillifera Salm. Dyck) cell cultures for textile dye degradation. *International Journal of Phytoremediation*, *14*(6), 554–569. doi:10.1080/15226514.2011.619226 PMID:22908626

Adnan, L. A., Sathishkumar, P., Mohd Yusoff, A. R., & Hadibarata, T. (2015). Metabolites characterisation of laccase mediated Reactive Black 5 biodegradation by fast growing ascomycete fungus *Trichoderma atroviride* F03. *International Biodeterioration & Biodegradation*, *104*, 274–282. doi:10.1016/j.ibiod.2015.05.019

Adrio, J. L., & Demain, A. L. (2014). Microbial enzymes: Tools for biotechnological processes. *Biomolecules*, *4*(1), 117–139. doi:10.3390/biom4010117 PMID:24970208

Aeron, A., Kumar, S., Pandey, P., & Maheshwari, D. K. (2011). Emerging Role of Plant Growth Promoting Rhizobacteria in Agrobiology. In D. Maheshwari (Ed.), *Bacteria in Agrobiology: Crop Ecosystems*. Springer. doi:10.1007/978-3-642-18357-7_1

Aftab, M. N., Hameed, A., & Ikram-ul-Haq, C. (2006). Biodegradation of leather waste by enzymatic treatment. *The Chinese Journal of Process Engineering*, *6*(3), 462–465.

Afzal, M. S., Ashraf, A., & Nabeel, M. (2018). Characterization of industrial effluents and groundwater of Hattar industrial estate, Haripur. *Advances in Agriculture and Environmental Science: Open Access*, *1*(2), 70–77.

Agamuthu, P., Tan, Y. S., & Fauziah, S. H. (2013). Bioremediation of hydrocarbon contaminated soil using selected organic wastes. *Procedia Environmental Sciences*, *18*, 694–702. doi:10.1016/j.proenv.2013.04.094

Agarwal, P., Giri, B. S. & Rani, R. (2020). Unraveling the role of rhizospheric plant-microbe synergy in phytoremediation: A genomic perspective. *Current Genentics*, 21.

Agbogidi, O. M., Dolor, E. D., & Okechukwu, E. M. (2007). Evaluation of *Tectona grandis* (Linn.) and *Gmelina arborea* (Roxb.) for phytoremediation in crude oil contaminated soils. *ACS. Agriculturae Conspectus Scientificus*, *72*(2), 149–152.

Agency for Toxic Substances and Disease Registry (ATSDR). (2002). *Toxicological Profile for Copper*. Centers for Disease Control.

Agency for Toxic Substances and Disease Registry. (1999). *Toxicological profile for total petroleum hydrocarbons (TPH). Toxicol. profile Total Pet. Hydrocarb.* TPH.

Aguelmous, A., El Fels, L., Souabi, S., Zamama, M., & Hafidi, M. (2019). The fate of total petroleum hydrocarbons during oily sludge composting: A critical review. *Reviews in Environmental Science and Biotechnology, 18*, 473 - 493.

Aguelmous, A., El Fels, L., Souabi, S., Zamama, M., & Hafidi, M. (2019). The fate of total petroleum hydrocarbons during oily sludge composting: A critical review. *Reviews in Environmental Science and Biotechnology*, *18*(3), 1–21. doi:10.100711157-019-09509-w

Ahalya, N., Ramachandra, T. V., & Kanamadi, R. D. (2003). Biosorption of heavy metals. *Research Journal of Chemistry and Environment*, *7*(4), 71–78.

Ahalya, N., & Ramachandra, T. Y. (2006). Phvtoremediation: Processes and mechanisms. *Journal of Ecobiology*, *18*(1), 33–38.

Ahemad Munees, M. (2013). Pesticides as Antagonists of Rhizobia and the Legume-Rhizobium Symbiosis: A Paradigmatic and Mechanistic Outlook. *Biochemistry and Molecular Biology*, *1*(4), 63–75. doi:10.12966/bmb.12.02.2013

Ahemad, M., & Khan, M. S. (2012a). Effect of fungicides on plant growth promoting activities of phosphate solubilizing Pseudomonas putida isolated from mustard (*Brassica compestris*) rhizosphere. *Chemosphere*, *86*(9), 945–950. doi:10.1016/j.chemosphere.2011.11.013 PMID:22133911

Ahemad, M., & Khan, M. S. (2012b). Evaluation of plant-growth-promoting activities of rhizobacterium *Pseudomonas putida* under herbicide stress. *Annals of Microbiology*, *62*(4), 1531–1540. doi:10.100713213-011-0407-2

Ahemad, M., & Kibret, M. (2014). Mechanisms and applications of plant growth promoting rhizobacteria: Current perspective. *Journal of King Saud University. Science*, *26*(1), 1–20. doi:10.1016/j.jksus.2013.05.001

Ahmad, F., Ahmad, I., & Khan, M. S. (2005). Indole acetic acid production by the indigenous isolates of *Azotobacter* and fluorescent *Pseudomonas* in the presence and absence of tryptophan. *Turkish Journal of Biology*, *29*, 29–34.

Ahmad, I., Pichtel, J., & Hayat, S. (Eds.). (2008). *Plant-bacteria interactions: strategies and techniques to promote plant growth*. John Wiley & Sons. doi:10.1002/9783527621989

Ahmad, P., Rasool, S., Gul, A., Sheikh, S. A., Akram, N. A., Ashraf, M., Kazi, A. M., & Gucel, S. (2016). Jasmonates: Multifunctional roles in stress tolerance. *Frontiers in Plant Science*, *7*, 813. doi:10.3389/fpls.2016.00813 PMID:27379115

Ahmed, A., & Hasnain, S. (2014). Auxins as one of the factors of plant growth improvement by plant growth promoting rhizobacteria. *Polish Journal of Microbiology*, *63*(3), 261–266. doi:10.33073/pjm-2014-035 PMID:25546935

Ahmed, E., & Holmström, S. J. (2014). Siderophores in environmental research: Roles and applications. *Microbial Biotechnology*, *7*(3), 196–208. doi:10.1111/1751-7915.12117 PMID:24576157

Ahmed, M. K. E., Parvin, M. M., Islam, M. S., Akter, S., Khan, S., & Al-Mamun, M. H. (2014). Lead- and cadmium-induced histopathological changes in gill, kidney and liver tissue of freshwater climbing perch *Anabas testudineus* (Bloch, 1792). *Chemistry and Ecology*, *30*(6), 532–540. doi:10.1080/02757540.2014.889123

Ahuja, S. K., Ferreira, G. M., & Moreira, A. R. (2004). Utilization of enzymes for environmental applications. *Critical Reviews in Biotechnology*, *24*(2-3), 125–154. doi:10.1080/07388550490493726 PMID:15493529

Akan, J. C., Abdulrahman, F. I., Ayodele, J. T., & Ogugbuaja, V. O. (2009). Impact of tannery and textile effluent on the chemical characteristics of Challawa River, Kano State, Nigeria. *Australian Journal of Basic and Applied Sciences*, *3*(3), 1933–1947.

Akansha, K., Chakraborty, D., & Sachan, S. G. (2019). Decolorization and degradation of methyl orange by *Bacillus stratosphericus* SCA1007. *Biocatalysis and Agricultural Biotechnology*, *18*, 101044. doi:10.1016/j.bcab.2019.101044

Akaraonye, E., Moreno, C., Knowles, J. C., Keshavarz, T., & Roy, I. (2012). Poly(3-hydroxybutyrate) production by *Bacillus cereus* SPV using sugarcane molasses as the main carbon source. *Biotechnology Journal*, *7*(2), 293–303. doi:10.1002/biot.201100122 PMID:22147642

Akar, T., Tunali, S., & Kiran, I. (2005). *Botrytis cinerea* as a new fungal biosorbent for removal of Pb(II) from aqueous solutions. *Biochemical Engineering Journal*, *5*(25), 227–235. doi:10.1016/j.bej.2005.05.006

Akbari, A., Gharanjik, S., Koobaz, P., & Sadeghi, A. (2020). Plant growth promoting Streptomyces strains are selectively interacting with the wheat cultivars especially in saline conditions. *Heliyon*, *6*(2), e03445. doi:10.1016/j.heliyon.2020.e03445 PMID:32095655

Akcil, A., Erust, C., Ozdemiroglu, S., Fonti, V., & Beolchini, F. (2015). A review of approaches and techniques used in aquatic contaminated sediments: Metal removal and stabilization by chemical and biotechnological processes. *Journal of Cleaner Production*, *86*, 24–36. doi:10.1016/j.jclepro.2014.08.009

Aken, B. V., Correa, P. A., & Schnoor, J. L. (2010). Phytoremediation of polychlorinated biphenyls: New trends and promises. *Environmental Science & Technology*, *44*(8), 276–2776. doi:10.1021/es902514d PMID:20384372

Akhtar, M. E., Sardar, A., Ashraf, M., Akhtar, M., & Khan, M. Z. (2003). Effect of potash application on seed cotton yield and yield components of selected cotton varieties–I. *Asian Journal of Plant Sciences*, *2*, 602–604. doi:10.3923/ajps.2003.602.604

Akhtar, M. S., & Siddiqui, Z. A. (2009). Use of plant growth promoting rhizobacteria for the biocontrol of root-rot disease complex of chickpea. *Australasian Plant Pathology*, *38*(1), 44–50. doi:10.1071/AP08075

Akif, M., Khan, A. R., & Sok, K., Min, Hussain, Z., Maal-Abrar, Khan, M., & Yousafjai, A. M. (2002). Textile effluents and their contribution towards aquatic pollution in the Kabul River (Pakistan). *Journal of the Chemical Society of Pakistan*, *24*, 106–111.

Aksu, Z. (2005). Application of biosorption for the removal of organic pollutants: A review. *Process Biochemistry*, *40*(3-4), 997–1026. doi:10.1016/j.procbio.2004.04.008

Aksu, Z., & Donmez, G. (2003). A Comparative Study on the Biosorption Characteristics of some Yeasts for Remazol Blue Reactive Dye. *Chemosphere*, *50*(8), 1075–1083. doi:10.1016/S0045-6535(02)00623-9 PMID:12531715

Al Hasin, A., Gurman, S. J., Murphy, L. M., Perry, A., Smith, T. J., & Gardiner, P. H. E. (2010). Remediation of Chromium (VI) by a Methane-Oxidizing Bacterium. *Environmental Science & Technology*, *44*(1), 400–405. doi:10.1021/es901723c PMID:20039753

Al Raisi, S. A. H., Sulaiman, H., Suliman, F. E., & Abdallah, O. (2014). Assessment of heavy metals in leachate of an unlined landfill in the Sultanate of Oman. *International Journal of Environmental Sciences and Development*, *5*(1), 60.

Alam, G., & Alam, G. (2000). *A study of biopesticides and biofertilisers in Haryana*. International Institute for Environment and Development.

Alamgir, M. (2016). The Effects of Soil Properties to the Extent of Soil Contamination with Metals. Environmental Remediation Technologies for Metal-Contaminated Soils. *Chapter*, *1*, 1–19.

Alaton, A., Balcioglu, I. A., & Bahnemann, D. W. (2002). Advanced Oxidation of a Reactive Dyebath Effluent: Comparison of O3, H2O2/UV-C and TiO2/UV-A Processes. *Water Research*, *36*(5), 1143–1154. doi:10.1016/S0043-1354(01)00335-9 PMID:11902771

Alavanja, M. C., & Bonner, M. R. (2012). Occupational pesticide exposures and cancer risk: A review. *Journal of Toxicology and Environmental Health. Part B, Critical Reviews*, *15*(4), 238–263. doi:10.1080/10937404.2012.632358 PMID:22571220

Albdaiwi, R. N., Khaymi-Horani, H., Ayad, J. Y., Alananbeh, K. M., Kholoud, M., & Al-Sayaydeh, R. (2019). Isolation and characterization of halotolerant plant growth promoting rhizobacteria from durum wheat (Triticum turgidum subsp. durum) cultivated in saline areas of the dead sea region. *Frontiers in Microbiology*, *10*, 1639. doi:10.3389/fmicb.2019.01639 PMID:31396175

Albertsen, A., Ravnskov, S., Green, H., Jensen, D. F., & Larsen, J. (2006). Interactions between the external mycelium of the mycorrhizal fungus *Glomus intraradices* and other soil microorganisms as affected by organic matter. *Soil Biology & Biochemistry, 38*(5), 1008–1014. doi:10.1016/j.soilbio.2005.08.015

Alburquerque, J. A., De La Fuente, C., & Bernal, M. P. (2011). Improvement of soil quality after "alperujo" compost application to two contaminated soils characterised by differing heavy metal solubility. *Journal of Environmental Management, 92*(3), 733–741. doi:10.1016/j.jenvman.2010.10.018 PMID:21035939

Aldor, I. S., & Keasling, J. D. (2003). Process design for microbial plastic factories: Metabolic engineering of polyhydroxyalkanoates. *Current Opinion in Biotechnology, 14*(5), 475–483. doi:10.1016/j.copbio.2003.09.002 PMID:14580576

Alexander, M. (1994). *Biodegradation and bioremediation.* Academic.

Alexandratos, N., & Bruinsma, J. (2012). *World agriculture towards 2030/2050: 2012 revision.* ESA.

Alexopoulos, C. J., Mims, C. W., & Blackwell, M. (1996). Introductory mycology (No. Ed. 4). John Wiley and Sons.

Algreen, M., Rein, A., Legind, C. N., Amundsen, C. E., Karlson, U. G., & Trapp, S. (2012). Test of tree core sampling for screening of toxic elements in soils from a Norwegian site. *International Journal of Phytoremediation, 14*(4), 305–319. doi:10.1080/15226514.2011.620648 PMID:22567713

Ali, H. (2010). Biodegradation of Synthetic Dyes—A Review. *Water, Air, and Soil Pollution, 213*(1-4), 251–273. doi:10.100711270-010-0382-4

Ali, H., & Khan, E. (2018). What are heavy metals? Long-standing controversy over the scientific use of the term' heavy metals'- proposal of a comprehensive definition. *Toxicological and Environmental Chemistry, 100*(1), 6–19. doi:10.1080/02772248.2017.1413652

Ali, H., Khan, E., & Sajad, M. A. (2013). Phytoremediation of heavy metals—Concepts and applications. *Chemosphere, 91*(7), 869–881. doi:10.1016/j.chemosphere.2013.01.075 PMID:23466085

Ali, S. M., Pervaiz, A., Afzal, B., Hamid, N., & Yasmin, A. (2014). Open dumping of municipal solid waste and its hazardous impacts on soil and vegetation diversity at waste dumping sites of Islamabad city. *Journal of King Saud University. Science, 26*(1), 59–65. doi:10.1016/j.jksus.2013.08.003

Ali, S., Hamid, N., Khan, D., & Malik, K. A. (1998). Use of *Azolla* as biofertilizer to enhance crop yield in a rice—wheat cropping system under mild climate. In *Nitrogen Fixation with Non-Legumes* (pp. 353–357). Springer. doi:10.1007/978-94-011-5232-7_41

Ali, W. A. (2019). Biodegradation and phytotoxicity of crude oil hydrocarbons in an agricultural soil. *Chilean Journal of Agricultural Research, 79*(2), 266–277. doi:10.4067/S0718-58392019000200266

Allen, R. L. M. (1971). Reactive dyes. In *Colour Chemistry. Studies in Modern Chemistry.* Springer. doi:10.1007/978-1-4615-6663-2_13

Allison, E., & Mandler, B. (2018). *Petroleum and the environment.* American Geoscience Institute.

Alloway, B. J. (2013). Sources of heavy metals and metalloids in soils. In *Heavy Metals in Soils: Trace Metals and Metalloids in Soils and their Bioavailability* (pp. 11–50). Springer. doi:10.1007/978-94-007-4470-7_2

Almaghrabi, O. A., Massoud, S. I., & Abdelmoneim, T. S. (2013). Influence of inoculation with plant growth promoting rhizobacteria (PGPR) on tomato plant growth and nematode reproduction under greenhouse conditions. *Saudi Journal of Biological Sciences, 20*(1), 57–61. doi:10.1016/j.sjbs.2012.10.004 PMID:23961220

Almeida, E. J. R., & Corso, C. R. (2014). Comparative study of toxicity of azo dye procion red MX-5B following biosorption and biodegradation treatments with the fungi *Aspergillus niger* and *Aspergillus terreus*. *Chemosphere, 112*, 317–322. doi:10.1016/j.chemosphere.2014.04.060 PMID:25048922

Alori, E. T., Glick, B. R., & Babalola, O. O. (2017). Microbial phosphorus solubilization and its potential for use in sustainable agriculture. *Frontiers in Microbiology, 8*, 971. doi:10.3389/fmicb.2017.00971 PMID:28626450

Al-Shammary, A. A. G., Kouzani, A. Z., Kaynak, A., Khoo, S. Y., Norton, M., & Gates, W. (2018). Soil Bulk Density Estimation Methods: A Review. *Pedosphere, 28*(4), 581–596. doi:10.10161002-0160(18)60034

Altieri, M. A., Nicholls, C. I., & Montalba, R. (2017). Technological approaches to sustainable agriculture at a crossroads: An agroecological perspective. *Sustainability, 9*(3), 349. doi:10.3390u9030349

Altimira, F., Yãez, C., Bravo, G., González, M., Rojas, L. A., & Seeger, M. (2012). Characterization of copper-resistant bacteria and bacterial communities from copper-polluted agricultural soils of central Chile. *BMC Microbiology, 12*(1), 193. doi:10.1186/1471-2180-12-193 PMID:22950448

Alwendawi, S. A. (2019). In vitro assessment the potential antioxidant and antitumor activities of Bifido bacterium derived bacteriocins. *International Journal of Drug Delivery Technology, 9*(02), 207–216. doi:10.25258/ijddt.9.2.15

Al-Zaban, M. I., Mahmoud, M. A., Al Harbi, M. A., & Bahatheq, A. M. (2020). Bioremediation of Crude Oil by Rhizosphere Fungal Isolates in the Presence of Silver Nanoparticles. *International Journal of Environmental Research and Public Health, 17*(18), 6564. doi:10.3390/ijerph17186564 PMID:32916946

Amadi, A., Dickson, A. A., & Maate, G. O. (1993). Remediation of oil polluted soils: Effects of organic and inorganic nutrient supplements on the performance of maize (*Zea mays* L.). *Water, Air, and Soil Pollution, 66*(1-2), 59–76. doi:10.1007/BF00477060

Amin, A., Naik, A. T. R., Azhar, M., & Nayak, H. (2013). Bioremediation of different waste waters—A review. *Canadian Journal of Fisheries and Aquatic Sciences, 7*(2), 7–17.

Amsellem, L., Brouat, C., Duron, O., Porter, S. S., & Facon, B. (2017). Networks of invasion: Empirical evidence and case studies. *Advances in Ecological Research, 57*, 99–146. doi:10.1016/bs.aecr.2016.10.005

Ananthashankar, R. (2012). *Treatment of textile effluent containing reactive red 120 dye using advanced oxidation* (M.Sc. Thesis). Dalhousie University, Halifax, Nova Scotia, Canada.

Andersen, R., Chapman, S. J., & Artz, R. R. E. (2013). Microbial communities in natural and disturbed peatlands: A review. *Soil Biology & Biochemistry, 57*, 979–994. doi:10.1016/j.soilbio.2012.10.003

Anderson, J. M. (1991). The effects of climate change on decomposition processes in grassland and coniferous forests. *Ecological Applications, 1*(3), 326–347. doi:10.2307/1941761 PMID:27755768

Anderson, R. F., & Jackson, R. W. (1958). Essential amino acids in microbial proteins. *Applied Microbiology, 6*(5), 369–373. doi:10.1128/AM.6.5.369-373.1958 PMID:13571982

Andrady, A. L. (2011). Microplastics in the marine environment. *Marine Pollution Bulletin, 62*(8), 1596–1605. doi:10.1016/j.marpolbul.2011.05.030 PMID:21742351

Andrews, G. K. (2000). Regulation of metallothionein gene expression by oxidative stress and metal ions. *Biochemical Pharmacology, 59*(1), 95–104. doi:10.1016/S0006-2952(99)00301-9 PMID:10605938

Andrews, S. C., Robinson, A. K., & Rodríguez-Quiñones, F. (2003). Bacterial iron homeostasis. *FEMS Microbiology Reviews, 27*(2-3), 215–237. doi:10.1016/S0168-6445(03)00055-X PMID:12829269

Anetor, J. I., Wanibuchi, H., & Fukushima, S. (2007). Arsenic exposure and its health effects and risk of cancer in developing countries: Micronutrients as host defence. *Asian Pacific Journal of Cancer Prevention*, *8*(1), 13. PMID:17477765

Angelucci, D., & Tomei, M. (2016). Ex-situ bioremediation of chlorophenol contaminated soil: Comparison of slurry and solid-phase bioreactors with the two-step polymer extraction-bioregeneration process. *Journal of Chemical Technology and Biotechnology (Oxford, Oxfordshire)*, *91*(6), 1577–1584. doi:10.1002/jctb.4882

Anikwe, M. A. N., & Nwobodo, K. C. A. (2002). Long-term effect of municipal waste disposal on soil properties and productivity of sites used for urban agriculture in Abakaliki, Nigeria. *Bioresource Technology*, *83*(3), 241–250. doi:10.1016/S0960-8524(01)00154-7 PMID:12094801

Anjaneyulu, Y., Sreedhara Chary, N., & Raj, S. S. D. (2005). Decolourization of Industrial Effluents-Available Methods and Emerging Technologies—A Review. *Reviews in Environmental Science and Biotechnology*, *4*(4), 245–273. doi:10.100711157-005-1246-z

Anjum, M., Miandad, R., Waqas, M., Gehany, F., & Barakat, M. A. (2016). Remediation of wastewater using various nanomaterials. *Arabian Journal of Chemistry*, *12*(8), 4897–4919. doi:10.1016/j.arabjc.2016.10.004

Anjum, S. A., Xie, X. Y., Wang, L. C., Saleem, M. F., Man, C., & Lei, W. (2011). Morphological, physiological and biochemical responses of plants to drought stress. *African Journal of Agricultural Research*, *6*(9), 2026–2032.

Antoun, H., & Kloepper, J. W. (2001). Plant growth promoting rhizobacteria. Encyclopedia of Genetics, 1477-1480.

Antoun, H., Beauchamp, C. J., Goussard, N., Chabot, R., & Lalande, R. (1998). Potential of Rhizobium and Bradyrhizobium species as plant growth promoting rhizobacteria on non-legumes: effect on radishes (*Raphanus sativus* L.). In *Molecular microbial ecology of the soil* (pp. 57–67). Springer. doi:10.1007/978-94-017-2321-3_5

Anum, W., Naeem, M., Tanveer, A., Ali, H. H., Nazir, M. Q., Hanif, Z., & Kashif, M. S. (2016). Effects of African purslane (*Zaleya pentandra* L.) on germination and seedling growth of maize. *Allelopathy Journal*, *39*(1), 83–91.

Anwar, M. S., Paliwal, A., Firdous, N., Verma, A., Kumar, A., & Pande, V. (2018). Co-culture development and bioformulation efficacy of psychrotrophic PGPRs to promote growth and development of Pea (Pisum sativum) plant. *The Journal of General and Applied Microbiology*. PMID:30381611

An, Y., Li, T., Jin, Z., Dong, M., Xia, H., & Wang, X. (2010). Effect of bimetallic and polymer coated Fe nanoparticles on biological denitrification. *Bioresource Technology*, *101*(24), 9825–9828. doi:10.1016/j.biortech.2010.07.110 PMID:20727742

Anyasi, R. O. & Atagana, H. I. (2018). Profiling of plants at petroleum contaminated site for 820 phytoremediation. *Int. J. Phytoremediation, 20*, 352–361. . doi:10.1080/15226514.2017.1393386

Anyasi, R. O., & Atagana, H. I. (2017). Assessment of plants at petroleum contaminated site for phytoremediation. *Proceedings of the international conference of recent trends in environmental science and engineering (RTESE'17).* 10.11159/rtese17.105

Anyasi, R. O., Atagana, H. I., & Sutherland, R. (2019). Identification and characterization of PAH-degrading endophytes isolated from plants growing around a sludge dam. *International Journal of Phytoremediation*, *21*(7), 672–682. doi:10.1080/15226514.2018.1556585 PMID:30942084

Apostol, L. C., Pereira, L., Pereira, R., Gavrilescu, M., & Alves, M. M. (2012). Biological decolorizartion of Xanthene dyes by anaerobic granular biomass. *Biodegradation*, *23*(5), 725–737. doi:10.100710532-012-9548-7 PMID:22437968

Appleton, J. D. (2007). Radon: Sources, health risks, and hazard mapping. *Ambio*, *36*(1), 85–89. doi:10.1579/0044-7447(2007)36[85:RSHRAH]2.0.CO;2 PMID:17408197

Aprill, W., & Sims, R. C. (1990). Evaluation of the use of prairie grasses for stimulating polycyclic aromatic hydrocarbon treatment in soil. *Chemosphere*, *20*(1-2), 253–265. doi:10.1016/0045-6535(90)90100-8

Aquilanti, L., Favilli, F., & Clementi, F. (2004). Comparison of different strategies for isolation and preliminary identification of *Azotobacter* from soil samples. *Soil Biology & Biochemistry*, *36*(9), 1475–1483. doi:10.1016/j.soilbio.2004.04.024

Arai, S., Yonezawa, Y., Ishibashi, M., Matsumoto, F., Adachi, M., Tamada, T., Tokunaga, H., Blaber, M., Tokunaga, M., & Kuroki, R. (2014). Structural characteristics of alkaline phosphatase from the moderately halophilic bacterium *Halomonas sp.* 593. *Acta Crystallographica. Section D, Biological Crystallography*, *70*(Pt 3), 811–820. doi:10.1107/S1399004713033609 PMID:24598750

Araki, A., Saito, I., Kanazawa, A., Morimoto, K., Nakayama, K., Shibata, E., ... Saijo, Y. (2014). Phosphorus flame retardants in indoor dust and their relation to asthma and allergies of inhabitants. *Indoor Air*, *24*(1), 3–15. doi:10.1111/ina.12054 PMID:23724807

Aranganathan, V., Kanimozhi, A. M., & Palvannan, T. (2013). Statistical optimization of synthetic azo dye (orange II) degradation by azoreductase from *Pseudomonas oleovorans* PAMD_1. *Preparative Biochemistry & Biotechnology*, *43*(7), 649–667. doi:10.1080/10826068.2013.772063 PMID:23768111

Aransiola, S. A., Ijah, U. J. J., Abioye, O. P., & Bala, J. D. (2019). Microbial-aided phytoremediation of heavy metals contaminated soil: A review. *European Journal of Biological Research*, *9*(2), 104–125.

Araujo, R., Gupta, V. V., Reith, F., Bissett, A., Mele, P., & Franco, C. M. (2020). Biogeography and emerging significance of *Actinobacteria* in Australia and Northern Antarctica soils. *Soil Biology & Biochemistry*, *146*, 107805. doi:10.1016/j.soilbio.2020.107805

Aravindhan, R., Rao, J. R., & Nair, B. U. (2007). Removal of basic yellow dye from aqueous solution by sorption on green alga Caulerpa scalpelliformis. *Journal of Hazardous Materials*, *142*(1-2), 68–76. doi:10.1016/j.jhazmat.2006.07.058 PMID:16938392

Ardakani, S. S., Heydari, A., Khorasani, N. A., Arjmandi, R., & Ehteshami, M. (2009). Preparation of new biofungicides using antagonistic bacteria and mineral compounds for controlling cotton seedling damping-off disease. *Journal of Plant Protection Research*, *49*(1). Advance online publication. doi:10.2478/v10045-009-0007-3

Argelis, D. T., Gonzala, D. A., Vizcaino, C., & Gartia, M. T. (1993). Biochemical mechanism of stone alteration carried out by filamentous fungi living in monuments. *Biogeochemistry*, *19*, 129–147.

Arikan, Ş., & Pirlak, L. (2016). Effects of plant growth promoting rhizobacteria (PGPR) on growth, yield and fruit quality of sour cherry (*Prunus cerasus* L.). *Erwerbs-Obstbau*, *58*(4), 221–226. doi:10.100710341-016-0278-6

Armengaud, J., Marie Hartmann, E., & Bland, C. (2013). Proteogenomics for environmental microbiology. *Proteomics*, *13*(18-19), 2731–2742. doi:10.1002/pmic.201200576 PMID:23636904

Armengaud, P., Breitling, R., & Amtmann, A. (2010). Coronatine-intensive 1 (COI1) mediates transcriptional responses of *Arabidopsis thaliana* to external potassium supply. *Molecular Plant*, *3*(2), 390–405. doi:10.1093/mpsq012 PMID:20339157

Arora, D. S., & Sharma, R. K. (2010). Ligninolytic fungal laccases and their biotechnological applications. *Applied Biochemistry and Biotechnology*, *160*(6), 1760–1788. doi:10.100712010-009-8676-y PMID:19513857

Arora, N. K. (Ed.). (2015). *Plant Microbes Symbiosis: Applied Facets*. Springer.

Arora, P. K., Kumar, M., Chauhan, A., Raghava, G. P., & Jain, R. K. (2009). OxDBase: A database of oxygenases involved in biodegradation. *BMC Research Notes*, *2*(1), 67. doi:10.1186/1756-0500-2-67 PMID:19405962

Arora, P. K., Srivastava, A., & Singh, V. P. (2010). Application of Monooxygenases in Dehalogenation, Desulphurization, Denitrification and Hydroxylation of Aromatic Compounds. *Journal of Bioremediation & Biodegradation, 1*(03), 112. doi:10.4172/2155-6199.1000112

Arrigo, K. R. (2005). Marine microorganisms and global nutrient cycles. *Nature, 437*(7057), 349–355. doi:10.1038/nature04159 PMID:16163345

Arseneault, T., Goyer, C., & Filion, M. (2013). Phenazine production by *Pseudomonas sp.* LBUM223 contributes to the biological control of potato common scab. *Phytopathology, 103*(10), 995–1000. doi:10.1094/PHYTO-01-13-0022-R PMID:23883153

Arshad, M., Shaharoona, B., & Mahmood, T. (2008). Inoculation with *Pseudomonas* spp. containing ACC-deaminase partially eliminates the effects of drought stress on growth, yield, and ripening of pea (*Pisum sativum* L.). *Pedosphere, 18*(5), 611–620. doi:10.1016/S1002-0160(08)60055-7

Artursson, V., Finlay, R. D., & Jansson, J. K. (2006). Interactions between arbuscular mycorrhizal fungi and bacteria and their potential for stimulating plant growth. *Environmental Microbiology, 8*(1), 1–10. doi:10.1111/j.1462-2920.2005.00942.x PMID:16343316

Arunrat, N., Pumijumnong, N., Sereenonchai, S., & Chareonwong, U. (2020). Factors Controlling Soil Organic Carbon Sequestration of Highland Agricultural Areas in the Mae Chaem Basin, Northern Thailand. *Agronomy (Basel), 10*(2), 305–330. doi:10.3390/agronomy10020305

Arzanesh, M. H., Alikhani, H. A., Khavazi, K., Rahimian, H. A., & Miransari, M. (2011). Wheat (Triticum aestivum L.) growth enhancement by Azospirillum sp. under drought stress. *World Journal of Microbiology & Biotechnology, 27*(2), 197–205. doi:10.100711274-010-0444-1

Asadgol, Z., Forootanfar, H., Rezaei, S., Mahvi, A. H., & Faramarzi, M. A. (2014). Removal of phenol and bisphenol-A catalyzed by laccase in aqueous solution. *Journal of Environmental Health Science & Engineering, 12*(1), 93. doi:10.1186/2052-336X-12-93 PMID:25031840

Asami, T., Nakano, T., & Fujioka, S. (2005). Plant brassinosteroid hormones. *Vitamins and Hormones, 72*, 479–504. doi:10.1016/S0083-6729(05)72014-8 PMID:16492480

Asari, S., Tarkowská, D., Rolčík, J., Novák, O., Palmero, D. V., Bejai, S., & Meijer, J. (2017). Analysis of plant growth-promoting properties of *Bacillus amyloliquefaciens* UCMB5113 using *Arabidopsis thaliana* as host plant. *Planta, 245*(1), 15–30. doi:10.100700425-016-2580-9 PMID:27541497

Ashokkumar, V., & Rengasamy, R. (2012). Mass culture of *Botryococcus braunii Kutz.* under open raceway pond for biofuel production. *Bioresource Technology, 104*, 394–399. doi:10.1016/j.biortech.2011.10.093 PMID:22115530

Ashraf, M. A., Asif, M., Zaheer, A., Malik, A., Ali, Q., & Rasool, M. (2013). Plant growth promoting rhizobacteria and sustainable agriculture: A review. *African Journal of Microbiological Research, 7*(9), 704–709. doi:10.5897/AJMR12.936

Asiabadi, F. I., Mirbagheri, S. A., Najafi, P., & Moatar, F. (2018). Concentrations of petroleum hydrocarbons at different depths of soil following phytoremediation. *Environmental Engineering and Management Journal, 17*(9), 2129–2135. doi:10.30638/eemj.2018.211

Atashgahi, S., Shetty, S. A., Smidt, H., & de Vos, W. M. (2018). Flux, impact, and fate of halogenated xenobiotic compounds in the gut. *Frontiers in Physiology, 9*, 888. doi:10.3389/fphys.2018.00888 PMID:30042695

Atkins, P., & Jones, L. (1997). Chemistry—Molecules, Matter and Change. W. H. Freeman.

Atlas, R. M. (1981). Microbial degradation of petroleum hydrocarbons: An environmental perspective. *Microbiological Reviews*, *45*(1), 180–209. doi:10.1128/MR.45.1.180-209.1981 PMID:7012571

Atlas, R. M., & Philp, J. (2005). *Bioremediation: Applied microbial solutions for real-world environmental cleanup*. ASM Press. doi:10.1128/9781555817596

Aulenta, F., Di Tomassi, C., Cupo, C., Papini, M. P., & Majone, M. (2006). Influence of hydrogen on the reductive dechlorination of tetrachloroethene (PCE) to ethane in a methanogenic biofilm reactor: Role of mass transport phenomena. *Journal of Chemical Technology and Biotechnology (Oxford, Oxfordshire)*, *81*(9), 1520–1529. doi:10.1002/jctb.1562

Awasthi, M. K., Wang, Q., Ren, X., Zhao, J., Huang, H., Awasthi, S. K., Lahori, A. H., Li, R., Zhou, L., & Zhang, Z. (2016). Role of biochar amendment in mitigation of nitrogen loss and greenhouse gas emission during sewage sludge composting. *Bioresource Technology*, *219*, 270–280. doi:10.1016/j.biortech.2016.07.128 PMID:27497088

Awuchi Godswill, C., & Awuchi Gospel, C. (2019a). Physiological Effects of Plastic Wastes on the Endocrine System (Bisphenol A, Phthalates, Bisphenol S, PBDEs, TBBPA). *International Journal of Bioinformatics and Computational Biology*, *4*(2), 11–29.

Awuchi Godswill, C., & Awuchi Gospel, C. (2019b). Impacts of Plastic Pollution on the Sustainability of Seafood Value Chain and Human Health. *International Journal of Advanced Academic Research*, *5*(11), 46–138.

Ayangbenro, A. S., & Babalola, O. O. (2017). A New Strategy for Heavy Metal Polluted Environments: A Review of Microbial Biosorbents. *International Journal of Environmental Research and Public Health*, *14*(1), 94. Advance online publication. doi:10.3390/ijerph14010094 PMID:28106848

Azam, F., Farooq, S., & Lodhi, A. (2003). Microbial biomass in agricultural soils-determination, synthesis, dynamics and role in plant nutrition. *Pakistan Journal of Biological Sciences*, *6*(7), 629–639. doi:10.3923/pjbs.2003.629.639

Aznar, A., Chen, N. W., Rigault, M., Riache, N., Joseph, D., Desmaële, D., Mouille, G., Boutet, S., Soubigou-Taconnat, L., Renou, J. P., Thomine, S., Expert, D., & Dellagi, A. (2014). Scavenging iron: A novel mechanism of plant immunity activation by microbial siderophores. *Plant Physiology*, *164*(4), 2167–2183. doi:10.1104/pp.113.233585 PMID:24501001

Azubuike, C. C., Chikere, C. B., & Okpokwasili, G. C. (2016). Bioremediation techniques–classification based on site of application: Principles, advantages, limitations and prospects. *World Journal of Microbiology & Biotechnology*, *32*(11), 180. doi:10.100711274-016-2137-x PMID:27638318

Azuka, C. V., Igu eacute, A. M., Diekkr uuml ger, B., & Igwe, C. A. (2015). Soil survey and soil classification of the Koupendri catchment in Benin, West Africa. *African Journal of Agricultural Research*, *10*(42), 3938–3951. doi:10.5897/AJAR2015.9904

Baakza, A., Vala, A. K., Dave, B. P., & Dube, H. C. (2004). A comparative study of siderophore production by fungi from marine and terrestrial habitats. *Journal of Experimental Marine Biology and Ecology*, *311*(1), 1–9. doi:10.1016/j.jembe.2003.12.028

Bååth, E. (1989). Effects of heavy metals in soil on microbial processes and populations (a review). *Water, Air, and Soil Pollution*, *47*(3-4), 335–379. doi:10.1007/BF00279331

Babu, G., & Reddy, M. S. (2011). *Aspergillus tubingensis* Improves the Growth and Native Mycorrhizal Colonization of Bermudagrass in Bauxite Residue. *Bioremediation Journal*, *15*(3), 157–164. doi:10.1080/10889868.2011.598486

Babu, S., Bidyarani, N., Chopra, P., Monga, D., Kumar, R., Radha, P., Kranthi, S., & … . (2015). Evaluating microbe-plant interactions and varietal differences for enhancing biocontrol efficacy in root rot challenged cotton crop. *European Journal of Plant Pathology*, *142*(2), 345–362. doi:10.100710658-015-0619-6

Backer, R., Rokem, J. S., Ilangumaran, G., Lamont, J., Praslickova, D., Ricci, E., Subramanian, S., & Smith, D. L. (2018). Plant Growth-Promoting Rhizobacteria: Context, Mechanisms of Action, and Roadmap to Commercialization of Biostimulants for Sustainable Agriculture. *Frontiers in Plant Science*, *9*, 1473. doi:10.3389/fpls.2018.01473 PMID:30405652

Bacon, C. W., Yates, I. E., Hinton, D. M., & Meredith, F. (2001). Biological control of *Fusarium moniliforme* in maize. *Environmental Health Perspectives*, *109*(suppl 2), 325–332. PMID:11359703

Badhan, K., Mehra, R., & Sonkawade, R. G. (2017). Natural Radioactivity Measurements in Soils of Jalandhar and Hoshiarpur Districts of Punjab, India. *International Journal of Pure and Applied Physics*, *13*, 232–237.

Badri, D. V., Chaparro, J. M., Manter, D. K., Martinoia, E., & Vivanco, J. M. (2012). Influence of ATP-binding cassette transporters in root exudation of phytoalexins, signals, and in disease resistance. *Frontiers in Plant Science*, *3*, 149. doi:10.3389/fpls.2012.00149 PMID:22783269

Badri, D. V., Weir, T. L., van der Lelie, D., & Vivanco, J. M. (2009). Rhizosphere chemical dialogues: Plant–microbe interactions. *Current Opinion in Biotechnology*, *20*(6), 642–650. doi:10.1016/j.copbio.2009.09.014 PMID:19875278

Baez-Rogelio, A., Morales-García, Y. E., Quintero-Hernández, V., & Muñoz-Rojas, J. (2017). Next generation of microbial inoculants for agriculture and bioremediation. *Microbial Biotechnology*, *10*(1), 19–21. doi:10.1111/1751-7915.12448 PMID:27790851

Bafana, A., Devi, S. S., & Chakrabarti, T. (2011). Azo dyes: past, present and the future. *Environmental Review*, *19*, 350–371.

Bafana, A., Chakrabarti, T., & Devi, S. S. (2008). Azoreductase and dye detoxification activities of *Bacillus velezensis* strain AB. *Applied Microbiology and Biotechnology*, *77*(5), 1139–1144. doi:10.100700253-007-1212-5 PMID:18034237

Bagade, A., Nandre, V., Paul, D., Patil, Y., Sharma, N., Giri, A., & Kodam, K. (2020). Characterisation of hyper tolerant *Bacillus firmus* L-148 for arsenic oxidation. *Environmental Pollution*, *261*, 1–13. doi:10.1016/j.envpol.2020.114124 PMID:32078878

Bagal-Kestwal, D., Karve, M., Kakade, B., & Pillai, V. (2008). Invertase inhibition based electrochemical sensor for the detection of heavy metal ions in aqueous system: Application of ultra-microelectrode to enhance sucrose biosensor's sensitivity. *Biosensors & Bioelectronics*, *24*(4), 657–664. doi:10.1016/j.bios.2008.06.027 PMID:18667298

Bageshwar, U. K., Srivastava, M., Pardha-Saradhi, P., Paul, S., Gothandapani, S., Jaat, R. S., Shankar, P., Yadav, R., Biswas, D. R., Kumar, P. A., Padaria, J. C., Mandal, P. K., Annapurna, K., & Dasa, H. K. (2017). An environmentally friendly engineered *Azotobacter* strain that replaces a substantial amount of urea fertilizer while sustaining the same wheat yield. *Applied and Environmental Microbiology*, *83*(15), e00590–e17. doi:10.1128/AEM.00590-17 PMID:28550063

Bagyalakshmi, B., Ponmurugan, P., & Balamurugan, A. (2012). Impact of different temperature, carbon and nitrogen sources on solubilization efficiency of native potassium solubilizing bacteria from tea (*Camellia sinensis*). *Journal of Biological Research (Thessaloniki)*, *3*, 36–42.

Bagyaraj, D. J., & Revanna, A. (2016). Effect of chemical fertilizers on the beneficial soil microorganisms. *Fertilizers and Environment News*, *2*, 10–11.

Bahar, M. M., Megharaj, M., & Naidu, R. (2013). Bioremediation of arsenic-contaminated water: Recent advances and future prospects. *Water, Air, and Soil Pollution*, *224*(12), 1722. doi:10.100711270-013-1722-y

Bai, L., Xu, H., Wang, C., Deng, J., & Jiang, H. (2016). Extracellular polymeric substances facilitate the biosorption of phenanthrene on cyanobacteria *Microcystis aeruginosa*. *Chemosphere*, *162*, 172–180. doi:10.1016/j.chemosphere.2016.07.063 PMID:27497347

Bailey, D. C., Bohac, T. J., Shapiro, J. A., Giblin, D. E., Wencewicz, T. A., & Gulick, A. M. (2018). Crystal Structure of the Siderophore Binding Protein BauB Bound to an Unusual 2:1 Complex Between Acinetobactin and Ferric Iron. *Biochemistry*, *57*(48), 6653–6661. doi:10.1021/acs.biochem.8b00986 PMID:30406986

Bailey-Serres, J., Parker, J. E., Ainsworth, E. A., Oldroyd, G., & Schroeder, J. I. (2019). Genetic strategies for improving crop yields. *Nature*, *575*(7781), 109–118. doi:10.103841586-019-1679-0 PMID:31695205

Bais, H. P., Loyola-Vargas, V. M., Flores, H. E., & Vivanco, J. M. (2001). Root-specific metabolism: The biology and biochemistry of underground organs. *In Vitro Cellular & Developmental Biology. Plant*, *37*(6), 730–741. doi:10.100711627-001-0122-y

Bais, H. P., Weir, T. L., Perry, L. G., Gilroy, S., & Vivanco, J. M. (2006). The role of root exudates in rhizosphere interactions with plants and other organisms. *Annual Review of Plant Biology*, *57*(1), 233–266. doi:10.1146/annurev.arplant.57.032905.105159 PMID:16669762

Baishya, K., & Samra, H. P. (2014). Effect of agrochemicals application on accumulation of heavy metals on soil of different land uses with respect to nutrient status. *Journal of Environmental Science Toxicology and Food Technology*, *8*(7), 46–54. doi:10.9790/2402-08724654

Bai, Z., Caspari, T., Gonzalez, M. R., Batjes, N. H., Mäder, P., Bünemann, E. K., & Tóth, Z. (2018). Effects of agricultural management practices on soil quality: A review of long-term experiments for Europe and China. *Agriculture, Ecosystems & Environment*, *265*, 1–7. doi:10.1016/j.agee.2018.05.028

Bajaj, S., Sagar, S., Khare, S., & Singh, D. K. (2017). Biodegradation of γhexachlorocyclohexane (lindane) by halophilic bacterium *Chromohalobacter* sp. LD2 isolated from HCH dumpsite. *International Biodeterioration & Biodegradation*, *122*, 23–28. doi:10.1016/j.ibiod.2017.04.014

Bajsa, O., Nair, J., Mathew, K., & Ho, G. E. (2004). *Pathogen Die-Off in Vermicomposting Process*. Paper presented at the International Conference on "Small Water and Wastewater Treatment Systems", Perth, Australia.

Bakermans, C., Tsapin, A. I., Souza-Egipsy, V., Gilichinsky, D. A., & Nealson, K. H. (2003). Reproduction and metabolism at− 10 C of bacteria isolated from Siberian permafrost. *Environmental Microbiology*, *5*(4), 321–326. doi:10.1046/j.1462-2920.2003.00419.x PMID:12662179

Bakker, P. A. H. M., Ran, L., & Mercado-Blanco, J. (2014). Rhizobacterial salicylate production provokes headaches. *Plant and Soil*, *382*(1-2), 1–16. doi:10.100711104-014-2102-0

Bakker, P. A., Pieterse, C. M., & Van Loon, L. C. (2007). Induced systemic resistance by fluorescent Pseudomonas spp. *Phytopathology*, *97*(2), 239–243. doi:10.1094/PHYTO-97-2-0239 PMID:18944381

Balaji, S., Kalaivani, T., Sushma, B., Pillai, C. V., Shalini, M., & Rajasekaran, C. (2016). Characterization of sorption sites and different stress response of microalgae isolates against application towards phytoremediation. *International Journal of Phytoremediation*, *18*(8), 747–753. doi:10.1080/15226514.2015.1115960 PMID:26587690

Balarak, D., Jaafari, J., Hassani, G., Mahdavi, Y., Tyagi, I., Agarwal, S., & Gupta, V. K. (2015). The use of low-cost adsorbent (Canola residues) for the adsorption of methylene blue from aqueous solution: Isotherm, kinetic and thermodynamic studies. *Colloid and Interface Science Communications*, *7*, 16–19. doi:10.1016/j.colcom.2015.11.004

Baldiris, R., Acosta-Tapia, N., Montes, A., Hernández, J., & Vivas-Reyes, R. (2018). Reduction of Hexavalent Chromium and Detection of Chromate Reductase (ChrR) in *Stenotrophomonas maltophilia. Molecules (Basel, Switzerland)*, *23*(2), 1–20. doi:10.3390/molecules23020406 PMID:29438314

Baldrian, P. (2006). Fungal laccases–occurrence and properties. *FEMS Microbiology Reviews*, *30*(2), 215–242. doi:10.1111/j.1574-4976.2005.00010.x PMID:16472305

Balogh-Brunstad, Z., Keller, C. K., Gill, R. A., Bormann, B. T., & Li, C. Y. (2008). The effect of bacteria and fungi on chemical weathering and chemical denudation fluxes in pine growth experiments. *Biogeochemistry*, *88*(2), 153–167. doi:10.100710533-008-9202-y

Balseiro-Romero, M., & Baveye, P. C. (2018). Book Review: Soil Pollution: A Hidden Danger Beneath our Feet. *Frontiers in Environmental Science*, *6*, 130–134. doi:10.3389/fenvs.2018.00130

Banat, I. M., Nigam, P., Singh, D., & Marchant, R. (1996). Microbial decolorization of textile-dyecontaining effluents: A review. *Bioresource Technology*, *58*(3), 217–227. doi:10.1016/S0960-8524(96)00113-7

Baoune, H., Aparicio, J. D., Acuña, A., El Hadj-khelil, A. O., Sanchez, L., Polti, M. A., & Alvarez, A. (2019). Effectiveness of the *Zea mays-Streptomyces* association for the phytoremediation of petroleum hydrocarbons impacted soils. *Ecotoxicology and Environmental Safety*, *184*, 109591. doi:10.1016/j.ecoenv.2019.109591 PMID:31514081

Baoune, H., El Hadj-Khelil, A. O., Pucci, G., Sineli, P., Loucif, L., & Polti, M. A. (2018). Petroleum degradation by endophytic *Streptomyces spp.* isolated from plants grown in contaminated soil of southern Algeria. *Ecotoxicology and Environmental Safety*, *147*, 602–609. doi:10.1016/j.ecoenv.2017.09.013 PMID:28923725

Barac, T., Taghavi, S., Borremans, B., Provoost, A., Oeyen, L., Colpaert, J. V., Vangronsveld, J., & van der Lelie, D. (2004). Engineered endophytic bacteria improve phytoremediation of water-soluble, volatile, organic pollutants. *Nature Biotechnology*, *22*(5), 583–588. doi:10.1038/nbt960 PMID:15077119

Baragaño, D., Forján, R., Welte, L., & Gallego, J. L. R. (2020). Nanoremediation of As and metals polluted soils by means of graphene oxide nanoparticles. *Scientific Reports*, *10*(1), 1896. doi:10.103841598-020-58852-4 PMID:32024880

Barbash, J. E., & Reinhard, M. (1989). Abiotic dehalogenation of 1, 2-dichloroethane and 1, 2-dibromoethane in aqueous solution containing hydrogen sulfide. *Environmental Science & Technology*, *23*(11), 1349–1358. doi:10.1021/es00069a004

Barber, M. S., Giesecke, U., Reichert, A., & Minas, W. (2004). Industrial enzymatic production of cephalosporin-based β-lactams. In *Molecular Biotechnolgy of Fungal beta-Lactam Antibiotics and Related Peptide Synthetases* (pp. 179–215). Springer. doi:10.1007/b99261

Barber, S. A. (1995). *Soil Nutrient Bioavailability A Mechanistic Approach* (2nd ed.). Wiley.

Barbez, E., Dünser, K., Gaidora, A., Lendl, T., & Busch, W. (2017). Auxin steers root cell expansion via apoplastic pH regulation in *Arabidopsis thaliana. Proceedings of the National Academy of Sciences of the United States of America*, *114*(24), E4884–E4893. doi:10.1073/pnas.1613499114 PMID:28559333

Bar, C., Patil, R., Doshi, J., Kulkarni, M. J., & Gade, W. N. (2007). Characterization of the proteins of bacterial strain isolated from contaminated site involved in heavy metal resistance-A proteomic approach. *Journal of Biotechnology*, *128*(3), 444–451. doi:10.1016/j.jbiotec.2006.11.010 PMID:17210198

Barea, J. M. (1997). *Mycorrhiza/bacteria interactions on plant growth promotion. In Plant growth-promoting rhizobacteria, present status and future prospects*. OECD.

Barea, J. M., & Brown, M. E. (1974). Effects on plant growth produced by *Azotobacter paspali* related to synthesis of plant growth regulating substances. *The Journal of Applied Bacteriology*, *37*(4), 583–593. doi:10.1111/j.1365-2672.1974.tb00483.x PMID:4611996

Bargaz, A., Lyamlouli, K., Chtouki, M., Zeroual, Y., & Dhiba, D. (2018). Soil microbial resources for improving fertilizers efficiency in an integrated plant nutrient management system. *Frontiers in Microbiology*, *9*, 1606. doi:10.3389/fmicb.2018.01606 PMID:30108553

Barka, E. A., Nowak, J., & Clément, C. (2006). Enhancement of chilling resistance of inoculated grapevine plantlets with a plant growth-promoting rhizobacterium, Burkholderiaphytofirmans strain PsJN. *Applied and Environmental Microbiology*, *72*(11), 7246–7252. doi:10.1128/AEM.01047-06 PMID:16980419

Barkay, T., Miller, S. M., & Summers, A. O. (2003). Bacterial mercury resistance from atoms to ecosystems. *FEMS Microbiology Reviews*, *27*(2-3), 355–384. doi:10.1016/S0168-6445(03)00046-9 PMID:12829275

Barrera, A. D., & Soto, E. (2010). Biotechnological uses of *Azotobacter vinelandii*: Current state limits and prospects. *African Journal of Biotechnology*, *9*(33), 5240–5250.

Baruah, R. (2018). Towards the Bioavailability of Zinc in Agricultural Soils. In *Role of Rhizospheric Microbes in Soil* (pp. 99–136). Springer. doi:10.1007/978-981-13-0044-8_4

Baruthio, F. (1992). Toxic effects of chromium and its compounds. *Biological Trace Element Research*, *32*(1-3), 145–153. doi:10.1007/BF02784599 PMID:1375051

Basak, B. B., Sarkar, B., Biswas, D. R., Sarkar, S., Sanderson, P., & Naidu, R. (2016). Bio-intervention of naturally occurring silicate minerals for alternative source of potassium: Challenges and opportunities. *Advances in Agronomy*, *141*, 115–145. doi:10.1016/bs.agron.2016.10.016

Basak, B., & Biswas, D. (2012). *Modification of waste mica for alternative source of potassium: evaluation of potassium release in soil from waste mica treated with potassium solubilizing bacteria (KSB)*. Lap Lambert Academic Publishing.

Bashagaluke, J. B., Logah, V., Opoku, A., Sarkodie-Addo, J., & Quansah, C. (2018). Soil nutrient loss through erosion: Impact of different cropping systems and soil amendments in Ghana. *PLoS One*, *13*(12), 1–17. doi:10.1371/journal.pone.0208250 PMID:30566517

Bashan, Y. (1998). Inoculants of plant growth-promoting bacteria for use in agriculture. *Biotechnology Advances*, *16*(4), 729–770. doi:10.1016/S0734-9750(98)00003-2

Bashan, Y., de Bashan, L. E., Prabhu, S. R., & Hernandez, J. P. (2014). Advances in plant growthpromoting bacterial inoculant technology: Formulations and practical perspectives (1998- 2013). *Plant and Soil*, *378*(1-2), 1–33. doi:10.100711104-013-1956-x

Bashan, Y., & Holguin, G. (1997). Azospirillum-plant relationships: Environmental and physiological advances (1990–1996). *Canadian Journal of Microbiology*, *43*(2), 103–121. doi:10.1139/m97-015

Bashan, Y., & Holguin, G. (1998). Proposal for the division of plant growth-promoting rhizobacteria into two classifications: biocontrol-PGPB (plant growth-promoting bacteria) and PGPB. *Soil Biology & Biochemistry*, *30*(8-9), 1225–1228. doi:10.1016/S0038-0717(97)00187-9

Bashan, Y., Menéndez, A., & Toledo, G. (1992). Responses of soybean and cowpea root membranes to inoculation with Azospirillum brasilense. *Symbiosis*, *13*, 217–228.

Basheer, T., & Umesh, M. (2018). Valorization of Tannery Solid Waste Materials Using Microbial Techniques: Microbes in Tannery Solid Waste Management. In *Handbook of Research on Microbial Tools for Environmental Waste Management* (pp. 127–145). IGI Global. doi:10.4018/978-1-5225-3540-9.ch007

Basumatary, B., & Bordoloi, S. (2016). Phytoremediation of crude oil-contaminated soil using *Cynodon dactylon* (L.) Pers. In A. Ansari, S. Gill, R. Gill, G. Lanza, & L. Newman (Eds.), *Phytoremediation*. Springer. doi:10.1007/978-3-319-41811-7_3

Basumatary, B., Saikia, R., Bordoloi, S., Das, H. C., & Sarma, H. P. (2012). Assessment of potential plant species for phytoremediation of hydrocarbon-contaminated areas of upper Assam, India. *Journal of Chemical Technology and Biotechnology (Oxford, Oxfordshire)*, *87*(9), 1329–1334. doi:10.1002/jctb.3773

Baudoin, E., Nazaret, S., Mougel, C., Ranjard, L., & Moënne-Loccoz, Y. (2009). Impact of inoculation with the phytostimulatory PGPR *Azospirillum lipoferum* CRT1 on the genetic structure of the rhizobacterial community of field-grown maize. *Soil Biology & Biochemistry*, *41*(2), 409–413. doi:10.1016/j.soilbio.2008.10.015

Baughman, G. L., & Weber, E. J. (1994). Transformation of dyes and related compounds in anoxic sediment: Kinetics and products. *Environmental Science & Technology*, *28*(2), 267–276. doi:10.1021/es00051a013 PMID:22176172

Beardmore, J. A., & Porter, J. S. (2003). *Genetically modified organisms and aquaculture*. Food and Agriculture Organization of the United Nations.

Beauregard, P. B., Chai, Y., Vlamakis, H., Losick, R., & Kolter, R. (2013). *Bacillus subtilis* biofilm induction by plant polysaccharides. *Proceedings of the National Academy of Sciences of the United States of America*, *110*(17), E1621–E1630. doi:10.1073/pnas.1218984110 PMID:23569226

Beauvais-Flück, R., Slaveykova, V. I., & Cosio, C. (2017). Cellular toxicity pathways of inorganic and methyl mercury in the green microalga *Chlamydomonas reinhardtii. Scientific Reports*, *7*(1), 1–12. doi:10.103841598-017-08515-8 PMID:28808314

Becard, G., Taylor, L. P., & Douds, D. D. (1995). Flavonoids are not necessary plant signal compounds in arbuscular mycorrhizal symbiosis. *Molecular Plant-Microbe Interactions*, *8*(2), 252–258. doi:10.1094/MPMI-8-0252

Becher, D., Bernhardt, J., Fuchs, S., & Riedel, K. (2013). Metaproteomics to unravel major microbial players in leaf litter and soil environments: C hallenges and perspectives. *Proteomics*, *13*(18-19), 2895–2909. doi:10.1002/pmic.201300095 PMID:23894095

Beckers, B., Op De Beeck, M., Thijs, S., Truyens, S., Weyens, N., Boerjan, W., & Vangronsveld, J. (2016). Performance of 16s rDNA primer pairs in the study of rhizosphere and endosphere bacterial microbiomes in metabarcoding studies. *Frontiers in Microbiology*, *7*, 650. doi:10.3389/fmicb.2016.00650 PMID:27242686

Bedard, D. L., & May, R. J. (1995). Characterization of the polychlorinated biphenyls in the sediments of Woods Pond: Evidence for microbial dechlorination of Aroclor 1260 *in situ. Environmental Science & Technology*, *30*(1), 237–245. doi:10.1021/es950262e

Behera, B. C., Yadav, H., Singh, S. K., Mishra, R. R., Sethi, B. K., Dutta, S. K., & Thatoi, H. N. (2017). Phosphate solubilization and acid phosphatase activity of *Serratia sp.* isolated from mangrove soil of Mahanadi river delta, Odisha, India. *Journal of Genetic Engineering and Biotechnology*, *15*(1), 169–178. doi:10.1016/j.jgeb.2017.01.003 PMID:30647653

Behie, S. W., & Bidochka, M. J. (2013). Insects as a nitrogen source for plants. *Insects*, *4*(3), 413–424. doi:10.3390/insects4030413 PMID:26462427

Behrman, J., Alderman, H., & Hoddinott, J. (2004). Hunger and malnutrition. *Global crises. Global Solutions*, *363*, 420.

Beijerinck, M. W. (1888). Die bacterien der papilionaceenknöllchen. *Botanische Zeitung*, *46*(46), 725–735.

Beiyuan, J., Awad, Y. M., Beckers, F., Tsang, D. C., Ok, Y. S., & Rinklebe, J. (2017). Mobility and phytoavailability of As and Pb in a contaminated soil using pine sawdust biochar under systematic change of redox conditions. *Chemosphere*, *178*, 110–118. doi:10.1016/j.chemosphere.2017.03.022 PMID:28319738

Bell, G. D. H. (1987). The history of wheat cultivation. In *Wheat breeding* (pp. 31–49). Springer. doi:10.1007/978-94-009-3131-2_2

Benderliev, K. M., Ivanova, N. I., & Pilarski, P. S. (2003). Singlet oxygen and other reactive oxygen species are involved in regulation of release of iron-binding chelators from *Scenedesmus* cells. *Biologia Plantarum*, *47*(4), 523–526. doi:10.1023/B:BIOP.0000041056.07819.df

Beneduzi, A., Ambrosini, A., & Passaglia, L. M. (2012). Plant growth-promoting rhizobacteria (PGPR): Their potential as antagonists and biocontrol agents. *Genetics and Molecular Biology*, *35*(4, suppl), 1044–1051. doi:10.1590/S1415-47572012000600020 PMID:23411488

Benka-Coker, M. O., & Olumagin, A. (1996). Effects of waste drilling fluid on bacterial isolates from a mangrove swamp oilfield location in the Niger Delta of Nigeria. *Bioresource Technology*, *55*(3), 175–179. doi:10.1016/0960-8524(95)00165-4

Bennett, P. C., Choi, W. J., & Rogera, J. R. (1998). Microbial destruction of feldspars. *Mineral Management*, *8*(1), 149–150. doi:10.1180/minmag.1998.62A.1.79

Benson, D. R., & Silvester, W. B. (1993). Biology of Frankia strains, actinomycete symbionts of actinorhizal plants. *Microbiological Reviews*, *57*(2), 293–319. doi:10.1128/MR.57.2.293-319.1993 PMID:8336669

Berditsch, M., Afonin, S., & Ulrich, A. S. (2007). The ability of *Aneurinibacillus migulanus* (*Bacillus brevis*) to produce the antibiotic gramicidin S is correlated with phenotype variation. *Applied and Environmental Microbiology*, *73*(20), 6620–6628. doi:10.1128/AEM.00881-07 PMID:17720841

Berdy, J. (2005). Bioactive microbial metabolites. *The Journal of Antibiotics*, *58*(1), 1–26. doi:10.1038/ja.2005.1 PMID:15813176

Berg, G. (2009). Plant–microbe interactions promoting plant growth and health: Perspectives for controlled use of microorganisms in agriculture. *Applied Microbiology and Biotechnology*, *84*(1), 11–18. doi:10.100700253-009-2092-7 PMID:19568745

Berg, G., Grube, M., Schloter, M., & Smalla, K. (2014). Unravelling the plant microbiome: Looking back and future perspectives. *Frontiers in Microbiology*, (5), 175–175.

Bertling, J., Bertling, R., & Hamann, L. (2018). Kunststoffe in der Umwelt. In Fraunhofer Institut für Umwelt, Sicherheits- und Energietechnik Umsicht (pp. 1–56). Oberhausen, Germany: Academic Press.

Bertrand, F. N. (2020). Phytoremediation of petroleum hydrocarbon-contaminated soils with two plant species: *Jatropha curcas* and *Vetiveria zizanioides* at Ghana Manganese Company Ltd. *International Journal of Phytoremediation*, 1–10. doi:10.1080/15226514.2020.1803204 PMID:32805144

Besis, A., & Samara, C. (2012). Polybrominated diphenyl ethers (PBDEs) in the indoor and outdoor environments–a review on occurrence and human exposure. *Environmental Pollution*, *169*, 217–229. doi:10.1016/j.envpol.2012.04.009 PMID:22578798

Bhagobaty, R. K., & Malik, A. (2008). Utilization of chlorpyrifos as a sole source of carbon by bacteria isolated from wastewater irrigated agricultural soils in an industrial area of western Uttar Pradesh, India. *Research Journal of Microbiology*, *3*(5), 293–307. doi:10.3923/jm.2008.293.307

Bhalerao, T. S. (2012). Bioremediation of endosulfancontaminated soil by using bioaugmentation treatment of fungal inoculant *Aspergillus niger. Turkish Journal of Biology*, *35*, 561–567.

Bhalerao, T. S., & Puranik, P. R. (2007). Biodegradation of organochlorine pesticide, endosulfan, by a fungal soil isolate, Aspergillus niger. *International Biodeterioration & Biodegradation*, *59*(4), 315–321. doi:10.1016/j.ibiod.2006.09.002

Bhalla, T. C., Sharma, N. N., & Sharma, M. (2007). *Production of metabolites, industrial enzymes, amino acid, organic acids, antibiotics, vitamins and single cell proteins*. Academic Press.

Bharadwaj, A. (2018). Green Chemistry in Environmental Sustainability and Chemical Education. In *Green Chemistry in Environmental Sustainability and Chemical Education*. Springer Singapore.

Bharagava, R. N., Purchase, D., Saxena, G., & Mulla, S. I. (2019). Applications of metagenomics in microbial bioremediation of pollutants: from genomics to environmental cleanup. In *Microbial diversity in the genomic era* (pp. 459–477). Academic Press. doi:10.1016/B978-0-12-814849-5.00026-5

Bharathiraja, B., Jayamuthunagai, J., Praveenkumar, R., & Iyyappan, J. (2018). Phytoremediation techniques for the removal of dye in wastewater. In *Bioremediation: applications for environmental protection and management* (pp. 243–252). Springer. doi:10.1007/978-981-10-7485-1_12

Bhardwaj, D., Ansari, M. W., Sahoo, R. K., & Tuteja, N. (2014). Biofertilizers function as key player in sustainable agriculture by improving soil fertility, plant tolerance and crop productivity. *Microbial Cell Factories*, *13*(1), 1–10. doi:10.1186/1475-2859-13-66 PMID:24885352

Bhargava, A., & Srivastava, S. (2019). *Participatory Plant Breeding: Concept and Applications*. Springer Singapore. doi:10.1007/978-981-13-7119-6

Bharti, A., Velmourougane, K., & Prasanna, R. (2017). Phototrophic biofilms: Diversity, ecology and applications. *Journal of Applied Phycology*, 1–16.

Bharti, N., Pandey, S. S., Barnawal, D., Patel, V. K., & Kalra, A. (2016). Plant growth promoting rhizobacteria *Dietzianatronolimnaea* modulates the expression of stress responsive genes providing protection of wheat from salinity stress. *Scientific Reports*, *6*(1), 1–16. doi:10.1038rep34768 PMID:28442746

Bhatia, M., Girdhar, A., Chandrakar, B., & Tiwari, A. (2013). Implicating nanoparticles as potential biodegradation enhancers: A review. *Journal of Nanomedicine & Nanotechnology*, *4*(175), 2. doi:10.4172/2157-7439.1000175

Bhatia, M., Girdhar, A., Tiwari, A., & Nayarisseri, A. (2014). Implications of a novel Pseudomonas species on low density polyethylene biodegradation: An in vitro to in silico approach. *SpringerPlus*, *3*(1), 497. doi:10.1186/2193-1801-3-497 PMID:25932357

Bhat, S. (2013). Ecotoxicology & Impact on Biodiversity. *Journal of Pharmacognosy and Phytochemistry*, *2*(2).

Bhattacharya, A., Goyal, N., & Gupta, A. (2017). Degradation of azo dye methyl red by alkaliphilic, halotolerant *Nesterenkonia lacusekhoensis* EMLA3: Application in alkaline and salt-rich dyeing effluent treatment. *Extremophiles*, *21*(3), 479–490. doi:10.100700792-017-0918-2 PMID:28255636

Bhattacharyya, P. N., & Jha, D. K. (2012). Plant growth-promoting rhizobacteria (PGPR): Emergence in agriculture. *World Journal of Microbiology & Biotechnology*, *28*(4), 1327–1350. doi:10.100711274-011-0979-9 PMID:22805914

Bhatti, S. S., Kumar, V., Singh, N., Sambyal, V., Singh, J., Katnoria, J. K., & Nagpal, A. K. (2016). Physico-chemical Properties and Heavy Metal Contents of Soils and Kharif Crops of Punjab, India. *Procedia Environmental Sciences*, *35*, 801–808. doi:10.1016/j.proenv.2016.07.096

Bhatti, S. S., Sambyal, V., & Nagpal, A. K. (2016). Heavy metals bioaccumulation in Berseem (*Trifolium alexandrinum*) cultivated in areas under intensive agriculture, Punjab, India. *SpringerPlus*, *5*(1), 173. doi:10.118640064-016-1777-5 PMID:27026870

Bhatt, P., Gangola, S., Chaudhary, P., Khati, P., Kumar, G., Sharma, A., & Srivastava, A. (2019). Pesticide induced up-regulation of esterase and aldehyde dehydrogenase in indigenous *Bacillus* spp. *Bioremediation Journal*, *23*(1), 42–52. doi:10.1080/10889868.2019.1569586

Bhatt, P., Huang, Y., Zhang, W., Sharma, A., & Chen, S. (2020). Enhanced cypermethrin degradation kinetics and metabolic pathway in *Bacillus thuringiensis* strain SG4. *Microorganisms*, *8*(2), 223. doi:10.3390/microorganisms8020223 PMID:32046050

Bhosale, H. J., Kadam, T. A., & Bobade, A. R. (2013). Identification and production of *Azotobacter vinelandii* and its antifungal activity against *Fusarium oxysporum*. *Journal of Environmental Biology*, *34*(2), 177–182. PMID:24620576

Bhosale, S., Saratale, G., & Govindwar, S. (2006). Biotransformation enzymes in *Cunninghamella blakesleeana* (NCIM-687). *Journal of Basic Microbiology*, *46*(6), 444–448. doi:10.1002/jobm.200510117 PMID:17139609

Bhunia, S. K., & Jana, N. R. (2014). Reduced graphene oxide-silver nanoparticle composite as visible light photocatalyst for degradation of colorless endocrine disruptors. *ACS Applied Materials & Interfaces*, *6*(22), 20085–20092. doi:10.1021/am505677x PMID:25296393

Bibi, I., Bhatti, H. N., & Asgher, M. (2011). Comparative study of natural and synthetic phenolic compounds as efficient laccase mediators for the transformation of cationic dye. *Biochemical Engineering Journal*, *56*(3), 225–231. doi:10.1016/j.bej.2011.07.002

Bidlan, R., Afsar, M., & Mononmani, H. K. (2004). Bioremediation of HCH-contaminated soil: Elimination of inhibitory effects of the insecticide on radish and green gram seed germination. *Chemosphere*, *56*(8), 803–811. doi:10.1016/j.chemosphere.2004.01.015 PMID:15251295

Bidyarani, N., Prasanna, R., Chawla, G., Babu, S., & Singh, R. (2015). Deciphering the factors associated with the colonization of rice plants by cyanobacteria. *Journal of Basic Microbiology*, *55*(4), 407–419. doi:10.1002/jobm.201400591 PMID:25515189

Bielicka, A., Ryłko, E., & Bojanowska, I. (2009). Zawartość pierwiastków metalicznych w glebach i warzywach z ogrodów działkowych Gdańska i okolic. *Ochrona Środowiska i Zasobów Naturalnych*, (40).

Biesalski Hans, K., & Jana, T. (2018). Micronutrients in the life cycle: Requirements and sufficient supply. *NFS Journal*, *11*, 1–11. doi:10.1016/j.nfs.2018.03.001

Bijtenhoorn, P., Schipper, C., Hornung, C., Quitschau, M., Grond, S., Weiland, N., & Streit, W. R. (2011). BpiB05: A novel metagenome-derived hydrolase acting on N-acylhomoserine lactones. *Journal of Biotechnology*, *155*(1), 86–94. doi:10.1016/j.jbiotec.2010.12.016 PMID:21215778

Bing-Cheng, Y. U. A. N., & Dong-Xia, Y. U. E. (2012). Soil microbial and enzymatic activities across a chronosequence of Chinese pine plantation development on the loess plateau of China. *Pedosphere*, *22*(1), 1–12. doi:10.1016/S1002-0160(11)60186-0

Biondi, N., Piccardi, R., Margheri, M. C., Rodolfi, L., Smith, G. D., & Tredici, M. R. (2004). Evaluation of *Nostoc* strain ATCC 53789 as a potential source of natural pesticides. *Applied and Environmental Microbiology*, *70*(6), 3313–3320. doi:10.1128/AEM.70.6.3313-3320.2004 PMID:15184126

Birolli, W. G., Alvarenga, N., Seleghim, M. H. R., & Porto, A. L. M. (2016). Biodegradation of the pyrethroid pesticide esfenvalerate by marine-derived fungi. *Marine Biotechnology (New York, N.Y.)*, *18*(4), 511–520. doi:10.100710126-016-9710-z PMID:27381569

Birolli, W. G., Arai, M. S., Nitschke, M., & Porto, A. L. M. (2019). The pyrethroid (±)-lambda-cyhalothrin enantioselective biodegradation by a bacterial consortium. *Pesticide Biochemistry and Physiology*, *156*, 129–137. doi:10.1016/j.pestbp.2019.02.014 PMID:31027572

Biswas, B., Qi, F., Biswas, J., Wijayawardena, A., Khan, M., & Naidu, R. (2018). The Fate of Chemical Pollutants with Soil Properties and Processes in the Climate Change Paradigm—A Review. *Soil Systems*, *2*(3), 51–71. doi:10.3390oilsystems2030051

Bloemberg, G. V., & Lugtenberg, B. J. J. (2001). Molecular basis of plant growth promotion and biocontrol by rhizobacteria. *Current Opinion in Plant Biology*, *4*(4), 343–350. doi:10.1016/S1369-5266(00)00183-7 PMID:11418345

Bobrov, A. G., Kirillina, O., Fetherston, J. D., Miller, M. C., Burlison, J. A., & Perry, R. D. (2014). The *Yersinia pestis* siderophore, yersiniabactin, and the ZnuABC system both contribute to zinc acquisition and the development of lethal septicaemic plague in mice. *Molecular Microbiology*, *93*(4), 759–775. doi:10.1111/mmi.12693 PMID:24979062

Boddey, R. M., Da Silva, L. G., Reis, V., Alves, B. J. R., & Urquiaga, S. (2000). Assessment of bacterial nitrogen fixation in grass species. In Prokaryotic nitrogen fixation: a model system for analysis of a biological process. Horizon Scientific Press.

Boddey, R. M., Baldani, V. L., Baldani, J. I., & Döbereiner, J. (1986). Effect of inoculation of *Azospirillum* spp. on nitrogen accumulation by field-grown wheat. *Plant and Soil*, *95*(1), 109–121. doi:10.1007/BF02378857

Boddey, R. M., Polidoro, J. C., Resende, A. S., Alves, B. J., & Urquiaga, S. (2001). Use of the15N natural abundance technique for the quantification of the contribution of N2 fixation to sugar cane and other grasses. *Functional Plant Biology*, *28*(9), 889–895. doi:10.1071/PP01058

Boeing, D. W. (2000). Ecological effects, transport, and fate of mercury: A general review. *Chemosphere*, *40*(12), 1335–1351. doi:10.1016/S0045-6535(99)00283-0 PMID:10789973

Boente, C., Sierra, C., Martínez-Blanco, D., Menéndez-Aguado, J. M., & Gallego, J. R. (2018). Nanoscalezero-valent iron-assisted soil washing for the removal of potentially toxic elements. *Journal of Hazardous Materials*, *350*, 55–65. doi:10.1016/j.jhazmat.2018.02.016 PMID:29448214

Bohlool, B., Ladha, J., Garrity, D., & George, T. (1992). Biological nitrogen fixation for sustainable agriculture: A perspective. *Plant and Soil*, *141*(1-2), 1–11. doi:10.1007/BF00011307

Bokare, V., Murugesan, K., Kim, J. H., Kim, E. J., & Chang, Y. S. (2012). Integrated hybrid treatment for theremediation of 2, 3, 7, 8-tetrachlorodibenzo-p-dioxin. *The Science of the Total Environment*, *435*, 563–566. doi:10.1016/j.scitotenv.2012.07.079 PMID:22909785

Bokare, V., Murugesan, K., Kim, Y. M., Jeon, J. R., Kim, E. J., & Chang, Y. S. (2010). Degradation of triclosan byan integrated nano-bio redox process. *Bioresource Technology*, *101*(16), 6354–6360. doi:10.1016/j.biortech.2010.03.062 PMID:20381343

Bolan, N. S., Choppala, G., Kunhikrishnan, A., Park, J., & Naidu, R. (2013). Microbial transformation of trace elements in soils in relation to bioavailability and remediation. *Reviews of Environmental Contamination and Toxicology*, *225*, 1–56. doi:10.1007/978-1-4614-6470-9_1 PMID:23494555

Bolton, H., Fredrickson, J. K., & Elliot, L. E. (1992). Microbial ecology of the rhizosphere. Marcel Dekker.

Bonartsev, A. P., Bonartseva, G. A., Reshetov, I. V., Kirpichnikov, M. P., & Shaitan, K. V. (2019). Application of Polyhydroxyalkanoates in Medicine and the Biological Activity of Natural Poly(3-Hydroxybutyrate). *Acta Naturae*, *11*(2), 4–16. doi:10.32607/20758251-2019-11-2-4-16 PMID:31413875

Bondarenko, O., Rolova, T., Kahru, A., & Ivask, A. (2008). Bioavailability of Cd, Zn and Hg in soil to nine recombinant luminescent metal sensor bacteria. *Sensors (Basel)*, *8*(11), 6899–6923. doi:10.33908116899 PMID:27873907

Bondoc, K. G., Heuschele, J., Gillard, J., Vyverman, W., & Pohnert, G. (2016). Selective silicate-directed motility in diatoms. *Nature Communications*, *7*(1), 10540. doi:10.1038/ncomms10540 PMID:26842428

Boonchai, R., Kaewsuk, J., & Seo, G. (2014). Effect of nutrient starvation on nutrient uptake and extracellular polymeric substance for microalgae cultivation and separation. *Desalination and Water Treatment*, *55*(2), 360–367. doi:10.1080/19443994.2014.939501

Booth, S. C., Workentine, M. L., Weljie, A. M., & Turner, R. J. (2011a). Metabolomics and its application to studying metal toxicity. *Metallomics*, *3*(11), 1142–1152. doi:10.1039/c1mt00070e PMID:21922109

Booth, S. C., Workentine, M. L., Wen, J., Shaykhutdinov, R., Vogel, H. J., Ceri, H., Turner, R. J., & Weljie, A. M. (2011b). Differences in metabolism between the biofilm and planktonic response to metal stress. *Journal of Proteome Research*, *10*(7), 3190–3199. doi:10.1021/pr2002353 PMID:21561166

Borah, B., Thakur, P. S., & Nigam, J. N. (2002). The influence of nutritional and environmental conditions on the accumulation of poly-beta-hydroxybutyrate in *Bacillus mycoides* RLJ B-017. *Journal of Applied Microbiology*, *92*(4), 776–783. doi:10.1046/j.1365-2672.2002.01590.x PMID:11966920

Borremans, B., Hobman, J. L., Provoost, A., Brown, N. L., & Van Der Lelie, D. (2001). Cloning and functional analysis of the *pbr* lead resistance determinant of *Ralstonia metallidurans* CH34. *Journal of Bacteriology*, *183*(19), 5651–5658. doi:10.1128/JB.183.19.5651-5658.2001 PMID:11544228

Borriss, R. (2011). Use of Plant-Associated Bacillus Strains as Biofertilizers and Biocontrol Agents in Agriculture. In D. K. Maheshwari (Ed.), *Bacteria in Agrobiology: Plant Growth Responses* (pp. 41–76). Springer Berlin Heidelberg. doi:10.1007/978-3-642-20332-9_3

Boschker, H. T., Vasquez-Cardenas, D., Bolhuis, H., Moerdijk-Poortvliet, T. W., & Moodley, L. (2014). Chemoautotrophic carbon fixation rates and active bacterial communities in intertidal marine sediments. *PLoS One*, *9*(7), e101443. doi:10.1371/journal.pone.0101443 PMID:25003508

Bosso, L., Scelza, R., Testa, A., Cristinzio, G., & Rao, M. A. (2015). Depletion of pentachlorophenol contamination in an agricultural soil treated with Byssochlamys nivea, Scopulariopsis brumptii and urban waste compost: A laboratory microcosm study. *Water, Air, and Soil Pollution*, *226*(6), 183. doi:10.100711270-015-2436-0

Botta, A. (2012). Enhancing plant tolerance to temperature stress with amino acids: an approach to their mode of action. *I World Congress on the Use of Biostimulants in Agriculture 1009*, 29-35.

Bouchard, M. F., Chevrier, J., Harley, K. G., Kogut, K., Vedar, M., Calderon, N., Trujillo, C., Johnson, C., Bradman, A., Barr, D. B., & Eskenazi, B. (2011). Prenatal exposure to organophosphate pesticides and IQ in 7-year-old children. *Environmental Health Perspectives*, *119*(8), 1189–1195. doi:10.1289/ehp.1003185 PMID:21507776

Bouffaud, M. L., Poirier, M. A., Muller, D., & Moënne-Loccoz, Y. (2014). Root microbiome relates to plant host evolution in maize and other Poaceae. *Environmental Microbiology*, *16*(9), 2804–2814. doi:10.1111/1462-2920.12442 PMID:24588973

Boukhalfa, H., Icopini, G. A., Reilly, S. D., & Neu, M. P. (2007). Plutonium (IV) reduction by the metal-reducing bacteria *Geobacter metallireducens* GS15 and Shewanella oneidensis MR1. *Applied and Environmental Microbiology, 73*(18), 5897–5903. doi:10.1128/AEM.00747-07 PMID:17644643

Boulter, J. I., Trevors, J. T., & Boland, G. J. (2002). Microbial studies of compost: Bacterial identification and their potential for turfgrass pathogen suppression. *World Journal of Microbiology & Biotechnology, 18*(7), 661–671. doi:10.1023/A:1016827929432

Bourguet, D., & Guillemaud, T. (2016). The hidden and external costs of pesticide use. In *Sustainable agriculture reviews* (pp. 35–120). Springer. doi:10.1007/978-3-319-26777-7_2

Boyd, E. S., & Peters, J. W. (2013). New insights into the evolutionary history of biological nitrogen fixation. *Frontiers in Microbiology, 4*, 201. doi:10.3389/fmicb.2013.00201 PMID:23935594

Bradl, H. (2002). *Heavy Metals in the Environment: Origin, Interaction and Remediation* (Vol. 6). Academic Press.

Brahmbhatt, N. H., & Jasrai, R. T. (2016). The Role of Algae in Bioremediation of Textile Effluent. *International Journal of Engineering Research and General Science, 4*(1), 443–451.

Brahushi, F., Kengara, F. O., & Yang, S. O. N. G. (2017). Fate processes of chlorobenzenes in soil and potential remediation strategies: A review. *Pedosphere, 27*(3), 407–420. doi:10.1016/S1002-0160(17)60338-2

Brandelli, A., Daroit, D. J., & Riffel, A. (2010). Biochemical features of microbial keratinases and their production and applications. *Applied Microbiology and Biotechnology, 85*(6), 1735–1750. doi:10.100700253-009-2398-5 PMID:20039036

Brandelli, A., Sala, L., & Kalil, S. J. (2015). Microbial enzymes for bioconversion of poultry waste into added-value products. *Food Research International, 73*, 3–12. doi:10.1016/j.foodres.2015.01.015

Braud, A., Hannauer, M., Mislin, G. L., & Schalk, I. J. (2009). The Pseudomonas aeruginosa pyochelin-iron uptake pathway and its metal specificity. *Journal of Bacteriology, 191*(11), 3517–3525. doi:10.1128/JB.00010-09 PMID:19329644

Brautaset, T., Jakobsen, Ø. M., Josefsen, K. D., Flickinger, M. C., & Ellingsen, T. E. (2007). *Bacillus methanolicus*: A candidate for industrial production of amino acids from methanol at 50 C. *Applied Microbiology and Biotechnology, 74*(1), 22–34. doi:10.100700253-006-0757-z PMID:17216461

Briat, J. F., Curie, C., & Gaymard, F. (2007). Iron utilization and metabolism in plants. *Current Opinion in Plant Biology, 10*(3), 276–282. doi:10.1016/j.pbi.2007.04.003 PMID:17434791

Brígido, C., Glick, B. R., & Oliveira, S. (2017). Survey of plant growth-promoting mechanisms in native Portuguese chickpea Mesorhizobium isolates. *Microbial Ecology, 73*(4), 900–915. doi:10.100700248-016-0891-9 PMID:27904921

Brillet, K., Ruffenach, F., Adams, H., Journet, L., Gasser, V., Hoegy, F., Guillon, L., Hannauer, M., Page, A., & Schalk, I. J. (2012). An ABC transporter with two periplasmic binding proteins involved in iron acquisition in *Pseudomonas aeruginosa. ACS Chemical Biology, 7*(12), 2036–2045. doi:10.1021/cb300330v PMID:23009327

Brown, L. D., Cologgi, D. L., Gee, K. F., & Ulrich, A. C. (2017). Bioremediation of Oil Spills on Land. In Oil Spill Science and Technology (2nd ed., pp. 699 - 729). Gulf Professional Publishing. doi:10.1016/B978-0-12-809413-6.00012-6

Bruenn, J. (2005). The Ustilago maydis killer toxins. In M. Schmitt, R. Schaffrath, & M. J. Schmitt (Eds.), *Microbial Protein Toxins* (pp. 157–174). Springer.

Buckley, D. H., & Schmidt, T. M. (2001). The structure of microbial communities in soil and the lasting impact of cultivation. *Microbial Ecology, 42*(1), 11–21. doi:10.1007002480000108 PMID:12035077

Bulgarelli, D., Schlaeppi, K., Spaepen, S., Van Themaat, E. V. L., & Schulze-Lefert, P. (2013). Structure and functions of the bacterial microbiota of plants. *Annual Review of Plant Biology*, *64*(1), 807–838. doi:10.1146/annurev-arplant-050312-120106 PMID:23373698

Bull, S., Burnett, K., Vassaux, K., Ashdown, L., Brown, T., & Rushton, L. (2014). Extensive literature search and provision of summaries of studies related to the oral toxicity of perfluoroalkylated substances (PFASs), their precursors and potential replacements in experimental animals and humans. Area 1: Data on toxicokinetics (absorption, distribution, metabolism, excretion) in in vitro studies, experimental animals and humans. Area 2: Data on toxicity in experimental animals. Area 3: Data on observations in humans. *EFSA Supporting Publications*, *11*(4), 572E. doi:10.2903p.efsa.2014.EN-572

Bumpus, J., Tien, M., Wright, D., & Aust, S. (1985). Oxidation of persistent environmental pollutants by a white rot fungus. *Science*, *228*(4706), 1434–1436. doi:10.1126cience.3925550 PMID:3925550

Bünemann, E. K., Bongiorno, G., Bai, Z., Creamer, R. E., De Deyn, G., de Goede, R., & Brussaard, L. (2018). Soil quality – A critical review. *Soil Biology & Biochemistry*, *120*, 105–125. doi:10.1016/j.soilbio.2018.01.030

Bünemann, E. K., Schwenke, G. D., & Van Zwieten, L. (2006). Impact of agricultural inputs on soil organisms—A review. *Australian Journal of Soil Research*, *44*(4), 379. doi:10.1071/SR05125

Burén, S., & Rubio, L. M. (2018). State of the art in eukaryotic nitrogenase engineering. *FEMS Microbiology Letters*, *365*(2), fnx274. doi:10.1093/femsle/fnx274 PMID:29240940

Burke, E. L. (1994). A survey of recent literature on medical waste. *Journal of Environmental Health*, 11–14.

Burkhead, K. D., Schisler, D. A., & Slininger, P. J. (1994). Pyrrolnitrin Production by Biological Control Agent *Pseudomonas cepacia* B37w in Culture and in Colonized Wounds of Potatoes. *Applied and Environmental Microbiology*, *60*(6), 2031–2039. doi:10.1128/AEM.60.6.2031-2039.1994 PMID:16349289

Burkinshaw, S. M. (1995). *Chemical Principles of Synthetic Fibre Dyeing*. Springer Science & Business Media. doi:10.1007/978-94-011-0593-4

Burkinshaw, S. M., Jeong, D. S., & Chun, T. I. (2013). The coloration of poly (lactic acid) fibres with indigoid dyes: Part 2: Wash fastness. *Dyes and Pigments*, *97*(2), 374–387. doi:10.1016/j.dyepig.2012.12.026

Burkinshaw, S. M., & Salihu, G. (2019). The role of auxiliaries in the immersion dyeing of textile fibres part 2: Analysis of conventional models that describe the manner by which inorganic electrolytes promote direct dye uptake on cellulosic fibres. *Dyes and Pigments*, *161*, 531–545. doi:10.1016/j.dyepig.2017.08.034

Burns, R. C., & Hardy, R. W. F. (1975). *Nitrogen Fixation in Bacteria and Higher Plants*. Springer-Verlag. doi:10.1007/978-3-642-80926-2

Bursle, E., & Robson, J. (2016). Non-culture methods for detecting infection. *Australian Prescriber*, *39*(5), 171–175. doi:10.18773/austprescr.2016.059 PMID:27789929

Buscot, F., & Varma, A. (2005). *Microorganisms in Soils: Roles in Genesis and Functions*. Springer.

Byers, J. T., Lucas, C., Salmond, G. P., & Welch, M. (2002). Nonenzymatic turnover of an Erwinia carotovora quorum-sensing signaling molecule. *Journal of Bacteriology*, *184*(4), 1163–1171. doi:10.1128/jb.184.4.1163-1171.2002 PMID:11807077

Byun, G. H., Moon, H. B., Choi, J. H., Hwang, J., & Kang, C. K. (2013). Biomagnification of persistent chlorinated and brominated contaminants in food web components of the Yellow Sea. *Marine Pollution Bulletin*, *73*(1), 210–219. doi:10.1016/j.marpolbul.2013.05.017 PMID:23768977

Cabugao, K. G., Timm, C. M., Carrell, A. A., Childs, J., Lu, T. S., Pelletier, D. A., Weston, D. J., & Norby, R. J. (2017). Root and Rhizosphere Bacterial Phosphatase Activity Varies with Tree Species and Soil Phosphorus Availability in Puerto Rico Tropical Forest. *Frontiers in Plant Science*, *8*, 1834. doi:10.3389/fpls.2017.01834 PMID:29163572

Cachada, A., Pato, P., Rocha-Santos, T., Da Silva, E. F., & Duarte, A. C. (2012). Levels, sources and potential human health risks of organic pollutants in urban soils. *The Science of the Total Environment*, *430*, 184–192. doi:10.1016/j.scitotenv.2012.04.075 PMID:22652008

Cakmakci, R., Dönmez, M. F., & Erdoğan, Ü. (2007). The effect of plant growth promoting rhizobacteria on barley seedling growth, nutrient uptake, some soil properties, and bacterial counts. *Turkish Journal of Agriculture and Forestry*, *31*(3), 189–199.

Callesen, I., Liski, J., Raulund-Rasmussen, K., Olsson, M. T., Tau-Strand, L., Vesterdal, L., & Westman, C. J. (2003). Soil carbon stores in Nordic well-drained forest soils—Relationships with climate and texture class. *Global Change Biology*, *9*(3), 358–370. doi:10.1046/j.1365-2486.2003.00587.x

Calvo, P., Nelson, L. M., & Kloeppe, J. W. (2014). Agricultural uses of plant biostimulants. *Plant and Soil*, *383*(1-2), 3–41. doi:10.100711104-014-2131-8

Calvo, P., Ormeño-Orrillo, E., Martínez-Romero, E., & Zúñiga, D. (2010). Characterization of *Bacillus* isolates of potato rhizosphere from andean soils of Peru and their potential PGPR characteristics. *Brazilian Journal of Microbiology*, *41*(4), 899–906. doi:10.1590/S1517-83822010000400008 PMID:24031569

Camachomorales, R. L., Karina, G. N., & José, E. S. (2017). Degradation of the herbicide paraquat by macromycetes isolated from southeastern Mexico. *Biotech, 7*, 324 - 334.

Camarero, S., Ibarra, D., Martínez, M. J., & Martínez, Á. T. (2005). Lignin-derived compounds as efficient laccase mediators for decolorization of different types of recalcitrant dyes. *Applied and Environmental Microbiology*, *71*(4), 1775–1784. doi:10.1128/AEM.71.4.1775-1784.2005 PMID:15812000

Cameron, A. D., Ridderström, M., Olin, B., & Mannervik, B. (1999). Crystal structure of human glyoxalase II and its complex with a glutathione thiolester substrate analogue. *Structure, 7*(9), 1067–1078.. doi:10.10160969-2126(99)80174-9

Cameron, D. D., Neal, A. L., van Wees, S. C., & Ton, J. (2013). Mycorrhiza-induced resistance: More than the sum of its parts? *Trends in Plant Science*, *18*(10), 539–545. doi:10.1016/j.tplants.2013.06.004 PMID:23871659

Cameselle, C., Chirakkara, R. A., & Reddy, K. R. (2013). Electrokinetic-enhanced phytoremediation of soils: Status and opportunities. *Chemosphere*, *93*(4), 626–636. doi:10.1016/j.chemosphere.2013.06.029 PMID:23835413

Camilios-Neto, D., Bonato, P., Wassem, R., Tadra-Sfeir, M. Z., Brusamarello-Santos, L. C., Valdameri, G., Donatti, L., Faoro, H., Weiss, V. A., Chubatsu, L. S., Pedrosa, F. O., & Souza, E. M. (2014). Dual RNA-seq transcriptional analysis of wheat roots colonized by Azospirillum brasilense reveals up-regulation of nutrient acquisition and cell cycle genes. *BMC Genomics*, *15*(1), 378. doi:10.1186/1471-2164-15-378 PMID:24886190

Campos, R., Kandelbauer, A., Robra, K., Cavaco-Paulo, A., & Gübitz, G. (2001). Indigo degradation with purified laccases from Trametes hirsuta and Sclerotium rolfsii. *Journal of Biotechnology*, *89*(2-3), 131–139. doi:10.1016/S0168-1656(01)00303-0 PMID:11500206

Canadian Council of Ministers of the Environment (CCME). (2008). *Canada-wide standard for petroleum hydrocarbons (PHC) in soil: Scientific rationale*. Supporting technical document. PN 1399, 1–383.

Canarini, A., Kaiser, C., Merchant, A., Richter, A., & Wanek, W. (2019). Root exudation of primary metabolites: Mechanisms and their roles in plant responses to environmental stimuli. *Frontiers in Plant Science*, *10*, 157. doi:10.3389/fpls.2019.00157 PMID:30881364

Canizares, M. C., Perez Artes, E., & Garcia-Pedrajas, M. D. (2014). The complete nucleotide sequence of a novel partitivirus isolated from the plant pathogenic fungus *Verticillium albo-atrum*. *Archives of Virology*, *159*(11), 3141–3144. doi:10.100700705-014-2156-6 PMID:24986717

Cao, J., Elliott, D., & Zhang, W.-X. (2005). Perchlorate reduction by nanoscale iron particles. *Journal of Nanoparticle Research*, *7*(4-5), 499–506. doi:10.100711051-005-4412-x

Cao, J., & Zhang, W. (2006). Stabilization of chromium ore processing residue (COPR) with nanoscale iron particles. *Journal of Hazardous Materials*, *132*(2-3), 213–219. doi:10.1016/j.jhazmat.2005.09.008 PMID:16621279

Cao, Y. F., Zhu, X. W., Xiang, Y., Li, D. Q., Yang, J. R., Mao, Q. Z., & Chen, J. S. (2011). Genomic characterization of a novel dsRNA virus detected in the phytopathogenic fungus *Verticillium dahliae* Kleb. *Virus Research*, *159*(1), 73–78. doi:10.1016/j.virusres.2011.04.029 PMID:21571013

Carlier, A., Uroz, S., Smadja, B., Fray, R., Latour, X., Dessaux, Y., & Faure, D. (2003). The Ti plasmid of Agrobacterium tumefaciens harbors an attM-paralogous gene, aiiB, also encoding N-Acyl homoserine lactonase activity. *Applied and Environmental Microbiology*, *69*(8), 4989–4993. doi:10.1128/AEM.69.8.4989-4993.2003 PMID:12902298

Carlin, A., Shi, W., Dey, S., & Rosen, B. P. (1995). The *ars* operon of *Escherichia coli* confers arsenical and antimonial resistance. *Journal of Bacteriology*, *177*(4), 177. doi:10.1128/JB.177.4.981-986.1995 PMID:7860609

Carmichael, S. L., Yang, W., Roberts, E., Kegley, S. E., Brown, T. J., English, P. B., & Shaw, G. M. (2016). Residential agricultural pesticide exposures and risks of selected birth defects among offspring in the San Joaquin Valley of California. *Birth Defects Research. Part A, Clinical and Molecular Teratology*, *106*(1), 27–35. doi:10.1002/bdra.23459 PMID:26689858

Carpenter, C., & Payne, S. M. (2014). Regulation of iron transport systems in Enterobacteriaceae in response to oxygen and iron availability. *Journal of Inorganic Biochemistry*, *133*, 110–117. doi:10.1016/j.jinorgbio.2014.01.007 PMID:24485010

Carrano, C. J., Jordan, M., Drechsel, H., Schmid, D. G., & Winkelmann, G. (2001). Heterobactins: A new class of siderophores from *Rhodococcus erythropolis* IGTS8 containing both hydroxamate and catecholate donor groups. *Biometals*, *14*(2), 119–125. doi:10.1023/A:1016633529461 PMID:11508844

Carter, R. A., Worsley, P. S., Sawers, G., Challis, G. L., Dilworth, M. J., Carson, K. C., Lawrence, J. A., Wexler, M., Johnston, A. W., & Yeoman, K. H. (2002). The vbs genes that direct synthesis of the siderophore vicibactin in *Rhizobium leguminosarum*: Their expression in other genera requires ECF sigma factor RpoI. *Molecular Microbiology*, *44*(5), 1153–1166. doi:10.1046/j.1365-2958.2002.02951.x PMID:12028377

Carvajal-Muñoz, J. S., & Carmona-Garcia, C. E. (2012). Benefits and limitations of biofertilization in agricultural practices. *Livestock Research for Rural Development*, *24*, 43.

Carvalhais, L. C., Dennis, P. G., Fan, B., Fedoseyenko, D., Kierul, K., Becker, A., von Wiren, N., & Borriss, R. (2013). Linking plant nutritional status to plant-microbe interactions. *PLoS One*, *8*(7), e68555. doi:10.1371/journal.pone.0068555 PMID:23874669

Carvalho, F. P. (2017). Mining industry and sustainable development: Time for change. *Food and Energy Security*, *6*(2), 61–77. doi:10.1002/fes3.109

Carvalho, F. P. (2017). Pesticides, environment, and food safety. *Food and Energy Security*, *6*(2), 48–60. doi:10.1002/fes3.108

Casas, N., Parella, T., Vicent, T., Caminal, G., & Sarrà, M. (2009). Metabolites from the biodegradation of triphenylmethane dyes by *Trametes versicolor* or laccase. *Chemosphere*, *75*(10), 1344–1349. doi:10.1016/j.chemosphere.2009.02.029 PMID:19298999

Cassan, F., Maiale, S., Masciarelli, O., Vidal, A., Luna, V., & Ruiz, O. (2009). Cadaverine production by *Azospirillum brasilense* and its possible role in plant growth promotion and osmotic stress mitigation. *European Journal of Soil Biology*, *45*(1), 12–19. doi:10.1016/j.ejsobi.2008.08.003

Cassidy, D. P., Srivastava, V. J., Dombrowski, F. J., & Lingle, J. W. (2015). Combining in situ chemical oxidation, stabilization, and anaerobic bioremediation in a single application to reduce contaminant mass and leachability in soil. *Journal of Hazardous Materials*, *297*, 347–355. doi:10.1016/j.jhazmat.2015.05.030 PMID:26093352

Castellane, T. C. L., Otoboni, A. M. M. B., & Lemos, E. G. D. M. (2015). Characterization of exopolysaccharides produced by rhizobia species. *Revista Brasileira de Ciência do Solo*, *39*(6), 1566–1575. doi:10.1590/01000683rbcs20150084

Cavaglieri, L., Orlando, J. R. M. I., Rodriguez, M. I., Chulze, S., & Etcheverry, M. (2005). Biocontrol of Bacillus subtilis against Fusarium verticillioides in vitro and at the maize root level. *Research in Microbiology*, *156*(5-6), 748–754. doi:10.1016/j.resmic.2005.03.001 PMID:15950130

Cavagnaro, T. R., Jackson, L. E., Six, J., Ferris, H., Goyal, S., Asami, D., & Scow, K. M. (2006). Arbuscular mycorrhizas, microbial communities, nutrient availability, and soil aggregates in organic tomato production. *Plant and Soil*, *282*(1-2), 209–225. doi:10.100711104-005-5847-7

Cecchin, I., Reddy, K. R., Thomé, A., Tessaro, E. F., & Schnaid, F. (2017). Nano-bioremediation: Integration of nanoparticles and bioremediation for sustainable remediation of chlorinated organic contaminants in soils. *International Biodeterioration & Biodegradation*, *119*, 419–428. doi:10.1016/j.ibiod.2016.09.027

Celia, H., Botos, I., Ni, X., Fox, T., De Val, N., Lloubes, R., Jiang, J., & Buchanan, S. K. (2019). Cryo-EM structure of the bacterial Ton motor subcomplex ExbB–ExbD provides information on structure and stoichiometry. *Communications Biology*, *2*(1), 1–6. doi:10.103842003-019-0604-2 PMID:31602407

Cempel, M., & Nikel, G. (2006). Nickel: A review of its sources and environmental toxicology. *Polish Journal of Environmental Studies*, *15*, 375–382.

Center for Disease Control. (2017). *Biomonitoring Program Activities: Environmental Chemicals*. National Biomonitoring Program. https://www.cdc.gov/biomonitoring/environmental_chemicals.html

Cephidian, A., Makhdoumi, A., Mashreghi, M., & Mahmudy Gharaie, M. H. (2016). Removal of anthropogenic lead pollutions by a potent *Bacillus* species AS2 isolated from geogenic contaminated site. *International Journal of Environmental Science and Technology*, *13*(9), 2135–2142. doi:10.100713762-016-1023-2

Cerda, A., Artola, A., Font, X., Barrena, R., Gea, T., & Sánchez, A. (2018). Composting of food wastes: Status and challenges. *Bioresource Technology*, *248*, 57–67. doi:10.1016/j.biortech.2017.06.133 PMID:28693949

Cerdán, M., Sánchez-Sánchez, A., Oliver, M., Juárez, M., & Sánchez-Andreu, J. J. (2008). Effect of foliar and root applications of amino acids on iron uptake by tomato plants. *IV Balkan Symposium on Vegetables and Potatoes 830*, 481-488.

Cervantes, C., & Campos-García, J. (2007). Reduction and efflux of chromate by bacteria. Molecular Microbiology of Heavy Metals, 407–419. doi:10.1007/7171_2006_087

Cesa-Luna, C., Baez, A., Quintero-Hernández, V., De la Cruz-Enríquez, J., Castañeda-Antonio, M. D., & Muñoz-Rojas, J. (2020). The importance of antimicrobial compounds produced by beneficial bacteria on the biocontrol of phytopathogens. *Acta Biologica Colombiana*, *25*(1), 140–154. doi:10.15446/abc.v25n1.76867

Chaîneau, C. H., Morel, J. L., & Oudot, J. (1997). Phytotoxicity and plant uptake of fuel oil hydrocarbons. *Journal of Environmental Quality*, *26*(6), 1478–1483. doi:10.2134/jeq1997.00472425002600060005x

Cha, J. S., & Cooksey, D. A. (1991). Copper resistance in Pseudomonas syringae mediated by periplasmic and outer membrane proteins. *Proceedings of the National Academy of Sciences of the United States of America*, *88*(20), 8915–8919. doi:10.1073/pnas.88.20.8915 PMID:1924351

Chakdar, H., Dastager, S. G., Khire, J. M., Rane, D., & Dharne, M. S. (2018). Characterization of mineral phosphate solubilizing and plant growth promoting bacteria from termite soil of arid region. *Biotech, 8*(11), 463.

Chakraborty, R., Wu, C. H., & Hazen, T. C. (2012). Systems biology approach to bioremediation. *Current Opinion in Biotechnology*, *23*(3), 483–490. doi:10.1016/j.copbio.2012.01.015 PMID:22342400

Chakraborty, S. K. (2019). Bioinvasion and Environmental Perturbation: Synergistic Impact on Coastal–Mangrove Ecosystems of West Bengal, India. In *Impacts of Invasive Species on Coastal Environments* (pp. 171–245). Springer. doi:10.1007/978-3-319-91382-7_6

Chakravarty, P., Bauddh, K., & Kumar, M. (2017). Phytoremediation: A multidimensional and ecologically viable practice for the cleanup of environmental contaminants. In *Phytoremediation potential of bioenergy plants* (pp. 1–46)., doi:10.1007/978-981-10-3084-0_1

Chalam, A. V., Sasikala, C., Ramana, C. V., Uma, N. R., & Rao, P. R. (1997). Effect of pesticides on the diazotrophic growth and nitrogenase activity of purple nonsulfur bacteria. *Bulletin of Environmental Contamination and Toxicology*, *58*(3), 463–468. doi:10.1007001289900357 PMID:9008058

Chandanshive, V. V., Rane, N. R., Tamboli, A. S., Gholave, A. R., Khandare, R. V., & Govindwar, S. P. (2017). Co-plantation of aquatic macrophytes *Typha angustifolia* and *Paspalum scrobiculatum* for effective treatment of textile industry effluent. *Journal of Hazardous Materials*, *338*, 47–56. doi:10.1016/j.jhazmat.2017.05.021 PMID:28531658

Chandel, S. T. (2009). Nematicidal activity of the Cyanobacterium, *Aulosira fertilissima* on the hatch of *Meloidogyne triticoryzae* and *Meloidogyne incognita. Archiv für Phytopathologie und Pflanzenschutz*, *42*(1), 32–38. doi:10.1080/03235400600914363

Chandra, D., & General, T. Nisha, & Chandra, S. (2019). Microorganisms: an asset for decontamination of soil. In Smart Bioremediation Technologies (pp. 319–345). Elsevier.

Chandra, R., & Chowdhary, P. (2015). Properties of bacterial laccases and their application in bioremediation of industrial wastes. *Environmental Science. Processes & Impacts*, *17*(2), 326–342. doi:10.1039/C4EM00627E PMID:25590782

Chandra, R., & Singh, S. k. (2009). *Fundamentals and management of soil quality*. Westville Publishing House.

Chang, C. Y., Yu, H. Y., Chen, J. J., Li, F. B., Zhang, H. H., & Liu, C. P. (2014). Accumulation of heavymetalsinleaf vegetables from agricultural soils and associated potential health risks in the Pearl River Delta, South China. *Environmental Monitoring and Assessment*, *186*(3), 1547–1560. doi:10.100710661-013-3472-0 PMID:24185814

Chang, J. S., Chen, B. Y., & Lin, Y. S. (2004). Stimulation of bacterial decolorization of an azo dye by extracellular metabolites from Escherichia coli strain NO3. *Bioresource Technology*, *91*(3), 243–248. doi:10.1016/S0960-8524(03)00196-2 PMID:14607483

Chang, L. W., Magos, L., & Suzuki, T. (Eds.). (1996). *Toxicology of Metals*. CRC Press.

Chanmugathas, P., & Bollag, J. M. (1988). A column study of the biological mobilization and speciation of cadmium in soil. *Archives of Environmental Contamination and Toxicology*, *17*(2), 229–237. doi:10.1007/BF01056029

Chaoua, S., Boussaa, S., El Gharmali, A., & Boumezzough, A. (2019). Impact of irrigation with wastewater on accumulation of heavy metals in soil and crops in the region of Marrakech in Morocco. *Journal of the Saudi Society of Agricultural Sciences*, *18*(4), 429–436. doi:10.1016/j.jssas.2018.02.003

Chaparro, J. M., Badri, D. V., & Vivanco, J. M. (2014). Rhizosphere microbiome assemblage is affected by plant development. *The ISME Journal*, *8*(4), 790–803. doi:10.1038/ismej.2013.196 PMID:24196324

Chaturvedi, M. K. (2011). Studies on chromate removal by chromium-resistant *Bacillus sp.* isolated from tannery effluent. *The Journal of Electronic Publishing: JEP*, *2*(1), 76–82.

Chaudhary, D., Narula, N., Sindhu, S. S., & Behl, R. K. (2013). Plant growth stimulation of wheat (*Triticum aestivum* L.) by inoculation of salinity tolerant Azotobacter strains. *Physiology and Molecular Biology of Plants: An International Journal of Functional Plant Biology*, *19*(4), 515–519.

Chaudhary, D. K., & Kim, J. (2019). New insights into bioremediation strategies for oil-contaminated soil in cold environments. *International Biodeterioration & Biodegradation*, *142*, 58–72. doi:10.1016/j.ibiod.2019.05.001

Chaudhary, D., Narula, N., Sindhu, S. S., & Behl, R. K. (2013). Plant growth stimulation of wheat (Triticum aestivum L.) by inoculation of salinity tolerant Azotobacter strains. *Physiology and Molecular Biology of Plants*, *19*(4), 515–519. doi:10.100712298-013-0178-2 PMID:24431520

Chaudhary, S., Shankar, A., Singh, A., & Prasad, V. (2018). Usefulness of Penicillium in Enhancing Plants Resistance to Abiotic Stresses: An Overview. In *New and Future Developments in Microbial Biotechnology and Bioengineering* (pp. 277–284). Elsevier. doi:10.1016/B978-0-444-63501-3.00017-X

Chaudhary, V., Prasanna, R., Nain, L., Dubey, S. C., Gupta, V., Singh, R., Jaggi, S., & Bhatnagar, A. K. (2012). Bioefficacy of novel cyanobacteria-amended formulations in suppressing damping off disease in tomato seedlings. *World Journal of Microbiology & Biotechnology*, *28*(12), 3301–3310. doi:10.100711274-012-1141-z PMID:22869418

Chaudhry, G. R., & Chapalamadugu, S. (1991). Biodegradation of halogenated organic compounds. *Microbiology and Molecular Biology Reviews*, *55*(1), 59–79. PMID:2030673

Chaudhry, Q., Blom-Zandstra, M., Gupta, S., & Joner, E. (2005). Utilising the synergy between plants and rhizosphere microorganisms to enhance breakdown of organic pollutants in the environment. *Environmental Science and Pollution Research International*, *12*(1), 34–48. doi:10.1065/espr2004.08.213 PMID:15768739

Chavan, R. S., & Chavan, S. R. (2011). Sourdough technology—a traditional way for wholesome foods: A review. *Comprehensive Reviews in Food Science and Food Safety*, *10*(3), 169–182. doi:10.1111/j.1541-4337.2011.00148.x

Chee, J. Y., Lakshmanan, M., Jeepery, I. F., Hairudin, N. H. M., & Sudesh, K. (2019). The potential application of *Cupriavidus necator* as polyhydroxyalkanoates producer and single cell protein: A review on scientific, cultural and religious perspectives. *Applied Food Biotechnology*, *6*(1), 19–34.

Chen, J. H. (2006). The combined use of chemical and organic fertilizers and/or biofertilizer for crop growth and soil fertility. In *International workshop on sustained management of the soil-rhizosphere system for efficient crop production and fertilizer use* (*Vol. 16*, No. 20, pp. 1-11). Land Development Department.

Chen, B., Stein, A. F., Castell, N., Gonzalez-Castanedo, Y., De La Campa, A. S., & De La Rosa, J. (2016). Modeling and evaluation of urban pollution events of atmospheric heavy metals from a large Cu-smelter. *The Science of the Total Environment*, *539*, 17–25. doi:10.1016/j.scitotenv.2015.08.117 PMID:26352643

Chen, C., Belanger, R. R., Benhamou, N., & Paulitz, T. C. (2000). Defense enzymes induced in cucumber roots by treatment with plant growth-promoting rhizobacteria (PGPR) and Pythium aphanidermatum. *Physiological and Molecular Plant Pathology*, *56*(1), 13–23. doi:10.1006/pmpp.1999.0243

Chen, G.-Q. (2009). A microbial polyhydroxyalkanoates (PHA) based bio- and materials industry. *Chemical Society Reviews*, *38*(8), 2434–2446. doi:10.1039/b812677c PMID:19623359

Cheng, Z., Park, E., & Glick, B. R. (2007). 1-Aminocyclopropane-1-carboxylate deaminase from *Pseudomonas putida*UW4 facilitates the growth of canola in the presence of salt. *Canadian Journal of Microbiology*, *53*(7), 912–918. doi:10.1139/W07-050 PMID:17898846

Chen, H. W., Xu, M., Ma, X. W., Tong, Z. H., & Liu, D. F. (2019). Isolation and characterization of a chlorate-reducing bacterium *Ochrobactrum anthropi* XM-1. *Journal of Hazardous Materials*, *380*, 120–873. doi:10.1016/j.jhazmat.2019.120873 PMID:31325697

Chen, M., Yang, G., Sheng, Y., Li, P., Qiu, H., Zhou, X., Huang, L., & Chao, Z. (2017). Glomus mosseae inoculation improves the root system architecture, photosynthetic efficiency and flavonoids accumulation of liquorice under nutrient stress. *Frontiers in Plant Science*, *8*, 931. doi:10.3389/fpls.2017.00931 PMID:28638391

Chennappa, G., Adkar-Purushothama, C. R., Suraj, U., Tamilvendan, K., & Sreenivasa, M. Y. (2014). Pesticide tolerant *Azotobacter* isolates from paddy growing areas of northern Karnataka, India. *World Journal of Microbiology & Biotechnology*, *30*(1), 1–7. doi:10.100711274-013-1412-3 PMID:23813305

Chen, Q., & Liu, S. (2019). Identification and Characterization of the Phosphate-Solubilizing Bacterium Pantoea sp. S32 in Reclamation Soil in Shanxi, China. *Frontiers in Microbiology*, *10*, 2171. doi:10.3389/fmicb.2019.02171 PMID:31608027

Chen, S., Chang, C., Deng, Y., An, S., Dong, Y. H., Zhou, J., Hu, M., Zhong, G., & Zhang, L. H. (2014). Fenpropathrin biodegradation pathway in Bacillus sp. DG-02 and its potential for bioremediation of pyrethroid-contaminated soils. *Journal of Agricultural and Food Chemistry*, *12*, 147–157. PMID:24576059

Chen, S., Chao, L., Sun, L. N., & Sun, T. H. (2013). Plant-microorganism combined remediation for sediments contaminated with heavy metals. *Advanced Materials Research*, *123*, 610–613. doi:10.4028/www.scientific.net/AMR.726-731.610

Chen, S., Geng, P., Xiao, Y., & Hu, M. (2012). Bioremediation of b-cypermethrin and 3- phenoxybenzaldehyde contaminated soils using Streptomyces aureus HP-S- 01. *Applied Microbiology and Biotechnology*, *94*(2), 505–515. doi:10.100700253-011-3640-5 PMID:22038248

Chen, S., Yang, L., Hu, M., & Liu, J. (2011). Biodegradation of fenvalerate and 3- phenoxybenzoic acid by a novel *Stenotrophomonas* sp. strain ZS-S-01 and its use in bioremediation of contaminated soils. *Applied Microbiology and Biotechnology*, *90*(2), 755–767. doi:10.100700253-010-3035-z PMID:21184062

Chen, X., Schauder, S., Potier, N., Van Dorsselaer, A., Pelczer, I., Bassler, B. L., & Hughson, F. M. (2002). Structural identification of a bacterial quorum-sensing signal containing boron. *Nature*, *415*(6871), 545–549. doi:10.1038/415545a PMID:11823863

Chen, X., Zhou, Q., Liu, F., Peng, Q., & Teng, P. (2019). Removal of nine pesticide residues from water and soil by biosorption coupled with degradation on biosorbent immobilized laccase. *Chemosphere*, *233*, 49–56. doi:10.1016/j.chemosphere.2019.05.144 PMID:31163308

Cheon, J., Fujioka, S., Dilkes, B. P., & Choe, S. (2013). Brassinosteroids regulate plant growth through distinct signaling pathways in *Selaginella* and *Arabidopsis*. *PLoS One*, *8*(12), e81938. doi:10.1371/journal.pone.0081938 PMID:24349155

Che, R., Deng, Y., Wang, F., Wang, W., Xu, Z., Hao, Y., Xue, K., Zhang, B., Tang, L., Zhou, H., & Cui, X. (2018). Autotrophic and symbiotic diazotrophs dominate nitrogen-fixing communities in Tibetan grassland soils. *The Science of the Total Environment*, *639*, 997–1006. doi:10.1016/j.scitotenv.2018.05.238 PMID:29929338

Cherubin, M. R., Tormena, C. A., & Karlen, D. L. (2017). Soil quality evaluation using the soil management assessment framework (SMAF) in Brazilian oxisols with contrasting texture. *Revista Brasileira de Ciência do Solo*, *41*(0), 41. doi:10.1590/18069657rbcs20160148

Chet, I., Ordentilich, A., Shapira, R., & Oppenheim, A. (1990). Mechanisms of biocontrol of soil borne plant pathogens by rhizobacteria. *Plant and Soil*, *129*(1), 85–92. doi:10.1007/BF00011694

Cheung, K. H., & Gu, J.-D. (2007). Mechanism of hexavalent chromium detoxification by microorganisms and bioremediation application potential: A review. *International Biodeterioration & Biodegradation*, *59*(1), 8–15. doi:10.1016/j.ibiod.2006.05.002

Chevrot, R., Rosen, R., Haudecoeur, E., Cirou, A., Shelp, B. J., Ron, E., & Faure, D. (2006). GABA controls the level of quorum-sensing signal in Agrobacterium tumefaciens. *Proceedings of the National Academy of Sciences of the United States of America*, *103*(19), 7460–7464. doi:10.1073/pnas.0600313103 PMID:16645034

Chiaiese, P., Corrado, G., Colla, G., Kyriacou, M. C., & Rouphael, Y. (2018). Renewable sources of plant biostimulation: Microalgae as a sustainable means to improve crop performance. *Frontiers in Plant Science*, *871*(12), 1–6. doi:10.3389/fpls.2018.01782 PMID:30581447

Chiani, M., Akbarzadeh, A., Farhangi, A., Mazinani, M., Saffari, Z., Emadzadeh, K., & Mehrabi, M. R. (2010). Optimization of culture medium to increase the production of desferrioxamine B (Desferal) in *Streptomyces pilosus*. *Pakistan Journal of Biological Sciences*, *13*(11), 546–550. doi:10.3923/pjbs.2010.546.550 PMID:21848068

Chiba, S., Salaipeth, L., Lin, Y. H., Sasaki, A., Kanematsu, S., & Suzuki, N. (2009). A novel bipartite double-stranded RNA Mycovirus from the white root rot Fungus *Rosellinia necatrix*: Molecular and biological characterization, taxonomic considerations, and potential for biological control. *Journal of Virology*, *83*(24), 12801–12812. doi:10.1128/JVI.01830-09 PMID:19828620

Chibata, I., & Tosa, T. (1981). Use of immobilized cells. *Annual Review of Biophysics and Bioengineering*, *10*(1), 197–216. doi:10.1146/annurev.bb.10.060181.001213 PMID:7020575

Chibuike, G. U., & Obiora, S. C. (2014). Heavy metal polluted soils: Effect on plants and bioremediation methods. *Applied and Environmental Soil Science*, *2014*, 2014. doi:10.1155/2014/752708

Chidambaram, D., Hennebel, T., Taghavi, S., Mast, J., Boon, N., Verstraete, W., van der Lelie, D., & Fitts, J. P. (2010). Concomitant microbial generation of palladium nanoparticles and hydrogen to immobilize chromate. *Environmental Science & Technology*, *44*(19), 7635–7640. doi:10.1021/es101559r PMID:20822130

Chien, C., Yumei, K. U. O., Changchieh, C. H. E. N., Chunwei, H. U. N. G., Chihwei, Y. E. H., & Weijen, Y. E. H. (2008). Microbial diversity of soil bacteria in agricultural field contaminated with heavy metals. *Journal of Environmental Sciences (China)*, *20*(3), 359–363. doi:10.1016/S1001-0742(08)60056-X PMID:18595405

Chivukula, M., & Renganathan, V. (1995). Phenolic azo dye oxidation by laccase from *Pyricularia oryzae*. *Applied and Environmental Microbiology*, *61*(12), 4374–4377. doi:10.1128/AEM.61.12.4374-4377.1995 PMID:16535191

Choudhary, R. B., Jana, A. K., & Jha, M. K. (2004). *Enzyme technology applications in leather processing*. Academic Press.

Choudhary, D. K., Johri, B. N., & Prakash, A. (2008). Volatiles as priming agents that initiate plant growth and defence responses. *Current Science*, *94*(5), 595–604.

Choudhary, D. K., Varma, A., & Tuteja, N. (Eds.). (2016). *Plant-microbe interaction: an approach to sustainable agriculture*. Springer Singapore. doi:10.1007/978-981-10-2854-0

Chovancová, E., Kosinski, J., Bujnicki, J. M., & Damborský, J. (2007). Phylogenetic analysis of haloalkane dehalogenases. *Proteins*, *67*(2), 305–316. doi:10.1002/prot.21313 PMID:17295320

Chovanec, P., Sparacino-Watkins, C. E., Zhang, N., Basu, P., & Stolz, J. (2012). Microbial reduction of chromate in the presence of nitrate by three nitrate respiring organisms. *Frontiers in Microbiology*, *3*, 416. doi:10.3389/fmicb.2012.00416 PMID:23251135

Chowdhary, P. K., Keshavan, N., Nguyen, H. Q., Peterson, J. A., González, J. E., & Haines, D. C. (2007). Bacillus megaterium CYP102A1 oxidation of acyl homoserine lactones and acyl homoserines. *Biochemistry*, *46*(50), 14429–14437. doi:10.1021/bi701945j PMID:18020460

Chudasama, K. S., & Thaker, V. S. (2017). Genome sequence of *Ochrobactrum anthropi* strain SUBG007, a plant pathogen and potential xenobiotic compounds degradation bacterium. *Genomics Data*, *11*, 116–117. doi:10.1016/j.gdata.2017.01.001 PMID:28119820

Chumthong, A., Kanjanamaneesathian, M., Pengnoo, A., & Wiwattanapatapee, R. (2008). Water-soluble granules containing *Bacillus megaterium* for biological control of rice sheath blight: Formulation, bacterial viability and efficacy testing. *World Journal of Microbiology & Biotechnology*, *24*(11), 2499–2507. doi:10.100711274-008-9774-7

Chun, C. K., Ozer, E. A., Welsh, M. J., Zabner, J., & Greenberg, E. P. (2004). Inactivation of a Pseudomonas aeruginosa quorum-sensing signal by human airway epithelia. *Proceedings of the National Academy of Sciences of the United States of America*, *101*(10), 3587–3590. doi:10.1073/pnas.0308750101 PMID:14970327

Chung, H., Park, M., Madhaiyan, M., Seshadri, S., Song, J., Cho, H., & Sa, T. (2005). Isolation and characterization of phosphate solubilizing bacteria from the rhizosphere of crop plants of Korea. *Soil Biology & Biochemistry*, *37*(10), 1970–1974. doi:10.1016/j.soilbio.2005.02.025

Chung, J. Y., Yu, S. D., & Hong, Y. S. (2014). Environmental source of arsenic exposure. *Journal of Preventive Medicine and Public Health*, *47*(5), 253–257. doi:10.3961/jpmph.14.036 PMID:25284196

Chu, Y. M., Jeon, J. J., Yea, S. J., Kim, Y. H., Yun, S. H., Lee, Y. W., & Kim, K. H. (2002). Double-stranded RNA mycovirus from *Fusarium graminearum*. *Applied and Environmental Microbiology*, *68*(5), 2529–2534. doi:10.1128/AEM.68.5.2529-2534.2002 PMID:11976130

Cirou, A., Diallo, S., Kurt, C., Latour, X., & Faure, D. (2007). Growth promotion of quorum-quenching bacteria in the rhizosphere of Solanum tuberosum. *Environmental Microbiology*, *9*(6), 1511–1522. doi:10.1111/j.1462-2920.2007.01270.x PMID:17504488

Cirou, A., Mondy, S., An, S., Charrier, A., Sarrazin, A., Thoison, O., DuBow, M., & Faure, D. (2012). Efficient bio stimulation of native and introduced quorum-quenching Rhodococcus erythropolis populations is revealed by a combination of analytical chemistry, microbiology, and pyrosequencing. *Applied and Environmental Microbiology*, *78*(2), 481–492. doi:10.1128/AEM.06159-11 PMID:22081576

Cirou, A., Raffoux, A., Diallo, S., Latour, X., Dessaux, Y., & Faure, D. (2011). Gamma- caprolactone stimulates growth of quorum-quenching Rhodococcus populations in a large-scale hydroponic system for culturing Solanum tuberosum. *Research in Microbiology*, *162*(9), 945–950. doi:10.1016/j.resmic.2011.01.010 PMID:21288487

Clark, M. (2011). *Handbook of Textile and Industrial Dyeing: Principles, Processes and Types of Dyes* (Vol. 1). Elsevier.

Claussen, M., Lüthe, H., Blatt, M., & Böttger, M. (1997). Auxin-induced growth and its linkage to potassium channels. *Planta*, *201*(2), 227–234. doi:10.1007/BF01007708

Clemens, S. (2006). Toxic metal accumulation, responses to exposure and mechanisms of tolerance in plants. *Biochimie*, *88*(11), 1707–1719. doi:10.1016/j.biochi.2006.07.003 PMID:16914250

Coates, R., Trentacoste, E., & Gerwick, W. (2013). Bioactive and novel chemicals from microalgae. In A. Richmond & Q. Hu (Eds.), *Handbook of microalgal culture: applied phycology and biotechnology* (2nd ed., pp. 504–531). Wiley. doi:10.1002/9781118567166.ch26

Coelho, L. M., Rezende, H. C., Coelho, L. M., de Sousa, P. A., Melo, D. F., & Coelho, N. M. (2015). Bioremediation of polluted waters using microorganisms. In N. Shiomi (Ed.), *Advances in Bioremediation of Wastewater and Polluted Soil*. InTech. doi:10.5772/60770

Cohen, Y. (2002). Bioremediation of oil by marine microbial mats. *International Microbiology*, *5*(4), 189–193. doi:10.100710123-002-0089-5 PMID:12497184

Colazo, J. C., & Buschiazzo, D. (2014). The Impact of Agriculture on Soil Texture Due to Wind Erosion. *Land Degradation & Development*, *26*(1), 62–70. doi:10.1002/ldr.2297

Colla, G., Nardi, S., Cardarelli, M., Ertani, A., Lucini, L., Canaguier, R., & Rouphael, Y. (2015a). Protein hydrolysates as biostimulants in horticulture. *Scientia Horticulturae*, *196*, 28–38. doi:10.1016/j.scienta.2015.08.037

Colla, G., Rouphael, Y., Canaguier, R., Svecova, E., & Cardarelli, M. (2014). Biostimulant action of a plant-derived protein hydrolysate produced through enzymatic hydrolysis. *Frontiers in Plant Science*, *5*, 448. doi:10.3389/fpls.2014.00448 PMID:25250039

Colla, G., Rouphael, Y., Di Mattia, E., El-Nakhel, C., & Cardarelli, M. (2015b). Co-inoculation of Glomus intraradices and Trichoderma atroviride acts as a biostimulant to promote growth, yield and nutrient uptake of vegetable crops. *Journal of the Science of Food and Agriculture*, *95*(8), 1706–1715. doi:10.1002/jsfa.6875 PMID:25123953

Colla, G., Rouphael, Y., Jawad, R., Kumar, P., Rea, E., & Cardarelli, M. (2013a). The effectiveness of grafting to improve NaCl and CaCl2 tolerance in cucumber. *Scientia Horticulturae*, *164*, 380–391. doi:10.1016/j.scienta.2013.09.023

Collins, D. P., & Jacobsen, B. J. (2003). Optimizing a *Bacillus subtilis* isolate for biological control of sugar beet Cercospora leaf spot. *Biological Control*, *26*(2), 153–161. doi:10.1016/S1049-9644(02)00132-9

Comba, S., Di Molfetta, A., & Sethi, R. (2011). A Comparison Between Field Applications of Nano-, Micro-, and Millimetric Zero-Valent Iron for the Remediation of Contaminated Aquifers. *Water, Air, and Soil Pollution*, *215*(1-4), 595–607. doi:10.100711270-010-0502-1

Compant, S., Duffy, B., Nowak, J., Clément, C., & Barka, E. A. (2005). Use of plant growth-promoting bacteria for biocontrol of plant diseases: Principles, mechanisms of action, and future prospects. *Applied and Environmental Microbiology*, *71*(9), 4951–4959. doi:10.1128/AEM.71.9.4951-4959.2005 PMID:16151072

Compant, S., Sessitsch, A., & Mathieu, F. (2012). The 125th anniversary of the first postulation of the soil origin of endophytic bacteria – a tribute to M.L.V. Galippe. *Plant and Soil*, *356*(1-2), 299–301. doi:10.100711104-012-1204-9

Conde, M. I. R., Ocampo, S. A., Castañeda, G. C., Hernández, A. R. G., & Aguilar, G. M. B. (2018). Efect of fuorescent Pseudomonas on tomato seed germination and seedling vigor. *Revista Chapingo Serie Horticultura*, *24*, 121–131. doi:10.5154/r.rchsh.2017.06.023

Contreras, A. R., Casals, E., Puntes, V., Komilis, D., Sánchez, A., & Font, X. (2015). Use of cerium oxide (CeO2) nanoparticles for the adsorption of dissolved cadmium (II), lead (II) and chromium (VI) at two different pHs in single and multi-component systems. *Global NEST Journal, 17*(3), 536–543. doi:10.30955/gnj.001687

Contreras-Cornejo, H. A., Macías-Rodríguez, L., López-Bucio, J. S., & López-Bucio, J. (2014). Enhanced plant immunity using Trichoderma. In *Biotechnology and Biology of Trichoderma* (pp. 495–504). Elsevier. doi:10.1016/B978-0-444-59576-8.00036-9

Cook, R. L., & Hesterberg, D. (2013). Comparison of trees and grasses for rhizoremediation of petroleum hydrocarbons. *International Journal of Phytoremediation, 15*(9), 844–860. doi:10.1080/15226514.2012.760518 PMID:23819280

Coppens, J., Grunert, O., Van Den Hende, S., Vanhoutte, I., Boon, N., Haesaert, G., & De Gelder, L. (2016). The use of microalgae as a high-value organic slow-release fertilizer results in tomatoes with increased carotenoid and sugar levels. *Journal of Applied Phycology, 28*(4), 2367–2377. doi:10.100710811-015-0775-2

Correa-Garcia, S., Armand, P. S., St-Arnaud, M., & Yergeau, E. (2018). Rhizoremediation of petroleum hydrocarbons: A model system for plant microbiome manipulation. *Microbial Biotechnology, 11*(5), 819–832. doi:10.1111/1751-7915.13303 PMID:30066464

Costa, J. A. V., Freitas, B. C. B., Cruz, C. G., Silveira, J., & Morais, M. G. (2019). Potential of microalgae as biopesticides to contribute to sustainable agriculture and environmental development. *Journal of Environmental Science and Health. Part B, Pesticides, Food Contaminants, and Agricultural Wastes, 54*(5), 366–375. doi:10.1080/03601234.2019.1571366 PMID:30729858

Costa, J. P., Santos, P. S. M., Duarte, A. C., & Rocha-Santos, T. (2016). Nanoplastics in the environment - sources, fates and effects. *The Science of the Total Environment, 567*, 15–26. doi:10.1016/j.scitotenv.2016.05.041 PMID:27213666

Courtney, R. G., & Timpson, J. P. (2005). Reclamation of Fine Fraction Bauxite Processing Residue (Red Mud) Amended with Coarse Fraction Residue and Gypsum. *Water, Air, and Soil Pollution, 164*(1-4), 91–102. doi:10.100711270-005-2251-0

Courtney, R., Mullen, G., & Harrington, T. (2009). An Evaluation of Revegetation Success on Bauxite Residue. *Restoration Ecology, 17*(3), 350–358. doi:10.1111/j.1526-100X.2008.00375.x

Cox, M. S., Bell, P. F., & Kovar, J. L. (1996). Differential tolerance of canola to arsenic when grown hydroponically or in soil. *Journal of Plant Nutrition, 19*(12), 1599–1610. doi:10.1080/01904169609365224

Craven, D., Isbell, F., Manning, P., Connolly, J., Bruelheide, H., Ebeling, A., ... Beierkuhnlein, C. (2016). Plant diversity effects on grassland productivity are robust to both nutrient enrichment and drought. *Philosophical Transactions of the Royal Society B: Biological Sciences, 371*(1694), 20150277.

Crichton, R. R., Wilmet, S., Legssyer, R., & Ward, R. J. (2002). Molecular and cellular mechanisms of iron homeostasis and toxicity in mammalian cells. *Journal of Inorganic Biochemistry, 91*(1), 9–18. doi:10.1016/S0162-0134(02)00461-0 PMID:12121757

Croes, S., Weyens, N., Colpaert, J., & Vangronsveld, J. (2015). Characterization of the cultivable bacterial populations associated with field grown *Brassica napus L.*: An evaluation of sampling and isolation protocols. *Environmental Microbiology, 17*(7), 2379–2392. doi:10.1111/1462-2920.12701 PMID:25367683

Crowley, D. E. (2006). Microbial Siderophores in the Plant Rhizosphere. In L. L. Barton & J. Abadia (Eds.), *Iron Nutrition in Plants and Rhizospheric Microorganisms*. Springer. doi:10.1007/1-4020-4743-6_8

Crowley, D. E., Wang, Y. C., Reid, C. P. P., & Szaniszlo, P. J. (1991). Mechanisms of iron acquisition from siderophores by microorganisms and plants. In *Iron Nutrition and Interactions in Plants* (pp. 213–232). Springer. doi:10.1007/978-94-011-3294-7_27

Cunningham, S. D., Anderson, T. A., Schwab, P. A., & Hsu, F. C. (1996). Phytoremediation of soils contaminated with organic pollutants. *Advances in Agronomy*, *56*, 55–114. doi:10.1016/S0065-2113(08)60179-0

Cvjetko, P., Zovko, M., & Balen, B. (2014). Proteomics of heavy metal toxicity in plants. *Archives of Industrial Hygiene and Toxicology*, *65*(1), 1-18.

Cycon, M. Z., Mijowska, A., Wojcik, M., & Piotrowska-Seget, Z. (2013). Biodegradation and bioremediation potential of diazinon-degrading Serratia marcescens to remove other organophosphorus pesticides from soils. *Journal of Environmental Management*, *117*, 7–16. doi:10.1016/j.jenvman.2012.12.031 PMID:23333465

Cycon, M., Mrozik, A., & Piotrowska-Seget, Z. (2017). Bioaugmentation as a strategy for the remediation of pesticide-polluted soil: A review. *Chemosphere*, *172*, 52–71. doi:10.1016/j.chemosphere.2016.12.129 PMID:28061345

Cycon, M., & Piotrowska-Seget, Z. (2016). Pyrethroid-degrading microorganisms and their potential for the bioremediation of contaminated soils: A review. *Frontiers in Microbiology*, *7*, 1463. doi:10.3389/fmicb.2016.01463 PMID:27695449

Cycon, M., Zmijowska, A., & Piotrowska-Seget, Z. (2014). Enhancement of deltamethrin degradation by soil bioaugmentation with two different strains of *Serratia marcescens*. *International Journal of Environmental Science and Technology*, *11*(5), 1305–1316. doi:10.100713762-013-0322-0

Czech, B., & Rubinowska, K. (2013). TiO_2-assisted photocatalytic degradation of diclofenac, metoprolol, estrone and chloramphenicol as endocrine disruptors in water. *Adsorption*, *19*(2-4), 619–630. doi:10.100710450-013-9485-8

D'Angelo-Picard, C., Chapelle, E., Ratet, P., Faure, D., & Dessaux, Y. (2011). Transgenic plants expressing the quorum quenching lactonase AttM do not significantly alter root-associated bacterial populations. *Research in Microbiology*, *162*(9), 951–958. doi:10.1016/j.resmic.2011.01.009 PMID:21315818

D'Angelo-Picard, C., Faure, D., Penot, I., & Dessaux, Y. (2005). Diversity of N-acyl homoserine lactone-producing and -degrading bacteria in soil and tobacco rhizosphere. *Environmental Microbiology*, *7*(11), 1796–1808. doi:10.1111/j.1462-2920.2005.00886.x PMID:16232294

D'Annibale, A., Rosetto, F., Leonardi, V., Federici, F., & Petruccioli, M. (2006). Role of autochthonous filamentous fungi in bioremediation of a soil historically contaminated with aromatic hydrocarbons. *Applied and Environmental Microbiology*, *72*(1), 28–36. doi:10.1128/AEM.72.1.28-36.2006 PMID:16391021

D'haeze, W., & Holsters, M. (2002). Nod factor structures, responses, and perception during initiation of nodule development. *Glycobiology*, *12*(6), 79R–105R. doi:10.1093/glycob/12.6.79R PMID:12107077

D'Souza, R., Varun, M., Lakhani, A., Singla, V., & Paul, M. S. (2015). PAH Contamination of urban soils and phytoremediation. In *Phytoremediation: Management of environmental contaminants* (Vol. 1). doi:10.1007/978-3-319-10395-2_15

da Costa, A. C. A., & Duta, F. P. (2001). Bioaccumulation of copper, zinc, cadmium and lead by *Bacillus sp., Bacillus cereus, Bacillus sphaericus* and *Bacillus subtilis. Brazilian Journal of Microbiology*, *32*(1), 1–5. doi:10.1590/S1517-83822001000100001 PMID:30637653

Dabrock, P. (2009). Playing God? Synthetic biology as a theological and ethical challenge. *Systems and Synthetic Biology*, *3*(1-4), 47–54. doi:10.100711693-009-9028-5 PMID:19816799

Daeshwar, N., Ayazloo, M., Khataee, A. R., & Pourhassan, M. (2007). Biological Decolorization of Dye Solution Containing Malachite Green by Microalgae *Cosmarium* sp. *Bioresource Technology*, *98*(6), 1176–1182. doi:10.1016/j.biortech.2006.05.025 PMID:16844368

Dai, Y., Liu, R., Zhou, Y., Li, N., Hou, L., Ma, Q., & Gao, B. (2020). Fire Phoenix facilitates phytoremediation of PAH-Cd co-contaminated soil through promotion of beneficial rhizosphere bacterial communities. *Environment International*, *136*, 105421. doi:10.1016/j.envint.2019.105421 PMID:31884414

Dal Cortivo, C., Barion, G., Visioli, G., Mattarozzi, M., Mosca, G., & Vamerali, T. (2017). Increased root growth and nitrogen accumulation in common wheat following PGPR inoculation: Assessment of plant-microbe interactions by ESEM. *Agriculture, Ecosystems & Environment*, *247*, 396–408. doi:10.1016/j.agee.2017.07.006

Dal Cortivo, C., Ferrari, M., Visioli, G., Lauro, M., Fornasier, F., Barion, G., & Vamerali, T. (2020). Effects of seed-applied biofertilizers on rhizosphere biodiversity and growth of common wheat (*Triticum aestivum* L.) in the *Field. Frontiers in Plant Science*, *11*, 72. doi:10.3389/fpls.2020.00072 PMID:32174929

DalCorso, G., Fasani, E., Manara, A., Visioli, G., & Furini, A. (2019). Heavy metal pollutions: State of the art and innovation in phytoremediation. *International Journal of Molecular Sciences*, *20*(14), 3412. doi:10.3390/ijms20143412 PMID:31336773

Dall'Asta, P., Velho, A. C., Pereira, T. P., Stadnik, M. J., & Arisi, A. (2019). *Herbaspirillum seropedicae* promotes maize growth but fails to control the maize leaf anthracnose. *Physiology and Molecular Biology of Plants: An International Journal of Functional Plant Biology*, *25*(1), 167–176.

Dalton, D. A., & Kramer, S. (2007). Nitrogen-fixing bacteria in non-legumes. In *Plant-Associated Bacteria* (pp. 105–130). Springer.

Damalas, C. A., & Koutroubas, S. D. (2016). Farmers' Exposure to Pesticides: Toxicity Types and Ways of Prevention. *Toxics*, *4*(1), 1. doi:10.3390/toxics4010001 PMID:29051407

Dangi, A. K., Sharma, B., Hill, R. T., & Shukla, P. (2019). Bioremediation through microbes: Systems biology and metabolic engineering approach. *Critical Reviews in Biotechnology*, *39*(1), 79–98. doi:10.1080/07388551.2018.1500997 PMID:30198342

Danish, S., Kiran, S., Fahad, S., Ahmad, N., Ali, M. A., Tahir, F. A., Rasheed, M. K., Shahzad, K., Li, X., Wang, D., Mubeen, M., Abbas, S., Munir, T. M., Hashmi, M. Z., Adnan, M., Saeed, B., Saud, S., Khan, M. N., Ullah, A., & Nasim, W. (2019). Alleviation of chromium toxicity in maize by Fe fortification and chromium tolerant ACC deaminase producing plant growth promoting rhizobacteria. *Ecotoxicology and Environmental Safety*, *185*, 109706. doi:10.1016/j.ecoenv.2019.109706 PMID:31561073

Danish, S., Zafar-ul-Hye, M., Mohsin, F., & Hussain, M. (2020). ACC-deaminase producing plant growth promoting rhizobacteria and biochar mitigate adverse effects of drought stress on maize growth. *PLoS One*, *15*(4), e0230615. doi:10.1371/journal.pone.0230615 PMID:32251430

Dar, A., Zahir, Z. A., Asghar, H. N., & Ahmad, R. (2020). Preliminary screening of rhizobacteria for biocontrol of little seed canary grass (Phalaris minor Retz.) and wild oat (*Avena fatua* L.) in wheat. *Canadian Journal of Microbiology*, *66*(5), 368–376. doi:10.1139/cjm-2019-0427 PMID:32040347

Dartsch, P. C., Hildenbrand, S., Kimmel, R., & Schmahl, F. W. (1998). Investigations on the nephrotoxicity and hepatotoxicity of trivalent and hexavalent chromium compounds. *International Archives of Occupational and Environmental Health*, *71*, S40–S45. PMID:9827879

Dary, M., Chamber-Pérez, M. A., Palomares, A. J., & Pajuelo, E. (2010). *In situ* phyto stabilisation of heavy metal polluted soils using *Lupinus luteus* inoculated with metal resistant plant-growth promoting rhizobacteria. *Journal of Hazardous Materials, 177*(1-3), 323–330. doi:10.1016/j.jhazmat.2009.12.035 PMID:20056325

Das, S. N., & Thakur, R. S. (1995). International series on environment, Red mud analysis and utilization. Publication and Information Directorate.

Das, H. K. (2019). Azotobacters as biofertilizer. *Advances in Applied Microbiology, 108*, 1–43. doi:10.1016/bs.aambs.2019.07.001 PMID:31495403

Dash, H. R., & Das, S. (2012). Bioremediation of mercury and the importance of bacterial *mer* genes. *International Biodeterioration & Biodegradation, 75*, 207–213. doi:10.1016/j.ibiod.2012.07.023

Dash, H. R., & Das, S. (2015). Bioremediation of inorganic mercury through volatilization and biosorption by transgenic *Bacillus cereus* BW-03(pPW-05). *International Biodeterioration & Biodegradation, 103*, 179–185. doi:10.1016/j.ibiod.2015.04.022

Dashti, N., Zhang, F., Hynes, R., & Smith, D. L. (1997). Application of plant growth-promoting rhizobacteria to soybean (*Glycine max* [L.] Merr.) increases protein and dry matter yield under short-season conditions. *Plant and Soil, 188*(1), 33–41. doi:10.1023/A:1004295827311

Das, K., Prasanna, R., & Saxena, A. K. (2017). Rhizobia: A potential biocontrol agent for soilborne fungal pathogens. *Folia Microbiologica, 62*(5), 425–435. doi:10.100712223-017-0513-z PMID:28285373

Das, M., & Adholeya, A. (2012). Role of microorganisms in remediation of contaminated soil. In *Microorganisms in environmental management* (pp. 81–111). Springer. doi:10.1007/978-94-007-2229-3_4

Das, N., & Chandran, P. (2011). Microbial degradation of petroleum hydrocarbon contaminants: An overview. *Biotechnology Research International, 2011*, 1–14. doi:10.4061/2011/941810 PMID:21350672

Dasola, A. M., Adeyemi, A. L., Tunbosun, L. A., Abidemi, O. O., & Razaq, S. O. (2014). Kinetic and equilibrium studies of the heavy metal remediation potential of *Helix pomentia. African Journal of Pure and Applied Chemistry, 8*, 123–133.

Das, P., Sinha, S., & Mukherjee, S. K. (2014). Nickel Bioremediation Potential of *Bacillus thuringiensis* KUNi1 and Some Environmental Factors in Nickel Removal. *Bioremediation Journal, 18*(2), 169–177. doi:10.1080/10889868.2014.889071

Das, S. K., & Varma, A. (2010). Role of enzymes in maintaining soil health. In *Soil enzymology* (pp. 25–42). Springer. doi:10.1007/978-3-642-14225-3_2

Das, S., & Dash, H. R. (2014). 1 - Microbial Bioremediation: *A Potential Tool for Restoration of Contaminated Areas.* In S. Das (Ed.), *Microbial Biodegradation and Bioremediation* (pp. 1–21). Elsevier. doi:10.1016/B978-0-12-800021-2.00001-7

Das, S., Dash, H. R., & Chakraborty, J. (2016). Genetic basis and importance of metal resistant genes in bacteria for bioremediation of contaminated environments with toxic metal pollutants. *Applied Microbiology and Biotechnology, 100*(7), 2967–2984. doi:10.100700253-016-7364-4 PMID:26860944

Datta, B., & Chakrabartty, P. K. (2014). Siderophore biosynthesis genes of *Rhizobium sp.* isolated from *Cicer arietinum* L. *Biotech, 4*(4), 391-401.

Datta, S., Rajnish, K. N., Samuel, M. S., Pugazlendhi, A., & Selvarajan, E. (2020). Metagenomic applications in microbial diversity, bioremediation, pollution monitoring, enzyme and drug discovery. A review. *Environmental Chemistry Letters, 18*(4), 1229–1241. doi:10.100710311-020-01010-z

Datta, S., Singh, S., Kumar, V., Dhanjal, D. S., Sidhu, G. K., Amin, D. S., & Singh, J. (2020). Endophytic bacteria in xenobiotic degradation. In *Microbial Endophytes* (pp. 125–156). Woodhead Publishing. doi:10.1016/B978-0-12-818734-0.00006-1

Daughton, C. G., & Ternes, T. A. (1999). Pharmaceuticals and personal care products in the environment: Agents of subtle change? *Environmental Health Perspectives*, *107*(suppl 6), 907–938. doi:10.1289/ehp.99107s6907 PMID:10592150

Dave, S., Maitri, D., & Tipre, D. (2010). Copper remediation by *Eichhornia* spp. and sulphate-reducing bacteria. *Journal of Hazardous Materials*, *173*(1-3), 231–235. doi:10.1016/j.jhazmat.2009.08.073 PMID:19747776

David, S. R., Ihiawakrim, D., Regis, R., & Geoffroy, V. A. (2020). Efficiency of pyoverdines in iron removal from flocking asbestos waste: An innovative bacterial bioremediation strategy. *Journal of Hazardous Materials*, *122532*. Advance online publication. doi:10.1016/j.jhazmat.2020.122532 PMID:32200235

Davin, M., Starren, A., Marit, E., Lefébure, K., Fauconnier, M.-L., & Colinet, G. (2019). Investigating the effect of *Medicago sativa* L. and *Trifolium pratense* L. root exudates on PAHs bioremediation in an aged-contaminated soil. *Water, Air, and Soil Pollution*, *230*(12), 296. doi:10.100711270-019-4341-4

Davis, A. S., Prakash, P., & Thamaraiselvi, K. (2017). Nanobioremediation technologies for sustainable environment. In *Bioremediation and Sustainable Technologies for Cleaner Environment* (pp. 13–33). Springer. doi:10.1007/978-3-319-48439-6_2

Dawson, J. O. (1986). Actinorhizal plants: Their use in forestry and agriculture. *Outlook on Agriculture*, *15*(4), 202–208. doi:10.1177/003072708601500406

Dawwam, G. E., Elbeltagy, A., Emara, H. M., Abbas, I. H., & Hassan, M. M. (2013). Beneficial effect of plant growth promoting bacteria isolated from the roots of potato plant. *Annals of Agricultural Science*, *58*(2), 195–201. doi:10.1016/j.aoas.2013.07.007

De Bach, P. (1964). *Biological control of insect pests and weeds*. Reihold.

De Curtis, F., Lima, G., Vitullo, D., & De Cicco, V. (2010). Biocontrol of Rhizoctonia solani and Sclerotium rolfsii on tomato by delivering antagonistic bacteria through a drip irrigation system. *Crop Protection (Guildford, Surrey)*, *29*(7), 663–670. doi:10.1016/j.cropro.2010.01.012

De Freitas, J. R., Banerjee, M. R., & Germida, J. J. (1997). Phosphate-solubilizing rhizobacteria enhance the growth and yield but not phosphorus uptake of canola (Brassica napus L.). *Biology and Fertility of Soils*, *24*(4), 358–364. doi:10.1007003740050258

de la Fuente-Núñez, C., Korolik, V., Bains, M., Nguyen, U., Breidenstein, E. B., Horsman, S., Lewenza, S., Burrows, L., & Hancock, R. E. (2012). Inhibition of bacterial biofilm formation and swarming motility by a small synthetic cationic peptide. *Antimicrobial Agents and Chemotherapy*, *56*(5), 2696–2704. doi:10.1128/AAC.00064-12 PMID:22354291

de los Santos-Villalobos, S., de Folter, S., Délano-Frier, J. P., Gómez-Lim, M. A., Guzmán-Ortiz, D. A., & Pena-Cabriales, J. J. (2013). Growth promotion and flowering induction in mango (*Mangifera indica* L. cv "*Ataulfo*") trees by Burkholderia and Rhizobium Inoculation: Morphometric, biochemical, and molecular events. *Journal of Plant Growth Regulation*, *32*(3), 615–627. doi:10.100700344-013-9329-5

de Salamone, I. G., Döbereiner, J., Urquiaga, S., & Boddey, R. M. (1996). Biological nitrogen fixation in Azospirillum strain-maize genotype associations as evaluated by the 15 N isotope dilution technique. *Biology and Fertility of Soils*, *23*(3), 249–256. doi:10.1007/BF00335952

De Souza Machado, A. A., Kloas, W., Zarfl, C., Hempel, S., & Rillig, M. C. (2018). Microplastics as an emerging threat to terrestrial ecosystems. *Global Change Biology, 24*(4), 1405–1416. doi:10.1111/gcb.14020 PMID:29245177

de Souza, J. T., Arnould, C., Deulvot, C., Lemanceau, P., Gianinazzi-Pearson, V., & Raaijmakers, J. M. (2003). Effect of 2,4-diacetylphloroglucinol on pythium: Cellular responses and variation in sensitivity among propagules and species. *Phytopathology, 93*(8), 966–975. doi:10.1094/PHYTO.2003.93.8.966 PMID:18943863

de Souza, R. B., Maziviero, T. G., Christofoletti, C. A., & Fontanetti, C. S. (2013). Soil contamination with heavy metals and petroleum derivatives: Impact on edaphic fauna and remediation strategies. In M. C. H. Soriano (Ed.), *Soil processes and current trend in quality assessment*. doi:10.5772/52868

De Wet, J., Bihon, W., Preisig, O., Wingfield, B. D., & Wingfield, M. J. (2011). Characterization of a novel dsRNA element in the pine endophytic fungus *Diplodia scrobiculata. Archives of Virology, 156*(7), 1199–1208. doi:10.100700705-011-0978-z PMID:21442227

De Wildeman, S., Nollet, H., Van Langenhove, H., Diekert, G., & Verstraete, W. (2002). Reductive biodegradation of 1, 2-dichloroethane by methanogenic granular sludge: Perspectives for in situ remediation. *Water Science and Technology, 45*(10), 43–48. doi:10.2166/wst.2002.0284 PMID:12188575

Debnath, S., Rawat, D., Mukherjee, A. K., Adhikary, S., & Kundu, R. (2019). Applications and Constraints of Plant Beneficial Microorganisms in Agriculture. In *Rhizosphere and Soil Microbes-Utilization in Agriculture and Industry Under Current Scenario*. IntechOpen.

Decho, A. W., Visscher, P. T., Ferry, J., Kawaguchi, T., He, L., Przekop, K. M., Norman, R. S., & Reid, R. P. (2009). Autoinducers extracted from microbial mats reveal a surprising diversity of N-acylhomoserine lactones (AHLs) and abundance changes that may relate to diel pH. *Environmental Microbiology, 11*(2), 409–420. doi:10.1111/j.1462-2920.2008.01780.x PMID:19196272

Deeken, R., Engelmann, J. C., Efetova, M., Czirjak, T., Müller, T., Kaiser, W. M., Tietz, O., Krischke, M., Mueller, M. J., Palme, K., Dandekar, T., & Hedrich, R. (2006). An integrated view of gene expression and solute profiles of Arabidopsis tumors: A genome-wide approach. *The Plant Cell, 18*(12), 3617–3634. doi:10.1105/tpc.106.044743 PMID:17172353

Del Buono, D., Terzano, R., Panfili, I., & Bartucca, M. L. (2020). Phytoremediation and detoxification of xenobiotics in plants: Herbicide-safeners as a tool to improve plant efficiency in the remediation of polluted environments. A mini-review. *International Journal of Phytoremediation, 22*(8), 789–803. doi:10.1080/15226514.2019.1710817 PMID:31960714

Delalande, L., Faure, D., Raffoux, A., Uroz, S., D'Angelo-Picard, C., Elasri, M., Carlier, A., Berruyer, R., Petit, A., Williams, P., & Dessaux, Y. (2005). N-hexanoyl-L-homoserine lactone, a mediator of bacterial quorum-sensing regulation, exhibits plant-dependent stability and may be inactivated by germinating Lotus corniculatus seedlings. *FEMS Microbiology Ecology, 52*(1), 13–20. doi:10.1016/j.femsec.2004.10.005 PMID:16329888

Delmont, T. O., Robe, P., Clark, I., Simonet, P., & Vogel, T. M. (2011). Metagenomic comparison of direct and indirect soil DNA extraction approaches. *Journal of Microbiological Methods, 86*(3), 397–400. doi:10.1016/j.mimet.2011.06.013 PMID:21723887

Demain, A. L. (2008). A new opportunity for industry. *Secondary Metabolites: Their Function and Evolution, 171*, 3. PMID:1302184

Demain, A. L., & Sanchez, S. (2009). Microbial drug discovery: 80 years of progress. *The Journal of Antibiotics, 62*(1), 5–16. doi:10.1038/ja.2008.16 PMID:19132062

Dematté, J. A. M., Nanni, M. R., da Silva, A. P., de Melo Filho, J. F., Dos Santos, W. C., & Campos, R. C. (2010). Soil density evaluated by spectral reflectance as an evidence of compaction effects. *International Journal of Remote Sensing*, *31*(2), 403–422. doi:10.1080/01431160902893469

Deng, A., Wu, J., Zhang, G., & Wen, T. (2011). Molecular and structural characterization of a surfactant-stable high-alkaline protease AprB with a novel structural feature unique to subtilisin family. *Biochimie*, *93*(4), 783–791. doi:10.1016/j.biochi.2011.01.011 PMID:21281692

Deng, F., Xu, R., & Boland, G. J. (2003). Hypovirulence-Associated Double-Stranded RNA from *Sclerotinia homoeocarpa* Is Conspecific with *Ophiostoma novoulmi* Mitovirus 3a-Ld. *Phytopathology*, *93*(11), 1407–1414. doi:10.1094/PHYTO.2003.93.11.1407 PMID:18944069

Deng, Z., & Cao, L. (2017). Fungal endophytes and their interactions with plants in phytoremediation: A review. *Chemosphere*, *168*, 1100–1106. doi:10.1016/j.chemosphere.2016.10.097 PMID:28029384

Derakhshan, Z., Ehrampoush, M. H., Mahvi, A. H., Dehghani, M., Faramarzian, M., & Eslami, H. (2019). A comparative study of hybrid membrane photobioreactor and membrane photobioreactor for simultaneous biological removal of atrazine and CNP from wastewater: A performance analysis and modeling. *Chemical Engineering Journal*, *355*, 428–438. doi:10.1016/j.cej.2018.08.155

Dermont, G., Bergeron, M., Mercier, G., & Richerlaflèche, M. (2008). Soil washing for metal removal: A review of physical/chemical technologies and field applications. *Journal of Hazardous Materials*, *152*(1), 1–31. doi:10.1016/j.jhazmat.2007.10.043 PMID:18036735

Desai, C., Jain, K., & Madamwar, D. (2008). Evaluation of in vitro Cr(VI) reduction potential in cytosolic extracts of three indigenous *Bacillus sp.* isolated from Cr(VI) polluted industrial landfill. *Bioresource Technology*, *99*(14), 6059–6069. doi:10.1016/j.biortech.2007.12.046 PMID:18255287

Desbrosses, G. J., & Stougaard, J. (2011). Root nodulation: A paradigm for how plant-microbe symbiosis influences host developmental pathways. *Cell Host & Microbe*, *10*(4), 348–358. doi:10.1016/j.chom.2011.09.005 PMID:22018235

Desbrosses, G., Contesto, C., Varoquaux, F., Galland, M., & Touraine, B. (2009). PGPR-Arabidopsis interactions is a useful system to study signaling pathways involved in plant developmental control. *Plant Signaling & Behavior*, *4*(4), 321–323. doi:10.4161/psb.4.4.8106 PMID:19794852

Deshmukh, R., Khardenavis, A. A., & Purohit, H. J. (2016). Diverse metabolic capacities of fungi for bioremediation. *Indian Journal of Microbiology*, *56*(3), 247–264. doi:10.100712088-016-0584-6 PMID:27407289

Dessaux, Y., Grandclément, C., & Faure, D. (2016). Engineering the rhizosphere. *Trends in Plant Science*, *21*(3), 266–278. doi:10.1016/j.tplants.2016.01.002 PMID:26818718

Devasia, S., & Nair, A. J. (2016). Screening of potent laccase producing organisms based on the oxidation pattern of different phenolic substrates. *International Journal of Current Microbiology and Applied Sciences*, *5*(5), 127–137. doi:10.20546/ijcmas.2016.505.014

Devescovi, G., Aguilar, C., Majolini, M. B., Marugg, J., Weisbeek, P., & Venturi, V. (2001). A siderophore peptide synthetase gene from plant-growth-promoting Pseudomonas putida WCS358. *Systematic and Applied Microbiology*, *24*(3), 321–330. doi:10.1078/0723-2020-00063 PMID:11822666

Dhal, P. K., & Sar, P. (2014). Microbial communities in uranium mine tailings and mine water sediment from JadugudaU mine, India: A culture independent analysis. *Journal of Environmental Science and Health. Part A, Toxic/Hazardous Substances & Environmental Engineering*, *49*(6), 694–709. doi:10.1080/10934529.2014.865458 PMID:24521415

Dhiman, K., & Shirkot, P. (2015). Bioprospecting and molecular characterization of laccase producing bacteria from paper mills of Himachal Pradesh. *Proceedings of the National Academy of Sciences. India. Section B, Biological Sciences*, *85*(4), 1095–1103. doi:10.100740011-015-0541-x

Dhir, B. (2017). Biofertilizers and biopesticides: eco-friendly biological agents. In *Advances in Environmental Biotechnology* (pp. 167–188). Springer. doi:10.1007/978-981-10-4041-2_10

Dhull, S., & Gera, R. (2018). Siderophore Production by Rhizobia Isolated from Cluster Bean [*Cyamopsis tetragonoloba* (L.) Taub.] Growing in Semi-Arid Regions of Haryana, India. *International Journal of Current Microbiology and Applied Sciences*, *7*(3), 3187–3191. doi:10.20546/ijcmas.2018.703.368

Di Benedetto, N. A., Corbo, M. R., Campaniello, D., Cataldi, M. P., Bevilacqua, A., Sinigaglia, M., & Flagella, Z. (2017). The role of plant growth promoting bacteria in improving nitrogen use efficiency for sustainable crop production: A focus on wheat. *AIMS Microbiology*, *3*(3), 413–434. doi:10.3934/microbiol.2017.3.413 PMID:31294169

Diab, E. (2008). Phytoremediation of oil contaminated desert soil using the rhizosphere effects. *Global Journal of Environmental Research*, *2*(2), 66–73. doi:10.1016/j.envpol.2020.114787

Diamantidis, G., Effosse, A., Potier, P., & Bally, R. (2000). Purification and characterization of the first bacterial laccase in the rhizospheric bacterium *Azospirillum lipoferum*. *Soil Biology & Biochemistry*, *32*(7), 919–927. doi:10.1016/S0038-0717(99)00221-7

Díaz-Magaña, A., Aguilar-Barajas, E., Moreno-Sánchez, R., Ramírez-Díaz, M. I., Riveros-Rosas, H., Vargas, E., & Cervantes, C. (2009). Short-chain chromate ion transporter proteins from *Bacillus subtilis* confer chromate resistance in *Escherichia coli*. *Journal of Bacteriology*, *191*(17), 5441–5445. doi:10.1128/JB.00625-09 PMID:19581367

Dick, R. P., Breakwell, D. P., & Turco, R. F. (1997). Soil enzyme activities and biodiversity measurements as integrative microbiological indicators. *Methods for Assessing Soil Quality*, *49*, 247-271.

Diep, C. N., & Hieu, T. N. (2013). Phosphate and potassium solubilizing bacteria from weathered materials of denatured rock mountain, Ha Tien, Kie^n Giang province Vietnam. *American Journal of Life Sciences*, *1*(3), 88–92. doi:10.11648/j.ajls.20130103.12

Dignac, M., Derrien, D., Barré, P., Barot, S., Cécillon, L., Chenu, C., Chevallier, T., Freschet, G. T., Garnier, P., Guenet, B., Hedde, M., Klumpp, K., Lashermes, G., Maron, P.-A., Nunan, N., Roumet, C., & Basile-Doelsch, I. (2017). Increasing soil carbon storage: Mechanisms, effects of agricultural practices and proxies A review. *Agronomy for Sustainable Development*, *37*(2), 14–44. doi:10.100713593-017-0421-2

Dijkhuizen, L., & Harder, W. (1984). Current views on the regulation of autotrophic carbon dioxide fixation via the Calvin cycle in bacteria. *Antonie van Leeuwenhoek*, *50*(5-6), 473–487. doi:10.1007/BF02386221 PMID:6099093

Dimkpa, C. O., Svatos, A., Dabrowska, P., Schmidt, A., Boland, W., & Kothe, E. (2008). Involvement of siderophores in the reduction of metal-induced inhibition of auxin synthesis in *Streptomyces* spp. *Chemosphere*, *74*(1), 19–25. doi:10.1016/j.chemosphere.2008.09.079 PMID:18986679

Dimkpa, C., Weinand, T., & Asch, F. (2009). Plant-rhizobacteria interactions alleviate abiotic stress conditions. *Plant, Cell & Environment*, *32*(12), 1682–1694. doi:10.1111/j.1365-3040.2009.02028.x PMID:19671096

Ding, L., Cheng, J., Xia, A., Jacob, A., Voelklein, M., & Murphy, J. D. (2016). Co-generation of biohydrogen and biomethane through two-stage batch co-fermentation of macro- and micro-algal biomass. *Bioresource Technology*, *218*, 224–231. doi:10.1016/j.biortech.2016.06.092 PMID:27371795

Dinglasan-Panlilio, M. J., Dworatzek, S., Mabury, S., & Edwards, E. (2006). Microbial oxidation of 1, 2-dichloroethane under anoxic conditions with nitrate as electron acceptor in mixed and pure cultures. *FEMS Microbiology Ecology*, *56*(3), 355–364. doi:10.1111/j.1574-6941.2006.00077.x PMID:16689868

Dixit, R., Wasiullah, E., Malaviya, D., Pandiyan, K., Singh, U., Sahu, A., Shukla, R., Singh, B., Rai, J., Sharma, P., Lade, H., & Paul, D. (2015). Bioremediation of heavy metals from soil and aquatic environment: An overview of principles and criteria of fundamental processes. *Sustainability*, *7*(2), 2189–2212. doi:10.3390u7022189

Dixon, M., Simonne, E., Obreza, T., & Liu, G. (2020). Crop Response to Low Phosphorus Bioavailability with a Focus on Tomato. *Agronomy (Basel)*, *10*(5), 617. doi:10.3390/agronomy10050617

Djoko, K. Y., Chong, L. X., Wedd, A. G., & Xiao, Z. (2010). Reaction mechanisms of the multicopper oxidase CuO from *Escherichia coli* support its functional role as a cuprous oxidase. *Journal of the American Chemical Society*, *132*(6), 2005–2015. doi:10.1021/ja9091903 PMID:20088522

Dobbelaere, S., Croonenborghs, A., Thys, A., Ptacek, D., Vanderleyden, J., Dutto, P., & Brener, S. (2001). Responses of agronomically important crops to inoculation with Azospirillum. *Functional Plant Biology*, *28*(9), 871–879. doi:10.1071/PP01074

Doble, M., & Kumar, A. (2007). *Biotreatment of Industrial Effluents. Resource Conservation Recycling*. Elsevier.

Doherty, M., Coutts, R. H. A., Brasier, C. M., & Buck, K. W. (2006). Sequence of RNA-dependent RNA polymerase genes provides evidence for three more distinct mitoviruses in *Ophiostoma novo-ulmi* isolate Ld. *Virus Genes*, *33*(1), 41–44. doi:10.100711262-005-0029-5 PMID:16791417

Domingues, V. S., de Souza Monteiro, A., Júlio, A. D. L., Queiroz, A. L. L., & dos Santos, V. L. (2020). Diversity of metal-resistant and tensoactive-producing culturable heterotrophic bacteria isolated from a copper mine in Brazilian Amazonia. *Scientific Reports*, *10*(1), 1–12. doi:10.103841598-020-62780-8 PMID:32277075

Dominguez-Nunez, J. A., Benito, B., Berrocal-Lobo, M., & Albanesi, A. (2016). Mycorrhizal fungi: role in the solubilization of potassium. In V. S. Meena, B. R. Maurya, J. P. Verma, & R. S. Meena (Eds.), *Potassium solubilizing microorganisms for sustainable agriculture* (pp. 77–98). Springer. doi:10.1007/978-81-322-2776-2_6

Dong, H., Jiang, Z., Deng, J., Zhang, C., Cheng, Y., Hou, K., Zhang, L., Tang, L., & Zeng, G. (2018). Physicochemical transformation of Fe/Ni bimetallic nanoparticles during aging in simulated groundwater and the consequent effect on contaminant removal. *Water Research*, *129*, 51–57. doi:10.1016/j.watres.2017.11.002 PMID:29128681

Dong, W. Y., Zhang, X. Y., Liu, X. Y., Fu, X. L., Chen, F. S., Wang, H. M., Sun, X. M., & Wen, X. F. (2015). Responses of soil microbial communities and enzyme activities to nitrogen and phosphorus additions in Chinese fir plantations of subtropical China. *Biogeosciences*, *12*(18), 5537–5546. doi:10.5194/bg-12-5537-2015

Dong, Y. H., Gusti, A. R., Zhang, Q., Xu, J. L., & Zhang, L. H. (2002). Identification of quorum-quenching N-acyl homoserine lactonases from Bacillus species. *Applied and Environmental Microbiology*, *68*(4), 1754–1759. doi:10.1128/AEM.68.4.1754-1759.2002 PMID:11916693

Dong, Y. H., Wang, L. H., Xu, J. L., Zhang, H. B., Zhang, X. F., & Zhang, L. H. (2001). Quenching quorum-sensing-dependent bacterial infection by an N-acyl homoserine lactonase. *Nature*, *411*(6839), 813–817. doi:10.1038/35081101 PMID:11459062

Dong, Y. H., Xu, J. L., Li, X. Z., & Zhang, L. H. (2000). aiiA, an enzyme that inactivates the acylhomoserine lactone quorum-sensing signal and attenuates the virulence of Erwinia carotovora. *Proceedings of the National Academy of Sciences of the United States of America*, *97*(7), 3526–3531. doi:10.1073/pnas.97.7.3526 PMID:10716724

Dong, Y. H., & Zhang, L. H. (2005). Quorum sensing and quorum-quenching enzymes. *Journal of Microbiology (Seoul, Korea)*, *43*(Spec No), 101–109. PMID:15765063

Dong, Y. H., Zhang, X. F., Xu, J. L., & Zhang, L. H. (2004). Insecticidal Bacillus thuringiensis silences Erwinia carotovora virulence by a new form of microbial antagonism, signal interference. *Applied and Environmental Microbiology*, *70*(2), 954–960. doi:10.1128/AEM.70.2.954-960.2004 PMID:14766576

Donova, M. V. (2007). Transformation of steroids by actinobacteria: A review. *Applied Biochemistry and Microbiology*, *43*(1), 1–14. doi:10.1134/S0003683807010012 PMID:17345852

Doran, J. W., Sarrantonio, M., & Liebig, M. (1996). Soil health and sustainability. In D. L. Sparks (Ed.), *Advances in Agronomy* (Vol. 56). Academic Press.

Doran, J. W., Sarrantonio, M., & Liebig, M. A. (1996). Soil health and sustainability. *Advances in Agronomy*, *56*, 1–54. doi:10.1016/S0065-2113(08)60178-9

Dos Santos, A. B., Cervantes, F. J., & van Lier, J. B. (2007). Review Paper on Current Technologies for Decolourisation of Textile Wastewaters: Perspectives for Anaerobic Biotechnology. *Bioresource Technology*, *98*(12), 2369–2385. doi:10.1016/j.biortech.2006.11.013 PMID:17204423

dos Santos, L. F., & Livingston, A. G. (1993). A novel bioreactor system for the destruction of volatile organic compounds. *Applied Microbiology and Biotechnology*, *40*(1), 151–157. doi:10.1007/BF00170444

dos Santos, L. F., & Livingston, A. G. (1994). Extraction and biodegradation of a toxic volatile organic compound (1, 2-dichloroethane) from waste-water in a membrane bioreactor. *Applied Microbiology and Biotechnology*, *42*(2-3), 421–431. doi:10.1007002530050273 PMID:7765782

dos Santos, L. F., & Livingston, A. G. (1995). Novel membrane bioreactor for detoxification of VOC wastewaters: Biodegradation of 1, 2-dichloroethane. *Water Research*, *29*(1), 179–194. doi:10.1016/0043-1354(94)00137-V

Dotaniya, M. L., Meena, H. M., & Lata, M. (2013b). Heavy metal toxicity: Pandora's box for human disease. *Read Shelf*, *9*(6), 5–6.

Dotaniya, M. L., Rajendiran, S., Meena, V. D., Saha, J. K., Coumar, M. V., Kundu, S., & Patra, A. K. (2016). Influence of chromium contamination on carbon mineralization and enzymatic activities in Vertisol. *Agricultural Research*, *6*(1), 91–96. doi:10.100740003-016-0242-6

Dotaniya, M. L., Thakur, J. K., Meena, D., Jajoria, D. K., & Rathor, G. (2014). Chromium pollution: A threat to environment. *Agricultural Reviews (Karnal)*, *35*(2), 153–157. doi:10.5958/0976-0741.2014.00094.4

Dott, W., Feidieker, D., Steiof, M., Becker, P. M., & Kämpfer, P. (1995). Comparison of ex situ and in situ techniques for bioremediation of hydrocarbon-polluted soils. *International Biodeterioration & Biodegradation*, *35*, 301–316. doi:10.1007/978-94-011-1312-0_4

Doty, S. L., Freeman, J. L., Cohu, C. M., Burken, J. G., Firrincieli, A., Simon, A., & Blaylock, M. J. (2017). Enhanced degradation of TCE on a superfund site using endophyte-assisted poplar tree phytoremediation. *Environmental Science & Technology*, *51*(17), 10050–10058. doi:10.1021/acs.est.7b01504 PMID:28737929

Doucette, W. J., Hall, A. J., & Gorder, K. A. (2010). Emissions of 1, 2-dichloroethane from holiday decorations as a source of indoor air contamination. *Ground Water Monitoring and Remediation*, *30*(1), 67–73. doi:10.1111/j.1745-6592.2009.01267.x

Doucette, W. J., Shunthirasingham, C., Dettenmaier, E. M., Zaleski, R. T., Fantke, P., & Arnot, J. A. (2018). A review of measured bioaccumulation data on terrestrial plants for organic chemicals: Metrics, variability, and the need for standardized measurement protocols. *Environmental Toxicology and Chemistry*, *37*(1), 21–33. doi:10.1002/etc.3992 PMID:28976607

Doudle, S., & Williams, W. (2010). Can we kick-start mining rehablitation with cyanobacterial crusts? In *Proceedings of the 16th Biennial Conference of the Australian Rangeland Society.* Australian Rangeland Society.

Douds, D. D., Gadkar, V., & Adholeya, A. (2000). Mass production of VAM fungus biofertilizer. In *Mycorrhizal Biology* (pp. 197–215). Springer. doi:10.1007/978-1-4615-4265-0_13

Douet, V., Expert, D., Barras, F., & Py, B. (2009). Erwinia chrysanthemi iron metabolism: The unexpected implication of the inner membrane platform within the type II secretion system. *Journal of Bacteriology*, *191*(3), 795–804. doi:10.1128/JB.00845-08 PMID:18978048

Dowling, D. N., & O'Gara, F. (1994). Metabolites of Pseudomonas involved in the biocontrol of plant disease. *Trends in Biotechnology*, *12*(4), 133–141. doi:10.1016/0167-7799(94)90091-4

Draganov, D. I., Teiber, J. F., Speelman, A., Osawa, Y., Sunahara, R., & La Du, B. N. (2005). Human paraoxonases (PON1, PON2, and PON3) are lactonases with overlapping and distinct substrate specificities. *Journal of Lipid Research*, *46*(6), 1239–1247. doi:10.1194/jlr.M400511-JLR200 PMID:15772423

Du Jardin, P. (2015). Plant biostimulants: Definition, concept, main categories and regulation. *Scientia Horticulturae*, *196*, 3–14. doi:10.1016/j.scienta.2015.09.021

Dua, M., & Singh, A. (2002). Biotechnology and bioremediation: Successes and limitations. *Applied Microbiology and Biotechnology*, *59*(2-3), 143–152. doi:10.100700253-002-1024-6 PMID:12111139

Duan, J., Muller, K. M., Charles, T. C., Vesely, S., & Glick, B. R. (2009). 1-Aminocyclopropane-1-carboxylate (ACC) deaminase genes in rhizobia from Southern Saskatchewan. *Microbial Ecology*, *57*(3), 423–436. doi:10.100700248-008-9407-6 PMID:18548183

Dubey, K. (2012). Bio-remediation of Bauxite residue (red mud) generated from Aluminium industry by using blue green algae / bio-inoculants. ICFRE Project Report.

Dubey, K. K., Kumar, P., Singh, P. K., & Shukla, P. (2014). Exploring Prospects of Monooxygenase-Based Biocatalysts in Xenobiotics. In *Microbial Biodegradation and Bioremediation* (pp. 577–614). Elsevier. doi:10.1016/B978-0-12-800021-2.00026-1

Dubey, K., & Dubey, K. P. (2011). A Study of the Effect of Red Mud Amendments on the Growth of Cyanobacterial Species. *Bioremediation Journal*, *15*(3), 133–139. doi:10.1080/10889868.2011.598483

Dubey, S. K., Dubey, J., Mehra, S., Tiwari, P., & Bishwas, A. J. (2011). Potential use of cyanobacterial species in bio-remediation of industrial effluents. *African Journal of Biotechnology*, *10*(7), 1125–1132.

Dudley, R. E., Gammal, L. M., & Klaassen, C. D. (1985). Cadmium-induced hepatic and renal injury in chronically exposed rats: Likely role of hepatic cadmium-metallothionein in nephrotoxicity. *Toxicology and Applied Pharmacology*, *77*(3), 414–426. doi:10.1016/0041-008X(85)90181-4 PMID:3975909

Duffner, A., Hoffland, E., & Temminghoff, E. J. (2012). Bioavailability of zinc and phosphorus in calcareous soils as affected by citrate exudation. *Plant and Soil*, *361*(1-2), 165–175. doi:10.100711104-012-1273-9

Duffy, B., Schouten, A., & Raaijmakers, J. M. (2003). Pathogen self-defense: Mechanisms to counteract microbial antagonism. *Annual Review of Phytopathology*, *41*(1), 501–538. doi:10.1146/annurev.phyto.41.052002.095606 PMID:12730392

Dukare, A. S., Prasanna, R., Chandra Dubey, S., Nain, L., Chaudhary, V., Singh, R., & Saxena, A. K. (2011). Evaluating novel microbe amended composts as biocontrol agents in tomato. *Crop Protection (Guildford, Surrey)*, *30*(4), 436–442. doi:10.1016/j.cropro.2010.12.017

Dulla, G. F., & Lindow, S. E. (2009). Acyl-homoserine lactone-mediated cross talk among epiphytic bacteria modulates behavior of Pseudomonas syringae on leaves. *The ISME Journal*, *3*(7), 825–834. doi:10.1038/ismej.2009.30 PMID:19340082

Dun-chun, H. E., Jia-sui, Z. H. A. N., & Lian-hui, X. I. E. (2016). Problems, challenges and future of plant disease management: From an ecological point of view. *Journal of Integrative Agriculture*, *154*(4), 705–715.

Dunne, E. J., Clark, M. W., Corstanje, R., & Reddy, K. R. (2011). Legacy phosphorus in subtropical wetland soils: Influence of dairy, improved and unimproved pasture land use. *Ecological Engineering*, *37*(10), 1481–1491. doi:10.1016/j.ecoleng.2011.04.003

Duran, N., & Esposito, E. (2000). Potential Applications of Oxidative Enzymes and Phenol oxidase-like Compounds in Wastewater and Soil Treatment. A Review. *Applied Catalysis B: Environmental*, *28*(2), 83–99. doi:10.1016/S0926-3373(00)00168-5

Dutta, S., Datta, J. K., & Mandal, N. C. (2017). Evaluation of indigenous rhizobacterial strains with reduced dose of chemical fertilizer towards growth and yield of mustard (Brassica campestris) under old alluvial soil zone of West Bengal, India. *Annals of Agrarian Science*, *15*(4), 447–452. doi:10.1016/j.aasci.2017.02.015

Dutta, S., & Podile, A. R. (2010). Plant growth promoting rhizobacteria (PGPR): The bugs to debug the root zone. *Critical Reviews in Microbiology*, *36*(3), 232–244. doi:10.3109/10408411003766806 PMID:20635858

Dwivedi, P., Mishra, P. K., Mondal, M. K., & Srivastava, N. (2019). Non-biodegradable polymeric waste pyrolysis for energy recovery. *Heliyon*, *5*(8), 1–15. doi:10.1016/j.heliyon.2019.e02198 PMID:32368634

Dwivedi, R. S. (1997). Perspectives of Microbial interactions. *Journal of the Indian Botanical Society*, *76*, 145–156.

Dymond, J. S. (2013). Explanatory chapter: Quantitative PCR. *Methods in Enzymology*, *529*, 279–289. doi:10.1016/B978-0-12-418687-3.00023-9 PMID:24011054

Dzionek, A., Wojcieszy, D., & Guzik, U. (2016). Natural carriers in bioremediation: A review. *Electronic Journal of Biotechnology*, *23*, 28–36. doi:10.1016/j.ejbt.2016.07.003

Eastmond, D. A., MacGregor, J. T., & Slesinski, R. S. (2008). Trivalent chromium: Assessing the genotoxic risk of an essential trace element and widely used human and animal nutritional supplement. *Critical Reviews in Toxicology*, *38*(3), 173–190. doi:10.1080/10408440701845401 PMID:18324515

East, R. (2013). Soil science comes to life: Plants may be getting a little help with their tolerance of drought and heat. *Nature*, *501*, 18–19. doi:10.1038/501S18a

Edberg, F., Kalinowski, B. E., Holmström, S. J., & Holm, K. (2010). Mobilization of metals from uranium mine waste: The role of pyoverdines produced by *Pseudomonas fluorescens*. *Geobiology*, *8*(4), 278–292. doi:10.1111/j.1472-4669.2010.00241.x PMID:20456501

Egamberdieva, D., Wirth, S. J., Shurigin, V. V., Hashem, A., & Abd-Allah, E. F. (2017). Endophytic bacteria improve plant growth, symbiotic performance of chickpea. (*Cicer arietinum* L.). and induce suppression of root rot caused by Fusarium under salt stress. *Frontiers in Microbiology*, *8*, 1887. doi:10.3389/fmicb.2017.01887 PMID:29033922

Egorova, D. O., Buzmakov, S. A., Nazarova, E. A., Andreev, D. N., Demakov, V. A., & Plotnikova, E. G. (2017). Bioremediation of hexachlorocyclohexane-contaminated soil by the new *Rhodococcuswratislaviensis* Strain Ch628. *Water, Air, and Soil Pollution*, *228*(5), 183. doi:10.100711270-017-3344-2

Eichlerova, I., Homolka, L., Benada, O., Kofroňová, O., Hubálek, T., & Nerud, F. (2007). Decolorization of Orange G and Remazol Brilliant Blue R by the white rot fungus *Dichomitus squalens*: Toxicological evaluation and morphological study. *Chemosphere*, *69*(5), 795–802. doi:10.1016/j.chemosphere.2007.04.083 PMID:17604080

El Bouraie, M., & El Din, W. S. (2016). Biodegradation of Reactive Black 5 by *Aeromonas hydrophila* strain isolated from dye-contaminated textile wastewater. *Sustainable Environment Research*, *26*(5), 209–216. doi:10.1016/j.serj.2016.04.014

El Fantroussi, S., & Agathos, S. N. (2005). Is bioaugmentation a feasible strategy for pollutant removal and site remediation? *Current Opinion in Microbiology*, *8*(3), 268–275. doi:10.1016/j.mib.2005.04.011 PMID:15939349

El-Akhdar, I., Elsakhawy, T., & Abo-Koura, H. A. (2020). Alleviation of Salt Stress on Wheat (*Triticum aestivum* L.) by Plant Growth Promoting Bacteria strains *Bacillus halotolerans* MSR-H4 and *Lelliottia amnigena* MSR-M49. *Journal of Advances in Microbiology*, 44–58. doi:10.9734/jamb/2020/v20i130208

Elango, G., Rathika, G., & Elango, S. (2016). Physico-Chemical Parameters of Textile Dyeing Effluent and Its Impacts with Case study. *International Journal of Research in Chemistry and Environment*, *7*(1), 17–24.

Elangovan, R., Abhipsa, S., Rohit, B., Ligy, P., & Chandraraj, K. (2006). Reduction of Cr(VI) by a *Bacillus sp. Biotechnology Letters*, *28*(4), 247–252. doi:10.100710529-005-5526-z PMID:16555008

El-Bestawy, E., El-Salam, Z., & Mansy, E. R. H. (2007). Potential Use of Environmental Cyanobacterial Species in Bioremediation of Lindane-Contaminated Effluents. *International Biodeterioration & Biodegradation*, *59*(3), 180–192. doi:10.1016/j.ibiod.2006.12.005

El-Bestawy, E., Sabir, J., Mansy, A. H., & Zabermawi, N. (2014). Comparison among the efficiency of different bioremediation technologies of Atrazine-contaminated soils. *Journal of Bioremediation & Biodegradation*, *5*(5), 237.

Elcossy, S. A. E., Abbas, M. H. H., Farid, I. M., Beheiry, G. G. S., Abou Yuossef, M. F., Abbas, H. H., Abdelhafez, A. A., & Mohamed, I. (2020). Dynamics of soil organic carbon in Typic Torripsamment soils irrigated with raw effluent sewage water. *Environmental Science and Pollution Research International*, *27*(8), 8188–8198. doi:10.100711356-019-07526-4 PMID:31900766

Eldridge, D. J. (1996). Cryptogamic soil crusts: fixers of the desert. In *Proc 1996 Workshop on rehabilitation of arid and semi-arid lands*. Goldfields Land Rehabilitation Group.

El-Howeity, M. A., & Asfour, M. M. (2012). Response of some varieties of canola plant (Brassica napus L.) cultivated in a newly reclaimed desert to plant growth promoting rhizobacteria and mineral nitrogen fertilizer. *Annals of Agricultural Science*, *57*(2), 129–136. doi:10.1016/j.aoas.2012.08.006

Elias, F., Woyessa, D., & Muleta, D. (2016). Phosphate Solubilization Potential of Rhizosphere Fungi Isolated from Plants in Jimma Zone, Southwest Ethiopia. *International Journal of Microbiology*, *5472601*, 1–11. Advance online publication. doi:10.1155/2016/5472601 PMID:27688771

Elliott, A., Hanby, W., & Malcolm, B. (1954). The near infra-red absorption spectra of natural and synthetic fibres. *British Journal of Applied Physics*, *5*(11), 337. doi:10.1088/0508-3443/5/11/301

El-Maraghy, S. S., Tohamy, T. A., & Hussein, K. A. (2020). Role of Plant-Growth Promoting Fungi (PGPF) in Defensive Genes Expression of Triticum aestivum against Wilt Disease. *Rhizosphere*, *15*, 100223. doi:10.1016/j.rhisph.2020.100223

El-Naggar, M. Y., El-Assar, S. A., & Abdul-Gawad, S. M. (2006). Meroparamycin production by newly isolated *Streptomyces sp.* strain MAR01: Taxonomy, fermentation, purification and structural elucidation. *Journal of Microbiology (Seoul, Korea)*, *44*(4), 432–438. PMID:16953179

El-Sheekh, M. M. & Mahmoud, Ya-G. (2017). Technological approach of bioremediation using microbial tools: bacteria, fungi, and algae. In Handbook of research on inventive bioremediation techniques. IGI Global.

El-Sheekh, M. M., Farghl, A. A., Galal, H. R., & Bayoumi, H. S. (2016). Bioremediation of different types of polluted water using microalgae. *Rend. Fis. Acc. Lincei.*, *27*(2), 401–410. doi:10.100712210-015-0495-1

El-Sheekh, M. M., Osman, M. E., Dyab, M. A., & Amer, M. S. (2006). Production and characterization of antimicrobial active substance from the cyanobacterium *Nostoc muscorum. Environmental Toxicology and Pharmacology*, *21*(1), 42–50. doi:10.1016/j.etap.2005.06.006 PMID:21783637

Elumalai, P., Parthipan, P., Karthikeyan, O. P., & Rajasekar, A. (2017). Enzyme-mediated biodegradation of long-chain n-alkanes (C 32 and C 40) by thermophilic bacteria. *Biotech, 7*(2), 1-10.

Elumalai, P., Lim, J.-M., Park, Y.-J., Cho, M., Shea, P. J., & Oh, B.-T. (2020). Agricultural waste materials enhance protease production by *Bacillus subtilis* B22 in submerged fermentation under blue light-emitting diodes. *Bioprocess and Biosystems Engineering*, *43*(5), 821–830. doi:10.100700449-019-02277-5 PMID:31919603

Emami, S., Alikhani, H. A., Pourbabaee, A. A., Etesami, H., Motasharezadeh, B., & Sarmadian, F. (2020). Consortium of endophyte and rhizosphere phosphate solubilizing bacteria improves phosphorous use efficiency in wheat cultivars in phosphorus deficient soils. *Rhizosphere*, *14*, 100196. doi:10.1016/j.rhisph.2020.100196

Embrandiri, A., Kiyasudeen, S. K., Rupani, P. F., & Ibrahim, M. H. (2016). Environmental xenobiotics and its effects on natural ecosystem. In *Plant Responses to Xenobiotics* (pp. 1–18). Springer. doi:10.1007/978-981-10-2860-1_1

Emerson, W. W., & McGarry, D. (2003). Organic carbon and soil porosity. *Australian Journal of Soil Research*, *41*(1), 107–118. doi:10.1071/SR01064

Emmert, E. A., & Handelsman, J. (1999). Biocontrol of plant disease: A (Gram-) positive perspective. *FEMS Microbiology Letters*, *171*(1), 1–9. doi:10.1111/j.1574-6968.1999.tb13405.x PMID:9987836

EPA. (2006). In Situ Treatment Technologies for Contaminated Soil. Engineering Issue. EPA 542-F-06-013.

Erable, B., Goubet, I., Lamare, S., Legoy, M. D., & Maugard, T. (2006). Bioremediation of halogenated compounds: Comparison of dehalogenating bacteria and improvement of catalyst stability. *Chemosphere*, *65*(7), 1146–1152. doi:10.1016/j.chemosphere.2006.04.007 PMID:16723151

Erturk, Y., Ercisli, S., Haznedar, A., & Cakmakci, R. (2010). Effects of plant growth promoting rhizobacteria (PGPR) on rooting and root growth of kiwifruit (Actinidia deliciosa) stem cuttings. *Biological Research*, *43*(1), 91–98. doi:10.4067/S0716-97602010000100011 PMID:21157636

Eshaghi, E., Nosrati, R., Owlia, P., Malboobi, M. A., Ghaseminejad, P., & Ganjali, M. R. (2019). Zinc solubilization characteristics of efficient siderophore-producing soil bacteria. *Iranian Journal of Microbiology*, *11*(5), 419. doi:10.18502/ijm.v11i5.1961 PMID:32148673

Esitken, A., Pirlak, L., Turan, M., & Sahin, F. (2006). Effects of floral and foliar application of plant growth promoting rhizobacteria (PGPR) on yield, growth and nutrition of sweet cherry. *Scientia Horticulturae*, *110*(4), 324–327. doi:10.1016/j.scienta.2006.07.023

Eskandari, F., Shahnavaz, B., & Mashreghi, M. (2019). Optimization of complete RB-5 azo dye decolorization using novel cold-adapted and mesophilic bacterial consortia. *Journal of Environmental Management*, *241*(March), 91–98. doi:10.1016/j.jenvman.2019.03.125 PMID:30986666

Eslami, H., Shariatifar, A., Rafiee, E., Shiranian, M., Salehi, F., Hosseini, S. S., & Ebrahimi, A. A. (2019). Decolorization and biodegradation of reactive Red 198 Azo dye by a new *Enterococcus faecalis–Klebsiella variicola* bacterial consortium isolated from textile wastewater sludge. *World Journal of Microbiology & Biotechnology, 35*(3), 38. doi:10.100711274-019-2608-y PMID:30739299

Esmaeili, A., Pourbabaee, A. A., Alikhani, H. A., Shabani, F., & Esmaeili, E. (2013). Biodegradation of Low-Density Polyethylene (LDPE) by Mixed Culture of *Lysinibacillus xylanilyticus* and *Aspergillus niger* in Soil. *PlosOne*, 8. doi:10.1371/journal.pone.0071720

Essumang, D. K. (2013). *Environmental Xenobiotics: PAHs In Soil (Heavy Metals), Indoor Air And Water Environment, Case Studies Of Ghana And Denmark* (Doctoral Dissertation). Luma Print.

Essumang, D. K. (2010). Distribution, levels, and risk assessment of polycyclic aromatic hydrocarbons (PAHs) in some water bodies along the coastal belt of Ghana. *TheScientificWorldJournal, 10*, 972–985. doi:10.1100/tsw.2010.96 PMID:20526527

Estabrook, E. M., & Yoder, J. I. (1998). Plant-plant communications: Rhizosphere signaling between parasitic angiosperms and their hosts. *Plant Physiology, 116*(1), 1–7. doi:10.1104/pp.116.1.1

Etesami, H., & Alikhani, H. A. (2016). Co-inoculation with endophytic and rhizosphere bacteria allows reduced application rates of N-fertilizer for rice plantquery id="q1">. *Rhizosphere, 2*, 5–12. doi:10.1016/j.rhisph.2016.09.003

Etesami, H., Emami, S., & Alikhani, H. A. (2017). Potassium solubilizing bacteria (KSB): Mechanisms, promotion of plant growth, and future prospects A review. *Journal of Soil Science and Plant Nutrition, 17*(4), 897–911. doi:10.4067/S0718-95162017000400005

Etesami, H., & Noori, F. (2019). Soil salinity as a challenge for sustainable agriculture and bacterial-mediated alleviation of salinity stress in crop plants. In *Saline Soil-based Agriculture by Halotolerant Microorganisms* (pp. 1–22). Springer. doi:10.1007/978-981-13-8335-9_1

Euliss, K., Ho, C. H., Schwab, A. P., Rock, S., & Banks, M. K. (2008). Greenhouse and field assessment of phytoremediation for petroleum contaminants in a riparian zone. *Bioresource Technology.*

Fabra, A., Duffard, R., & De Duffard, A. E. (1997). Toxicity of 2, 4-dichlorophenoxyacetic acid to *Rhizobium sp* in pure culture. *Bulletin of Environmental Contamination and Toxicology, 59*(4), 645–652. doi:10.1007001289900528 PMID:9307432

Fajardo, C., Ortiz, L. T., Rodriguez-Membibre, M. L., Nande, M., Lobo, M. C., & Martin, M. (2012). Assessing the impact of zero-valent iron (ZVI) nanotechnology on soil microbial structure and functionality: A molecular approach. *Chemosphere, 86*(8), 802–808. doi:10.1016/j.chemosphere.2011.11.041 PMID:22169206

Fakruddin, M., & Mannan, K. S. (2013). Methods for analyzing diversity of microbial communities in natural environments. *Ceylon Journal of Science. Biological Sciences, 42*(1), 19. doi:10.4038/cjsbs.v42i1.5896

Fang, L., Huang, Q., Wei, X., Liang, W., Rong, X., Chen, W., & Cai, P. (2010). Microcalorimetric and potentiometric titration studies on the adsorption of copper by extracellular polymeric substances (EPS), minerals and their composites. *Bioresource Technology, 101*(15), 5774–5779. doi:10.1016/j.biortech.2010.02.075 PMID:20227874

Fang, Q., Fan, Z., Xie, Y., Wang, X., Li, K., & Liu, Y. (2016). Screening and Evaluation of the Bioremediation Potential of Cu/Zn-Resistant, Autochthonous Acinetobacter sp. FQ-44 from *Sonchus oleraceus* L. *Frontiers in Plant Science, 7*, 1487. doi:10.3389/fpls.2016.01487 PMID:27746807

Fan, J., Yang, G., Zhao, H., Shi, G., Geng, Y., Hou, T., & Tao, K. (2012). Isolation, identification and characterization of a glyphosate-degrading bacterium, *Bacillus cereus* CB4, from soil. *The Journal of General and Applied Microbiology, 58*(4), 263–271. doi:10.2323/jgam.58.263 PMID:22990486

FAO. (2002). International code of conduct on the distribution and use of pesticides. In *Hundred and Twenty-third Session of the FAO Council in November 2002*. Publishing Management Service, Information Division, FAO.

FAO. (2020). *Agriculturalpollution: pesticides*. documents.worldbank.org/curated/en/689281521218090562/pdf/124345-BRI-p153343-PUBLIC-march-22-9-pm-WB-Knowledge-Pesticides.pdf

Farahat, E., & Linderholm, H. W. (2015). The effect of long-term wastewater irrigation on accumulation and transfer of heavy metals in Cupressus sempervirens leaves and adjacent soils. *The Science of the Total Environment, 51*, 1–7. doi:10.1016/j.scitotenv.2015.01.032 PMID:25613764

Farmer, E. E. (2001). Surface-to-air signals. *Nature, 411*(6839), 854–856. doi:10.1038/35081189 PMID:11459069

Farrell, R. E., & Germida, J. J. (2002). *Phytotechnologies: Plant-based Systems for remediation of oil impacted soils*. Available from https://www.esaa.org/wp-content/uploads/2015/06/02-09FarrellPaper

Fasani, E., Manara, A., Martini, F., Furini, A., & DalCorso, G. (2018). The potential of genetic engineering of plants for the remediation of soils contaminated with heavy metals. *Plant, Cell & Environment, 41*(5), 1201-1232.

Fasani, E., Manara, A., Martini, F., Furini, A., & DalCorso, G. (2018). The potential of genetic engineering of plants for the remediation of soils contaminated with heavy metals. *Plant, Cell & Environment, 41*(5), 1201–1232. doi:10.1111/pce.12963 PMID:28386947

Fatima, K., Imran, A., Amin, I., Khan, Q. M., & Afzal, M. (2018). Successful phytoremediation of crude-oil contaminated soil at an oil exploration and production company by plants-bacterial synergism. *International Journal of Phytoremediation, 20*(7), 675–681. doi:10.1080/15226514.2017.1413331 PMID:29723052

Fatta-Kassinos, D., Meric, S., & Nikolaou, A. (2011). Pharmaceutical residues in environmental waters and wastewater: Current state of knowledge and future research. *Analytical and Bioanalytical Chemistry, 399*(1), 251–275. doi:10.100700216-010-4300-9 PMID:21063687

Fauquet, C. M., & Fargette, D. (2005). International Committee on Taxonomy of Viruses and the unassigned species. *Virology Journal, 2*(1), 64. doi:10.1186/1743-422X-2-64 PMID:16105179

Faure, D., & Dessaux, Y. (2007). Quorum sensing as a target for developing control strategies for the plant pathogen *Pectobacterium. European Journal of Plant Pathology, 119*(3), 353–365. doi:10.100710658-007-9149-1

Favas, P. J., Pratas, J., Varun, M., D'Souza, R., & Paul, M. S. (2014). Phytoremediation of soils contaminated with metals and metalloids at mining areas: potential of native flora. *Environmental Risk Assessment of Soil Contamination, 3*, 485-516..

Fay, P. A., Prober, S. M., Harpole, W. S., Knops, J. M., Bakker, J. D., Borer, E. T., ... Adler, P. B. (2015). Grassland productivity limited by multiple nutrients. *Nature Plants, 1*(7), 1–5. doi:10.1038/nplants.2015.80 PMID:27250253

Feng, L., Wang, W., Cheng, J., Ren, Y., Zhao, G., Gao, C., Gao, C., Tang, Y., Liu, X., Han, W., Peng, X., & Liu, R. (2007). Genome and proteome of long-chain alkane degrading *Geobacillus thermodenitrificans* NG80-2 isolated from a deep-subsurface oil reservoir. *Proceedings of the National Academy of Sciences of the United States of America, 104*(13), 5602–5607. doi:10.1073/pnas.0609650104 PMID:17372208

Feng, N. X., Yu, J., Zhao, H. M., Cheng, Y. T., Mo, C. H., Cai, Q. Y., & Wong, M. H. (2017). Efficient phytoremediation of organic contaminants in soils using plant–endophyte partnerships. *The Science of the Total Environment, 583*, 352–368. doi:10.1016/j.scitotenv.2017.01.075 PMID:28117167

Fenibo, E. O., Ijoma, G. N., Selvarajan, R., & Chikere, C. B. (2019). Microbial Surfactants: The Next Generation Multifunctional Biomolecules for Applications in the Petroleum Industry and Its Associated Environmental Remediation. *Microorganisms*, *7*(11), 581. doi:10.3390/microorganisms7110581 PMID:31752381

Fergusson, J. E. (1990). *The Heavy Elements: Chemistry, Environmental Impact and Health Effects*. Pergamon Press.

Fernández-Fuego, D., Bertrand, A., & González, A. (2017). Metal accumulation and detoxification mechanisms in mycorrhizal *Betula pubescens*. *Environmental Pollution*, *231*, 1153–1162. doi:10.1016/j.envpol.2017.07.072 PMID:28941719

Fernández, H., Prandoni, N., Fernández-Pascual, M., Fajardo, S., Morcillo, C., Díaz, E., & Carmona, M. (2014). Azoarcus sp. CIB, an anaerobic biodegrader of aromatic compounds shows an endophytic lifestyle. *PLoS One*, *9*(10), e110771. doi:10.1371/journal.pone.0110771 PMID:25340341

Fernando, W. D., Nakkeeran, S., & Zhang, Y. (2005). Biosynthesis of antibiotics by PGPR and its relation in biocontrol of plant diseases. In *PGPR: biocontrol and biofertilization* (pp. 67–109). Springer.

FESLM. (1993). *An International Framework for Evaluating Sustainable Land Management Report 73*. Retrieved june 05, 2019, from http://www.fao.org/3/T1079E/t1079e04.htm#nature%20of%20sustainable%20land%20management%20(slm)

Fetzner, S. (2002). Biodegradation of xenobiotics. *Biotechnology*, *10*, 215–246.

Fiedler, H. P., Krastel, P., Müller, J., Gebhardt, K., & Zeeck, A. (2001). Enterobactin: The characteristic catecholate siderophore of Enterobacteriaceae is produced by Streptomyces species. *FEMS Microbiology Letters*, *196*(2), 147–151. doi:10.1111/j.1574-6968.2001.tb10556.x PMID:11267771

Fiedurek, J., Trytek, M., & Szczodrak, J. (2017). Strain improvement of industrially important microorganisms based on resistance to toxic metabolites and abiotic stress. *Journal of Basic Microbiology*, *57*(6), 445–459. doi:10.1002/jobm.201600710 PMID:28370185

Fierer, N. (2017). Embracing the unknown: Disentangling the complexities of the soil microbiome. *Nature Reviews. Microbiology*, *15*(10), 579–590. doi:10.1038/nrmicro.2017.87 PMID:28824177

Fingas, M. F., Fieldhouse, B., Sigouin, L., Wang, Z., & Mullin, J. V. (2001). Dispersant effectiveness testing: laboratory studies of fresh and weathered oils. *Proceedings of the 24th Arctic and Marine Oil spill Program Technical Seminar*, 551-566.

Finlay, R. D., Ek, H., Odham, G., & Söderström, B. (1990). Mycelial uptake, translocation and assimilation of 15N-labelled nitrogen by ectomycorrhizal Pinus sylvestris plants. *Agriculture, Ecosystems & Environment*, *28*(1-4), 133–137. doi:10.1016/0167-8809(90)90028-C

Firáková, S., Šturdíková, M., & Múčková, M. (2007). Bioactive secondary metabolites produced by microorganisms associated with plants. *Biologia.*, *62*(3), 251–257. doi:10.247811756-007-0044-1

Fischer, A. R., Werner, P., & Goss, K.-U. (2011). Photodegradation of malachite green and malachite green carbinol under irradiation with different wavelength ranges. *Chemosphere*, *82*(2), 210–214. doi:10.1016/j.chemosphere.2010.10.019 PMID:21035831

Fisk, M. C., Ruether, K. F., & Yavitt, J. B. (2003). Microbial activity and functional composition among northern peatland ecosystems. *Soil Biology & Biochemistry*, *35*(4), 591–602. doi:10.1016/S0038-0717(03)00053-1

Fiyadh, S. S., AlSaadi, M. A., Jaafar, W. Z., AlOmar, M. K., Fayaed, S. S., Mohd, N. S., Hin, L. S., & El-Shafie, A. (2019). Review on heavy metal adsorption processes by carbon nanotubes. *Journal of Cleaner Production*, *230*, 783–793. doi:10.1016/j.jclepro.2019.05.154

Flemming, H.-C., Wingender, J., Szewzyk, U., Steinberg, P., Rice, S. A., & Kjelleberg, S. (2016). Biofilms: An emergent form of bacterial life. *Nature Reviews. Microbiology*, *14*(9), 563–575. doi:10.1038/nrmicro.2016.94 PMID:27510863

Flint, M. L., & Dreistadt, S. H. (1998). *Natural Enemies Handbook: The Illustrated Guide to Biological Pest Control* (J. K. Clark, Ed.). University of California Press.

Flores, H. E., Vivanco, J. M., & Loyola-Vargas, V. M. (1999). Radicle biochemistry: The biology of root-specific metabolism. *Trends in Plant Science*, *4*(6), 220–226. doi:10.1016/S1360-1385(99)01411-9 PMID:10366878

Flores, O., Alcaíno, J., Fernandez-Lobato, M., Cifuentes, V., & Baeza, M. (2015). Characterization of virus-like particles and identification of capsid proteins in *Xanthophyllomyces dendrorhous*. *Virus Genes*, *50*(2), 253–259. doi:10.100711262-015-1171-3 PMID:25663143

Folch, A., Vilaplana, M., Amado, L., Vicent, R., & Caminal, G. (2013). Fungal permeable reactive barrier to remediate groundwater in an artificial aquifer. *Journal of Hazardous Materials*, *262*, 554–560. doi:10.1016/j.jhazmat.2013.09.004 PMID:24095995

Fomina, M., & Gadd, G. M. (2014). Biosorption: Current perspectives on concept, definition and application. *Bioresource Technology*, *160*, 3–14. doi:10.1016/j.biortech.2013.12.102 PMID:24468322

Fondi, M., Orlandini, V., Emiliani, G., Papaleo, M. C., Maida, I., Perrin, E., ... Fani, R. (2012). *Draft genome sequence of the hydrocarbon-degrading and emulsan-producing strain* Acinetobacter venetianus *RAG-1T*. Academic Press.

Forgacs, E., Cserháti, T., & Oros, G. (2004). Removal of synthetic dyes from wastewaters: A review. *Environment International*, *30*(7), 953–971. doi:10.1016/j.envint.2004.02.001 PMID:15196844

Forlani, G. M., Mantelli, M., & Nielsen, E. (1999). Biochemical evidence for multiple acetoin-forming enzymes in cultured plant cells. *Phytochemistry*, *50*(2), 255–262. doi:10.1016/S0031-9422(98)00550-0

Forootanfar, H., Moezzi, A., Aghaie-Khozani, M., Mahmoudjanlou, Y., Ameri, A., Niknejad, F., & Faramarzi, M. A. (2012). Synthetic dye decolorization by three sources of fungal laccase. *Iranian Journal of Environmental Health Sciences & Engineering*, *9*(1), 27. doi:10.1186/1735-2746-9-27 PMID:23369690

Fox, J. E., Starcevic, M., Kow, K. Y., Burow, M. E., & McLachlan, J. A. (2001). Endocrine disrupters and flavonoid signalling. *Nature*, *413*(6852), 128–129. doi:10.1038/35093163 PMID:11557969

Francis, D. (2011). A commentary on the G2/M transition of the plant cell cycle. *Annals of Botany*, *107*(7), 1065–1070. doi:10.1093/aob/mcr055 PMID:21558458

Frascari, D., Zanaroli, G., & Danko, A. S. (2015). *In situ* aerobic cometabolism of chlorinated solvents: A review. *Journal of Hazardous Materials*, *283*, 382–399. doi:10.1016/j.jhazmat.2014.09.041 PMID:25306537

Frick, C. M., Farrell, R. E., & Germida, J. J. (1999). *Assessment of phytoremediation as an in-situ technique for cleaning oil-contaminated sites. Petroleum Technology Alliance of Canada*. PTAC.

Friedel, J. K., Langer, T., Siebe, C., & Stahr, K. (2000). Effects of long-term waste water irrigation on soil organic matter, soil microbial biomass and its activities in central Mexico. *Biology and Fertility of Soils*, *31*(5), 414–421. doi:10.1007003749900188

Friedlova, M. (2010). The influence of heavy metals on soil biological and chemical properties. *Soil and Water Research*, *5*(1), 21–27. doi:10.17221/11/2009-SWR

Frutos, F. J. G., Escolano, O., Garcı'a, S., Mar Babın, M., & Ferna'ndez, M. D. (2010). Bioventing remediation and ecotoxicity evaluation of phenanthrene-contaminated soil. *Journal of Hazardous Materials*, *183*(1-3), 806–813. doi:10.1016/j.jhazmat.2010.07.098 PMID:20800967

Frutos, F. J. G., Pe'rez, R., Escolano, O., Rubio, A., Gimeno, A., Fernandez, M. D., Carbonell, G., Perucha, C., & Laguna, J. (2012). Remediation trials for hydrocarbon-contaminated sludge from a soil washing process: Evaluation of bioremediation technologies. *Journal of Hazardous Materials*, *199*, 262–271. doi:10.1016/j.jhazmat.2011.11.017 PMID:22118850

Fulekar, M. H. (2009). Bioremediation of fenvalerate by Pseudomonas aeruginosa in a scale upbioreactor. *Romanian Biotechnological Letters*, *14*(6), 4900–4905.

Fuller, S., & Gautam, A. (2016). A procedure for measuring microplasticsusing pressurized fluid extraction. *Environmental Science & Technology*, *50*(11), 5774–5780. doi:10.1021/acs.est.6b00816 PMID:27172172

Funhoff, E. G., & Van Beilen, J. B. (2007). Alkane activation by P450 oxygenases. *Biocatalysis and Biotransformation*, *25*(2-4), 186–193. doi:10.1080/10242420701379254

Fu, Q., Liao, C., Du, X., Schlenk, D., & Gan, J. (2018). Back conversion from product to parent: Methyl triclosan to triclosan in plants. *Environmental Science & Technology Letters*, *5*(3), 181–185. doi:10.1021/acs.estlett.8b00071

Fuqua, C., Burbea, M., & Winans, S. C. (1995). Activity of the Agrobacterium Ti plasmid conjugal transfer regulator TraR is inhibited by the product of the traM gene. *Journal of Bacteriology*, *177*(5), 1367–1373. doi:10.1128/JB.177.5.1367-1373.1995 PMID:7868612

Fu, Y., & Viraraghavan, T. (2001). Fungal decolorization of dye wastewaters: A review. *Bioresource Technology*, *79*(3), 251–262. doi:10.1016/S0960-8524(01)00028-1 PMID:11499579

Gadd, G. M. (2007). Geomycology: Biogeochemical transformations of rocks, minerals, metals and radionuclides by fungi, bioweathering and bioremediation. *Mycological Research*, *11*(1), 3–49. doi:10.1016/j.mycres.2006.12.001 PMID:17307120

Gadd, G. M. (2010). Metals, minerals and microbes: Geomicrobiology and bioremediation. *Microbiology*, *156*(3), 609–643. doi:10.1099/mic.0.037143-0 PMID:20019082

Gadhave, K. R., Hourston, J. E., & Gange, A. C. (2016). Developing soil microbial inoculants for pest management: Can one have too much of a good thing? *Journal of Chemical Ecology*, *42*(4), 348–356. doi:10.100710886-016-0689-8 PMID:27059329

Galdames, A., Mendoza, A., Orueta, M., de Soto García, I. S., Sánchez, M., Virto, I., & Vilas, J. L. (2017). Development of new remediation technologies for contaminated soils based on the applicationof zero-valent iron nanoparticles and bioremediation with compost. *Resour Effic Technol.*, *3*(2), 166–176. doi:10.1016/j.reffit.2017.03.008

Galgani, F., Fleet, J., Van Franeker, S., Kastanevakis, T., Maes, J., Mouat, L., & Oosterban, I. (2010). Marine Strategy Framework Directive. Task Group 10 Report. Marine Litter; EU Commission, Joint Research Centre.

Gallego, J. L., Loredo, J., Llamas, J., Vázquez, F., & Sánchez, J. (2001). Bioremediation of diesel-contaminated soils: Evaluation of potential in situ techniques by study of bacterial degradation. *Biodegradation*, *12*(5), 325–335. doi:10.1023/A:1014397732435 PMID:11995826

Gallon, J. R. (2001). N_2 fixation in phototrophs: Adaptation to a specialized way of life. *Plant and Soil*, *230*(1), 39–48. doi:10.1023/A:1004640219659

Gamalero, E., Berta, G., Massa, N., Glick, B. R., & Lingua, G. (2008). Synergistic interactions between the ACC deaminase-producing bacterium Pseudomonas putida UW4 and the AM fungus Gigasporarosea positively affect cucumber plant growth. *FEMS Microbiology Ecology*, *64*(3), 459–467. doi:10.1111/j.1574-6941.2008.00485.x PMID:18400004

Gamalero, E., Martinotti, M. G., Trotta, A., Lemanceau, P., & Berta, G. (2002). Morphogenetic modifications induced by Pseudomonas fluorescens A6RI and Glomus mosseae BEG12 in the root system of tomato differ according to plant growth conditions. *The New Phytologist*, *155*(2), 293–300. doi:10.1046/j.1469-8137.2002.00460.x

Game, B. C., Ilhe, B. M., Pawar, V. S., & Khandagale, P. P. (2020). Effect of Azotobacter, Phosphate Solubilising Bacteria and Potash Mobilising Bacteria Inoculants on Productivity of Wheat (Triticum aestivum L.). *International Journal of Current Microbiology and Applied Sciences*, *9*(3), 2800–2807. doi:10.20546/ijcmas.2020.903.322

Gandhi, N. U., & Chandra, S. B. (2012). A comparative analysis of three classes of bacterial non-specific acid phosphatases and archaeal phosphoesterases: Evolutionary perspective. *Acta Informatica Medica*, *20*(3), 167. doi:10.5455/aim.2012.20.167-173 PMID:23322973

Gandini, A. (2008). Polymers from renewable resources: A challenge for the future of macromolecular materials. *Macromolecules*, *41*(24), 9491–9504. doi:10.1021/ma801735u

Gangola, S., Joshi, S., Kumar, S., & Pandey, S. C. (2019). Comparative analysis of fungal and bacterial enzymes in biodegradation of xenobiotic compounds. *Smart Bioremediation Technologies: Microbial Enzymes*, *10*, 169–189. doi:10.1016/B978-0-12-818307-6.00010-X

Ganzert, L., Lipski, A., Hubberten, H. W., & Wagner, D. (2011). The impact of different soil parameters on the community structure of dominant bacteria from nine different soils located on Livingston Island, South Shetland Archipelago, Antarctica. *FEMS Microbiology Ecology*, *76*(3), 476–491. doi:10.1111/j.1574-6941.2011.01068.x PMID:21314705

Gao, P., Qin, J., Li, D., & Zhou, S. (2018). Inhibitory effect and possible mechanism of a Pseudomonas strain QBA5 against gray mold on tomato leaves and fruits caused by Botrytis cinerea. *PLoS One*, *13*(1), e0190932. doi:10.1371/journal.pone.0190932 PMID:29320571

Garbisu, C., & Alkorta, I. (2001). Phytoextraction: A cost-effective plant-based technology for the removal of metals from the environment. *Bioresource Technology*, *77*(3), 229–236. doi:10.1016/S0960-8524(00)00108-5 PMID:11272009

García de Salamone, I. E., Hynes, R. K., & Nelson, L. M. (2001). Cytokinin production by plant growth promoting rhizobacteria and selected mutants. *Canadian Journal of Microbiology*, *47*(5), 404–411. doi:10.1139/w01-029 PMID:11400730

García-Delgado, C., Alfaro-Barta, I., & Eymar, E. (2015). Combination of biochar amendment and mycoremediation for polycyclic aromatic hydrocarbons immobilization and biodegradation in creosote-contaminated soil. *Journal of Hazardous Materials*, *285*, 259–266. doi:10.1016/j.jhazmat.2014.12.002 PMID:25506817

Garcia-Gonzalez, J., & Sommerfeld, M. (2016). Biofertilizer and biostimulant properties of the microalga Acutodesmus dimorphus. *Journal of Applied Phycology*, *28*(2), 1051–1061. doi:10.100710811-015-0625-2 PMID:27057088

Garnham, C. P., Gilbert, J. A., Hartman, C. P., Campbell, R. L., Laybourn-Parry, J., & Davies, P. L. (2008). A Ca2+-dependent bacterial antifreeze protein domain has a novel β-helical ice-binding fold. *The Biochemical Journal*, *411*(1), 171–180. doi:10.1042/BJ20071372 PMID:18095937

Gaspar, P., Carvalho, A. L., Vinga, S., Santos, H., & Neves, A. R. (2013). From physiology to systems metabolic engineering for the production of biochemicals by lactic acid bacteria. *Biotechnology Advances*, *31*(6), 764–788. doi:10.1016/j.biotechadv.2013.03.011 PMID:23567148

Gaur, N., Flora, G., Yadav, M., & Tiwari, A. (2014). A review with recent advancements on bioremediation-based abolition of heavy metals. *Environmental Science. Processes & Impacts*, *16*(2), 180–193. doi:10.1039/C3EM00491K PMID:24362580

Gaur, N., Narasimhulu, K., & y, P. S. (2018). Recent advances in the bio-remediation of persistent organic pollutants and its effect on environment. *Journal of Cleaner Production, 198*, 1602–1631. doi:10.1016/j.jclepro.2018.07.076

Gautam, S. P., Bundela, P. S., Pandey, A. K., Jamaluddin, M. K., Awasthi, M. K., & Sarsaiya, S. (2012). Diversity of cellulolytic microbes and the biodegradation of municipal solid waste by a potential strain. *International Journal of Microbiology, 2012*, 1–12. doi:10.1155/2012/325907 PMID:22518141

Gavrilescu, M. (2004). Removal of heavy metals from the environment by biosorption. *Engineering in Life Sciences, 4*(3), 219–232. doi:10.1002/elsc.200420026

Gayathri, M., Kumar, P. S., Prabha, A. M. L., & Muralitharan, G. (2015). In vitro regeneration of *Arachis hypogaea L.* and *Moringa oleifera Lam.* using extracellular phytohormones from *Aphanothece* sp. MBDU 515. *Algal Research, 7*, 100–105. doi:10.1016/j.algal.2014.12.009

Ge, B., Liu, B., Nwet, T. T., Zhao, W., Shi, L., & Zhang, K. (2016). Bacillus methylotrophicus strain NKG-1, isolated from Changbai Mountain, China, has potential applications as a biofertilizer or biocontrol agent. *PLoS One, 11*(11), e0166079. doi:10.1371/journal.pone.0166079 PMID:27832162

Gentili, F., & Jumpponen, A. (2006). Bacterial and Fungal Biofertilizers. Handbook of microbial biofertilizers, 1, 25-89.

Gentili, F. G., & Fick, J. (2017). Algal cultivation in urban wastewater: An efficient way to reduce pharmaceutical pollutants. *Journal of Applied Phycology, 29*(1), 255–262. doi:10.100710811-016-0950-0 PMID:28344390

Gentili, R., Ambrosini, R., Montagnani, C., Caronni, S., & Citterio, S. (2018). Effect of Soil pH on the Growth, Reproductive Investment and Pollen Allergenicity of Ambrosia artemisiifolia L. *Frontiers in Plant Science, 9*, 1335–1347. doi:10.3389/fpls.2018.01335 PMID:30294333

Genuini, M., Bidet, P., Benoist, J. F., Schlemmer, D., Lemaitre, C., Birgy, A., & Bonacorsi, S. (2019). ShiF acts as an auxiliary factor of aerobactin secretion in meningitis *Escherichia coli* strain S88. *BMC Microbiology, 19*(1), 298. doi:10.118612866-019-1677-2 PMID:31847813

Georgiou, D., Hatiras, J., & Aivasidis, A. (2005). Microbial immobilization in a two-stage fixed-bed-reactor pilot plant for on-site anaerobic decolorization of textile wastewater. *Enzyme and Microbial Technology, 37*(6), 597–605. doi:10.1016/j.enzmictec.2005.03.019

Gerhardt, K. E., Huang, X. D., Glick, B. R., Bruce, M., & Greenberg, B. M. (2009). Phytoremediation and rhizoremediation of organic soil contaminants: Potential and challenges. *Plant Science, 176*(1), 20–30. doi:10.1016/j.plantsci.2008.09.014

Germaine, K. J., Byrne, J., Liu, X., Keohane, J., Culhane, J., Lally, R. D., Kiwanuka, S., Ryan, D., & Dowling, D. N. (2015). Ecopiling: A combined phytoremediation and passive biopiling system for remediating hydrocarbon impacted soils at field scale. *Front. Plant Sci., 5*, 756. doi:10.3389/fpls.2014.00756 PMID:25601875

Germaine, K. J., Keogh, E., Ryan, D., & Dowling, D. N. (2009). Bacterial endophyte mediated naphthalene phytoprotection and phytoremediation. *Fems. Microbiol., 296*(2), 226–234. doi:10.1111/j.1574-6968.2009.01637.x PMID:19459954

Germanier, R. (Ed.). (2012). *Bacterial vaccines*. Academic Press.

Germinda, J. J., Frick, C. M., & Farrell, R. E. (2002). Phytoremediation of oil-contaminated soils. *Developments. Soil Science, 28B*, 169–186.

GESAMP. (2015). Sources, fate and effects of microplastics in the marine environment: a global assessment. In IMO/FAO/ UNESCOIOC/ UNIDO/WMO/IAEA/UN/UNEP/UNDP Joint Group of Experts on the Scientific Aspects of Marine Environmental Protection. Rep Stud GESAMP.

Ghabrial, S. A., Caston, J. R., Jiang, D., Nibert, M. L., & Suzuki, N. (2015). 50-plus years of fungal viruses. *Virology*, *479*, 356–368. doi:10.1016/j.virol.2015.02.034 PMID:25771805

Ghabrial, S. A., Soldevila, A. I., & Havens, W. M. (2002). Molecular genetics of the viruses infecting the plant pathogenic fungus *Helminthosporium victoriae*. In S. Tavantzis (Ed.), *Molecular Biology of Double-Stranded RNA: Concepts and Applications in Agriculture* (pp. 213–236). Forestry and Medicine.

Ghabrial, S., & Suzuki, N. (2009). Viruses of plant pathogenic fungi. *Annual Review of Phytopathology*, *47*(1), 353–384. doi:10.1146/annurev-phyto-080508-081932 PMID:19400634

Ghareeb, B. (2009). Genetic engineering and transgenesis: advantages, assessment of risks and ethics. *Al-Quds Open University Research Journal*.

Ghorbanpour, M., Hatami, M., & Khavazi, K. (2013). Role of plant growth promoting rhizobacteria on antioxidant enzyme activities and tropane alkaloid production of *Hyoscyamus niger* under water deficit stress. *Turkish Journal of Biology*, *37*(3), 350–360. doi:10.3906/biy-1209-12

Ghosh, M., & Singh, S. P. (2005). A review on phytoremediation of heavy metals and utilization of It's by products. *Aust J Energy Environ*, *6*(4), 214–231.

Ghosh, M., & Singh, S. P. (2005). Comparative uptake and phytoextraction study of soil induced chromium by accumulator and high biomass weed species. *Applied Ecology and Environmental Research*, *3*(2), 67–79. doi:10.15666/aeer/0302_067079

Ghosh, N. (2004). Promoting biofertilisers in Indian agriculture. *Economic and Political Weekly*, 5617–5625.

Ghosh, P. K., De, T. K., & Maiti, T. K. (2018). Role of ACC Deaminase as a Stress Ameliorating Enzyme of Plant Growth-Promoting Rhizobacteria Useful in Stress Agriculture: A Review. In V. Meena (Ed.), *Role of Rhizospheric Microbes in Soil*. Springer., doi:10.1007/978-981-10-8402-7_3

Ghosh, P. K., Maiti, T. K., Pramanik, K., Ghosh, S. K., Mitra, S., & De, T. K. (2018). The role of arsenic resistant Bacillus aryabhattai MCC3374 in promotion of rice seedlings growth and alleviation of arsenic phytotoxicity. *Chemosphere*, *211*, 407–419. doi:10.1016/j.chemosphere.2018.07.148 PMID:30077937

Ghosh, S. J. N., Penterman, R. D., Little, R., Chavez, R., & Glick, B. R. (2003). Three newly isolated plant growth-promoting bacilli facilitate the seedling growth of canola, Brassica campestris. *Plant Physiology and Biochemistry*, *41*(3), 277–281. doi:10.1016/S0981-9428(03)00019-6

Gianfreda, L., Sannino, F., Rao, M. A., & Bollag, J. M. (2003). Oxidative transformation of phenols in aqueous mixtures. *Water Research*, *37*(13), 3205–3215.

Gidarakos, E., & Aivalioti, M. (2007). Large scale and long term application of bioslurping: The case of a Greek petroleum refinery site. *Journal of Hazardous Materials*, *149*(3), 574–581. doi:10.1016/j.jhazmat.2007.06.110 PMID:17709182

Gilichinsky, D. A. (2002). Permafrost model of extraterrestrial habitat. In *Astrobiology* (pp. 125–142). Springer. doi:10.1007/978-3-642-59381-9_9

Giller, K. E., Andersson, J. A., Corbeels, M., Kirkegaard, J., Mortensen, D., Erenstein, O., & Vanlauwe, B. (2015). Beyond conservation agriculture. *Frontiers in Plant Science*, *6*, 870. doi:10.3389/fpls.2015.00870 PMID:26579139

Giller, K. E., Witter, E., & Mcgrath, S. P. (1998). Toxicity of heavy metals to microorganisms and microbial processes in agricultural soils: A review. *Soil Biology & Biochemistry*, *30*(10-11), 1389–1414. doi:10.1016/S0038-0717(97)00270-8

Gill, S. S., Cowles, E. A., & Pietrantonio, P. V. (1992). The mode of action of *Bacillus thuringiensis* endotoxins. *Annual Review of Entomology*, *37*(1), 615–634. doi:10.1146/annurev.en.37.010192.003151 PMID:1311541

Giovannetti, M., Fortuna, P., Citernesi, A. S., Morini, S., & Nuti, M. P. (2001). The occurrence of anastomosis formation and nuclear exchange in intact arbuscular mycorrhizal networks. *The New Phytologist*, *151*(3), 717–724. doi:10.1046/j.0028-646x.2001.00216.x

Girgis, M. G. Z. (2006). Response of wheat to inoculation with phosphate and potassium mobilizers and organic amendment. *Annals of Agricultural Sciences Shams Univ Cairo*, *51*(1), 85–100.

Giri, B., Prasad, R., Wu, Q. S., & Varma, A. (Eds.). (2019). *Biofertilizers for sustainable agriculture and environment* (Vol. 55). Springer. doi:10.1007/978-3-030-18933-4

Givskov, M., de Nys, R., Manefield, M., Gram, L., Maximilien, R., Eberl, L., Molin, S., Steinberg, P. D., & Kjelleberg, S. (1996). Eukaryotic interference with homoserine lactone-mediated prokaryotic signalling. *Journal of Bacteriology*, *178*(22), 6618–6622. doi:10.1128/JB.178.22.6618-6622.1996 PMID:8932319

Gizaw, B., Tsegay, Z., Tefera, G., Aynalem, E., Wassie, M., & Abatneh, E. (2017). Phosphate Solubilizing Fungi Isolated and Characterized from Teff Rhizosphere Soil Collected from North Showa and Gojam, Ethiopia. *J Fertil Pestic*, *8*(02), 180. doi:10.4172/2471-2728.1000180

Glazer, A. N., & Nikaido, H. (2007). *Microbial biotechnology: Fundamentals of applied Microbiology*. Cambridge University Press. doi:10.1017/CBO9780511811227

Glick, B., Patten, C., Holguin, G., & Penrose, D. (1999). Overview of plant growth-promoting bacteria. Biochemical and Genetic Mechanisms Used by Plant Growth Promoting Bacteria, 1-13.

Glick, B. R. (2010). Using soil bacteria to facilitate phytoremediation. *Biotechnology Advances*, *28*(3), 367–374. doi:10.1016/j.biotechadv.2010.02.001 PMID:20149857

Glick, B. R. (2012). Plant growth-promoting bacteria: Mechanisms and applications. *Scientifica*, *2012*, 1–15. doi:10.6064/2012/963401 PMID:24278762

Glick, B. R. (2018). Soil microbes and sustainable agriculture. *Pedosphere*, *28*(2), 167–169. doi:10.1016/S1002-0160(18)60020-7

Glick, B. R., Cheng, Z., Czarny, J., & Duan, J. (2013). Promotion of plant growth by ACC deaminase-producing soil bacteria. In *New perspectives and approaches in plant growth-promoting Rhizobacteria research* (pp. 329–339). Springer.

Glick, B. R., Patten, C. L., Holguin, G., & Penrose, D. M. (1999). *Biochemical and Genetic Mechanisms Used by Plant Growth Promoting Bacteria*. London Imperial College Press. doi:10.1142/p130

Gliński, J., Horabik, J., & Lipiec, J. (Eds.). (2011). *Encyclopedia of agrophysics* (pp. 264–267). Springer. doi:10.1007/978-90-481-3585-1

Gmach, M. R., & Cherubin, M. R. (2020). Processes that influence dissolved organic matter in the soil: A review. *Scientia Agrícola*, *77*(3), 1–10. doi:10.1590/1678-992X-2018-0164

Godfray, H. C. J., Beddington, J. R., Crute, I. R., Haddad, L., Lawrence, D., Muir, J. F., & Toulmin, C. (2010). Food security: The challenge of feeding 9 billion people. *Science*, *327*(5967), 812–818. doi:10.1126cience.1185383 PMID:20110467

Godheja, J., Modi, D. R., Kolla, V., Pereira, A. M., Bajpai, R., Mishra, M., Sharma, S. V., Sinha, K., & Shekhar, S. K. (2019). Environmental remediation: Microbial and nonmicrobial prospects. In *Microbial Interventions in Agriculture and Environment* (Vol. 1, pp. 379–409). Springer. doi:10.1007/978-981-13-8383-0_13

Goering, P. L., Aposhian, H. V., Mass, M. J., Cebrián, M., Beck, B. D., & Waalkes, M. P. (1999). The enigma of arsenic carcinogenesis: role of metabolism. *Toxicological Sciences: An Official Journal of the Society of Toxicology, 49*(1), 5-14.

Goglio, P., Smith, W. N., Grant, B. B., Desjardins, R. L., McConkey, B. G., Campbell, C. A., & Nemecek, T. (2015). Accounting for soil carbon changes in agricultural life cycle assessment (LCA): A review. *Journal of Cleaner Production, 104*, 23–39. doi:10.1016/j.jclepro.2015.05.040

Göhre, V., & Paszkowski, U. (2006). Contribution of the arbuscular mycorrhizal symbiosis to heavy metal phytoremediation. *Planta, 223*(6), 1115–1122. doi:10.100700425-006-0225-0 PMID:16555102

Goker, M., Scheuner, C., Klenk, H. P., Stielow, J. B., & Menzel, W. (2014). Codivergence of mycoviruses with their hosts. *PLoS One, 6*(7), 22252. doi:10.1371/journal.pone.0022252 PMID:21829452

Gola, D., Malik, A., Shaikh, Z. A., & Sreekrishnan, T. R. (2016). Impact of Heavy Metal Containing Wastewater on Agricultural Soil and Produce: Relevance of Biological Treatment. *Environ. Process., 3*(4), 1063–1080. doi:10.100740710-016-0176-9

Gold, M. H., Glenn, J. K., & Alic, M. (1988). Use of polymeric dyes in lignin biodegradation assays. *Biomass Part B: Lignin, Pectin, and Chitin, 181*, 74–78. doi:10.1016/0076-6879(88)61011-1

Goldman, S., Lammers, P., Berman, M., & Sanders-Loehr, J. (1983). Siderophore-mediated iron uptake in different strains of *Anabaena* sp. *Journal of Bacteriology, 156*(3), 1144–1150. doi:10.1128/JB.156.3.1144-1150.1983 PMID:6227608

Goldstein, A. H. (1994). Involvement of the quinoprotein glucose dehydrogenase in the solubilization of exogenous mineral phosphates by gram-negative bacteria. In A. Torriani-Gorni, E. Yagil, & S. Silver (Eds.), *Phosphate in microorganisms: Cellular and molecular biology* (pp. 197–203). ASM Press.

Golovchenko, A. V., Tikhonova, E. Y., & Zvyagintsev, D. G. (2007). Abundance, biomass, structure, and activity of the microbial complexes of minerotrophic and ombrotrophic peatlands. *Microbiology, 76*(5), 630–637. doi:10.1134/S0026261707050177 PMID:18069333

Gomes, H. I., Dias-Ferreira, C., & Ribeiro, A. B. (2013). Overview of in situ and ex situ remediation technologies for PCB-contaminated soils and sediments and obstacles for full-scale application. *The Science of the Total Environment, 445-446*, 237–260. doi:10.1016/j.scitotenv.2012.11.098 PMID:23334318

Gong, X., Huang, D., Liu, Y., Peng, Z., Zeng, G., Xu, P., Cheng, M., Wang, R., & Wan, J. (2018). Remediation of contaminated soils by biotechnology with nanomaterials: Bio-behavior, applications, and perspectives. *Critical Reviews in Biotechnology, 38*(3), 455–468. doi:10.1080/07388551.2017.1368446 PMID:28903604

Gonzalez, C. F., Aekerley, D. F., Lynch, S. V., & Matin, A. (2005). ChrR, a soluble quinone reductase of *Pseudomonas putida* that defends against H_2O_2. *The Journal of Biological Chemistry, 280*(24), 22590–22595. doi:10.1074/jbc.M501654200 PMID:15840577

Gopalakrishnan, S., Sathya, A., Vijayabharathi, R., Varshney, R.K., Gowda, C.L.L., & Krishnamurthy, L. (2015). Plant growth promoting rhizobia Challenges and opportunities. *Biotech, 5*(4), 355-377.

Gopalakrishnan, S., Srinivas, V., Alekhya, G., & Prakash, B. (2015). Effect of plant growth-promoting *Streptomyces sp.* on growth promotion and grain yield in chickpea (Cicer arietinum L). *Biotech, 5*(5), 799-806.

Gopalakrishnan, S., Srinivas, V., Prakash, B., Sathya, A., & Vijayabharathi, R. (2015). Plant growth-promoting traits of Pseudomonas geniculata isolated from chickpea nodules. *Biotech, 5*(5), 653-661.

Gopalakrishnan, S., Srinivas, V., Prakash, B., Sathya, A., & Vijayabharathi, R. (2015). Plant growth-promoting traits of Pseudomonas geniculata isolated from chickpea nodules. *Biotech, 5*(5), 653–661. . doi:10.100713205-014-0263-4

Gopalakrishnan, S. V., Srinivas, V., Vemula, A. K., Samineni, S., & Rathore, A. (2018). Influence of diazotrophic bacteria on nodulation, nitrogen fixation, growth promotion and yield traits in five cultivars of chickpea. *Biocatalysis and Agricultural Biotechnology*, *15*, 35–42. doi:10.1016/j.bcab.2018.05.006

Gopalakrishnan, S., Srinivas, V., & Samineni, S. (2017). Nitrogen fixation, plant growth and yield enhancements by diazotrophic growth-promoting bacteria in two cultivars of chickpea. (Cicero arietinum L.). *Biocatalysis and Agricultural Biotechnology*, *11*, 116–123. doi:10.1016/j.bcab.2017.06.012

Gornall, J., Betts, R., Burke, E., Clark, R., Camp, J., Willett, K., & Wiltshire, A. (2010). Implications of climate change for agricultural productivity in the early twenty-first century. *Philosophical Transactions of the Royal Society of London. Series B, Biological Sciences*, *365*(1554), 2973–2989. doi:10.1098/rstb.2010.0158 PMID:20713397

Gorzelak, M. A., Asay, A. K., Pickles, B. J., & Simard, S. W. (2015). Inter-plant communication through mycorrhizal networks mediates complex adaptive behaviour in plant communities. *AoB Plants*, *7*, 7. doi:10.1093/aobpla/plv050 PMID:25979966

Goswami, D., Thakker, J. N., & Dhandhukia, P. C. (2016). Portraying mechanics of plant growth promoting rhizobacteria (PGPR): A review. *Cogent Food & Agriculture*, *2*(1), 1127500. doi:10.1080/23311932.2015.1127500

Götz, C., Fekete, A., Gebefuegi, I., Forczek, S. T., Fuksová, K., Li, X., Englmann, M., Gryndler, M., Hartmann, A., Matucha, M., Schmitt-Kopplin, P., & Schröder, P. (2007). Uptake, degradation and chiral discrimination of N-acyl-D/L-homoserine lactones by barley (Hordeum vulgare) and yam bean (Pachyrhizuserosus) plants. *Analytical and Bioanalytical Chemistry*, *389*(5), 1447–1457. doi:10.100700216-007-1579-2 PMID:17899036

Gouda, S., Kerry, R. G., Samal, D., Mahapatra, G. P., Das, G., & Patra, J. K. (2018). Application of Plant Growth Promoting Rhizobacteria in Agriculture. *Advances in Microbial Biotechnology,* 74-86.

Gouda, S., Kerry, R. G., Das, G., Paramithiotis, S., Shin, H. S., & Patra, J. K. (2018). Revitalization of plant growth promoting rhizobacteria for sustainable development in agriculture. *Microbiological Research*, *206*, 131–140. doi:10.1016/j.micres.2017.08.016 PMID:29146250

Gouka, R. J., Punt, P. J., & Van Den Hondel, C. A. M. J. J. (1997). Efficient production of secreted proteins by *Aspergillus*: Progress, limitations and prospects. *Applied Microbiology and Biotechnology*, *47*(1), 1–11. doi:10.1007002530050880 PMID:9035405

Goulding, K. W. T. (2016). Soil acidification and the importance of liming agricultural soils with particular reference to the United Kingdom. *Soil Use and Management*, *32*(3), 390–399. doi:10.1111um.12270 PMID:27708478

Govindasamy, V., Senthilkumar, M., Gaikwad, K., & Annapurna, K. (2008). Isolation and characterization of ACC deaminase gene from two plant growth-promoting rhizobacteria. *Current Microbiology*, *57*(4), 312–317. doi:10.100700284-008-9195-8 PMID:18654819

Gowda, V., & Shivakumar, S. (2014). Agrowaste-based Polyhydroxyalkanoate (PHA) production using hydrolytic potential of *Bacillus thuringiensis* IAM 12077. *Brazilian Archives of Biology and Technology*, *57*(1), 55–61. doi:10.1590/S1516-89132014000100009

Goyer, R. A. (2001). Toxic effects of metals. In C. D. Klaassen (Ed.), *Cassarett and Doull's Toxicology: The Basic Science of Poisons* (pp. 811–867). McGraw-Hill Publisher.

Graham, D. E., Wallenstein, M. D., Vishnivetskaya, T. A., Waldrop, M. P., Phelps, T. J., Pfiffner, S. M., ... Elias, D. A. (2012). Microbes in thawing permafrost: The unknown variable in the climate change equation. *The ISME Journal*, *6*(4), 709–712. doi:10.1038/ismej.2011.163 PMID:22094350

Graham, P. H. (1992). Stress tolerance in Rhizobium and Bradyrhizobium, and nodulation under adverse soil conditions. *Canadian Journal of Microbiology*, *38*(6), 475–484. doi:10.1139/m92-079

Graham, P. H., & Vance, C. P. (2000). Nitrogen fixation in perspective: An overview of research and extension needs. *Field Crops Research*, *65*(2-3), 93–106. doi:10.1016/S0378-4290(99)00080-5

Graham, P. H., & Vance, C. P. (2003). Legumes: Importance and constraints to greater use. *Plant Physiology*, *131*(3), 872–877. doi:10.1104/pp.017004 PMID:12644639

Grandclément, C., Tannières, M., Moréra, S., Dessaux, Y., & Faure, D. (2016). Quorum quenching: Role in nature and applied developments. *FEMS Microbiology Reviews*, *40*(1), 86–116. doi:10.1093/femsre/fuv038 PMID:26432822

Grasserová, A., Hanc, A., Innemanová, P., & Cajthaml, T. (2020). Composting and vermicomposting used to break down and remove pollutants from organic waste: A mini review. *European Journal of Environmental Sciences*, *10*(1), 9–14. doi:10.14712/23361964.2020.2

Grass, G., Fricke, B., & Nies, D. H. (2005). Control of expression of a periplasmic nickel efflux pump by periplasmic nickel concentrations. *Biometals*, *18*(4), 437–448. doi:10.100710534-005-3718-6 PMID:16158236

Grass, G., Große, C., & Nies, D. H. (2000). Regulation of the cnr cobalt and nickel resistance determinant from *Ralstonia* sp. strain CH34. *Journal of Bacteriology*, *182*(5), 1390–1398. doi:10.1128/JB.182.5.1390-1398.2000 PMID:10671463

Gray, E. J., & Smith, D. L. (2005). Intracellular and extracellular PGPR: Commonalities and distinctions in the plant-bacterium signaling processes. *Soil Biology & Biochemistry*, *37*(3), 395–412. doi:10.1016/j.soilbio.2004.08.030

Gregory, M. R. (2009). Environmental implications of plastic debris in marine settings–entanglement, ingestion, smothering, hangers-on, hitch-hiking and alien invasions. *Philosophical Transactions of the Royal Society of London. Series B, Biological Sciences*, *364*(1526), 2013–2025. doi:10.1098/rstb.2008.0265 PMID:19528053

Gregson, B. H., Metodieva, G., Metodiev, M. V., Golyshin, P. N., & McKew, B. A. (2018). Differential protein expression during growth on medium versus long-Chain alkanes in the obligate marine hydrocarbon-degrading bacterium *Thalassolituus oleivorans* MIL-1. *Frontiers in Microbiology*, *9*, 3130. doi:10.3389/fmicb.2018.03130 PMID:30619200

Grobelak, A., Kokot, P., Hutchison, D., Grosser, A., & Kacprzak, M. (2018). Plant growth-promoting rhizobacteria as an alternative to mineral fertilizers in assisted bioremediation-sustainable land and waste management. *Journal of Environmental Management*, *227*, 1–9. doi:10.1016/j.jenvman.2018.08.075 PMID:30170232

Grotz, N., & Guerinot, M. L. (2006). Molecular aspects of Cu, Fe and Zn homeostasis in plants. *Biochimica et Biophysica Acta (BBA)-. Molecular Cell Research*, *1763*(7), 595–608.

Grzesik, M., Romanowska-Duda, Z., & Kalaji, H. M. (2017). Effectiveness of cyanobacteria and green algae in enhancing the photosynthetic performance and growth of willow (*Salix viminalis L.*) plants under limited synthetic fertilizers application. *Photosynthetica*, *55*(3), 1–12. doi:10.100711099-017-0716-1

Guan, Z.-B., Luo, Q., Wang, H.-R., Chen, Y., & Liao, X.-R. (2018). Bacterial laccases: Promising biological green tools for industrial applications. *Cellular and Molecular Life Sciences: CMLS*, *75*(19), 3569–3592. doi:10.100700018-018-2883-z PMID:30046841

Guauque-Torres, M. P., & Bustos, A. Y. (2019). Laccases for Soil Bioremediation. In A. Kumar & S. Sharma (Eds.), *Microbes and Enzymes in Soil Health and Bioremediation* (Vol. 16, pp. 165–209). Springer Singapore. doi:10.1007/978-981-13-9117-0_8

Guéguen, P., & Bard, P.-Y. (2005). Soil-Structure and Soil-Structure-Soil Interaction: Experimental Evidence at the Volvi Test Site. *Journal of Earthquake Engineering*, *9*(5), 657–693. doi:10.1080/13632460509350561

Guerrieri, M. C., Fanfoni, E., Fiorini, A., Trevisan, M., & Puglisi, E. (2020). Isolation and Screening of Extracellular PGPR from the Rhizosphere of Tomato Plants after Long-Term Reduced Tillage and Cover Crops. *Plants*, *9*(5), 668. doi:10.3390/plants9050668 PMID:32466288

Guinel, F. C. (2015). Ethylene, a hormone at the center-stage of nodulation. *Frontiers in Plant Science*, *6*, 1121. doi:10.3389/fpls.2015.01121 PMID:26834752

Gulati, A., Sharma, N., Vyas, P., Sood, S., Rahi, P., Pathania, V., & Prasad, R. (2010). Organic acid production and plant growth promotion as a function of phosphate solubilization by *Acinetobacter rhizosphaerae* strain BIHB 723 isolated from the cold deserts of the trans-Himalayas. *Archives of Microbiology*, *192*(11), 975–983. doi:10.100700203-010-0615-3 PMID:20821196

Gunther, T., Dornberger, U., & Fritsche, W. (1996). Effects of ryegrasss on biodegradation of hydrocarbons in soil. *Chemosphere*, *33*(2), 203–215. doi:10.1016/0045-6535(96)00164-6 PMID:8696773

Guo, H., Yang, Y., Liu, K., Xu, W., Gao, J., Duan, H., Du, B., Ding, Y., & Wang, C. (2016). Comparative Genomic Analysis of *Delftia tsuruhatensis* MTQ3 and the Identification of Functional NRPS Genes for Siderophore Production. *BioMed Research International*, *2016*, 3687619. doi:10.1155/2016/3687619 PMID:27847812

Guo, K., & Liu, X. (2019). Effect of initial soil water content and bulk density on the infiltration and desalination of melting saline ice water in coastal saline soil. *European Journal of Soil Science*, *70*(6), 1249–1266. doi:10.1111/ejss.12816

Guo, R. F., Yuan, G. F., & Wang, Q. M. (2013). Effect of NaCl treatments on glucosinolate metabolism in broccoli sprouts. *Journal of Zhejiang University. Science. B.*, *14*(2), 124–131. doi:10.1631/jzus.B1200096 PMID:23365011

Gupta, G., Parihar, S. S., Ahirwar, N. K., Snehi, S. K., & Singh, V. (2015). Plant growth promoting rhizobacteria (PGPR): current and future prospects for development of sustainable agriculture. *Journal of Microbial and Biochemical Technology, 7*(2), 96-102.

Gupta, A., Gopal, M., Thomas, G. V., Manikandan, V., Gajewski, J., Thomas, G., Seshagiri, S., Schuster, S. C., Rajesh, P., & Gupta, R. (2014). Whole genome sequencing and analysis of plant growth promoting bacteria isolated from the rhizosphere of plantation crops coconut, cocoa and arecanut. *PLoS One*, *9*(8), e104259. doi:10.1371/journal.pone.0104259 PMID:25162593

Gupta, A., Patel, A. K., Gupta, D., Singh, G., & Mishra, V. K. (2020). Rhizospheric remediation of organic pollutants from the soil; a green and sustainable technology for soil clean up. In *Abatement of Environmental Pollutants* (pp. 263–286). Elsevier. doi:10.1016/B978-0-12-818095-2.00013-8

Gupta, D. K., Huang, H. G., & Corpas, F. J. (2013). Lead tolerance in plants: Strategies for phytoremediation. *Environmental Science and Pollution Research International*, *20*(4), 2150–2161. doi:10.100711356-013-1485-4 PMID:23338995

Gupta, P., & Diwan, B. (2017). Bacterial exopolysaccharide mediated heavy metal removal: A review on biosynthesis, mechanism and remediation strategies. *Biotechnology Reports (Amsterdam, Netherlands)*, *13*, 58–71. doi:10.1016/j.btre.2016.12.006 PMID:28352564

Gupta, S. K., & Shukla, P. (2017). Microbial platform technology for recombinant antibody fragment production: A review. *Critical Reviews in Microbiology*, *43*(1), 31–42. doi:10.3109/1040841X.2016.1150959 PMID:27387055

Gupta, S., & Dikshit, A. K. (2010). Biopesticides: An ecofriendly approach for pest control. *Journal of Biopesticides*, *3*, 186.

Gupta, S., & Pandey, S. (2019). ACC Deaminase Producing Bacteria With Multifarious Plant Growth Promoting Traits Alleviates Salinity Stress in French Bean (*Phaseolus vulgaris*) Plants. *Frontiers in Microbiology, 10*, 1506. doi:10.3389/fmicb.2019.01506 PMID:31338077

Gupta, V. K., Nayak, A., & Agarwal, S. (2015). Bioadsorbents for remediation of heavy metals: Current status and their future prospects. *Environmental Engineering Research, 20*(1), 1–18. doi:10.4491/eer.2015.018

Gupta, V., Moradi, O., Tyagi, I., Agarwal, S., Sadegh, H., Shahryari-Ghoshekandi, R., Makhlouf, A., Goodarzi, M., & Garshasbi, A. (2016). Study on the removal of heavy metal ions from industry waste by carbon nanotubes: effect of the surface modification: A review. *Critical Reviews in Environmental Science and Technology, 46*(2), 93–118. doi:10.1080/10643389.2015.1061874

Gupta, V., Ratha, S. K., Sood, A., Chaudhary, V., & Prasanna, R. (2013). New insights into the biodiversity and applications of cyanobacteria (blue-green algae)-Prospects and challenges. *Algal Research, 2*(2), 79–97. doi:10.1016/j.algal.2013.01.006

Gürses, A., Acikyildiz, M., Günes, K., & Gürses, M.S. (2016). Dyes and pigments VIII. *Springer Brief in Green Chemistry for Sustainability, 83*, 23.

Gusain, Y. S., Singh, U. S., & Sharma, A. K. (2015). Bacterial mediated amelioration of drought stress in drought tolerant and susceptible cultivars of rice (*Oryza sativa* L.). *African Journal of Biotechnology, 14*(9), 764–773. doi:10.5897/AJB2015.14405

Gwinn, M. R., Johns, D. O., Bateson, T. F., & Guyton, K. Z. (2011). A review of the genotoxicity of 1, 2-dichloroethane (EDC). *Mutation Research/Reviews in Mutation Research, 727*(1-2), 42–53. doi:10.1016/j.mrrev.2011.01.001 PMID:21255676

Gyaneshwar, P., Naresh Kumar, G., Parekh, L. J., & Poole, P. S. (2002). Role of soil microorganisms in improving P nutrition of plants. *Plant and Soil, 245*(1), 83–93. doi:10.1023/A:1020663916259

Haas, D., & Défago, G. (2005). Biological control of soil-borne pathogens by fluorescent pseudomonads. *Nature Reviews. Microbiology, 3*(4), 307–319. doi:10.1038/nrmicro1129 PMID:15759041

Haba, E., Pinazo, A., Jauregui, O., Espuny, M. J., Infante, M. R., & Manresa, A. (2003). Physicochemical characterization and antimicrobial properties of rhamnolipids produced by *Pseudomonas aeruginosa* 47T2 NCBIM 40044. *Biotechnology and Bioengineering, 81*(3), 316–322. doi:10.1002/bit.10474 PMID:12474254

Habibi, S., Djedidi, S., Ohkama-Ohtsu, N., Sarhadi, W. A., Kojima, K., Rallos, R. V., Ramirez, M., Yamaya, H., Sekimoto, H., & Yokoyama, T. (2019). Isolation and Screening of Indigenous Plant Growth-promoting Rhizobacteria from Different Rice Cultivars in Afghanistan Soils. *Microbes and Environments, 34*(4), 347–355. doi:10.1264/jsme2.ME18168 PMID:31527341

Hadad, D., Geresh, S., & Sivan, A. (2005). Biodegradation of polyethylene by the thermophilic bacterium *Brevibacillus borstelensis*. *Journal of Applied Microbiology, 98*(5), 1093–1100. doi:10.1111/j.1365-2672.2005.02553.x PMID:15836478

Haddock, B. A., & Jones, C. W. (1977). Bacterial respiration. *Bacteriological Reviews, 41*(1), 47–99. doi:10.1128/BR.41.1.47-99.1977 PMID:140652

Hadi, F., Mousavi, A., Noghabi, K. A., Tabar, H. G., & Salmanian, A. H. (2013). New bacterial strain of the genus *Ochrobactrum* with glyphosate-degrading activity. *Journal of Environmental Science and Health. Part B, Pesticides, Food Contaminants, and Agricultural Wastes, 48*(3), 208–213. doi:10.1080/03601234.2013.730319 PMID:23356342

Hafeez, F. Y., Yasmin, S., Ariani, D., Zafar, Y., & Malik, K. A. (2006). *Plant growth-promoting bacteria as biofertilizer*. Academic Press.

Hage, J. C., & Hartmans, S. (1999). Monooxygenase-mediated 1, 2-dichloroethane degradation by Pseudomonas sp. strain DCA1. *Applied and Environmental Microbiology, 65*(6), 2466–2470. doi:10.1128/AEM.65.6.2466-2470.1999 PMID:10347028

Hagerberg, D., Thelin, G., & Wallander, H. (2003). The production of ectomycorrhizal mycelium in forests: Relation between forest nutrient status and local mineral sources. *Plant and Soil, 252*(2), 279–290. doi:10.1023/A:1024719607740

Haidar, B., Ferdous, M., Fatema, B., Ferdous, A. S., Islam, M. R., & Khan, H. (2018). Population diversity of bacterial endophytes from jute (*Corchorus olitorius*) and evaluation of their potential role as bioinoculants. *Microbiological Research, 208*, 43–53. doi:10.1016/j.micres.2018.01.008 PMID:29551211

Håkansson, I., & Lipiec, J. (2000). A review of the usefulness of relative bulk density values in studies of soil structure and compaction. *Soil & Tillage Research, 53*(2), 71–85. doi:10.1016/S0167-1987(99)00095-1

Hakeem, K. R., Sabir, M., Ozturk, M., Mermut, A. R., Surriya, O., Saleem, S. S., & Kazi, A. G. (2015). *Soil remediation and plants*. Soil Remediation. Plants.

Haling, R. E., Simpson, R. J., Culvenor, R. A., Lambers, H., & Richardson, A. E. (2011). Effect of soil acidity, soil strength and macropores on root growth and morphology of perennial grass species differing in acid-soil resistance. *Plant, Cell & Environment, 34*(3), 444–456. doi:10.1111/j.1365-3040.2010.02254.x PMID:21062319

Hall, D. O., Markov, S. A., Watanabe, Y., & Rao, K. K. (1995). The potential applications of cyanobacterial photosynthesis for clean technology. *Photosynthesis Research, 46*(1-2), 159–167. doi:10.1007/BF00020426 PMID:24301578

Hall, J., Soole, K., & Bentham, R. (2011). Hydrocarbon phytoremediation in the family fabacea—A review. *International Journal of Phytoremediation, 13*(4), 317–332. doi:10.1080/15226514.2010.495143 PMID:21598795

Hamed, S. M., Selim, S., Klöck, G., & AbdElgawad, H. (2017). Sensitivity of two green microalgae to copper stress: Growth, oxidative and antioxidants analyses. *Ecotoxicology and Environmental Safety, 144*, 19–25. doi:10.1016/j.ecoenv.2017.05.048 PMID:28599127

Hamid, A. A. A., Wong, E. L., Joyce-Tan, K. H., Shamsir, M. S., Hamid, T. H. T. A., & Huyop, F. (2013). Molecular modelling and functional studies of the non-stereospecific α-haloalkanoic acid Dehalogenase (DehE) from Rhizobium sp. RC1 and its association with 3-chloropropionic acid (β-chlorinated aliphatic acid). *Biotechnology, Biotechnological Equipment, 27*(2), 3725–3736. doi:10.5504/BBEQ.2012.0142

Hamlett, N. V., Landale, E. C., Davis, B. H., & Summers, A. O. (1992). Roles of the Tn21 merT, merP, and merC gene products in mercury resistance and mercury binding. *Journal of Bacteriology, 174*(20), 6377–6385. doi:10.1128/JB.174.20.6377-6385.1992 PMID:1328156

Hammel, K. E., Gai, W. Z., Green, B., & Moen, M. A. (1992). Oxidative degradation of phenanthrene by the ligninolytic fungus *Phanerochaete chrysosporium. Applied and Environmental Microbiology, 58*(6), 1832–1838. doi:10.1128/AEM.58.6.1832-1838.1992 PMID:1622259

Hamzah, A., Hapsari, R. I., & Wisnubroto, E. I. (2016). Phytoremediation of Cadmium-contaminated agricultural land using indigenous plants. *International Journal of Environmental & Agriculture Research, 2*, 8–14.

Han, B., Zhang, M., Zhao, D., & Feng, Y. (2015). Degradation of aqueous and soil-sorbed estradiol using a new class of stabilized manganese oxide nanoparticles. *Water Research, 70*, 288–299. doi:10.1016/j.watres.2014.12.017 PMID:25543239

Handelsman, J., Rondon, M. R., Brady, S. F., Clardy, J., & Goodman, R. M. (1998). Molecular biological access to the chemistry of unknown soil microbes: A new frontier for natural products. *Chemistry & Biology*, *5*(10), R245–R249. doi:10.1016/S1074-5521(98)90108-9 PMID:9818143

Handelsman, J., & Stabb, E. V. (1996). Bio-control of soil borne plant pathogens. *The Plant Cell*, *8*(10), 1855–1869. doi:10.2307/3870235 PMID:12239367

Hang, N. T. T., Oh, S. O., Kim, G. H., Hur, J. S., & Koh, Y. J. (2005). Bacillus subtilis S1-0210 as a biocontrol agent against Botrytis cinerea in strawberries. *The Plant Pathology Journal*, *21*(1), 59–63. doi:10.5423/PPJ.2005.21.1.059

Han, H. S., Supanjani, & Lee, K. D. (2006). Effect of co-inoculation with phosphate and potassium solubilizing bacteria on mineral uptake and growth of pepper and cucumber. *Plant, Soil and Environment*, *52*(3), 130–136. doi:10.17221/3356-PSE

Hansda, A., & Kumar, V. (2017). Cu-resistant Kocuria sp. CRB15: a potential PGPR isolated from the dry tailing of Rakha copper mine. *Biotech, 7*(2), 132.

Hansen, V., Bonnichsen, L., Nunes, I., Sexlinger, K., Lopez, S. R., van der Bom, F. J. T., & Jensen, L. S. (2020). Seed inoculation with *Penicillium bilaiae* and *Bacillus simplex* affects the nutrient status of winter wheat. *Biology and Fertility of Soils*, *56*(1), 97–109. doi:10.100700374-019-01401-7

Han, X., Yuan, R., Wang, G. Q., & Zhang, C. J. (2014). Isolation of paraquat degrading bacteria and identification of degradation characteristics. *Anhui Agricultural Science Bulletin*, *20*, 38–39.

Haque, F., Sajid, M., Cameotra, S. S., & Battacharyya, M. S. (2017). Anti-biofilm activity of a sophorolipid-amphotericin B niosomal formulation against *Candida albicans*. *Biofouling*, *33*(9), 768–779. doi:10.1080/08927014.2017.1363191 PMID:28946803

Hardoim, P. R., Van Overbeek, L. S., Berg, G., Pirttilä, A. M., Compant, S., Campisano, A., Döring, M., & Sessitsch, A. (2015). The hidden world within plants: Ecological and evolutionary considerations for defining functioning of microbial endophytes. *Microbiology and Molecular Biology Reviews*, *79*(3), 293–320. doi:10.1128/MMBR.00050-14 PMID:26136581

Hardoim, P. R., van Overbeek, L. S., & Elsas, J. D. (2008). Properties of bacterial endophytes and their proposed role in plant growth. *Trends in Microbiology*, *16*(10), 463–471. doi:10.1016/j.tim.2008.07.008 PMID:18789693

Harish, S., Kavino, M., Kumar, N., Balasubramanian, P., & Samiyappan, R. (2009a). Induction of defense-related proteins by mixtures of plant growth promoting endophytic bacteria against Banana bunchy top virus. *Biological Control*, *51*(1), 16–25. doi:10.1016/j.biocontrol.2009.06.002

Harish, S., Kavino, M., Kumar, N., & Samiyappan, R. (2009b). Differential expression of pathogenesis-related proteins and defense enzymes in banana: Interaction between endophytic bacteria, Banana bunchy top virus and Pentaloniani-gronervosa. *Biocontrol Science and Technology*, *19*(8), 843–857. doi:10.1080/09583150903145000

Harman, G. E. (2011). Trichoderma—Not just for biocontrol anymore. *Phytoparasitica*, *39*(2), 103–108. doi:10.100712600-011-0151-y

Hartemink, A. E. (2015). The use of soil classification in journal papers between 1975 and 2014. *Geoderma Regional*, *5*, 127–139. doi:10.1016/j.geodrs.2015.05.002

Hartwig, A., & Schwerdtle, T. (2002). Interactions by carcinogenic metal compounds with DNA repair processes: Toxicological implications. *Toxicology Letters*, *127*(1-3), 47–54. doi:10.1016/S0378-4274(01)00482-9 PMID:12052640

Harvey, L. J., & McArdle, H. J. (2008). Biomarkers of copper status: A brief update. *British Journal of Nutrition*, *99*(S3), S10–S13. doi:10.1017/S0007114508006806 PMID:18598583

Harvey, P. J., Campanella, B. F., Castro, P. M. L., Harms, H., Lichtfouse, E., Schäffner, A. R., Smrcek, S., & Werck-Reichhart, D. (2002). Phytoremediation of polyaromatic hydrocarbons, anilines and phenols. *Environmental Science and Pollution Research International*, *9*(1), 29–47. doi:10.1007/BF02987315 PMID:11885416

Hasan, F., Shah, A. A., & Hameed, A. (2006). Industrial applications of microbial lipases. *Enzyme and Microbial Technology*, *39*(2), 235–251. doi:10.1016/j.enzmictec.2005.10.016

Hasanuzzaman, M., Bhuyan, M. H. M., Nahar, K., Hossain, M., Mahmud, J. A., Hossen, M., Masud, A. A., & Fujita, M. (2018). Potassium: A vital regulator of plant responses and tolerance to abiotic stresses. *Agronomy (Basel)*, *8*(3), 31. doi:10.3390/agronomy8030031

Hashem, M. A. (2001). Problems and prospects of cyanobacterial biofertilizer for rice cultivation. *Australian Journal of Plant Physiology*, *28*, 881–888.

Hassaan & Ahmed. (2017). Health and Environmental Impacts of Dyes: Mini Review. *American Journal of Environmental Science and Engineering*, *1*(3), 64–67.

Hassaan, M. A. (2016). *Advanced oxidation processes of some organic pollutants in fresh and seawater* (PhD Thesis). Faculty of Science, Port Said University.

Hassan, M. M., & Carr, C. M. (2018). A critical review on recentadvancements of the removal of reactive dyes from dyehouseeffluent by ion-exchange adsorbents. *Chemosphere*, *209*(1), 201–219. doi:10.1016/j.chemosphere.2018.06.043 PMID:29933158

Hassan, M. U., Nayab, H., Rehman, T. U., Williamson, M. P., Haq, K. U., Shafi, N., & Shafique, F. (2020). Characterization of bacteriocins produced by *Lactobacillus spp.* isolated from the traditional Pakistani yoghurt and their antimicrobial activity against common foodborne pathogens. *BioMed Research International*.

Hassan, S. E. D., Boon, E., St-Arnaud, M., & Hijri, M. (2011). Molecular biodiversity of arbuscular mycorrhizal fungi in trace metal-polluted soils. *Molecular Ecology*, *20*(16), 3469–3483. doi:10.1111/j.1365-294X.2011.05142.x PMID:21668808

Hassan, T. U., Bano, A., & Naz, I. (2017). Alleviation of heavy metals toxicity by the application of plant growth promoting rhizobacteria and effects on wheat grown in saline sodic field. *International Journal of Phytoremediation*, *19*(6), 522–529. doi:10.1080/15226514.2016.1267696 PMID:27936865

Hassen, W., Neifar, M., Cherif, H., Najjari, A., Chouchane, H., Driouich, R. C., Salah, A., Naili, F., Mosbah, A., Souissi, Y., Raddadi, N., Ouzari, H. I., Fava, F., & Cherif, A. (2018). *Pseudomonas rhizophila* S211, a New Plant Growth-Promoting Rhizobacterium with Potential in Pesticide-Bioremediation. *Frontiers in Microbiology*, *9*, 34. doi:10.3389/fmicb.2018.00034 PMID:29527191

Hatvani, N., & Mecs, I. (2001). Production of Laccase and Manganese Peroxidase by *Lentinus edodes* on Malt Containing by Product of the Brewing Process. *Process Biochemistry*, *37*(5), 491–496. doi:10.1016/S0032-9592(01)00236-9

Haudecoeur, E., Planamente, S., Cirou, A., Tannières, M., Shelp, B. J., Moréra, S., & Faure, D. (2009). Proline antagonizes GABA-induced quenching of quorum-sensing in Agrobacterium tumefaciens. *Proceedings of the National Academy of Sciences of the United States of America*, *106*(34), 14587–14592. doi:10.1073/pnas.0808005106 PMID:19706545

Havugimana, E., Bhople, B. S., Kumar, A., Byiringiro, E., Mugabo, J. P., & Kumar, A. (2015). Soil pollution–major sources and types of soil pollutants. *Environmental Science and Engineering, 11*, 53-86.

Head, I., Singleton, I., & Milner, M. (2003). *Bioremediation; a critical review*. Caister Academic Press.

He, D., Luo, Y., Lu, S., Liu, M., Song, Y., & Lei, L. (2018). Microplasticsin soils: Analytical methods, pollution characteristics and ecologicalrisks. *Trends in Analytical Chemistry*, *109*, 163–172. doi:10.1016/j.trac.2018.10.006

Hedin, L. O., Brookshire, E. J., Menge, D. N., & Barron, A. R. (2009). The nitrogen paradox in tropical forest ecosystems. *Annual Review of Ecology, Evolution, and Systematics*, *40*(1), 613–635. doi:10.1146/annurev.ecolsys.37.091305.110246

Heera, S., & Rajor, A. (2014). Bacterial treatment and metal characterization of biomedical waste ash. *Journal of Waste Management*.

Heera, S., Kunal, & Rajor, A. (2014). Bacterial treatment and metal characterization of biomedical waste ash. *Journal of Waste Management*, *2014*, 1–7. doi:10.1155/2014/956316

He, F., & Zhao, D. (2005). Preparation and characterization of a new class of starch-stabilized bimetallic nanoparticles for degradation of chlorinated hydrocarbons in water. *Environmental Science & Technology*, *39*(9), 3314–3320. doi:10.1021/es048743y PMID:15926584

He, F., & Zhao, D. (2007). Manipulating the size and dispersibility of zerovalent iron nanoparticles by use of carboxymethyl cellulose stabilizers. *Environmental Science & Technology*, *41*(17), 6216–6221. doi:10.1021/es0705543 PMID:17937305

Heil, M., & Silva, B. J. C. (2007). Within-plant signaling by volatiles leads to induction and priming of an indirect plant defense in nature. *Proceedings of the National Academy of Sciences of the United States of America*, *104*(103), 5467–5472. doi:10.1073/pnas.0610266104 PMID:17360371

Heinfling, A., Bergbauer, M., & Szewzyk, U. (1997). Biodegradation of azo and phthalocyanine dyes by *Trametes versicolour* and *Bjerkandera adusta*. *Applied Microbiology and Biotechnology*, *48*(2), 261–266. doi:10.1007002530051048

Heinze, J., Gensch, S., Weber, E., & Joshi, J. (2016). Soil temperature modifies effects of soil biota on plant growth. *Journal of Plant Ecology*, *10*(5), 898–821. doi:10.1093/jpe/rtw097

Hell, R., & Stephan, U. W. (2003). Iron uptake, trafficking and homeostasis in plants. *Planta*, *216*(4), 541–551. doi:10.100700425-002-0920-4 PMID:12569395

Hemlata, P., Jambhulkar, P., & Juwarkar, A. (2009). Assessment of bioaccumulation of heavy metals by different plant species grown on fly ash dump. *Ecotoxicology and Environmental Safety*, *72*(4), 1122–1128. doi:10.1016/j.ecoenv.2008.11.002 PMID:19171381

Hemmat-Jou, M. H., Safari-Sinegani, A. A., Mirzaie-Asl, A., & Tahmourespour, A. (2018). Analysis of microbial communities in heavy metals-contaminated soils using the metagenomic approach. *Ecotoxicology (London, England)*, *27*(9), 1281–1291. doi:10.100710646-018-1981-x PMID:30242595

He, N., Li, P., Zhou, Y., Fan, S., & Ren, W. (2006). Degradation of pentachlorobiphenyl by a sequential treatment using Pd coating iron and aerobic bacterium (H1). *Chemosphere*, *76*(11), 1491–1497. doi:10.1016/j.chemosphere.2009.06.046 PMID:19596135

Henn, K. W., & Waddill, D. W. (2006). Utilization of nanoscale zerovalent iron for source remediation—A case study. *Remediation*, *16*(2), 57–77. doi:10.1002/rem.20081

Hentzer, M., Wu, H., Andersen, J. B., Riedel, K., Rasmussen, T. B., Bagge, N., Kumar, N., Schembri, M. A., Song, Z., Kristoffersen, P., Manefield, M., Costerton, J. W., Molin, S., Eberl, L., Steinberg, P., Kjelleberg, S., Høiby, N., & Givskov, M. (2003). Attenuation of Pseudomonas aeruginosa virulence by quorum sensing inhibitors. *The EMBO Journal*, *22*(15), 3803–3815. doi:10.1093/emboj/cdg366 PMID:12881415

Herawati, N., Suzuki, S., Hayashi, K., Rivai, I. F., & Koyoma, H. (2000). Cadmium, copper and zinc levels in rice and soil of Japan, Indonesia and China by soil type. *Bulletin of Environmental Contamination and Toxicology*, *64*, 33–39. doi:10.1007001289910006 PMID:10606690

Hermosa, R., Viterbo, A., Chet, I., & Monte, E. (2012). Plant-beneficial effects of Trichoderma and of its genes. *Microbiology, 158*(1), 17–25. doi:10.1099/mic.0.052274-0 PMID:21998166

Hernandez-Carlos, B., & Gamboa-Angulo, M. M. (2011). Metabolites from freshwater aquatic microalgae and fungi as potential natural pesticides. *Phytochemistry Reviews, 10*(2), 261–286. doi:10.100711101-010-9192-y

Hernández, L. E., Garate, A., & Carpena-Ruiz, R. (1997). Effects of cadmium on the uptake, distribution and assimilation of nitrate in *Pisum sativum. Plant and Soil, 189*(1), 97–106. doi:10.1023/A:1004252816355

Hesseler, M., Bogdanović, X., Hidalgo, A., Berenguer, J., Palm, G. J., Hinrichs, W., & Bornscheuer, U. T. (2011). Cloning, functional expression, biochemical characterization, and structural analysis of a haloalkane dehalogenase from Plesiocystis pacifica SIR-1. *Applied Microbiology and Biotechnology, 91*(4), 1049–1060. doi:10.100700253-011-3328-x PMID:21603934

Hewitt, L., Kasche, V., Lummer, K., Lewis, R. J., Murshudov, G. N., Verma, C. S., Dodson, G. G., & Wilson, K. S. (2000). Structure of a slow processing precursor penicillin acylase from Escherichia coli reveals the linker peptide blocking the active-site cleft. *Journal of Molecular Biology, 302*(4), 887–898. doi:10.1006/jmbi.2000.4105 PMID:10993730

He, Z. L., & Yang, X. E. (2007). Role of soil rhizobacteria in phytoremediation of heavy metal contaminated soils. *Journal of Zhejiang University. Science. B., 8*(3), 192–207. doi:10.1631/jzus.2007.B0192 PMID:17323432

He, Z. L., Yang, X. E., & Stoffella, P. J. (2005). Trace elements in agro-ecosystems and impacts on the environment. *Journal of Trace Elements in Medicine and Biology, 19*(2–3), 125–140. doi:10.1016/j.jtemb.2005.02.010 PMID:16325528

Hicks, B. (2019). *Agricultural pesticides and human health, department of earth sciences.* Montana State University. serc.carleton.edu/NAGTWorkshops/health/case_studies/pesticides.html

Higham, D. P., Sadler, P. J., & Scawen, M. D. (1986). Cadmium-binding proteins in *Pseudomonas putida*: Pseudothioneins. *Environmental Health Perspectives, 65*, 5–11. PMID:3709466

Hill, D. S., Stein, J. I., Torkewitz, N. R., Morse, A. M., Howell, C. R., Pachlatko, J. P., Becker, J. O., & Ligon, J. M. (1994). Cloning of Genes Involved in the Synthesis of Pyrrolnitrin from *Pseudomonas fluorescens* and Role of Pyrrolnitrin Synthesis in Biological Control of Plant Disease. *Applied and Environmental Microbiology, 60*(1), 78–85. doi:10.1128/AEM.60.1.78-85.1994 PMID:16349167

Hillel, D., & Hatfield, J. L. (Eds.). (2005). *Encyclopedia of Soils in the Environment* (Vol. 3). Elsevier.

Hillman, B. I., & Suzuki, N. (2004). Viruses of the Chestnut bligth fungus, *Cryphonectria parasitica. Advances in Virus Research, 63*, 423–472. doi:10.1016/S0065-3527(04)63007-7 PMID:15530566

Himanen, K., Boucheron, E., Vanneste, S., de Almeida Engler, J., Inzé, D., & Beeckman, T. (2002). Auxin-mediated cell cycle activation during early lateral root initiation. *The Plant Cell, 14*(10), 2339–2351. doi:10.1105/tpc.004960 PMID:12368490

Hinchee, R. E., & Leeson, A. (1996). *Soil bioventing: Principles and practice.* CRC Press.

Hiner, A. N., Ruiz, J. H., López, J. N. R., Cánovas, F. G., Brisset, N. C., Smith, A. T., Arnao, M. B., & Acosta, M. (2002). Reactions of the class II peroxidases, lignin peroxidase andArthromyces ramosus peroxidase, with hydrogen peroxide catalase-like activity, compound III formation, and enzyme inactivation. *The Journal of Biological Chemistry, 277*(30), 26879–26885. doi:10.1074/jbc.M200002200 PMID:11983689

Hintz, W. E., Carneiro, J. S., Kassatenko, I., Varga, A., & James, D. (2013). Two novel mitoviruses from a Canadian isolate of the Dutch elm pathogen *Ophiostoma novoulmi* (93-1224). *Virology Journal, 10*(1), 252. doi:10.1186/1743-422X-10-252 PMID:23924036

Hiraishi, T., & Taguchi, S. (2013). Protein Engineering of Enzymes Involved in Bioplastic Metabolism. In T. Ogawa (Ed.), *Protein Engineering - Technology and Application* (pp. 133–165). InTech. doi:10.5772/55552

Hirasawa, T., & Shimizu, H. (2016). Recent advances in amino acid production by microbial cells. *Current Opinion in Biotechnology*, *42*, 133–146. doi:10.1016/j.copbio.2016.04.017 PMID:27151315

Hlihor, R. M., Figueiredo, H., Tavares, T., & Gavrilescu, M. (2017). Biosorption potential of dead and living *Arthrobacter viscosus* biomass in the removal of Cr (VI): Batch and column studies. *Process Safety and Environmental Protection*, *108*, 44–56. doi:10.1016/j.psep.2016.06.016

Hoanga, S. A., Lamba, D., Seshadria, B., Sarkarb, B., Choppalaa, G., Kirkhamc, M. B., & Bolana, N. S. (2021). Rhizoremediation as a green technology for the remediation of petroleum hydrocarbon-contaminated soils. *Journal of Hazardous Materials*, *401*, 123282. doi:10.1016/j.jhazmat.2020.123282 PMID:32634659

Hobbie, E. A., & Högberg, P. (2012). Nitrogen isotopes link mycorrhizal fungi and plants to nitrogen dynamics. *The New Phytologist*, *196*(2), 367–382. doi:10.1111/j.1469-8137.2012.04300.x PMID:22963677

Hobbie, S. E., & Gough, L. (2004). Litter decomposition in moist acidic and non-acidic tundra with different glacial histories. *Oecologia*, *140*(1), 113–124. doi:10.100700442-004-1556-9 PMID:15164284

Hoffman, B. M., Lukoyanov, D., Yang, Z. Y., Dean, D. R., & Seefeldt, L. C. (2014). Mechanism of nitrogen fixation by nitrogenase: The next stage. *Chemical Reviews*, *114*(8), 4041–4062. doi:10.1021/cr400641x PMID:24467365

Höhener, P., & Ponsin, V. (2014). In situ vadose zone bioremediation. *Current Opinion in Biotechnology*, *27*, 1–7. doi:10.1016/j.copbio.2013.08.018 PMID:24863890

Holden, M. T., Ram Chhabra, S., de Nys, R., Stead, P., Bainton, N. J., Hill, P. J., Manefield, M., Kumar, N., Labatte, M., England, D., Rice, S., Givskov, M., Salmond, G. P., Stewart, G. S., Bycroft, B. W., Kjelleberg, S., & Williams, P. (1999). Quorum-sensing cross talk: Isolation and chemical characterization of cyclic dipeptides from Pseudomonas aeruginosa and other gram-negative bacteria. *Molecular Microbiology*, *33*(6), 1254–1266. doi:10.1046/j.1365-2958.1999.01577.x PMID:10510239

Hollings, M. (1962). Viruses associated with a die-back disease of cultivated mushroom. *Nature*, *196*(4858), 962–965. doi:10.1038/196962a0

Holme, I. (2002). Recent developments in colorants for textile applications. *Surface Coatings International. Part B, Coatings Transactions*, *85*(4), 243–264. doi:10.1007/BF02699548

Holme, I. (2006). Sir William Henry Perkin: A review of his life, work and legacy. *Coloration Technology*, *122*(5), 235–251. doi:10.1111/j.1478-4408.2006.00041.x

Holopainen, J. K., & Gershenzon, J. (2010). Multiple stress factors and the emission of plant VOCs. *Trends in Plant Science*, *15*(3), 176–184. doi:10.1016/j.tplants.2010.01.006 PMID:20144557

Hong, B., Phornphisutthimas, S., Tilley, E., Baumberg, S., & McDowall, K. J. (2007). Streptomycin production by *Streptomyces griseus* can be modulated by a mechanism not associated with change in the adpA component of the A-factor cascade. *Biotechnology Letters*, *29*(1), 57–64. doi:10.100710529-006-9216-2 PMID:17120093

Horne, I., Sutherland, T. D., Harcourt, R. L., Russell, R. J., & Oakeshott, J. G. (2002). Identification of an opd (organophosphate degradation) gene in an Agrobacterium isolate. *Applied and Environmental Microbiology*, *68*(7), 3371–3376. doi:10.1128/AEM.68.7.3371-3376.2002 PMID:12089017

Horne, J. E., Kalevitch, A. E., & Filimonova, M. V. (1996). Soil acidity effect on initial wheat growth and development. *Journal of Sustainable Agriculture*, *7*(2-3), 5–13. doi:10.1300/J064v07n02_03

Hossain, A., Islam Masum, M. M., Wu, X., Abdallah, Y., Ogunyemi, S. O., Wang, Y., Sun, G., Li, B., & An, Q. (2020). Screening of Bacillus strains in biocontrol of pathogen Dickeya dadantii causing stem and root rot disease of sweet potato. *Biocontrol Science and Technology*, *30*(11), 1180–1198. doi:10.1080/09583157.2020.1798356

Hossen, M. Z., Hussain, M. E., Hakim, A., Islam, K., Uddin, M. N., & Azad, A. K. (2019). Biodegradation of reactive textile dye Novacron Super Black G by free cells of newly isolated *Alcaligenes faecalis AZ26* and *Bacillus spp* obtained from textile effluents. *Heliyon*, *5*(7), e02068. doi:10.1016/j.heliyon.2019.e02068 PMID:31338473

Hou, J., Liu, W., Wang, B., Wang, Q., Luo, Y., & Franks, A. E. (2015). PGPR enhanced phytoremediation of petroleum contaminated soil and rhizosphere microbial community response. *Chemosphere*, *138*, 592–598. doi:10.1016/j.chemosphere.2015.07.025 PMID:26210024

Hou, X., Yu, M., Liu, A., Li, Y., Ruan, T., Liu, J., Schnoor, J. L., & Jiang, G. (2018). Biotransformation of tetrabromobisphenol A dimethyl ether back to tetrabromobisphenol A in whole pumpkin plants. *Environmental Pollution*, *241*, 331–338. doi:10.1016/j.envpol.2018.05.075 PMID:29843015

Hrynkiewicz, K., Baum, C., Niedojadło, J., & Dahm, H. (2009). Promotion of mycorrhiza formation and growth of willows by the bacterial strain *Sphingomonas* sp. 23L on Fly Ash. *Biology and Fertility of Soils*, *45*(4), 385–394. doi:10.100700374-008-0346-7

Hsieh, P., Lin, H., Lin, T., & Wang, J. (2010). CadC regulates *cad* and *tdc* operons in response to gastrointestinal stresses and enhances intestinal colonization of *Klebsiella pneumoniae*. *The Journal of Infectious Diseases*, *202*(1), 52–64. doi:10.1086/653079 PMID:20497056

Hu¨ckelhoven, R. (2007). Transport and secretion in plant_microbe interactions. *Current Opinion in Plant Biology*, *10*(6), 573–579. doi:10.1016/j.pbi.2007.08.002 PMID:17875397

Huang, H., Liu, B., Liu, L., & Song, S. (2017). Jasmonate action in plant growth and development. *Journal of Experimental Botany*, *68*(6), 1349–1359. doi:10.1093/jxb/erw495 PMID:28158849

Huang, J. J., Han, J. I., Zhang, L. H., & Leadbetter, J. R. (2003). Utilization of acyl-homoserine lactone quorum signals for growth by a soil pseudomonad and Pseudomonas aeruginosa PAO1. *Applied and Environmental Microbiology*, *69*(10), 5941–5949. doi:10.1128/AEM.69.10.5941-5949.2003 PMID:14532048

Huang, J. J., Petersen, A., Whiteley, M., & Leadbetter, J. R. (2006). Identification of QuiP, the product of gene PA1032, as the second acyl-homoserine lactone acylase of Pseudomonas aeruginosa PAO1. *Applied and Environmental Microbiology*, *72*(2), 1190–1197. doi:10.1128/AEM.72.2.1190-1197.2006 PMID:16461666

Huang, R. (2018). Research progress on plant tolerance to soil salinity and alkalinity in sorghum. *Journal of Integrative Agriculture*, *17*(4), 739–746. doi:10.1016/S2095-3119(17)61728-3

Huang, X. D., El-Alawi, Y., Penrose, D. M., Glick, B. R., & Greenberg, B. M. (2004). A multi-process phytoremediation system for removal of polycyclic aromatic hydrocarbons from contaminated soils. *Environmental Pollution*, *130*(3), 465–476. doi:10.1016/j.envpol.2003.09.031 PMID:15182977

Huang, Y., Fulton, A. N., & Keller, A. A. (2016). Simultaneous removal of PAHs and metal contaminants from water using magnetic nanoparticle adsorbents. *The Science of the Total Environment*, *571*, 1029–1036. doi:10.1016/j.scitotenv.2016.07.093 PMID:27450251

Huang, Z., He, L., Sheng, X., & He, Z. (2013). Weathering of potash feldspar by *Bacillus sp.* L11. Wei sheng wu xue bao. *Acta Microbiologica Sinica*, *53*, 1172–1178. PMID:24617258

Hu, B., Chen, S., Hu, J., Xia, F., Xu, J., Li, Y., & Shi, Z. (2017). Application of portable XRF and VNIR sensors for rapid assessment of soil heavy metal pollution. *PLoS One*, *12*(2), 1–13. doi:10.1371/journal.pone.0172438 PMID:28234944

Huerta Lwanga, E., Gertsen, H., Gooren, H., Peters, P., Salánki, T., van der Ploeg, M., Besseling, E., Koelmans, A. A., & Geissen, V. (2016). Microplastics in the terrestrial ecosystem: Implications for Lumbricus terrestris (Oligochaeta, Lumbricidae). *Environmental Science & Technology*, *50*(5), 2685–2691. doi:10.1021/acs.est.5b05478 PMID:26852875

Hu, J. Y., Fan, Y., Lin, Y. H., Zhang, H. B., Ong, S. L., Dong, N., Xu, J. L., Ng, W. J., & Zhang, L. H. (2003). Microbial diversity and prevalence of virulent pathogens in biofilms developed in a water reclamation system. *Research in Microbiology*, *154*(9), 623–629. doi:10.1016/j.resmic.2003.09.004 PMID:14596899

Hu, K., Deng, W., Zhu, Y., Yao, K., Li, J., Liu, A., Ao, X., Zou, L., Zhou, K., & He, L. (2018). Simultaneous degradation of β-cypermethrin and 3-phenoxybenzoic acid by Eurotium cristatum ET1, a novel"golden flower fungus" strain isolated from Fu Brick Tea. *MicrobiologyOpen*, *8*, 7. PMID:30548839

Hungria, M., Nogueira, M. A., & Araujo, R. S. (2013). Co-inoculation of soybeans and common beans with rhizobia and azospirilla: Strategies to improve sustainability. *Biology and Fertility of Soils*, *49*(7), 791–801. doi:10.100700374-012-0771-5

Hunt, L. J., Duca, D., Dan, T., & Knopper, L. D. (2018). Petroleum hydrocarbon (PHC) uptake in plants: A literature review. *Environmental Pollution*, *245*, 472–484. doi:10.1016/j.envpol.2018.11.012 PMID:30458377

Hu, Q. H., Weng, J. Q., & Wang, J. S. (2008). Sources of anthropogenic radionuclides in the environment: A review. *Journal of Environmental Radioactivity*, *101*(6), 426–437. doi:10.1016/j.jenvrad.2008.08.004 PMID:18819734

Hu, S., Chapin, F. S. III, Firestone, M. K., Field, C. B., & Chiariello, N. R. (2001). Nitrogen limitation of microbial decomposition in a grassland under elevated CO 2. *Nature*, *409*(6817), 188–191. doi:10.1038/35051576 PMID:11196641

Husain, Q. (2006). Potential applications of the oxidoreductive enzymes in the decolorization and detoxification of textile and other synthetic dyes from polluted water: A review. *Critical Reviews in Biotechnology*, *26*(4), 201–221. doi:10.1080/07388550600969936 PMID:17095432

Hussain, A., Abbas, N., Arshad, F., Akram, M., Khan, Z. I., Ahmad, K., Mansha, M., & Mirzaei, F. (2013). Effects of diverse doses of lead (Pb) on different growth attributes of *Zea mays* L. *Agricultural Sciences*, *4*(5), 262–265. doi:10.4236/as.2013.45037

Hu, T. L. (2001). Kinetics of Azoreductase and Assessment of Toxicity of Metabolic Products from Azo Dyes by *Pseudomonas luteola. Water Science and Technology*, *43*(2), 261–269. doi:10.2166/wst.2001.0098 PMID:11380189

Hu, W., Lu, Q., Zhong, G., Hu, M., & Yi, X. (2019). Biodegradation of pyrethroids by a hydrolyzing carboxylesterase EstA from *Bacillus cereus* BCC01. *Applied Sciences (Basel, Switzerland)*, *9*(3), 477. doi:10.3390/app9030477

Hu, Y., & Ribbe, M. W. (2015). Nitrogenase and homologs. *JBIC Journal of Biological Inorganic Chemistry*, *20*(2), 435–445. doi:10.100700775-014-1225-3 PMID:25491285

Hu, Y., Xie, Q., & Chua, N. H. (2003). The Arabidopsis auxin-inducible gene ARGOS controls lateral organ size. *The Plant Cell*, *15*(9), 1951–1961. doi:10.1105/tpc.013557 PMID:12953103

Hu, Z., Guo, J., Da Gao, B., & Zhong, J. (2020). A novel mycovirus isolated from the plant-pathogenic fungus *Alternaria dianthicola. Archives of Virology*, *165*(9), 2105–2109. doi:10.100700705-020-04700-9 PMID:32556598

Hwangbo, H., Park, R. D., Kim, Y. W., Rim, Y. S., Park, K. H., Kim, T. H., Sun, J.S., & Kim, K. Y. (2003). 2-Ketogluconic acid production and phosphate solubilization by *Enterobacter intermedium. Current Microbiology, 47*(2), 87-92.

Hwang, I., Li, P. L., Zhang, L., Piper, K. R., Cook, D. M., Tate, M. E., & Farrand, S. K. (1994). TraI, a LuxI homologue, is responsible for production of conjugation factor, the Ti plasmid N-acylhomoserine lactone autoinducer. *Proceedings of the National Academy of Sciences of the United States of America*, *91*(11), 4639–4643. doi:10.1073/pnas.91.11.4639 PMID:8197112

Hyde, F. W., Hunt, G. R., & Errede, L. A. (1991). Immobilization of bacteria and *Saccharomyces cerevisiae* in poly (tetrafluoroethylene) membranes. *Applied and Environmental Microbiology*, *57*(1), 219–222. doi:10.1128/AEM.57.1.219-222.1991 PMID:2036008

Hyder, R., Pennanen, T., Hamberg, L., Vainio, E. J., Piri, T., & Hantula, J. (2013). Two viruses of *Heterobasidion* confer beneficial, cryptic or detrimental effects to their hosts in different situations. *Fungal Ecology*, *6*(5), 387–396. doi:10.1016/j.funeco.2013.05.005

Hyman, M., & Dupont, R. R. (2001). *Groundwater and soil remediation: process design and cost estimating of proven technologies* (Vol. 137). ASCE Press. doi:10.1061/9780784404270

Hynninen, A., Touzé, T., Pitkänen, L., Mengin-Lecreulx, D., & Virta, M. (2009). An efflux transporter PbrA and a phosphatase PbrB cooperate in a lead-resistance mechanism in bacteria. *Molecular Microbiology*, *74*(2), 384–394. doi:10.1111/j.1365-2958.2009.06868.x PMID:19737357

IAEA (International Atomic Energy Agency). (1970). *Standardization of radioactive waste categories; TRS no. 101*. IAEA.

IAEA. (1999). *Review of the factors affecting the selection and implementation of waste management technologies; IAEA-TECDOC-1096*. IAEA.

Iavicoli, A., Boutet, E., Buchala, A., & Métraux, J. P. (2003). Induced systemic resistance in Arabidopsis thaliana in response to root inoculation with Pseudomonas fluorescens CHA0. *Molecular Plant-Microbe Interactions*, *16*(10), 851–858. doi:10.1094/MPMI.2003.16.10.851 PMID:14558686

Ibrahim, A. S., Elbadawi, Y. B., El-Tayeb, M. A., & Al-Salamah, A. A. (2012). Hexavalent chromium reduction by novel chromate resistant alkaliphilic *Bacillus* sp. strain KSUCr9a. *African Journal of Biotechnology*, *11*, 3832–3841.

ICRP Recommendations of the International Commission on Radiological Protection (Users Edition). (2007). ICRP Publication 103 (Users Edition). *Annals of the ICRP*, *37*(2-4), 1–332. PMID:18082557

ICTV. (2014). *International Committee on Taxonomy of Viruses*. http://ictvonline.org/index.asp

ICTVdB. (2002). *The Universal Virus Database, version 4*. https://www.ncbi.nlm.nih.gov/ICTVdb/ICTVdB/

Igiehon, N. O., Babalola, O. O., & Aremu, B. R. (2019). Genomic insights into plant growth promoting rhizobia capable of enhancing soybean germination under drought stress. *BMC Microbiology*, *19*(1), 159. doi:10.118612866-019-1536-1 PMID:31296165

Igiri, B. E., Okoduwa, S. I., Idoko, G. O., Akabuogu, E. P., Adeyi, A. O., & Ejiogu, I. K. (2018). Toxicity and bioremediation of heavy metals contaminated ecosystem from tannery wastewater: A review. *Journal of Toxicology*, *2018*, 2018. doi:10.1155/2018/2568038 PMID:30363677

Ihrmark, K., Johannesson, H., Stenstrom, E., & Stenlid, J. (2002). Transmission of double-stranded RNA in *Heterobasidion annosum. Fungal Genetics and Biology*, *36*(2), 147–154. doi:10.1016/S1087-1845(02)00011-7 PMID:12081468

Imam, J., Singh, P. K., & Shukla, P. (2016). Plant microbe interactions in post genomic era: Perspectives and applications. *Frontiers in Microbiology*, *7*, 1488. doi:10.3389/fmicb.2016.01488 PMID:27725809

Imran, M. A., Ch, M. N., Khan, R. M., Ali, Z., & Mahmood, T. (2013). Toxicity of arsenic (As) on seed germination of sunflower (*Helianthus annuus L.*). *International Journal of Physical Sciences*, *8*(17), 840–847. doi:10.5897/IJPS2013.3894

Innerebner, G., Knief, C., & Vorholt, J. A. (2011). Protection of Arabidopsis thaliana against leaf-pathogenic Pseudomonas syringae by Sphingomonas strains in a controlled model system. *Applied and Environmental Microbiology*, *77*(10), 3202–3210. doi:10.1128/AEM.00133-11 PMID:21421777

Inoue, S., & Ito, S. (1982). Sophorolipids from *Torulopsis bombicola* as microbial surfactants in alkane fermentations. *Biotechnology Letters*, *4*(1), 3–8. doi:10.1007/BF00139273

International Fertilizer Development Center (1980). *Fertilizer manual* (No. 13). UN.

International Tanker Owners Pollution Federation (ITOPF). (2019). *Oil Tanker Spill Statistics*. Available at http://www.itopf.org

Inthorn, D., Sidtitoon, N., Silapanuntakul, S., & Incharoensakdi, A. (2002). Sorption of mercury, cadmium and lead by microalgae. *Science Asia*, *28*(3), 253–261. doi:10.2306cienceasia1513-1874.2002.28.253

Iqbal, A., & Hasnain, S. (2013). Aeromonas punctata PNS-1: A promising candidate to change the root morphogenesis of Arabidopsis thaliana in MS and sand system. *Acta Physiologiae Plantarum*, *35*(3), 657–665. doi:10.100711738-012-1106-8

Iqbal, M., Iqbal, N., Bhatti, I. A., Ahmad, N., & Zahid, M. (2016). Response surface methodology application in optimization of cadmium adsorption by shoe waste: A good option of waste mitigation by waste. *Ecological Engineering*, *88*, 265–275. doi:10.1016/j.ecoleng.2015.12.041

Isaac, P., Alessandrello, M. J., Macedo, A. J., Estévez, M. C., & Ferrero, M. A. (2017). Pre-exposition to polycyclic aromatic hydrocarbons (PAHs) enhance biofilm formation and hydrocarbon removal by native multi-species consortium. *Journal of Environmental Chemical Engineering*, *5*(2), 1372–1378. doi:10.1016/j.jece.2017.02.031

Isaka, K., Yoshie, S., Sumino, T., Inamori, Y., & Tsuneda, S. (2007). Nitrification of landfill leachate using immobilized nitrifying bacteria at low temperatures. *Biochemical Engineering Journal*, *37*(1), 49–55. doi:10.1016/j.bej.2007.03.008

Islami, F., Torre, L. A., & Jemal, A. (2015). Global trends of lung cancer mortality and smoking prevalence. *Translational Lung Cancer Research*, *4*(4), 327. PMID:26380174

Islam, M. M., Mahmud, K., Faruk, O., & Billah, M. S. (2011). Textile Dyeing Industries in Bangladesh for Sustainable Development. *International Journal of Environmental Sciences and Development*, *2*(6), 428–436. doi:10.7763/IJESD.2011.V2.164

Ismail, H. Y., Ijah, U. J. J., Riskuwa, M. L., & Allamin, I. I. (2014). Biodegradation of spent engine oil by bacteria isolated from the rhizosphere of legumes grown in contaminated soil. *International Journal of Environment*, *3*(2), 63–75. doi:10.3126/ije.v3i2.10515

Ismail, H. Y., Riskuwa-Shehu, M. L., Allamin, I. A., Farouq, A. A., & Abakwak, C. S. (2019). Biostimulation Potentials of Vigna Species (L.) in Hydrocarbon Impacted Soil. *American Journal of Bioscience and Bioengineering*, *7*(1), 22–27. doi:10.11648/j.bio.20190701.15

Ite, A. E., Ibok, U. J., Ite, M. U., & Petters, S. W. (2013). Petroleum exploration and production: Past and present environmental issues in the Nigeria's Niger Delta. *Nature*, *1*, 78–90.

Itelima, J. U., Bang, W. J., Onyimba, I. A., & Oj, E. (2018). A review: Biofertilizer; a key player in enhancing soil fertility and crop productivity. *Journal of Microbiology and Biotechnology Research*, *2*(1), 22–28.

Iturbe, R., Flores, C., Castro, A., & Torres, L. G. (2007). Sub-soil contamination due to oil spills in zones surrounding oil pipeline-pump stations and oil pipeline right-of-ways in Southwest-Mexico. *Environmental Monitoring and Assessment, 133*(1-3), 387–398. doi:10.100710661-006-9593-y PMID:17286169

Izrael- Zivkoví, L., Rikaloví, M., Gojgí Cvijoví, G., Kazazí, S., Vrví, M., Bř ceski, I., Bě skoski, V., Loň Careví, B., GopčevíGopˇGopčeví, K., & KaradžíKaradží, I. (2018). Cadmium specific proteomic responses of a highly resistant Pseudomonas aeruginosa san ai. *RSC Advances, 8*, 10549–10560.

Jackson, M. L. (1964). Chemical composition of soils. In Chemistry of the Soil (pp. 71-141). Reinhold.

Jackson, L. E., Bowles, T. M., Hodson, A. K., & Lazcano, C. (2012). Soil microbial-root and microbial-rhizosphere processes to increase nitrogen availability and retention in agroecosystems. *Current Opinion in Environmental Sustainability, 4*(5), 517–522. doi:10.1016/j.cosust.2012.08.003

Jacoby, R., Peukert, M., Succurro, A., Koprivova, A., & Kopriva, S. (2017). The Role of Soil Microorganisms in Plant Mineral Nutrition—Current Knowledge and Future Directions. *Frontiers in Plant Science, 8*, 1617–1636. doi:10.3389/fpls.2017.01617 PMID:28974956

Jacoud, C., Faure, D., Wadoux, P., & Bally, R. (1998). Development of a strain-specific probe to follow inoculated Azospirillum lipoferum CRT1 under field conditions and enhancement of maize root development by inoculation. *FEMS Microbiology Ecology, 27*(1), 43–51. doi:10.1111/j.1574-6941.1998.tb00524.x

Jadhav, J. P., Phugare, S. S., Dhanve, R. S., & Jadhav, S. B. (2010). Rapid biodegradation and decolorization of Direct Orange 39 (Orange TGLL) by an isolated bacterium *Pseudomonas aeruginosa* strain BCH. *Biodegradation, 21*(3), 453–463. doi:10.100710532-009-9315-6 PMID:19937265

Jadhav, S. U., Jadhav, M. U., Kagalkar, A. N., & Govindwar, S. P. (2008). Decolorization of Brilliant Blue G dye mediated by degradation of the microbial consortium of *Galactomyces geotrichum* and *Bacillus* sp. *Journal of the Chinese Institute of Chemical Engineers, 39*(6), 563–570. doi:10.1016/j.jcice.2008.06.003

Jafaria, N., Soudib, M. R., & Kasra-Kermanshahi, R. (2014). Biodegradation Perspectives of Azo Dyes by Yeasts. *Microbiology, 83*(5), 484–497. doi:10.1134/S0026261714050130

Jafra, S., Przysowa, J., Czajkowski, R., Michta, A., Garbeva, P., & van der Wolf, J. M. (2006). Detection and characterization of bacteria from the potato rhizosphere degrading N-acyl-homoserine lactone. *Canadian Journal of Microbiology, 52*(10), 1006–1015. doi:10.1139/w06-062 PMID:17110970

Jagtap, U. B., & Bapat, V. A. (2015). Genetic engineering of plants for heavy metal removal from soil. In *Heavy Metal Contamination of Soils* (pp. 433–470). Springer. doi:10.1007/978-3-319-14526-6_22

Jahan, S., Khan, M., Ahmed, S., & Ullah, H. (2014). Comparative analysis of antioxidants against cadmium induced reproductive toxicity in adult male rats. *Systems Biology in Reproductive Medicine, 60*(1), 28–34. doi:10.3109/19396368.2013.843039 PMID:24156729

Jain, S., & Bhatt, A. (2013). Proteomic analysis of diversified extremophilic strains of *Pseudomonas* in the presence of cadmium. *Agricultural Research, 2*(4), 354–359.

Jambon, I., Thijs, S., Weyens, N., & Vangronsveld, J. (2018). Harnessing plant-bacteria-fungi interactions to improve plant growth and degradation of organic pollutants. *Journal of Plant Interactions, 13*(1), 119–130. doi:10.1080/17429145.2018.1441450

James, E. K., Olivares, F. L., Baldani, J. I., & Döbereiner, J. (1997). Herbaspirillum, an endophytic diazotroph colonizing vascular tissue in leaves of Sorghum bicolor L. *Moench J Exp Bot, 48*(3), 785–797. doi:10.1093/jxb/48.3.785

Jan, A. T., Azam, M., Ali, A., & Haq, Q. M. R. (2014). Prospects for exploiting bacteria for bioremediation of metal pollution. *Critical Reviews in Environmental Science and Technology*, *44*(5), 519–560. doi:10.1080/10643389.2012.728811

Jan, A. T., Azam, M., Choi, I., Ali, A., & Haq, Q. M. R. (2016). Analysis for the presence of determinants involved in the transport of mercury across bacterial membrane from polluted water bodies of India. *Brazilian Journal of Microbiology*, *47*(1), 55–62. doi:10.1016/j.bjm.2015.11.023 PMID:26887227

Jan, A. T., Azam, M., Siddiqui, K., Ali, A., Choi, I., & Haq, Q. M. (2015). Heavy metals and human health: Mechanistic insight into toxicity and counter defense system of antioxidants. *International Journal of Molecular Sciences*, *16*(12), 29592–29630. doi:10.3390/ijms161226183 PMID:26690422

Jan, A. T., Murtaza, I., Ali, A., & Rizwanul Haq, Q. M. (2009). Mercury pollution: An emerging problem and potential bacterial remediation strategies. *World Journal of Microbiology & Biotechnology*, *25*(9), 1529–1537. doi:10.100711274-009-0050-2

Janas, M., & Zawadzka, A. (2017). The impact of waste landfill on the environment. *Inzynieria Ekologiczna*, *18*(3), 64–73. doi:10.12912/23920629/70259

Jangid, K., Williams, M. A., Franzluebbers, A. J., Sanderlin, J. S., Reeves, J. H., Jenkins, M. B., Endale, D. M., Coleman, D. C., & Whitman, W. B. (2008). Relative impacts of land-use, management intensity and fertilization upon soil microbial community structure in agricultural systems. *Soil Biology & Biochemistry*, *40*(11), 2843–2853. doi:10.1016/j.soilbio.2008.07.030

Jan, M., Shah, G., Masood, S., Iqbal Shinwari, K., Hameed, R., Rha, E. S., & Jamil, M. (2019). Bacillus cereus enhanced phytoremediation ability of rice seedlings under cadmium toxicity. *BioMed Research International*, *2019*, 2019. doi:10.1155/2019/8134651 PMID:31428647

Janssen, D. B. (2004). Evolving haloalkane dehalogenases. *Current Opinion in Chemical Biology*, *8*(2), 150–159. doi:10.1016/j.cbpa.2004.02.012 PMID:15062775

Janssen, D. B. (2007). Biocatalysis by dehalogenating enzymes. *Advances in Applied Microbiology*, *61*, 233–252. doi:10.1016/S0065-2164(06)61006-X PMID:17448791

Janssen, D. B., Oppentocht, J. E., & Poelarends, G. J. (2001). Microbial dehalogenation. *Current Opinion in Biotechnology*, *12*(3), 254–258. doi:10.1016/S0958-1669(00)00208-1 PMID:11404103

Janssen, D. B., Pries, F., & van der Ploeg, J. R. (1994). Genetics and biochemistry of dehalogenating enzymes. *Annual Review of Microbiology*, *48*(1), 163–191. doi:10.1146/annurev.mi.48.100194.001115 PMID:7826004

Janssen, D. B., Scheper, A., Dijkhuizen, L., & Witholt, B. (1985). Degradation of halogenated aliphatic compounds by Xanthobacter autotrophicus GJ10. *Applied and Environmental Microbiology*, *49*(3), 673–677. doi:10.1128/AEM.49.3.673-677.1985 PMID:3994371

Janssen, D. B., Scheper, A., & Witholt, B. (1984). Biodegradation of 2-chloroethanol and 1, 2-dichloroethane by pure bacterial cultures. *Progress in Industrial Microbiology*, *20*, 169–178.

Janssen, D. B., & Stucki, G. (2020). Perspectives of genetically engineered microbes for groundwater bioremediation. *Environmental Science. Processes & Impacts*, *22*(3), 487–499. doi:10.1039/C9EM00601J PMID:32095798

Janssen, D. B., van der Ploeg, J. R., & Pries, F. (1994). Genetics and biochemistry of 1, 2-dichloroethane degradation. *Biodegradation*, *5*(3-4), 249–257. doi:10.1007/BF00696463 PMID:7765836

Jansson, J. K., & Baker, E. S. (2016). A multi-omic future for microbiome studies. *Nature Microbiology*, *1*(5), 1–3. doi:10.1038/nmicrobiol.2016.49 PMID:27572648

Jaramillo, M. F., & Restrepo, I. (2017). Wastewater Reuse in Agriculture: A Review about Its Limitations and Benefits. *Sustainability*, *9*(10), 1734–1753. doi:10.3390u9101734

Jarosiewicz, M., Duchnowicz, P., Włuka, A., & Bukowska, B. (2017). Evaluation of the effect of brominated flame retardants on hemoglobin oxidation and hemolysis in human erythrocytes. *Food and Chemical Toxicology*, *109*, 264–271. doi:10.1016/j.fct.2017.09.016 PMID:28893619

Järup, L. (2003). Hazards of heavy metal contamination. *British Medical Bulletin*, *68*(1), 167–182. doi:10.1093/bmb/ldg032 PMID:14757716

Jat, H. S., Datta, A., Choudhary, M., Yadav, A. K., Choudhary, V., Sharma, P. C., & McDonald, A. (2019). Effects of tillage, crop establishment and diversification on soil organic carbon, aggregation, aggregate associated carbon and productivity in cereal systems of semi-arid Northwest India. *Soil & Tillage Research*, *190*, 128–138. doi:10.1016/j.still.2019.03.005 PMID:32055081

Jat, H. S., Datta, A., Sharma, P. C., Kumar, V., Yadav, A. K., Choudhary, M., & McDonald, A. (2017). Assessing soil properties and nutrient availability under conservation agriculture practices in a reclaimed sodic soil in cereal-based systems of North-West India. *Archives of Agronomy and Soil Science*, *64*(4), 531–545. doi:10.1080/03650340.2017.1359415 PMID:30363929

Jayaraj, R., Megha, P., & Sreedev, P. (2016). Organochlorine pesticides, their toxic effects on living organisms and their fate in the environment. *Interdisciplinary Toxicology*, *9*(3-4), 90–100. doi:10.1515/intox-2016-0012 PMID:28652852

Jayasiri, H. B., Purushothaman, C. S., & Vennila, A. (2013). Plastic litter accumulation on high-water strandline of urban beaches in Mumbai, India. *Environmental Monitoring and Assessment*, *185*(9), 7709–7719. doi:10.100710661-013-3129-z PMID:23430068

Jeganathan, K. (2006). *Bioremediation studies on oil refinery industry effluent using* Oscillatoria earli Gartner (M.Phil dissertation). Bharathidasan University, Tiruchirapalli, India.

Jem, K. J., van der Pol, J. F., & de Vos, S. (2010). Microbial lactic acid, its polymer poly (lactic acid), and their industrial applications. In *Plastics from bacteria* (pp. 323–346). Springer. doi:10.1007/978-3-642-03287-5_13

Jeong, S. W., Jeong, J., & Kim, J. (2015). Simple surface foam application enhances bioremediation of oil-contaminated soil in cold conditions. *Journal of Hazardous Materials*, *286*, 164–170. doi:10.1016/j.jhazmat.2014.12.058 PMID:25577318

Jesubunmi, C. O. (2014). Isolation of oil-degrading microorganisms in spent engine oil-contaminated soil. *Journal of Biology, Agriculture and Healthcare*, *4*(24), 191–195.

Jha, Y., & Subramanian, R. B. (2013). Paddy plants inoculated with PGPR show better growth physiology and nutrient content under saline condition. *Chilean Journal of Agricultural Research*, *73*(3), 213–219. doi:10.4067/S0718-58392013000300002

Ji, C., Tian, H., Wang, X., Hao, L., Wang, C., Zhou, Y., & Liu, X. (2020). Bacillus subtilis *HG-15, a Halotolerant Rhizoplane Bacterium, Promotes Growth and Salinity Tolerance in Wheat (*Triticum aestivum*)*. Preprint.

Jiang, Q. Y., Zhuo, F., Long, S. H., Zhao, H. D., Yang, D. J., Ye, Z. H., & Jing, Y. X. (2016). Can arbuscular mycorrhizal fungi reduce Cd uptake and alleviate Cd toxicity of *Lonicera japonica* grown in Cd-added soils? *Scientific Reports*, *6*(1), 1–9. doi:10.1038rep21805 PMID:26892768

Ji, C., Wang, X., Tian, H., Hao, L., Wang, C., Zhou, Y., & Liu, X. (2020). Effects of Bacillus methylotrophicus M4-1 on physiological and biochemical traits of wheat under salinity stress. *Journal of Applied Microbiology*, *129*(3), 695–711. doi:10.1111/jam.14644 PMID:32215987

Ji, G., & Silver, S. (1992). Reduction of arsenate to arsenite by the ArsC protein of the arsenic resistance operon of *Staphylococcus aureus* plasmid pI258. *Proceedings of the National Academy of Sciences of the United States of America*, *89*(20), 9474–9478.

Jiménez, D. J., Montaña, J. S., & Martínez, M. M. (2011). Characterization of free nitrogen fixing bacteria of the genus Azotobacter in organic vegetable-grown Colombian soils. *Brazilian Journal of Microbiology*, *42*(3), 846–858. doi:10.1590/S1517-83822011000300003 PMID:24031700

Jin, X.-C., Liu, G.-Q., Xu, Z.-H., & Tao, W.-Y. (2007). Decolorization of a dye industry effluent by *Aspergillus fumigatus* XC6. *Applied Microbiology and Biotechnology*, *74*(1), 239–243. doi:10.100700253-006-0658-1 PMID:17086413

Jin, Z. P., Luo, K., Zhang, S., Zheng, Q., & Yang, H. (2012). Bioaccumulation and catabolism of prometryne in green algae. *Chemosphere*, *87*(3), 278–284. doi:10.1016/j.chemosphere.2011.12.071 PMID:22273183

Ji, Y., Mao, G., Wang, Y., & Bartlam, M. (2013). Structural insights into diversity and n-alkane biodegradation mechanisms of alkane hydroxylases. *Frontiers in Microbiology*, *4*, 58. doi:10.3389/fmicb.2013.00058 PMID:23519435

Joardar, J. C., & Rahman, M. M. (2018). Poultry feather waste management and effects on plant growth. *International Journal of Recycling of Organic Waste in Agriculture*, *7*(3), 183–188. doi:10.100740093-018-0204-z

Jobby, R., Jha, P., Gupta, A., Gupte, A., & Desai, N. (2019). Biotransformation of chromium by root nodule bacteria *Sinorhizobium sp.* SAR1. *PLoS One*, *14*(7), e0219387. doi:10.1371/journal.pone.0219387 PMID:31361751

Jobling, S., Nolan, M., Tyler, C. R., Brighty, G., & Sumpter, J. P. (1998). Widespread sexual disruption in wild fish. *Environmental Science & Technology*, *32*(17), 2498–2506. doi:10.1021/es9710870

Jogaiah, S., Shivanna, R. K., Gnanaprakash, P. H., & Hunthrike, S. S. (2010). Evaluation of plant growth-promoting rhizobacteria for their efficiency to promote growth and induce systemic resistance in pearl millet against downy mildew disease. *Archiv für Phytopathologie und Pflanzenschutz*, *43*(4), 368–378. doi:10.1080/03235400701806377

John Sundar, V., Gnanamani, A., Muralidharan, C., Chandrababu, N. K., & Mandal, A. B. (2011). Recovery and utilization of proteinous wastes of leather making: A review. *Reviews in Environmental Science and Biotechnology*, *10*(2), 151–163. doi:10.100711157-010-9223-6

Johns, C. (2017). Living soils: the role of microorganisms in soil health. *Fut Direct Intl*, 1-7.

Johnson, D. L., Ambrose, S. H., Bassett, T. J., Bowen, M. L., Crummey, D. E., Isaacson, J. S., & Winter-Nelson, A. E. (1997). Meanings of environmental terms. *Journal of Environmental Quality*, *26*(3), 581–589. doi:10.2134/jeq1997.00472425002600030002x

Johnson, D. L., Thompson, J. L., Brinkmann, S. M., Schuller, K. A., & Martin, L. L. (2003). Electrochemical Characterization of Purified *Rhus vernicifera* Laccase: Voltammetric Evidence for a Sequential Four-Electron Transfer. *Biochemistry*, *42*(34), 10229–10237. doi:10.1021/bi034268p PMID:12939151

Jones, J. P., O'Hare, E. J., & Wong, L. L. (2001). Oxidation of polychlorinated benzenes by genetically engineered CYP101 (cytochrome P450cam). *European Journal of Biochemistry*, *268*(5), 1460–1467. doi:10.1046/j.1432-1327.2001.02018.x PMID:11231299

Jones, K. M., Kobayashi, H., Davies, B. W., Taga, M. E., & Walker, G. C. (2007). How rhizobial symbionts invade plants: The Sinorhizobium–Medicago model. *Nature Reviews. Microbiology*, *5*(8), 619–633. doi:10.1038/nrmicro1705 PMID:17632573

Jones, S. W., Karpol, A., Friedman, S., Maru, B. T., & Tracy, B. P. (2020). Recent advances in single cell protein use as a feed ingredient in aquaculture. *Current Opinion in Biotechnology*, *61*, 189–197. doi:10.1016/j.copbio.2019.12.026 PMID:31991311

Joosten, H., & Clarke, D. (2002). Wise use of mires and peatlands. International Mire Conservation Group and International Peat Society, 304.

Joshi, B., Kabariya, K., Nakrani, S., Khan, A., Parabia, F. M., Doshi, H. V., & Thakur, M. C. (2013). Biodegradation of turquoise blue dye by *Bacillus megaterium* isolated from industrial effluent. *American Journal of Environmental Protection*, *1*(2), 41–46. doi:10.12691/env-1-2-5

Joshi, H., Shourie, A., & Singh, A. (2020). Cyanobacteria as a source of biofertilizers for sustainable agriculture. In *Advances in Cyanobacterial Biology* (pp. 385–396). Academic Press. doi:10.1016/B978-0-12-819311-2.00025-5

Joshi, P. A., Jaybhaye, S., & Mhatre, K. (2015). Biodegradation of dyes using consortium of bacterial strains isolated from textile effluent. *European Journal of Experimental Biology*, *5*, 36–40.

Joshi, T., Iyengar, L., Singh, K., & Garg, S. (2008). Isolation, identification and application of novel bacterial consortium TJ-1 for the decolourization of structurally different azo dyes. *Bioresource Technology*, *99*(15), 7115–7121. doi:10.1016/j.biortech.2007.12.074 PMID:18289845

Joutey, N. T., Bahafid, W., Sayel, H., & El Ghachtouli, N. (2013). Biodegradation: involved microorganisms and genetically engineered microorganisms. *Biodegradation-Life of Science*, 289-320.

Juárez-Hernández, R. E., Franzblau, S. G., & Miller, M. J. (2012). Syntheses of mycobactin analogs as potent and selective inhibitors of *Mycobacterium tuberculosis*. *Organic & Biomolecular Chemistry*, *10*(37), 7584–7593. doi:10.1039/c2ob26077h PMID:22895786

Ju, M., Navarreto-Lugo, M., Wickramasinghe, S., Milbrandt, N. B., McWhorter, A., & Samia, A. C. S. (2019). Exploring the chelation-based plant strategy for iron oxide nanoparticle uptake in garden cress (*Lepidium sativum*) using magnetic particle spectrometry. *Nanoscale*, *11*(40), 18582–18594. doi:10.1039/C9NR05477D PMID:31528944

Junghare, M., Spiteller, D., & Schink, B. (2019). Anaerobic degradation of xenobiotic isophthalate by the fermenting bacterium *Syntrophorhabdus aromaticivorans*. *The ISME Journal*, *13*(5), 1252–1268. doi:10.103841396-019-0348-5 PMID:30647456

Ju, W., Liu, L., Fang, L., Cui, Y., Duan, C., & Wu, H. (2019). Impact of co-inoculation with plant-growth-promoting rhizobacteria and *Rhizobium* on the biochemical responses of alfalfa-soil system in copper contaminated soil. *Ecotoxicology and Environmental Safety*, *167*, 218–226. doi:10.1016/j.ecoenv.2018.10.016 PMID:30342354

Juwarkar, A. A., Nair, A., Dubey, K. V., Singh, S. K., & Devotta, S. (2007). Biosurfactant technology for remediation of cadmium and lead contaminated soils. *Chemosphere*, *68*(10), 1996–2002. doi:10.1016/j.chemosphere.2007.02.027 PMID:17399765

Kabbashi, N. A., Atiehc, M. A., Al-Mamuna, A., Mirghamia, M. E. S., Alama, M. D. Z., & Yahyaa, N. (2009). Kinetic adsorption of application ofcarbon nanotubes for Pb(II) removal from aqueous solution. *Journal of Environmental Sciences (China)*, *21*(4), 539–544. doi:10.1016/S1001-0742(08)62305-0 PMID:19634432

Kabra, A. N., Khandare, R. V., & Govindwar, S. P. (2013). Development of a bioreactor for remediation of textile effluent and dye mixture: A plant–bacterial synergistic strategy. *Water Research*, *47*(3), 1035–1048. doi:10.1016/j.watres.2012.11.007 PMID:23245543

Kabra, A. N., Khandare, R. V., Kurade, M. B., & Govindwar, S. P. (2011). Phytoremediation of a sulphonated azo dye Green HE4B by *Glandularia pulchella* (Sweet) Tronc.(Moss Verbena). *Environmental Science and Pollution Research International*, *18*(8), 1360–1373. doi:10.100711356-011-0491-7 PMID:21465161

Kadam, A. A., Kulkarni, A. N., Lade, H. S., & Govindwar, S. P. (2014). Exploiting the potential of plant growth promoting bacteria in decolorization of dye Disperse Red 73 adsorbed on milled sugarcane bagasse under solid state fermentation. *International Biodeterioration & Biodegradation*, *86*, 364–371. doi:10.1016/j.ibiod.2013.10.012

Kaddafa, A. A. (2012). Oil exploration and spillage in the Niger Delta of Nigeria. *Civ. Environ. Res.*, *2*, 38–51.

Kaewlaoyoong, A., Cheng, C. Y., Lin, C., Chen, J. R., Huang, W. Y., & Sriprom, P. (2020). White rot fungus Pleurotus pulmonarius enhanced bioremediation of highly PCDD/F-contaminated field soil via solid state fermentation. *The Science of the Total Environment*, *738*, 139670. doi:10.1016/j.scitotenv.2020.139670 PMID:32534283

Kagalkar, A. N., Khandare, R. V., & Govindwar, S. P. (2015). Textile dye degradation potential of plant laccase significantly enhances upon augmentation with redox mediators. *RSC Advances*, *5*(98), 80505–80517. doi:10.1039/C5RA12454A

Kalam, S. U., Naushin, F., Khan, F. A., & Rajakaruna, N. (2019). Long-term phytoremediating abilities of *Dalbergia sissoo* Roxb. (Fabaceae). *SN Applied Sciences*, *1*(501), 1–8. doi:10.100742452-019-0510-8

Kalavathy, S. (2004). The Multidisciplinary nature of environmental studies. *Environmental Studies, 1.*

Kalayu, G. (2019). Phosphate solubilizing microorganisms: promising approach as biofertilizers. *International Journal of Agronomy*.

Kalme, S., Jadhav, S., Jadhav, M., & Govindwar, S. (2009). Textile dye degrading laccase from *Pseudomonas desmolyticum* NCIM 2112. *Enzyme and Microbial Technology*, *44*(2), 65–71. doi:10.1016/j.enzmictec.2008.10.005

Kalyani, P., Sailaja, B., & Hemalatha, K. P. J. (2017). Degradation of textile dyes by *Aspergillus fumigatus* strain and their culture optimization. *International Journal of Current Research*, *9*(12), 62229–62232.

Kaminsky, L. M., Trexler, R. V., Malik, R. J., Hockett, K. L., & Bell, T. H. (2019). The Inherent Conflicts in Developing Soil Microbial Inoculants. In Trends in Biotechnology (Vol. 37, Issue 2, pp. 140–151). doi:10.1016/j.tibtech.2018.11.011

Kandeler, E. (2007). Physiological and biochemical methods for studying soil biota and their function. In *Soil microbiology, ecology and biochemistry* (pp. 53–83). Academic Press. doi:10.1016/B978-0-08-047514-1.50007-X

Kanel, S. R., Greeneche, J. M., & Choi, H. (2006). Arsenic (V) removal from groundwater using nano scale zerovalent iron as a colloidal reactive barrier material. *Environmental Science & Technology*, *40*(6), 2045–2050. doi:10.1021/es0520924 PMID:16570634

Kang, B. G., Kim, W. T., Yun, H. S., & Chang, S. C. (2010). Use of plant growth-promoting rhizobacteria to control stress responses of plant roots. *Plant Biotechnology Reports*, *4*(3), 179–183. doi:10.100711816-010-0136-1

Kang, B. R., Lee, J. H., Ko, S. J., Lee, Y. H., Cha, J. S., Cho, B. H., & Kim, Y. C. (2004). Degradation of acyl-homoserine lactone molecules by Acinetobacter sp. strain C1010. *Canadian Journal of Microbiology*, *50*(11), 935–941. doi:10.1139/w04-083 PMID:15644910

Kang, C.-H., Kwon, Y.-J., & So, J.-S. (2016). Bioremediation of heavy metals by using bacterial mixtures. *Ecological Engineering*, *89*, 64–69.

Kang, J. P., Huo, Y., Kim, Y. J., Ahn, J. C., Hurh, J., Yang, D. U., & Yang, D. C. (2019). *Rhizobium panacihumi* sp. *nov*., an isolate from ginseng-cultivated soil, as a potential plant growth promoting bacterium. *Archives of Microbiology*, *201*(1), 99–105. doi:10.100700203-018-1578-z PMID:30259064

Kang, S. M., Khan, A. L., You, Y. H., Kim, J. G., Kamran, M., & Lee, I. J. (2014). Gibberellin production by newly isolated strain Leifsonia soli SE134 and its potential to promote plant growth. *Journal of Microbiology and Biotechnology, 24*(1), 106–112. doi:10.4014/jmb.1304.04015 PMID:24100624

Kang, S. M., Radhakrishnan, R., Khan, A. L., Kim, M. J., Park, J. M., Kim, B. R., Shin, D.-H., & Lee, I. J. (2014). gibberellin secreting rhizobacterium, Pseudomonas putida H-2-3 modulates the hormonal and stress physiology of soybean to improve the plant growth under saline and drought conditions. *Plant Physiology and Biochemistry, 84*, 115–124. doi:10.1016/j.plaphy.2014.09.001 PMID:25270162

Kang, S. M., Shahzad, R., Bilal, S., Khan, A. L., Park, Y. G., Lee, K. E., Asaf, S., Khan, M. A., & Lee, I. J. (2019). Indole-3-acetic-acid and ACC deaminase producing Leclercia adecarboxylata MO1 improves Solanum lycopersicum L. growth and salinity stress tolerance by endogenous secondary metabolites regulation. *BMC Microbiology, 19*(1), 80. doi:10.118612866-019-1450-6 PMID:31023221

Kanhayuwa, L., Kotta-Loizou, I., Ozkan, S., Gunning, A. P., & Coutts, R. H. (2015). A novel mycovirus from *Aspergillus fumigatus* contains four unique dsRNAs as its genome and is infectious as dsRNA. *Proceedings of the National Academy of Sciences of the United States of America, 112*(29), 9100–9105. doi:10.1073/pnas.1419225112 PMID:26139522

Kanianska, R. (2016). Agriculture and Its Impact on Land-Use, Environment,and Ecosystem Services. In A. Almusaed (Eds.), Landscape Ecology. The Influences of Land Use and Anthropogenic Impacts of Landscape Creation. IntechOpen. doi:10.5772/63719

Kanmani, S., & Gandhimathi, R. (2013). Assessment of heavy metal contamination in soil due to leachate migration from an open dumping site. *Applied Water Science, 3*(1), 193–205. doi:10.100713201-012-0072-z

Kannaiyan, S. (2002). Biotechnology of biofertilizers. Alpha Science Int'l Ltd.

Kapahi, M., & Sachdeva, S. (2019). Bioremediation Options for Heavy Metal Pollution. *Journal of Health & Pollution, 9*(24), 1–20. doi:10.5696/2156-9614-9.24.191203 PMID:31893164

Kapoor, A., Viraraghavan, T., & Cullimore, D. R. (1999). Removal of heavy metals using the fungus *Aspergillus niger. Bioresource Technology, 70*(1), 95–104. doi:10.1016/S0960-8524(98)00192-8

Karadeniz, A., Topcuoğlu, Ş. F., & Inan, S. (2006). Auxin, gibberellin, cytokinin and abscisic acid production in some bacteria. *World Journal of Microbiology & Biotechnology, 22*(10), 1061–1064. doi:10.100711274-005-4561-1

Karcher, S., Kornmüller, A., & Jekel, M. (2001). Screening of commercial sorbents for the removal of reactive dyes. *Dyes and Pigments, 51*(2-3), 111–125. doi:10.1016/S0143-7208(01)00066-3

Kardos, N., & Demain, A. L. (2011). Penicillin: The medicine with the greatest impact on therapeutic outcomes. *Applied Microbiology and Biotechnology, 92*(4), 677–687. doi:10.100700253-011-3587-6 PMID:21964640

Karigar, C. S., & Rao, S. S. (2011). Role of microbial enzymes in the bioremediation of pollutants: A review. *Enzyme Research, 2011*, 2011. doi:10.4061/2011/805187 PMID:21912739

Karlidag, H., Esitken, A., Turan, M., & Sahin, F. (2007). Effects of root inoculation of plant growth promoting rhizobacteria (PGPR) on yield, growth and nutrient element contents of leaves of apple. *Scientia Horticulturae, 114*(1), 16–20. doi:10.1016/j.scienta.2007.04.013

Karpouzas, D. G., & Walker, A. (2000). Factors influencing the ability of Pseudomonas putida epI to degrade ethoprophos in soil. *Soil Biology & Biochemistry, 32*(11-12), 1753–1762. doi:10.1016/S0038-0717(00)00093-6

Karthik, C., Oves, M., Thangabalu, R., Sharma, R., Santhosh, S. B., & Arulselvi, P. I. (2016). *Cellulosimicrobium funkei-like* enhances the growth of *Phaseolus vulgaris* by modulating oxidative damage under Chromium (VI) toxicity. *Journal of Advanced Research*, *7*(6), 839–850. doi:10.1016/j.jare.2016.08.007 PMID:27668092

Karthikeyan, S., Balasubramanian, R., & Iyer, C. (2007). Evaluation of the marine algae *Ulva fasciata* and *Sargassum* sp. for the biosorption of Cu (II) from aqueous solutions. *Bioresource Technology*, *98*(2), 452–455. doi:10.1016/j.biortech.2006.01.010 PMID:16530408

Kasno, A., Adiningsih, S., & Subowo, S. (2000). Pollution status of Lead and Cadmium in lowland rice intensifies the Pantura route of West Java. *Journal of Soil and Environmental Sciences*, *3*(2), 25–32.

Kästner, M., & Miltner, A. (2016). Application of compost for effective bioremediation of organic contaminants and pollutants in soil. *Applied Microbiology and Biotechnology*, *100*(8), 3433–3449. doi:10.100700253-016-7378-y PMID:26921182

Kataoka, R., Takagi, K., & Sakakibara, F. (2011). Biodegradation of endosulfan by *Mortieralla* sp. strain W8 in soil: Influence of different substrates on biodegradation. *Chemosphere*, *85*(3), 548–552. doi:10.1016/j.chemosphere.2011.08.021 PMID:21893334

Kathi, S., & Khan, A. B. (2011). Phytoremediation approaches to PAH contaminated soil. *Indian Journal of Science and Technology*, *4*(1), 56–63. doi:10.17485/ijst/2011/v4i1.15

Kaur, H., Rajor, A., & Singh Kaleka, A. S. (2019). Role of Phycoremediation to Remove Heavy Metals from Sewage Water: Review Article. *Journal of Environmental Science and Technology*, *12*(1), 1–9. doi:10.3923/jest.2019.1.9

Kaur, P., Singh, S., Kumar, V., Singh, N., & Singh, J. (2018). Effect of rhizobacteria on arsenic uptake by macrophyte *Eichhornia crassipes* (Mart.) Solms. *International Journal of Phytoremediation*, *20*(2), 114–120. doi:10.1080/15226514.2017.1337071 PMID:28613914

Kaushik, B. D. (1989). Reclamative potential of cyanobacteria in salt-affected soils. *Phykos (Algiers)*, *28*, 101–109.

Kaushik, B. D., & Krishnamurti, G. S. R. (1981). Effect of blue-green algae and gypsum application on physico-chemical properties of alkali soils. *Phykos (Algiers)*, *20*, 91–94.

Kaushik, B. D., & Subhashini, D. (1985). Amelioration of salt-affected soils with blue-green algae. II Improvement in soil properties. *Proc Indian Natl Sci Acad Part B*, *51*, 386–389.

Kavino, M., Harish, S., Kumar, N., Saravanakumar, D., Damodaran, T., Soorianathasundaram, K., & Samiyappan, R. (2007). Rhizosphere and endophytic bacteria for induction of systemic resistance of banana plantlets against bunchy top virus. *Soil Biology & Biochemistry*, *39*(5), 1087–1098. doi:10.1016/j.soilbio.2006.11.020

Kaya, A. G. A., Dogmus Lehtijarvi, H. T., & Lehtijarvi, A. (2015). The usage of mycoviruses in biological control against tree pathogenic fungi. *Istanbul Üniversitesi Orman Fakültesi Dergisi*, *1*, 60–71.

Kdasi, A., Idris, A., Saed, K., & Guan, C. (2004). Treatment of textile wastewater by advanced oxidation processes: A review. *Global Nest: The International Journal*, *6*(3), 222–230.

Keceli, S. A. (2017). Mycoviruses and importance in mycology. *Mikrobiyoloji Bulteni*, *51*(4), 404–412. PMID:29153071

Keharia, H., & Madamwar, D. (2003). Bioremediation concepts for treatment of dye containing waste water: A review. *Indian Journal of Experimental Biology*, *41*(9), 1068–1075. PMID:15242298

Keller, A. A., & Fox, J. (2019). Giving credit to reforestation for water quality benefits. *PLoS One*, *14*(6), 1–18. doi:10.1371/journal.pone.0217756 PMID:31163057

Kembel, S. W., O'Connor, T. K., Arnold, H. K., Hubbell, S. P., Wright, S. J., & Green, J. L. (2014). Relationships between phyllosphere bacterial communities and plant functional traits in a neotropical forest. *Proceedings of the National Academy of Sciences of the United States of America*, *111*(38), 13715–13720. doi:10.1073/pnas.1216057111 PMID:25225376

Kenawy, A., Dailin, D. J., Abo-Zaid, G. A., Abd Malek, R., Ambehabati, K. K., Zakaria, K. H. N., Sayyed, R. Z., & El Enshasy, H. A. (2019). Biosynthesis of antibiotics by PGPR and their roles in biocontrol of plant diseases. In *Plant Growth Promoting Rhizobacteria for Sustainable Stress Management* (pp. 1–35). Springer. doi:10.1007/978-981-13-6986-5_1

Kennedy, C., Rudnick, P., MacDonald, T., & Melton, T. (2005). Genus Azotobacter. In G. M. Garirity. Bergey's Manual of Systematic Bacteriology, 2, 384–401.

Khaledian, Y., Kiani, F., & Ebrahimi, S. (2012). The effect of land use change on soil and water quality in northern Iran. *Journal of Mountain Science*, *9*(6), 798–816. doi:10.100711629-012-2301-1

Khalid, S., Shahid, M., Natasha, Bibi, I., Sarwar, T., Shah, A., & Niazi, N. (2018). A Review of Environmental Contamination and Health Risk Assessment of Wastewater Use for Crop Irrigation with a Focus on Low and High-Income Countries. *International Journal of Environmental Research and Public Health*, *15*(5), 895–931. doi:10.3390/ijerph15050895 PMID:29724015

Khan, S. M., & Gomes, J. (2018). *An interdisciplinary population health approach to the radon health risk management in Canada*. Academic Press.

Khandare, R. N., Chandra, R., Pareek, N., & Raverkar, K. P. (2020). Carrier-based and liquid bioinoculants of Azotobacter and PSB saved chemical fertilizers in wheat (Triticum aestivum L.) and enhanced soil biological properties in Mollisols. *Journal of Plant Nutrition*, *43*(1), 36–50. doi:10.1080/01904167.2019.1659333

Khandare, R. V., Kabra, A. N., Awate, A. V., & Govindwar, S. P. (2013). Synergistic degradation of diazo dye Direct Red 5B by *Portulaca grandiflora* and *Pseudomonas putida*. *International Journal of Environmental Science and Technology*, *10*(5), 1039–1050. doi:10.100713762-013-0244-x

Khan, M. I. R., Saha, R. K., & Saha, H. (2018). Muli bamboo (*Melocanna baccifera*) leaves ethanolic extract a non-toxic phyto-prophylactic against low pH stress and saprolegniasis in *Labeo rohita* fingerlings. *Fish & Shellfish Immunology*, *74*, 609–619. doi:10.1016/j.fsi.2017.11.047 PMID:29183812

Khan, M. I. R., Singh, M., & Monsang, S. J. (2019). Prophylactic potential of ethanolic leaves extract of Muli bamboo (*Melocanna baccifera*) against *Aeromonas hydrophila* infection in *Labeo rohita*. *Journal of Entomology and Zoology Studies*, *7*(2), 288–291.

Khan, M. S., Zaidi, A., & Ahmad, E. (2014). Mechanism of Phosphate Solubilization and Physiological Functions of Phosphate-Solubilizing Microorganisms. In M. Khan, A. Zaidi, & J. Musarrat (Eds.), *Phosphate Solubilizing Microorganisms*. Springer., doi:10.1007/978-3-319-08216-5_2

Khan, M. S., Zaidi, A., & Wani, P. A. (2007). Role of phosphate-solubilizing microorganisms in sustainable agriculture—A review. *Agronomy for Sustainable Development*, *27*(1), 29–43. doi:10.1051/agro:2006011

Khan, N., Bano, A., & Babar, M. A. (2019). The stimulatory effects of plant growth promoting rhizobacteria and plant growth regulators on wheat physiology grown in sandy soil. *Archives of Microbiology*, *201*(6), 769–785. doi:10.100700203-019-01644-w PMID:30843087

Khanna, S., & Gharpure, A. S. (2017). Petroleum Carcinogenicity and Aerodigestive Tract: In Context of Developing Nations. *Cureus*, *9*(4), e1202. doi:10.7759/cureus.1202 PMID:28573078

Khan, R., Bhawana, P., & Fulekar, M. H. (2013). Microbial decolorization and degradation of synthetic dyes: A review. *Reviews in Environmental Science and Biotechnology, 12*(1), 75–97. doi:10.100711157-012-9287-6

Khan, S. A., Rashmi, Hussain, M. Z., Prasad, S., & Banerjee, U. C. (2009). Prospects of biodiesel production from microalgae in India. *Renewable & Sustainable Energy Reviews, 13*(9), 2361–2372. doi:10.1016/j.rser.2009.04.005

Khan, S. R., & Farrand, S. K. (2009). The BlcC (AttM) lactonase of Agrobacterium tumefaciens does not quench the quorum-sensing system that regulates Ti plasmid conjugative transfer. *Journal of Bacteriology, 191*(4), 1320–1329. doi:10.1128/JB.01304-08 PMID:19011037

Khan, S., Cao, Q., Zheng, Y. M., Huang, Y. Z., & Zhu, Y. G. (2008). Health risks of heavy metals in contaminated soils and food crops irrigated with wastewater in Beijing, China. *Environmental Pollution, 152*(3), 686–692. doi:10.1016/j.envpol.2007.06.056 PMID:17720286

Khan, Z., Kim, S. G., Jeon, Y. H., Khan, H. U., Son, S. H., & Kim, Y. H. (2008). A plant growth promoting rhizobacterium, *Paenibacillus polymyxa* strain GBR-1, suppresses root-knot nematode. *Bioresource Technology, 99*(8), 3016–3023. doi:10.1016/j.biortech.2007.06.031 PMID:17706411

Khan, Z., Kim, Y. H., Kim, S. G., & Kim, H. W. (2007). Observations on the suppression of root knot nematode (*Meloidogyne arenari*a) on tomato by incorporation of cyanobacterial powder (*Oscillatoria chlorina*) into potting field soil. *Bioresource Technology, 98*(1), 69–73. doi:10.1016/j.biortech.2005.11.029 PMID:16458501

Khan, Z., & Park, S. D. (1999). Effects of inoculum level and time of *Microcoleus vaginatus* on control of *Meloidogyne incognita* on tomato. *Journal of Asia-Pacific Entomology, 2*(2), 93–96. doi:10.1016/S1226-8615(08)60036-9

Khasa, Y. P. (2017). Microbes as biocontrol agents. In *Probiotics and Plant Health* (pp. 507–552). Springer.

Khehra, M. S., Saini, H. S., Sharma, D. K., Chadha, B. S., & Chimni, S. S. (2006). Biodegradation of azo dye C.I. Acid Red 88 by an anoxic–aerobic sequential bioreactor. *Dyes and Pigments, 70*(1), 1–7. doi:10.1016/j.dyepig.2004.12.021

Kheyrodin, H. (2014). Methodology for measurement of enzyme activity in soil. *World J Biol Med Science, 1*(1), 18–25.

Kholssi, R., Marks, E. A., Miñón, J., Montero, O., Debdoubi, A., & Rad, C. (2019). Biofertilizing effect of Chlorella sorokiniana suspensions on wheat growth. *Journal of Plant Growth Regulation, 38*(2), 644–649. doi:10.100700344-018-9879-7

Khosravi-Darani, K., & Bucci, D. Z. (2015). Application of poly (hydroxyalkanoate) in food packaging: Improvements by nanotechnology. *Chemical and Biochemical Engineering Quarterly, 29*(2), 275–285. doi:10.15255/CABEQ.2014.2260

Khuong, N. Q., Kantachote, D., Nookongbut, P., Xuan, L. N. T., Nhan, T. C., Xuan, N. T. T., & Tantirungkij, M. (2020). Potential of Mn 2+-Resistant Purple Nonsulfur Bacteria Isolated from Acid Sulfate Soils to Act as Bioremediators and Plant Growth Promoters via Mechanisms of Resistance. *Journal of Soil Science and Plant Nutrition, 20*(4), 1–15. doi:10.100742729-020-00303-0

Kiba, T., & Krapp, A. (2016). Plant Nitrogen Acquisition Under Low Availability: Regulation of Uptake and Root Architecture. *Plant & Cell Physiology, 57*(4), 707–714. doi:10.1093/pcp/pcw052 PMID:27025887

Kibblewhite, M. G., Ritz, K., & Swift, M. J. (2008). Soil health in agricultural systems. *Philosophical Transactions of the Royal Society of London. Series B, Biological Sciences, 363*(1492), 685–701. doi:10.1098/rstb.2007.2178 PMID:17785275

Kiflu, A., & Beyene, S. (2013). Effects of different land use systems on selected soil properties in south Ethiopia. *Journal of Soil Science and Environmental Management, 4*(5), 100–107. doi:10.5897/JSSEM2013.0380

Kim, H. S., Kim, Y. J., & Seo, Y. R. (2015). An overview of carcinogenic heavy metal: Molecular toxicity mechanism and prevention. *Journal of Cancer Prevention*, *20*(4), 232–240. doi:10.15430/JCP.2015.20.4.232 PMID:26734585

Kim, H., Hong, H.-J., Jung, J., Kim, S.-H., & Yang, J.-W. (2010). Degradation of trichloroethylene (TCE) by nanoscale zero-valent iron (nZVI) immobilized in alginate bead. *Journal of Hazardous Materials*, *176*(1-3), 1038–1043. doi:10.1016/j.jhazmat.2009.11.145 PMID:20042289

Kim, I. H., Choi, J. H., Joo, J. O., Kim, Y. K., Choi, J. W., & Oh, B. K. (2015). Development of a microbe-zeolite carrier for the efective elimination of heavy metals from seawater. *Journal of Microbiology and Biotechnology*, *25*(9), 1542–1546. doi:10.4014/jmb.1504.04067 PMID:26032363

Kim, J., & Rees, D. C. (1994). Nitrogenase and biological nitrogen fixation. *Biochemistry*, *33*(2), 389–397. doi:10.1021/bi00168a001 PMID:8286368

Kim, J.-Y., Cohen, C., Shuler, M. L., & Lion, L. W. (2000). Use of Amphiphilic Polymer Particles for In Situ Extraction of Sorbed Phenanthrene from a Contaminated Aquifer Material. *Environmental Science & Technology*, *34*(19), 4133–4139. doi:10.1021/es001021w

Kim, K., Jordan, D., & McDonald, G. (1997). Effect of phosphate-solubilizing bacteria and vesicular-arbuscular mycorrhizae on tomato growth and soil microbial activity. *Biology and Fertility of Soils*, *26*(2), 79–87. doi:10.1007003740050347

Kim, M. H., Choi, W. C., Kang, H. O., Lee, J. S., Kang, B. S., Kim, K. J., Derewenda, Z. S., Oh, T. K., Lee, C. H., & Lee, J. K. (2005). The molecular structure and catalytic mechanism of a quorum-quenching N-acyl-L-homoserine lactone hydrolase. *Proceedings of the National Academy of Sciences of the United States of America*, *102*(49), 17606–17611. doi:10.1073/pnas.0504996102 PMID:16314577

Kim, S., Krajmalnik-Brown, R., Kim, J. O., & Chung, J. (2014). Remediation of petroleum hydrocarbon-contaminated sites by DNA diagnosis-based bioslurping technology. *The Science of the Total Environment*, *497*, 250–259. doi:10.1016/j.scitotenv.2014.08.002 PMID:25129160

Kim, Y. M., Murugesan, K., Chang, Y. Y., Kim, E. J., & Chang, Y. S. (2011). Degradation of polybrominated diphenyl ethers by a sequential treatment with nanoscale zero valent iron and aerobic biodegradation. *Journal of Chemical Technology and Biotechnology (Oxford, Oxfordshire)*, *87*(2), 216–224. doi:10.1002/jctb.2699

King, A. M. Q., Lefkowitz, E. M., Adams, J., & Carstens, E. B. (2011). *Virus Taxonomy: Ninth Report of the International Committee on Taxonomy of Viruses (ICTV)*. Elsevier Academic.

Kinoshita, S., Nakayama, K., & Kitada, S. (1958). Production of aspartic acid from fumaric acid by microorganism. *Hakko Kyokaishi*, *16*, 517–520.

Kinuthia, G. K., Ngure, V., Beti, D., Lugalia, R., Wangila, A., & Kamau, L. (2020). Levels of heavy metals in wastewater and soil samples from open drainage channels in Nairobi, Kenya: Community health implication. *Scientific Reports*, *10*(1), 8434–8447. doi:10.103841598-020-65359-5 PMID:32439896

Kiran, B., Kaushik, A., & Kaushik, C. P. (2008). Metal-salt co-tolerance and metal removal by indigenous cyanobacterial strains. *Process Biochemistry*, *43*(6), 598–604. doi:10.1016/j.procbio.2008.01.019

Kirpichtchikova, T. A., Manceau, A., Spadini, L., Panfili, F., Marcus, M. A., & Jacquet, T. (2006). Speciation and solubility of heavy metals in contaminated soil using X-ray microfluorescence, EXAFS spectroscopy, chemical extraction, and thermodynamic modeling. *Geochimica et Cosmochimica Acta*, *70*(9), 2163–2190. doi:10.1016/j.gca.2006.02.006

Kirschbaum, M. U. (2000). Will changes in soil organic carbon act as a positive or negative feedback on global warming? *Biogeochemistry*, *48*(1), 21–51. doi:10.1023/A:1006238902976

Kirschling, T. L., Gregory, K. B., Minkley, E. G. Jr, Lowry, G. V., & Tilton, R. D. (2010). Impact of Nanoscale Zero Valent Iron on Geochemistry and Microbial Populations in Trichloroethylene Contaminated Aquifer Materials. *Environmental Science & Technology*, *44*(9), 3474–3480. doi:10.1021/es903744f PMID:20350000

Kishi, R. N. I., Júnior, R. F. G., Val-Moraes, S. P., & Kishi, L. T. (2017). Soil Microbiome and Their Effects on Nutrient Management for Plants. In Probiotics in Agroecosystem (pp. 117-143). Academic Press.

Kisiala, A., Laffont, C., Emery, R. N., & Frugier, F. (2013). Bioactive cytokinins are selectively secreted by Sinorhizobium meliloti nodulating and nonnodulating strains. *Molecular Plant-Microbe Interactions*, *26*(10), 1225–1231. doi:10.1094/MPMI-02-13-0054-R PMID:24001254

Kiyono, M., & Pan-Hou, H. (1999). The merG gene product is involved in phenylmercury resistance in Pseudomonas strain K-62. *Journal of Bacteriology*, *181*(3), 726–730.

Kjerstadius, H., Saraiva, A. B., & Spångberg, J. (2016). Can source separation increase sustainability of sanitation management. *Impact on Nutrient Recovery, Climate Change and Eutrophication of Two Sanitation Systems for a Hypothetical Urban Area in Southern Sweden Using Life Cycle Assessment (opublicerat)*. https://va-tekniksodra.se/2016/10/can-source-separation-increase-sustainability-of-sanitation-management/

Klangsin, P., & Harding, A. K. (1998). Medical waste treatment and disposal methods used by hospitals in Oregon, Washington, and Idaho. *Journal of the Air & Waste Management Association*, *48*(6), 516–526. doi:10.1080/10473289.1998.10463706 PMID:9949738

Klironomos, J. N. (2003). Variation in plant response to native and exotic arbuscular mycorrhizal fungi. *Ecology*, *84*(9), 2292–2301. doi:10.1890/02-0413

Kloepper, J. W., Lifshitz, R., & Zablotowicz, R. M. (1989). Free-living bacterial inocula for enhancing crop productivity. *Trends in Biotechnology*, *7*(2), 39–43. doi:10.1016/0167-7799(89)90057-7

Klonowska, A., Gaudin, C., Fournel, A., Asso, M., Le Petit, J., Giorgi, M., & Tron, T. (2002). Characterization of a low redox potential laccase from the basidiomycete C30. *European Journal of Biochemistry*, *269*(24), 6119–6125. doi:10.1046/j.1432-1033.2002.03324.x PMID:12473107

Knapp, J. S., & Newby, P. S. (1995). The microbiological decolorization of an industrial effluent containing a diazo-linked chromophore. *Water Research*, *29*(7), 1807–1809. doi:10.1016/0043-1354(94)00341-4

Knapp, J. S., Newby, P. S., & Reece, L. P. (1995). Decolorization of dyes by wood-rotting basidiomycete fungi. *Enzyme and Microbial Technology*, *17*(7), 664–668. doi:10.1016/0141-0229(94)00112-5

Kobayashi, D. Y., Reedy, R. M., Bick, J., & Oudemans, P. V. (2002). Characterization of a chitinase gene from Stenotrophomonas maltophilia strain 34S1 and its involvement in biological control. *Applied and Environmental Microbiology*, *68*(3), 1047–1054. doi:10.1128/AEM.68.3.1047-1054.2002 PMID:11872449

Kobayashi, T., & Nishizawa, N. K. (2012). Iron uptake, translocation, and regulation in higher plants. *Annual Review of Plant Biology*, *63*(1), 131–152. doi:10.1146/annurev-arplant-042811-105522 PMID:22404471

Kobayashi, T., Ogo, Y., Aung, M. S., Nozoye, T., Itai, R. N., Nakanishi, H., Yamakawa, T., & Nishizawa, N. K. (2010). The spatial expression and regulation of transcription factors IDEF1 and IDEF2. *Annals of Botany*, *105*(7), 1109–1117. doi:10.1093/aob/mcq002 PMID:20197292

Koenig, J. C., Boparai, H. K., Lee, M. J., O'Carroll, D. M., Barnes, R. J., & Manefield, M. J. (2016). Particles and enzymes: Combining nanoscale zero valent iron and organochlorine respiring bacteria for the detoxification of chloroethane mixtures. *Journal of Hazardous Materials*, *308*, 106–112. doi:10.1016/j.jhazmat.2015.12.036 PMID:26808236

Koh, H., Kwon, S., & Thomson, M. (2015). *Current Technologies in Plant Molecular Breeding: A Guide Book of Plant Molecular Breeding for Researchers*. Springer. doi:10.1007/978-94-017-9996-6

Kohler, J., Hernaındez, J. A., Caravaca, F., & Roldaın, A. (2008). Plant-growth promoting rhizobacteria and arbuscular mycorrhizal fungi modify alleviation biochemical mechanisms in water stressed plants. *Functional Plant Biology*, *35*(2), 141–151. doi:10.1071/FP07218 PMID:32688765

Köhl, J., Kolnaar, R., & Ravensberg, W. J. (2019). Mode of action of microbial biological control agents against plant diseases: Relevance beyond efficacy. *Frontiers in Plant Science*, *10*, 845. doi:10.3389/fpls.2019.00845 PMID:31379891

Koller, M. (2014). Poly (hydroxyalkanoates) for food packaging: Application and attempts towards implementation. *Applied Food Biotechnology, 1*(1), 3-15.

Komancová, M., Jurčová, I., Kochánková, L., & Burkhard, J. (2003). Metabolic pathways of polychlorinated biphenyls degradation by *Pseudomonas* sp. 2. *Chemosphere*, *50*(4), 537–543. doi:10.1016/S0045-6535(02)00374-0 PMID:12685753

Komossa, D., Langebartels, C., & Sandermann, H. Jr. (1995). Metabolic processes for organic chemicals in plants. In *Plant contamination: Modeling and simulation of organic chemical processes* (pp. 60–103). CRC Press.

Koo, Y. M., Heo, A. Y., & Choi, H. W. (2020). Salicylic acid as a safe plant protector and growth regulator. *The Plant Pathology Journal*, *36*(1), 1–10. doi:10.5423/PPJ.RW.12.2019.0295 PMID:32089657

Koppel, N., Rekdal, V. M., & Balskus, E. P. (2017). Chemical transformation of xenobiotics by the human gut microbiota. *Science*, *356*(6344), eaag2770. doi:10.1126cience.aag2770 PMID:28642381

Koshlaf, E., & Ball, A. S. (2017). Soil bioremediation approaches for petroleum hydrocarbon polluted environments. *AIMS Microbiology*, *3*(1), 25–49. doi:10.3934/microbiol.2017.1.25 PMID:31294147

Kotasthane, A. S., Agrawal, T., Zaidi, N. W., & Singh, U. S. (2017). Identification of siderophore producing and cyno-genic fluorescent Pseudomonas and a simple confrontation assay to identify potential bio-control agent for collar rot of chickpea. *Biotech, 7*(2), 137.

Kothawala, D. N., Roehm, C., Blodau, C., & Moore, T. R. (2012). Selective adsorption of dissolved organic matter to mineral soils. *Geoderma*, *189-190*, 334–342. doi:10.1016/j.geoderma.2012.07.001

Kotoky, R., & Pandey, P. (2019). Rhizosphere mediated biodegradation of benzo (A) pyrene by surfactin producing soil bacilli applied through *Melia azadirachta* rhizosphere. *International Journal of Phytoremediation*, **•••**, 1–10. PMID:31522524

Kotoky, R., Rajkumari, J., & Pandey, P. (2018). The rhizosphere microbiome: Significance in rhizoremediation of polyaromatic hydrocarbon contaminated soil. *Journal of Environmental Management*, *217*, 858–870. doi:10.1016/j.jenvman.2018.04.022 PMID:29660711

Kotta-Loizou, I., & Coutts, R. H. A. (2017). Mycoviruses in Aspergilli: A comprehensive review. *Frontiers in Microbiology*, *8*, 1699. doi:10.3389/fmicb.2017.01699 PMID:28932216

Koua, D., Cerutti, L., Falquet, L., Sigrist, C. J., Theiler, G., Hulo, N., & Dunand, C. (2009). PeroxiBase: A database with new tools for peroxidase family classification. *Nucleic Acids Research*, *37*(Database, suppl_1), D261–D266. doi:10.1093/nar/gkn680 PMID:18948296

Koudelakova, T., Chovancova, E., Brezovsky, J., Monincova, M., Fortova, A., Jarkovsky, J., & Damborsky, J. (2011). Substrate specificity of haloalkane dehalogenases. *The Biochemical Journal*, *435*(2), 345–354. doi:10.1042/BJ20101405 PMID:21294712

Kour, R., Jain, D., Bhojiya, A. A., Sukhwal, A., Sanadhya, S., Saheewala, H., Jat, G., Singh, A., & Mohanty, S. R. (2019). Zinc biosorption, biochemical and molecular characterization of plant growth-promoting zinc-tolerant bacteria. *Biotech, 9*(11), 421.

Kourmentza, C., Plácido, J., Venetsaneas, N., Burniol-Figols, A., Varrone, C., Gavala, H. N., & Reis, M. A. M. (2017). Recent Advances and Challenges towards Sustainable Polyhydroxyalkanoate (PHA) Production. *Bioengineering (Basel, Switzerland), 4*(2), 55. Advance online publication. doi:10.3390/bioengineering4020055 PMID:28952534

Ko, Y. S., Kim, J. W., Lee, J. A., Han, T., Kim, G. B., Park, J. E., & Lee, S. Y. (2020). Tools and strategies of systems metabolic engineering for the development of microbial cell factories for chemical production. *Chemical Society Reviews, 49*(14), 4615–4636. doi:10.1039/D0CS00155D PMID:32567619

Kraepiel, A. M., Bellenger, J. P., Wichard, T., & Morel, F. M. (2009). Multiple roles of siderophores in free-living nitrogen-fixing bacteria. *Biometals: An International Journal on the Role of Metal Ions in Biology, Biochemistry, and Medicine, 22*(4), 573–581. doi:10.100710534-009-9222-7

Krewulak, K. D., & Vogel, H. J. (2008). Structural biology of bacterial iron uptake. *Biochimica et Biophysica Acta (BBA)- Biomembranes, 1778*(9), 1781–1804. doi:10.1016/j.bbamem.2007.07.026

Krings, M., Hass, H., Kerp, H., Taylor, T. N., Agerer, R., & Dotzler, N. (2009). Endophytic cyanobacteria in a 400-million-yr-old land plant: A scenario for the origin of a symbiosis? *Review of Palaeobotany and Palynology, 153*(1-2), 62–69. doi:10.1016/j.revpalbo.2008.06.006

Krishna, P. (2003). *Bioremediation of bauxite residue (red mud) using microbes* (Dissertation). Thapar Institute of Engineering and Technology, Patiala, Punjab, India

Krishnan, J., Kishore, A. A., Suresh, A., Madhumeetha, B., & Prakash, D. G. (2017). Effect of pH, inoculum dose and initial dye concentration on the removal of azo dye mixture under aerobic conditions. *International Biodeterioration & Biodegradation, 119*, 16–27. doi:10.1016/j.ibiod.2016.11.024

Krishna, P., Reddy, M. S., & Patnaik, S. K. (2005). *Aspergillus Tubingensis* Reduces the pH of the Bauxite Residue (Red Mud) Amended Soils. *Water, Air, and Soil Pollution, 167*(1-4), 201–209. doi:10.100711270-005-0242-9

Kruus, K., Kiiskinen, L. L., Saloheimo, M., Haklinen, N., Rouvinen, J., Paananen, A., Linder, M., & Viikari, L. (2002). A novel laccase from the ascomycete Melanocarpus albomyces. *Applied Microbiology and Biotechnology, 59*, 198–204. doi:10.100700253-002-1012-x PMID:12111146

Kshetri, L., Pandey, P., & Sharma, G. D. (2018). Rhizosphere mediated nutrient management in *Allium hookeri* Thwaites by using phosphate solubilizing rhizobacteria and tricalcium phosphate amended soil. *Journal of Plant Interactions, 13*(1), 256–269. doi:10.1080/17429145.2018.1472307

Kuan, K. B., Othman, R., Abdul Rahim, K., & Shamsuddin, Z. H. (2016). Plant growth-promoting rhizobacteria inoculation to enhance vegetative growth, nitrogen fixation and nitrogen remobilisation of maize under greenhouse conditions. *PLoS One, 11*(3), e0152478. doi:10.1371/journal.pone.0152478 PMID:27011317

Kube, M., Beck, A., Zinder, S. H., Kuhl, H., Reinhardt, R., & Adrian, L. (2005). Genome sequence of the chlorinated compound respiring bacterium *Dehalococcoides* species strain CBDB1. *Nature Biotechnology, 23*(10), 1269–1273. doi:10.1038/nbt1131 PMID:16116419

Kudlich, M., Keck, A., Klein, J., & Stolz, A. (1997). Localization of the enzyme system involved in anaerobic reduction of azo dyes by *Sphingomonas* sp. strain BN6 and effect of artificial redox mediators on the rate of azo dye reduction. *Applied and Environmental Microbiology, 63*(9), 3691–3694. doi:10.1128/AEM.63.9.3691-3694.1997 PMID:16535698

Kues, U. (2015). Fungal enzymes for environmental management. *Current Opinion in Biotechnology, 33*, 268–278. doi:10.1016/j.copbio.2015.03.006 PMID:25867110

Kuhad, R. C., Singh, S., & Singh, A. (2011). Phosphate-solubilizing microorganisms. In *Bioaugmentation, Biostimulation and Biocontrol* (pp. 65–84). Springer. doi:10.1007/978-3-642-19769-7_4

Kulkarni, A. N., Kadam, A. A., Kachole, M. S., & Govindwar, S. P. (2014). Lichen *Permelia perlata*: A novel system for biodegradation and detoxification of disperse dye Solvent Red 24. *Journal of Hazardous Materials, 276*, 461–468. doi:10.1016/j.jhazmat.2014.05.055 PMID:24929306

Kulshreshtha, S., Mathur, N., & Bhatnagar, P. (2014). Mushroom as a product and their role in mycoremediation. *AMB Express, 4*(1), 29. doi:10.118613568-014-0029-8 PMID:24949264

Kumar, A., Singh, A. K., Kaushik, M. S., Mishra, S. K., Raj, P., Singh, P. K., & Pandey, K. D. (2017a). Interaction of turmeric (*Curcuma longa* L.) with beneficial microbes: a review. *Biotech, 7*(6), 357. doi:10.100713205-017-0971-7

Kumar, K., & Dubey, K. (2009). Microbial-amelioration of Sodic/Alkaline Soil. In *National Seminar on Frontiers in Biotechnology* (NSFB-2009). Department of Biotechnology, Bharathiar University.

Kumar, A., Bahadur, I., Maurya, B. R., Raghuwanshi, R., Meena, V. S., Singh, D. K., & Dixit, J. (2015). Does a plant growth promoting rhizobacteria enhance agricultural sustainability? *Journal of Pure & Applied Microbiology, 9*, 715–724.

Kumar, A., Singh, R., Yadav, A., Giri, D. D., Singh, P. K., & Pandey, K. D. (2016). Isolation and characterization of bacterial endophytes of *Curcuma longa* L. *Biotech, 6*(1), 60. PMID:28330130

Kumar, A., Tripathi, R. D., Singh, N., Rai, U. N., & Singh, S. N. (2002). Biochemical Responses of *Cassia siamea* Lamk. Grown on Coal Combustion Residue Fly-ash. *Bulletin of Environmental Contamination and Toxicology, 68*(5), 675–683. doi:10.1007001280307 PMID:12068933

Kumar, A., Verma, H., Singh, V. K., Singh, P. P., Singh, S. K., & Ansari, W. A. (2017b). Role of *Pseudomonas* sp. in Sustainable Agriculture and Disease Management. In V. Meena, P. Mishra, J. Bisht, & A. Pattanayak (Eds.), *Agriculturally Important Microbes for Sustainable Agriculture*. Springer. doi:10.1007/978-981-10-5343-6_7

Kumar, B. (2016). Biocontrol of insect pests. In *Ecofriendly pest management for food security* (pp. 25–61). Academic Press.

Kumar, D. (2018). Biodegradation of γ-Hexachlorocyclohexane by Burkholderia sp. IPL04. *Biocatalysis and Agricultural Biotechnology, 16*, 331–339. doi:10.1016/j.bcab.2018.09.001

Kumar, D., Kumar, A., & Sharma, J. (2016). Degradation study of lindane by novel strains Kocuria sp. DAB-1Y and *Staphylococcus* sp. DAB-1W. *Bioresources and Bioprocessing, 3*(1), 53. doi:10.118640643-016-0130-8 PMID:28090433

Kumari, V., & Srivastava, J. (1999). Molecular and biochemical aspects of rhizobacterial ecology with emphasis on biological control. *World Journal of Microbiology & Biotechnology, 15*(5), 535–543. doi:10.1023/A:1008958912647

Kumar, M., Prasad, R., Goyal, P., Teotia, P., Tuteja, N., Varma, A., & Kumar, V. (2017). Environmental biodegradation of xenobiotics: role of potential microflora. In *Xenobiotics in the Soil Environment* (pp. 319–334). Springer. doi:10.1007/978-3-319-47744-2_21

Kumar, M., Prasanna, R., Bidyarani, N., Babu, S., Mishra, B. K., Kumar, A., Adak, A., Jauhari, S., Yadav, K., Singh, R., & Saxena, A. K. (2013). Evaluating the plant growth promoting ability of thermotolerant bacteria and cyanobacteria and their interactions with seed spice crops. *Scientia Horticulturae, 164*, 94–101. doi:10.1016/j.scienta.2013.09.014

Kumar, P. S. (2019). Soil bioremediation techniques. In *Advanced Treatment Techniques for Industrial Wastewater* (pp. 35–50). IGI Global. doi:10.4018/978-1-5225-5754-8.ch003

Kumar, R. R., & Prasad, S. (2011). Metabolic engineering of bacteria. *Indian Journal of Microbiology*, *51*(3), 403–409. doi:10.100712088-011-0172-8 PMID:22754024

Kumar, S. N., Siji, J. V., Ramya, R., Nambisan, B., & Mohandas, C. (2020). Improvement of antimicrobial activity of compounds produced by *Bacillus sp.* associated with a *Rhabditid sp.* (entomopathogenic nematode) by changing carbon and nitrogen sources in fermentation media. *Journal of Microbiology, Biotechnology and Food Sciences*, *9*(4), 1424–1438.

Kumar, S. R., Arumugam, T., Anandakumar, C. R., Balakrishnan, S., & Rajave, D. S. (2013). Use of Plant Species in Controlling Environmental Pollution- A Review. *Bull. Env. Pharmacol. Life Sci.*, *2*(2), 52–63.

Kumar, S., Chaudhuri, S., & Maiti, S. K. (2013). Soil dehydrogenase enzyme activity in natural and mine soil-a review. *Middle East Journal of Scientific Research*, *13*(7), 898–906.

Kumar, V., Behl, R. K., & Narula, N. (2001). Effect of phosphate solubilizing strains of *Azotobacter chroococcum* on yield traits and their survival in the rhizosphere of wheat genotypes under field conditions. *Acta Agronomica Hungarica*, *49*(2), 141–149. doi:10.1556/AAgr.49.2001.2.4

Kundan, R., Pant, G., Jadon, N., & Agrawal, P. K. (2015). Plant growth promoting rhizobacteria: Mechanism and current prospective. *Journal of Fertilizers and Pesticides*, *6*(2), 9. doi:10.4172/2471-2728.1000155

Kundu, D. K., & Ladha, J. K. (1995). Efficient management of soil and biologically fixed N2 in intensively cultivated rice fields. *Soil Biology & Biochemistry*, *27*(4-5), 431–439. doi:10.1016/0038-0717(95)98615-U

Kuppusamy, S., Maddela, N. R., Megharaj, M., Venkateswarlu, K., Kuppusamy, S., Megharaj, M., & Venkateswarlu, K. (2020). Approaches for remediation of sites contaminated with total petroleum hydrocarbons. In *Total Petroleum Hydrocarbons* (pp. 167–205). Springer International Publishing. doi:10.1007/978-3-030-24035-6_7

Kurasvili, M. V., Adamia, G. S., Ananiasvili, T. I., Varazi, T. G., Pruidze, M. V., Gordeziani, M. S., & Khatisashvili, G. A. (2014). Plants as tools for control and remediation of the environment polluted by organochlorine toxicants. *Annals of Agrarian Science*, *12*(3), 84–87.

Kurihara, T., & Esaki, N. (2008). Bacterial hydrolytic dehalogenases and related enzymes: Occurrences, reaction mechanisms, and applications. *Chemical Record (New York, N.Y.)*, *8*(2), 67–74. doi:10.1002/tcr.20141 PMID:18366103

Kuritz, T., & Wolk, C. P. (1995). Use of filamentous cyanobacteria for biodegradation of organic pollutants. *Applied and Environmental Microbiology*, *61*(1), 234–238. doi:10.1128/AEM.61.1.234-238.1995 PMID:7534052

Kurnia, Sutono, Anda, Sulaeman, Kurniawansyah, & Tala'ohu. (2000). *Assessment of soil quality standards on agricultural land.* Final Report on Research Collaboration between Bapedal and Puslitbangtanak.

Kurth, C., Kage, H., & Nett, M. (2016). Siderophores as molecular tools in medical and environmental applications. *Organic & Biomolecular Chemistry*, *14*(35), 8212–8227. doi:10.1039/C6OB01400C PMID:27492756

Ku, Y., Xu, G., Tian, X., Xie, H., Yang, X., & Cao, C. (2018). Root colonization and growth promotion of soybean, wheat and Chinese cabbage by *Bacillus cereus* YL6. *PLoS One*, *13*(11), e0200181. doi:10.1371/journal.pone.0200181 PMID:30462642

Kuzyakov, Y., & Razavi, B. S. (2019). Rhizosphere size and shape: Temporal dynamics and spatial stationarity. *Soil Biology & Biochemistry*, *135*, 343–360. doi:10.1016/j.soilbio.2019.05.011

Kvesitadze, E., Sadunishvili, T., & Kvesitadze, G. (2009). Mechanisms of organic contaminants uptake and degradation in plants. *World Academy of Science, Engineering and Technology*, *55*, 458–468.

Kyei-Boahen, S., Slinkard, A. E., & Walley, F. L. (2001). Rhizobial survival and nodulation of chickpea as influenced by fungicide seed treatment. *Canadian Journal of Microbiology*, *47*(6), 585–589. doi:10.1139/w01-038 PMID:11467735

Laakso, H. A., Marolda, C. L., Pinter, T. B., Stillman, M. J., & Heinrichs, D. E. (2016). A heme-responsive regulator controls synthesis of staphyloferrin B in *Staphylococcus aureus*. *The Journal of Biological Chemistry*, *291*(1), 29–40. doi:10.1074/jbc.M115.696625 PMID:26534960

Lacina, P., Dvorak, V., Vodickova, E., Barson, P., Kalivoda, J., & Goold, S. (2015). The application of nanosizedzero-valent iron for in situ remediation of chlorinated ethylenes in groundwater: A field case study. *Water Environment Research*, *87*(4), 326–333. doi:10.2175/106143015X14212658613596 PMID:26462077

Lakshmidevi, R., & Muthukumar, K. (2010). Enzymatic saccharification and fermentation of paper and pulp industry effluent for biohydrogen production. *International Journal of Hydrogen Energy*, *35*(8), 3389–3400. doi:10.1016/j.ijhydene.2009.12.165

Lale, G., Jogdand, V. V., & Gadre, R. V. (2006). Morphological mutants of *Gibberella fujikuroi* for enhanced production of gibberellic acid. *Journal of Applied Microbiology*, *100*(1), 65–72. doi:10.1111/j.1365-2672.2005.02754.x PMID:16405686

Lalnunhlimi, S., & Krishnaswamy, V. (2016). Decolorization of azo dyes (Direct Blue 151 and Direct Red 31) by moderately alkaliphilic bacterial consortium. *Brazilian Journal of Microbiology, 47*(1), 39-46.

Lal, R. (2016). Soil health and carbon management. *Food and Energy Security*, *5*(4), 212–222. doi:10.1002/fes3.96

Lambers, H., Raven, J. A., Shaver, G. R., & Smith, S. E. (2008). Plant nutrient-acquisition strategies change with soil age. *Trends in Ecology & Evolution*, *23*(2), 95–103. doi:10.1016/j.tree.2007.10.008 PMID:18191280

Lamb, R. L. (2003). Fertilizer use, risk, and off-farm labor markets in the semi-arid tropics of India. *American Journal of Agricultural Economics*, *85*(2), 359–371. doi:10.1111/1467-8276.00125

Lampis, S., Santi, C., Ciurli, A., Andreolli, M., & Vallini, G. (2015). Promotion of arsenic phytoextraction efficiency in the fern Pteris vittata by the inoculation of As-resistant bacteria: A soil bioremediation perspective. *Frontiers in Plant Science*, *6*, 80. doi:10.3389/fpls.2015.00080 PMID:25741356

Lanfranconi, M. P., Alvarez, H. M., & Studdert, C. A. (2003). A strain isolated from gas oil-contaminated soil displays chemotaxis towards gas oil and hexadecane. *Environmental Microbiology*, *5*(10), 1002–1008. doi:10.1046/j.1462-2920.2003.00507.x PMID:14510854

Lang, J., Planamente, S., Mondy, S., Dessaux, Y., Moréra, S., & Faure, D. (2013). Concerted transfer of the virulence Ti plasmid and companion at plasmid in the Agrobacterium tumefaciens-induced plant tumour. *Molecular Microbiology*, *90*(6), 1178–1189. doi:10.1111/mmi.12423 PMID:24118167

LaSarre, B., & Federle, M. J. (2013). Exploiting quorum sensing to confuse bacterial pathogens. *Microbiology and molecular biology reviews. Microbiology and Molecular Biology Reviews*, *77*(1), 73–111. doi:10.1128/MMBR.00046-12 PMID:23471618

Latif Khan, A., Ahmed Halo, B., Elyassi, A., Ali, S., Al-Hosni, K., Hussain, J., Al-Harrasi, A., & Lee, I. J. (2016). Indole acetic acid and ACC deaminase from endophytic bacteria improves the growth of Solarium lycopersicum. *Electronic Journal of Biotechnology*, *19*(3), 58–64. doi:10.1016/j.ejbt.2016.02.001

Latifi, A. M., Khodi, S., Mirzaei, M., Miresmaeili, M., & Babavalian, H. (2012). Isolation and characterization of five chlorpyrifos degrading bacteria. *African Journal of Biotechnology*, *11*, 3140–3146.

Latinwo, L. M., Donald, C., Ikediobi, C., & Silver, S. (1998). Effects of intracellular glutathione on sensitivity of *Escherichia coli* to mercury and arsenite. *Biochemical and Biophysical Research Communications*, *242*(1), 67–70. doi:10.1006/bbrc.1997.7911 PMID:9439611

Lau, E. T., Tani, A., Khew, C. Y., Chua, Y. Q., & San Hwang, S. (2020). Plant growth-promoting bacteria as potential bio-inoculants and biocontrol agents to promote black pepper plant cultivation. *Microbiological Research*, *240*, 126549. doi:10.1016/j.micres.2020.126549 PMID:32688172

Leadbetter, J. R., & Greenberg, E. P. (2000). Metabolism of acyl-homoserine lactone quorum-sensing signals by Variovorax paradoxus. *Journal of Bacteriology*, *182*(24), 6921–6926. doi:10.1128/JB.182.24.6921-6926.2000 PMID:11092851

Leake, J., Johnson, D., Donnelly, D., Muckle, G., Boddy, L., & Read, D. (2004). Networks of power and influence: The role of mycorrhizal mycelium in controlling plant communities and agroecosystem functioning. *Canadian Journal of Botany*, *82*(8), 1016–1045. doi:10.1139/b04-060

Łebkowska, M., Zborowska, E., Karwowska, E., Mia'skiewicz-Peska, E., Muszy'nski, A., Tabernacka, A., Naumczyk, J., & Jeczalik, M. (2011). Bioremediation of soil polluted with fuels by sequential multiple injection of native microorganisms: Field-scale processes in Poland. *Ecological Engineering*, *37*(11), 1895–1900. doi:10.1016/j.ecoleng.2011.06.047

Lech, M., Skutnik, Z., Bajda, M., & Markowska-Lech, K. (2020). Applications of Electrical Resistivity Surveys in Solving Selected Geotechnical and Environmental Problems. *Applied Sciences (Basel, Switzerland)*, *10*(7), 2263–2282. doi:10.3390/app10072263

Ledgard, S. F., & Steele, K. W. (1992). Biological nitrogen fixation in mixed legume/grass pastures. *Plant and Soil*, *141*(1-2), 137–153. doi:10.1007/BF00011314

Lee Chang, K. J., Nichols, C. M., Blackburn, S. I., Dunstan, G. A., Koutoulis, A., & Nichols, P. D. (2014). Comparison of thraustochytrids *Aurantiochytrium* sp., *Schizochytrium* sp., *Thraustochytrium* sp., and *Ulkenia* sp. for production of biodiesel, long-chain omega-3 oils, and exopolysaccharide. *Marine Biotechnology (New York, N.Y.)*, *16*, 396–411.

Lee, G., Bigham, J. M., & Faure, G. (2002). Removal of trace metals by coprecipitation with Fe, Al and Mn from natural waters contaminated with acid mine drainage in the Ducktown Mining District, Tennessee. *Applied Geochemistry*, *17*(5), 569–581.

Lee, S. J., Park, S. Y., Lee, J. J., Yum, D. Y., Koo, B. T., & Lee, J. K. (2002). Genes encoding the N-acyl homoserine lactone-degrading enzyme are widespread in many subspecies of Bacillus thuringiensis. *Applied and Environmental Microbiology*, *68*(8), 3919–3924. doi:10.1128/AEM.68.8.3919-3924.2002 PMID:12147491

Lee, S. M., Grass, G., Rensing, C., Barrett, S. R., Yates, C. J. D., Stoyanov, J. V., & Brown, N. L. (2002). The Pco proteins are involved in periplasmic copper handling in *Escherichia coli. Biochemical and Biophysical Research Communications*, *295*(3), 616–620.

Lee, S. Y. (1996a). Bacterial polyhydroxyalkanoates. *Biotechnology and Bioengineering*, *49*(1), 1–14. doi:10.1002/(SICI)1097-0290(19960105)49:1<1::AID-BIT1>3.0.CO;2-P PMID:18623547

Lee, S. Y. (1996b). Plastic bacteria? Progress and prospects for polyhydroxyalkanoate production in bacteria. *Trends in Biotechnology*, *14*(11), 431–438. doi:10.1016/0167-7799(96)10061-5

Lee, S. Y., & Kim, H. U. (2015). Systems strategies for developing industrial microbial strains. *Nature Biotechnology*, *33*(10), 1061–1072. doi:10.1038/nbt.3365 PMID:26448090

Lees, Z. M., & Senior, E. (1995). Bioremediation: a practical solution to land pollution. In *Clean technology and the environment*. Springer.

Lefebvre, D. D., Kelly, D., & Budd, K. (2007). Biotransformation of Hg (II) by cyanobacteria. *Applied and Environmental Microbiology, 73*(1), 243–249. doi:10.1128/AEM.01794-06 PMID:17071784

Lefkowitz, E. J., Dempsey, D. M., Hendrickson, R. C., Orton, R. J., Siddell, S. G., & Smith, D. B. (2018). Virus taxonomy: The database of the Internatinal Committee on Taxonomy of Viruses (ICTV). *Nucleic Acids Research, 46*(D1), D708–D717. doi:10.1093/nar/gkx932 PMID:29040670

Lemes, A. P., Montanheiro, T. L. A., Passador, F. R., & Durán, N. (2015). Nanocomposites of polyhydroxyalkanoates reinforced with carbon nanotubes: chemical and biological properties. In V. K. Thakur & M. K. Thakur (Eds.), *Eco-friendly Polymer Nanocomposites: Processing and Properties* (pp. 79–108). Springer India. doi:10.1007/978-81-322-2470-9_3

Leng, L., Wei, L., Xiong, Q., Xu, S., Li, W., Lv, S., Lu, Q., Wan, L., Wen, Z., & Zhou, W. (2020). Use of microalgae based technology for the removal of antibiotics from wastewater: A review. *Chemosphere, 238*, 124680. Advance online publication. doi:10.1016/j.chemosphere.2019.124680 PMID:31545213

Le, T. T., Nguyen, K. H., Jeon, J. R., Francis, A. J., & Chang, Y. S. (2015). Nano/bio treatment of polychlorinated-biphenyls with evaluation of comparative toxicity. *Journal of Hazardous Materials, 287*, 335–341. doi:10.1016/j.jhazmat.2015.02.001 PMID:25679799

Leung, M. (2004). Bioremediation: Techniques for cleaning up a mess. *BioTeach Journal, 2*, 18–22.

Lewis, D. R., Olex, A. L., Lundy, S. R., Turkett, W. H., Fetrow, J. S., & Muday, G. K. (2013). A kinetic analysis of the auxin transcriptome reveals cell wall remodeling proteins that modulate lateral root development in *Arabidopsis. The Plant Cell, 25*(9), 3329–3346. doi:10.1105/tpc.113.114868 PMID:24045021

Leyval, C., & Berthelin, J. (1989). Interactions between *Laccaria laccata, Agrobacterium radiobacter* and beech roots: Influence on P, K, Mg, and Fe mobilization from minerals and plant growth. *Plant and Soil, 117*(1), 103–110. doi:10.1007/BF02206262

Li de la Sierra-Gallay, I., Pellegrini, O., & Condon, C. (2005). Structural basis for substrate binding, cleavage and allostery in the tRNA maturase RNase Z. *Nature, 433*(7026), 657–661. doi:10.1038/nature03284 PMID:15654328

Li, Y., Ge, X. Z., Wang, X. Y., & Gao, R. (2017). *The Invention Discloses a Compound Bacterial Agent used to Degrade Paraquat and a Preparation Method. China*. Patent No CN 106520618 A. Beijing: National Intellectual Property Administration.

Li, A., & Shao, Z. (2014). Biochemical characterization of a haloalkane dehalogenase DadB from Alcanivorax dieselolei B-5. *PLoS One, 9*(2), e89144. doi:10.1371/journal.pone.0089144 PMID:24586552

Lian, B., Fu, P. Q., Mo, D. M., & Liu, C. Q. (2002). A comprehensive review of the mechanism of potassium release by silicate bacteria. *Acta Mineralogica Sinica, 22*, 179–182.

Lian, B., Wang, B., Pan, M., Liu, C., & Teng, H. H. (2008). Microbial release of potassium from K-bearing minerals by thermophilic fungus *Aspergillus fumigatus. Geochimica et Cosmochimica Acta, 72*(1), 87–98. doi:10.1016/j.gca.2007.10.005

Liang, Q., & Zhao, D. (2014). Immobilization of arsenate in a sandy loam soil using starch-stabilized magnetite nanoparticles. *Journal of Hazardous Materials, 271*, 16–23. doi:10.1016/j.jhazmat.2014.01.055 PMID:24584068

Liao, J., Jiang, J., Xue, S., Qingyu, C., Wu, H., Manikandan, R., Hartley, W., & Huang, L. (2018). A novel acid-producing fungus isolated from bauxite residue: The potential to reduce the alkalinity. *Geomicrobiology Journal, 35*(10), 840–847. doi:10.1080/01490451.2018.1479807

Liba, C. M., Ferrara, F. I. S., Manfio, G. P., Fantinatti-Garboggini, F., Albuquerque, R. C., Pavan, C., Ramos, P. L., Moreira-Filho, C. A., & Barbosa, H. R. (2006). Nitrogen-fixing chemo-organotrophic bacteria isolated from cyanobacteria-deprived lichens and their ability to solubilize phosphate and to release amino acids and phytohormones. *Journal of Applied Microbiology*, *101*(5), 1076–1086. doi:10.1111/j.1365-2672.2006.03010.x PMID:17040231

Li, C., Li, Q., Wang, Z., Ji, G., Zhao, H., Gao, F., Su, M., Jiao, J., Li, H., & Li, H. (2019). Environmental fungi and bacteria facilitate lecithin decomposition and the transformation of phosphorus to apatite. *Scientific Reports*, *9*(1), 1–8. doi:10.103841598-019-51804-7 PMID:31653926

Li, C., Zhou, K., Qin, W., Tian, C., Qi, M., Yan, X., & Han, W. (2019). A review on heavy metals contamination in soil: Effects, sources, and remediation techniques. *Soil and Sediment Contamination: An International Journal*, *28*(4), 380–394. doi:10.1080/15320383.2019.1592108

Lien, H. L., & Zhang, W.-X. (2001). Nanoscale iron particles for complete reduction of chlorinated ethenes. *Colloids and Surfaces. A, Physicochemical and Engineering Aspects*, *191*(1-2), 97–105. doi:10.1016/S0927-7757(01)00767-1

Li, G., Ma, J., Tan, M., Mao, J., An, N., Sha, G., Zhang, D., Zhao, C., & Han, M. (2016). Transcriptome analysis reveals the effects of sugar metabolism and auxin and cytokinin signaling pathways on root growth and development of grafted apple. *BMC Genomics*, *17*(1), 150. doi:10.118612864-016-2484-x PMID:26923909

Li, H. B., Singh, R. K., Singh, P., Song, Q. Q., Xing, Y. X., Yang, L. T., & Li, Y. R. (2017). Genetic diversity of nitrogen-fixing and plant growth promoting Pseudomonas species isolated from sugarcane rhizosphere. *Frontiers in Microbiology*, *8*, 1268. doi:10.3389/fmicb.2017.01268 PMID:28769881

Li, M., Guo, R., Yu, F., Chen, X., Zhao, H., Li, H., & Wu, J. (2018). Indole-3-acetic acid biosynthesis pathways in the plant-beneficial bacterium *Arthrobacter pascens* ZZ21. *International Journal of Molecular Sciences*, *19*(2), 443. doi:10.3390/ijms19020443 PMID:29389906

Li, M.-H., Gao, X.-Y., Li, C., Yang, C.-L., Fu, C.-A., Liu, J., Wang, R., Chen, L.-X., Lin, J.-Q., Liu, X.-M., Lin, J.-Q., & Pang, X. (2020). Isolation and identification of chromium reducing *Bacillus Cereus* species from chromium-contaminated soil for the biological detoxification of chromium. *International Journal of Environmental Research and Public Health*, *17*(6), 1–13. doi:10.3390/ijerph17062118 PMID:32209989

Lima, A. I. G., Corticeiro, S. C., & Figueira, E. M. D. A. P. (2006). Glutathione-mediated cadmium sequestration in *Rhizobium leguminosarum. Enzyme and Microbial Technology*, *39*(4), 763–769. doi:10.1016/j.enzmictec.2005.12.009

Lim, J. L., Wilhelmus, M. M., deVries, H. E., Drukarch, B., Hoozemans, J. J., & van Horssen, J. (2014). Antioxidative defense mechanisms controlled by Nrf2: State of the art and clinical perspectives in neurodegenerative diseases. *Archives of Toxicology*, *88*, 1773–1786.

Limmer, M., & Burken, J. (2016). Phytovolatilization of organic contaminants. *Environmental Science & Technology*, *50*(13), 6632–6643. doi:10.1021/acs.est.5b04113 PMID:27249664

Ling, L. L., Schneider, T., Peoples, A. J., Spoering, A. L., Engels, I., Conlon, B. P., ... Jones, M. (2015). A new antibiotic kills pathogens without detectable resistance. *Nature*, *517*(7535), 455–459. doi:10.1038/nature14098 PMID:25561178

Lin, Q., Rao, Z., Sun, Y., Yao, J., & Xing, L. (2002). Identification and practical application of silicate-dissolving bacteria. *Agricultural Sciences in China*, *1*, 81–85.

Lin, T.-S., & Cheng, F.-Y. (2016). Impact of Soil Moisture Initialization and Soil Texture on Simulated Land–Atmosphere Interaction in Taiwan. *Journal of Hydrometeorology*, *17*(5), 1337–1355. doi:10.1175/JHM-D-15-0024.1

Lin, W., Lin, M., Zhou, H., Wu, H., Li, Z., & Lin, W. (2019). The effects of chemical and organic fertilizer usage on rhizosphere soil in tea orchards. *PLoS One*, *14*(5), e0217018. doi:10.1371/journal.pone.0217018 PMID:31136614

Lin, Y. H., Xu, J. L., Hu, J., Wang, L. H., Ong, S. L., Leadbetter, J. R., & Zhang, L. H. (2003). Acyl-homoserine lactone acylase from Ralstonia strain XJ12B represents a novel and potent class of quorum-quenching enzymes. *Molecular Microbiology*, *47*(3), 849–860. doi:10.1046/j.1365-2958.2003.03351.x PMID:12535081

Lin, Y.-F., & Aarts, M. G. M. (2012). The molecular mechanism of zinc and cadmium stress response in plants. *Cellular and Molecular Life Sciences: CMLS*, *69*(19), 3187–3206. doi:10.100700018-012-1089-z PMID:22903262

Li, P., Bhattacharjee, P., Wang, S., Zhang, L., Ahmed, I., & Guo, L. (2019). Mycoviruses in *Fusarium* Species: An Update. *Frontiers in Cellular and Infection Microbiology*, *9*, 257. doi:10.3389/fcimb.2019.00257 PMID:31380300

Lipczynska-Kochany, E. (2018). Humic substances, their microbial interactions and effects on biological transformations of organic pollutants in water and soil: A review. *Chemosphere*, *202*, 420–437. doi:10.1016/j.chemosphere.2018.03.104 PMID:29579677

Li, S., Wang, Z., Miao, Y., & Li, S. (2014). Soil Organic Nitrogen and Its Contribution to Crop Production. *Journal of Integrative Agriculture*, *13*(10), 2061–2080. doi:10.1016/S2095-3119(14)60847-9

Lisk, D. J. (1988). Environmental implications of incineration of municipal solid waste and ash disposal. *The Science of the Total Environment*, *74*, 39–66. doi:10.1016/0048-9697(88)90128-3 PMID:3065938

Li, T., Liu, J., Bai, R., Ohandja, D. G., & Wong, F. S. (2007). Biodegradation of organonitriles by adapted activated sludge consortium with acetonitrile-degrading microorganisms. *Water Research*, *41*(15), 3465–3473. doi:10.1016/j.watres.2007.04.033 PMID:17544472

Liu, C., Li, C., Zheng, F., Zhang, H. & Yu, H. (2019). Composition identification and allelopathic effect of root exudates of ginseng in different continuous cropping years. *Acta Microsc, 28*.

Liu, D., Yang, Q., Ge, K., Hu, X., Qi, G., Du, B., Liu, K., & Ding, Y. (2017). Promotion of iron nutrition and growth on peanut by Paenibacillus illinoisensis and Bacillus sp. strains in calcareous soil. *Brazilian Journal of Microbiology, 48*(4), 656-670. doi:10.1016/j.bjm.2017.02.006

Liu, A., Wang, H., Gao, P., & Xu, H. (2013). Chemical fractionation of Cu and Zn and their impacts on microbial properties in slightly contaminated soils. *International Journal of Agricultural Research, Innovation and Technology*, *3*(1), 20–25. doi:10.3329/ijarit.v3i1.16045

Liu, C. H., Chen, X., Liu, T. T., Lian, B., Gu, Y., Caer, V., Xue, Y. R., & Wang, B. T. (2007). Study of the antifungal activity of Acinetobacter baumannii LCH001 in vitro and identification of its antifungal components. *Applied Microbiology and Biotechnology*, *76*(2), 459–466. doi:10.100700253-007-1010-0 PMID:17534613

Liu, D., Lepore, B. W., Petsko, G. A., Thomas, P. W., Stone, E. M., Fast, W., & Ringe, D. (2005). Three-dimensional structure of the quorum-quenching N-acyl homoserine lactone hydrolase from Bacillus thuringiensis. *Proceedings of the National Academy of Sciences of the United States of America*, *102*(33), 11882–11887. doi:10.1073/pnas.0505255102 PMID:16087890

Liu, D., Lian, B., & Dong, H. (2012). Isolation of *Paenibacillus sp.* and assessment of its potential for enhancing mineral weathering. *Geomicrobiology Journal*, *29*(5), 413–421. doi:10.1080/01490451.2011.576602

Liu, D., Thomas, P. W., Momb, J., Hoang, Q. Q., Petsko, G. A., Ringe, D., & Fast, W. (2007). Structure and specificity of a quorum-quenching lactonase (AiiB) from Agrobacterium tumefaciens. *Biochemistry*, *46*(42), 11789–11799. doi:10.1021/bi7012849 PMID:17900178

Liu, F., Chi, Y., Wu, S., Jia, J., & Yao, K. (2014). Simultaneous degradation of cypermethrin and its metabolite, 3-phenoxy-benzoic acid, by the cooperation of *Bacillus licheniformis* B-1 and *Sphingomonas* sp. SC-1. *Journal of Agricultural and Food Chemistry*, *62*(33), 8256–8262. doi:10.1021/jf502835n PMID:25068244

Liu, H. F., Qian, T. W., & Zhao, D. Y. (2013). Reductive immobilization of perrhenate in soil and groundwater using starch stabilized ZVI nanoparticles. *Chinese Science Bulletin*, *58*(2), 275–281. doi:10.100711434-012-5425-3

Liu, H., Probst, A., & Liao, B. (2005). Metal contamination of soils and crops affected by the Chenzhoulead/zincmine-spill (Hunan, China). *The Science of the Total Environment*, *339*(1-3), 153–166. doi:10.1016/j.scitotenv.2004.07.030 PMID:15740766

Liu, L., Li, W., Song, W., & Guo, M. (2018). Remediation techniques for heavy metal-contaminated soils: Principles and applicability. *The Science of the Total Environment*, *633*, 206–219. doi:10.1016/j.scitotenv.2018.03.161 PMID:29573687

Liu, M., Lu, S., Song, Y., Lei, L., Hu, J., Lv, W., & He, D. (2018). Microplasticand mesoplastic pollution in farmland soils in suburbs of Shanghai,China. *Environmental Pollution*, *242*, 855–862. doi:10.1016/j.envpol.2018.07.051 PMID:30036839

Liu, Q., Liu, C., Yu, J., Yan, J., & Qi, X. (2012). Analysis of the ketosynthase genes in *Streptomyces* and its implications for preventing reinvestigation of polyketides with bioactivities. *The Journal of Agricultural Science*, *4*(7), 262–270.

Liu, R., & Zhao, D. (2007). Reducing leachability and bio-accessibility of lead in soils using a new class of stabilized iron phosphate nanoparticles. *Water Research*, *41*(12), 2491–2502. doi:10.1016/j.watres.2007.03.026 PMID:17482234

Liu, S., Zhang, F., Chen, J., & Sun, G. X. (2011). Arsenic removal from contaminated soil via biovolatilization by genetically engineered bacteria under laboratory conditions. *Journal of Environmental Sciences (China)*, *23*(9), 1544–1550. doi:10.1016/S1001-0742(10)60570-0 PMID:22432292

Liu, X., & Kokare, C. (2017). Microbial enzymes of use in industry. In *Biotechnology of Microbial Enzymes* (pp. 267–298). Academic Press. doi:10.1016/B978-0-12-803725-6.00011-X

Liu, Y. Q., Majetich, S. A., Tilton, R. D., Sholl, D. S., & Lowry, G. V. (2005). TCE dechlorination rates, pathways, and efficiency of nanoscale iron particles with different properties. *Environmental Science & Technology*, *39*(5), 1338–1345. doi:10.1021/es049195r PMID:15787375

Liu, Y., Li, S., Chen, Z., Megharaj, M., & Naidu, R. (2014). Influence of zero-valent iron nanoparticles on nitrate removal by *Paracoccus* sp. *Chemosphere*, *108*, 426–432. doi:10.1016/j.chemosphere.2014.02.045 PMID:24630453

Liu, Y., & Lowry, G. V. (2006). Effect of particle age (Fe0 content) and solution pH on NZVI reactivity: H_2 evolution and TCE dechlorination. *Environmental Science & Technology*, *40*(19), 6085–6090. doi:10.1021/es060685o PMID:17051804

Liu, Y., Phenrat, T., & Lowry, G. V. (2007). Effect of TCE concentration and dissolved groundwater solutes on NZVI-promoted TCE dechlorination and H_2 evolution. *Environmental Science & Technology*, *41*(22), 7881–7887. doi:10.1021/es0711967 PMID:18075103

Liu, Y., Serrano, A., Wyman, V., Marcellin, E., Southam, G., Vaughan, J., & Villa-Gomez, D. (2020). Nickel complexation as an innovative approach for nickel-cobalt selective recovery using sulfate-reducing bacteria. *Journal of Hazardous Materials*, *402*, 123506.

Liu, Z., Dugan, B., Masiello, C. A., & Gonnermann, H. M. (2017). Biochar particle size, shape, and porosity act together to influence soil water properties. *PLoS One*, *12*(6), 1–28. doi:10.1371/journal.pone.0179079 PMID:28598988

Liu, Z., Rong, Q., Zhou, W., & Liang, G. (2017). Effects of inorganic and organic amendment on soil chemical properties, enzyme activities, microbial community and soil quality in yellow clayey soil. *PLoS One*, *12*(3), 1–20. doi:10.1371/journal.pone.0172767 PMID:28263999

Li, X. Q., & Zhang, W. (2006). Iron nanoparticles: The core-shell structure and unique properties for Ni(II) sequestration. *Langmuir*, *22*(10), 4638–4642. doi:10.1021/la060057k PMID:16649775

Li, X., Elliott, D. W., & Zhang, W. (2006). Zero-Valent Iron Nanoparticles for Abatement of Environmental Pollutants: Materials and Engineering Aspects. *Critical Reviews in Solid State and Material Sciences*, *31*(4), 111–122. doi:10.1080/10408430601057611

Li, X., He, J., & Li, S. (2007). Isolation of a chlorpyrifos-degrading bacterium, *Sphingomonas* sp. strain Dsp-2, and cloning of the mpd gene. *Research in Microbiology*, *158*(2), 143–149. doi:10.1016/j.resmic.2006.11.007 PMID:17306510

Li, X., Li, D., Yan, Z., & Ao, Y. (2018). Biosorption and bioaccumulation characteristics of cadmium by plant growth-promoting rhizobacteria. *RSC Advances*, *8*(54), 30902–30911. doi:10.1039/C8RA06270F

Li, Y., Chuang, Y. H., Sallach, J. B., Zhang, W., Boyd, S. A., & Li, H. (2018). Potential metabolism of pharmaceuticals in radish: Comparison of in vivo and in vitro exposure. *Environmental Pollution*, *242*, 962–969. doi:10.1016/j.envpol.2018.07.060 PMID:30373041

Li, Y., Du, X., Wu, C., Liu, X., Wang, X., & Xu, P. (2013). An efficient magnetically modified microbial cell biocomposite for carbazole biodegradation. *Nanoscale Research Letters*, *8*(1), 522. doi:10.1186/1556-276X-8-522 PMID:24330511

Li, Y., Li, Q., Guan, G., & Chen, S. (2020). Phosphate solubilizing bacteria stimulate wheat rhizosphere and endosphere biological nitrogen fixation by improving phosphorus content. *PeerJ*, *8*, e9062. doi:10.7717/peerj.9062 PMID:32411531

Li, Z., Greden, K., Alvarez, P. J. J., Gregory, K. B., & Lowry, G. V. (2010). Adsorbed polymer and NOM limits adhesion and toxicity of nano scale zerovalent iron to *E. coli*. *Environmental Science & Technology*, *44*(9), 3462–3467. doi:10.1021/es9031198 PMID:20355703

Lloyd, J. R. (2002). Bioremediation of metals; the application of micro-organisms that make and break minerals. *Microbiology Today*, *29*, 67–69.

Loaces, I., Ferrando, L., & Scavino, A. F. (2011). Dynamics, diversity and function of endophytic siderophore-producing bacteria in rice. *Microbial Ecology*, *61*(3), 606–618. doi:10.100700248-010-9780-9 PMID:21128071

Lobell, D. B., & Field, C. B. (2007). Global scale climate–crop yield relationships and the impacts of recent warming. *Environmental Research Letters*, *2*(1), 014002. doi:10.1088/1748-9326/2/1/014002

Loeffler, W., Katzer, W., Kremer, S., Kugler, M., Petersen, F., Jung, G., & Tschen, J. S. M. (1990). Gegen Pilze wirksame antibiotika der Bacillus subtilis-gruppe. Forum Mikrobiol, 90, 156-163.

Loick, N., Hobbs, P. J., Hale, M. D. C., & Jones, D. L. (2009). Bioremediation of poly-aromatic hydrocarbon (PAH)-contaminated soil by composting. *Critical Reviews in Environmental Science and Technology*, *39*(4), 271–332. doi:10.1080/10643380701413682

Longkumer, I. Y., & Ahmad, M. (2020). Potential of Mycovirus in the Biological Control of Fungal Plant Pathogens: A Review. *Agricultural Reviews (Karnal)*, *41*(3). Advance online publication. doi:10.18805/ag.R-1896

Loper, J. E., & Henkels, M. D. (1999). Utilization of heterologous siderophores enhances levels of iron available to *Pseudomonas putida* in the rhizosphere. *Applied and Environmental Microbiology*, *65*(12), 5357–5363. doi:10.1128/AEM.65.12.5357-5363.1999 PMID:10583989

López-Bucio, J., Pelagio-Flores, R., & Herrera-Estrella, A. (2015). Trichoderma as biostimulant: Exploiting the multilevel properties of a plant beneficial fungus. *Scientia Horticulturae*, *196*, 109–123. doi:10.1016/j.scienta.2015.08.043

Loreto, F., & Schnitzler, J. P. (2010). Abiotic stresses and induced BVOCs. *Trends in Plant Science*, *15*(3), 154–166. doi:10.1016/j.tplants.2009.12.006 PMID:20133178

Lovley, D. R. (2003). Cleaning up with genomics: Applying molecular biology to bioremediation. *Nature Reviews. Microbiology*, *1*(1), 35–44.

Lowry, G. V., & Johnson, K. M. (2004). Congener-specific dechlorination of dissolved PCBs by microscale and nanoscale zerovalent iron in a water/methanol solution. *Environmental Science & Technology*, *38*(19), 5208–5216. doi:10.1021/es049835q PMID:15506219

Lozowicka, B., Jankowska, M., Hrynko, I., & Kaczynski, P. (2016). Removal of 16 pesticide residues from strawberries by washing with tap and ozone water, ultrasonic cleaning and boiling. *Environmental Monitoring and Assessment*, *188*(1), 51. doi:10.100710661-015-4850-6 PMID:26694708

Luan, H., Gao, W., Huang, S., Tang, J., Li, M., Zhang, H., Chen, X., & Masiliūnas, D. (2020). Substitution of manure for chemical fertilizer affects soil microbial community diversity, structure and function in greenhouse vegetable production systems. *PLoS One*, *15*(2), e0214041. doi:10.1371/journal.pone.0214041 PMID:32084129

Lucas, M. S., Amaral, C., Sampaio, A., Peres, J. A., & Dias, A. A. (2006). Biodegradation of the diazo dye Reactive Black 5 by a wild isolate of *Candida oleophila. Enzyme and Microbial Technology*, *39*(1), 51–55. doi:10.1016/j.enzmictec.2005.09.004

Lucini, L., Rouphael, Y., Cardarelli, M., Canaguier, R., Kumar, P., & Colla, G. (2015). The effect of a plant-derived biostimulant on metabolic profiling and crop performance of lettuce grown under saline conditions. *Scientia Horticulturae*, *182*, 124–133. doi:10.1016/j.scienta.2014.11.022

Lugtenberg, B., & Kamilova, F. (2009). Plant-growth-promoting rhizobacteria. *Annual Review of Microbiology*, *63*(1), 541–556. doi:10.1146/annurev.micro.62.081307.162918 PMID:19575558

Lu, H., Wanga, W., Li, F., & Zhu, L. (2019). Mixed-surfactant-enhanced phytoremediation of PAHs in soil: Bioavailability of PAHs and responses of microbial community structure. *The Science of the Total Environment*, *653*, 658–666. doi:10.1016/j.scitotenv.2018.10.385 PMID:30759591

Lu, L., Zhao, M., Wang, T. N., Zhao, L. Y., Du, M. H., Li, T. L., & Li, D. B. (2012). Characterization and dye decolorization ability of an alkaline resistant and organic solvents tolerant laccase from Bacillus licheniformis LS04. *Bioresource Technology*, *115*, 35–40. doi:10.1016/j.biortech.2011.07.111 PMID:21868217

Lu, M., Zhang, Z., Sun, S., Wei, X., Wang, Q., & Su, Y. (2010). The use of goosegrass (*Eleusine indica*) to remediate soil contaminated with petroleum. *Water, Air, and Soil Pollution*, *209*(1-4), 181–189. doi:10.100711270-009-0190-x

Lundberg, D. S., Lebeis, S. L., Paredes, S. H., Yourstone, S., Gehring, J., Malfatti, S., & Edgar, R. C. (2012). Defining the core Arabidopsis thaliana root microbiome. *Nature*, *488*(7409), 86–90. doi:10.1038/nature11237 PMID:22859206

Luo, X. S., Yu, S., Zhu, Y. G., & Li, X. D. (2012). Trace metal contamination in urban soils of China. *The Science of the Total Environment*, *421*, 17–30. doi:10.1016/j.scitotenv.2011.04.020 PMID:21575982

Lupton, F. G. H. (1987). History of wheat breeding. In Wheat breeding (pp. 51-70). doi:10.1007/978-94-009-3131-2_3

Lurthy, T., Cantat, C., Jeudy, C., Declerck, P., Gallardo, K., Barraud, C., Leroy, F., Ourry, A., Lemanceau, P., Salon, C., & Mazurier, S. (2020). Impact of Bacterial Siderophores on Iron Status and Ionome in Pea. *Frontiers in Plant Science*, *11*, 730. doi:10.3389/fpls.2020.00730 PMID:32595663

Lu, S., Teng, Y., Wang, J., & Sun, Z. (2010). Enhancement of pyrene removed from contaminated soils by *Bidens maximowicziana. Chemosphere*, *81*(5), 645–650. doi:10.1016/j.chemosphere.2010.08.022 PMID:20832842

Lushchak, V. I., Matviishyn, T. M., Husak, V. V., Storey, J. M., & Storey, K. B. (2018). Pesticide toxicity: A mechanistic approach. *EXCLI Journal*, *17*, 1101. PMID:30564086

Lu, Y., & Xu, J. (2015). Phytohormones in microalgae: A new opportunity for microalgal biotechnology? *Trends in Plant Science*, *20*(5), 273–282. doi:10.1016/j.tplants.2015.01.006 PMID:25697753

Lynch, J. M., & Whipps, J. M. (1990). Substrate flow in the rhizosphere. *Plant and Soil*, *129*(1), 1–10. doi:10.1007/BF00011685

Lyon, G. J., Mayville, P., Muir, T. W., & Novick, R. P. (2000). Rational design of a global inhibitor of the virulence response in Staphylococcus aureus, based in part on localization of the site of inhibition to the receptor-histidine kinase, Agr C. *Proceedings of the National Academy of Sciences of the United States of America*, *97*(24), 13330–13335. doi:10.1073/pnas.97.24.13330 PMID:11087872

Macaulay, B. M., & Rees, D. (2014). Bioremediation of oil spills. A review of challenges for resesrch advancement. *Annals of Environmental Science (Boston, Mass.)*, *8*, 9–37.

Macé, C., Desrocher, S., Gheorghiu, F., Kane, A., Pupeza, M., Cernik, M., Kvapil, P., Venkatakrishnan, R., & Zhang, W.-X. (2006). Nanotechnology and groundwater remediation: A step forward in technology understanding. *Remediation*, *16*(2), 23–33. doi:10.1002/rem.20079

Maceiras, R. (2016). Emerging technologies for soil remediation of hydrocarbons. *Pharm Anal Chem*, *2*(01), 102. doi:10.4172/2471-2698.1000e102

Mącik, M., Gryta, A., & Frą, M. (2020). Biofertilizers in agriculture: An overview on concepts, strategies and effects on soil microorganisms. *Advances in Agronomy*, *162*, 31–87. doi:10.1016/bs.agron.2020.02.001

MacMillan, J. (2001). Occurrence of gibberellins in vascular plants, fungi, and bacteria. *Journal of Plant Growth Regulation*, *20*(4), 387–442. doi:10.1007003440010038 PMID:11986764

Madar, R., Singh, Y. V., Meena, M. C., Das, T. K., Gaind, S., & Verma, R. K. (2020). Potassium and Residue Management Options to Enhance Productivity and Soil Quality in Zero till Maize–Wheat Rotation. *CLEAN–Soil, Air. Water (Basel)*, *48*(3), 1900316.

Maddela, N. R., Scalvenzi, L., & Venkateswarlu, K. (2017). Microbial degradation of total petroleum hydrocarbons in crude oil: A field-scale study at the low-land rainforest of Ecuador. *Environmental Technology*, *38*(20), 2543–2550. doi:10.1080/09593330.2016.1270356 PMID:27928937

Madhavi, V., & Lele, S. S. (2009). Laccase: Properties and applications. *BioResources*, *4*(4), 1694–1717.

Madhuri, R. J., & Vijayalakshmi, G. (2014). Biodegradation of diazodye, trypan blue by Aspergillus species from dye contaminated sites. *International Journal of Research Studies in Biosciences*, *2*(11), 49–61.

Mager, D. M., & Thomas, A. D. (2011). Extracellular polysaccharides from cyanobacterial soil crusts: A review of their role in dryland soil processes. *Journal of Arid Environments*, *75*(2), 91–97. doi:10.1016/j.jaridenv.2010.10.001

Maguire, R., Alley, M. M., & Flowers, W. (2019). *Fertilizer types and calculating application rates*. Academic Press.

Mahaffee, W. F., & Backman, P. A. (1993). Effects of seed factors on spermosphere and rhizosphere colonization of cotton by Bacillus subtilis GB03. *Phytopathology*, *83*(10), 1120–1125. doi:10.1094/Phyto-83-1120

Mahalik, S., Sharma, A. K., & Mukherjee, K. J. (2014). Genome engineering for improved recombinant protein expression in *Escherichia coli*. *Microbial Cell Factories*, *13*(1), 177. doi:10.118612934-014-0177-1 PMID:25523647

Mahanta, D., Rai, R. K., Mishra, S. D., Raja, A. J., & Varghese, E. (2014). Influence of phosphorus and biofertilizers on soybean and wheat root growth and properties. *Field Crops Research, 166*, 1–9. doi:10.1016/j.fcr.2014.06.016

Mahanty, T., Bhattacharjee, S., Goswami, M., Bhattacharyya, P., Das, B., Ghosh, A., & Tribedi, P. (2017). Biofertilizers: A potential approach for sustainable agriculture development. *Environmental Science and Pollution Research International, 24*(4), 3315–3335. doi:10.100711356-016-8104-0 PMID:27888482

Mahapatra, N. N. (2016). Textile dyes. Boca Raton: CRC Press.

Mahar, A., Wang, P., Ali, A., Awasthi, M. K., Lahori, A. H., Wang, Q., & Zhang, Z. (2016). Challenges and opportunities in the phytoremediation of heavy metals contaminated soils: A review. *Ecotoxicology and Environmental Safety, 126*, 111–121. doi:10.1016/j.ecoenv.2015.12.023 PMID:26741880

Maharani, V., Vijayalakshmi, S., & Balasubramanian, T. (2013). Degradation and detoxification of reactive azo dyes by native bacterial communities. *African Journal of Microbiological Research, 7*(20), 2274–2282. doi:10.5897/AJMR12.1539

Mahbub, K. R., Ferdouse, J., & Anwar, M. N. (2011). Demonstration of Decolorization of Various Dyes by Some Bacterial Isolates Recovered from Textile Effluents. *Bangladesh Journal of Scientific and Industrial Research, 46*(3), 323–328. doi:10.3329/bjsir.v46i3.9037

Mahdi, I., Fahsi, N., Hafidi, M., Allaoui, A., & Biskri, L. (2020). Plant Growth Enhancement using Rhizospheric Halotolerant Phosphate Solubilizing Bacterium Bacillus licheniformis QA1 and Enterobacter asburiae QF11 Isolated from *Chenopodium quinoa* Willd. *Microorganisms, 8*(6), 948. doi:10.3390/microorganisms8060948 PMID:32599701

Mahmoud, A., & Abd-Alla, M. (2001). Siderophore production by some microorganisms and their effect on *Bradyrhizobium*-Mung Bean symbiosis. *International Journal of Agriculture and Biology, 3*(2), 157–162.

Mahmoud, S. (2001). Nutritional Status and Growth of Maize Plants as Affected by Green Microalgae as Soil Additives. *The Journal of Biological Sciences, 1*(6), 475–479. doi:10.3923/jbs.2001.475.479

Mahmud, K., Makaju, S., Ibrahim, R., & Missaoui, A. (2020). Current Progress in Nitrogen Fixing Plants and Microbiome Research. *Plants (Basel, Switzerland), 9*(1), 97. doi:10.3390/plants9010097 PMID:31940996

Maila, M. P., & Cloete, T. E. (2004). Bioremediation of petroleum hydrocarbons through landfarming: Are simplicity and costeffectiveness the only advantages? *Reviews in Environmental Science and Biotechnology, 3*(4), 49–360. doi:10.100711157-004-6653-z

Maimaiti, Y., Ding, L., Chai, M., Jing, X., Yang, D., Han, S., Sun, L., & Chen, W. (2020). Identification and analysis of new mycoviruses from melon powdery mildew. *Journal of Plant Pathology*, 1–6.

Majda, M., & Robert, S. (2018). The role of auxin in cell wall expansion. *International Journal of Molecular Sciences, 19*(4), 951. doi:10.3390/ijms19040951 PMID:29565829

Maksimov, I. V., Abizgil'Dina, R. R., & Pusenkova, L. I. (2011). Plant growth promoting rhizobacteria as alternative to chemical crop protectors from pathogens. *Applied Biochemistry and Microbiology, 47*(4), 333–345. doi:10.1134/S0003683811040090

Malhi, Y. A., Baldocchi, D. D., & Jarvis, P. G. (1999). The carbon balance of tropical, temperate and boreal forests. *Plant, Cell & Environment, 22*(6), 715–740. doi:10.1046/j.1365-3040.1999.00453.x

Malik, A. (2004). Metal bioremediation through growing cells. *Environment International, 30*, 261–278.

Malik, K. A., Bilal, R., Mehnaz, S., Rasul, G., Mirza, M. S., & Ali, S. (1997). Association of nitrogen-fixing, plant-growth-promoting rhizobacteria (PGPR) with kallar grass and rice. In *Opportunities for Biological Nitrogen Fixation in Rice and Other Non-Legumes* (pp. 37–44). Springer. doi:10.1007/978-94-011-5744-5_5

Malinovskaya, I. M., Kosenko, L. V., Votselko, S. K., & Podgorskii, V. S. (1990). Role of *Bacillus mucilaginosus* polysaccharide in degradation of silicate minerals. *Microbiology*, *59*, 49–55.

Malla, M. A., Dubey, A., Yadav, S., Kumar, A., Hashem, A., & Abd-Allah, E. F. (2018). Understanding and designing the strategies for the microbe-mediated remediation of environmental contaminants using omics approaches. *Frontiers in Microbiology*, *9*, 1132.

Mall, I. D. (2007). *Petrochemical process technology*. Sanat Pronters.

Malusá, E., Sas-Paszt, L., & Ciesielska, J. (2012). Technologies for beneficial microorganisms inocula used as biofertilizers. *TheScientificWorldJournal*, *2012*, 1–13. doi:10.1100/2012/491206 PMID:22547984

Malusá, E., & Vassilev, N. (2014). A contribution to set a legal framework for biofertilisers. *Applied Microbiology and Biotechnology*, *98*(15), 6599–6607. doi:10.100700253-014-5828-y PMID:24903811

Mancinelli, R. L. (1996). The nature of nitrogen: An overview. *Life Support & Biosphere Science*, *3*(1-2), 17–24. PMID:11539154

Manfred, R., Wilhelm, L., Gerhard, P., Adolf, T. L., & Ernest, L. (2006). Chlorinated hydrocarbons Ullmann's Encyclopedia of Industrial Chemistry.

Mani, A., Hameed, S. S., Ramalingam, S., & Narayanan, M. (2012). Assessment of Quorum Quenching Activity of Bacillus Species Against Pseudomonas aeruginosa MTCC 2297. *Global Journal of Pharmacology*, *6*(2), 118–125.

Mani, D., & Kumar, C. (2014). Biotechnological advances in bioremediation of heavy metals contaminated ecosystems: An overview with special reference to phytoremediation. *International Journal of Environmental Science and Technology*, *11*, 843–872.

Manigandan, P. K., & Shekar, C. (2014). Evaluation of radionuclides in the terrestrial environment of Western Ghats. *Journal of Radiation Research and Applied. The Sciences*, *7*, 310–316.

Manikandan, M., Kannan, V., Mahalingam, K., Vimala, A., & Chun, S. (2016). Phytoremediation potential of chromium containing tannery effluent-contaminated soil by native Indian timber-yielding tree species. *Preparative Biochemistry & Biotechnology*, *46*(1), 100–108. doi:10.1080/10826068.2015.1045607 PMID:26177918

Manjunath, M., Kanchan, A., Ranjan, K., Venkatachalam, S., Prasanna, R., Ramakrishnan, B., Hossain, F., Nain, L., Shivay, Y. S., Rai, A. B., & Singh, B. (2016). Beneficial cyanobacteria and eubacteria synergistically enhance bioavailability of soil nutrients and yield of okra. *Heliyon*, *2*(2), e00066. doi:10.1016/j.heliyon.2016.e00066 PMID:27441245

Manjunath, M., Prasanna, R., Nain, L., Dureja, P., Singh, R., Kumar, A., Jaggi, S., & Kaushik, B. D. (2010). Biocontrol potential of cyanobacterial metabolites against damping off disease caused by *Pythium aphanidermatum* in solanaceous vegetables. *Archiv für Phytopathologie und Pflanzenschutz*, *43*(7), 666–677. doi:10.1080/03235400802075815

Man, L. Y., Cao, X. Y., & Sun, D. S. (2014). Effect of potassium-solubilizing bacteria-mineral contact mode on decomposition behavior of potassium-rich shale. *Zhongguo Youse Jinshu Xuebao*, *24*, 48–52.

Manoharachary, C., & Mukerji, K. G. (2006). Rhizosphere biology – an Overview. In K. G. Mukerji, C. Manoharachary, & J. Singh (Eds.), *Microbial activity in the rhizosphere soil biology*. Springer-Verlag Berlin Heidelberg. doi:10.1007/3-540-29420-1_1

Manu, B. (2003). *Decolourization of indigo and azo dye in semicontinous reactors with long hydraulic retention time* (PhD Thesis). IIT Bombay.

Manzoor, J., & Sharma, M. (2019). Impact of Biomedical Waste on Environment and Human Health. *Environmental Claims Journal*, *31*(4), 311–334. doi:10.1080/10406026.2019.1619265

Mao, J., & Guan, W. (2016). Fungal degradation of polycyclic aromatic hydrocarbons (PAHs) by Scopulariopsis brevicaulis and its application in bioremediation of PAH-contaminated soil. *Acta Agriculturæ Scandinavica. Section B, Soil and Plant Science*, *66*(5), 399–405. doi:10.1080/09064710.2015.1137629

Marek, J., Vévodová, J., Smatanová, I. K., Nagata, Y., Svensson, L. A., Newman, J., Takagi, M., & Damborský, J. (2000). Crystal structure of the haloalkane dehalogenase from Sphingomonas paucimobilis UT26. *Biochemistry*, *39*(46), 14082–14086. doi:10.1021/bi001539c PMID:11087355

Margesin, R. (2012). Psychrophilic microorganisms in alpine soils. In *Plants in Alpine Regions* (pp. 187–198). Springer. doi:10.1007/978-3-7091-0136-0_14

Margesin, R., Hammerle, M., & Tscherko, D. (2007). Microbial activity and community composition during bioremediation of diesel-oilcontaminated soil: Effects of hydrocarbon concentration, fertilizers, and incubation time. *Microbial Ecology*, *53*(2), 259–269. doi:10.100700248-006-9136-7 PMID:17265002

Margesin, R., & Schinner, F. (2001). Bioremediation (natural attenuation and biostimulation) of diesel-oil-contaminated soil in an alpine glacier skiing area. *Applied and Environmental Microbiology*, *67*(7), 3127–3133. doi:10.1128/AEM.67.7.3127-3133.2001 PMID:11425732

Markowicz, A., Plaza, G., & Piotrowskaseget, Z. (2016). Activity and functional diversity of microbial communities in long-term hydrocarbon and potentially toxic element contaminated soils. *Archives of Environmental Protection*, *42*, 3–11. doi:10.1515/aep-2016-0041

Maron, P. A., Ranjard, L., Mougel, C., & Lemanceau, P. (2007). Metaproteomics: A new approach for studying functional microbial ecology. *Microbial Ecology*, *53*(3), 486–493. doi:10.100700248-006-9196-8 PMID:17431707

Marques, A. P., Rangel, A. O., & Castro, P. M. (2009). Remediation of heavy metal contaminated soils: Phytoremediation as a potentially promising clean-up technology. *Critical Reviews in Environmental Science and Technology*, *39*(8), 622–654. doi:10.1080/10643380701798272

Marrone, P. G. (2002). An effective biofungicide with novel modes of action. *Pesticide Outlook*, *13*(5), 193–194. doi:10.1039/b209431m

Marschner, H., Römheld, V., Horst, W. J., & Martin, P. (1986). Root-induced changes in the rhizosphere: Importance for the mineral nutrition of plants. *ZeitschriftfürPflanzenernährung und Bodenkunde*, *149*(4), 441–456. doi:10.1002/jpln.19861490408

Martin, B. C., George, S. J., Price, C. A., Ryan, M. H., & Tibbett, M. (2014). The role of root exuded low molecular weight organic anions in facilitating petroleum hydrocarbon degradation: Current knowledge and future directions. *The Science of the Total Environment*, *472*, 642–653. doi:10.1016/j.scitotenv.2013.11.050 PMID:24317170

Martinez-Alvarez, P., Pando, V., & Diez, J. J. (2014). Alternative species to replace Monterey pine plantations affected by pitch canker caused by *Fusarium circinatum* in northern Spain. *Plant Pathology*, *63*(5), 1086–1094. doi:10.1111/ppa.12187

Martínez-Pascual, E., Grotenhuis, T., Solanas, A. M., & Viñas, M. (2015). Coupling chemical oxidation and biostimulation: Effects on the natural attenuation capacity and resilience of the native microbial community in alkylbenzene-polluted soil. *Journal of Hazardous Materials*, *300*, 135–143. doi:10.1016/j.jhazmat.2015.06.061 PMID:26177489

Martins, D. A. B., Prado, H. F. A., Leite, R. S. R., Ferreira, H., Moretti, M. M. S., Silva, R., & Gomes, E. (2011). Agroindustrial wastes as substrates for microbial enzymes production and source of sugar for bioethanol production. *Integrated Waste Management*, *2*, 319–361.

Martins, M. A. M., Cardoso, M. H., Queiroz, M. J., Ramalho, M. T., & Campus, A. M. O. (1999). Biodegradation of azo dyes by the yeast *Candida zeylanoides* in batch aerated cultures. *Chemosphere*, *38*(11), 2455–2460. doi:10.1016/S0045-6535(98)00448-2 PMID:10204232

Marwa, N., Singh, N., Srivastava, S., Saxena, G., Pandey, V., & Singh, N. (2019). Characterizing the hypertolerance potential of two indigenous bacterial strains (*Bacillus flexus* and *Acinetobacter junii*) and their efficacy in arsenic bioremediation. *Journal of Applied Microbiology*, *126*(4), 1117–1127. doi:10.1111/jam.14179 PMID:30556924

Mary Mangaiyarkarasi, M. S., Vincent, S., Janarthanan, S., Subba Rao, T., & Tata, B. V. R. (2011). Bioreduction of Cr(VI) by alkaliphilic *Bacillus subtilis* and interaction of the membrane groups. *Saudi Journal of Biological Sciences*, *18*(2), 157–167. doi:10.1016/j.sjbs.2010.12.003 PMID:23961119

Masciarelli, O., Llanes, A., & Luna, V. (2014). A new PGPR co-inoculated with Bradyrhizobium japonicum enhances soybean nodulation. *Microbiological Research*, *169*(7-8), 609–615. doi:10.1016/j.micres.2013.10.001 PMID:24280513

Mashjoor, S., Yousefzadi, M., Esmaeili, M. A., & Rafiee, R. (2016). Cytotoxicity and antimicrobial activity of marine macro algae (*Dictyotaceae* and *Ulvaceae*) from the Persian Gulf. *Cytotechnology*, *68*(5), 1717–1726. doi:10.100710616-015-9921-6 PMID:26507649

Masindi, V., & Muedi, K. L. (2018). Environmental Contamination by Heavy Metals. In H. E.-D. M. Saleh & R. F. Aglan (Eds.), *Heavy Metals*. InTech. doi:10.5772/intechopen.76082

Masood, F., & Malik, A. (2011). Hexavalent chromium reduction by *Bacillus sp.* strain FM1 isolated from heavy-metal contaminated soil. *Bulletin of Environmental Contamination and Toxicology*, *86*(1), 114–119. doi:10.100700128-010-0181-z PMID:21181113

Masu, S., Cojocariu, L., Horablaga, N. M., Bordean, D. M., & Borozan, A. B. (2013). The Effects of Triticum aestivum species for the phytoremediation of petroleum-contaminated soil. International Multidisciplinary Scientific GeoConference, 1, 963-970.

Matassa, S., Boon, N., Pikaar, I., & Verstraete, W. (2016). Microbial protein: Future sustainable food supply route with low environmental footprint. *Microbial Biotechnology*, *9*(5), 568–575. doi:10.1111/1751-7915.12369 PMID:27389856

Mathew, B. B., Biju, V. G., & Nideghatta Beeregowda, K. (2019). Accumulation of lead (Pb II) metal ions by *Bacillus toyonensis* SCE1 species, innate to industrial-area ground water and nanoparticle synthesis. *Applied Nanoscience*, *9*(1), 49–66. doi:10.100713204-018-0892-8

Mathur, N., Bhatnagar, P., & Bakre, P. (2006). Assessing mutagenicity of textile dyes from Pali (Rajasthan) using Ames bioassay. *Applied Ecology and Environmental Research*, *4*(1), 111–118. doi:10.15666/aeer/0401_111118

Matos, M. (2016). *Soil decontamination using nanomaterials* (MSc Thesis). University of Coimbra, Coimbra, Portugal.

Matos, M. P. S. R., Correia, A. A. S., & Rasteiro, M. G. (2017). Application of carbon nanotubes to immobilize heavy metals in contaminated soils. *Journal of Nanoparticle Research*, *19*(4), 126. doi:10.100711051-017-3830-x

Matson, S. W., Bean, D. W., & George, J. W. (1994). DNA helicases: Enzymes with essential roles in all aspects of DNA metabolism. *BioEssays*, *16*(1), 13–22. doi:10.1002/bies.950160103 PMID:8141804

Maurhofer, M., Keel, C., Haas, D., & Défago, G. (1994). Pyoluteorin production by *Pseudomonas fluorescens* strain CHA0 is involved in the suppression of *Pythium* damping-off of cress but not of cucumber. *European Journal of Plant Pathology*, *100*(3-4), 221–232. doi:10.1007/BF01876237

Maurya, B. R., Meena, V. S., & Meena, O. P. (2014). Influence of Inceptisol and Alfisol's potassium solubilizing bacteria (KSB) isolates on release of K from waste mica. *Vegetos*, *27*(1), 181–187. doi:10.5958/j.2229-4473.27.1.028

Ma, X., Li, X., Liu, J., Cheng, Y., Zhai, F., Sun, Z., & Han, L. (2020). Enhancing Salix viminalis L.–mediated phytoremediation of polycyclic aromatic hydrocarbon–contaminated soil by inoculation with Crucibulum laeve (white-rot fungus). *Environmental Science and Pollution Research International*, *27*(33), 41326–41341. doi:10.100711356-020-10125-3 PMID:32681334

Ma, Y., Prasad, M. N. V., Rajkumar, M., & Freitas, H. (2011). Plant growth promoting rhizobacteria and endophytes accelerate phytoremediation of metalliferous soils. *Biotechnology Advances*, *29*(2), 248–258. doi:10.1016/j.biotechadv.2010.12.001 PMID:21147211

Ma, Y., Rajkumar, M., Luo, Y., & Freitas, H. (2011). Inoculation of endophytic bacteria on host and non-host plants—Effects on plant growth and Ni uptake. *Journal of Hazardous Materials*, *195*, 230–237. doi:10.1016/j.jhazmat.2011.08.034 PMID:21872991

Maya, K., Singh, R. S., Upadhyay, S. N., & Dubey, S. K. (2011). Kinetic analysis reveals bacterial efficacy for biodegradation of chlorpyrifos and its hydrolyzing metabolite TCP. *Process Biochemistry*, *46*(11), 2130–2136. doi:10.1016/j.procbio.2011.08.012

Mayak, S., Tirosh, T., & Glick, B. R. (2004). Plant growth-promoting bacteria that confer resistance to water stress in tomatoes and peppers. *Plant Science*, *166*(2), 525–530. doi:10.1016/j.plantsci.2003.10.025

Mayville, P., Ji, G., Beavis, R., Yang, H., Goger, M., Novick, R. P., & Muir, T. W. (1999). Structure-activity analysis of synthetic autoinducing thiolactone peptides from Staphylococcus aureus responsible for virulence. *Proceedings of the National Academy of Sciences of the United States of America*, *96*(4), 1218–1223. doi:10.1073/pnas.96.4.1218 PMID:9990004

Mazen, M. M., El-Batanony, N. H., Abd El-Monium, M. M., & Massoud, O. N. (2008). Cultural filtrate of Rhizobium spp. and arbuscular mycorrhiza are potential biological control agents against root rot fungal diseases of faba bean. *Global Journal of Biotechnology and Biochemistry*, *3*, 32–41.

Mbachu, A. E., Chukwura, E. I., & Mbachu, N. A. (2020). Role of Microorganisms in the Degradation of Organic Pollutants: A Review. *Energy and Environmental Engineering*, *7*(1), 1–11. doi:10.13189/eee.2020.070101

McBride, M. B., Shayler, H. A., Russell-Anelli, J. M., Spliethoff, H. M., & Marquez-Bravo, L. G. (2015). Arsenic and lead uptake by vegetable crops grown on an old orchard site amended with compost. *Water, Air, and Soil Pollution*, *226*(8), 265–279. doi:10.100711270-015-2529-9 PMID:26900187

McCann, A. E., & Cullimore, D. R. (1979). Influence of pesticides on the soil algal flora. *Residue Reviews*, *72*, 1–31. doi:10.1007/978-1-4612-6214-5_1

Mcdaniel, L. E., Schaffner, C. P., & Bailey, E. G. (1965). *U.S. Patent No. 3,182,004*. Washington, DC: U.S. Patent and Trademark Office.

McGrath, S. P., Zhao, F. J., & Lombi, E. (2001). Plant and rhizosphere process involved in phytoremediation of metal-contaminated soils. *Plant and Soil*, *232*(1/2), 207–214. doi:10.1023/A:1010358708525

McIntosh, P., Schulthess, C. P., Kuzovkina, Y. A., & Guillard, K. (2017). Bioremediation and phytoremediation of total petroleum hydrocarbons (TPH) under various conditions. *International Journal of Phytoremediation*, *19*(8), 755–764. doi:10.1080/15226514.2017.1284753 PMID:28165761

McLaren, T. I., Smernik, R. J., McLaughlin, M. J., McBeath, T. M., Kirby, J. K., Simpson, R. J., Guppy, C. N., Doolette, A. L., & Richardson, A. E. (2015). Complex Forms of Soil Organic Phosphorus-A Major Component of Soil Phosphorus. *Environmental Science & Technology*, *49*(22), 13238–13245. doi:10.1021/acs.est.5b02948 PMID:26492192

McNear, D. H. Jr. (2013). The rhizosphere-roots, soil and everything in between. *Nature Education Knowledge*, *4*(3), 1.

McRose, D. L., Baars, O., Morel, F. M. M., & Kraepiel, A. M. L. (2017). Siderophore production in *Azotobacter vinelandii* in response to Fe-, Mo- and V-limitation. *Environmental Microbiology*, *19*(9), 3595–3605. doi:10.1111/1462-2920.13857 PMID:28703469

McWilliams, A. (2015). *Microbial products: technologies, applications and global markets*. Academic Press.

Medeiros, E. V., Krystal, A. N., Jamilly, A. B., Wendson, S. M., Aline, O. S., & Keila, A. M. (2015). Absolute and specific enzymatic activities of sandy entisol from tropical dry forest, monoculture and intercropping areas. *Soil & Tillage Research*, *145*, 208–215. doi:10.1016/j.still.2014.09.013

Medfu Tarekegn, M., Zewdu Salilih, F., & Ishetu, A. I. (2020). Microbes used as a tool for bioremediation of heavy metal from the environment. *Cogent Food & Agriculture*, *6*(1), 74. doi:10.1080/23311932.2020.1783174

Meehan, C., Banat, I., McMullan, G., Nigam, P., Smyth, F., & Marchant, R. (2000). Decolorization of Remazol Black-B using a thermotolerant yeast, *Kluyveromyces marxianus* IMB3. *Environment International*, *26*(1-2), 75–79. doi:10.1016/S0160-4120(00)00084-2 PMID:11345742

Meeks, J. C., & Elhai, J. (2002). Regulation of cellular differentiation in filamentous cyanobacteria in free-living and plant-associated symbiotic growth states. *Microbiology and Molecular Biology Reviews*, *66*(1), 94–121. doi:10.1128/MMBR.66.1.94-121.2002 PMID:11875129

Meena, H., Meena, R. S., Rajput, B. S., & Kumar, S. (2016). Response of bio-regulators to morphology and yield of clusterbean [*Cyamopsistetragonoloba* (L.) Taub.] under different sowing environments. *Journal of Applied and Natural Science*, *8*(2), 715–718. doi:10.31018/jans.v8i2.863

Meena, R. S., Kumar, S., Datta, R., Lal, R., Vijayakumar, V., Brtnicky, M., ... Pathan, S. I. (2020). Impact of agrochemicals on soil microbiota and management: A review. *Land (Basel)*, *9*(2), 34. doi:10.3390/land9020034

Meena, R. S., Meena, P. D., Yadav, G. S., & Yadav, S. S. (2017). Phosphate solubilizing microorganisms, principles and application of microphos technology. *Journal of Cleaner Production*, *145*, 157–158. doi:10.1016/j.jclepro.2017.01.024

Meena, R. S., Meena, V. S., Meena, S. K., & Verma, J. P. (2015b). The needs of healthy soils for a healthy world. *Journal of Cleaner Production*, *102*, 560–561. doi:10.1016/j.jclepro.2015.04.045

Meena, R. S., Vijayakumar, V., Yadav, G. S., & Mitran, T. (2018). Response and interaction of *Bradyrhizobium japonicum* and arbuscular mycorrhizal fungi in the soybean rhizosphere. *Plant Growth Regulation*, *84*(2), 207–223. doi:10.100710725-017-0334-8

Meena, V. D., Dotaniya, M. L., Meena, B. P., & Das, H. (2013a). Organic food safer but not healthy: Truth in myth. *Indian Farmers Dig*, *46*(8), 43–44.

Meena, V. S., Maurya, B. R., & Bahadur, I. (2014). Potassium solubilization by bacterial strain in waste mica. *Bangladesh Journal of Botany*, *43*(2), 235–237. doi:10.3329/bjb.v43i2.21680

Meena, V. S., Maurya, B. R., & Meena, R. S. (2015). Residual impact of wellgrow formulation and NPK on growth and yield of wheat (*Triticum aestivum*L.). *Bangladesh Journal of Botany*, *44*(1), 143–146. doi:10.3329/bjb.v44i1.22738

Meena, V. S., Maurya, B. R., Verma, J. P., Aeron, A., Kumar, A., Kim, K., & Bajpai, V. K. (2015). Potassium solubilizing rhizobacteria (KSR): Isolation, identification, and K-release dynamics from waste mica. *Ecological Engineering*, *81*, 340–347. doi:10.1016/j.ecoleng.2015.04.065

Meena, V. S., Maurya, B. R., Verma, J. P., & Meena, R. S. (2016). *Potassium solubilizing microorganisms for sustainable agriculture*. Springer India., doi:10.1007/978-81-322-2776-2

Meenavilli, H., Potumarthi, R., & Jetty, A. (2008). Gentamicin production by *Micromonospora echinospora* (Me-22) in stirred tank reactor: Effect of various parameters. *Journal of Basic Microbiology*, *48*(1), 53–58. doi:10.1002/jobm.200700116 PMID:18247396

Megharaj, M., Ramakrishnan, B., Venkateswarlu, K., Sethunathan, N., & Naidu, R. (2011). Bioremediation approaches for organic pollutants: A critical perspective. *Environment International*, *37*(8), 1362–1375. doi:10.1016/j.envint.2011.06.003 PMID:21722961

Megharaj, M., Venkateswarlu, K., & Rao, A. S. (1987). Metabolism of monocrotophos and quinalphos by algae isolated from soil. *Bulletin of Environmental Contamination and Toxicology*, *39*(2), 251–256. doi:10.1007/BF01689414 PMID:3663978

Mehta, P. K., & Sehgal, S. (2019). Microbial enzymes in food processing. In *Biocatalysis* (pp. 255–275). Springer. doi:10.1007/978-3-030-25023-2_13

Mena-Benitez, G. L., Gandia-Herrero, F., Graham, S., Larson, T. R., McQueen-Mason, S. J., French, C. E., Rylott, E. L., & Bruce, N. C. (2008). Engineering a catabolic pathway in plants for the degradation of 1, 2-dichloroethane. *Plant Physiology*, *147*(3), 1192–1198. doi:10.1104/pp.108.119008 PMID:18467461

Mendes, R., Garbeva, P., & Raaijmakers, J. M. (2013). The rhizosphere microbiome: Significance of plant beneficial, plant pathogenic, and human pathogenic microorganisms. *FEMS Microbiology Reviews*, *37*(5), 634–663. doi:10.1111/1574-6976.12028 PMID:23790204

Menendez-Vega, D., Gallego, J. L. R., Pelaez, A. I., de Cordoba, G. F., Moreno, J., Munoz, D., & Sanchez, J. (2007). Engineered in situ bioremediation of soil and groundwater polluted with weathered hydrocarbons. *European Journal of Soil Biology*, *43*(5-6), 310–321. doi:10.1016/j.ejsobi.2007.03.005

Meneses, C., Gonçalves, T., Alquéres, S., Rouws, L., Serrato, R., Vidal, M., & Baldani, J. I. (2017). Gluconacetobacter diazotrophicus exopolysaccharide protects bacterial cells against oxidative stress in vitro and during rice plant colonization. *Plant and Soil*, *416*(1-2), 133–147. doi:10.100711104-017-3201-5

Menge, J. A. (1983). Utilization of vesicular arbuscular mycorrhizal fungi in agriculture. *The New Phytologist*, *81*, 553–559. doi:10.1111/j.1469-8137.1978.tb01628.x

Mengel, K., & Kirkby, E. A. (2001). *Principles of plant nutrition* (5th ed.). Kluwer Acad Publishers. doi:10.1007/978-94-010-1009-2

Meng, Q., Yan, W., Yu, M., & Huang, D. (2003). A study of third-order nonlinear optical properties for anthraquinone derivatives. *Dyes and Pigments*, *56*(2), 145–149. doi:10.1016/S0143-7208(02)00123-7

Menn, F. M., Easter, J. P., & Sayler, G. S. (2008). Genetically engineered microorganisms and bioremediation. In H. J. Rehm & B. Reed (Eds.), *Biotechnology set* (pp. 441–463). Wiley. doi:10.1002/9783527620951.ch21

Menn, F.-M., Easter, J. P., & Sayler, G. S. (2008). Genetically engineered microorganisms and bioremediation. *Biotechnology (Faisalabad), 11*, 441–463.

Menzies, N. W., Fulton, I. M., & Morrell, W. J. (2004). Seawater Neutralization of Alkaline Bauxite Residue and Implications for Revegetation. *Journal of Environmental Quality, 33*(5), 1877–1884. doi:10.2134/jeq2004.1877 PMID:15356249

Merian, E. (1984). Introduction on environmental chemistry and global cycles of chromium, nickel, cobalt beryllium, arsenic, cadmium and selenium, and their derivatives. *Toxicological and Environmental Chemistry, 8*, 9–38.

Metcalf, E. (2003). *Waste water Engineering: Treatment and Reuse* (4th ed.). McGraw-Hill.

Metting, B. (1981). The systematics and ecology of soil algae. The Botanical Review, 47(2), 195 – 312.

Meudec, A., Poupart, N., Dussauze, J., & Deslandes, E. (2007). Relationship between heavy fuel oil phytotoxicity and polycyclic aromatic hydrocarbon contamination in Mishra, V. K., Singh, G., & Shukla, R. (2019). Impact of Xenobiotics under a Changing Climate Scenario. In *Climate Change and Agricultural Ecosystems* (pp. 133–151). Woodhead Publishing.

Mhlongo, M. I., Piater, L. A., Madala, N. E., Labuschagne, N., & Dubery, I. A. (2018). The chemistry of plant–microbe interactions in the rhizosphere and the potential for metabolomics to reveal signaling related to defense priming and induced systemic resistance. *Front. Plant Sci., 9*, 112. doi:10.3389/fpls.2018.00112 PMID:29479360

Mia, M. B., & Shamsuddin, Z. H. (2010). Rhizobium as a crop enhancer and biofertilizer for increased cereal production. *African Journal of Biotechnology, 9*(37), 6001–6009.

Michael Beman, J., Arrigo, K. R., & Matson, P. A. (2005). Agricultural runoff fuels large phytoplankton blooms in vulnerable areas of the ocean. *Nature, 434*(7030), 211–214. doi:10.1038/nature03370 PMID:15758999

Michalak, I., & Chojnacka, K. (2015). Algae as production systems of bioactive compounds. *Engineering in Life Sciences, 15*(2), 160–176. doi:10.1002/elsc.201400191

Michalik, B. (2017). NORM contaminated area identification using radionuclides activity concentration pattern in a soil profile. *Journal of Environmental Radioactivity, 169*, 9–18. doi:10.1016/j.jenvrad.2016.11.035 PMID:28408134

Michaud, A. M., Cambier, P., Sappin-Didier, V., Deltreil, V., Mercier, V., Rampon, J.-N., & Houot, S. (2020). Mass balance and long-term soil accumulation of trace elements in arable crop systems amended with urban composts or cattle manure during 17 years. *Aurélia Marcelline Michaud, 27*(5), 5367–5386. doi:10.100711356-019-07166-8 PMID:31848970

Michel, J., & Fingas, M. (2016). Oil spills: Causes, consequences, prevention, and countermeasures. Fossil Fuels, 159-201.

Michels, C. A. (2002). Genetic Techniques for Biological Research: A Case Study Approach. John Wiley & Sons.

Michelsen, C. F., & Stougaard, P. (2012). Hydrogen cyanide synthesis and antifungal activity of the biocontrol strain Pseudomonas fluorescens In5 from Greenland is highly dependent on growth medium. *Canadian Journal of Microbiology, 58*(4), 381–390. doi:10.1139/w2012-004 PMID:22417387

Middleton, B. A. (2008). Invasive Species. In S. E. Jørgensen & B. D. Fath (Eds.), *Encyclopedia of Ecology* (pp. 2020–2028). Academic Press. doi:10.1016/B978-008045405-4.00060-4

Miethke, M., & Marahiel, M. A. (2007). Siderophore-based iron acquisition and pathogen control. *Microbiology and Molecular Biology Reviews, 71*(3), 413–451. doi:10.1128/MMBR.00012-07 PMID:17804665

Milgroom, M. G., & Hillman, B. I. (2011). The ecology and evolution of fungal viruses. In *Studies in viral ecology: microbial and botanical host systems* (pp. 217–253). John Wiley and Sons. doi:10.1002/9781118025666.ch9

Miller, G., Suzuki, N., Ciftci-Yilmaz, S., & Mittler, R. (2010). Reactive oxygen species homeostasis and signalling during drought and salinity stresses. *Plant, Cell & Environment*, *33*(4), 453–467. doi:10.1111/j.1365-3040.2009.02041.x PMID:19712065

Miller, M. G. (2007). Environmental metabolomics: A SWOT analysis (strengths, weaknesses, opportunities, and threats). *Journal of Proteome Research*, *6*, 540–545.

Millow, C. J., Mackintosh, S. A., Lewison, R. L., Dodder, N. G., & Hoh, E. (2015). Identifying bioaccumulative halogenated organic compounds using a nontargeted analytical approach: Seabirds as sentinels. *PLoS One*, *10*(5), e0127205. doi:10.1371/journal.pone.0127205 PMID:26020245

Milošević, N. A., & Govedarica, M. M. (2002). Effect of herbicides on microbiological properties of soil. *Zbornik Matice Srpske za Prirodne Nauke*, (102), 5–21. doi:10.2298/ZMSPN0201005M

Mire, C. E., Tourjee, J. A., O'Brien, W. F., Ramanujachary, K. V., & Hecht, G. B. (2004). Lead precipitation by *Vibrio harveyi*: Evidence for novel quorum-sensing interactions. *Applied and Environmental Microbiology*, *70*(2), 855–864. doi:10.1128/AEM.70.2.855-864.2004 PMID:14766565

Mirminachi, F., Zhang, A., & Roehr, M. (2002). Citric acid fermentation and heavy metal ions. *Acta Biotechnologica*, *22*(3-4), 363–373. doi:10.1002/1521-3846(200207)22:3/4<363::AID-ABIO363>3.0.CO;2-A

Mirsal, I. A. (2008). Sources of soil pollution. In *Soil Pollution* (pp. 137–173). Springer. doi:10.1007/978-3-540-70777-6_7

Mirshad, P. P., & Puthur, J. T. (2017). Drought tolerance of bioenergy grass Saccharum spontaneum L. enhanced by arbuscular mycorrhizae. *Rhizosphere*, *3*, 1–8. doi:10.1016/j.rhisph.2016.09.004

Misal, S. A., Lingojwar, D. P., Shinde, R. M., & Gawai, K. R. (2011). Purification and characterization of azoreductase from alkaliphilic strain *Bacillus badius*. *Process Biochemistry*, *46*(6), 1264–1269. doi:10.1016/j.procbio.2011.02.013

Mishra, A., & Malik, A. (2013). Recent Advances in Microbial Metal Bioaccumulation. *Critical Reviews in Environmental Science and Technology*, *43*(11), 1162–1222. doi:10.1080/10934529.2011.627044

Mishra, A., & Malik, A. (2013). Recent advances in microbial metal bioaccumulation. *Critical Reviews in Environmental Science and Technology*, *43*, 1162–1222.

Mishra, J., Prakash, J., & Arora, N. K. (2016). Role of beneficial soil microbes in sustainable agriculture and environmental management. *Climate Change and Environmental Sustainability*, *4*(2), 137–149. doi:10.5958/2320-642X.2016.00015.6

Mishra, J., Singh, R., & Arora, N. K. (2017). Alleviation of heavy metal stress in plants and remediation of soil by rhizosphere microorganisms. *Frontiers in Microbiology*, *8*, 1706. doi:10.3389/fmicb.2017.01706 PMID:28932218

Mishra, J., Singh, R., & Arora, N. K. (2017). Plant growth-promoting microbes: diverse roles in agriculture and environmental sustainability. In *Probiotics and plant health* (pp. 71–111). Springer. doi:10.1007/978-981-10-3473-2_4

Mishra, P., Kaur, S., Sharma, A. N., & Jolly, R. S. (2016). Characterization of an Indole-3-Acetamide Hydrolase from *Alcaligenes faecalis* subsp. *parafaecalis* and Its Application in Efficient Preparation of Both Enantiomers of Chiral Building Block 2, 3-Dihydro-1, 4-Benzodioxin-2-Carboxylic Acid. *PLoS One*, *11*(7), e0159009. doi:10.1371/journal.pone.0159009 PMID:27391673

Mishra, S., Singh, A., Keswani, C., Saxena, A., Sarma, B. K., & Singh, H. B. (2015). Harnessing Plant-Microbe Interactions for Enhanced Protection Against Phytopathogens. In N. Arora (Ed.), *Plant Microbes Symbiosis: Applied Facets*. Springer. doi:10.1007/978-81-322-2068-8_5

Mitra, A., Chatterjee, S., & Gupta, D. K. (2017). Potential role of microbes in bioremediation of arsenic. In *Arsenic Contamination in the Environment* (pp. 195–213). Springer. doi:10.1007/978-3-319-54356-7_10

Mitra, S., Pramanik, K., Sarkar, A., Ghosh, P. K., Soren, T., & Maiti, T. K. (2018). Bioaccumulation of cadmium by *Enterobacter sp.* and enhancement of rice seedling growth under cadmium stress. *Ecotoxicology and Environmental Safety*, *156*, 183–196. doi:10.1016/j.ecoenv.2018.03.001 PMID:29550436

Mitter, B., Brader, G., Afzal, M., Compant, S., Naveed, M., Trognitz, F., & Sessitsch, A. (2013). Advances in elucidating beneficial interactions between plants, soil and bacteria. In *Advances in Agronomy* (Vol. 121, pp. 381–445). Academic Press.

Mitter, E. K., Kataoka, R., Renato de Freitas, J., & Germida, J. J. (2019). Potential use of endophytic root bacteria and host plants to degrade hydrocarbons. *International Journal of Phytoremediation*, *21*(9), 928–938. doi:10.1080/15226514.2019.1583637 PMID:30907105

Miyatake, M., & Hayashi, S. (2011). Characteristics of Arsenic Removal by Bacillus cereus Strain W2. Resources Processing, 58(3), 101–107. doi:10.4144/rpsj.58.101

Moghaddam, M. R., Fatemi, S., & Keshtkar, A. (2013). Adsorption of lead (Pb2þ) and uranium (UO22þ) cations by brown algae; experimental and thermodynamic modeling. *Chemical Engineering Journal*, *231*, 294–303. doi:10.1016/j.cej.2013.07.037

Mogharabi, M., & Faramarzi, M. A. (2014). Laccase and laccase-mediated systems in the synthesis of organic compounds. *Advanced Synthesis & Catalysis*, *356*(5), 897–927. doi:10.1002/adsc.201300960

Mohamed, M. F., Thalooth, A. T., Elewa, T. A., & Ahmed, A. G. (2019). Yield and nutrient status of wheat plants (Triticum aestivum) as affected by sludge, compost, and biofertilizers under newly reclaimed soil. *Bulletin of the National Research Center*, *43*(1), 31. doi:10.118642269-019-0069-y

Mohammadi, K., & Sohrabi, Y. (2012). Bacterial biofertilizers for sustainable crop production: A review. *American Journal of Agricultural and Biological Sciences*, *7*, 307–316.

Mohan, V., Rao, C., & Karthikeyan, J. (2002). Adsorptive Removal of Direct Azo Dye from Aqueous Phase into Coal Based Sorbents: A Kinetic and Mechanistic Study. *Journal of Hazardous Materials*, *90*(2), 189–204. doi:10.1016/S0304-3894(01)00348-X PMID:11827721

Mohapatra, R. K., Srichandan, H., Mishra, S., & Parhi, P. K. (2019). Native Soil Bacteria: Potential Agent for Bioremediation. *Soil Microenvironment for Bioremediation and Polymer Production*, 17-34.

Mohapatra, S., Maity, S., Dash, H. R., Das, S., Pattnaik, S., Rath, C. C., & Samantaray, D. (2017). Bacillus and biopolymer: Prospects and challenges. In Biochemistry and Biophysics Reports (Vol. 12, pp. 206–213). doi:10.1016/j.bbrep.2017.10.001

Mohapatra, R. K., Parhi, P. K., Patra, J. K., Panda, C. R., & Thatoi, H. N. (2017). Biodetoxification of toxic heavy metals by marine metal resistant bacteria - a novel approach for bioremediation of the polluted saline environment. In J. Patra, C. Vishnuprasad, & G. Das (Eds.), *Microbial Biotechnology* (pp. 343–376). Springer.

Mohapatra, S., Sarkar, B., Samantaray, D. P., Daware, A., Maity, S., Pattnaik, S., & Bhattacharjee, S. (2017). Bioconversion of fish solid waste into PHB using *Bacillus subtilis* based submerged fermentation process. *Environmental Technology*, *38*(24), 3201–3208. doi:10.1080/09593330.2017.1291759 PMID:28162048

Moharrery, L., Otadi, M., Miraly, N., Rezaei Zangeneh, M. M., & Amiri, R. (2019). Degradation of toluidine red, an oil soluble azo dye by *Halomonas strain IP8* at alkaline condition. *Chemical Engineering Communications*, *206*(1), 61–68. doi:10.1080/00986445.2018.1472587

Mohidin, H., Hanafi, M. M., Rafii, Y. M., Abdullah, S. N. A., Idris, A. S., Man, S., Idris, J., & Sahebi, M. (2015). Determination of optimum levels of nitrogen, phosphorus and potassium of oil palm seedlings in solution culture. *Bragantia, 74*(3), 247–254. doi:10.1590/1678-4499.0408

Mole, B. M., Baltrus, D. A., Dangl, J. L., & Grant, S. R. (2007). Global virulence regulation networks in phytopathogenic bacteria. *Trends in Microbiology, 15*(8), 363–371. doi:10.1016/j.tim.2007.06.005 PMID:17627825

Molina, L., Rezzonico, F., Défago, G., & Duffy, B. (2005). Autoinduction in Erwinia amylovora: Evidence of an acyl-homoserine lactone signal in the fire blight pathogen. *Journal of Bacteriology, 187*(9), 3206–3213. doi:10.1128/JB.187.9.3206-3213.2005 PMID:15838048

Monchy, S., Benotmane, M. A., Janssen, P., Vallaeys, T., Taghavi, S., Van Der Lelie, D., & Mergeay, M. (2007). Plasmids pMOL28 and pMOL30 of *Cupriavidus metallidurans* are specialized in the maximal viable response to heavy metals. *Journal of Bacteriology, 189*(20), 7417–7425.

Mondal, T., Datta, J. K., & Mondal, N. K. (2017). Chemical fertilizer in conjunction with biofertilizer and vermicompost induced changes in morpho-physiological and bio-chemical traits of mustard crop. *Journal of the Saudi Society of Agricultural Sciences, 16*(2), 135–144. doi:10.1016/j.jssas.2015.05.001

Mongkolthanaruk, W., Tongbopit, S., & Bhoonobtong, A. (2012). Independent behavior of bacterial laccases to inducers and metal ions during production and activity. *African Journal of Biotechnology, 11*(39), 9391–9398. doi:10.5897/AJB11.3042

Monica, S., Karthik, L., Mythili, S., & Sathiavelu, A. (2011). Formulation of effective microbial consortia and its application for sewage treatment. *J Microbial Biochem Technol, 3*, 51-55.

Monincová, M., Prokop, Z., Vévodová, J., Nagata, Y., & Damborský, J. (2007). Weak activity of haloalkane dehalogenase LinB with 1, 2, 3-trichloropropane revealed by X-ray crystallography and microcalorimetry. *Applied and Environmental Microbiology, 73*(6), 2005–2008. doi:10.1128/AEM.02416-06 PMID:17259360

Monsan, P. (1982). Les methodes immobilisation enzymes. In G. Durand & P. Monsan (Eds.), *Les enzymes, productions utilizations industrielles* (pp. 81–118). Gauthier-Villards.

Montalbán, B., Thijs, S., Lobo, M. C., Weyens, N., Ameloot, M., Vangronsveld, J., & Pérez-Sanz, A. (2017). Cultivar and metal-specific effects of endophytic bacteria in Helianthus tuberosus exposed to Cd and Zn. *International Journal of Molecular Sciences, 18*(10), 20–26. doi:10.3390/ijms18102026 PMID:28934107

Montaya. (2015). decolourisation of dyes with different molecular properties using free and immobilized laccases from *Trametes versicolor. Journal of Molecular Liquids, 212*, 30–37. doi:10.1016/j.molliq.2015.08.040

Monteiro, C. M., Marques, A. P., Castro, P. M., & Xavier Malcata, F. (2009). Characterization of *Desmodesmuspleiomorphus* isolated from a heavy metal-contaminated site: Biosorption of zinc. *Biodegradation, 20*(5), 629–641. doi:10.100710532-009-9250-6 PMID:19225897

Montemurro, F., Maiorana, M., Convertini, G., & Fornaro, F. (2005). Improvement of soil properties and nitrogen utilization of sunflower by amending municipal solid waste compost. *Agronomy for Sustainable Development, 25*(3), 369–375. doi:10.1051/agro:2005038

Montesinos, E., & Bonaterra, A. (2009). *Microbial pesticides. Encyclopedia of microbiology* (3rd ed.). Elsevier Inc.

Moore, C. H., Foster, L. A., Gerbig, D. G. Jr, Dyer, D. W., & Gibson, B. W. (1995). Identification of alcaligin as the siderophore produced by *Bordetella pertussis* and *B. bronchiseptica. Journal of Bacteriology, 177*(4), 1116–1118. doi:10.1128/JB.177.4.1116-1118.1995 PMID:7860593

Moore, M. J., Distefano, M. D., Zydowsky, L. D., Cummings, R. T., & Walsh, C. T. (1990). Organomercurial lyase and mercuric ion reductase: Nature's mercury detoxification catalysts. *Accounts of Chemical Research*, *23*(9), 301–308.

Morais, T. G., Teixeira, R. F. M., & Domingos, T. (2019). Detailed global modelling of soil organic carbon in cropland, grassland and forest soils. *PLoS One*, *14*(9), 1–27. doi:10.1371/journal.pone.0222604 PMID:31536571

Morales, S. E., Mouser, P. J., Ward, N., Hudman, S. P., Gotelli, N. J., Ross, D. S., & Lewis, T. A. (2006). Comparison of bacterial communities in New England Sphagnum bogs using terminal restriction fragment length polymorphism (T-RFLP). *Microbial Ecology*, *52*(1), 34–44. doi:10.100700248-005-0264-2 PMID:16729225

Moran, M. A. (2009). Metatranscriptomics: Eavesdropping on complex microbial communities-large-scale sequencing of mRNAs retrieved from natural communities provides insights into microbial activities and how they are regulated. *Microbe*, *4*(7), 329.

Moritsuka, N., Matsuoka, K., Katsura, K., Sano, S., & Yanai, J. (2014). Soil color analysis for statistically estimating total carbon, total nitrogen and active iron contents in Japanese agricultural soils. *Soil Science and Plant Nutrition*, *60*(4), 475–485. doi:10.1080/00380768.2014.906295

Morohoshi, T., Nakazawa, S., Ebata, A., Kato, N., & Ikeda, T. (2008). Identification and characterization of N-acylhomoserine lactone-acylase from the fish intestinal Shewanella sp. strain MIB015. *Bioscience, Biotechnology, and Biochemistry*, *72*(7), 1887–1893. doi:10.1271/bbb.80139 PMID:18603799

Morozova, O. V., Shumakovich, G. P., Shleev, S. V., & Yaropolov, Y. I. (2007). Laccase-mediator systems and their applications: A review. *Applied Biochemistry and Microbiology*, *43*(5), 523–535. doi:10.1134/S0003683807050055 PMID:18038679

Morris, C. F., Wrigley, C., Corke, H., Seetharaman, K., & Faubion, J. (2016). *Cereals: Overview of uses: Accent on wheat grain. Encyclopedia of food grains* (2nd ed.). Academic Press.

Morrison, C. K., Arseneault, T., Novinscak, A., & Filion, M. (2017). Phenazine-1-Carboxylic Acid Production by *Pseudomonas fluorescens* LBUM636 Alters *Phytophthora infestans* Growth and Late Blight Development. *Phytopathology*, *107*(3), 273–279. doi:10.1094/PHYTO-06-16-0247-R PMID:27827009

Morrissey, J., & Guerinot, M. L. (2009). Iron uptake and transport in plants: The good, the bad, and the ionome. *Chemical Reviews*, *109*(10), 4553–4567. doi:10.1021/cr900112r PMID:19754138

Mosa, K. A., Saadoun, I., Kumar, K., Helmy, M., & Dhankher, O. P. (2016). Potential Biotechnological Strategies for the Cleanup of Heavy Metals and Metalloids. *Frontiers in Plant Science*, *7*, 1–14. doi:10.3389/fpls.2016.00303 PMID:27014323

Mostafa, S. (2013). Microalgal biotechnology: Prospects and applications. *Plant Science*.

Mousa, W. K., & Raizada, M. N. (2013). The diversity of anti-microbial secondary metabolites produced by fungal endophytes: An interdisciplinary perspective. *Frontiers in Microbiology*, *4*, 65. doi:10.3389/fmicb.2013.00065 PMID:23543048

Mrkovacki, N., & Milic, V. (2001). Use of Azotobacter chroococcum as potentially useful in agricultural application. *Annals of Microbiology*, *51*, 145–158.

Mubeen, F., Aslam, A., Sheikh, M., Iqbal, T., Hameed, S., Malik, K. A., & Hafeez, F. Y. (2006). Response of wheat yield under combine use of Fungicide and Biofertilizer. *International Journal of Agriculture and Biology*, *8*(5), 580–582.

Muhammad, S., Muhammad, S., & Sarfraz, H. (2008). Perspectives of bacterial ACC deaminase in phytoremediation. *Trends in Biotechnology*, *25*, 356–362. PMID:17573137

Mukherjee, A. K., & Bordoloi, N. K. (2011). Bioremediation and reclamation of soil contaminated withpetroleum oil hydrocarbons by exogenously seeded bacterial consortium: A pilot-scale study. *Environmental Science and Pollution Research International, 18*(3), 471–478. doi:10.100711356-010-0391-2 PMID:20835890

Mullen, M. D., Wolf, D. C., Ferris, F. G., Beveridge, T. J., Flemming, C. A., & Bailey, G. W. (1989). Bacterial sorption of heavy metals. *Applied and Environmental Microbiology, 55*(12), 3143–3149. doi:10.1128/AEM.55.12.3143-3149.1989 PMID:2515800

Mulligana, C. N., & Yongb, R. N. (2004). Natural attenuation of contaminatedsoils. *Environment International, 30*(4), 587–601. doi:10.1016/j.envint.2003.11.001 PMID:15031019

Mulligan, C. N., Young, R. N., & Gibbs, B. F. (2001). Remediation technologies for metal-contaminated soils and groundwater: An evaluation. *Engineering Geology, 60*(1-4), 193–207. doi:10.1016/S0013-7952(00)00101-0

Muminah, B., Hazarin, S., & Fahruddin, D. B. (2015). Isolation and screening of exopolysaccharide producing bacterial (EPS) from potato rhizosphere for soil aggregation. *International Journal of Current Microbiology and Applied Sciences, 4*(6), 341–349.

Munson, G. P., Lam, D. L., Outten, F. W., & O'Halloran, T. V. (2000). Identification of a copper-responsive two-component system on the chromosome of *Escherichia coli* K-12. *Journal of Bacteriology, 182*(20), 5864–5871.

Muradoglu, F., Gundogdu, M., Ercisli, S., Encu, T., Balta, F., Jaafar, H. Z., & Zia-Ul-Haq, M. (2015). Cadmium toxicity affects chlorophyll a and b content, antioxidant enzyme activities and mineral nutrient accumulation in strawberry. *Biological Research, 48*(1), 11. doi:10.118640659-015-0001-3 PMID:25762051

Murray, E. W., Greenberg, B. M., Cryer, K., Poltorak, B., McKeown, J., Spies, J., & Gerwing, P. D. (2019). Kinetics of phytoremediation of petroleum hydrocarbon contaminated soil. *International Journal of Phytoremediation, 21*(1), 27–33. doi:10.1080/15226514.2018.1523870 PMID:30701992

Murray, J. D. (2011). Invasion by invitation: Rhizobial infection in legumes. *Molecular Plant-Microbe Interactions, 24*(6), 631–639. doi:10.1094/MPMI-08-10-0181 PMID:21542766

Mus, F., Crook, M. B., Garcia, K., Garcia Costas, A., Geddes, B. A., Kouri, E. D., Paramasivan, P., Ryu, M. H., Oldroyd, G., Poole, P. S., Udvardi, M. K., Voigt, C. A., Ané, J. M., & Peters, J. W. (2016). Symbiotic Nitrogen Fixation and the Challenges to Its Extension to Nonlegumes. *Applied and Environmental Microbiology, 82*(13), 3698–3710. doi:10.1128/AEM.01055-16 PMID:27084023

Musilova, J., Arvay, J., Vollmannova, A., Toth, T., & Tomas, J. (2016). Environmental contamination by heavy metals in region with previous mining activity. *Bulletin of Environmental Contamination and Toxicology, 97*, 569–575.

Muthu, S. S. (2017). Introduction. In S. S. Muthu (Ed.), *Sustainabilityin the textile industry* (pp. 1–8). Springer. doi:10.1007/978-981-10-2639-3_1

Muthusamy, K., Gopalakrishnan, S., Ravi, T. K., & Sivachidambaram, P. (2008). Biosurfactants: Properties, commercial production and application. *Current Science*, 736–747.

Muyzer, G. (1999). Genetic fingerprinting of microbial communities-present status and future perspectives. Methods of microbial community analysis. In *Proceedings of 8th International symposium on microbial ecology*. Atlantic Canada Society for Microbial Ecology.

Myresiotis, C. K., Vryzas, Z., & Papadopoulou-Mourkidou, E. (2012). Biodegradation of soil-applied pesticides by selected strains of plant growth-promoting rhizobacteria (PGPR) and their effects on bacterial growth. *Biodegradation, 23*(2), 297–310. doi:10.100710532-011-9509-6 PMID:21870159

Nabti, E., Jha, B., & Hartmann, A. (2017). Impact of seaweeds on agricultural crop production as biofertilizer. *International Journal of Environmental Science and Technology*, *14*(5), 1119–1134. doi:10.100713762-016-1202-1

Nagaraju, Y., Triveni, S., Subhashreddy, R., & Jhansi, P. (2017). Biofilm formation of zinc solubilizing, potassium releasing bacteria on the surface of fungi. *International Journal of Current Microbiology and Applied Sciences*, *6*(4), 2037–2047. doi:10.20546/ijcmas.2017.604.241

Nagargade, M., Tyagi, V., & Singh, M. K. (2018). Plant Growth-Promoting Rhizobacteria: A Biological Approach Toward the Production of Sustainable Agriculture. In V. Meena (Ed.), *Role of Rhizospheric Microbes in Soil*. Springer. doi:10.1007/978-981-10-8402-7_8

Naidoo, G., & Naidoo, K. (2016). Uptake of polycyclic aromatic hydrocarbons and their cellular effects inthe mangrove *Bruguiera gymnorrhiza*. *Marine Pollution Bulletin*, *113*(1-2), 193–199. doi:10.1016/j.marpolbul.2016.09.012 PMID:27634737

Naik, K., Mishra, S., Srichandan, H., Singh, P. K., & Sarangi, P. K. (2019). Plant growth promoting microbes: Potential link to sustainable agriculture and environment. *Biocatalysis and Agricultural Biotechnology*, *21*, 101326. doi:10.1016/j.bcab.2019.101326

Naik, M. M., Pandey, A., & Dubey, S. K. (2012a). *Pseudomonas aeruginosa* strain WI-1 from Mandovi estuary possesses metallothionein to alleviate lead toxicity and promotes plant growth. *Ecotoxicology and Environmental Safety*, *79*, 129–133.

Naik, M. M., Shamim, K., & Dubey, S. K. (2012b). Biological characterization of lead-resistant bacteria to explore role of bacterial metallothionein in lead resistance. *Current Science*, *103*(4), 426–429.

Nam, S., & Renganathan, V. (2000). Non-enzymatic reduction of azo dyes by NADH. *Chemosphere*, *40*(4), 351–357. doi:10.1016/S0045-6535(99)00226-X PMID:10665399

Namvar, A., & Khandan, T. (2013). Response of wheat to mineral nitrogen fertilizer and biofertilizer (Azotobacter sp. and Azospirillum sp.) inoculation under different levels of weed interference. *Ekologija (Lietuvos Mokslu Akademija)*, *59*(2), 2. doi:10.6001/ekologija.v59i2.2711

Nandakumar, R., Babu, S., Viswanathan, R., Raguchander, Y., & Samiyappan, R. (2001). Induction of systemic resistance in rice against sheath blight disease by *Pseudomonas fluorescens*. *Soil Biology & Biochemistry*, *33*(4-5), 603–612. doi:10.1016/S0038-0717(00)00202-9

Nangul, A., & Bhatia, R. (2020). Microorganisms: A marvelous source of single cell proteins. *Journal of Microbiology, Biotechnology and Food Sciences*, *9*(4), 15–18.

Nannipieri, P., Badalucco, L., Landi, L., & Pietramellara, G. (1997). Measurement in assessing the risk of chemicals to the soil ecosystem. *Ecotoxicology: Responses, biomarkers and risk assessment. SOS Publications. Fair Haven, NJ*, *7704*, 507–534.

Nannipieri, P., Giagnoni, L., Landi, L., & Renella, G. (2011). Role of Phosphatase Enzymes in Soil. In E. Bünemann, A. Oberson, & E. Frossard (Eds.), *Phosphorus in Action. Soil Biology* (Vol. 26). Springer. doi:10.1007/978-3-642-15271-9_9

Naqqash, T., Hameed, S., Imran, A., Hanif, M. K., Majeed, A., & van Elsas, J. D. (2016). Differential response of potato toward inoculation with taxonomically diverse plant growth promoting rhizobacteria. *Frontiers in Plant Science*, *7*, 144. doi:10.3389/fpls.2016.00144 PMID:26925072

Narayan Chadar, S. (2018). Composting as an eco-friendly method to recycle organic waste. *Progress in Petrochemical Science*, *2*(5), 252–254. doi:10.31031/PPS.2018.02.000548

Narayanan, M. P., Murugan, S., Eva, A. S., Devina, S. U., & Kalidass, S. (2015). Application of immobilized laccase from Bacillus subtilis MTCC 2414 on decolourization of synthetic dyes. *Res J Microbiol*, *10*(9), 421–432. doi:10.3923/jm.2015.421.432

Nardi, S., Concheri, G., Pizzeghello, D., Sturaro, A., Rella, R., & Parvoli, G. (2000). Soil organic matter mobilization by root exudates. *Chemosphere*, *41*(5), 653–658. doi:10.1016/S0045-6535(99)00488-9 PMID:10834364

Nargund, V. B., Amaresh, Y. S., Sreenivas, A. G., & Nadagouda, S. (2007). *Trichoderma harzianum*–a potential bioagent for seed and soil borne diseases management in Upper Krishna project command area of Karnataka, India. *International Journal of Agricultural Sciences*, *3*, 158–160.

Naseri, R., Maleki, A., Naserirad, H., Shebibi, S., & Omidian, A. (2013). Effect of plant growth promoting rhizobacteria (PGPR) on reduction nitrogen fertilizer application in rapeseed (Brassica napus L.). *Middle East Journal of Scientific Research*, *14*(2), 213–220.

Nataraja, K. N., Suryanarayanan, T. S., Shaanker, R. U., Senthil-Kumar, M., & Oelmuller, R. (2019). Plant–microbe interaction: Prospects for crop improvement and management. *Plant Physiology Reproduction*, *24*(4), 461–462. doi:10.100740502-019-00494-4

Natarajan, C., Prasanna, R., Gupta, V., Dureja, P., & Nain, L. (2012). Characterization of the fungicidal activity of *Calothrix elenkinii* using chemical methods and microscopy. *Applied Biochemistry and Microbiology*, *48*(1), 51–57. doi:10.1134/S0003683812010115 PMID:22567886

Navarro, F., Forján, E., Vázquez, M., Toimil, A., Montero, Z., Ruiz-Domínguez, M., Garbayo, I., Castaño, M. Á., Vílchez, C., & Vega, J. M. (2017). Antimicrobial activity of the acidophilic eukaryotic microalga *Coccomyxaonubensis*. *Phycological Research*, *65*(1), 38–43. doi:10.1111/pre.12158

Nawrot, T., Plusquin, M., Hogervorst, J., Roels, H. A., Celis, H., Thijs, L., & Staessen, J. A. (2006). Environmental exposure to cadmium and risk of cancer: A prospective population-based study. *The Lancet. Oncology*, *7*(2), 119–126. doi:10.1016/S1470-2045(06)70545-9 PMID:16455475

Nayak, S., & Mukherjee, A. K. (2015). Management of Agricultural Wastes Using Microbial Agents. In Singh & Sarkar (Eds.), Waste Management: Challenges, Threats and Opportunities (pp. 65-91). Nova Scientific.

Nayak, T., Panda, A. N., Adhya, T. K., Das, B., & Raina, V. (2019). Biodegradation of Chlorpyrifos and 3, 5, 6-trichloro-2-pyridinol (TCP) by *Ochrobactrum* sp. CPD-03: Insights from genome analysis on organophosphorus pesticides degradation, chemotaxis and PGPR activity. *bioRxiv*. doi:10.1101/2019.12.12.866210

NCBI. (2014). *National Center for Biotechnology Information*. https://www.ncbi.nlm.nih.gov

Ndeddy Aka, R. J., & Babalola, O. O. (2016). Effect of bacterial inoculation of strains of *Pseudomonas aeruginosa, Alcaligenes feacalis* and *Bacillus subtilis* on germination, growth and heavy metal (Cd, Cr, and Ni) uptake of Brassica juncea. *International Journal of Phytoremediation*, *18*(2), 200–209. doi:10.1080/15226514.2015.1073671 PMID:26503637

Ndukwu, B. N., Osujieke, D. N., Ahukaemere, C. M., & Umeh, M. O. (2016). Micronutrients and Physicochemical Properties Of Soils Affected By Municipal Solid Wastes In Ekwulobia Southeastern Nigeria. *Journal of Multidisciplinary Engineering Science Studies*, *2*(11), 1032–1040.

Nedunuri, K., Lowell, C., Meade, W., Vonderheide, A., & Shann, J. (2009). Management practices and phytoremediation by native grasses. *International Journal of Phytoremediation*, *12*(2), 200–214. doi:10.1080/15226510903213928 PMID:20734616

Neerja, G. (2016). Biodegradation of 1, 1, 1-trichloro-2, 2-bis (4-chlorophenyl) ethane (DDT) by using *Serratia marcescens* NCIM 2919. *Journal of Environmental Science and Health. Part B, Pesticides, Food Contaminants, and Agricultural Wastes*, *51*(12), 809–816. doi:10.1080/03601234.2016.1208455 PMID:27494385

Neff, J., Lee, K., & Deblois, E. M. (2011). Produced water: overview of composition, fates, and effects. In K. Lee & J. Neff (Eds.), *Produced Water: Environmental Risks and Advances in Mitigation Technologies* (pp. 3–54). Springer. doi:10.1007/978-1-4614-0046-2_1

Negráo, S., Schmöckel, S. M., & Tester, M. (2017). Evaluating physiological responses of plants to salinity stress. *Annals of Botany*, *119*(1), 1–11. doi:10.1093/aob/mcw191 PMID:27707746

Neha, S., Tuhina, V., & Rajeeva, G. (2013). Detoxifcation of hexavalent chromium by an indigenous facultative anaerobic *Bacillus cereus* strain isolated from tannery effluent. *African Journal of Biotechnology*, *12*(10), 1091–1103.

Neifar, M., Chouchane, H., Mahjoubi, M., Jaouani, A., & Cherif, A. (2016). Pseudomonasextremorientalis BU118: a new salt-tolerant laccase-secreting bacterium with biotechnological potential in textile azo dye decolourization. *Biotech*, *6*(1), 107.

Neilands, J. B. (1995). Siderophores: Structure and function of microbial Iron transport compounds. *The Journal of Biological Chemistry*, *270*(45), 26723–26726. doi:10.1074/jbc.270.45.26723 PMID:7592901

Nelson, L. M. (2004). Plant Growth Promoting Rhizobacteria (PGPR): Prospects for New Inoculants. *Crop Management*, *3*(1), 1–7. doi:10.1094/CM-2004-0301-05-RV

Nelson, M. N., & Sorenson, J. (1999). Chitinolytic activity of *Pseudomonas fluorescens* isolates from barley and sugar beet rhizosphere. *FEMS Microbiology Ecology*, *30*(3), 217–227. doi:10.1111/j.1574-6941.1999.tb00650.x PMID:10525178

Nematshahi, N., Lahouti, M., & Ganjeali, A. (2012). Accumulation of chromium and its effect on growth of (*Allium cepa* cv. Hybrid). *European Journal of Experimental Biology*, *2*(4), 969–974.

Nemecek, J., Pokorný, P., Lhotský, O., Knytl, V., Najmanov, P., Steinov, J., Cerník, M., Filipov, A., Filip, J., & Cajthaml, T. (2016). Combined nano-biotechnology for in-situ remediation of mixed contamination of groundwater by hexavalent chromium and chlorinated solvents. *The Science of the Total Environment*, *563-564*, 822–834. doi:10.1016/j.scitotenv.2016.01.019 PMID:26850861

Němeček, J., Pokorný, T. P., Lacinová, L., Černík, M., Masopustová, Z., Lhotský, O., Filipová, A., & Cajthaml, T. (2015). Combined abiotic and biotic in-situ reduction of hexavalent chromium in groundwater using nZVI and whey: A remedial pilot test. *Journal of Hazardous Materials*, *300*, 670–679. doi:10.1016/j.jhazmat.2015.07.056 PMID:26292054

Nendel, C., Melzer, D., & Thorburn, P. J. (2019). The nitrogen nutrition potential of arable soils. *Scientific Reports*, *9*(1), 5851–5860. doi:10.103841598-019-42274-y PMID:30971710

Nese, T., Sivri, N., & Toroz, I. (2007). Pollutants of textile industry wastewater an assessment of its discharge limits by water quality standards. *Turkish Journal*, *7*, 97–103.

Newman, K. L., Chatterjee, S., Ho, K. A., & Lindow, S. E. (2008). Virulence of plant pathogenic bacteria attenuated by degradation of fatty acid cell-to-cell signaling factors. *Molecular plant-microbe interactions. Molecular Plant-Microbe Interactions*, *21*(3), 326–334. doi:10.1094/MPMI-21-3-0326 PMID:18257682

Ney, L., Franklin, D., Mahmud, K., Cabrera, M., Hancock, D., Habteselassie, M., Newcomer, Q., & Fatzinger, B. (2019). Rebuilding Soil Ecosystems for Improved Productivity in Biosolarized Soils. *International Journal of Agronomy*, *2019*, 5827585. Advance online publication. doi:10.1155/2019/5827585

Ngivprom, U., Milintawisamai, N., & Reungsang, A. (2020). *Reductive dechlorination of 1, 2-dichloroethane to ethylene by anaerobic enrichment culture containing Vulcanibacillus spp. Walailak Journal of Science and Technology.*

Ng, S. P., Davis, B., & Polombo, E. A. (2009). Tn5051 like mer containing transposon identified in a heavy metal tolerant strain *Achromobacter* sp. AO22. *BMC Research Notes*, *7*(1), 2–38. doi:10.1186/1756-0500-2-38 PMID:19284535

Nibert, M. L., Ghabrial, S. A., Maiss, E., Lesker, T., Vainio, E. J., Jiang, D., & Suzuki, N. (2014). Taxonomic reorganization of family Partitiviridae and other recent progress in partitivirus research. *Virus Research*, *188*, 128–141. doi:10.1016/j.virusres.2014.04.007 PMID:24768846

Nicholl, S. T. (2008). *An Introduction to Genetic Engineering*. Cambridge University Press. doi:10.1017/CBO9780511800986

Nicolopoulou-Stamati, P., Maipas, S., Kotampasi, C., Stamatis, P., & Hens, L. (2016). Chemical pesticides and human health: The urgent need for a new concept in agriculture. *Frontiers in Public Health*, *4*, 148. doi:10.3389/fpubh.2016.00148 PMID:27486573

Nie, M., Wang, Y., Yu, J., Xiao, M., Jiang, L., Yang, J., Fang, C., Chen, J., & Li, B. (2011). Understanding plant-microbe interactions for phytoremediation of petroleum-polluted soil. *PLoS One*, *6*(3), e17961. doi:10.1371/journal.pone.0017961 PMID:21437257

Nie, M., Yang, Q., Jiang, L. F., Fang, C. M., Chen, J. K., & Li, B. (2010). Do plants modulate biomass allocation in response to petroleum pollution? *Biology Letters*, *6*(6), 811–814. doi:10.1098/rsbl.2010.0261 PMID:20484231

Nie, P., Li, X., Wang, S., Guo, J., Zhao, H., & Niu, D. (2017). Induced systemic resistance against Botrytis cinerea by Bacillus cereus AR156 through a JA/ET-and NPR1-dependent signaling pathway and activates PAMP-triggered immunity in Arabidopsis. *Frontiers in Plant Science*, *8*, 238. doi:10.3389/fpls.2017.00238 PMID:28293243

Niewiadomska, A. (2004). Effect of Carbendazim, Imazetapir and Thiram on nitrogenase activity, the number of microorganisms in soil and yield of Red Clover (*Trifolium pratense* L.). *Polish Journal of Environmental Studies*, *13*(4).

Niinemets, Ü. (2010). Mild versus severe stress and BVOCs: Thresholds, priming and consequences. *Trends in Plant Science*, *15*(3), 145–153. doi:10.1016/j.tplants.2009.11.008 PMID:20006534

Nion, Y. A., & Toyota, K. (2015). Recent trends in control methods for bacterial wilt diseases caused by Ralstonia solanacearum. *Microbes and Environments*. PMID:25762345

Nisha, R., Kaushik, A., & Kaushik, C. P. (2007). Effect of indigenous cyanobacterial application on structural stability and productivity of an organically poor semi-arid soil. *Geoderma*, *138*(1–2), 49–56. doi:10.1016/j.geoderma.2006.10.007

Nishimoto, R. (2019). Global trends in the crop protection industry. *Journal of pesticide science*, D19-101.Nwoko, C. O. (2010). Trends in phytoremediation of toxic elemental and organic pollutants. *African Journal of Biotechnology*, *9*(37), 6010–6016.

Niu, B., Vater, J., Rueckert, C., Blom, J., Lehmann, M., Ru, J. J., Chen, X. H., Wang, Q., & Borriss, R. (2013). Polymyxin P is the active principle in suppressing phytopathogenic *Erwinia spp.* by the biocontrol rhizobacterium *Paenibacillus polymyxa* M-1. *BMC Microbiology*, *13*(1), 137. doi:10.1186/1471-2180-13-137 PMID:23773687

Niu, Y., Yuan, Y., Mao, J., Yang, Z., Cao, Q., Zhang, T., Wang, S., & Liu, D. (2018). Characterization of two novel mycoviruses from *Penicillium digitatum* and the related fungicide resistance analysis. *Scientific Reports*, *8*(1), 1–12. doi:10.103841598-018-23807-3 PMID:29615698

Nizzetto, L., Futter, M., & Langaas, S. (2016). Are agricultural soils dumpsfor microplastics of urban origin? *Environmental Science & Technology*, *50*(20), 10777–10779. doi:10.1021/acs.est.6b04140 PMID:27682621

Njoku, K. L., Nomba, E. U., & Olatunde, A. M. (2017). Vermiremediation of Crude Oil Contaminated Soil Using *Eudrillus euginae* and *Lumbricus terrestris. Journal of Biological and Environmental Sciences, 11*(31), 43–50.

Nocelli, N., Bogino, P. C., Banchio, E., & Giordano, W. (2016). Roles of extracellular polysaccharides and biofilm formation in heavy metal resistance of rhizobia. *Materials (Basel), 9*(6), 418. doi:10.3390/ma9060418 PMID:28773540

Novinscak, A., Surette, C., Allain, C., & Filion, M. (2008). Application of molecular technologies to monitor the microbial content of biosolids and composted biosolids. *Water Science and Technology: A Journal of the International Association on Water Pollution Research, 57*(4), 471–477.

Nriagu, J. O. (1989). A global assessment of natural sources of atmospheric trace metals. *Nature, 338*, 47–49.

Nunn, D. M. (1979). *The Dyeing of Synthetic-Polymer and Acetate Fibres*. Dyers Co. Publications Trust.

Nuss, D. L. (2005). Hypovirulence: Mycoviruses at the fungal-plant interface. *Nature Reviews. Microbiology, 3*(8), 632–642. doi:10.1038/nrmicro1206 PMID:16064055

Nutaratat, P., Monprasit, A., Srisuk, N.(2017). High-yield production of indole-3-acetic acid by Enterobacter sp. DMKU-RP206, a rice phyllosphere bacterium that possesses plant growth-promoting traits. *Biotech, 7*(5), 305.

Nutt, M. O., Hughes, J. B., & Wong, M. S. (2005). Designing Pd-on-Au Bimetallic Nanoparticle Catalysts for Trichloroethene Hydrodechlorination. *Environmental Science & Technology, 39*(5), 1346–1353. doi:10.1021/es048560b PMID:15787376

Nwaaichi, E. O., Frac, M., Nwoha, P. A., & Eragbor, P. (2015). Enhanced phytoremediation of crude oil-polluted soil by four plant species: Effect of inorganic and organic bioaugumentation. *International Journal of Phytoremediation, 17*(12), 1253–1261. doi:10.1080/15226514.2015.1058324 PMID:26090948

O'Banion, B. S., O'Neal, L., Alexander, G., & Lebeis, S. L. (2020). Bridging the gap between single-strain and community-level plant-microbe chemical interaction. *Molecular Plant-Microbe Interactions, 3*(2), 124–134. doi:10.1094/MPMI-04-19-0115-CR PMID:31687914

O'Hara, S., Krug, T., Quinn, J., Clausen, C., & Geiger, C. (2006). Field andlaboratory evaluation of the treatment of DNAPL sourcezones using emulsified zero-valent iron. *Remediation, 16*(2), 35–56. doi:10.1002/rem.20080

O'sullivan, D. J., & O'gara, F. (1992). Traits of fluorescent *Pseudomonas* spp. involved in suppression of plant root pathogens. *Microbiological Reviews, 56*(4), 662–676. doi:10.1128/MR.56.4.662-676.1992 PMID:1480114

O'Callaghan-Gordo, C., Orta-Martínez, M., & Kogevinas, M. (2016). Health effects of non-occupational exposure to oil extraction. *Environmental Health: A Global Access Science Source, 15*, 56.

Ochoa-Hueso, R., Delgado-Baquerizo, M., King, P. T. A., Benham, M., Arca, V., & Power, S. A. (2019). Ecosystem type and resource quality are more important than global change drivers in regulating early stages of litter decomposition. *Soil Biology & Biochemistry, 129*, 144–152. doi:10.1016/j.soilbio.2018.11.009

Oden, K. L., Gladysheva, T. B., & Rosen, B. P. (1994). Arsenate reduction mediated by the plasmid-encoded ArsC protein is coupled to glutathione. *Molecular Microbiology, 12*(2), 301–306.

Odjegba, V. J., & Sadiq, A. O. (2002). Effects of spent engine oil on growth parameters, chlorophyll and protein level of *Amaranthus hybrious. The Environmentalist, 22*(1), 23–28. doi:10.1023/A:1014515924037

Odukkathil, G., & Vasudevan, M. (2013). Toxicity and bioremediation of pesticides in agricultural soil. *Reviews in Environmental Science and Biotechnology, 12*(4), 421–444. doi:10.100711157-013-9320-4

Oetiker, J. H., Lee, D. H., & Kato, A. (1999). Molecular analysis of a tryptophan-2-monooxygenase gene (IaaM) of *Agrobacterium vitis*. *DNA Sequence*, *10*(4-5), 349–354. doi:10.3109/10425179909033963 PMID:10727091

Ogut, M., Er, F., & Kandemir, N. (2010). Phosphate solubilization potentials of soil Acinetobacter strains. *Biology and Fertility of Soils*, *46*(7), 707–715. doi:10.100700374-010-0475-7

Oh, B. T., Just, C. L., & Alvarez, P. J. (2001). Hexahydro-1, 3, 5-trinitro-1, 3, 5-triazine mineralization byzerovalent iron and mixed anaerobic cultures. *Environmental Science & Technology*, *35*(21), 4341–4346. doi:10.1021/es010852e PMID:11718353

Ojuederie, O. B., & Babalola, O. O. (2017). Microbial and plant-assisted bioremediation of heavy metal polluted environments: A review. *International Journal of Environmental Research and Public Health*, *14*(12), 1504. doi:10.3390/ijerph14121504 PMID:29207531

Ojumu, T. V., Yu, J., & Solomon, B. O. (2004). Production of Polyhydroxyalkanoates, a bacterial biodegradable polymer. In African Journal of Biotechnology (Vol. 3, Issue 1, pp. 18–24). doi:10.5897/AJB2004.000-2004

Okereke, J. N., Ogidi, O. I., & Obasi, K. O. (2016). Environmental and health impact of industrial wastewater effluents in Nigeria-A Review. *International Journal of Advanced Research in Biological Sciences*, *3*(6), 55-67.

Okino-Delgado, C. H., Zanutto-Elgui, M. R., do Prado, D. Z., Pereira, M. S., & Fleuri, L. F. (2019). Enzymatic Bioremediation: Current Status, Challenges of Obtaining Process, and Applications. In P. Arora (Ed.), *Microbial Metabolism ofXenobiotic Compounds. Microorganisms for Sustainability* (pp. 79–101). Springer. doi:10.1007/978-981-13-7462-3_4

Okoduwa, S. I. R., Igiri, B., Udeh, C. B., Edenta, C., & Gauje, B. (2017). Tannery effluent treatment by yeast species isolates from watermelon. *Toxics*, *5*(1), 6. doi:10.3390/toxics5010006 PMID:29051437

Okon, Y., & Labandera-Gonzalez, C. A. (1994). Agronomic applications of Azospirillum: An evaluation of 20 years worldwide field inoculation. *Soil Biology & Biochemistry*, *26*(12), 1591–1601. doi:10.1016/0038-0717(94)90311-5

Okumura, M., Inoue, S. I., Kuwata, K., & Kinoshita, T. (2016). Photosynthesis activates plasma membrane H+-ATPase via sugar accumulation. *Plant Physiology*, *171*(1), 580–589. doi:10.1104/pp.16.00355 PMID:27016447

Olaniran, A. O., Balgobind, A., & Pillay, B. (2013). Review bioavailability of heavy metal in soil: Impact on microbial biodegradation of organic compounds and possible improvement strategies. *International Journal of Molecular Sciences*, *14*(5), 10197–10228. doi:10.3390/ijms140510197 PMID:23676353

Olaniran, A. O., Naidoo, S., Masango, M. G., & Pillay, B. (2007). Aerobic biodegradation of 1, 2-dichloroethane and 1, 3-dichloropropene by bacteria isolated from a pulp mill wastewater effluent in South Africa. *Biotechnology and Bioprocess Engineering; BBE*, *12*(3), 276–281. doi:10.1007/BF02931104

Olanrewaju, O. S., Glick, B. R., & Babalola, O. O. (2017). Mechanisms of action of plant growth promoting bacteria. *World Journal of Microbiology & Biotechnology*, *33*(11), 197. doi:10.100711274-017-2364-9 PMID:28986676

Oldroyd, G. E., & Downie, J. A. (2008). Coordinating nodule morphogenesis with rhizobial infection in legumes. *Annual Review of Plant Biology*, *59*(1), 519–546. doi:10.1146/annurev.arplant.59.032607.092839 PMID:18444906

Oliver, D. P., Bramley, R. G. V., Riches, D., Porter, I., & Edwards, J. (2013). Review: Soil physical and chemical properties as indicators of soil quality in Australian viticulture. *Australian Journal of Grape and Wine Research*, *19*(2), 129–139. doi:10.1111/ajgw.12016

Ollikka, P., Harjunpää, T., Palmu, K., Mäntsälä, P., & Suominen, I. (1998). Oxidation of crocein orange G by lignin peroxidase isoenzymes. *Applied Biochemistry and Biotechnology*, *75*(2-3), 307–321. doi:10.1007/BF02787783 PMID:10230025

Olorunfemi, I., Fasinmirin, J., & Ojo, A. (2016). Modeling cation exchange capacity and soil water holding capacity from basic soil properties. *Eurasian Journal of Soil Science*, *5*(4), 266–274. doi:10.18393/ejss.2016.4.266-274

Olowoyo, J. O., & Mugivhisa, L. L. (2019). Evidence of uptake of different pollutants in plants harvested from soil treated and fertilized with organic materials as source of soil nutrients from developing countries. *Chemical and Biological Technologies in Agriculture*, *6*(1), 28–39. doi:10.118640538-019-0165-0

Onakpa, M. M., Njan, A. A., & Kalu, O. C. (2018). A Review of Heavy Metal Contamination of Food Crops in Nigeria. *Annals of Global Health*, *84*(3), 488–494. doi:10.29024/aogh.2314 PMID:30835390

Ona, O., Van Impe, J., Prinsen, E., & Vanderleyden, J. (2005). Growth and indole-3-acetic acid biosynthesis of Azospirillum brasilense Sp245 is environmentally controlled. *FEMS Microbiology Letters*, *246*(1), 125–132. doi:10.1016/j.femsle.2005.03.048 PMID:15869971

Ongena, M., Daayf, F., Jacques, P., Thonart, P., Benhamou, N., Paulitz, T.C., Cornélis, P., Koedam, N., & Bélanger, R.R, (1999). Protection of cucumber against Pythium root rot by fluorescent pseudomonads: predominant role of induced resistance over siderophores and antibiosis. *Plant Pathology, 48*, 66-76. doi:10.1046/j.1365-3059.1999.00315.x

Onneby, K., Håkansson, S., Pizzul, L., & Stenstrom, J. (2014). Reduced leaching of the herbicide MCPA after bioaugmentation with formulated and stored *Sphingobium* sp. *Biodegradation*, *25*(2), 291–300. doi:10.100710532-013-9660-3 PMID:23982656

Onofre-Lemus, J., Hernández-Lucas, I., Girard, L., & Caballero-Mellado, J. (2009). ACC (1-aminocyclopropane-1-carboxylate) deaminase activity, a widespread trait in Burkholderia species, and its growth-promoting effect on tomato plants. *Applied and Environmental Microbiology*, *75*(20), 6581–6590. doi:10.1128/AEM.01240-09 PMID:19700546

Onwuka, B. M. (2016). Effects of soil temperature on Some Soil properties and plant growth. *The Journal of Agricultural Science*, *6*(3), 89–93.

Ordinioha, B., & Brisibe, S. (2013). The human health implications of crude oil spills in the Niger Delta, Nigeria: An interpretation of published studies. *Nigerian Medical Journal*, *54*(1), 10–16. doi:10.4103/0300-1652.108887 PMID:23661893

Oren, A. (2014). Halophilic archaea on Earth and in space: Growth and survival under extreme conditions. *Philosophical Transactions - Royal Society. Mathematical, Physical, and Engineering Sciences*, *372*(2030), 20140194. doi:10.1098/rsta.2014.0194 PMID:25368347

Organization of the Petroleum Exporting Countries (OPEC). (2019). OPEC Share of World Crude Oil Reserves. *OPEC Annual Statistical Bulletin, 17*, 52. Available at: https://www.opec.org/opec_web/en/76.html

Organum, N., & Bacon, F. (2006). Bioremediation technologies. In P. J. J. Alvarez & W. A. Illman (Eds.), *Bioremediation and natural attenuation* (pp. 351–455). John Wiley & Sons.

Orhan, E., Esitken, A., Ercisli, S., Turan, M., & Sahin, F. (2006). Effects of plant growth promoting rhizobacteria (PGPR) on yield, growth and nutrient contents in organically growing raspberry. *Scientia Horticulturae*, *111*(1), 38–43. doi:10.1016/j.scienta.2006.09.002

Orhorhoro, E., & Oghoghorie, O. (2019). Review on solid waste generation and management in sub-Saharan Africa: A case study of Nigeria. *Journal of Applied Science & Environmental Management*, *23*(9), 1729–1737. doi:10.4314/jasem.v23i9.19

Ortíz-Castro, R., Contreras-Cornejo, H. A., Macías-Rodríguez, L., & López-Bucio, J. (2009). The role of microbial signals in plant growth and development. *Plant Signaling & Behavior*, *4*(8), 701–712. doi:10.4161/psb.4.8.9047 PMID:19820333

Ortiz-Hernández, M. L., Sánchez-Salinas, E., Dantán-González, E., & Castrejón-Godínez, M. L. (2013). Pesticide biodegradation: mechanisms, genetics and strategies to enhance the process. *Biodegradation-Life of Science*, 251-287.

Oses, R., Valenzuela, S., Freer, J., Sanfuentes, E., & Rodriguez, J. (2008). Fungal endophytes in xylem of healthy chilean trees and their possible role in early wood decay. *Fungal Diversity*, *33*, 77–86.

Osma, J. F., Toca-Herrera, J. L., & Rodriguez-Couto, S. (2010). Transformation pathway of Remazol Brilliant Blue R by immobilised laccase. *Bioresource Technology*, *101*(22), 8509–8514. doi:10.1016/j.biortech.2010.06.074 PMID:20609582

Osman, N. I., & Yin, S. (2018). Isolation and characterization of pea plant (Pisum sativum L.) growth-promoting Rhizobacteria. *African Journal of Microbiological Research*, *12*(34), 820–828. doi:10.5897/AJMR2018.8859

Otieno, N., Lally, R. D., Kiwanuka, S., Lloyd, A., Ryan, D., Germaine, K. J., & Dowling, D. N. (2015). Plant growth promotion induced by phosphate solubilizing endophytic *Pseudomonas* isolates. *Frontiers in Microbiology*, *6*, 745. doi:10.3389/fmicb.2015.00745 PMID:26257721

Otto, M., Floyd, M., & Bajpai, S. (2008). Nanotechnology for site remediation. *Remediation*, *19*(1), 99–108. doi:10.1002/rem.20194

Otyepka, M., & Damborský, J. (2002). Functionally relevant motions of haloalkane dehalogenases occur in the specificity-modulating cap domains. *Protein Science*, *11*(5), 1206–1217. doi:10.1110/ps.ps3830102 PMID:11967377

Ouni, L., Ramazani, A., & Taghavi Fardood, S. (2019). An overview of carbon nanotubes role in heavy metals removal from wastewater. *Frontiers of Chemical Science and Engineering*, *13*(2), 274–295. doi:10.100711705-018-1765-0

Ovaa, H. (2014). Unnatural amino acid incorporation in *E. coli*: Current and future applications in the design of therapeutic proteins. *Frontiers in Chemistry*, *2*, 15. PMID:24790983

Owens, P. R., & Rutledge, E. M. (2005). Morphology. Encyclopedia of Soils in the Environment, 511–520. doi:10.1016/B0-12-348530-4/00002-3

Oyewole, O. A., Raji, R. O., Musa, I. O., Enemanna, C. E., Abdulsalam, O. N., & Yakubu, J. G. (2019). Enhanced degradation of Crude Oil with *Alcaligenes faecalis* ADY25 and Iron oxide Nanoparticle. *International Journal of Applied Biological Research*, *10*(2), 62–72.

Ozer, E. A., Pezzulo, A., Shih, D. M., Chun, C., Furlong, C., Lusis, A. J., Greenberg, E. P., & Zabner, J. (2005). Human and murine paraoxonase 1 are host modulators of Pseudomonas aeruginosa quorum-sensing. *FEMS Microbiology Letters*, *253*(1), 29–37. doi:10.1016/j.femsle.2005.09.023 PMID:16260097

Ozkan-Kotiloglu, S., & Coutts, R. H. A. (2018). Multiplex detection of *Aspergillus fumigatus* Mycoviruses. *Viruses*, *10*(5), 247. doi:10.3390/v10050247 PMID:29738445

Ozsoy, G., & Aksoy, E. (2015). Estimation of soil erosion risk within an important agricultural sub-watershed in Bursa, Turkey, in relation to rapid urbanization. *Environmental Monitoring and Assessment*, *187*(7), 419. doi:10.100710661-015-4653-9 PMID:26059559

Padma, S. D., & Sukumar, J. (2015). Response of mulber- ry to inoculation of potash mobilizing bacterial isolate and other bio-inoculants. *Global Journal of Bio-Science and BioTechnology*, *4*, 50–53.

Paes, B. G., & Almeida, J. R. (2014). Genetic improvement of microorganisms for applications in biorefineries. *Chemical and Biological Technologies in Agriculture*, *1*(1), 21. doi:10.118640538-014-0021-1

Page, S. E., Rieley, J. O., & Banks, C. J. (2011). Global and regional importance of the tropical peatland carbon pool. *Global Change Biology*, *17*(2), 798–818. doi:10.1111/j.1365-2486.2010.02279.x

Pahari, A., Pradhan, A., Nayak, S. K., & Mishra, B. B. (2017). Bacterial siderophore as a plant growth promoter. In *Microbial Biotechnology* (pp. 163–180). Springer. doi:10.1007/978-981-10-6847-8_7

Pailan, S., Gupta, D., Apte, S., Krishnamurthi, S., & Saha, P. (2015). Degradation of organophosphate insecticide by a novel Bacillus aryabhattai strain SanPS1, isolated from soil of agricultural field in Burdwan, West Bengal, India. *International Biodeterioration & Biodegradation, 103*, 191–195. doi:10.1016/j.ibiod.2015.05.006

Pal, A. K., & Sengupta, C. (2019). Isolation of Cadmium and Lead Tolerant Plant Growth Promoting Rhizobacteria: *Lysinibacillus varians* and *Pseudomonas putida* from Indian Agricultural Soil. *Soil and Sediment Contamination: An International Journal, 28*(7), 601–629. doi:10.1080/15320383.2019.1637398

Palmer-Brown, W., de Melo Souza, P. L., & Murphy, C. D. (2018). Cyhalothrin biodegradation in Cunninghamella elegans. *Environmental Science and Pollution Research International, 26*(2), 1414–1421. doi:10.100711356-018-3689-0 PMID:30426373

Palmroth, M. R. T., Pichtel, J., & Puhakka, J. A. (2002). Phytoremediation of subarctic soil contaminated with diesel fuel. *Bioresource Technology, 84*(3), 221–228. doi:10.1016/S0960-8524(02)00055-X PMID:12118697

Paltseva, A., Cheng, Z., Deeb, M., Groffman, P. M., Shaw, R. K., & Maddaloni, M. (2018). Accumulation of arsenic and lead in garden-grown vegetables: Factors and mitigation strategies. *The Science of the Total Environment, 640-641*, 640–641. doi:10.1016/j.scitotenv.2018.05.296 PMID:29859443

Palva, I. (1982). Molecular cloning of α-amylase gene from *Bacillus amyloliquefaciens* and its expression in *B. subtilis. Gene, 19*(1), 81–87. doi:10.1016/0378-1119(82)90191-3 PMID:6183169

Pan, I., Dam, B., & Sen, S. K. (2012). Composting of common organic wastes using microbial inoculants. *Biotech, 2*(2), 127–134.

Panchenko, L., Muratova, A., & Turkovskaya, O. (2017). Comparison of the phytoremediation potentials of Medicago falcata L. And Medicago sativa L. in aged oil-sludge-contaminated soil. *Environmental Science and Pollution Research International, 24*(3), 3117–3130. doi:10.100711356-016-8025-y PMID:27858273

Panda, H. (2011). *Manufacture of biofertilizer and organic farming*. ASIA PACIFIC BUSINESS PRESS Inc.

Panday, S. C., Choudhary, M., Singh, S., Meena, V. S., Mahanta, D., Yadav, R. P., Pattanayak, A., & Bisht, J. K. (2018). Increasing farmer's income and water use efficiency as affected by long-term fertilization under a rainfed and supplementary irrigation in a soybean-wheat cropping system of Indian mid-Himalaya. *Field Crops Research, 219*, 214–221. doi:10.1016/j.fcr.2018.02.004

Pande, V., Pandey, S. C., Joshi, T., Sati, D., Gangola, S., Kumar, S., & Samant, M. (2019). Biodegradation of toxic dyes: a comparative study of enzyme action in a microbial system. In *Smart Bioremediation Technologies* (pp. 255–287). Elsevier. doi:10.1016/B978-0-12-818307-6.00014-7

Pandey, V. C., & Bajpai, O. (2019). Phytoremediation: From theory toward practice. In Phytomanagement of polluted sites (pp. 1-49). Elsevier. doi:10.1016/B978-0-12-813912-7.00001-6

Pandey, A. K., Sarada, D. V., & Kumar, A. (2016). Microbial Decolorization and Degradation of Reactive Red 198. *Proceedings of the National Academy of Sciences*, 805–815.

Pandey, G. (2018). Prospects of Nanobioremediation in Environmental cleanup. *Oriental Journal of Chemistry, 34*(6), 2838–2850. doi:10.13005/ojc/340622

Pandey, J., Verma, R. K., & Singh, S. (2019). Suitability of aromatic plants for phytoremediation of heavy metal contaminated areas: A review. *International Journal of Phytoremediation*, *21*(5), 405–418. doi:10.1080/15226514.2018.1540546 PMID:30656974

Pandey, S., Ghosh, P. K., Ghosh, S., De, T. K., & Maiti, T. K. (2013). Role of heavy metal resistant *Ochrobactrum sp.* and *Bacillus spp.* strains in bioremediation of a rice cultivar and their PGPR like activities. *Journal of Microbiology (Seoul, Korea)*, *51*(1), 11–17. doi:10.100712275-013-2330-7 PMID:23456706

Pandey, V. C., Pandey, D. N., & Singh, N. (2015). Sustainable phytoremediation based on naturally colonizing and economically valuable plants. *Journal of Cleaner Production*, *86*, 37–39. doi:10.1016/j.jclepro.2014.08.030

Panfili, I., Bartucca, M. L., Ballerini, E., & Del Buono, D. (2017). Combination of aquatic species and safeners improves the remediation of copper polluted water. *The Science of the Total Environment*, *601*, 1263–1270. doi:10.1016/j.scitotenv.2017.06.003 PMID:28605844

Panfili, I., Bartucca, M. L., & Del Buono, D. (2019). The treatment of duckweed with a plant biostimulant or a safener improves the plant capacity to clean water polluted by terbuthylazine. *The Science of the Total Environment*, *646*, 832–840. doi:10.1016/j.scitotenv.2018.07.356 PMID:30064109

Pang, Y., Zeng, G. M., Tang, L., Zhang, Y., Liu, Y. Y., Lei, X. X., Wu, M. S., Li, Z., & Liu, C. (2011). Cr(VI) reduction by Pseudomonas aeruginosa immobilized in a polyvinyl alcohol/sodium alginatematrix containing multi-walled carbon nanotubes. *Bioresource Technology*, *102*(22), 10733–10736. doi:10.1016/j.biortech.2011.08.078 PMID:21937224

Pant, D., & Adholeya, A. (2009). Concentration of fungal ligninolytic enzymes by ultrafiltration and their use in distillery effluent decolorization. *World Journal of Microbiology & Biotechnology*, *25*(10), 1793–1800. doi:10.100711274-009-0079-2

Pan, X., Chen, Z., Chen, F., Cheng, Y., Lin, Z., & Guan, X. (2015). The mechanism of uranium transformation from U (VI) into nano-uramphite by two indigenous *Bacillus thuringiensis* strains. *Journal of Hazardous Materials*, *297*, 313–319. doi:10.1016/j.jhazmat.2015.05.019 PMID:26026850

Pappas, K. M., & Winans, S. C. (2003). A LuxR-type regulator from Agrobacterium tumefaciens elevates Ti plasmid copy number by activating transcription of plasmid replication genes. *Molecular Microbiology*, *48*(4), 1059–1073. doi:10.1046/j.1365-2958.2003.03488.x PMID:12753196

Parales, R. E., Ditty, J. L., & Harwood, C. S. (2000). Toluene-degrading bacteria are chemotactic towards the environmental pollutants benzene, toluene, and trichloroethylene. *Applied and Environmental Microbiology*, *66*(9), 4098–4104. doi:10.1128/AEM.66.9.4098-4104.2000 PMID:10966434

Parikh, A., & Madamwar, D. (2006). Partial characterization of extracellular polysaccharides from cyanobacteria. *Bioresource Technology*, *97*, 1822–1827.

Park, C. M., Banerjee, N., Koltin, Y., & Bruenn, J. A. (1996). The *Ustilago maydis* virally encoded KP1 killer toxin. *Molecular Microbiology*, *20*(5), 957–963. doi:10.1111/j.1365-2958.1996.tb02537.x PMID:8809749

Park, C., Lee, M., Lee, B., Kim, S.-W., Chase, H. A., Lee, J., & Kim, S. (2007). Biodegradation and biosorption for decolorization of synthetic dyes by *Funalia trogii*. *Biochemical Engineering Journal*, *36*(1), 59–65. doi:10.1016/j.bej.2006.06.007

Park, C., Li, X. R., Zhao, Y., Jia, R. L., & Hur, J. (2017). Rapid development of cyanobacterial crust in the field for combating desertification. *PLoS One*, *12*(6), e0179903. doi:10.1371/journal.pone.0179903 PMID:28644849

Park, J. W., Park, B. K., & Kim, J. E. (2006). Remediation of soil contaminated with 2, 4-dichlorophenol by treatment of minced shepherd's purse roots. *Archives of Environmental Contamination and Toxicology*, *50*(2), 191–195. doi:10.100700244-004-0119-8 PMID:16392021

Park, S. R., Yoon, Y. J., Pham, J. V., Yilma, M. A., Feliz, A., Majid, M. T., ... Song, M. C. (2019). A review of the microbial production of bioactive natural products and biologics. *Frontiers in Microbiology*, *10*, 1404. doi:10.3389/fmicb.2019.01404 PMID:31281299

Park, S. Y., Binkley, R. M., Kim, W. J., Lee, M. H., & Lee, S. Y. (2018). Metabolic engineering of *Escherichia coli* for high-level astaxanthin production with high productivity. *Metabolic Engineering*, *49*, 105–115. doi:10.1016/j.ymben.2018.08.002 PMID:30096424

Park, S. Y., Hwang, B. J., Shin, M. H., Kim, J. A., Kim, H. K., & Lee, J. K. (2006). N-acylhomoserine lactonase producing Rhodococcus spp. with different AHL-degrading activities. *FEMS Microbiology Letters*, *261*(1), 102–108. doi:10.1111/j.1574-6968.2006.00336.x PMID:16842366

Park, S. Y., Kang, H. O., Jang, H. S., Lee, J. K., Koo, B. T., & Yum, D. Y. (2005). Identification of extracellular N-acylhomoserine lactone acylase from a Streptomyces sp. and its application to quorum quenching. *Applied and Environmental Microbiology*, *71*(5), 2632–2641. doi:10.1128/AEM.71.5.2632-2641.2005 PMID:15870355

Park, S. Y., Lee, S. J., Oh, T. K., Oh, J. W., Koo, B. T., Yum, D. Y., & Lee, J. K. (2003). AhlD, an N-acylhomoserine lactonase in Arthrobacter sp., and predicted homologues in other bacteria. *Microbiology (Reading, England)*, *149*(Pt 6), 1541–1550. doi:10.1099/mic.0.26269-0 PMID:12777494

Park, Y., James, D., & Punja, Z. K. (2005). Co-infection by two distinct totivirus-like double-stranded RNA elements in *Chalara elegans* (*Thielaviopsis basicola*). *Virus Research*, *109*(1), 71–85. doi:10.1016/j.virusres.2004.10.011 PMID:15826915

Parmar, P., & Sindhu, S. S. (2013). Potassium solubilization by rhizosphere bacteria: Influence of nutritional and environmental conditions. *Journal of Microbiology Research (Rosemead, Calif.)*, *3*, 25–31.

Parnell, J. J., Park, J., Denef, V., Tsoi, T., Hashsham, S., Quensen, J. I. III, & Tiedje, J. M. (2006). Coping with polychlorinated biphenyl (PCB) toxicity: Physiological and genomewide responses of Burkholderia xenovorans LB400 to PCB-mediated stress. *Applied and Environmental Microbiology*, *72*(10), 6607–6614. doi:10.1128/AEM.01129-06 PMID:17021212

Parrish, Z. D., Banks, M. K., & Schwab, A. P. (2005). Effect of root death and decay on dissipation of polycyclic aromatic hydrocarbons in the rhizosphere of yellow sweet clover and tall fescue. *Journal of Environmental Quality*, *34*(1), 207–216. doi:10.2134/jeq2005.0207 PMID:15647551

Parveda, M., Kiran, B., Punita, D. L., & Kishor, P. K. (2017). Overexpression of SbAP37 in rice alleviates concurrent imposition of combination stresses and modulates different sets of leaf protein profiles. *Plant Cell Reports*, *36*(5), 773–786. doi:10.100700299-017-2134-z PMID:28393269

Pasquini, M. W., & Alexander, M. J. (2004). Chemical properties of urban waste ash produced by open burning on the Jos Plateau: Implications for agriculture. *The Science of the Total Environment*, *319*(1-3), 225–240. doi:10.1016/S0048-9697(03)00434-0 PMID:14967513

Paszczynski, A., & Crawford, R. L. (1991). Degradation of azo compounds by ligninase from *Phanerochaete chrysosporium*: Involvement of veratryl alcohol. *Biochemical and Biophysical Research Communications*, *178*(3), 1056–1063. doi:10.1016/0006-291X(91)90999-N PMID:1872828

Patel, P. R., Shaikh, S. S., & Sayyed, R. Z. (2016). Dynamism of PGPR in bioremediation and plant growth promotion in heavy metal contaminated soil. *Indian Journal of Experimental Biology*, *54*(4), 286–290. PMID:27295926

Patel, P., Trivedi, G., & Saraf, M. (2018). Iron biofortification in mungbean using siderophore producing plant growth promoting bacteria. *Environmental Sustainability*, *1*(4), 357–365. doi:10.100742398-018-00031-3

Patel, T., & Saraf, M. (2017). Biosynthesis of phytohormones from novel rhizobacterial isolates and their in vitro plant growth-promoting efficacy. *Journal of Plant Interactions*, *12*(1), 480–487. doi:10.1080/17429145.2017.1392625

Patowary, R., Patowary, K., Devi, A., Kalita, M. C., & Deka, S. (2017). Uptake of total petroleum hydrocarbon (TPH) and polycyclic aromatic hydrocarbons (PAHs) by *Oryza sativa* L. grown in soil contaminated with crude oil. *Bulletin of Environmental Contamination and Toxicology*, *98*(1), 120–126. doi:10.100700128-016-1990-5 PMID:27896384

Patra, S., Mishra, P., Mahapatra, S. C., & Mithun, S. K. (2016). Modelling impacts of chemical fertilizer on agricultural production: A case study on Hooghly district, West Bengal, India. *Modeling Earth Systems and Environment*, *2*(4), 1–11. doi:10.100740808-016-0223-6

Paul, A., & Dubey, R. (2014). Characterization of Protein involved in Nitrogen Fixation and Estimation of Co-Factor. *International Journal of Advanced Biotechnology and Research*, *5*(4), 582–597.

Paul, T., Mandal, A., & Mondal, K. C. (2018). *Waste to value aided fertilizer: an alternative cleaning technique for poultry feathers waste disposal*. Ann Microbiol Immunol.

Paungfoo-Lonhienne, C., Rentsch, D., Robatzek, S., Webb, R. I., Sagulenko, E., Näsholm, T., Schmidt, S., & Lonhienne, T. G. A. (2010). Turning the table: Plants consume microbes as a source of nutrients. *PLoS One*, *5*(7), e11915. doi:10.1371/journal.pone.0011915 PMID:20689833

Pavlova, M., Klvana, M., Prokop, Z., Chaloupkova, R., Banas, P., Otyepka, M., & Damborsky, J. (2009). Redesigning dehalogenase access tunnels as a strategy for degrading an anthropogenic substrate. *Nature Chemical Biology*, *5*(10), 727–733. doi:10.1038/nchembio.205 PMID:19701186

Pawlik, M., Cania, B., Thijs, S., Vangronsveld, J., & Piotrowska-Seget, Z. (2017). Hydrocarbon degradation potential and plant growth-promoting activity of culturable endophytic bacteria of *Lotus corniculatus* and *Oenothera biennis* from a long-term polluted site. *Environmental Science and Pollution Research International*, *24*(24), 19640–19652. doi:10.100711356-017-9496-1 PMID:28681302

Paz-Alberto, A. M., & Sigua, G. C. (2013). Phytoremediation: A green technology to remove environmental pollutants. *American Journal of Climate Change*, *2*(01), 71–86. doi:10.4236/ajcc.2013.21008

Paz-Ferreiro, J., Gascó, G., Méndez, A., & Reichman, S. (2018). Soil Pollution and Remediation. *International Journal of Environmental Research and Public Health*, *15*(8), 1657–1660. doi:10.3390/ijerph15081657 PMID:30081583

Pazirandeh, M., Chrisey, L. A., Mauro, J. M., Campbell, J. R., & Gaber, B. P. (1995). Expression of the Neurospora crassa metallothionein gene in Escherichia coli and its effect on heavy-metal uptake. *Applied Microbiology and Biotechnology*, *43*(6), 1112–1117. doi:10.1007/BF00166934 PMID:8590662

Pearce, C. I., Lloyd, J. R., & Guthrie, J. T. (2003). The removal of color from textile wastewater using whole bacterial cells: A review. *Dyes and Pigments*, *58*(3), 179–196.

Pearson, M. N., Beever, R. E., Boine, B., & Arthur, K. (2009). Mycoviruses of filamentous fungi and their relevance to plant pathology. *Molecular Plant Pathology*, *10*(1), 115–128. doi:10.1111/j.1364-3703.2008.00503.x PMID:19161358

Peer, W. A., Baxter, I. R., Richards, E. L., Freeman, J. L., & Murphy, A. S. (2006). Phytoremediation and hyperaccumulator plants. In *Molecular Biology of Metal Homeostasis and Detoxification* (pp. 299–340). Springer.

Peixoto, R. S., Vermelho, A. B., & Rosado, A. S. (2011). Petroleum-degrading enzymes: Bioremediation and new prospects. *Enzyme Research*, *2011*, 475193. doi:10.4061/2011/475193 PMID:21811673

Peña-Montenegro, T. D., Lozano, L., & Dussán, J. (2015). Genome sequence and description of the mosquitocidal and heavy metal tolerant strain *Lysinibacillussphaericus* CBAM5. *Standards in Genomic Sciences*, *10*(1), 2. doi:10.1186/1944-3277-10-2 PMID:25685257

Peng, Y., Joseph, J. P., Minori, U., & Jason, C. W. (2015). Hetero aggregation of cerium oxide nanoparticles and nanoparticles of pyrolyzed biomass. *Environmental Science & Technology*, *49*(22), 13294–13303. doi:10.1021/acs.est.5b03541 PMID:26461459

Penkhrue, W., Jendrossek, D., Khanongnuch, C., Pathom-Aree, W., Aizawa, T., Behrens, R. L., & Lumyong, S. (2020). Response surface method for polyhydroxybutyrate (PHB) bioplastic accumulation in *Bacillus drentensis* BP17 using pineapple peel. *PLoS One*, *15*(3), 1–21. doi:10.1371/journal.pone.0230443 PMID:32191752

Penn, C. J., & Camberato, J. J. (2019). A Critical Review on Soil Chemical Processes that Control How Soil pH Affects Phosphorus Availability to Plants. *Agriculture*, *9*(6), 120–138. doi:10.3390/agriculture9060120

Peoples, M. B., Hauggaard-Nielsen, H., & Jensen, E. S. (2009). The Potential Environmental Benefits and Risks Derived from Legumes in Rotations. In D. W. Emerich & H. B. Krishnan (Eds.), *Nitrogen Fixation in Crop Production*. John Wiley & Sons. doi:10.2134/agronmonogr52.c13

Percival, D. C., Abbey, J., Lu, H., & Harris, L. (2016). Use of biofungicides to address conventional Botrytis blight control challenges in wild blueberry production. In *XI International Vaccinium Symposium 1180* (pp. 241-248). Academic Press.

Pereda Reyes, I., & Sárvári Horváth, I. (2015). *Anaerobic Biodegradation of Solid Substrates from Agroindustrial Activities—Slaughterhouse Wastes and Agrowastes*. Academic Press.

Pereg, L., & McMillan, M. (2015). Scoping the potential uses of beneficial microorganisms for increasing productivity in cotton cropping systems. *Soil Biology & Biochemistry*, *80*, 349–358. doi:10.1016/j.soilbio.2014.10.020

Pereira, A. R. B., & de Freitas, D. A. F. (2012). Uso de micro-organismos para a biorremediação de ambientes impactados. *Revista Eletrônica em Gestão. Educação e Tecnologia Ambiental*, *6*(6), 995–1006.

Pereira, L., Coelho, A. V., Viegas, C. A., Santos, M. M. C. D., Robalo, M. P., & Martins, L. O. (2009). Enzymatic biotransformation of the azo dye Sudan Orange G with bacterial CotA-laccase. *Journal of Biotechnology*, *139*(1), 68–77. doi:10.1016/j.jbiotec.2008.09.001 PMID:18938200

Pérez-Montaño, F., Alías-Villegas, C., Bellogín, R. A., Del Cerro, P., Espuny, M. R., Jiménez-Guerrero, I., & Cubo, T. (2014). Plant growth promotion in cereal and leguminous agricultural important plants: From microorganism capacities to crop production. *Microbiological Research*, *169*(5-6), 325–336. doi:10.1016/j.micres.2013.09.011 PMID:24144612

Pérez-Rodriguez, M. M., Piccoli, P., Anzuay, M. S., Baraldi, R., Neri, L., Taurian, T., Lobato Ureche, M. A., Segura, D. M., & Cohen, A. C. (2020). Native bacteria isolated from roots and rhizosphere of *Solanum lycopersicum* L. increase tomato seedling growth under a reduced fertilization regime. *Scientific Reports*, *10*(1), 15642. doi:10.103841598-020-72507-4 PMID:32973225

Perissini-Lopes, B., Egea, T. C., Monteiro, D. A., Vici, A. C., & Da Silva, D. G. H. (2016). Lisboa DCO, De Almeida EA, Parsons JR, Da Silva R, Gomes E:Evaluation of diuron tolerance and biotransformation byfungi from a sugar cane plantation sandy-loam soil. *Journal of Agricultural and Food Chemistry*, *64*, 268–9275. doi:10.1021/acs.jafc.6b03247

Perret, X., Staehelin, C., & Broughton, W. J. (2000). Molecular basis of symbiotic promiscuity. *Microbiology and Molecular Biology Reviews: MMBR*, *64*(1), 180–201. doi:10.1128/MMBR.64.1.180-201.2000 PMID:10704479

Perrot-Rechenmann, C. (2010). Cellular responses to auxin: Division versus expansion. *Cold Spring Harbor Perspectives in Biology*, *2*(5), a001446. doi:10.1101/cshperspect.a001446 PMID:20452959

Petatán-Sagahón, I., Anducho-Reyes, M. A., Silva-Rojas, H. V., Arana-Cuenca, A., Tellez-Jurado, A., Cárdenas-Álvarez, I. O., & Mercado-Flores, Y. (2011). Isolation of bacteria with antifungal activity against the phytopathogenic fungi Stenocarpella maydis and Stenocarpella macrospora. *International Journal of Molecular Sciences*, *12*(9), 5522–5537. doi:10.3390/ijms12095522 PMID:22016606

Peters, J. W., Boyd, E. S., Hamilton, T., & Rubio, L. M. (2011). Biochemistry of Mo-nitrogenase. In J. W. B. Moir (Ed.), *Nitrogen cycling in bacteria: Molecular Analysis*. Caister Academic Press.

Petrozza, A., Santaniello, A., Summerer, S., Di Tommaso, G., Di Tommaso, D., Paparelli, E., & Cellini, F. (2014). Physiological responses to Megafol® treatments in tomato plants under drought stress: A phenomic and molecular approach. *Scientia Horticulturae*, *174*, 185–192. doi:10.1016/j.scienta.2014.05.023

Philippar, K., Fuchs, I., Luthen, H., Hoth, S., Bauer, C. S., Haga, K., Thiel, G., Ljung, K., Sandberg, G., Bottger, M., Becker, D., & Hedrich, R. (1999). Auxin-induced K+ channel expression represents an essential step in coleoptile growth and gravitropism. *Proceedings of the National Academy of Sciences of the United States of America*, *96*(21), 12186–12191. doi:10.1073/pnas.96.21.12186 PMID:10518597

Philip, S., Keshavarz, T., & Roy, I. (2007). Polyhydroxyalkanoates: Biodegradable polymers with a range of applications. *Journal of Chemical Technology and Biotechnology (Oxford, Oxfordshire)*, *82*(3), 233–247. doi:10.1002/jctb.1667

Phillips, A. L. (2008). *The Relationship between plants and their root-associated microbial communities in hydrocarbon phytoremediation systems* (Ph.D. Thesis). College of Graduate Studies and Research University of Saskatchewan, Saskatoon, Canada.

Philp, J. C., & Atlas, R. M. (2005). Bioremediation of contaminated soils and aquifers. In *Bioremediation* (pp. 139–236). American Society of Microbiology.

Picarelli, M. A. S., Forgia, M., Rivas, E. B., Nerva, L., Chiapello, M., Turina, M., & Colariccio, A. (2019). Extreme diversity of mycoviruses present in isolates of *Rhizoctonia solani* AG2-2 LP from *Zoysia japonica* from Brazil. *Frontiers in Cellular and Infection Microbiology*, *9*, 244. doi:10.3389/fcimb.2019.00244 PMID:31355150

Pichtel, J. (2016). Oil and gas production wastewater: Soil contamination and pollution prevention. *Applied and Environmental Soil Science*, *2707989*, 1–24. Advance online publication. doi:10.1155/2016/2707989

Pierzynski, G. M., Sims, J. T., & Vance, G. F. (1994). *Soils and environmental quality*. Lewis Publishers.

Pieterse, C. M., Zamioudis, C., Does, D. V. der & Van Wees, S.C. (2014). Signalling Networks Involved in Induced Resistance. In Induced Resistance for Plant Defense. John Wiley & Sons. doi:10.1002/9781118371848.ch4

Pieterse, C. M., Zamioudis, C., Berendsen, R. L., Weller, D. M., Van Wees, S. C., & Bakker, P. A. (2014). Induced systemic resistance by beneficial microbes. *Annual Review of Phytopathology*, 52. PMID:24906124

Pilon-Smits, E. A. H. (2005). Phytoremediation. *Annual Review of Plant Biology*, *56*(1), 15–39. doi:10.1146/annurev.arplant.56.032604.144214 PMID:15862088

Pimda, W., & Bunnag, S. (2012). Biodegradation of used motor oil by single and mixed cultures of cyanobacteria. *African Journal of Biotechnology*, *11*(37), 9074–9078.

Pimmata, P., Reungsang, A., & Plangklang, P. (2013). Comparative bioremediation of carbofuran contaminated soil by natural attenuation, bioaugmentation and biostimulation. *International Biodeterioration & Biodegradation*, *85*, 196–204. doi:10.1016/j.ibiod.2013.07.009

Pingali, P. L. (2012). Green revolution: Impacts, limits, and the path ahead. *Proceedings of the National Academy of Sciences of the United States of America, 109*(31), 12302–12308. doi:10.1073/pnas.0912953109 PMID:22826253

Piotrowska-Długosz, A. (2017). The use of enzymes in bioremediation of soil xenobiotics. In M. Z. Hashmi, V. Kumar, & A. Varma (Eds.), *Xenobiotics in the Soil Environment* (Vol. 49, pp. 243–265). Springer International Publishing. doi:10.1007/978-3-319-47744-2_17

Piper, C. S. (1944). Soil and Plant Analysis. Interscience Publishers.

Pırlak, L., & Köse, M. (2009). Effects of plant growth promoting rhizobacteria on yield and some fruit properties of strawberry. *Journal of Plant Nutrition, 32*(7), 1173–1184. doi:10.1080/01904160902943197

Pirlak, L., Turan, M., Sahin, F., & Esitken, A. (2007). Floral and foliar application of plant growth promoting rhizobacteria (PGPR) to apples increases yield, growth, and nutrient element contents of leaves. *Journal of Sustainable Agriculture, 30*(4), 145–155. doi:10.1300/J064v30n04_11

Pivetz, B. E. (2001). *Ground water issue: phytoremediation of contaminated soil and ground water at hazardous waste sites*. National Risk Management Research Lab ADA OK.

Plaza, B. M., Gómez-Serrano, C., Acién-Fernández, F. G., & Jimenez-Becker, S. (2018). Effects of microalgae hydrolysate foliar application (*Arthrospira platensis* and *Scenedesmus* sp.) on Petunia x hybrida growth. *Journal of Applied Phycology, 30*(4), 2359–2365. doi:10.100710811-018-1427-0

Płociniczak, T., Fic, E., Pacwa-Płociniczak, M., Pawlik, M., & Piotrowska-Seget, Z. (2017). Improvement of phytoremediation of an aged petroleum hydrocarbon-contaminated soil by *Rhodococcus erythropolis* CD 106 strain. *International Journal of Phytoremediation, 19*(7), 614–620. doi:10.1080/15226514.2016.1278420 PMID:28103078

Podgornik, H., Grgić, I., & Perdih, A. (1999). Decolorization rate of dyes using lignin peroxidases of *Phanerochaete chrysosporium. Chemosphere, 38*(6), 1353–1359. doi:10.1016/S0045-6535(98)00537-2

Podile, A. R., & Kishore, G. K. (2007). Plant growth-promoting rhizobacteria. In *Plant-associated bacteria* (pp. 195–230). Springer.

Poeplau, C., & Don, A. (2015). Carbon sequestration in agricultural soils via cultivation of cover crops–A meta-analysis. *Agriculture, Ecosystems & Environment, 200*, 33–41. doi:10.1016/j.agee.2014.10.024

Polak-Juszczak, L. (2009). Temporal trends in the bioaccumulation of trace metals in herring, sprat, and cod from the southern Baltic Sea in the 1994-2003 period. *Chemosphere, 76*(10), 1334–1339. doi:10.1016/j.chemosphere.2009.06.030 PMID:19580989

Polyanskaya, L. M., Vedina, O. T., Lysak, L. V., & Zvyagintsev, D. G. (2002). The growth-promoting effects of Beijerinckiamobilis and Clostridium sp. cultures on some agricultural crops. *Microbiology, 71*(1), 109–115. doi:10.1023/A:1017914803544

Ponder, S. M., Darab, J. G., & Mallouk, T. E. (2000). Remediation of Cr(VI) and Pb(II) aqueous solutions using supported, nanoscale zero-valent iron. *Environmental Science & Technology, 34*(12), 2564–2569. doi:10.1021/es9911420

Ponraj, M., Jamunarani, P., & Zambare, V. (2011). Isolation and optimization of culture condition for decolourization of true blue using dye decolorizing fungi. *Asian Journal of Experimental Biological Sciences, 2*, 270–277.

Popova-Kroumova, P., Vasileva, E., & Beschkov, V. (2015). Modelling of 1, 2-dichloroethane Biodegradation, Stimulated by Constant Electric Field. *Biomath Communications, 2*(1).

Postma, J., Os, E. V., & Bonants, P. J. M. (2008). *Pathogen detection and management strategies in soilless plant growing systems*. Academic Press.

Poupin, M. J., Timmermann, T., Vega, A., Zuñiga, A., & González, B. (2013). Effects of the plant growth-promoting bacterium Burkholderia phytofirmans PsJN throughout the life cycle of *Arabidopsis thaliana. PLoS One*, *8*(7), e69435. doi:10.1371/journal.pone.0069435 PMID:23869243

Pourbabaee, A., Bahmani, E., Alikhani, H., & Emami, S. (2016). Promotion of wheat growth under salt stress by halo tolerant bacteria containing ACC deaminase. *Journal of Agricultural Science and Technology*, *18*, 855–864.

Power, I. M., Wilson, S. A., Thom, J. M., Dipple, G. M., & Southam, G. (2007). Biologically induced mineralization of dypingite by cyanobacteria from an alkaline wetland near Atlin, British Columbia, Canada. *Geochemical Transactions*, *8*(1), 13. doi:10.1186/1467-4866-8-13 PMID:18053262

Prabhu, M., Chemodanov, A., Gottlieb, R., Kazir, M., Nahor, O., Gozin, M., Israel, A., Livney, Y. D., & Golberg, A. (2019). Starch from the sea: The green macroalga *Ulva ohnoi* as a potential source for sustainable starch production in the marine biorefinery. *Algal Research*, *37*, 215–227. doi:10.1016/j.algal.2018.11.007

Pradhan, G. K., & Parida, K. M. (2010). Fabrication of iron-cerium mixed oxide: An efficient photo catalyst for dye degradation. *International Journal of Engineering Science and Technology*, *2*, 9.

Prajapati, J. B., & Nair, B. M. (2008). The history of fermented foods. Handbook of Fermented Functional Foods, 1-24.

Prajapati, S. K., & Meravi, N. (2014). Potentially toxic element speciation of soil and calotropis procera from thermal power plant area. *Proceedings of the International Academy of Ecology and Environmental Sciences*, *4*, 68–71.

Prakash, S., & Verma, J. P. (2016). Global perspective of potash for fertilizer production. In V. S. Meena, B. R. Maurya, J. P. Verma, & R. S. Meena (Eds.), *Potassium Solubilizing Microorganisms for Sustainable Agriculture* (pp. 327–331). Springer. doi:10.1007/978-81-322-2776-2_23

Pramanik, A., Zhang, F., Schwarz, H., Schreiber, F., & Braun, V. (2010). ExbB protein in the cytoplasmic membrane of Escherichia coli forms a stable oligomer. *Biochemistry*, *49*(40), 8721–8728. doi:10.1021/bi101143y PMID:20799747

Prasad, M. N. V., & Tewari, J. C. (2016). Prosopis juliflora (Sw) DC: Potential for Bioremediation and Bioeconomy. In Bioremediation and Bioeconomy. Elsevier. doi:10.1016/B978-0-12-802830-8.00003-4

Prasad, M., Srinivasan, R., Chaudhary, M., Choudhary, M., & Jat, L. K. (2019). Plant growth promoting rhizobacteria (PGPR) for sustainable agriculture: perspectives and challenges. In PGPR Amelioration in Sustainable Agriculture (pp. 129-157). Woodhead Publishing.

Prasad, M. N. V., & Freitas, H. M. O. (2003). Metal hyperaccumulation in plants biodiversity prospecting for phytoremediation technology. *Electronic Journal of Biotechnology*, *6*(3), 285–321. doi:10.2225/vol6-issue3-fulltext-6

Prasad, M. P., Bhakat, P., & Chatterjee, S. (2013). Optimization of textile dye degradation by bacterial species isolated from natural sources. *Journal of Ecology and Environmental Sciences*, *4*(1), 97–99.

Prasad, R. (2009). Efficient fertilizer use: The key to food security and better environment. *Journal of Tropical Agriculture*, *47*(1-2), 1–17.

Prasanna, R., Bidyarani, N., Babu, S., Hossain, F., Shivay, Y. S., Nain, L., & Moral, M. T. (2015). Cyanobacterial inoculation elicits plant defense response and enhanced Zn mobilization in maize hybrids. *Cogent Food & Agriculture*, *1*(1), 998507. doi:10.1080/23311932.2014.998507

Prasanna, R., Joshi, M., Rana, A., Shivay, Y. S., & Nain, L. (2012). Influence of co-inoculation of bacteria-cyanobacteria on crop yield and C–N sequestration in soil under rice crop. *World Journal of Microbiology & Biotechnology*, *28*(3), 1223–1235. doi:10.100711274-011-0926-9 PMID:22805842

Prasanna, R., Kanchan, A., Kaur, S., Ramakrishnan, B., Ranjan, K., Singh, M. C., Hasan, M., Saxena, A. K., & Shivay, Y. S. (2016). *Chrysanthemum* Growth Gains from Beneficial Microbial Interactions and Fertility Improvements in Soil Under Protected Cultivation. *Horticultural Plant Journal*, *2*(4), 229–239. doi:10.1016/j.hpj.2016.08.008

Prasanna, R., Nain, L., Tripathi, R., Gupta, V., Chaudhary, V., Middha, S., Joshi, M., Ancha, R., & Kaushik, B. D. (2008). Evaluation of fungicidal activity of extracellular filtrates of cyanobacteria – possible role of hydrolytic enzymes. *Journal of Basic Microbiology*, *48*(3), 186–194. doi:10.1002/jobm.200700199 PMID:18506903

Prasanna, R., Ramakrishnan, B., Simranjit, K., Ranjan, K., Kanchan, A., Hossain, F., & Nain, L. (2017). Cyanobacterial and rhizobial inoculation modulates the plant physiological attributes and nodule microbial communities of chickpea. *Archives of Microbiology*, *199*(9), 1311–1323. doi:10.100700203-017-1405-y PMID:28669069

Preethi, K., Anand, M., & Thazeem, B. (2015). Isolation and identification of keratinolytic bacteria from tannery effluent: A study on their biodegradative and dehairing activity. *International Journal of Multidisciplinary Research and Development*, *2*(10), 227–234.

Preethi, K., & Vineetha, U. M. (2015). Water hyacinth: A potential substrate for bioplastic (PHA) production using *Pseudomonas aeruginosa. International Journal of Applied Research in Veterinary Medicine*, *1*(11), 349–354.

Preisig, O., Wingfield, B. D., & Wingfield, M. J. (1998). Coinfection of a fungal pathogen by two distinct double-stranded RNA viruses. *Virology*, *252*(2), 399–406. doi:10.1006/viro.1998.9480 PMID:9878619

Prescott, M. I., Harle, J. D., & Klein, D. A. (2002). *Microbiology of Food* (5th ed.). McGraw-Hill Ltd.

Pricelius, S., Held, C., Murkovic, M., Bozic, M., Kokol, V., Cavaco-Paulo, A., & Guebitz, G. M. (2007). Enzymatic reduction of azo and indigoid compounds. *Applied Microbiology and Biotechnology*, *77*(2), 321–327. doi:10.100700253-007-1165-8 PMID:17891390

Priester, J. H., Olson, S. G., Webb, S. M., Neu, M. P., Hersman, L. E., & Holden, P. A. (2006). Enhanced exopolymer production and chromium stabilization in Pseudomonas putida unsaturated biofilms. *Applied and Environmental Microbiology*, *72*(3), 1988–1996. doi:10.1128/AEM.72.3.1988-1996.2006 PMID:16517647

Prieto, P., Schilirò, E., Maldonado-González, M. M., Valderrama, R., Barroso-Albarracín, J. B., & Mercado-Blanco, J. (2011). Root hairs play a key role in the endophytic colonization of olive roots by Pseudomonas spp. with biocontrol activity. *Microbial Ecology*, *62*(2), 435–445. doi:10.100700248-011-9827-6 PMID:21347721

Priyadarshani, I., & Rath, B. (2012). Commercial and industrial applications of micro algae –. *RE:view*, *3*(4), 89–100.

Priya, H., Prasanna, R., Ramakrishnan, B., Bidyarani, N., Babu, S., Thapa, S., & Renuka, N. (2015). Influence of cyanobacterial inoculation on the culturable microbiome and growth of rice. *Microbiological Research*, *171*, 78–89. doi:10.1016/j.micres.2014.12.011 PMID:25644956

Priyanka, T. A., Kotasthane, A. S., Kosharia, A., Kushwah, R., Zaidi, N. W., & Singh, U. S. (2017). Crop specific plant growth promoting effects of ACCd enzyme and siderophore producing and cynogenic fluorescent *Pseudomonas. Biotech, 7*(1).

Priyanka, K., Umesh, M., Thazeem, B., & Preethi, K. (2020). Polyhydroxyalkanoate biosynthesis and characterization from optimized medium utilizing distillery effluent using *Bacillus endophyticus* MTCC 9021: A statistical approach. *Biocatalysis and Biotransformation*, 1–13. doi:10.1080/10242422.2020.1789112

Prosekov, A. Y., & Ivanova, S. A. (2018). Food security: The challenge of the present. *Geoforum*, *91*, 73–77. doi:10.1016/j.geoforum.2018.02.030

Prüss-Üstün, A., Giroult, E., Rushbrook, P., & World Health Organization. (1999). *Safe management of wastes from health-care activities*. World Health Organization.

Prusty, J. S., Rath, B. P., & Thatoi, H. (2019). Production optimization and application of extracellular chromate reductase from *Bacillus sp.* for bioremediation of hexavalent Chromium. In R. Kundu, R. Narula, R. Paul, & S. Mukherjee (Eds.), *Environmental Biotechnology For Soil and Wastewater Implications on Ecosystems* (Vol. 36, pp. 103–108). Springer Singapore. doi:10.1007/978-981-13-6846-2_13

Przydatek, G., & Kanownik, W. (2019). Impact of small municipal solid waste landfill on groundwater quality. *Environmental Monitoring and Assessment*, *191*(3), 169–183. doi:10.100710661-019-7279-5 PMID:30778777

Puente, M. E., Li, C. Y., & Bashan, Y. (2009). Endophytic bacteria in cacti seeds can improve the development of cactus seedlings. *Environmental and Experimental Botany*, *66*(3), 402–408. doi:10.1016/j.envexpbot.2009.04.007

Pugazhendhi, A., Ranganathan, K., & Kaliannan, T. (2018). biosorptive removal of copper(ii) by *Bacillus cereus* isolated from contaminated soil of electroplating industry in India. *Water, Air, and Soil Pollution*, *229*(3), 1780. doi:10.100711270-018-3734-0

Pulz, O. (2001). Photobioreactors: Production systems for phototrophic microorganisms. *Applied Microbiology and Biotechnology*, *57*(3), 287–293. doi:10.1007002530100702 PMID:11759675

Pulz, O., & Gross, W. (2004). Valuable products from biotechnology of microalgae. *Applied Microbiology and Biotechnology*, *65*(6), 635–648. doi:10.100700253-004-1647-x PMID:15300417

Purdom, C. E., Hardiman, P. A., Bye, V. V. J., Eno, N. C., Tyler, C. R., & Sumpter, J. P. (1994). Estrogenic effects of effluents from sewage treatment works. *Chemistry and Ecology*, *8*(4), 275–285. doi:10.1080/02757549408038554

Puvaneswari, N., Muthukrishnan, J., & Gunasekaran, P. (2006). *Toxicity assessment and microbial degradation of azo dyes*. Academic Press.

Puzon, G. J., Roberts, A. G., Kramer, D. M., & Xun, L. (2005). Formation of soluble organo-chromium(III) complexes after chromate reduction in the presence of cellular organics. *Environmental Science & Technology*, *39*(8), 2811–2817. doi:10.1021/es048967g PMID:15884380

Pyae, H. A., Win, W. A., Yossapol, C., & Dararatana, S. (2019). Micro-particle ZVI inhibition threshold in cassava pulp bio-methanation. *Environment Asia*, *12*(Special Issue), 64–73.

Qaswar, M., Ahmed, W., Jing, H., Hongzhu, F., Xiaojun, S., Xianjun, J., & Zhang, H. (2019). Soil carbon (C), nitrogen (N) and phosphorus (P) stoichiometry drives phosphorus lability in paddy soil under long-term fertilization: A fractionation and path analysis study. *PLoS One*, *14*(6), 1–20. doi:10.1371/journal.pone.0218195 PMID:31233510

Qessaoui, R., Bouharroud, R., Furze, J. N., El Aalaoui, M., Akroud, H., Amarraque, A., Vaerenbergh, J. V., Tahzima, R., Mayad, E. H., & Chebli, B. (2019). Applications of New Rhizobacteria Pseudomonas Isolates in Agroecology via Fundamental Processes Complementing Plant Growth. *Scientific Reports*, *9*(1), 12832. doi:10.103841598-019-49216-8 PMID:31492898

Qiao, K., Takano, T., & Liu, S. (2015). Discovery of two novel highly tolerant $NaHCO_3$ Trebouxiophytes: Identification and characterization of microalgae from extreme saline–alkali soil. *Algal Research*, *9*, 245–253. doi:10.1016/j.algal.2015.03.023

Qi, F., & Zhang, F. (2020). Cell Cycle Regulation in the Plant Response to Stress. *Frontiers in Plant Science*, *10*, 1765. doi:10.3389/fpls.2019.01765 PMID:32082337

Qin, Y., Su, S., & Farrand, S. K. (2007). Molecular basis of transcriptional antiactivation. TraM disrupts the TraR-DNA complex through stepwise interactions. *The Journal of Biological Chemistry*, *282*(27), 19979–19991. doi:10.1074/jbc.M703332200 PMID:17475619

Qiu, X., Leland, T. W., Shah, S. I., Sorensen, D. L., & Kendall, E. W. (1997). Field study: grass remediation for clay soil contaminated with polycyclic aromatic hydrocarbons. In *Phytoremediation of soil and water contaminants*. American Chemcial Society. 10.1021/bk-1997-0664.ch014

Quero, G. M., Cassin, D., Botter, M., Perini, L., & Luna, G. M. (2015). Patterns of benthic bacterialdiversity in coastal areas contaminated by heavy metals,polycyclic aromatic hydrocarbons (PAHs) and polychlorinatedbiphenyls (PCBs). *Frontiers in Microbiology*, *6*, 1–15. doi:10.3389/fmicb.2015.01053 PMID:26528247

Quesada, C. A., Paz, C., Oblitas Mendoza, E., Phillips, O. L., Saiz, G., & Lloyd, J. (2020). Variations in soil chemical and physical properties explain basin-wide Amazon forest soil carbon concentrations. *Soil (Göttingen)*, *6*(1), 53–88. doi:10.5194oil-6-53-2020

Quinn, L. D., Straker, K. C., Guo, J., Kim, S., Thapa, S., Kling, G., & Voigt, T. B. (2015). Stress-tolerant feedstocks for sustainable bioenergy production on marginal land. *BioEnergy Research*, *8*(3), 1081–1100. doi:10.100712155-014-9557-y

Quinones, B., Pujol, C. J., & Lindow, S. E. (2004). Regulation of AHL production and its contribution to epiphytic fitness in Pseudomonassyringae. *Molecular Plant-Microbe Interactions*, *17*(5), 521–531. doi:10.1094/MPMI.2004.17.5.521 PMID:15141956

Qu, J., Xu, Y., Ai, G. M., Liu, Y., & Liu, Z. P. (2015). Novel Chryseobacterium sp. PYR2 degrades various organochlorine pesticides (OCPs) and achieves enhancing removal and complete degradation of DDT in highly contaminated soil. *Journal of Environmental Management*, *161*, 350–357. doi:10.1016/j.jenvman.2015.07.025 PMID:26203874

Raddadi, N., Crotti, E., Rolli, E., Marasco, R., Fava, F., & Daffonchio, D. (2012). The most important Bacillus species in biotechnology. In Bacillus thuringiensis Biotechnology (pp. 329–345). doi:10.1007/978-94-007-3021-2_17

Raddadi, N., Cherif, A., Boudabous, A., & Daffonchio, D. (2008). Screening of plant growth promoting traits of Bacillus thuringiensis. *Annals of Microbiology*, *58*(1), 47–52. doi:10.1007/BF03179444

Raddadi, N., Cherif, A., Ouzari, H., Marzorati, M., Brusetti, L., Boudabous, A., & Daffonchio, D. (2007). Bacillus thuringiensis beyond insect biocontrol: Plant growth promotion and biosafety of polyvalent strains. *Annals of Microbiology*, *57*(4), 481–494. doi:10.1007/BF03175344

Radhakrishnan, R., Hashem, A., & Abd Allah, E. F. (2017). Bacillus: A biological tool for crop improvement through bio-molecular changes in adverse environments. *Frontiers in Physiology*, *8*, 1–14. doi:10.3389/fphys.2017.00667 PMID:28932199

Radjacommare, R., Kandan, A., Nandakumar, R., & Samiyappan, R. (2004). Association of the Hydrolytic Enzyme Chitinase against *Rhizoctonia solani* in Rhizobacteria-treated Rice Plants. *Journal of Phytopathology*, *152*(6), 365–370. doi:10.1111/j.1439-0434.2004.00857.x

Radwan, S. S., Al-Awadhi, H., & El-Nemr, I. M. (2000). Cropping as a phytoremediation practice for oily desert soil with reference to crop safety as food. *International Journal of Phytoremediation*, *2*(4), 383–396. doi:10.1080/15226510008500046

Radwan, S., Sorkhoh, N. A., & El-Nemr, I. (1995). Oil biodegradation around roots. *Nature*, *376*(6538), 302. doi:10.1038/376302a0 PMID:7630395

Radzki, W., Gutierrez Mañero, F. J., Algar, E., Lucas García, J. A., García-Villaraco, A., & Ramos Solano, B. (2013). Bacterial siderophores efficiently provide iron to iron-starved tomato plants in hydroponics culture. *Antonie van Leeuwenhoek*, *104*(3), 321–330. doi:10.100710482-013-9954-9 PMID:23812968

Rafique, N., & Tariq, S. R. (2016). Distribution and source apportionment studies of heavy metals in soil of cotton/wheat fields. *Environmental Monitoring and Assessment*, *188*(5), 309. doi:10.100710661-016-5309-0 PMID:27115422

Rahbar, M. H., Samms-Vaughan, M., Hessabi, M., Dickerson, A. S., Lee, M., Bressler, J., & Shakespeare-Pellington, S. (2016). Concentrations of polychlorinated biphenyls and organochlorine pesticides in umbilical cord blood serum of newborns in Kingston, Jamaica. *International Journal of Environmental Research and Public Health*, *13*(10), 1032. doi:10.3390/ijerph13101032 PMID:27775677

Rahman, R. O. (2011). Abdel et al., "Liquid Radioactive Wastes Treatment: A Review". *Water*, 551-565.

Rai, A. N., Söderbäck, E., & Bergman, B. (2000). Cyanobacterium-plant symbioses. *The New Phytologist*, *147*(3), 449–481. doi:10.1046/j.1469-8137.2000.00720.x

Rai, H., Bhattacharya, M., Singh, J., Bansal, T. K., Vats, P., & Banerjee, U. C. (2005). Removal of Dyes from the Effluent of Textile and Dyestuff Manufacturing Industry: A Review of Emerging Techniques with Reference to Biological Treatment. *Critical Reviews in Environmental Science and Technology*, *35*(3), 219–238. doi:10.1080/10643380590917932

Rai, M. (Ed.). (2006). *Handbook of microbial biofertilizers*. CRC Press. doi:10.1201/9781482277760

Rai, N., Rai, R., & Venkatesh, K. V. (2015). Quorum sensing biosensors. In V. C. Kalia (Ed.), *Quorum Sensing vs Quorum quenching: A Battle with No End in Sight* (pp. 173–183). Springer India.

Rai, U., Tripathi, R., Singh, N., Kumar, A., Ali, M. B., Pal, A., & Singh, S. N. (2000). Amelioration of Fly-Ash by Selected Nitrogen Fixing Blue Green Algae. *Bulletin of Environmental Contamination and Toxicology*, *64*(2), 294–301. doi:10.1007001289910043 PMID:10656898

Rajaei, G., Mansouri, B., Jahantigh, H., & Hamidian, A. H. (2012). Metal concentrations in the water of Chah nimeh reservoirs in Zabol, Iran. *Bulletin of Environmental Contamination and Toxicology*, *89*(3), 495–500.

Rajaguru, P., Vidya, L., Baskarasethupathi, B., Kumar, P. A., Palanivel, M., & Kalaiselvi, K. (2002). Genotoxicity evaluation of polluted ground water in human peripheral blood lymphocytes using the comet assay. *Mutation Research/Genetic Toxicology and Environmental Mutagenesis*, *517*(1-2), 29–37. doi:10.1016/S1383-5718(02)00025-6 PMID:12034306

Rajan, M. R. D. N. S. (2014). Impact of Dyeing Industry Effluent on Groundwater Quality by Water Quality Index and Correlation Analysis. *Journal of Pollution Effects and Control*, *2*(2), 1–4. doi:10.4172/2375-4397.1000126

Rajawat, M. V. S., Singh, S., Singh, G., & Saxena, A. K. (2012). Isolation and characterization of K-solubilizing bacteria isolated from different rhizospheric soil. In *Proceeding of 53rd annual conference of association of microbiologists of India* (p. 124). Academic Press.

Rajawat, M. V. S., Singh, S., Tyagi, S. P., & Saxena, A. K. (2016). A modified plate assay for rapid screening of potassium-solubilizing bacteria. *Pedosphere*, *26*(5), 768–773. doi:10.1016/S1002-0160(15)60080-7

Rajendran, K., & Sen, S. (2018). Adsorptive removal of carbamazepine using biosynthesized hematite nanoparticles. *Environmental Nanotechnology, Monitoring & Management*, *9*, 122–127. doi:10.1016/j.enmm.2018.01.001

Rajesha, J. B., Ramasami, A. K., Nagaraju, G., & Balakrishna, G. R. (2017). Photochemical elimination of Endocrine Disrupting Chemical (EDC) by ZnO nanoparticles, synthesized by gel combustion. *Water Environment Research*, *89*(5), 396–405. doi:10.2175/106143016X14733681696086 PMID:27779923

Raju, M. N., Leo, R., Herminia, S. S., Morán, R. E. B., Venkateswarlu, K., & Laura, S. (2017). Biodegradation of Diesel, Crude Oil and Spent Lubricating Oil by Soil Isolates of *Bacillus spp. Bulletin of Environmental Contamination and Toxicology*, *98*(5), 698–705. doi:10.100700128-017-2039-0 PMID:28210752

Ramadas, N. V., Soccol, C. R., & Pandey, A. (2010). A statistical approach for optimization of polyhydroxybutyrate production by *Bacillus sphaericus* NCIM 5149 under submerged fermentation using central composite design. *Applied Biochemistry and Biotechnology*, *162*(4), 996–1007. doi:10.100712010-009-8807-5 PMID:19812909

Ramakrishna, K. R., & Viraraghavan, T. (1997). Dye Removal Using Low Cost Adsorbents. *Water Science and Technology*, *36*(2-3), 189–196. doi:10.2166/wst.1997.0516

Ramakrishnan, B., Kaur, S., Prasanna, R., Ranjan, K., Kanchan, A., Hossain, F., Shivay, Y. S., & Nain, L. (2017). Microbial inoculation of seeds characteristically shapes the rhizosphere microbiome in desi and kabuli chickpea types. *Journal of Soils and Sediments*, *17*(8), 2040–2053. doi:10.100711368-017-1685-5

Ramalho, P. A., Cardoso, M. H., Cavaco-Paulo, A., & Ramalho, M. T. (2004). Characterization of Azo Reduction Activity in a Novel Ascomycete Yeast Strain. *Applied and Environmental Microbiology*, *70*(4), 2279–2288. doi:10.1128/AEM.70.4.2279-2288.2004 PMID:15066823

Ramalho, P. A., Paiva, S., Cavaco-Paulo, A., Casal, M., Cardoso, M. H., & Ramalho, M. T. (2005). Azo Reductase Activity of Intact Saccharomyces cerevisiae Cells Is Dependent on the Fre1p Component of Plasma Membrane Ferric Reductase. *Applied and Environmental Microbiology*, *71*(7), 3882–3888. doi:10.1128/AEM.71.7.3882-3888.2005 PMID:16000801

Ramalho, P. A., Scholze, H., Cardoso, M. H., Ramalho, M. T., & Oliveira-Campos, A. (2002). Improved conditions for the aerobic reductive decolourisation of azo dyes by *Candida zeylanoides. Enzyme and Microbial Technology*, *31*(6), 848–854. doi:10.1016/S0141-0229(02)00189-8

Ramamoorthy, V., Viswanathan, R., Raguchander, T., Prakasam, V., & Samayapan, R. (2001). Induction of systemic resistance by plant growth promoting rhizobacteria in crop plants against pests and diseases. *Crop Protection (Guildford, Surrey)*, *20*(1), 1–11. doi:10.1016/S0261-2194(00)00056-9

Ramankutty, N., Evan, A. T., Monfreda, C., & Foley, J. A. (2008). Farming the planet: 1. Geographic distribution of global agricultural lands in the year 2000. *Global Biogeochemical Cycles*, *22*(1), n/a. doi:10.1029/2007GB002952

Ramasamy, K., Kamaludeen, S., & Parwin, B. (2006). Bioremediation of metals microbial processes and techniques. In Environmental Bioremediation Technologies (pp. 173–187). Springer Publication.

Ramasamy, K., Kamaludeen, S., & Parwin, B. (2006). Bioremediation of metals microbial processes and techniques. In S. N. Singh & R. D. Tripathi (Eds.), *Environmental Bioremediation Technologies* (pp. 173–187). Springer Publication.

Ramírez, V., Baez, A., López, P., Bustillos, R., Villalobos, M. Á., Carreño, R., Contreras, J. L., Muñoz-Rojas, J., Fuentes, L. E., Martínez, J., & Munive, J. A. (2019). Chromium hyper-tolerant Bacillus sp. MH778713 assists phytoremediation of heavy metals by mesquite trees (Prosopis laevigata). *Frontiers in Microbiology*, *10*, 1833. doi:10.3389/fmicb.2019.01833 PMID:31456770

Ram, N. M., David, H. B., Robert, F., & Maureen, L. (1993). A decision framework for selecting remediation technologies at hydrocarbon- contaminated sites. *Journal of Soil Contamination*, *2*(2), 1–24. doi:10.1080/15320389309383436

Ramos, L., Hernandez, L. M., & Gonzalez, M. J. (1994). Sequential fractionation of copper, lead, cadmium and zinc in soils from or near Donana National Park. *Journal of Environmental Quality*, *23*(1), 50–57. doi:10.2134/jeq1994.00472425002300010009x

Rana, S., Jindal, V., Mandal, K., Kaur, G., & Gupta, V. K. (2015). Thiamethoxam degradation by Pseudomonas and Bacillus strains isolated from agricultural soils. *Environmental Monitoring and Assessment, 187*(5), 4532. doi:10.100710661-015-4532-4 PMID:25917187

Ranaweera, I., Shrestha, U., Ranjana, K. C., Kakarla, P., Willmon, T. M., Hernandez, A. J., Mukherjee, M. M., Barr, S. R., & Varela, M. F. (2015). Structural comparison of bacterial multidrug efflux pumps of the major facilitator superfamily. *Trends in Cell & Molecular Biology, 10*, 131–140.

Rane, N. R., Patil, S. M., Chandanshive, V. V., Kadam, S. K., Khandare, R. V., Jadhav, J. P., & Govindwar, S. P. (2016). Ipomoea hederifolia rooted soil bed and Ipomoea aquatica rhizofiltration coupled phytoreactors for efficient treatment of textile wastewater. *Water Research, 96*, 1–11. doi:10.1016/j.watres.2016.03.029 PMID:27016633

Ranghar, S., Agrawal, S., & Agrawal, P. K. (2019). Microbial products: protein, enzyme, secondary metabolites and chemicals. In *Microbial Interventions in Agriculture and Environment* (pp. 347–384). Springer. doi:10.1007/978-981-32-9084-6_17

Rani, B., Kumar, V., Singh, J., Bisht, S., Teotia, P., Sharma, S., & Kela, R. (2014). Bioremediation of dyes by fungi isolated from contaminated dye effluent sites for bio-usability. *Brazilian Journal of Microbiology, 45*(3), 1055–1063. doi:10.1590/S1517-83822014000300039 PMID:25477943

Rani, R., & Juwarkar, A. (2012). Biodegradation of phorate in soil and rhizosphere of *Brassica juncea* L. (Indian Mustard) by a microbial consortium. *International Biodeterioration & Biodegradation, 71*, 36–42. doi:10.1016/j.ibiod.2012.04.004

Rani, R., & Kumar, V. (2017). Endosulfan degradation by selected strains of plant growth promoting rhizobacteria. *Bulletin of Environmental Contamination and Toxicology, 99*(1), 138–145. doi:10.100700128-017-2102-x PMID:28484804

Ranjan, K., Priya, H., Ramakrishnan, B., Prasanna, R., Venkatachalam, S., Thapa, S., Tiwari, R., Nain, L., Singh, R., & Shivay, Y. S. (2016). Cyanobacterial inoculation modifies the rhizosphere microbiome of rice planted to a tropical alluvial soil. *Applied Soil Ecology, 108*, 195–203. doi:10.1016/j.apsoil.2016.08.010

Ranjbar, R., Behzadi, P., Najafi, A., & Roudi, R. (2017). DNA microarray for rapid detection and identification of food and water borne bacteria: From dry to wet lab. *The Open Microbiology Journal, 11*(1), 330–338.

Rao, H. C. Y., Mohana, N. C., & Satish, S. (2020). *Biocommercial aspects of microbial endophytes for sustainable agriculture*. Microbial Endophytes, Functional Biology and Applications. doi:10.1016/B978-0-12-819654-0.00013-2

Rao, P. M., Anitha, Y., & Satyaprasad, K. (2016). Combined effect of Bacillus cereus CPOU13 and *B. subtilis* SPC14 on polycyclic aromatic hydrocarbons degradation in vitro. *International Journal of Bioassays, 5*(4), 505–4511.

Rashidipour, M., Maleki, A., Kordi, S., Birjandi, M., Pajouhi, N., Mohammadi, E., Heydari, R., Rezaee, R., Rasoulian, B., & Davari, B. (2019). Pectin/chitosan/tripolyphosphate nanoparticles: Efficient carriers for reducing soil sorption, cytotoxicity, and mutagenicity of paraquat and enhancing its herbicide activity. *Journal of Agricultural and Food Chemistry, 67*, 5736–5745. doi:10.1021/acs.jafc.9b01106 PMID:31042035

Rasmussen, L. D., Sørensen, S. J., Turner, R. R., & Barkay, T. (2000). Application of a mer-lux biosensor for estimating bioavailable mercury in soil. *Soil Biology & Biochemistry, 32*, 639–646.

Rasolomanana, J. L., & Balandreau, J. (1987). Role de la rhizosphere dans la biodegradation decomposes recalcitrants: Cas d'une riziere polluee par des residus petroliers. *Revue D'Ecologie et de Biologie du Sol., 24*(3), 443–457.

Rastogi, G., Sbodio, A., Tech, J. J., Suslow, T. V., Coaker, G. L., & Leveau, J. H. (2012). Leaf microbiota in an agroecosystem: Spatiotemporal variation in bacterial community composition on field-grown lettuce. *The ISME Journal, 6*(10), 1812–1822. doi:10.1038/ismej.2012.32 PMID:22534606

Rastogi, M., Nandal, M., & Khosla, B. (2020). Microbes as vital additives for solid waste composting. *Heliyon*, *6*(2), 1–11. doi:10.1016/j.heliyon.2020.e03343 PMID:32095647

Raupach, G. S., & Kloepper, J. W. (1998). Mixtures of plant growth-promoting rhizobacteria enhance biological control of multiple cucumber pathogens. *Phytopathology*, *88*(11), 1158–1164. doi:10.1094/PHYTO.1998.88.11.1158 PMID:18944848

Ravanbakhsh, M., Sasidharan, R., Voesenek, L. A. C. J., Kowalchuk, G. A., & Jousset, A. (2017). ACC deaminase-producing rhizosphere bacteria modulate plant responses to flooding. *Journal of Ecology*, *105*(4), 979–986. doi:10.1111/1365-2745.12721

Ravikumar, K. V. G., Kumar, D., Kumar, G., Mrudula, P., Natarajan, C., & Mukherjee, A. (2016). Enhanced Cr(VI) removal by nano zerovalent iron-immobilized alginate beads in the presence of a biofilm ina continuous-flow reactor. *Industrial & Engineering Chemistry Research*, *55*(20), 5973–5982. doi:10.1021/acs.iecr.6b01006

Ravindran, B., Kumar, A. G., Bhavani, P. S. A., & Sekaran, G. (2011). Solid-state fermentation for the production of alkaline protease by *Bacillus cereus* 1173900 using proteinaceous tannery solid waste. *Current Science*, *100*(5), 726–730.

Ray, D. K., Mueller, N. D., West, P. C., & Foley, J. A. (2013). Yield Trends Are Insufficient to Double Global Crop Production by 2050. *PLoS One*, *8*(6), e66428. doi:10.1371/journal.pone.0066428 PMID:23840465

Rayu, S., Karpouzas, D. G., & Singh, B. K. (2012). Emerging technologies in bioremediation: Constraints and opportunities. *Biodegradation*, *23*(6), 917–926. doi:10.100710532-012-9576-3 PMID:22836784

Raza, A., Razzaq, A., Mehmood, S. S., Zou, X., Zhang, X., Lv, Y., & Xu, J. (2019). Impact of Climate Change on Crops Adaptation and Strategies to Tackle Its Outcome: A Review. *Plants (Basel, Switzerland)*, *8*(2), 34. doi:10.3390/plants8020034 PMID:30704089

Raza, W., & Shen, Q. (2010). Growth, Fe^{3+} reductase activity, and siderophore production by Paenibacillus polymyxa SQR-21 under differential iron conditions. *Current Microbiology*, *61*(5), 390–395. doi:10.100700284-010-9624-3 PMID:20358373

Recio, E., Colinas, Á., Rumbero, Á., Aparicio, J. F., & Martín, J. F. (2004). PI factor, a novel type quorum-sensing inducer elicits pimaricin production in *Streptomyces natalensis*. *The Journal of Biological Chemistry*, *279*(40), 41586–41593. doi:10.1074/jbc.M402340200 PMID:15231842

Reddy, G. C., Goyal, R. K., Puranik, S., Waghmar, V., Vikram, K. V., & Sruthy, K. S. (2020). *Biofertilizers Toward Sustainable Agricultural Development*. Plant Microbe Symbiosis. doi:10.1007/978-3-030-36248-5_7

Reddy, K. J., McDonald, K. J., & King, H. (2013). A novel arsenic removal process for water using cupric oxide nanoparticles. *Journal of Colloid and Interface Science*, *397*, 96–102. doi:10.1016/j.jcis.2013.01.041 PMID:23452518

Reddy, K. R., Darnault, C. J., & Darko-Kagya, K. (2014). Transport of lactate-modified nanoscale iron particles in porous media. *Journal of Geotechnical and Geoenvironmental Engineering*, *140*(2), 04013013. doi:10.1061/(ASCE)GT.1943-5606.0001015

Reddy, K. R., Khodadoust, A. P., & Darko-Kagya, K. (2014). Transport and reactivity of lactate-modified nanoscale iron particles for remediation of DNT in subsurface soils. *Journal of Environmental Engineering*, *140*(12), 04014042. doi:10.1061/(ASCE)EE.1943-7870.0000870

Reddy, M. S., Kour, M., Aggarwal, S., Ahuja, S., Marmeisse, R., & Fraissinet-Tachet, L. (2016). Metal induction of a P isolithus albus metallothionein and its potential involvement in heavy metal tolerance during mycorrhizal symbiosis. *Environmental Microbiology*, *18*(8), 2446–2454. doi:10.1111/1462-2920.13149 PMID:26626627

Reddy, P. P. (2012). Plant growth-promoting rhizobacteria (PGPR). In *Recent advances in crop protection* (pp. 131–158). Springer. doi:10.1007/978-81-322-0723-8_10

Reddy, S. M., Basha, S., Adimurthy, S., & Ramachandraiah, G. (2006). Description of small plastics fragments in marine sediments along the Alang–Sosiyaship-breaking yard, India. *Estuarine, Coastal and Shelf Science*, *68*(3-4), 656–660. doi:10.1016/j.ecss.2006.03.018

Reguera, M., Peleg, Z., & Blumwald, E. (2012). Targeting metabolic pathways for genetic engineering abiotic stress-tolerance in crops. *Biochimica et Biophysica Acta (BBA)-. Gene Regulatory Mechanisms*, *1819*(2), 186–194. PMID:21867784

Rehman, Z. U., Ali, H. H., & Akbar, A. (2020). Production of polyhydroxybutyrate (PHB) from soil bacterium (*Bacillus megaterium* TISTR 1814) with Cantaloupe waste extract as potential carbon source. *Pak-Euro Journal of*. https://readersinsight.net/PJMLS/article/view/1293

Rehm, B. H. A. (2003). Polyester synthases: Natural catalysts for plastics. *The Biochemical Journal*, *376*(1), 15–33. doi:10.1042/bj20031254 PMID:12954080

Rehm, G., & Schmitt, M. (2002). *Potassium for Crop Production*. University of Minnesota.

Reimmann, C., Ginet, N., Michel, L., Keel, C., Michaux, P., Krishnapillai, V., Zala, M., Heurlier, K., Triandafillu, K., Harms, H., Défago, G., & Haas, D. (2002). Genetically programmed autoinducer destruction reduces virulence gene expression and swarming motility in *Pseudomonas aeruginosa* PAO1. *Microbiology (Reading, England)*, *148*(Pt 4), 923–932. doi:10.1099/00221287-148-4-923 PMID:11932439

Reineke, W., & Knackmuss, H. J. (1988). Microbial degradation of haloaromatics. *Annual Review of Microbiology*, *42*(1), 263–287. doi:10.1146/annurev.mi.42.100188.001403 PMID:3059995

Reinsch, S., Koller, E., Sowerby, A., De Dato, G., Estiarte, M., Guidolotti, G., ... Liberati, D. (2017). Shrubland primary production and soil respiration diverge along European climate gradient. *Scientific Reports*, *7*(1), 43952. doi:10.1038rep43952 PMID:28256623

Reiss, R., Ihssen, J., & Thöny-Meyer, L. (2011). *Bacillus pumilus* laccase: A heat stable enzyme with a wide substrate spectrum. *BMC Biotechnology*, *11*(1), 1–11. doi:10.1186/1472-6750-11-9 PMID:21266052

Rekik, H., Zaraî Jaouadi, N., Bouacem, K., Zenati, B., Kourdali, S., Badis, A., Annane, R., Bouanane-Darenfed, A., Bejar, S., & Jaouadi, B. (2019). Physical and enzymatic properties of a new manganese peroxidase from the white-rot fungus *Trametes pubescens* strain i8 for lignin biodegradation and textile-dyes biodecolorization. *International Journal of Biological Macromolecules*, *15*, 514–525. doi:10.1016/j.ijbiomac.2018.12.053 PMID:30528991

Reller, L., Weinstein, M., & Petti, C. (2007). Detection and identification of microorganisms by gene amplification and sequencing. *Clinical Infectious Diseases*, *44*, 1108–1114.

Ren, D., Cheng, Y., Huang, C., Wang, Z., Zhang, S., Zhang, X., & Gong, X. (2020). Study on Remediation-improvement of 2,4-Dichlorophenol Contaminated Soil by Organic Fertilizer Immobilized Laccase. *Soil and Sediment Contamination: An International Journal*, *30*(2), 201–215. doi:10.1080/15320383.2020.1828266

Ren, D., Sims, J. J., & Wood, T. K. (2001). Inhibition of biofilm formation and swarming of *Escherichia coli* by (*5Z*)-4-bromo-5-(bromomethylene)-3-butyl-2(5H)-furanone. *Environmental Microbiology*, *3*(11), 731–736. doi:10.1046/j.1462-2920.2001.00249.x PMID:11846763

Renuka, N., Guldhe, A., Prasanna, R., Singh, P., & Bux, F. (2018). Microalgae as multi-functional options in modern agriculture: Current trends, prospects and challenges. *Biotechnology Advances*, *36*(4), 1255–1273. doi:10.1016/j.biotechadv.2018.04.004 PMID:29673972

Renuka, N., Prasanna, R., Sood, A., Ahluwalia, A. S., Bansal, R., Babu, S., Singh, R., Shivay, Y. S., & Nain, L. (2016). Exploring the efficacy of wastewater-grown microalgal biomass as a biofertilizer for wheat. *Environmental Science and Pollution Research International*, *23*(7), 6608–6620. doi:10.100711356-015-5884-6 PMID:26638970

Renuka, N., Prasanna, R., Sood, A., Bansal, R., Bidyarani, N., Singh, R., Shivay, Y. S., Nain, L., & Ahluwalia, A. S. (2017). Wastewater grown microalgal biomass as inoculants for improving micronutrient availability in wheat. *Rhizosphere*, *1*, 150–159. doi:10.1016/j.rhisph.2017.04.005

Requena, B. N., Jimenez, I., Toro, M., & Barea, J. M. (1997). Interactions between plant growth promoting rhizobacteria (PGPR), arbuscular mycorrhizal fungi and *Rhizobium spp*.in the rhizosphere of *Anthyllis cytiisoides*, a model legume for revegetation in Mediterranean semi-arid ecosystem. *The New Phytologist*, *136*(4), 667–677. doi:10.1046/j.1469-8137.1997.00786.x

Reverter, M., Bontemps, N., Lecchini, D., Banaigs, B., & Sasal, P. (2014). Use of plant extracts in fish aquaculture as an alternative to chemotherapy: Current status and future perspectives. *Aquaculture (Amsterdam, Netherlands)*, *433*, 50–61. doi:10.1016/j.aquaculture.2014.05.048

Revillas, J., Rodelas, B., Pozo, C., Martínez-Toledo, M., & González-López, J. (2000). Production of B-group vitamins by two *Azotobacter* strains with phenolic compounds as sole carbon source under diazotrophic and adiazotrophic conditions. *Journal of Applied Microbiology*, *89*(3), 486–493. doi:10.1046/j.1365-2672.2000.01139.x PMID:11021581

Rex Consortium. (2013). Heterogeneity of selection and the evolution of resistance. *Trends in Ecology & Evolution*, *28*(2), 110–118. doi:10.1016/j.tree.2012.09.001 PMID:23040463

Reyes, I., Bernier, L., & Antoun, H. (2002). Rock phosphate solubilization and colonization of maize rhizosphere by wild and genetically modified strains of Penicillium rugulosum. *Microbial Ecology*, *44*(1), 39–48. doi:10.100700248-002-1001-8 PMID:12019460

Rezaei, S., Shahverdi, A. R., & Faramarzi, M. A. (2017). Isolation, one-step affinity purification, and characterization of a polyextremotolerant laccase from the halophilic bacterium Aquisalibacillus elongatus and its application in the delignification of sugar beet pulp. *Bioresource Technology*, *230*, 67–75. doi:10.1016/j.biortech.2017.01.036 PMID:28161622

Rezania, S., Park, J., Md Din, M. F., Mat Taib, S., Talaiekhozani, A., Kumar Yadav, K., & Kamyab, H. (2018). Microplastics pollution in differentaquatic environments and biota: A review of recent studies. *Marine Pollution Bulletin*, *133*, 191–208. doi:10.1016/j.marpolbul.2018.05.022 PMID:30041307

Rezania, S., Taib, S. M., Md Din, M. F., Dahalan, F. A., & Kamyab, H. (2016). Comprehensive review on phytotechnology: Heavy metals removal by diverse aquatic plants species from wastewater. *Journal of Hazardous Materials*, *318*, 587–599.

Rezzonico, F., Binder, C., Défago, G., & Moënne-Loccoz, Y. (2005). The type III secretion system of biocontrol Pseudomonas fluorescens KD targets the phytopathogenic Chromista Pythium ultimum and promotes cucumber protection. *Molecular Plant-Microbe Interactions*, *18*(9), 991–1001. doi:10.1094/MPMI-18-0991 PMID:16167769

Rfaki, A., Zennouhi, O., Aliyat, F. Z., Nassiri, L., & Ibijbijen, J. (2020). Isolation, selection and characterization of root-associated rock phosphate solubilizing bacteria in moroccan wheat (Triticum aestivum L.). *Geomicrobiology Journal*, *37*(3), 230–241. doi:10.1080/01490451.2019.1694106

Riah, W., Laval, K., Laroche-Ajzenberg, E., Mougin, C., Latour, X., & Trinsoutrot-Gattin, I. (2014). Effects of pesticides on soil enzymes: A review. *Environmental Chemistry Letters*, *12*(2), 257–273. doi:10.100710311-014-0458-2

Riaz, K., Elmerich, C., Moreira, D., Raffoux, A., Dessaux, Y., & Faure, D. (2008). A metagenomic analysis of rhizospheric bacteria extends the diversity of quorum-quenching lactonases. *Environmental Microbiology*, *10*(3), 560–570. doi:10.1111/j.1462-2920.2007.01475.x PMID:18201196

Riaz, U., Aziz, H., Anum, W., Mehdi, S. M., Murtaza, G., & Jamil, M. (2020a). Biofortification Technologies Used in Agriculture in Relation to Micronutrients. In *Plant Micronutrients* (pp. 225–239). Springer. doi:10.1007/978-3-030-49856-6_9

Riaz, U., Mehdi, S. M., Iqbal, S., Khalid, H. I., Qadir, A. A., Anum, W., & Murtaza, G. (2020b). Bio-fertilizers: Eco-Friendly Approach for Plant and Soil Environment. In *Bioremediation and Biotechnology* (pp. 189–213). Springer. doi:10.1007/978-3-030-35691-0_9

Richardson, A. E., & Simpson, R. J. (2011). Soil Microorganisms Mediating Phosphorus Availability: Phosphorus Plant Physiology. *Plant physiology (Bethesda), 156*(3), 989–996. doi:10.1104/pp.111.175448

Richardson, A. E. (2001). Prospects for using soil microorganisms to improve the acquisition of phosphorus by plants. *Australian Journal of Plant Physiology, 28*, 897–906.

Riesenfeld, C. S., Schloss, P. D., & Handelsman, J. (2004). Metagenomics: Genomic analysis of microbial communities. *Annual Review of Genetics, 38*, 525–552.

Rillig, M. C., Wright, S. F., & Eviner, V. T. (2002). The role of arbuscular mycorrhizal fungi and glomalin in soil aggregation: Comparing effects of five plant species. *Plant and Soil, 238*(2), 325–333. doi:10.1023/A:1014483303813

Rilling, J. I., Acuña, J. J., Sadowsky, M. J., & Jorquera, M. A. (2018). Putative Nitrogen-Fixing Bacteria Associated With the Rhizosphere and Root Endosphere of Wheat Plants Grown in an Andisol From Southern Chile. *Frontiers in Microbiology, 9*, 2710. doi:10.3389/fmicb.2018.02710 PMID:30524385

Rippka, R., Deruelles, J., Waterbery, J. B., Herdman, M., & Stainer, R. Y. (1979). Generic assignments, strain histories and properties, pure cultures of cyanobacteria. *Journal of General Microbiology, 111*, 1–61.

Riskuwa-Shehu, M. L., & Ismail, H. Y. (2018). Isolation of endophytic bacteria and phytoremediation of soil contaminated with polycyclic aromatic hydrocarbons using *Cajanus cajan* and *Lablab purpereus. Bioremediation Science and Technology Research, 6*(1), 26-30. https://journal.hibiscuspublisher.com/index.php/BSTR/issue/view/42

Rizwan, M., Singh, M., Mitra, C. K., & Morve, R. K. (2014). Ecofriendly application of nanomaterials: nanobioremediation. *Journal of Nanoparticles*.

Rober-Kleber, N., Albrechtová, J. T., Fleig, S., Huck, N., Michalke, W., Wagner, E., Speth, V., Neuhaus, G., & Fischer-Iglesias, C. (2003). Plasma membrane H+-ATPase is involved in auxin-mediated cell elongation during wheat embryo development. *Plant Physiology, 131*(3), 1302–1312. doi:10.1104/pp.013466 PMID:12644680

Robinson, N. J., Gupta, A., Fordham-Skelton, A. P., Croy, R. R. D., Whitton, B. A., & Huckle, J. W. (1990). Prokaryotic metallothionein gene characterization and expression: Chromosome crawling by ligation-mediated PCR. *Proceedings of the Royal Society of London. Series B, Biological Sciences, 242*(1305), 241–247.

Robinson, T., McMullan, G., Marchant, R., & Nigam, P. (2001). Remediation of dyes in textile effluent: A critical review on current treatment technologies with a proposed alternative. *Bioresource Technology, 77*(3), 247–255. doi:10.1016/S0960-8524(00)00080-8 PMID:11272011

Rocha-Martins, M., Cavalheiro, G. R., Matos-Rodrigues, G. E., & Martins, R. P. (2015). From Gene Targeting to Genome Editing: Transgenic animals applications and beyond. *Anais da Academia Brasileira de Ciências, 87*(2), 1323–1348. doi:10.1590/0001-3765201520140710 PMID:26397828

Rodrigues, E. P., Rodrigues, L. S., de Oliveira, A. L. M., Baldani, V. L. D., dos Santos Teixeira, K. R., Urquiaga, S., & Reis, V. M. (2008). Azospirillum amazonense inoculation: Effects on growth, yield and N 2 fixation of rice (Oryza sativa L.). *Plant and Soil, 302*(1-2), 249–261. doi:10.100711104-007-9476-1

Rodrigues, J. L. M., Kachel, C. A., Aiello, M. R., Quensen, J. F. III, Maltseva, O. V., Tsoi, T. V., & Tiedje, J. M. (2006). Degradation of Aroclor 1242 Dechlorination Products in Sediments by Burkholderiaxenovorans LB400(Ohb) and *Rhodococcus* sp. Strain RHA1(Fcb). *Applied and Environmental Microbiology, 72*(4), 2476–2482. doi:10.1128/AEM.72.4.2476-2482.2006 PMID:16597946

Rodriguez Couto, S., & Toca Herrera, J. L. (2006). Industrial and biotechnological applications of laccases: A review. *Biotechnology Advances, 24*(5), 500–513. doi:10.1016/j.biotechadv.2006.04.003 PMID:16716556

Rodríguez Eugenio, N., McLaughlin, M., & Pennock, D. (2018). *Soil Pollution: a hidden reality. Food and Agriculture Organization of the United Nations edition.* FAO.

Rodriguez-Campos, J., Perales-Garcia, A., Hernandez-Carballo, J., Martinez-Rabelo, F., Hernández-Castellanos, B., Barois, I., & Contreras-Ramos, S. M. (2018). Bioremediation of soil contaminated by hydrocarbons with the combination of three technologies: Bioaugmentation, phytoremediation, and vermiremediation. *Journal of Soils and Sediments, 19*(4), 1981–1994. doi:10.100711368-018-2213-y

Rodriguez, L., Lopez-Bellido, F. J., Carnicer, A., Recreo, F., Tallos, A., & Monteagudo, J. M. (2005). Mercury recovery from soils by phytoremediation. In *Book of environmental chemistry* (pp. 197–204). Springer. doi:10.1007/3-540-26531-7_18

Rodríguez, M., Torres, M., Blanco, L., Béjar, V., Sampedro, I., & Llamas, I. (2020). Plant growth-promoting activity and quorum quenching-mediated biocontrol of bacterial phytopathogens by Pseudomonas segetis strain P6. *Scientific Reports, 10*(1), 1–2. doi:10.103841598-020-61084-1 PMID:32139754

Rodriguez, R. J., Henson, J., Van Volkenburgh, E., Hoy, M., Wright, L., Beckwith, F., & Redman, R. S. (2008). Stress tolerance in plants via habitat-adapted symbiosis. *The ISME Journal, 2*(4), 404–416. doi:10.1038/ismej.2007.106 PMID:18256707

Rodríguez-Rojas, F., Tapia, P., Castro-Nallar, E., Undabarrena, A., Muñoz-Díaz, P., Arenas-Salinas, M., Díaz-Vásquez, W., Valdés, J., & Vásquez, C. (2016). Draft genome sequence of a multi-metal resistant bacterium *Pseudomonas putida* ATH-43 isolated from Greenwich Island, Antarctica. *Frontiers in Microbiology, 7*, 1777.

Roesch, L. F., Fulthorpe, R. R., Riva, A., Casella, G., Hadwin, A. K., Kent, A. D., Daroub, S. H., Camargo, F. A. O., Farmerie, W. G., & Triplett, E. W. (2007). Pyrosequencing enumerates and contrasts soil microbial diversity. *The ISME Journal, 1*(4), 283–290. doi:10.1038/ismej.2007.53 PMID:18043639

Roeselers, G., Van Loosdrecht, M., & Muyzer, G. (2008). Phototrophic biofilms and their potential applications. *Journal of Applied Phycology, 20*(3), 227–235. doi:10.100710811-007-9223-2 PMID:19396356

Rohrbacher, F., & St-Arnaud, M. (2016). Root exudation: The ecological driver of hydrocarbon rhizoremediation. *Agronomy (Basel), 6*(1), 19. doi:10.3390/agronomy6010019

Rojas-Tapias, D., Moreno-Galván, A., Pardo-Díaz, S., Obando, M., Rivera, D., & Bonilla, R. (2012). Effect of inoculation with plant growth-promoting bacteria (PGPB) on amelioration of saline stress in maize (*Zea mays*). *Applied Soil Ecology, 61*, 264–272. doi:10.1016/j.apsoil.2012.01.006

Romanenko, E. A., Kosakovskaya, I. V., & Romanenko, P. A. (2015). Phytohormones of microalgae: Biological role and involvement in the regulation of physiological processes. Pt I. auxins, abscisic acid, ethylene. *International Journal on Algae, 17*(3), 275–289. doi:10.1615/InterJAlgae.v17.i3.80

Romano, I., Ventorino, V., & Pepe, O. (2020). Effectiveness of plant beneficial microbes: Overview of the methodological approaches for the assessment of root colonization and persistence. *Frontiers in Plant Science, 11*, 6. doi:10.3389/fpls.2020.00006 PMID:32076431

Romera, F. J., García, M. J., Lucena, C., Martínez-Medina, A., Aparicio, M. A., Ramos, J., Alcántara, E., Angulo, M., & Pérez-Vicente, R. (2019). Induced systemic resistance (ISR) and Fe deficiency responses in dicot plants. *Frontiers in Plant Science*, *10*, 287. doi:10.3389/fpls.2019.00287 PMID:30915094

Romero, D., de Vicente, A., Rakotoaly, R. H., Dufour, S. E., Veening, J. W., Arrebola, E., Cazorla, F. M., Kuipers, O. P., Paquot, M., & Pérez-García, A. (2007). The iturin and fengycin families of lipopeptides are key factors in antagonism of *Bacillus subtilis* toward *Podosphaera fusca. Molecular plant-microbe interactions. Molecular Plant-Microbe Interactions*, *20*(4), 430–440. doi:10.1094/MPMI-20-4-0430 PMID:17427813

Romero, M., Diggle, S. P., Heeb, S., Cámara, M., & Otero, A. (2008). Quorum quenching activity in Anabaena sp. PCC 7120: Identification of AiiC, a novel AHL-acylase. *FEMS Microbiology Letters*, *280*(1), 73–80. doi:10.1111/j.1574-6968.2007.01046.x PMID:18194337

Romheld, V., & Kirkby, E. A. (2010). Research on potassium in agriculture: Needs and prospects. *Plant and Soil*, *335*(1-2), 155–180. doi:10.100711104-010-0520-1

Ron, E. Z., & Rosenberg, E. (2002). Biosurfactants and oil bioremediation. *Current Opinion in Biotechnology*, *13*(3), 249–252. doi:10.1016/S0958-1669(02)00316-6 PMID:12180101

Ron, E. Z., & Rosenberg, E. (2014). Enhanced bioremediation of oil spills in the sea. *Current Opinion in Biotechnology*, *27*, 191–194. doi:10.1016/j.copbio.2014.02.004 PMID:24657912

Ronga, D., Biazzi, E., Parati, K., Carminati, D., Carminati, E., & Tava, A. (2019). Microalgal biostimulants and biofertilisers in crop productions. *Agronomy (Basel)*, *9*(4), 1–22. doi:10.3390/agronomy9040192

Roper, M. M., Gault, R. R., & Smith, N. A. (1995). Contribution to the N status of soil by free-living N2-fixing bacteria in a Lucerne stand. *Soil Biology & Biochemistry*, *27*(4-5), 467–471. doi:10.1016/0038-0717(95)98621-T

Rorat, A., Wloka, D., Grobelak, A., Grosser, A., Sosnecka, A., Milczarek, M., Jelonek, P., Vandenbulcke, F., & Kacprzak, M. (2017). Vermi remediation of polycyclic aromatic hydrocarbons and heavy metals in sewage sludge composting process. *Journal of Environmental Management*, *187*, 347–353. doi:10.1016/j.jenvman.2016.10.062 PMID:27836561

Rosca, M., Hlihor, R. M., & Gavrilescu, M. (2019). Bioremediation of Persistent Toxic Substances: From conventional to new approaches in using microorganisms and plants. In *Microbial Technology for the Welfare of Society* (pp. 289–312). Springer. doi:10.1007/978-981-13-8844-6_14

Rosi-Marshall, E. (2013). *Streams stressed by pharmaceutical pollution*. www.environmentalchange.nd.edu/events/2

Rossi, F., Li, H., Liu, Y., & De Philippis, R. (2017). Cyanobacterial inoculation (cyanobacterisation): Perspectives for the development of a standardized multifunctional technology for soil fertilization and desertification reversal. *Earth-Science Reviews*, *171*, 28–43. doi:10.1016/j.earscirev.2017.05.006

Rouphael, Y., Cardarelli, M., & Colla, G. (2015a). Role of arbuscular mycorrhizal fungi in alleviating the adverse effects of acidity and aluminium toxicity in zucchini squash. *Scientia Horticulturae*, *188*, 97–105. doi:10.1016/j.scienta.2015.03.031

Rouphael, Y., Franken, P., Schneider, C., Schwarz, D., Giovannetti, M., Agnolucci, M., & Colla, G. (2015b). Arbuscular mycorrhizal fungi act as biostimulants in horticultural crops. *Scientia Horticulturae*, *196*, 91–108. doi:10.1016/j.scienta.2015.09.002

Rout, G. R., & Sahoo, S. (2015). Role of iron in plant growth and metabolism. *Reviews in Agricultural Science*, *3*(0), 1–24. doi:10.7831/ras.3.1

Rovira, A. D. (1969). Plant root exudates. *Botanical Review*, *35*(1), 35–57. doi:10.1007/BF02859887

Roy, M., & McDonald, L. M. (2015). Metal uptake in plants and health risk assessments in metal-contaminated smelter soils. *Land Degradation & Development, 26,* 785-792.

Roy, M., Giri, A. K., Dutta, S., & Mukherjee, P. (2015). Integrated phytobial remediation for sustainable management of arsenic in soil and water. *Environment International, 75,* 180–198. doi:10.1016/j.envint.2014.11.010 PMID:25481297

Rubio, L. M., & Ludden, P. W. (2008). Biosynthesis of the iron-molybdenum cofactor of nitrogenase. *Annual Review of Microbiology, 62*(1), 93–111. doi:10.1146/annurev.micro.62.081307.162737 PMID:18429691

Rui, Q. C., Hao, X., & William, W. (2020). Ecological Stoichiometry of Microbial Biomass Carbon, Nitrogen and Phosphorus on Bauxite Residue Disposal Areas. *Geomicrobiology Journal, 37*(5), 467–474. doi:10.1080/01490451.2020.1722768

Ruley, J. A., Tumuhairwe, J. B., Amoding, A., Opolot, E., Oryem-Origa, H., & Basamba, T. (2019). Assessment of plants for phytoremediation of hydrocarbon-contaminated soils in the Sudd Wetland of South Sudan. *Plant, Soil and Environment, 65*(9), 463–469. doi:10.17221/322/2019-PSE

Rware, H., Kayuki, C., Macharia, M., & Oduor, G. (2016). Fertilizer use optimization approach: An innovation to increase agricultural profitability for African farmers. *African Journal of Agricultural Research, 11*(38), 3587–3597. doi:10.5897/AJAR2016.11408

Ryder, M. H., Yan, Z., Terrace, T. E., Rovira, A. D., Tang, W., & Correll, R. L. (1999). Use of Bacillus isolated in China to suppress take-all and *rhizoctonia* root rot, and promote seedling growth of glass house grown wheat in Australian soils. *Soil Biology & Biochemistry, 31*(1), 19–29. doi:10.1016/S0038-0717(98)00095-9

Rylott, E. L., & Bruce, N. C. (2009). Plants disarm soil: Engineering plants for the phytoremediation of explosives. *Trends in Biotechnology, 27*(2), 73–81. doi:10.1016/j.tibtech.2008.11.001 PMID:19110329

Ryu, C. M., Farag, M. A., Hu, C. H., Reddy, M. S., Wei, H. X., Paré, P. W., & Kloepper, J. W. (2003). Bacterial volatiles promote growth in Arabidopsis. *Proceedings of the National Academy of Sciences of the United States of America, 100*(8), 4927–4932. doi:10.1073/pnas.0730845100 PMID:12684534

Saad, M. M., Kandil, M., & Mohammed, Y. M. (2020). Isolation and Identification of Plant Growth-Promoting Bacteria Highly Effective in Suppressing Root Rot in Fava Beans. *Current Microbiology, 77*(9), 1–11. doi:10.100700284-020-02015-1 PMID:32372106

Saba, B., Jabeen, M., Khalid, A., Aziz, I., & Christy, A. D. (2015). Effectiveness of rice agricultural waste, microbes and wetland plants in the removal of reactive black-5 azo dye in microcosm constructed wetlands. *International Journal of Phytoremediation, 17*(11), 1060–1067. doi:10.1080/15226514.2014.1003787 PMID:25849115

Sabirova, J. S., Becker, A., Lünsdorf, H., Nicaud, J. M., Timmis, K. N., & Golyshin, P. N. (2011). Transcriptional profiling of the marine oil-degrading bacterium *Alcanivorax borkumensis* during growth on n-alkanes. *FEMS Microbiology Letters, 319*(2), 160–168. doi:10.1111/j.1574-6968.2011.02279.x PMID:21470299

Sachdev, S., & Singh, R. P. (2018). Root colonization: Imperative mechanism for efficient plant protection and growth. *MOJ Ecology & Environmental Sciences, 3,* 240–242.

Sachidanand, B., Mitra, N., Kumar, V., Roy, R., & Mishra, B. (2019). Soil as a huge laboratory for microorganisms. *Agricultural Research & Technology: Open Access Journal, 22*(4).

Sadettin, S., & Donmez, G. (2006). Simultaneous bioaccumulation of reactive dye and chromium (VI) by using thermophilic *Phormidium* sp. *Enzyme and Microbial Technology, 41*(1-2), 175–180. doi:10.1016/j.enzmictec.2006.12.015

Sadfi, N., Cherif, M., Fliss, I., Boudabbous, A., & Antoun, H. (2001). Evaluation of bacterial isolates from salty soils and *Bacillus thuringiensis* strains for the biocontrol of *Fusarium* dry rot of potato tubers. *Journal of Plant Pathology*, *83*, 101–118.

Sadh, P. K., Duhan, S., & Duhan, J. S. (2018). Agro-industrial wastes and their utilization using solid state fermentation: A review. *Bioresources and Bioprocessing*, *5*(1), 1–15. doi:10.118640643-017-0187-z

Saeid, A., Prochownik, E., & Dobrowolska-Iwanek, J. (2018). Phosphorus solubilization by *Bacillus* species. *Molecules (Basel, Switzerland)*, *23*(11), 2897. doi:10.3390/molecules23112897 PMID:30404208

Saez, J. M., Alvarez, A., Benimelli, C. S., & Amorosso, M. J. (2014). Enhanced lindane removal &from soil slurry by immobilized Streptomyces consortium. *International Biodeterioration & Biodegradation*, *93*, 63–69. doi:10.1016/j.ibiod.2014.05.013

Safari, M., Ahmadi, A. S., & Soltani, N. (2016). *The Potential Of Cyanobacterium Fischerella Ambigua Isc67↑in Biodegradation Of Crude Oil*. Academic Press.

Safarikova, M., Ptackova, L., Kibrikova, I., & Safarik, I. (2005). Biosorption of water-soluble dyes on magnetically modified *Saccharomyces cerevisiae* subsp. uvarum cells. *Chemosphere*, *59*(6), 831–835. doi:10.1016/j.chemosphere.2004.10.062 PMID:15811411

Safronova, V. I., Stepanok, V. V., Engqvist, G. L., Alekseyev, Y. V., & Belimov, A. A. (2006). Root-associated bacteria containing 1- minocyclopropane- 1-carboxylate deaminase improve growth and nutrient uptake by pea genotypes cultivated in cadmium supplemented soil. *Biology and Fertility of Soils*, *42*(3), 267–272. doi:10.100700374-005-0024-y

Sagarkar, S., Mukherjee, S., Nousiainen, A., Björklöf, K., Purohit, H. J., Jørgensen, K. S., & Kapley, A. (2013). Monitoring bioremediation of atrazine in soil microcosms using molecular tools. *Environmental Pollution*, *172*, 108–115. doi:10.1016/j.envpol.2012.07.048 PMID:23022948

Saha, S., Loganathan, M., Rai, A. B., Singh, A., & Garg, R. (2016). Role of microbes in soil health improvement. *SATSA Mukhapatra Ann Tech*, (20), 53–62.

Sahoo, B., Ningthoujam, R., & Chaudhuri, S. (2019). Isolation and characterization of a lindane degrading bacteria Paracoccus sp. NITDBR1 and evaluation of its plant growth promoting traits. *International Microbiology*, *22*(1), 155–167. doi:10.100710123-018-00037-1 PMID:30810939

Sahu, J., Vaishnav, A., & Singh, H. B. (2020). Insights in plant-microbe interaction through genomics approach (Part 1). *Current Genetics*, *21*(3), 155–156. PMID:33071608

Saikia, J., Sarma, R. K., Dhandia, R., Yadav, A., Bharali, R., Gupta, V. K., & Saikia, R. (2018). Alleviation of drought stress in pulse crops with ACC deaminase producing rhizobacteria isolated from acidic soil of Northeast India. *Scientific Reports*, *8*(1), 3560. doi:10.103841598-018-21921-w PMID:29476114

Saiyad, S. A., Jhala, Y. K., & Vyas, R. V. (2015). Comparative efficiency of five potash and phosphate solubilizing bacteria and their key enzymes useful for enhancing and improvement of soil fertility. *International Journal of Scientific Research*, *5*, 1–6.

Salokhe, M. D., & Govindwar, S. P. (1999). Effect of carbon source on the biotransformation enzymes in Serratia marcescens. *World Journal of Microbiology & Biotechnology*, *15*(2), 229–232. doi:10.1023/A:1008875404889

Salomons, W., Forstner, U., & Mader, P. (1995). *Heavy Metals: Problems and Solutions*. Springer-Verlag.

Salunkhe, V. P., Sawant, I. S., Banerjee, K., Rajguru, Y. R., Wadkar, P. N., Oulkar, D. P., Naik, D. G., & Sawant, S. D. (2013). Biodegradation of profenofos by *Bacillus subtilis* isolated from grapevines (*Vitis vinifera*). *Journal of Agricultural and Food Chemistry*, *61*(30), 7195–7202. doi:10.1021/jf400528d PMID:23806113

Samir, D., Mohcem, R., Selma, O., & Asma, S. (2020). The Effect of Herbicide Metribuzin on Environment and Human: A Systematic Review. *Pharmaceutical and Biosciences Journal*. Available at www.ukjpb.com

Sam, K., Coulon, F., & Prpich, G. (2017). Management of petroleum hydrocarbon contaminated sites in Nigeria: Current challenges and future direction. *Land Use Policy*, *64*, 133–144. doi:10.1016/j.landusepol.2017.01.051

Sammauria, R., & Kumawat, S. (2018). Legume Plant Growth-Promoting Rhizobacteria (PGPRs): Role in Soil Sustainability. In *Legumes for Soil Health and Sustainable Management* (pp. 409–443). Springer. doi:10.1007/978-981-13-0253-4_13

Sampaio, C., de Souza, J., Damião, A. O., Bahiense, T. C., & Roque, M. (2019). Biodegradation of polycyclic aromatic hydrocarbons (PAHs) in a diesel oil-contaminated mangrove by plant growth-promoting rhizobacteria. *Biotech, 9*(4), 155.

Samuelson, P., Wernérus, H., Svedberg, M., & Ståhl, S. (2000). Staphylococcal surface display of metal-binding polyhistidyl peptides. *Applied and Environmental Microbiology*, *66*(3), 1243–1248.

San Miguel, A., Ravanel, P., & Raveton, M. (2013). A comparative study on the uptake and translocation of organochlorines by *Phragmites australis*. *Journal of Hazardous Materials*, *244*, 60–69. doi:10.1016/j.jhazmat.2012.11.025 PMID:23246941

Sánchez-Porro, C., Martin, S., Mellado, E., & Ventosa, A. (2003). Diversity of moderately halophilic bacteria producing extracellular hydrolytic enzymes. *Journal of Applied Microbiology*, *94*(2), 295–300. doi:10.1046/j.1365-2672.2003.01834.x PMID:12534822

Sanchez, S., & Demain, A. L. (2011). Enzymes and bioconversions of industrial, pharmaceutical, and biotechnological significance. *Organic Process Research & Development*, *15*(1), 224–230. doi:10.1021/op100302x

Sandaa, R. A., Torsvik, V., Enger, O., Daae, F. L., Castberg, T., & Hahn, D. (1999). Analysis of bacterial communities in heavy metal-contaminated soils at different levels of resolution. *FEMS Microbiology Ecology*, *30*(3), 237–251. doi:10.1111/j.1574-6941.1999.tb00652.x PMID:10525180

Sandana Mala, J. G., Sujatha, D., & Rose, C. (2015). Inducible chromate reductase exhibiting extracellular activity in *Bacillus methylotrophicus* for chromium bioremediation. *Microbiological Research*, *170*, 235–241. doi:10.1016/j.micres.2014.06.001 PMID:24985094

Sanderlin, R. S., & Ghabrial, S. A. (1978). Physicochemical properties of two distinct types of virus-like particles from *Helminthosporium victoriae*. *Virology*, *87*(1), 142–151. doi:10.1016/0042-6822(78)90166-6 PMID:664249

Sandermann, H. (1994). Higher plant metabolism of xenobiotics: the "green liver" concept. *Pharmacogenetics, 4*, 225-241.

Sandhya, V., & Ali, S. Z. (2015). The production of exopolysaccharide by Pseudomonas putida GAP-P45 under various abiotic stress conditions and its role in soil aggregation. *Microbiology*, *84*(4), 512–519. doi:10.1134/S0026261715040153

Sangeeth, K. P., Bhai, R. S., & Srinivasan, V. (2012). *Paenibacillus glucanolyticus*, a promising potassium solubilizing bacterium isolated from black pepper (Piper nigrum L.) rhizosphere. *Journal of Spices and Aromatic Crops*, *21*(2), 118–124.

Sani, R. K., & Banerjee, U. C. (1999). Decolorization of triphenylmethane dyes and textile and dye-stuff effluent by Kurthia sp. *Enzyme and Microbial Technology*, *24*(7), 433–437. doi:10.1016/S0141-0229(98)00159-8

Sannino, F., & Gianfreda, L. (2001). Pesticide influence on soil enzymatic activities. *Chemosphere*, *45*(4-5), 417–425. doi:10.1016/S0045-6535(01)00045-5 PMID:11680737

Santi, C., Bogusz, D., & Franche, C. (2013). Biological nitrogen fixation in non-legume plants. *Annals of Botany*, *111*(5), 743–767. doi:10.1093/aob/mct048 PMID:23478942

Santornchot, P., Satapanajaru, T., & Comfort, S. D. (2010). Application of nano-zero valent iron for treating metolachlor in aqueous solution. *World Academy of Science, Engineering and Technology*, *48*, 625–628.

Santos, D. K. F., Rufino, R. D., Luna, J. M., Santos, V. A., & Sarubbo, L. A. (2016). Biosurfactants: Multifunctional biomolecules of the 21st century. *International Journal of Molecular Sciences*, *17*(3), 401. doi:10.3390/ijms17030401 PMID:26999123

Santos, R. M., Kandasamy, S., & Rigobelo, E. C. (2018). Sugarcane growth and nutrition levels are differentially affected by the application of PGPR and cane waste. *MicrobiologyOpen*, *7*(6), e00617. doi:10.1002/mbo3.617 PMID:29653035

Sanz, L., Dewitte, W., Forzani, C., Patell, F., Nieuwland, J., Wen, B., Quelhas, P., De Jager, S., Titmus, C., Campilho, A., Ren, H., Estelle, M., Wang, H., & Murray, J. A. (2011). The *Arabidopsis* D-type cyclin CYCD2; 1 and the inhibitor ICK2/KRP2 modulate auxin-induced lateral root formation. *The Plant Cell*, *23*(2), 641–660. doi:10.1105/tpc.110.080002 PMID:21357490

Saran, A., Fernandez, L., Cora, F., Savio, M., Thijs, S., Vangronsveld, J., & Merini, L. J. (2020). Phytostabilization of Pb and Cd polluted soils using *Helianthus petiolaris* as pioneer aromatic plant species. *International Journal of Phytoremediation*, *22*(5), 459–467. doi:10.1080/15226514.2019.1675140 PMID:31602996

Saranraj, P., Sumathi, V., Reetha, D., & Stella, D. (2010). Fungal decolourization of direct Azo dyes and biodegradation of textile dye effluent. *Journal of Ecobiotechnology*, *2*(7), 12–16.

Saranya, K., Sundaramanickam, A., Shekhar, S., & Swaminathan, S. (2019). Biosorption of mercury by *Bacillus thuringiensis* (CASKS3) isolated from mangrove sediments of southeast coast India. *Indian Journal of Geo-Marine Sciences*, *48*(2), 143–150.

Saratale, G. D., Humnabadkar, R. P., & Govindwar, S. P. (2007). Study of mixed function oxidase system in Aspergillus ochraceus (NCIM 1146). *Indian Journal of Microbiology*, *47*(4), 304–309. doi:10.100712088-007-0056-0 PMID:23100682

Saratale, R. G., Saratale, G. D., Chang, J. S., & Govindwar, S. P. (2009a). Decolorization and biodegradation of textile dye Navy blue HER by Trichosporon beigelii NCIM-3326. *Journal of Hazardous Materials*, *166*(2-3), 1421–1428. doi:10.1016/j.jhazmat.2008.12.068 PMID:19157708

Saratale, R. G., Saratale, G. D., Chang, J. S., & Govindwar, S. P. (2009c). Ecofriendly degradation of sulfonated diazo dye C.I. Reactive Green 19A using *Micrococcus glutamicus* NCIM-2168. *Bioresource Technology*, *100*(17), 3897–3905. doi:10.1016/j.biortech.2009.03.051 PMID:19375909

Saratale, R. G., Saratale, G. D., Chang, J. S., & Govindwar, S. P. (2011). Bacterial decolorization and degradation of azo dyes: A review. *Journal of the Taiwan Institute of Chemical Engineers*, *42*(1), 138–157. doi:10.1016/j.jtice.2010.06.006

Saratale, R. G., Saratale, G. D., Kalyani, D. C., Chang, J. S., & Govindwar, S. P. (2009b). Enhanced decolorization and biodegradation of textile azo dye Scarlet R by using developed microbial consortium-GR. *Bioresource Technology*, *100*(9), 2493–2500. doi:10.1016/j.biortech.2008.12.013 PMID:19157864

Sardar, K. H. A. N., Qing, C. A. O., Hesham, A. E. L., Yue, X., & He, J. Z. (2007). Soil enzymatic activities and microbial community structure with different application rates of Cd and Pb. *Journal of Environmental Sciences (China)*, *19*(7), 834–840. doi:10.1016/S1001-0742(07)60139-9 PMID:17966871

Sardrood, B. P., Goltapeh, E. M., & Varma, A. (2013). An introduction to bioremediation. In Funghi as bioremediators (pp. 3-29). Springer Science and Business Media. doi:10.1007/978-3-642-33811-3_1

Sari, A. A., Tachibana, S., Muryanto, & Hadibarata, T. (2016). Development of bioreactor systems for decolorization of reactive green 19 using white rot fungus. *Desalination and Water Treatment*, *57*(15), 7029–7039. doi:10.1080/19443994.2015.1012121

Saritha, M., & Tollamadugu, N. V. K. V. P. (2019). The Status of Research and Application of Biofertilizers and Biopesticides: Global Scenario. In V. Buddolla (Ed.), *Recent Developments in Applied Microbiology and Biochemistry*. Academic Press., doi:10.1016/B978-0-12-816328-3.00015-5

Sarkar, A., Ghosh, P. K., Pramanik, K., Mitra, S., Soren, T., Pandey, S., Mondal, M. H., & Maiti, T. K. (2018). A halotolerant Enterobacter sp. displaying ACC deaminase activity promotes rice seedling growth under salt stress. *Research in Microbiology*, *169*(1), 20–32. doi:10.1016/j.resmic.2017.08.005 PMID:28893659

Sarkar, S., Banerjee, A., Halder, U., Biswas, R., & Bandopadhyay, R. (2017). Degradation of synthetic azo dyes of textile industry: A sustainable approach using microbial enzymes. *Water Conservation Science and Engineering*, *2*(4), 121–131. doi:10.100741101-017-0031-5

Sasec, V., & Cajthaml, T. (2014). *Mycoremediation*: Current status and perspectives. *International Journal of Medicinal Mushrooms*, *7*(3), 360–361. doi:10.1615/IntJMedMushr.v7.i3.200

Sasikala, C., Jiwal, S., Rout, P., & Ramya, M. (2012). Biodegradation of chlorpyrifos by bacterial consortium isolated from agriculture soil. *World Journal of Microbiology & Biotechnology*, *28*(3), 1301–1308. doi:10.100711274-011-0879-z PMID:22805851

Sasikumar, C. S., & Papinazath, T. (2003). Environmental management: bioremediation of polluted environment. In *Proceedings of the third international conference on environment and health* (pp 465-469). Chennai, India: Department of Geography, University of Madras and Faculty of Environmental Studies, York University, Chennai.

Sato, T., Usui, S., Tsuchiya, Y., & Kondo, Y. (2006). Invention of outdoor closed type photobioreactor for microalgae. *Energy Conversion and Management*, *47*(6), 791–799. doi:10.1016/j.enconman.2005.06.010

Satpathy, R. (2019). Computational Tools and Techniques to Predict Aquatic Toxicity of Some Halogenated Pollutants. In *Handbook of Research on the Adverse Effects of Pesticide Pollution in Aquatic Ecosystems* (pp. 318–337). IGI Global. doi:10.4018/978-1-5225-6111-8.ch018

Satpathy, R., Konkimalla, V. B., & Ratha, J. (2015). Application of bioinformatics tools and databases in microbial dehalogenation research: A review. *Applied Biochemistry and Microbiology*, *51*(1), 11–20. doi:10.1134/S0003683815010147 PMID:25842899

Satpathy, R., Konkimalla, V. B., & Ratha, J. (2016). In silico phylogenetic analysis and molecular modelling study of 2-haloalkanoic acid dehalogenase enzymes from bacterial and fungal origin. *Advances in Bioinformatics*, *2016*, 2016. doi:10.1155/2016/8701201 PMID:26880911

Satpathy, R., Konkimalla, V. B., & Ratha, J. (2017). Microbial dehalogenation: 3-chloropropanoic acid (3-CPA) degradation as a case study. *Microbiology*, *86*(1), 32–41. doi:10.1134/S0026261716060175

Satpathy, R., Konkimalla, V. S. B., & Ratha, J. (2015). Dehalobase: A database of dehalogenase and other allied enzymes. *International Journal of Applied Research on Information Technology and Computing*, *6*(1), 33–37. doi:10.5958/0975-8089.2015.00004.4

Satyapal, G. K., Mishra, S. K., Srivastava, A., Ranjan, R. K., Prakash, K., Haque, R., & Kumar, N. (2018). Possible bioremediation of arsenic toxicity by isolating indigenous bacteria from the middle Gangetic plain of Bihar, India. *Biotechnology Reports (Amsterdam, Netherlands)*, *17*, 117–125. doi:10.1016/j.btre.2018.02.002 PMID:29541605

Savci, S. (2012). An agricultural pollutant: Chemical fertilizer. *International Journal of Environmental Sciences and Development*, *3*(1), 73–80. doi:10.7763/IJESD.2012.V3.191

Savin, I. I., & Butnaru, R. (2008). Wastewater characteristics in textile finishing mills. [EEMJ]. *Environmental Engineering and Management Journal*, *7*(6).

Savin, I. I., & Butnaru, R. (2008). Wastewater Characteristics in Textile Finishing Mills. *Environmental Engineering and Management Journal*, *7*(6), 859–864. doi:10.30638/eemj.2008.113

Saxena, G., & Bharagava, R. N. (2020). Bioremediation of Industrial Waste for Environmental Safety: Volume I: Industrial Waste and Its Management. Springer.

Saxena, M., Chauhan, A., & Asokan, P. (1998). Flyash vermicompost from non-eco-friendly organic wastes. *Pollution Research*, *17*, 5–11.

Saxena, S. (2015). Strategies of strain improvement of industrial microbes. In *Applied Microbiology* (pp. 155–171). Springer.

Sayara, T., Borràs, E., Caminal, G., Sarrà, M., & Sánchez, A. (2011). Bioremediation of PAHs-contaminated soil through composting: Influence of bioaugmentation and biostimulation on contaminant biodegradation. *International Biodeterioration & Biodegradation*, *65*(6), 859–865. doi:10.1016/j.ibiod.2011.05.006

Saykhedkar, S. S., & Singhal, R. S. (2004). Solid-state fermentation for production of griseofulvin on rice bran using *Penicillium griseofulvum. Biotechnology Progress*, *20*(4), 1280–1284. doi:10.1021/bp0343662 PMID:15296463

Scavino, A. F., & Pedraza, R. O. (2013). The role of siderophores in plant growth-promoting bacteria. In *Bacteria in agrobiology: crop productivity* (pp. 265–285). Springer. doi:10.1007/978-3-642-37241-4_11

Schachtman, D. P., Reid, R. J., & Ayling, S. M. (1998). Phosphorus uptake by plants: From soil to cell. *Plant Physiology*, *116*(2), 447–453. doi:10.1104/pp.116.2.447 PMID:9490752

Schaefer, J. K., Rocks, S. S., Zheng, W., Liang, L., Gu, B., & Morel, F. M. M. (2011). Active transport, substrate specificity, and methylation of Hg(II) in anaerobic bacteria. *Proceedings of the National Academy of Sciences of the United States of America*, *108*(21), 8714–8719.

Schandry, N., & Becker, C. (2019). Allelopathic plants: Models for studying plant–interkingdom interactions. *Trends in Plant Science*, *6*, 234–256. PMID:31837955

Schenk, P. M., Thomas-Hall, S. R., Stephens, E., Marx, U. C., Mussgnug, J. H., Posten, C., Kruse, O., & Hankamer, B. (2008). Second Generation Biofuels: High-Efficiency Microalgae for Biodiesel Production. *BioEnergy Research*, *1*(1), 20–43. doi:10.100712155-008-9008-8

Schipper, C., Hornung, C., Bijtenhoorn, P., Quitschau, M., Grond, S., & Streit, W. R. (2009). Metagenome derived clones encoding two novel lactonase family proteins involved in biofilm inhibition in *Pseudomonas aeruginosa. Applied and Environmental Microbiology*, *75*(1), 224–233. doi:10.1128/AEM.01389-08 PMID:18997026

Schisler, D. A., Slininger, P. J., Behle, R. W., & Jackson, M. A. (2004). Formulation of *Bacillus spp.* for biological control of plant diseases. *Phytopathology*, *94*(11), 1267–1271. doi:10.1094/PHYTO.2004.94.11.1267 PMID:18944465

Schmid, M., & HartMann, A. (2007). Molecular Phylogeny and Ecology of Root Associated Diazotrophic α- and β-Proteobacteria. In *Associative and Endophytic Nitrogen-fixing Bacteria and Cyanobacterial Associations. Nitrogen Fixation: Origins, Applications, and Research Progress* (Vol. 5). Springer. doi:10.1007/1-4020-3546-2_2

Schmidt, O. (2006). *Wood and tree fungi*. Springer-Verlag.

Schmidt, U. (2003). Enhancing phytoremediation: The effect of chemical soil manipulation on mobility, plant accumulation, and leaching of heavy metals. *Journal of Environmental Quality*, *32*, 1939–1954. doi:10.2134/jeq2003.1939 PMID:14674516

Schmidt, W. (1999). Mechanisms and regulation of reduction-based iron uptake in plants. *The New Phytologist*, *141*(1), 1–26. doi:10.1046/j.1469-8137.1999.00331.x

Schmiedeknecht, G., Bochow, H., & Junge, H. (1998). Use of Bacillus subtilis as biocontrol agent. II. Biological control of potato diseases/Anwendung von *Bacillus subtilis* als Mittel für den biologischen Pflanzenschutz. II. Biologische Bekämpfung von Kartoffelkrankheiten. Zeitschrift für Pflanzenkrankheiten und Pflanzenschutz. *Journal of Plant Diseases and Protection*, 376–386.

Schneider, P., Caspersen, M. B., Mondorf, K., Halkier, T., Skov, L. K., Østergaard, P. R., Brown, K. M., Brown, S. H., & Xu, F. (1999). Characterization of a Coprinus cinereus laccase. *Enzyme and Microbial Technology*, *25*(6), 502–508. doi:10.1016/S0141-0229(99)00085-X

Schneiker, S., Keller, M., Dröge, M., Lanka, E., Pühler, A., & Selbitschka, W. (2001). The genetic organization and evolution of the broad host range mercury resistance plasmid pSB102 isolated from a microbial population residing in the rhizosphere of alfalfa. *Nucleic Acids Research*, *29*(24), 5169–5181.

Schnider-Keel, U., Seematter, A., Maurhofer, M., Blumer, C., Duffy, B., Gigot-Bonnefoy, C., Reimmann, C., Notz, R., Défago, G., Haas, D., & Keel, C. (2000). Autoinduction of 2, 4-diacetylphloroglucinol biosynthesis in the biocontrol agent *Pseudomonas fluorescens* CHA0 and repression by the bacterial metabolites salicylate and pyoluteorin. *Journal of Bacteriology*, *182*(5), 1215–1225. doi:10.1128/JB.182.5.1215-1225.2000 PMID:10671440

Schnoor, J. L., Light, L. A., McCutcheon, S. C., Wolfe, N. L., & Carreia, L. H. (1995). Phytoremediation of organic and nutrient contaminants. *Environmental Science & Technology*, *29*(7), 318A–323A. doi:10.1021/es00007a747 PMID:22667744

Schoebel, C. N., Zoller, S., & Rigling, D. (2014). Detection and genetic characterisation of a novel mycovirus in *Hymenoscyphus fraxineus*, the causal agent of ash dieback. *Infection, Genetics and Evolution*, *28*, 78–86. doi:10.1016/j.meegid.2014.09.001 PMID:25219345

Schoenborn, L., Yates, P. S., Grinton, B. E., Hugenholtz, P., & Janssen, P. H. (2004). Liquid serial dilution is inferior to solid media for isolation of cultures representative of the phylum-level diversity of soil bacteria. *Applied and Environmental Microbiology*, *70*(7), 4363–4366. doi:10.1128/AEM.70.7.4363-4366.2004 PMID:15240320

Schoonover, J. E., & Crim, J. F. (2015). An Introduction to Soil Concepts and the Role of Soils in Watershed Management. *Journal of Contemporary Water Research & Education*, *154*(1), 21–47. doi:10.1111/j.1936-704X.2015.03186.x

Schrick, B., Hydutsky, B. W., Blough, J. L., & Mallouk, T. E. (2004). Delivery vehicles for zerovalent metal nanoparticles in soil and groundwater. *Chemistry of Materials*, *16*(11), 2187–2193. doi:10.1021/cm0218108

Schröder, I., Johnson, E., & De Vries, S. (2003). Microbial ferric iron reductases. *FEMS Microbiology Reviews*, *27*(2-3), 427–447. doi:10.1016/S0168-6445(03)00043-3 PMID:12829278

Schue, M., Dover, L. G., Besra, G. S., Parkhill, J., & Brown, N. L. (2009). Sequence and analysis of a plasmid encoded mercury resistance operon from *Mycobacterium marinum* identifies MerH, a new mercuric ion transporter. *Journal of Bacteriology*, *19*(1), 439–444. doi:10.1128/JB.01063-08 PMID:18931130

Schultze, M., & Kondorosi, A. (1998). Regulation of symbiotic root nodule development. *Annual Review of Genetics*, *32*(1), 33–57. doi:10.1146/annurev.genet.32.1.33 PMID:9928474

Schumpp, O., & Deakin, W. J. (2010). How inefficient rhizobia prolong their existence within nodules. *Trends in Plant Science*, *15*(4), 189–195. doi:10.1016/j.tplants.2010.01.001 PMID:20117958

Schwartz, C., Echevarria, G., & Morel, J. L. (2003). Phytoextraction of cadmium with *Thlaspi caerulescens*. *Plant and Soil*, *249*(1), 27–35. doi:10.1023/A:1022584220411

Schwarzott, D., & Walker, C. (2001). A new fungal phylum, the Glomermycota: Phylogeny and evolution. *Mycological Research*, *105*(12), 1413–1421. doi:10.1017/S0953756201005196

Schwitzgue'bel, J. P. (2017). Phytoremediation of soils contaminated by organic compounds: Hype, hope and facts. *Journal of Soils and Sediments*, *17*(5), 1492–1502. doi:10.100711368-015-1253-9

Scott, M. F., Botigué, L. R., Brace, S., Stevens, C. J., Mullin, V. E., Stevenson, A., & Mott, R. (2019). A 3,000-year-old Egyptian emmer wheat genome reveals dispersal and domestication history. *Nature Plants*, *5*(11), 1120–1128. doi:10.103841477-019-0534-5 PMID:31685951

Sebastián, F. A., Méndez, V., Aguila, P., & Seeger, M. (2014). Bioremediation of petroleum hydrocarbons: Catabolic genes, microbial communities, and applications. *Applied Microbiology and Biotechnology*, *98*(11), 4781–4794. Advance online publication. doi:10.100700253-014-5684-9 PMID:24691868

Seema, D., Dahiya, R., Phogat, V. K., & Sheoran, H. S. (2019). Hydraulic Properties and Their Dependence on Physico-chemical Properties of Soils: A Review. *Current Journal of Applied Science and Technology*, *38*(2), 1–7. doi:10.9734/cjast/2019/v38i230355

Sekine, M., Watanabe, K., & Syono, K. (1989). Molecular cloning of a gene for indole-3-acetamide hydrolase from *Bradyrhizobium japonicum*. *Journal of Bacteriology*, *171*(3), 1718–1724. doi:10.1128/JB.171.3.1718-1724.1989 PMID:2646294

Selvakumar, G., Panneerselvam, P., & Ganeshamurthy, A. N. (2012). Bacterial Mediated Alleviation of Abiotic Stress in Crops. In D. Maheshwari (Ed.), *Bacteria in Agrobiology: Stress Management*. Springer. doi:10.1007/978-3-662-45795-5_10

Selvakumar, G., Panneerselvam, P., & Ganeshamurthy, A. N. (2014). Biosafety of novel bioinoculants. *Journal of Biofertilizers & Biopesticides*, *5*(2), 145.

Semu, E., & Singh, B. R. (1996). Accumulation of heavy metals in soils and plants after long-term use of fertilizers and fungicides in Tanzania. *Fertilizer Research*, *44*(3), 241–248.

Senthil Kumar, P., & Gunasundari, E. (2018). Bioremediation of Heavy Metals. In S. J. Varjani, A. K. Agarwal, E. Gnansounou, & B. Gurunathan (Eds.), *Bioremediation: Applications for Environmental Protection and Management* (Vol. 201–202, pp. 165–195). Springer Singapore. doi:10.1007/978-981-10-7485-1_9

Sergeeva, E., Shah, S., & Glick, B. R. (2006). Growth of transgenic canola (Brassica napus cv. Westar) expressing a bacterial 1-aminocyclopropane-1-carboxylate (ACC) deaminase gene on high concentrations of salt. *World Journal of Microbiology & Biotechnology*, *22*(3), 277–282. doi:10.100711274-005-9032-1

Serteyn, L., Quaghebeur, C., Ongena, M., Cabrera, N., Barrera, A., Molina-Montenegro, M. A., Francis, F., & Ramírez, C. C. (2020). Induced Systemic Resistance by a Plant Growth-Promoting Rhizobacterium Impacts Development and Feeding Behavior of Aphids. *Insects*, *11*(4), 234. doi:10.3390/insects11040234 PMID:32276327

Sessitsch, A., Hardoim, P., Döring, J., Weilharter, A., Krause, A., Woyke, T., & Reinhold-Hurek, B. (2012). Functional characteristics of an endophyte community colonizing rice roots as revealed by metagenomic analysis. *Molecular Plant-Microbe Interactions*, *25*(1), 28–36. doi:10.1094/MPMI-08-11-0204 PMID:21970692

Sessitsch, A., Reiter, B., Pfeifer, U., & Wilhelm, E. (2002). Cultivation-independent population analysis of bacterial endophytes in three potato varieties based on eubacterial and Actinomycetes-specific PCR of 16S rRNA genes. *FEMS Microbiology Ecology*, *39*(1), 23–32. doi:10.1111/j.1574-6941.2002.tb00903.x PMID:19709181

Sethi, S. K., Sahu J. K., Adhikary S. P. (2014). Microbial biofertilizers and their pilot-scale production. *Microbial Biotechnol. Progr. Trends,* 297.

Sethy, S. K., & Ghosh, S. (2013). Effect of heavy metals on germination of seeds. *Journal of Natural Science, Biology, and Medicine*, *4*(2), 272. doi:10.4103/0976-9668.116964 PMID:24082715

Setubal, J. C., dos Santos, P., Goldman, B. S., Ertesvåg, H., Espin, G., Rubio, L. M., Valla, S., Almeida, N. F., Balasubramanian, D., Cromes, L., Curatti, L., Du, Z., Godsy, E., Goodner, B., Hellner-Burris, K., Hernandez, J. A., Houmiel, K., Imperial, J., Kennedy, C., ... Wood, D. (2009). Genome sequence of Azotobacter vinelandii, an obligate aerobe specialized to support diverse anaerobic metabolic processes. *Journal of Bacteriology*, *191*(14), 4534–4545. doi:10.1128/JB.00504-09 PMID:19429624

Shaaban, M. (2001). Green Microalgae Water Extract as Foliar Feeding to Wheat Plants. *Pakistan Journal of Biological Sciences*, *4*(6), 628–632. doi:10.3923/pjbs.2001.628.632

Shabana, E. F., Senousy, H. H., & Khourshid, E. B. (2019). Pharmaceutical wastewater treatment using free and immobilized Cyanobacteria. *Egyptian Journal of Phycology*, *20*(1), 123–154. doi:10.21608/egyjs.2019.116025

Shafi, J., Tian, H., & Ji, M. (2017). *Bacillus* species as versatile weapons for plant pathogens: A review. *Biotechnology, Biotechnological Equipment*, *31*(3), 446–459. doi:10.1080/13102818.2017.1286950

Shaheen, S., & Sundari, K. (2013). Exploring the applicability of PGPR to remediate residual organophosphate and carbamate pesticides used in agriculture fields. *International Journal of Agriculture and Food Science Technology*, *4*(10), 947–954.

Shahid, I., Rizwan, M., Baig, D. N., Saleem, R. S., Malik, K. A., & Mehnaz, S. (2017). Secondary Metabolites Production and Plant Growth Promotion by *Pseudomonas chlororaphis* and *P. aurantiaca* Strains Isolated from Cactus, Cotton, and Para Grass. *Journal of Microbiology and Biotechnology*, *27*(3), 480–491. doi:10.4014/jmb.1601.01021 PMID:27974729

Shahid, M., Khalid, S., Abbas, G., Shahid, N., Nadeem, M., & Sabir, M. (2015). Heavy Metal Stress and Crop Productivity. In K. Hakeem (Ed.), *Crop Production and Global Environmental Issues*. Springer. doi:10.1007/978-3-319-23162-4_1

Shah, M. (2014). Effective treatment systems for azo dye degradation: A joint venture between physico – chemical and microbiological process. *International Journal of Environmental Bioremediation & Biodegradation*, *2*(5), 231–242.

Shah, U. A., Kotta-Loizou, I., Fitt, B. D., & Coutts, R. H. (2020). Mycovirus-induced hypervirulence of *Leptosphaeria biglobosa* enhances systemic acquired resistance to *Leptosphaeria maculans* in *Brassica napus*. *Molecular Plant-Microbe Interactions*, *33*(1), 98–107. doi:10.1094/MPMI-09-19-0254-R PMID:31652089

Shahzadi, A., Saddiqui, S., & Bano, A. (2016). The response of maize (*Zea mays* L.) plant assisted with bacterial consortium and fertilizer under oily sludge. *International Journal of Phytoremediation*, *18*(5), 521–526. doi:10.1080/15226514.2015.1115964 PMID:26587972

Shakeel, M., Rais, A., Hassan, M. N., & Hafeez, F. Y. (2015). Root Associated Bacillus sp. Improves Growth, Yield and Zinc Translocation for Basmati Rice (Oryza sativa) Varieties. *Frontiers in Microbiology*, *6*, 1286. doi:10.3389/fmicb.2015.01286 PMID:26635754

Shakir, E., Zahraw, Z., & Al-Obaidy, A. H. M. J. (2017). Environmental and health risks associated with reuse of wastewater for irrigation. *Egyptian Journal of Petroleum*, *26*(1), 95–102. doi:10.1016/j.ejpe.2016.01.003

Shallari, S., Schwartz, C., Hasko, A., & Morel, J. L. (1998). Heavy metals in soils and plants of serpentine and industrial sites of Albania. *The Science of the Total Environment*, *19209*, 133–142.

Shamraiz, U., Hussain, R. A., Badshah, A., Raza, B., & Saba, S. (2016). Functional metal sulfides and selenides for the removal of hazardous dyes from water. *Journal of Photochemistry and Photobiology*, *159*, 33–41. doi:10.1016/j.jphotobiol.2016.03.013 PMID:27010842

Shan, G., Xing, J., Zhang, H., & Liu, H. (2005). Bio-desulfurization of dibenzothiophene by microbial cells coated with magnetite nanoparticles. *Applied and Environmental Microbiology*, *71*(8), 4497–4502. doi:10.1128/AEM.71.8.4497-4502.2005 PMID:16085841

Shankar, S., Kansrajh, C., Dinesh, M. G., Satyan, R. S., Kiruthika, S., & Tharanipriya, A. (2014). Application of indigenous microbial consortiain bioremediation of oil-contaminated soils. *International Journal of Environmental Science and Technology*, *11*(2), 367–376. doi:10.100713762-013-0366-1

Shanker, A. K., Cervantes, C., Loza-Tavera, H., & Avudainayagam, S. (2005). Chromium toxicity in plants. *Environment International*, *31*(5), 739–753. doi:10.1016/j.envint.2005.02.003 PMID:15878200

Shanmugam, L., Ahire, M., & Nikam, T. (2020). *Bacopa monnieri (L.) Pennell*, a potential plant species for degradation of textile azo dyes. *Environmental Science and Pollution Research International*, *27*(9), 9349–9363. doi:10.100711356-019-07430-x PMID:31912399

Shannon, T. P., Ahler, S. J., Mathers, A., Ziter, C. D., & Dugan, H. A. (2020). Road salt impact on soil electrical conductivity across an urban landscape. *Journal of Urban Economics*, *6*(1), juaa006. Advance online publication. doi:10.1093/jue/juaa006

Sharma, M., Sudheer, S., Usmani, Z., Rani, R. & Gupta, P. (2020). Deciphering the omics of plant- microbe interaction: perspectives and new insights. *Curr. Gen.,* 21.

Sharma, S., Tiwari, S., Hasan, A., Saxena, V., & Pandey, L. M. (2018). Recent advances in conventional and contemporary methods for remediation of heavy metal-contaminated soils. *Biotech, 8*(4), 216.

Sharma, A., Gupta, A. K., & Ganguly, R. (2018). Impact of open dumping of municipal solid waste on soil properties in mountainous region. *Journal of Rock Mechanics and Geotechnical Engineering*, *10*(4), 725–739. doi:10.1016/j.jrmge.2017.12.009

Sharma, A., & Johri, B. N. (2003). Growth promoting influence of siderophore-producing Pseudomonas strains GRP3A and PRS9 in maize (Zea mays L.) under iron limiting conditions. *Microbiological Research*, *158*(3), 243–248. doi:10.1078/0944-5013-00197 PMID:14521234

Sharma, A., Johri, B. N., Sharma, A. K., & Glick, B. R. (2003). Plant growth-promoting bacterium Pseudomonas sp. strain GRP3 influences iron acquisition in mung bean (Vigna radiata L. Wilzeck). *Soil Biology & Biochemistry*, *35*(7), 887–894. doi:10.1016/S0038-0717(03)00119-6

Sharma, B., Dangi, A. K., & Shukla, P. (2018). Contemporary enzyme based technologies for bioremediation: A review. *Journal of Environmental Management*, *210*, 10–22. doi:10.1016/j.jenvman.2017.12.075 PMID:29329004

Sharma, J. K., & Juwarkar, A. A. (2015). Phytoremediation: General account and its application. In B. Bahadur (Ed.), *Plant Biology and Biotechnology: Plant Genomics and Biotechnology* (Vol. 2, pp. 469–497). Springer India. doi:10.1007/978-81-322-2283-5_34

Sharma, K. K., Shrivastava, B., Sastry, V. R. B., Sehgal, N., & Kuhad, R. C. (2013). Middle-redox potential laccase from *Ganoderma* sp.: Its application in improvement of feed for monogastric animals. *Scientific Reports*, *3*(1), 1299. doi:10.1038rep01299 PMID:23416696

Sharma, M., Guleria, S., Singh, K., Chauhan, A., & Kulshrestha, S. (2018). Mycovirus associated hypovirulence, a potential method for biological control of Fusarium species. *Virusdisease*, *29*(2), 134–140. doi:10.100713337-018-0438-4 PMID:29911145

Sharma, M., Mishra, V., Rau, N., & Sharma, R. S. (2016). Increased iron-stress resilience of maize through inoculation of siderophore-producing *Arthrobacter globiformis* from mine. *Journal of Basic Microbiology*, *56*(7), 719–735. doi:10.1002/jobm.201500450 PMID:26632776

Sharma, P. K., Balkwill, D. L., Frenkel, A., & Vairavamurthy, M. A. (2000). A new Klebsiella planticola strain (Cd-1) grows anaerobically at high cadmium concentrations and precipitates cadmium sulfide. *Applied and Environmental Microbiology*, *66*(7), 3083–3087. doi:10.1128/AEM.66.7.3083-3087.2000 PMID:10877810

Sharma, P., Khanna, V., & Kumari, P. (2013a). Efficacy of aminocyclopropane-1-carboxylic acid (ACC)-deaminase-producing rhizobacteria in ameliorating water stress in chickpea under axenic conditions. *African Journal of Microbiological Research*, *7*(50), 5749–5757. doi:10.5897/AJMR2013.5918

Sharma, R., Sindhu, S., & Sindhu, S. S. (2018). Bioinoculation of mustard (*Brassica juncea* L.) with beneficial rhizobacteria: A sustainable alternative to improve crop growth. *International Journal of Current Microbiology and Applied Sciences*, *7*(5), 1375–1386. doi:10.20546/ijcmas.2018.705.163

Sharma, S. B., Sayyed, R. Z., Trivedi, M. H., & Gobi, T. A. (2013). Phosphate solubilizing microbes: Sustainable approach for managing phosphorus deficiency in agricultural soils. *SpringerPlus*, *2*(1), 587. doi:10.1186/2193-1801-2-587 PMID:25674415

Sharma, S., & Adholeya, A. (2011). Detoxification and accumulation of chromium from tannery effluent and spent chrome effluent by Paecilomyces lilacinus fungi. *International Biodeterioration & Biodegradation*, *65*(2), 309–317. doi:10.1016/j.ibiod.2010.12.003

Sharma, S., Kulkarni, J., & Jha, B. (2016). Halotolerant rhizobacteria promote growth and enhance salinity tolerance in peanut. *Frontiers in Microbiology*, *7*, 1600. doi:10.3389/fmicb.2016.01600 PMID:27790198

Sharma, S., Kumar, V., & Tripathi, R. B. (2011). Isolation of phosphate solubilizing microorganism (PSMs) from soil. *Journal of Microbiology and Biotechnology Research*, *1*(2), 90–95.

Sharma, S., & Malik, P. (2012). Biopesticides: Types and applications. *International Journal of Advances in Pharmacy Biological Chemistry*, *1*(4), 2277–4688.

Sharma, S., Singh, B., & Gupta, V. K. (2014). Assessment of imidacloprid degradation bysoilisolated *Bacillus alkalinitrilicus*. *Environmental Monitoring and Assessment*, *186*(11), 7183–7193. doi:10.100710661-014-3919-y PMID:25052329

Sharma, S., Szele, Z., Schilling, R., Munch, J. C., & Schloter, M. (2006). Influence of freeze-thaw stress on the structure and function of microbial communities and denitrifying populations in soil. *Applied and Environmental Microbiology*, *72*(3), 2148–2154. doi:10.1128/AEM.72.3.2148-2154.2006 PMID:16517665

Shedbalkar, U., Dhanve, R., & Jadhav, J. (2008). Biodegradation of triphenylmethane dye cotton blue by *Penicillium ochrochloron* MTCC 517. *Journal of Hazardous Materials*, *157*(2-3), 472–479. doi:10.1016/j.jhazmat.2008.01.023 PMID:18282658

Shehzadi, M., Afzal, M., Khan, M. U., Islam, E., Mobin, A., Anwar, S., & Khan, Q. M. (2014). Enhanced degradation of textile effluent in constructed wetland system using *Typha domingensis* and textile effluent-degrading endophytic bacteria. *Water Research*, *58*, 152–159. doi:10.1016/j.watres.2014.03.064 PMID:24755300

Sheikh, L. I., Dawar, S., Zaki, M. J., & Ghaffar, A. (2006). Efficacy of *Bacillus thuringiensis* and *Rhizobium meliloti* with nursery fertilizers in the control of root infecting fungi on mung bean and okra plants. *Pakistan Journal of Botany*, *38*, 465–473.

Shekar, C. H. C., Sammaiah, D., Shasthree, T., & Reddy, K. J. (2011). Effect of mercury on tomato growth and yield attributes. *International Journal of Pharma and Bio Sciences*, *2*(2), B358–B364.

Shekher, R., Sehgal, S., Kamthania, M., & Kumar, A. (2011). Laccase: Microbial sources, production, purification, and potential biotechnological applications. *Enzyme Research*. PMID:21755038

Sheng, M. M., Jia, H. K., Zhang, G. Y., Zeng, L. N., Zhang, T. T., Long, Y. H., Lan, J., Hu, Z. Q., Zeng, Z., Wang, B., & Liu, H. M. (2020). Siderophore Production by Rhizosphere Biological Control Bacteria *Brevibacillus brevis* GZDF3 of *Pinellia ternata* and Its Antifungal Effects on *Candida albicans*. *Journal of Microbiology and Biotechnology*, *30*(5), 689–699. doi:10.4014/jmb.1910.10066 PMID:32482934

Sheng, X. F., & He, L. Y. (2006). Solubilization of potassium-bearing minerals by a wild-type strain of *Bacillus edaphicus* and its mutants and increased potassium uptake by wheat. *Canadian Journal of Microbiology*, *52*(1), 66–72. doi:10.1139/w05-117 PMID:16541160

Sheng, X. F., & Huang, W. Y. (2002). Mechanism of potassium release from feldspar affected by the strain NBT of silicate bacterium. *Turang Xuebao*, *39*, 863–871.

Sheng, X. F., Zhao, F., He, H., Qiu, G., & Chen, L. (2008). Isolation, characterization of silicate mineral solubilizing *Bacillus globisporus* Q12 from the surface of weathered feldspar. *Canadian Journal of Microbiology*, *54*(12), 1064–1068. doi:10.1139/W08-089 PMID:19096461

Shen, J., Yuan, L., Zhang, J., Li, H., Bai, Z., Chen, X., Zhang, W., & Zhang, F. (2011). Phosphorus dynamics: From soil to plant. *Plant Physiology*, *156*(3), 997–1005. doi:10.1104/pp.111.175232 PMID:21571668

Shen, X., Hu, H., Peng, H., Wang, W., & Zhang, X. (2013). Comparative genomic analysis of four representative plant growth-promoting rhizobacteria in Pseudomonas. *BMC Genomics*, *14*(1), 271. doi:10.1186/1471-2164-14-271 PMID:23607266

Shi, H., Shi, X., & Liu, K. J. (2004). Oxidative mechanism of arsenic toxicity and carcinogenesis. *Molecular and Cellular Biochemistry*, *255*(1-2), 67–78. doi:10.1023/B:MCBI.0000007262.26044.e8 PMID:14971647

Shinohara, M., Nakajima, N., & Uehara, Y. (2007). Purification and characterization of a novel esterase (beta-hydroxypalmitate methyl ester hydrolase) and prevention of the expression of virulence by Ralstonia solanacearum. *Journal of Applied Microbiology, 103*(1), 152–. doi:10.1111/j.1365-2672.2006.03222.x

Shinozaki, K., Uemura, M., Bailey-Serres, J., Bray, E. A., & Weretilnyk, E. (2015). *Biochemistry and Molecular Biology of Plants* (B. B. Buchanan, W. Gruissem, & R. L. Jones, Eds.). Wiley.

Shipley, H. J., Engates, K. E., & Guettner, A. M. (2011). Study of iron oxide nanoparticles in soil for remediation of arsenic. *Journal of Nanoparticle Research*, *13*(6), 2387–2397. doi:10.100711051-010-9999-x

Shi, R., Hong, Z., Li, J., Jiang, J., Baquy, M. A.-A., Xu, R., & Qian, W. (2017). Mechanisms for Increasing the pH Buffering Capacity of an Acidic Ultisol by Crop Residue-Derived Biochars. *Journal of Agricultural and Food Chemistry*, *65*(37), 8111–8119. doi:10.1021/acs.jafc.7b02266 PMID:28846405

Shi, W., & Ma, X. (2017). Effects of heavy metal Cd pollution on microbial activities in soil. *Annals of Agricultural and Environmental Medicine*, *24*(4), 722–725. doi:10.26444/aaem/80920 PMID:29284254

Shleev, S., Jarosz-Wilkolazka, A., Khalunina, A., Morozova, O., Yaropolov, A., Ruzgas, T., & Gorton, L. (2005). Direct electron transfer reactions of laccases from different origins on carbon electrodes. *Bioelectrochemistry (Amsterdam, Netherlands)*, *67*(1), 115–124. doi:10.1016/j.bioelechem.2005.02.004 PMID:15941673

Shleev, S., Nikitina, O., Christenson, A., Reimann, C. T., Yaropolov, A. I., Ruzgas, T., & Gorton, L. (2007). Characterization of two new multiforms of *Trametes pubescens* laccase. *Bioorganic Chemistry*, *35*(1), 35–49. doi:10.1016/j.bioorg.2006.08.001 PMID:16989887

Shore, J. (1995). Dyeing with reactive dyes. *Cellulosics Dyeing*, 189-245.

Show, P. L., Oladele, K. O., Siew, Q. Y., Aziz Zakry, F. A., Lan, J. C. W., & Ling, T. C. (2015). Overview of citric acid production from *Aspergillus niger. Frontiers in Life Science*, *8*(3), 271–283. doi:10.1080/21553769.2015.1033653

Shrestha, G., & Clair, L. L. S. (2013). Lichens: A promising source of antibiotic and anticancer drugs. *Phytochemistry Reviews*, *12*(1), 229–244. doi:10.100711101-013-9283-7

Shrestha, P., Bellitürk, K., & Görres, J. H. (2019). Phytoremediation of heavy metal-contaminated soil by switchgrass: A comparative study utilizing different composts and coir fiber on pollution remediation, plant productivity, and nutrient leaching. *International Journal of Environmental Research and Public Health*, *16*(7), 1261. doi:10.3390/ijerph16071261 PMID:30970575

Shridhar, B. S. (2012). nitrogen fixing microorganisms. *International Journal of Microbiology Research*, *3*(1), 46–52.

Shrivastava, P., & Kumar, R. (2015). Soil salinity: A serious environmental issue and plant growth promoting bacteria as one of the tools for its alleviation. *Saudi Journal of Biological Sciences*, *22*(2), 123–131. doi:10.1016/j.sjbs.2014.12.001 PMID:25737642

Shukla, S. K., Mangwani, N., Karley, D., & Rao, T. S. (2017). Bacterial biofilms and genetic regulation for metal detoxification. Handbook of Metal-Microbe Interactions and Bioremediation, 317.

Shukla, S. K., Mangwani, N., Rao, T. S., & Das, S. (2014). Biofilm-mediated bioremediation of polycyclic aromatic hydrocarbons. Microbial Biodegradation and Bioremediation, 203–232.

Shukla, K. P., Sharma, S., Singh, N. K., Singh, V., Tiwari, K., & Singh, A. S. (2011). Nature and role of root exudates: Efficacy in bioremediation. *African Journal of Biotechnology*, *10*, 9717–9724.

Siciliano, S. D., & Germida, J. J. (1998). Mechanisms of phytoremediation: Biochemical and ecological interactions between plants and bacteria. *Environmental Reviews*, *6*(1), 65–79. doi:10.1139/a98-005

Siciliano, S. D., Germida, J. J., Banks, K., & Greer, C. W. (2003). Changes in microbial community composition and function during a polyaromatic hydrocarbon phytoremediation field trial. *Applied and Environmental Microbiology*, *69*(1), 483–489. doi:10.1128/AEM.69.1.483-489.2003 PMID:12514031

Siddiqui, Z. A. (2005). PGPR: Prospective Biocontrol Agents of Plant Pathogens. In Z. A. Siddiqui (Ed.), *PGPR: Biocontrol and Biofertilization*. Springer. doi:10.1007/1-4020-4152-7_4

Siddiqui, Z. A., Iqbal, A., & Mahmood, I. (2001). Efects of Pseudomonas fuorescens and fertilizers on the reproduction of Meloidogyne incognita and growth of tomato. *Applied Soil Ecology*, *16*(2), 179–185. doi:10.1016/S0929-1393(00)00083-4

Sidhu, G. P. S. (2016). Heavy metal toxicity in soils: Sources, remediation technologies and challenges. *Advances in Plants & Agriculture Research*, *5*(1), 445–446.

Si, J., Peng, F., & Cui, B. (2013). Purification, biochemical characterization and dye decolorization capacity of an alkali-resistant and metal-tolerant laccase from Trametes pubescens. *Bioresource Technology*, *128*, 49–57. doi:10.1016/j.biortech.2012.10.085 PMID:23196221

Silo-Suh, L. A., Lethbridge, B. J., Raffel, S. J., He, H., Clardy, J., & Handelsman, J. (1994). Biological activities of two fungistatic antibiotics produced by *Bacillus cereus* UW85. *Applied and Environmental Microbiology*, *60*(6), 2023–2030. doi:10.1128/AEM.60.6.2023-2030.1994 PMID:8031096

Silva-Bailão, M. G., Bailão, E. F. L. C., Lechner, B. E., Gauthier, G. M., Lindner, H., Bailão, A. M., Haas, H., & de Almeida Soares, C. M. (2014). Hydroxamate production as a high affinity iron acquisition mechanism in *Paracoccidioides* spp. *PLoS One*, *9*(8), e105805. doi:10.1371/journal.pone.0105805 PMID:25157575

Silva, H. S. A., Romeiro, R., Macagnan, D., Halfeld-Vieira, B. A., Pereira, M. C. B., & Mounteer, A. (2004). Rhizobacterial induction of systemic resistance in tomato plants: Non-specific protection and increase in enzyme activities. *Biological Control*, *29*(2), 288–295. doi:10.1016/S1049-9644(03)00163-4

Silva, V. P., Moreira-Santos, M., Mateus, C., Teixeira, T., Ribeiro, R., & Viegas, C. (2015). Evaluation of *Arthrobacter aurescens* strain TC1 as bioaugmentation bacterium in soils contaminated with the herbicidal substance terbuthylazine. *PLoS One*, *10*(12), e0144978. doi:10.1371/journal.pone.0144978 PMID:26662024

Silveira, E., Marques, P. P., Silva, S. S., Lima-Filho, J. L., Porto, A. L. F., & Tambourgi, E. B. (2009). Selection of Pseudomonas for industrial textile dyes decolourization. *International Biodeterioration & Biodegradation*, *63*(2), 230–235. doi:10.1016/j.ibiod.2008.09.007

Simarro, R., González, N., Bautista, L., & Molina, M. (2013). Assessment of the efficiency of in situ bioremediation techniques in a creosote polluted soil: Change in bacterial community. *Journal of Hazardous Materials*, *262*, 158–167. doi:10.1016/j.jhazmat.2013.08.025 PMID:24025312

Simon, C., & Daniel, R. (2011). Metagenomic analyses: Past and future trends. *Applied and Environmental Microbiology*, *77*(4), 1153–1161. doi:10.1128/AEM.02345-10 PMID:21169428

Simon, J. C., Marchesi, J. R., Mougel, C., & Selosse, M. A. (2019). Host-microbiota interactions: From holobiont theory to analysis. *Microbiome*, *7*(1), 1–5. doi:10.118640168-019-0619-4 PMID:30635058

Sims, G. K., & Cupples, A. M. (1999). Factors controlling degradation of pesticides in soil. *Pesticide Science*, *55*(5), 598–601. doi:10.1002/(SICI)1096-9063(199905)55:5<598::AID-PS962>3.0.CO;2-N

Sineli, P. E., Tortella, G., Costa, J. D., Benimeli, C. S., & Cuozzo, S. A. (2016). Evidence of α-β-and γ-HCH mixture aerobic degradation by the native actinobacteria *Streptomyces* sp. M7. *World Journal of Microbiology & Biotechnology*, *32*(5), 81. doi:10.100711274-016-2037-0 PMID:27038951

Singer, A. C., Crowley, D. E., & Thompson, I. P. (2003). Secondary plant metabolites in phytoremediation and biotransformation. *Trends in Biotechnology*, *21*(3), 123–130. doi:10.1016/S0167-7799(02)00041-0 PMID:12628369

Singh, L., & Singh, V. P. (2015). Textile dyes degradation: A microbial approach for biodegradation of pollutants. In Microbial degradation of synthetic dyes in wastewaters (pp. 187-204). Springer. doi:10.1007/978-3-319-10942-8_9

Singh, M., Kumar, A., Singh, R., & Pandey, K. D. (2017). Endophytic bacteria: a new source of bioactive compounds. *Biotech, 7*(5), 315. doi:10.100713205-017-0942-z

Singh, P. P., Kujur, A., Yadav, A., Kumar, A., Singh, S. K., & Prakash, B. (2019). Mechanisms of Plant-Microbe Interactions and its Significance for Sustainable Agriculture. In PGPR Amelioration in Sustainable Agriculture. doi:10.1016/B978-0-12-815879-1.00002-1

Singh, R., Kumar, M., Mittal, A., & Mehta, P. K. (2016). Microbial enzymes: industrial progress in 21st century. *Biotech, 6*(2), 174.

Singh, V.K., Singh, A.K. & Kumar, A. (2017c). Disease management of tomato through PGPB: current trends and future perspective. *Biotech, 7*, 255. doi:10.100713205-017-0896-1

Singh, A., & Gauba, P. (2014). Mycoremediation: A treatment for heavy metal pollution of soil. *Journal of Civil and Engineeringand Environmental Technology, 1*, 59–61.

Singh, A., Jain, A., Sarma, B. K., Upadhyay, R. S., & Singh, H. B. (2014). Rhizosphere competent microbial consortium mediates rapid changes in phenolic profiles in chickpea during *Sclerotium rolfsii* infection. *Microbiological Research, 169*(5-6), 353–360. doi:10.1016/j.micres.2013.09.014 PMID:24168925

Singh, B. (2018). Are Nitrogen Fertilizers Deleterious to Soil Health? *Agronomy (Basel), 8*(4), 48–67. doi:10.3390/agronomy8040048

Singh, B. K., Quince, C., Macdonald, C. A., Khachane, A., Thomas, N., AlSoud, W. A., Sørensen, S. J., He, Z., White, D., Sinclair, A., Crooks, B., Zhou, J., & Campbell, C. D. (2014). Loss of microbial diversity in soils is coincident with reductions in some specialized functions. *Environmental Microbiology, 16*(8), 2408–2420. doi:10.1111/1462-2920.12353 PMID:24422656

Singh, B. K., Walker, A., Morgan, J. A. W., & Wright, D. J. (2004). Biodegradation of chlorpyrifos by *Enterobacter* strain B-14 and its use in bioremediation of contaminated soils. *Applied and Environmental Microbiology, 70*(8), 4855–4863. doi:10.1128/AEM.70.8.4855-4863.2004 PMID:15294824

Singh, B. K., Walker, A., & Wright, D. J. (2006). Bioremedial potential of fenamiphos and chlorpyrifos degrading isolates: Influence of different environmental conditions. *Soil Biology & Biochemistry, 38*(9), 2682–2693. doi:10.1016/j.soilbio.2006.04.019

Singh, B., & Satyanarayana, T. (2011). Microbial phytases in phosphorus acquisition and plant growth promotion. *Physiology and Molecular Biology of Plants, 17*(2), 93–103. doi:10.100712298-011-0062-x PMID:23572999

Singh, G., Bhalla, A., Capalash, N., & Sharma, P. (2010). Characterization of immobilized laccase from γ-proteobacterium JB: Approach towards the development of biosensor for the detection of phenolic compounds. *Indian Journal of Science and Technology, 2*(1), 48–53. doi:10.17485/ijst/2010/v3i1.8

Singh, G., Biswas, D. R., & Marwaha, T. S. (2010). Mobilization of potassium from waste mica by plant growth promoting rhizobacteria and its assimilation by maize (*Zea mays*) and wheat (*Triticum aestivum* L.): A hydroponics study under phytotron growth chamber. *Journal of Plant Nutrition, 33*(8), 1236–1251. doi:10.1080/01904161003765760

Singh, J. S., Pandey, V. C., & Singh, D. P. (2011). Efficient soil microorganisms: A new dimension for sustainable agriculture and environmental development. *Agriculture, Ecosystems & Environment, 140*(3-4), 339–353. doi:10.1016/j.agee.2011.01.017

Singh, L. (2003). *Microbial degradation of hazardous dyes: A fungal approach.* University of Delhi.

Singh, L., & Singh, V. P. (2010). Biodegradation of textiles dyes, Bromophenol blue and congo red by fungus-*Aspergillus flavus. Environment & We: An International Journal of Science and Technology, 5*(4), 235–242.

Singh, L., & Singh, V. P. (2010). Microbial degradation and decolourization of dyes in semi-solid medium by the fungus-*Trichoderma harzianum*. Environment &*We. International Journal of Science and Technology, 5*(3), 147–153.

Singh, L., & Singh, V. P. (2017). Decolourization of azo (acid red) and anthraquinonic (basic blue) dyes by the fungus *Aspergillus flavus. International Journal of Biomedical Engineering and Clinical Science*, *3*(1), 1. doi:10.11648/j.ijbecs.20160301.11

Singh, M. J., & Sedhuraman, P. (2015). Biosurfactant, polythene, plastic, and diesel biodegradation activity of endophytic *Nocardiopsis sp.* mrinalini9 isolated from *Hibiscus rosasinensis* leaves. *Bioresources and Bioprocessing*, *2*(1), 2. doi:10.118640643-014-0034-4

Singh, P., & Cameotra, S. S. (2004). Enhancement of metal bioremediation by use of microbial surfactants. *Biochemical and Biophysical Research Communications*, *319*(2), 291–297.

Singh, P., Jain, R., Srivastava, N., Borthakur, A., Pal, D. B., Singh, R., & Mishra, P. K. (2017). Current and emerging trends in bioremediation of petrochemical waste: A review. *Critical Reviews in Environmental Science and Technology*, *47*(3), 155–201. doi:10.1080/10643389.2017.1318616

Singh, R. (2013). Exploring Flavin as Catalyst for the Remediation of Halogenated Compounds. In *New and Future Developments in Catalysis*. doi:10.1016/B978-0-444-53870-3.00015-0

Singh, R. (2014). Microorganism as a tool of bioremediation technology for cleaning environment: A review. *Proceedings of the International Academy of Ecology and Environmental Sciences*, *4*(1), 1.

Singh, R. L., Singh, P. K., & Singh, R. P. (2015). Enzymatic decolorization and degradation of azo dyes – A review. *International Biodeterioration & Biodegradation*, *104*, 21–31. doi:10.1016/j.ibiod.2015.04.027

Singh, R. P., & Jha, P. N. (2016). The multifarious PGPR *Serratia marcescens* CDP-13 augments induced systemic resistance and enhanced salinity tolerance of wheat (Triticum aestivum L.). *PLoS One*, *11*(6), e0155026. doi:10.1371/journal.pone.0155026 PMID:27322827

Singh, R. P., Singh, P. K., & Singh, R. L. (2014). Bacterial decolorization of textile azo dye acid orange by *Staphylococcus hominis* RMLRT03. *Toxicology International*, *21*(2), 160–166. doi:10.4103/0971-6580.139797 PMID:25253925

Singh, R. R., & Prasad, K. (2011). Effect of bio-fertilizers on growth and productivity of wheat (*Triticum aestivum*). *International Journal of Farm Sciences*, *1*(1), 1–8.

Singh, R., Behera, M., & Kumar, S. (2020). Nano-bioremediation: An innovative remediation technology for treatment and management of contaminated sites. In *Bioremediation of Industrial Waste for Environmental Safety* (pp. 165–182). Springer. doi:10.1007/978-981-13-3426-9_7

Singh, R., Manickam, N., Mudiam, M. K. R., Murthy, R. C., & Misra, V. (2013). An integrated (nano-bio)technique for degradation of γ-HCH contaminated soil. *Journal of Hazardous Materials*, *258*, 35–41. doi:10.1016/j.jhazmat.2013.04.016 PMID:23692681

Singh, R., Pandey, D. K., Kumar, A., & Singh, M. (2017b). PGPR isolates from the rhizosphere of vegetable crop *Momordica charantia*: Characterization and application as biofertilizer. *International Journal of Current Microbiology and Applied Sciences*, *6*(3), 1789–1802. doi:10.20546/ijcmas.2017.603.205

Singh, S. (2014). A review on possible elicitor molecules of cyanobacteria: Their role in improving plant growth and providing tolerance against biotic or abiotic stress. *Journal of Applied Microbiology*, *117*(5), 1221–1244. doi:10.1111/jam.12612 PMID:25069397

Singh, S. P., Shrivastava, S. K., Kolhe, S. S., Patel, J. R., & Bargali, S. S. (2007). Prospects of biofertilizers and organic manure utilization: A case study in Durg district. *Agricultural Science Digest*, *27*(3), 157–161.

Singh, S., Adak, A., Saritha, M., Sharma, S., Tiwari, R., Rana, S., & Nain, L. (2017). Bioethanol production scenario in India: Potential and policy perspective. In *Sustainable Biofuels Development in India* (pp. 21–37). Springer. doi:10.1007/978-3-319-50219-9_2

Singh, T., & Singh, D. K. (2019). Rhizospheric Microbacterium sp. P27 showing potential of lindane degradation and plant growth promoting traits. *Current Microbiology*, *76*(7), 888–895. doi:10.100700284-019-01703-x PMID:31093691

Singh, T., & Singh, D. K. (2019a). Lindane degradation by root epiphytic bacterium *Achromobacter* sp. strain A3 from Acorus calamus and characterization of associated proteins. *International Journal of Phytoremediation*, *21*(5), 419–424. doi:10.1080/15226514.2018.1524835 PMID:30648424

Singh, U. V., Abhishek, A., Bhaskar, M., Tandan, N., Ansari, N. G., & Singh, N. P. (2015). Phyto-extraction of heavy metals and biochemical changes with Brassica nigra L. grown in rayon grade paper mill effluent irrigated soil. *Bioinformation*, *11*(3), 138–144. doi:10.6026/97320630011138 PMID:25914448

Singh, V. K., Singh, A. K., Singh, P. P., & Kumar, A. (2018). Interaction of plant growth promoting bacteria with tomato under abiotic stress: A review. *Agriculture, Ecosystems & Environment*, *267*, 129–140. doi:10.1016/j.agee.2018.08.020

Singh, V. P., Singh, L., Singh, I., & Kumar, R. (2006). Microbial Degradation of Hazardous Dyes. In *Current Concepts in Botany* (pp. 273–285). IK International Publishing House Pvt Ltd.

Singh, V., & Sood, A. K. (2016). Plant Nutrition: A tool for the management of hemipteran insect-pests-A review. *Agricultural Reviews (Karnal)*, *38*(04), 260–270. doi:10.18805/ag.R-1637

Sivasakthi, S., Usharani, G., & Saranraj, P. (2014). Biocontrol potentiality of plant growth promoting bacteria (PGPR)-Pseudomonas fluorescens and Bacillus subtilis: A review. *African Journal of Agricultural Research*, *9*(16), 1265–1277.

Slattery, B., & Hollier, C. (2002). *The impact of acid soils in Victoria*. Agriculture Victoria.

Slotton, D. G., Goldman, C. R., & Frank, A. (1989). Commercially Grown *Spirulina* Found to Contain Low Levels of Mercury and Lead. *Nutrition Reports International*, *40*(2), 1165–1172.

Smith, L. A., Means, J. L., Chen, A., Alleman, B., Chapma, C. C., Tixier, J. R., Brauning, S. E., Gavaskar, A. R., & Royer, M. D. (1995). Remedial options for metal contaminated sites. Academic Press.

Smith, E., Thavamani, P., Ramadass, K., Naidu, R., Srivastava, P., & Megharaj, M. (2015). Remediation trials for hydrocarbon-contaminated soils in arid environments: Evaluation of bioslurry and biopiling techniques. *International Biodeterioration & Biodegradation*, *101*, 56–65. doi:10.1016/j.ibiod.2015.03.029

Smith, K. M., Bu, Y., & Suga, H. (2003). Induction and inhibition of Pseudomonas aeruginosa quorum sensing by synthetic autoinducer analogs. *Chemistry & Biology*, *10*(1), 81–89. doi:10.1016/S1074-5521(03)00002-4 PMID:12573701

Smith, K., & Novick, R. P. (1972). Genetic studies on plasmid-linked cadmium resistance in *Staphylococcus aureus*. *Journal of Bacteriology*, *112*(2), 761–772.

Smith, M. J., Flowers, T. H., Duncan, H. J., & Alder, J. (2006). Effects of polycyclic aromatic hydrocarbons on germination and subsequent growth of grasses and legumes in freshly contaminated soil and soil with aged PAHs residues. *Environmental Pollution*, *141*(3), 519–525. doi:10.1016/j.envpol.2005.08.061 PMID:16246476

Smith, P., Lutfalla, S., Riley, W. J., Torn, M. S., Schmidt, M. W. I., & Soussana, J. F. (2017). The changing faces of soil organic matter research. *European Journal of Soil Science*, *69*(1), 23–30. doi:10.1111/ejss.12500

Smith, S. E., & Gianinazzi-Pearson, V. (1988). Physiological interactions between symbionts in vesicular-arbuscular mycorrhizal plants. *Annual Review of Plant Physiology and Plant Molecular Biology*, *39*(1), 221–244. doi:10.1146/annurev.pp.39.060188.001253

Smith, S. E., & Read, D. J. (1997). *Mycorrhizal symbiosis*. Academic Press.

Smith, S. E., & Read, D. J. (2008). *Mycorrhizal Symbiosis* (3rd ed.). Academic Press.

Smith, S. E., Smith, F. A., & Jakobsen, I. (2003). Mycorrhizal fungi can dominate phosphate supply to plants irrespective of growth responses. *Plant Physiology*, *133*(1), 16–20. doi:10.1104/pp.103.024380 PMID:12970469

Smits, T. H., Witholt, B., & van Beilen, J. B. (2003). Functional characterization of genes involved in alkane oxidation by *Pseudomonas aeruginosa*. *Antonie van Leeuwenhoek*, *84*(3), 193–200. doi:10.1023/A:1026000622765 PMID:14574114

Smolander, A., & Sarsa, M. L. (1990). Frankia strains of soil under Betula pendula: Behaviour in soil and in pure culture. *Plant and Soil*, *122*(1), 129–136. doi:10.1007/BF02851920

Sohn, K., Kang, S. W., Ahn, S., Woo, M., & Yang, S. K. (2006). Fe(0) nanoparticles for nitrate reduction: Stability, reactivity, and transformation. *Environmental Science & Technology*, *40*(17), 5514–5519. doi:10.1021/es0525758 PMID:16999133

Soleimani, M., Amini, N., Sadeghian, B., Wang, D., & Fang, L. (2018). Heavy metals and their source identification in particulate matter (PM 2.5) in Isfahan City, Iran. *Journal of Environmental Sciences (China)*, 72.

Solly, E. F., Schöning, I., Boch, S., Kandeler, E., Marhan, S., Michalzik, B., Müller, J., Zscheischler, J., Trumbore, S. E., & Schrumpf, M. (2014). Factors controlling decomposition rates of fine root litter in temperate forests and grasslands. *Plant and Soil*, *382*(1-2), 203–218. doi:10.100711104-014-2151-4

Someya, N., Kataoka, N., Komagata, T., Hirayae, K., Hibi, T., & Akutsu, K. (2000). Biological Control of Cyclamen Soilborne Diseases by *Serratia marcescens* Strain B2. *Plant Disease*, *84*(3), 334–340. doi:10.1094/PDIS.2000.84.3.334 PMID:30841252

Somtrakoona, K., & Chouychai, W. (2013). Phytotoxicity of single and combined polycyclic aromatic hydrocarbons toward economic crops. *Russian Journal of Plant Physiology: a Comprehensive Russian Journal on Modern Phytophysiology*, *60*(1), 139–148. doi:10.1134/S1021443712060155

Sondhi, S., Sharma, P., Saini, S., Puri, N., & Gupta, N. (2014). Purification and characterization of an extracellular, thermo-alkali-stable, metal tolerant laccase from Bacillus tequilensis SN4. *PLoS One*, *9*(5), e96951. doi:10.1371/journal.pone.0096951 PMID:24871763

Song, B., Zeng, G., Gong, J., Liang, J., Xu, P., Liu, Z., Zhang, Y., Zhang, C., Cheng, M., Liu, Y., Ye, S., Yi, H., & Ren, X. (2017). Evaluation methods for assessing effectiveness of in situ remediation of soil and sediment contaminated with organic pollutants and heavy metals. *Environment International*, *105*, 43–55. doi:10.1016/j.envint.2017.05.001 PMID:28500873

Song, H., & Carraway, E. R. (2005). Reduction of chlorinated ethanes by nanosized zerovalent iron: Kinetics, pathways, and effects of reaction conditions. *Environmental Science & Technology*, *39*(16), 6237–6245. doi:10.1021/es048262e PMID:16173587

Song, Z., Song, L., Shao, Y., & Tan, L. (2018). Degradation and detoxification of azo dyes by a salt-tolerant yeast Cyberlindnera samutprakarnensis S4 under high-salt conditions. *World Journal of Microbiology & Biotechnology*, *34*(9), 131. doi:10.100711274-018-2515-7 PMID:30105649

Son, M., Yu, J., & Kim, K. H. (2015). Five Questions about Mycoviruses. *PLoS Pathogens*, *11*(11), e1005172. doi:10.1371/journal.ppat.1005172 PMID:26539725

Sorkau, E., Boch, S., Boeddinghaus, R. S., Bonkowski, M., Fischer, M., Kandeler, E., & Oelmann, Y. (2017). The role of soil chemical properties, land use and plant diversity for microbial phosphorus in forest and grassland soils. *Journal of Plant Nutrition and Soil Science*, *181*(2), 185–197. doi:10.1002/jpln.201700082

Sottorff, I., Wiese, J., Lipfert, M., Preußke, N., Sönnichsen, F. D., & Imhoff, J. F. (2019). Different secondary metabolite profiles of phylogenetically almost identical *Streptomyces griseus* strains originating from geographically remote locations. *Microorganisms*, *7*(6), 166. doi:10.3390/microorganisms7060166 PMID:31174336

Soumare, A., Diedhiou, A. G., Thuita, M., Hafidi, M., Ouhdouch, Y., Gopalakrishnan, S., & Kouisni, L. (2020). Exploiting Biological Nitrogen Fixation: A Route Towards a Sustainable Agriculture. *Plants*, *9*(8), 10–11. doi:10.3390/plants9081011 PMID:32796519

Souza, E. C., Vessoni-Penna, T. C., & de Souza Oliveira, R. P. (2014). Biosurfactant-enhanced hydrocarbon bioremediation: An overview. *International Biodeterioration & Biodegradation*, *89*, 88–94. doi:10.1016/j.ibiod.2014.01.007

Souza, R. D., Ambrosini, A., & Passaglia, L. M. (2015). Plant growth-promoting bacteria as inoculants in agricultural soils. *Genetics and Molecular Biology*, *38*(4), 401–419. doi:10.1590/S1415-475738420150053 PMID:26537605

Spaepen, S., Bossuyt, S., Engelen, K., Marchal, K., & Vanderleyden, J. (2014). Phenotypical and molecular responses of *Arabidopsis thaliana* roots as a result of inoculation with the auxin-producing bacterium *Azospirillum brasilense. The New Phytologist*, *201*(3), 850–861. doi:10.1111/nph.12590 PMID:24219779

Spaepen, S., & Vanderleyden, J. (2011). Auxin and plant-microbe interactions. *Cold Spring Harbor Perspectives in Biology*, *3*(4), a001438. doi:10.1101/cshperspect.a001438 PMID:21084388

Spaepen, S., Vanderleyden, J., & Remans, R. (2007). Indole-3-acetic acid in microbial and microorganism-plant signaling. *FEMS Microbiology Reviews*, *31*(4), 425–448. doi:10.1111/j.1574-6976.2007.00072.x PMID:17509086

Spaepen, S., Versées, W., Gocke, D., Pohl, M., Steyaert, J., & Vanderleyden, J. (2007). Characterization of phenylpyruvate decarboxylase, involved in auxin production of *Azospirillum brasilense. Journal of Bacteriology*, *189*(21), 7626–7633. doi:10.1128/JB.00830-07 PMID:17766418

Spaink, H. P., Kondorosi, A., & Hooykaas, P. J. J. (Eds.). (1998). *The Rhizobiaceae*. Kluwer Academic Publishers. doi:10.1007/978-94-011-5060-6

Sparks, D. L. (2000). Bioavailability of soil potassium. In M. E. Sumner (Ed.), *Handbook of soil science*. CRC Press.

Speight, J. G. (2006). *The Chemistry and Technology of Petroleum*. CRC Press. doi:10.1201/9781420008388

Spolaore, P., Joannis-Cassan, C., Duran, E., & Isambert, A. (2006). Optimization of *Nannochloropsis oculata* growth using the response surface method. *Journal of Chemical Technology & Biotechnology: International Research in Process, Environmental & Clean Technology*, *81*(6), 1049–1056.

Srikantan, C., Suraishkumar, G. K., & Srivastava, S. (2018). Effect of light on the kinetics and equilibrium of the textile dye (Reactive Red 120) adsorption by *Helianthus annuus* hairy roots. *Bioresource Technology*, *257*, 84–91. doi:10.1016/j.biortech.2018.02.075 PMID:29486410

Srinivasan, A., & Viraraghavan, T. (2010). Decolorization of dye wastewaters by biosorbents: A review. *Journal of Environmental Management*, *91*(10), 1915–1929. doi:10.1016/j.jenvman.2010.05.003 PMID:20627542

Srivastava, N. (2020). Phytoremediation of Toxic Metals/Metalloids and Pollutants by *Brassicaceae* Plants. In *The Plant Family Brassicaceae* (pp. 409–435). Springer. doi:10.1007/978-981-15-6345-4_14

Srivastava, N. K., Jha, M. K., Mall, I. D., & Singh, D. (2010). Application of Genetic Engineering for Chromium Removal from Industrial Wastewater. *International Journal of Chemical and Biological Engineering*, *3*, 3.

Srivastava, N. K., & Majumder, C. B. (2008). Novel biofiltration methods for the treatment of heavy metals from industrial waste water. *Journal of Hazardous Materials*, *151*, 18.

Srivastava, S., & Thakur, I. S. (2006). Isolation and process parameter optimization of *Aspergillus* sp. for removal of chromium from tannery effluent. *Bioresource Technology*, *97*(10), 1167–1173.

Srivastava, V., de Araujo, A. S. F., Vaish, B., Bartelt-Hunt, S., Singh, P., & Singh, R. P. (2016). Biological response of using municipal solid waste compost in agriculture as fertilizer supplement. *Reviews in Environmental Science and Biotechnology*, *15*(4), 677–696. doi:10.100711157-016-9407-9 PMID:32214923

Sruthy, S., & Ramasamy, E. V. (2017). *Microplastic pollution in Vembanad Lake, Kerala, India: The first report of microplastics in lake and estuarine sediments in India*. Academic Press.

Sruthy, S., & Ramasamy, E. V. (2016). Microplastic pollution in Vembanad Lake, Kerala, India: The first report of microplastics in lake and estuarine sediments in India. *Environmental Pollution*, 1–8. PMID:28041839

Stamenković, S., Beškoski, V., Karabegović, I., Lazić, M., & Nikolić, N. (2018). Microbial fertilizers: A comprehensive review of current findings and future perspectives. *Spanish Journal of Agricultural Research*, *16*(1), 1–18. doi:10.5424jar/2018161-12117

Stamenković-Stojanović, S., Karabegović, I., Beškoski, V., Nikolić, N., & Lazić, M. (2019). Bacillus based microbial formulations: Optimization of the production process. *Hemijska Industrija*, *73*(3), 169–182. doi:10.2298/HEMIND190214014S

Stamford, N. P., Ortega, A. D., Temprano, F., & Santos, D. R. (1997). Effects of phosphorus fertilization and inoculation of Bradyrhizobium and mycorrhizal fungi on growth of Mimosacaesalpiniaefolia in an acid soil. *Soil Biology & Biochemistry*, *29*(5-6), 959–964. doi:10.1016/S0038-0717(96)00240-4

Steffan, J. J., Brevik, E. C., Burgess, L. C., & Cerdà, A. (2017). The effect of soil on human health: An overview. *European Journal of Soil Science*, *69*(1), 159–171. doi:10.1111/ejss.12451 PMID:29430209

Stein, T., Hayen-Schneg, N., & Fendrik, I. (1997). Contribution of BNF by Azoarcus sp. BH72 in Sorghum vulgare. *Soil Biology & Biochemistry*, *29*(5-6), 969–971. doi:10.1016/S0038-0717(96)00211-8

Stenehjem, J. S., Robsahm, T. E., Bråtveit, M., Samuelsen, S. O., Kirkeleit, J., & Grimsrud, T. K. (2017). Aromatic hydrocarbons and risk of skin cancer by anatomical site in 25 000 male offshore petroleum workers. *American Journal of Industrial Medicine*, *60*(8), 679–688. doi:10.1002/ajim.22741 PMID:28692192

Stern, B. R. (2010). Essentiality and toxicity in copper health risk assessment: Overview, update and regulatory considerations. *Toxicology Environment and Health*, *73*(2), 114–127.

Stewart, Z. P., Pierzynski, G. M., Middendorf, B. J., & Vara Prasad, P. V. (2020). Approaches to improve soil fertility in sub-Saharan Africa. *Journal of Experimental Botany*, *71*(2), 632–641. doi:10.1093/jxb/erz446 PMID:31586430

Stoebner, J. A., Butterton, J. R., Calderwood, S. B., & Payne, S. M. (1992). Identification of the vibriobactin receptor of *Vibrio cholerae*. *Journal of Bacteriology*, *174*(10), 3270–3274. doi:10.1128/JB.174.10.3270-3274.1992 PMID:1315733

Stolz, A. (2001). Basic and applied aspects in the microbial degradation of azo dyes. *Applied Microbiology and Biotechnology*, *56*(1-2), 69–80. doi:10.1007002530100686 PMID:11499949

Straube, W. L., Nestler, C. C., Hansen, L. D., Ringleberg, D., Pritchard, P. H., & Jones-Meehan, J. (2003). Remediation of polyaromatic hydrocarbons (PAHs) through landfarming with biostimulation and bioaugmentation. *Acta Biotechnologica*, *23*(23), 179–196. doi:10.1002/abio.200390025

Streche, C., Cocârţă, D. M., Istrate, I., & Badea, A. A. (2018). Decontamination of petroleum-contaminated soils using the electrochemical technique: Remediation degree and energy consumption. *Scientific Reports*, *8*(1), 3272. doi:10.103841598-018-21606-4 PMID:29459642

Strobel, G., & Daisy, B. (2003). Bioprospecting for microbial endophytes and their natural products. *Microbiology and Molecular Biology Reviews*, *67*(4), 491–502. doi:10.1128/MMBR.67.4.491-502.2003 PMID:14665674

Strong, P. J., & Burgess, J. E. (2008). Treatment methods for wine-related ad distillery wastewaters: A review. *Bioremediation Journal*, *12*(2), 70–87. doi:10.1080/10889860802060063

Stroo, H. F., Leeson, A., & Ward, C. H. (Eds.). (2012). *Bioaugmentation for groundwater remediation* (Vol. 5). Springer Science & Business Media.

Stroud, J., Paton, G., & Semple, K. T. (2007). Microbe-aliphatic hydrocarbon interactions in soil: Implications for biodegradation and bioremediation. *Journal of Applied Microbiology*, *102*(5), 1239–1253. doi:10.1111/j.1365-2672.2007.03401.x PMID:17448159

Stucki, G., Krebser, U., & Leisinger, T. (1983). Bacterial growth on 1, 2-dichloroethane. *Experientia*, *39*(11), 1271–1273. doi:10.1007/BF01990365 PMID:6641901

Stucki, G., Thüer, M., & Bentz, R. (1992). Biological degradation of 1, 2-dichloroethane under groundwater conditions. *Water Research*, *26*(3), 273–278. doi:10.1016/0043-1354(92)90023-W

Stumm, W., & Morgan, J. J. (1996). *Aquatic Chemistry: Chemical Equilibria and Rates in Natural Waters* (3rd ed.). JohnWiley & Sons, Inc.

Štursová, M., Žifčáková, L., Leigh, M. B., Burgess, R., & Baldrian, P. (2012). Cellulose utilization in forest litter and soil: Identification of bacterial and fungal decomposers. *FEMS Microbiology Ecology*, *80*(3), 735–746. doi:10.1111/j.1574-6941.2012.01343.x PMID:22379979

Sturz, A. V., & Nowak, J. (2000). Endophytic communities of rhizobacteria and the strategies required to create yield enhancing associations with crops. *Applied Soil Ecology*, *15*(2), 183–190. doi:10.1016/S0929-1393(00)00094-9

Suárez, R., Wong, A., Ramírez, M., Barraza, A., Orozco, M. D. C., Cevallos, M. A., Lara, M., Hernández, G., & Iturriaga, G. (2008). Improvement of drought tolerance and grain yield in common bean by overexpressing trehalose-6-phosphate synthase in rhizobia. *Molecular Plant-Microbe Interactions*, *21*(7), 958–966. doi:10.1094/MPMI-21-7-0958 PMID:18533836

Subashchandrabose, S. R., Ramakrishnan, B., Megharaj, M., Venkateswarlu, K., & Naidu, R. (2013). Mixotrophic Cyanobacteria and Microalgae as Distinctive Biological Agents for Organic Pollutant Degradation. *Environment International*, *51*, 59–72. doi:10.1016/j.envint.2012.10.007 PMID:23201778

Subba Rao, N. S. (2017). *Bio-fertilizers in Agriculture and Forestry* (4th ed.). Oxford & IBH Publishing Co. Pvt. Ltd.

Subhashini, D., & Kaushik, B. D. (1981). Amelioration of sodic soils with blue green algae. *Australian Journal of Soil Research*, *19*(3), 361–367. doi:10.1071/SR9810361

Subramaniyam, V., Subashchandrabose, S. R., Thavamani, P., Chen, Z., Krishnamurti, G. S. R., Naidu, R., & Megharaj, M. (2016). Toxicity and bioaccumulation of iron in soil microalgae. *Journal of Applied Phycology*, *28*(5), 2767–2776. doi:10.100710811-016-0837-0

Su, C., LiQin, J., Zhang, W. (2014). A review on heavy metal contamination in the soil worldwide: Situation, impact and remediation techniques. *Environmental Skeptics and Critics*, *3*(2), 24–38.

Sudha, V., Govindaraj, R., Baskar, K., Al-Dhabi, N. A., & Duraipandiyan, V. (2016). Biological properties of endophytic fungi. *Brazilian Archives of Biology and Technology*, *59*(0), e16150436. doi:10.1590/1678-4324-2016150436

Suganya, T., Varman, M., Masjuki, H. H., & Renganathan, S. (2016). Macroalgae and microalgae as a potential source for commercial applications along with biofuels production: A biorefinery approach. *Renewable & Sustainable Energy Reviews*, *55*, 909–941. doi:10.1016/j.rser.2015.11.026

Sugawara, M., Okazaki, S., Nukui, N., Ezura, H., Mitsui, H., & Minamisawa, K. (2006). Rhizobitoxine modulates plant–microbe interactions by ethylene inhibition. *Biotechnology Advances*, *24*(4), 382–388. doi:10.1016/j.biotechadv.2006.01.004 PMID:16516430

Sugumaran, P., & Janarthanam, B. (2007). Solubilization of potassium containing minerals by bacteria and their effect on plant growth. *World Journal of Agricultural Sciences*, *3*, 350–355.

Sulieman, S., & Tran, L. P. (2016). *Legume Nitrogen Fixation in a Changing Environment- Achievements and Challenges*. Springer Nature Switzerland AG.

Suman, G., Nupur, M., Anuradha, S., & Pradeep, B. (2015). Single cell protein production: A review. *International Journal of Current Microbiology and Applied Sciences*, *4*(9), 251–262.

Suman, J., Uhlik, O., Viktorova, J., & Macek, T. (2018). Phytoextraction of heavy metals: A promising tool for clean-up of polluted environment? *Frontiers in Plant Science*, *9*, 1476. doi:10.3389/fpls.2018.01476 PMID:30459775

Sumiahadi, A., & Acar, R. (2018). A review of phytoremediation technology: Heavy metals uptake by plants. *IOP Conference Series. Earth and Environmental Science*, *142*, 12–23. doi:10.1088/1755-1315/142/1/012023

Sun, Y. P. (2006). *Dispersion of nanoscale iron particles* (Doctoral dissertation). Department of Civil and Environmental Engineering, Lehigh University, Bethlehem, PA.

Sundara, B., Natarajan, V., & Hari, K. (2002). Influence of phosphorus solubilizing bacteria on the changes in soil available phosphorus and sugarcane and sugar yields. *Field Crops Research*, *77*(1), 43–49. doi:10.1016/S0378-4290(02)00048-5

Sundarram, A., & Murthy, T. P. K. (2014). α-amylase production and applications: A review. *Journal of Applied & Environmental Microbiology*, *2*(4), 166–175.

Sun, H., Tao, J., Gu, P., Xu, G., & Zhang, Y. (2016). The role of strigolactones in root development. *Plant Signaling & Behavior*, *11*(1), e1110662. doi:10.1080/15592324.2015.1110662 PMID:26515106

Sun, J., Chen, Q., Qian, Z., Zheng, Y., Yu, S., & Zhang, A. (2018). Plant uptake and metabolism of 2, 4-dibromophenol in carrot: In vitro enzymatic direct conjugation. *Journal of Agricultural and Food Chemistry*, *66*(17), 4328–4335. doi:10.1021/acs.jafc.8b00543 PMID:29656645

Sun, J., Wu, X., & Gan, J. (2015). Uptake and metabolism of phthalate esters by edible plants. *Environmental Science & Technology*, *49*(14), 8471–8478. doi:10.1021/acs.est.5b01233 PMID:26090545

Sun, M., Fu, D., Teng, Y., Shen, Y., Luo, Y., Li, Z., & Christie, P. (2011). In situ phytoremediation of PAH-contaminated soil by intercropping alfalfa (*Medicago sativa* L.) with tall fescue (*Festuca arundinacea* Schreb.) and associated soil microbial activity. *Journal of Soils and Sediments*, *11*(6), 980–989. doi:10.100711368-011-0382-z

Sun, S. L., Yang, W. L., Guo, J. J., Zhou, Y. N., Rui, X., Chen, C., Ge, F., & Dai, Y. J. (2017). Biodegradation of the neonicotinoid insecticide acetamiprid in surface water by the bacterium Variovorax boronicumulans CGMCC 4969 and its enzymatic mechanism. *Royal Society of Chemistry Advances*, *7*(41), 25387–25397. doi:10.1039/C7RA01501A

Supanjani, H. H. S., Jung, S. J., & Lee, K. D. (2006). Rock phosphate and potassium rock solubilizing bacteria as alternative sustainable fertilizers. *Agronomy for Sustainable Development*, *26*(4), 233–340. doi:10.1051/agro:2006020

Su, Q., Guan, T., & Lv, H. (2016). Siderophore biosynthesis coordinately modulated the virulence-associated interactive metabolome of uropathogenic Escherichia coli and human urine. *Scientific Reports*, *6*(1), 1–11. doi:10.1038rep24099 PMID:27076285

Suryadi, Y., Susilowati, D. N., & Fauziah, F. (2019). Management of Plant Diseases by PGPR-Mediated Induced Resistance with Special Reference to Tea and Rice Crops. In *Plant Growth Promoting Rhizobacteria for Sustainable Stress Management* (pp. 65–110). Springer. doi:10.1007/978-981-13-6986-5_4

Susarla, S., Medina, V. F., & McCutcheon, S. C. (2002). Phytoremediation: An ecological solution to organic chemical contamination. *Ecological Engineering*, *18*(5), 647–658. doi:10.1016/S0925-8574(02)00026-5

Susilowati, D. N., Riyanti, E. I., Setyowati, M., & Mulya, K. (2018). Indole-3-acetic acid producing bacteria and its application on the growth of rice. In AIP Conference Proceedings: 2018. AIP Publishing LLC. doi:10.1063/1.5050112

Sutar, H., & Das, C. K. (2012). A review on: Bioremediation. *International Journal of Research in Chemistry and Environment*, *2*(1), 3–21.

Sutar, H., Mishra, S. C., Sahoo, S. K., Chakraverty, A. P., & Mahrana, H. S. (2014). Progress of Red Mud Utilization: An Overview. *American Chemical Science Journal*, *4*(3), 255–279. doi:10.9734/ACSJ/2014/7258

Sutherland, T. D., Horne, I., Harcourt, R. L., Russell, R. J., & Oakeshott, J. G. (2002). Isolation and characterization of a *Mycobacterium* strain that metabolizes the insecticide endosulfan. *Journal of Applied Microbiology*, *93*(3), 380–389. doi:10.1046/j.1365-2672.2002.01728.x PMID:12174035

Sutton, M. A., Oenema, O., Erisman, J. W., Leip, A., van Grinsven, H., & Winiwarter, W. (2011). Too much of a good thing. *Nature*, *472*(7342), 159–161. doi:10.1038/472159a PMID:21478874

Suzaki, T., Yoro, E., & Kawaguchi, M. (2015). Leguminous plants: Inventors of root nodules to accommodate symbiotic bacteria. *International Review of Cell and Molecular Biology*, *316*, 111–158. doi:10.1016/bs.ircmb.2015.01.004 PMID:25805123

Suzuki, N., Supyani, S., Maruyama, K., & Hillman, B. I. (2004). Complete genome sequence of Mycoreovirus-1/Cp9B21., a member of a novel genus within the family *Reoviridae,* isolated from the chestnut blight fungus *Cryphonectria parasitica. The Journal of General Virology*, *85*(11), 3437–3448. doi:10.1099/vir.0.80293-0 PMID:15483262

Swain, S. S., Paidesetty, S. K., & Padhy, R. N. (2017). Antibacterial, antifungal and antimycobacterial compounds from cyanobacteria. *Biomedicine and Pharmacotherapy*, *90*, 760–776. doi:10.1016/j.biopha.2017.04.030 PMID:28419973

Swarnalakshmi, K., Prasanna, R., Kumar, A., Pattnaik, S., Chakravarty, K., Shivay, Y. S., Singh, R., & Saxena, A. K. (2013). Evaluating the influence of novel cyanobacterial biofilmed biofertilizers on soil fertility and plant nutrition in wheat. *European Journal of Soil Biology*, *55*, 107–116. doi:10.1016/j.ejsobi.2012.12.008

Syed, B. A., & Patel, B. (2014). Investigation and correlation of soil biotic and abiotic factors affecting agricultural productivity in semi-arid regions of north Gujarat, India. *International Journal of Research Studies in Biosciences*, *2*, 18–29.

Syers, J. K. (2003). Potassium in soils: Current concepts. In *Proceedings of the IPI Golden Jubilee Congress 1952–2002 on Feed the soil to feed the people: The role of potash in sustainable agriculture*. (pp. 301–10). International Potash Institute.

Sylvia, A. (2019). Introduction of petroleum hydrocarbons contaminants and its human effects. *Journal of Environmental Science and Public Health, 3*, 1-9.

Sylvia, D. M. (1990). Inoculation of native woody plants with vesicular–arbuscular fungi for phosphate mine land reclamation. *Agriculture, Ecosystems & Environment, 31*(3), 847–897. doi:10.1016/0167-8809(90)90224-2

Sytsma, K. J., Morawetz, J., Pires, J. C., Nepokroeff, M., Conti, E., Zjhra, M., Hall, J. C., & Chase, M. W. (2002). *Urticalean rosids*: Circumscription, rosid ancestry, and phylogenetics based on rbcL, trnL-F, and ndhF sequences. *American Journal of Botany, 89*(9), 1531–1546. doi:10.3732/ajb.89.9.1531 PMID:21665755

Taccari, M., Milanovic, V., Comitini, F., Casucci, C., & Ciani, M. (2012). Effects of biostimulation and bioaugmentation on diesel removal and bacterial community. *International Biodeterioration & Biodegradation, 66*(1), 39–46. doi:10.1016/j.ibiod.2011.09.012

Tagele, S. B., Kim, S. W., Lee, H. G., Kim, H. S., & Lee, Y. S. (2018). Effectiveness of multi-trait Burkholderia contaminans KNU17BI1 in growth promotion and management of banded leaf and sheath blight in maize seedling. *Microbiological Research, 214*, 8–18. doi:10.1016/j.micres.2018.05.004 PMID:30031484

Taghavi, S., Barac, T., Greenberg, B., Borremans, B., Vangronsveld, J., & van der Lelie, D. (2005). Horizontal gene transfer to endogenous endophytic bacteria from poplar improves phytoremediation of toluene. *Applied and Environmental Microbiology, 71*(12), 8500–8505. doi:10.1128/AEM.71.12.8500-8505.2005 PMID:16332840

Tahat, M., Alananbeh, K., Othman, Y., & Leskovar, D. (2020). Soil Health and Sustainable Agriculture. *Sustainability, 12*(12), 4859. doi:10.3390u12124859

Tahri, N., Bahafid, W., Sayel, H., & El Ghachtouli, N. (2013). Biodegradation: involved microorganisms and genetically engineered microorganisms. In Biodegradation - Life of Science. IntechOpen.

Tailor, A. J., & Joshi, B. H. (2014). Harnessing plant growth promoting rhizobacteria beyond nature: A review. *Journal of Plant Nutrition, 37*(9), 1534–1571. doi:10.1080/01904167.2014.911319

Takagi, S. I., Nomoto, K., & Takemoto, T. (1984). Physiological aspect of mugineic acid, a possible phytosiderophore of graminaceous plants. *Journal of Plant Nutrition, 7*(1-5), 469–477. doi:10.1080/01904168409363213

Takahashi-Nakaguchi, A., Shishido, E., Yahara, M., Urayama, S. I., Sakai, K., Chibana, H., Kamei, K., Moriyama, H., & Gonoi, T. (2020). Analysis of an intrinsic mycovirus associated with reduced virulence of the human pathogenic fungus *Aspergillus fumigatus. Frontiers in Microbiology, 10*, 3045. doi:10.3389/fmicb.2019.03045 PMID:32010101

Talboys, P. J., Owen, D. W., Healey, J. R., Withers, P. J., & Jones, D. L. (2014). Auxin secretion by *Bacillus amyloliquefaciens* FZB42 both stimulates root exudation and limits phosphorus uptake in *Triticum aestivum. BMC Plant Biology, 14*(1), 51. doi:10.1186/1471-2229-14-51 PMID:24558978

Tamás, L., Mistrík, I., Huttová, J., Halusková, L., Valentovicová, K., & Zelinová, V. (2010). Role of reactive oxygen species-generating enzymes and hydrogen peroxide during cadmium, mercury and osmotic stresses in barley root tip. *Planta, 231*(2), 221–231. doi:10.100700425-009-1042-z PMID:19898864

Tanaka, Y., Higashi, T., Rakwal, R., Wakida, S., & Iwahashi, H. (2007). Quantitative analysis of sulfur-related metabolites during cadmium stress response in yeast by capillary electrophoresis–mass spectrometry. *Journal of Pharmaceutical and Biomedical Analysis, 44*(2), 608–613.

Tanaka, Y., & Omura, S. (1993). Agroactive compounds of microbial origin. *Annual Review of Microbiology, 47*(1), 57–87. doi:10.1146/annurev.mi.47.100193.000421 PMID:8257109

Tanamool, V., Imai, T., Danvirutai, P., & Kaewkannetra, P. (2013). An alternative approach to the fermentation of sweet sorghum juice into biopolymer of poly-β-hydroxyalkanoates (PHAs) by newly isolated, *Bacillus aryabhattai* PKV01. *Biotechnology and Bioprocess Engineering; BBE, 18*(1), 65–74. doi:10.100712257-012-0315-8

Tandon, P. K., & Singh, S. B. (2016). Redox processes in water remediation. *Environmental Chemistry Letters, 14*(1), 15–25. doi:10.100710311-015-0540-4

Tang, C. Y., Kwon, Y. N., & Leckie, J. O. (2007). Fouling of reverse osmosis and nanofiltration membranes by humic acid—Effects of solution composition and hydrodynamic conditions. *Journal of Membrane Science, 290*(1-2), 86–94. doi:10.1016/j.memsci.2006.12.017

Tan, L., He, M., Song, L., Fu, X., & Shi, S. (2016). Aerobic decolorization, degradation and detoxification of azo dyes by a newly isolated salt-tolerant yeast *Scheffersomyces spartinae TLHS-SF1. Bioresource Technology, 203*, 287–294. doi:10.1016/j.biortech.2015.12.058 PMID:26744802

Tano, Z. J. (2011). *7 Ecological Effects of Pesticides*. Academic Press.

Tawate, S., Gupta, R., & Jain, K. (2018). Technology Commercialization in Bio-fertilizer Firm: An Indian Case. *International Journal of Global Business and Competitiveness, 13*(1), 65–74.

Taylor, J. A. (2000). Recent developments in reactive dyes. *Review of Progress in Coloration and Related Topics, 30*(1), 93–108. doi:10.1111/j.1478-4408.2000.tb03785.x

Tchounwou, P. B., Yedjou, C. G., Patlolla, A. K., & Sutton, D. J. (2012). Heavy metal toxicity and the environment. In *Molecular, clinical and environmental toxicology* (pp. 133–164). Springer. doi:10.1007/978-3-7643-8340-4_6

Tchounwou, P., Newsome, C., Williams, J., & Glass, K. (2008). Copper-induced cytotoxicity and transcriptional activation of stress genes in human liver carcinoma cells. *Metal Ions in Biology and Medicine, 10*, 285–290.

Tehrani-Bagha, A. R., & Holmberg, K. (2013). Solubilization of Hydrophobic Dyes in Surfactant Solutions. *Materials (Basel), 6*(2), 580–608. doi:10.3390/ma6020580 PMID:28809328

Telke, A. A., Ghodake, G. S., Kalyani, D. C., Dhanve, R. S., & Govindwar, S. P. (2011). Biochemical characteristics of a textile dye degrading extracellular laccase from a *Bacillus* sp. ADR. *Bioresource Technology, 102*(2), 1752–1756. doi:10.1016/j.biortech.2010.08.086 PMID:20855194

Telke, A. A., Kadam, A. A., Jagtap, S. S., Jadhav, J. P., & Govindwar, S. P. (2010). Biochemical characterization and potential for textile dye degradation of blue laccase from *Aspergillus ochraceus* NCIM-1146. *Biotechnology and Bioprocess Engineering; BBE, 15*(4), 696–703. doi:10.100712257-009-3126-9

Telke, A. A., Kalyani, D. C., Dawkar, V. V., & Govindwar, S. P. (2009). Influence of organic and inorganic compounds on oxidoreductive decolorization of sulfonated azo dye C.I. Reactive Orange 16. *Journal of Hazardous Materials, 172*(1), 298–309. doi:10.1016/j.jhazmat.2009.07.008 PMID:19640646

Teng, Y., & Chen, W. (2019). Soil Microbiomes—a Promising Strategy for Contaminated Soil Remediation: A Review. *Pedosphere, 29*(3), 283–297. doi:10.1016/S1002-0160(18)60061-X

Teplitski, M., Robinson, J. B., & Bauer, W. D. (2000). Plants secrete substances that mimic bacterial N-acyl homoserine lactone signal activities and affect population density-dependent behaviors in associated bacteria. *Molecular Plant-Microbe Interactions, 13*(6), 637–648. doi:10.1094/MPMI.2000.13.6.637 PMID:10830263

Terpe, K. (2006). Overview of bacterial expression systems for heterologous protein production: From molecular and biochemical fundamentals to commercial systems. *Applied Microbiology and Biotechnology, 72*(2), 211–222. doi:10.100700253-006-0465-8 PMID:16791589

Tesfahunegn, G. B., & Gebru, T. A. (2020). Variation in soil properties under different cropping and other land-use systems in Dura catchment, Northern Ethiopia. *PLoS One*, *15*(2), 1–27. doi:10.1371/journal.pone.0222476 PMID:32023243

Thakur, P., & Singh, I. (2018). Biocontrol of soilborne root pathogens: An Overview. *Root Biology*, 181-220.

Thakur, R. S., & Das, S. N. (2003). Red mud Analysis and utilisation of metal values. Publication and Information Directorate (CSIR) and Willy Eastern Ltd.

Thatoi, H., Das, S., Mishra, J., Rath, B. P., & Das, N. (2014). Bacterial chromate reductase, a potential enzyme for bioremediation of hexavalent chromium: A review. *Journal of Environmental Management*, *146*, 383–399. doi:10.1016/j.jenvman.2014.07.014 PMID:25199606

Thazeem, B., Umesh, M., & Vikas, O. V. (2016b). Isolation and identification of proteolytic *Bacillus pumilus* from tannery lime effluent and its dehairing activity. *International Journal of Academic Research and Development, 1*(2), 5–7.

Thazeem, B., Beryl, G. P., & Umesh, M. (2017). A comparative study on alkaline protease production from *Bacillus spp*. and their biodegradative, dehairing and destaining activity. *International Journal of Academic Research and Development*, *2*(2), 74–79.

Thazeem, B., Preethi, K., & Umesh, M. (2015). Characterization and fermentative utilization of tannery fleshings using *Lactobacillus plantarum. International Journal of Recent Scientific Research*, *6*(3), 3037–3041.

Thazeem, B., Preethi, K., Umesh, M., & Radhakrishnan, S. (2018). Nutritive characterization of delimed bovine tannery fleshings for their possible use as a proteinaceous aqua feed ingredient. *Waste and Biomass Valorization*, *9*(8), 1289–1301. doi:10.100712649-017-9922-0

Thazeem, B., Umesh, M., Mani, V. M., Beryl, G. P., & Preethi, K. (2020). Biotransformation of bovine tannery fleshing into utilizable product with multifunctionalities. *Biocatalysis and Biotransformation*, 1–19. doi:10.1080/10242422.2020.1786071

Thazeem, B., Umesh, M., & Vikas, O. V. (2016a). Bioconversion of poultry feather into feather meal using proteolytic *Bacillus* Species – A comparative study. *International Journal of Advances in Scientific Research, 1*(1), 14–16.

Theerachat, M., Guieysse, D., Morel, S., Remaud-Siméon, M., & Chulalaksananukul, W. (2019). Laccases from marine organisms and their applications in the biodegradation of toxic and environmental pollutants: A review. *Applied Biochemistry and Biotechnology*, *187*(2), 583–611. doi:10.100712010-018-2829-9 PMID:30009326

Thenmozhi, R., Arumugam, K., Nagasathya, A., Thajuddin, N., & Paneerselvam, A. (2013). Studies on mycoremediation of used engine oil contaminated soil samples. *Advances in Applied Science Research*, *4*, 110–118.

Thierry, L., Armelle, B., & Karine, J. (2008). Performance of bioaugmentation–assisted phytoextraction applied to metal contaminated soils: A review. *Environmental Pollution*, *153*(3), 497–522. doi:10.1016/j.envpol.2007.09.015 PMID:17981382

Thijs, S., Sillen, W., Rineau, F., Weyens, N., & Vangronsveld, J. (2016). Towards an enhanced understanding of plant-microbiome interactions to improve phytoremediation: Engineering the metaorganism. *Frontiers in Microbiology*, *7*, 341. doi:10.3389/fmicb.2016.00341 PMID:27014254

Thomas, L., & Singh, I. (2019). Microbial biofertilizers: types and applications. In *Biofertilizers for Sustainable Agriculture and Environment* (pp. 1–19). Springer. doi:10.1007/978-3-030-18933-4_1

Thomas, P. W., Stone, E. M., Costello, A. L., Tierney, D. L., & Fast, W. (2005). The quorum-quenching lactonase from *Bacillus thuringiensis* is a metalloprotein. *Biochemistry*, *44*(20), 7559–7569. doi:10.1021/bi050050m PMID:15895999

Thomas, T., Gilbert, J., & Meyer, F. (2012). Metagenomics - a guide from sampling to data analysis. *Microbial Informatics and Experimentation*, *2*(1), 3.

Thompson, I. P., & Christopher, J. (2005). Bioaugmentation for bioremediation: The challenge of strain selection. *Environmental Microbiology*, *7*(7), 909–915. doi:10.1111/j.1462-2920.2005.00804.x PMID:15946288

Thompson, L. A., & Darwish, W. S. (2019). Environmental Chemical Contaminants in Food: Review of a Global Problem. *Journal of Toxicology*, *2019*, 2345283. doi:10.1155/2019/2345283 PMID:30693025

Tigini, V., Giansanti, P., Mangiavillano, A., Pannocchiam, A., & Varese, G. C. (2011). Evaluation of toxicity, genotoxicity and environmental risk of simulated textile and tannery wastewaters with a battery of biotests. *Ecotoxicology and Environmental Safety*, *74*(4), 866–873. doi:10.1016/j.ecoenv.2010.12.001 PMID:21176963

Timmusk, S., Behers, L., Muthoni, J., Muraya, A., & Aronsson, A. C. (2017). Perspectives and challenges of microbial application for crop improvement. *Frontiers in Plant Science*, *8*, 49. doi:10.3389/fpls.2017.00049 PMID:28232839

Timmusk, S., Islam, A., Abd El, D., Lucian, C., Tanilas, T., & Kannaste, A. (2014). Drought-tolerance of wheat improved by rhizosphere bacteria from harsh environments: Enhanced biomass production and reduced emissions of stress volatiles. *PLoS One*, *9*(5), 1–13. doi:10.1371/journal.pone.0096086 PMID:24811199

Titah, H. S., Abdullah, S., Idris, M., Anuar, N., Basri, H., Mukhlisin, M., Tangahu, B. V., Purwanti, I. F., & Kurniawan, S. B. (2018). Arsenic Resistance and Biosorption by Isolated Rhizobacteria from the Roots of *Ludwigia octovalvis. International Journal of Microbiology*, *2018*, 3101498. doi:10.1155/2018/3101498 PMID:30723505

Tittabutr, P., Sripakdi, S., Boonkerd, N., Tanthanuch, W., Minamisawa, K., & Teaumroong, N. (2015). Possible role of 1-aminocyclopropane-1-carboxylate (ACC) deaminase activity of *Sinorhizobium sp.* BL3 on symbiosis with mung bean and determinate nodule senescence. *Microbes and Environments*, *30*(4), ME15120. doi:10.1264/jsme2.ME15120 PMID:26657304

Tiwari, S., Lata, C., Chauhan, P. S., & Nautiyal, C. S. (2016). *Pseudomonas putida* attunes morphophysiological, biochemical and molecular responses in *Cicer arietinum* L. during drought stress and recovery. *Plant Physiology and Biochemistry*, *99*, 108–117. doi:10.1016/j.plaphy.2015.11.001 PMID:26744996

Tiwari, S., Lata, C., Chauhan, P. S., Prasad, V., & Prasad, M. (2017). A Functional Genomic Perspective on Drought Signalling and its Crosstalk with Phytohormone-mediated Signalling Pathways in Plants. *Current Genomics*, *18*(6), 469–482. doi:10.2174/1389202918666170605083319 PMID:29204077

Tomei, M. C., & Daugulis, A. J. (2013). Feasibility of operating a solid–liquid bioreactor with used automobile tires as the sequestering phase for the biodegradation of inhibitory compounds. *Journal of Environmental Management*, *125*, 7–11. doi:10.1016/j.jenvman.2013.03.047 PMID:23629012

Tomotada, I., & Masao, N. (2001). Current Bioremediation Practice and Perspective. Journal of Bioscience and Bioengineering, 92.

Tony, B. D., Goyal, D., & Khanna, S. (2009a). Decolorization of textile azo dyes by aerobic bacterial consortium. *International Biodeterioration & Biodegradation*, *63*(4), 462–469. doi:10.1016/j.ibiod.2009.01.003

Tony, B. D., Goyal, D., & Khanna, S. (2009b). Decolorization of Direct Red 28 by mixed bacterial culture in an up-flow immobilized bioreactor. *Journal of Industrial Microbiology & Biotechnology*, *36*(7), 955–960. doi:10.100710295-009-0574-3 PMID:19390882

Tonziello, G., Caraffa, E., Pinchera, B., Granata, G., & Petrosillo, N. (2019). Present and future of siderophore-based therapeutic and diagnostic approaches in infectious diseases. *Infectious Disease Reports, 11*(2). Advance online publication. doi:10.4081/idr.2019.8208 PMID:31649808

Torrey, J. G. (1978). Nitrogen fixation by actinomycete-nodulated angiosperms. *Bioscience, 28*(9), 586–592. doi:10.2307/1307515

Torsvik, V., & Øvreås, L. (2002). Microbial diversity and function in soil: From genes to ecosystems. *Current Opinion in Microbiology, 5*(3), 240–245. doi:10.1016/S1369-5274(02)00324-7 PMID:12057676

Tosco, T., Papini, M. P., Viggi, C. C., & Sethi, R. (2014). Nanoscale zerovalent iron particles for groundwater remediation: A review. *Journal of Cleaner Production, 77*, 10–21. doi:10.1016/j.jclepro.2013.12.026

Toxics Link. (2014). *Plastics and the Environment Assessing the Impact of the Complete Ban on Plastic Carry Bag.* Central Pollution Control Board, CPCB.

Trabelsi, D., & Mhamdi, R. (2013). Microbial inoculants and their impact on soil microbial communities: A review. *BioMed Research International, 2013*, 1–11. doi:10.1155/2013/863240 PMID:23957006

Trace Elements in Human Nutrition and Health. (1996). WHO/FAO/IAEA. World Health Organization.

Tran, H. T., Lin, C., Bui, X. T., Ngo, H. H., Cheruiyot, N. K., Hoang, H. G., & Vu, C. T. (1920). Aerobic composting remediation of petroleum hydrocarbon-contaminated soil. Current and future perspectives. *The Science of the Total Environment, 753*, 142250. doi:10.1016/j.scitotenv.2020.142250 PMID:33207468

Trejo, A., De-Bashan, L. E., Hartmann, A., Hernandez, J.-P., Rothballer, M., Schmid, M., & Bashan, Y. (2012). Recycling waste debris of immobilized microalgae and plant growth-promoting bacteria from wastewater treatment as a resource to improve fertility of eroded desert soil. *Environmental and Experimental Botany, 75*, 65–73. doi:10.1016/j.envexpbot.2011.08.007

Trentacoste, E. M., Martinez, A. M., & Zenk, T. (2015). The place of algae in agriculture: Policies for algal biomass production. *Photosynthesis Research, 123*(3), 305–315. doi:10.100711120-014-9985-8 PMID:24599393

Trinder, C. J., Johnson, D., & Artz, R. R. (2008). Interactions among fungal community structure, litter decomposition and depth of water table in a cutover peatland. *FEMS Microbiology Ecology, 64*(3), 433–448. doi:10.1111/j.1574-6941.2008.00487.x PMID:18430005

Tripathi, S., Sanjeevi, R., Anuradha, J., Chauhan, D. S., & Rathoure, A. K. (2018). Nano-bioremediation: nanotechnology and bioremediation. In Biostimulation Remediation Technologies for Groundwater Contaminants (pp. 202-219). IGI Global. doi:10.4018/978-1-5225-4162-2.ch012

Tripathi, S., Srivastava, P., Devi, R. S., & Bhadouria, R. (2020). Influence of synthetic fertilizers and pesticides on soil health and soil microbiology. In Agrochemicals Detection, Treatment and Remediation (pp. 25-54). Butterworth-Heinemann. doi:10.1016/B978-0-08-103017-2.00002-7

Tripathi, D. K., Singh, S., Gaur, S., Singh, S., Yadav, V., Liu, S., ... Dubey, N. K. (2018). Acquisition and homeostasis of iron in higher plants and their probable role in abiotic stress tolerance. *Frontiers in Environmental Science, 5*, 86. doi:10.3389/fenvs.2017.00086

Triplett, E. (1996). Diazotrophic endophytes: Progress and prospects for nitrogen fixation in monocots. *Plant and Soil, 186*(1), 29–38. doi:10.1007/BF00035052

Trivedi, P., Schenk, P. M., Wallenstein, M. D., & Singh, B. K. (2017). Tiny microbes, big yields: Enhancing food crop production with biological solutions. *Microbial Biotechnology*, *10*(5), 999–1003. doi:10.1111/1751-7915.12804 PMID:28840959

Trujillo-Tapia, M. N., & Ramírez-Fuentes, E. (2016). Bio-fertilizer: An alternative to reduce chemical fertilizer in agriculture. *Journal of Global Agriculture and Ecology*, *4*(2), 99–103.

Truskewycz, A., Gundry, T. D., Khudur, L. S., Kolobaric, A., Taha, M., Aburto-Medina, A., Ball, A. S., & Shahsavari, E. (2019). Petroleum Hydrocarbon Contamination in Terrestrial Ecosystems-Fate and Microbial Responses. *Molecules (Basel, Switzerland)*, *24*(18), 3400. doi:10.3390/molecules24183400 PMID:31546774

Tsao, D. T. (2003). Overview of phytotechnologies. In T. Scheper & D. T. Tsao (Eds.), *Advances in Biochemical Engineering Biotechnology* (Vol. 78, pp. 1–50). Springer.

Tsukanova, K. A., Meyer, J. J. M., & Bibikova, T. N. (2017). Effect of plant growth-promoting Rhizobacteria on plant hormone homeostasis. *South African Journal of Botany*, *113*, 91–102. doi:10.1016/j.sajb.2017.07.007

Tsygankov, V. Y., Lukyanova, O., Boyarova, M., Gumovskiy, A., Donets, M., Lyakh, V., Korchagin, V. P., & Prikhodko, Y. V. (2019). Organochlorine pesticides in commercial Pacific salmon in the Russian Far Eastern seas: Food safety and human health risk assessment. *Marine Pollution Bulletin*, *140*, 503–508. doi:10.1016/j.marpolbul.2019.02.008 PMID:30803671

Tuğrul, K. M. (2019). Soil Management in Sustainable Agriculture. In *Soil Management and Plant Nutrition for Sustainable Crop Production*. IntechOpen.

Tully, K. L., & McAskill, C. (2020). Promoting soil health in organically managed systems: A review. *Organic Agriculture*, *10*(3), 339–358. doi:10.100713165-019-00275-1

Tungittiplakorn, W., Cohen, C., & Lion, L. W. (2005). Engineeredpolymeric nanoparticles for bioremediation of hydrophobic contaminants. *Environmental Science & Technology*, *39*(5), 1354–1358. doi:10.1021/es049031a PMID:15787377

Tungittiplakorn, W., Lion, L. W., Cohen, C., & Kim, J. Y. (2004). Engineered polymeric nanoparticles for soil remediation. *Environmental Science & Technology*, *38*(5), 1605–1610. doi:10.1021/es0348997 PMID:15046367

TuomistoH. L.ScheelbeekP. F.ChalabiZ.GreenR.SmithR. D.HainesA.DangourA. D. (2017). Effects of environmental change on agriculture, nutrition and health: A framework with a focus on fruits and vegetables. Wellcome open research, *2*. doi:10.12688/wellcomeopenres.11190.2

Tuomivirta, T. T., & Hantula, J. (2003a). *Gremmeniella abietina* mitochondrial RNA virus S1 is phylogenetically related to the members of the genus *Mitovirus*. *Archives of Virology*, *148*(12), 2429–2436. doi:10.100700705-003-0195-5 PMID:14648296

Tuomivirta, T. T., & Hantula, J. (2003b). Two unrelated double-stranded RNA molecule patterns in *Gremmeniella abietina* type A code for putative viruses of the families *Totiviridae* and *Partitiviridae*. *Archives of Virology*, *148*(12), 2293–2305. doi:10.100700705-003-0194-6 PMID:14648287

Turan, M., Yildirim, E., Kitir, N., Unek, C., Nikerel, E., & Ozdemir, B. S. (2017). Beneficial Role of Plant Growth-Promoting Bacteria in Vegetable Production under Abiotic Stress. In A. Zaidi & M. Khan (Eds.), *Microbial Strategies for Vegetable Production*. Springer. doi:10.1007/978-3-319-54401-4_7

Türker, M. (2014). Yeast biotechnology: diversity and applications. In *27th VH yeast conference: advances in science and industrial production of baker's yeast, Istanbul* (pp. 14-15). Academic Press.

Turner, B. L., Papházy, M. J., Haygarth, P. M., & McKelvie, I. D. (2002). Inositol phosphates in the environment. *Philosophical Transactions of the Royal Society of London. Series B, Biological Sciences*, *357*(1420), 449–469. doi:10.1098/rstb.2001.0837 PMID:12028785

Tzvetkov, M., Klopprogge, C., Zelder, O., & Liebl, W. (2003). Genetic dissection of trehalose biosynthesis in *Corynebacterium glutamicum*: Inactivation of trehalose production leads to impaired growth and an altered cell wall lipid composition. *Microbiology*, *149*(7), 1659–1673. doi:10.1099/mic.0.26205-0 PMID:12855718

Ubani, O., Atagana, I. H., & Thantsha, S. M. (2013). Biological degradation of oil sludge: A review of the current state of development. *African Journal of Biotechnology*, *12*(47), 6544–6567. doi:10.5897/AJB11.1139

Ubogu, M., Odokuma, L. O., & Akponah, E. (2019). Enhanced rhizoremediation of crude oil–contaminated mangrove swamp soil using two wetland plants (Phragmites australis and Eichhornia crassipes). *Brazilian Journal of Microbiology*, *50*(3), 715–728. doi:10.100742770-019-00077-3 PMID:30993597

Ulery, B. D., Nair, L. S., & Laurencin, C. T. (2011). Biomedical applications of biodegradable polymers. *Journal of Polymer Science. Part B, Polymer Physics*, *49*(12), 832–864. doi:10.1002/polb.22259 PMID:21769165

Ullah, S., & Bano, A. (2015). Isolation of plant-growth-promoting rhizobacteria from rhizospheric soil of halophytes and their impact on maize (*Zea mays* L.) under induced soil salinity. *Canadian Journal of Microbiology*, *61*(4), 307–313. doi:10.1139/cjm-2014-0668 PMID:25776270

Ulrich, R. L. (2004). Quorum quenching: Enzymatic disruption of N-acylhomoserine lactone-mediated bacterial communication in *Burkholderia thailandensis*. *Applied and Environmental Microbiology*, *70*(10), 6173–6180. doi:10.1128/AEM.70.10.6173-6180.2004 PMID:15466564

Umali, G. M., & Krishnapillay, B. (2002). Biological nitrogen fixation in tree species. *Basic principles of biotechnology and their application in forestry*, 145-148.

Umesh, M., Mani, V. M., Thazeem, B., & Preethi, K. (2018). Statistical optimization of process parameters for bioplastic (PHA) Production by *Bacillus subtilis* NCDC0671 using orange peel-based medium. *Iranian Journal of Science and Technology. Transaction A, Science*, *42*(4), 1947–1955. doi:10.100740995-017-0457-9

Umesh, M., & Preethi, K. (2017). Fabrication of antibacterial bioplastic sheet using orange peel medium and its antagonistic effect against common clinical pathogens. *Research Journal of Biotechnology*, *12*(7), 67–74.

Umesh, M., Priyanka, K., Thazeem, B., & Preethi, K. (2017). Production of Single Cell Protein and Polyhydroxyalkanoate from *Carica papaya* Waste. *Arabian Journal for Science and Engineering*, *42*(6), 2361–2369. doi:10.100713369-017-2519-x

Umesh, M., Priyanka, K., Thazeem, B., & Preethi, K. (2018). Biogenic PHA nanoparticle synthesis and characterization from Bacillus subtilis NCDC0671 using orange peel medium. *International Journal of Polymeric Materials and Polymeric Biomaterials*, *67*(17), 996–1004. doi:10.1080/00914037.2017.1417284

Umesh, M., & Thazeem, B. (2019). Biodegradation Studies of Polyhydroxyalkanoates extracted from *Bacillus Subtilis* NCDC 0671. *Research Journal of Chemistry and Environment*, *23*(6), 107–114.

Umsakul, K., Dissara, Y., & Srimuang, N. (2010). Chemical, physical and microbiological changes during composting of the water hyacinth. *Pakistan Journal of Biological Sciences*, *13*(20), 985–992. doi:10.3923/pjbs.2010.985.992 PMID:21319457

UN DESA (United Nations, Department of Economic and Social Affairs). (2014). World Urbanization Prospects. Population Department, United Nations.

UNEP. (2009). *Stockholm Convention on Persistent Organic Pollutants (POPs)*. http://irptc.unep.ch/pops

Unkovich, M. J., & Pate, J. S. (2000). An appraisal of recent field measurements of symbiotic N_2 fixation by annual legumes. *Field Crops Research*, *65*(2-3), 211–228. doi:10.1016/S0378-4290(99)00088-X

Unkovich, M. J., Pate, J. S., & Sanford, P. (1997). Nitrogen fixation by annual legumes in Australian Mediterranean agriculture. *Australian Journal of Agricultural Research*, *48*(3), 267–293. doi:10.1071/A96099

Upadhyay, N., Vishwakarma, K., Singh, J., Mishra, M., Kumar, V., Rani, R., Mishra, R. K., Chauhan, D. K., Tripathi, D. K., & Sharma, S. (2017). Tolerance and Reduction of Chromium (VI) by *Bacillus* sp. MNU16 Isolated from Contaminated Coal Mining Soil. *Frontiers in Plant Science*, *8*, 778. doi:10.3389/fpls.2017.00778 PMID:28588589

Upadhyay, S. K., Singh, J. S., & Singh, D. P. (2011). Exopolysaccharide-producing plant growth-promoting rhizobacteria under salinity condition. *Pedosphere*, *21*(2), 214–222. doi:10.1016/S1002-0160(11)60120-3

Uren, N. C. (2000). Types, amounts, and possible functions of compounds released into the rhizosphere by soil-grown plants. In *The Rhizosphere - Biochemistry and Organic Substances at the Soil-Plant Interface* (2nd ed.). CRC Press.

Üreyen Esertaş, Ü. Z., Uzunalioğlu, E., Güzel, Ş., Bozdeveci, A., & Alpay Karaoğlu, Ş. (2020). Determination of bioremediation properties of soil-borne *Bacillus sp.* 5O5Y11 and its effect on the development of *Zea mays* in the presence of copper. *Archives of Microbiology*, *202*(7), 1817–1829. doi:10.100700203-020-01900-4 PMID:32440759

Uroz, S., Calvaruso, C., Turpault, M. P., & Frey-Klett, P. (2009). Mineral weathering by bacteria: Ecology, actors and mechanisms. *Trends in Microbiology*, *17*(8), 378–387. doi:10.1016/j.tim.2009.05.004 PMID:19660952

Uroz, S., Chhabra, S. R., Camara, M., Williams, P., Oger, P., & Dessaux, Y. (2005). N-Acylhomoserine lactone quorum-sensing molecules are modified and degraded by *Rhodococcus erythropolis* W2 by both amidolytic and novel oxidoreductase activities. *Microbiology*, *151*(10), 3313–3322. doi:10.1099/mic.0.27961-0 PMID:16207914

Uroz, S., & Heinonsalo, J. (2008). Degradation of N-acyl homoserine lactone quorum sensing signal molecules by forest root-associated fungi. *FEMS Microbiology Ecology*, *65*(2), 271–278. doi:10.1111/j.1574-6941.2008.00477.x PMID:18400006

Uroz, S., Oger, P. M., Chapelle, E., Adeline, M. T., Faure, D., & Dessaux, Y. (2008). A *Rhodococcus* qsdA-encoded enzyme defines a novel class of large-spectrum quorum-quenching lactonases. *Applied and Environmental Microbiology*, *74*(5), 1357–1366. doi:10.1128/AEM.02014-07 PMID:18192419

Uroz, S., Oger, P., Chhabra, S. R., Cámara, M., Williams, P., & Dessaux, Y. (2007). N-acyl homoserine lactones are degraded via an amidolytic activity in Comamonas sp. strain D1. *Archives of Microbiology*, *187*(3), 249–256. doi:10.100700203-006-0186-5 PMID:17136382

USEPA (United States Environmental Protection Agency). (2006). *A Citizen's Guide to Bioremediation*. USEPA.

USEPA. (2011). *Petroleum refining*. OAR, Office of Air Quality Planning and Standards (OAQPS). Available on: https://www3.epa.gov/ttnchie1/ap42/ch05/final/c05s01.pdf

Usman, S. (2018). *Technology of bioorganic fertilizer production: Treasures for north-west Nigeria*. Furtunate Print.

Uysal, A. K., & Gunal, S. (2014). The impact of preprocessing on text classification. *Information Processing & Management*, *50*(1), 104–112. doi:10.1016/j.ipm.2013.08.006

Uysal, O., Uysal, F. O., & Ekinci, K. (2015). Evaluation of microalgae as microbial fertilizer. *European Journal of Sustainable Development*, *4*(2), 77. doi:10.14207/ejsd.2015.v4n2p77

Vacheron, J., Desbrosses, G., Bouffaud, M. L., Touraine, B., Moënne-Loccoz, Y., Muller, D., Legendre, L., Wisniewski-Dyé, F., & Prigent-Combaret, C. (2013). Plant growth-promoting rhizobacteria and root system functioning. *Frontiers in Plant Science*, *4*, 356. doi:10.3389/fpls.2013.00356 PMID:24062756

Vaikuntapu, P. R., Dutta, S., Samudrala, R. B., Rao, V. R. V. N., Kalam, S., & Podile, A. R. (2014). Preferential Promotion of *Lycopersicon esculentum* (Tomato) Growth by Plant Growth Promoting Bacteria Associated with Tomato. *Indian Journal of Microbiology*, *54*(4), 403–412. doi:10.100712088-014-0470-z PMID:25320438

Vainio, E. J., Müller, M. M., Korhonen, K., Piri, T., & Hantula, J. (2014). Viruses accumulate in aging infection centers of a fungal forest pathogen. *The ISME Journal*, *9*(2), 497–507. doi:10.1038/ismej.2014.145 PMID:25126757

Valarie, E. (1999). Bioremediation of bauxite residue using indigenous bacteria. Minerals council of Australia Environmental Workshop, 311.

Vallet, I., Diggle, S. P., Stacey, R. E., Cámara, M., Ventre, I., Lory, S., Lazdunski, A., Williams, P., & Filloux, A. (2004). Biofilm formation in Pseudomonas aeruginosa: Fimbrial cup gene clusters are controlled by the transcriptional regulator MvaT. *Journal of Bacteriology*, *186*(9), 2880–2890. doi:10.1128/JB.186.9.2880-2890.2004 PMID:15090530

Valls, M., Atrian, S., de Lorenzo, V., & Fernandez, L. A. (2000). Engineering a mouse metallothionein on the cell surface of Ralstonia eutropha CH34 for immobilization of heavy metals in soil. *Nature Biotechnology*, *18*, 661–665.

Valls, M., & de Lorenzo, V. (2002). Exploiting the genetic and biochemical capacities of bacteria for the remediation of heavy metal pollution. *FEMS Microbiology Reviews*, *26*(4), 327–338.

Van Den Berg, M. A., Albang, R., Albermann, K., Badger, J. H., Daran, J. M., Driessen, A. J., ... Joardar, V. (2008). Genome sequencing and analysis of the filamentous fungus *Penicillium chrysogenum. Nature Biotechnology*, *26*(10), 1161–1168. doi:10.1038/nbt.1498 PMID:18820685

Van den Wijngaard, A. J., Van der Kamp, K. W., Van der Ploeg, J., Pries, F., Kazemier, B., & Janssen, D. B. (1992). Degradation of 1, 2-dichloroethane by Ancylobacter aquaticus and other facultative methylotrophs. *Applied and Environmental Microbiology*, *58*(3), 976–983. doi:10.1128/AEM.58.3.976-983.1992 PMID:1575500

van den Wijngaard, A. J., Wind, R. D., & Janssen, D. B. (1993). Kinetics of bacterial growth on chlorinated aliphatic compounds. *Applied and Environmental Microbiology*, *59*(7), 2041–2048. doi:10.1128/AEM.59.7.2041-2048.1993 PMID:16348981

Van Der Heijden, M. G., & Horton, T. R. (2009). Socialism in soil? The importance of mycorrhizal fungal networks for facilitation in natural ecosystems. *Journal of Ecology*, *97*(6), 1139–1150. doi:10.1111/j.1365-2745.2009.01570.x

van der Zaan, B., de Weert, J., Rijnaarts, H., de Vos, W. M., Smidt, H., & Gerritse, J. (2009). Degradation of 1, 2-dichloroethane by microbial communities from river sediment at various redox conditions. *Water Research*, *43*(13), 3207–3216. doi:10.1016/j.watres.2009.04.042 PMID:19501382

Van der Zee, F. P., & Villaverde, S. (2005). Combined anaerobic–aerobic treatment of azo dyes—A short review of bioreactor studies. *Water Research*, *39*(8), 1425–1440. doi:10.1016/j.watres.2005.03.007 PMID:15878014

Van Epps, A. (2006). *Phytoremediation of Petroleum Hydrocarbons*. U.S. Environmental Protection Agency.

Van Loon, L. C., Bakker, P. A. H. M., & Pieterse, C. M. J. (1998). Systemic resistance induced by rhizosphere bacteria. *Annual Review of Phytopathology*, *36*(1), 453–483. doi:10.1146/annurev.phyto.36.1.453 PMID:15012509

Van Schöll, L., Kuyper, T. W., Smits, M. M., Landeweert, R., Hoffland, E., & Van Breemen, N. (2008). Rock-eating mycorrhizas: Their role in plant nutrition and biogeochemical cycles. *Plant and Soil*, *303*(1-2), 35–47. doi:10.100711104-007-9513-0

Vandevivere, P. C., Bianchi, R., & Verstraete, W. (1998). Treatment and Reuse of Wastewater from the Textile Wet-processing Industry: Review of Emerging Technologies. *Journal of Industrial Microbiology & Biotechnology*, *72*, 289.

Vangronsveld, J., Weyens, N., Thijs, S., Dubin, D., Clemmens, M., Van Geert, K., van den Eeckhaut, M., van den Bossche, P., van Gestel, G., Bruneel, N., Crauwels, L. & Lemmens, C. (2019). Phytoremediation – Code of Good Practice. *Ovam*, 1-132.

Vardhan, V., & Dubey, K. (2009). Bioremediation of Soil Contaminated with Petroleum Crude Oil. In *National Seminar on Frontiers in Biotechnology (NSFB-2009)*. Department of Biotechnology, Bharathiar University.

Varga, J., Rigo, K., Molnar, J., Toth, B., Szencz, S., & Teren, J. (2003). Mycotoxin production and evolutionary relationships among species of *Aspergillus* section Clavati. *Antonie van Leeuwenhoek*, *83*(2), 191–200. doi:10.1023/A:1023355707646 PMID:12785313

Varinderpal-Singh, Sharma, S., Kunal, Gosal, S. K., Choudhary, R., Singh, R., Adholeya, A., & Bijay-Singh. (2020). Synergistic Use of Plant Growth-Promoting Rhizobacteria, Arbuscular Mycorrhizal Fungi, and Spectral Properties for Improving Nutrient Use Efficiencies in Wheat (*Triticum aestivum* L.). *Communications in Soil Science and Plant Analysis*, *51*(1), 14–27. doi:10.1080/00103624.2019.1689259

Varjani, S. J. (2017). Microbial degradation of petroleum hydrocarbons. *Bioresource Technology*, *223*, 277–286.

Varma, P. K., Uppala, S., Pavuluri, K., Chandra, K. J., Chapala, M. M., & Kumar, K. V. K. (2017). Endophytes: Role and functions in crop health. In *Plant-Microbe Interactions in Agro-Ecological Perspectives*. doi:10.1007/978-981-10-5813-4_15

Vasavi, A., Usha, R., & Swamy, P. M. (2010). Phytoremediation–an overview review. *J Ind Pollut Control*, *26*(1), 83–88.

Vassileva, M., Azcon, R., Barea, J., & Vassilev, N. (2000). Rock phosphate solubilization by free and encapsulated cells of *Yarrowia lipolytica*. *Process Biochemistry*, *35*(7), 693–697. doi:10.1016/S0032-9592(99)00132-6

Vaziri, A., Panahpour, E., & Beni, M. H. M. (2013). Phytoremediation, a method for treatment of petroleum hydrocarbon contaminated soils. *International Journal of Farm and Allied Sciences*, *2*(21), 909–913.

Vázquez-Núñez, E., Molina-Guerrero, C. E., Peña-Castro, J. M., Fernández-Luqueño, F., & de la Rosa-Álvarez, M. (2020). Use of Nanotechnology for the Bioremediation of Contaminants: A Review. *Processes (Basel, Switzerland)*, *8*(7), 826. doi:10.3390/pr8070826

Vejan, P., Abdullah, R., Khadiran, T., Ismail, S., & Nasrulhaq Boyce, A. (2016). Role of Plant Growth Promoting Rhizobacteria in Agricultural Sustainability-A Review. *Molecules (Basel, Switzerland)*, *21*(5), 573. doi:10.3390/molecules21050573 PMID:27136521

Velmourougane, K., Saxena, G., & Prasanna, R. (2017). Plant-microbe interactions in the rhizosphere: Mechanisms and their ecological benefits. In *Plant-microbe interactions in agro-ecological perspectives*. doi:10.1007/978-981-10-6593-4_7

Venkata Mohan, S., Chandrasekhar Rao, N., Krishna Prasad, K., & Karthikeyan, J. (2002). Treatment of simulated Reactive Yellow 22 (Azo) dye effluents using *Spirogyra* species. *Waste Management (New York, N.Y.)*, *22*(6), 575–582. doi:10.1016/S0956-053X(02)00030-2 PMID:12214968

Ventura, L. M. B., Mateus, V. L., de Almeida, A. C. S. L., Wanderley, K. B., Taira, F. T., Saint'Pierre, T. D., & Gioda, A. (2017). Chemical composition of fine particles (PM2.5): Water-soluble organic fraction and trace metals. *Air Quality, Atmosphere & Health*, *10*, 845–852.

Venturi, V., & Keel, C. (2016). Signaling in the rhizosphere. *Trends in Plant Science*, *21*(3), 187–198. doi:10.1016/j.tplants.2016.01.005 PMID:26832945

Verbon, E. H., & Liberman, L. M. (2016). Beneficial microbes affect endogenous mechanisms controlling root development. *Trends in Plant Science*, *21*(3), 218–229. doi:10.1016/j.tplants.2016.01.013 PMID:26875056

Verma, S. K., & Gond, S. K. (2017). Fungal endophytes representing diverse habitats and their role in plant protection. *Developments in Fungal Biology and Applied Mycology*, 135-157.

Verma, J. P., Yadav, J., Tiwari, K. N., & Jaiswal, D. K. (2014). Evaluation of plant growth promoting activities of microbial strains and their effect on growth and yield of chickpea (*Cicer arietinum* L.) in India. *Soil Biology & Biochemistry*, *70*, 33–37. doi:10.1016/j.soilbio.2013.12.001

Verma, J. P., Yadav, J., Tiwari, K. N., Lavakush, & Singh, V. (2010). Impact of plant growth promoting rhizobacteria on crop production. *International Journal of Agricultural Research*, *5*(11), 954–983. doi:10.3923/ijar.2010.954.983

Verma, N., & Sharma, R. (2017). Bioremediation of Toxic Heavy Metals: A Patent Review. *Recent Patents on Biotechnology*, *11*(3), 171–187. doi:10.2174/1872208311666170111111631 PMID:28078980

Verma, P., & Madamwar, D. (2003). Decolourization of synthetic dyes by a newly isolated strain of Serratia marcescens. *World Journal of Microbiology & Biotechnology*, *19*(6), 615–618. doi:10.1023/A:1025115801331

Verma, S. K., & White, J. F. (2019). *Seed Endophytes*. Springer. doi:10.1007/978-3-030-10504-4

Verma, S., & Kuila, A. (2019). Bioremediation of heavy metals by microbial process. *Environmental Technology & Innovation*, *14*, 100369. doi:10.1016/j.eti.2019.100369

Verma, S., & Kuila, A. (2019). Bioremediation of heavy metals by microbial process. *Environmental Technology and Innovation*, *14*, 100369.

Verslues, P. E. (2017). Time to grow: Factors that control plant growth during mild to moderate drought stress. *Plant, Cell & Environment*, *40*(2), 177–179. doi:10.1111/pce.12827 PMID:27588960

Vessey, J. K. (2003). Plant growth promoting rhizobacteria as biofertilizers. *Plant and Soil*, *255*(2), 571–586. doi:10.1023/A:1026037216893

Vessey, J. K., & Buss, T. J. (2002). Bacillus cereus UW85 inoculation effects on growth, nodulation, and N accumulation in grain legumes: Controlled-environment studies. *Canadian Journal of Plant Science*, *82*(2), 282–290. doi:10.4141/P01-047

Vicente-Serrano, S. M., Gouveia, C., Camarero, J. J., Beguería, S., Trigo, R., López-Moreno, J. I., ... Morán-Tejeda, E. (2013). Response of vegetation to drought time-scales across global land biomes. *Proceedings of the National Academy of Sciences of the United States of America*, *110*(1), 52–57. doi:10.1073/pnas.1207068110 PMID:23248309

Vidali, M. (2001). Bioremediation. an overview. *Pure and Applied Chemistry*, *73*(7), 1163–1172. doi:10.1351/pac200173071163

Vidali, M. (2001). Bioremediation. An overview. *Pure and Applied Chemistry*, *73*, 1163–1172.

Vidya Lakshmi, C., Kumar, M., & Khanna, S. (2008). Biotransformation of chlorpyrifos and bioremediation of contaminated soil. *International Biodeterioration & Biodegradation*, *62*, 204–209.

Vieira, S., Sikorski, J., Dietz, S., Herz, K., Schrumpf, M., Bruelheide, H., Scheel, D., Friedrich, M. W., & Overmann, J. (2020). Drivers of the composition of active rhizosphere bacterial communities in temperate grasslands. *The ISME Journal*, *14*(2), 463–475. doi:10.103841396-019-0543-4 PMID:31659233

Vijayakumar, S. (2012). Potential Applications of Cyanobacteria in Industrial Effluents-A Review. *Journal of Bioremediation & Biodegradation*, *3*(154), 1–4. doi:10.4172/2155-6199.1000154

Vijayakumar, S., & Manoharan, C. (2012). Treatment of dye industry effluent using free and immobilized cyanobacteria. *Bioremed. Biodeg., 3*(165), 1–6. doi:10.4172/2155-6199.1000165

Vijayakumar, S., Thajuddin, N., & Manoharan, C. (2005). Role of cyanobacteria in the treatment of dye industry effluent. *Pollution Research, 24*(1), 69–74.

Vijayakumar, S., Thajuddin, N., & Manoharan, C. (2007). Biodiversity of cyanobacteria in industrial effluents. *Acta Botanica Malacitana, 32*, 27–34. doi:10.24310/abm.v32i0.7026

Vijayaraghavan, K., & Yun, Y. S. (2007a). Chemical Modification and Immobilization of *Corynebacterium glutamicum* for Biosorption of Reactive Black 5 from Aqueous Solution. *Industrial & Engineering Chemistry Research, 46*(2), 608–617. doi:10.1021/ie061158g

Vijayaraghavan, K., & Yun, Y. S. (2007b). Utilization of fermentation waste (*Corynebacterium glutamicum*) for biosorption of Reactive Black 5 from aqueous solution. *Journal of Hazardous Materials, 141*(1), 45–52. doi:10.1016/j.jhazmat.2006.06.081 PMID:16879915

Vimala, P. P., & Mathew, L. (2016). Biodegradation of Polyethylene using *Bacillus subtilis. Procedia Technology, 24*, 232–239.

Vinay, B. R., Uzma, M., Govindappa, M., Vasantha, R. A., & Lokesh, S. (2016). Screening and Identification of Polyurethane (PU) and low density poly-ethene (LDPE) degrading soil fungi isolated from municipal solid waste. *International Journal of Current Research, 8*, 34752–34761.

Virág, D., Naár, Z., & Kiss, A. (2007). Microbial toxicity of pesticide derivatives produced with UV-photodegradation. *Bulletin of Environmental Contamination and Toxicology, 79*(3), 356–359. doi:10.100700128-007-9230-7 PMID:17639315

Visconti, D., Álvarez-Robles, M. J., Fiorentino, N., Fagnano, M., & Clemente, R. (2020). Use of *Brassica juncea* and *Dactylis glomerata* for the phytostabilization of mine soils amended with compost or biochar. *Chemosphere, 260*, 127661. doi:10.1016/j.chemosphere.2020.127661 PMID:32688327

Vishnivetskaya, T. A., Petrova, M. A., Urbance, J., Ponder, M., Moyer, C. L., Gilichinsky, D. A., & Tiedje, J. M. (2006). Bacterial community in ancient Siberian permafrost as characterized by culture and culture-independent methods. *Astrobiology, 6*(3), 400–414. doi:10.1089/ast.2006.6.400 PMID:16805696

Vitor, S. L., Eliana, F. C. S., & Fernando, J. S. O. (2018). Biosurfactant- assisted phytoremediation of multi-contaminated industrial soil using sunflower (*Helianthus annuus* L.). *Journal of Environmental Science and Health. Part A, Toxic/Hazardous Substances & Environmental Engineering, 53*(7), 609–616. doi:10.1080/10934529.2018.1429726 PMID:29388890

Vitor, V., & Corso, C. R. (2008). Decolorization of textile dye by Candida albicans isolated from industrial effluents. *Journal of Industrial Microbiology & Biotechnology, 35*(11), 1353–1357. doi:10.100710295-008-0435-5 PMID:18712543

Vitousek, P. M., & Howarth, R. W. (1991). Nitrogen limitation on land and in the sea: How can it occur? *Biogeochemistry, 13*(2), 87–115. doi:10.1007/BF00002772

Vivekananthan, V., Selvapriya, A., Janani, D., & Narendhar, C. (2014). Synthesis of mixed oxides of cerium-iron nanostructures for effective removal of heavy metals from wastewater. *Research Journal of Recent Sciences, 3*, 212–217.

Voberkova, S., Vaverkova, M. D., Buresova, A., Adamcova, D., Vrsanska, M., Kynicky, J., Brtnicky, M., & Adam, V. (2017). Effect of inoculation with white-rot fungi and fungal consortium on the composting efficiency of municipal solid waste. *Waste Management (New York, N.Y.), 61*, 157–164.

Vogt, C., & Richnow, H. (2014). Bioremediation via in situ microbial degradations of organic pollutants. *Advances in Biochemical Engineering/Biotechnology, 142*, 123–146.

Vos, I. A., Pieterse, C. M., & Van Wees, S. C. (2013). Costs and benefits of hormone-regulated plant defences. *Plant Pathology*, *62*, 43–55. doi:10.1111/ppa.12105

Vurukonda, S. S. K. P., Vardharajula, S., Shrivastava, M., & SkZ, A. (2016a). Multifunctional Pseudomonas putida strain *FBKV2* from arid rhizosphere soil and its growth promotional effects on maize under drought stress. *Rhizosphere*, *1*, 4–13. doi:10.1016/j.rhisph.2016.07.005

Vurukonda, S. S. K. P., Vardharajula, S., Shrivastava, M., & SkZ, A. (2016b). Enhancement of drought stress tolerance in crops by plant growth promoting rhizobacteria. *Microbiological Research*, *184*, 13–24. doi:10.1016/j.micres.2015.12.003 PMID:26856449

Vyas, B. R., & Molitoris, H. P. (1995). Involvement of an extracellular H_2O_2-dependent ligninolytic activity of the white-rot fungus *Pleurotus ostrus* in the decolorization of remazol brilliant blue R. *Applied and Environmental Microbiology*, *61*(11), 3919–3927. doi:10.1128/AEM.61.11.3919-3927.1995 PMID:8526504

Wagner, D., Gattinger, A., Embacher, A., Pfeiffer, E.-M., Schloter, M., & Lipski, A. (2007). Methanogenic activity and biomass in Holocene permafrost deposits of the Lena Delta, Siberian Arctic and its implication for the global methane budget. *Global Change Biology*, *13*(5), 1089–1099. doi:10.1111/j.1365-2486.2007.01331.x

Walker, V., Bertrand, C., Bellvert, F., Moënne-Loccoz, Y., Bally, R., & Comte, G. (2011). Host plant secondary metabolite profiling shows a complex, strain-dependent response of maize to plant growth-promoting rhizobacteria of the genus Azospirillum. *The New Phytologist*, *189*(2), 494–506. doi:10.1111/j.1469-8137.2010.03484.x PMID:20946131

Wallace, R. L., Hirkala, D. L., & Nelson, L. M. (2018). Mechanisms of action of three isolates of Pseudomonas fluorescens active against postharvest grey mold decay of apple during commercial storage. *Biological Control*, *117*, 13–20. doi:10.1016/j.biocontrol.2017.08.019

Walpola, B., & Yoon, M. (2012). Prospectus of phosphate solubilizing microorganisms and phosphorus availability in agricultural soils. A review. *African Journal of Microbiological Research*, *6*(37), 6600–6605.

Walsh, U. F., Morrissey, J. P., & O'Gara, F. (2001). *Pseudomonas* for biocontrol of phytopathogens: From functional genomics to commercial exploitation. *Current Opinion in Biotechnology*, *12*(3), 289–295. doi:10.1016/S0958-1669(00)00212-3 PMID:11404107

Walton, B. T., Hoylman, A. M., Perez, M. M., Anderson, T. A., Johnson, T. R., Guthrie, E. A., & Christman, R. F. (1994). Rhizosphere microbial communitites as a plant defense against toxic substances in soils. In T. A. Anderson & J. R. Coats (Eds.), *Bioremediation Through Rhizosphere Technology* (pp. 82–92). American Chemical Society. doi:10.1021/bk-1994-0563.ch007

Wandersman, C., & Delepelaire, P. (2004). Bacterial iron sources: From siderophores to hemophores. *Annual Review of Microbiology*, *58*(1), 611–647. doi:10.1146/annurev.micro.58.030603.123811 PMID:15487950

Wang, B., Liu, L., O'Leary, G. J., Asseng, S., Macadam, I., Lines-Kelly, R., Yang, X., Clark, A., Crean, J., Sides, T., Xing, H., Mi, C., & Yu, Q. (2018). Australian wheat production expected to decrease by the late 21st century. *Global Change Biology*, *24*(6), 2403–2415. doi:10.1111/gcb.14034 PMID:29284201

Wang, C. B., & Zhang, W. X. (1997). Synthesizing nanoscale iron particles for rapid and complete dechlorination of TCE and PCBs. *Environmental Science & Technology*, *31*(7), 2154–2156. doi:10.1021/es970039c

Wang, C. L., Michels, P. C., Dawson, S. C., Kitisakkul, S., Baross, J. A., Keasling, J. D., & Clark, D. S. (1997). Cadmium removal by a new strain of *Pseudomonas aeruginosa* in aerobic culture. *Applied and Environmental Microbiology*, *63*(10), 4075–4078. doi:10.1128/AEM.63.10.4075-4078.1997 PMID:9327571

Wang, C. L., Zhao, M., Wei, X. D., Li, T. L., & Lu, L. (2010). Characteristics of spore-bound laccase from *Bacillus subtilis* WD23 and its use in dye decolorization. *Advanced Materials Research, 113–116*, 226–230. . doi:10.4028/www.scientific.net/AMR.113-116.226

Wang, C., Zhao, M., Lu, L., Wei, X., & Li, T. (2011). Characterization of spore laccase from Bacillus subtilis WD23 and its use in dye decolorization. *African Journal of Biotechnology*, *10*(11), 2186–2192.

Wang, J. S., & Chiu, K. (2009). Destruction of pentachlorobiphenyl in soil by supercritical CO2 extraction coupled with polymer-stabilized palladium nanoparticles. *Chemosphere*, *75*(5), 629–633. doi:10.1016/j.chemosphere.2009.01.018 PMID:19211124

Wang, L. H., Weng, L. X., Dong, Y. H., & Zhang, L. H. (2004). Specificity and enzyme kinetics of the quorum-quenching N-Acyl homoserine lactone lactonase (AHL-lactonase). *The Journal of Biological Chemistry*, *279*(14), 13645–13651. doi:10.1074/jbc.M311194200 PMID:14734559

Wang, L., Chi, X. Q., Zhang, J. J., Sun, D. L., & Zhou, N. Y. (2014). Bioaugmentation of a methyl parathion contaminated soil with Pseudomonas sp. strain WBC-3. *International Biodeterioration & Biodegradation*, *87*, 116–121.

Wang, L., Dong, X. P., Zhang, W., Zhang, G. S., Liu, G. X., & Feng, H. Y. (2011). Quantitative characters of microorganisms in permafrost at different depths and their relation to soil physicochemical properties. *Bingchuan Dongtu*, *33*, 436–441.

Wang, L., Jiang, J., Wang, Y., Hong, N., Zhang, F., Xu, W., & Wang, G. (2014). Hypovirulence of the phytopathogenic fungus *Botryosphaeria dothidea*: Association with a coinfecting chrysovirus and a partitivirus. *Journal of Virology*, *88*(13), 7517–7527. doi:10.1128/JVI.00538-14 PMID:24760881

Wang, L., Ji, B., Hu, Y., Liu, R., & Sun, W. (2017). A review on in situ phytoremediation of mine tailings. *Chemosphere*, *184*, 594–600. doi:10.1016/j.chemosphere.2017.06.025 PMID:28623832

Wang, M., & Jin, H. (2017). Spray-induced gene silencing: A powerful innovative strategy for crop protection. *Trends in Microbiology*, *25*(1), 4–6. doi:10.1016/j.tim.2016.11.011 PMID:27923542

Wang, N., Hua, H., Eneji, A. E., Li, Z., Duan, L., & Tian, X. (2012). Genotypic variation in photosynthetic and physiological adjustment to potassium deficiency in cotton (*Gossypium hirsutum*). *Journal of Photochemistry and Photobiology*, *110*, 1–8. doi:10.1016/j.jphotobiol.2012.02.002 PMID:22387141

Wang, Q., Xie, S., & Hu, R. (2013). Bioaugmentation with Arthrobacter sp. strain DAT1 for remediation of heavily atrazine-contaminated soil. *International Biodeterioration & Biodegradation*, *77*, 63–67.

Wang, Q., Xiong, D., Zhao, P., Yu, X., Tu, B., & Wang, G. (2011). Effect of applying an arsenic-resistant and plant growth–promoting rhizobacterium to enhance soil arsenic phytoremediation by Populus deltoides LH05-17. *Journal of Applied Microbiology*, *111*(5), 1065–1074. doi:10.1111/j.1365-2672.2011.05142.x PMID:21895895

Wang, S., Kondo, H., Liu, L., Guo, L., & Qiu, D. (2013). A novel virus in the family Hypoviridae from the plant pathogenic fungus *Fusarium graminearum. Virus Research*, *174*(1-2), 69–77. doi:10.1016/j.virusres.2013.03.002 PMID:23499998

Wang, S., Xu, Y., Lin, Z., Zhang, J., Norbu, N., & Liu, W. (2017). The harm of petroleum-polluted soil and its remediation research. *AIP Conference Proceedings*, *1864*, 020222. doi:10.1063/1.4993039

Wang, T., Weissman, J., Ramesh, G., Varadarajan, R., & Benemann, J. R. (1998). Heavy Metal Binding and Removal by Phormidium. *Bulletin of Environmental Contamination and Toxicology*, *60*(5), 739–744. doi:10.1007001289900688 PMID:9595189

Wang, W., & Shao, Z. (2012). Diversity of flavin-binding monooxygenase genes (almA) in marine bacteria capable of degradation long-chain alkanes. *FEMS Microbiology Ecology*, *80*(3), 523–533. doi:10.1111/j.1574-6941.2012.01322.x PMID:22304419

Wang, W., & Shao, Z. (2013). Enzymes and genes involved in aerobic alkane degradation. *Frontiers in Microbiology*, *4*, 116. doi:10.3389/fmicb.2013.00116 PMID:23755043

Wang, X., Cheng, X., Sun, D., & Qi, H. (2008). Biodecolorization and partial mineralization of Reactive Black 5 by a strain of *Rhodopseudomonas palustris*. *Journal of Environmental Sciences (China)*, *20*(10), 1218–1225. doi:10.1016/S1001-0742(08)62212-3 PMID:19143346

Wang, X., Gai, Z., Yu, B., Feng, J., Xu, C., Yuan, Y., Lin, Z., & Xu, P. (2007). Degradation of carbazole bymicrobial cells immobilized in magnetic gellan gum gel beads. *Applied and Environmental Microbiology*, *73*(20), 6421–6428. doi:10.1128/AEM.01051-07 PMID:17827304

Wang, Y. H., & Irving, H. R. (2011). Developing a model of plant hormone interactions. *Plant Signaling & Behavior*, *6*(4), 494–500. doi:10.4161/psb.6.4.14558 PMID:21406974

Wang, Y., Guo, J., & Liu, R. (2001). Biosorption of heavy metals by bacteria isolated from activated sludge. *Applied Biochemistry and Biotechnology*, *91–93*, 171–184.

Wang, Y., Jiao, X., & Song, L. (2014). Soil and soil environmental quality monitoring in China: A review. *Environment International*, *69*, 177–199. doi:10.1016/j.envint.2014.04.014 PMID:24875802

Wang, Y., Pan, C., Chu, W., Vipin, A. K., & Sun, L. (2019). Environmental Remediation Applications of Carbon Nanotubes and Graphene Oxide: Adsorption and Catalysis. *Nanomaterials (Basel, Switzerland)*, *9*(3), 439. doi:10.3390/nano9030439 PMID:30875970

Wang, Y., Shi, J., Lin, Q., Chen, X., & Chen, Y. (2007). Heavy metal availability and impact on activity of soil microorganisms along a Cu/Zn contamination gradient. *Journal of Environmental Sciences (China)*, *19*(7), 848–853. doi:10.1016/S1001-0742(07)60141-7 PMID:17966873

Wang, Y., Zhao, H., Xue, C., Xu, C., Geng, Y., & Zang, R., Guo, Y., Wu, H., & Zhang, M. (2020). Complete genome sequence of a novel mycovirus isolated from the phytopathogenic fungus *Corynespora cassiicola* in China. *Archives of Virology*, 1–4. PMID:32757057

Wang, Z., Ren, D., Kang, C., Zhang, S., Zhang, X., Deng, Z., Huang, C., & Guo, H. (2020). Migration of heavy metals and migration-degradation of phenanthrene in soil using electro kinetic-laccase combined remediation system. *Journal of Environmental Science and Health. Part B, Pesticides, Food Contaminants, and Agricultural Wastes*, *55*(8), 704–711. doi:10.1080/03601234.2020.1773719 PMID:32500809

Wang, Z., Wang, Y., Gong, F., Zhang, J., Hong, Q., & Li, S. (2010). Biodegradation of carbendazim by a novel actinobacterium *Rhodococcus* jialingiae djl-6-2. *Chemosphere*, *81*, 639–644.

Wani, S. H., Kumar, V., Shriram, V., & Sah, S. K. (2016). Phytohormones and their metabolic engineering for abiotic stress tolerance in crop plants. *The Crop Journal*, *4*(3), 162–176. doi:10.1016/j.cj.2016.01.010

Wankhade, R., & Ghugal, Y. M. (2016). Study on Soil-Structure Interaction: A Review. *International Journal of Engine Research*, *5*(3), 737–741. doi:10.17950/ijer/v5i3/047

Wan, W., Qin, Y., Wu, H., Zuo, W., He, H., Tan, J., Wang, Y., & He, D. (2020). Isolation and Characterization of Phosphorus Solubilizing Bacteria With Multiple Phosphorus Sources Utilizing Capability and Their Potential for Lead Immobilization in Soil. *Frontiers in Microbiology*, *11*, 752. doi:10.3389/fmicb.2020.00752 PMID:32390988

Wan, X., Yang, J., & Song, W. (2018). Pollution Status of Agricultural Land in China: Impact of Land Use and Geographical Position. *Soil and Water Research*, *13*, 234–242.

Ward, M. H., Jones, R. R., Brender, J. D., de Kok, T. M., Weyer, P. J., Nolan, B. T., Villanueva, C. M., & van Breda, S. G. (2018). Drinking Water Nitrate and Human Health: An Updated Review. *International Journal of Environmental Research and Public Health*, *15*(7), 1557. doi:10.3390/ijerph15071557 PMID:30041450

Ward, M. H., Lubin, J., Giglierano, J., Colt, J. S., Wolter, C., Bekiroglu, N., Camann, D., Hartge, P., & Nuckols, J. R. (2006). Proximity to crops and residential exposure to agricultural herbicides in Iowa. *Environmental Health Perspectives*, *114*(6), 893–897. doi:10.1289/ehp.8770 PMID:16759991

Ware, G. W. (1988). 1, 2-Dichloroethane. In *Reviews of Environmental Contamination and Toxicology* (pp. 69–79). Springer. doi:10.1007/978-1-4612-3922-2_6

Watharkar, A. D., & Jadhav, J. P. (2014). Detoxification and decolorization of a simulated textile dye mixture by phytoremediation using Petunia grandiflora and, Gailardia grandiflora: A plant–plant consortial strategy. *Ecotoxicology and Environmental Safety*, *103*, 1–8. doi:10.1016/j.ecoenv.2014.01.033 PMID:24561240

Watson, J. G. (1996). Physical/chemical treatment of organically contaminated soils and sediments. *Journal of the Air & Waste Management Association*, *46*(10), 993–1003. doi:10.1080/10473289.1996.10467536 PMID:28065141

Weathers, L. J., Parkin, G. F., & Alvarez, P. J. (1997). Utilization of cathodic hydrogen as electron donor for chloroform cometabolism by a mixed, methanogenic culture. *Environmental Science & Technology*, *31*(3), 880–885. doi:10.1021/es960582d

Weekley, J., Gabbard, J., & Nowak, J. (2012). Micro-level management of agricultural inputs: Emerging approaches. *Agronomy (Basel)*, *2*(4), 321–357. doi:10.3390/agronomy2040321

Wehr, J. B., Fulton, I., & Menzies, N. W. (2006). Revegetation strategies for bauxite refinery residue: A case study of Alcan Gove in Northern Territory, Australia. *Environmental Management*, *37*(3), 297–306. doi:10.100700267-004-0385-2 PMID:16456629

Wei, C. Z., Osaki, H., Iwanami, T., Matsumoto, N., & Ohtsu, Y. (2003). Molecular characterization of dsRNA segments 2 and 5 and electron microscopy of a novel reovirus from a hypovirulent isolate of the plant pathogen *Rosellina necatrix*. *The Journal of General Virology*, *84*(9), 2431–2437. doi:10.1099/vir.0.19098-0 PMID:12917464

Wei, H., Liu, Y., Xiang, H., Zhang, J., Li, S., & Yang, J. (2019). Soil pH Responses to Simulated Acid Rain Leaching in Three Agricultural Soils. *Sustainability*, *12*(1), 280–292. doi:10.3390u12010280

Weir, K. M., Sutherland, T. D., Horne, I., Russell, R. J., & Oakeshott, J. G. (2006). A single monooxygenase, ese, is involved in the metabolism of the organochlorides endosulfan and endosulfate in an *Arthrobacter sp. Applied and Environmental Microbiology*, *72*(5), 3524–3530. doi:10.1128/AEM.72.5.3524-3530.2006 PMID:16672499

Weisman, W. (1998). Total Petroleum Hydrocarbon Criteria Working Group Series: Vol. 1. *Analysis of petroleum hydrocarbons in environmental media*. Amherst Scientific Publishers.

Welch, S. A., Taunton, A. E., & Banfield, J. F. (2002). Effect of microorganisms and microbial metabolites on apatite dissolution. *Geophysical Journal of the Royal Astronomical Society*, *19*, 343–367.

Welch, S. A., & Vandevivere, P. (1994). Effect of microbial and other naturally occurring polymers on mineral dissolution. *Geomicrobiology Journal*, *12*(4), 227–238. doi:10.1080/01490459409377991

Wentzel, A., Ellingsen, T. E., Kotlar, H., Zotchev, S. B., & Throne-Holst, M. (2007). Bacterial metabolism of long-chain n-alkanes. *Applied Microbiology and Biotechnology*, *76*(6), 1209–1221. doi:10.100700253-007-1119-1 PMID:17673997

Wenzel, N., van der Lelie, D., Taghavi, S., & Vangronsveld, J. (2009). Phytoremediation: Plant-endophyte partnerships take the challenge. *Current Opinion in Biotechnology*, *20*(2), 248–254. doi:10.1016/j.copbio.2009.02.012 PMID:19327979

Wenzel, W. W. (2009). Rhizosphere processes and management in plant-assisted bioremediation (phytoremediation) of soils. *Plant and Soil*, *321*(1-2), 385–408. doi:10.100711104-008-9686-1

Wesenberg, D., Kyriakides, I., & Agathos, S. N. (2003). White-rot fungi and their enzymes for the treatment of industrial dye effluents. *Biotechnology Advances*, *22*(1-2), 161–187. doi:10.1016/j.biotechadv.2003.08.011 PMID:14623049

Weyens, N., Croes, S., Dupae, J., Newman, L., van der Lelie, D., Carleer, R., & Vangronsveld, J. (2010). Endophytic bacteria improve phytoremediation of Ni and TCE co- contamination. *Environmental Pollution*, *158*(7), 2422–2427. doi:10.1016/j.envpol.2010.04.004 PMID:20462680

Whelan, M. J., Coulon, F., Hince, G., Rayner, J., McWatters, R., Spedding, T., & Snape, I. (2015). Fate and transport of petroleum hydrocarbons in engineered biopiles in polar regions. *Chemosphere*, *131*, 232–240. doi:10.1016/j.chemosphere.2014.10.088 PMID:25563162

Whipps, J. M. (2001). Microbial interactions and biocontrol in the rhizosphere. *Journal of Experimental Botany*, *52*(suppl_1), 487–511. doi:10.1093/jxb/52.suppl_1.487 PMID:11326055

White, A. F. (2003). Natural weathering rates of silicate minerals. *Treatise on Geochemistry*, *5*, 133–168. doi:10.1016/B0-08-043751-6/05076-3

White, D. P. (1941). Prairie soil as a medium for tree growth. *Ecology*, *22*(4), 398–407. doi:10.2307/1930714

Whitehead, N. A., Barnard, A. M., Slater, H., Simpson, N. J., & Salmond, G. P. (2001). Quorum-sensing in Gram-negative bacteria. *FEMS Microbiology Reviews*, *25*(4), 365–404. doi:10.1111/j.1574-6976.2001.tb00583.x PMID:11524130

Whiteley, C. G., & Lee, D. J. (2006). Enzyme technology and biological remediation. *Enzyme and Microbial Technology*, *38*(3-4), 291–316. doi:10.1016/j.enzmictec.2005.10.010

White, P. J., & Karley, A. J. (2010). "Potassium," in Plant Cell Monographs 17. In R. Hell & R. R. Mendel (Eds.), *Cell Biology of Metals and Nutrients* (pp. 199–224). Springer-Verlag. doi:10.1007/978-3-642-10613-2_9

WHO. (1995). *1,2-Dichloroethane* (2nd ed.). World Health Organization. https://apps.who.int/iris/handle/10665/37243

WHO. (2016). *News Release*. https://www.who.int/news-room/detail/15-03-2016-an-estimated-12-6-million-deaths-each-year-are-attributable-to-unhealthyenvironments

Wiegel, J., & Wu, Q. (2000). Microbial reductive dehalogenation of polychlorinated biphenyls. *FEMS Microbiology Ecology*, *32*, 1–15.

Wildberger, P., Pfeiffer, M., Brecker, L., Rechberger, G. N., Birner-Gruenberger, R., & Nidetzky, B. (2015). Phosphoryl transfer from α-d-glucose 1-phosphate catalyzed by *Escherichia coli* sugar-phosphate phosphatases of two protein superfamily types. *Applied and Environmental Microbiology*, *81*(5), 1559–1572. doi:10.1128/AEM.03314-14 PMID:25527541

Wild, E., Dent, J., Thomas, G. O., & Jones, K. C. (2005). Direct observation of organic contaminant uptake, storage, and metabolism within plant roots. *Environmental Science & Technology*, *39*(10), 3695–3702. doi:10.1021/es048136a PMID:15952374

Wilde, E. J., Hughes, A., Blagova, E. V., Moroz, O. V., Thomas, R. P., Turkenburg, J. P., Raines, D. J., Duhme-Klair, A. K., & Wilson, K. S. (2017). Interactions of the periplasmic binding protein CeuE with Fe(III) n-LICAM^{4-} siderophore analogues of varied linker length. *Scientific Reports*, *7*(1), 45941. doi:10.1038rcp45941 PMID:28383577

Wilde, S. A. (1944). Mycorrhizae and silviculture. *Journal of Forestry*, *42*, 290.

Williams, P. P. (1977). Metabolism of synthetic organic pesticides by anaerobic microorganisms. In *Residue reviews* (pp. 63–135). Springer. doi:10.1007/978-1-4612-6352-4_3

Wilmes, P., & Bond, P. L. (2006). Metaproteomics: Studying functional gene expression in microbial ecosystems. *Trends in Microbiology*, *14*, 92–97.

Wilson, B. R., Bogdan, A. R., Miyazawa, M., Hashimoto, K., & Tsuji, Y. (2016). Siderophores in iron metabolism: From mechanism to therapy potential. *Trends in Molecular Medicine*, *22*(12), 1077–1090. doi:10.1016/j.molmed.2016.10.005 PMID:27825668

Wilson, M. K., Abergel, R. J., Arceneaux, J. E., Raymond, K. N., & Byers, B. R. (2010). Temporal production of the two *Bacillus anthracis* siderophores, petrobactin and bacillibactin. *Biometals*, *23*(1), 129–134. doi:10.100710534-009-9272-x PMID:19816776

Windt, W. D., Aelterman, P., & Verstraete, W. (2005). Bioreductive deposition of palladium(0) nanoparticles on *Shewanella oneidensis* with catalytic activity towards reductive dechlorination of polychlorinated biphenyls. *Environmental Microbiology*, *7*(3), 314–325. doi:10.1111/j.1462-2920.2005.00696.x PMID:15683392

Winkelmann, G. (2002). Microbial siderophore-mediated transport. *Biochemical Society Transactions*, *30*(4), 691–696. doi:10.1042/bst0300691 PMID:12196166

Winsborough, C., & Basiliko, N. (2010). Fungal and bacterial activity in northern peatlands. *Geomicrobiology Journal*, *27*(4), 315–320. doi:10.1080/01490450903424432

Wiszniewska, A., Hanus-Fajerska, E., Muszyńska, E., & Ciarkowska, K. (2016). Natural organic amendments for improved phytoremediation of polluted soils: A review of recent progress. *Pedosphere*, *26*(1), 1–12. doi:10.1016/S1002-0160(15)60017-0

Wong, M. H. (2012). *Environmental Contamination: Health Risks and Ecological Restoration*. CRC Press. doi:10.1201/b12531

Wong, W. C., & Ho, G. E. (1993). Use of Waste Gypsum in the Revegetation on Red Mud Deposits: A Greenhouse Study. *Waste Management & Research*, *11*(3), 249–256. doi:10.1177/0734242X9301100306

Wong, Y. (1999). Laccase-catalyzed decolorization of synthetic dyes. *Water Research*, *33*(16), 3512–3520. doi:10.1016/S0043-1354(99)00066-4

Wood, N. (2001). Nodulation by numbers: The role of ethylene in symbiotic nitrogen fixation. *Trends in Plant Science*, *6*(11), 501–502. doi:10.1016/S1360-1385(01)02128-8 PMID:11701355

Wood, T., & Cummings, B. (1992). Biotechnology and the future of VAM commercialization. In M. F. Allen (Ed.), *Mycorrhizal functioning* (pp. 468–487). Chapman and Hall.

Woo, S. L., & Pepe, O. (2018). Microbial consortia: Promising probiotics as plant biostimulants for sustainable agriculture. *Frontiers in Plant Science*, *9*, 1801. doi:10.3389/fpls.2018.01801 PMID:30564264

Wuana, R. A., & Okieimen, F. E. (2011). Heavy Metals in Contaminated Soils: A Review of Sources, Chemistry, Risks and Best Available Strategies for Remediation. *International Scholarly Research Network Ecology*, *2011*, 1–20. doi:10.5402/2011/402647

Wu, J. Y., Hwang, S. C. J., Chen, C. T., & Chen, K. C. (2005). Decolorization of azo dye in a FBR reactor using immobilized bacteria. *Enzyme and Microbial Technology*, *37*(1), 102–112. doi:10.1016/j.enzmictec.2005.02.012

Wu, J., Kamal, N., Hao, H., Qian, C., Liu, Z., Shao, Y., Zhong, X., & Xu, B. (2019). Endophytic *Bacillus megaterium* BM18-2 mutated for cadmium accumulation and improving plant growth in Hybrid *Pennisetum*. *Biotechnology Reports (Amsterdam, Netherlands)*, *24*, 1–9. doi:10.1016/j.btre.2019.e00374 PMID:31763195

Wu, K., Dumat, C., Li, H., Xia, H., Li, Z., & Wu, J. (2019). Responses of soil microbial community and enzymes during plant-assisted biodegradation of di-(2-ethylhexyl) phthalate and pyrene. *International Journal of Phytoremediation*, *21*(7), 683–692. doi:10.1080/15226514.2018.1556586 PMID:30924369

Wu, L., Jiang, Y., Zhao, F., He, X., Liu, H., & Yu, K. (2020). Increased organic fertilizer application and reduced chemical fertilizer application affect the soil properties and bacterial communities of grape rhizosphere soil. *Scientific Reports*, *10*(1), 1–10. doi:10.103841598-020-66648-9 PMID:32533037

Wu, M., Chen, L., Tian, Y., Ding, Y., & Dick, W. A. (2013). Degradation of polycyclic aromatic hydrocarbons by microbial consortia enriched from three soils using two different culture media. *Environmental Pollution*, *178*, 152–158.

Wu, M., Dick, W. A., Li, W., Wang, X., Yang, Q., & Wang, T. (2016). Bioaugmentation and biostimulation of hydrocarbon degradation and the microbial community in a petroleum-contaminated soil. *International Biodeterioration & Biodegradation*, *107*, 158–164.

Wu, M., Li, W., Dick, W. A., Ye, X., Chen, K., Kost, D., & Chen, L. (2017). Bioremediation of hydrocarbon degradation in a petroleum-contaminated soil and microbial population and activity determination. *Chemosphere*, *169*, 124–130. doi:10.1016/j.chemosphere.2016.11.059 PMID:27870933

Wu, M., Zhang, L., Li, G., Jiang, D., & Ghabrial, S. A. (2010). Genome characterization of a debilitation-associated mitovirus infecting the phytopathogenic fungus *Botrytis cinerea*. *Virology*, *406*(1), 117–126. doi:10.1016/j.virol.2010.07.010 PMID:20674953

Xiao, A. W., Li, Z., Li, W. C., & Ye, Z. H. (2020). The effect of plant growth-promoting rhizobacteria (PGPR) on arsenic accumulation and the growth of rice plants (*Oryza sativa* L.). *Chemosphere*, *242*, 125136. doi:10.1016/j.chemosphere.2019.125136 PMID:31654806

Xiao, B., Lian, B., Sun, L., & Shao, W. (2012). Gene transcription response to weathering of K-bearing minerals by *Aspergillus fumigatus*. *Chemical Geology*, *306-307*, 1–9. doi:10.1016/j.chemgeo.2012.02.014

Xiao, C., Liu, T., Guang, X., & Ruan, C. (2018). Characteristics and Mechanisms of Biosolubilization of Rock Phosphate by *Aspergillus japonicas*. *Brazilian Archives of Biology and Technology*, *60*(0), 1–21. doi:10.1590/1678-4324-2017160541

Xiao, P., Mori, T., Kamei, I., Kiyota, H., Takagi, K., & Kondo, R. (2011). Novel metabolic pathways of organochlorine pesticides dieldrin and aldrin by the white rot fungi of the genus Phlebia. *Chemosphere*, *85*(2), 218–224.

Xiao, R., & Zheng, Y. (2016). Overview of microalgal extracellular polymeric substances (EPS) and their applications. *Biotechnology Advances*, *34*(7), 1225–1244. doi:10.1016/j.biotechadv.2016.08.004 PMID:27576096

Xiao, W., Ye, X., Yang, X., Zhu, Z., Sun, C., & Zhang, Q. (2017). Isolation and characterization of chromium (VI)-reducing Bacillus sp. FY1 and Arthrobacter sp. WZ2 and their bioremediation potential. *Bioremediation Journal*, *2*, 100–108.

Xiao, Y., Wang, X., Chen, W., & Huang, Q. (2017). Isolation and identification of three potassium solubilizing bacteria from rape rhizospheric soil and their effects on ryegrass. *Geomicrobiology Journal*, *34*(10), 873–880. doi:10.1080/01490451.2017.1286416

Xia, X. X., Qian, Z. G., Ki, C. S., Park, Y. H., Kaplan, D. L., & Lee, S. Y. (2010). Native-sized recombinant spider silk protein produced in metabolically engineered Escherichia coli results in a strong fiber. *Proceedings of the National Academy of Sciences of the United States of America*, *107*(32), 14059–14063. doi:10.1073/pnas.1003366107 PMID:20660779

Xie, J., & Jiang, D. (2014). New insights into mycoviruses and exploration for the biological Control of Crop Fungal Diseases. *Annual Review of Phytopathology, 52*, 3.1–3.24.

Xie, Y., Fan, J., Zhu, W., Amombo, E., Lou, Y., Chen, L., & Fu, J. (2016). Effect of Heavy Metals Pollution on Soil Microbial Diversity and Bermudagrass Genetic Variation. *Frontiers in Plant Science*, *7*, 755.

Xiu, Z. M., Gregory, K. B., Lowry, G. V., & Alvarez, P. J. (2010). Effect of bare and coated nanoscale zerovalent iron on tceA and vcrA gene expression in *Dehalococcoides* spp. *Environmental Science & Technology*, *44*(19), 7647–7651. doi:10.1021/es101786y PMID:20804135

Xiu, Z., Jin, Z., Li, T., Mahendra, S., Lowry, G. V., & Alvarez, P. J. J. (2010). Effects of nano-scale zero-valent iron particles on a mixed culture dechlorinating trichloroethylene. *Bioresource Technology*, *101*(4), 1141–1146. doi:10.1016/j.biortech.2009.09.057 PMID:19819128

Xu, X., Wen, B., Huang, H., Wang, S., Han, R., & Zhang, S. (2016). Uptake, translocation and biotransformation kinetics of BDE-47, 6-OH-BDE-47 and 6-MeO-BDE-47 in maize (Zea mays L.). *Environmental Pollution, 208*(Pt B), 714–722.

Xu, D., & Pei, J. (2011). Construction and characterization of a photosynthetic bacterium ge- netically engineered for Hg^{2+} uptake. *Bioresource Technology*, *102*, 3083–3088.

Xuezhi, D., Anum, A. A., Ishaq, M., Tariq, S., & Qudratullah, K. (2020). Remediation methods of crude oil contaminated soil. *World J Agri & Soil Sci.*, *4*(3), 34–46. doi:10.33552/WJASS.2020.04.000595

Xu, F., Byun, T., Deussen, H. J., & Duke, K. R. (2003). Degradation of N-acylhomoserine lactones, the bacterial quorum-sensing molecules, by acylase. *Journal of Biotechnology*, *101*(1), 89–96. doi:10.1016/S0168-1656(02)00305-X PMID:12523973

Xu, F., Mou, Z., Geng, J., Zhang, X., & Li, C. (2016). Azo dye decolorization by a halotolerant exoelectrogenic decolorizer isolated frommarine sediment. *Chemosphere*, *158*, 30–36.

Xu, J., Dozier, A., & Bhattacharyya, D. (2005). Synthesis of Nanoscale Bimetallic Particles in Polyelectrolyte Membrane Matrix for Reductive Transformation of Halogenated Organic Compounds. *Journal of Nanoparticle Research*, *7*(4-5), 449–467. doi:10.100711051-005-4273-3

Xu, J., Morris, P. J., Liu, J., & Holden, J. (2018). PEATMAP: Refining estimates of global peatland distribution based on a meta-analysis. *Catena*, *160*, 134–140. doi:10.1016/j.catena.2017.09.010

Xu, Y., Rossi, F., Colica, G., Deng, S., De Philippis, R., & Chen, L. (2013). Use of cyanobacterial polysaccharides to promote shrub performances in desert soils: A potential approach for the restoration of desertified areas. *Biology and Fertility of Soils*, *49*(2), 143–152. doi:10.100700374-012-0707-0

Xu, Y., & Zhao, D. (2007). Reductive immobilization of chromate in water and soil using stabilized iron nanoparticles. *Water Research*, *41*(10), 2101–2108. doi:10.1016/j.watres.2007.02.037 PMID:17412389

Yadav, R., Ror, P., Rathore, P., & Ramakrishna, W. (2020). Bacteria from native soil in combination with arbuscular mycorrhizal fungi augment wheat yield and biofortification. *Plant Physiology and Biochemistry*, *150*, 222–233. doi:10.1016/j.plaphy.2020.02.039 PMID:32155450

Yadegari, M., & Rahmani, H. A. (2010). Evaluation of bean (Phaseolus vulgaris) seeds inoculation with *Rhizobium phaseoli* and plant growth promoting Rhizobacteria (PGPR) on yield and yield components. *African Journal of Agricultural Research*, *5*(9), 792–799.

Yadegari, M., Rahmani, H. A., Noormohammadi, G., & Ayneband, A. (2010). Plant growth promoting rhizobacteria increase growth, yield and nitrogen fixation in Phaseolus vulgaris. *Journal of Plant Nutrition*, *33*(12), 1733–1743. doi:10.1080/01904167.2010.503776

Yamada, Y., & Nihira, T. (1998). Microbial hormones and microbial chemical ecology. In D. H. R. Barton & K. Nakanishi (Eds.), *Comprehensive natural products chemistry*. Elsevier Sciences.

Yan, L. (2012). *The use of plants, including trees, to remediate oil contaminated soils: a review and empirical study* (MSc thesis). Department of Forestry, University of Helsinki. Available on: https://core.ac.uk/download/pdf/14926191.pdf

Yanai, K., Murakami, T., & Bibb, M. (2006). Amplification of the entire kanamycin biosynthetic gene cluster during empirical strain improvement of *Streptomyces kanamyceticus. Proceedings of the National Academy of Sciences of the United States of America*, *103*(25), 9661–9666. doi:10.1073/pnas.0603251103 PMID:16766657

Yandigeri, M. S., Yadav, A. K., Meena, K. K., & Pabbi, S. (2010). Effect of mineral phosphates on growth and nitrogen fixation of diazotrophic cyanobacteria Anabaena variabilis and Westiellopsis prolifica. *Antonie van Leeuwenhoek*, *97*(3), 297–306. doi:10.100710482-009-9411-y PMID:20069361

Yan, F. F., Wu, C., Cheng, Y. Y., He, Y. R., Li, W. W., & Yu, H. Q. (2013). Carbon nanotubes promote Cr(VI) reduction by alginate-immobilized *Shewanella oneidensis* MR-1. *Biochemical Engineering Journal*, *77*, 183–189. doi:10.1016/j.bej.2013.06.009

Yan, G. Y., & Viraraghavan, T. (2000). Effect of pretreatment on the biosorption of heavy metals on *Mucor rouxii. Water S.A.*, *26*, 119–123.

Yang, C., Song, C., Mulchandani, A., & Qiao, C. (2010). Genetic engineering of Stenotrophomonas strain YC-1 to possess a broader substrate range for organophosphates. *Journal of Agricultural and Food Chemistry*, *58*, 6762–6766.

Yang, F., Wang, L. H., Wang, J., Dong, Y. H., Hu, J. Y., & Zhang, L. H. (2005). Quorum quenching enzyme activity is widely conserved in the sera of mammalian species. *FEBS Letters*, *579*(17), 3713–3717. doi:10.1016/j.febslet.2005.05.060 PMID:15963993

Yang, J., Li, W., Ng, T. B., Deng, X., Lin, J., & Ye, X. (2017). Laccases: Production, expression regulation, and applications in pharmaceutical biodegradation. *Frontiers in Microbiology*, *8*, 832. doi:10.3389/fmicb.2017.00832 PMID:28559880

Yang, J., Yang, X., Lin, Y., Ng, T. B., Lin, J., & Ye, X. (2015). Laccase-Catalyzed Decolorization of Malachite Green: Performance Optimization and Degradation Mechanism. *PLoS One*, *10*(5), e0127714. doi:10.1371/journal.pone.0127714 PMID:26020270

Yang, M., Yang, D., & Yu, X. (2018). Soil microbial communities and enzyme activities in sea-buckthorn (Hippophae rhamnoides) plantation at different ages. *PLoS One*, *13*(1), e0190959. doi:10.1371/journal.pone.0190959 PMID:29324845

Yang, M., Zhou, X., Zhai, L., Xiao, F., Hong, N., & Wang, G. (2020). Molecular characterization of a novel mycovirus infecting the phytopathogenic fungus *Botryosphaeria dothidea. Archives of Virology*, *165*(7), 1667–1670. doi:10.100700705-020-04629-z PMID:32328855

Yang, O., Kim, H. L., Weon, J. I., & Seo, Y. R. (2015). Endocrine-disrupting chemicals: Review of toxicological mechanisms using molecular pathway analysis. *Journal of Cancer Prevention*, *20*(1), 12–24. doi:10.15430/JCP.2015.20.1.12 PMID:25853100

Yang, P., Zhou, X. F., Wang, L. L., Li, Q. S., Zhou, T., Chen, Y. K., Zhao, Z. Y., & He, B. Y. (2018). Effect of Phosphate-Solubilizing Bacteria on the Mobility of Insoluble Cadmium and Metabolic Analysis. *International Journal of Environmental Research and Public Health*, *15*(7), 1330. doi:10.3390/ijerph15071330 PMID:29941813

Yang, Q., Yang, M., Pritsch, K., Yediler, A., Hagn, A., Schloter, M., & Kettrup, A. (2003). Decolorization of synthetic dyes and production of manganese-dependent peroxidase by new fungal isolates. *Biotechnology Letters*, *25*(9), 709–713. doi:10.1023/A:1023454513952 PMID:12882171

Yang, Y. T., Bennett, G. N., & San, K. Y. (1998). Genetic and metabolic engineering. *Electronic Journal of Biotechnology*, *1*, 49–60.

Yang, Y. Y., Du, L. N., Wang, G., Jia, X. M., & Zhao, Y. H. (2011). The decolorisation capacity and mechanism of *Shewanella oneidensis* MR-1 for Methyl Orange and Acid Yellow 199 under microaerophilic conditions. *Water Science and Technology*, *63*(5), 956–963. doi:10.2166/wst.2011.275 PMID:21411946

Yan, H., & Pan, G. (2004). Increase in Biodegradation of Dimethyl Phthalate by *Closterium lunula* Using Inorganic Carbon. *Chemosphere*, *55*(9), 1281–1285. doi:10.1016/j.chemosphere.2003.12.019 PMID:15081769

Yanni, Y. G., Rizk, R. Y., Corich, V., Squartini, A., Ninke, K., Philip-Hollingsworth, S., Orgambide, G., de Bruijn, F., Stoltzfus, J., Buckley, D., Schmidt, T. M., Mateos, P. F., Ladha, J. K., & Dazzo, F. B. (1997). Natural endophytic association between *Rhizobium leguminosarum* bv. trifolii and rice roots and assessment of its potential to promote rice growth. *Plant and Soil*, *194*(1/2), 99–114. doi:10.1023/A:1004269902246

Yao, Z., Li, J., Xie, H., & Yu, C. (2012). Review on Remediation Technologies of Soil Contaminated by Heavy Metals. *Procedia Environmental Sciences*, *16*, 722–729. doi:10.1016/j.proenv.2012.10.099

Yap, C. K., & Penge, S. H. T. (2019). Cleaning Up of Contaminated Soils by Using Microbial Remediation: A Review and Challenges to the Weaknesses. *American Journal of Biomedical Science & Research*, *2*(3), 126–128.

Yaqoob, A., Nasim, F. H., Sumreen, A., & Munawar, N. (2019). Current scenario of phytoremediation: Progresses and limitations. *International Journal of Biosciences*, *14*(3), 191–206. doi:10.12692/ijb/14.3.191-206

Yateem, A., Balba, M. T., El-Nawawy, A. S. N., & AlAwadhi, N. (2000). Plants-associated microflora and the remediation of oil contaminated soil. *International Journal of Phytoremediation*, *2*(3), 183–191. doi:10.1080/15226510009359031

Yates, E. A., Philipp, B., Buckley, C., Atkinson, S., Chhabra, S. R., Sockett, R. E., Goldner, M., Dessaux, Y., Cámara, M., Smith, H., & Williams, P. (2002). N-acylhomoserine lactones undergo lactonolysis in a pH-, temperature-, and acyl chain length-dependent manner during growth of *Yersinia pseudotuberculosis* and *Pseudomonas aeruginosa. Infection and Immunity*, *70*(10), 5635–5646. doi:10.1128/IAI.70.10.5635-5646.2002 PMID:12228292

Ye, X., Dong, F., & Lei, X. (2018). Microbial resources and ecology-microbial degradation of pesticides. *Natural Resources Conservation and Research, 1*(1).

Ye, L., Zhao, X., Bao, E., Li, J., Zou, Z., & Cao, K. (2020). Bio-organic fertilizer with reduced rates of chemical fertilization improves soil fertility and enhances tomato yield and quality. *Scientific Reports*, *10*(1), 177. Advance online publication. doi:10.103841598-019-56954-2 PMID:31932626

Yeung, A. T. (2009). Remediation technologies for contaminated sites. In Y. Chen, X. Tang, & L. Zhan (Eds.), *Advances in Environmental Geotechnics* (pp. 328–369). Springer.

Yilmaz, E. I., & Ensari, N. Y. (2005). Cadmium biosorption by *Bacillus circulans* strain EB1. *World Journal of Microbiology & Biotechnology*, *21*(5), 777–779. doi:10.100711274-004-7258-y

Yilmaz, E., & Sonmez, M. (2017). The role of organic/bio–fertilizer amendment on aggregate stability and organic carbon content in different aggregate scales. *Soil & Tillage Research*, *168*, 118–124. doi:10.1016/j.still.2017.01.003

Ying, X., Yizhi, S., David, M. J., Meng, L., Huigang, L., & Yingping, H. (2018). Se enhanced phytoremediation of diesel in soil by *Trifolium repens. Ecotoxicology and Environmental Safety, 154*, 137-144. doi:10.1016/j.ecoenv.2018.01.061

Yin, K., Wang, Q., Lu, M., & Chen, L. (2019). Microorganism remediation strategies towards heavy metals. *Chemical Engineering Journal*, 360.

Yin, P., Yu, Q., Jin, B., & Ling, Z. (1999). Biosorption removal of cadmium from aqueous solution by using pretreated fungal biomass cultured from starch wastewater. *Water Research*, *33*(8), 1960–1963. doi:10.1016/S0043-1354(98)00400-X

Yin, W. F., Tung, H. J., Sam, C. K., Koh, C. L., & Chan, K. G. (2012). Quorum quenching *Bacillus sonorensis* isolated from soya sauce fermentation brine. *Sensors (Basel)*, *12*(4), 4065–4073. doi:10.3390120404065 PMID:22666018

Yousaf, S., Khan, S., & Aslam, M. T. (2013). Effect of pesticides on the soil microbial activity. *Pakistan Journal of Zoology*, *45*(4).

Yousefi, N., Chehregani, A., Malayeri, B., Lorestani, B., & Cheraghi, M. (2011). Investigating the effect of heavy metals on developmental stages of anther and pollen in *Chenopodium botrys* L. (Chenopodiaceae). *Biological Trace Element Research*, *140*(3), 368–376. doi:10.100712011-010-8701-6 PMID:20499206

Youssef, M. M. A., & Eissa, M. F. M. (2014). Biofertilizers and their role in management of plant parasitic nematodes. A review. *Journal of Biotechnology and Pharmaceutical Research*, *5*(1), 1–6.

Youssef, N., Sheik, C. S., Krumholz, L. R., Najar, F. Z., Roe, B. A., & Elshahed, M. S. (2009). Comparison of species richness estimates obtained using nearly complete fragments and simulated pyrosequencing-generated fragments in 16S rRNA gene-based environmental surveys. *Applied and Environmental Microbiology*, *75*(16), 5227–5236.

Yuan, Y., Liu, Z. Q., Jin, H., Sun, S., Liu, T. J., Wang, X., Fan, H.-J., Hou, S.-K., & Ding, H. (2017). Photodynamic antimicrobial chemotherapy with the novel amino acid-porphyrin conjugate 4I: In vitro and in vivo studies. *PLoS One*, *12*(5), e0176529. doi:10.1371/journal.pone.0176529 PMID:28493985

Yu, G., Ran, W., & Shen, Q. (2016). Compost process and organic fertilizers application in China. In *Organic fertilizers – From Basic concepts to applied outcomes*. IntechOpen. doi:10.5772/62324

Yuksel, O., Kavdr, Y., & Bahtiyar, M. (2004). The effect of municipal waste compost on physical characteristics of clay soils. *Fresenius Environmental Bulletin*, *13*(11a), 1094–1098.

Yuniati, M. D. (2018). Bioremediation of petroleum-contaminated soil: A Review. I*OP Conf. Series: Earth and Environmental Science, 118*. doi:10.1088/1755-1315/118/1/012063

Yu, P., Yuan, J., Deng, X., Ma, M., & Zhang, H. (2014). Subcellular targeting of bacterial CusF enhances Cu accumulation and alters root to shoot Cu translocation in *Arabidopsis. Plant & Cell Physiology*, *55*(9), 1568–1581.

Yu, R., Peethambaram, H. S., Falta, R. W., Verce, M. F., Henderson, J. K., Bagwell, C. E., Brigmon, R. L., & Freedman, D. L. (2013). Kinetics of 1, 2-dichloroethane and 1, 2-dibromoethane biodegradation in anaerobic enrichment cultures. *Applied and Environmental Microbiology*, *79*(4), 1359–1367. doi:10.1128/AEM.02163-12 PMID:23263950

Yürekli, F., Yesilada, O., Yürekli, M., & Topcuoglu, S. F. (1999). Plant growth hormone production from olive oil mill and alcohol factory wastewaters by white rot fungi. *World Journal of Microbiology & Biotechnology*, *15*(4), 503–505. doi:10.1023/A:1008952732015

Yu, S., Teng, C., Bai, X., Liang, J., Song, T., Dong, L., Jin, Y., & Qu, J. (2017). Optimization of siderophore production by Bacillus sp. PZ-1 and its potential enhancement of phytoextration of Pb from soil. *Journal of Microbiology and Biotechnology*, *27*(8), 1500–1512. doi:10.4014/jmb.1705.05021 PMID:28633518

Yu, X., Li, B., Fu, Y., Xie, J., Cheng, J., Ghabrial, S. A., Li, G., Yi, X., & Jiang, D. (2013). Extracellular transmission of a DNA mycovirus and its use as a natural fungicide. *Proceedings of the National Academy of Sciences of the United States of America*, *110*(4), 1452–1457. doi:10.1073/pnas.1213755110 PMID:23297222

Yu, Y. (2016). *Study on Rapid Screening of Pesticides and Antibiotics in Soil*. Chinese Academy of Agricultural Sciences.

Yu, Y. L., Fang, H., Wang, X., Wu, X. M., & Shan, M. (2011). Characteristics of fungal a fungal strain capable of degrading chlorpyrifos and its use in detoxification of the insecticide on vegetables. *Biodegradation*, *17*, 487–494.

Zabel, F., Putzenlechner, B., & Mauser, W. (2014). Global agricultural land resources—A high resolution suitability evaluation and its perspectives until 2100 under climate change conditions. *PLoS One*, *9*(9), e107522. doi:10.1371/journal.pone.0107522 PMID:25229634

Zaccardelli, M., Sorrentino, R., Caputo, M., Scotti, R., De Falco, E., & Pane, C. (2020). Stepwise-selected *Bacillus amyloliquefaciens* and *B. subtilis* strains from composted aromatic plant waste able to control soil-borne diseases. *Collection FAO: Agriculture*, *10*(2), 1–15. doi:10.3390/agriculture10020030

Zafar-Ul-Hye, M., Tahzeeb-Ul-Hassan, M., Abid, M., Fahad, S., Brtnicky, M., Dokulilova, T., Datta, R., & Danish, S. (2020). Potential role of compost mixed biochar with rhizobacteria in mitigating lead toxicity in spinach. *Scientific Reports*, *10*(1), 12159. doi:10.103841598-020-69183-9 PMID:32699323

Zaidi, S., Usmani, S., Singh, B. R., & Musarrat, J. (2006). Significance of *Bacillus subtilis* strain SJ-101 as a bioinoculant for concurrent plant growth promotion and nickel accumulation in *Brassica juncea*. *Chemosphere*, *64*(6), 991–997. doi:10.1016/j.chemosphere.2005.12.057 PMID:16487570

Zaller, J. G., Heigl, F., Ruess, L., & Grabmaier, A. (2014). Glyphosate herbicide affects belowground interactions between earthworms and symbiotic mycorrhizal fungi in a model ecosystem. *Scientific Reports*, *4*(1), 5634. doi:10.1038rep05634 PMID:25005713

Zand, A. D., Bidhendi, G. N., & Mehrdadi, M. (2009). Phytoremediation of total petroleum hydrocarbons (TPHs) using plant species in Iran. *Turkish Journal of Agriculture and Forestry*, *34*, 429–438. doi:10.3906/tar-0903-2

Zarjani, J. K., Aliasgharzad, N., Oustan, S., Emadi, M., & Ahmadi, A. (2013). Isolation and characterization of potassium solubilizing bacteria in some Iranian soils. *Archives of Agronomy and Soil Science*, *59*(12), 1713–1723. doi:10.1080/03650340.2012.756977

Zawadzka, A. M., Abergel, R. J., Nichiporuk, R., Andersen, U. N., & Raymond, K. N. (2009). Siderophore-mediated iron acquisition systems in *Bacillus cereus*: Identification of receptors for anthrax virulence-associated petrobactin. *Biochemistry*, *48*(16), 3645–3657. doi:10.1021/bi8018674 PMID:19254027

Zeffa, D. M., Fantin, L. H., Koltun, A., de Oliveira, A. L., Nunes, M. P., Canteri, M. G., & Gonçalves, L. S. (2020). Effects of plant growth-promoting rhizobacteria on co-inoculation with Bradyrhizobium in soybean crop: A meta-analysis of studies from 1987 to 2018. *PeerJ*, *8*, e7905. doi:10.7717/peerj.7905 PMID:31942248

Zeffa, D. M., Perini, L. J., Silva, M. B., de Sousa, N. V., Scapim, C. A., Oliveira, A., Amaral Júnior, A., & Azeredo Gonçalves, L. S. (2019). Azospirillum brasilense promotes increases in growth and nitrogen use efficiency of maize genotypes. *PLoS One*, *14*(4). doi:10.1371/journal.pone.0215332 PMID:30998695

Zeng, Q., Xie, J., Li, Y., Gao, T., Xu, C., & Wang, Q. (2018). Comparative genomic and functional analyses of four sequenced Bacillus cereus genomes reveal conservation of genes relevant to plant-growth-promoting traits. *Scientific Reports*, *8*(1), 1–10. doi:10.103841598-018-35300-y PMID:30451927

Zeng, X., Cai, Y., Liao, X., Zeng, X., Li, W., & Zhang, D. (2011). Decolorization of synthetic dyes by crude laccase from a newly isolated *Trametes trogii* strain cultivated on solid agro-industrial residue. *Journal of Hazardous Materials*, *187*(1-3), 517–525. doi:10.1016/j.jhazmat.2011.01.068 PMID:21315513

Zeng, X., Liu, X., Tang, J., Hu, S., Jiang, P., Li, W., & Xu, L. (2012). Characterization and potassium-solubilizing ability of *Bacillus Circulans* Z 1–3. *Advanced Science Letters*, *10*(1), 173–176. doi:10.1166/asl.2012.3726

Zerdani, I., Faid, M., & Malki, A. (2004). Digestion of solid tannery wastes by strains of *Bacillus sp.* isolated from compost in morocco. *International Journal of Agriculture and Biology*, *6*(5), 758–761.

Zeriouh, H., Romero, D., Garcia-Gutierrez, L., Cazorla, F. M., de Vicente, A., & Perez-Garcia, A. (2011). The iturin-like lipopeptides are essential components in the biological control arsenal of *Bacillus subtilis* against bacterial diseases of cucurbits. *Molecular plant-microbe interactions. Molecular Plant-Microbe Interactions*, *24*(12), 1540–1552. doi:10.1094/MPMI-06-11-0162 PMID:22066902

Zerkle, A. L., & Mikhail, S. (2017). The geobiological nitrogen cycle: From microbes to the mantle. *Geobiology*, *15*(3), 343–352. doi:10.1111/gbi.12228 PMID:28158920

Zeroual, Y., Moutaouakkil, A., & Blaghen, M. (2001). Volatilization of mercury by immobilized bacteria (Klebsiella pneumoniae) in different support by using fluidized bed bioreactor. *Current Microbiology*, *43*(5), 322–327. doi:10.1007002840010310 PMID:11688795

Zhai, Y., Li, K., Song, J., Shi, Y., & Yan, Y. (2012). Molecular cloning, purification and biochemical characterization of a novel pyrethroid-hydrolyzing carboxylesterase gene from *Ochrobactrumanthropi* YZ-1. *Journal of Hazardous Materials*, *221*, 206–212.

Zhang, C., Meng, X., Li, N., Wang, W., Sun, Y., Jiang, W., Guan, G., & Li, Y. (2013). Two bifunctional enzymes with ferric reduction ability play complementary roles during magnetosome synthesis in *Magnetospirillum gryphiswaldense* MSR-1. *Journal of Bacteriology*, *195*(4), 876–885. doi:10.1128/JB.01750-12 PMID:23243303

Zhang, C., Wu, D., & Ren, H. (2020). Bioremediation of oil contaminated soil using agricultural wastes via microbial consortium. *Scientific Reports*, *10*(1), 1–8. PMID:32513982

Zhang, H. B., Wang, L. H., & Zhang, L. H. (2002). Genetic control of quorum-sensing signal turnover in Agrobacterium tumefaciens. *Proceedings of the National Academy of Sciences of the United States of America*, *99*(7), 4638–4643. doi:10.1073/pnas.022056699 PMID:11930013

Zhang, J., Howell, C. R., & Starr, J. L. (1996). Suppression of Fusarium colonization of cotton roots and Fusarium wilt by seed treatments with *Gliocladium virens* and *Bacillus subtilis. Biocontrol Science and Technology*, *6*(2), 175–188. doi:10.1080/09583159650039377

Zhang, J., Lu, L., Chen, F., Chen, L., Yin, J., & Huang, X. (2018). Detoxification of diphenyl ether herbicide lactofen by *Bacillus* sp. Za and enantioselective characteristics of an esterase gene lacE. *Journal of Hazardous Materials*, *341*, 336–345.

Zhang, L. H., & Dong, Y. H. (2004). Quorum sensing and signal interference: Diverse implications. *Molecular Microbiology*, *53*(6), 1563–1571. doi:10.1111/j.1365-2958.2004.04234.x PMID:15341639

Zhang, L., Guo, Z., Gao, H., Peng, X., Li, Y., Sun, S., Lee, J. K., & Lin, W. (2016). Interaction of *Pseudostellaria heterophylla* with Quorum Sensing and Quorum Quenching Bacteria Mediated by Root Exudates in a Consecutive Monoculture System. *Journal of Microbiology and Biotechnology*, *26*(12), 2159–2170. doi:10.4014/jmb.1607.07073 PMID:27666992

Zhang, L., Ruan, L., Hu, C., Wu, H., Chen, S., Yu, Z., & Sun, M. (2007). Fusion of the genes for AHL-lactonase and S-layer protein in *Bacillus thuringiensis* increases its ability to inhibit soft rot caused by Erwinia carotovora. *Applied Microbiology and Biotechnology*, *74*(3), 667–675. doi:10.100700253-006-0696-8 PMID:17216466

Zhang, M., Liu, G. H., Song, K., Wang, Z., Zhao, Q., Li, S., & Ye, Z. (2015). Biological treatment of 2,4,6-trinitrotoluene (TNT) red water by immobilized anaerobic and aerobic microbial filters. *Chemical Engineering Journal*, *259*, 876–884. doi:10.1016/j.cej.2014.08.041

Zhang, N., Wang, D., Liu, Y., Li, S., Shen, Q., & Zhang, R. (2014). Effects of different plant root exudates and their organic acid components on chemotaxis, biofilm formation and colonization by beneficial rhizosphere-associated bacterial strains. *Plant and Soil*, *374*(1-2), 689–700. doi:10.100711104-013-1915-6

Zhang, Q., Kong, W., Wei, L., Wang, Y., Luo, Y., Wang, P., & Jiang, G. (2020). Uptake, phytovolatilization, and interconversion of 2, 4-dibromophenol and 2, 4-dibromoanisole in rice plants. *Environment International*, *142*, 105888. doi:10.1016/j.envint.2020.105888 PMID:32593840

Zhang, Q., Liu, Y., Lin, Y., Kong, W., Zhao, X., Ruan, T., & Jiang, G. (2019). Multiple metabolic pathways of 2, 4, 6-tribromophenol in rice plants. *Environmental Science & Technology*, *53*(13), 7473–7482. doi:10.1021/acs.est.9b01514 PMID:31244074

Zhang, W. F., Dou, Z.-X., He, P., Ju, X.-T., Powlson, D., Chadwick, D., Norse, D., Lu, Y.-L., Zhang, Y., Wu, L., Chen, X.-P., Cassman, K. G., & Zhang, F.-S. (2013). New technologies reduce greenhouse gas emissions from nitrogenous fertilizer in China. *Proceedings of the National Academy of Sciences of the United States of America*, *110*(21), 8375–8380. doi:10.1073/pnas.1210447110 PMID:23671096

Zhang, W. X., & Elliott, D. W. (2006). Applications of iron nanoparticlesfor groundwater remediation. *Remediation*, *16*(2), 7–21. doi:10.1002/rem.20078

Zhang, W.-X. (2003). Nanoscale iron particles for environmental remediation: An overview. *Journal of Nanoparticle Research*, *5*(3/4), 323–332. doi:10.1023/A:1025520116015

Zhang, X., & Flurkey, W. (1997). Phenol oxidases in Portabella Mushrooms. *Journal of Food Science*, *62*(1), 97–100. doi:10.1111/j.1365-2621.1997.tb04376.x

Zhang, X., Zhang, R., Gao, J., Wang, X., Fan, F., Ma, X., Yin, H., Zhang, C., Feng, K., & Deng, Y. (2017). Thirtyone years of rice-rice-green manure rotations shape the rhizosphere microbial community and enrich beneficial bacteria. *Soil Biology & Biochemistry*, *104*, 208–217. doi:10.1016/j.soilbio.2016.10.023

Zhang, Y., Xue, R., He, X., Cheng, Q., Hartley, W., & Xue, S. (2020). Effect of acid production by Penicillium oxalicum on physicochemical properties of bauxite residue. *Geomicrobiology Journal*, *37*(10), 929–936. doi:10.1080/01490451.2020.1801907

Zhang, Y., Yu, Z., Fu, X., & Liang, C. (2002). Noc3p, a bHLH protein, plays an integral role in the initiation of DNA replication in budding yeast. *Cell*, *109*(7), 849–860. doi:10.1016/S0092-8674(02)00805-X PMID:12110182

Zhang, Z., Hong, Q., Xu, J., Zhang, X., & Li, S. (2006). Isolation of fenitrothion-degrading strain Burkholderia sp. FDS-1 and cloning of mpd gene. *Biodegradation*, *17*, 275–283.

Zhao, B., & Poh, C. L. (2008). Insights into environmental bioremediation by microorganisms through functional genomics and proteomics. *Proteomics*, *8*, 874–881.

Zhao, C., Liu, B., Piao, S., Wang, X., Lobell, D. B., Huang, Y., Huang, M., Yao, Y., Bassu, S., Ciais, P., Durand, J. L., Elliott, J., Ewert, F., Janssens, I. A., Li, T., Lin, E., Liu, Q., Martre, P., Müller, C., ... Asseng, S. (2017). Temperature increase reduces global yields of major crops in four independent estimates. *Proceedings of the National Academy of Sciences of the United States of America*, *114*(35), 9326–9331. doi:10.1073/pnas.1701762114 PMID:28811375

Zhao, D., Oosterhuis, D. M., & Bednarz, C. W. (2001). Influence of potassium deficiency on photosynthesis, chlorophyll content, and chloroplast ultrastructure of cotton plants. *Photosynthetica, 39*(1), 103–109. doi:10.1023/A:1012404204910

Zhao, H., Tao, K., Zhu, J., Liu, S., Gao, H., & Zhou, X. (2015). Bioremediation potential of glyphosate-degrading *Pseudomonas spp.* strains isolated from contaminated soil. *The Journal of General and Applied Microbiology, 61*(5), 165–170. doi:10.2323/jgam.61.165 PMID:26582285

Zhao, J., Jia, D., Du, J., Chi, Y., & Yao, K. (2019). Substrate regulation on cometabolic degradation of β-cypermethrin by *Bacillus licheniformis* B-1. *AMB Express, 9*, 83.

Zhao, J., Zhao, X., Wang, J., Gong, Q., Zhang, X., & Zhang, G. (2020). Isolation, Identification and Characterization of Endophytic Bacterium *Rhizobium oryzihabitans* sp. *nov*., from Rice Root with Biotechnological Potential in Agriculture. *Microorganisms, 8*(4), 608. doi:10.3390/microorganisms8040608 PMID:32331293

Zhao, Y. (2014). Auxin biosynthesis. *The Arabidopsis Book / American Society of Plant Biologists, 12*, e0173. doi:10.1199/tab.0173 PMID:24955076

Zheng, B. X., Ibrahim, M., Zhang, D. P., Bi, Q. F., Li, H. Z., Zhou, G. W., Ding, K., Peñuelas, J., Zhu, Y. G., & Yang, X. R. (2018). Identification and characterization of inorganic-phosphate-solubilizing bacteria from agricultural fields with a rapid isolation method. *AMB Express, 8*(1), 47. doi:10.118613568-018-0575-6 PMID:29589217

Zheng, F., Cui, B.-K., Wu, X.-J., Meng, G., Liu, H.-X., & Si, J. (2016). Immobilization of laccase onto chitosan beads to enhance its capability to degrade synthetic dyes. *International Biodeterioration & Biodegradation, 110*, 69–78. doi:10.1016/j.ibiod.2016.03.004

Zheng, J., Li, R., Zhu, J., Zhang, J., He, J., Li, S., & Jiang, J. (2012). Degradation of the chloroacetamide herbicide butachlor by *Catellibacterium caeni* sp. nov DCA-1T. *International Biodeterioration & Biodegradation, 73*, 16–22.

Zhou, C., Guo, J., Zhu, L., Xiao, X., Xie, Y., Zhu, J., Ma, Z., & Wang, J. (2016). *Paenibacillus polymyxa* BFKC01 enhances plant iron absorption via improved root systems and activated iron acquisition mechanisms. *Plant physiology and biochemistry. PPB, 105*, 162–173. PMID:27105423

Zhou, X., & Xiang, X. (2013). Effect of different plants on azo-dye wastewater biodecolorization. *Procedia Environmental Sciences, 18*, 540–546. doi:10.1016/j.proenv.2013.04.073

Zhu, Y., Boye, A., Body-Malapel, M., & Herkovits, J. (2017). *The toxic effects of xenobiotics on the health of humans and animals*. Academic Press.

Zhuang, P., Li, Z.-a., McBride, M. B., Zou, B., & Wang, G. (2013). Health risk assessment for consumption of fish originating from ponds near Dabaoshan mine, South China. *Environmental Science and Pollution Research International, 20*(8), 5844–5854.

Zhu, D., Chen, Q. L., An, X. L., Yang, X. R., Christie, P., Ke, X., & Zhu, Y. G. (2018). Exposure of Soil collembolans to microplastics perturbs their gut microbiota and alters their isotopic composition. *Soil Biology & Biochemistry, 116*, 302–310.

Zhu, F., Qu, L., Hong, X., & Sun, X. (2011). Isolation and characterization of a phosphate-solubilizing halophilic bacterium Kushneria sp. YCWA18 from Daqiao Saltern on the coast of Yellow Sea of China. *Evidence-Based Complementary and Alternative Medicine*. PMID:21716683

Zhu, J. K. (2016). Abiotic stress signalling and responses in plants. *Cell, 167*(2), 313–324. doi:10.1016/j.cell.2016.08.029 PMID:27716505

Zhu, J., & Winans, S. C. (2001). The quorum-sensing transcriptional regulator TraR requires its cognate signaling ligand for protein folding, protease resistance, and dimerization. *Proceedings of the National Academy of Sciences of the United States of America*, *98*(4), 1507–1512. doi:10.1073/pnas.98.4.1507 PMID:11171981

Zia-ul-Hassan & Arshad, M. (2010). *Cotton growth under potassium deficiency stress is influenced by photos*. Academic Press.

Zimbardi, A., Camargo, P., Carli, S., Aquino Neto, S., Meleiro, L., Rosa, J., De Andrade, A. R., Jorge, J. A., & Furriel, R. (2016). A High Redox Potential Laccase from Pycnoporus sanguineus RP15: Potential Application for Dye Decolorization. *International Journal of Molecular Sciences*, *17*(5), 672. doi:10.3390/ijms17050672 PMID:27164083

Zojiali, F., Hassani, A. H., & Sayedi, M. H. (2014). Bioaccumulation of chromium by Zea mays in a waste water-irrigated soil. An experimental study. *Proceeding of the International Academy of Ecology and Environmental Sciences*, *4*, 62-67.

Zok, S., Boutonnet, J. C., De Rooij, C., Garny, V., Lecloux, A., Papp, R., Thompson, R. S., & Van Wijk, D. (1998). Euro Chlor risk assessment for the marine environment OSPARCOM region: North sea-Chloroform. *Environmental Monitoring and Assessment*, *53*(3), 401–424. doi:10.1023/A:1006010515371

Zolgharnein, H., Karami, K., Assadi, M.M., & Sohrab, A.D. (2010). Molecular characterization and phylogenetic analyses of heavy metal removal bacteria from the Persian *Gulf Biotechnology, 29*, 1-8.

Zolnerciks, J. K., Andress, E. J., Nicolaou, M., & Linton, K. J. (2011). Structure of ABC transporters. *Essays in Biochemistry*, *50*(1), 43. PMID:21967051

Zouboulis, A. I., & Moussas, P. A. (2011). Groundwater and Soil Pollution: Bioremediation. Encyclopedia of Environmental Health, 1037–1044. doi:10.1016/B978-0-444-52272-6.00035-0

Zubair, M., Shakir, M., Ali, Q., Rani, N., Fatima, N., Farooq, S., Shafiq, S., Kanwal, N., Ali, F., & Nasir, I. A. (2016). Rhizobacteria and phytoremediation of heavy metals. *Environmental Technology Reviews*, *5*(1), 112–119. doi:10.1080/21622515.2016.1259358

Zwanenburg, B., & Blanco-Ania, D. (2018). Strigolactones: New plant hormones in the spotlight. *Journal of Experimental Botany*, *69*(9), 2205–2218. doi:10.1093/jxb/erx487 PMID:29385517

Zwolak, A., Sarzyńska, M., Szpyrka, E., & Stawarczyk, K. (2019). Sources of soil pollution by heavy metals and their accumulation in vegetables: A review. *Water, Air, and Soil Pollution*, *230*(7), 164. doi:10.100711270-019-4221-y

About the Contributors

Junaid Ahmad Malik received B.Sc. (2008) Science from the University of Kashmir, Srinagar, J&K; M.Sc. (2010) in Zoology from Barkatullah University, Bhopal, Madhya Pradesh; and PhD (2015) in Zoology from the same university. He completed his B.Ed program in 2017 from the University of Kashmir, Srinagar, J&K. He started his career as Lecturer in School Education Department, Govt. of J&K for 2 years. Dr. Malik is now working as a Lecturer in Department of Zoology, Govt. Degree College, Bijbehara, Kashmir (J&K) and actively involved in teaching and research activities. He has more than 8 years of research experience. His areas of interest are ecology, soil macrofauna, wildlife biology, conservation biology etc. Dr. Malik has published 21 research articles and technical papers in International peer reviewed journals like Springer, Elsevier, and T&F etc. He has authored 3 books, edited 2 books, 11 book chapters and more than 10 popular editorial articles. He is also serving as editor and reviewer of several journals with a reasonable reputation. He has participated in several State, National, and International conferences, seminars, workshops, and symposia and more than 20 conference papers are to his credit. He is the life member of SBBS (Society for Bioinformatics and Biological Sciences) with membership id LMJ-243.

* * *

Ankita Agarwal is a scholar at Amity Institute of Biotechnology, Amity University Jharkhand. She is pursuing B. Tech Biotechnology. She is working under guidance of Dr. Sumira Malik and Dr. Shilpa Prasad, Assistant Professor, Amity institute of Biotechnology, Amity University Jharkhand. She has presented posters and research papers in various International and National conferences.

Swati Agarwal is working as Research Associate in DBT (BIG) BIRAC project from the Banasthali Vidyapith, She has an expertise in Plant molecular biology and biofuel production. She had earned her Doctor of Philosophy in BIotechnology in 2018 from Banasthali Vidyapith. She has several research articles, chapters and review papers in peer reviewed journals.

I. A. Allamin has a doctorate in Environmental Biotechnology from Universiti Putra Malaysia. He is currently the head of the Microbiology Department University of Maiduguri where he has been teaching for the last ten years. His major interest is in phytoremediation, Bioprospecting of natural products, and environmental management. He has over twenty research papers to his credit.

Wajiha Anum is scientific officer at regional agricultural research institute, Bahawalpur, Pakistan. She is a researcher with over 5 years of experience. This institute is dedicated for varietal development of various field crops like wheat, pulses, sorghum and millets. As she is working solely under agronomy department so the basic research interests are improvements in the production technologies, testing of newly developed crop varieties under agro- ecological conditions of the area. Recently more emphasis is on the testing of crop growth model (Decision support system for agro technology transfer) for making adaptation packages and yield prediction under uncertain climate conditions. Apart from the this, she has great interest in research writing and like to produce contents related to dimensional areas of agricultural research.

Thazeem Basheer originally from Tamil Nadu, India. An active researcher and advisor with six years of research experience, who earned her Post Graduation in Industrial Biotechnology and PhD in Microbial Biotechnology, from Bharathiar University, Coimbatore, Tamil Nadu. She was a recipient of INSPIRE Fellowship from the Department of Science and Technology, New Delhi, India (2014-2018) during her PhD. She is also recipient of Gold Medals during her UG and PG (University First Rank Holder). As a need of the hour, her thrust areas involve tannery and poultry waste management, aqua feed formulations, bio-polymer production, bio-fertilizer production, and enzyme applications.

Chanda Berde is presently working as Assistant Professor in Marine Microbiology, School of Earth, Ocean and Atmospheric Sciences, Goa University, Goa . She has completed doctoral and post doctoral studies from Department of Microbiology, Goa University. Thereafter worked as Scientist at Vidya Pratishthan's School of Biotechnology, Baramati, Maharashtra, India, for 5 years and as Assistant Professor, Department of Microbiology, Gogate Jogalekar College, Ratnagiri for 10 years. Having guided more than 60 post graduate research projects in the subjects of Biotechnology and Microiology, she has to her credit more than 35 publications in national and international journals and has authored several books and chapters.

Vikrant B. Berde is presently working as Head and Assistant Professor, Department of Zoology, Arts Commerce and Science College, Lanja. He has been in the teaching field for the last 20 years. He has worked as Scientist at Vidya Pratishthan's School of Biotechnology, Baramati, wherein he involved in the major components of a socialite project "Biovillage", including sericulture, aquaculture, vermiculture and apiculture. The thrust area of his research being freshwater limnology, entomology, bioremediation, and biodiversity.

Brij Bhushan is a researcher involved in recovery of resources by integration of membrane technology from wastewater, activated carbon development and its use in wastewater treatment. Actively involved in research since the last 8 years. PhD from IIT Roorkee and postdoc from UPC, Barcelona, Spain. Currently working as Associate Professor at Graphic Era University, Dehradun.

Leena Merlin Biju has completed her PG and MPhil in Microbiology. She has teaching experience of 15 years in Microbiology and published two papers at national conferences. She is currently a research scholar at Stella Maris College (Autonomous), Chennai, India and pursuing her teaching career at Kumararani Meena Muthiah College of Arts and Science (Co-Ed), Chennai, India.

Trinath Biswal is presently working as Associate Professor in the department of Chemistry, Veer Surendra Sai University of Technology, Burla, and Odisha, India. He completed his M.Sc. and M.Phil. from Ravenshaw University Cuttack, India and Ph.D. in polymer composite from Utkal University, Bhubaneswar, India. He has more than twenty (20) years of teaching experience in different reputed institutions of Odisha, India and more than 15 years of research experience. His area of interest includes polymer biocomposite, polymer nanocomposite, environmental monitoring, and environmental analysis. He has published 35 research papers in various national and international journals, ten book chapters and is the author of two books on environmental science and one book on Materials science. He is a member of several national and international organizations and is a reviewer of various journals .He guided many M.Sc. and M.Phil. Students and presently five research scholars are working under his guidance for Ph.D. work. He also organized different workshops and a number of environmental awareness programmer in the state Odisha, India.

U. A. Bukar is a Staff of Microbiology Department, University of Maiduguri. He has a Diploma in Agricultural Technology, a Bachelor's degree in Microbiology, and currently studying Environmental Microbiology at Bayero University Kano, Nigeria as a Graduate student.

A. A. Farouq is a Professor of Environmental Microbiology and holds a Ph. D in Applied Microbiology from Universiti Putra Malaysia. He was formerly the Head, Department of Microbiology, Acting Dean and immediate past Deputy Dean, Faculty of Science, UDU Sokoto. His interest is in the generation of sustainable renewable energy through microbial biotechnological processes, reclamation of polluted soil, and public and environmental health. He has won research grants and has published many research articles and conference papers in the last two decades. He is a member of many professional academic bodies like ASM, NSM, MSM, MSPTM, BSN, etc.

Allwyn Vyas G. originally from Kerala, India. Earned his MSc. degree in Microbiology from Garden City College, Bangalore affiliated to Bangalore University. He has received his undergraduate degree in Industrial Microbiology & Zoology from St. Berchmans college, Changanacherry affiliated to Mahatma Gandhi University, Kottayam, Kerala. He is currently pursuing Ph.D degree in Biotechnology at SRM Institute of Science and Technology, Tamil Nadu. His area of interests are Environmental Microbiology & Microbial Quorum sensing.

Shreya Ghoshal is a scholar at Amity Institute of Biotechnology, Amity University Jharkhand. She is pursuing B. Tech Biotechnology. She is working under guidance of Dr. Sumira Malik and Dr. Shilpa Prasad, Assistant Professor, Amity institute of Biotechnology, Amity University Jharkhand. She has presented posters/e posters in various International and National conferences.

Asha Giriyan is currently pursuing PhD degree in Marine Microbiology at Goa University and Holds a Master's Degree in Environmental Science . she joined Coastal Ecology and Marine Resources Centre, regional office of TERI working on projects, which cut boundaries between the two disciplines. Worked on projects that addressed the interface of environment and development and pollution impacts particularly in coastal and marine sector. Her interest was on Community base resource management, particularly in the field of coastal and marine resources She is involved in research on documentation of traditional knowledge in managing ecosystems, using participatory approaches and suggesting alternative livelihood opportunities, especially for Fishermen and farmers. Her work includes evaluating ecological footprints, assessing impacts of various development activities, recommending better policies, and developing indicators and tools for better ecosystem management and demonstrating Aquaculture practices.

Mir Zahoor Gul is currently working as Post-Doctoral Fellow at the Department of Biochemistry, University College of Science, Osmania University, Hyderabad, India. He has earned his Ph.D. degree at School of Life Sciences, University of Hyderabad, India in the year 2015. His research interests lie in the area of Nano-phytomedicine, Drug delivery, Nano-biocomposites, Nano-bioremdiation of synthetic wastes, to name a few. Dr. Gul has over 30 research publications in peer reviewed International journals, including 12 book chapters in edited volumes with reputed international publishers. Dr. Gul is a recipient of several national fellowships for his research endeavors.

U. B. Ibrahim is currently an Environmental Microbiology Ph.D. scholar at Bayero University Kano in Nigeria. He received his B.Sc and M. Sc Degrees in Microbiology from Usmanu Danfodiyo University, Sokoto-Nigeria. He is an avid researcher, with several works published in reputable journals. He teaches at Department of Microbiology of Usmanu Danfodiyo University, Sokoto, Nigeria. He is a member of several professional bodies which include the Nigerian Society for Microbiology (NSM), Science Association of Nigeria (SAN), Nigerian Society of Biochemistry and Molecular Biology (NSBMB).

H. Y. Ismail is with the Department of Microbiology, University of Maiduguri, and currently a Ph.D. Scholar at Usman Danfodiyo University Sokoto, Nigeria. He has an interest in the bioremediation of environmental pollutants and has published more than 20 research articles in referred journals in the last decade.

Anvi Jain is a research scholar under Dr. Suphiya Khan at the Department of Bioscience and Biotechnology, Banasthali Vidyapith, Newai, District - Tonk, Rajasthan, India.

Meghna Jindal is M.Sc. Botany qualified (2019) from Department of Botany, University of Delhi, India. She is also a Joint CSIR-UGC NET L.S. (June 2020) and GATE Life Sciences (2020) qualified fellow. She has completed her graduation, B.Sc. (Hons.) Botany (2017), from Deen Dayal Upadhyaya College, University of Delhi, India.

Md. Idrish Khan has completed his PhD in the discipline Aquatic Animal Health and his area of expertise includes fish diseases and their management, immunostimulants, probiotics, bacteriophages, etc.

Suphiya Khan is Associate Professor at Banasthali Vidyapith which is the world largest residential NAAC A++ WOMEN University. She is a scientist/Professor by passion and Bioentrepreneur by choice to nurture women scientist. She is the recipient of many national and International awards for contribution in science and innovation. She has been selected as Chevening Research Science and Innovation Policy (CRISP) fellow at University of Oxford for her contribution in Science and Innovation. Recently she got TIE-BIRAC WInER award (5 lakh cash Prize) and innovation grant from BIRAC-BIG (50 Lakhs) for the development of affordable Defluoridation Technology. She has attracted several crores of funding from different funding agencies viz MHRD, DST, DBT, UGC etc. She has successfully run the MHRD Centre for Excellence (COE) on "Water and Energy (2.5 Crores). She has founded her own spinoff company

Drumlins Water Technologies Pvt Ltd from this COE and currently 8 technologies are being ready for the commercialization, four has been patented. This company also got many innovation awards. She is recently awarded with Niti Ayog Women Transforming (WTI) Awardtop30, her success story was released as a Women Entrepreneur leading change in India by NITI AYOG through Hon'ble Defence Minister Shri Rajnath Singh. Currently, she is running multiinsitutional and transdisiplinary research grant on Water Innovation Centre (5.65 Crores). Her strong background mainly relates to water research which special focus on Nanomaterial synthesis Fluoride remediation, DNA fingerprinting, chemoprofiling and Fluoride (F) phytoremediation technology. She was the finalist in the National Bioentrepreneurship competition 2017 conducted by BIRAC-C-CAMP. She has received various awards viz DBT-research Associateship, young scientist award By ISGBRD, ICAR, recognition award for research and teaching and Indian National Academy of Sciences (INSA) international visiting scientist fellowship. She has selected as INSA visiting scientist for Turkey and widely travelled as visiting scientist and speaker to different countries. Her vision is to develop technologies for affordable safe water and create employment for women. As a scientist, teacher and entrepreneur she is following the new path nurturing women scientist to become an entrepreneur, the path which is less travelled.

Veena Gayathri Krishnaswamy is currently an Assistant Professor, Department of Biotechnology, Stella Maris College, Chennai. She has a patent and has completed six projects from UGC, DBT, TNSCST, and SEED Grant from Stella Maris College. She has published her research findings in 50 international journals. She has guided 53 M.Sc. Projects and currently guiding 4 Ph.D. Scholars. She has organized about 40 National workshops.

Amrendra Kumar received his PhD in Biosciences and Bioengineering from Indian Institute of Technology Guwahati and currently is senior research fellow at ICAR Research Center for NEH region, Umiam, Meghalaya, India. His research focuses on the protein biochemistry, biophysics and spectroscopy to understand the complex structural and functional nature of biological macromolecules.

Sonu Kumari is a passionate research scholar who completed PhD from Banasthali Vidyapith and developed a cost effective product to remove toxicants from soil and water.

Tanvi Kumari is a scholar at Amity Institute of Biotechnology, Amity University Jharkhand. She is pursuing B. Tech Biotechnology. Dr. Sumira Malik and Dr. Shilpa Prasad, Assistant Professor, Amity institute of Biotechnology, Amity University Jharkhand. She is working under guidance of Dr. Sumira Malik, Assistant Professor, Amity University Jharkhand. She has presented posters in various International and National conferences.

A. Madhavi obtained M.Sc in Applied Microbiology (2006) from VIT University, Vellore and having the project experience in Dr. Reddy's Laboratories, Hyderabad. She had joined as project fellow (UGC – MRP) in the year 2016, under the guidance of Prof. V. Rangaswamy (Rtd.), Department of Microbiology, S.K. University, Anantapuramu, Andhra Pradesh, India, entitled "Impact of Xenobiotics on Microbial Diversity, Enzymes and Biodegradation In Agricultural Soils In Anantapuramu District Of Andhra Pradesh. At present, she is a research scholar and worked as academic consultant (2018 – 2019) in the Department of Microbiology, S.K University, Anantapuramu. To her credit, she published 5 papers in national and international journals.

Dakshayani Mahapatra has completed her graduation from Presidency College in Physiology. She has completed her post graduation and PhD from University of Calcutta. She is interested in female reproductive physiology, endocrinology and related studies. She has publications in peer reviewed journals and has received couple of awards as a student and research fellow. Presently she is Assistant Professor in the Department of Physiology in Government General Degree College in Mohanpur, Paschim Medinipur, West Bengal, India.

Sumira Malik is currently working as an Assistant Professor in Amity institute of biotechnology, Amity University Jharkhand, Ranchi, Jharkhand. She has total experience of 09 years, with 06 years of teaching and 08 years of research experience. She has completed her Ph.D. at School of Biological sciences and Biotechnology, Chonnam national university, Gwangju, South Korea. She has worked intensively as research fellow and pursued postdoctoral studies from lab of stem cell biology, Hormone research centre, Chonnam National university, South Korea. She had 31 research and review articles in SCI, Peer-Reviewed International, UGC approved and NAAS rated journals, 5 book chapters, 2 books, 2 published patents, six awards for research excellence and two awards for young scientist award to her credit. She is a Life Member in various Professional Societies and reviewer in SCI and international journals. She has been awarded with competitive research fund for working upon Uttarakhand technical university state government based project as co-Principal investigator.

A. B. Muhammad is currently a professor of Applied & Petroleum Chemistry. He has B.Sc. and M.Sc. degrees (Applied Chemistry) from Usmanu Danfodiyo University, as well as M.Sc. and Ph.D. degrees (Petroleum Geochemistry) from Newcastle University, UK. He was also a postdoctoral research fellow during which he worked on Geochemical control of organic matter turnover in peatlands under Dr. Geoff D Abbott at Newcastle University. His research interests cover upstream and downstream petroleum chemistry. His current research is mainly on development of sustainable chemical technology for biofuels and biochemicals production from biomass as a subset of bioeconomy. He has over 35 publications in peer reviewed journals.

Suchetana Mukherjee completed Ph.D. in Botany in 2014 from University of Kalyani, West Bengal, India. PhD work was on "Induced Mutagenesis in Nigella sativa with Special Emphasis to Genetic Male Sterility." Field of interest: Genetics, Cytology, Plant Breeding, recent trends of sustainable agriculture, therapeutic options of COVID 19 with special emphasis to medicinal plants & Ayurveda. Currently working as an Assistant professor in Sripat Singh College, Jiaganj, Murshidabad, West Bengal, India, which is affiliated to the University of Kalyani and a member of U.G. Board of Studies, Botany, University of Kalyani. Former Head of the Department, Botany, Sripat Singh College (01.11.2017 to 10.11.2019). Publications Some of the important publications include 1. Induced genetic male sterility in Nigella sativa L. (black cumin). Cytologia. 2. Mitotic and Meiotic consequences of gamma irradiation on dry seeds of Nigella sativa L. (Black Cumin). Journal of plant Developmet Sciences 3. A note on non-shattering mutant in Nigella sativa L. (black cumin). International research journal of pharmacy. 4. Pollination events in Nigella sativa L. (black cumin). Int. J. Res. Ayurveda Pharm. 5. Seasonal influence on the chromosome behaviour of diploid (Solanum nigrum L.) and hexaploidy (S. americanum Mill.) species of Solanum. Asian J. Exp. Biol. Sci. 6. An overview on Withania somnifera (L.) Dunal- The 'Indian ginseng'. Medicinal and aromatic Plant Science and Biotechnology.

Anandkumar Naorem, Ph.D, is a soil scientist in ICAR-Central Arid Zone Research Institute, RRS-Bhuj, Gujarat, India. His field of specialisation is soil fertility, dryland soil, cactus research and soil microbiology.

Arunima Nayak is a researcher involved in the development and advancement in the knowledge, tools and processes in the field of different biorefinery and valorisation strategies for wastes and wastewater treatment via 1. Contribution to the area of recovery of biofuels, nutrients, and value added components from the waste matrices. 2. Contribution to the development of materials/method for use as cheap adsorbents for water treatment 3. Contribution to the recovery of recycled water via wastewater monitoring and development of sustainable treatment technologies. 8 years of active research. PhD from IIT, Roorkee; Research Associate from IIT Roorkee and Marie Curie Experienced Researcher from Spain. Currently associated as Associate Professor at Graphic Era University, Dehradun, India.

Joan Nyika is an Environmental Science PhD student at the University of South Africa and a Tutorial Fellow at the Technical University of Kenya, Geoscience and the Environment Department. She obtained a Bsc. degree in Biochemistry and Molecular Biology from the Jomo Kenyatta University of Agriculture and Technology in 2011. She also obtained an Msc degree in Land and Water Management from the University of Nairobi and, Kenya in 2017. Her PhD work revolves around solid waste management and the effects of landfill leachate on soil and groundwater. She has also done research on the use of microbes for pollutant bioremediation and phytoremediation. She has interests in sustainable environmental management, water management, ecological and ecosystem management. Joan has published more than 15 articles and 6 book chapters in peer reviewed journals and books, respectively. She has and is mentoring both undergraduate and diploma students and has more than 5 years experience teaching and training in environmental science.

Sachin Patel is working as a project associate in ICAR-CAZRI, RRS-Bhuj, His field of specialization is soil biotechnology and cactus nutrient management.

Elroy Pereira has done his doctoral thesis on the bioremediation of chromium metal by marine bacteria.

Shilpa Prasad is presently working as an Assistant Professor at Amity Institute of Biotechnology, Amity University Jharkhand. She has been working on the factors that are responsible to determine egg quality. Her research area includes cell physiology. She has completed her Post-Doctoral research work from University of the Free States, Bloemfontein, South Africa. She has completed her Ph. D from Banaras Hindu University, Varanasi, India. She has published various research/review articles in International and National Journals of repute.

Kumaresan Priyanka originally from Tamil Nadu, India. She earned her undergraduate degree from RVS college of Arts and Science in Microbiology and obtained her post graduate degree in Industrial Biotechnology from Bharathiar University, Tamil Nadu. She also obtained her M.Phil degree in Microbial biotechnology and is currently pursuing Ph.D in the same avenue from Bharathiar University, Coimbatore, Tamil Nadu India. Her research area includes biopolymer production from microbial sources and published articles for the same in reputed journals.

A. B. Rabah is an Associate Professor of Environmental Microbiology In the Department of Microbiology, UDU Sokoto. He has authored over 35 research articles in local and international peer-reviewed journals. He has been teaching environmental microbiology courses to undergraduate and postgraduate students since 2006 in the same department where he is currently the HOD. His expertise is in bioremediation, bioenergy, and biotechnology.

Karthikeyan Ramalingam, Dean, Student Affairs and Associate professor, BSA Crescent Institute of Science and Technology with over 20 years of laboratory knowledge, including academic, government, and industrial projects, handling experience with excellent supervisory skills and a strong record of scientific accomplishments and publications. In nanoemulsion sciences experienced from the University of Texas Health Science Center at San Antonio, Texas, USA, US Naval Medical Research Unit, San Antonio, Texas, USA, New York University, USA, Nottingham University, UK and received funds for nanoemulsions research from Department of Defense, Navy, USA ($105,900), DST-SERB (Rs. 23 lakhs) and ICMR (15 lakhs) and USA-NSF-SBIR ($475,766) as principle investigator.

V. Rangaswamy born in 1957 June 28th. He did M.Sc., M.Phil., B.Ed., and Ph.D. He is Professor (Rtd.) in Microbiology, Department of Microbiology, Sri Krihnadevaraya University, Anantapur- 515003, A.P. India. He did Post Doctoral Research work in Penn State, Pennsylvania State Univeristy, USA for two years (1994-96), and Second Post Doctoral Research work in University of Wales, U.K. for One Year (1999-2000) and also worked as a Visiting Professor in Ethiopia (Africa) for period of 1 and 1/2 year (2005-2007). He completed one Minor Research Project (UGC) and Two Major Research Projects (UGC). Three Ph.D., s and 10 MPhil.,s awarded under his guidance. He worked as a Head of the Department for 3 Years and Chairman of BOS for 3 Years. His main research area is "Bio-remediation and Detoxification of toxic chemicals/Xenobiotics in environment". He published a quite good number of National and International publications in reputed journals. He attended quite good number of National and International workshops. He had quite good number of administrative experiences (10 Nos) in the University. He received A.P. State Meritorious Best Teacher Award in September 2011.

Shabir A. Rather is working as a Postdoc fellow at the College of Life Sciences, Northwest A&F University, Yangling, P.R. China.

Umair Riaz is a young scientist working as a Scientific Officer at Soil and Water Testing Laboratory for Research, Bahawalpur (ISO 17025 Accredited Lab). He obtained his B.Sc. (Hons.) from Islamia University of Bahawalpur while M.Sc. (Hons.) and Ph.D. (Soil & Environmental Sciences) from the University of Agriculture, Faisalabad, Punjab, Pakistan. He has a good record of publications and working on different projects to put his technical share in developing advanced technologies for researchers. Major areas of his research are potassium release kinetics from mica minerals, solid waste management, sewage sludge management, toxicity of heavy metals in rice, wheat, maize, etc. abiotic stresses and pore water chemistry.

Adhithya Sankar S. originally from Kerala, India. He has completed his undergraduate studies in triple major Biotechnology, Chemistry and Zoology from CHRIST (Deemed to be University), Bangalore, India. He is currently pursuing Postgraduate studies in Biotechnology at CHRIST (Deemed to be University), Bangalore, India. His area of interests are microbiology and biopolymers.

Bijaya Samal is a scholar at Amity Institute of Biotechnology, Amity University Jharkhand. She is pursuing B. Tech Biotechnology. Dr. Sumira Malik and Dr. Shilpa Prasad, Assistant Professor, Amity institute of Biotechnology, Amity University Jharkhand. She is working under guidance of Dr. Sumira Malik, Assistant Professor, Amity University Jharkhand. She has presented posters/e posters in various International and National conferences.

Raghunath Satpathy received his M.Sc in Botany (with specialization in Biotechnology) from Berhampur University science, India; (Post M.Sc.) Advanced P.G Diploma in Bioinformatics from University of Hyderabad and M. Tech. degree in Biotechnology from VIT University, Vellore India. He was awarded the degree for the doctor of philosophy (PhD) in Biotechnology by Sambalpur University, Odisha India. Currently he is working as an Assistant professor in the School of Biotechnology, Gangadhar Meher University, Samabalpur, Odisha, India. He is having more than 11 years of experiences in both teaching and research. He has authored many national and international journal papers and book chapters to his credit. In addition to this Dr. Satpathy is associated with many professional bodies and he is the recipient of many academic and research awards.

Ann Mary Sebastian originally from Kerala, India. She pursued her undergraduate degree in Chemistry, Botany and Zoology (CBZ) and completed her post graduate degree in Botany from CHRIST (Deemed to be University), Bangalore. She has ample research experience in the study of plant beneficial microorganisms and has worked closely with Bacillus sp. Her research interest includes formulation of microbial inoculants, plant growth regulation and plant biotechnology

Laila Shahzad is working as an Assistant Professor at Sustainable Development Study Center, GC University Lahore. Dr. Laila has over a decade of development and research experience in the areas of disaster risk reduction, community based natural resource management, climate change adaptation, and public health issues. With growing needs of research in Pakistan, environmental research and awareness among youth is what she focused on. Laila have produced several research papers which highlighted the need of climate change adaptation, studies and analyses disaster risk. She has launched a new curriculum on sustainable consumption and production which will be of graduate and post graduate level and much needed theme in era of SDGs.

Shashank Shekhar is a scholar at Department of Agriculture and pursuing B. Sc Agriculture from Shivalik Institute of Professional Studies, Dehradun. He is working under guidance of Dr. Sumira Malik, Assistant Professor, Amity University Jharkhand. She has published research/ review papers in International and National Journals of repute. He has presented his work in various International and National platforms.

P. Sindhura is faculty with 9 years of teaching experience at UG and PG level. She is presently working as Assistant professor at TSWRDC(W), Bhupalpally, Telangana, India. She has completed her M.Sc. (Microbiology) from Osmania University, Hyderabad, India. She has Qualified state level eligibility test for Lecturership - (TS SET). Her areas of interest include Microbial physiology, Medical microbiology, Microbial biochemistry and Immunology.

Neha Singh is a research scholar under Dr. Suphiya khan at the Department of Bioscience and Biotechnology, Banasthali Vidyapith, Newai, District - Tonk, Rajasthan, India.

Dwaipayan Sinha is an Assistant Professor in Botany, West Bengal Education Service, Government of West Bengal and has years of teaching experience in Botany and 15 years of Research Experience. He has done his graduation with honours in Botany from the esteemed Presidency College, Kolkata and then Masters Degree in Botany from Delhi University North Campus with specializations in crop genetics and plant molecular biology. He obtained his Doctoral Degree from CSIR-Indian Institute of Toxicology Research, Lucknow India and enrolled in Department of Botany, Lucknow University. He is having a number of publication in reputed peer reviewed journal and is a life member of International Academy Of Science And Research (IASR), Kolkata and International Foundation for Environment and Ecology, Kolkata. His research interests are mineral nutrition in plants and improvement in yield of agricultural crops using beneficial bacteria. He also has keen interest on ethnobotany and ethnopharmacology. ORCID iD: https://orcid.org/0000-0001-7870-8998.

Rajalakshmi Sridharan is currently studying Ph.D in Stella Maris College (Autonomous), Chennai, India. She completed M.Sc Biotechnology in Stella Maris College (Autonomous), Chennai, India and scored 79.5% and won best outgoing student award. She did her B.Sc Advanced Zoology and Biotechnology in Meenakshi College for Women, Chennai, India and scored 89.5%. She has published 3 research articles and 3 chapters in international journals and publishers. She has presented Papers and posters in many national and International conferences and won best paper and best poster awards. She worked as a Project Assistant under the INDO-US DBT project entitled "Maximal usage of foldscope to appraise South and North – East Indian farmers on the relative abundance of earthworm species, associated microflora and resistance to pesticides to preserve biodiversity".

M. Srinivasulu born in Peddakammavaripalli, Puttaparthi mandal, Anantapur District, Andhra Pradesh, India on 10-05-1980. Awarded M.Phil and Ph.D from the Department of Microbiology, Sri Krishnadevaraya University, Anantapur. Currently working as Academic Consultant in the Department of Biotechnoloy, Yogi Venmana University, Kadapa. He had worked as Post-Doctoral Fellow in Pohang Univeristy of Sceince and Technology (POSTECH), South Korea; Univesridad De las Fuerzas Armadas-ESPE, Ecuador, South America; and Nanjing Forestry University, Nanjing, China. He also worked as Academic Consultant in the Department of Microbiology, Sri Krishnadevaraya University, Anantapur. To his credit, he published 45 papers in national and international journals, one book and 4 book chapters with international publishers. He received Dr. K.V. Rao Research Award for the year 2010 in Biological Sciences. Awarded Prometeo Research Project, PI, Ecuador, South America in 2014. Awarded Senior Research Fellowship funded by ICAR Govt. of India in 2007. Received UGC Project Fellowship in 2008.

Uzma Sultana is currently working as a Assistant Professor at TSWRDC (W) Kamareddy, Telangana. She has done her Ph.D. from Central Research Institute for Dryland Agriculture, affiliated to Osmania University, Hyderabad, Telangana, India. Dr. Uzma Sultana has published her research work in peer reviewed international journals. Her area of interest are Microbial physiology, Medical microbiology, Microbial biochemistry, and Immunology.

Shiva Kumar Udayana, Ph.D, is working as a research associate in ICAR-KVK, Andhra Pradesh. His field of specialization is soil heavy metal toxicity.

Mridul Umesh originally from Kerala, India. An active researcher in the field of biopolymers and waste management, with 6 years of research experience, Mridul pursued his Post Graduation in Industrial Biotechnology, MPhil in Microbial Biotechnology and PhD in Microbial Biotechnology, from Bharathiar University, Coimbatore, Tamil Nadu. He is currently working as Assistant Professor in the Department of Life Sciences, CHRIST(Deemed to be University), Karnataka, India. His trust area of research includes, bioplastic production, single cell protein, organic acids, enzyme production, aqua feed formulation and waste management.

P. Vanamala is a senior faculty with 14 years of teaching experience at UG and PG level. She is currently working as Assistant Professor at Telangana Social Welfare Residential Degree College for Women, (TSWRDCW), Kamareddy, Telangana, India. She did her M.Sc. in Applied Microbiology from Padmavati women's university, Tirupati. She has Qualified state level eligibility test for Lectureship-AP-TS SET. Her area of interest is Molecular Biology, medical Virology and Applied Microbiology.

Himashi Verma is MSc. (Botany) qualified from Department of Botany, University of Delhi, India. She is also a joint CSIR-UGC NET JRF (2020) holder, DBT-JRF (2020) holder and joint CSIR-UGC NET L.S (2019) qualified fellow. She has completed her graduation, B.Sc. (Hons.) Botany (2017), from Miranda House, University of Delhi, India.

Anam Waheed is a student at Government College University Lahore in postgraduate program. She is majoring in Environmental Sciences programme at Sustainable Development Study Centre. She has worked in flood affected communities of Muzaffargarh and has proposed climate adaptations for local farmers. Anam has also researched about environmental impacts of orange train as a developmental project.

Monika Yadav is working as research scholar in department of Bioscience and Biotechnology, Banasthali Vidyapith for "Development of Defluoridation Technology" under the supervision of Dr. Suphiya Khan, Associate Professor, Department of Bioscience and Biotechnology, Banasthali Vidyapith, India. Her research interest focuses on sensing -nanotechnology, plant stress, phytoremediation, polymers, and rare earth metal nano-composite applications in water treatment. She has published in several reputed journals such as sustainable Chemistry and pharmacy and filed one patents on the development of colirimetric detector for detection of fluoride.

Index

C

D

E

F

G

H

T

V

W

X

www.ingramcontent.com/pod-product-compliance
Lightning Source LLC
LaVergne TN
LVHW080616300926
849162LV00005B/1183
* 9 7 8 1 7 9 9 8 7 0 6 2 3 *